Representations of Algebraic Groups

Second Edition

Mathematical Surveys and Monographs

Volume 107

Representations of Algebraic Groups

Second Edition

Jens Carsten Jantzen

AMERICAN MATHEMATICAL SOCIETY

Providence, Rhode Island

2000 *Mathematics Subject Classification.* Primary 20-02, 20G05; Secondary 17B10, 17B45, 17B56, 22E45.

For additional information and updates on this book, visit
www.ams.org/bookpages/surv-107

Library of Congress Cataloging-in-Publication Data

Jantzen, Jens Carsten
Representations of algebraic groups / Jens Carsten Jantzen. — 2nd ed.
p. cm. — (Mathematical surveys and monographs, ISSN 0076-5376 ; v. 107)
Includes bibliographical references and index.
ISBN 0-8218-3527-0 (alk. paper)
1. Representations of groups. 2. Linear algebraic groups. I. Title. II. Mathematical surveys and monographs ; no. 107.

QA176.J37 2003
512′.22–dc22 2003058381

AMS softcover ISBN 978-0-8218-4377-2

♾ The paper used in this book is acid-free and falls within the guidelines
established to ensure permanence and durability.
Visit the AMS home page at http://www.ams.org/

13 12 11 10 9 8 7 6 5 4 29 28 27 26 25 24

Contents

Introduction

I This book is meant to give its reader an introduction to the representation theory of such groups as the general linear groups $GL_n(k)$, the special linear groups $SL_n(k)$, the special orthogonal groups $SO_n(k)$, and the symplectic groups $Sp_{2n}(k)$ over an algebraically closed field k. These groups are algebraic groups, and we shall look only at representations $G \to GL(V)$ that are homomorphisms of algebraic groups. So any G–module (vector space with a representation of G) will be a space over the same ground field k.

Many different techniques have been introduced into the theory, especially during the last thirty years. Therefore, it is necessary (in my opinion) to start with a general introduction to the representation theory of algebraic group schemes. This is the aim of Part I of this book, whereas Part II then deals with the representations of reductive groups.

II The book begins with an introduction to schemes (Chapter I.1) and to (affine) group schemes and their representations (Chapter I.2). We adopt the "functorial" point of view for schemes. For example, the group scheme SL_n over $\mathbf{Z}$ is the functor mapping each commutative ring A to the group $SL_n(A)$. Almost everything about these matters can also be found in the first two chapters of [DG]. I have tried to enable the reader to understand the basic definitions and constructions independently of [DG]. However, I refer to [DG] for some results that I feel the reader might be inclined to accept without going through the proof. Let me add that the reader (of Part I) is supposed to have a reasonably good knowledge of varieties and algebraic groups. For example, he or she should know [Bo] up to Chapter III, or the first seventeen chapters of [Hu2], or the first six ones of [Sp2]. (There are additional prerequisites for Part II mentioned below.)

In Chapter I.3, induction functors are defined in the context of group schemes, their elementary properties are proved, and they are used to construct injective modules and injective resolutions. These in turn are applied in Chapter I.4 to the construction of derived functors, especially to that of the Hochschild cohomology groups and of the derived functors of induction. In contrast to the situation for finite groups, the induction from a subgroup scheme H to the whole group scheme G is (usually) not exact, only left exact. The values of the derived functors of induction can also be interpreted (and are so in Chapter I.5) as cohomology groups of certain associated bundles on the quotient G/H (at least for algebraic schemes over a field). Before doing that, we have to understand the construction of the quotient G/H. The situation gets simpler and has some additional features if H is normal in G. This is discussed in Chapter I.6.

One can associate to any group scheme G an (associative) algebra $\mathrm{Dist}(G)$ of distributions on G (called the hyperalgebra of G by some authors). When working over a field of characteristic 0, it is just the universal enveloping algebra of the Lie algebra Lie(G) of G. In general, it reflects the properties of G much better than Lie(G) does. This is described in Chapter I.7.

A group scheme G (say over a field) is called finite if the algebra of regular functions on G is finite dimensional. For such G the representation theory is equivalent to that of a certain finite dimensional algebra and has additional features (Chapter I.8). For us, the most important cases of finite group schemes arise as Frobenius kernels (Chapter I.9) of algebraic groups over an algebraically closed field k of characteristic $p \neq 0$. For example, for $G = GL_n(k)$ the map $F : G \to G$ sending any matrix (a_{ij}) to (a_{ij}^p) is a Frobenius endomorphism. The kernel of F^r (in the sense of group schemes) is the rth Frobenius kernel G_r of G. The representation theory of G_1 (for any G) is equivalent to that of Lie(G) regarded as a p–Lie algebra.

In order to apply our rather extensive knowledge of the representation theory of groups like $SL_n(\mathbf{C})$ to that of $SL_n(k)$, where k is a field of prime characteristic, one uses the group scheme SL_n over $\mathbf{Z}$. One chooses SL_n–stable lattices in $SL_n(\mathbf{C})$–modules and tensors with k in order to get $SL_n(k)$–modules. Some general properties of this procedure are proved in Chapter I.10.

From Part I, the contents of Chapters 1 (until 1.6), 2, 3, 4 (until 4.18), 5 (mainly 5.8–5.13), and 6 (until 6.9) are fundamental for everything to follow. The other sections are used less often.

In Part II, the reader is supposed to know the structure theory of reductive algebraic groups (over an algebraically closed field) as to be found in [Bo], [Hu2], [Sp2]. The reader is invited (in Chapter II.1) to believe that there is for each possible root datum a (unique) group scheme over $\mathbf{Z}$ that yields for every field k (by extension of the base ring) a split reductive group defined over k having the prescribed root datum. Furthermore, he or she has to accept that all "standard" constructions (like root subgroups, parabolic subgroups, etc.) can be carried out over $\mathbf{Z}$. (The sceptical reader should turn to [SGA 3] for proof.) I have included a proof (following Takeuchi) of the uniqueness of an algebraic group with a given root datum (over an algebraically closed field) that does not use case-by-case considerations.

III Let me describe a selection of the contents of the remaining chapters in more detail. Assume from now on (in this introduction) that k is an algebraically closed field and that G is a (connected) reductive algebraic group over k with a Borel subgroup $B \subset G$ and a maximal torus $T \subset B$. Let $X(T)$ denote the group of characters of T.

In case char(k) $= 0$ the representation theory of G is well understood. Each G–module is semi-simple. The simple G–modules are classified (as in the case of compact Lie groups or of complex semi-simple Lie algebras) by their highest weights. Furthermore, one has a character formula for these simple modules. In fact, Weyl's formula for the compact groups holds when interpreted in the right way. (For us, the character of a finite dimensional G–module will always be the family of the dimensions of its weight spaces with respect to T. As the semi-simple elements in G are dense in G and as each semi-simple element is conjugate to one in T, the character determines the trace of any $g \in G$ on the G–module.)

The situation in prime characteristic is much worse. Except for the case of a torus, there are non-semi-simple G–modules. Except for a few low rank cases, we do not know a character formula for the simple modules, and Weyl's formula will certainly not carry over. Only one property survives: The simple modules are still classified by their highest weights, and the possible highest weights are the "dominant" weights in $X(T)$. (The notion of dominant depends on the choice of an ordering of $X(T)$. We shall always work with an ordering for which the weights of T on Lie(B) are negative.) This classification is due to Chevalley, cf. [SC]. Let $L(\lambda)$ denote the simple module with highest weight λ.

The difference of the situations in zero and prime characteristic can be observed already in the case $G = SL_2(k)$. Let $H(n)$ be the nth symmetric power of the natural representation of G on k^2. If $\text{char}(k) = 0$, then $H(n) = L(n)$ for all $n \in \mathbf{N}$. (For SL_2 we identify $X(T) \simeq \mathbf{Z}$ in such a way that the dominant weights correspond to $\mathbf{N}$.) If $\text{char}(k) = p \neq 0$, then obviously not all $H(n)$ can be simple: For all positive $r, n \in \mathbf{N}$ the image of the map $f \mapsto f^{p^r}$ from $H(n)$ to $H(p^r n)$ is a proper submodule of $H(p^r n)$, so $H(p^r n)$ is not simple. It is not too difficult to show for any n that $H(n)$ contains $L(n)$ as its unique simple submodule, and that $H(n) = L(n)$ if and only if $n = ap^r - 1$ for some $a, r \in \mathbf{N}$ with $0 < a \leq p$. So for all other n the module $H(n)$ is not semi-simple.

For arbitrary G one gets $L(\lambda)$ as the unique simple submodule of an induced module $H^0(\lambda)$: One extends $\lambda \in X(T)$ to a one dimensional representation of B such that the unipotent radical of B acts trivially. Then $H^0(\lambda)$ is the G–module induced by this B–module. It is nonzero if and only if λ is dominant. (In the case $G = SL_2(k)$ the $H^0(\lambda)$ are just the $H(n)$ from above.) This is the main content of Chapter II.2.

The case $G = SL_2(k)$ with $\text{char}(k) = p \neq 0$ can serve to illustrate other general results also. For any vector space V over k let $V^{(r)}$ be the vector space that is equal to V as an additive group and where any $a \in k$ acts as $a^{p^{-r}}$ does on V. Then the map $f \mapsto f^{p^r}$ is linear when regarded as a map $H(n)^{(r)} \to H(p^r n)$, hence a homomorphism of G–modules. It is not difficult to show: If $n = \sum_{i=0}^r a_i p^i$ with $0 \leq a_i < p$ for all i, then $f_0 \otimes f_1 \otimes \cdots \otimes f_r \mapsto \prod_{i=0}^r f_i^{p^i}$ is an isomorphism

$$H(a_0) \otimes H(a_1)^{(1)} \otimes \cdots \otimes H(a_r)^{(r)} \xrightarrow{\sim} L(n).$$

This result was generalised in [Steinberg 2] to all G: A suitable p–adic expansion of the highest weight λ leads to a decomposition of $L(\lambda)$ into a tensor product of the form $L(\lambda_0) \otimes L(\lambda_1)^{(1)} \otimes \cdots \otimes L(\lambda_r)^{(r)}$. This tensor product theorem reduces the problem of calculating the characters of all simple G–modules to a finite problem (for each G). Steinberg's proof relied on a theorem from [Curtis 1] on the representations of Lie(G). In the special case of $G = SL_2(k)$, this theorem says: Each $L(n)$ with $n < p$ remains simple for Lie(G), and each simple module of Lie(G) as a p–Lie algebra is isomorphic to exactly one $L(n)$ with $n < p$. More generally, each $L(n)$ with $n < p^r$ is simple for the rth Frobenius kernel of $SL_2(k)$, and we get thus each simple module for this infinitesimal group scheme. This result again has an extension to all G and then leads to a rather simple proof of Steinberg's tensor product theorem, discovered by Cline, Parshall, and Scott. (All this is done in Chapter II.3.)

The choice of the notation $H^0(\lambda)$ for the induced module has been influenced by the fact that $H^0(\lambda)$ is the zeroth cohomology group of a line bundle on G/B associated to λ. Let $H^i(\lambda)$ denote the ith cohomology group (for any $\lambda \in X(T)$, not only for dominant ones). We could have constructed $H^i(\lambda)$ also by applying the ith

derived functor of induction from B to G to the one-dimensional B–module defined by λ. Another result from characteristic zero that does not carry over to prime characteristic is the Borel-Bott-Weil theorem. It describes explicitly all $H^i(\mu)$ with $i \in \mathbf{N}$ and $\mu \in X(T)$: For each μ there is at most one i with $H^i(\mu) \neq 0$, and this $H^i(\mu)$ can then be identified with a specific $L(\lambda)$. We observed already that we cannot expect the $H^i(\mu)$ to be simple in prime characteristic. But, even worse, there can be more than one i for a given μ with $H^i(\mu) \neq 0$, and the character of $H^i(\mu)$ will depend on the field. (This was first discovered by Mumford.) It is crucial for the representation theory that one special case of the Borel-Bott-Weil theorem holds over any k: If λ is dominant, then $H^i(\lambda) = 0$ for all $i > 0$. This is Kempf's vanishing theorem from [Kempf 1]. The proof given here in Chapter II.4 is due to Haboush and Andersen (independently). In Chapter II.5, we give Demazure's proof of the Borel-Bott-Weil theorem in case $\operatorname{char}(k) = 0$. Furthermore we prove (following Donkin) that Weyl's character formula yields the alternating sum (over i) of the characters of all $H^i(\mu)$.

Assume from now on that $\operatorname{char}(k) = p \neq 0$. Kempf's vanishing theorem implies that one can construct for any k the modules $H^0(\lambda)$ with λ dominant by starting with the similar object over $\mathbf{C}$, taking a suitable lattice stable under a $\mathbf{Z}$–form of G, and then tensoring with k. To construct representations in this way has the advantage that one can carry out specific computations more easily. Several examples computed especially by Braden then led Verma in the late 1960s to several conjectures (cf. [Verma]) that had a great influence on the further development of the theory. One conjecture is the linkage principle (Chapter II.6): If $L(\mu)$ is a composition factor of $H^0(\lambda)$ (or, more generally, if $L(\mu)$ and $L(\lambda)$ are both composition factors of a given indecomposable G–module), then $\mu \in W_p \bullet \lambda$. Here W_p is the group generated by the Weyl group W and by all translations by $p\alpha$ with α a root. The dot is to indicate a shift in the action by ρ, the half sum of the positive roots (i.e., $w \bullet \lambda = w(\lambda+\rho) - \rho$). For large p this principle was proved in [Humphreys 1]. The result was then extended by several people to almost all cases, but a general proof appeared only in 1980 (in [Andersen 4]). It relies on an analysis of the failure of Demazure's proof (of the Borel-Bott-Weil theorem) in prime characteristic.

Another conjecture of Verma was a special case of the translation principle (Chapter II.7): If two dominant weights λ, μ belong to the same "facet" with respect to the affine Weyl group W_p, then the multiplicity of any $L(w \bullet \lambda)$ with $w \in W_p$ as a composition factor of $H^0(\lambda)$ should be equal to that of $L(w \bullet \mu)$ in $H^0(\mu)$. This was proved (modulo the linkage principle) in [Jantzen 2].

The approach to the $H^0(\lambda)$ via representations over $\mathbf{Z}$ also has the advantage that it allows the construction of a certain filtration (Chapter II.8) of $H^0(\lambda)$. One can compute the sum of the characters of the terms in the filtration ([Jantzen 3] for large p, [Andersen 12] in general) and use this "sum formula" to get information about composition factors. For example, it leads to a computation of the characters of all simple modules for $G = SL_4(k)$ or for G of type G_2.

If λ and $\lambda + p\nu$ are weights that are "small" with respect to p^2 and that are "sufficiently dominant" (see II.9.17/18 for a more precise condition), then one gets the composition factors of $H^0(\lambda + p\nu)$ from those of $H^0(\lambda)$ by adding $p\nu$ to the highest weights. This was proved first in [Jantzen 4] using involved computations. Later on it was realised that it follows rather easily if one develops the representation theory of the group scheme G_rT. For λ as above experimental evidence (cf.

[Humphreys 10]) indicated that the $H^i(w \bullet \lambda)$ with $w \in W$ satisfy a weak version of the Borel-Bott-Weil theorem ($H^i(w \bullet \lambda) \neq 0$ for at most one i). This was then proved in [Cline, Parshall, and Scott 10] using the representation theory of the group scheme G_rB. All this is described in Chapter II.9.

Let us assume that G is semi-simple and simply connected. There is for each positive integer r a unique simple G–module that is simple and injective for G_r. It is called the rth Steinberg module and was first discovered by Steinberg within the representation theory of finite Chevalley groups. We do not look at its great importance there, but discuss some applications to the representation theory of G (Chapter II.10). It plays a crucial role in Haboush's proof that G is geometrically reductive. One may wonder whether any injective G_r–module can be extended to a G–module. For large p this was proved by Ballard. We discuss this (with some applications to the representation theory of G) in Chapter II.11.

One can write down the character of a simple G–module $L(\lambda)$ if one knows all extension groups $\mathrm{Ext}^n_G(L(\lambda), H^0(\mu))$, see II.6.21. Unfortunately, rather little is known about these groups. There has been a considerable amount of work (especially by Cline, Parshall, and Scott) to understand better the Hochschild cohomology groups $H^n(G, M) \simeq \mathrm{Ext}^n_G(k, M)$. One has $H^n(G, M) \simeq \varprojlim H^n(G_r, M)$ if $\dim M < \infty$. So one may hope to get information on G–cohomology from information on G_r–cohomology. Here the most remarkable result is due to Friedlander and Parshall: For large p the cohomology ring $H^\bullet(G_1, k)$ is isomorphic to the ring of regular functions on the nilpotent cone in Lie(G). This can be found in Chapter II.12.

The orbits of B on G/B are isomorphic to affine spaces. They are called Bruhat cells, while their closures are called Schubert varieties. For example, G/B itself is a Schubert variety. One can extend Kempf's vanishing theorem to any Schubert variety $Y \subset G/B$: If one restricts to Y the line bundle on G/B corresponding to a dominant weight λ, then all higher cohomology groups vanish. As an application one can prove the normality of Y and a character formula for the space of global sections. These results were proved by Mehta, Ramanathan, Seshadri, Ramanan, and Andersen. One can find this in Chapter II.14, whereas Chapter II.13 provides the necessary background on Schubert varieties.

The last seven chapters mentioned above can be divided into three groups (II.8, II.9–12, II.13–14), which are independent of each other. Also, the logical interdependence of Chapters II.10–12 is rather weak.

IV So far this introduction has been copied (with minor modifications) from the introduction of the first edition. For this new edition I have added a few chapters that I shall discuss in a moment.

As far as the old chapters are concerned, I have tried to correct mistakes and misprints. I have added several remarks and in a few cases rearranged things. In doing so, I have tried to avoid renumbering subsections and equations so that references to the first edition would also work with the second one. However, in a few cases (in particular in Chapter II.9) this turned out to be impossible. In these cases I have summed up the changes at the end of the introductions to the chapters (see II.7–9, 11, 12).

V The new chapters were added to Part II. They are not identified by numbers, but by capital letters so to indicate the break between the old and the new.

Keep the general assumptions from above (III). Let π be a finite set of dominant weights that is "saturated". This means that for each $\mu \in \pi$ also all dominant weights $\nu < \mu$ belong to π. Then it makes sense to consider the "truncated" category of all G–modules having only composition factors with a highest weight in π. Such categories are studied in Chapter II.A. Each of them is equivalent to the category of all modules over a suitable finite dimensional algebra. This allows the application of techniques from the representation theory of finite dimensional algebras to the theory of G–modules.

The categories of homogeneous polynomial GL_n–modules are special cases of truncated categories for $G = GL_n$. They connect the representation theory of GL_n with that of Schur algebras and of symmetric groups as well as with the theory of polynomial functors.

In Chapter II.B several cohomological results for G–modules are generalised from the case of a ground field to the case where one works over a principal ideal domain. For some of these proofs we have to use results from Chapter II.A.

In Chapters II.C and II.D we describe some consequences of Lusztig's conjecture leading to the calculation of Ext groups and to information about submodule structures, e.g., on the layers in the radical filtration of "baby Verma modules" (induced modules for G_1). One gets also that some of these consequences in turn imply Lusztig's conjecture.

Tilting modules (discussed in Chapter II.E) are G–modules that have a filtration with factors of the form $H^0(\lambda)$ as well as a filtration with factors of the form $H^0(\mu)^*$. The indecomposable tilting modules are classified by the dominant weights (like the simple G–modules) and as for the simple G–modules the computation of the characters of indecomposable tilting modules is a major open problem. In the case of $G = GL_n$ these tilting modules lead to yet another connection between the representation theory of GL_n and that of the symmetric groups.

The technique of "Frobenius splitting" is a powerful method to prove vanishing results for varieties in prime characteristics. We describe this in Chapter II.F and then use it to give alternative approaches to results from Chapter II.14. In Chapter II.G we use then Frobenius splitting techniques to prove the main properties of modules with a good filtration (announced in Chapter II.4).

The final chapter II.H surveys certain parts of the representation theory of quantum groups. Using these groups one can construct a representation theory in characteristic 0 that is similar to that of G in prime characteristic. However, one can prove stronger results on the quantum groups side, e.g., on characters of simple modules or of indecomposable tilting modules. This has then applications to the characteristic p theory.

VI Suppose that $\mathbf{F}_q$ is a finite field and that k is an algebraically closed extension of $\mathbf{F}_q$. The representation theory of groups like $GL_n(k)$ or $Sp_{2n}(k)$ has been developed in close interaction with that of groups like $GL_n(\mathbf{F}_q)$ or $Sp_{2n}(\mathbf{F}_q)$. It would therefore have been desirable to have a third part of the book dealing with representations of finite Chevalley groups (mainly over fields of the same characteristic as that over which the groups are defined). In fact, I promised to write such a part in a preliminary foreword to a preprint version of Part I. However, I hope to be forgiven for breaking this promise, as otherwise the book would have grown to an unreasonable size. Furthermore, I suspect that people most interested in these finite groups would prefer another book where they would not have to devour at first all of Parts I and II. Now (2003) a book on this topic is under preparation by Jim Humphreys.

VII In the summer of 1984, I gave a series of lectures on some topics discussed in this book at the East China Normal University in Shanghai. I had been asked in advance to provide the audience with some notes. When doing so, I was still undecided about the precise contents of my lectures. I therefore included more material than I could possibly cover in my lectures. The first edition of this book has grown out of those notes.

I should like to use this opportunity to thank the mathematicians I met in Shanghai, especially Professor Cao Xihua, for their hospitality during my stay and for the patience with which they listened to my lectures.

Thanks are also due to Henning Haahr Andersen, Rolf Farnsteiner, Burkhard Haastert, Jim Humphreys, Niels Lauritzen, Zongzhu Lin, and Jesper Funch Thomsen for useful comments on my manuscript and for providing lists of misprints, before and after the publication of the first edition and during the preparation of the second edition.

Part I

General Theory

CHAPTER 1

Schemes

It is the purpose of this first chapter to give the necessary introduction to schemes following the functorial approach of [DG]. This approach appears to be most suitable when dealing with group schemes later on. After trying to motivate the definitions in 1.1, we discuss affine schemes in 1.2–1.6. What is done there is fundamental for the understanding of everything to follow.

As far as arbitrary schemes are concerned, it is most of the time enough to know that they form a certain class of functors that includes the affine schemes and so that over an algebraically closed field any variety gives rise to a scheme in a canonical way. Sometimes, e.g., when dealing with quotients, it is useful to know more. So we give the appropriate definitions in 1.7–1.9 and mention the comparison with other approaches to schemes and with varieties in 1.11. The elementary discussion of a base change in 1.10 is again necessary for many parts later on.

There is also a discussion of closed subfunctors and of closures (1.12–1.14). Finally, we describe the functor of morphisms between two functors (1.15) and prove some of its properties. These results are used only in few places.

A ring or an associative algebra will always be assumed to have a 1, and homomorphisms are assumed to respect this 1. Let k be a fixed commutative ring. Notations of linear algebra (like Hom, $\otimes$) without special reference to a ground ring always refer to structures as k–modules. A k–algebra is always assumed to be commutative and associative. (For noncommutative algebras we shall use the terminology: algebras over k.)

1.1. Before giving the definitions, I want to point out how functors arise naturally in algebraic geometry. Assume for the moment that k is an algebraically closed field.

Consider a Zariski closed subset X of some k^n and denote by I the ideal of all polynomials $f \in k[T_1, T_2, \dots, T_n]$ (over k in n variables $T_1, T_2, \dots, T_n$) with $f(X) = 0$. Instead of looking at the zeroes of I only over k, we can also look at the zeroes over any k–algebra A, i.e., at $\mathcal{X}(A) = \{x \in A^n \mid f(x) = 0 \text{ for all } f \in I\}$. The map $A \mapsto \mathcal{X}(A)$ from $\{\, k\text{–algebras} \,\}$ to $\{\, \text{sets} \,\}$ is a functor: Any homomorphism $\varphi : A \to A'$ of k–algebras induces a map

$$\varphi^n : A^n \to (A')^n, \qquad (a_1, a_2, \dots, a_n) \mapsto (\varphi(a_1), \varphi(a_2), \dots, \varphi(a_n))$$

with $f(\varphi^n(x)) = \varphi(f(x))$ for all $x \in A^n$ and $f \in k[T_1, T_2, \dots, T_n]$. Therefore φ^n maps $\mathcal{X}(A)$ to $\mathcal{X}(A')$. Denote its restriction by $\mathcal{X}(\varphi) : \mathcal{X}(A) \to \mathcal{X}(A')$. For another homomorphism $\varphi' : A' \to A''$ of k–algebras, one has obviously $\mathcal{X}(\varphi') \circ \mathcal{X}(\varphi) = \mathcal{X}(\varphi' \circ \varphi)$, proving that $\mathcal{X}$ is indeed a functor.

A regular map f from X to a Zariski closed subset Y of some k^m is given by m polynomials $f_1, f_2, \dots, f_m \in k[T_1, T_2, \dots, T_n]$ as

$$f : X \to Y, \qquad x \mapsto (f_1(x), f_2(x), \dots, f_m(x)).$$

The f_i define for each k–algebra A a map $f(A) : A^n \to A^m$. The comorphism $f^* : k[T_1, T_2, \ldots, T_m] \to k[T_1, T_2, \ldots, T_n]$ maps the ideal defining Y into the ideal I defining X. This implies that any $f(A)$ maps $\mathcal{X}(A)$ to $\mathcal{Y}(A)$. The family of all $f(A)$ defines a morphism $\mathfrak{f} : \mathcal{X} \to \mathcal{Y}$ of functors, i.e., a natural transformation. The more general discussion in 1.3 (cf. 1.3(2)) shows that the map $f \mapsto \mathfrak{f}$ is bijective (from $\{$ regular maps $X \to Y \}$ to $\{$ natural transformations $\mathcal{X} \to \mathcal{Y} \}$).

Taking this for granted, we have embedded the category of all affine algebraic varieties over k into the category of all functors from $\{$ k–algebras $\}$ to $\{$ sets $\}$ as a full subcategory. This embedding can be extended to the category of all algebraic varieties, see 1.11.

One advantage of working with functors instead of varieties (i.e., of working with $\mathcal{X}$ instead of X) will be that it gives a natural way to work with "varieties" over other fields and over rings. Furthermore, we get new objects over k (algebraically closed) in a natural way. Instead of working with I, we might have taken any ideal $I' \subset k[T_1, T_2, \ldots, T_n]$ defining X, i.e., with $X = \{x \in k^n \mid f(x) = 0 \text{ for all } f \in I'\}$ or, equivalently by Hilbert's Nullstellensatz, with $\sqrt{I'} = I$. Replacing I by I' in the definition of $\mathcal{X}$, we get a functor, say $\mathcal{X}'$, with $\mathcal{X}'(A) = \mathcal{X}(A)$ whenever A is a field (or just an integral domain), but with $\mathcal{X}'(A) \neq \mathcal{X}(A)$ for some A if $I \neq I'$. Such functors arise in a natural way even when we deal with varieties, and they play an important role in representation theory.

Before giving the proper definitions, let us describe the functor $\mathcal{X}$ without using the embedding of X into k^n. For each k–algebra A, we have a bijection $\mathrm{Hom}_{k-\mathrm{alg}}(k[T_1, T_2, \ldots, T_n], A) \to A^n$, sending any α to $(\alpha(T_1), \alpha(T_2), \ldots, \alpha(T_n))$. The inverse image of $\mathcal{X}(A)$ consists of those α with $0 = f(\alpha(T_1), \alpha(T_2), \ldots, \alpha(T_n))$ $= \alpha(f)$ for all $f \in I$, hence can be identified with $\mathrm{Hom}_{k-\mathrm{alg}}(k[T_1, T_2, \ldots, T_n]/I, A)$. As $k[T_1, T_2, \ldots, T_n]/I$ is the algebra $k[X]$ of regular functions on X, we have thus a bijection $\mathcal{X}(A) \simeq \mathrm{Hom}_{k-\mathrm{alg}}(k[X], A)$. If $\varphi : A \to A'$ is a homomorphism of k–algebras, the $\mathcal{X}(\varphi)$ corresponds to the map

$$\mathrm{Hom}_{k-\mathrm{alg}}(k[X], A) \to \mathrm{Hom}_{k-\mathrm{alg}}(k[X], A'), \qquad \alpha \mapsto \varphi \circ \alpha.$$

A morphism $f : X \to Y$ is given by its comorphism $f^* : k[Y] \to k[X]$. Then the morphism $\mathfrak{f} : \mathcal{X} \to \mathcal{Y}$ is given by $\mathfrak{f}(A) : \mathrm{Hom}_{k-\mathrm{alg}}(k[X], A) \to \mathrm{Hom}_{k-\mathrm{alg}}(k[Y], A)$ with $\alpha \mapsto \alpha \circ f^*$ for any k–algebra A.

1.2. (k–functors) Let us assume k to be arbitrary again. In the definitions to follow, we shall be rather careless about the foundations of mathematics. Instead of working with "all" k–algebras, we should (as in [DG]) take only those in some universe. We leave the appropriate modifications to the interested reader.

A *k–functor* is a functor from the category of k–algebras to the category of sets.

Let X be a k–functor. A *subfunctor* of X is a k–functor Y with $Y(A) \subset X(A)$ and $Y(\varphi) = X(\varphi)_{|Y(A)}$ for all k–algebras A, A' and all $\varphi \in \mathrm{Hom}_{k-\mathrm{alg}}(A, A')$.

Obviously, a map Y that associates to each k–algebra A a subset $Y(A) \subset X(A)$ is a subfunctor if and only if $X(\varphi)\, Y(A) \subset Y(A')$ for each homomorphism $\varphi : A \to A'$ of k–algebras.

For any family $(Y_i)_{i \in I}$ of subfunctors of X, we define their *intersection* $\bigcap_{i\in I} Y_i$ through $(\bigcap_{i \in I} Y_i)(A) = \bigcap_{i \in I} Y_i(A)$ for each k–algebra A. This is again a subfunctor. The obvious definition of a union is not the useful one, so we shall not denote it by $\bigcup_{i \in I} Y_i$.

For any two k–functors X, X', we denote by $\mathrm{Mor}(X, X')$ the set of all morphisms (i.e., natural transformations) from X to X'. For any $f \in \mathrm{Mor}(X, X')$ and any subfunctor Y' of X', we define the *inverse image* $f^{-1}(Y')$ of Y' under f through $f^{-1}(Y')(A) = f(A)^{-1}(Y'(A))$ for each k–algebra A. Clearly $f^{-1}(Y')$ is a subfunctor of X. (The obvious definition of an image of a subfunctor is not the useful one.) Obviously, f^{-1} commutes with intersections.

For two k–functors X_1, X_2, the *direct product* $X_1 \times X_2$ is defined through $(X_1 \times X_2)(A) = X_1(A) \times X_2(A)$ for all A. The projections $p_i : X_1 \times X_2 \to X_i$ are morphisms and $(X_1 \times X_2, p_1, p_2)$ has the usual universal property of a direct product.

For three k–functors X_1, X_2, S and two morphisms $f_1 : X_1 \to S$, $f_2 : X_2 \to S$, the *fibre product* $X_1 \times_S X_2$ (relative to f_1, f_2) is defined through

$$
\begin{aligned}
(X_1 \times_S X_2)(A) &= X_1(A) \times_{S(A)} X_2(A) \\
&= \{\, (x_1, x_2) \in X_1(A) \times X_2(A) \mid f_1(A)(x_1) = f_2(A)(x_2) \,\}.
\end{aligned}
$$

The projections from $X_1 \times_S X_2$ to X_1 and X_2 are morphisms and $X_1 \times_S X_2$, together with these projections, has the usual universal property of a fibre product. Of course, we may also regard $X_1 \times_S X_2$ as the inverse image of the *diagonal* subfunctor $D_S \subset S \times S$ (with $D_S(A) = \{(s, s) \mid s \in S(A)\}$ for all A) under the (obvious) morphism $(f_1, f_2) : X_1 \times X_2 \to S \times S$. (On the other hand, inverse images and intersections can also be regarded as special cases of fibre products.)

1.3. (Affine Schemes) For any $n \in \mathbf{N}$ the functor $\mathbf{A}^n$ with $\mathbf{A}^n(A) = A^n$ for all A and $\mathbf{A}^n(\varphi) = \varphi^n : (a_1, a_2, \dots, a_n) \mapsto (\varphi(a_1), \varphi(a_2), \dots, \varphi(a_n))$ for all $\varphi : A \to A'$ is called the *affine n–space* over k. (We also sometimes use the notation $\mathbf{A}_k^n$ when it may be doubtful which k we consider.) Note that $\mathbf{A}^0$ is the functor with $\mathbf{A}^0(A) = \{0\}$ for all A. Hence there is for each k–functor X exactly one morphism from X to $\mathbf{A}^0$ (i.e., $\mathbf{A}^0$ is a final object in the category of k functors), and we can regard any direct product as a fibre product over $\mathbf{A}^0$.

For any k–algebra R, we can define a k–functor $Sp_k R$ through $(Sp_k R)(A) = \mathrm{Hom}_{k\text{–alg}}(R, A)$ for all A and

$$
(Sp_k R)(\varphi) : \mathrm{Hom}_{k\text{–alg}}(R, A) \to \mathrm{Hom}_{k\text{–alg}}(R, A'), \qquad \alpha \mapsto \varphi \circ \varphi
$$

for all homomorphisms $\varphi : A \to A'$. We call $Sp_k R$ the *spectrum* of R. Any k–functor isomorphic to some $Sp_k R$ is called an *affine scheme* over k. (Note that the $Sp_k R$ generalise the functors $\mathcal{X}$ considered in 1.1.) For example, the affine n–space $\mathbf{A}^n$ is isomorphic to $Sp_k k[T_1, T_2, \dots, T_n]$ (and will usually be identified with it), where $k[T_1, T_2, \dots, T_n]$ is the polynomial ring over k in n variables $T_1, T_2, \dots, T_n$.

We can recover R from $Sp_k R$. This follows from:

Yoneda's Lemma: *For any k–algebra R and any k–functor X, the map $f \mapsto f(R)(\mathrm{id}_R)$ is a bijection* $\mathrm{Mor}(Sp_k R, X) \xrightarrow{\sim} X(R)$.

Indeed, take any k–algebra A and any $\alpha \in \mathrm{Hom}_{k\text{–alg}}(R, A) = (Sp_k R)(A)$. As f is a natural transformation, we have $X(\alpha) \circ f(R) = f(A) \circ (Sp_k R)(\alpha)$. Let us abbreviate $x_f = f(R)(\mathrm{id}_R)$. As $(Sp_k R)(\alpha)(\mathrm{id}_R) = \alpha \circ \mathrm{id}_R = \alpha$, we get

$$
f(A)(\alpha) = X(\alpha)(x_f). \tag{1}
$$

This shows that f is uniquely determined by x_f and indicates how to construct an inverse map. For each $x \in X(R)$ and any k–algebra A, let $f_x(A) : (Sp_k R)(A) \to X(A)$ be the map $\alpha \mapsto X(\alpha)(x)$. Then one may check that $f_x \in \mathrm{Mor}(Sp_k R, X)$ and that $x \mapsto f_x$ is inverse to $f \mapsto x_f$.

An immediate consequence of Yoneda's lemma is

$$\mathrm{Mor}(Sp_k R, Sp_k R') \xrightarrow{\sim} \mathrm{Hom}_{k-\mathrm{alg}}(R', R) \tag{2}$$

for all k–algebras R, R'. We denote this bijection by $f \mapsto f^*$ and call f^* the *comorphism* corresponding to f. As $\mathrm{Hom}_{k-\mathrm{alg}}(k[T_1], R) \xrightarrow{\sim} R$ under $\alpha \mapsto \alpha(T_1)$ we get especially

$$\mathrm{Mor}(Sp_k R, \mathbf{A}^1) \xrightarrow{\sim} R. \tag{3}$$

For any k–functor X, we denote $\mathrm{Mor}(X, \mathbf{A}^1)$ by $k[X]$. This set has a natural structure as a k–algebra and (3) is an isomorphism $k[Sp_k R] \xrightarrow{\sim} R$ of k–algebras. (For $f_1, f_2 \in k[X]$, define $f_1 + f_2$ through $(f_1 + f_2)(A)(x) = f_1(A)(x) + f_2(A)(x)$ for all A and all $x \in X(A)$. Similarly, $f_1 f_2$ and $a f_1$ for $a \in k$ are defined.) We shall usually write $f(x) = f(A)(x)$ for $x \in X(A)$ and $f \in k[X]$. Note that for $X = Sp_k R$ and $f \in R \simeq k[X]$ we have $f(x) = x(f)$ for $f \in (Sp_k R)(A) = \mathrm{Hom}_{k-\mathrm{alg}}(R, A)$.

The universal property of the tensor product implies immediately that a direct product $X_1 \times X_2$ of affine schemes over k is again an affine scheme over k with $k[X_1 \times X_2] \simeq k[X_1] \otimes k[X_2]$. More generally, a fibre product $X_1 \times_S X_2$ with X_1, X_2, S affine schemes over k is an affine scheme over k with

$$k[X_1 \times_S X_2] \simeq k[X_1] \otimes_{k[S]} k[X_2]. \tag{4}$$

1.4. (Closed Subfunctors of Affine Schemes) Let X be an affine scheme over k.

For any subset $I \subset k[X]$, we define a subfunctor $V(I)$ of X through

$$\begin{aligned} V(I)(A) &= \{\, x \in X(A) \mid f(x) = 0 \text{ for all } f \in I \,\} \\ &\simeq \{\, \alpha \in \mathrm{Hom}_{k-\mathrm{alg}}(k[X], A) \mid \alpha(I) = 0 \,\} \end{aligned} \tag{1}$$

for all A. (One can check immediately that this is indeed a subfunctor, i.e., that $X(\varphi)\, V(I)(A) \subset V(I)(A')$ for any homomorphism $\varphi : A \to A'$.)

Of course, $V(I)$ depends only on the ideal generated by I in $k[X]$. We claim:

(2) *The map $I \mapsto V(I)$ from $\{$ideals in $k[X]\}$ to $\{$subfunctors of $X\}$ is injective.*

More precisely, we claim that

$$I = \{\, f \in k[X] \mid f(V(I)(A)) = 0 \text{ for all } A \,\}.$$

Here the inclusion "$\subset$" holds by the definition of $V(I)$. On the other hand, consider the canonical map $\alpha : k[X] \to k[X]/I$, which we regard as an element of $V(I)(k[X]/I)$. Now $f(V(I)(A)) = 0$ for all A yields $0 = f(\alpha) = \alpha(f)$, hence $f \in I$.

Of course, this more precise statement implies immediately for all ideals I, I' of $k[X]$:

$$I \subset I' \iff V(I) \supset V(I'). \tag{3}$$

We call a subfunctor Y of X *closed* if it is of the form $Y = V(I)$ for some ideal $I \subset k[X]$. Obviously, any closed subfunctor is again an affine scheme over k as

$$V(I) \simeq Sp_k(k[X]/I). \tag{4}$$

In particular X itself and the "empty subfunctor" (mapping each $A \neq (0)$ to $\emptyset$) are closed subfunctors: Take $I = \{0\}$ and $I = k[X]$ respectively.

Each $x \in X(k)$ defines a closed subfunctor of X as follows: For each k–algebra A let $i_A : k \to A$ denote the homomorphism that makes A into a k–algebra. Set $x_A = X(i_A)x$. Then $A \mapsto \{x_A\}$ is a closed subfunctor of X isomorphic to $\mathbf{A}^0$. (If $X = Sp_k R$, then x corresponds to an algebra homomorphism $f_x : R \to k$ and our subfunctor is equal to $V(\ker f_x)$.)

For any family $(I_j)_{j\in J}$ of ideals in $k[X]$, one checks easily

$$\bigcap_{j\in J} V(I_j) = V(\sum_{j\in J} I_j). \tag{5}$$

Thus the intersection of closed subfunctors is closed again.

For each subfunctor Y of X, there is a smallest closed subfunctor $\overline{Y}$ of X with $Y(A) \subset \overline{Y}(A)$ for all A. (Take the intersection of all closed subfunctors with the last property.) This subfunctor $\overline{Y}$ is called the *closure* of Y. We really do not have to assume here that Y is a subfunctor: Any map Y will do that associates to each A a subset $Y(A) \subset X(A)$. We can, for example, fix an A and consider a subset $M \subset X(A)$. Then the closure $\overline{M}$ of M is the smallest closed subfunctor of X with $M \subset \overline{M}(A)$.

Let I_1, I_2 be ideals in $k[X]$. Because of (3), the closure of the subfunctor $A \mapsto V(I_1)(A) \cup V(I_2)(A)$ is equal to $V(I_1 \cap I_2)$. If A is an integral domain, then one checks easily that $V(I_1)(A) \cup V(I_2)(A) = V(I_1 \cap I_2)(A)$. For arbitrary A, this equality can be false. Still, we *define* the union as $V(I_1) \cup V(I_2) = V(I_1 \cap I_2)$.

Let $f : X' \to X$ be a morphism of affine schemes over k. One checks easily for any ideal I of $k[X]$ that

$$f^{-1}V(I) = V(k[X']\, f^*(I)). \tag{6}$$

Thus the inverse image of a closed subfunctor is again a closed subfunctor. For any ideal $I' \subset k[X']$, the closure of the subfunctor $A \mapsto f(A)\,(V(I')(A))$ is $V((f^*)^{-1}I')$. This functor is also denoted as $\overline{f(V(I'))}$, but we do not want to define $f(V(I'))$ here.

For two affine schemes X_1, X_2 over k and ideals $I_1 \subset k[X_1]$, $I_2 \subset k[X_2]$, one checks easily

$$V(I_1) \times V(I_2) \simeq V(I_1 \otimes k[X_2] + k[X_1] \otimes I_2). \tag{7}$$

If S is another affine scheme and if morphisms $X_1 \to S$, $X_2 \to S$ are fixed, then one gets

$$V(I_1) \times_S V(I_2) \simeq V(I_1 \otimes_{k[S]} k[X_2] + k[X_1] \otimes_{k[S]} I_2). \tag{8}$$

(Use, e.g., that $V(I_1) \times_S V(I_2) = p_1^{-1}V(I_1) \cap p_2^{-1}V(I_2)$ together with (5), (6), where $p_i : X_1 \times_S X_2 \to X_i$ for $i = 1, 2$ are the canonical projections.)

1.5. (Open Subfunctors of Affine Schemes) Let X be an affine scheme over k.

A subfunctor Y of X is called *open* if there is a subset $I \subset k[X]$ with $Y = D(I)$ where we set for all k–algebras A:

$$D(I)(A) = \{\, x \in X(A) \mid \sum_{f\in I} Af(x) = A \,\} \tag{1}$$
$$= \{\, \alpha \in \mathrm{Hom}_{k-\mathrm{alg}}(k[X], A) \mid A\alpha(I) = A \,\}.$$

Note that (1) defines for each subset I a subfunctor: For each $\varphi \in \mathrm{Hom}_{k-\mathrm{alg}}(A, A')$ and each $x \in D(I)(A)$ one has

$$\sum_{f\in I} A' f(X(\varphi)x) = \sum_{f\in I} A'\varphi(f(x)) = A'\varphi(\sum_{f\in I} Af(x)) = A'\varphi(A) = A',$$

hence $X(\varphi)x \in D(I)(A')$. One has clearly:

(2) *If A is a field, then* $D(I)(A) = \bigcup_{f\in I}\{x \in X(A) \mid f(x) \neq 0\}$.

Of course, the right hand side in (2) would be the obvious choice for something open. But it does not define a subfunctor, as homomorphisms between k–algebras are not injective in general. Therefore we have to take (1) as the appropriate generalisation to rings.

Note that $D(k[X]) = X$ and that $D(\{0\})$ is the empty subfunctor of X. For I of the form $I = \{f\}$ for some $f \in k[X]$, one writes $X_f = D(f) = D(\{f\})$ and gets

$$X_f(A) = \{\, \alpha \in \mathrm{Hom}_{k-\mathrm{alg}}(k[X], A) \mid \alpha(f) \in A^\times \,\} \tag{3}$$

(where $A^\times$ denotes the group of units in A), hence

$$X_f \simeq Sp_k(k[X]_f) \tag{4}$$

where $k[X]_f = k[X][f^{-1}]$ is the localisation of $k[X]$ at f. So the open subfunctors of the form X_f are again affine schemes. For arbitrary I, however, $D(I)$ may be no longer an affine scheme.

Obviously, $D(I)$ depends only on the ideal of $k[X]$ generated by I. As any proper ideal in any ring is contained in a maximal ideal, we have for any A

$$D(I)(A) = \{\, \alpha \in \mathrm{Hom}_{k-\mathrm{alg}}(k[X], A) \mid \alpha(I) \not\subset \mathfrak{m} \text{ for all } \mathfrak{m} \in \mathrm{Max}(A) \,\}$$
$$= \{\, \alpha \in \mathrm{Hom}_{k-\mathrm{alg}}(k[X], A) \mid \alpha_{\mathfrak{m}} \in D(I)(A/\mathfrak{m}) \text{ for all } \mathfrak{m} \in \mathrm{Max}(A) \,\}$$

where $\mathrm{Max}(A)$ is the set of all maximal ideals of A and $\alpha_{\mathfrak{m}}$ is the composed map $k[X] \xrightarrow{\alpha} A \xrightarrow{\mathrm{can}} A/\mathfrak{m}$. This shows that $D(I)$ is uniquely determined by its values over fields and especially that $D(I) = D(\sqrt{I})$ for any ideal $I \subset k[X]$.

Denote for each prime ideal $P \subset k[X]$ the fraction field of $k[X]/P$ by Q_P and the canonical homomorphism $k[X] \to k[X]/P \to Q_P$ by α_P. Then

$$\alpha_P \notin D(I)(Q_P) \iff \alpha_P(I) = 0 \iff P \supset I.$$

As $\sqrt{I}$ is the intersection of all prime ideals $P \supset I$ of $k[X]$, we get for any two ideals I, I' of $k[X]$

$$D(I) \subset D(I') \iff \sqrt{I} \subset \sqrt{I'}. \tag{5}$$

Thus $I \mapsto D(I)$ is a bijection $\{$ ideals I of $k[X]$ with $I = \sqrt{I}\,\} \to \{$ open subfunctors of $X\,\}$.

For two ideals I, I' in $k[X]$, one checks easily

$$D(I) \cap D(I') = D(I \cap I') = D(I \cdot I') \tag{6}$$

and one gets for two elements $f, f' \in k[X]$

$$X_f \cap X_{f'} = X_{ff'}. \tag{7}$$

For any ideal I in $k[X]$ one has:

(8) *If A is a field, then $X(A)$ is the disjoint union of $D(I)(A)$ and $V(I)(A)$.*

For arbitrary A, the union may be smaller. Also, the next statement may be false for arbitrary A: Consider a family $(I_j)_{j\in J}$ of ideals in $k[X]$. Then obviously:

(9) *If A is a field, then $\bigcup_{j\in J} D(I_j)(A) = D(\sum_{j\in J} I_j)(A)$.*

For a morphism $f : X' \to X$ of affine schemes over k, one has

$$f^{-1}D(I) = D(k[X']f^*(I)) \tag{10}$$

for any ideal $I \subset k[X]$ as one may check easily. We get especially for any $f' \in k[X]$

$$f^{-1}(X_{f'}) = X'_{f^*(f')}. \tag{11}$$

For any fibre product $X_1 \times_S X_2$ of affine scheme over k (with respect to suitable morphisms) and any ideals $I_1 \subset k[X_1]$, $I_2 \subset k[X_2]$, one has

$$D(I_1) \times_S D(I_2) = D(I_1 \otimes_{k[S]} I_2) \tag{12}$$

(Argue as for 1.4(8).)

1.6. (Affine Varieties and Affine Schemes) An affine scheme X over k is called *algebraic* if $k[X]$ is isomorphic to a k–algebra of the form $k[T_1, T_2, \dots, T_n]/I$ for some $n \in \mathbf{N}$ and some finitely generated ideal I in the polynomial ring $k[T_1, T_2, \dots, T_n]$. It is called *reduced* if $k[X]$ does not contain any nilpotent element other than 0.

Assume until the end of 1.6 that k is an algebraically closed field. Any affine variety X over k defines as in 1.1 a k–functor $\mathcal{X}$ which we may identify with $Sp_k k[X]$. One gets in this way exactly all reduced algebraic affine schemes over k. For two affine varieties X, X', one has $\mathrm{Mor}(X, X') \simeq \mathrm{Hom}_{k-\mathrm{alg}}(k[X'], k[X]) \simeq \mathrm{Mor}(\mathcal{X}, \mathcal{X}')$. So we have indeed embedded the category of affine varieties as a full subcategory into the category of affine schemes.

When doing this, one has to be aware of several points. Any closed subset Y of an affine variety X is itself an affine variety. The corresponding functor $\mathcal{Y}$ is the closed subfunctor $V(I) \subset \mathcal{X}$, where $I = \{f \in k[X] \mid f(Y) = 0\}$. In this way one gets an embedding $\{$ closed subsets of $X\,\} \to \{$ closed subfunctors of $\mathcal{X}\,\}$. On the level of ideals (cf. 1.4(2)), it corresponds to the inclusion $\{$ ideals I of $k[X]$ with $I = \sqrt{I}\,\} \subset \{$ ideals of $k[X]\,\}$. The embedding is certainly compatible with inclusions (i.e., $Y \subset Y' \iff \mathcal{Y} \subset \mathcal{Y}'$), but in general not with intersections: It may happen that $\mathcal{Y} \cap \mathcal{Y}'$ is strictly larger than the functor corresponding to $Y \cap Y'$.

Take for example in $X = k^2$ (where $k[X] = k[T_1, T_2]$) the line $Y = \{(a, 0) \mid a \in k\}$ and the parabola $Y' = \{(a, a^2) \mid a \in k\}$. Then $Y \cap Y' = \{(0, 0)\}$. The ideals I, I' of Y, Y' are $I = (T_2)$ and $I' = (T_1^2 - T_2)$; it follows that $I + I' = (T_1^2, T_2) \neq (T_1, T_2)$ and $\mathcal{Y} \cap \mathcal{Y}' = V(I) \cap V(I') = V(I + I')$ differs from the subfunctor $V(T_1, T_2)$ corresponding to $Y \cap Y'$.

So, when regarding affine varieties as (special) affine schemes, we have to be careful, whether intersections are taken as varieties or as schemes. The same is true for inverse images and (more generally) for fibre products.

Similar problems do not arise with open subsets. To any open $Y \subset X$ we can associate the ideal $I = \{f \in k[X] \mid f(X \setminus Y) = 0\}$ and then the open subfunctor $\mathcal{Y} = D(I)$. Because of 1.5(5), the map $Y \mapsto \mathcal{Y}$ is a bijection from $\{$ open subsets of $X\}$ to $\{$ open subfunctors of $\mathcal{X}\}$ that is compatible with intersections. It follows from 1.5(10), (12) that this bijection is also compatible with inverse images and fibre products. (In case Y is affine, the notation $\mathcal{Y}$ is compatible with the earlier one.)

1.7. (Open Subfunctors) Let k again be arbitrary. We shall define arbitrary schemes over k in 1.9 as "local" k–functors with "open coverings" by affine schemes. We first have to explain these terms and begin with the open coverings.

Let X be a k–functor. A subfunctor $Y \subset X$ is called *open* if for any affine scheme X' over k and any morphism $f : X' \to X$ there is an ideal $I \subset k[X']$ with $f^{-1}(Y) = D(I)$.

Note that this definition is compatible with the one at the beginning of 1.5 because of 1.5(10). From 1.5(6) one gets:

(1) *If Y, Y' are open subfunctors of X, then so is $Y \cap Y'$.*

Let $f : X' \to X$ be a morphism of k–functors. Then one has, obviously:

(2) *If Y is an open subfunctor of X, then $f^{-1}(Y)$ is an open subfunctor of X'.*

Applied to an inclusion $Z \hookrightarrow X$ this yields: Let Y, Z be subfunctors of a k–functor X. If Y is open in X, then $Y \cap Z$ is open in Z.

Let X_1,X_2, S be k–functors and suppose that $X_1 \times_S X_2$ is defined with respect to some morphisms. Then one gets (using $Y_1 \times_S Y_2 = p_1^{-1}(Y_1) \cap p_2^{-1}(Y_2)$):

(3) *If $Y_1 \subset X_1$ and $Y_2 \subset X_2$ are open subfunctors, then $Y_1 \times_S Y_2$ is an open subfunctor of $X_1 \times_S X_2$.*

Let Y, Y' be open subfunctors of X. Then:

(4) $Y = Y' \iff Y(A) = Y'(A)$ *for each k–algebra A that is a field.*

(Of course "$\Rightarrow$" is trivial. In order to show "$\Leftarrow$", suppose that $Y \neq Y'$. Then there is some k–algebra A with $Y(A) \neq Y'(A)$. Assume that there exists $x \in Y(A)$ with $x \notin Y'(A)$. Via $Y(A) \simeq \mathrm{Mor}(Sp_k A, Y) \subset \mathrm{Mor}(Sp_k A, X)$, regard x as a morphism $Sp_k A \to X$. Then $\mathrm{id}_A \in x^{-1}(Y)(A)$, $\notin x^{-1}(Y')(A)$, hence $x^{-1}(Y) \neq x^{-1}(Y')$. Now the result follows from the discussion preceding 1.5(5).)

A family $(Y_j)_{j \in J}$ of open subfunctors of X is called an *open covering* of X, if $X(A) = \bigcup_{j \in J} Y_j(A)$ for each k–algebra A *which is a field.*

If X is affine and if $Y_j = D(I_j)$ for some ideal I_j for some ideal $I_j \subset k[X]$, then 1.5(9) implies that the $D(I_j)$ form an open covering of X if and only if $k[X] = \sum_{j \in J} I_j$. A comparison with the case of an affine variety shows that the present definition of open coverings is the appropriate generalisation of the definition for varieties. Note that especially:

(5) *Let X be affine and $f_1, f_2, \ldots, f_r \in k[X]$. Then the X_{f_i} form an open covering of X if and only if $k[X] = \sum_{i=1}^r k[X]f_i$.*

Let $Y \subset X$ be an open subfunctor, and let $\varphi : A \to A'$ be a homomorphism of k–algebras. Then:

(6) *If A' is a faithfully flat A–module via φ, then $Y(A) = X(\varphi)^{-1}Y(A')$.*

We have to prove only "$\supset$". Suppose at first that X is affine. Then $Y = D(I)$ for some ideal $I \subset k[X]$. Consider some $\alpha \in X(A) = \mathrm{Hom}_{k-\mathrm{alg}}(k[X], A)$ with $\varphi \circ \alpha = X(\varphi)(\alpha) \in Y(A')$, i.e., with $A' = A'\varphi(\alpha(I))$. The isomorphism $A \otimes_A A' \xrightarrow{\sim} A'$, $a \otimes a' \mapsto \varphi(a)a'$ induces an isomorphism $A\alpha(I) \otimes_A A' \xrightarrow{\sim} A'\varphi(\alpha(I))$. Therefore $A' = A'\varphi(\alpha(I))$ together with the flatness of A' implies $(A/A\alpha(I)) \otimes_A A' = 0$, hence $A = A\alpha(I)$ by the faithful flatness.

For arbitrary X, we regard $x \in X(A)$ as a morphism $x : Sp_k A \to X$ with $x(A)(\mathrm{id}_A) = x$, hence with $X(\varphi)x = x(A')Sp_k(\varphi)(\mathrm{id}_A)$. So, if $x \in X(\varphi)^{-1}Y(A')$, then $\mathrm{id}_A \in Sp_k(\varphi)^{-1}(x^{-1}(Y)(A'))$, hence $\mathrm{id}_A \in x^{-1}(Y)(A)$, as $x^{-1}(Y)$ is an open subfunctor of the affine scheme $Sp_k A$. Now we get $x = x(A)(\mathrm{id}_A) \in Y(A)$ as desired.

Of course, (6) implies that we can restrict to algebraically closed fields in (4). Also, a family $(Y_j)_{j\in J}$ of open subfunctors of X is an open covering of X if and only if $X(A) = \bigcup_{j\in J} Y_j(A)$ for all k–algebras A that are *algebraically closed* fields.

1.8. (Local Functors) Consider two k–functors X, Y and an open covering $(Y_j)_{j\in J}$ of Y. If X, Y correspond to geometric objects, then a morphism $f : Y \to X$ ought to be determined by its restrictions $f_{|Y_j}$ to all Y_j. Furthermore, given for each j a morphism $f_j : Y_j \to X$ such that $f_{j\,|Y_j\cap Y_{j'}} = f_{j'\,|Y_j \cap Y_{j'}}$ for all $j, j' \in J$, then there ought to be a (unique) morphism $f : X \to Y$ with $f_{|Y_j} = f_j$ for all j. In other words, the sequence

$$\mathrm{Mor}(Y, X) \longrightarrow \prod_{j\in J} \mathrm{Mor}(Y_j, X) \rightrightarrows \prod_{j,j'\in J} \mathrm{Mor}(Y_j \cap Y_{j'}, X) \tag{1}$$

ought to be exact where the map on the left takes any $f \in \mathrm{Mor}(Y, X)$ to the family of all $f_{|Y_j}$ and the upper (resp. lower) map on the right takes a family $(f_j)_{j\in J}$ to the family with (j, j')–component $f_{j\,|Y_j\cap Y_{j'}}$ (resp. $f_{j'\,|Y_j\cap Y_{j'}}$).

For arbitrary X, Y, (Y_j), the sequence (1) will not be exact. So we define a k–functor X to be *local* if the sequence (1) is exact for all k–functors Y and all open coverings $(Y_j)_{j\in J}$ of Y. (One can express this as saying that the functor $\mathrm{Mor}(\,?\,, X)$ is a sheaf in a suitable sense.)

For any k–algebra R and any $f_1, f_2, \ldots, f_r \in R$ with $\sum_{i=1}^r Rf_i = R$, the $Y_i = Sp_k(R_{f_i})$ form an open covering of the affine scheme $Y = Sp_k R$. In this case the sequence (1) takes (because of Yoneda's lemma) the form

$$X(R) \longrightarrow \prod_{i=1}^{r} X(R_{f_i}) \rightrightarrows \prod_{1\le i,j\le r} X(R_{f_i f_j}) \tag{2}$$

where the maps have components of the form $X(\alpha)$ with α one of the canonical maps $R \to R_{f_i}$ or $R_{f_i} \to R_{f_i f_j}$. Now one can prove (cf. [DG], I, §1, 4.13):

Proposition: *A k–functor X is local if and only if for any k–algebra R and any $f_1, f_2, \ldots, f_r \in R$ with $\sum_{i=1}^r Rf_i = R$ the sequence* (2) *is exact.*

(Note that in [DG] the second property is taken as the definition of local.)

For R and $f_1, f_2, \ldots, f_r$ as in (2) the sequence

$$(3) \qquad R \longrightarrow \prod_{i=1}^{r} R_{f_i} \rightrightarrows \prod_{1\le i,j\le r} R_{f_i f_j}$$

(induced by the natural maps $R \to R_{f_i}$ and $R_{f_i} \to R_{f_i f_j}$) is exact. (This is really the description of the structure sheaf on Spec R, e.g., in [Ha], II, 2.2.) For an affine scheme X over k the exactness property of $\mathrm{Hom}_{k-\mathrm{alg}}(k[X], ?) = X(?)$ shows that the exactness of (3) implies the exactness of (2). Thus we get:

(4) *Any affine scheme over k is a local k–functor.*

We can apply this to the affine scheme $\mathbf{A}^1$ and get for any open covering $(Y_j)_{j\in J}$ of a k–functor Y an exact sequence

$$k[Y] \longrightarrow \prod_{j\in J} k[Y_j] \rightrightarrows \prod_{j,j'\in J} k[Y_j \cap Y_{j'}].$$

Consider k–algebras $A_1, A_2, \ldots, A_n$ and the projections $p_j : \prod_{i=1}^n A_i \to A_j$. If we apply (2) to $R = \prod_{i=1}^n A_i$ and to $f_i = (0, \ldots, 0, 1, 0, \ldots, 0)$, then we get:

(5) *If X is a local functor, then $X(\prod_{i=1}^n A_i) \xrightarrow{\sim} \prod_{i=1}^n X(A_i)$ for all k–algebras $A_1, A_2, \ldots, A_n$.*

(The bijection maps any x to $(X(p_i)x)_{1\le i\le n}$.)

1.9. (Schemes) A k–functor is called a *scheme* (over k) if it is local and if it admits an open covering by affine schemes. Obviously, 1.8(4) implies:

(1) *Any affine scheme over k is a scheme over k.*

The category of schemes over k (a full subcategory of { k–functors }) is closed under important operations:

(2) *If X is a local k–functor (resp. a scheme over k) and if X' is an open subfunctor of X, then X' is local (resp. a scheme).*

In the situation of 1.8(1) the injectivity of the map on the left for X implies its injectivity for X'. In order to show the exactness for X', one has to show for any $f \in \mathrm{Mor}(Y, X)$ such that each $f_{|Y_j}$ factors through X', that also f factors through X'. The assumption implies that $Y_j \subset f^{-1}(X')$ for each j, hence by the definition of an open covering $f^{-1}(X')(A) = Y(A)$ for each k–algebra A that is a field. Then 1.7(4) implies $Y = f^{-1}(X')$ and f factors through X'. In order to get the affine covering of X' in case X is a scheme, one can restrict to the case where X is affine, hence $X' = D(I)$ for some ideal I in $k[X]$. Then the $(X_f)_{f\in I}$ form an open affine covering.

Let X_1, X_2, S be k–functors and form $X_1 \times_S X_2$ with respect to suitable morphisms. Then:

(3) *If X_1, X_2, S are local (resp. schemes), then so is $X_1 \times_S X_2$.*

The proof is left to the reader.

The most important non-affine schemes are the projective spaces and, more generally, the *Grassmann schemes* $\mathcal{G}_{r,n}$ for each $r, n \in \mathbf{N}$. For any k–algebra A, one sets $\mathcal{G}_{r,n}(A)$ equal to the set of direct summands of the A–module A^{r+n} having rank r. (Then $\mathbf{P}^n = \mathcal{G}_{1,n}$ is the *projective n–space*.) In [DG], I.1.3.9/13, there is a proof that all $\mathcal{G}_{r,n}$ are schemes.

1.10. (Base Change) Let k' be a k–algebra. Any k'–algebra A is in a natural way a k–algebra, just by combining the structural homomorphisms $k \to k'$ and $k' \to A$. We can therefore associate to each k–functor X a k'–functor $X_{k'}$ by $X_{k'}(A) = X(A)$ for any k'–algebra A. For any morphism $f : X \to X'$ of k–functors, we get a morphism $f_{k'} : X_{k'} \to X'_{k'}$ of k'–functors simply by $f_{k'}(A) = f(A)$ for any k'–algebra A. In this way we get a functor $X \mapsto X_{k'}$, $f \mapsto f_{k'}$ from $\{ k\text{–functors} \}$ to $\{ k'\text{–functors} \}$, which we call *base change* from k to k'.

For any subfunctor Y of a k–functor X, the k'–functor $Y_{k'}$ is a subfunctor of $X_{k'}$. Furthermore, the base change commutes with taking inverse images under morphisms, with taking intersections of subfunctors, and with forming fibre products.

The universal property of the tensor product implies $(Sp_k R)_{k'} \simeq Sp_{k'}(R \otimes k')$ for any k–algebra R. In other words, if X is an affine scheme over k, then $X_{k'}$ is an affine scheme over k' with $k'[X_{k'}] \simeq k[X] \otimes k'$. For any ideal I of $k[X]$, one gets then $V(I)_{k'} = V(I \otimes k')$ and $D(I)_{k'} = D(I \otimes k')$. (We really ought to replace $I \otimes k'$ in these formulas by its canonical image in $k[X] \otimes k'$, but for once we indulge in some abuse of notation.)

For any k'–algebras A, R one has

$$(Sp_{k'} R)(A) = \mathrm{Hom}_{k'\text{–alg}}(R, A) \subset \mathrm{Hom}_{k\text{–alg}}(R, A) = (Sp_k R)_{k'}(A).$$

Thus we have embedded $Sp_{k'} R$ as a subfunctor into $(Sp_k R)_{k'}$. For any ideal I of R, denote the corresponding subfunctors as in 1.4/5 by $V(I), D(I) \subset Sp_k R$ and $V_{k'}(I), D_{k'}(I) \subset Sp_{k'} R$. Then one sees immediately $D_{k'}(I) = (Sp_{k'} R) \cap D(I)_{k'}$ and $V_{k'}(I) = (Sp_{k'} R) \cap V(I)_{k'}$.

Using the last results, it is easy to prove for any open subfunctor Y of a k–functor X that $Y_{k'}$ is open in $X_{k'}$. (We have to show for any k'–algebra R and any morphism $f : Sp_{k'} R \to X_{k'}$ that $f^{-1}(Y_{k'})$ is open in $Sp_{k'} R$. Via Yoneda's lemma we can regard f as an element in $X_{k'}(R)$. Since $X_{k'}(R) = X(R)$, there corresponds to f also a morphism $f' : Sp_k R \to X$. One checks that $(f')_{k'} : (Sp_k R)_{k'} \to X_{k'}$ restricts to f on $Sp_{k'} R \subset (Sp_k R)_{k'}$. There is an ideal $I \subset R$ with $(f')^{-1}(Y') = D(I)$. One gets now that $f^{-1}(Y_{k'}) = D_{k'}(I)$.)

If X is a local functor, then obviously $X_{k'}$ is a local functor. Now it is easy to show that $X_{k'}$ is a scheme over k' if X is one over k.

Let k_1 be a subring of k. A k–functor *defined over* k_1 is k–functor X together with a fixed k_1–functor X_1 such that $X = (X_1)_k$.

1.11. ("Schemes") In text books on algebraic geometry (like that by Hartshorne to which I shall usually refer in such matters) another notion of scheme is introduced that I shall denote by "schemes" in case a distinction is useful. A "scheme" over k

is a topological space together with a sheaf of k–algebras and an open covering by "affine schemes" over k. An "affine scheme" is the prime spectrum Spec(R) of some k–algebra R endowed with the Zariski topology and with a sheaf having sections R_f on $\mathrm{Spec}(R_f) \subset \mathrm{Spec}(R)$ for each $f \in R$.

To each "scheme" X' one associates a k–functor $\mathcal{X}'$ via $\mathcal{X}'(A) = \mathrm{Mor}(\mathrm{Spec}\, A, X')$ for all A. On the other hand, one can associate in a functorial way to each k–functor X a topological space $|X|$ together with a sheaf of k–algebras such that $|Sp_k\, R| = \mathrm{Spec}\,(R)$ for each k–algebra R. It turns out that $|X|$ is a "scheme" if and only if X is a scheme and that $X \mapsto |X|$ and $X' \mapsto \mathcal{X}'$ are quasi-inverse equivalences of categories. (This is the content of the comparison theorem [DG], I, §1, 4.4.)

One property of this construction is that the open subfunctors of any k–functor X correspond bijectively to the open subsets of $|X|$, cf. [DG], I, §1, 4.12. More precisely, if Y is an open subfunctor of X, then $|Y|$ can be identified with an open subset of $|X|$, and the k–algebra of sections in $|Y|$ of the structure sheaf of $|X|$ is isomorphic to $\mathrm{Mor}(Y, \mathbf{A}^1)$, see ibid. 4.14/15.

Suppose that k is an algebraically closed field. Consider a scheme X over k that has an open covering by algebraic affine schemes. We can define on $X(k)$ a topology such that the open subsets are the $Y(k)$ for open subfunctors $Y \subset X$. The map $Y \mapsto Y(k)$ turns out to be injective ([DG], I, §3, 6.8). We can define a sheaf $\mathcal{O}_{X(k)}$ on $X(k)$ through $\mathcal{O}_{X(k)}(Y(k)) = \mathrm{Mor}(Y, \mathbf{A}^1)$. Then $X \mapsto (X(k), \mathcal{O}_{X(k)})$ is a faithful functor and its image contains all varieties over k in the usual sense.

There are some fundamental notions of algebraic geometry (like smoothness and dimension) that we have to consider only in a few places. The necessary definitions and the main properties from the point of view of k–functors are contained in [DG]. I do not want to repeat what is done there in order to keep the length of this book down. Any reader who is familiar with those notions in the context of "schemes" (e.g., from [Ha]) can use the correspondence of X and $|X|$ as above to translate. For example, a scheme X is smooth if and only if $|X|$ is so.

1.12. (Closed Subfunctors) A subfunctor Y of a k–functor X is called *closed* if and only if for each affine functor X' and any morphism $f : X' \to X$ of k–functors the subfunctor $f^{-1}(Y)$ of X' is closed in the old sense (as in 1.4). Because of 1.4(6), this is compatible with the old definition in case X is affine. (The definition in [DG], I, §2, 4.1, is equivalent to the version here by the final statement in that subsection.)

The following statements are clear from 1.4(5) or by definition:

(1) *If $(Y_i)_{i\in I}$ is a family of closed subfunctors of a k–functor X, then $\bigcap_{i\in I} Y_i$ is closed in X.*

(2) *Let $f : X \to X'$ be a morphism of k–functors. If $Y' \subset X'$ is a closed subfunctor, then $f^{-1}(Y') \subset X$ is closed.*

(3) *Let X_1, X_2, S be k–functors with fixed morphisms $X_1 \to S$ and $X_2 \to S$. If $Y_1 \subset X_1$ and $Y_2 \subset X_2$ are closed subfunctors, then also $Y_1 \times_S Y_2 \subset X_1 \times_S X_2$ is closed.*

Note that (2) applied to an inclusion $Z \hookrightarrow X$ yields: Let Y, Z be subfunctors of a k–functor X. If Y is closed in X, then $Y \cap Z$ is closed in Z.

Because of (1), we can define the closure $\overline{Y}$ of any subfunctor Y of X as the intersection of all closed subfunctors containing Y. In order to get some deeper results we need:

(4) *Let X be an affine scheme and $(X_j)_{j\in J}$ an open covering of X. If Y, Y' are local subfunctors of X with $Y \cap X_j = Y' \cap X_j$ for all $j \in J$, then $Y = Y'$.*

If $X_j = D(I_j)$ for some ideal $I_j \subset k[X]$, then $\sum_{j\in J} I_j = k[X]$, cf. 1.7. We can choose a finite subset $J_0 \subset J$ and $f_j \in I_j$ for all $j \in J_0$ such that $k[X] = \sum_{j\in J_0} k[X]f_j$. Then also the $D(f_j) \subset X_j$ with $j \in J_0$ form an open covering of X (refining the original one). We have also $Y \cap D(f_j) = Y' \cap D(f_j)$ for all $j \in J_0$, so we may as well assume that $J = \{1, 2, \dots, r\}$ and $X_j = X_{f_j}$ for some $f_j \in k[X]$ with $k[X] = \sum_{j=1}^r k[X]f_j$.

Consider now $x \in X(A) = \operatorname{Hom}_{k-\mathrm{alg}}(k[X], A)$ for some k–algebra A; set $f'_i = x(f_i) \in A$ and let $x_i \in X(A_{f'_i})$ correspond to the composed homomorphism $k[X] \xrightarrow{x} A \xrightarrow{\mathrm{can}} A_{f'_i}$. Now $\sum_{i=1}^r k[X]f_i = k[X]$ implies $A = \sum_{i=1}^r Af'_i$, so the local property of Y and Y' yields

$$\begin{aligned} x \in Y(A) &\iff x_i \in Y(A_{f'_i}) && \text{for all } i \\ &\iff x_i \in (Y \cap X_i)(A_{f'_i}) && \text{for all } i \\ &\iff x_i \in (Y' \cap X_i)(A_{f'_i}) && \text{for all } i \\ &\iff x \in Y'(A). \end{aligned}$$

In the affine case, any closed subfunctor is again an affine scheme, cf. 1.4(4), hence local, so we can apply (4) to it.

(5) *Let $(X_j)_{j\subset J}$ be an open covering of some k–functor X. If Y is a closed subfunctor with $Y \supset X_j$ for all j, then $Y = X$.*

Indeed, consider $x \in X(R)$ for some k–algebra R and let $f : Sp_k R \to X$ be the morphism with $f(R)(\mathrm{id}_R) = x$, cf. 1.3. We can apply (4) to the closed, hence local, subfunctors $f^{-1}(Y)$ and $f^{-1}(X) = Sp_k R$ of $Sp_k R$ and the open covering $(f^{-1}(X_j))_{j\in J}$. We get $f^{-1}(Y) = f^{-1}(X)$, hence $\mathrm{id}_R \in f^{-1}(Y)(R)$ and $x \in Y(R)$.

(6) *Any closed subfunctor Y of a local functor X (resp. a scheme X) is again local (resp. a scheme).*

Indeed, consider a morphism $f : X' \to X$ and an open covering $(X'_j)_{j\in J}$ of X' such that each $f_{|X'_j}$ factors through Y, i.e., with $X'_j \subset f^{-1}(Y)$. As $f^{-1}(Y)$ is closed, (5) yields $f^{-1}(Y) = X'$, hence f factors through Y. This together with the local property of X implies easily that Y is local. If X is a scheme and if $(X_j)_{j\in J}$ is an open covering by affine schemes, then $(Y \cap X_j)_{j\in J}$ is an open covering of Y by closed subschemes of affine schemes, hence by affine schemes.

The argument used in 1.10 for open subfunctors yields now:

(7) *For any closed subfunctor Y of a k–functor X and any k–algebra k' the subfunctor $Y_{k'}$ of $X_{k'}$ is closed.*

1.13. Lemma: *Let X be a local functor and $Y \subset X$ a local subfunctor. Let $(X_j)_{j\in J}$ be an open covering of X. Then Y is closed in X if and only if each $Y \cap X_j$ is closed in X_j.*

Proof: One direction being obvious, let us suppose that each $Y \cap X_j$ is closed in X_j. For each morphism $f : X' \to X$ with X' affine, also $f^{-1}(Y) \simeq X' \times_X Y$ is local, the $f^{-1}(X_j)$ are an open covering of X', and each $f^{-1}(Y) \cap f^{-1}(X_j) = f^{-1}(Y \cap X_j)$ is closed in $f^{-1}(X_j)$. So we may as well assume that X is affine.

As in the proof of 1.12(4), we can assume $J = \{1, 2, \dots, r\}$ and $X_j = X_{f_j}$ for some $f_j \in k[X]$. Let I resp. I_j denote the kernel of the restriction map $k[X] \to k[Y]$ resp. $k[X] \to k[Y \cap X_j]$. Then $\overline{Y} = V(I)$. The $Y \cap X_j$ form an open covering of Y. Since $\mathbf{A}^1$ is local, the restriction induces an injective map $k[Y] \to \prod_{j=1}^r k[Y \cap X_j]$, hence $I = \bigcap_{j=1}^r I_j$.

The kernel of each restriction map $k[X_j] = k[X]_{f_j} \to k[Y \cap X_j]$ is equal to $(I_j)_{f_j}$; as $Y \cap X_j$ is closed in X_j, we have $Y \cap X_j = V((I_j)_{f_j})$. For all i and j we can identify $k[X_i \cap X_j]$ with $(k[X]_{f_j})_{f_i} \simeq k[X]_{f_j f_i} \simeq (k[X]_{f_i})_{f_j}$. The kernel of the restriction map $k[X_i \cap X_j] \to k[Y \cap X_i \cap X_j]$ identifies with $((I_j)_{f_j})_{f_i} \simeq (I_j)_{f_j f_i}$ and with $((I_i)_{f_i})_{f_j} \simeq (I_i)_{f_j f_i}$. It follows that $(I_j)_{f_j f_i} = (I_i)_{f_j f_i}$. Given i and $a \in I_i$ we can now find some n with $(f_i f_j)^n a \in I_j$ for all j, hence with $f_i^n a \in I_j$ for all j and thus $f_i^n a \in I = \bigcap_{j=1}^r I_j$. This implies that $I_{f_i} = (I_i)_{f_i}$ for all i, hence $\overline{Y} \cap X_i = Y \cap X_i$. Now apply 1.12(4) to the local subfunctors Y and $\overline{Y}$ of X and get $Y = \overline{Y}$.

1.14. (Closures and Direct Products) Let us assume in this subsection that k is noetherian (in order to simplify the following definition). A scheme X over k is called *algebraic* if it admits a finite open covering by affine subschemes which are algebraic in the sense of 1.6. (One can check that this yields the old definition in the affine case.)

Let Y, Z be schemes and X a subscheme of Y such that X and Y are algebraic and such that Z is *flat*. (This means that Z admits an open and affine covering $(Z_i)_i$ such that each $k[Z_i]$ is a flat k–module.) Then we have in $Y \times Z$

(1) $$\overline{X \times Z} = \overline{X} \times Z.$$

This follows, e.g., by applying [DG], I, §2, 4.14 to $Y' = Y \times Z$ and the projection $Y \times Z \to Y$.

1.15. (Functors of Morphisms) For any k–functors X, Y, we can define a k–functor $\mathfrak{Mor}(X, Y)$ through

(1) $$\mathfrak{Mor}(X, Y)(A) = \mathrm{Mor}(X_A, Y_A) \qquad \text{for any } k\text{–algebra } A.$$

For any homomorphism $\varphi : A \to A'$ of k–algebras $\mathfrak{Mor}(X, Y)(\varphi)$ maps any morphism $f : X_A \to Y_A$ to the morphisms $f_{A'} : X_{A'} \simeq (X_A)_{A'} \to (Y_A)_{A'} \simeq Y_{A'}$ using the structure of A' as an A–algebra via φ.

The construction of $\mathfrak{Mor}(X, Y)$ is clearly functorial: To each morphism $X' \to X$ resp. $Y' \to Y$ of k–functors there corresponds an obvious morphism $\mathfrak{Mor}(X, Y) \to \mathfrak{Mor}(X', Y)$ resp. $\mathfrak{Mor}(X, Y') \to \mathfrak{Mor}(X, Y)$. If Y' is a subfunctor of Y, then we shall always regard $\mathfrak{Mor}(X, Y')$ as a subfunctor of $\mathfrak{Mor}(X, Y)$.

Our goal in this subsection is to show that $\mathfrak{Mor}(X, Y')$ is closed in $\mathfrak{Mor}(X, Y)$ if Y' is closed in Y and if some additional condition holds. We need some preparation:

Consider an open covering $(X_j)_{j\in J}$ of X and a *closed* subfunctor Y' of Y. Let $\rho_j : \mathfrak{Mor}(X, Y) \to \mathfrak{Mor}(X_j, Y)$ denote the obvious restriction map. We claim:

$$\mathfrak{Mor}(X, Y') = \bigcap_{j\in J} \rho_j^{-1} \mathfrak{Mor}(X_j, Y'). \tag{2}$$

Of course, one inclusion ("$\subset$") is trivial. Consider on the other hand $f \in \mathfrak{Mor}(X, Y)(A) = \mathrm{Mor}(X_A, Y_A)$ for some k–algebra A with $\rho_j(A)f \in \mathrm{Mor}(X_{j\,A}, Y'_A)$ for all $j \in J$, i.e., with $X_{j\,A} \subset f^{-1}(Y'_A)$ for all j. Now the $(X_{j\,A})_{j\in J}$ are an open covering of X_A and $f^{-1}(Y'_A)$ is a closed subfunctor of X_A, so 1.12(5) yields $f^{-1}(Y'_A) = X_A$, hence $f \in \mathfrak{Mor}(X, Y')(A)$.

(3) *Let X and Y be k–functors and $Y' \subset Y$ a closed subfunctor. If X admits an open covering $(X_j)_{j\in J}$ with affine schemes such that each $k[X_j]$ is free as a k–module, then $\mathfrak{Mor}(X, Y')$ is closed in $\mathfrak{Mor}(X, Y)$.*

(If X is a scheme, then X is called *locally free* if and only if there is an open covering as above.)

One sees using (2) that it suffices to prove (3) in the case where $X = Sp_k R$ for some k–algebra R that is free as a k–module. We now have to show for each k–algebra A and each morphism $f : Sp_k A \to \mathfrak{Mor}(X, Y)$ that $f^{-1}\mathfrak{Mor}(X, Y')$ is closed.

We have for each k–algebra B natural bijections

$$\begin{aligned} \mathrm{Mor}(Sp_k B, \mathfrak{Mor}(Sp_k R, Y)) &\simeq \mathfrak{Mor}(Sp_k R, Y)(B) \simeq \mathrm{Mor}((Sp_k R)_B, Y_B) \\ &\simeq \mathrm{Mor}(Sp_B(R \otimes B), Y_B) \simeq Y(R \otimes B) \end{aligned}$$

arising from Yoneda's lemma, the definitions, or the universal property of the tensor product.

Taking $B = A$, we see that any f as above corresponds to some $y \in Y(R \otimes A)$. One checks then for all B that

$$f(B) : (Sp_k A)(B) = \mathrm{Hom}_{k\text{–alg}}(A, B) \to \mathfrak{Mor}(Sp_k R, Y)(B) \simeq Y(R \otimes B)$$

maps any $\beta \in \mathrm{Hom}_{k\text{–alg}}(A, B)$ to $Y(\mathrm{id}_R \otimes \beta)(y) \in Y(R \otimes B)$.

On the other hand $y \in Y(R \otimes A)$ defines also a morphism $f' : Sp_k(R \otimes A) \to Y$ that maps for all B any $\gamma \in \mathrm{Hom}_{k\text{–alg}}(R \otimes A, B)$ to $Y(\gamma)(y)$. So we get above

$$f(B)(\beta) = f'(R \otimes B)(\mathrm{id}_R \otimes \beta).$$

So $f^{-1}\mathfrak{Mor}(Sp_k R, Y')(B)$ consists of all β with $f'(R \otimes B)(\mathrm{id}_R \otimes \beta) \in Y'(R \otimes B)$.

Since Y' is closed in Y, there is an ideal I' in $R \otimes A$ with $(f')^{-1}(Y') = V(I')$. We get now for all B

$$f^{-1}\mathfrak{Mor}(Sp_k R, Y')(B) = \{\, \beta \in \mathrm{Hom}_{k\text{–alg}}(A, B) \mid I' \subset \ker(\mathrm{id}_R \otimes \beta) \,\}.$$

Now we use for the first time that R is free as a k–module. It implies that $\ker(\mathrm{id}_R \otimes \beta) = R \otimes \ker(\beta)$, hence

$$f^{-1}\mathfrak{Mor}(Sp_k R, Y')(B) = \{\, \beta \in \mathrm{Hom}_{k\text{–alg}}(A, B) \mid I' \subset R \otimes \ker(\beta) \,\}.$$

The freeness implies also $R \otimes \bigcap_i I_i = \bigcap_i R \otimes I_i$ for any family $(I_i)_i$ of ideals in A. If we take as the I_i all ideals with $R \otimes I_i \supset I'$, then $I = \bigcap_i I_i$ is the smallest ideal of A with $I' \subset R \otimes I$. Then

$$I' \subset R \otimes \ker(\beta) \iff I \subset \ker(\beta) \iff \beta \in V(I)(B),$$

so $f^{-1}\mathfrak{Mor}(Sp_k R, Y') = V(I)$ is closed.

CHAPTER 2

Group Schemes and Representations

In this chapter we define group schemes and modules over these objects and discuss their fundamental properties. As in Chapter 1 we more or less follow [DG].

After making the definitions of k–group functors and k–group schemes in 2.1, we describe some examples in 2.2. The relationship between algebraic groups and Hopf algebras generalises to group schemes. This is done in 2.3–2.4. (We always assume our group schemes to be affine.) We then discuss the class of diagonalisable group schemes in 2.5 and group operations in 2.6.

We then go on to define representations (2.7) and to discuss the relationship between G–modules and $k[G]$–comodules (2.8). We generalise standard notions of representation theory to G–modules: submodules (2.9), fixed points (2.10), centralisers and stabilisers (2.12), and simple modules (2.14). The definition of a submodule has an unpleasant aspect that disappears only when G is a flat group scheme (i.e., a group scheme such that $k[G]$ is a flat k–module). This is the reason why we restrict ourselves to such groups later on.

We show that representations of group schemes are locally finite (2.13). Furthermore, we describe representations of diagonalisable group schemes (2.11) and mention results about modules for trigonalisable and unipotent groups over fields (in 2.14). One can twist a given representation of an abstract group over k by composing with a group endomorphism. One can also construct a new representation by changing the operation of k on the module by a ring endomorphism. This is generalised to group schemes in 2.15–2.16.

2.1. (Definitions) A *k–group functor* is a functor from the category of all k–algebras to the category of groups. We can regard any k–group functor also as a k–functor by composing it with the forgetful functor from $\{\text{groups}\}$ to $\{\text{sets}\}$. In this way we can and shall apply all ideas and notions from Chapter 1 also to k–group functors. For two group functors G, H, we shall denote by $\mathrm{Mor}(G, H)$ the set of all morphisms (= natural transformations) from G to H considered as k–functors and by $\mathrm{Hom}(G, H)$ the set of all morphisms from G to H considered as k–group functors. So $\mathrm{Hom}(G, H)$ consists of all those $f \in \mathrm{Mor}(G, H)$ with $f(A)$ a group homomorphism for each k–algebra A. These elements are called *homomorphisms* from G to H. Let $\mathrm{Aut}(G)$ be the group of all automorphisms of the k–group functor G.

A *k–group scheme* is a k–group functor that is an *affine* scheme over k when considered as a k–functor. (Of course, we really ought to call such an object an affine k–group scheme and drop the word "affine" in the definition of a k–group scheme. But we shall consider only affine group schemes, and then it is more economical to call them group schemes.) An *algebraic k–group* is a k–group scheme that is algebraic as an affine scheme. A k–group scheme is called *reduced* if it is so as an affine scheme. Over an algebraically closed field, the category of algebraic groups as in [Hu2] or [Sp2] can be identified with the subcategory of all reduced algebraic k–groups.

Let G be a k–group functor. A *subgroup functor* of G is a subfunctor H of G such that each $H(A)$ is a subgroup of $G(A)$. The intersection of subgroup functors is again a subgroup functor. The inverse image of a subgroup functor under a homomorphism is again one. A direct product of k–group functors is again a k–group functor; so is a fibre product if the morphisms used in its construction are homomorphisms of k–group functors.

A subgroup functor H of G is called *normal* (resp. *central*) if each $H(A)$ is a normal (resp. central) subgroup of $G(A)$. Again, normality is preserved under taking intersections and inverse images under homomorphisms. The kernel $\ker\varphi$ of a homomorphism $\varphi : G \to G'$ is always a normal subgroup functor.

A *closed subgroup scheme* of a k–group scheme is a subgroup functor which is closed if considered as a subfunctor of the affine scheme G over k. If G and H are algebraic k–groups, we simply call H a *closed subgroup* of G. For each k–group scheme the subgroup functor that maps each A to $\{1\}$ is closed, see the remark following 1.4(4). Hence the kernel of a homomorphism between k–group schemes is always closed.

A k–group functor G is called *commutative* if all $G(A)$ are commutative.

2.2. (Examples) The notations introduced here for special group functors G and their algebras $k[G]$ will be used always. The *additive group* over k is the k–group functor G_a with $G_a(A) = (A, +)$ for all k–algebras A. It is an algebraic k–group with $k[G_a]$ isomorphic to (and usually identified with) the polynomial ring $k[T]$ in one variable.

Any k–module M defines a k–group functor M_a with $M_a(A) = (M \otimes A, +)$ for all A. (So we have $G_a \simeq k_a$.) If M is projective of finite rank as a k–module, then M_a is an algebraic k–group with $k[M_a] = S(M^*)$, the symmetric algebra over the dual k–module M^*. In case $M = k^n$ for some $n \in \mathbf{N}$, we may identify M_a with $G_a \times G_a \times \cdots \times G_a$ (n factors) and $k[M_a]$ with the polynomial ring $k[T_1, T_2, \dots, T_n]$.

The *multiplicative group* over k is the k–group functor G_m with $G_m(A) = A^\times = \{\text{ units of } A\}$ for all A. It is an algebraic k–group with $k[G_a] \simeq k[T, T^{-1}]$.

Any k–module M leads to a k–group functor $GL(M)$ defined by $GL(M)(A) = (\mathrm{End}_A(M \otimes A))^\times$ called the *general linear group* of M. In case $M = k^n$, we may identify $GL(M)$ with GL_n where $GL_n(A)$ is the group of all invertible $(n \times n)$–matrices over A. Obviously, GL_n is an algebraic k–group with $k[GL_n]$ isomorphic to the localisation of the polynomial ring $k[T_{ij},\ 1 \le i, j \le n]$ with respect to $\{(\det)^n \mid n \in \mathbf{N}\}$. More generally, if M is projective of finite rank as a k–module, then the k–functor $A \mapsto \mathrm{End}(M \otimes A)$ can be identified with the affine scheme $(M^* \otimes M)_a$ from above and $GL(M)$ with the open subfunctor $D(\det)$, cf. [B2], ch. II, §5, exerc. 9. For such M (projective of finite rank), the determinant defines a homomorphism of algebraic k–groups $GL(M) \to G_m$. Its kernel is denoted by $SL(M)$ and is called the *special linear group* of M. It is an algebraic k–group. Similarly, we define $SL_n \subset GL_n$. Note that $GL_1 = G_m$ and $SL_1 = 1 =$ the group functor associating to each A the trivial group $\{1\}$.

For each $n \in \mathbf{N}$, let T_n be the algebraic k–group such that $T_n(A)$ is the group of all invertible upper-triangular $(n \times n)$–matrices over A, i.e., of all upper-triangular matrices such that all diagonal entries belong to $A^\times$. One may identify

$k[T_n] \simeq k[T_{ij} \mid 1 \le i \le j \le n, T_{ii}^{-1} \mid 1 \le i \le n]$. Furthermore, let U_n be the algebraic k–group such that each $U_n(A)$ consists of all $g \in T_n(A)$ having all diagonal entries equal to 1. We may identify $k[U_n] \simeq k[T_{ij} \mid 1 \le i < j \le n]$.

For any $n \in \mathbf{N}$ we denote by $\mu_{(n)}$ the group functor with $\mu_{(n)}(A) = \{a \in A \mid a^n = 1\}$ for all A. This an algebraic k–functor with $k[\mu_{(n)}] \simeq k[T]/(T^n - 1)$ and a closed subgroup of G_m.

Let p be a prime number; assume that $p1 = 0$ in k. Then we can define for each $r \in \mathbf{N}$ a closed subgroup $G_{a,r}$ of G_a through $G_{a,r}(A) = \{a \in A \mid a^{p^r} = 0\}$.

2.3. (Group Schemes and Hopf Algebras) Let G be a k–group functor. The group structures on all $G(A)$ define morphisms of k–functors $m_G : G \times G \to G$ (such that each $m_G(A) : G(A) \times G(A) \to G(A)$ is the multiplication), and $1_G : Sp_k k \to G$ (such that $1_G(A)$ maps the unique element of $(Sp_k k)(A)$ to the 1 of $G(A)$), and $i_G : G \to G$ (inducing on each $G(A)$ the map $g \mapsto g^{-1}$).

Now assume G to be a k–group scheme. Then these morphisms correspond uniquely to their comorphisms $\Delta_G = m_G^* : k[G] \to k[G] \otimes k[G]$ (called *comultiplication*), $\varepsilon_G = 1_G^* : k[G] \to k$ (called *counit* or *augmentation*), and $\sigma_G = i_G^* : k[G] \to k[G]$ (called *coinverse* or *antipode*). So, if $\Delta_G(f) = \sum_{i=1}^r f_i \otimes f_i'$ for some $f \in k[G]$, then $f(g_1 g_2) = \sum_{i=1}^r f_i(g_1) f_i'(g_2)$ for all $g_1, g_2 \in G(A)$ and all A. Furthermore, we have $\varepsilon_G(f) = f(1)$ and $\sigma_G(f)(g) = f(g^{-1})$ for all $g \in G(A)$ and all A.

We shall drop the index G in our notation whenever no confusion is likely.

As in the case of algebraic groups (cf. [Bo], 1.5, or [Hu2], 7.6, or [Sp2], 2.1.2) the group axioms imply that Δ, ε, σ satisfy

$$(\mathrm{id} \otimes \Delta) \circ \Delta = (\Delta \otimes \mathrm{id}) \circ \Delta, \tag{1}$$

$$(\varepsilon \mathbin{\overline{\otimes}} \mathrm{id}) \circ \Delta = \mathrm{id} = (\mathrm{id} \mathbin{\overline{\otimes}} \varepsilon) \circ \Delta, \tag{2}$$

$$(\sigma \mathbin{\overline{\otimes}} \mathrm{id}) \circ \Delta = \overline{\varepsilon} = (\mathrm{id} \mathbin{\overline{\otimes}} \sigma) \circ \Delta. \tag{3}$$

(Here we denote by $\varphi \mathbin{\overline{\otimes}} \psi$ the map $a \otimes a' \mapsto \varphi(a)\psi(a')$ in contrast to $\varphi \otimes \psi : a \otimes a' \mapsto \varphi(a) \otimes \psi(a')$ and by $\overline{\varepsilon}$ the endomorphism $a \mapsto \varepsilon(a)1$ of $k[G]$.)

A morphism $\varphi : G \to G'$ between two k–group schemes is a homomorphism if and only if its comorphism $\varphi^* : k[G'] \to k[G]$ satisfies

$$\Delta_G \circ \varphi^* = (\varphi^* \otimes \varphi^*) \circ \Delta_{G'}. \tag{4}$$

If so, then one has automatically

$$\varepsilon_G \circ \varphi^* = \varepsilon_{G'} \tag{5}$$

and

$$\sigma_G \circ \varphi^* = \varphi^* \circ \sigma_{G'}. \tag{6}$$

A *Hopf algebra* over k is an associative (not necessarily commutative) algebra R over k together with homomorphisms of algebras $\Delta : R \to R \otimes R$, $\varepsilon : R \to k$, and $\sigma : R \to R$ satisfying (1)–(3). A homomorphism between two Hopf algebras is a homomorphism of algebras satisfying additionally (4)–(6) (with the appropriate changes in the notation). We call R commutative if it is so as an algebra, and *cocommutative* if $s \circ \Delta = \Delta$ where $s : R \otimes R \to R \otimes R$ is the map $a \otimes b \mapsto b \otimes a$.

Let R be a commutative Hopf algebra over k. Then we can define on each $(Sp_k R)(A) = \mathrm{Hom}_{k-\mathrm{alg}}(R, A)$ a multiplication via $\alpha\beta = (\alpha \overline{\otimes} \beta) \circ \Delta$. In this way we get on $Sp_k R$ a structure as a k–group scheme. It is elementary to see that we get in this way a functor $\{$ commutative Hopf algebras over $k \} \to \{ k$–group schemes $\}$ that is quasi-inverse to $G \mapsto k[G]$. Thus these categories are anti-equivalent.

Note that G is commutative if and only if $k[G]$ is cocommutative.

2.4. (Properties of the Hopf Algebra Structure) Let us look at the Hopf algebra structures on $k[G]$ in our examples in 2.2. In the case of G_a, one has $\Delta(T) = 1 \otimes T + T \otimes 1$, $\varepsilon(T) = 0$, and $\sigma(T) = -T$. Similar formulas hold for the $G_{a,r}$. In the case of G_m, one has $\Delta(T) = T \otimes T$, $\varepsilon(T) = 1$, and $\sigma(T) = T^{-1}$. For GL_n, one has $\Delta(T_{ij}) = \sum_{m=1}^n T_{im} \otimes T_{mj}$ and $\varepsilon(T_{ij}) = \delta_{ij}$ (the Kronecker delta); the formula for $\sigma(T_{ij})$ is more complicated. Furthermore, one has $\Delta(\det) = \det \otimes \det$, $\varepsilon(\det) = 1$, and $\sigma(\det) = \det^{-1}$.

Let G be a k–group scheme and set $I_1 = \ker \varepsilon$, the *augmentation ideal* in $k[G]$. One has $k[G] = k1 \oplus I_1$, and $a \mapsto a1$ is a bijection $k \to k1$. This implies $k[G] \otimes k[G] = k(1 \otimes 1) \oplus (k \otimes I_1) \oplus (I_1 \otimes k) \oplus (I_1 \otimes I_1)$. The formula 2.3(2) implies

(1) $$\Delta(f) \in f \otimes 1 + 1 \otimes f + I_1 \otimes I_1 \qquad \text{for all } f \in I_1.$$

We have $\sigma(I_1) = I_1$ (since $\sigma(f)(1) = f(1)$); therefore 2.3(3) implies now

(2) $$\sigma(f) \in -f + I_1^2 \qquad \text{for all } f \in I_1.$$

Set

(3) $$X(G) = \mathrm{Hom}(G, G_m).$$

This is a commutative group in a natural way, called the *character group* of G. The embedding of affine schemes $G_m \subset G_a = \mathbf{A}^1$ yields an embedding

$$X(G) \subset \mathrm{Mor}(G, G_m) \subset \mathrm{Mor}(G, G_a) \simeq k[G]$$

that is compatible with the multiplication. Take $f \in k[G]$. One has $f^*(T) = f$. Therefore 2.3(4) implies easily

(4) $$X(G) \simeq \{ f \in k[G] \mid f(1) = 1,\ \Delta_G(f) = f \otimes f \}.$$

Of course, $\Delta_G(f) = f \otimes f$ implies $f(1)^2 = f(1)$. If $f(1) = 0$, then $f(g) = f(g \cdot 1) = f(g)f(1) = 0$ for all $g \in G(A)$ and all A, hence:

(4′) $$k \text{ integral domain} \implies X(G) \simeq \{ f \in k[G] \mid \Delta_G(f) = f \otimes f,\ f \neq 0 \}.$$

Let me refer to [DG], II, §1, 2.9 for the proof of:

(5) *If k is a field, then $X(G)$ is linearly independent.*

(This is just another variation on the theme "linear independence of characters".) *Usually* we shall write the group law in $X(G)$ *additively.*

Let I be an ideal in $k[G]$. Using 1.4(6), (7) one checks easily that $V(I)$ is a subgroup functor if and only if

(6) $$\Delta(I) \subset I \otimes k[G] + k[G] \otimes I,\ \varepsilon(I) = 0,\ \sigma(I) \subset I.$$

If so, it will be a normal subgroup if and only if

(7) $$c^*(I) \subset k[G] \otimes I$$

where c^* is the comorphism of the *conjugation map* $c : G \times G \to G$ with $c(A)(g_1, g_2) = g_1 g_2 g_1^{-1}$ for all A and all $g_1, g_2 \in G(A)$. One may check that

(8) $$c^* = t \circ (\Delta \otimes \mathrm{id}) \circ \Delta$$

where $t(f_1 \otimes f_2 \otimes f_3) = f_1 \sigma(f_3) \otimes f_2$.

2.5. (Diagonalisable Groups) Let Λ be a commutative group (written multiplicatively); identify Λ with the standard basis of the group algebra $k[\Lambda]$. We make $k[\Lambda]$ into a commutative and cocommutative Hopf algebra via $\Delta(\lambda) = \lambda \otimes \lambda$ and $\varepsilon(\lambda) = 1$ and $\sigma(\lambda) = \lambda^{-1}$ for all $\lambda \in \Lambda$. By 2.3 we get now on $Sp_k k[\Lambda]$ a structure as a k–group scheme; we denote this group scheme by $\mathrm{Diag}(\Lambda)$. If Λ is finitely generated, then $\mathrm{Diag}(\Lambda)$ is an algebraic k–group.

Now $\mathrm{Diag}(\Lambda)(A)$ consists (for each k–algebra A) of all group homomorphisms $\varphi : \Lambda \to A^\times$. The multiplication is the obvious one: $(\varphi\varphi')(\lambda) = \varphi(\lambda)\varphi'(\lambda)$ for all $\lambda \in \Lambda$.

We call a k–group scheme *diagonalisable* if it is isomorphic to $\mathrm{Diag}(\Lambda)$ for some commutative group Λ. For example, $G_m \simeq \mathrm{Diag}(\mathbf{Z})$ and $\mu_{(n)} \simeq \mathrm{Diag}(\mathbf{Z}/(n))$ are diagonalisable. We get also direct products of these groups as $\mathrm{Diag}(\Lambda_1 \times \Lambda_2) \simeq \mathrm{Diag}(\Lambda_1) \times \mathrm{Diag}(\Lambda_2)$ for all commutative groups Λ_1, Λ_2.

Any group homomorphism $\alpha : \Lambda_1 \to \Lambda_2$ induces a homomorphism of group algebras $\alpha_* : k[\Lambda_1] \to k[\Lambda_2]$ which is a homomorphism of Hopf algebras. Hence we get a homomorphism $\mathrm{Diag}(\alpha) : \mathrm{Diag}(\Lambda_2) \to \mathrm{Diag}(\Lambda_1)$ of k–group schemes. Thus $\Lambda \mapsto \mathrm{Diag}(\Lambda)$ is a contravariant functor from { commutative groups } to { k–group schemes } that maps finitely generated commutative groups to algebraic k–groups.

If α is surjective, then $\mathrm{Diag}(\alpha)(A)$ is injective for each A, and we can regard $\mathrm{Diag}(\Lambda_2)$ as a subfunctor of $\mathrm{Diag}(\Lambda_1)$. If α is not surjective, then there is a non-trivial homomorphism $\Lambda_1/\alpha(\Lambda_2) \to A^\times$ for some k–algebra A, hence $\mathrm{Diag}(\alpha)(A)$ is not injective.

Suppose that k is an integral domain. Then an easy computation shows (cf. [DG], II, §1, 2.11) for all Λ, Λ'

(1) $$X(\mathrm{Diag}(\Lambda)) \simeq \Lambda \qquad (k \text{ integral})$$

and

(2) $$\mathrm{Hom}_{gp}(\Lambda, \Lambda') \xrightarrow{\sim} \mathrm{Hom}(\mathrm{Diag}(\Lambda'), \mathrm{Diag}(\Lambda)) \qquad (k \text{ integral}).$$

Thus in this case $\mathrm{Diag}(?)$ is an anti-equivalence of categories from { commutative groups } to { diagonalisable k–group schemes }. Furthermore, Λ is finitely generated if and only if $\mathrm{Diag}(\Lambda)$ is an algebraic k–group. We get from (1) that a k–group scheme G is diagonalisable if and only if $X(G)$ is a basis of $k[G]$ (for k integral).

2.6. (Actions) Let G be a k–group functor. A left *action* of G on a k–functor X is a morphism $\alpha : G \times X \to X$ such that for each k–algebra A the map $\alpha(A) : G(A) \times X(A) \to X(A)$ is a (left) action of the group $G(A)$ on the set $X(A)$. We usually write $g\,x$ instead of $\alpha(A)(g,x)$ for $g \in G(A)$ and $x \in X(A)$. We can similarly define right actions.

For example, the conjugation map c in 2.4 is an action of G on itself. Other actions of G on itself are by left ($\alpha(A)(g,g') = gg'$) and right ($\alpha(A)(g,g') = g'g^{-1}$) multiplication.

Let k' be a k–algebra. Then any action of G on a k–functor X defines in a natural way an action of $G_{k'}$ on $X_{k'}$.

For any action α as above we set

$$X^G(k) = \{\, x \in X(k) \mid gx = x \text{ for all } g \in G(A) \text{ and all } A \,\}. \tag{1}$$

(This is done by some abuse of notation. The x in $gx = x$ is really the image of x under the map $X(k) \to X(A)$ corresponding to the structural homomorphism $k \to A$. We shall stick to this abuse.) We can define a subfunctor X^G of X, the *fixed point functor* via

$$X^G(A) = (X_A)^{G_A}(A), \tag{2}$$

i.e., via $X^G(A) = \{\, x \in X(A) \mid gx = x \text{ for all } g \in G(A') \text{ and all } A\text{–algebras } A' \,\}$.

If Y is a subfunctor of X, then its *stabiliser* in G is the subgroup functor $\mathrm{Stab}_G(Y)$ with

$$\mathrm{Stab}_G(Y)(A) = \{\, g \in G(A) \mid gY(A') = Y(A') \text{ for all } A\text{–algebras } A' \,\} \tag{3}$$

for all A, and its *centraliser* is the subgroup functor $\mathrm{Cent}_G(Y)$ with

$$\mathrm{Cent}_G(Y)(A) = \{\, g \in G(A) \mid gy = y \text{ for all } y \in Y(A') \text{ and all } A\text{–algebras } A' \,\} \tag{4}$$

for all A.

These constructions can also be described using the k–functors $\mathfrak{Mor}(X_1, X_2)$ as in 1.15. The action of G defines a morphism $\gamma : G \to \mathfrak{Mor}(X, X)$ associating to each $g \in G(A)$ the morphism $X_A \to X_A$ defined by the action of g. For each subfunctor Y of X we get by restriction a morphism $\gamma_Y : G \to \mathfrak{Mor}(Y, X)$. Recall that $i_G : G \to G$ was the morphism $g \mapsto g^{-1}$. Now we have obviously

$$\mathrm{Stab}_G(Y) = \gamma_Y^{-1}\,\mathfrak{Mor}(Y,Y) \cap i_G\,(\gamma_Y^{-1}\,\mathfrak{Mor}(Y,Y)). \tag{5}$$

Let $\varphi : G \to \mathfrak{Mor}(Y, X \times X)$ denote the morphism associating to each g the morphism $y \mapsto (gy, y)$. Using the notation D_X for the diagonal subfunctor of $X \times X$ as in 1.2 we see

$$\mathrm{Cent}_G(Y) = \varphi^{-1}\,\mathfrak{Mor}(Y, D_X). \tag{6}$$

Let $\psi : X \to \mathfrak{Mor}(G, X \times X)$ denote the morphism associating to each x the morphism $g \mapsto (gx, x)$. Then

$$X^G = \psi^{-1}\,\mathfrak{Mor}(G, D_X). \tag{7}$$

Now 1.15(3) implies (using that i_G is an automorphism of G as a k–functor):

(8) *If Y is a closed subfunctor of X and a locally free k–scheme, then* $\mathrm{Stab}_G(Y)$ *is closed in G.*

(9) *If Y is a locally free k–scheme and if D_X is closed in $X \times X$, then* $\mathrm{Cent}_G(Y)$ *is closed in G.*

(10) *If G is a locally free k–scheme and if D_X is closed in $X \times X$, then X^G is closed in X.*

(One calls X *separate* if D_X is closed in $X \times X$. Any affine scheme is separate.)

In case $X = G$ with G acting via conjugation, one usually calls $\mathrm{Stab}_G(Y)$ the *normaliser* of Y and denotes it by $N_G(Y)$. Furthermore, we then usually write $C_G(Y)$ instead of $\mathrm{Cent}_G(Y)$, and $Z(G)$ instead of $C_G(G)$. Of course, $Z(G)$ is just the *centre* of G.

Consider a k–algebra k' which is an algebraically closed field. Suppose that $X_{k'}$ is an *algebraic* and *separate scheme.* The map $Z \mapsto Z_{k'}$ is a bijection from $\{$ closed and reduced subfunctors of $X_{k'}$ $\}$ to $\{$ closed subsets of $X(k')$ $\}$, cf. 1.6 for X affine. We claim for any *closed subfunctor* Y of X such that $Y_{k'}$ is *reduced*:

$$\mathrm{Stab}_G(Y)(k') = \mathrm{Stab}_{G(k')}(Y(k')), \tag{11}$$

$$\mathrm{Cent}_G(Y)(k') = \mathrm{Cent}_{G(k')}(Y(k')). \tag{12}$$

Indeed, if $g \in G(k')$, then $gY_{k'}$ and $Y_{k'}$ are two closed and reduced subfunctors of $X_{k'}$, hence $gY_{k'} = Y_{k'}$ if and only if $(gY_{k'})(k') = gY(k')$ is equal to $Y(k')$, i.e., if $g \in \mathrm{Stab}_{G(k')}(Y(k'))$. This yields (11). In order to get (12), we embed $Y_{k'}$ via $y \mapsto (y, y)$ and via $y \mapsto (y, gy)$ into $X_{k'} \times X_{k'}$. The images Y_1 and Y_2 are closed subfunctors of $X_{k'} \times X_{k'}$, both isomorphic to $Y_{k'}$, hence reduced. Therefore $Y_1 = Y_2$ if and only if $Y_1(k') = Y_2(k')$. On the other hand, $g \in \mathrm{Cent}_G(Y)(k')$ resp. $g \in \mathrm{Cent}_{G(k')}(Y(k'))$ if and only if $Y_1 = Y_2$ resp. $Y_1(k') = Y_2(k')$. This implies (12).

Suppose that G acts on another k–group functor H such that each $G(A)$ acts on $H(A)$ through group automorphisms. The we can form the *semi-direct product* $H \rtimes G$ where each $(H \rtimes G)(A)$ is the usual semi-direct product $H(A) \rtimes G(A)$. As a k–functor $H \rtimes G$ is of course the direct product of G and H.

Let H, N be subgroup functors of G such that H normalises N, i.e., that each $H(A)$ normalises $N(A)$. We can then construct $N \rtimes H$ as above and get a homomorphism $\varphi : N \rtimes H \to G$ via $(n, h) \mapsto nh$ for all $h \in H(A)$, $n \in N(A)$ and all A. Its kernel is isomorphic to $H \cap N$ under $h \mapsto (h, h^{-1})$ for all $h \in H(A) \cap N(A)$ and all A. If φ is an isomorphism, then we say that G is the semi-direct product of N and H and write $G = N \rtimes H$. (If G is a k–group scheme and $G = N \rtimes H$, then necessarily H and N are closed subgroup schemes.)

2.7. (Representations) Let G be a k–group functor and M a k–module. A *representation* of G on M (or: a G–*module* structure on M) is an action of G on the k–functor M_a (as in 2.2) such that each $G(A)$ acts on $M_a(A) = M \otimes A$ through A–linear maps. Such a representation gives for each A a group homomorphism $G(A) \to \mathrm{End}_A(M \otimes A)^\times$, leading to a homomorphism $G \to GL(M)$ of group functors. Vice versa, any such homomorphism defines a representation of G on M.

There is an obvious notion of a G–module homomorphism (or G–equivariant map) between two G–modules M and M'. The k–module of all such homomorphisms is denoted by $\operatorname{Hom}_G(M, M')$.

The representations of G on the k–module k, for example, correspond bijectively to the group homomorphisms from G to $GL_1 = G_m$, i.e., to the elements of $X(G)$. For each $\lambda \in X(G)$ we denote k considered as a G–module via λ by k_λ. In case $\lambda(g) = 1$ for all $g \in G(A)$ and all A we simply write k.

Given one or several G–modules, we can construct in a natural way other G–modules. For example:

(1) *Any direct sum of G–modules is a G–module in a natural way.*

(2) *The tensor product of two G–modules is a G–module in a natural way.*

(3) *Symmetric and exterior powers of a G–module are G–modules in a natural way.*

In (3), for example, we consider for each commutative ring R the functor F_R from R–modules to itself with $F_R(M) = S^n M$. We have for each R–algebra R' canonical isomorphisms $F_R(M) \otimes_R R' \xrightarrow{\sim} F_{R'}(M \otimes_R R')$ for all R–modules M, i.e., the functors $M \mapsto F_R(M) \otimes_R R'$ and $M \mapsto F_{R'}(M \otimes_R R')$ are isomorphic. If M is a G–module, then G acts on the k–functor $A \mapsto F_A(M \otimes A)$, each $g \in G(A)$ via F_A applied to the action of g on $M \otimes A$. By our observation this functor is isomorphic to $F_k(M)_a$, hence we get a G–module structure on $F_k(M)$. The functor $M \mapsto \Lambda^n M$ has the same property, hence we can argue as above. Our reasoning can easily be extended to functors in several variables and then yields (1) and (2).

If we deal with contravariant functors $(F_R)_R$ in our situation above, we ought to let $g \in G(A)$ act via $F_A(g^{-1})$. This applies to the functor $M \mapsto M^*$ which, however, will "commute with ring extensions" only when restricted to finitely generated projective modules. Thus we get:

(4) *Let M be a G–module which is finitely generated and projective over k. Then M^* is a G–module in a natural way.*

For M as in (4) one has canonically $M^* \otimes M' \simeq \operatorname{Hom}(M, M')$ for any k–module M'. Combining (2) and (4) we get:

(5) *Let M, M' be G–modules with M is finitely generated and projective over k. Then* $\operatorname{Hom}(M, M')$ *is a G–module in a natural way.*

The next statement is obvious from the definitions:

(6) *Let k' be a k–algebra and M a G–module. Then $M_{k'}$ is a $G_{k'}$–module in a natural way.*

Another way, how representations arise, is from an action of G on an affine scheme. We get then a G–module structure on $k[X]$: If $g \in G(A)$ and $f \in A[X_A] = k[X] \otimes A$, then $gf \in A[X_A]$ is defined through $(gf)(x) = f(g^{-1}x)$ resp. $= f(xg)$ (for a left resp. right action) for all $x \in X(A')$ and all A–algebras A'. (Again, the g in $g^{-1}x$ or xg is really the image of g under $G(A) \to G(A') \dots$)

In case G is a k–group scheme, we get thus the *left* and *right regular representations* of G on $k[G]$ derived from the action of G on itself by left and right multiplications. We shall always denote the corresponding homomorphisms $G \to GL(k[G])$ by ρ_l and ρ_r. The coinverse σ_G is an isomorphism of G–modules from $k[G]$ with ρ_r to $k[G]$ with ρ_l. Finally, the conjugation action of G on itself gives rise to the *conjugation representation* of G on $k[G]$.

2.8. (The Comodule Map) Let G be a k–group scheme. If M is a G–module, then $\mathrm{id}_{k[G]} \in G(k[G]) = \mathrm{End}_{k-\mathrm{alg}}(k[G])$ acts on $M \otimes k[G]$. We get thus a k–linear map $\Delta_M : M \to M \otimes k[G]$ with $\Delta_M(m) = \mathrm{id}_{k[G]}(m \otimes 1)$ for all $m \in M$. We call Δ_M the *comodule map* of the G–module M. It determines the representation of G on M completely: For any k–algebra A and any $g \in G(A) = \mathrm{Hom}_{k-\mathrm{alg}}(k[G], A)$ we have a commutative diagram

$$\begin{array}{ccc} G(k[G]) \times (M \otimes k[G]) & \longrightarrow & M \otimes k[G] \\ {\scriptstyle G(g)\times(\mathrm{id}_M \otimes g)}\big\downarrow & & \big\downarrow{\scriptstyle \mathrm{id}_M \otimes g} \\ G(A) \times (M \otimes A) & \longrightarrow & M \otimes A \end{array}$$

by the functorial property of an action. As $G(g)\varphi = g \circ \varphi$ for any $\varphi \in G(k[G])$, we have $g = G(g)\,\mathrm{id}_{k[G]}$, hence

$$g(m \otimes 1) = (\mathrm{id}_M \otimes g) \circ \Delta_M(m) \qquad \text{for all } m \in M.$$

More explicitly, if $\Delta_M(m) = \sum_{i=1}^r m_i \otimes f_i$, then

$$g\,(m \otimes 1) = \sum_{i=1}^r m_i \otimes f_i(g). \tag{1}$$

The fact that each $G(A)$ acts on $M(A)$ (i.e., that $g(g'm) = (gg')m$ and $1m = m$) yields easily:

$$(\Delta_M \otimes \mathrm{id}_{k[G]}) \circ \Delta_M = (\mathrm{id}_M \otimes \Delta_G) \circ \Delta_M \tag{2}$$

and

$$(\mathrm{id}_M \overline{\otimes}\, \varepsilon_G) \circ \Delta_M = \mathrm{id}_M\,. \tag{3}$$

If M' is another G–module, then a linear map $\varphi : M \to M'$ is a homomorphism of G–modules if and only if

$$\Delta_{M'} \circ \varphi = (\varphi \otimes \mathrm{id}_{k[G]}) \circ \Delta_M. \tag{4}$$

A *comodule* over the Hopf algebra $k[G]$ is a k–module M together with a linear map $\Delta_M : M \to M \otimes k[G]$ such that (2) and (3) are satisfied. A homomorphism between two comodules is a linear map satisfying (4). So we have defined a faithful functor from $\{\, G\text{–modules} \,\}$ to $\{\, k[G]\text{–comodules} \,\}$. On the other hand, any $k[G]$–comodule gives rise to a G–module: Just take (1) as a definition. In this way we can see that the two categories of G–modules and of $k[G]$–comodules are equivalent.

Let $\alpha : X \times G \to X$ be a right action of G on an affine scheme X over k. Then $k[X]$ is a G–module in a natural way (see 2.7) and the comodule map $\Delta_{k[X]} : k[X] \to k[X] \otimes k[G]$ is easily checked to be equal to the comorphism α^*. If we take $X = G$ and the action by right multiplication, we get

$$\Delta_{\rho_r} = \Delta_G. \tag{5}$$

(We write Δ_{ρ_r} and also Δ_{ρ_l} below instead of $\Delta_{k[G]}$ in order to indicate which representation is considered.) For the left regular action we get

$$\Delta_{\rho_l} = s \circ (\sigma_G \otimes \mathrm{id}_{k[G]}) \circ \Delta_G \tag{6}$$

where $s(f \otimes f') = f' \otimes f$ for all f, f'. For the conjugation representation on $k[G]$ the comodule map is equal to

$$t' \circ (\mathrm{id}_{k[G]} \otimes \Delta_G) \circ \Delta_G \tag{7}$$

where $t'(f_1 \otimes f_2 \otimes f_3) = f_2 \otimes \sigma_G(f_1) f_3$.

Remark: Suppose for the moment that k is an algebraically closed field and that G is a reduced algebraic k–group. There is a natural notion of representations of $G(k)$ as an algebraic group (or of a *rational $G(k)$–module*), cf. [Hu2], p. 60. If $(v_i)_{i\in I}$ is a basis for such a rational $G(k)$–module M, then there are $f_{ji} \in k[G(k)] = k[G]$ with $g\,v_i = \sum_{j\in J} f_{ji}(g)v_j$ for all $g \in G(k)$; for each i almost all f_{ji} are 0. Then the linear map $\Delta_M : M \to M \otimes k[G]$ with $\Delta_M(v_i) = \sum_{j\in J} v_j \otimes f_{ji}$ for all i and j can be checked to satisfy (2) and (3). In this way M is turned into a comodule over $k[G]$, hence into a G–module by the discussion above. We get thus an equivalence of categories between { rational $G(k)$–modules } and { G–modules }. The inverse map associates to any G–module M the action of $G(k)$ on M given by the definition of a G–module. One gets $\mathrm{Hom}_G(M, M') = \mathrm{Hom}_{G(k)}(M, M')$ for any G–modules M, M' from (4) above.

Furthermore, one can now show that the notions of G–submodules (to be defined in 2.9) and of $G(k)$–submodules coincide, using 2.9(1), and that $M^{G(k)} = M^G$ (to be defined in 2.10), using 2.10(2). One can also get these claims from the density of $G(k)$ in G and from 2.12(4), (5) below.

2.9. (Submodules) Let G be a k–group functor. If k is a field, we can define a submodule of a G–module M as a subspace $N \subset M$ such that $N \otimes A$ is a $G(A)$–submodule of $M \otimes A$ for each k–algebra A. Then N is a G–module in a natural way. For arbitrary k this works out well as long as the natural map $N \otimes A \to M \otimes A$ is injective for all A, e.g., if N is a direct summand of M. Taking only such "pure" submodules (as in [DG], II, §2, 1.3) would be too restrictive and not allow kernels and images of all homomorphisms.

So let us define a *submodule* of a G–module M to be a k–submodule N of M that has itself a G–module structure such that the inclusion of N into M is a homomorphism of G–modules. If so, then M/N has a natural structure as G–module: We have for each A an exact sequence of $G(A)$–modules $N \otimes A \to M \otimes A \to (M/N) \otimes A \to 0$. We call M/N the *factor module* of M by N. It has the usual property of a factor module.

Still, our definition of a submodule has one disadvantage: A given k–submodule N of M may conceivably carry more than one structure as a G–module. In order to prevent this we shall prefer to make special assumptions on our group and not on the modules.

An affine scheme X over k is called *flat* if $k[X]$ is a flat k–module. A k–group scheme is called flat, if it is so as an affine scheme. This property is obviously preserved under base change.

Assume now that G is a flat k–group scheme. If N is a submodule of a G–module M, then $N \otimes k[G]$ is a $G(k[G])$–stable submodule of $M \otimes k[G]$ (by our assumption of flatness). Then we obviously get

$$\Delta_M(N) \subset N \otimes k[G] \tag{1}$$

and

$$\Delta_N = (\Delta_M)_{|N}. \tag{2}$$

Together with 2.8 the second equality implies that the G–module structure on N is unique. On the other hand, if N is a k–submodule of M satisfying (1), then

(2) defines a G–module structure on N and N is a G–submodule of M. *So the G–submodules of M are precisely the k–submodules of M satisfying* (1).

Using 2.8(4) one easily checks:

(3) *Let G be a flat k–group scheme. The kernel* $\ker(\varphi)$ *and the image* $\mathrm{im}(\varphi)$ *of a G–module homomorphism $\varphi : M \to M'$ are G–submodules of M resp. of M'.*

We get from this that the G–modules form an abelian category (for G flat). Under the same assumption *intersections* and *sums* of submodules are again submodules. Note that *inductive limits* exist in the category of G–modules (for G flat): Just take the inductive limit as a k–module. This is a factor module of the direct sum (which is O.K. by 2.7(1)) where we divide by the sum of images of homomorphisms.

2.10. (Fixed Points) Let G be a k–group scheme and M a G–module. Set

(1) $$M^G = \{\, m \in M \mid g(m \otimes 1) = m \otimes 1 \text{ for all } g \in G(A) \text{ and all } A \,\}.$$

This is a k–submodule of M and its elements are called the *fixed points* of G on M. We call M a *trivial* G–module if $M = M^G$. In the notations of 2.6 one has $M^G = (M_a)^G(k)$. If we take $g = \mathrm{id}_{k[G]} \in G(k[G])$ in (1), then we get

(2) $$M^G = \{\, m \in M \mid \Delta_M(m) = m \otimes 1 \,\}.$$

This description of M^G as kernel of $\Delta_M - \mathrm{id}_M \otimes 1$ yields for each k–algebra k':

(3) *If k' is flat as a k–module, then $(M \otimes k')^{G_{k'}} = M^G \otimes k'$.*

In case k is a field, this implies $(M_a)^G = (M^G)_a$. (See [DG], II, §2, 1.6 for a generalisation to k–group functors.)

If $\varphi : M \to M'$ is a homomorphism of G–modules, then obviously $\varphi(M^G) \subset (M')^G$. In this way $M \mapsto M^G$ is a functor from $\{\, G\text{-modules} \,\}$ to $\{\, k\text{–modules} \,\}$ that we call *fixed point functor* (relative to G). It is certainly additive. We get from (2):

(4) *If G is flat, then the fixed point functor is left exact.*

Furthermore it commutes with taking direct sums, intersections of submodules, and direct limits (but in general not with arbitrary inductive limits).

If we consider $k[G]$ as a G–module via the left or right regular representation, then the definition immediately yields:

(5) $$k[G]^G = k\,1 \qquad (\text{for } \rho_l \text{ and for } \rho_r).$$

Let M' be another G–module; suppose that M is finitely generated and projective over k. We can then regard $\mathrm{Hom}(M, M')$ as a G–module and get easily:

(6) $$\mathrm{Hom}(M, M')^G = \mathrm{Hom}_G(M, M').$$

Therefore (3) implies:

(7) *Let k' be a flat k–algebra and let M be finitely generated and projective as a k–module. Then the canonical map*

$$\mathrm{Hom}_G(M, M') \otimes k' \longrightarrow \mathrm{Hom}_{G_{k'}}(M \otimes k', M' \otimes k')$$

is an isomorphism for all G–modules M'.

Suppose that k' is finitely generated and projective as a k–module. Then $\varprojlim$ (and $\varinjlim$) commute with $? \otimes k'$. So if M is a direct limit of G–modules M_i that are finitely generated and projective over k, then

$$\begin{aligned}\mathrm{Hom}_{G_{k'}}(M \otimes k', M' \otimes k') &= \mathrm{Hom}_{G_{k'}}(\varinjlim M_i \otimes k', M' \otimes k')\\ &\simeq \varprojlim \mathrm{Hom}_{G_{k'}}(M_i \otimes k', M' \otimes k')\\ &\simeq \varprojlim\,(\mathrm{Hom}_G(M_i, M') \otimes k')\\ &\simeq (\varprojlim \mathrm{Hom}_G(M_i, M')) \otimes k'\\ &\simeq \mathrm{Hom}_G(\varinjlim M_i, M') \otimes k' = \mathrm{Hom}_G(M, M') \otimes k'.\end{aligned}$$

So (7) extends to any such M. It will follow from 2.13(3) that we can take any M, if k is a field, or any torsion free M, if k is a Dedekind ring and G is flat.

We can generalise (1)–(4) as follows. For each $\lambda \in X(G)$ set

(1′) $\quad M_\lambda = \{\, m \in M \mid g(m \otimes 1) = m \otimes \lambda(g) \text{ for all } g \in G(A) \text{ and all } A\,\}.$

Then:

(2′) $$M_\lambda = \{\, m \in M \mid \Delta_M(m) = m \otimes \lambda\,\}.$$

(3′) *For k' as in (3) we have $(M \otimes k')_{\lambda \otimes 1} = M_\lambda \otimes k'$.*

(4′) *If G is flat, then the functor $M \mapsto M_\lambda$ is exact.*

We have furthermore:

(8) *If k is a field, then the sum of all M_λ is direct.*

(If $\sum_\lambda m_\lambda = 0$ where each $m_\lambda \in M_\lambda$, then $0 = \Delta_M(\sum_\lambda m_\lambda) = \sum_\lambda m_\lambda \otimes \lambda$. Now apply 2.4(5).)

2.11. (Representations of Diagonalisable Group Schemes) Let Λ be a commutative group and consider $G = \mathrm{Diag}(\Lambda)$ as in 2.5. As $k[G]$ is a free k–module with basis Λ, we can write the comodule map Δ_M for any G–module M as

(1) $$\Delta_M(m) = \sum_{\lambda \in \Lambda} p_\lambda(m) \otimes \lambda$$

for suitable $p_\lambda \in \mathrm{End}(M)$. Using the description of Δ_G, ε_G in 2.5 and the formulas 2.8(2), (3) one easily checks (cf. [DG], II, §2, 2.5) that $\sum_{\lambda \in \Lambda} p_\lambda = \mathrm{id}_M$ and $p_\lambda p_{\lambda'} = 0$ for $\lambda \neq \lambda'$ and $p_\lambda^2 = p_\lambda$ for all λ. This implies that M is the direct sum of all $p_\lambda(M)$, that

(2) $$p_\lambda(M) = \{\, m \in M \mid \Delta_M(m) = m \otimes \lambda\,\} = M_\lambda$$

(using 2.10(2′)), and

(3) $$M = \bigoplus_{\lambda \in \Lambda} M_\lambda.$$

It follows easily that for all G–modules M, M'

(4) $$\operatorname{Hom}_G(M, M') = \prod_{\lambda \in \Lambda} \operatorname{Hom}(M_\lambda, M'_\lambda)$$

and that the functor $M \mapsto M_\lambda$ is exact for each λ.

If we consider for example $k[G]$ and ρ_r, then we get

(5) $$k[G]_\lambda = k\,\lambda \qquad \text{for all } \lambda \in \Lambda.$$

Let $(e(\lambda))_{\lambda \in \Lambda}$ denote the standard basis of the group ring $\mathbf{Z}[\Lambda]$ over $\mathbf{Z}$. We write Λ additively and have then $e(\lambda)\,e(\lambda') = e(\lambda + \lambda')$ for all $\lambda, \lambda' \in \Lambda$. If M is a G–module such that each M_λ is a projective module of finite rank and such that $M_\lambda \neq 0$ only for finitely many λ, then we define the *formal character* of M as

(6) $$\operatorname{ch} M = \sum_{\lambda \in \Lambda} \operatorname{rk}(M_\lambda)\, e(\lambda) \in \mathbf{Z}[\Lambda].$$

For an exact sequence $0 \to M' \to M \to M'' \to 0$ of G–modules of this type one has

(7) $$\operatorname{ch} M = \operatorname{ch} M' + \operatorname{ch} M''.$$

For two G–modules M_1, M_2 of this type, also $M_1 \otimes M_2$ has this property, and one has

(8) $$\operatorname{ch}(M_1 \otimes M_2) = (\operatorname{ch} M_1)\,(\operatorname{ch} M_2).$$

One uses for (8) that for arbitrary M_1 and M_2 and all λ, $\lambda' \in \Lambda$

(9) $$(M_1)_\lambda \otimes (M_2)_{\lambda'} \subset (M_1 \otimes M_2)_{\lambda+\lambda'}.$$

(One can generalise (6) to the case where the M_λ are only assumed to be finitely generated over k and where we replace $\mathbf{Z}$ by the Grothendieck group of such modules.)

If k' is a k–algebra, then one has obviously for all λ

(10) $$(M \otimes k')_\lambda = M_\lambda \otimes k'.$$

If $\operatorname{ch}(M)$ is defined, then so is $\operatorname{ch}(M \otimes k')$ and it is equal to $\operatorname{ch}(M)$.

Suppose that M is a G–module finitely generated and projective over k. Then also M^* is a G–module, cf. 2.7(4). The number of λ with $M_\lambda \neq 0$ is finite, so M^* is the direct sum of all $\{f \in M^* \mid f(M_\mu) = 0 \text{ for all } \mu \neq \lambda\} \simeq (M_\lambda)^*$. This subspace is contained in $(M^*)_{-\lambda}$ as one easily checks. Since the sum of the $(M^*)_\nu$ is direct, this implies

(11) $$(M^*)_\nu \simeq (M_{-\nu})^* \qquad \text{for all } \nu \in \Lambda.$$

If all M_λ are projective of finite rank, so are the $(M_\lambda)^*$; it follows that in this case

(12) $$\mathrm{ch}(M^*) = \sum_{\lambda\in\Lambda} \mathrm{rk}(M_\lambda)\, e(-\lambda).$$

2.12. (Centralisers and Stabilisers) Let G be a k–group scheme and M a G–module.

For any subset $S \subset M$, we define its *centraliser* $Z_G(S)$ as the subgroup functor of G with

(1) $$Z_G(S)(A) = \{\, g \in G(A) \mid g(m \otimes 1) = m \otimes 1 \text{ for all } m \in S\,\}.$$

Obviously $Z_G(S)$ depends only on the k–module generated by S. It is equal to the intersection of all $Z_G(m)$ with $m \in S$.

For any k–submodule $N \subset M$, we define its *stabiliser* $\mathrm{Stab}_G(N)$ in G as the subgroup functor of G with

(2) $$\mathrm{Stab}_G(N)(A) = \{\, g \in G(A) \mid g\,(\,\overline{N \otimes A}\,) = \overline{N \otimes A}\,\},$$

where $\overline{N \otimes A}$ denotes the canonical image of $N \otimes A$ in $M \otimes A$. (If N is finitely generated over k and a direct summand of M as a k–module, then it suffices to demand $g\,(\,\overline{N \otimes A}\,) \subset \overline{N \otimes A}$ in (2), cf. [DG], II, §2, 1.3.)

For two k–submodules $N' \subset N$ of M, we define a subgroup functor $G_{N',N}$ of G through

(3) $$G_{N',N}(A) = \{\, g \in \mathrm{Stab}_G(N') \mid g(n \otimes 1) - n \otimes 1 \in \overline{N' \otimes A} \text{ for all } n \in N\,\}.$$

Obviously $G_{0,N} = Z_G(N)$ and $G_{N,N} = \mathrm{Stab}_G(N)$.

Suppose that M/N' is a projective k–module. Then N' is a direct summand of $M \simeq N' \oplus M/N'$. By adding to M a suitable k–module with trivial G–action, we can assume that M/N' is free (without changing $G_{N',N}$). Choose a basis $(e_j)_{j\in J}$ of some complement to N' in M. For any $m \in M$ there are $a_j(m) \in k$ and $f_{j,m} \in k[G]$, almost all equal to 0 in both cases, such that $m \in \sum_{j\in J} a_j(m)e_j + N'$ and $\Delta_M(m) \in \sum_{j\in J} e_j \otimes f_{j,m} + N' \otimes k[G]$. Then

$$g(m \otimes 1) - m \otimes 1 = \sum_{j\in J} e_j \otimes (f_{j,m}(g) - a_j(m)) + N' \otimes A$$

for any A and any $g \in G(A)$. So we have $g \in G_{N',N}(A)$ if and only if $f_{j,n}(g) = a_j(n) = f_{j,n}(g^{-1})$ for all $n \in N$ and $j \in J$. (The condition involving g^{-1} is needed to ensure that g maps $\overline{N' \otimes A}$ onto itself.) Since $f_{j,n}(g^{-1}) = \sigma_G(f_{j,n})(g)$ and $\sigma_G(f_{j,n}) \in k[G]$, this shows that $G_{N',N}$ is a closed subgroup scheme of G. It follows that:

(4) *If M is a projective module over k, then $Z_G(S)$ is a closed subgroup scheme of G for each subset $S \subset M$.*

(5) *If N is a k–submodule of M such that M/N is projective over k, then $\mathrm{Stab}_G(N)$ is a closed subgroup scheme of G.*

(6) *If $N' \subset N$ are k–submodules of M such that M/N' is projective over k, then $G_{N',N}$ is a closed subgroup scheme of G.*

2.13. (Local Finiteness) Let G be a *flat* k–group scheme and M a G–module.

We know that an intersection of G–submodules of M is again a G–submodule. So for each subset S of M there is a smallest G–submodule of M containing S. It is called the G–submodule *generated* by S and usually denoted by kGS. (Note that in general $kGS \neq kG(k)S$, the k–$G(k)$–submodule generated by S.)

Now take $m \in M$ and write $\Delta_M(m) = \sum_{i=1}^r m_i \otimes f_i$ with $m_i \in M$ and $f_i \in k[G]$. We claim that

(1) $$kGm \subset \sum_{i=1}^{r} k\, m_i.$$

Let us write $M' = \sum_{i=1}^r k\, m_i$. As $1m = m$, we have $m = \sum_{i=1}^r f_i(1) m_i \in M'$. The same argument proves more generally that $N \subset M'$ where $N = \{n \in M \mid \Delta_M(n) \in M' \otimes k[G]\}$. Since obviously $m \in N$, it will be enough to show that N is a submodule of M, i.e., that $\Delta_M(N) \subset N \otimes k[G]$.

By definition $N = \Delta_M^{-1}(M' \otimes k[G])$. Using the flatness of $k[G]$ we get $N \otimes k[G] = (\Delta_M \otimes \mathrm{id}_{k[G]})^{-1}(M' \otimes k[G] \otimes k[G])$. Therefore it is enough to show $(\Delta_M \otimes \mathrm{id}_{k[G]})\Delta_M(N) \subset M' \otimes k[G] \otimes k[G]$. By 2.8(2) the left hand side is equal to $(\mathrm{id}_M \otimes \Delta_G)\Delta_M(N) \subset (\mathrm{id}_M \otimes \Delta_G)(M' \otimes k[G]) \subset M' \otimes k[G] \otimes k[G]$.

As kGm is a G–submodule, we have $\Delta_M(m) \in kGm \otimes k[G]$. We therefore may choose the m_i above all in kGm. Then (1) yields $kGm = \sum_{i=1}^r km_i$. This shows:

(2) *Each kGm with $m \in M$ is a finitely generated k–module*

and:

(3) *Each finitely generated k–submodule of M is contained in a G–submodule of M that is a finitely generated over k.*

This property is usually stated as: *Any G–module is locally finite.*

In the case of a field one can show:

(4) *If k is a field and if $\Delta_M(m) = \sum_{i=1}^r m_i \otimes f_i$ with $(f_i)_{1 \le i \le r}$ linearly independent, then $kGm = \sum_{i=1}^r km_i$.*

(We may assume that also the m_i are linearly independent. If $(m'_j)_{1\le j\le s}$ is a basis of kGm, then there are $a_{ji} \in k$ with $m'_j = \sum_{i=1}^r a_{ji} m_i$ for all j [by (1)] and there are $f'_j \in k[G]$ with $\Delta_M(m) = \sum_{j=1}^s m'_j \otimes f'_j = \sum_{i=1}^r m_i \otimes (\sum_{j=1}^s a_{ji} f'_j)$, hence $f_i = \sum_{j=1}^s a_{ji} f'_j$. This yields $r = s$ and the claim.)

2.14. (Simple Modules) In this subsection we assume that k is a *field*. Let G be a k–group scheme.

As usual, a G–module M is called *simple* (and the corresponding representation is called *irreducible*) if $M \neq 0$ and if M has no G–submodules other than 0 and M. It is called *semi-simple* if it is a direct sum of simple G–submodules.

For any M, the sum of all its simple submodules is called the *socle* of M and denoted by $\mathrm{soc}_G M$ (or simply by $\mathrm{soc}\, M$ if it is clear which G is considered). It is the largest semi-simple G–submodule of M. For a given simple G–module E, the sum of all simple G–submodules of M isomorphic to E is called the *E–isotypic component* of $\mathrm{soc}_G M$ (or the isotypic component of type E) and denoted by $(\mathrm{soc}_G M)_E$.

By 2.13(1) each element in a G–module is contained in a finite dimensional submodule. This implies:

(1) *Each simple G–module is finite dimensional.*

(2) *If M is a non-zero G–module, then* $\operatorname{soc}_G M \neq 0$.

For any G–module M and any simple G–module E, the map $\varphi \otimes e \mapsto \varphi(e)$ is an isomorphism

(3) $$\operatorname{Hom}_G(E, M) \otimes_D E \xrightarrow{\sim} (\operatorname{soc}_G M)_E \qquad \text{where } D = \operatorname{End}_G(E).$$

(Of course D is a finite dimensional algebra over k and a skew field by Schur's lemma. If k is algebraically closed, then $D = k$.)

Each representation of dimension 1 is irreducible. For $\lambda \in X(G)$, the k_λ–isotypic component of $\operatorname{soc}_G M$ is just M_λ. We get in particular $M^G = (\operatorname{soc}_G M)_k$. The discussion in 2.11 shows:

(4) *If G is a diagonalisable k–group scheme, then each G–module is semi-simple.*

The *socle series* or (ascending) *Loewy series* of M

$$0 \subset \operatorname{soc}_1 M = \operatorname{soc}_G M \subset \operatorname{soc}_2 M \subset \operatorname{soc}_3 M \subset \cdots$$

is defined iteratively through $\operatorname{soc}(M/\operatorname{soc}_{i-1} M) = \operatorname{soc}_i M/\operatorname{soc}_{i-1} M$. Again because of 2.13(3) one has

(5) $$\bigcup_{i>0} \operatorname{soc}_i M = M.$$

Any finite dimensional G–module M has a composition series (or Jordan-Hölder series). The number of factors isomorphic to a given simple G–module E is independent of the choice of the series. It is called the *multiplicity* of E as a *composition factor* of M and usually denoted by $[M : E]$ or $[M : E]_G$.

For arbitrary M we say that a simple G–module E is a composition factor of M if E is a direct summand of some $\operatorname{soc}_i M/\operatorname{soc}_{i-1} M$. This is by 2.13(3) equivalent to the condition that E is a composition factor of some finite dimensional submodule of M. We can define $[M : E] \in \mathbf{N} \cup \{\infty\}$ as the sum of the multiplicities of E in all $\operatorname{soc}_i M/\operatorname{soc}_{i-1} M$ with $i \geq 0$.

The *radical* $\operatorname{rad}_G M$ of a G–module M is the intersection of all maximal submodules of M. If $\dim M < \infty$, then $\operatorname{rad}_G M$ is the smallest submodule of M with $M/\operatorname{rad}_G M$ semi-simple. The higher radicals of M are defined inductively by $\operatorname{rad}_G^{i+1}(M) = \operatorname{rad}_G(\operatorname{rad}_G^i M)$ and $\operatorname{rad}_G^1 M = \operatorname{rad}_G M$.

If G is an algebraic k–group, then it is called *trigonalisable* (resp. *unipotent*) if it is isomorphic to a closed subgroup of T_n (resp. U_n) for some $n \in \mathbf{N}$, cf. 2.2. One can show ([DG], IV, §2, 2.5 and 3.4):

(6) *G trigonalisable $\iff$ Each simple G–module has dimension one.*

(7) *G unipotent $\iff$ Up to isomorphism k is the only simple G–module.*

For arbitrary k–group schemes one takes these results as definitions. For unipotent G we deduce $\operatorname{soc}_G M = M^G$ for each G–module M. We get now from (2):

(8) *G unipotent $\iff$ $M^G \neq 0$ for each G–module $M \neq 0$.*

If $\operatorname{soc}_G M$ is simple, then it is contained in every non-zero submodule of M; so M has to be indecomposable. Therefore (8) and 2.10(5) imply:

(9) *If G is unipotent, then $k[G]$ is indecomposable (for ρ_l and ρ_r).*

2.15. (Twisting Representations) Any homomorphism $\alpha : G \to G'$ of k–group functors leads to a functor α^* from $\{\,G'\text{–modules}\,\}$ to $\{\,G\text{–modules}\,\}$: If M is a G'–module, then set $\alpha^*(M)$ equal to M as a k–module, and let any $G(A)$ act on $M \otimes A$ via the homomorphism $\alpha(A) : G(A) \to G'(A)$ and the given action of $G'(A)$. Obviously α^* is exact and commutes with elementary constructions as in 2.7(1)–(6).

Suppose that G and G' are group schemes. The comodule map $\Delta_{\alpha^*(M)} : M \to M \otimes k[G]$ is the composition of $\Delta_M : M \to M \otimes k[G']$ with $\mathrm{id}_M \otimes \alpha^*$ where this α^* is the comorphism corresponding to α.

For any $\lambda \in X(G')$ one has obviously

$$\alpha^*(k_\lambda) \simeq k_{\alpha^*(\lambda)} \tag{1}$$

where on the right hand side α^* denotes the homomorphism $X(G') \to X(G)$ with $\lambda \mapsto \lambda \circ \alpha$. If k is integral, if G, G' are diagonalisable, if $\alpha^* : X(G') \to X(G)$ has finite kernel, and if M is a G'–module such that $\mathrm{ch}(M)$ is defined (cf. 2.11(6)), then

$$\mathrm{ch}(\alpha^* M) = \alpha^*(\mathrm{ch}\, M) \tag{2}$$

where α^* on the right hand side now denotes the ring homomorphism $\mathbf{Z}[X(G')] \to \mathbf{Z}[X(G)]$ with $e(\lambda) \mapsto e(\lambda \circ \alpha)$ for all $\lambda \in X(G')$.

In case $G = G'$ (so if α is a group endomorphism) then we say that $\alpha^*(M)$ arises from M by *twisting* with α. If $\alpha \in \mathrm{Aut}(G)$, then we usually write ${}^{\alpha}M = (\alpha^{-1})^* M$. Then ${}^{\alpha}({}^{\beta}M) = {}^{(\alpha\beta)}M$ for all $\alpha, \beta \in \mathrm{Aut}(G)$. Note that (both for the left and right regular representations)

$${}^{\alpha}k[G] \simeq k[G] \tag{3}$$

via $f \mapsto f \circ \alpha$, i.e., the comorphism of α.

If G is a normal subgroup scheme of some k–group scheme H, then each $h \in H(k)$ induces by conjugation an automorphism $\mathrm{Int}(h)$ of G. We usually write ${}^{h}M = {}^{\mathrm{Int}(h)}M$ and get then ${}^{h}({}^{h'}M) = {}^{(hh')}M$ for all $h, h' \in H(k)$. We usually say that ${}^{h}M$ arises from M by twisting with h. If M extends to an H–module, then ${}^{h}M \simeq M$ for all $h \in H(k)$; the isomorphism is given by the action of h on M.

2.16. (Twisting with Ring Endomorphisms) A representation over k of an abstract group can also be twisted by a ring endomorphism of k: If M is a k–module, then let $M^{(\varphi)}$ denote the k–module that coincides with M as an abelian group, but where $a \in k$ acts as $\varphi(a)$ does on M. Then $\mathrm{End}_k(M) \subset \mathrm{End}_k(M^{(\varphi)})$, so any group acting linearly on M automatically acts linearly on $M^{(\varphi)}$.

Suppose for the moment that k is algebraically closed and that φ is bijective. Consider a group of the form $G(k)$ with G a reduced algebraic k–group. Let M be a $G(k)$–module and let $(m_i)_{i \in I}$ be a basis of M. There are functions c_{ij} on $G(k)$ such that $gm_j = \sum_i c_{ij}(g) m_i$ for all j. Then M is a rational $G(k)$–module (cf. 2.8, remark) if and only if $c_{ij} \in k[G]$ for all i, j. As we assume φ to be bijective, $(m_i)_{i \in I}$ is also a basis of $M^{(\varphi)}$. In $M^{(\varphi)}$ one has $gm_j = \sum_i \varphi^{-1}(c_{ij}(g)) m_i$, so $M^{(\varphi)}$ is a rational $G(k)$–module if and only if $\varphi^{-1} \circ c_{ij} \in k[G]$ for all i, j. So $M \mapsto M^{(\varphi)}$ will map rational $G(k)$–modules to rational $G(k)$–modules if $f \mapsto \varphi^{-1} \circ f$ maps $k[G]$ to $k[G]$, e.g., if φ^{-1} is a polynomial.

This can be generalised to arbitrary k and k–group schemes as follows: Let $\psi : \mathbf{A}^1 \to \mathbf{A}^1$ be a morphism such that each $\psi(A)$ is a ring endomorphism on $\mathbf{A}^1(A) = A$ and such that $\psi(k)$ is bijective. Set $\varphi = \psi(k)^{-1}$. Then $\psi_* : f \mapsto \psi \circ f$ is a ring endomorphism on $k[G] = \mathrm{Mor}(G, \mathbf{A}^1)$, but not, in general, k–linear. If we change the k–structure on $k[G]$ to that of $k[G]^{(\varphi)}$, then ψ_* is a homomorphism of k–algebras $k[G]^{(\varphi)} \to k[G]$. If M is a G–module, then the comodule map $\Delta_M : M \to M \otimes k[G]$ can also be regarded as a k–linear map

$$M^{(\varphi)} \to (M \otimes k[G])^{(\varphi)} \xrightarrow{\sim} M^{(\varphi)} \otimes k[G]^{(\varphi)}.$$

If we compose with $\mathrm{id}_M \otimes \psi_*$, we get a k–linear map $M^{(\varphi)} \to M^{(\varphi)} \otimes k[G]$ that can be checked to be a comodule map for $M^{(\varphi)}$ (one uses $(\psi_* \otimes \psi_*) \circ \Delta_G = \Delta_G \circ \psi_*$ and $\varepsilon_G \circ \psi_* = \varphi^{-1} \circ \varepsilon_G$). We get thus a structure as a G–module on $M^{(\varphi)}$.

For any k–algebra A with $\psi(A)$ bijective, we have an isomorphism of A–modules

$$M^{(\varphi)} \otimes A \xrightarrow{\sim} (M \otimes A)^{(\psi(A)^{-1})}, \qquad m \otimes a \mapsto m \otimes \psi(A)^{-1}(a).$$

Under this identification any $g \in G(A)$ acts on $M^{(\varphi)} \otimes A$ as on $M \otimes A$.

If M has a finite basis $(m_i)_{1 \le i \le r}$ and if $c_{ij} \in k[G]$ are the matrix coefficients for this basis and the given action (i.e., any $g \in G(A)$ has matrix $(c_{ij}(A)(g))_{i,j}$ with respect to the basis $(m_i \otimes 1)_i$ of $M \otimes A$), then $(m_i)_i$ is also a basis of $M^{(\varphi)}$ and here the matrix coefficients are the $\psi \circ c_{ij}$.

We shall apply twists of this second kind only in the following situation: Let p be a prime number and suppose that $p1 = 0$ in k. Choose $r \in \mathbf{N}$ and let $\psi : \mathbf{A}^1 \to \mathbf{A}^1$ with $\psi(a) = a^{p^r}$ for all $a \in A$ and all A. Then our assumptions are satisfied if $a \mapsto a^p$ is bijective on k (e.g., if k is a perfect field of characteristic p). In this case we usually write $M^{(r)}$ instead of $M^{(\varphi)}$. So $M^{(r)}$ is M with any $a \in k$ acting as $a^{p^{-r}}$ on M.

CHAPTER 3

Induction and Injective Modules

In the representation theory of finite groups or of Lie groups the process of inducing representations from a subgroup to the whole group is an important technique. The same holds for algebraic group schemes. So we start this chapter with the necessary definitions (3.3), prove elementary properties (3.4–3.6), and describe some easy special cases (3.7/8). All this is a more or less straightforward generalisation of what is done in the finite group case or the Lie group case. We have, however, to assume that the group G and its subgroup are flat.

We then use the induction functor to show that the category of G–modules contains enough injective objects, i.e., that each G–module can be embedded into an injective one (3.9).

In the case where our ground ring k is a field we can be more precise. Then the injective G–modules are determined up to isomorphism by their socles and any semi-simple G–module M occurs as the socle of such an injective G–module: the injective hull of M. The indecomposable injective G–modules are just the injective hulls of the simple G–modules. We get especially a decomposition of $k[G]$ generalising the decomposition of the regular representation of a finite group into principal indecomposable modules. (These results are proved in 3.10–3.17.)

Let me mention as a source [Green 1] for the last part (3.12–3.17). For the first part one may compare [Haboush 2], [Oberst], [Cline, Parshall and Scott 3], or [Donkin 1]. (There is not much point in attributing priorities for these generalisations.)

We assume from 3.3 on that G is a *flat* k–group scheme and from 3.10 on that k is a field.

3.1. (Restriction) Let G be a k–group functor and H a subgroup functor of G. Each G–module is an H–module in a natural way: Restrict the action of $G(A)$ for each k–algebra to $H(A)$. In this way we get a functor

$$\operatorname{res}_H^G : \{\, G\text{–modules} \,\} \to \{\, H\text{–modules} \,\},$$

which is obviously exact. It is a special case of an α^* as in 2.15 and commutes (as any α^*) with the constructions described in 2.7(1)–(6).

If G and H are group schemes, then we get the comodule map for $\operatorname{res}_H^G M$ from Δ_M as $(\mathrm{id}_M \otimes \gamma) \circ \Delta_M$ where $\gamma : k[G] \to k[H]$ is the restriction of functions.

3.2. Consider a k–module M and a group scheme H over k. We have $M \otimes k[H] = M_a(k[H]) \simeq \operatorname{Mor}(H, M_a)$ by 1.3, and more generally $(M \otimes k[H]) \otimes A \simeq (M \otimes A) \otimes_A (k[H] \otimes A) = (M \otimes A) \otimes_A A[H_A] \simeq \operatorname{Mor}(H_A, (M \otimes A)_a)$ for each k–algebra A. If M is an H–module with comodule map Δ_M, then this isomorphism maps any $(\Delta_M \otimes \mathrm{id}_A)(v) \in M \otimes k[H] \otimes A$ with $v \in M \otimes A$ to the morphism $H_A \to (M \otimes A)_a$ with $h \mapsto h(v \otimes 1)$ for any A–algebra A' and all $h \in H(A')$. (Here $v \otimes 1 \in (M \otimes A) \otimes_A A'$ is identified with its natural image in $M \otimes A'$.)

Lemma: *Let H, H' be subgroup schemes of a k–group functor G such that H' normalises H and is flat. Let M be a G–module. Then M^H is an H'–submodule of M.*

Proof: Let $\Delta_M : M \to M \otimes k[H]$ denote the comodule map of M, considered as an H–module. We regard $M \otimes k[H]$ as an H'–module via the given action on M and the conjugation action on $k[H]$. Under the isomorphism $(M \otimes k[H]) \otimes A \simeq \mathrm{Mor}(H_A, (M \otimes A)_a)$ as above we get an action of $H'(A)$ on $\mathrm{Mor}(H_A, (M \otimes A)_a)$ such that for all $h' \in H'(A)$ and $f : H_A \to (M \otimes A)_a$ we have

$$(h'f)(h) = h'\, f((h')^{-1}hh')$$

for all $h \in H(A')$ and each A–algebra A'. (More rigorously we should here replace h' on the right hand side by its image under $H'(A) \to H'(A')$ and we should write $f(A')$ instead of f; similarly for $h'f$.)

Using this description, it is easy to check that Δ_M is a homomorphism of H'–modules. The same holds for the map $m \mapsto \Delta_M(m) - m \otimes 1$. Therefore its kernel M^H is an H'–submodule by 2.9(3).

3.3. (Induction) Let H be a subgroup scheme of G. For each H–module M there is a natural $(G \times H)$–module structure on $M \otimes k[G]$: Let G act trivially on M and via the left regular representation on $k[G]$, let H act as given on M and via the right regular representation on $k[G]$, and then take tensor products. Now $(M \otimes k[G])^H$ is a G–submodule of $M \otimes k[G]$ by Lemma 3.2. We denote this G–module by $\mathrm{ind}_H^G M$ and call it the *induced module* of M from H to G. Obviously

$$\mathrm{ind}_H^G : \{\, H\text{–modules}\,\} \to \{\, G\text{–modules}\,\}$$

is a functor.

We can interpret the action of $G \times H$ on $M \otimes k[G]$ in a different way. Identify $(M \otimes k[G]) \otimes A \simeq \mathrm{Mor}(G_A, (M \otimes A)_a)$ for each k–algebra A as in 3.2. Any $(g, h) \in G(A) \times H(A)$ acts on some $f \in \mathrm{Mor}(G_A, (M \otimes A)_a)$ through

$$((g,h)f)(x) = h(f(g^{-1}xh)) \tag{1}$$

for all $x \in G(A')$ and all A–algebras A'. (Let me remind you that there is some abuse of notation going on: We really ought to write $((g,h)f)(A')(x) = h_{A'}\, f(A')(g_{A'}^{-1}xh_{A'})$ with $g_{A'} \in G(A')$ the image of g under the map $G(A) \to G(A')$ defined by the structural map $A \to A'$, and similarly for $h_{A'}$.) In this interpretation we have

$$\begin{aligned} \mathrm{ind}_H^G M = \{\, f \in \mathrm{Mor}(G, M_a) \mid f(gh) = h^{-1}f(g) \text{ for all } g \in G(A),\\ h \in H(A) \text{ and all } k\text{–algebras } A \,\} \end{aligned} \tag{2}$$

and G acts by left translation (in a natural sense).

Proposition: *Let H be a flat subgroup scheme of G.*

a) *The functor* ind_H^G *is left exact.*

b) *The functor* ind_H^G *commutes with forming direct sums, intersections of submodules, and direct limits.*

Proof: a) As we assume G to be flat, the functor $M \mapsto M \otimes k[G]$ is exact. Therefore the claim follows from 2.10(4).

b) All these constructions commute with tensoring with a flat k–module and with the fixed point functor, cf. 2.10.

Remark: If the fixed point functor $?^H$ is exact, then obviously also ind_H^G is exact. So ind_H^G is certainly exact whenever H is diagonalisable (by 2.11).

3.4. For any k–module M let $\varepsilon_M : M \otimes k[G] \to M$ be the linear map $\varepsilon_M = \operatorname{id}_M \overline{\otimes} \varepsilon_G$. If we make the identification $M \otimes k[G] \simeq \operatorname{Mor}(G, M_a)$, then we get $\varepsilon_M(f) = f(1)$. We shall also use the notation ε_M for the restriction of ε_M to various submodules of $M \otimes k[G]$.

Proposition (Frobenius Reciprocity): *Let H be a flat subgroup scheme of G and M an H–module.*

a) $\varepsilon_M : \operatorname{ind}_H^G M \to M$ *is a homomorphism of H–modules.*

b) *For each G–module N the map $\varphi \mapsto \varepsilon_M \circ \varphi$ is an isomorphism*

$$\operatorname{Hom}_G(N, \operatorname{ind}_H^G M) \xrightarrow{\sim} \operatorname{Hom}_H(\operatorname{res}_H^G N, M).$$

Proof: a) We have for all A, all $h \in H(A)$ and $f \in \operatorname{ind}_H^G M$

$$(\varepsilon_M \otimes \operatorname{id}_A)(hf) = (hf)(1) = f(h^{-1}) = h(f(1)) = h(\varepsilon_M(f) \otimes 1).$$

b) In order to define an inverse, consider for each $\psi \in \operatorname{Hom}_H(N, M)$ and any $x \in N$ the morphism $\widetilde{\psi}(x) \in \operatorname{Mor}(G, M_a)$ with $\widetilde{\psi}(x)(g) = (\psi \otimes \operatorname{id}_A)(g^{-1}(x \otimes 1))$ for all A and all $g \in G(A)$. Using the description in 3.3(2), one easily checks that $\widetilde{\psi}(x) \in \operatorname{ind}_H^G M \subset \operatorname{Mor}(G, M_a)$. Another straightforward calculation shows that $\widetilde{\psi} \in \operatorname{Hom}_G(N, \operatorname{ind}_H^G M)$ and that the maps $\psi \mapsto \widetilde{\psi}$ and $\varphi \mapsto \varepsilon_M \circ \varphi$ are inverse to each other.

3.5. (Transitivity of Induction) The last result implies (for G, H as above):

(1) *The functor* ind_H^G *is right adjoint to* res_H^G.

This of course determines ind_H^G uniquely up to isomorphism. (One can also say that the pair $(\operatorname{ind}_H^G M, \varepsilon_M)$ is uniquely determined up to isomorphism by 3.4.b.)

Let H' be another flat subgroup scheme of G with $H \subset H'$. We have obviously $\operatorname{res}_H^{H'} \circ \operatorname{res}_{H'}^G = \operatorname{res}_H^G$. Therefore (1) yields:

(2) *There is an isomorphism of functor* $\operatorname{ind}_{H'}^G \circ \operatorname{ind}_H^{H'} \simeq \operatorname{ind}_H^G$.

We can express this also in this way: *Induction is transitive.* For any H–module M we can write down isomorphisms between $\operatorname{ind}_H^G M$ and $\operatorname{ind}_{H'}^G \circ \operatorname{ind}_H^{H'} M$ explicitly.

To any $f \in \operatorname{ind}_H^G M$ we associate $\widetilde{f} \in \operatorname{Mor}(G, (\operatorname{ind}_H^{H'} M)_a)$ with $\widetilde{f}(g)(h') = f(gh')$ for all $g \in G(A)$, $h' \in H'(A)$ and all A. To any $f \in \operatorname{ind}_{H'}^G(\operatorname{ind}_H^{H'} M)$ we associate $\overline{f} \in \operatorname{Mor}(G, M_a)$ with $\overline{f}(g) = f(g)(1)$ for all $g \in G(A)$ and all A. The maps $f \mapsto \widetilde{f}$ and $f \mapsto \overline{f}$ turn out to be inverse isomorphisms.

Observe that 2.10(3) implies:

Let k' be a flat k–algebra. Then we have for each H–module M a canonical isomorphism

$$(3) \qquad (\operatorname{ind}_H^G M) \otimes k' \simeq \operatorname{ind}_{H_{k'}}^{G_{k'}}(M \otimes k').$$

Let σ be an automorphism of G. For any H–module M let ${}^\sigma M$ denote the $\sigma(H)$–module which coincides as a k–module with M, and where any $g \in \sigma(G)(A)$ acts on $M \otimes A$ as $\sigma^{-1}(g) \in H(A)$ does (for each k–algebra A). Then $\operatorname{Hom}_{\sigma(H)}({}^\sigma V, {}^\sigma M) \simeq \operatorname{Hom}_H(V, M)$ for any G–module V. So (1) easily yields

$$(4) \qquad {}^\sigma(\operatorname{ind}_H^G M) \simeq \operatorname{ind}_{\sigma(H)}^G({}^\sigma M) \qquad \text{for each } H\text{–module } M.$$

3.6. Proposition (The Tensor Identity): *Let N be a G-module that is flat over k. For any flat subgroup scheme H of G and any H-module M there is a canonical isomorphism of G–modules*

$$\operatorname{ind}_H^G(M \otimes \operatorname{res}_H^G N) \xrightarrow{\sim} (\operatorname{ind}_H^G M) \otimes N.$$

Proof: Both sides may be embedded into $\operatorname{Mor}(G, (M \otimes N)_a) \simeq M \otimes N \otimes k[G]$ using 3.3(2), the left hand side as

$$L = \{ f : G \to (M \otimes N)_a \mid f(gh) = (h^{-1} \otimes h^{-1}) f(g) \text{ for all } g,\, h \}$$

and the right hand side as

$$R = \{ f : G \to (M \otimes N)_a \mid f(gh) = (h^{-1} \otimes 1) f(g) \text{ for all } g,\, h \}.$$

Here "for all g, h" means "for all $g \in G(A)$, $h \in H(A)$ and all A".

We define two endomorphisms α, β of $\operatorname{Mor}(G, (M \otimes N)_a)$ through $(\alpha f)(g) = (1 \otimes g) f(g)$ and $(\beta f)(g) = (1 \otimes g^{-1}) f(g)$ for all g. Obviously, these are isomorphisms inverse to each other. A straightforward calculation shows that $\alpha(L) \subset R$ and $\beta(R) \subset L$, and that α, β are G–equivariant for the two actions of G we consider. (On L we have $gf = f(g^{-1}?)$ and on R we have $gf = (1 \otimes g) f(g^{-1}?)$.) This implies the proposition.

Remark: The flatness of N is needed for the obvious map from $(\operatorname{ind}_H^G M) \otimes N = (M \otimes k[G])^H \otimes N$ to $M \otimes k[G] \otimes N$ to be injective. In case $H = 1$ this injectivity holds automatically; so Proposition 3.6 extends for $H = 1$ to arbitrary G–modules N.

3.7. (Trivial Examples) We can apply all this especially to the subgroup schemes $H = 1$ and $H = G$. The first case yields

$$(1) \qquad \operatorname{ind}_1^G M = M \otimes k[G] \qquad \text{for all } k\text{–modules } M$$

(where M is considered as a trivial G–module on the right hand side), especially

(2) $$\operatorname{ind}_1^G k = k[G].$$

(Here and below $k[G]$ is considered as a G–module via ρ_l.) Combining (2) with 3.4.b (Frobenius reciprocity) we get for each G–module M

(3) $$\operatorname{Hom}_G(M, k[G]) \simeq M^*.$$

(This can also be shown directly using matrix coefficients, cf. [DG], §2, 2.3.) Taking $M = k$ in 3.6 we get for each G–module N an isomorphism

(4) $$N \otimes k[G] \xrightarrow{\sim} \operatorname{ind}_1^G N = N_{tr} \otimes k[G]$$

where N_{tr} denotes the k–module N considered as a trivial G–module. (Note that N need not be flat over k by the remark in 3.6.) Going back to the proof and the definitions one checks that this isomorphism is given by

$$x \otimes f \mapsto (1 \otimes f) \cdot (\mathrm{id}_N \otimes \sigma_G) \circ \Delta_N(x).$$

If we restrict this map to the G–submodule $N \otimes k1 \simeq N$ of $N \otimes k[G]$, then we get (as σ_G intertwines between ρ_r and ρ_l):

(5) $\Delta_N : N \to N_{tr} \otimes k[G]$ *is an injective homomorphism of G–modules if we consider $k[G]$ as a G–module via ρ_r.*

(This can also be checked directly or using 2.8(2), (4), (5).)

As $\operatorname{res}_G^G M = M$ for each G–module M, we have also by 3.5(1) (canonically)

(6) $$M \xrightarrow{\sim} \operatorname{ind}_G^G M \qquad \text{for each } G\text{–module } M.$$

This isomorphism $M \xrightarrow{\sim} (M \otimes k[G])^G \subset M \otimes k[G]$ is given by $(\mathrm{id}_M \otimes \sigma_G) \circ \Delta_M$. (In other words, any $m \in M$ is mapped to the morphism $G \to M_a$ with $g \mapsto g^{-1}(m \otimes 1)$ for all $g \in G(A)$ and all A.)

3.8. (Induction and Semi-direct Products) Let G' be a flat k–group scheme acting on G through automorphisms and let H be a flat k–subgroup scheme of G stable under G'. We can then form the semi-direct products $H \rtimes G'$ and $G \rtimes G'$ (which are again flat k–group schemes). We regard G, G', H, and $H \rtimes G'$ as subgroup schemes of $G \rtimes G'$; so the action of G' on G is now by conjugation inside $G \rtimes G'$.

Let M be an $(H \rtimes G')$–module, i.e., a k–module that is simultaneously an H–module and a G'–module so that these two actions are compatible: $g'(hm) = (g'hg'^{-1})(g'm)$. Then G' acts naturally on $\operatorname{Mor}(G, M_a) \simeq k[G] \otimes M$ via $(g'f)(g) = g'(f(g'^{-1}gg'))$, i.e., through the tensor product of the conjugation action on $k[G]$ with the given action on M. This leads to structures as $(H \rtimes G')$–module and as $(G \rtimes G')$–module on $\operatorname{Mor}(G, M_a)$ where H, G act as in the construction of $\operatorname{ind}_H^G M$. As G' normalises H, it stabilises the subspace $\operatorname{ind}_H^G M$ of $\operatorname{Mor}(G, M_a)$, cf. 3.2. We get thus a structure as a $(G \rtimes G')$–module on $\operatorname{ind}_H^G M$. We claim that we have an isomorphism of $(G \rtimes G')$–modules

(1) $$\operatorname{ind}_H^G M \xrightarrow{\sim} \operatorname{ind}_{H \rtimes G'}^{G \rtimes G'} M.$$

We simply associate to $f \in \operatorname{ind}_H^G M \subset \operatorname{Mor}(G, M_a)$ the map $F \in \operatorname{ind}_{H\rtimes G'}^{G\rtimes G'} M \subset \operatorname{Mor}(G \rtimes G', M_a)$ with $F(g,g') = {g'}^{-1}F(g)$, and to any F the map f with $f(g) = F(g,1)$. The claim follows now from elementary calculations.

Taking $H = 1$ we get especially for each G'–module M

(2) $$\operatorname{ind}_{G'}^{G\rtimes G'} M \simeq k[G] \otimes M \simeq \operatorname{Mor}(G, M_a)$$

with G acting via ρ_l on $k[G]$ and trivially on M, and with G' acting via the conjugation action on $k[G]$ and as given on M.

We can also describe $\operatorname{ind}_G^{G\rtimes G'} N$ for any G–module N. There is an isomorphism

(3) $$\operatorname{ind}_G^{G\rtimes G'} N \xrightarrow{\sim} \operatorname{Mor}(G', N_a) \simeq k[G'] \otimes N$$

mapping any $F \in \operatorname{ind}_G^{G\rtimes G'} N \subset \operatorname{Mor}(G \rtimes G', N_a)$ to $f : G' \to N_a$ with $f(g') = F(1,g')$; its inverse takes any f to F with $F(g,g') = {g'}^{-1}gg'f(g')$. This isomorphism is compatible with the G'–action if we let G' act on $k[G']$ via ρ_l and trivially on N. The action of G on some $f : G' \to N_a$ is given by $(gf)(g') = (g'g{g'}^{-1})f(g')$. This implies:

(4) *If N is a trivial G–module, then G acts trivially on $\operatorname{ind}_G^{G\rtimes G'} N$.*

Lemma: *Let G, G' be flat k–group schemes, let $H \subset G$ and $H' \subset G'$ be flat k–subgroup schemes. Let M be an H–module and M' an H'–module. If M and $\operatorname{ind}_{H'}^{G'} M'$ are flat over k, then there is an isomorphism of $(G \times G')$–modules*

(5) $$\operatorname{ind}_H^G M \otimes \operatorname{ind}_{H'}^{G'} M' \simeq \operatorname{ind}_{H\times H'}^{G\times G'}(M \otimes M').$$

Proof: Note that any G–module V can be regarded as a $(G \times G')$–module such that any $(g,g') \in (G \times G')(A)$ acts on $V \otimes A$ as $g \in G(A)$ does. Similarly, any G'–module V' can be turned into a $(G \times G')$–module and we then get a $(G \times G')$–module structure on $V \otimes V'$. In this way the left hand side of (5) is to be interpreted as a $(G \times G')$–module.

The same observation holds for the direct products $H \times H'$, $H \times G'$, and $G \times H'$. Applying (1) to the case of direct products we get isomorphisms of $(G \times G')$–modules

$$\operatorname{ind}_H^G M \simeq \operatorname{ind}_{H\times G'}^{G\times G'} M \qquad \text{and} \qquad \operatorname{ind}_{H'}^{G'} M' \simeq \operatorname{ind}_{G\times H'}^{G\times G'} M'.$$

Now the transitivity of induction and the tensor identity (where we need the flatness assumptions on the modules) yield

$$\begin{aligned}\operatorname{ind}_{H\times H'}^{G\times G'}(M \otimes M') &\simeq \operatorname{ind}_{H\times G'}^{G\times G'} \operatorname{ind}_{H\times H'}^{H\times G'}(M \otimes M') \simeq \operatorname{ind}_{H\times G'}^{G\times G'}(M \otimes \operatorname{ind}_{G\times H'}^{G\times G'} M') \\ &\simeq \operatorname{ind}_{H\times G'}^{G\times G'}(M \otimes \operatorname{ind}_{H'}^{G'} M') \simeq \operatorname{ind}_{H\times G'}^{G\times G'}(M) \otimes \operatorname{ind}_{H'}^{G'} M' \\ &\simeq \operatorname{ind}_H^G M \otimes \operatorname{ind}_{H'}^{G'} M'.\end{aligned}$$

Remark: Reversing the roles of G and G' one gets the same result when M' and $\operatorname{ind}_H^G M$ are flat over k. If k is a principal ideal domain (where flat means the same as torsion free), then we can replace the assumption with the more symmetric: *If M and M' are flat over k, ...* since then the flatness of G' and M' implies that of $\operatorname{ind}_{H'}^{G'} M' \subset M' \otimes k[G']$.

3.9. We define an *injective* G–module to be an injective object in the category of all G–modules.

Proposition: a) *For each flat subgroup scheme H of G the functor ind_H^G maps injective H–modules to injective G–modules.*

b) *Any G–module can be embedded into an injective G–module.*

c) *A G–module M is injective if and only if there is an injective k–module I such that M is isomorphic to a direct summand of $I \otimes k[G]$ with I regarded as a trivial G–module.*

Proof: a) This is obvious as ind_H^G is right adjoint to the exact functor res_H^G.

b) Let M be G–module. We can embed M as a k–submodule into an injective k–module I. Then $I \otimes k[G] \simeq \operatorname{ind}_1^G I$ is injective by (a) and $\operatorname{ind}_1^G M \simeq M_{tr} \otimes k[G]$ is a submodule of $I \otimes k[G]$. Now combine this with the embedding of M into $M_{tr} \otimes k[G]$ via $(\mathrm{id}_M \otimes \sigma_G) \circ \Delta_M$, cf. 3.7.

c) If M is injective, then the embedding $M \hookrightarrow I \otimes k[G]$ constructed in the proof of (b) has to split. This gives one direction in (c). The other is obvious, as $I \otimes k[G]$ is injective by (a), hence so is each direct summand.

3.10. Let us assume from now on in Chapter 3 that k is a *field.* Then we can simplify the preceding result:

Proposition: a) *A G–module M is injective if and only if there is a vector space V over k such that M is isomorphic to a direct summand of $V \otimes k[G]$ with V regarded as a trivial G–module.*

b) *Any direct sum of injective G–modules is injective.*

c) *If M, Q are G–modules with Q injective, then $M \otimes Q$ is injective.*

Proof: (a) is just 3.9.c and (b) is an immediate consequence of (a). If Q is a direct summand of $V \otimes k[G]$ as in (a), then $M \otimes Q$ is a direct summand of $M \otimes V \otimes k[G]$, which is isomorphic to $M_{tr} \otimes V \otimes k[G]$ by 3.7(4). This yields (c).

3.11. Before looking at indecomposable injective G–modules in general, let us treat an important example.

Suppose $G = G' \rtimes H$ with H diagonalisable and G' a unipotent group scheme. We set for each $\lambda \in X(H)$

(1) $$Y_\lambda = \operatorname{ind}_H^G k_\lambda.$$

We have $k[G] \simeq \operatorname{ind}_1^G k \simeq \operatorname{ind}_H^G \operatorname{ind}_1^H k \simeq \operatorname{ind}_H^G k[H]$ by the transitivity of induction, and $k[H] \simeq \bigoplus_{\lambda \in X(H)} k_\lambda$ by 2.11(5) (also with respect to ρ_l, of course), hence

(2) $$k[G] \simeq \bigoplus_{\lambda \in X(H)} Y_\lambda.$$

We know by 3.8 that Y_λ is isomorphic to $k[G']$ when considered as a G'–module. Therefore 2.14(9) implies:

(3) *Each Y_λ is an indecomposable and injective G–module.*

Each $\lambda \in X(H)$ can be extended to an element in $X(G)$ with G' in the kernel. We denote also this extension by λ, and the corresponding G–module by k_λ. For each G–module M, the subspace $M^{G'}$ is a G–submodule by 3.2. Because of 2.11 it is a direct sum of one dimensional G–submodules of the form k_λ with $\lambda \in X(H)$. This shows especially that $M^{G'}$ is a semi-simple G–module. On the other hand, we have $M^{G'} \neq 0$ for any simple G–module because of 2.14(8). Therefore the k_λ with $\lambda \in X(H)$ are all simple G–modules (up to isomorphism) and we have

$$\operatorname{soc}_G M = M^{G'} \tag{4}$$

for each G–module M. The discussion in 3.8 shows that $Y_\lambda \simeq k_\lambda \otimes k[G']$ where H acts on $k[G']$ via the conjugation action. Then $(Y_\lambda)^{G'} \simeq k_\lambda \otimes (k[G']^{G'}) = k_\lambda \otimes k\,1 \simeq k_\lambda$, hence by (4)

$$\operatorname{soc}_G Y_\lambda \simeq k_\lambda. \tag{5}$$

This shows that in this case there is, for each simple G–module E, an indecomposable and injective G–module with socle isomorphic to E. We want to generalise this result. We first prove the uniqueness of such a module (up to isomorphism).

3.12. Proposition: *Let M, M' be injective G–modules and $\varphi \in \operatorname{Hom}_G(M, M')$. Then φ is an isomorphism if and only if φ induces an isomorphism $\operatorname{soc}_G M \xrightarrow{\sim} \operatorname{soc}_G M'$.*

Proof: The "only if" part is obvious, so let us look at the "if". We know by 2.14(2) that

$$\ker \varphi \neq 0 \implies 0 \neq \operatorname{soc}_G(\ker \varphi) = \ker(\varphi_{|\operatorname{soc}_G M}).$$

If φ induces an isomorphism of the socles, then we get $\ker \varphi = 0$ and the injectivity of φ. Therefore $\varphi(M) \simeq M$ is an injective G–module, hence a direct summand of M'. If M_1 is a G–stable complement, then $M' = \varphi(M) \oplus M_1$ implies $\operatorname{soc}_G(M') = \operatorname{soc}_G \varphi(M) \oplus \operatorname{soc}_G M_1$. The assumption $\operatorname{soc}_G M' = \varphi(\operatorname{soc}_G M)$ yields $\operatorname{soc}_G M_1 = 0$, hence $M_1 = 0$ by 2.14(2). Therefore φ is bijective.

3.13. Corollary: *Two injective G–modules are isomorphic if and only if their socles are isomorphic.*

Proof: Because of the injectivity, any isomorphism of the socles can be extended to a homomorphism of the whole modules. Then apply 3.12.

3.14. Proposition: *Let M be an injective G–module and $\varphi_1 \in \operatorname{End}_G(\operatorname{soc}_G M)$ be idempotent. Then there is $\varphi \in \operatorname{End}_G(M)$ idempotent with $\varphi_{|\operatorname{soc}_G M} = \varphi_1$.*

Proof: Consider the socle series of M as in 2.14(5). Let us abbreviate $M_i = \operatorname{soc}_i M$. Each endomorphism of M has to preserve all M_i. Therefore the injectivity of M yields, for each i, an exact sequence

$$0 \to \mathfrak{m}_i \longrightarrow \operatorname{End}_G(M) \xrightarrow{\text{res}} \operatorname{End}_G(M_i) \to 0 \tag{1}$$

where $\mathfrak{m}_i$ is the two-sided ideal

(2) $$\mathfrak{m}_i = \{\varphi \in \operatorname{End}_G(M) \mid \varphi(M_i) = 0\}.$$

Any $\varphi \in \mathfrak{m}_i$ maps M_j to M_{j-i} for all $j > i$. This implies

(3) $$\mathfrak{m}_i\,\mathfrak{m}_j \subset \mathfrak{m}_{i+j} \qquad \text{for all } i, j \geq 1.$$

We deduce from $M = \bigcup_{i\geq 1} M_i$ that

(4) $$\operatorname{End}_G(M) = \varprojlim \operatorname{End}_G(M_i).$$

Therefore the proposition follows from a version of Hensel's lemma proved in the next subsection.

3.15. Proposition: *Let R be a ring and let $\mathfrak{m}_1 \supset \mathfrak{m}_2 \supset \cdots$ be a chain of two-sided ideals of R with $\mathfrak{m}_i\mathfrak{m}_j \subset \mathfrak{m}_{i+j}$ for all $i, j \geq 1$ and $R \simeq \varprojlim R/\mathfrak{m}_i$ naturally. Then there is for each idempotent element $\overline{e} \in R/\mathfrak{m}_1$ an idempotent element $e \in R$ with $\overline{e} = e + \mathfrak{m}_1$.*

Proof: Because of $R \simeq \varprojlim R/\mathfrak{m}_i$ is suffices to construct $e_1, e_2, \ldots \in R$ such that each $e_i + \mathfrak{m}_i \in R/\mathfrak{m}_i$ is idempotent and such that $e_i + \mathfrak{m}_{i-1} = e_{i-1} + \mathfrak{m}_{i-1}$ for all $i > 1$ and $e_1 + \mathfrak{m}_1 = \overline{e}$. We choose as e_1 an arbitrary representative in R of $\overline{e} \in R/\mathfrak{m}_1$. We then define iteratively $e_{i+1} = 2e_i(e_i - e_i^2) + e_i^2$ for each $i \geq 1$. As $e_i + \mathfrak{m}_i$ is supposed to be idempotent, we have $(e_i - e_i^2)^2 \in \mathfrak{m}_i^2 \subset \mathfrak{m}_{i+1}$, hence $e_i^4 \equiv 2e_i^3 - e_i^2 \pmod{\mathfrak{m}_{i+1}}$. Using this one checks that $e_{i+1}^2 \in 2e_i(e_i - e_i^2) + e_i^2 + \mathfrak{m}_{i+1}$, i.e., that $e_{i+1} + \mathfrak{m}_{i+1}$ is idempotent. Therefore our iterative construction works.

3.16. Proposition: a) *For each simple G–module E there is an injective G–module Q_E (unique up to isomorphism) with $E \simeq \operatorname{soc}_G Q_E$.*

b) *An injective G–module is indecomposable if and only if it is isomorphic to Q_E for some simple G–module E.*

c) *Any injective G–module Q is a direct sum of indecomposable submodules. For each simple G–module E the number of summands isomorphic to Q_E is equal to the multiplicity of E in $\operatorname{soc}_G Q$.*

Proof: Let Q be an injective G–module. Any decomposition $\operatorname{soc}_G Q = M_1 \oplus M_2$ leads by 3.14 to a decomposition $Q_1 \oplus Q_2$. As we can embed any G–module into an injective G–module by 3.9.b, we get the existence of the Q_E in (a) immediately. The uniqueness follows from 3.13. The other parts of the proposition are now obvious. (Note that we admit here infinite multiplicities.)

3.17. The module Q_E from 3.16.a is called the *injective hull* of E. More generally, we can find for each G–module M an injective G–module Q_M (unique up to isomorphism) with $\operatorname{soc}_G Q_M = \operatorname{soc}_G M$. The embedding of $\operatorname{soc}_G M$ into Q_M can be extended to an embedding of M into Q_M. We call Q_M the injective hull of M. It is clear that this is compatible with the general definition, e.g., in [B1], ch. X, §1, n° 9.

In the situation of 3.16.c, the (possibly infinite) number of summands isomorphic to Q_E is equal to

$$\dim \operatorname{Hom}_G(E,Q)/\dim \operatorname{End}_G(E),$$

cf. 2.14(3). If we take especially $Q = k[G]$, then we get from 3.7(3)

$$k[G] \simeq \bigoplus_E Q_E^{d(E)} \tag{1}$$

where

$$d(E) = \dim(E)/\dim(\operatorname{End}_G(E)) \tag{2}$$

and where the direct sum is taken over a system of representatives of all simple G–modules. (If k is algebraically closed, then $d(E) = \dim(E)$ of course.)

In the situation of 3.11 we have obviously $Y_\lambda = Q_{k_\lambda}$, and 3.11(2) illustrates (1) very well. In the case of a unipotent group one has $k[G] = Q_k$, cf. 2.14(9).

Let me mention one standard property of injective hulls: Let E be a simple G–module and M a finite dimensional G–module. Then

$$[M : E]_G = \dim \operatorname{Hom}_G(M, Q_E)/\dim \operatorname{End}_G(E). \tag{3}$$

(For the notation cf. 2.14.)

3.18. (Projective and Injective Modules) We call a projective object in the category of all G–modules simply a *projective G–module.* Note that, for example, by 2.11(4) any G–module is projective if G is diagonalisable. (Recall that we assume k to be a field; otherwise we should require the G–module to be projective over k.) In general, 0 may be the only the only projective G–module. [Donkin 27] (with a correction in [Donkin 31], p. 218) contains a classification of all G that have non-zero projective G–modules. Here we shall meet important examples in 8.10 and II.9.3. At this point we prove (following [Donkin 27], Lemma 1):

Lemma: a) *Any projective G–module is injective.*

b) *If there exists a non-zero projective G–module, then any injective G–module is projective.*

Proof: a) Let P be a projective G–module. For each finite dimensional G–module V also $P \otimes V$ is projective since we have for each G–module M a canonical isomorphism $\operatorname{Hom}_G(P \otimes V, M) \xrightarrow{\sim} \operatorname{Hom}_G(P, V^* \otimes M)$, cf. 4.4 below. Any short exact sequence $0 \to V_1 \to V \to V_2 \to 0$ of finite dimensional G–modules leads to a short exact sequence

$$0 \to V_2^* \otimes P \longrightarrow V^* \otimes P \longrightarrow V_1^* \otimes P \to 0$$

with all terms projective. So the latter sequence splits and the map on the fixed points $(V^* \otimes P)^G \to (V_1^* \otimes P)^G$ is surjective. But this map identifies with the natural map $\operatorname{Hom}_G(V, P) \to \operatorname{Hom}_G(V_1, P)$. Now one gets for arbitrary injective homomorphisms $N \to M$ that the induced map $\operatorname{Hom}_G(M, P) \to \operatorname{Hom}_G(N, P)$ is surjective using the local finiteness of M and Zorn's lemma. So P is injective.

b) Let P be a non-zero projective G–module. Choose a non-zero finite dimensional submodule P_1 of P. The embedding of the trivial G–module k as $k \cdot \mathrm{id}$ into $\mathrm{End}(P_1) \simeq P_1^* \otimes P_1 \subset P_1^* \otimes P$ induces for each G–module M an embedding into $M \otimes P_1^* \otimes P$. If M is simple, hence finite dimensional, then $M \otimes P_1^* \otimes P$ is projective (as above), hence injective by a). By 3.16.c the injective hull Q_M of M is isomorphic to a direct summand of $M \otimes P_1^* \otimes P$, hence also projective. Now the claim follows from 3.16.b/c.

CHAPTER 4

Cohomology

Throughout this chapter let G be a flat k–group scheme.

We have shown in the preceding chapter that each G–module has a resolution by injective G–modules. Therefore we can define (right) derived functors of left exact functors on the category of G–modules. We can, for example, describe the Ext-functors as derived from the Hom-functor, and we can introduce the cohomology functors $H^n(G, ?)$ as derived from the fixed point functor. Furthermore, there are for each flat subgroup scheme H of G the derived functors $R^n \operatorname{ind}_H^G$ of the induction functor.

After recalling some general facts about derived functors (4.1), and making the definitions (4.2) we prove many elementary properties of the derived functors mentioned above (4.3–4.13, 4.17). We prove equalities between certain derived functors and mention several spectral sequences. We show that the cohomology can be computed using an explicit complex: the Hochschild complex (4.14–4.16). This complex is used for the proof of a universal coefficient theorem (4.18) and for the computation of the cohomology of the additive group over a field (4.20–4.27). Because of later applications we formulate the results at once not for G_a but for direct products $G_a \times G_a \times \cdots \times G_a$.

As in the preceding chapter there is not much point in attributing priorities for generalities. In addition to the papers listed there, one ought to mention [Andersen 12] where some results were extended to the case of an arbitrary ground ring (instead of a field). When discussing the Hochschild complex I follow [DG] more or less. The computation of $H^\bullet(G_a, k)$ is due to [Cline, Parshall, Scott, and van der Kallen].

4.1. (Derived Functors) Let $\mathcal{C}$ be an abelian category containing enough injectives, i.e., such that each object can be embedded into an injective object. Then certainly each object admits an injective resolution. We can then define the (right) *derived functors* $R^n\mathcal{F}$ of any additive (covariant) functor $\mathcal{F}$ from $\mathcal{C}$ to some other abelian category $\mathcal{C}'$. We have $R^0\mathcal{F} = \mathcal{F}$ if and only if $\mathcal{F}$ is left exact. An object M in $\mathcal{C}$ is called *acyclic* for $\mathcal{F}$ if $R^n\mathcal{F}(M) = 0$ for all $n > 0$. Any short exact sequence in $\mathcal{C}$ gives rise to a long exact sequence in $\mathcal{C}'$.

Suppose now that $\mathcal{F} : \mathcal{C} \to \mathcal{C}'$ and $\mathcal{F}' : \mathcal{C}' \to \mathcal{C}''$ are additive (covariant) functors where $\mathcal{C}$, $\mathcal{C}'$, $\mathcal{C}''$ are abelian categories with $\mathcal{C}$ and $\mathcal{C}'$ having enough injectives.

Proposition (Grothendieck's Spectral Sequence): *If $\mathcal{F}'$ is left exact and if $\mathcal{F}$ maps injective objects in $\mathcal{C}$ to objects acyclic for $\mathcal{F}'$, then there is a spectral sequence for each object M in $\mathcal{C}$ with differentials d_r of bidegree $(r, 1-r)$ and*

(1) $$E_2^{n,m} = (R^n\mathcal{F}')(R^m\mathcal{F})M \;\Rightarrow\; R^{n+m}(\mathcal{F}' \circ \mathcal{F})M.$$

One can find a proof (and more background material) in S. Lang's *Algebra* (from the second edition) or in [Ro].

Here are two trivial special cases:

(2) *If $\mathcal{F}'$ is exact, then $\mathcal{F}' \circ R^m\mathcal{F} \simeq R^m(\mathcal{F}' \circ \mathcal{F})$ for all $m \in \mathbf{N}$.*

(This is obvious.)

(3) *If $\mathcal{F}$ is exact and maps injective objects in $\mathcal{C}$ to objects acyclic for $\mathcal{F}'$, then $R^n\mathcal{F}' \circ \mathcal{F} \simeq R^n(\mathcal{F}' \circ \mathcal{F})$ for all $m \in \mathbf{N}$.*

(This can be proved by degree shifting, i.e., by induction on n using the long exact sequence.)

Consider an arbitrary spectral sequence $(E_r^{n,m})$ with differentials $d_r^{n,m} : E_r^{n,m} \to E_r^{n+r,m+1-r}$ such that $E_2^{n,m} = 0$ for $n < 0$ or $m < 0$, converging to some abutment (E^r). As $d_r^{n,0} = 0$ for all n and r (resp. as $d_r^{-r,m} = 0$ for all m and r) we get epimorphisms $E_2^{n,0} \to E_\infty^{n,0}$ (resp. monomorphisms $E_\infty^{0,m} \to E_2^{0,m}$). Combining this with the monomorphisms $E_\infty^{n,0} \to E^n$ resp. the epimorphisms $E^m \to E_\infty^{0,m}$ we get natural maps

$$E_2^{n,0} \to E^n \qquad \text{and} \qquad E^m \to E_2^{0,m}$$

called the *base maps* of the spectral sequence. In the situation as in (1) they have the form

$$(R^n\mathcal{F}')(\mathcal{F}M) \to R^n(\mathcal{F}' \circ \mathcal{F})M \qquad \text{and} \qquad R^m(\mathcal{F}' \circ \mathcal{F})M \to \mathcal{F}'((R^m\mathcal{F})M).$$

If we take the base maps for $n \in \{1, 2\}$ and for $m = 1$ together with $d_2^{0,1}$, then we get an exact sequence (cf. [B1], ch. X, §2, exerc. 15c)

(4) $$0 \to E_2^{1,0} \longrightarrow E^1 \longrightarrow E_2^{0,1} \longrightarrow E_2^{2,0} \longrightarrow E^2$$

called the *five term exact sequence.*

4.2. Throughout this chapter let G be a flat group scheme over k and H a flat subgroup scheme of G.

We know by 2.9 and 3.9.b that the G–modules form an abelian category containing enough injective objects. So we can apply the general principles from 4.1. For example, the fixed point functor from $\{G\text{–modules}\}$ to $\{k\text{–modules}\}$ is left exact. We denote its derived functors by $M \mapsto H^n(G, M)$, and call $H^n(G, M)$ the nth (rational) *cohomology* group of M.

For any G–module M, the functor $\operatorname{Hom}_G(M, ?)$ is left exact. Its derived functors are denoted (as usual) by $\operatorname{Ext}_G^n(M, ?)$. They can (as always) also be defined using equivalence classes of exact sequences of G–modules.

For the trivial module k the functor $\operatorname{Hom}_G(k, ?)$ is isomorphic to the fixed point functor: For each G–module M we have an isomorphism $\operatorname{Hom}_G(k, M) \xrightarrow{\sim} M^G$ with $\varphi \mapsto \varphi(1)$. We get therefore isomorphisms of derived functors

(1) $$\operatorname{Ext}_G^n(k, ?) \simeq H^n(G, ?).$$

The induction functor from H to G is left exact. We can therefore define also its derived functors $R^n \operatorname{ind}_H^G$.

4.3. Lemma: *Suppose that G is diagonalisable. Let Λ be an abelian group with $G \simeq \mathrm{Diag}(\Lambda)$. Then one has for all G–modules M, N:*

a) $\mathrm{Ext}^n_G(M,N) \simeq \prod_{\lambda\in\Lambda} \mathrm{Ext}^n_k(M_\lambda, N_\lambda)$ *for all* $n \in \mathbf{N}$.

b) $H^n(G,M) = 0$ *for all* $n \in \mathbf{N}$, $n > 0$.

c) *If k is a field, then* $\mathrm{Ext}^n_G(M,N) = 0$ *for all* $n \in \mathbf{N}$, $n > 0$.

Proof: We get from 2.11(4) that a G–module Q is injective if and only if all Q_λ are injective k–modules. (In order to see that this condition is necessary, note that given λ each k–module V can be regarded as a G–module with $V = V_\lambda$.) Another application of 2.11(4) yields then a). The other statements are immediate consequences.

4.4. Lemma: *Let M, N, V be G–modules. If V is finitely generated and projective as a k–module, then we have for all $n \in \mathbf{N}$ a canonical isomorphism*

$$\mathrm{Ext}^n_G(M, V\otimes N) \xrightarrow{\sim} \mathrm{Ext}^n_G(M\otimes V^*, N).$$

Proof: We have a canonical isomorphism

$$\mathrm{Hom}(M, V\otimes N) \xrightarrow{\sim} \mathrm{Hom}(M\otimes V^*, N)$$

sending any φ to the map $m\otimes\alpha \mapsto (\alpha\,\overline{\otimes}\,\mathrm{id}_N)(\varphi(m))$. It is easy to check that this induces an isomorphism

(1) $$\mathrm{Hom}_G(M, V\otimes N) \xrightarrow{\sim} \mathrm{Hom}_G(M\otimes V^*, N).$$

This is functorial in N and can be interpreted as an isomorphism of functors

$$\mathrm{Hom}_G(M,?)\circ(V\otimes ?) \xrightarrow{\sim} \mathrm{Hom}_G(M\otimes V^*,?).$$

The functor $V\otimes ?$ is exact and maps injective G–modules to injective G–modules [by (1), for example]. We can therefore apply 4.1(3).

Remark: If V_1 and V_2 are G–modules that both are finitely generated and projective as k–modules, then we get

(2) $$\mathrm{Ext}^n_G(V_1,V_2) \simeq \mathrm{Ext}^n_G(V_2^*, V_1^*) \qquad \text{for all } n\in\mathbf{N}.$$

(One applies the lemma twice and uses $(V_i^*)^* \simeq V_i$.)

4.5. Proposition: *Let M be an H–module.*

a) *For each G–module N we have a spectral sequence*

$$E_2^{n,m} = \mathrm{Ext}^n_G(N, R^m \mathrm{ind}_H^G M) \Rightarrow \mathrm{Ext}^{n+m}_H(N,M).$$

b) *There is a spectral sequence with*

$$E_2^{n,m} = H^n(G, R^m \mathrm{ind}_H^G M) \Rightarrow H^{n+m}(H,M).$$

c) *Let H' be a flat subgroup scheme of G with $H \subset H'$. Then there is a spectral sequence with*

$$E_2^{n,m} = (R^n \operatorname{ind}_{H'}^G)(R^m \operatorname{ind}_H^{H'})M \Rightarrow (R^{n+m} \operatorname{ind}_H^G)M.$$

Proof: a) The Frobenius reciprocity in 3.4 can be interpreted as an isomorphism of functors

$$\operatorname{Hom}_G(N, ?) \circ \operatorname{ind}_H^G \simeq \operatorname{Hom}_H(N, ?).$$

As ind_H^G maps injective H–modules to injective G–modules by 3.9.a, we can apply 4.1(1).

b) This is the special case $N = k$ of a).

c) Take the isomorphism in 3.5(2) and argue as in the proof of a).

4.6. We call H *exact in* G if ind_H^G is an exact functor. For example, any diagonalisable subgroup scheme of G is exact in G, see the remark in 3.3. The preceding proposition implies obviously:

Corollary: *Suppose that H is exact in G. Let M be an H–module.*

a) *For each G–module N and each $n \in \mathbf{N}$ there is an isomorphism*

$$\operatorname{Ext}_G^n(N, \operatorname{ind}_H^G M) \simeq \operatorname{Ext}_H^n(N, M).$$

b) *For each $n \in \mathbf{N}$ there is an isomorphism*

$$H^n(G, \operatorname{ind}_H^G M) \simeq H^n(H, M).$$

Remark: These results are also known as *generalised Frobenius reciprocity* and *Shapiro's lemma.*

4.7. When we regard $k[G]$ as a G–module and do not mention the representation explicitly, we will deal with ρ_l or ρ_r. As both structures are equivalent, it is most of the time not necessary to specify which of these two we consider. The same applies to H instead of G.

Lemma: *Let $n \in \mathbf{N}$.*

a) *We have for each G–module N*

$$H^n(G, N \otimes k[G]) \simeq \begin{cases} N, & \text{if } n = 0, \\ 0, & \text{if } n > 0. \end{cases}$$

b) *We have for each H–module M*

$$R^n \operatorname{ind}_H^G(M \otimes k[H]) \simeq \begin{cases} M \otimes k[G], & \text{if } n = 0, \\ 0, & \text{if } n > 0. \end{cases}$$

Proof: a) The trivial subgroup 1 of G is exact in G as it is diagonalisable (or even more trivially, as $\mathrm{ind}_1^G = k[G]\otimes ?$ is obviously exact). Therefore a) is an immediate consequence of 4.6.b (applied to $H = 1$) and of 3.7(4).

b) Apply the spectral sequence 4.5.c to $(H, 1)$ instead of (H', H). As 1 is exact in H, the spectral sequence together with 3.7(4) yields isomorphisms

$$R^n \mathrm{ind}_H^G(M \otimes k[H]) \simeq R^n \mathrm{ind}_1^G(M).$$

As 1 is exact in G, the right hand side is 0 for $n > 0$, and equal to $M \otimes k[G]$ for $n = 0$ by 3.7(4). This implies b).

Remark: If k is a field, then $N \otimes k[G]$ is an injective G–module by 3.10. Similarly, $M \otimes k[H]$ is an injective H–module. So the claims for $n > 0$ are obvious in this case.

4.8. Proposition (The Generalised Tensor Identity): *Let N be a G–module that is flat over k. Then we have for each H–module M and each $n \in \mathbf{N}$ an isomorphism*

$$R^n \mathrm{ind}_H^G(M \otimes N) \simeq (R^n \mathrm{ind}_H^G M) \otimes N.$$

Proof: The tensor identity may be interpreted as an isomorphism of functors

$$\mathrm{ind}_H^G \circ (\mathrm{res}_H^G N \otimes ?) \;\simeq\; (N \otimes ?) \circ \mathrm{ind}_H^G .$$

Tensoring with N is exact and maps, because of 3.9.c and 4.7.b, injective H–modules to modules acyclic for ind_H^G. So we can apply 4.1(2), (3).

4.9. (Semi-direct Products) Let G' be a flat k–group scheme that acts on G via automorphisms. We can therefore form the semi-direct product $G \rtimes G'$.

By 3.2, we may regard the fixed point functor $?^G$ as a functor from $\{(G \rtimes G')\text{–modules}\}$ to $\{G'\text{–modules}\}$. There is an obvious isomorphism $\mathrm{res}_1^{G'} \circ ?^G \simeq ?^G \circ \mathrm{res}_G^{G\rtimes G'}$ of functors. The isomorphism of k–algebras $k[G \rtimes G'] \simeq k[G] \otimes k[G']$ is compatible with the action of G via ρ_l on $k[G \rtimes G']$ and $k[G]$, and with the trivial action on $k[G']$. Therefore, 3.9.c and 4.7.a imply that $\mathrm{res}_G^{G\rtimes G'}$ maps injective modules to modules acyclic for the fixed point functor. We get therefore isomorphisms of derived functors by 4.1(2), (3). So we have for each $n \in \mathbf{N}$ and each $(G \rtimes G')$–module M a natural structure as a G'–module on $H^n(G, M)$.

Suppose now that G' stabilises our subgroup scheme H of G. We can interpret 3.8(1) as an isomorphism $\mathrm{res}_G^{G\rtimes G'} \circ \mathrm{ind}_{H\rtimes G'}^{G\rtimes G'} \simeq \mathrm{ind}_H^G \circ \mathrm{res}_H^{H\rtimes G'}$ of functors. As above, 3.9.c and 4.7.b imply that $\mathrm{res}_H^{H\rtimes G'}$ maps injective modules to modules acyclic for ind_H^G. Therefore 4.1(2), (3) yield isomorphisms of derived functors (for all $n \in \mathbf{N}$)

$$\text{(1)} \qquad \mathrm{res}_G^{G\rtimes G'} \circ R^n \mathrm{ind}_{H\rtimes G'}^{G\rtimes G'} \;\simeq\; R^n \mathrm{ind}_H^G \circ \mathrm{res}_H^{H\rtimes G'} .$$

For $H = 1$ this shows that G' is exact in $G \rtimes G'$, which is already clear by 3.8(2). Similarly, G is exact in $G \rtimes G'$ by 3.8(3).

4.10. Proposition: *We have for each H–module M and each $n \in \mathbf{N}$ an isomorphism of k–modules*

$$H^n(H, M \otimes k[G]) \xrightarrow{\sim} (R^n \operatorname{ind}_H^G)M.$$

Proof: The definition of ind_H^G yields an isomorphism of functors

$$\mathcal{F} \circ \operatorname{ind}_H^G \simeq ?^H \circ (k[G] \otimes ?)$$

where $\mathcal{F}$ is the forgetful functor from { G–modules } to { k–modules }. As $k[G]\otimes ?$ is exact and maps injective H–modules to modules acyclic for the fixed point functor (by 4.7.a), we can apply 4.1(2), (3).

4.11. Corollary: *If $k[G]$ is an injective H–module, then H is exact in G.*

Proof: Under our assumption $k[G]$ is a direct summand of $M_1 \otimes k[H]$ for a suitable H–module M_1. For each H–module M we can then find an H–module M_2 such that $M \otimes k[G]$ is a direct summand of $M_2 \otimes k[H]$. Now 4.10 and 4.7.a imply the claim.

Remarks: 1) Suppose that k is a field. Then the corollary can be proved directly as follows. If $0 \to M_1 \to M_2 \to M_3 \to 0$ is an exact sequence of H–modules, then $0 \to M_1 \otimes k[G] \to M_2 \otimes k[G] \to M_3 \otimes k[G] \to 0$ is an exact sequence of injective H–modules (by 3.10), hence split as a sequence of H–modules. Therefore the sequence of all $(M_i \otimes k[G])^H = \operatorname{ind}_H^G(M_i)$ also has to be exact.

2) The example $H = 1$ shows that the converse will not hold in general. However:

4.12. Proposition: *Suppose that k is a field. Then H is exact in G if and only if $k[G]$ is an injective H–module.*

Proof: Because of 4.11 we have to prove one direction only. Suppose that H is exact in G. We have for each finite dimensional H–module V, by 4.4, 4.2(1), and 4.10,

$$\operatorname{Ext}_H^n(V, k[G]) \simeq \operatorname{Ext}_H^n(k, V^* \otimes k[G]) \simeq H^n(H, V^* \otimes k[G]) = 0$$

for all $n > 0$. Therefore the functor $\operatorname{Hom}_H(?, k[G])$ is exact when restricted to finite dimensional H–modules. This easily implies the exactness on all H–modules (i.e., the injectivity of $k[G]$) because each H–module is the direct limit of finite dimensional H–modules.

Remark: Now Proposition 3.10.a shows: If k is a field and if H is exact in G, then all injective G–modules are injective for H.

4.13. Proposition: *Let k' be a flat k–algebra. Let $n \in \mathbf{N}$.*

a) *For each G–module N there is an isomorphism*

$$H^n(G, N) \otimes k' \simeq H^n(G_{k'}, N \otimes k').$$

b) *For each H–module M there is an isomorphism*

$$(R^n \operatorname{ind}_H^G M) \otimes k' \simeq (R^n \operatorname{ind}_{H_{k'}}^{G_{k'}})(M \otimes k').$$

Proof: We get from 2.10(3) and 3.5(3) isomorphisms of functors to which we want to apply 4.1(2), (3). This is possible as $? \otimes k'$ is exact and maps injective G–modules to modules acyclic for the $G_{k'}$–fixed point functor (by 3.9 and 4.7.a), and also maps injective H–modules to modules acyclic for the induction from $H_{k'}$ to $G_{k'}$ (by 3.9 and 4.7.b).

4.14. Let M be a G–module. The cohomology $H^\bullet(G, M)$ can be computed using the *Hochschild complex* $C^\bullet(G, M)$ that we are going to describe now.

We set $C^n(G, M) = M \otimes \bigotimes^n k[G]$ for all $n \in \mathbf{N}$, and define boundary maps $\partial^n : C^n(G, M) \to C^{n+1}(G, M)$ of the form $\partial^n = \sum_{i=0}^{n+1} (-1)^i \partial_i^n$ where

$$\begin{aligned}
\partial_0^n(m \otimes f_1 \otimes \cdots \otimes f_n) &= \Delta_M(m) \otimes f_1 \otimes \cdots \otimes f_n,\\
\partial_i^n(m \otimes f_1 \otimes \cdots \otimes f_n) &= m \otimes f_1 \otimes \cdots \otimes f_{i-1} \otimes \Delta_G(f_i) \otimes f_{i+1} \otimes \cdots \otimes f_n\\
&\qquad \text{for } 1 \le i \le n,\\
\partial_{n+1}^n(m \otimes f_1 \otimes \cdots \otimes f_n) &= m \otimes f_1 \otimes \cdots \otimes f_n \otimes 1.
\end{aligned}$$

We can interpret $C^n(G, M)$ as $\mathrm{Mor}(G^n, M_a)$ where G^n is the direct product of n copies of G, cf. 3.3. Then the ∂_i^n look like

$$\begin{aligned}
\partial_0^n f(g_1, g_2, \dots, g_{n+1}) &= g_1 f(g_2, \dots, g_{n+1}),\\
\partial_i^n f(g_1, g_2, \dots, g_{n+1}) &= f(g_1, \dots, g_{i-1}, g_i g_{i+1}, g_{i+2}, \dots, g_{n+1})\\
&\qquad \text{for } 1 \le i \le n,\\
\partial_{n+1}^n f(g_1, g_2, \dots, g_{n+1}) &= f(g_1, g_2, \dots, g_n).
\end{aligned}$$

It is easy to check that $\partial^n \partial^{n-1} = 0$ for all n. Therefore $(C^\bullet(G, M), \partial^\bullet)$ is a complex. We want to prove that its cohomology is just $H^\bullet(G, M)$.

4.15. If our last claim is correct, then $C^\bullet(G, k[G])$ ought to be exact except in degree 0, see 4.7.a. Let us consider $k[G]$ as a G–module via ρ_r so that $\Delta_{k[G]} = \Delta_G$. We define for each n a linear map

$$s^n : C^{n+1}(G, k[G]) = \bigotimes^{n+2} k[G] \longrightarrow \bigotimes^{n+1} k[G] = C^n(G, k[G])$$

through $s^n = \varepsilon_G \,\overline{\otimes}\, \bigotimes^{n+1} \mathrm{id}_{k[G]}$. An elementary calculation using 2.3(2) shows $s^n \partial^n = \mathrm{id} - \partial^{n-1} s^{n-1}$ for all $n > 0$. This implies the exactness of $C^\bullet(G, k[G])$ in each degree $n > 0$ whereas

$$\partial^0 : C^0(G, k[G]) = k[G] \longrightarrow C^1(G, k[G]) = k[G] \otimes k[G]$$

maps f to $\Delta_G(f) - f \otimes 1$, hence has kernel $k1 = k[G]^G$. Therefore we have an exact sequence

$$(1) \qquad 0 \to k \longrightarrow k[G] \longrightarrow \bigotimes^2 k[G] \longrightarrow \bigotimes^3 k[G] \longrightarrow \cdots.$$

This sequence can be regarded as a sequence of homomorphisms of G–modules when we let G act on $\bigotimes^n k[G]$ via ρ_l on the first factor and trivially on all the other factors. It is for this action that $k[G] \to k[G] \otimes k[G], f \mapsto \Delta_G(f) - f \otimes 1$ is G–equivariant. If we now tensor (1) with M, then we get an exact sequence

$$(2) \qquad 0 \to M \longrightarrow M \otimes k[G] \longrightarrow M \otimes \bigotimes{}^2 k[G] \longrightarrow \cdots.$$

Furthermore, by 3.7(4) we can make the action of G on the factor M in each $M \otimes \bigotimes^i k[G]$ with $i > 0$ trivial, hence get a resolution

$$(3) \qquad 0 \to M \longrightarrow M_{tr} \otimes k[G] \longrightarrow M_{tr} \otimes \bigotimes{}^2 k[G] \longrightarrow \cdots$$

of M by acyclic modules (cf. 4.7.a) using the notation M_{tr} as in 3.7(4). (If k is a field, then this is an injective resolution.) Therefore $H^\bullet(G, M)$ is the cohomology of the complex

$$(4) \qquad 0 \to (M_{tr} \otimes k[G])^G \longrightarrow (M_{tr} \otimes \bigotimes{}^2 k[G])^G \longrightarrow \cdots.$$

As G acts trivially on all but one factor, and as $k[G]^G = k$, the nth term in (4) is equal to $(M_{tr} \otimes \bigotimes^{n+1} k[G])^G \simeq M_{tr} \otimes \bigotimes^n k[G] \simeq C^n(G, M)$. Furthermore, tracing back the maps one finds that ∂^n is just the map from $C^n(G, M)$ to $C^{n+1}(G, M)$ occurring in (4). [The shortest way of doing it is via the interpretation of functions $G^n \to M_a$.] This proves our claim.

4.16. Let us state our result explicitly:

Proposition: *For each G–module M the cohomology of the complex $C^\bullet(G, M)$ is equal to $H^\bullet(G, M)$.*

Remarks: 1) In [DG], II, §3, the case of arbitrary group functors (instead of our flat group schemes) is treated and more general coefficients are considered.

2) It is sometimes convenient to replace the Hochschild complex $C^\bullet(G, M)$ by a "normalised" Hochschild complex $C_0^\bullet(G, M)$. Let $I_1 = \{f \in k[G] \mid f(1) = 0\}$; we have then $k[G] = k1 \oplus I_1$ as k–modules. Set $C_0^n(G, M) = M \otimes \bigotimes^n I_1$ for each $n \in \mathbf{N}$; this is a direct summand of $C^n(G, M)$. Note that 2.8(3) implies $\Delta_M(m) \in m \otimes 1 + M \otimes I_1$ for all $m \in M$. Using this and 2.4(1) one checks easily that $\partial^n C_0^n(G, M) \subset C_0^{n+1}(G, M)$ for all n. So $C_0^\bullet(G, M)$ is a subcomplex of $C^\bullet(G, M)$. The point is now that $H^\bullet(G, M)$ *is also the cohomology of* $C_0^\bullet(G, M)$.

This is easily seen in the first degrees: There is clearly no problem with $H^0(G, M)$ since $C_0^0(G, M) = M = C^0(G, M)$. The image of $\partial^0 : M \mapsto \Delta_M(m) - m \otimes 1$ is contained in $M \otimes I_1 = C_0^1(G, M)$. So is the kernel of ∂^1 as $C^1(G, M) = C_0^1(G, M) \oplus M \otimes k1$ and $\partial^1(m \otimes 1) = \Delta_M(m) \otimes 1$ while $\partial^1(C_0^1(G, M)) \subset C_0^2(G, M)$. (Note that $\Delta_M(m) \neq 0$ if $m \neq 0$ by 2.8(3).)

In general, one has to show for all $f \in \ker(\partial^n)$ that there exists $f' \in C_0^n(G, M)$ with $f' - f \in \partial^{n-1} C^{n-1}(G, M)$ and that for any $h \in C_0^n(G, M) \cap \partial^{n-1} C^{n-1}(G, M)$ that already $h \in \partial^{n-1} C_0^{n-1}(G, M)$. The first part yields that the natural map from the cohomology of $C_0^\bullet(G, M)$ to the cohomology of $C^\bullet(G, M)$ is surjective, the second one that this map is injective.

One can now proceed as for abstract groups, say, as in [Hall], 15.7: If we interpret $C^n(G, M)$ as $\text{Mor}(G^n, M_a)$, then $C_0^n(G, M)$ consists of all $f : G^n \to M_a$

with $f(g_1, g_2, \dots, g_n) = 0$ whenever $g_i = 1$ for some i. For any $f \in C^n(G, M)$ one now constructs a sequence $f_0 = f,\ f_1, \dots, f_n$ in $C^n(G, M)$ where we define inductively $f_{i+1} = f_i - \partial^{n-1}(h_{i+1})$ with

$$h_{i+1}(g_1, g_2, \dots, g_{n-1}) = (-1)^i f_i(g_1, \dots, g_i, 1, g_{i+1}, \dots, g_{n-1}).$$

One then checks inductively: If $\partial^n(f) \in C_0^{n+1}(G, M)$, then $f_i \in M \otimes \bigotimes^i I_1 \otimes \bigotimes^{n-i} k[G]$, in particular $f_n \in C_0^n(G, M)$. This now solves the two problems in the last paragraph: In the first one we take $f' = f_n$. In the second one we apply this procedure (with n replaced by $n-1$) to any $f \in C^{n-1}(G, M)$ with $h = \partial^{n-1}(f)$ and get then $h = \partial^{n-1}(f_{n-1}) \in \partial^{n-1} C_0^{n-1}(G, M)$.

4.17. Lemma: *Let $(M_i)_{i \in I}$ be a directed system of G–modules. Then there are natural isomorphisms for all $n \in \mathbf{N}$*

$$\varinjlim H^n(G, M_i) \xrightarrow{\sim} H^n(G, \varinjlim M_i).$$

Proof: We have natural isomorphisms $\varinjlim C^n(G, M_i) \xrightarrow{\sim} C^n(G, \varinjlim M_i)$ for all $n \in \mathbf{N}$ as tensoring with $k[G]$ is exact. So the claim follows from the exactness of $\varinjlim$, cf. [B1], ch. II, §6, prop. 3.

Remark: For any directed system $(N_i)_{i \in I}$ of H–modules one similarly gets isomorphisms of G–modules

$$\varinjlim R^n \operatorname{ind}_H^G(N_i) \xrightarrow{\sim} R^n \operatorname{ind}_H^G(\varinjlim N_i), \tag{1}$$

cf. also [Donkin 9], 1.1.1.

4.18. We can identify $C^\bullet(G_{k'}, M \otimes k')$ for each k–algebra k' with $C^\bullet(G, M) \otimes k'$. Suppose that M is a flat k–module. Then all $C^n(G, M)$ are also flat. If k has the property that each submodule of a flat module is flat, then we get a universal coefficient theorem, e.g., by [B1], ch. X, §4, cor. 1 du th. 3 (after re-indexing). Any Dedekind ring has this property, as for such a ring the notions "flat" and "torsion free" coincide, cf. [B2], ch. VII, §4, prop. 22. We therefore get the first part of:

Proposition: *Suppose that k is a Dedekind ring. Let k' be a k–algebra and let $n \in \mathbf{N}$.*

a) *There is for each G–module N which is flat over k an exact sequence*

$$0 \to H^n(G, N) \otimes k' \longrightarrow H^n(G_{k'}, N \otimes k') \longrightarrow \operatorname{Tor}_1^k(H^{n+1}(G, N), k') \to 0.$$

b) *There is for each H–module M which is flat over k an exact sequence of $G_{k'}$–modules*

$$0 \to (R^n \operatorname{ind}_H^G M) \otimes k' \longrightarrow R^n \operatorname{ind}_{H_{k'}}^{G_{k'}}(M \otimes k') \longrightarrow \operatorname{Tor}_1^k(R^{n+1} \operatorname{ind}_H^G M, k') \to 0.$$

Note that b) follows on the level of k'–modules from a) and 4.10. It is left to the reader to find the $G_{k'}$–module structure on the Tor-group and to prove the equivariance of the maps.

Remark: If k' is flat over k, then we get from part a) that $H^0(G,N)\otimes k' \simeq H^0(G_{k'},N\otimes k')$, which we know already from 2.10(3) to hold for all N. If k' is not flat, however, such a statement will not be true, even for flat N (in spite of Lemma 1.17 in [Andersen 12]). Take, e.g., $G=G_a$ and its representation $a\mapsto \left(\begin{smallmatrix}1&2a\\0&1\end{smallmatrix}\right)$ on k^2 and get a contradiction for $k=\mathbf{Z}$ and $k'=\mathbf{F}_2$.

Such a formula will, however, hold for acyclic modules as the last term in part a) is then 0. We can, for example, take for N a direct summand of some $E\otimes k[G]$ where E is a flat k–module regarded as a trivial G–module. If N' is another G–module which is finitely generated and projective over k, then $\operatorname{Hom}(N',N)\simeq (N')^*\otimes N$ is again of this type by 3.7(4). This shows (for any Dedekind ring k):

Let N, N' be G–modules such that N' is finitely generated and projective over k, and such that N is isomorphic to a direct summand of some G–module $E\otimes k[G]$ with E flat over k. Then we have for all $n\in\mathbf{N}$ and for each k–algebra k' a natural isomorphism

$$\operatorname{Ext}^n_G(N',N)\otimes k' \simeq \operatorname{Ext}^n_{G_{k'}}(N'\otimes k', N\otimes k'). \tag{1}$$

4.19. For any k there is on $H^\bullet(G,k)=\bigoplus_{i\geq 0} H^i(G,k)$ a structure as an (associative) algebra over k. The multiplication is called the *cup product*; it is compatible with the grading and satisfies the usual anti-commutativity formula: If $a\in H^i(G,k)$ and $b\in H^j(G,k)$, then $ab\in H^{i+j}(G,k)$ and $ab=(-1)^{ij}ba$. Furthermore, there is (for each G–module N) a natural structure as an $H^\bullet(G,k)$–right module on $H^\bullet(G,N)=\bigoplus_{i\geq 0} H^i(G,N)$.

Let us describe these structures using the Hochschild complexes for k and N. We can obviously identify $C^n(G,k)=\bigotimes^n k[G]$ and then have to write $\partial_0^n(x)=1\otimes x$. Furthermore, we identify $C^n(G,N)\otimes C^m(G,k)$ and $C^{n+m}(G,N)$ for all $n,m\in\mathbf{N}$. For all $a\in C^n(G,N)$ and $b\in C^m(G,k)$ one easily checks $\partial^{n+m}(a\otimes b)=(\partial^n a)\otimes b+(-1)^n a\otimes\partial^m(b)$. Hence $a\otimes b$ is a cocycle if a and b are. Another simple computation shows, then, that the cohomology class $[a\otimes b]$ of $a\otimes b$ depends only on the classes $[a]$ of a and $[b]$ of b. Then the action of $[b]\in H^m(G,k)$ on $[a]\in H^n(G,N)$ is defined through $[a][b]=[a\otimes b]$. In case $N=k$ we get thus the cup product on $H^\bullet(G,k)$.

Let G' be a flat group scheme acting on G through group automorphisms. If N is a $(G\rtimes G')$–module (e.g., if $N=k$), then G' acts on each $H^n(G,N)$, cf. 4.9. This action can be described using the Hochschild complex. The discussion above shows that G' acts on $H^\bullet(G,k)$ through algebra automorphisms, and that the action of $H^\bullet(G,k)$ on an arbitrary $H^\bullet(G,N)$ is compatible with the G'–action, i.e., that $H^\bullet(G,N)\otimes H^\bullet(G,k)\to H^\bullet(G,N)$ is a homomorphism of G'–modules.

4.20. We want to describe $H^\bullet(G_a,k)$ or (more generally) $H^\bullet(V_a,k)$ for a free k–module V of finite rank, say $\operatorname{rk}(V)=n$. Of course, there is a Künneth formula, reducing the second problem to the first one. But we shall prefer to formulate our results at once for V in order to keep track of the $GL(V)$–action on the cohomology groups (as in 4.19).

Choosing a basis we identify $k[V_a]$ with the polynomial ring $k[T_1,T_2,\dots,T_n]$. We get then an $\mathbf{N}^n$–grading and an $\mathbf{N}$–grading on the complex $C^\bullet(V_a,k)$: For each $\alpha=(\alpha_1,\alpha_2,\dots,\alpha_n)\in\mathbf{N}^n$ let $C^r(V_a,k)_\alpha$ denote the span of all tensor products of monomials such that the degrees of T_i in the factors add up to α_i

for each i. Set $C^r(V_a,k)_m$ equal to the sum of all $C^r(V_a,k)_\alpha$ with $m = |\alpha|$ (where $|(\alpha_1,\alpha_2,\dots,\alpha_n)| = \sum_{i=1}^n \alpha_i$). Obviously, the $C^r(V_a,k)_m$ are $GL(V)$–stable whereas the $C^r(V_a,k)_\alpha$ are not (for $n>1$). As the comultiplication on $k[V_a]$ is given by $\Delta(T_j) = 1\otimes T_j + T_j\otimes 1$ for all j, the formulae for the ∂^i in 4.14 show that $\partial^i C^i(V_a,k)_\alpha \subset C^{i+1}(V_a,k)_\alpha$ for all α, and $\partial^i C^i(V_a,k)_m \subset C^{i+1}(V_a,k)_m$. Therefore we also get gradings for the cohomology groups

$$H^i(V_a,k) = \bigoplus_{\alpha\in\mathbf{N}^n} H^i(V_a,k)_\alpha = \bigoplus_{m\in\mathbf{N}} H^i(V_a,k)_m. \tag{1}$$

(Note that these gradings simply describe the representations on the cohomology of the diagonal subgroup of $GL(V)$ resp. of the subgroup of scalar matrices.)

4.21. We can now easily compute $H^1(V_a,k)$.

Lemma: *Suppose that k is an integral domain.*

a) *If* $\operatorname{char}(k)=0$, *then* $H^1(V_a,k) = \sum_{i=1}^n kT_i \simeq V$ *as $GL(V)$–module.*

b) *If* $\operatorname{char}(k)=p\neq 0$, *then* $H^1(V_a,k) = \sum_{i=1}^n \sum_{r=0}^\infty kT_i^{p^r}$.

Proof: We have obviously $H^0(V_a,k)=k$ and $\partial^0=0$, hence $H^1(V_a,k)=\ker(\partial^1)$. This map is given by $\partial^1(f) = 1\otimes f - \Delta(f) + f\otimes 1$. Because of 4.20(1), the monomials $\prod_{i=1}^n T_i^{r(i)}$ with $\partial^1(\prod_{i=1}^n T_i^{r(i)})=0$ form a basis of $\ker(\partial^1)$. If there exist two indices $h\neq j$ with $r(h), r(j)>0$, then $T_h^{r(h)}\otimes \prod_{i\neq h} T_i^{r(i)}$ occurs with coefficient -1 in $\partial^1(\prod_{i=1}^n T_i^{r(i)})$ so that this image is not equal to 0. As $\partial^1(1)=1\otimes 1$, we have to look only at all

$$\partial^1(T_i^r) = -\sum_{j=1}^{r-1}\binom{r}{j} T_i^j\otimes T_i^{r-j} \tag{1}$$

with $r>0$. This is certainly 0 if $r=1$. We then have to determine all $r>1$ such that all those binomial coefficients are 0 in k. The result is well known and implies the lemma.

4.22. Keep the assumption of Lemma 4.21. The cup product induces a homomorphism of $GL(V)$–modules

$$H^1(V_a,k)\otimes H^1(V_a,k) \longrightarrow H^2(V_a,k).$$

Because of the anti-commutativity of the cup product (i.e., because of $f\otimes f' + f'\otimes f = -\partial^1(ff')$) this map has to factor through $\Lambda^2 H^1(V_a,k)$ if $\operatorname{char}(k)\neq 2$, and through $S^2H^1(V_a,k)$ if $\operatorname{char}(k)=2$.

Let us denote the image of this map by M. We want to show that

$$M \simeq \begin{cases} \Lambda^2 H^1(V_a,k), & \text{if } \operatorname{char}(k)\neq 2,\\ S^2H^1(V_a,k), & \text{if } \operatorname{char}(k)=2.\end{cases} \tag{1}$$

The image of ∂^1 in $C^2(V_a,k) = k[V_a]\otimes k[V_a]$ consists of symmetric elements, i.e., of elements stable under $f\otimes f'\mapsto f'\otimes f$. If we take two different basis elements f,

f' in Lemma 4.21, then $f \otimes f'$ is not symmetric, hence the class $[f][f'] = [f \otimes f'] \in H^2(V_a, k)$ is non-zero. In order to get their linear independence we just have to observe that these tensor products are homogeneous of pairwise different degrees (except for the trivial equality $[f \otimes f'] = -[f' \otimes f]$).

This proves (1) for $\operatorname{char}(k) \neq 2$. For $\operatorname{char}(k) = 2$ we still have to show $f \otimes f \notin \operatorname{im}(\partial^1)$ for each basis element f in Lemma 4.21. We can do something more general. Suppose $\operatorname{char}(k) = p \neq 0$, set $\left\{ {p \atop i} \right\} = \frac{1}{p}\binom{p}{i}$ for $1 \le i \le p-1$ and

$$\beta(f) = \sum_{i=1}^{p-1} \left\{ {p \atop i} \right\} f^i \otimes f^{p-i} \tag{2}$$

for all $f \in k[V_a]$. (So $\beta(f) = f \otimes f$ if $\operatorname{char}(k) = 2$.) This map is of course induced from the map $f \mapsto ((1 \otimes f + f \otimes 1)^p - 1 \otimes f^p - f^p \otimes 1)/p$ on $\mathbf{Z}[T_1, T_2, \dots, T_n]$. Using this fact (or a direct calculation) we get that β maps $\ker(\partial^1) = H^1(V_a, k)$ into $\ker(\partial^2)$, hence leads to a map $\overline{\beta} : H^1(V_a, k) \to H^2(V_a, k)$. A simple computation shows

$$\beta(f_1 + f_2) = \beta(f_1) + \beta(f_2) - \partial^1 \sum_{i=1}^{p-1} \left\{ {p \atop i} \right\} f_1^i f_2^{p-i}$$

for all $f_1, f_2 \in H^1(V_a, k)$. Therefore $\overline{\beta}$ is additive. Obviously, $\overline{\beta}$ is $GL(V)$–equivariant and satisfies $\overline{\beta}(af) = a^p \overline{\beta}(f)$ for all $a \in k$.

Take now for f a basis element from Lemma 4.21. Then $\beta(f)$ is homogeneous of degree equal to p times the degree of f. The homogeneous component of this degree in $k[V_a]$ is spanned by f^p. As $\partial^1(f^p) = 0$, we get $\beta(f) \notin \operatorname{im}(\partial^1)$. This concludes the proof of (1) in the remaining case $\operatorname{char}(k) = 2$. For $p \neq 2$ we see that the $\overline{\beta}(T_i^{p^r})$ with $1 \le i \le n$ and $r \in \mathbf{N}$ are a basis for a $GL(V)$–submodule of $H^2(V_a, k)$ intersecting M in 0.

We claim that we have found all of $H^2(V_a, k)$ in case k is a field. We refer to [DG], II, §3, 4.6 for the proof and just state the result:

Lemma: *Suppose that k is a field.*

a) *If $\operatorname{char}(k) = 0$, then $H^2(V_a, k) \simeq \Lambda^2 H^1(V_a, k)$.*

b) *If $\operatorname{char}(k) = 2$, then $H^2(V_a, k) \simeq S^2 H^1(V_a, k)$.*

c) *If $\operatorname{char}(k) \neq 2, 0$, then $H^2(V_a, k) \simeq \Lambda^2 H^1(V_a, k) \oplus k\overline{\beta} H^1(V_a, k)$.*

4.23. In order to get all of $H^\bullet(V_a, k)$, we shall reduce its computation to that of the cohomology of finite cyclic groups. This is done using a filtration of the Hochschild complex.

Set $k[V_a, m]$ for each $m \in \mathbf{N}$ equal to the span of all monomials

$$T_1^{r(1)} T_2^{r(2)} \dots T_n^{r(n)}$$

with $r(i) < m$ for all i. Then the formula $\Delta(T_i) = 1 \otimes T_i + T_i \otimes 1$ implies $\Delta(k[V_a, m]) \subset k[V_a, m] \otimes k[V_a, m]$. Set

$$C^j(V_a, k, m) = \bigotimes^j k[V_a, m] \subset \bigotimes^j k[V_a] = C^j(V_a, k).$$

Then we see that

$$\partial^j C^j(V_a,k,m) \subset C^{j+1}(V_a,k,m).$$

Hence $C^\bullet(V_a,k,m) = \bigoplus_{j\geq 0} C^j(V_a,k,m)$ is a subcomplex of $C^\bullet(V_a,k)$. Let us denote the cohomology of this subcomplex by $H^\bullet(V_a,k,m) = \bigoplus_{i\geq 0} H^i(V_a,k,m)$.

For all $m, m' \in \mathbf{N}$ with $m' \leq m$ the inclusion $C^\bullet(V_a,k,m') \subset C^\bullet(V_a,k,m)$ leads to a homomorphism $\alpha_{m,m'} : H^\bullet(V_a,k,m') \to H^\bullet(V_a,k,m)$. We have obviously $\alpha_{m,m'} \circ \alpha_{m',m''} = \alpha_{m,m''}$ for any $m'' \leq m'$. Similarly, the inclusion $C^\bullet(V_a,k,m) \subset C^\bullet(V_a,k)$ induces a homomorphism $\alpha_m : H^\bullet(V_a,k,m) \to H^\bullet(V_a,k)$ with $\alpha_m \circ \alpha_{m,m'} = \alpha_{m'}$. We get thus a homomorphism $\alpha : \varinjlim H^\bullet(V_a,k,m) \to H^\bullet(V_a,k)$. Obviously, $H^\bullet(V_a,k)$ is the union of all $\alpha_m(H^\bullet(V_a,k,m))$, and for each $f \in \ker(\alpha_m)$ there is $m' \geq m$ with $f \in \ker(\alpha_{m',m})$. This implies

$$\varinjlim H^\bullet(V_a,k,m) \xrightarrow{\sim} H^\bullet(V_a,k). \tag{1}$$

Note that $C^i(V_a,k,m) \otimes C^j(V_a,k,m) = C^{i+j}(V_a,k,m)$. Therefore we can define a cup product on each $H^\bullet(V_a,k,m)$, and the α_m are homomorphisms of algebras. Hence so is the isomorphism (1).

Let me point out that this construction can be generalised to any V_a–module M that is finitely generated over k. For such an M there is some $r(M) \in \mathbf{N}$ with $\Delta_M(M) \subset M \otimes k[V_a, r(M)]$. Then we get for all $m \geq r(M)$ a subcomplex $C^\bullet(V_a,M,m)$ of $C^\bullet(V_a,M)$ leading as above to an isomorphism

$$\varinjlim H^\bullet(V_a,M,m) \xrightarrow{\sim} H^\bullet(V_a,M). \tag{2}$$

4.24. Obviously, we can define a complement $C^j(V_a,k,m)^c$ to $C^j(V_a,k,m)$ in $C^j(V_a,k)$: Take the span of all tensor products of monomials not belonging to $C^j(V_a,k,m)$, i.e., where in at least one factor some T_i occurs with an exponent $\geq m$. In general the $C^j(V_a,k,m)^c$ do not form a subcomplex.

Suppose, however, that p is a prime number with $p1 = 0$ in k. Then $\Delta(T_i^{p^r}) = 1 \otimes T_i^{p^r} + T_i^{p^r} \otimes 1$ for all i and r. This implies that all $C^\bullet(V_a,k,p^r)^c$ are subcomplexes and that $H^j(V_a,k,p^r)$ is a direct summand of $H^j(V_a,k)$. We may write 4.23(1) in the form

$$H^\bullet(V_a,k) = \bigcup_{r>0} H^\bullet(V_a,k,p^r) \qquad (\text{if } p\,k = 0). \tag{1}$$

(We can restate 4.23(2) in a similar way.)

Of course, our calculations in 4.21/22 are compatible with this formula. In the situation of 4.21.b we have

$$H^1(V_a,k,p^r) = \sum_{i=1}^{n} \sum_{j=0}^{r-1} kT_i^{p^j}, \tag{2}$$

in 4.22.c:

$$H^2(V_a,k,p^r) \simeq \Lambda^2 H^1(V_a,k,p^r) \oplus k\overline{\beta} H^1(V_a,k,p^r), \tag{3}$$

and in 4.22.b:

$$H^2(V_a, k, 2^r) \simeq S^2 H^1(V_a, k, 2^r). \tag{4}$$

4.25. The groups $H^\bullet(V_a, k, p^r)$ in 4.24 have a different interpretation. Let p still be a prime number with $p1 = 0$ in k. Identify V with k^n via the T_i, and consider the (Frobenius) endomorphism F of V_a with $F(a_1, \dots, a_n) = (a_1^p, \dots, a_n^p)$ for all $(a_1, \dots, a_n) \in A^n = k^n \otimes A \simeq V_a(A)$ and all k–algebras A. This is an endomorphism of algebraic k–groups with $F^*(T_i) = T_i^p$ for all i. The kernel $V_{a,r}$ of F^r is also an algebraic k–group; we have $k[V_{a,r}] \simeq k[T_1, \dots, T_n]/(T_1^{p^r}, \dots, T_n^{p^r})$. [Obviously, $V_{a,r}$ is independent of the choice of the identification $V \simeq k^n$. Note that $V_{a,r}$ is isomorphic to the direct product of n copies of the algebraic k–group $G_{a,r}$ introduced in 2.2.]

Clearly, the restriction of functions $k[V_a] \to k[V_{a,r}]$ induces an isomorphism $k[V_a, p^r] \to k[V_{a,r}]$ of vector spaces. It is compatible with the comultiplication and induces an isomorphism of complexes $C^\bullet(V_a, k, p^r) \xrightarrow{\sim} C^\bullet(V_{a,r}, k)$ and an isomorphism of algebras

$$H^\bullet(V_a, k, p^r) \xrightarrow{\sim} H^\bullet(V_{a,r}, k). \tag{1}$$

Any $C^j(V_a, k, p^r)^c$ is the kernel of the restriction map $C^j(V_a, k) \to C^j(V_{a,r}, k)$. This gives a better reason for $\bigoplus_j C^j(V_a, k, p^r)^c$ to form a subcomplex and hence for the injectivity of the map $H^\bullet(V_a, k, p^r) \to H^\bullet(V_a, k)$.

We can generalise (1) to any V_a–module M, finitely generated over k, and get

$$H^\bullet(V_a, M, p^r) \xrightarrow{\sim} H^\bullet(V_{a,r}, M) \qquad \text{if } p^r > r(M). \tag{2}$$

Note that the gradings on $H^\bullet(V_a, k)$ considered in 4.20 induce similar gradings on $H^\bullet(V_{a,r}, k)$.

4.26. Let us assume that k is a field of characteristic $p \neq 0$. It will be convenient to assume for the moment that k contains an algebraic closure of $\mathbf{F}_p$.

Consider the endomorphism F of V_a as in 4.25 and define for each $r \in \mathbf{N}$, $r > 0$ a closed subgroup $V_a(p^r)$ of V_a via

$$V_a(p^r)(A) = \{ v \in V_a(A) \mid F^r(v) = v \}.$$

It is defined by the ideal generated by all $T_i^{p^r} - T_i$ with $1 \le i \le n$. Therefore the restriction of functions induces also an isomorphism $k[V_a, p^r] \to k[V_a(p^r)]$ of vector spaces. It is compatible with the comultiplication and leads to an isomorphism of algebras

$$H^\bullet(V_a, k, p^r) \xrightarrow{\sim} H^\bullet(V_a(p^r), k). \tag{1}$$

If A is an extension field of k, then $V_a(p^r)(A)$ is simply the group of all points in A^n having all coordinates in $\mathbf{F}_{p^r}$. Let us denote this group by $V(p^r)$. It is an elementary abelian p–group of order p^{rn}. We may regard $k[V_a(p^r)]$ as the algebra of all functions from $V(p^r)$ to k. The comultiplication on $k[V_a(p^r)]$ is given by the group law on the finite group $V(p^r)$. Therefore the Hochschild complex for $V_a(p^r)$

computes the cohomology of the finite group $V(p^r)$. (Equivalently one can say that the category of $V_a(p^r)$–modules is "the same" as the category of k-$V(p^r)$–modules.)

Now the cohomology of a cyclic group is well known (see, e.g., [HS]) and the cohomology of an elementary abelian group follows using the Künneth formula. The result can be formulated as follows:

(2) *If $p = 2$, then $H^\bullet(V_a, k, p^r) \simeq SH^1(V_a, k, p^r)$.*

We denote here by $S(M)$ resp. $\Lambda(M)$ the symmetric resp. exterior algebra of a k–module M given its natural grading. If we put each element of S^iM in degree $2i$, then we write $S'(M)$.

If $p \neq 2$, then

$$H^\bullet(V_a, k, p^r) \simeq \Lambda H^1(V_a, k, p^r) \otimes S'(V') \tag{3}$$

where $V' \simeq H^2(V_a, k, p^r)/\Lambda^2 H^1(V_a, k, p^r)$.

These results are certainly also true if k is finite, e.g., by 4.13.a.

4.27. Combining 4.26(2), (3) with 4.24(1) we get a complete description of $H^\bullet(V_a, k)$. Before stating the result, we want to introduce some notation to describe the action of $GL(V)$ on the spaces $\sum_{i=1}^n kT_i^{p^r}$ and $\sum_{i=1}^n k\overline{\beta}(T_i^{p^r})$.

Let us assume for the sake of simplicity that k is a perfect field of characteristic p. We shall use the notation $M^{(r)}$ as in 2.16. The map $\gamma : f \mapsto f^{p^r}$ is an isomorphism $V^{*(r)} \to \sum_{i=1}^n kT_i^{p^r}$ of vector spaces over k. The map $\gamma \otimes \mathrm{id}_{\overline{k}}$ on $V^{*(r)} \otimes \overline{k} \simeq (V^* \otimes \overline{k})^{(r)}$ for some algebraic closure $\overline{k}$ of k commutes with $GL(V)(\overline{k})$. As $GL(V)$ is reduced, this implies that γ is an isomorphism of $GL(V)$–modules. Similarly, one checks that $f \mapsto \overline{\beta}(f^{p^r})$ is an isomorphism $V^{*(r+1)} \to \sum_{i=1}^n k\overline{\beta}(T_i^{p^r})$ of $GL(V)$–modules. This shows:

Proposition: *Suppose that k is a perfect field of characteristic $p \neq 0$.*

a) *If $p = 2$, then*

$$H^\bullet(V_a, k) \simeq S\Big(\bigoplus_{j\geq 0} V^{*(j)}\Big)$$

and for all $r > 0$

$$H^\bullet(V_{a,r}, k) \simeq S\Big(\bigoplus_{j=0}^{r-1} V^{*(j)}\Big).$$

b) *If $p \neq 2$, then*

$$H^\bullet(V_a, k) \simeq \Lambda\Big(\bigoplus_{j\geq 0} V^{*(j)}\Big) \otimes S'\Big(\bigoplus_{j\geq 1} V^{*(j)}\Big)$$

and for all $r > 0$

$$H^\bullet(V_{a,r}, k) \simeq \Lambda\Big(\bigoplus_{j=0}^{r-1} V^{*(j)}\Big) \otimes S'\Big(\bigoplus_{j=1}^{r} V^{*(j)}\Big).$$

Remarks: 1) The explicit description of H^1 and H^2 also gives the degrees [as in 4.20] of the generators of $H^\bullet(V_a, k)$ and $H^\bullet(V_{a,r}, k)$. All elements in $V^{*(j)}$ are homogeneous of degree p^j with respect to the $\mathbf{N}$–grading.

2) If k is a field of characteristic 0, then $H^\bullet(V_a, k) \simeq \Lambda(V^*)$. This follows, e.g., from the proposition applying the universal coefficient theorem to $\mathbf{Z}^n$.

CHAPTER 5

Quotients and Associated Sheaves

Some properties of the derived functors of induction can be proved only by interpreting the $R^n \operatorname{ind}_H^G M$ as cohomology groups $H^n(G/H, \mathcal{L}(M))$ of certain quasi-coherent sheaves on G/H. Before we can define these "associated sheaves" (5.8/9), and prove the equality $R^n \operatorname{ind}_H^G M = H^n(G/H, \mathcal{L}(M))$ in 5.12, we have to introduce the quotients G/H.

This is a nontrivial problem. Assuming G to be a (flat) group scheme and H a (flat) subgroup scheme we want G/H to be a scheme. The choice at first sight, the functor $A \mapsto G(A)/H(A)$, will in general not be a scheme. On the other hand, there is an obvious definition of a quotient scheme via a universal property (cf. 5.1) that, however, gives no information about existence and what the quotient looks like, if it happens to exist.

It has turned out to be useful to construct quotients not at once in the category of schemes over k but in the larger category of all k–faisceaux. These are the k–functors having a sheaf property with respect to the faithfully flat, finitely presented (Grothendieck) topology, cf. 5.2/3. The quotient faisceau G/H has a not too complicated description (5.4/5). In the most important cases (e.g., over a field) the quotient faisceau is a scheme (hence the quotient scheme) and has nice properties (5.6/7). It is only in this case that we can prove the relation between sheaf cohomology and the derived functors of induction mentioned above.

One consequence of this relation is that ind_H^G is an exact functor if G/H is an affine scheme. One can find direct proofs in [Oberst] and [Cline, Parshall, and Scott 3]. The converse is also proved in these papers.

The contents of the subsections 5.14–5.21 are applied in only a few places later on. They deal mainly with a description of $\mathcal{L}(M)$ as the sheaf of sections of a certain fibrations $G \times^H M_a \to G/H$ and with computing inverse and (also higher) direct images of $\mathcal{L}(M)$ in special situations.

I more or less follow [DG] in the subsections 5.1–5.7. Proposition 5.12 was first proved in [Haboush 2]. Let me add that closely related material is treated in [Cline, Parshall, and Scott 9].

5.1. (Quotients) For a linear algebraic group H over an algebraically closed field and a closed subgroup H of G, it is well known how to make the coset space G/H into a variety. We should like to have a generalisation to the case where G is a k–group scheme and H a closed subgroup scheme. Unfortunately, the "obvious" choice, i.e., the functor $A \mapsto G(A)/H(A)$ turns out to be wrong one (in general) as it in general will not be a scheme.

Let us instead define a quotient via a universal property. This can be done in the more general situation of a k–group scheme G acting on a scheme X over k. A *quotient scheme* of X by G is a pair (Y, π) where Y is a scheme and $\pi : X \to Y$ is a morphism such that π is constant on G–orbits, and such that for each morphism $f : X \to Y'$ of schemes constant on G–orbits there is exactly one morphism $f' : Y \to Y'$ with $f = f' \circ \pi$. ("Constant on G–orbits" means, e.g., for π that each $\pi(A) : X(A) \to Y(A)$ is constant on $G(A)$–orbits.) Of course, such a quotient scheme is unique up to unique isomorphism — if it exists (and that is the problem).

Let me give another formulation of this definition. We want to assume that G acts on the right. (The necessary changes for left actions will be obvious.) Consider the two morphisms $\alpha, \alpha' : X \times G \to X$ with $\alpha(x, g) = xg$ and $\alpha'(x, g) = x$. Then a morphism $f : X \to Y'$ will be constant on G–orbits if and only if $f \circ \alpha = f \circ \alpha'$. So (Y, π) is a quotient scheme if and only if $\pi \circ \alpha = \pi \circ \alpha'$ and if for all morphisms $f : X \to Y'$ with $f \circ \alpha = f \circ \alpha'$ there is a unique morphism $f' : Y \to Y'$ with $f = f' \circ \pi$. (We assume Y, Y' are schemes.) So a quotient scheme of X by G is (in categorical language) the cokernel of the pair (α, α') in the category of schemes over k.

This way of formulating the universal property allows for generalisations. Take, for example, a "schematic" equivalence relation on X, i.e., a subscheme $R \subset X \times X$ such that each $R(A)$ is an equivalence relation on $X(A)$. Then a quotient scheme of X by R is the cokernel in the category of schemes of the pair of projections from R to X. There is a generalisation of these two situations (i.e., of group actions and of equivalence relations) called *groupoid*. This is discussed, e.g., in [DG], III, §2, n° 1.

5.2. (The fppf-topology) Of course, we can define quotients by group actions in larger categories that $\{$ schemes over $k \}$ using the same type of universal property as before, but allowing any Y, Y' in that larger category. If we take, e.g., the category of all k–functors, then certainly $A \mapsto X(A)/G(A)$ is the quotient. If we now had a functor from $\{ k$–functors $\}$ to $\{$ schemes over $k \}$ left adjoint to the inclusion, then it would map $A \mapsto X(A)/G(A)$ to the quotient scheme. But we do not have such a functor. It has been proved to be useful in this situation to replace the category $\{$ schemes over $k \}$ by a larger one for which there is such a functor with nice properties.

Any scheme X is by definition local (cf. 1.8), i.e., $Y \mapsto \mathrm{Mor}(Y, X)$ is a sheaf in some sense: If $(Y_j)_j$ is an open covering of Y, then any $\alpha \in \mathrm{Mor}(Y, X)$ is uniquely determined by its restrictions to the Y_j, and we can glue together morphisms $\alpha_j \in \mathrm{Mor}(Y_j, X)$ if they coincide on intersections. The open coverings were defined using the Zariski topology.

One can now consider more general topologies, called *Grothendieck topologies*, where the property "open" is no longer attached to subsets (or rather subfunctors) but to certain morphisms. We shall consider only the *faithfully flat, finitely presented* topology (for short "fppf" as the French is much more symmetric in this case), and the k–functors with the sheaf property for this topology will be called faisceaux (reserving the term "sheaf" for objects related to the Zariski topology).

As in 1.8, it is enough to consider open coverings of affine schemes by affine schemes. Let R be a k–algebra. An *fppf-open covering* of R is a finite family R_1, $R_2, \dots, R_n$ of R–algebras such that each R_i is a finitely presented R–algebra, and such that $R_1 \times R_2 \times \cdots \times R_n$ is a faithfully flat R–module. (An R–algebra is finitely presented if it is the quotient of a polynomial ring in a finite number of variables by a finitely generated ideal.) Note that $R_1 \times R_2 \times \cdots \times R_n$ is faithfully flat over

R if and only if each R_i is a flat R–module and $\operatorname{Spec}(R)$ is the union of the images of all $\operatorname{Spec}(R_i)$, cf. [B2], ch. II, §2, cor. 4 de la prop. 4.

For each $f \in R$ the R–algebra R_f is finitely presented (as $R_f \simeq R[T]/(Tf-1)$) and a flat R–module. If $f_1, f_2, \dots, f_n \in R$ satisfy $\sum_{i=1}^n Rf_i = R$, then $\prod_{i=1}^n R_{f_i}$ is faithfully flat, cf. [B2], ch. II, §5, prop. 3, so the R_{f_i} form an fppf-open covering of R.

5.3. (Faisceaux) A k–functor X is called a *faisceau* if for each k–algebra R and each fppf-open covering $R_1, R_2, \dots, R_n$ of R the sequence

$$X(R) \longrightarrow \prod_i X(R_i) \rightrightarrows \prod_{i,j} X(R_i \otimes_R R_j) \tag{1}$$

is exact. (The maps are the obvious ones, induced by the structural maps $R \to R_i$ and by $R_i \to R_i \otimes_R R_j$ resp. $R_i \to R_j \otimes_R R_i$ with $a \mapsto a \otimes 1$ resp. $a \mapsto 1 \otimes a$.) A k–functor that is a faisceau is also called a k–faisceau.

For any k–algebras $R_1, R_2, \dots, R_n$ we can regard each R_i as a $(\prod_{i=1}^n R_i)$–algebra via the projection. The R_i obviously form an fppf-open covering of $R = \prod_{i=1}^n R_i$. As $R_i \otimes_R R_j = 0$ for $i \neq j$, the exactness of (1) amounts in this case to:

(2) *The projections induce for all k–algebras $R_1, R_2, \dots, R_n$ a bijection*

$$X(R_1 \times R_2 \times \cdots \times R_n) \xrightarrow{\sim} X(R_1) \times X(R_2) \times \cdots \times X(R_n).$$

A single R-algebra R' is an fppf-open covering of R if and only if it is a faithfully flat R–module and a finitely presented R–algebra. Let us call such an object an *fppf–R–algebra.* Now the exactness of (1) implies:

(3) *If R is a k–algebra and if R' is an fppf–R–algebra, then*

$$X(R) \longrightarrow X(R') \rightrightarrows X(R' \otimes_R R')$$

is exact.

We have thus proved one direction of:

(4) *A k–functor X is faisceau if and only if it satisfies* (2) *and* (3).

For the converse apply (3) to $R' = \prod_{i=1}^n R_i$ and (2) to $\prod_i R_i$ and $\prod_i R_i \otimes_R \prod_j R_j$.

Let $f_1, \dots, f_n \in R$ with $\sum_{i=1}^n Rf_i = R$. Then the R_{f_i} $(1 \le i \le n)$ form an fppf-open covering of R, cf. 5.2. Therefore a comparison of the definitions above and in 1.8 yields:

(5) *Each faisceau is a local functor.*

Suppose that R' is a faithfully flat R–algebra. We have then an exact sequence

$$0 \longrightarrow R \longrightarrow R' \longrightarrow R' \otimes_R R'$$

where $R \to R'$ is the structural map, and where any $a \in R'$ is mapped to $a \otimes 1 - 1 \otimes a$. (This is only the beginning of a long exact sequence, see [DG], I, §1, 2.7. It is enough to show the exactness of $0 \to R \otimes_R R' \to R' \otimes_R R' \to R' \otimes_R R' \otimes_R R'$. The last map sends $a \otimes a'$ to $a \otimes 1 \otimes a' - 1 \otimes a \otimes a'$. If this is 0, then $0 = a \otimes a' - 1 \otimes aa'$, hence $a \otimes a'$ is in the image of the previous map.) We can express this exactness also as saying that

$$(6) \qquad R \longrightarrow R' \rightrightarrows R' \otimes_R R'$$

is exact. (Here the two maps are $a \mapsto a \otimes 1$ and $a \mapsto 1 \otimes a$.) Now the left exactness of $\mathrm{Hom}_{k-\mathrm{alg}}(A, ?)$ shows that each affine scheme $Sp_k A$ over k is a faisceau. More generally, one can show (see [DG], III, §1, 1.3):

(7) *Any scheme over k is a faisceau.*

Let M be a k–module and k' a faithfully flat k–algebra. Then the same argument as above gives us an exact sequence

$$0 \to M \longrightarrow M \otimes k' \longrightarrow M \otimes k' \otimes k'$$

with maps $m \mapsto m \otimes 1$ and $m \otimes b \mapsto m \otimes b \otimes 1 - m \otimes 1 \otimes b$. Applying this to all $M \otimes A$ we get:

(8) *For each k–module M the functor M_a is a faisceau.*

The following property is obvious:

(9) *Let X be a k–functor and k' a k–algebra. If X is a faisceau, then so is $X_{k'}$.*

Let X be a k–faisceau and Y a subfaisceau of X. For each k–algebra R and each fppf–R–algebra R' we can by (3) identify $X(R)$ with a subset of $X(R')$ and $Y(R)$ with a subset of $Y(R')$. The exactness in (3) yields then

$$(10) \qquad Y(R) = Y(R') \cap X(R).$$

5.4. (Associated Faisceaux) There is a natural construction that associates to each k–functor X a k–faisceau $\widetilde{X}$ (called the *associated faisceau*) together with a morphism $i : X \to \widetilde{X}$ such that for all k–faisceau Y the map $f \mapsto f \circ i$ is a bijection $\mathrm{Mor}(\widetilde{X}, Y) \to \mathrm{Mor}(X, Y)$. We get thus a functor $X \mapsto \widetilde{X}$ from $\{k\text{–functors}\}$ to $\{k\text{–faisceaux}\}$ left adjoint to the inclusion of $\{k\text{–faisceaux}\}$ into $\{k\text{–functors}\}$. This construction should be regarded as an analogue of the construction of a sheaf associated to a presheaf. The details may be found in [DG], III, §1, 1.8–1.12. I shall describe $\widetilde{X}$ only in a particularly simple case where X is already close to being a faisceau. To be more precise, I want to assume the following:

(1) *X satisfies 5.3(2) and $X(R) \to X(R')$ is injective for each k–algebra R and each fppf–R–algebra R'.*

Under this assumption $\widetilde{X}$ has the following form. Take a k–algebra A and consider for each fppf–A–algebra B the kernel $X(B, A)$ of $X(B) \rightrightarrows X(B \otimes_A B)$. If B' is an fppf–B–algebra, then B' is also fppf over A, and the natural inclusion from $X(B)$ into $X(B')$ maps $X(B, A)$ into $X(B', A)$. More precisely, $B' \otimes_A B'$ is fppf over $B \otimes_A B$, hence the standard map $X(B \otimes_A B) \to X(B' \otimes_A B')$ is injective, and we can identify $X(B,A)$ with the intersection of $X(B',A)$ and $X(B)$. The $X(B,A)$ with B fppf over A form a direct system. (If B_1, B_2 are fppf over A, then $B_1 \otimes_A B_2$ is fppf over B_1 and B_2.) So we can form the direct limit of these $X(B,A)$ and this will be our $\widetilde{X}(A)$:

$$\widetilde{X}(A) = \varinjlim X(B,A). \tag{2}$$

As all maps $X(B,A) \to X(B',A)$ are injective, so are all maps $X(B,A) \to \widetilde{X}(A)$, we can identify $X(B,A)$ with its image in $\widetilde{X}(A)$ and regard $\widetilde{X}(A)$ as the union of all $X(B,A)$. We get in particular for $X(A,A) = X(A)$:

(3) *For X as in* (1) *each $X(A) \to \widetilde{X}(A)$ is injective.*

(For arbitrary X this will not be the case.)

We now check that $\widetilde{X}$ indeed is a k–faisceau. If $A \to A'$ is a homomorphism of k–algebras, then $B \otimes_A A'$ is fppf over A' for each fppf–A–algebra B, and the natural map $X(B) \to X(B \otimes_A A')$ maps $X(B,A)$ to $X'(B \otimes_A A', A')$. Taking direct limits we get a map $\widetilde{X}(A) \to \widetilde{X}(A')$, which is easily checked to be functorial. In this way $\widetilde{X}$ is a k–functor.

It is rather obvious that $\widetilde{X}$ inherits the property (1) from X. Consider any element x in the kernel of $\widetilde{X}(B) \rightrightarrows \widetilde{X}(B \otimes_A B)$ for some fppf–A–algebra B. We have to show that x belongs to the image of $\widetilde{X}(A)$. By construction $x \in X(B',B)$ for some fppf–B–algebra B'. The restrictions to $X(B',B)$ of the two maps from $\widetilde{X}(B)$ to $\widetilde{X}(B \otimes_A B)$ are induced by the maps

$$\begin{aligned} X(B',B) \to X(B' \otimes_B (B \otimes_A B), B \otimes_A B) &\xrightarrow{\sim} X(B' \otimes_A B, B \otimes_A B) \\ \subset X(B' \otimes_A B', B \otimes_A B) &\subset \widetilde{X}(B \otimes_A B) \end{aligned}$$

and

$$\begin{aligned} X(B',B) \to X(B' \otimes_B (B \otimes_A B), B \otimes_A B) &\xrightarrow{\sim} X(B \otimes_A B', B \otimes_A B) \\ \subset X(B' \otimes_A B', B \otimes_A B) &\subset \widetilde{X}(B \otimes_A B), \end{aligned}$$

where the isomorphism in the second step is induced by $b' \otimes (b_1 \otimes b_2) \mapsto (b'b_2) \otimes b_1$ in the first case, by $b' \otimes (b_1 \otimes b_2) \mapsto b_1 \otimes (b'b_2)$ in the second case. (We use here that $B' \otimes_A B' \simeq B' \otimes_B (B \otimes_A B') \simeq (B' \otimes_A B) \otimes_B B'$ is fppf over $B \otimes_A B'$ and $B' \otimes_A B$.) Therefore the intersection of $\ker(\widetilde{X}(B) \rightrightarrows \widetilde{X}(B \otimes_A B))$ with $X(B',B)$ is equal to

$$\begin{aligned} \ker(\widetilde{X}(B',B) \rightrightarrows X(B' \otimes_A B', B \otimes_A B)) &= \ker(X(B') \rightrightarrows X(B' \otimes_A B')) \\ &= X(B',A), \end{aligned}$$

hence contained in $\widetilde{X}(A)$. This shows that $\widetilde{X}(A)$ is a faisceau.

For any morphism $f : X \to Y$ into a k–faisceau Y, any $f(B)x$ with $x \in X(B,A)$ as above has to belong to $Y(A) \subset Y(B)$; so we can define $\widetilde{f} : \widetilde{X} \to Y$ by

$\widetilde{f}(A)x = f(B)x \in Y(A)$. This is easily checked to be a morphism and to be unique with $\widetilde{f}_{|X} = f$. So $\widetilde{X}$ has the universal property we wanted.

Note: If each $f(A)$ is injective, then so is each $\widetilde{f}(A)$. So we can then regard $\widetilde{X}$ as a subfunctor of Y. One easily gets the following:

Let X be a subfunctor of a k–faisceau Y such that X satisfies 5.3(2). *Then $\widetilde{X}$ is a subfunctor of Y. One has*

(4) $\quad \widetilde{X}(A) = \{\, x \in Y(A) \mid$ *there is an fppf–A–algebra B with* $x \in X(B)\,\}$.

It is clear in a situation as in (1), but can be proved also in the general case, that taking the associated faisceau commutes with base change:

(5) *Let X be a k–functor and k' a k–algebra. Then $(\widetilde{X})_{k'}$ is the faisceau associated to $X_{k'}$.*

5.5. (Images and Quotients) Let $f : X \to Y$ be a morphism of k–faisceaux. The subfunctor $A \mapsto \mathrm{im}(f(A)) = f(A)X(A)$ of Y clearly satisfies 5.3(2). So, 5.4(4) yields a rather precise description of the associated faisceau, which is called the *image faisceau* of f. We shall usually denote this associated faisceau by $f(X)$ or $\mathrm{im}(f)$. So in general, $f(A)X(A)$ is properly contained in $f(X)(A)$.

Let G be a k–group faisceau acting on a k–faisceau X, say from the right. We define the *quotient faisceau* X/G as the associated faisceau of the k–functor Z with $Z(A) = X(A)/G(A)$. We denote by $\pi : X \to X/G$ the canonical map that is the composition of the morphism $Z \to \widetilde{Z} = X/G$ from the definition of an associated faisceau with the morphism $X \to Z$ that takes any $x \in X(A)$ to $xG(A)$.

The universal property of the associated faisceau implies that $(X/G, \pi)$ is a quotient of X by G in the category of k–faisceaux. If X is a scheme and if X/G also happens to be a scheme, then X/G is a quotient in the category of k–schemes.

In general the functor $A \mapsto X(A)/G(A)$ will not satisfy 5.4(1), so, in general, the description of X/G is more complicated than what is done in 5.4. However, let us assume that G acts freely on X, i.e., that each $G(A)$ acts freely on $X(A)$. We claim:

(1) *If G acts freely on X, then the functor $A \mapsto X(A)/G(A)$ satisfies* 5.4(1).

The part about direct products is obvious (for any action). Consider now some k–algebra A and an fppf–A–algebra B. In order to prove the injectivity of $X(A)/G(A) \to X(B)/G(B)$, we have to take $x, x' \in X(A)$ with $xG(B) = x'G(B)$ in $X(B)$ and show that $xG(A) = x'G(A)$. [We use here and below the same notation for x and its canonical image in $X(B)$.] So there is $g \in G(B)$ with $x = x'g$ in $X(B)$. Let g_1, g_2 be the images of g in $G(B \otimes_A B)$ under the maps derived from the two homomorphisms $B \to B \otimes_A B$ with $b \mapsto b \otimes 1$ and $b \mapsto 1 \otimes b$. We have then $x = x'g_1 = x'g_2$ in $X(B \otimes_A B)$, hence $g_1 = g_2$ as $G(B \otimes_A B)$ acts freely. The faisceau property of G yields that we may assume $g \in G(A)$. The faisceau property of X then implies that $x = x'g$ already in $X(A)$, hence that $xG(A) = x'G(A)$ as claimed.

We get now from 5.4(3) that the natural map $X(A)/G(A) \to (X/G)(A)$ is injective in the case of a free action. This inclusion will in general be strict. We can express this injectivity as follows: The maps $X(A) \times G(A) \to X(A) \times X(A)$ with $(x, g) \mapsto (x, xg)$ induce an isomorphism

$$X \times G \xrightarrow{\sim} X \times_{X/G} X \qquad \text{(for a free action).} \tag{2}$$

(The fibre product is taken with respect to the canonical projection taken twice.)

Let me describe an example. Denote by $\mathbf{A}_*^{n+1}$ the open subscheme of $\mathbf{A}^{n+1}$ that associates to each A the set of all $(a_0, a_1, \ldots, a_n) \in A^{n+1}$ with $\sum_{i=0}^n Aa_i = A$. For each such $\alpha = (a_0, a_1, \ldots, a_n)$ there is a linear map $\varphi : A^{n+1} \to A$ with $\varphi(\alpha) = 1$. Then $A^{n+1} = A\alpha \oplus \ker(\varphi)$ and so α defines a point $A\alpha \in \mathbf{P}^n(A)$, cf. 1.9. Two such tuples α and α' define the same submodule $A\alpha = A\alpha'$ if and only if they differ by a factor in $A^\times = G_m(A)$. So $\alpha \mapsto A\alpha$ yields an embedding $\mathbf{A}_*^{n+1}(A)/G_m(A) \hookrightarrow \mathbf{P}^n(A)$. We claim that this induces an isomorphism (for the obvious, free action of G_m on $\mathbf{A}_*^{n+1}$)

$$\mathbf{A}_*^{n+1}/G_m \xrightarrow{\sim} \mathbf{P}^n. \tag{3}$$

Because of 5.4(4) we only have to show the following: If $M \in \mathbf{P}^n(A)$ for some A, then there is some fppf–A–algebra B and some $\beta \in \mathbf{A}_*^{n+1}(B)$ with $B\beta = M \otimes_A B$, where we identify $M \otimes_A B \subset A^{n+1} \otimes_A B$ with its image in B^{n+1}. Equivalently, we have to find B, fppf over A, such that $M \otimes_A B$ is free over B. But by [B2], ch. II, §5, thm. 1, there are $f_1, \ldots, f_r \in A$ such that $A = \sum_{i=1}^r Af_i$ and such that each M_{f_i} is free over A_{f_i}. (One can even assume that $(A^{n+1}/M)_{f_i}$ is also free. That will be used when proving 5.6(3).) Now $= \prod_{i=1}^r A_{f_i}$ will do, cf. 5.2.

If $X' \subset X$ is a G–stable subfaisceau, then $A \mapsto X'(A)/G(A)$ is a subfunctor of $A \mapsto X(A)/G(A)$. This leads to a morphism $X'/G \to X/G$. In the case of a free action, our construction shows that this morphism is an embedding; in fact, X'/G is then identified with the image faisceau $\pi(X')$ where $\pi : X \to X/G$ is the canonical map.

Note that 5.4(5) implies for any k–algebra k':

$$X_{k'}/G_{k'} = (X/G)_{k'}. \tag{4}$$

If a direct product $G_1 \times G_2$ of k–group faisceaux acts on a k–faisceau X, then G_1 acts in a natural way on X/G_2 and we get a natural isomorphism

$$(X/G_2)/G_1 \xrightarrow{\sim} X/(G_1 \times G_2). \tag{5}$$

This follows from the universal property of a quotient faisceau. In case $G_1 \times G_2$ acts freely, one also gets (5) from the explicit construction in 5.4. In that case the action of G_1 on X/G_2 is also free.

Suppose for the moment that k is an algebraically closed field. One can show that $(X/G)(k) = X(k)/G(k)$, see [DG], III, §1, 1.15. Suppose that X and G are reduced and algebraic (hence correspond to varieties). There will be, in general, orbits of $G(k)$ on $X(k)$ that are not closed. If so, then $X(k)/G(k) = (X/G)(k)$ cannot have a topology where all points are closed and for which the natural map $X(k) \to (X/G)(k)$ is continuous. Therefore X/G is not a scheme. It is only for very nice group actions [as that in (3)] that the quotient faisceau leads to the quotient scheme. Let me mention without proof one case (now again for arbitrary k) where this happens.

We call an affine group scheme G *finite* if $k[G]$ is a finitely generated and projective k–module.

(6) *Let X be an affine scheme and G an affine k–group scheme acting freely on X. If G is finite, then X/G is an affine scheme isomorphic to $Sp_k(k[X]^G)$.*

Though not stated in this way, this follows easily by combining [DG], III, §2, n° 4 and §1, 2.10. The results at the first place imply that $k[X]$ is finitely generated and projective as a $k[X]^G$–module and that X/G is a subfunctor of $Sp_k(k[X]^G)$. The second result quoted implies that the inclusion $k[X]^G \subset k[X]$ induces an epimorphism $X \to Sp_k(k[X]^G)$ in the category of k–faisceaux, while on the other hand the image faisceau is equal to X/G. (In [DG], III, §2, the case where X is not affine or where G does not act freely is also discussed.)

5.6. (Quotients by Subgroups and Orbit Faisceaux) Let G be a k–group faisceau.

The action on G of any subgroup faisceau H of G by right translation is obviously free. So we can apply the explicit construction in 5.4 in order to get G/H, cf. 5.5(1). We see especially that each map $G(A)/H(A) \to (G/H)(A)$ is injective, and that we get an isomorphism [cf. 5.5(2)]:

$$G \times H \xrightarrow{\sim} G \times_{G/H} G. \tag{1}$$

Let X be a k–faisceau on which G acts (on the left). Take $x \in X(k)$ and consider its stabiliser $\mathrm{Stab}_G(x)$ in G, i.e., the subgroup functor of G with

$$\mathrm{Stab}_G(x)(A) = \{\, g \in G(A) \mid gx = x \,\}$$

for all A. (In the notation of 2.6(3) this is $\mathrm{Stab}_G(Y)$ where $Y \subset X$ is the subfunctor with $Y(A) = \{x\}$ for all A.) It is easy to check that $\mathrm{Stab}_G(x)$ is a faisceau. We can identify $A \mapsto G(A)/\mathrm{Stab}_G(x)(A)$ with a subfunctor of X, hence also its associated faisceau $G/\mathrm{Stab}_G(x)$. More precisely, the morphism $\pi_x : G \to X$, $g \mapsto gx$ factors through $G/\mathrm{Stab}_G(x)$ and induces an isomorphism

$$G/\mathrm{Stab}_G(x) \xrightarrow{\sim} \mathrm{im}(\pi_x) \tag{2}$$

onto the image faisceau of π_x, which is also called the *orbit faisceau* of x.

Take, for example, $G = GL_{r+n}$ for some $r, n \in \mathbf{N}$, $\neq 0$. The natural action of G on $\mathbf{A}^{r+n} = (k^{r+n})_a$ induces an action on the Grassmann scheme $\mathcal{G}_{r,n}$, cf. 1.9. Let $x \in \mathcal{G}_{r,n}(k)$ denote the direct summand $\{(x_1, x_2, \dots, x_r, 0, 0, \dots, 0) \mid x_1, x_2, \dots, x_r \in k\}$ of k^{r+n}. Set $P_r = \mathrm{Stab}_G(x)$. Then $G(A)x$ consists of all direct summands M of A^{r+n} such that M is free of rank r and A^{r+n}/M is free of rank n. Arguing as in 5.5(3) one shows

$$GL_{r+n}/P_r \xrightarrow{\sim} \mathcal{G}_{r,n}. \tag{3}$$

Let us consider another example. Let Λ be a finitely generated abelian group and Λ' a subgroup of Λ. We can regard (cf. 2.5) $\mathrm{Diag}(\Lambda/\Lambda')$ as a subgroup functor of $\mathrm{Diag}(\Lambda)$:

$$\begin{aligned}\mathrm{Diag}(\Lambda/\Lambda')(A) &= \{\, \varphi \in \mathrm{Hom}_{gp}(\Lambda, A^\times) \mid \varphi(\Lambda') = \{1\} \,\}\\ &\subset \mathrm{Hom}_{gp}(\Lambda, A^\times) = \mathrm{Diag}(\Lambda)(A)\end{aligned}$$

for each k–algebra A. Then $\varphi \mapsto \varphi_{|\Lambda'}$ defines an embedding

$$\mathrm{Diag}(\Lambda)(A)/\mathrm{Diag}(\Lambda/\Lambda')(A) \hookrightarrow \mathrm{Diag}(\Lambda')(A),$$

which we claim to yield an isomorphism

(4) $$\mathrm{Diag}(\Lambda)/\mathrm{Diag}(\Lambda/\Lambda') \xrightarrow{\sim} \mathrm{Diag}(\Lambda').$$

We have to show for any k–algebra A and any group homomorphism $\varphi : \Lambda' \to A^\times$ that there exists an fppf–A–algebra B and a group homomorphism $\overline{\varphi} : \Lambda \to B^\times$ extending φ. By induction we can assume that $\Lambda = \mathbf{Z}\lambda + \Lambda'$ for some $\lambda \in \Lambda$. There is $n \in \mathbf{N}$ with $\mathbf{Z}\lambda \cap \Lambda' = \mathbf{Z}n\lambda$. Then take the polynomial algebra $A[t]$ and its homomorphic image $B = A[t]/(t^n - \varphi(n\lambda))$ and map λ to the residue class of t.

More generally, consider a homomorphism $\alpha : \Lambda_2 \to \Lambda_1$ of finitely generated abelian groups, and the corresponding homomorphism $\mathrm{Diag}(\alpha) : \mathrm{Diag}(\Lambda_1) \to \mathrm{Diag}(\Lambda_2)$ of diagonalisable group schemes. One can obviously identify

(5) $$\ker \mathrm{Diag}(\alpha) \simeq \mathrm{Diag}(\Lambda_1/\alpha(\Lambda_2))$$

and the discussion above yields

(6) $$\mathrm{im}\ \mathrm{Diag}(\alpha) \simeq \mathrm{Diag}(\alpha(\Lambda_2)) \simeq \mathrm{Diag}(\Lambda_2/\ker(\alpha)).$$

In this situation we can define a cokernel of $\mathrm{Diag}(\alpha)$ as the quotient of $\mathrm{Diag}(\Lambda_2)$ by the image of $\mathrm{Diag}(\alpha)$ and we get

(7) $$\mathrm{coker}\ \mathrm{Diag}(\alpha) \simeq \mathrm{Diag}(\ker(\alpha)).$$

This object has the usual properties of a cokernel.

Recall that an affine scheme X over k is called algebraic if $k[X]$ is a finitely presented k–algebra.

(8) *Suppose that G is an algebraic k–group. If k is a field, then G/H is a scheme for each closed subgroup scheme H of G.*

One can find a proof in [DG], III, §3, 5.4. It uses a generalisation of 5.5(6) in order reduce to the case of reduced group schemes. If G is reduced, then so is G/H, in fact, it is a smooth variety. If in addition H is reduced, then G/H is the same object as constructed, e.g., in [Bo], 6.8 using a different language. The construction in Section 17 of [Bo] is an example for a quotient with G reduced and H not reduced.

For arbitrary k, G, H the quotient faisceau G/H will not be a scheme, cf. the counter examples in [DG], III, §3, 3.3, and in [Ra], p. 157. Let me however mention:

(9) *Let G be an algebraic k–group. If k is a Dedekind ring, then G/H is a scheme for each closed and flat subgroup scheme H of G.*

We shall use this result only in cases where it is clear for other reasons. A proof is contained in [A], Thm. 4.C.

5.7. (Flatness of Quotients) Let X be a k–scheme and let G be a k–group scheme acting freely (on the right) on X. Let $\pi : X \to X/G$ denote the canonical map. Then:

(1) *Suppose that X/G is a scheme. Let Y be an open and affine subscheme Y of X/G. Then $\pi^{-1}(Y)$ is open and affine in X. If G is flat (resp. flat and algebraic), then $k[\pi^{-1}(Y)]$ is faithfully flat over $k[Y]$ (resp. an fppf–$k[Y]$–algebra).*

One usually expresses this result in the form: π is affine, flat (for G flat) and finitely presented (for G flat and algebraic). In [DG], III, §3, 2.5/6 one can find a proof of (1) in the special case where a subgroup acts by right multiplication on the whole group. Let me go through the main steps of the proof in order to show that it generalises to our situation.

To start with, $\pi^{-1}(Y)$ is a G–stable and open subscheme of X with $\pi^{-1}(Y)/G \simeq Y$. So we can replace X by $\pi^{-1}(Y)$ and assume that $Y = X/G$. Set $A = k[Y]$ and regard id_Y as an element of $Y(A)$. By the construction of X/G there is an fppf–A–algebra B such that id_Y is in the image of $X(B)$. Set $X' = Sp_k B$ for such a B. Using the identification $\mathrm{Mor}(X', X) = X(B)$ we get a morphism $\alpha : X' \to X$ such that $\pi \circ \alpha : X' = Sp_k B \to Sp_k A = Y$ is given by the structural homomorphism $A \to B$. The isomorphism $X \times G \to X \times_Y X$ from 5.5(2) induces an isomorphism $X' \times G \to X' \times_Y X$ given by $(x', g) \mapsto (x', \alpha(x')g)$. So the faithfully flat base change $X' \to Y$ applied to $\pi : X \to Y$ yields the first projection $X' \times G \to X'$. As G is affine, so is this projection, hence also X, cf. [DG], I, §2, 3.9. Furthermore, we get

$$B \otimes_A k[X] \simeq k[X' \times_Y X] \simeq k[X' \times G] \simeq k[X'] \otimes k[G].$$

So, if $k[G]$ is flat over k (hence faithfully flat as k is a direct summand of $k[G]$), then so is $B \otimes_A k[X]$ over B, hence $k[X]$ over A, cf. [DG], I, §3, 1.4; the same type of argument works in case $k[G]$ is fppf over k.

Here is one consequence of (1):

(2) *If X and G are flat and if X/G is a scheme, then X/G is flat.*

Indeed, for any Y as above $k[\pi^{-1}(Y)]$ is flat over k and faithfully flat over $k[Y]$. So $k[Y]$ is flat over k.

Another corollary to the flatness is the openness of π:

(3) *Suppose that X/G is a scheme and that G is flat and algebraic. If Y is an open subfunctor in X, then $\pi(Y)$ is open in X/G.*

Let me refer to [DG], I, §3, 3.11 combined with the proof of [DG], III, §3, 2.5 for this result. (In the noetherian case one can also use [Ha], III, exerc. 9.1.)

5.8. (Associated Sheaves) As mentioned in 1.11, there corresponds to each k–scheme X a topological space $|X|$ together with a sheaf of k–algebras on $|X|$. The open subsets of $|X|$ correspond bijectively to the open subfunctors of X. So we can describe a sheaf on $|X|$ as a contravariant functor from { open subfunctors of X } (with inclusions as morphisms) to some other category, having the usual sheaf property with respect to open coverings of open subfunctors (defined in 1.7). For example, the structure sheaf $\mathcal{O}_X$ associates to each open subfunctor Y the k–algebra $\mathcal{O}_X(Y) = \mathrm{Mor}(Y, \mathbf{A}^1) = k[Y]$.

Let G be a flat k–subgroup scheme acting freely (from the right) on a flat k–scheme X such that X/G is a scheme. Let us denote the canonical morphism by $\pi : X \to X/G$.

We now associate to each G–module M a sheaf $\mathcal{L}(M) = \mathcal{L}_{X/G}(M)$ on X/G. We set for each open subfunctor $U \subset X/G$

$$\mathcal{L}(M)(U) = \{f \in \operatorname{Mor}(\pi^{-1}U, M_a) \mid f(xg) = g^{-1}f(x) \text{ for all } x \in (\pi^{-1}U)(A), \text{ all } g \in G(A), \text{ and all } A\}. \tag{1}$$

If $\pi^{-1}U$ is affine, then $\operatorname{Mor}(\pi^{-1}U, M_a) \simeq M_a(k[\pi^{-1}U]) = M \otimes k[\pi^{-1}U]$. This is a G–module via the given representation on M and the action on $k[\pi^{-1}U]$ derived from the action on $\pi^{-1}U \subset X$. Then obviously

$$\mathcal{L}(M)(U) = (M \otimes k[\pi^{-1}U])^G \qquad (\text{for } \pi^{-1}U \text{ affine}). \tag{2}$$

Any $f_1 \in \mathcal{O}_{X/G}(U)$ (for arbitrary U) can be regarded as a G–invariant element in $\mathcal{O}_X(\pi^{-1}U)$. So we can multiply any $f \in \mathcal{L}(M)(U)$ with f_1 and get still an element in $\mathcal{L}(M)(U)$. Thus $\mathcal{L}(M)(U)$ is an $\mathcal{O}_{X/G}(U)$–module. For open subfunctors $U' \subset U$ of X/G we have an obvious restriction map $\mathcal{L}(M)(U) \to \mathcal{L}(M)(U')$. Thus $\mathcal{L}(M)$ is a presheaf of $\mathcal{O}_{X/G}$–modules.

Consider the morphisms $\gamma_U : (\pi^1 U) \times G \to \pi^1 U$, $(x,g) \mapsto xg$ and $\beta : M_a \times G \to M_a$, $(m,g) \mapsto g^{-1}m$. A morphism $f \in \operatorname{Mor}(\pi^{-1}U, M_a)$ belongs to $\mathcal{L}(M)(U)$ if and only if $f \circ \gamma_U = \beta \circ (f \times \mathrm{id}_G)$. So we have an exact sequence

$$\mathcal{L}(M)(U) \longrightarrow \operatorname{Mor}(\pi^{-1}U, M_a) \rightrightarrows \operatorname{Mor}(\pi^{-1}U \times G, M_a). \tag{3}$$

Because of 5.3(8) the functors $U \mapsto \operatorname{Mor}(\pi^{-1}U, M_a)$ and $U \mapsto \operatorname{Mor}(\pi^{-1}U \times G, M_a)$ are sheaves. An elementary argument shows now that $\mathcal{L}(M)$ is also a sheaf (of $\mathcal{O}_{X/G}$–modules). It is called the *associated sheaf* (to M on X/G).

5.9. Let X, G, π be as in 5.8. Any homomorphism $\varphi : M \to M'$ of G–modules induces a homomorphism of $\mathcal{O}_{X/G}$–modules:

$$\mathcal{L}(\varphi) : \mathcal{L}(M) \longrightarrow \mathcal{L}(M'), \quad f \mapsto \varphi_a \circ f. \tag{1}$$

So $\mathcal{L}$ is a functor from $\{$ G–modules $\}$ to $\{$ $\mathcal{O}_{X/G}$–modules $\}$.

Proposition: a) *The functor $\mathcal{L}$ is exact.*

b) *For each G–module M the $\mathcal{O}_{X/G}$–module $\mathcal{L}(M)$ is quasi-coherent.*

c) *If a G–module M is finitely generated over k, then $\mathcal{L}(M)$ is a coherent $\mathcal{O}_{X/G}$–module.*

Proof: a) It is enough to show that $M \mapsto \mathcal{L}(M)(U)$ is exact for each open and affine $U \subset X/G$. For such U also $U' = \pi^{-1}U$ is affine and $k[U']$ is faithfully flat over $k[U]$, cf. 5.7(1). It is therefore enough to show that

$$M \mapsto k[U'] \otimes_{k[U]} \mathcal{L}(M)(U) = k[U'] \otimes_{k[U]} (k[U'] \otimes M)^G \tag{2}$$

is exact. The isomorphism in 5.5(2) induces an isomorphism $U' \times G \to U' \times_U U'$ compatible with the actions of G (on the second factors). So the corresponding isomorphism $k[U'] \otimes_{k[U]} k[U'] \simeq k[U'] \otimes k[G]$ is also G–equivariant. As $(k[G] \otimes M)^G \simeq M$, cf. 3.7(6), the functor in (2) can be identified with $M \mapsto k[U'] \otimes M$. It is exact as X is assumed to be flat.

b) For each scheme Y and each k–module M, the sheaf $U \mapsto \mathrm{Mor}(U, M_a)$ is quasi-coherent. (If Y is affine, then it is the quasi-coherent sheaf associated to the $k[Y]$–module $k[Y] \otimes M$, cf. Yoneda's lemma 1.3.) The sheaves $U \mapsto \mathrm{Mor}(\pi^{-1}U, M_a)$ and $U \mapsto \mathrm{Mor}((\pi^{-1}U) \times G, M_a)$ on X/G are direct images of such sheaves, hence quasi-coherent, cf. [Ha], II, 5.2.d. (We can reduce to the affine case using 5.7(1).) By 5.8(3) we can now regard $\mathcal{L}(M)$ as a kernel of a homomorphism between quasi-coherent $\mathcal{O}_{X/G}$–modules. So it is quasi-coherent itself, cf. [Ha], II, 5.7.

c) We have to show for all $U \subset X/G$ (open and affine) that $\mathcal{L}(M)(U)$ is finitely generated over $k[U]$. As $k[\pi^{-1}U]$ is faithfully flat over $k[U]$, cf. 5.7(1), it is enough to show that $k[\pi^{-1}U] \otimes_{k[U]} \mathcal{L}(M)(U)$ is finitely generated over $k[\pi^{-1}U]$, cf. [B2], ch. I, §3, prop. 11. This module is isomorphic to $k[\pi^{-1}U] \otimes M$ as we saw in the proof of a), so the claim is now obvious.

5.10. (Examples) Here are a few simple cases. For the trivial G–module k one has $\mathcal{L}(k)(U) = \mathrm{Mor}(\pi^{-1}U, \mathbf{A}^1)^G = \mathrm{Mor}(\pi^{-1}U/G, \mathbf{A}^1) = \mathrm{Mor}(U, \mathbf{A}^1)$, hence

(1) $$\mathcal{L}(k) = \mathcal{O}_{X/G}.$$

Consider, on the other hand, the G–module $k[G]$ under ρ_l or, more generally, $M \otimes k[G]$ for any k–module M regarded as a trivial G–module. For any $U \subset X/G$ open and affine, we can identify $\mathrm{Mor}(\pi^{-1}U, (M \otimes k[G])_a) \simeq \mathrm{Mor}(\pi^{-1}U \times G, M_a)$. Then the G–invariance condition translates into $f(x, g') = f(xg, g^{-1}g')$. The map $(x, g) \mapsto (xg, g)$ is an automorphism of $\pi^{-1}U \times G$ and transfers the condition into $f(x, gg') = f(x, g')$. In this way $\mathcal{L}(M)(U)$ is identified with $\mathrm{Mor}(\pi^{-1}U, M_a) = \mathcal{L}_{X/1}(M)(\pi^{-1}U) = (\pi_* \mathcal{L}_{X/1}(M))(U)$. This implies

(2) $$\mathcal{L}(M \otimes k[G]) \simeq \pi_* \mathcal{L}_{X/1}(M).$$

5.11. Let X, G, π be as in 5.8.

Proposition: *The functor $M \mapsto H^0(X/G, \mathcal{L}(M))$ from {G–modules} to {k–modules} is left exact. If X is affine, then the $M \mapsto H^n(X/G, \mathcal{L}(M))$ are its derived functors.*

Proof: The first claim is clear from 5.9.a and the left exactness of $H^0(X/G, ?)$. In order to get the second claim, it is enough to show that $\mathcal{L}$ maps injective G–modules to acyclic sheaves, cf. 4.1(3). By 3.9.c it is enough to consider G–modules of the form $M \otimes k[G]$ with G acting trivially on M. As π is affine (cf. 5.7(1)) we get from 5.10(2) and [Ha], III, exerc. 4.1

$$H^n(X/G, \mathcal{L}(M \otimes k[G])) \simeq H^n(X, \mathcal{L}_{X/1}(M)).$$

If X is affine, these cohomology groups vanish for $n > 0$, cf. [Ha], III, thm. 3.5.

5.12. Proposition: *Let G be a flat k–group scheme and H a flat subgroup scheme of G such that G/H is a scheme.*

a) *There is for each H–module M and each $n \in \mathbf{N}$ a canonical isomorphism of k–modules*

$$R^n \operatorname{ind}_H^G M \simeq H^n(G/H, \mathcal{L}_{G/H}(M)). \tag{1}$$

b) *If G/H is noetherian, then $R^n \operatorname{ind}_H^G = 0$ for all $n > \dim(G/H)$.*

c) *Suppose that k is noetherian and that G/H is a projective scheme. For each H–module M that is finitely generated over k, all $R^n \operatorname{ind}_H^G M$ are finitely generated over k.*

Proof: As the forgetful functor from { G–modules } to { k–modules } is exact, the $R^n \operatorname{ind}_H^G$ are also the derived functors of ind_H^G considered as a functor from { H–modules } to { k–modules }, cf. 4.1(2). Comparing 5.8(1) and 3.3(2) we get

$$\operatorname{ind}_H^G M = \mathcal{L}_{G/H}(M)(G/H) = H^0(G/H, \mathcal{L}_{G/H}(M)). \tag{2}$$

So (1) follows from 5.11. Now b) and c) are translations of well known results on sheaf cohomology, cf. [Ha], III, 2.7 and 5.2.a.

Remarks: 1) We get from (1) a G–module structure on each $H^n(G/H, \mathcal{L}_{G/H}(M))$. The existence of such a structure [satisfying (1)] is, however, already clear for other reasons: Each $\mathcal{L}_{G/H}(M)$ is a G–linearised sheaf in the sense, say, of [MF], p. 30. For such sheaves one gets the G–module structure on the cohomology using the functoriality of the cohomology.

2) Let M be an H–module and N a G–module such that N is flat over k. Then we get now a sheaf version of the tensor identity:

$$\mathcal{L}_{G/H}(M \otimes N) \simeq \mathcal{L}_{G/H}(M) \otimes N \tag{3}$$

where the right hand side denotes the sheaf $U \mapsto \mathcal{L}_{G/H}(M)(U) \otimes N$. Let $\pi : G \to G/H$ denote the canonical map. Then the sections on U of both sides in (3) can be embedded into

$$\operatorname{Mor}(\pi^{-1}U, (M \otimes N)_a) \simeq M \otimes N \otimes k[\pi^{-1}U]$$

and one can argue as in the proof of Proposition 3.6.

3) Let H' be a flat subgroup scheme of G with $H' \supset H$. Let M be an H–module. We get now a sheaf version of the transitivity of induction:

$$\pi_* \mathcal{L}_{G/H}(M) \simeq \mathcal{L}_{G/H'}(\operatorname{ind}_H^{H'} M) \tag{4}$$

where $\pi : G/H \to G/H'$ is the canonical map. We have $\pi \circ \pi_2 = \pi_1$ where $\pi_1 : G \to G/H'$ and $\pi_2 : G \to G/H$ are the canonical maps. For each open U in G/H' the sections on U of the left hand side in (4) are contained in $\operatorname{Mor}(\pi_1^{-1}(U), M_a)$, those of the right hand side in $\operatorname{Mor}(\pi_1^{-1}(U), (\operatorname{ind}_H^{H'} M)_a)$. One now proceeds as in the proof of 3.5(2) replacing G by $\pi_1^{-1}(U)$. Alternatively one can get (4) as a special case of 5.18(5), cf. Remark 5.19.

5.13. Corollary: *Let G be a flat k–group scheme and H a flat subgroup scheme of G.*

a) *If G/H is an affine scheme, then ind_H^G is exact.*

b) *If H is a finite group scheme, then ind_H^G is exact.*

Proof: To get a) combine 5.12 with [Ha], III, 3.7. The claim in b) is a special case of a) because of 5.5(6).

Remarks: 1) One can prove a) directly using 5.7(1) without going through the construction of associated sheaves, cf. [Oberst] or [Cline, Parshall, and Scott 3].

2) If k is a field, then the converse of a) also holds, i.e., if ind_H^G is exact, then G/H is affine. This is proved in [Oberst]. In case k is algebraically closed and the groups are reduced, this was proved independently in [Cline, Parshall, and Scott 3]. Another proof was later given in [Donkin 9], 12.4.

5.14. (Associated Fibrations) Let G be a k–group faisceau acting freely (from the right) on a k–faisceau X. Let Y be a k–faisceau with a left G–action (not necessarily free). Then G acts via $(x,y)g = (xg, g^{-1}y)$ freely on $X \times Y$. We denote the quotient $(X \times Y)/G$ for this action by $X \times^G Y$.

Let $\pi : X \to X/G$ denote the canonical map. Then the map $X \times Y \to X/G$, $(x,y) \mapsto \pi(x)$ is constant on G–orbits, hence induces a morphism $\pi_Y : X \times^G Y \to X/G$ with $(x,y)G(A) \mapsto xG(A)$. For any k–algebra A and any $x \in X(A)$ the map $y \mapsto (x,y)G(A)$ is a bijection from $Y(A)$ onto the inverse image $\pi_Y^{-1}(\pi(x))$ in $(X \times^G Y)(A)$. That is why we call π_Y a fibration with fibre Y, or more precisely, the *fibration* over X/G *associated* to the faisceau with G–action Y.

Here are two trivial cases: If $Y = G$ (resp. $X = G$) with the G–action via left (resp. right) multiplication, then $(x,g) \mapsto xg$ (resp. $(g,y) \mapsto gy$) induces an isomorphism

(1) $$X \times^G G \xrightarrow{\sim} X$$

resp.

(2) $$G \times^G Y \xrightarrow{\sim} Y.$$

The inverse maps are $x \mapsto (x,1)G$ resp. $y \mapsto (1,y)G$. If we let G act on the left hand side in (1) (resp. in (2)) via right (resp. left) multiplication on G, then the maps in (1) and (2) are G–equivariant.

Back to the general case! The map $(x,y) \mapsto (x,(x,y)G)$ induces (for all X, Y) an isomorphism

(3) $$X \times Y \xrightarrow{\sim} X \times_{X/G} (X \times^G Y).$$

The inverse maps any $(x',(x,y)G)$ to (x,gy) where g is the unique element in G with $x' = xg$. (In the special case (1) we get just 5.5(2).)

So, if we apply the base change π to π_Y, then we get the first projection $X \times Y \to X$. If Y is an affine scheme, then we see now that π_Y is an affine morphism, cf. [DG], III, §1, 2.12. This means by definition ([DG], I, §2, 3.1) for any affine scheme Z and any morphism $Z \to X/G$ that $Z \times_{X/G} (X \times^G Y)$ is also an affine scheme. So if $(U_i)_i$ is an open covering of X/G by affine subschemes, then $(\pi_Y^{-1}U_i)_i$ is an open covering of $X \times^G Y$ by affine schemes. As $X \times^G Y$ is a faisceau, hence a local functor, we get now the first claims in:

(4) *Suppose that X/G is scheme and that Y is an affine scheme. Then $X \times^G Y$ is a scheme. For any $U \subset X/G$ open and affine, $\pi_Y^{-1}(U) \subset X \times^G Y$ is open and affine. If Y is flat (resp. flat and algebraic), then $k[\pi_Y^{-1}(U)]$ is flat over $k[U]$ (resp. an fppf-$k[U]$-algebra).*

For the proof of the last claims one has to argue as in the proof of 5.7(1); this is left to the reader.

As in 5.7, we get from (4) that $X \times^G Y$ is flat if X/G and Y are so (and if the other assumptions in (4) are satisfied). We recover the results in 5.7 by taking $Y = \mathbf{A}^0$.

5.15. (Sheaves and Fibrations) Let G be a flat k-group scheme acting freely (from the right) on a scheme X such that X/G is a scheme. We can apply the construction in 5.14 for any G-module M to $Y = M_a$ with the obvious G-action. We shall write (by abuse of notation) $X \times^G M = X \times^G M_a$ and $\pi_M = \pi_{M_a}$.

A *section* of π_M on an open subscheme U of X/G is a morphism $s : U \to X \times^G M$ with $\pi_M \circ s = \mathrm{id}_U$. Let us denote the set of all such s by $\Gamma(U, X \times^G M)$. We claim that there is a canonical bijection (for each U)

$$\mathcal{L}(M)(U) \xrightarrow{\sim} \Gamma(U, X \times^G M). \tag{1}$$

Indeed, each $f \in \mathcal{L}(M)(U) \subset \mathrm{Mor}(\pi^{-1}U, M_a)$ defines a morphism $f_1 : \pi^{-1}U \to X \times^G M$ via $x \mapsto (x, f(x))G$. Then $\pi_M \circ f_1 = f$. As f_1 is obviously constant on G-orbits, it induces a morphism $s_f : U \to X \times^G M$ with $\pi_M \circ s_f = \mathrm{id}_U$, i.e., with $s_f \in \Gamma(U, X \times^G M)$.

On the other hand, take any $s \in \Gamma(U, X \times^G M)$. We compose the morphism $x \mapsto (x, s(\pi(x)))$ from $\pi^{-1}U$ to $\pi^{-1}U \times_{X/G} (X \times^G M)$ with the isomorphism $\gamma : X \times_{X/G} (X \times^G M) \xrightarrow{\sim} X \times M_a$ inverse to the map in 5.14(3), and then take the second projection $X \times M_a \to M_a$. This yields a morphism $f_s : \pi^{-1}U \to M_a$ that can be checked to be in $\mathcal{L}(M)(U)$. In more down to earth terms: If $x \in \pi^{-1}(U)(A)$, then $f_s(x)$ is the unique element in $M \otimes A$ such that $s(xG(A)) = (x, f_s(x))G(A)$. Now it is elementary to check that $s \mapsto f_s$ and $f \mapsto s_f$ are inverse maps. For more details one can consult [Cline, Parshall, and Scott 9], 1.3.

We get via (1) a structure as an $\mathcal{O}_{X/G}(U)$-module on $\Gamma(U, X \times^G M)$. It is left to the reader to construct this structure directly.

5.16. (Local Triviality) Let G be a k-group faisceau acting freely (from the right) on a k-faisceau X and let $\pi : X \to X/G$ be the canonical map. We call π *locally trivial* if there is an open covering $(U_i)_i$ of X/G such that there exist morphisms $\sigma_i : U_i \to X$ with $\pi \circ \sigma_i = \mathrm{id}_{U_i}$ for all i. (So the σ_i are *local sections* of π.) Then $(u, g) \mapsto \sigma_i(u)g$ is an isomorphism $U_i \times G \xrightarrow{\sim} \pi^{-1}(U_i)$ for each i. It is compatible with the action of G if we let G act on $U_i \times G$ via right multiplication on G. The inverse map is given by $x \mapsto (\pi(x), g)$ where g is the unique element in G with $\sigma_i(\pi(x))g = x$.

Suppose from now on that π is locally trivial and fix U_i, σ_i as above. Let Y be a k–faisceau with a G–action from the left. We want to consider $X \times^G Y$ and $\pi_Y : X \times^G Y \to X/G$ as in 5.14. We now get for each i an isomorphism

$$\pi_Y^{-1}(U_i) = \pi^{-1}(U_i) \times^G Y \simeq (U_i \times G) \times^G Y \simeq U_i \times (G \times^G Y) \simeq U_i \times Y$$

such that $U_i \times Y \to \pi_Y^{-1}(U_i)$ is given by $(x, y) \mapsto (\sigma_i(x), y)G$. Under this identification the restriction of π_Y to $\pi_Y^{-1}(U_i)$ is just the first projection $U_i \times Y \to U_i$. This is why we call π_Y in this case a *locally trivial fibration.* As the $\pi_Y^{-1}(U_i)$ form an open covering of $X \times^G Y$, we can conclude:

(1) *Suppose that X/G and Y are schemes and that π is locally trivial. Then $X \times^G Y$ is a scheme. If both X/G and Y are flat (resp. smooth, resp. algebraic), then so is $X \times^G Y$.*

Suppose now that G is a flat k–group scheme and consider the special case where $Y = M_a$ for some G–module M. For all i and j the two identifications $(U_i \cap U_j) \times M_a \xrightarrow{\sim} \pi_M^{-1}(U_i \cap U_j)$ derived from σ_i and σ_j differ by the map $(x, m) \mapsto (x, gm)$ where g is the unique element in G with $\sigma_i(x)g = \sigma_j(x)$. This shows: If M is a projective k–module of rank $n < \infty$ (and if π is locally free), then $X \times^G M$ is a vector bundle of rank n over X/G (in the sense, say, of [Ha], II, exerc. 5.18.) It is called the *bundle associated* to the G–module M. Furthermore, 5.15(1) shows that that $\mathcal{L}(M)$ is the locally free sheaf associated to this vector bundle in the usual way. Let us state explicitly:

(2) *If π is locally trivial and if M is a projective k–module of rank $n < \infty$, then $\mathcal{L}(M)$ is a locally free sheaf of $\mathcal{O}_{X/G}$–modules of rank n.*

5.17. (Inverse Images) Let H be a flat subgroup scheme of a flat k–group scheme G. Let X (resp. X') be k–schemes with free actions (on the right) by G (resp. H) such that X/G and X'/H are schemes. Denote the canonical maps by $\pi : X \to X/G$ and $\pi' : X' \to X'/H$.

Suppose that we have an H–equivariant morphism $\varphi : X' \to X$. Then φ induces a morphism $\overline{\varphi} : X'/H \to X/G$ with $x'H \mapsto \varphi(x')G$. For any G–module M we can look at the inverse image $\overline{\varphi}^* \mathcal{L}_{X/G}(M)$ of the $\mathcal{O}_{X/G}$–module $\mathcal{L}_{X/G}(M)$. We want to compare it with the $\mathcal{O}_{X'/H}$–module $\mathcal{L}_{X'/H}(M)$ where we regard M as an H–module (via res_H^G).

Consider open subschemes $U \subset X/G$ and $U' \subset X'/H$ with $\overline{\varphi}(U') \subset U$. Then $\varphi(\pi'^{-1}U') \subset \pi^{-1}U$; the restriction to $\pi'^{-1}U'$ of $f \circ \varphi$ for any $f \in \mathcal{L}_{X/G}(M)(U)$ belongs to $\mathcal{L}_{X'/H}(M)(U')$. This leads to a homomorphism from $\overline{\varphi}^* \mathcal{L}_{X/G}(M)$ to $\mathcal{L}_{X'/H}(M)$. We claim:

(1) *If π is locally trivial, then $\overline{\varphi}^* \mathcal{L}_{X/G}(M) \xrightarrow{\sim} \mathcal{L}_{X'/H}(M)$.*

As this is a local problem, we may assume that X'/H and X/G are affine and that $X \simeq X/G \times G$ (compatible with the G–action). By 5.7(1) X' is also affine. Now $\mathcal{L}_{X/G}(M)$ is the unique quasi-coherent sheaf on X/G with global sections

$$(k[X] \otimes M)^G \simeq \mathrm{Mor}(X/G \times G, M_a)^G \simeq \mathrm{Mor}(X/G, M_a) \simeq k[X/G] \otimes M.$$

Therefore, $\overline{\varphi}^*\mathcal{L}_{X/G}(M)$ is the unique quasi-coherent sheaf on X'/H with global sections

$$k[X'/H] \otimes_{k[X/G]} (k[X/G] \otimes M) \simeq k[X'/H] \otimes M \simeq \mathrm{Mor}(X'/H, M_a) \\ \simeq \mathrm{Mor}(X', M_a)^H = \mathcal{L}_{X'/H}(M)(X'/H).$$

It is left to the reader to check that these identifications yield the same homomorphism as constructed above.

Remark: One can generalise (1) to the case where H is no longer a subgroup scheme of G, but where one has a homomorphism $\alpha : H \to G$ and where φ is compatible with α (i.e., $\varphi(x'h) = \varphi(x')\alpha(h)$). Then one has to replace the right hand side in (1) by $\mathcal{L}_{X'/H}(\alpha^*(M))$, with $\alpha^*(M)$ as in 2.15.

5.18. (Direct Images) Let G, H be flat k–group schemes and X, Y be flat k–schemes together with free actions (on the right) by G (resp. H) on X (resp. Y). Furthermore, assume that G also acts on the left on Y and that this action commutes with that of H (so that $(gy)h = g(yh)$ always). Then $(x,y)(g,h) = (xg, g^{-1}yh)$ defines a free action of $G \times H$ on $X \times Y$. Set $Z = (X \times Y)/(G \times H)$. As we can take the quotient in steps (cf. 5.5(5)), we have isomorphisms

$$(1) \qquad Z \simeq (X \times^G Y)/H \simeq X \times^G (Y/H).$$

We want to assume that Z, X/G, and Y/H are schemes and that Y is affine. Let $\pi : X \to X/G$ denote the canonical map, and let $\pi' = \pi_{Y/H} : Z \to X/G$ denote the fibration with fibre Y/H arising from the identification $Z \simeq X \times^G (Y/H)$ as in (1). For any open subscheme $U \subset X/G$ the inverse image of $\pi'^{-1}U \subset Z$ in $X \times Y$ is equal to $\pi^{-1}U \times Y$. So we have for any $(G \times H)$–module M:

$$(2) \qquad \pi'_*\mathcal{L}_Z(M)(U) = \mathrm{Mor}(\pi^{-1}U \times Y, M_a)^{G \times H}.$$

We have a natural action of G on

$$H^0(Y/H, \mathcal{L}_{Y/H}(M)) = \mathrm{Mor}(Y, M_a)^H \simeq (k[Y] \otimes M)^H$$

derived from the representation of $G \times H$ on $k[Y] \otimes M$, see 3.2. (Recall that Yoneda's lemma implies $\mathrm{Mor}(Y', M_a) \simeq k[Y'] \otimes M$ for all affine schemes Y'.) The associated sheaf on X/G satisfies, for any U as above,

$$(3) \qquad \mathcal{L}_{X/G}(H^0(Y/H, \mathcal{L}_{Y/H}(M)))(U) = \mathrm{Mor}(\pi^{-1}U, (\mathrm{Mor}(Y, M_a)^H)_a)^G.$$

We can associate to any $f \in \mathrm{Mor}(\pi^{-1}U, (\mathrm{Mor}(Y, M_a)^H)_a)$ a morphism $F : \pi^{-1}U \times Y \to M_a$ via $F(x,y) = f(x)(y)$. One easily checks that F is H–equivariant and that $f \mapsto F$ is G–equivariant. So, for $f \in \mathrm{Mor}(\pi^{-1}U, (\mathrm{Mor}(Y, M_a)^H)_a)^G$ we get $F \in \mathrm{Mor}(\pi^{-1}U \times Y, M_a)^{G \times H}$. Hence $f \mapsto F$ defines a homomorphism

$$(4) \qquad \mathcal{L}_{X/G}(H^0(Y/H, \mathcal{L}_{Y/H}(M))) \xrightarrow{\sim} \pi'_*\mathcal{L}_Z(M)$$

that we claim to be an isomorphism. In order to check this, we can restrict ourselves to affine U. Then $\pi^{-1}U \subset X$ is also affine by 5.7(1). So, the right hand side in (3) is isomorphic to

$$\begin{aligned}(k[\pi^{-1}U] \otimes \mathrm{Mor}(Y, M_a)^H)^G &\simeq (k[\pi^{-1}U] \otimes (k[Y] \otimes M)^H)^G \\ &\simeq (k[\pi^{-1}U] \otimes k[Y] \otimes M)^{G\times H} \\ &\simeq (k[\pi^{-1}U \times Y] \otimes M)^{G\times H} \\ &\simeq \mathrm{Mor}(\pi^{-1}U \times Y, M_a)^{G\times H}\end{aligned}$$

using the flatness of $k[\pi^{-1}U]$ for the second step. This yields the isomorphism in (4).

It is clear that the construction is functorial in M, i.e., that we have an isomorphism of functors

$$(5) \qquad \mathcal{L}_{X/G} \circ H^0(Y/H, ?) \circ \mathcal{L}_{Y/H} \xrightarrow{\sim} \pi'_* \circ \mathcal{L}_Z.$$

5.19. (Higher Direct Images) Keep all the assumptions and notations from 5.18. We want to discuss the derived functors of those in 5.18(5). As Y is affine, the functor $H^0(Y/H, ?) \circ \mathcal{L}_{Y/H}$ has, by 5.11, as derived functors the $H^i(Y/H, ?) \circ \mathcal{L}_{Y/H}$. Arguing as in the proof of 5.12.a, one checks that this is still true if we regard $H^0(Y/H, ?) \circ \mathcal{L}_{Y/H}$ as a functor from $\{ (G \times H)$–modules$\}$ to $\{ G$–modules$\}$. So the left hand side in 5.18(5) has derived functors $\mathcal{L}_{X/G} \circ H^i(Y/H, ?) \circ \mathcal{L}_{Y/H}$ as $\mathcal{L}_{X/G}$ is exact, cf. 4.1(2).

We want to show for all $i \in \mathbf{N}$:

$$(1) \qquad \mathcal{L}_{X/G} \circ H^i(Y/H, ?) \circ \mathcal{L}_{Y/H} \xrightarrow{\sim} R^i\pi'_* \circ \mathcal{L}_Z.$$

By the remarks above, by the exactness of $\mathcal{L}_Z$, and by 4.1(3) we just have to show that $\mathcal{L}_Z$ maps injective $(G \times H)$–modules to sheaves acyclic for π'_*.

Let $\pi_Z : X \times Y \to Z = (X \times Y)/(G \times H)$ denote the canonical map. Then $\pi' \circ \pi_Z = \pi \circ \mathrm{pr}_1$ where $\mathrm{pr}_1 : X \times Y \to X$ is the first projection. By 5.7(1) both π_Z and π are affine. So is pr_1 as Y is affine. Hence,

$$0 = R^i(\pi \circ \mathrm{pr}_1)_* = R^i(\pi' \circ \pi_Z)_* = R^i\pi'_* \circ \pi_Z$$

for all $i > 0$ because of [Ha], III, 8.3 and [G], Théorème II.3.1.1.

For any injective $(G \times H)$–module M, the sheaf $\mathcal{L}_Z(M)$ is a direct summand of $(\pi_Z)_*\mathcal{F}$ for some sheaf $\mathcal{F}$ on $X \times Y$, cf. 5.10(2). It follows that $R^i\pi'_*(\mathcal{L}_Z(M)) = 0$ for all $i > 0$. This concludes the proof of (1).

If we apply the Leray spectral sequence $H^j(X/G, R^i\pi'_*?) \Rightarrow H^{i+j}(Z, ?)$, then we get for any $(G \times H)$–module M a spectral sequence

$$(2) \qquad H^j(X/G, \mathcal{L}_{X/G}(H^i(Y/H, \mathcal{L}_{Y/H}(M)))) \Rightarrow H^{i+j}(Z, \mathcal{L}_Z(M)).$$

Remark: Let G' be a flat k–group scheme acting on X from the left such that the actions of G' and G commute (i.e., such that $g'(xg) = (g'x)g$ always holds). Then G' also acts on $X \times Y$ (compatibly with $G \times H$), on X/G, and on Z. The maps in 5.18(1) are G'–equivariant. Assume also that X is affine. Then G' acts on $H^0(X/G, \mathcal{L}_{X/G}(M'))$ and $H^0(Z, \mathcal{L}_Z(M))$ for each G–module M' and each $(G \times H)$–module M. We can regard the functors in 5.18(5) as functors with values in 5.18(5) as functors with values in { G'–linearised sheaves } and then still get an isomorphism, similarly for 5.19(1). In particular, the isomorphism

$$H^0(X/G, \mathcal{L}_{X/G}(H^0(Y/H, \mathcal{L}_{Y/H}(M)))) \simeq H^0(Z, \mathcal{L}_Z(M))$$

is G'–equivariant.

In the special case where $H \subset G$ are subgroups of G' and where $Y = G$ and $X = G'$ (with the action by left or right multiplication), we get back the transitivity of induction (3.5(2)) and the corresponding spectral sequence (4.5.c).

5.20. Keep all the assumptions and notations from 5.18. Suppose in addition that we have a homomorphism $\alpha : G \to H$ and $y_0 \in Y(k)$ with $gy_0 = y_0\alpha(g)$ for all $g \in G(A)$ and all A. Embed G into $G \times H$ via $\alpha' : g \mapsto (g, \alpha(g))$. Then the map $\varphi : X \to X \times Y$, $x \mapsto (x, y_0)$ is G–equivariant. We can therefore apply 5.17(1) to each $(G \times H)$–module M and get:

(1) *If $X \times Y \to Z$ is locally trivial, then $\overline{\varphi}^* \mathcal{L}_Z(M) \simeq \mathcal{L}_{X/G}(\alpha'^* M)$.*

If we are in the situation of the remark to 5.19, then this is an isomorphism of G'–linearised sheaves.

5.21. Assume that k is noetherian. Consider an algebraic k–group G acting freely (from the right) on a k–scheme such that X/G is an algebraic scheme and such the canonical map $\pi : X \to X/G$ is locally trivial. (Some of these assumptions are made only to simplify some proofs.)

Let $X' \subset X$ be a G–stable subscheme. We claim:

(1) *If X' is closed in X, then so is X'/G in X/G.*

Using 1.13 we can assume that $X \simeq X/G \times G$, hence $X' \simeq X'/G \times G$ and $\overline{X'} = \overline{X'/G} \times G$, cf. 1.14. So $X' = \overline{X'}$ implies $X'/G = \overline{X'/G}$.

In general we get:

(2) *$\overline{X'}$ is G–stable with $\overline{X'}/G = \overline{X'/G}$.*

Indeed, the action morphism $X \times G \to X$ is continuous and maps $X' \times G$ to X', hence $\overline{X' \times G} = \overline{X'} \times G$ to $\overline{X'}$. So $\overline{X'}$ is G–stable. Furthermore, $\overline{X'}/G \supset \overline{X'/G}$ is obvious. In order to prove equality it is enough to do so on an open covering, cf. 1.12(5). So we may assume $X = X/G \times G$, hence $X' = X'/G \times G$ and $\overline{X'} = \overline{X'/G} \times G = \overline{X'}/G \times G$ and, finally, $\overline{X'}/G = \overline{X'/G}$.

Furthermore, 5.7(3) implies:

(3) *If X' is open in $\overline{X'}$, then X'/G is open in $\overline{X'}/G$.*

If one wants to prove similar results without assuming π to be locally trivial, then one has to apply results like [DG], I, §2, 5.3, III, §1, 2.12, and I, §2, 3.3.

CHAPTER 6

Factor Groups

If G is a k–group faisceau and N a normal subgroup faisceau of G, then G/N is again a k–group faisceau and has the universal property of a factor group. This and related things are described in 6.1/2 following [DG].

In this chapter we discuss the relation between the representation theories of G, N, and G/N under the assumption that they all are flat group schemes. The results are usually generalisations of known theorems from the representation theory of abstract groups such as the Lyndon-Hochschild-Serre spectral sequence in 6.6 or the Clifford theory in 6.15/16.

More or less all necessary references have been given before. Let me add that 6.12 generalises 3.1 in [Andersen and Jantzen].

6.1. (Factor Groups) Let G be a k–group faisceau and N a normal subgroup faisceau of G. Obviously $A \mapsto G(A)/N(A)$ is a k–group functor. Then so is the associated faisceau G/N. This follows (on the one hand) from the universal property (cf. [DG], III, §3, 1.2) and is (on the other hand) clear from the construction in 5.4/5: For any $g, g' \in (G/N)(A)$ there is an fppf–A–algebra B with g, g' both in the kernel of $G(B)/N(B) \rightrightarrows G(B \otimes_A B)/N(B \otimes_A B)$. As these maps are group homomorphisms, gg' and g^{-1} also belong to the kernel. This easily yields the group structure on each $(G/N)(A)$. Furthermore, it is easy to see that all maps $(G/N)(A) \to (G/N)(A')$ and $G(A) \to (G/N)(A)$ are group homomorphisms. Hence G/N is a k–group faisceau and the canonical map $\pi : G \to G/N$ is a group homomorphism. We call G/N the *factor group* of G by N.

Note that G/N has the universal property of a factor group: If $\varphi : G \to G'$ is a homomorphism of k–group faisceaux with $N \subset \ker(\varphi)$, then there is a unique group homomorphism $\overline{\varphi} : G/N \to G'$ with $\overline{\varphi} \circ \pi = \varphi$. (As φ is constant on the N–cosets, the universal property of G/N as a quotient faisceau gives the existence and uniqueness of $\overline{\varphi}$ as a morphism. It is immediately seen from the construction that $\overline{\varphi}$ is a group homomorphism. This can also be deduced from the uniqueness of $\overline{\varphi}$.)

For any homomorphism $\varphi : G \to G'$ of k–group faisceaux the kernel $\ker(\varphi)$ is a normal subgroup faisceau of G. We can identify $G/\ker(\varphi)$ with the image faisceau $\operatorname{im}(\varphi)$, which is a subgroup faisceau of G'. This is really a special case of an orbit faisceau as we can make any $g \in G(A)$ act on $G'(A)$ as multiplication with $\varphi(g)$.

6.2. (Product Subgroups) Let G be a k–group faisceau and let H, N be subgroup faisceaux of G such that H normalises N. We can then form the semi-direct product $N \rtimes H$, and have a natural homomorphism $N \rtimes H \to G$, $(n, h) \mapsto nh$ with kernel isomorphic to the intersection $N \cap H$, cf. 2.6. Both $N \rtimes H$ and $N \cap H$ are k–group faisceaux. We denote the image faisceau of the homomorphism $N \rtimes H \to G$ by NH and call it the product of N and H. It is a subgroup faisceau of G with

$$(N \rtimes H)/(N \cap H) \simeq NH. \tag{1}$$

The construction in 5.4 yields for any k–algebra A

$$(NH)(A) = \{\, g \in G(A) \mid \text{there are an fppf–}A\text{–algebra } B \text{ and } h \in H(B), \; n \in N(B) \text{ with } g = nh \text{ in } G(B) \,\}. \tag{2}$$

Obviously N is a normal subgroup faisceau of NH. The canonical homomorphism $NH \to (NH)/N$ has kernel N, hence its restriction to H has kernel $N \cap H$. We get thus an embedding $H/(N \cap H) \to (NH)/N$ that turns out to be an isomorphism: For all g, h, n as in (2) the element $h(N(B) \cap H(B))$ defines an element in $(H/(N \cap H))(A)$ that is mapped to $gN(A)$. Therefore $(NH)(A)/N(A)$ belongs to the image for each A, hence $((NH)/N)(A)$ to the image faisceau. So, we get the isomorphism theorem

$$H/(N \cap H) \xrightarrow{\sim} (NH)/N. \tag{3}$$

Suppose now that N is normal in G; let $\pi : G \to G/N$ denote the canonical map. We write $\pi(H)$ for the image faisceau of $\pi_{|H}$. Then

$$NH = \pi^{-1}(\pi(H)). \tag{4}$$

Indeed, if $g \in \pi^{-1}(\pi(H))(A)$, then there is some B (fppf over A) with $\pi(g) \in \pi(H(B))$, hence some $h \in H(B)$ with $h^{-1}g \in \ker(\pi)(B) = N(B)$ and $g \in (NH)(A)$ by (2). The other inclusion is even more obvious.

If $H \supset N$, then obviously $NH = N$ and $H = \pi^{-1}(\pi(N))$. So we have for normal N the usual bijection between $\{$ subgroup faisceaux of G containing $N \}$ and $\{$ subgroup faisceaux of $G/N \}$. Furthermore, one can then show (for $H \supset N$) that H is normal in G if and only if H/N is normal in G/N, and that one has a canonical isomorphism $(G/N)/(H/N) \xrightarrow{\sim} G/H$ of faisceaux, which is a group isomorphism if H is normal, cf. [DG], III, §3, 3.7.

6.3. (G/N–modules) Let us assume from now on until the end of this chapter that G is a flat group scheme over k, and that N is a normal and flat subgroup scheme of G.

The general construction from 2.15 applied to the canonical homomorphism $\pi : G \to G/N$ yields for any G/N–module M an action of G on M. As in 2.15 we use the notation π^*M for this G–module. (If there is no confusion possible, then we simply write M instead of π^*M.)

Obviously π^* is a functor from $\{ G/N\text{–modules} \}$ to $\{ G\text{–modules} \}$. This functor is faithful: We have

$$\operatorname{Hom}_{G/N}(M, M') = \operatorname{Hom}_G(\pi^*M, \pi^*M') \tag{1}$$

for all G/N–modules M, M'. (Any $\overline{g} \in (G/N)(A)$ has a representative $g \in G(B)$ with B fppf over A. If $\varphi \in \mathrm{Hom}_G(\pi^* M, \pi^* M')$, then $\varphi \otimes \mathrm{id}_B$ commutes with g, hence $\varphi \otimes \mathrm{id}_A$ with $\overline{g}$ as $M \otimes A$ is mapped injectively into $M \otimes B$.)

Clearly N acts trivially on each G–module of the form $\pi^* M$. In fact, the image of π^* consists of all G–modules V on which N acts trivially: For such V the k–group functor $A \mapsto G(A)/N(A)$ acts naturally on V_a, and this action can be extended uniquely to the associated faisceau G/N as V_a is itself a faisceau. This follows from the universal property of G/N and also from its explicit description in 5.4/5.

The full subcategory of all G–modules on which N acts trivially is obviously an abelian category. So we see that the category of all G/N–modules is an abelian category even without knowing whether G/N is a flat group scheme (what we needed in 2.9) or not. The functor π^* is now obviously exact.

6.4. For any G–module V, the subspace V^N is a G–submodule of V, by 3.2, on which N acts trivially. We therefore can regard V^N as a G/N–module and $V \mapsto V^N$ as a (clearly: left exact) functor from $\{\,G\text{–modules}\,\}$ to $\{\,G/N\text{–modules}\,\}$.

Lemma: *The functor $V \mapsto V^N$ from $\{\,G\text{–modules}\,\}$ to $\{\,G/N\text{–modules}\,\}$ is right adjoint to π^*. It maps injective G–modules to injective G/N–modules. The category of G/N–modules contains enough injective objects.*

Proof: We have for any G/N–module M and any G–module V

$$\mathrm{Hom}_G(\pi^* M, V) \simeq \mathrm{Hom}_G(\pi^* M, V^N) = \mathrm{Hom}_{G/N}(M, V^N),$$

by 6.3(1), where the first isomorphism is induced by the inclusion $V^N \subset V$. This shows that $V \mapsto V^N$ is right adjoint to the exact functor π^*, hence that injective objects are mapped to injective objects. Any embedding of $\pi^* M$ into an injective G–module Q induces an embedding of M into the injective G/N–module Q^N. Therefore $\{\,G/N\text{–modules}\,\}$ contains enough injective objects.

Remark: We can generalise the above as follows. Let E be a G–module that is finitely generated and projective over k. Then $M \mapsto \pi^*(M) \otimes E$ is an exact functor from $\{\,G/N\text{–modules}\,\}$ to $\{\,G\text{–modules}\,\}$. The functor $V \mapsto \mathrm{Hom}_N(E, V) \simeq (E^* \otimes V)^N$, cf. 2.10(6), is right adjoint to it. It is therefore left exact and maps injective G–modules to injective G/N–modules. Indeed, one has for any G–module V and any G/N–module M

$$\begin{aligned}\mathrm{Hom}_{G/N}(M, (E^* \otimes V)^N) = \mathrm{Hom}_G(M, (E^* \otimes V)^N) &\simeq \mathrm{Hom}_G(M, E^* \otimes V)\\ &\simeq \mathrm{Hom}_G(M \otimes E, V),\end{aligned}$$

using 4.4(1) for the last step. Notice that we can also regard this as an isomorphism of functors

$$\text{(1)} \qquad \mathrm{Hom}_{G/N}(M, ?) \circ \mathrm{Hom}_N(E, ?) \simeq \mathrm{Hom}_G(M \otimes E, ?).$$

6.5. (Factor Groups as Affine Schemes) Let me quote the following result from [DG], III, §3, 5.6:

(1) *If k is a field and if G, N are algebraic k–groups, then G/N is an algebraic k–group.*

(Recall that in our convention an algebraic k–group is assumed to be affine.) Another case where we know G/N to be affine is when N is a finite group scheme, see 5.5(6).

Now recall from 5.7(2) and 5.13:

(2) *If G/N is an affine scheme, then it is flat and N is exact in G.*

Of course, in this case we do not need 6.3/4 to see that $\{\,G/N\text{–modules}\,\}$ is an abelian category and has enough injective objects. The functor $V \mapsto V^N$ maps $M \otimes k[G]$ for any k–module M to $M \otimes k[G]^N = M \otimes k[G/N]$ if k is a field. Therefore we can also use 3.9.c to show that this functor maps injective G–modules to injective G/N–modules (in that case).

6.6. Proposition: *Suppose N is exact in G. Let E be a G–module that is finitely generated and projective over k. Then the derived functors of $V \mapsto \mathrm{Hom}_N(E,V)$ from $\{\,G\text{–modules}\,\}$ to $\{\,G/N\text{–modules}\,\}$ can be identified with $V \mapsto \mathrm{Ext}^n_N(E,V)$. There are for each G/N–module M and each G–module V spectral sequences*

$$E_2^{n,m} = \mathrm{Ext}^n_{G/N}(M, \mathrm{Ext}^m_N(E,V)) \Rightarrow \mathrm{Ext}^{n+m}_G(M \otimes E, V), \tag{1}$$

$$E_2^{n,m} = \mathrm{Ext}^n_{G/N}(M, H^m(N,V)) \Rightarrow \mathrm{Ext}^{n+m}_G(M, V), \tag{2}$$

$$E_2^{n,m} = H^n(G/N, H^m(N,V)) \Rightarrow H^{n+m}(G, V). \tag{3}$$

Proof: As N is exact in G, the functor res^G_N maps injective G–modules to modules acyclic for the fixed point functor $?^N$. (Use 3.9.c and 4.10.) The composition of $?^N$ (taken from $\{\,N\text{–modules}\,\}$ to $\{\,k\text{–modules}\,\}$) with res^G_N is isomorphic to the composition of $\mathrm{res}^{G/N}_1$ with $?^N$ (taken from $\{\,G\text{–modules}\,\}$ to $\{\,G/N\text{–modules}\,\}$). Therefore 4.1(2), (3) imply that the $V \mapsto H^n(N,V)$ can be regarded as the derived functors of $V \mapsto V^N$ from $\{\,G\text{–modules}\,\}$ to $\{\,G/N\text{–modules}\,\}$. The same is true for $V \mapsto H^n(N, E^* \otimes V) \simeq \mathrm{Ext}^n_N(E,V)$, cf. 4.4, and $V \mapsto (E^* \otimes V)^N \simeq \mathrm{Hom}_N(E,V)$.

As $\mathrm{Hom}_N(E,?)$ maps injective G–modules to injective G/N–modules, we can apply 4.1(1) to 6.4(1) and get the spectral sequence in (1). Taking $E = k$ we get (2), and setting $M = k$ yields (3).

Remark: The spectral sequence in (3) is known as the *Lyndon-Hochschild-Serre* spectral sequence. Let us be a bit more precise about its construction.

There is a standard injective resolution

$$0 \longrightarrow V \longrightarrow V \otimes k[G] \longrightarrow V \otimes \bigotimes{}^2 k[G] \longrightarrow V \otimes \bigotimes{}^3 k[G] \longrightarrow \cdots$$

of V as a G–module, cf. 4.15(2). Then $H^\bullet(N,V)$ is the cohomology of the complex $(V \otimes \bigotimes^\bullet k[G])^N$, because the $V \otimes \bigotimes^i k[G]$ are acyclic for $?^N$, see the beginning of the proof above. Constructing the standard injective resolution of the G/N–modules $(V \otimes \bigotimes^i k[G])^N$ we get a double complex $\bigoplus_{i,j} (V \otimes \bigotimes^i k[G])^N \otimes \bigotimes^j k[G/N]$. The spectral sequence associated to this double complex is just the one in (3).

Using this, the edge maps $H^i(G,V) = E^i \to E_2^{0,i} = H^i(N,V)^{G/N}$ can be checked to be the restriction maps $H^i(G,V) \to H^i(N,V)$ that take values in the G/N–fixed points, cf. [Mac], XI.9.1.

6.7. In the special case $E = k$, Proposition 6.6 implies that each $H^n(N, V)$ for any G–module V has a natural structure as a G/N–module. This structure can also be constructed using the Hochschild complex. We make G act on each $C^n(N, V) \simeq V \otimes \bigotimes^n k[N]$ via the given representation on V and via the conjugation action on each factor $k[N]$. Then each ∂^n is a homomorphism of G–modules as Δ_V and Δ_N are so. This makes each $H^n(N, V)$ into a G–module.

For any exact sequence $0 \to V' \to V \to V'' \to 0$ of G–modules one can now check that all connecting homomorphisms $H^n(N, V'') \to H^{n+1}(N, V')$ are homomorphisms of G–modules. (See [Sullivan 3], 4.1 for the case of a field.) The universal property of derived functors (via δ–functors) shows, then, that the G–modules $H^n(N, V)$ constructed in this way yield the derived functors of $V \mapsto V^N$ from $\{ G\text{–modules} \}$ to $\{ G\text{–modules} \}$. This functor can be regarded as the composition of $V \mapsto V^N$ from $\{ G\text{–modules} \}$ to $\{ G/N\text{–modules} \}$ with the natural inclusion of $\{ G/N\text{–modules} \}$ into $\{ G\text{–modules} \}$. The last functor being exact implies that the G/N–action on $H^n(N, V)$ given by Proposition 6.6 must lead to the same G–action as the construction using the Hochschild complex.

Note that this implies, in the case $G = N$, that the action of G on the $H^n(G, V)$ constructed with the conjugation action on the Hochschild complex is trivial.

6.8. Corollary: *Suppose that N is diagonalisable. Then we have for all G–modules V and E with E finitely generated and projective over k, for all G/N–modules M, and all $n \in \mathbf{N}$ isomorphisms*

$$\text{(1)} \qquad \operatorname{Ext}^n_{G/N}(M, \operatorname{Hom}_N(E, V)) \simeq \operatorname{Ext}^n_G(M \otimes E, V),$$

$$\text{(2)} \qquad \operatorname{Ext}^n_{G/N}(M, V^N) \simeq \operatorname{Ext}^n_G(M, V),$$

$$\text{(3)} \qquad H^n(G/N, V^N) \simeq H^n(G, V).$$

Proof: All this follows immediately from 6.6 and 4.3 as each E_λ is a projective k–module and as N is exact in G, cf. 4.6.

Remark: If we apply (3) to the G–module $\pi^* M$, then we get

$$\text{(4)} \qquad H^n(G/N, M) \simeq H^n(G, M).$$

6.9. Corollary: *Suppose that G/N is a diagonalisable group scheme. Then we have for all G–modules V and E with E finitely generated and projective over k, for all G/N–modules M projective over k, and all $n \in \mathbf{N}$ isomorphisms*

$$\text{(1)} \qquad \operatorname{Hom}_{G/N}(M, \operatorname{Ext}^n_N(E, V)) \simeq \operatorname{Ext}^n_G(M \otimes E, V),$$

$$\text{(2)} \qquad \operatorname{Hom}_{G/N}(M, H^n(N, V)) \simeq \operatorname{Ext}^n_G(M, V),$$

$$\text{(3)} \qquad H^n(N, V)^{G/N} \simeq H^n(G, V).$$

Proof: As G/N is affine, hence N exact in G, we can apply 6.6. The formulas follow now immediately from 4.3.

Remark: Suppose $G/N \simeq \operatorname{Diag}(\Lambda)$ for some abelian group Λ. We have by 2.11(3) decompositions

$$\text{(4)} \qquad \operatorname{Ext}^n_N(E, V) = \bigoplus_{\lambda \in \Lambda} \operatorname{Ext}^n_N(E, V)_\lambda$$

(for all n). The map $\varphi \mapsto \varphi(1)$ is for any G/N–module M' an isomorphism $\operatorname{Hom}_{G/N}(k_\lambda, M') \xrightarrow{\sim} M'_\lambda$, cf. 2.11(4). We can therefore identify the direct summands in (4) using (1) and 4.4 as follows:

$$\operatorname{Ext}^n_N(E,V)_\lambda \simeq \operatorname{Ext}^n_G(E\otimes\lambda, V) \simeq \operatorname{Ext}^n_G(E, V\otimes(-\lambda)). \tag{5}$$

(We use the convention $E\otimes\lambda = E\otimes k_\lambda$ etc.) In the special case $E=k$ we get (for all $\lambda\in\Lambda$ and $n\in\mathbf{N}$)

$$H^n(N,V)_\lambda \simeq H^n(G, V\otimes(-\lambda)). \tag{6}$$

6.10. Let M be a G/N–module. The spectral sequence 6.6(3) yields base maps of the form

$$H^n(G/N,M)\to H^n(G,M).$$

They are induced by the map $C^\bullet(G/N,M)\to C^\bullet(G,M)$ of the Hochschild complexes arising from the inclusion $k[G/N]\simeq k[G]^N \hookrightarrow k[G]$. This is just the usual "inflation" map in group cohomology, cf. the proof of [Mac], XI, 10.2.

We shall show that these (and more general) base maps are injective provided certain conditions are satisfied. At first the conditions may look artificial, but we shall see later on (II.10.11/12/18) that they are satisfied in some important cases.

Proposition: *Let N be exact in G. Let Q be a G–module and E a G–submodule of Q such that E is finitely generated and projective over k. If Q is injective as an N–module and if $\operatorname{Hom}_N(E,Q)=k$, then the base maps*

$$\operatorname{Ext}^n_{G/N}(M_1,M_2)\longrightarrow \operatorname{Ext}^n_G(E\otimes M_1, E\otimes M_2) \tag{1}$$

are injective for all $n\in\mathbf{N}$ and all G/N–modules M_1, M_2. Furthermore, we get an exact sequence

$$\begin{aligned} 0\to \operatorname{Ext}^1_{G/N}(M_1,M_2) &\longrightarrow \operatorname{Ext}^1_G(E\otimes M_1, E\otimes M_2)\\ &\longrightarrow \operatorname{Hom}_{G/N}(M_1, \operatorname{Ext}^1_N(E,E)\otimes M_2)\to 0. \end{aligned} \tag{2}$$

Proof: From 6.6(1) we get two spectral sequences:

$$E_2^{n,m} = \operatorname{Ext}^n_{G/N}(M_1, \operatorname{Ext}^m_N(E,E)\otimes M_2) \Rightarrow \operatorname{Ext}^{n+m}_G(E\otimes M_1, E\otimes M_2) \tag{3}$$

and

$$E_2'^{\,n,m} = \operatorname{Ext}^n_{G/N}(M_1, \operatorname{Ext}^m_N(E,Q)\otimes M_2) \Rightarrow \operatorname{Ext}^{n+m}_G(E\otimes M_1, Q\otimes M_2). \tag{4}$$

As $\operatorname{Hom}_N(E,E)\subset \operatorname{Hom}_N(E,Q)=k$, we have $E_2^{n,0}\simeq \operatorname{Ext}^n_{G/N}(M_1,M_2)\simeq E_2'^{\,n,0}$. So our first claim says that the base maps for (3) are injective. If so, then (2) will follow from 4.1(4) [the five term exact sequence], as the map $E_2^{2,0}\to E^2$ there is one of our base maps.

As Q is injective for N, the spectral sequence (4) degenerates and yields isomorphisms $E_2'^{\,n,0}\simeq E'^{\,n,0}_\infty \simeq E'^n$. The embedding of E into Q induces homomorphisms $E_r^{n,m}\to E_r'^{\,n,m}$ for all r, n, m. These maps are compatible with the differentials of the spectral sequences. Thus we get a commutative diagram

$$\begin{array}{ccc} E_2^{n,0} & \longrightarrow & E_2'^{\,n,0}\\ \downarrow & & \downarrow\\ E_\infty^{n,0} & \longrightarrow & E'^{\,n,0}_\infty \end{array}$$

We know $E_2^{n,0}\to E_2'^{\,n,0}$ and $E_2'^{\,n,0}\to E'^{\,n,0}_\infty$ to be isomorphisms. This implies that $E_2^{n,0}\to E_\infty^{n,0}$ is injective, hence so is the base map $E_2^{n,0}\to E^n$.

Remarks: 1) The special case $E = M_1 = k$ yields: *If there is a G–module Q injective for N such that $Q^G = Q^N \simeq k$, then the base maps*

$$H^n(G/N, M) \to H^n(G, M) \tag{5}$$

are injective for each G/N–module M. Furthermore, we have an exact sequence

$$0 \to H^1(G/N, M) \longrightarrow H^1(G, M) \longrightarrow (H^1(N, k) \otimes M)^{G/N} \to 0. \tag{6}$$

2) One can generalise the proposition as follows: Let V be a G–module that is finitely generated and projective over k. Drop the similar assumption on E and that $\operatorname{Hom}_N(E, Q) \simeq k$, but assume that $\operatorname{Hom}_G(V, E) = \operatorname{Hom}_N(V, Q) \simeq k$. Then the spectral sequence

$$\operatorname{Ext}^n_{G/N}(M_1, \operatorname{Ext}^m_N(V, E) \otimes M_2) \Rightarrow \operatorname{Ext}^{n+m}_G(V \otimes M_1, E \otimes M_2)$$

will yield injective base maps

$$\operatorname{Ext}^n_{G/N}(M_1, M_2) \longrightarrow \operatorname{Ext}^n_G(V \otimes M_1, E \otimes M_2) \tag{7}$$

and an exact sequence

$$\begin{aligned} 0 \to \operatorname{Ext}^1_{G/N}(M_1, M_2) &\longrightarrow \operatorname{Ext}^1_G(V \otimes M_1, E \otimes M_2) \\ &\longrightarrow \operatorname{Hom}_{G/N}(M_1, \operatorname{Ext}^1_N(V, E) \otimes M_2) \to 0. \end{aligned} \tag{8}$$

6.11. Proposition: *Let H be a flat subgroup scheme of G with $N \subset H$. Suppose that both G/N and H/N are affine. The one has for each H/N–module M and each $n \in \mathbf{N}$ an isomorphism of G–modules*

$$(R^n \operatorname{ind}_H^G)M \simeq (R^n \operatorname{ind}_{H/N}^{G/N})M. \tag{1}$$

Proof: Let $\pi : G \to G/N$ and $\pi' : H \to H/N$ be the canonical maps. Our claim ought to be formulated as an isomorphism of functors:

$$(R^n \operatorname{ind}_H^G) \circ \pi'^* \simeq \pi^* \circ (R^n \operatorname{ind}_{H/N}^{G/N}). \tag{1'}$$

Let us first consider the case $n = 0$, i.e., get an isomorphism

$$\operatorname{ind}_H^G M \simeq \operatorname{ind}_{H/N}^{G/N} M. \tag{2}$$

The right hand side is a subset of $\operatorname{Mor}(G/N, M_a)$, which we may identify with $\operatorname{Mor}(G, M_a)^N$ because of the universal property of G/N. (Recall that M_a is a faisceau.) Any $f \in \operatorname{Mor}(G, M_a)$ will belong to $\operatorname{ind}_{H/N}^{G/N} M$ if and only if $f(gn) = f(g)$ and $f(gh) = h^{-1}f(g)$ for all $g \in G(A)$, $h \in H(A)$, $n \in N(A)$, and all A. As $N(A) \subset H(A)$ acts trivially on $M \otimes A$, we can drop the first part of the condition. The second one alone describes just $\operatorname{ind}_H^G M$ so that we get (2). As above, we ought to have formulated this as an isomorphism of functors

$$\operatorname{ind}_H^G \circ \pi'^* \simeq \pi^* \circ \operatorname{ind}_{H/N}^{G/N}. \tag{2'}$$

This formula implies (1′) using 4.1(2), (3) as soon as we can show that π'^* maps injective H/N–modules to H–modules acyclic for ind_H^G. By 3.9.c it is enough to look at H/N–modules of the form $Q \otimes k[H/N] \simeq \operatorname{ind}_1^{H/N} Q$ for injective k–modules Q. Applying (2) to (H, N) instead of (G, N) we can identify $\pi'^*(\operatorname{ind}_1^{H/N} Q)$ with $\operatorname{ind}_N^H Q$ where we regard Q as a trivial N–module.

By our assumption N is exact in H and G. The spectral sequence 4.5.c yields therefore

$$(R^n \operatorname{ind}_H^G) \circ \operatorname{ind}_N^H = 0 \qquad \text{for all } n > 0. \tag{3}$$

This certainly implies the needed acyclicity of $\operatorname{ind}_N^H Q$ above, hence the claim.

Remark: We often use only the following part of the proposition: Let M be an H–module. If N acts trivially on M, then it acts trivially also on $\operatorname{ind}_H^G M$ and even on all $R^n \operatorname{ind}_H^G M$.

6.12. The isomorphism in 6.11(2) can be regarded as a special case of a more general statement that we are going to prove now.

Proposition: *Let H be a flat subgroup scheme of G. Suppose both G/N and $H/(H \cap N)$ are affine schemes.*

a) *The functors $\mathcal{F}_1$, $\mathcal{F}_2$ from $\{H\text{–modules}\}$ to $\{G/N\text{–modules}\}$ with*

$$\mathcal{F}_1(M) = (\operatorname{ind}_H^G M)^N \qquad \text{and} \qquad \mathcal{F}_2(M) = \operatorname{ind}_{H/(H\cap N)}^{G/N}(M^{H\cap N})$$

are isomorphic.

b) *For each H–module M there are spectral sequences*

$$E_2^{n,m} = H^n(N, R^m \operatorname{ind}_H^G M) \Rightarrow (R^{n+m}\mathcal{F}_1)M \tag{1}$$

and

$$E_2^{n,m} = (R^n \operatorname{ind}_{H/(H\cap N)}^{G/N}) H^m(H \cap N, M) \Rightarrow (R^{n+m}\mathcal{F}_2)M. \tag{2}$$

Proof: a) Let $\pi : G \to G/N$ and $\pi' : H \to H/(H \cap N)$ be the canonical maps. Obviously $\operatorname{res}_H^G \circ \pi^* = \pi'^* \circ \operatorname{res}_{H/(H\cap N)}^{G/N}$. This yields an isomorphism of the adjoint functors, i.e., of $\mathcal{F}_1$ and $\mathcal{F}_2$.

b) Both $\mathcal{F}_1$ and $\mathcal{F}_2$ are compositions of two left exact functors. It is enough to show that the first one maps injective objects to acyclic objects with respect to the second one. Then we can apply 4.1(1).

The functor ind_H^G maps injective H–modules to injective G–modules (3.9). This give the claim for $\mathcal{F}_1$. Note that we have to apply 6.6 in order to regard the $H^n(N, ?)$ as derived functors on the category of G–modules.

In the second case we have to apply 6.4 to $(H, H \cap N)$ instead of (G, N).

Remark: Note that a) implies $R^n\mathcal{F}_1 \simeq R^n\mathcal{F}_2$ for all n, so the two spectral sequences in (1) and (2) have the same abutment.

6.13. Proposition: *Let H be a flat subgroup scheme of G such that NH is an affine scheme. Then there are isomorphisms of functors*

$$\operatorname{res}_N^{NH} \circ \operatorname{ind}_H^{NH} \simeq \operatorname{ind}_{N\cap H}^N \circ \operatorname{res}_{N\cap H}^H \tag{1}$$

and

$$\operatorname{res}_H^{NH} \circ \operatorname{ind}_N^{NH} \simeq \operatorname{ind}_{N\cap H}^H \circ \operatorname{res}_{N\cap H}^N . \tag{2}$$

Proof: Let H' denote the kernel of the obvious homomorphism $N \rtimes H \to G$; we can identify H' with $N \cap H$ via $(h, h^{-1}) \mapsto h$, cf. 2.6. We have an isomorphism $(N \rtimes H)/H' \simeq NH$, so NH is flat by our assumption and 6.5(2).

Let M be an H–module and M' an N–module. Because of $H' \cap N = 1 = H' \cap H$ (in $N \rtimes H$) we get from 6.12.a isomorphisms

$$\operatorname{ind}_H^{NH} M \simeq (\operatorname{ind}_H^{N \rtimes H} M)^{H'} \qquad \text{and} \qquad \operatorname{ind}_N^{NH} M' \simeq (\operatorname{ind}_N^{N \rtimes H} M')^{H'}.$$

Now 3.8(2), (3) yield

$$\operatorname{ind}_H^{NH} M \simeq (k[N] \otimes M)^{H'} \qquad \text{and} \qquad \operatorname{ind}_N^{NH} M' \simeq (k[H] \otimes M')^{H'}.$$

Here any $h \in H(A)$ and $n \in N(A)$ act via $\rho_c(h) \otimes h$ resp. $\rho_l(n) \otimes 1$ on $k[N] \otimes M$, where ρ_c denotes the conjugation action. If $h \in N(A) \cap H(A)$, then $(h, h^{-1}) \in H'(A)$ acts therefore as $\rho_r(h) \otimes h$. So N and $N \cap H$ act on $k[N] \otimes M$ as in the definition of $\operatorname{ind}_{N \cap H}^N M$. This yields (1).

Similarly, any $h \in H(A)$ acts on $k[H] \otimes M'$ as in the definition of $\operatorname{ind}_{N \cap H}^H M'$. Some $(n, n^{-1}) \in H'(A)$ with $n \in N(A) \cap H(A)$ will not act in that way, but the set of fixed points will be the same. (Regard $f \in k[H] \otimes M'$ as morphism $H \to M_a$. Then $((n, n^{-1})f)(h) = (h^{-1}n^{-1}h)\, f(h \cdot h^{-1}n^{-1}h)$.) From this we get (2).

Remark: Suppose also that $H/(N \cap H)$ is affine. Then $\operatorname{res}_{N \cap H}^H$ maps injective H–modules to modules acyclic for $\operatorname{ind}_{N \cap H}^N$ as (cf. 4.10 and 5.13)

$$\begin{aligned} (R^n \operatorname{ind}_{N \cap H}^N)(Q \otimes k[H]) &\simeq H^n(N \cap H, Q \otimes k[H] \otimes k[N]) \\ &\simeq (R^n \operatorname{ind}_{N \cap H}^H)(Q \otimes k[H]) = 0 \end{aligned}$$

for all $n > 0$ and all k–modules Q. We therefore get from (1) and 4.1(2), (3) isomorphisms of derived functors (for each $n \in \mathbf{N}$)

(3) $$\operatorname{res}_N^{NH} \circ R^n \operatorname{ind}_H^{NH} \simeq R^n \operatorname{ind}_{N \cap H}^N \circ \operatorname{res}_{N \cap H}^H.$$

In (2) the higher derived functors are 0 (for $H/(N \cap H) \simeq (NH)/N$ affine).

6.14. Keep the notations of 6.12. The inclusion of H into NH induces by 6.2(3) an isomorphism $H/(N \cap H) \simeq (NH)/N$. Similarly, one can show that the inclusion of N into NH induces an isomorphism of faisceaux $N/(N \cap H) \simeq (NH)/H$. [One can regard $(NH)/H$ as an orbit faisceau of N, cf. 5.6(2).]

Suppose now that these quotient faisceaux are schemes. Then any H–module M resp. any N–module M' defines a sheaf $\mathcal{L}_{(NH)/H}(M)$ resp. $\mathcal{L}_{(NH)/N}(M')$ as in Chapter 5. The isomorphisms above identify this sheaf with $\mathcal{L}_{N/(N \cap H)}(\operatorname{res}_{N \cap H}^H M)$ resp. with $\mathcal{L}_{H/(N \cap H)}(\operatorname{res}_{N \cap H}^N M')$; this is a consequence of 5.17(1). Using 5.12 one gets another approach to 6.13(3) and the symmetric statement with H and N interchanged.

This can be generalised as follows: Let H, H' be flat subgroup schemes such that the multiplication map $m : H \times H' \to G$ has image faisceau equal to G. Then one gets an isomorphism of faisceaux $H/(H \cap H') \xrightarrow{\sim} G/H'$. If these quotient faisceaux are schemes, then one gets as above

(1) $$\operatorname{res}_H^G \circ R^n \operatorname{ind}_{H'}^G \simeq R^n \operatorname{ind}_{H \cap H'}^H \circ \operatorname{res}_{H \cap H'}^{H'}.$$

A (slightly) more general result is proved in [Cline, Parshall, and Scott 9], 4.1.

6.15. As explained in 2.15, we can twist any N–module V with any $g \in G(k)$. As there, we denote the twisted module by ${}^gV$. Recall that ${}^nV \simeq V$ for all $n \in N(k)$. More generally, if V is an N–submodule of a G–module M, then gV is also an N–submodule of M and gV is isomorphic to ${}^gV$.

Suppose from now on that k is a field. Any N–module V is simple (resp. semi-simple) if and only if ${}^gV$ is so. This implies:

(1) *If M is a G–module, then $\operatorname{soc}_N M$ is $G(k)$–stable.*

Let L, M be G–modules with $\dim(L) < \infty$. Then $\operatorname{Hom}(L, M) \simeq L^* \otimes M$ is also a G–module and $\operatorname{Hom}_N(L, M) \simeq (L^* \otimes M)^N$ is a G–submodule, cf. 6.3/4. The map $\varphi \otimes x \mapsto \varphi(x)$ from $\operatorname{Hom}(L, M) \otimes L$ to M is easily seen to be a homomorphism of G–modules. Therefore 2.14(3) implies:

(2) *If L is simple as an N–module with $\operatorname{End}_N(L) \simeq k$, then we have an isomorphism $\operatorname{Hom}_N(L, M) \otimes L \simeq (\operatorname{soc}_N M)_L$ of G–modules.*

6.16. We call $G(k)$ dense in G if there is no closed subfunctor $X \subset G$ with $X(k) \supset G(k)$ and $X \neq G$, cf. the definition of closures in 1.4. If k is an algebraically closed field and G a reduced algebraic k–group, then $G(k)$ is dense in G (by Hilbert's Nullstellensatz). The same is true for G reduced connected and algebraic over any infinite perfect field, see [Bo], 18.3. For reductive G one may even drop the assumption "perfect".

Proposition: *Suppose that k is a field and that $G(k)$ is dense in G. Then for any G–module M, its N–socle $\operatorname{soc}_N M$ is a G–submodule of M with $\operatorname{soc}_G M \subset \operatorname{soc}_N M$.*

Proof: As $G(k)$ is dense in G, any subspace of M is a G–submodule if and only if it is $G(k)$–stable, cf. 2.12(5). So 6.15(1) implies that $\operatorname{soc}_N M$ is a G–submodule.

If M is a simple G–module, then $\operatorname{soc}_N M$ is a G–submodule that is non-zero by 2.14(2), hence equal to M. This implies that every semi-simple G–module is also semi-simple for N, hence that $\operatorname{soc}_G M \subset \operatorname{soc}_N M$ for any M.

Remark: This proposition generalises to the following more general situation. Suppose that there is a subgroup scheme H of G such that $G = NH$ and such that $H(k)$ is dense in H. Then an N–submodule of M is a G–submodule if and only if it is an H–submodule, hence if and only if it is $H(k)$–stable. Therefore we can argue as in the proof above.

6.17. Any homomorphism $\alpha : G \to G'$ of flat k–group schemes gives rise (by 2.15) to an exact functor $\alpha^* : \{\, G'\text{–modules}\,\} \to \{\, G\text{–modules}\,\}$. In case α is an inclusion (resp. α induces an isomorphism $G/\ker(\alpha) \xrightarrow{\sim} G'$) we have constructed a right adjoint functor $\alpha_* = \operatorname{ind}_G^{G'}$ (resp. $\alpha_* = ?^{\ker(\alpha)}$). In general, α is a composition of maps of this type, cf. [DG], III, §3, 3.2, so we get always a right adjoint. See [Donkin 1], section 3 or [Cline, Parshall, and Scott 6], 1.2 for a unified treatment.

CHAPTER 7

Algebras of Distributions

Over a field of characteristic 0 the representation theory of a connected algebraic group G is very well reflected by the representation theory of its Lie algebra $\mathfrak{g}$. Any representation of G gives rise to a representation of $\mathfrak{g}$. Then the notions of "submodule", "fixed point", or "module homomorphism" yield the same result whether applied to G–modules or to $\mathfrak{g}$–modules.

This is no longer true in characteristic $p \neq 0$. Any G–module still leads to a $\mathfrak{g}$–module in a natural way, but now there may be $\mathfrak{g}$–submodules that are not G–submodules, or $\mathfrak{g}$–homomorphisms that are not G–homomorphisms, etc.

It is, however, still possible to save some of the advantages of the linearisation process (of going from G to $\mathfrak{g}$) by looking not only at $\mathfrak{g}$, but at the algebra $\operatorname{Dist}(G)$ of all distributions on G with support in the origin. (See 7.1 and 7.7 for the definition.)

In characteristic 0 there is not more information contained in $\operatorname{Dist}(G)$ than in $\mathfrak{g}$, as in this case $\operatorname{Dist}(G)$ is isomorphic to the universal enveloping algebra of $\mathfrak{g}$. This changes in prime characteristic and there $\operatorname{Dist}(G)$ will do everything that $\mathfrak{g}$ does not do (7.14–7.17).

In this chapter we give at first the definitions of distributions with support in a rational point on an affine scheme, prove elementary properties, and then go over to distributions on group schemes with support in the origin.

The definitions and results are more or less contained in [DG], [Ta], and [Y]. In [Ta] and [Y] there are many more results on distributions on schemes over a field than I could include here. In some cases it was necessary to extend their results from fields to rings. There [Haboush 3] was very useful.

7.1. (Distributions with Support in a Point) Let X be an affine scheme over k and $x \in X(k)$. Set $I_x = \{f \in k[X] \mid f(x) = 0\}$. Then $k[X] = k1 \oplus I_x \simeq k \oplus I_x$.

A *distribution* of order $\leq n$ on X with support in x is a linear map $\mu : k[X] \to k$ with $\mu(I_x^{n+1}) = 0$. These distributions form a k–module that we denote by $\operatorname{Dist}_n(X, x)$. We have

$$(1) \qquad (k[X]/I_x^{n+1})^* \simeq \operatorname{Dist}_n(X, x) \subset k[X]^*.$$

Obviously $\operatorname{Dist}_0(X, x) \simeq k^* \simeq k$, and for any n

$$(2) \qquad \operatorname{Dist}_n(X, x) \simeq k \oplus \operatorname{Dist}_n^+(X, x)$$

where

$$(3) \qquad \operatorname{Dist}_n^+(X, x) = \{\, \mu \in \operatorname{Dist}_n(X, x) \mid \mu(1) = 0 \,\} \simeq (I_x/I_x^{n+1})^*.$$

For each $\mu \in \mathrm{Dist}_n(X,x)$ we call $\mu(1)$ its *constant term*; elements in $\mathrm{Dist}_n^+(X,x)$ are called distributions *without* constant term. The k–module $\mathrm{Dist}_1^+(X,x) \simeq (I_x/I_x^2)^*$ is called the *tangent space* to X at x and is denoted by T_xX. (Cf. [DG], II, §4, 3.3 for another description.)

The union of all $\mathrm{Dist}_n(X,x)$ in $k[X]^*$ is denoted by $\mathrm{Dist}(X,x)$; its elements are called distributions on X with support in x:

$$\mathrm{Dist}(X,x) = \{\mu \in k[X]^* \mid \exists\, n \in \mathbf{N} : \mu(I_x^{n+1}) = 0\} = \bigcup_{n \geq 0} \mathrm{Dist}_n(X,x). \tag{4}$$

This is obviously a k–module. Similarly, $\mathrm{Dist}^+(X,x) = \bigcup_{n\geq 0} \mathrm{Dist}_n^+(X,x)$ is a k–module.

For each $f \in k[X]$ and $\mu \in k[X]^*$ define $f\mu \in k[X]^*$ by $(f\mu)(f_1) = \mu(ff_1)$ for all $f_1 \in k[X]$. In this way $k[X]^*$ is a $k[X]$–module. As each I_x^{n+1} is an ideal in $k[X]$, obviously each $\mathrm{Dist}_n(X,x)$ and hence also $\mathrm{Dist}(X,x)$ is a $k[X]$–submodule of $k[X]^*$.

We have restricted ourselves above to the case of affine schemes. There is, however, a definition available for all schemes. In general one defines distributions as special deviations ([DG], II, §4, 5.2), shows that all these deviations form a k–module ([DG], II, §4, 5.4), and uses [DG], II, §4, 5.7 in order to prove that in the affine case one gets the same definition as above.

In the case of a ground field, however, we can easily give another description that works for all schemes. Suppose that k is field. Then we can associate to $x \in X(k)$ the local ring $\mathcal{O}_{X,x}$ and its maximal ideal $\mathfrak{m}_x$. In the affine case these are localisations $\mathcal{O}_{X,x} = k[X]_x$ and $\mathfrak{m}_x = (I_x)_x$. Furthermore, the natural map $k[X] \to \mathcal{O}_{X,x}$ induces in the affine case isomorphisms $k[X]/I_x^{n+1} \simeq \mathcal{O}_{X,x}/\mathfrak{m}_x^{n+1}$ for all n. So we can in general define $\mathrm{Dist}_n(X,x)$ as $(\mathcal{O}_{X,x}/\mathfrak{m}_x^{n+1})^*$. Similarly, we get $\mathrm{Dist}(X,x)$, $\mathrm{Dist}_n^+(X,x)$, and $\mathrm{Dist}^+(X,x)$.

7.2. (Elementary Properties) Let $\varphi : X \to Y$ be a morphism of affine schemes over k and let $\varphi^* : k[Y] \to k[X]$ be its comorphism. Then $(\varphi^*)^{-1}I_x = I_{\varphi(x)}$ for all $x \in X(k)$, hence $\varphi^*(I_{\varphi(x)}^{n+1}) \subset I_x^{n+1}$ and φ^* induces a linear map $k[Y]/I_{\varphi(x)}^{n+1} \to k[X]/I_x^{n+1}$. The transposed maps for all n yield a linear map

$$(d\varphi)_x : \mathrm{Dist}(X,x) \to \mathrm{Dist}(Y,\varphi(x)) \tag{1}$$

with $(d\varphi)_x(\mathrm{Dist}_n(X,x)) \subset \mathrm{Dist}_n(Y,\varphi(x))$ and

$$(d\varphi)_x(\mathrm{Dist}_n^+(X,x)) \subset \mathrm{Dist}_n^+(Y,\varphi(x))$$

for all n. We call $(d\varphi)_x$ the *tangent map* of φ in x; it is the usual tangent map on $T_xX = \mathrm{Dist}_1^+(X,x)$. One easily checks $d(\psi\circ\varphi)_x = d(\psi)_{\varphi(x)} \circ (d\varphi)_x$ for any morphism $\psi : Y \to Z$ into another affine scheme.

Let X be an affine scheme over k and $x \in X(k)$. Suppose I is an ideal in $k[X]$ with $x \in V(I)(k)$, i.e., with $I \subset I_x$, cf. 1.4 for the notation. We can then apply the construction above to the inclusion of $V(I)$ into X. We have $k[V(I)] = k[X]/I$.

The ideal of x is I_x/I, and its nth power is $(I_x^n + I)/I$. This implies that the inclusion yields isomorphisms

$$\text{(2)} \qquad \operatorname{Dist}_n(V(I), x) \simeq \{\, \mu \in \operatorname{Dist}_n(X, x) \mid \mu(I) = 0 \,\}$$

and

$$\text{(3)} \qquad \operatorname{Dist}(V(I), x) \simeq \{\, \mu \in \operatorname{Dist}(X, x) \mid \mu(I) = 0 \,\},$$

similarly for Dist_n^+ and Dist^+. We shall usually identify both sides in (2) and (3). If I' is another ideal with $x \in V(I')(k)$, then 1.4(5) implies

$$\text{(4)} \qquad \operatorname{Dist}(V(I) \cap V(I'), x) = \operatorname{Dist}(V(I), x) \cap \operatorname{Dist}(V(I'), x),$$

similarly for Dist_n, Dist^+, and Dist_n^+.

If $x \in D(f)(k)$ for some $f \in k[X]$, then the canonical map $k[X] \to k[X]_f$ induces an isomorphism of each $k[X]/I_x^{n+1}$ onto the corresponding object for $D(f)$. Therefore the inclusion of $D(f)$ into X induces an isomorphism

$$\text{(5)} \qquad \operatorname{Dist}_n(D(f), x) \simeq \operatorname{Dist}_n(X, x),$$

similarly for Dist, etc.

Let $\varphi : X \to Y$ again be a morphism of affine schemes over k, let $Y' \subset Y$ be a closed subscheme defined by an ideal $J \subset k[Y]$, and let $x \in X(k)$ with $\varphi(x) \in Y'(k)$. Then $\varphi^{-1}(Y')$ is the closed subscheme $V(k[X]\varphi^*(J))$ of X. As $x \in \varphi^{-1}(Y')(k)$, we have $I_x \supset k[X]\varphi^*(J)$. We get therefore from $k[X] = k1 \oplus I_x$ that

$$k[X]\varphi^*(J) + I_x^2 = \varphi^*(J) + I_x\varphi^*(J) + I_x^2 = \varphi^*(J) + I_x^2.$$

Any $\mu \in T_x(X) = \operatorname{Dist}_1^+(X, x)$ belongs to $T_x(\varphi^{-1}(Y'))$ if and only if $\mu(k[X]\varphi^*(J))$ $= 0$, hence if and only if $\mu(\varphi^*(J)) = 0$, i.e., if $((d\varphi)_x\mu)(J) = 0$. This yields

$$\text{(6)} \qquad T_x(\varphi^{-1}(Y')) = (d\varphi)_x^{-1} T_{\varphi(x)}(Y').$$

Here and in (7) below we regard $(d\varphi)_x$ as a map defined on T_xX. The reader who is more familiar with varieties than with schemes should be aware that $\varphi^{-1}(Y')$ has to be taken as a scheme, even if X, Y, Y' happen to be varieties; so $T_x(\varphi^{-1}(Y'))$ can be larger than expected.

We can always apply (6) to $Y' = V(I_{\varphi(x)})$. This closed subscheme is isomorphic to $\mathbf{A}^0$; the unique element in any $Y'(A)$ is just the canonical image of $\varphi(x)$. Obviously $T_{\varphi(x)}(Y') = 0$ for this Y', hence

$$\text{(7)} \qquad \ker\,(d\varphi)_x = T_x(\varphi^{-1}(\varphi(x))).$$

The constructions and results above have generalisations to the case where the schemes are not affine. This is particularly obvious when k is a field and we can work with $\mathcal{O}_{X,x}$. One can also generalise (5) to $\operatorname{Dist}(Y, x) = \operatorname{Dist}(X, x)$ for any open subscheme Y of X with $x \in Y(k)$.

7.3. (Distributions on $\mathbf{A}^m$) First let us consider as an example $X = \mathbf{A}^1 = Sp_k k[T]$ and $x = 0$, hence $I_x = (T)$. The k–module $k[X]/I_x^{n+1}$ is free and has the residue classes of $1 = T^0, T = T^1, T^2, \dots, T^n$ as a basis. Define $\gamma_r \in k[T]^* = k[\mathbf{A}^1]^*$ through $\gamma_r(T^n) = 0$ for $n \neq r$ and $\gamma_r(T^r) = 1$. Then obviously $\operatorname{Dist}(\mathbf{A}^1, 0)$ is a free k–module with basis $(\gamma_r)_{r\in\mathbf{N}}$ and each $\operatorname{Dist}_n(\mathbf{A}^1, 0)$ is a free k–module with basis $(\gamma_r)_{0\leq r\leq n}$. If k is a field of characteristic 0, then obviously

$$\gamma_r(f) = \frac{1}{r!}\Big(\Big(\frac{\partial}{\partial T}\Big)^r f\Big)(0).$$

This description can easily be generalised to $\mathbf{A}^m = Sp_k\, k[T_1, T_2, \dots, T_m]$ for all m. For each multi-index $a = (a(1), a(2), \dots, a(m)) \in \mathbf{N}^m$ set $T^a = T_1^{a(1)}T_2^{a(2)}\dots T_m^{a(m)}$ and denote by γ_a the linear map with $\gamma_a(T^a) = 1$ and $\gamma_a(T^b) = 0$ for all $b \in \mathbf{N}^m$, $b \neq a$. One easily checks that $\operatorname{Dist}(\mathbf{A}^m, 0)$ is free over k with all γ_a as a basis, and that $\operatorname{Dist}_n(\mathbf{A}^m, 0)$ is free over k with all γ_a with $|a| \leq n$ as a basis. (For a as above set $|a| = \sum_{i=1}^m a(i)$.) If k is a field of characteristic 0, then

$$\gamma_a(f) = \frac{1}{\prod a(i)!}\Big(\Big(\frac{\partial}{\partial T_1}\Big)^{a(1)}\Big(\frac{\partial}{\partial T_2}\Big)^{a(2)}\cdots\Big(\frac{\partial}{\partial T_m}\Big)^{a(m)} f\Big)(0).$$

If k is a field, then any $\operatorname{Dist}(X, x)$ (for arbitrary X) will only depend on the $\mathfrak{m}_x$–adic completion of $\mathcal{O}_{X,x}$. So, for a simple point x all $\operatorname{Dist}_n(X, x)$ and $\operatorname{Dist}(X, x)$ will look like $\operatorname{Dist}_n(\mathbf{A}^m, 0)$ and $\operatorname{Dist}(\mathbf{A}^m, 0)$ where $m = \dim_x X$, cf. [DG], I, §4, 4.2.

7.4. (Infinitesimal Flatness) Let X be an affine scheme over k and $x \in X(k)$. We call X *infinitesimally flat* at x if each $k[X]/I_x^{n+1}$ with $n \in \mathbf{N}$ is a finitely presented and flat (or, equivalently, projective, cf. [B2], ch. II, §5, cor. 2 du th. 1) k–module. (In [Haboush 3] this property is called "infinitesimally smooth". As obviously over a field any algebraic scheme (cf. 1.6) has this property, I think that name is not appropriate.)

If X is infinitesimally flat at x, then each I_x^n/I_x^m with $n \leq m$ is also finitely presented and projective over k, and each I_x^n is a direct k–summand of $k[X]$.

Let k' be a k–algebra. Any $x \in X(k)$ defines a point in $X(k') = X_{k'}(k')$ with ideal $I_x \otimes k' \subset k[X] \otimes k' \simeq k'[X_{k'}]$. Then $k'[X_{k'}]/(I_x \otimes k')^{n+1} \simeq (k[X]/I_x^{n+1}) \otimes k'$. Now ring extension commutes with taking the dual module as long as the module is finitely generated and projective. So we get:

(1) *If X is infinitesimally flat at x, then $X_{k'}$ is infinitesimally flat at x for each k–algebra k' and there are natural isomorphisms* $\operatorname{Dist}_n(X, x) \otimes k' \simeq \operatorname{Dist}_n(X_{k'}, x)$ *and* $\operatorname{Dist}(X, x) \otimes k' \simeq \operatorname{Dist}(X_{k'}, x)$.

[We use here the letter x also for the image of x in $X_{k'}(k') = X(k')$.]

Consider two affine schemes X, X' and points $x \in X(k)$ and $x' \in X'(k)$. The the ideal of (x, x') in $k[X \times X'] \simeq k[X] \otimes k[X']$ is $I_{(x,x')} = I_x \otimes k[X'] + k[X] \otimes I_{x'}$. If X and X' are infinitesimally flat at x resp. x', then $I_{(x,x')}^{n+1}$ can be identified with $\sum_{j=0}^{n+1} I_x^j \otimes I_{x'}^{n+1-j}$, and then with $\bigcap_{j=0}^n (k[X] \otimes I_{x'}^{n+1-j} + I_x^{j+1} \otimes k[X'])$. Now elementary considerations yield:

(2) *Let X and X' be infinitesimally flat at x resp. x'. Then $X \times X'$ is infinitesimally flat at (x, x') and there is an isomorphism* $\operatorname{Dist}(X, x) \otimes \operatorname{Dist}(X', x') \simeq$

$\mathrm{Dist}(X \times X', (x, x'))$ *mapping each* $\sum_{m=0}^{n} \mathrm{Dist}_m(X, x) \otimes \mathrm{Dist}_{n-m}(X', x')$ *onto* $\mathrm{Dist}_n(X \times X', (x, x'))$.

Assume that X is infinitesimally flat at x. We can apply (2) to $X' = X$. Consider the diagonal morphism $\delta_X : X \to X \times X$, $x \mapsto (x, x)$. Let us regard the tangent map $(d\delta_X)_x$ as a map $\Delta'_{X,x} : \mathrm{Dist}(X, x) \to \mathrm{Dist}(X, x) \otimes \mathrm{Dist}(X, x)$. It makes $\mathrm{Dist}(X, x)$ into a coalgebra, i.e., satisfies 2.3(1) as $(\mathrm{id} \times \delta_X) \circ \delta_X = (\delta_X \times \mathrm{id}) \circ \delta_X$. This coalgebra is cocommutative, i.e., we have $s \circ \Delta'_{X,x} = \Delta'_{X,x}$ where $s(\mu_1 \otimes \mu_2) = \mu_2 \otimes \mu_1$. The map $\varepsilon'_x : \mu \mapsto \mu(1)$ is a counit, i.e., satisfies 2.3(2). If $\varphi : X \to Y$ is a morphism, then $(d\varphi)_x$ is a homomorphism of coalgebras, as $(\varphi \times \varphi) \circ \delta_X = \delta_Y \circ \varphi$. So we have seen:

(3) *If X is infinitesimally flat at x, then* $\mathrm{Dist}(X, x)$ *has a natural structure as a cocommutative coalgebra with a counit. Tangent maps are homomorphisms for these structures.*

Remark: There is another case besides (1) where $\mathrm{Dist}_n(X, x) \otimes k' \simeq \mathrm{Dist}_n(X_{k'}, x)$ for all n and $\mathrm{Dist}(X, x) \otimes k' \simeq \mathrm{Dist}(X_{k'}, x)$. Assume that k is a noetherian integral domain and that k' is its field of fractions. In this case tensoring with k' commutes with taking the dual module for all finitely generated k–modules, cf. [B1], ch. II, §7, exerc. 29b. If X is an algebraic scheme over k, then each $k[X]/I_x^{n+1}$ is finitely generated over k. So we get in this case the desired isomorphisms.

7.5. An affine k–scheme X is called *noetherian* if $k[X]$ is a noetherian ring, and it is called *integral* if $k[X]$ is an integral domain.

Proposition: *Let X be an affine scheme over k and $x \in X(k)$. Let I, I' be ideals in $k[X]$ with $x \in V(I)(k) \cap V(I')(k)$. If $V(I)$ is integral, noetherian, and infinitesimally flat at x, then:*

$$V(I) \subset V(I') \iff \mathrm{Dist}(V(I), x) \subset \mathrm{Dist}(V(I'), x).$$

Proof: If $V(I) \subset V(I')$, then $I' \subset I$ by 1.4(3), hence $\mathrm{Dist}(V(I), x) \subset \mathrm{Dist}(V(I'), x)$ by 7.2(3).

Suppose conversely that $\mathrm{Dist}(V(I), x) \subset \mathrm{Dist}(V(I'), x)$. We want to show

$$I' \subset I_x^{n+1} + I \tag{1}$$

for all $n \in \mathbf{N}$. If not, then $(I' + I_x^{n+1} + I)/(I_x^{n+1} + I) \neq 0$ for some n. Now I_x/I is the ideal of x in $k[V(I)] = k[X]/I$ and its $(n+1)^{\mathrm{st}}$ power is $(I_x^{n+1} + I)/I$. So $k[X]/(I_x^{n+1} + I)$ is a finitely presented and projective k–module. For any $a \in (I' + I_x^{n+1} + I)/(I_x^{n+1} + I)$, $a \neq 0$, there is some $\mu \in (k[X]/(I_x^{n+1} + I))^* = \mathrm{Dist}_n(V(I), x)$ with $\mu(a) \neq 0$. Regarding μ as an element in $k[X]^*$, we get $\mu(I') \neq 0$, hence $\mu \notin \mathrm{Dist}(V(I'), x)$. This contradicts our assumption and we have extablished (1).

We can now apply Krull's intersection theorem to $k[V(I)] = k[X]/I$ and get $I = \bigcap_{n \geq 0}(I_x^{n+1} + I) \supset I'$, hence $V(I) \subset V(I')$.

Remark: This obviously generalises to the case where $V(I)$ is no longer integral, but where I_x contains all associated prime ideals of I.

7.6. Proposition: *Suppose that k is a field. Let $\varphi : X \to Y$ be a morphism of algebraic schemes over k and let $x \in X(k)$. If φ is flat at x, then $(d\varphi)_x : \mathrm{Dist}(X,x) \to \mathrm{Dist}(Y,\varphi(x))$ is surjective.*

Proof: Set $A = \mathcal{O}_{Y,\varphi(x)}$ and $B = \mathcal{O}_{X,x}$. The flatness of φ amounts to the following: Using the comorphism (we may assume X and Y to be affine) we may regard A as a subalgebra of B such that B is a faithfully flat A–module. This faithful flatness implies $\mathfrak{m}_{\varphi(x)}^{n+1} = A \cap B\mathfrak{m}_{\varphi(x)}^{n+1}$ for all $n \in \mathbf{N}$, cf. [B2], ch. I, §3, prop. 9. As we assume our schemes to be algebraic, the rings A, B are noetherian and each $A/\mathfrak{m}_{\varphi(x)}^{n+1}$ is finite dimensional. So Krull's intersection theorem yields

$$B\mathfrak{m}_{\varphi(x)}^{n+1} = \bigcap_{r \geq 0} (\mathfrak{m}_x^{r+1} + B\mathfrak{m}_{\varphi(x)}^{n+1}),$$

hence

$$\mathfrak{m}_{\varphi(x)}^{n+1} = \bigcap_{r \geq 0} ((\mathfrak{m}_x^{r+1} + B\mathfrak{m}_{\varphi(x)}^{n+1}) \cap A).$$

Now $\dim(A/\mathfrak{m}_{\varphi(x)}^{n+1}) < \infty$ implies that there is some r with $A \cap (\mathfrak{m}_x^{r+1} + B\mathfrak{m}_{\varphi(x)}^{n+1}) = \mathfrak{m}_{\varphi(x)}^{n+1}$. We can therefore embed $A/\mathfrak{m}_{\varphi(x)}^{n+1} \simeq (A + B\mathfrak{m}_{\varphi(x)}^{n+1} + \mathfrak{m}_x^{r+1})/(B\mathfrak{m}_{\varphi(x)}^{n+1} + \mathfrak{m}_x^{r+1})$ into $B/(B\mathfrak{m}_{\varphi(x)}^{n+1} + \mathfrak{m}_x^{r+1})$. As k is a field, any $\mu \in \mathrm{Dist}_n(Y,\varphi(x)) = (A/\mathfrak{m}_{\varphi(x)}^{n+1})^*$ has an extension to $B/(B\mathfrak{m}_{\varphi(x)}^{n+1} + \mathfrak{m}_x^{r+1})$, which gives some $\mu' \in (B/\mathfrak{m}_x^{r+1})^* \simeq \mathrm{Dist}_r(X,x)$. Then obviously $(d\varphi)_x\mu' = \mu$. Therefore $(d\varphi)_x$ is surjective.

Remark: Note that we do not claim that $\mathrm{Dist}_n(X,x)$ is mapped onto $\mathrm{Dist}_n(Y,\varphi(x))$ for each n. Indeed, it is well known that, e.g., the "classical" tangent map $T_xX = \mathrm{Dist}_1^+(X,x) \to \mathrm{Dist}_1^+(Y,\varphi(x)) = T_{\varphi(x)}Y$ will not be surjective in general.

7.7. (Distributions on a Group Scheme) Let G be a group scheme over k. In this case we set

$$\mathrm{Dist}(G) = \mathrm{Dist}(G,1)$$

and $\mathrm{Dist}_n(G) = \mathrm{Dist}_n(G,1)$ for all n.

We can make $\mathrm{Dist}(G)$ into an associative algebra over k. For any $\mu, \nu \in k[G]^*$ we can define a product $\mu\nu$ as

$$\mu\nu : k[G] \xrightarrow{\Delta} k[G] \otimes k[G] \xrightarrow{\mu\otimes\nu} k \otimes k \xrightarrow{\sim} k. \tag{1}$$

It is clear that $\mu\nu \in k[G]^*$ and that $(\mu,\nu) \mapsto \mu\nu$ is bilinear. Furthermore, 2.3(1) implies that this multiplication is associative, and 2.3(2) that ε_G is a neutral element. So $k[G]^*$ has a structure as an associative algebra over k with one. It will in general be non-commutative.

Now $\mathrm{Dist}(G)$ is a subalgebra of $k[G]^*$ with

$$\mathrm{Dist}_n(G)\,\mathrm{Dist}_m(G) \subset \mathrm{Dist}_{n+m}(G) \tag{2}$$

for all n, m. This follows from the formula $\Delta(I_1^n) \subset \sum_{r=0}^n I_1^r \otimes I_1^{n-r}$, cf. 2.4(1). (Here $I_1^r \otimes I_1^{n-r}$ is some abuse of notation; we really mean the image of this tensor product in $k[G] \otimes k[G]$.) More precisely, 2.4(1) implies

$$\Delta(f_1 f_2 \dots f_n) \in \prod_{i=1}^n (1 \otimes f_i + f_i \otimes 1) + \sum_{r=1}^n I_1^r \otimes I_1^{n+1-r}$$

for all $f_1, f_2, \dots, f_n \in I_1$ and $n \in \mathbf{N}$. We get therefore:

If $\mu \in \mathrm{Dist}_n(G)$ *and* $\nu \in \mathrm{Dist}_m(G)$, *then*

(3) $$[\mu, \nu] = \mu\nu - \nu\mu \in \mathrm{Dist}_{n+m-1}(G).$$

So $\mathrm{Dist}(G)$ has a structure as a filtered associative algebra over k such that the associated graded algebra is commutative. We call $\mathrm{Dist}(G)$ the *algebra of distributions on* G, dropping the addendum "with support in the origin". (Some people call $\mathrm{Dist}(G)$ the *hyperalgebra* of G.)

Because of $\Delta(1) = 1 \otimes 1$, the subspace $\mathrm{Dist}^+(G)$ is a two-sided ideal in $\mathrm{Dist}(G)$. Therefore (3) implies $[\mathrm{Dist}_n^+(G), \mathrm{Dist}_m^+(G)] \subset \mathrm{Dist}_{n+m-1}^+(G)$. This shows in particular that $\mathrm{Dist}_1^+(G)$ is a Lie algebra, which we denote by $\mathrm{Lie}(G)$ and call *the Lie algebra of* G. Note that $\mathrm{Lie}(G) = T_1 G$ as a k–module, cf. 7.1. It can be shown that we have constructed the usual structure as a Lie algebra on $T_1 G$.

7.8. (Examples) Let us look first at the additive group $G = G_a$. As a scheme we may identify $G_a = Sp_k k[T]$ with $\mathbf{A}^1$. Therefore we have described $\mathrm{Dist}(G_a)$ as a k–module already in 7.3. As before let γ_n be the element with $\gamma_n(T^n) = 1$ and $\gamma_n(T^m) = 0$ for $m \neq n$. We have $\Delta(T) = 1 \otimes T + T \otimes 1$, hence $\Delta(T^m) = \sum_{i=0}^n \binom{n}{i} T^i \otimes T^{n-i}$. This implies easily

(1) $$\gamma_n\, \gamma_m = \binom{n+m}{n} \gamma_{n+m},$$

hence

(2) $$\gamma_1^n = n!\, \gamma_n.$$

So $\mathrm{Dist}(G_{a,\mathbf{C}})$ can be identified with the polynomial ring $\mathbf{C}[\gamma_1]$, and $\mathrm{Dist}(G_{a,\mathbf{Z}})$ with the $\mathbf{Z}$–lattice in $\mathbf{C}[\gamma_1]$ spanned by all $\gamma_1^n/n!$. In general, we have $\mathrm{Dist}(G_a) = \mathrm{Dist}(G_{a,\mathbf{Z}}) \otimes_{\mathbf{Z}} k$.

Let us consider next the multiplicative group $G_m = Sp_k k[T, T^{-1}]$. Then I_1 is generated by $T - 1$. The residue classes of 1, $T-1$, $(T-1)^2, \dots, (T-1)^n$ form a basis of $k[G_m]/I_1^{n+1}$. There is a unique $\delta_n \in \mathrm{Dist}(G_m)$ with $\delta_n(I_1^{n+1}) = 0$ and $\delta_n((T-1)^n) = 1$ and $\delta_n((T-1)^i) = 0$ for $0 \leq i < n$. From this and the binomial expansion of $T^n = ((T-1)+1)^n$ one gets $\delta_r(T^n) = \binom{n}{r}$ for all $n \in \mathbf{Z}$ and $r \in \mathbf{N}$. If k is a $\mathbf{Q}$–algebra, then obviously

$$\delta_r\, f = \frac{1}{r!}\Big(\Big(\frac{\partial}{\partial T}\Big)^r f\Big)(1).$$

All δ_r with $r \in \mathbf{N}$ form a basis of $\mathrm{Dist}(G_m)$, all δ_r with $r \leq n$ one of $\mathrm{Dist}_n(G_m)$. We have $\Delta(T) = T \otimes T$, hence $\Delta(T-1) = (T-1) \otimes (T-1) + (T-1) \otimes 1 + 1 \otimes (T-1)$, hence

(3) $$\delta_r\, \delta_s = \sum_{i=0}^{\min(r,s)} \frac{(r+s-i)!}{(r-i)!\,(s-i)!\,i!}\, \delta_{r+s-i}.$$

We get as a special case $\delta_1 \delta_r = (r+1)\delta_{r+1} + r\delta_r$, hence $(\delta_1 - r)\delta_r = (r+1)\delta_{r+1}$ and inductively

(4) $$r!\, \delta_r = \delta_1(\delta_1 - 1) \dots (\delta_1 - r + 1).$$

If k is a $\mathbf{Q}$–algebra, then $\delta_r = \binom{\delta_1}{r}$. Therefore $\mathrm{Dist}(G_{m,\mathbf{C}}) \simeq \mathbf{C}[\delta_1]$, and $\mathrm{Dist}(G_{m,\mathbf{Z}})$ is the $\mathbf{Z}$–lattice in $\mathrm{Dist}(G_{m,\mathbf{C}})$ generated by all $\binom{\delta_1}{r}$. In general, $\mathrm{Dist}(G_m) = \mathrm{Dist}(G_{m,\mathbf{Z}}) \otimes_{\mathbf{Z}} k$.

7.9. (Elementary Properties) If $\alpha : G \to G'$ is a homomorphism of group schemes over k, then

(1) $$d\alpha = (d\alpha)_1 : \mathrm{Dist}(G) \to \mathrm{Dist}(G')$$

is a homomorphism of algebras. This follows easily from the definition of the multiplication. On $\mathrm{Lie}(G) = \mathrm{Dist}_1^+(G)$ we get the usual tangent map $\mathrm{Lie}(G) \to \mathrm{Lie}(G')$, which is a homomorphism of Lie algebras.

If H, H' are closed subgroup schemes of a group scheme G, then the inclusions of $\mathrm{Dist}(H)$ and $\mathrm{Dist}(H')$ into $\mathrm{Dist}(G)$, cf. 7.2(3), are homomorphisms of algebras, and 7.2(4) implies

(2) $$\mathrm{Dist}(H \cap H') = \mathrm{Dist}(H) \cap \mathrm{Dist}(H'),$$

similarly $\mathrm{Lie}(H \cap H') = \mathrm{Lie}(H) \cap \mathrm{Lie}(H')$. (The same statement for linear algebraic groups is known to be false in general. There the intersection as varieties is considered, not as schemes as we do here.)

We call G *infinitesimally flat* if it is so at 1. Now 7.4(2) easily implies:

If G_1, G_2 are infinitesimally flat group schemes, then $G_1 \times G_2$ is infinitesimally flat and there is an isomorphism of algebras

(3) $$\mathrm{Dist}(G_1) \otimes \mathrm{Dist}(G_2) \xrightarrow{\sim} \mathrm{Dist}(G_1 \times G_2).$$

In the case of a semi-direct product we get here still an isomorphism of k–modules.

If we take $G_1 = G_2 = G$ and consider the multiplication map $m_G : G \times G \to G$, then we see easily:

(4) *If G is an infinitesimally flat group scheme over k, then $dm_G : \mathrm{Dist}(G) \otimes \mathrm{Dist}(G) \xrightarrow{\sim} \mathrm{Dist}(G)$ is given by $dm_G(\mu \otimes \nu) = \mu\nu$ for all $\mu, \nu \in \mathrm{Dist}(G)$.*

For G as in (4) and any k–algebra k', the isomorphisms $\mathrm{Lie}(G) \otimes k' \xrightarrow{\sim} \mathrm{Lie}(G_{k'})$ resp. $\mathrm{Dist}(G) \otimes k' \xrightarrow{\sim} \mathrm{Dist}(G_{k'})$, cf. 7.4(1), are isomorphisms of Lie algebras resp. of associative algebras. Furthermore, the comultiplication $\Delta'_G = \Delta'_{G,1} : \mathrm{Dist}(G) \to \mathrm{Dist}(G) \otimes \mathrm{Dist}(G)$ can be checked to be a homomorphism of algebras over k.

The map $i_G : G \to G$ with $g \mapsto g^{-1}$ has as a tangent map (cf. 2.3)

(5) $$\sigma'_G = di_G : \mu \mapsto \mu \circ \sigma_G.$$

One easily checks that σ'_G is an anti-automorphism of $\mathrm{Dist}(G)$, i.e., satisfies $\sigma'_G(\mu\nu) = \sigma'_G(\nu)\sigma'_G(\mu)$ for all μ, ν. If G is infinitesimally flat, then σ'_G is a coinverse for the coalgebra structure, i.e., 2.3(3) is satisfied by $(\Delta'_G, \sigma'_G, \varepsilon_G)$ instead of $(\Delta, \sigma, \varepsilon)$.

7.10. (Distributions and the Enveloping Algebra) To each Lie algebra $\mathfrak{g}$ over k one can associate its universal enveloping algebra $U(\mathfrak{g})$. One may consult [B3], ch. I, §2, or [Dix], ch. 2 for the definition and the elementary properties of this object. It has a natural filtration $U_0(\mathfrak{g}) = k1 \subset U_1(\mathfrak{g}) = k1 \oplus \mathfrak{g} \subset U_2(\mathfrak{g}) \subset \cdots$ where $U_n(\mathfrak{g})$ is spanned over k by all products $x_1 x_2 \dots x_r$ with $r \le n$ and all $x_i \in \mathfrak{g}$.

Let G be a group scheme over k. As $\mathrm{Lie}(G) = \mathrm{Dist}_1^+(G)$ is a Lie subalgebra of $\mathrm{Dist}(G)$, the universal property of $U(\mathrm{Lie}(G))$ yields a homomorphism

$\gamma : U(\text{Lie}(G)) \to \text{Dist}(G)$ of algebras that induces the identity on $\text{Lie}(G)$. It maps $U_n(\text{Lie}(G))$ to $\text{Dist}_n(G)$ by 7.7(2).

It is not very difficult to prove (cf. [DG], II, §6, n° 1):

(1) *If k is a field of characteristic* 0 *and G an algebraic k–group, then γ is an isomorphism* $U(\text{Lie}(G)) \xrightarrow{\sim} \text{Dist}(G)$ *and maps each* $U_n(\text{Lie}(G))$ *bijectively to* $\text{Dist}_n(G)$.

Using this one can then show that an algebraic k–group over a field of characteristic 0 is smooth and reduced, cf. [DG], II, §6, 1.1.

If k is a field of characteristic $p \neq 0$, then the situation is completely different. In this case for each $x \in \text{Lie}(G) = \text{Dist}_1^+(G)$ its pth power in $\text{Dist}(G)$ also belongs to $\text{Lie}(G)$. This is more easily seen by identifying $\text{Dist}(G)$ with the algebra of left or of right invariant derivations of $k[G]$ as in 7.18 below. Let us denote this p^{th} power by $x^{[p]}$ in order to distinguish it from the pth power x^p in $U(\text{Lie}(G))$. The pair $(\text{Lie}(G), x \mapsto x^{[p]})$ is an example of what is called a p–Lie algebra. (One can find the general definition in [DG], II, §7, n° 3.) For any p–Lie algebra $(\mathfrak{g}, x \mapsto x^{[p]})$ set $U^{[p]}(\mathfrak{g})$ equal to the quotient of $U(\mathfrak{g})$ by the two-sided ideal generated by all $x^p - x^{[p]}$ with $x \in \mathfrak{g}$. This algebra is called the *restricted enveloping algebra* of $\mathfrak{g}$. We can still regard $\mathfrak{g}$ as a subspace of $U^{[p]}(\mathfrak{g})$. If $x_1, x_2, \dots, x_m$ is a basis of $\mathfrak{g}$, then all $x_1^{a(1)} x_2^{a(2)} \cdots x_m^{a(m)}$ with $0 \le a(i) < p$ for all i form a basis of $U^{[p]}(\mathfrak{g})$, cf. [DG], II, §7, 3.6. So $\dim U^{[p]}(\mathfrak{g}) = p^{\dim \mathfrak{g}}$.

By the definition of $x^{[p]}$ for $x \in \text{Lie}(G)$ it is clear that γ has to factor through $U^{[p]}(\mathfrak{g})$. One can show:

(2) *If k is a field with* $\text{char}(k) = p \neq 0$ *and G an algebraic k–group, then γ induces an injective homomorphism* $U^{[p]}(\mathfrak{g}) \to \text{Dist}(G)$.

For this and for more details one may consult [DG], II, §7, n° 2–4.

7.11. (G–modules and Dist(G)–modules) Let G be a group scheme over k. Then any G–module M carries a natural structure as a $\text{Dist}(G)$–module: One sets for each $\mu \in \text{Dist}(G)$ and $m \in M$

$$\mu m = (\text{id}_M \mathbin{\overline{\otimes}} \mu)\, \Delta_M(m), \tag{1}$$

i.e., the action of μ on M is given by

$$M \xrightarrow{\Delta_M} M \otimes k[G] \xrightarrow{\text{id}_M \otimes \mu} M \otimes k \xrightarrow{\sim} M. \tag{2}$$

It is obvious that $(\mu, m) \mapsto \mu m$ is bilinear, and it is easy to see that $\mu(\nu m) = (\mu\nu)m$ and $\varepsilon_G m = m$ for all $m \in M$ and $\mu, \nu \in \text{Dist}(G)$, using 2.8(2), (3) and 7.7(1).

Obviously, 2.8(4) implies for all G–modules M, M':

$$\text{Hom}_G(M, M') \subset \text{Hom}_{\text{Dist}(G)}(M, M'). \tag{3}$$

Applying this to inclusions we get

(4) *Any G–submodule of a G–module M is also a* $\text{Dist}(G)$*–submodule of M.*

Of course, on a factor module the structure as a $\text{Dist}(G)$–module coming from the G–module structure is equal to the structure as a factor module for $\text{Dist}(G)$.

The Dist(G)–module structure on a direct sum of G–modules is the one as a direct sum of Dist(G)–modules.

We get from 2.10(2):

(5) *If $m \in M^G$, then $\mu m = \mu(1)m$ for all $\mu \in \operatorname{Dist}(G)$.*

More generally, 2.10(2′) implies for each $\lambda \in X(G) \subset k[G]$:

(6) *If $m \in M_\lambda$, then $\mu m = \mu(\lambda)m$ for all $\mu \in \operatorname{Dist}(G)$.*

For any G–module M and any $\lambda \in X(G)$ we can construct the G–module $M \otimes k_\lambda$, which we usually denote by $M \otimes \lambda$. We can identify $M \otimes \lambda$ as a k–module with M. If $\Delta_M(m) = \sum_i m_i \otimes f_i$, then $\Delta_{M\otimes\lambda}(m) = \sum_i m_i \otimes \lambda f_i$. This implies (cf. 7.1 for the $k[G]$–module structure on Dist(G)):

(7) *Any $\mu \in \operatorname{Dist}(G)$ acts on $M \otimes \lambda$ as $\lambda\mu$ acts on M.*

If G is infinitesimally flat, then any $\mu \in \operatorname{Dist}(G)$ acts on a tensor product of two G–modules through $\Delta'_G(\mu) \in \operatorname{Dist}(G) \otimes \operatorname{Dist}(G)$.

Let M be a G–module that is finitely generated and projective over k. Then M^* is a G–module in a canonical way, cf. 2.7(4). The action of Dist(G) on M^* is then given by

$$(\mu\varphi)(m) = \varphi(\sigma'_G(\mu)m) \tag{8}$$

for all $\mu \in \operatorname{Dist}(G)$, $\varphi \in M^*$, and $m \in M$.

If G is flat, then 2.13(2) implies that each $m \in M$ is contained in a Dist(G)–submodule of M finitely generated over k. In this sense M is a *locally finite* Dist(G)–*module.*

When we restrict the action of Dist(G) on M to Lie$(G) = \operatorname{Dist}_1^+(G)$, then we get a representation of Lie(G) as a Lie algebra. In case k is a field of characteristic $p \neq 0$, then M is in this way even a restricted module for the p–Lie algebra Lie(G).

Consider k–submodules $N' \subset N \subset M$ such that M/N' is a projective k–module and recall the definition of the closed subgroup scheme $G_{N',N}$ from 2.12. We claim:

$$\operatorname{Lie}(G_{N',N}) = \{\, \mu \in \operatorname{Lie}(G) \mid \mu N \subset N' \,\}. \tag{9}$$

As in 2.12 we may assume that M/N' is free and choose a basis $(e_j)_{j\in J}$ of a complement to N' in M. There are $a_j(m) \in k$ and $f_{j,m} \in k[G]$ for all $m \in M$ as in 2.12. We have then $G_{N',N} = V(k[G]I)$ where

$$I = \sum_{n\in N}\sum_{j\in J} k(f_{j,n} - a_j(n)1) + \sum_{n\in N}\sum_{j\in J} k(\sigma_G(f_{j,n}) - a_j(n)1).$$

We have $I \subset k[G]I \subset I_1$ since $1 \in G_{N',N}$, hence $k[G]I + I_1^2 = I + I_1^2$. Any $\mu \in \operatorname{Lie}(G)$ belongs to Lie$(G_{N',N})$ if and only if $\mu(k[G]I) = 0$, hence by the last equality if and only if $\mu(I) = 0$. Using $\mu(1) = 0$ and 2.4(2) we get that $\mu \in \operatorname{Lie}(G_{N',N})$ if and only if $\mu(f_{j,n}) = 0$ for all $n \in N$ and $j \in J$. Given the definition of the $f_{j,n}$ this means that $\mu n \in N'$ for all $n \in N$, hence (9).

Similarly, one gets for each subset $S \subset M$ if M is projective over k:

(10) $$\operatorname{Lie} Z_G(S) = \{\, \mu \in \operatorname{Lie}(G) \mid \mu\, m = 0 \text{ for all } m \in S \,\},$$

and for each k–submodule N of M such that M/N is projective over k:

(11) $$\operatorname{Lie} \operatorname{Stab}_G(N) = \{\, \mu \in \operatorname{Lie}(G) \mid \mu\, N \subset N \,\}.$$

The reader should be aware that we always regard $Z_G(S)$ and $\operatorname{Stab}_G(N)$ as schemes, even when we deal with a reduced algebraic group G over an algebraically closed field.

7.12. (The Case $G = G_a$) Let us use the basis $(\gamma_n)_{n\in \mathbf{N}}$ of $\operatorname{Dist}(G_a)$ as in 7.3 and 7.8. As $k[G_a] = k[T]$ is free with basis $(T^i)_{i\geq 0}$, we can write uniquely $\Delta_M(m) = \sum_{i\geq 0} m_i \otimes T^i$ for any G_a–module M and $m \in M$ with almost all $m_i = 0$. Then obviously $\gamma_n m = m_n$ for all n, i.e., $\Delta_M(m) = \sum_{n\geq 0} (\gamma_n m) \otimes T^n$. So the structure as a $\operatorname{Dist}(G_a)$–module determines the comodule map uniquely, hence also the structure as a G_a–module. This implies for $G = G_a$ that there is equality in 7.11(3) and that the converse holds in 7.11(4), (5).

In general, not all locally finite $\operatorname{Dist}(G_a)$–modules arise from G_a–modules. If, e.g., k is a field of characteristic 0, then one can define for each $b \in k$ a structure as a $\operatorname{Dist}(G_a)$–module on k, where each γ_r acts as multiplication by $b^r/(r!)$. For $b \neq 0$ this module does not come from a G_a–module. If k is a field of characteristic $p \neq 0$, then one can make k^2 into a $\operatorname{Dist}(G_a)$–module letting γ_i act as $\left(\begin{smallmatrix}0&1\\0&0\end{smallmatrix}\right)$ if $i = p^r$ for some $r \in \mathbf{N}$, $r > 0$, as 1 if $i = 0$, and as 0 otherwise. This structure does not come from G_a.

7.13. (The Case $G = G_m$) Let us use the basis $(\delta_r)_{r\geq 0}$ of $\operatorname{Dist}(G_m)$ as in 7.8. If M is a G_m–module and $m \in M$, then $\Delta_M(m) = \sum_{i\in\mathbf{Z}} m_i \otimes T^i$ with uniquely determined $m_i \in M$, almost all zero. Then

(1) $$\delta_n\, m = \sum_{i\in \mathbf{Z}} \binom{i}{n} m_i \qquad \text{for all } n \in \mathbf{N}.$$

Recall that $M = \bigoplus_{i\in\mathbf{Z}} M_i$ where $M_i = \{m' \in M \mid \Delta_M(m') = m' \otimes T^i\}$, and that $m_i \in M_i$ in the situation above, see 2.11.

For $a_1, a_2, \dots, a_r \in \mathbf{Z}$ pairwise distinct, there is $f \in \mathbf{Q}[T]$ with $f(a_1) = 1$, $f(a_2) = \cdots = f(a_r) = 0$, and $f(\mathbf{Z}) \subset \mathbf{Z}$. There are integers $b_j \in \mathbf{Z}$ with $f = \sum_{j\geq 0} b_j \binom{T}{j}$, cf. [St1], p. 16. Denote then by $\widetilde{f}$ the element $\sum_{j\geq 0} b_j \delta_j \in \operatorname{Dist}(G_m)$. If we apply this construction to $\{a_1, a_2, \dots, a_r\} = \{i \in \mathbf{Z} \mid m_i \neq 0\}$, then we get $\widetilde{f_i} \in \operatorname{Dist}(G_m)$ with $\widetilde{f_i}\, m = m_i$.

This shows for any $\operatorname{Dist}(G_m)$–submodule N of M that $N = \bigoplus (N \cap M_i)$, hence that N is also a G_m–submodule. So the converse of 7.11(4) holds for $G = G_m$. Also the converse of 7.11(5), (6) is true in this case: We have for all $j \in \mathbf{Z}$

(2) $$M_j = \{\, m \in M \mid \delta_n\, m = \binom{j}{n} m \text{ for all } n \in \mathbf{N} \,\}.$$

Indeed, consider any m as on the right hand side. Take the m_i as above. Then $\binom{j}{n} m_i = \binom{i}{n} m_i$ for all $n \in \mathbf{N}$. For $i \neq j$ we take f as above with $f(i) = 1$ and $f(j) = 0$ and get $m_i = \widetilde{f}\, m_i = 0$. Hence $m \in M_j$.

Note that (2) implies that the $\mathrm{Dist}(G_m)$–module structure on M determines the G_a–module structure, especially that we have equality in 7.11(3) for $G = G_m$.

In general, not every locally finite $\mathrm{Dist}(G_m)$–module arises from a G_m–module. If k is a field of characteristic 0 and if $a \in k$, then we make k into a $\mathrm{Dist}(G_m)$–module on k letting any δ_i act as multiplication by $\binom{a}{i}$. For $a \notin \mathbf{Z}$ this structure does not come from G_m–module. If k is a field of characteristic $p \neq 0$, then one can make a similar construction with a p–adic integer a.

7.14. Lemma: *Let G be an infinitesimally flat, noetherian, and integral group scheme over k. If M is a G–module that is projective over k, then for all $\lambda \in X(G)$*

$$M_\lambda = \{\, m \in M \mid \mu m = \mu(\lambda)\, m \text{ for all } \mu \in \mathrm{Dist}(G) \,\}.$$

Proof: Note: for each $x \in M \otimes k[G]$ with $x \notin M \otimes I_1^{n+1}$ for some n, there exist $\mu \in \mathrm{Dist}_n(G)$ with $(\mathrm{id}_M \otimes \mu)(x) \neq 0$. (Use embeddings of M and $k[G]/I_1^{n+1}$ as direct summands into free k–modules.)

If now $\mu m = \mu(\lambda) m$ for all $\mu \in \mathrm{Dist}(G)$, then $(\mathrm{id}_M \otimes \mu)(\Delta_M(m) - m \otimes \lambda) = 0$ for all μ, hence $\Delta_M(m) - m \otimes \lambda \in M \otimes I_1^{n+1}$ for all n by the argument above, hence $\Delta_M(m) - m \otimes \lambda \in \bigcap_{n>0}(M \otimes I_1^{n+1}) = M \otimes \bigcap_{n>0} I_1^{n+1}$. (Use a split embedding M into a free k–module for the last equality.) Now Krull's intersection theorem shows that the last term is 0, hence $\Delta_M(m) = m \otimes \lambda$ and $m \in M_\lambda$.

7.15. Lemma: *Let G be an infinitesimally flat, noetherian, and integral group scheme over k. Let M be a G–module and M' a k–submodule of M such that M/M' is projective over k. Then M' is a G–submodule of M if and only if it is a* $\mathrm{Dist}(G)$*–submodule.*

Proof: As M/M' is projective, the k–submodule M' is a direct summand of M and we can identify $M' \otimes k[G]$ with the kernel of $M \otimes k[G] \to (M/M') \otimes k[G]$. We have to show: If M' is a $\mathrm{Dist}(G)$–submodule, then $\Delta_M(M') \subset M' \otimes k[G]$, i.e., the image N of $\Delta_M(M')$ in $(M/M') \otimes k[G]$ is 0. Now $\mathrm{Dist}(G)M' \subset M'$ is equivalent to $(\mathrm{id}_M \overline{\otimes}\, \mu)\Delta_M(M') \subset M'$ for all $\mu \in \mathrm{Dist}(G)$, hence implies $(\mathrm{id}_{M/M'} \overline{\otimes}\, \mu)N = 0$. As in the preceding proof this implies

$$N \subset \bigcap_{n>0} (M/M') \otimes I_1^{n+1} = M/M' \otimes \bigcap_{n>0} I_1^{n+1} = 0,$$

hence the lemma.

7.16. Lemma: *Let G be an infinitesimally flat, noetherian, and integral group scheme over k. Then one has for all G–modules M, M' such that M' is projective over k:*

$$\mathrm{Hom}_G(M, M') = \mathrm{Hom}_{\mathrm{Dist}(G)}(M, M').$$

Proof: For any $\varphi \in \mathrm{Hom}(M, M')$ set

$$\nu(\varphi) = \Delta_{M'} \circ \varphi - (\varphi \otimes \mathrm{id}_{k[G]}) \circ \Delta_M .$$

So $\nu : \mathrm{Hom}(M, M') \to \mathrm{Hom}(M, M' \otimes k[G])$ is linear and has kernel equal to $\mathrm{Hom}_G(M, M')$, see 2.8(4). On the other hand, if $\varphi \in \mathrm{Hom}_{\mathrm{Dist}(G)}(M, M')$, then $(\mathrm{id}_{M'} \overline{\otimes}\, \mu)\, \nu(\varphi) = 0$ for all $\mu \in \mathrm{Dist}(G)$, hence (as in the proof of 7.14) for all $m \in M$

$$\nu(\varphi)\, m \in \bigcap_{n>0} M' \otimes I_1^{n+1} = 0.$$

7.17. (The Case of a Ground Field) Let us assume in this section that k is a field and that G is an algebraic k–group. One knows (cf. [DG], II, §5, 2.1):

(1) G *smooth* $\iff \dim G = \dim \mathrm{Lie}(G)$

and (cf. [DG], II, §6, 1.1):

(2) *If* $\mathrm{char}(k) = 0$, *then* G *is smooth.*

In the situation of 7.11(10), (11) we get therefore (using the notations as there)

(3) $Z_G(S)$ *smooth* $\iff \dim Z_G(S) = \dim\{\mu \in \mathrm{Lie}(G) \mid \mu m = 0$ *for all* $m \in S\}$,

(4) $\mathrm{Stab}_G(N)$ *smooth* $\iff \dim \mathrm{Stab}_G(S) = \dim \mathrm{Stab}_{\mathrm{Lie}(G)}(N)$.

If G acts on a scheme X and if $x \in X(k)$, then the morphism $\varphi : G \to X$, $g \mapsto g\,x$ satisfies $\varphi^{-1}(\varphi(1)) = \mathrm{Stab}_G(x)$. So 7.2(7) yields:

(5) $\mathrm{Stab}_G(x)$ *smooth* $\iff \dim \mathrm{Stab}_G(x) = \dim \ker(d\varphi)_1$.

Recall that an affine scheme Y is called *irreducible* if and only if $\sqrt{0}$ is a prime ideal in $k[Y]$. It is integral if and only if it is irreducible and reduced. In case $\mathrm{char}(k) = 0$, then (2) implies that G is integral if and only if it is irreducible as any smooth scheme is reduced.

Suppose now that k is a perfect field of characteristic $p > 0$. If G is irreducible, then there is by [DG], III, §3, 6.4 an isomorphism $G \simeq X \times Y$ of affine schemes with Y integral, and where $k[X]$ is a finite dimensional local k–algebra. The only maximal ideal of $k[X]$ is nilpotent. This implies that we have $\bigcap_{n>0} I_1^{n+1} = 0$ in $k[G]$. Now the proofs of the preceding lemmas show:

(6) *Suppose that k is a perfect field. Then the claims of* 7.14–7.16 *hold for any irreducible algebraic k–group.*

Similarly, the arguments in 7.5 yield now:

(7) *Suppose that k is a perfect field. Let G be an algebraic k–group and H, H' closed subgroups of G. If H is irreducible, then*

$$H \subset H' \iff \mathrm{Dist}(H) \subset \mathrm{Dist}(H').$$

Let K be an algebraic closure of k. If we no longer assume that H is irreducible, then we can still say:

(8) $H \subset H' \iff H(K) \subset H'(K)$ *and* $\mathrm{Dist}(H) \subset \mathrm{Dist}(H')$.

Indeed, we have to prove only one direction ("$\Leftarrow$"). Suppose $H = V(I)$ and $H' = V(I')$. As $I \supset I'$ if and only if $I \otimes K \supset I' \otimes K$ (and similarly for $\mathrm{Dist}(H)$, $\mathrm{Dist}(H')$), we may assume $k = K$. Decompose $I = \bigcap_{j=0}^{r} I_j$ such that the $V(I_j)$ are the irreducible components of H and $1 \in V(I_0)(k)$. If $\mathrm{Dist}(H) \subset \mathrm{Dist}(H')$, then (7) yields $I_0 \supset I'$. For any j we can choose $g_j \in V(I_j)(k)$ as $k = K$. Then $\rho_l(g_j)I_0 = I_j$. If $H(k) \subset H'(k)$, then

$$I' = \rho_l(g_j)I' \subset \rho_l(g_j)I_0 = I_j$$

for all j, hence $I' \subset \bigcap_{j=0}^{r} I_j = I$ and $H' \supset H$.

7.18. (Distributions as Differential Operators) Let G be a group scheme over k. Any action of G on an affine scheme X leads (cf. 2.7) to a representation of G on $k[X]$, hence makes $k[X]$ into a Dist(G)–module. When dealing with a right action $\alpha : X \times G \to X$ (resp. a left action $\beta : G \times X \to X$) then the action of $\mu \in \mathrm{Dist}(G)$ on $k[X]$ is given by $(\mathrm{id}_{k[X]} \overline{\otimes} \mu) \circ \alpha^*$ (resp. $(\sigma'_G(\mu) \overline{\otimes} \mathrm{id}_{k[X]}) \circ \beta^*$).

There is a notion of *differential operators* on a scheme, cf. [DG], II, §4, 5.3. In the case of an affine scheme X they can be described as follows ([DG], II, §4, 5.7): Each $f \in k[X]$ defines $\mathrm{ad}(f) : \mathrm{End}(k[X]) \to \mathrm{End}(k[X])$ through $(\mathrm{ad}(f)\varphi)(f_1) = f\varphi(f_1) - \varphi(ff_1)$, in other words, $\mathrm{ad}(f)\varphi$ is the commutator of the left multiplication by f and of φ. Then a differential operator on X of order $\leq n$ is some $D \in \mathrm{End}(k[X])$ with $\mathrm{ad}(f_0)\,\mathrm{ad}(f_1)\dots\mathrm{ad}(f_n)\,D = 0$ for all $f_0, f_1, \dots, f_n \in k[X]$. A differential operator on X is then defined as a differential operator of order $\leq n$ for some $n \in \mathbf{N}$. The differential operators on X form a subalgebra of $\mathrm{End}(k[X])$.

For G acting on X as above, any $\mu \in \mathrm{Dist}_n(G)$ acts on $k[X]$ as a differential operator of order $\leq n$; this follows from an elementary argument, cf. [DG], II, §4, 6.3.

When dealing with the action of G on itself by left resp. right translation, then we get get an action of any $\mu \in \mathrm{Dist}(G)$ as a differential operator on G that commutes with the action of G by multiplication on the other side. This construction turns out to yield an isomorphism of Dist(G) onto the algebra of all differential operators on G that are right resp. left invariant (i.e., that commute with the action of G by right resp. left translation), cf. [DG], II, §4, 6.5.

The conjugation action of G on itself yields a representation of G on $k[G]$ that stabilises I_1, hence also all I_1^{n+1}. We get thus G–structures on all $k[G]/I_1^{n+1}$, hence also on all $\mathrm{Dist}_n(G) = (k[G]/I_1^{n+1})^*$ — provided that G is infinitesimally flat. If so, then we also get a representation of G on the direct limit Dist(G). The representation of G on $\mathrm{Lie}(G) = \mathrm{Dist}_1^+(G)$ constructed thus is the *adjoint* representation of G. We use the notation "Ad" for the representation of G on Dist(G) and all $\mathrm{Dist}_n(G)$, $\mathrm{Dist}_n^+(G)$, and the notation "ad" for the corresponding actions of Dist(G) in itself and its submodules.

Suppose that G is infinitesimally flat. An elementary calculation shows that the adjoint representation on Dist(G) and the action of Dist(G) on any G–module M are related by the formula

$$(1) \qquad (\mathrm{Ad}(g)\varphi)m = g(\varphi(g^{-1}m)),$$

for all $\varphi \in \mathrm{Dist}(G) \otimes A \simeq \mathrm{Dist}(G_A)$, $m \in M \otimes A$, $g \in G(A)$, and all A.

Let us write down explicitly how any $\mu \in \mathrm{Dist}(G)$ acts on $k[G]$ and Dist(G) under the conjugation resp. adjoint action (for G infinitesimally flat). Suppose $\Delta'_G(\mu) = \sum_i \mu_i \otimes \mu'_i$. Then, because of 2.8(7), the conjugation action of μ is given by

$$(2) \qquad \sum_i (\sigma'_G(\mu_i) \overline{\otimes}\, \mathrm{id}_{k[G]} \overline{\otimes}\, \mu'_i) \circ (\mathrm{id}_{k[G]} \otimes \Delta_G) \circ \Delta_G.$$

As $\Delta'_G \circ \sigma'_G = (\sigma'_G \otimes \sigma'_G) \circ \Delta'_G$, the adjoint action is given by (using 7.11(8) and 7.7(1))

$$(3) \qquad \mathrm{ad}(\mu)\mu' = \sum_i (\mu_i \overline{\otimes} \mu' \overline{\otimes} \sigma'_G(\mu'_i)) \circ (\mathrm{id}_{k[G]} \otimes \Delta_G) \circ \Delta_G = \sum_i \mu_i \mu' \sigma'_G(\mu'_i).$$

Suppose now that k is a field and that G is algebraic. For any closed subgroup scheme H of G, its normaliser $N_G(H)$ and its centraliser $C_G(H)$ are closed subgroup schemes of G, cf. 2.6(8), (9). One has obviously $N_G(H) \subset \mathrm{Stab}_G(\mathrm{Lie}\, H)$ and $C_G(H) \subset \mathrm{Cent}_G(\mathrm{Lie}\, H)$, hence (cf. 7.11(10), (11)) $\mathrm{Lie}\, N_G(H) \subset \mathrm{Stab}_{\mathrm{Lie}(G)}(\mathrm{Lie}\, H)$, the normaliser of $\mathrm{Lie}(H)$ in $\mathrm{Lie}(G)$, and $\mathrm{Lie}\, C_G(H) \subset \mathrm{Cent}_{\mathrm{Lie}(G)}(\mathrm{Lie}\, H)$. So 7.17(1) implies:

(4) *If* $\dim N_G(H) = \dim \mathrm{Stab}_{\mathrm{Lie}(G)}(\mathrm{Lie}\, H)$, *then* $N_G(H)$ *is smooth.*

(5) *If* $\dim C_G(H) = \dim \mathrm{Cent}_{\mathrm{Lie}(G)}(\mathrm{Lie}\, H)$, *then* $C_G(H)$ *is smooth.*

Of course, one should expect this condition to work only when H is irreducible. One can show (cf. [DG], II, §5, 5.7) that $\mathrm{Lie}(C_G(H)) = (\mathrm{Lie}\, G)^H$ and that

$$\mathrm{Lie}(N_G(H))/\mathrm{Lie}(H)) = (\mathrm{Lie}(G)/\mathrm{Lie}(H))^H.$$

7.19. For any family $(X_j)_{j\in J}$ of subfunctors of a group scheme G there is a smallest closed subgroup scheme H of G containing all X_j. (Take the intersection of all closed subgroup schemes containing all X_j.) We call H the closed subgroup of G generated by all X_j.

Proposition: *Suppose that k is an algebraically closed field. Let G be an algebraic k–group and let $(H_j)_{j\in J}$ be a family of integral, closed subgroups of G. Let H be the closed subgroup of G generated by all H_j. Then H is integral and* $\mathrm{Dist}(H)$ *is the subalgebra of* $\mathrm{Dist}(G)$ *generated by all* $\mathrm{Dist}(H_j)$.

Proof: The reduced subgroup of G defined by $H(k)$ contains all H_j, hence H is reduced. We can assume (by [DG], II, §5, 4.6 or [Bo], 2.2) that $(H_j)_{j\in J} = \{H_1, H_2, \dots, H_r\}$ and that the multiplication map $\alpha : H_1 \times H_2 \times \cdots \times H_r \to H$ is surjective on points over k. This implies that H is irreducible, hence integral. Furthermore, the theorem of generic flatness ([DG], I, §3, 3.7) provides us with a point over k where α is flat, hence $d\alpha$, by 7.6, surjective on the distributions with support in that point. As $d\alpha$ in $(1, 1, \dots, 1)$ is multiplication, the same argument as in [Bo], 7.5 yields

(1) $$\mathrm{Dist}(H) = (\mathrm{Ad}(h_1)\, \mathrm{Dist}(H_1))(\mathrm{Ad}(h_2)\, \mathrm{Dist}(H_2)) \dots (\mathrm{Ad}(h_r)\, \mathrm{Dist}(H_r))$$

for suitable $h_1, h_2, \dots, h_r \in H(k)$.

Let R be the subalgebra of $\mathrm{Dist}(G)$ generated by all $\mathrm{Dist}(H_i)$. As $H_i \subset H$ for all i, also $R \subset \mathrm{Dist}(H)$. Because of (1) we have to show that R is stable under all $\mathrm{Ad}(h)$ with $h \in H(k)$, or, by the surjectivity of $\alpha(k)$, that R is an H_i–submodule of $\mathrm{Dist}(G)$ for all i. By 7.15 it is enough to show stability under each $\mathrm{Dist}(H_i)$ for the adjoint action. This is now clear by 7.18(3) as $\Delta'_G(\mathrm{Dist}(H_i)) \subset \mathrm{Dist}(H_i) \otimes \mathrm{Dist}(H_i)$ and $\sigma'_G(\mathrm{Dist}(H_i)) = \mathrm{Dist}(H_i)$ for all i. Indeed, Δ'_G and σ'_G restrict to Δ'_{H_i} resp. σ'_{H_i} on $\mathrm{Dist}(H_i)$.

Remarks: 1) There is another proof in [Y], 10.10. The proof above follows the one in [Bo], 7.6 that $\mathrm{Lie}(H)$ is generated as a Lie algebra by all $\mathrm{Lie}(H_i)$ provided $\mathrm{char}(k) = 0$.

2) Drop the assumption that k is algebraically closed. Let K be an algebraic closure of k. If each $(H_j)_K$ is still integral, then the claim of the proposition still holds: We get from [Bo], 2.2 that H_K is the closed subgroup generated by all $(H_j)_K$. With

R as in the proof, we get $R \otimes K = \mathrm{Dist}(H_K) = \mathrm{Dist}(H) \otimes K$, hence $R = \mathrm{Dist}(H)$ using 7.4(1).

Now $(H_j)_K$ is integral if and only if it is reduced, cf. [DG], II, §5, 1.1. This will certainly be satisfied if k is perfect, cf. [Bo], AG 2.2.

CHAPTER 8

Representations of Finite Algebraic Groups

Let us suppose throughout this chapter that k is a field.

A k–group scheme G is called a *finite algebraic group* if $\dim k[G] < \infty$. We have met already some examples: $\mu_{(n)}$, $G_{a,r}$. One can associate to each finite abstract group a finite algebraic group in a natural way (8.5.a). The examples that are most important for us will be introduced in Chapter 9: the Frobenius kernels.

In this chapter we look at some special features of the representation theory of such finite G. Let me mention right away that one can find in [Voigt 2] many more results that we do not look at here.

One of the special features is that injective G–modules are also projective as in the representation theory of abstract finite groups. Whereas in that case (abstract finite groups) the injective hull of a simple module is also its projective cover, this is no longer true in our situation (in general). Here the simple head and the simple socle of an injective indecomposable module differ by a character of G that we call the modular function of G (8.13).

Another special feature is seen when dealing with a closed subgroup H of G. We have not only the right adjoint ind_H^G to the restriction functor res_H^G but also a left adjoint coind_H^G, the coinduction. Both functors are exact and they are related by dualising (8.14–8.16). In fact, one can get from one to the other by first tensoring with a character depending on the modular functions of H and G (8.17).

One main ingredient in the proof of these results is the fact that $k[G]$ and $k[G]^*$ are isomorphic as G–modules (8.7 and 8.12). This is a special case of a more general theorem of Larson and Sweedler (cf. [Sw]). As a source for the other nontrivial results let me mention [Oberst and Schneider] and [Voigt 1].

When working not over a field but over an arbitrary commutative ring (say R), then one should define a finite algebraic group over R as an R–group scheme such that $R[G]$ is finitely generated and *projective* as an R–module. It is elementary how to generalise 8.1–8.6 to this more general situation. For an extension of 8.12 and 8.17 to this situation one may consult [Voigt 1], cf. also [Oberst and Schneider].

8.1. (Finite Algebraic Groups) A k–group scheme G is called *finite* (hence: a finite algebraic group) if $\dim k[G] < \infty$. It is called *infinitesimal* if it is finite and if the ideal $I_1 = \{f \in k[G] \mid f(1) = 0\}$ is nilpotent.

If k' is an extension field of k, then obviously G is finite (resp. infinitesimal) if and only if $G_{k'}$ is so.

The closed subgroups $G_{a,r}$ of the additive group (cf. the end of 2.2) are infinitesimal groups. They are examples of Frobenius kernels of reduced groups, the (for us) most important class of infinitesimal groups, which will be introduced in Chapter 9.

The groups $\mu_{(n)}$ for each $n \in \mathbf{N}$ are finite (cf. 2.2). If $\mathrm{char}(k) = p \neq 0$ and if n is a power of p, then $\mu_{(n)}$ is infinitesimal.

8.2. Lemma: *Let G be an algebraic k–group.*

a) *G is finite if and only if $G(K)$ is finite for each extension field K of k.*

b) *G is infinitesimal if and only if $G(K) = 1$ for each extension field K of k.*

Proof: a) If $\dim k[G] < \infty$, then each element in $k[G]$ is algebraic over k, hence has only a finite number of possible images in any K (under an element of $G(K) = \mathrm{Hom}_{k-\mathrm{alg}}(k[G], K)$). As any $g \in G(K)$ is given by its values on a basis of $k[G]$, there are only finitely many possibilities for g.

Consider on the other hand an algebraic closure K of k, and suppose that $G(K)$ is finite. We can replace G by G_K, hence suppose $k = K$. We can write $k[G]$ in the form $k[T_1, T_2, \dots, T_n]/I$ for some ideal I. Then any prime ideal containing I has to be a maximal ideal. The same is true for any associated prime ideal of I. This easily implies that $\dim k[G] = \dim k[T_1, T_2, \dots, T_n]/I < \infty$.

b) If I_1 is nilpotent, then it has to be annihilated by any homomorphism of k–algebras $k[G] \to K$ into an extension field. As $k[G] = k1 \oplus I_1$, there is only one such homomorphism, hence $G(K) = 1$.

Suppose on the other hand $G(K) = 1$ for an algebraic closure K of k. We may assume $k = K$ and can identify $k[G]/\sqrt{0}$ with an algebra of functions from $G(K)$ to K. This implies $I_1 = \sqrt{0}$, hence that I_1 is nilpotent.

8.3. (Duality of Finite Dimensional Hopf Algebras) For any finite dimensional vector space V (over k) the canonical map $V \to (V^*)^*$ is an isomorphism. Mapping any linear map $\varphi : V_1 \to V_2$ between two finite dimensional vector spaces to its transposed map $\varphi^* : V_2^* \to V_1^*$ is therefore a bijection $\mathrm{Hom}(V_1, V_2) \xrightarrow{\sim} \mathrm{Hom}(V_2^*, V_1^*)$.

Let R be a finite dimensional vector space over k. We get from above isomorphisms $\mathrm{Hom}(k, R) \xrightarrow{\sim} \mathrm{Hom}(R^*, k)$ and $\mathrm{End}(R) \xrightarrow{\sim} \mathrm{End}(R^*)$ and $\mathrm{Hom}(R \otimes R, R) \xrightarrow{\sim} \mathrm{Hom}(R^*, R^* \otimes R^*)$ using the isomorphism $R^* \otimes R^* \xrightarrow{\sim} (R \otimes R)^*$. So multiplications on R (i.e., bilinear maps $R \times R \to R$ or, equivalently, linear maps $m : R \otimes R \to R$) correspond bijectively to comultiplications on R^* (i.e., linear maps $m^* : R^* \to R^* \otimes R^*$). Similarly, comultiplications $\Delta : R \to R \otimes R$ correspond bijectively to multiplications $\Delta^* : R^* \otimes R^* \to R^*$. Furthermore m is associative (resp. Δ is coassociative, i.e., satisfies 2.3(1)) if and only if m^* is coassociative (resp. Δ^* is associative). An element $a \in R$ is a 1 for the multiplication m if and only if the map $\varepsilon_a : R^* \to k$, $\varphi \mapsto \varphi(a)$ is a counit for m^* (i.e., satisfies 2.3(2) with the appropriate modifications in the notation). Similarly, $\varepsilon \in R^*$ is a counit for Δ if and only if it is a 1 for Δ^*.

If we have on R both a multiplication m and a comultiplication Δ, then Δ is a homomorphism of algebras (with respect to m) if and only if m^* is a homomorphism of algebras (with respect to Δ^*). If so, then some $\sigma \in \mathrm{End}(R)$ is an antipode for Δ and m (i.e., satisfies 2.3(3) and $\sigma(ab) = \sigma(b)\sigma(a)$ for all $a, b \in R$) if and only if σ^* is an antipode for m^* and Δ^*. This shows: If R is a Hopf algebra, then so is R^* is a natural way. For two such finite dimensional Hopf algebras R_1, R_2, a linear map $\psi : R_1 \to R_2$ is a homomorphism of Hopf algebras if and only if $\psi^* : R_2^* \to R_1^*$ is a homomorphism of Hopf algebras. Thus we get:

(1) *The functor $R \mapsto R^*$, $\psi \mapsto \psi^*$ is a self-duality on the category of all finite dimensional Hopf algebras.*

This anti-equivalence obviously has the property that R is commutative if and only if R^* is cocommutative (cf. 2.3).

8.4. (Finite Algebraic Groups and Hopf Algebras) We have by 2.3 an anti-equivalence of categories $\{$ group schemes over $k \} \to \{$ commutative Hopf algebras over $k \}$. Combining this with 8.3(1) we get an equivalence of categories:

(1) $\{$ *finite algebraic k–groups* $\} \to \{$ *finite dimensional cocommutative Hopf algebras over k* $\}$.

Each finite algebraic k–group G is mapped to $k[G]^*$. We denote this Hopf algebra by $M(G)$ and call it the algebra of all *measures* on G. We usually denote its comultiplication by Δ'_G, its antipode by $\sigma'_G = \sigma^*_G$, and its counit by $\varepsilon'_G : \mu \mapsto \mu(1)$.

We have an obvious embedding

$$G(k) = \mathrm{Hom}_{k-\mathrm{alg}}(k[G], k) \hookrightarrow M(G) = \mathrm{Hom}(k[G], k)$$

where each $g \in G(k)$ is mapped to the (Dirac) measure $\delta_g : f \mapsto f(g)$. An element $\mu \in M(G) = k[G]^*$ is a homomorphism of algebras if and only if $\Delta'_G(\mu) = \mu \otimes \mu$ and $\varepsilon'_G(\mu) = 1$. The multiplication on $G(k)$ is just the multiplication in $M(G)$. More generally, one can identify

$$G(A) = \mathrm{Hom}_{k-\mathrm{alg}}(k[G], A) \hookrightarrow \mathrm{Hom}(k[G], A) \simeq k[G]^* \otimes A = M(G) \otimes A$$

for any k–algebra A with

$$\{ \mu \in M(G) \otimes A \mid (\Delta'_G \otimes \mathrm{id}_A)(\mu) = \mu \otimes \mu,\ \varepsilon'_G(\mu) = 1 \}.$$

In Chapter 7 we have associated to each group scheme G the algebra $\mathrm{Dist}(G)$, cf. 7.1 and 7.7. If G is finite, then obviously $\mathrm{Dist}(G)$ is a subalgebra of $M(G)$, and G is infinitesimal if and only if $M(G) = \mathrm{Dist}(G)$. One easily checks that

(2) $$\mathrm{Lie}(G) = \mathrm{Dist}_1^+(G) = \{ \mu \in M(G) \mid \Delta'_G(\mu) = \mu \otimes 1 + 1 \otimes \mu \}.$$

8.5. (Examples) a) If Γ is an abstract finite group, then its group algebra $k\Gamma$ is a cocommutative Hopf algebra in a natural way. Considered as a vector space $k\Gamma$ has a basis that we can identify with Γ. These basis elements multiply as in Γ and the comultiplication is defined via $\gamma \mapsto \gamma \otimes \gamma$, the counit via $\gamma \mapsto 1$, and the antipode via $\gamma \mapsto \gamma^{-1}$ for all $\gamma \in \Gamma$. Hence there is a finite algebraic group G with $M(G) \simeq k\Gamma$. For any k–algebra A the group $G(A)$ can be identified with the set of all $\sum_{\gamma \in \Gamma} a_\gamma \gamma \in A\Gamma \simeq k\Gamma \otimes A$ with $\sum_{\gamma \in \Gamma} a_\gamma (\gamma \otimes \gamma) = \sum_{\gamma, \gamma' \in \Gamma} a_\gamma a_{\gamma'} (\gamma \otimes \gamma')$ and $\sum_{\gamma \in \Gamma} a_\gamma = 1$. If A is an integral domain (or, more generally, has no idempotents $\neq 0, 1$), then $G(A) \simeq \Gamma$. (This construction can obviously be carried out over any ring, not only over a field.)

b) Suppose that $\mathrm{char}(k) = p \neq 0$ and let $\mathfrak{g}$ be a finite dimensional p–Lie algebra, cf. 7.10. Then its restricted enveloping algebra $U^{[p]}(\mathfrak{g})$ is a cocommutative Hopf algebra. Any $x \in \mathfrak{g}$ is mapped to $x \otimes 1 + 1 \otimes x$ under the comultiplication, to 0 under the counit, and to $-x$ under the antipode. So there is a finite algebraic group G with $M(G) \simeq U^{[p]}(\mathfrak{g})$. One obviously gets $\mathfrak{g} \subset \mathrm{Lie}(G)$ from 8.4(2). The embedding of $U^{[p]}(\mathrm{Lie}\, G)$ into $\mathrm{Dist}(G) \subset M(G) \simeq U^{[p]}(\mathfrak{g})$ (cf. 7.10(2)) therefore has to be an isomorphism. We get $\mathrm{Lie}(G) = \mathfrak{g}$ and $M(G) = \mathrm{Dist}(G)$ so that G is infinitesimal. See [DG], II, §7, 3.9–3.12 for more details.

8.6. (Modules for G and $M(G)$) Let R be a finite dimensional Hopf algebra. If M is an R–module, then M is an R^*–comodule in a natural way: Define the comodule map $M \to M \otimes R^* \simeq \mathrm{Hom}(R, M)$ by mapping m to $a \mapsto am$. If M is an R–comodule, then M is an R^*–module in a natural way: Define the action of any $\mu \in R^*$ as $(\mathrm{id}_M \overline{\otimes} \mu) \circ \Delta_M$ if Δ_M is the comodule map $M \to M \otimes R$. For two such comodules M_1, M_2 a linear map $\psi : M_1 \to M_2$ is a homomorphism of R–comodules if and only if it is a homomorphism of R^*–modules. In this way we get an equivalence of categories

(1) $\{\, R\text{–}comodules \,\} \xrightarrow{\sim} \{\, R^*\text{–}modules \,\}$.

Let G be a finite algebraic k–group. Then the categories of G–modules and of $k[G]$–comodules are equivalent by 2.8. Combining this with (1) we get an equivalence of categories

(2) $\{\, G\text{–}modules \,\} \xrightarrow{\sim} \{\, M(G)\text{–}modules \,\}$.

This equivalence takes a G–module M to the $M(G)$–module M with $\mu \in M(G)$ acting as $(\mathrm{id}_M \overline{\otimes} \mu) \circ \Delta_M$. We recover the action of $G(k)$ via the embedding $G(k) \subset M(G)^\times$ and, more generally, the action of any $G(A)$ via the embedding $G(A) \subset (M(G) \otimes A)^\times$ and the action of $M(G) \otimes A$ on $M \otimes A$.

It is clear that we get on $\mathrm{Dist}(G) \subset M(G)$ the same action as in 7.11. Furthermore, all the statements in 7.11 generalise to $M(G)$. The claims in 7.14–7.17 obviously hold for any finite algebraic group with $\mathrm{Dist}(G)$ replaced by $M(G)$.

The representations of G on $k[G]$ via ρ_l and ρ_r lead to two (contragredient) representations of G on $M(G)$, hence to two structures as an $M(G)$–module on $M(G)$. Using the generalisation of 7.11(8) one checks that any $\mu \in M(G)$ acts on $M(G)$ as left multiplication by μ when we deal with ρ_l, and as right multiplication by $\sigma'_G(\mu)$ when we deal with ρ_r.

For G corresponding to a finite abstract group Γ as in 8.5.a the theory of G–modules is the same as that of $k\Gamma$–modules, hence equal to the representation theory of Γ over k.

For G corresponding to a p–Lie algebra $\mathfrak{g}$ as in 8.5.b the theory of G–modules is the same as that of $U^{[p]}(\mathfrak{g})$–modules, hence equal to the representation theory of $\mathfrak{g}$ considered as a p–Lie algebra.

8.7. From now on until the end of this chapter let G be a finite algebraic k–group.

We denote $k[G]$ resp. $M(G)$ regarded as a G–module with respect to ρ_l resp. the contragredient representation of ρ_l by $k[G]_l$ and $M(G)_l$. We write $k[G]_r$ and $M(G)_r$ when we instead consider ρ_r and its contragredient representation. Recall from 2.7 that $k[G]_l$ is isomorphic to $k[G]_r$; so there is also an isomorphism $M(G)_r \simeq M(G)_l$. We drop sometimes the index "l" or "r" when it does not matter which of the two structures we consider.

Lemma: *The G-modules $M(G)$ and $k[G]$ are isomorphic to each other; we have* $\dim M(G)^G = 1$.

Proof: By 3.7(4) we have $M(G) \otimes k[G] \simeq k[G]^n$ where $n = \dim k[G]$. On the other hand, $M(G) \otimes k[G] \simeq k[G]^* \otimes k[G]$ is self-dual as a G–module, hence also isomorphic to $(k[G]^*)^n$. Using the Krull-Schmidt theorem about unique decomposition into (finite dimensional) indecomposable modules we deduce from $k[G]^n \simeq (k[G]^*)^n$ that $k[G] \simeq k[G]^* = M(G)$. The claim on $M(G)^G$ follows now from 2.10(5).

8.8. (Invariant Measures) We call an element in $M(G)_l^G$ (resp. $M(G)_r^G$) a *left* (resp. *right*) *invariant measure* on G. (In [Sw] such elements are called "integrals", in [Haboush 3] "norm forms".)

The description of the left and right regular representations of $M(G)$ on itself in 8.6 implies

(1) $$M(G)_l^G = \{\, \mu_0 \in M(G) \mid \mu\mu_0 = \mu(1)\mu_0 \text{ for all } \mu \in M(G) \,\}$$

and

(2) $$M(G)_r^G = \{\, \mu_0 \in M(G) \mid \mu_0\mu = \mu(1)\mu_0 \text{ for all } \mu \in M(G) \,\}$$

as $\sigma'_G(\mu)(1) = \mu(1)$ for all $\mu \in M(G)$. Furthermore, we have

(3) $$\sigma'_G(M(G)_r^G) = M(G)_l^G$$

as σ'_G intertwines the left and right regular representations (or, using (1) and (2), as it is an anti-automorphism of $M(G)$ considered as an algebra).

Obviously $M(G)_l^G$ is stable under right multiplication by elements of $M(G)$, hence an $M(G)$– and G–submodule of $M(G)$ with respect to the right regular representation. (This can also be seen directly.) As $\dim M(G)_l^G = 1$ the representation of G on $M(G)_l^G$ is given by some $\delta_G \in X(G) \subset k[G]$. So one has for all $g \in G(A)$ and all A

(4) $$\rho_r(g)(\mu_0 \otimes 1) = \mu_0 \otimes \delta_G(g) \qquad \text{for all } \mu_0 \in M(G)_l^G,$$

and, equivalently, for all $\mu \in M(G)$

(5) $$\mu_0\mu = \sigma'_G(\mu)(\delta_G)\mu_0 = \mu(\delta_G^{-1})\mu_0 \qquad \text{for all } \mu_0 \in M(G)_l^G.$$

(Observe that $\sigma_G(\chi) = \chi^{-1}$ for all $\chi \in X(G)$.) This character δ_G is called the *modular function* of G. We call G *unimodular* if $\delta_G = 1$. (In the examples in 8.9 each G will be unimodular. We shall meet a case where $\delta_G \neq 1$ later on in II.3.4.)

We could have defined δ_G also via $M(G)_r^G$ as (3) implies for all $\mu \in M(G)$

(6) $$\mu\mu_0 = \mu(\delta_G)\mu_0 \qquad \text{for all } \mu_0 \in M(G)_r^G.$$

and, equivalently, for all $g \in G(A)$ and all A

(7) $$\rho_l(g)(\mu_0 \otimes 1) = \mu_0 \otimes \delta_G(g) \qquad \text{for all } \mu_0 \in M(G)_r^G.$$

8.9. (Examples) If G corresponds to an abstract finite group Γ as in 8.5.a, then

(1) $$M(G)_l^G = M(G)_r^G = k\sum_{\gamma\in\Gamma} \gamma.$$

Consider as another example $G = G_{a,r}$ with $r \in \mathbf{N}$, $r > 0$ assuming $\mathrm{char}(k) = p \neq 0$. Set $q = p^r$. As $G_{a,r}$ is a subgroup of $G_a = Sp_k k[T]$, we can identify $M(G_{a,r}) = \mathrm{Dist}(G_{a,r})$ with the subalgebra of $\mathrm{Dist}(G_a)$ spanned by all μ with $\mu(T^{q+i}) = 0$ for all $i \geq 0$. Using the basis $(\gamma_n)_{n\in\mathbf{N}}$ of $\mathrm{Dist}(G_a)$ as in 7.8 we get

$$M(G_{a,r}) = \sum_{n=0}^{q-1} k\,\gamma_n.$$

As $\gamma_0(1) = 1$ and $\gamma_n(1) = 0$ for $n > 0$, as $\gamma_n\gamma_{q-1} = \binom{q+n-1}{q-1}\gamma_{n+q-1} = 0$ for $0 < n \leq q-1$ and $\gamma_0\gamma_{q-1} = \gamma_{q-1}$, we see that γ_{q-1} is an invariant measure on $G_{a,r}$. Using dim $M(G)^G = 1$ (or some additional computations) we get

$$M(G)_l^G = M(G)_r^G = k\,\gamma_{q-1} \qquad \text{for } G = G_{a,r}. \tag{2}$$

Assume again char$(k) = p \neq 0$, let $r \in \mathbf{N}$, $r > 0$, and set $q = p^r$. Let us consider $G = \mu_{(q)}$ and determine $M(G)^G$. As $\mu_{(q)}$ is an infinitesimal and closed subgroup of G_m, we can identify $M(\mu_{(q)}) = \text{Dist}(\mu_{(q)})$ with a subalgebra of Dist(G_m). Let us use the notations of 7.8. Then $M(\mu_{(q)})$ consists of all $\nu \in \text{Dist}(G_m)$ with $\nu(T^i(T^q - 1)) = 0$ for all $i \in \mathbf{Z}$. Obviously

$$\delta_n(T^i(T^q - 1)) = \binom{q+i}{n} - \binom{i}{n}$$

for all $i \in \mathbf{Z}$. The standard formula for binomial coefficients mod p (cf., e.g., [Haboush 3], 5.1) shows $\delta_n(T^i(T^q - 1)) = 0$ for all $i \in \mathbf{Z}$ if $0 \leq n < q$. As dim $M(\mu_{(q)}) = \dim k[\mu_{(q)}] = q$, we get

$$M(\mu_{(q)}) = \sum_{n=0}^{q-1} k\,\delta_n.$$

We claim

$$M(G)_l^G = M(G)_r^G = k\sum_{i=0}^{q-1}(-1)^i\delta_i \qquad \text{for } G = \mu_{(q)}. \tag{3}$$

Set $\mu_0 = \sum_{i=0}^{q-1}(-1)^i\delta_i$. As δ_0 is the 1 in $M(G)$ and $\delta_0(1) = 1$, we have to show $\delta_n\mu_0 = 0$ for all n with $0 < n < q$. We have by 7.8(3)

$$\delta_n\mu_0 = \sum_{i=0}^{q-1}\sum_{j=0}^{\min(i,n)}(-1)^i\binom{n+i-j}{i-j}\binom{n}{j}\delta_{n+1-j}.$$

If $n+i-j > q-1$, then $\binom{n+i-j}{i-j} = 0$ and we can delete the corresponding summand. Substituting $s = i - j$ we get

$$\begin{aligned}\delta_n\mu_0 &= \sum_{i=0}^{q-1}(-1)^i\sum_{s=\max(0,i-n)}^{\min(i,q-1-n)}\binom{n+s}{s}\binom{n}{i-s}\delta_{n+s}\\ &= \sum_{s=0}^{q-1-n}\Big(\sum_{i=s}^{n+s}(-1)^i\binom{n}{i-s}\Big)\binom{n+s}{s}\delta_{n+s} = 0.\end{aligned}$$

8.10. (Projective and Injective Modules) Recall (3.18) that we call a projective object in the category of all G–modules a projective G–module. They correspond under the equivalence of categories to the projective $M(G)$–modules. This shows that each G–module is a homomorphic image of a projective G–module, hence that projective resolutions exist in the category of G–modules.

The representation theory of finite dimensional algebras shows that the indecomposable projective G–modules are (up to isomorphism) the indecomposable direct summands of $M(G)$. For each simple G–module E there is a unique (up to isomorphism) projective G–module Q with $Q/\operatorname{rad}(Q) \simeq E$. It is called the *projective cover* of E. In this way one gets a bijection between the isomorphism classes of simple G–modules and of indecomposable projective G–modules.

Now the isomorphism $M(G) \simeq k[G]$ from 8.7 together with 3.10 shows that a finite dimensional G–module is projective if and only if it is injective. (This follows of course also from 3.18.) The indecomposable injective G–modules are exactly the indecomposable projective G–modules. There is a bijection $E \mapsto E'$ on the set of isomorphism classes of simple G–modules such that the injective hull Q_E of E (cf. 3.16) is the projective cover of E', i.e.,

$$(1) \qquad Q_E/\operatorname{rad}(Q_E) \simeq E'.$$

We want to describe this bijection and have to be more precise about the isomorphism $M(G) \simeq k[G]$ first.

8.11. ($M(G)$ as a module over $k[G]$) There is a natural structure as a $k[G]$–module on $M(G)$: For any $f \in k[G]$ and $\mu \in M(G)$ we define $f\,\mu$ through

$$(1) \qquad (f\,\mu)(f_1) = \mu(f f_1) \qquad \text{for all } f_1 \in k[G].$$

The following properties follow from straightforward calculations that are left to the reader.

$$(2) \qquad f\varepsilon_G = f(1)\,\varepsilon_G \qquad \text{for all } f \in k[G],$$

$$(3) \qquad \sigma'_G(f\,\mu) = \sigma_G(f)\,\sigma'_G(\mu) \qquad \text{for all } f \in k[G],\ \mu \in M(G),$$

(4) *If $\mu_1, \mu_2 \in M(G)$ and $f \in k[G]$ with $\Delta_G(f) = \sum_{i=1}^r f_i \otimes f'_i$, then $f \cdot \mu_1\mu_2 = \sum_i (f_i\,\mu_1)(f'_i\,\mu_2)$.*

We have $\Delta_G(\chi) = \chi \otimes \chi$ and $\chi(1) = 1$ for all $\chi \in X(G) \subset k[G]$. Therefore (2) and (4) imply:

(5) *If $\chi \in X(G)$, then $\mu \mapsto \chi\,\mu$ is an algebra automorphism of $M(G)$. Its inverse is $\mu \mapsto \chi^{-1}\mu$.*

We claim, furthermore, for any $f \in k[G]$, $\mu \in M(G)$, and $g \in G(A)$ (for all A):

$$(6) \qquad \rho_l(g)(f\,\mu) = \rho_l(g)(f) \cdot \rho_l(g)(\mu)$$

and

$$(7) \qquad \rho_r(g)(f\,\mu) = \rho_r(g)(f) \cdot \rho_r(g)(\mu).$$

(We really ought to write $\rho_l(g)(f\,\mu \otimes 1)$ etc.) Indeed, we have

$$\begin{aligned} \rho_l(g)(f\,\mu) &= (f\,\mu) \circ \rho_l(g^{-1}) = \mu \circ (f\,?) \circ \rho_l(g^{-1}) \\ &= \mu \circ \rho_l(g^{-1}) \circ (\rho_l(g)(f)\,?) = (\rho_l(g)(\mu)) \circ (\rho_l(g)(f)\,?) \\ &= \rho_l(g)(f) \cdot \rho_l(g)(\mu). \end{aligned}$$

The proof of (7) is similar.

8.12. If M is a G–module, then we denote by M^l the $(G \times G)$–module that is equal to M as a vector space and where the first factor G acts as on M and the second factor acts trivially. Similarly M^r is defined. For $\lambda \in X(G)$ we shall usually write λ^l and λ^r instead of $(k_\lambda)^l$ and $(k_\lambda)^r$. We regard $k[G]$ and $M(G)$ as $(G \times G)$–modules with the first factor acting via ρ_l and the second one via ρ_r.

Proposition: *Let $\mu_0 \in M(G)_l^G$, $\mu_0 \neq 0$. Then $f \mapsto f\,\mu_0$ is an isomorphism of $k[G]$–modules and of $(G \times G)$–modules:*

$$k[G] \otimes (\delta_G)^r \xrightarrow{\sim} M(G).$$

Proof: It is obvious from the definitions and from 8.11(6), (7) that the map considered is a homomorphism of $k[G]$– and of $(G \times G)$–modules. We have only to prove its bijectivity. As both sides have the same dimension, it is enough to show injectivity.

Consider the endomorphism γ of $M(G) \otimes k[G]$ that is the composite of the map $\mathrm{id}_{M(G)} \otimes \Delta_G : M(G) \otimes k[G] \to M(G) \otimes k[G] \otimes k[G]$ with the map $M(G) \otimes k[G] \otimes k[G] \to M(G) \otimes k[G]$, $\mu \otimes f_1 \otimes f_2 \mapsto f_1\,\mu \otimes f_2$. We can identify $M(G) \otimes k[G]$ with $\mathrm{Mor}(G, M(G)_a)$ associating to each $\mu \otimes f$ the map $g \mapsto f(g)\,\mu$. Then $\gamma(\mu \otimes f)$ is easily checked to be the map $g \mapsto (\rho_r(g) f)\,\mu$.

Let us fix now $f \in k[G]$ and consider $F \in \mathrm{Mor}(G, M(G)_a)$ with $F(g) = (\rho_l(g) f)\,\mu_0 = \rho_l(g)(f\,\mu_0)$ for all $g \in G(A)$ and all A. If $\Delta_G(f) = \sum_{i=1}^r f_i \otimes f_i'$, then $\rho_l(g) f = \sum_{i=1}^r f_i(g^{-1}) f_i'$, hence F corresponds to $\sum_{i=1}^r (f_i'\,\mu_0) \otimes \sigma_G(f_i) \in M(G) \otimes k[G]$. Its image under γ is therefore the morphism $g \mapsto (\sum_{i=1}^r (\rho_r(g)\sigma_G(f_i))\, f_i')\,\mu_0$. Now $\sum_{i=1}^r (\rho_r(g)\sigma_G(f_i))\, f_i'$ maps any g' to $\sum_{i=1}^r f_i((g'g)^{-1}) f_i'(g') = f((g'g)^{-1} g')$ $= f(g^{-1})$. This implies $\gamma(F) = \mu_0 \otimes \sigma_G(f)$. If $f\,\mu_0 = 0$, then $F = 0$, hence $\mu_0 \otimes \sigma_G(f) = 0$. As $\mu_0 \neq 0$, this implies $f = 0$. So the map considered is injective.

Remarks: 1) If we take $\mu_0 \in M(G)_r^G$, $\mu_0 \neq 0$, then $f \mapsto f\,\mu_0$ is an isomorphism of $k[G]$– and of $(G \times G)$–modules:

$$k[G] \otimes (\delta_G)^l \xrightarrow{\sim} M(G).$$

2) The affine and finite scheme G is also a projective scheme of dimension 0. It has therefore a dualising sheaf, cf. [Ha], p. 241. This is easily seen to be the coherent sheaf with global sections equal to $M(G) = k[G]^*$: We have for each finite dimensional $k[G]$–module M a non-degenerate pairing $\mathrm{Hom}_{k[G]}(M, k[G]^*) \times M \to k[G]^* \to k$ mapping at first (φ, m) to $\varphi(m)$ and then μ to $\mu(1)$. (Use $\mathrm{Hom}_{k[G]}(M, k[G]^*) \simeq \mathrm{Hom}_{k[G]}(k[G], M^*) \simeq M^*$ with the obvious structure as a $k[G]$–module on M^*.)

In [Kempf 5], 5.1 the proposition is proved using the interpretation of $M(G)$ as dualising sheaf.

8.13. Proposition: *Let E be a simple G–module and Q a projective cover of E. Then*

$$\operatorname{soc} Q \simeq E \otimes \delta_G.$$

Proof: Choose a complete set $e_1, e_2, \dots, e_r$ of primitive, orthogonal idempotents in $M(G)$, hence a decomposition

$$M(G) = \bigoplus_{i=1}^{r} M(G) e_i$$

into indecomposable (projective and injective) modules. There are simple G–modules E_i and E_i' $(1 \le i \le r)$ with $M(G)e_i/\operatorname{rad} M(G)e_i \simeq E_i$ and $\operatorname{soc} M(G)e_i \simeq E_i'$. We have to show that $E_i' \simeq E_i \otimes \delta_G$ for all i.

The map $\varphi \mapsto \varphi(e_i)$ is an isomorphism $\operatorname{Hom}_G(M(G)e_i, M) \xrightarrow{\sim} e_i M$ for any G–module M. If M is simple, then $M \simeq E_i$ if and only if $e_i M \neq 0$. Any $\mu \in M(G)$ acts on M^* through $\mu\varphi = \varphi \circ \sigma_G'(\mu)$, and on $M \otimes \chi$ for $\chi \in X(G)$ as $\chi\mu$ acts on M. Therefore (for M simple)

$$\text{(1)} \qquad \begin{aligned} M \simeq E_i \iff e_i M \neq 0 \iff \sigma_G'(e_i)M^* \neq 0 \\ \iff (\chi\sigma_G'(e_i))(M^* \otimes \chi^{-1}) \neq 0. \end{aligned}$$

Because of 8.11(5), the $\chi\sigma_G'(e_i)$ form also a complete orthogonal set of primitive idempotents in $M(G)$. We get from (1)

$$\text{(2)} \qquad E_i^* \otimes \chi^{-1} \simeq M(G)\chi\sigma_G'(e_i)/\operatorname{rad} M(G)\chi\sigma_G'(e_i)$$

for all i.

Choose μ_0 as in 8.12 and let $\psi : M(G) \to k[G]$ be inverse to the map $f \mapsto f\,\mu_0$ from 8.12. The $(G \times G)$–homomorphism property of 8.12 and the description in 8.6 of ρ_l and ρ_r on $M(G)$ imply for all $\mu, \mu' \in M(G)$

$$\begin{aligned} \psi(\mu\mu') &= \psi(\rho_l(\mu)\mu') = \rho_l(\mu)\,\psi(\mu') \\ &= \psi(\rho_r(\sigma_G'(\mu'))\mu) = \rho_r(\delta_G\sigma_G'(\mu'))\,\psi(\mu). \end{aligned}$$

Therefore $\psi(M(G)e_i) = \rho_r(\delta_G\sigma_G'(e_i))\,\psi(M(G))$ is orthogonal to $M(G)\delta_G\sigma_G'(e_j)$ for all i and j with $i \neq j$. As ψ is an isomorphism for ρ_l, we get for all i

$$M(G)e_i \simeq \psi(M(G)e_i) \simeq (M(G)\delta_G\sigma_G'(e_i))^*,$$

hence

$$\text{(3)} \qquad \operatorname{soc} M(G)e_i \simeq (M(G)\delta_G\sigma_G'(e_i)/\operatorname{rad} M(G)\delta_G\sigma_G'(e_i))^*.$$

Now (2) and (3) imply

$$E_i' \simeq (E_i^* \otimes \delta_G^{-1})^* \simeq E_i \otimes \delta_G.$$

Remark: If $\delta_G = 1$ (i.e., if G is unimodular), then the projective cover and the injective hull of every simple G–module coincide. If we apply the proposition to the trivial G–module k, then we get that the converse also holds.

One can show for unimodular G that $M(G)$ is a symmetric algebra in the sense of [CR], ch. IX, cf. [Humphreys 9]. In general $M(G)$ is only a Frobenius algebra.

8.14. (Coinduced Modules) Any closed subgroup H of G is itself a finite algebraic k–group. We can identify $M(H)$ with the subalgebra $\{\mu \in M(G) \mid \mu(I(H)) = 0\}$ where $I(H) \subset k[G]$ is the ideal of H, cf. the corresponding result for $\operatorname{Dist}(H)$ in 7.2(3).

The equivalence of categories 8.6(2) enables us to define a functor coind_H^G from $\{$ H–modules $\}$ to $\{$ G–modules $\}$ through

$$\text{(1)} \qquad \operatorname{coind}_H^G M = M(G) \otimes_{M(H)} M$$

for any H–module M. We call this functor the *coinduction* from H to G. (When comparing this to what is done for Lie algebras, e.g., in [Dix], ch. 5 one has to observe that there the terms induction and coinduction have just the opposite meanings. Also in [Voigt 2] our $\operatorname{coind}_H^G M$ is called an induced module.)

We have obviously:

(2) *The functor* coind_H^G *is right exact.*

For any H–module M the map $i_M : M \to \operatorname{coind}_H^G M$ with $i_M(m) = 1 \otimes m$ is a homomorphism of H–modules. The universal property of the tensor product implies that for each G–module V we get an isomorphism

(3) $$\operatorname{Hom}_G(\operatorname{coind}_H^G M, V) \xrightarrow{\sim} \operatorname{Hom}_H(M, \operatorname{res}_H^G V), \qquad \varphi \mapsto \varphi \circ i_M.$$

Hence

(4) *The functor* coind_H^G *is left adjoint to* res_H^G.

Furthermore:

(5) *The functor* coind_H^G *maps projective H–modules to projective G–modules.*

8.15. Lemma: *Let H be a closed subgroup of G and M a finite dimensional H–module. Then there is an isomorphism of G–modules*

$$\operatorname{coind}_H^G M \simeq (\operatorname{ind}_H^G(M^*))^*.$$

Proof: For all finite dimensional G–modules V_1, V_2 the bijection $\operatorname{Hom}(V_1, V_2) \xrightarrow{\sim} \operatorname{Hom}(V_2^*, V_1^*)$ mapping each φ to its transpose φ^* induces a bijection $\operatorname{Hom}_G(V_1, V_2) \xrightarrow{\sim} \operatorname{Hom}_G(V_2^*, V_1^*)$.

Using this and 8.14(3) we get for each finite dimensional G–module V canonical isomorphisms

$$\begin{aligned} \operatorname{Hom}_G(V, (\operatorname{coind}_H^G M)^*) &\xrightarrow{\sim} \operatorname{Hom}_G(\operatorname{coind}_H^G M, V^*) \\ &\xrightarrow{\sim} \operatorname{Hom}_H(M, V^*) \xrightarrow{\sim} \operatorname{Hom}_H(V, M^*) \end{aligned}$$

mapping any ψ to $(i_M)^* \circ \psi$. This generalises to all V by taking direct limits. Therefore $(\operatorname{coind}_H^G M)^*$ has the universal property of $\operatorname{ind}_H^G(M^*)$ as in 3.5, hence is isomorphic to $\operatorname{ind}_H^G(M^*)$.

8.16. (Exactness of Induction) Let H be a closed subgroup of G. As H is a finite algebraic k–group, 5.13.b implies:

(1) ind_H^G *is exact.*

We get now from 4.12:

(2) $k[G]$ *is an injective H–module.*

Hence:

(3) $M(G)$ *is a projective left and right $M(H)$–module,*

and:

(4) coind_H^G *is exact.*

Of course, (4) follows also directly from (1) and 8.15. One can improve (3) and show that $M(G)$ is a free module over $M(H)$, cf. [Oberst and Schneider], 2.4. We do not have to go into this.

If M' is a projective and finite dimensional right $M(H)$–module, then we have for each H–module M an isomorphism

$$M' \otimes_{M(H)} M \xrightarrow{\sim} \operatorname{Hom}_{M(H)}(\operatorname{Hom}_{M(H)}(M', M(H)), M) \tag{5}$$

mapping any $m' \otimes m$ with $m' \in M'$ and $m \in M$ to the map $\varphi \mapsto \varphi(m')m$. Here we form $\operatorname{Hom}_{M(H)}(M', M(H))$ via the action of $M(H)$ on itself by right multiplication, and we consider it as an $M(H)$–module via the left multiplication on $M(H)$. In order to prove bijectivity in (5) one restricts to the case $M' = M(H)^n$ for some n where both sides are isomorphic to M^n.

Because of (3) we can apply this to $M(G)$ considered as an $M(H)$–module under right multiplication. The map in (5) is now easily checked to be an isomorphism of G–modules

$$\operatorname{coind}_H^G M \xrightarrow{\sim} \operatorname{Hom}_H(\operatorname{Hom}_H(M(G), M(H)), M), \tag{6}$$

where the action of G on the right hand side is derived from the left regular representation on $M(G)$.

8.17. Proposition: *Let H be a closed subgroup of G. Then we have for each H–module M an isomorphism:*

$$\operatorname{coind}_H^G M \simeq \operatorname{ind}_H^G(M \otimes (\delta_G)_{|H}\delta_H^{-1}).$$

Proof: We have isomorphisms of $(G \times H)$–modules

$$\begin{aligned}\operatorname{Hom}_H(M(G), M(H)) &\simeq (M(G)^* \otimes M(H))^H \\ &= (k[G] \otimes M(H))^H \simeq (k[G] \otimes k[H] \otimes \delta_H)^H \\ &= \operatorname{ind}_H^H(k[G] \otimes \delta_H) \simeq k[G] \otimes \delta_H.\end{aligned}$$

Here the last term is regarded as a G–module via the left regular representation on $k[G]$, and as an H–module via the tensor product of δ_H with the right regular representation on $k[G]$.

We get now from 8.16(6) isomorphisms of G–modules

$$\begin{aligned}\operatorname{coind}_H^G M &\simeq \operatorname{Hom}_H(k[G] \otimes \delta_H, M) \simeq (M(G) \otimes \delta_H^{-1} \otimes M)^H \\ &\simeq (k[G] \otimes (\delta_G)_{|H}\delta_H^{-1} \otimes M)^H = \operatorname{ind}_H^G(M \otimes (\delta_G)_{|H}\delta_H^{-1}).\end{aligned}$$

8.18. Corollary: *Let H be a closed subgroup of G and M a finite dimensional H–module. Then:*

$$(\operatorname{ind}_H^G M)^* \simeq \operatorname{ind}_H^G(M^* \otimes (\delta_G)_{|H}\delta_H^{-1}).$$

Proof: This follows from 8.17 and 8.15.

Remark: One can interpret this result as Serre duality for the sheaf cohomology of $\mathcal{L}_{G/H}(M)$, cf. 5.12.

8.19. Proposition: *Let G' be a k–group scheme acting on G through group automorphisms. Then G' acts naturally on $k[G]$ and $M(G)$. The space $M(G)_l^G$ is a G'–submodule of $M(G)$ and the action of G' on $M(G)_l^G$ is given by some $\chi \in X(G')$. If $\mu_0 \in M(G)_l^G$, $\mu_0 \neq 0$, then the map $f \mapsto f\,\mu_0$ is an isomorphism $k[G] \otimes \chi \xrightarrow{\sim} M(G)$ of G'–modules. If G is a closed normal subgroup of G' and if we consider the action of G' on G by conjugation, then $\chi_{|G} = \delta_G$.*

Proof: We can form the semi-direct product $G \rtimes G'$ and make it act on G such that G acts through left multiplication and G' as given. This yields representations of $G \rtimes G'$ on $k[G]$ and $M(G) = k[G]^*$ that yield the actions considered in the proposition when restricted to G', and that yield the left regular representations when restricted to G. Hence $M(G)_l^G$ are the fixed points of the normal subgroup G of $G \rtimes G'$, hence a G'–submodule by 3.2.

It is now obvious that G' acts through some $\chi \in X(G')$ on $M(G)_l^G$. Proposition 8.12 and an analogue to 8.11(6) for the G'–action show then that $f \mapsto f\,\mu_0$ is an isomorphism $k[G] \otimes \chi \xrightarrow{\sim} M(G)$ of G'–modules.

Suppose finally that G is a normal subgroup of G' and that we consider the conjugation action of G' on G. Then each $g \in G(A) \subset G'(A)$ acts through the composition of $\rho_l(g)$ and $\rho_r(g)$ on $M(G) \otimes A$, hence through $\rho_r(g)$ on $\mu_0 \otimes 1$. Now the definitions yield $\chi(g) = \delta_G(g)$.

8.20. Proposition: *Let G' be a k–group scheme containing G as a closed normal subgroup. Let H' be a closed subgroup of G' and set $H = H' \cap G$. Let M be an H'–module. Then $\operatorname{coind}_H^G M$ has a natural structure as an GH'–module extending the structure as a G–module. For each GH'–module V there is a canonical isomorphism*

$$\text{(1)} \qquad \operatorname{Hom}_{GH'}(\operatorname{coind}_H^G M, V) \xrightarrow{\sim} \operatorname{Hom}_{H'}(M, V).$$

If $\chi \in X(G')$ resp. $\chi' \in X(H')$ is the character through which G' resp. H' acts on $M(G)_l^G$ resp. $M(H)_l^H$, then we have an isomorphism of GH'–modules

$$\text{(2)} \qquad \operatorname{coind}_H^G M \xrightarrow{\sim} \operatorname{ind}_{H'}^{GH'}(M \otimes \chi_{|H'}(\chi')^{-1}).$$

If $\dim M < \infty$, then we have an isomorphism of GH'–modules

$$\text{(3)} \qquad (\operatorname{ind}_{H'}^{GH'} M)^* \xrightarrow{\sim} \operatorname{ind}_{H'}^{GH'}(M \otimes \chi_{|H'}(\chi')^{-1}).$$

Proof: Let us work with the description of $\operatorname{coind}_H^G M$ as in 8.16(6). We make H' act on $M(G)$ and $M(H)$ via the conjugation action on G and H. We get thus a representation of H' on $\operatorname{Hom}(M(G), M(H))$ that extends to $H \rtimes H'$ if we let H act through the two right regular representations. By 3.2 the subspace $\operatorname{Hom}_H(M(G), M(H))$ is an H'–module. Together with the given action of H' on M this makes $\operatorname{Hom}(\operatorname{Hom}_H(M(G), M(H)), M)$ into an H'–module. This action of H' can be extended to $H \rtimes H'$ with H acting via ρ_l on $M(H)$ and through the restriction of the H'–action on M. Again $\operatorname{Hom}_H(\operatorname{Hom}_H(M(G), M(H)), M)$ is an H'–module. We can extend the action of H' on this space to $G \rtimes H'$ letting G act through ρ_l on $M(G)$, i.e., inducing the action of G on $\operatorname{coind}_H^G M$.

For any $h \in H(A)$ for some A, the element $(h, h^{-1}) \in (G \rtimes H')(A)$ acts trivially. (This follows easily from the construction.) Therefore we get a representation of $(G \rtimes H')/H \simeq GH'$, cf. 6.2(1), on $\operatorname{coind}_H^G M$ extending the given one on G.

Using this structure, the isomorphism in 8.14(3) is easily checked to be an isomorphism of H'–modules (provided V is an H'–module). It therefore has to induce an isomorphism of the H'–fixed points. This implies (1).

We get (2) by examining the proof of 8.17. After replacing δ_G by χ and δ_H by χ' all isomorphisms there are also compatible with the H'–action, hence with the structure as GH'–module. Similarly, 8.15 generalises from G to GH' and together with (2) yields (3) as in 8.18.

Remark: We denote $\mathrm{coind}_H^G M$ when considered as an GH'–module by $\mathrm{coind}_{H'}^{GH'} M$. Obviously $\mathrm{coind}_{H'}^{GH'}$ is a functor from $\{\, H'\text{–modules}\}$ to $\{\, GH'\text{–modules}\}$. The claims in 8.12(2)–(4) and 8.16(4) generalise to this functor. Note that we have by construction an isomorphism of functors

$$\mathrm{res}_G^{GH'} \circ \mathrm{coind}_{H'}^{GH'} \simeq \mathrm{coind}_H^G \circ \mathrm{res}_H^{H'}$$

which is dual to 6.13.

8.21. If $\mathrm{char}(k) = 0$, then each finite algebraic group G over k is associated to an abstract finite group Γ as in 8.5.a (cf. [DG], II, §6 and §5). Then $M(G) \simeq k\Gamma$ is a semi-simple algebra by the theorem of Maschke.

Suppose now that $\mathrm{char}(k) = p > 0$ and that k is algebraically closed. Then G is the semi-direct product of the infinitesimal normal subgroup G^0 and a subgroup G_{red} that is associated to an abstract finite group as in 8.5.a, see [DG], II, §5, 2.4. Then a theorem of Nagata (see [DG], IV, §3, 3.6) says that $M(G)$ is semi-simple if and only if G^0 is diagonalisable (see 2.5) and p does not divide the order of $G_{\mathrm{red}}(k) = G(k)$.

Keep the assumption on k. In case G is associated to an abstract finite group Γ as in 8.5.a, a theorem of D. G. Higman (see [CR], §64) says that $M(G) \simeq k\Gamma$ has finite representation type (i.e., only finitely many equivalence classes of indecomposable modules) if and only if the Sylow p–subgroups of Γ are cyclic. In general, the question of finite representation type is investigated in [Pfautsch and Voigt] and [Farnsteiner and Voigt]. (For G associated to a p–Lie algebra as in 8.5.b, see also [Feldvoss and Strade].) It turns out that $M(G)$ has finite representation type if and only if both $M(G^0)$ and $M(G_{\mathrm{red}})$ have this type and at least one of theses algebras is semi-simple. And $M(G^0)$ has finite representation type only if G^0 is solvable of a very special form.

CHAPTER 9

Representations of Frobenius Kernels

Throughout this chapter let p be a prime number. We shall always assume that k is a perfect field with $\mathrm{char}(k) = p$.

Let G be an algebraic k–group. If $k = \mathbf{F}_p$, then the map $f \mapsto f^p$ on $k[G]$ is an endomorphism of k–algebras and defines a morphism $F_G : G \to G$ that is a group endomorphism. The kernels $G_r = \ker(F_G^r)$ are called the *Frobenius kernels* of G. They are infinitesimal algebraic k–groups. One can generalise this construction to all k (as above) by replacing F_G^r as above by some group homomorphism $G \to G^{(r)}$ into a suitable k–group $G^{(r)}$.

In this chapter we give first the definitions and elementary properties (9.1–9.6, 9.8). We compute the modular function of the G_r in the case of reduced groups (9.7 combined with 8.19) and prove that $H^i(G, M) = \varprojlim H^i(G_r, M)$ under special assumptions on G and M (9.9).

The representation theory of the first Frobenius kernel G_1 of G is equivalent to that of $\mathrm{Lie}(G)$ as a p–Lie algebra. Therefore each cohomology group $H^i(G_1, M)$ is equal to the corresponding "restricted Lie algebra cohomology group" in the sense of [Hochschild 3]. In that paper these groups are compared to the ordinary Lie algebra cohomology groups (cf. 9.17), especially in low degrees.

One of his main results can now be interpreted as a "six term exact sequence" arising from a spectral sequence (9.19/20). This spectral sequence was found for $p \neq 2, 3$ and G reductive in [Friedlander and Parshall 3]. (But also compare the remark at the end of section 3 in [Hochschild 3].) Their results were generalised somewhat in [Andersen and Jantzen]. The present version of Proposition 9.20 is the same as in my lectures in Shanghai and was also proved in [Friedlander and Parshall 4].

9.1. (The Frobenius Morphism on an Affine Variety) Before defining Frobenius morphisms in general we want to motivate the definitions by an example. In this subsection let us assume k to be algebraically closed.

Let X be an affine variety over k (as in 1.1). We can embed X as a Zariski closed subset into some k^n. The map

$$F : k^n \to k^n, \qquad (a_1, a_2, \dots, a_n) \mapsto (a_1^p, a_2^p, \dots, a_n^p)$$

is a bijective morphism of varieties. It is also a closed map. (Using that $f^p \in \mathrm{im}(F^*)$ for all $f \in k[T_1, T_2, \dots, T_n]$ one shows $\sqrt{k[T_1, T_2, \dots, T_n]F^*(F^*)^{-1}I} = \sqrt{I}$ for each ideal $I \subset k[T_1, T_2, \dots, T_n]$.) Therefore each $F^r(X)$ with $r \in \mathbf{N}$ is a closed subset of k^n and F^r induces a bijective morphism $X \to F^r(X)$. We want to show that the pair $(F^r(X), F^r : X \to F^r(X))$ has an intrinsic meaning, i.e., is independent (up to isomorphism) of the embedding of X into k^n.

Define for each $f \in k[X]$ a map $\varphi_r(f) : F^r(X) \to k$ through $\varphi_r(f)(x') = f(F^{-r}(x))^{p^r}$ for all $x' \in F^r(X)$. Obviously φ_r is an injective ring homomorphism from $k[X]$ to the algebra of all functions from $F^r(X)$ to k and satisfies $\varphi_r(af) = a^{p^r}\varphi(f)$ for all $a \in k$ and $f \in k[X]$. If f is the ith coordinate function on k^n restricted to X, then $\varphi_r(f)$ is the ith coordinate function restricted to $F^r(X)$. Therefore φ_r induces a bijection from $k[X]$ to $k[F^r(X)]$.

Denote by $k[X]^{(r)}$ the k–algebra that coincides as a ring with $k[X]$, but where each $a \in k$ acts as $a^{p^{-r}}$ does on $k[X]$. Then we can regard φ_r as an isomorphism of k–algebras $k[X]^{(r)} \xrightarrow{\sim} k[F^r(X)]$. This shows that $F^r(X)$ as a variety has an intrinsic meaning. If we identify $k[F^r(X)]$ with $k[X]^{(r)}$ via φ_r, then the comorphism of F^r is the map $k[X]^{(r)} \to k[X]$ with $f \mapsto f^{p^r}$ for all f, hence also F^r has a description independent of the embedding of X into k^n.

9.2. (The Frobenius Morphism on a Scheme) From now on let k again be an arbitrary perfect field of characteristic p.

For each k–algebra A and each $m \in \mathbf{Z}$ we define $A^{(m)}$ as the k–algebra that coincides with A as a ring but where each $b \in k$ acts as $b^{p^{-m}}$ does on A. Trivially $A^{(0)} = A$. One obviously has isomorphisms

(1) $$(A^{(m)})^{(n)} = A^{(m+n)} \qquad \text{for all } m, n \in \mathbf{Z},$$

and (for all k–algebras A, A')

(2) $$\mathrm{Hom}_{k-\mathrm{alg}}(A^{(-m)}, A') \simeq \mathrm{Hom}_{k-\mathrm{alg}}(A, A'^{(m)}) \qquad \text{for all } m \in \mathbf{Z}.$$

(This is the identity map.) For each k–algebra A, each $m \in \mathbf{Z}$ and $r \in \mathbf{N}$ the map

(3) $$\gamma_r : A^{(m)} \to A^{(m-r)}, \qquad a \mapsto a^{p^r}$$

is a homomorphism of k–algebras.

We now define for each k–functor X and each $r \in \mathbf{N}$ a new k–functor $X^{(r)}$ through

(4) $$X^{(r)}(A) = X(A^{(-r)}) \qquad \text{for all } k\text{–algebras } A.$$

Furthermore, we define a morphism $F_X^r : X \to X^{(r)}$ through

(5) $$F_X^r(A) = X(\gamma_r) : X(A) \to X(A^{(-r)}) = X^{(r)}(A)$$

for all A. We call F_X^r the rth *Frobenius morphism* on X. Obviously $X \mapsto X^{(r)}$ is a faithful functor from { k–functors } to itself.

One gets from (1) for all $r, s \in \mathbf{N}$ and all X

(6) $$(X^{(r)})^{(s)} = X^{(r+s)} \qquad \text{and} \qquad F^s_{X^{(r)}} \circ F^r_X = F^{r+s}_X.$$

If we consider an affine scheme $X = Sp_k R$ for some k–algebra R, then (2) implies for all $r \in \mathbf{N}$

(7) $$(Sp_k R)^{(r)} \simeq Sp_k(R^{(r)}).$$

Furthermore, F_X^r has as comorphism $R^{(r)} \to R$, $f \mapsto f^{p^r}$. So the construction of $X^{(r)}$ and F_X^r generalises the situation considered in 9.1.

We can interpret the definition (4) as saying that $X^{(r)}$ arises from X through base change from k to $k^{(-r)}$ which then is identified with k as a ring. We can therefore apply the general remarks about base changes in 1.10. So the functor $X \mapsto X^{(r)}$ maps subfunctors to subfunctors, commutes with taking intersections and inverse images of subfunctors and with taking direct and fibre products. It maps local functors to local functors, schemes to schemes, and faisceaux to faisceaux (cf. 5.3(9)). If X is an affine scheme and I an ideal in $k[X]$, then $V(I)^{(r)} = V(I^{(r)})$ and $D(I)^{(r)} = D(I^{(r)})$ where $I^{(r)} \subset k[X]^{(r)}$ is just I with the new action of k.

If $k = \mathbf{F}_p$, then obviously $X^{(r)} = X$ for all r and any k–functor X. If X is affine and if F_X is the endomorphism of X with $F_X^*(f) = f^p$ for all $f \in k[X]$, then obviously $F_X^r = (F_X)^r$. More generally, if k is again arbitrary, but if X has an $\mathbf{F}_p$–structure (i.e., there is some $\mathbf{F}_p$–functor X' with $X = (X')_k$), then we can identify $X^{(r)}$ with X. In the affine case one has $k[X] = \mathbf{F}_p[X'] \otimes_{\mathbf{F}_p} k$, and the map $f \otimes a \mapsto f \otimes a^{p^r}$ (for all $f \in \mathbf{F}_p[X']$ and $a \in k$) induces an isomorphism $k[X^{(r)}] = k[X]^{(r)} \xrightarrow{\sim} k[X]$. For $r = 1$ this map is called the *arithmetic* Frobenius endomorphism of $k[X]$. Taking this identification, F_X^r is the endomorphism of X with comorphism $f \otimes a \mapsto f^{p^r} \otimes a$ (for all f, a as above). For $r = 1$ this map is called the *geometric* Frobenius endomorphism of $k[X]$.

Remark: If we replace k by an arbitrary $\mathbf{F}_p$–algebra, then we can still define $A^{(r)}$ for all $r \leq 0$, hence $X^{(r)}$ as in (4) for all $r \geq 0$. We can also take the interpretation via base change in that situation. It is left to the reader to find out later on how much generalises to this case.

9.3. (Fibres of the Frobenius Morphism) Let X be an affine scheme over k. Consider a point $x \in X(k)$ and let us denote its ideal by $I_x = \{f \in k[X] \mid f(x) = 0\}$. Then the ideal of $F_X^r(x) \subset X^{(r)}(k)$ in $k[X^{(r)}] = k[X]^{(r)}$ is $I_x^{(r)}$ (i.e., I_x with the new scalar multiplication) as $f(F_X^r(x)) = f(x)^{p^r}$ for all f. This implies (for all $r \in \mathbf{N}$)

$$(1) \qquad (F_X^r)^{-1}(F_X^r(x)) = V\Big(\sum_{f \in I_x} k[X] f^{p^r}\Big).$$

So the $(F_X^r)^{-1}(F_X^r(x))$ form an ascending chain of closed subschemes of X.

Suppose now that X is algebraic. Then I_x is a finitely generated ideal, say $I_x = \sum_{i=1}^m k[X] f_i$. Then

$$(F_X^r)^{-1}(F_X^r(x)) = V\Big(\sum_{i=1}^m k[X] f_i^{p^r}\Big)$$

for all r. The ideal defining $(F_X^r)^{-1}(F_X^r(x))$ is contained in $I_x^{p^r}$ and contains $I_x^{mp^r}$. This implies (cf. 7.1, 7.2(2))

$$(2) \qquad \operatorname{Dist}(X, x) = \bigcup_{r > 0} \operatorname{Dist}((F_X^r)^{-1}(F_X^r(x)), x).$$

We can choose the f_i such that the $f_i + I_x^2$ $(1 \leq i \leq m)$ form a basis for I_x/I_x^2. If x is a simple point of X, then $m = \dim_x X$ and the f_i $(1 \leq i \leq m)$ are algebraically independent. Therefore the residue classes of all $f_1^{n(1)} f_2^{n(2)} \dots f_m^{n(m)}$ with all $n(i) < p^r$ form a basis of $k[(F_X^r)^{-1}(F_X^r(x))]$. This shows:

(3) *If x is a simple point of X, then* $\dim k[(F_X^r)^{-1}(F_X^r(x))] = p^{rm}$ *where* $m = \dim_x X$.

Let me add that (1) generalises to

(4) $$(F_X^r)^{-1} V(I^{(r)}) = V\Big(\sum_{f \in I} k[X] f^{p^r}\Big)$$

for all ideals I in $k[X]$ (and any affine X), whereas

(5) $$(F_X^r)^{-1} D(I^{(r)}) = D(I).$$

(Use that $\sqrt{I} = \sqrt{\sum_{f \in I} k[X] f^{p^r}}$ and 1.5(5), (10).)

9.4. (Frobenius Kernels) Let G be a k–group functor. Then obviously each $G^{(r)}$ is also a k–group functor and $F_G^r : G \to G^{(r)}$ is a homomorphism of k–group functors. Its kernel $G_r = \ker(F_G^r)$ is a normal subgroup functor of G that we call the rth *Frobenius kernel* of G. The factorisation in 9.2(6) implies that we get an ascending chain

(1) $$G_1 \subset G_2 \subset G_3 \subset \cdots$$

of normal subgroup functors of G. (We have $G_0(A) = \{1\}$ for all A.)

If H is a subgroup functor of G, then $H^{(r)}$ is a subgroup functor of $G^{(r)}$ and F_H^r is the restriction of F_G^r to H. This implies

(2) $$H_r = H \cap G_r,$$

especially for all $r, r' \in \mathbf{N}$

(3) $$(G_r)_{r'} = \begin{cases} G_{r'} & \text{for } r' \le r, \\ G_r & \text{for } r' \ge r. \end{cases}$$

If $k = \mathbf{F}_p$ or if G is defined over $\mathbf{F}_p$, then we can identify each $G^{(r)}$ with G and interpret F_G^r as the rth power of some Frobenius endomorphism $F_G : G \to G$ (which is F_G^i after the identification $G \simeq G^{(1)}$). This is true, e.g., for $G = G_a$ and $G = G_m$. In these cases $(F_G)^*(T) = T^p$ in the notations of 2.2. Therefore $G_{m,r} = \mu_{(p^r)}$ for all r, and the $G_{a,r}$ from 2.2 are the Frobenius kernels of G_a. (So our new notation is compatible with the old one.)

9.5. Let G be a k–group scheme. The image faisceau (cf. 5.5) of F_G^r in $G^{(r)}$ is isomorphic to G/G_r (by 6.1) as $G_r = \ker(F_G^r)$. For each subgroup scheme H of G we can identify

(1) $$F_G^r(H) \simeq F_H^r(H) \simeq H/H_r$$

by 9.4(2), and $(F_G^r)^{-1} F_G^r(H)$ with $G_r H$, cf. 6.2(4).

The factorisation $F_G^{r'} = F_{G^{(r)}}^{r'-r} \circ F_G^r$ yields

(2) $$G_{r'} = (F_G^r)^{-1}((G^{(r)})_{r'-r})$$

for all $r' \ge r$.

Proposition: *If G is a reduced algebraic k–group, then each F_G^r induces isomorphisms $G/G_r \simeq G^{(r)}$ and $G_{r'}/G_r \simeq (G^{(r)})_{r'-r}$ for all $r' \geq r$.*

Proof: By [DG], II, §5, 5.1.b the embedding of $F_G^r(G) \simeq G/G_r$ into $G^{(r)}$ is a closed immersion. Therefore G/G_r is identified with the closed subgroup of $G^{(r)}$ defined by the kernel of the comorphism $(F_G^r)^* : k[G]^{(r)} \to k[G]$ that maps each f to f^{p^r}, i.e., we get

$$F_G^r(G) = V(\{f \in k[G] \mid f^{p^r} = 0\}^{(r)}). \tag{3}$$

If G is reduced, i.e., if $k[G]$ does not contain nilpotent elements $\neq 0$, then obviously $F_G^r(G) = G^{(r)}$.

As we have shown F_G^r to be an epimorphism of faisceaux, each subfaisceau Y of $G^{(r)}$ is equal to the image faisceau $F_G^r((F_G^r)^{-1}Y)$. Therefore the last claim follows from (1) and (2).

Remark: If G is defined over $\mathbf{F}_p$, we can express the results as $G/G_r \simeq G$ and $G_{r'}/G_r \simeq G_{r'-r}$.

9.6. (The G_r and Lie(G)) Let G be an algebraic k–group scheme and I_1 the ideal in $k[G]$ defining 1. Keep this assumption and convention until the end of this chapter.

Obviously G_r is the closed subscheme of G defined by $\sum_{f \in I_1} k[G]f^{p^r}$. Therefore $k[G_r]$ is finite dimensional and the ideal of 1 in $k[G_r]$ is nilpotent. Hence (cf. 8.1):

(1) *Each G_r is an infinitesimal k–group.*

Choose $f_1, f_2, \dots, f_m \in I_1$ such that the $f_i + I_1^2$ form a basis of I_1/I_1^2. Then $m = \dim \mathrm{Lie}(G)$ and the f_i generate I_1 as an ideal. One has obviously $\dim k[G_r] \leq p^{rm}$ for all r, and equality holds if 1 is a simple point of G (cf. 9.3(3)). So we get (e.g., by [DG], II, §5, 2.1/3)

(2) *If G is reduced, then* $\dim k[G_r] = p^{r\dim(G)}$ *for all* $r \in \mathbf{N}$.

We obviously have for all $r > 0$ (and any G)

$$\mathrm{Lie}(G_r) = \mathrm{Lie}(G). \tag{3}$$

The subalgebra $U^{[p]}(\mathrm{Lie}(G)) = U^{[p]}(\mathrm{Lie}(G_1))$ of $\mathrm{Dist}(G_1) \subset \mathrm{Dist}(G)$, cf. 7.10(2), has dimension p^m, whereas $\dim \mathrm{Dist}(G_1) = \dim k[G_1] \leq p^m$. This implies

$$U^{[p]}(\mathrm{Lie}(G)) \simeq \mathrm{Dist}(G_1). \tag{4}$$

This shows that G_1 is the infinitesimal k–group corresponding to the p–Lie algebra $\mathrm{Lie}(G)$ as in 8.5.b and that the representation theory of G_1 is equivalent to that of $\mathrm{Lie}(G)$ as a p–Lie algebra (cf. 8.6).

9.7. Proposition: *Let G be a reduced algebraic k–group and $r \in \mathbf{N}$. Then G acts on* $\mathrm{Dist}(G_r)_l^{G_r}$ *through the character*

$$g \mapsto \det(\mathrm{Ad}(g))^{p^r-1}$$

where Ad *denotes the adjoint representation of G on* $\mathrm{Lie}(G)$.

Proof: Recall from 8.19 that the conjugation action of G on G_r leads to representations of G on $k[G_r]$ and $M(G_r) = \mathrm{Dist}(G_r)$ and that $M(G_r)_l^{G_r}$ is a one dimensional submodule on which G has to act through some character $\chi \in X(G)$.

Set $q = p^r$ and choose $f_1, f_2, \dots, f_m \in I_1$ such that the $f_i + I_1^2$ form a basis of I_1/I_1^2. Let $\overline{f_i}$ denote the image of f_i in $k[G_r]$. As G is reduced, hence 1 a simple point, the monomials $\overline{f_1}^{a(1)}\overline{f_2}^{a(2)}\dots\overline{f_m}^{a(m)}$ with $0 \le a(i) < q$ for all i form a basis of $k[G_r]$.

We can identify $k[G_r]$ with the factor ring $k[T_1, T_2, \dots, T_m]/(T_1^q, T_2^q, \dots, T_m^q)$ of the polynomial ring $k[T_1, T_2, \dots, T_m]$. It is therefore a graded ring in a natural way. Any endomorphism φ of the vector space $\sum_{i=1}^m k\,\overline{f_i}$ induces an endomorphism of the graded algebra $k[G_r]$. As $F = \prod_{i=1}^m \overline{f_i}^{q-1}$ is the only basis element of degree $m(q-1)$, it has to be mapped under φ to a multiple $c(\varphi)F$ of itself. Obviously $\varphi \mapsto c(\varphi)$ has to be multiplicative. This implies $c(\varphi) = \det(\varphi)^{q-1}$, as this is clearly true for φ in upper or lower triangular form (with respect to the $\overline{f_i}$), hence for all i by multiplicativity. This extends easily to any k–algebra A and any endomorphism of $\sum_{i=1}^m k\,\overline{f_i} \otimes A$ as $c(\varphi)$ is obviously a polynomial in the matrix coefficients of φ.

This can be applied especially to the action of any $g \in G(A)$ for any k–algebra A on $k[G_r] \otimes A$ derived from the conjugation action on G_r. Then the action of g on $\sum_{i=1}^m k\,\overline{f_i} \otimes A \simeq (I_1/I_1^2) \otimes A \simeq \mathrm{Lie}(G)^* \otimes A$ is dual to the adjoint action on $\mathrm{Lie}(G) \otimes A$, hence has determinant equal to $\det(\mathrm{Ad}(g))^{-1}$. So this implies

$$g\,F = \det(\mathrm{Ad}(g))^{-(q-1)}\,F.$$

Consider now $\mu_0 \in \mathrm{Dist}(G_r)_l^{G_r}$, $\mu_0 \neq 0$. If $\mu_0(F) = 0$, then $\mu_0(Fk[G_r]) = 0$ as $k[G_r]F = kF$, hence $F\mu_0 = 0$ by the definition of the $k[G_r]$–module structure on $M(G_r)$ in 8.11, contradicting 8.12. So we must have $\mu_0(F) \neq 0$. Then

$$\chi(g)\mu_0(F) = (g\,\mu_0)(F) = \mu_0(g^{-1}F) = \det(\mathrm{Ad}(g))^{q-1}\,\mu_0(F)$$

implies $\chi(g) = \det(\mathrm{Ad}(g))^{q-1}$ as $\mu_0(F)$ is a unit in A.

Remark: The same proof works for any algebraic k–group G if $r = 1$, by 9.6(4). So we can take any p–Lie algebra $\mathfrak{g}$ over k and consider the infinitesimal k–group G corresponding to $\mathfrak{g}$ as in 8.5.b. Then $G = G_1$ and $\mathrm{Dist}(G) = U^{[p]}(\mathfrak{g})$. Then the proposition implies that the modular function δ_G is given by $\delta_G(g) = \det(\mathrm{Ad}(g))^{p-1}$. The representation of $\mathfrak{g}$ on $\mathrm{Dist}(G)_l^G$ is then given by the differential, i.e., by $(p-1)\,\mathrm{tr}\,(\mathrm{ad}(?)) = -\mathrm{tr}\,(\mathrm{ad}(?))$. As the action of $\mathfrak{g}$ determines that of G in this case, we see that G is unimodular if and only if $\mathrm{tr}(\mathrm{ad}(x)) = 0$ for all $x \in \mathfrak{g}$. This is a theorem of Larson and Sweedler, cf. the discussion in [Humphreys 9].

9.8. (G and the G_r for irreducible G) Because of 9.4(1) the $\mathrm{Dist}(G_r)$ form an ascending chain of subalgebras of G_r, and one has by 9.3(2):

(1) $$\mathrm{Dist}(G) = \bigcup_{r>0} \mathrm{Dist}(G_r).$$

Therefore 7.14–7.17 imply *if G is irreducible*:

(2) *If M is a G–module, then* $M^G = \bigcap_{r>0} M^{G_r}$.

(3) *If M, M' are G–modules, then* $\operatorname{Hom}_G(M, M') = \bigcap_{r>0} \operatorname{Hom}_{G_r}(M, M')$.

(4) *Let M be a G–module and N a subspace of M. Then N is a G–submodule if and only if it is a G_r–submodule for all $r \in \mathbf{N}$, $r > 0$.*

In (2) and (3) we have descending chains $M^{G_1} \supset M^{G_2} \supset M^{G_3} \supset \cdots$ and $\operatorname{Hom}_{G_1}(M, M') \supset \operatorname{Hom}_{G_2}(M, M') \supset \operatorname{Hom}_{G_3}(M, M') \supset \cdots$. If $\dim M < \infty$ resp. if $\dim M \otimes M' < \infty$, then these chains have to stabilise. So we get (still for G irreducible)

(5) *If M is a G–module with* $\dim M < \infty$, *then there is an $n \in \mathbf{N}$ with $M^G = M^{G_r}$ for all $r > n$.*

(6) *If M, M' are G–modules with* $\dim(M \otimes M') < \infty$, *then there is an $n \in \mathbf{N}$ with* $\operatorname{Hom}_G(M, M') = \operatorname{Hom}_{G_r}(M, M')$ *for all $r > n$.*

9.9. Given a G–module M, there are natural restriction maps $H^j(G, M) \to H^j(G_r, M)$ and (for $r' \geq r$) $H^j(G_{r'}, M) \to H^j(G_r, M)$ that induce a natural map $H^j(G, M) \to \varprojlim H^j(G_r, M)$ For $j = 0$ and G irreducible this is an isomorphism by 9.8(2). For arbitrary j such a result has been proved only under additional assumptions. We shall see in II.4.12 that they are satisfied in an important case.

Proposition: *Suppose that G is irreducible and reduced, that $H^i(G, k) = 0$ for all $i > 0$, and that* $\dim H^j(G, M) < \infty$ *for any G–module M with* $\dim M < \infty$ *and for any $j \in \mathbf{N}$. Then the natural map $H^j(G, V) \to \varprojlim H^j(G_r, V)$ is an isomorphism for all finite dimensional G–modules V.*

Proof: Consider for each $r \geq 1$ the Hochschild-Serre spectral sequence $E_l^{i,j}(r)$ converging to $H^\bullet(G, V)$ with E_2–terms $E_2^{i,j}(r) = H^i(G/G_r, H^j(G_r, V))$. We have $\dim k[G_r] < \infty$, hence also $\dim H^j(G_r, V) < \infty$ for all j. So $G/G_r \simeq G^{(r)}$ [since G is reduced, see 9.5] and our assumptions on G imply that all $E_l^{i,j}(r)$ have finite dimension. For all $s \geq r$ the natural maps $G_r \hookrightarrow G_s$ and $G/G_r \to G/G_s$ induce a morphism of spectral sequences $E_l^{i,j}(s) \to E_l^{i,j}(r)$ that is compatible with the identity on the abutment, cf. the remark to 6.6.

We get thus a directed system of spectral sequences where all terms are finite dimensional vector spaces. Now $\varprojlim$ is exact on directed systems of finite dimensional vector spaces, cf. [Roos], prop. 7, so we get a spectral sequence with $E_l^{i,j} = \varprojlim E_l^{i,j}(r)$ still converging to $H^\bullet(G, V)$.

For $s \geq r$ the map on the E_2–level factors as follows:

$$\begin{aligned} E_2^{i,j}(s) = H^i(G/G_s, H^j(G_s, V)) &\to H^i(G/G_r, H^j(G_s, V)) \\ &\to H^i(G/G_r, H^j(G_r, V)^{G_s}) \to H^i(G/G_r, H^j(G_r, V)) = E_2^{i,j}(r). \end{aligned}$$

One gets the first map from the restriction from G/G_s to G/G_r, and the other ones by observing that the restriction map $H^j(G_s, V) \to H^j(G_r, V)$ takes values in the G_s–fixed points of $H^j(G_r, V)$. For any r and j there is $s_0(r, j) \in \mathbf{N}$ such that $H^j(G_r, V)^{G_s} = H^j(G_r, V)^G$ for all $s > s_0(r, j)$, cf. 9.8(5). If $i > 0$, then $H^i(G/G_r, H^j(G_r, V)^G) \simeq H^i(G, k) \otimes H^j(G_r, V)^G = 0$ by our assumption. So

$E_2^{i,j}(s) \to E_2^{i,j}(r)$ is the zero map for all $i > 0$ if $s > s_0(r,j)$. Therefore $E_2^{i,j} = \varprojlim E_2^{i,j}(r) = 0$ for all $i > 0$, the spectral sequence $E_l^{i,j}$ degenerates and yields isomorphisms

$$H^i(G,V) \xrightarrow{\sim} \varprojlim E_2^{0,i}(r) = \varprojlim H^i(G_r,V)^G. \tag{1}$$

On the other hand, the restriction map $H^i(G_s,V) \to H^i(G_r,V)$ takes values in $H^i(G_r,V)^G$ for all $s > s_0(r,i)$. So the inverse limit of the $H^i(G_r,V)$ is equal to that of the subsystem of all $H^i(G_r,V)^G$. Now the claim follows from (1).

Remark: This result and its proof have been communicated to me by W. van der Kallen. Cf. II.4.12 for additional historical remarks.

9.10. (Frobenius Twists of Representations) For any vector space M over k and any $r \in \mathbf{N}$ we denote by $M^{(r)}$ the vector space that is equal to M as an abelian group and where any $a \in k$ acts as $a^{p^{-r}}$ does on M. If M is a G–module, then we have a natural structure as a G–module on each $M^{(r)}$ with $r \geq 0$, cf. 2.16. If $(v_i)_{i \in I}$ is a basis for M and if $\Delta_M(v_i) = \sum_{j \in I} v_j \otimes f_{ji}$, then $\Delta_{M^{(r)}}(v_i) = \sum_{j \in I} v_j \otimes f_{ji}^{p^r}$.

Suppose now that M has a fixed $\mathbf{F}_p$–structure, i.e., an $\mathbf{F}_p$–subspace $M' \subset M$ with $M' \otimes_{\mathbf{F}_p} k = M$. We get then a Frobenius endomorphism F_M on M through $F_M(m' \otimes a) = m' \otimes a^p$ for all $m' \in M'$ and $a \in k$. Let us assume that the v_i above are also a basis for M' over $\mathbf{F}_p$; then we have $F_M(\sum_{i \in I} a_i v_i) = \sum_{i \in I} a_i^p v_i$ for all $a_i \in k$. Each $(F_M)^r$ is an isomorphism $M^{(r)} \xrightarrow{\sim} M$ of vector spaces over k.

Suppose that G is defined over $\mathbf{F}_p$ and denote the corresponding Frobenius endomorphism by $F_G : G \to G$. The representation of G on M is defined over $\mathbf{F}_p$ if the f_{ji} above are functions "over $\mathbf{F}_p$", i.e., if $(F_G)^*(f_{ji}) = (f_{ji})^p$ for all i and j. In that case we can rewrite our formula above as $\Delta_{M^{(r)}}(v_i) = \sum_{j \in I} v_j \otimes (F_G^r)^*(f_{ji})$.

We can also define a new representation of G on M by composing the given $G \to GL(M)$ with $(F_G)^r : G \to G$. Denote the G–module we get thus by $M^{[r]}$. Then we have clearly $\Delta_{M^{[r]}}(v_i) = \sum_{j \in I} v_j \otimes (F_G^r)^*(f_{ji})$. So $(F_M)^r$ is an isomorphism of G–modules $M^{(r)} \to M^{[r]}$ if the original module was defined over $\mathbf{F}_p$.

9.11. (The Associated Graded Group) The powers of I_1 define a filtration of $k[G]$ and we can form the associated graded algebra $\operatorname{gr} k[G] = \bigoplus_{n \geq 0} I_1^n / I_1^{n+1}$. There is obviously a surjection from the symmetric algebra $S(I_1/I_1^2)$ onto $\operatorname{gr} k[G]$ compatible with the grading.

The formulas 2.4(1), (2) show that Δ_G and σ_G also induce a comultiplication and an antipode on $\operatorname{gr} k[G]$ making (together with the obvious augmentation) $\operatorname{gr} k[G]$ into a (commutative and cocommutative) Hopf algebra. So there is a k–group scheme $\operatorname{gr}(G)$ with $k[\operatorname{gr}(G)] \simeq \operatorname{gr} k[G]$ (the *associated graded group* of G).

We can interpret $S(I_1/I_1^2)$ as $k[((I_1/I_1^2)^*)_a] = k[(\operatorname{Lie} G)_a]$. Then the surjection $S(I_1/I_1^2) \to \operatorname{gr} k[G] = k[\operatorname{gr}(G)]$ is compatible with the Hopf algebra structure (again because of 2.4(1), (2)). Thus:

(1) *$\operatorname{gr}(G)$ is canonically isomorphic to a closed subgroup scheme of* $\operatorname{Lie}(G)_a$.

The same arguments as in 9.6(2) imply:

(2) *If G is reduced, then* $\operatorname{gr}(G) \simeq \operatorname{Lie}(G)_a$

and:

(3) *If G is reduced, then* $\operatorname{gr}(G_r) \simeq (\operatorname{Lie}(G)_a)_r$ *for all* r.

9.12. (A Filtration of the Hochschild Complex) The filtration of $k[G]$ as in 9.11 leads also to a filtration of the Hochschild complex $C^\bullet(G,M)$ for each G–module M. We set for all $i,n \in \mathbf{N}$

$$C^i(G,M)_{(n)} = \sum M \otimes I_1^{a(1)} \otimes I_1^{a(2)} \otimes \cdots \otimes I_1^{a(i)}, \tag{1}$$

where we sum over all i–tuples $(a(1),a(2),\dots,a(i)) \in \mathbf{N}^i$ with $\sum_j a(j) \geq n$. Because of 2.4(1), (2) and as $\Delta_M(m) - m \otimes 1 \in M \otimes I_1$ for all $m \in M$, the definition of the boundary operators in 4.14 shows

$$\partial^i\, C^i(G,M)_{(n)} \subset C^{i+1}(G,M)_{(n)} \tag{2}$$

for all i and n.

Each quotient $C^i(G,M)_{(n)}/C^i(G,M)_{(n+1)}$ can be identified with the direct sum of all

$$M \otimes (I_1^{a(1)}/I_1^{a(1)+1}) \otimes \cdots \otimes (I_1^{a(i)}/I_1^{a(i)+1}),$$

over all i–tuples $(a(1),a(2),\dots,a(i))$ with $\sum_j a(j) = n$. We can on the other hand regard M as a trivial $\mathrm{gr}(G)$–module and form $C^\bullet(\mathrm{gr}(G),M)$. The grading on $k[\mathrm{gr}(G)]$ leads in a natural way (cf. 4.20) to a grading on each $C^i(\mathrm{gr}(G),M)$. We denote the homogeneous part of degree n by $C^i(\mathrm{gr}(G),M)_n$. Then

$$C^i(G,M)_{(n)}/C^i(G,M)_{(n+1)} \simeq C^i(\mathrm{gr}(G),M)_n \tag{3}$$

for all i, n. These identifications are easily checked to be compatible with the boundary operators so that the associated graded complex of $C^\bullet(G,M)$ is isomorphic to the graded complex $C^\bullet(\mathrm{gr}(G),M) = C^\bullet(\mathrm{gr}(G),k) \otimes M$.

The general theory of filtered complexes (consult, e.g., [G], I.4) shows that there is a spectral sequence with E_1–terms $E_1^{i,j} = H^{i+j}(\mathrm{gr}(G),k)_j \otimes M$. If G is irreducible, then $\bigcap_{n>0} I_1^{n+1} = 0$ by the observation preceding 7.17(6), hence $\bigcap_{n>0} C^i(G,M)_{(n)} = 0$ for all i. Therefore in this case the spectral sequence converges to the cohomology of the original complex.

9.13. Proposition: *Suppose G is irreducible. Then there is for each G–module M a spectral sequence with*

$$E_1^{i,j} = H^{i+j}(\mathrm{gr}(G),k)_i \otimes M \;\Rightarrow\; H^{i+j}(G,M). \tag{1}$$

This is what we proved in the last subsection. Let me add that the spectral sequence is compatible with the cup product in case $M = k$ resp. with the $H^\bullet(G,k)$–module structure on $H^\bullet(G,M)$ in the general case.

If some other group H acts on G through group automorphisms, then it acts on $C^\bullet(G,k)$ preserving the filtration. Then we get a natural action of H on each term of the spectral sequence such that all differentials are homomorphisms of H–modules. Also the filtration on the abutment is compatible with the action of H. This generalises to an arbitrary G–module M if we also have an action of H on M compatible with the action of G (i.e., defining a representation of $G \rtimes H$).

9.14. Proposition: *Let G be reduced and irreducible. Set $\mathfrak{g} = \mathrm{Lie}(G)$.*

a) *There is for each G–module M a spectral sequence converging to $H^\bullet(G,M)$ with the following E_1–terms: If $p \neq 2$, then*

$$E_1^{i,j} = \bigoplus M \otimes (S^{a(1)}\mathfrak{g}^*)^{(1)} \otimes (S^{a(2)}\mathfrak{g}^*)^{(2)} \otimes \cdots \otimes \Lambda^{b(1)}\mathfrak{g}^* \otimes (\Lambda^{b(2)}\mathfrak{g}^*)^{(1)} \otimes \cdots$$

where we sum over all finite sequences $(a(n))_{n\geq 1}$ and $(b(n))_{n\geq 1}$ in $\mathbf{N}$ with

$$i+j = \sum_{n\geq 1}(2a(n)+b(n)) \qquad \text{and} \qquad i = \sum_{n\geq 1}(a(n)p^n + b(n)p^{n-1}).$$

If $p = 2$, then

$$E_1^{i,j} = \bigoplus M \otimes S^{a(1)}\mathfrak{g}^* \otimes (S^{a(2)}\mathfrak{g}^*)^{(1)} \otimes \cdots$$

where we sum over all finite sequences $(a(n))_{n\geq 1}$ in $\mathbf{N}$ with

$$i+j = \sum_{n\geq 1} a(n) \qquad \text{and} \qquad i = \sum_{n\geq 1} a(n)p^{n-1}.$$

b) *Let $r \in \mathbf{N}$, $r > 0$ and let M be a G_r–module. If we take above only r–tuples $(a(n))_{1\leq n\leq r}$ and (for $p \neq 2$) $(b(n))_{1\leq n\leq r}$, then we get the $E_1^{i,j}$–terms in a spectral sequence converging to $H^\bullet(G_r, M)$.*

Proof: This follows from 9.13 and 4.27 using 9.11(2), (3).

Remark: Again these spectral sequences are compatible with the action of some group H on G or G_r through automorphisms if H acts also on M in a compatible way (e.g., always for the trivial module $M = k$). This follows from the fact that H then acts on $\mathrm{gr}(G) = \mathfrak{g}_a$ or $\mathrm{gr}(G_r) = (\mathfrak{g}_a)_r$ through a representation on $\mathfrak{g}$ so that the isomorphisms in 4.27 are compatible with the action of H. (This applies especially to the action of G on G_r through conjugation.) The exponents (1), (2) etc. denote a twist of the H–action as in 9.10.

9.15. The spectral sequence in 9.14.b is especially simple for $r = 1$.

Lemma: *If $p = 2$, then we can compute $H^\bullet(G_1, M)$ for any G_1–module as the cohomology of a complex*

$$0 \to M \to M \otimes \mathfrak{g}^* \to M \otimes S^2\mathfrak{g}^* \to M \otimes S^3\mathfrak{g}^* \to \cdots$$

where $\mathfrak{g} = \mathrm{Lie}(G)$.

Proof: We have by 9.14 that $E_1^{i,0} = M \otimes S^i\mathfrak{g}^*$ whereas $E_1^{i,j} = 0$ for $j \neq 0$ or $i < 0$. So the only non-zero differentials in the spectral sequence are $d_1^{i,0} : E_1^{i,0} \to E_1^{i+1,0}$. They provide the maps in the complex above; its cohomology groups are the $E_2^{i,0}$, which are equal to the abutment of the spectral sequence.

Remark: Note that we do not have to assume G to be reduced and irreducible when dealing with G_1 (here and below). The assumption of irreducibility was needed to make the spectral sequence in 9.13 converge to the G–cohomology. As each G_r is irreducible, we do not need the irreducibility of G in 9.14.b. The assumption of reducedness was needed to get 9.11(3). But we have $\mathrm{gr}(G_1) = (\mathrm{Lie}(G)_a)_1$ for any G by 9.6(4).

9.16. Lemma: *Let M be a G_1–module. If $p \neq 2$, then there is a spectral sequence with*

$$E_0^{i,j} = M \otimes (S^i \mathfrak{g}^*)^{(1)} \otimes \Lambda^{j-i} \mathfrak{g}^* \;\Rightarrow\; H^{i+j}(G_1, M)$$

where $\mathfrak{g} = \mathrm{Lie}(G)$.

Proof: We have in 9.14.b as E_1–terms $E_1^{a(p-1)+b,-(p-2)a} = M \otimes (S^a \mathfrak{g}^*)^{(1)} \otimes \Lambda^{b-a} \mathfrak{g}^*$ for all $b \geq a \geq 0$; all other $E_1^{i,j}$ are 0. So $E_1^{i,j} = 0$ for $(p-2) \nmid j$, hence $d_r^{i,j} = 0$ for $r \not\equiv 1 \bmod (p-2)$ as d_r has bidegree $(r, 1-r)$. We can therefore re-index the spectral sequence by now calling $E_r^{i,j}$ the old $E_{(p-2)r+1}^{i+j,-(p-2)i}$. This then gives $E_0^{i,j}$ as above.

9.17. (Lie Algebra Cohomology) In order to compute the E_1–terms of the spectral sequence from 9.16 it will be necessary to deal with (ordinary) Lie algebra cohomology, cf., e.g., [B3], ch. I, §3, exerc. 12.

If $\mathfrak{g}$ is a finite dimensional Lie algebra over any field and if M is a $\mathfrak{g}$–module, then the Lie algebra cohomology $H^\bullet(\mathfrak{g}, M)$ of M can be computed using a complex $M \otimes \Lambda \mathfrak{g}^*$ where we take the standard grading of $\Lambda \mathfrak{g}^*$. The map $d_0 : M \to M \otimes \mathfrak{g}^*$ takes any $m \in M$ to the unique element $\sum_{j=1}^r m_j \otimes \varphi_j \in M \otimes \mathfrak{g}^*$ with $xm = \sum_{j=1}^r \varphi_j(x) m_j$ for all $x \in \mathfrak{g}$. (It is something like a comodule map.) In general, one has for any $m \in M$ and $\psi \in \Lambda^i \mathfrak{g}^*$

$$d_i(m \otimes \psi) = \sum_j m_j \otimes (\varphi_j \wedge \psi) \,+\, m \otimes d_i'(\psi) \tag{1}$$

with m_j, φ_j as above and where $d_i' : \Lambda^i \mathfrak{g}^* \to \Lambda^{i+1} \mathfrak{g}^*$ is the boundary operator in the case of the trivial module. This in turn is uniquely determined by $d_1' : \mathfrak{g}^* \to \Lambda^2 \mathfrak{g}^* \simeq (\Lambda^2 \mathfrak{g})^*$ which is the transpose of $\Lambda^2 \mathfrak{g} \to \mathfrak{g}$, $x \wedge y \mapsto -[x,y]$, and by the derivation property

$$d_{i+j}'(\varphi \wedge \psi) = d_i'(\varphi) \wedge \psi + (-1)^i \varphi \wedge d_j'(\psi) \tag{2}$$

for all $\varphi \in \Lambda^i \mathfrak{g}^*$ and $\psi \in \Lambda^j \mathfrak{g}^*$.

9.18. Lemma: *Let M be a G_1–module. If $p \neq 2$, then one has in 9.16*

$$E_1^{0,j} = H^j(\mathfrak{g}, M) \qquad \textit{for all } j \in \mathbf{N}$$

where $\mathfrak{g} = \mathrm{Lie}(G)$.

Proof: We have $E_0^{0,j} = M \otimes \Lambda^j \mathfrak{g}^*$ and $d_0^{0,j}$ maps $M \otimes \Lambda^j \mathfrak{g}^*$ to $M \otimes \Lambda^{j+1} \mathfrak{g}^*$ for all $j \in \mathbf{N}$. So we have to show that the complex $(E_0^{0,\bullet}, d_0^{0,\bullet})$ is the same as the one computing the Lie algebra cohomology.

The compatibility of the spectral sequence with the cup product in the case $k = M$ and with the corresponding module structures in general implies that the $d_0^{0,i}$ have derivation properties analogous to 9.17(1), (2). It is therefore enough to prove that $d_0^{0,0} : M \to M \otimes \mathfrak{g}^*$ and $d_0^{0,1} : \mathfrak{g}^* \to \Lambda^2 \mathfrak{g}^*$ in the case $M = k$ are the same maps as in 9.17.

In the original notation of 9.14 our present $E_0^{0,i}$ was called $E_1^{i,0}$ and arose as a subquotient of $C^i(G_1, M)_{(i)}/C^i(G_1, M)_{(i+1)}$. Any $e \in E_0^{0,i}$ has a representative

$\widetilde{e} \in C^i(G_1, M)_{(i)}$ with $\partial^i \widetilde{e} \in C^{i+1}(G_1, M)_{(i+1)}$ and $d_0^{0,i}(e)$ is the class of $\partial^i \widetilde{e}$ in the subquotient $E_0^{0,i+1}$ of $C^{i+1}(G_1, M)_{(i+1)}/C^{i+1}(G_1, M)_{(i+2)}$.

In the case $i = 0$ we have $\partial^0 : M \to M \otimes k[G_1]$, $m \mapsto \Delta_M(m) - m \otimes 1$. Fix m. We can write $\Delta_M(m) = m \otimes 1 + \sum_{i=1}^s m_i \otimes f_i$ where $f_i \in I_1 = \{f \in k[G_1] \mid f(1) = 0\}$. We have $C^0(G_1, M)_{(n)} = M$ and $C^1(G_1, M)_{(n)} = M \otimes I_1^n$ for all $n \in \mathbf{N}$, hence $E_0^{0,1} = M \otimes I_1/I_1^2 = C^1(G_1, M)_{(1)}/C^1(G_1, M)_{(2)}$. Therefore $d_0^{0,0}(m) = \sum_{i=1}^s m_i \otimes \overline{f_i}$ where $\overline{f_i} = f_i + I_1^2$. The action of any $x \in \mathfrak{g} = (I_1/I_1^2)^*$ is given by $x\,m = \sum_{i=1}^s \overline{f_i}(x) m_i$. This shows that $d_0^{0,0}$ is the same map as in 9.17.

Consider now $d_0^{0,1}$ for $M = k$. It maps $\mathfrak{g}^* = I_1/I_1^2 = C^1(G_1, k)_{(1)}/C^1(G_1, k)_{(2)}$ into a subquotient of $C^2(G_1, k)_{(2)}/C^2(G_1, k)_{(3)}$. For any $f \in I_1$ we can write $\Delta_G(f) = 1 \otimes f + f \otimes 1 + \sum_{i=1}^s f_i \otimes f_i'$ with $f_i, f_i' \in I_1$, cf. 2.4(1). Then $\partial^1 f = -\sum_{i=1}^s f_i \otimes f_i'$. So $\overline{f} = f + I_1^2 \in I_1/I_1^2$ is mapped to the class of $-\sum_{i=1}^s f_i \otimes f_i'$ in the subquotient $H^2(\operatorname{gr} G_1, k)_2$ of $C^2(G_1, k)_{(2)}/C^2(G_1, k)_{(3)} = C^2(\operatorname{gr} G_1, k)_2$. By the definition of the cup product, this is the sum of the products of $\overline{f_i} = f_i + I_1^2$ and $\overline{f_i}' = f_i' + I_1^2$ in $H^\bullet(\operatorname{gr} G_1, k)$. It belongs to the subalgebra generated by $H^1(\operatorname{gr} G_1, k) \simeq \mathfrak{g}^*$ which is identified with $\Lambda \mathfrak{g}^*$. So $d_0^{0,1}\overline{f} = -\sum_{i=1}^s f_i \wedge f_i'$. As the Lie algebra structure on $(I_1/I_1^2)^* = \operatorname{Dist}_1^+(G)$ is defined through $[x, y] = (x \otimes y - y \otimes x) \circ \Delta_G$, we see that $d_0^{0,1}$ is transposed to $x \wedge y \mapsto -[x, y]$ as claimed.

Remark: Note that this computation also gives the boundary maps in the complex of Lemma 9.15.

9.19. (Ordinary and Restricted Cohomology) If M, M' are G_1–modules, then we can interpret each $\operatorname{Ext}_{G_1}^i(M', M)$ resp. $\operatorname{Ext}_{\mathfrak{g}}^i(M', M)$ as set of equivalence classes of exact sequences

$$0 \to M \to M_1 \to M_2 \to \cdots \to M_i \to M' \to 0$$

of homomorphisms of G_1–modules (resp. $\mathfrak{g}$–modules). So there is a natural map $\operatorname{Ext}_{G_1}^i(M', M) \to \operatorname{Ext}_{\mathfrak{g}}^i(M', M)$. Taking $M' = k$ we get a natural map $H^i(G_1, M) \to H^i(\mathfrak{g}, M)$. Let us describe it explicitly for $i = 1$.

Each 1–cocycle $\varphi : \mathfrak{g} \to M$ defines an extension of $\mathfrak{g}$–modules $0 \to M \to M(\varphi) \to k \to 0$ where $M(\varphi) = M \oplus k$ as a vector space with $x \in \mathfrak{g}$ acting through $x(m, a) = (x\,m + a\varphi(x), 0)$ for all $a \in k$ and $m \in M$. One checks easily that this is an extension of G_1–modules if and only if $\varphi(x^{[p]}) = x^{p-1}\varphi(x)$ for all $x \in \mathfrak{g}$. This condition is certainly satisfied if φ is a coboundary, i.e., of the form $x \mapsto xm$ for some $m \in M$. So we get an embedding $H^1(G_1, M) \hookrightarrow H^1(\mathfrak{g}, M)$. More precisely, the image is exactly the kernel of the map associating to the class of φ as above (in $H^1(\mathfrak{g}, M)$) the map $x \mapsto \varphi(x^{[p]}) - x^{p-1}\varphi(x)$ from $\mathfrak{g}$ to M. This map is semilinear, i.e., it is additive and satisfies $\varphi(ax) = a^p \varphi(x)$ for all $a \in k$ and $x \in \mathfrak{g}$. Let $\operatorname{Hom}^s(\mathfrak{g}, M)$ denote the space of all such maps.

We have so far constructed an exact sequence

$$0 \to H^1(G_1, M) \longrightarrow H^1(\mathfrak{g}, M) \longrightarrow \operatorname{Hom}^s(\mathfrak{g}, M).$$

We can be more precise. An elementary calculation using the cocycle property $\varphi([x, y]) = x\varphi(y) - y\varphi(x)$ for all $x, y \in \mathfrak{g}$ shows $\varphi(x^{[p]}) - x^{p-1}\varphi(x) \in M^{\mathfrak{g}}$ for all $x \in \mathfrak{g}$. So we can replace $\operatorname{Hom}^s(\mathfrak{g}, M)$ by $\operatorname{Hom}^s(\mathfrak{g}, M^{\mathfrak{g}})$.

We can now go on and associate to any $\psi \in \operatorname{Hom}^s(\mathfrak{g}, M^{\mathfrak{g}})$ a p–Lie algebra $\mathfrak{g}(\psi)$ that is an extension

$$0 \to M \longrightarrow \mathfrak{g}(\psi) \longrightarrow \mathfrak{g} \to 0$$

of p–Lie algebras, where we regard M as a commutative p–Lie algebra with $m^{[p]} = 0$ for all $m \in M$. We take $\mathfrak{g}(\psi) = M \oplus \mathfrak{g}$ with Lie bracket $[(m,x),(m',x')] = (xm' - x'm, [x,x'])$ and p^{th} power $(m,x)^{[p]} = (x^{p-1}m + \psi(x), x^{[p]})$ for all $m, m' \in M$ and $x, x' \in \mathfrak{g}$. (It is left to the reader to check that this is indeed a pth power map on the semi-direct product.)

Now $\mathfrak{g}(\psi)$ and $\mathfrak{g}(0)$ are equivalent extensions if and only if there is an isomorphism $\mathfrak{g}(0) \xrightarrow{\sim} \mathfrak{g}(\psi)$ of p–Lie algebras of the form $(m,x) \mapsto (m + \varphi(x), x)$ for some $\varphi \in \operatorname{Hom}(\mathfrak{g}, M)$. Such a map is a homomorphism of Lie algebras if and only if φ is a 1–cocycle, and it is compatible with the pth power map if and only if $\psi(x) = \varphi(x^{[p]}) - x^{p-1}\varphi(x)$ for all $x \in \mathfrak{g}$. So $\mathfrak{g}(\psi)$, $\mathfrak{g}(0)$ are equivalent if and only if ψ is in the image of $H^1(\mathfrak{g}, M) \to \operatorname{Hom}^s(\mathfrak{g}, M^{\mathfrak{g}})$.

The set of all equivalence classes of all central extensions of p–Lie algebras (resp. of Lie algebras) $0 \to M \to \mathfrak{h} \to \mathfrak{g} \to 0$ such that the adjoint action of $\mathfrak{g} \simeq \mathfrak{h}/M$ on M is the given action, is a vector space in a natural way with $\mathfrak{g}(0)$ as *zero*. One can identify this space with $H^2(G_1, M)$ resp. $H^2(\mathfrak{g}, M)$, and one can show that $\psi \mapsto \mathfrak{g}(\psi)$ induces a linear map $\operatorname{Hom}^s(\mathfrak{g}, M^{\mathfrak{g}}) \to H^2(G_1, M)$. Furthermore, one can show that the image is exactly the kernel of the forgetful map $H^2(G_1, M) \to H^2(\mathfrak{g}, M)$. In this way we get an exact sequence

$$\begin{aligned} 0 &\longrightarrow H^1(G_1, M) \longrightarrow H^1(\mathfrak{g}, M) \longrightarrow \operatorname{Hom}^s(\mathfrak{g}, M^{\mathfrak{g}}) \\ &\longrightarrow H^2(G_1, M) \longrightarrow H^2(\mathfrak{g}, M) \longrightarrow \operatorname{Hom}^s(\mathfrak{g}, H^1(\mathfrak{g}, M)), \end{aligned} \tag{1}$$

where I want to refer to the original proof in [Hochschild 1] (cf. p. 575) for the last map and the exactness at the last two places to be looked at.

9.20. Proposition: *The spectral sequence in* 9.16 *has the following* E_1 *terms:*

$$E_1^{i,j} \simeq H^{j-i}(\mathfrak{g}, M) \otimes (S^i\mathfrak{g}^*)^{(1)}.$$

Proof: The derivation property of the differential

$$d_0^{i,j} : M \otimes \Lambda^{j-i}\mathfrak{g}^* \otimes (S^i\mathfrak{g}^*)^{(1)} \longrightarrow M \otimes \Lambda^{j-i+1}\mathfrak{g}^* \otimes (S^i\mathfrak{g}^*)^{(1)}$$

implies $d_0^{i,j}(m \otimes \varphi \otimes \psi) = d_0^{0,j-i}(m \otimes \varphi) \otimes \psi + (m \otimes \varphi \otimes 1)(1 \otimes d_{0,k}^{i,i}(\psi))$ for all $m \in M$, $\varphi \in \Lambda^{j-i+1}\mathfrak{g}^*$, $\psi \in (S^i\mathfrak{g}^*)^{(1)}$, where $d_{0,k}^{i,i}$ is the differential in the case $M = k$. Therefore it is, by 9.18, enough to show $d_{0,k}^{i,i} = 0$ for all i. Again the derivation property shows that it is enough to show $d_{0,k}^{1,1} = 0$.

Suppose for the moment that $G = SL_n$ for some $n \in \mathbf{N}$ with $p \nmid n$; this condition implies that $\mathfrak{g} = [\mathfrak{g}, \mathfrak{g}]$. The map $d_{0,k}^{1,1} : (\mathfrak{g}^*)^{(1)} \to \mathfrak{g}^* \otimes (\mathfrak{g}^*)^{(1)}$ is a homomorphism of G–modules where the G–action is induced by the conjugation action, see Remark 9.14. Since G is defined over $\mathbf{F}_p$, we have $(\mathfrak{g}^*)^{(1)} \simeq (\mathfrak{g}^*)^{[1]}$. This implies that G_1 (and hence $\mathfrak{g}$) acts trivially on $(\mathfrak{g}^*)^{(1)}$. Therefore the image of $d_{0,k}^{1,1}$ is contained in $(\mathfrak{g}^*)^{\mathfrak{g}} \otimes (\mathfrak{g}^*)^{(1)} \simeq (\mathfrak{g}/[\mathfrak{g},\mathfrak{g}])^* \otimes (\mathfrak{g}^*)^{(1)} = 0$. So we get $d_{0,k}^{1,1} = 0$ in this case.

A homomorphism $f : G \to H$ of algebraic k–group schemes (with G again arbitrary) induces a homomorphism $f : G_1 \to H_1$ that is just the tangent map $df : \mathfrak{g} \to \mathfrak{h} = \mathrm{Lie}(H)$ when we regard these p–Lie algebras as infinitesimal group schemes. The comorphism $f^* : k[H_1] \to k[G_1]$ induces a morphism of complexes $f^* : C^\bullet(H_1, k) \to C^\bullet(G_1, k)$ satisfying $f^*(C^i(H_1, k)_{(n)}) \subset C^i(G_1, k)_{(n)}$ for all i and n. It induces a morphism of the spectral sequences $E_r^{i,j}(H_1) \to E_r^{i,j}(G_1)$ that we consider. The maps on the present E_0–level (the E_1–level in 9.13) comes from $H^\bullet(\mathrm{gr}(H_1), k) \to H^\bullet(\mathrm{gr}(G_1), k)$ induced by $\mathrm{gr}(f^*) : k[\mathrm{gr}(H_1)] = \mathrm{gr}\, k[H_1] \to \mathrm{gr}\, k[G_1] = k[\mathrm{gr}(G_1)]$ which in turn is induced by $(df)_a : \mathrm{gr}(G_1) = \mathfrak{g}_a \to \mathfrak{h}_a = \mathrm{gr}(H_1)$. So each map $E_0^{i,j}(H_1) = \Lambda^{j-i}\mathfrak{h}^* \otimes (S^i\mathfrak{h}^*)^{(1)} \to \Lambda^{j-i}\mathfrak{g}^* \otimes (S^i\mathfrak{g}^*)^{(1)} = E_0^{i,j}(G_1)$ is induced by the transposed of $df : \mathfrak{g} \to \mathfrak{h}$. We see in particular: If df is injective, then each $E_0^{i,j}(H_1)$ maps onto $E_0^{i,j}(G_1)$; if in this case $d_{0,k}^{i,j}$ is zero for H_1, then it is also 0 for G_1.

Now the vanishing of $d_{0,k}^{1,1}$ will follow if we can embed G as a closed subgroup into some SL_n with $p \nmid n$. Since we can embed GL_n into SL_{n+1} and SL_{n+1} into SL_{n+2}, it suffices to embed G into some GL_n. The existence of such embeddings is well known, cf. [DG], II, §2, 3.4. (In our situation we may actually assume that $G = G_1$; then it suffices to find an embedding of $\mathfrak{g}$ as a p–Lie algebra into a p–Lie algebra of the form $\mathrm{End}(V)$ for some finite dimensional vector space V. This can, e.g., be done taking $V = U^{[p]}(\mathfrak{g})$ with $\mathfrak{g}$ acting through the left regular representation.)

Remark: In [Friedlander and Parshall 4] another normalisation of this spectral sequence is used. If we denote the terms in their version by $\mathbf{E}_r^{i,j}$ and the differentials by $\mathbf{d}_r^{i,j}$, then they are related as follows to the numbering used here: One has $\mathbf{E}_{2r}^{2i,j} = \mathbf{E}_{2r-1}^{2i,j} = E_r^{i,i+j}$ and $\mathbf{E}_{2r}^{2i+1,j} = \mathbf{E}_{2r-1}^{2i+1,j} = 0$ for all $i, j, r \in \mathbf{Z}$ with $r > 0$. The differentials are given by $\mathbf{d}_{2r}^{2i,j} = d_r^{i,i+j}$ and $\mathbf{d}_{2r}^{2i+1,j} = 0 = \mathbf{d}_{2r-1}^{i,j}$. We still have that $\mathbf{E}_r^{i,j} \Rightarrow H^{i+j}(G_1, M)$. In this set-up one gets $\mathbf{E}_2^{2i,j} \simeq H^j(\mathfrak{g}, M) \otimes S^i(\mathfrak{g}^*)^{(1)}$.

9.21. Using the spectral sequence from 9.16 and Proposition 9.20 one can show that $H^\bullet(G_1, k)$ is a finitely generated algebra over k (under the cup-product), and that $H^\bullet(G_1, M)$ is a finitely generated module over $H^\bullet(G_1, k)$ for any finite dimensional G_1–module M, cf. [Friedlander and Parshall 4].

More precisely: The isomorphisms $(S^i\mathfrak{g}^*)^{(1)} \simeq E_1^{i,i}$ in 9.20 for $M = k$ induce a homomorphism $(S\mathfrak{g}^*)^{(1)} \to H^\bullet(G_1, k)$ of algebras over k. (Note that all $d_r^{i,i}$ with $r > 0$ are zero; so we have maps $E_1^{i,i} \to H^{2i}(G_1, k)$.) This homomorphism turns then any $H^\bullet(G_1, M)$ as above into a finitely generated $(S^i\mathfrak{g}^*)^{(1)}$–module.

This finiteness result was generalised in [Friedlander and Suslin]. Here the result is:

Proposition: *Let G be a finite group scheme. Then $H^\bullet(G, k)$ is a finitely generated algebra over k. If M is a finite dimensional G–module, then $H^\bullet(G, M)$ is a finitely generated module over $H^\bullet(G, k)$.*

This theorem generalises not only the one mentioned above for G_1, but also an older result for abstract finite groups due independently to B. Venkov and L. Evens. Their theorem (for abstract groups) is used by Friedlander and Suslin to reduce to the case where G is infinitesimal.

If G is infinitesimal, then one can embed G as a closed subgroup scheme into some $(GL_n)_r$ and we can choose n such that $p \nmid n$. Then Friedlander's and Suslin's theorem follows from the more precise statement that there is an algebra homomorphism from

$$C_r = S(\mathfrak{gl}_n(k)^*\langle 2\rangle)^{(1)} \otimes S(\mathfrak{gl}_n(k)^*\langle 2p\rangle)^{(2)} \otimes \cdots \otimes S(\mathfrak{gl}_n(k)^*\langle 2p^{r-1}\rangle)^{(r)}$$

to $H^\bullet(G,k)$ that makes $H^\bullet(G,M)$ into a finitely generated C_r–module. Here $S(\mathfrak{gl}_n(k)^*\langle 2p^i\rangle)$ denotes $S(\mathfrak{gl}_n(k)^*)$ graded such that $\mathfrak{gl}_n(k)^*$ is the homogeneous part of degree $2p^i$. The homomorphism from C_r to $H^\bullet(G,k)$ then respects the grading.

The construction of the homomorphism $C_r \to H^\bullet(G,k)$ involves certain cohomology classes in

$$H^{2p^{i-1}}((GL_n)_r, \mathfrak{gl}_n(k)^{(i)}) \simeq \operatorname{Hom}_k(\mathfrak{gl}_n(k)^{*\,(i)}, H^{2p^{i-1}}((GL_n)_r, k))$$

discovered by Friedlander and Suslin via the theory of polynomial functors.

Remark: The finiteness results (first for G_1 and then in general) imply that the maximal spectrum of the even part of $H^\bullet(G,k)$ is a finite dimensional variety, called the *cohomology variety* of G. One can now associate to each finite dimensional G_1–module M a subvariety (the *support variety* of M) of that cohomology variety. For results on these objects one may consult [Friedlander and Parshall 5, 6], [Jantzen 12, 13, 14], [Suslin, Friedlander, and Bendel 1, 2].

CHAPTER 10

Reduction mod p

Let G be a flat group scheme over $\mathbf{Z}$. If V is a finite dimensional $G_{\mathbf{Q}}$–module, then we can find a G–module $V_{\mathbf{Z}}$, free over $\mathbf{Z}$, with $V_{\mathbf{Z}} \otimes_{\mathbf{Z}} \mathbf{Q} \simeq V$, cf. 10.4. We can then form the G_k–module $V_k = V_{\mathbf{Z}} \otimes_{\mathbf{Z}} k$ for any k. Assume that p is a prime number and that k is $\mathbf{F}_p$ (or an algebraic closure of $\mathbf{F}_p$); we say now that we get V_k from V through reduction mod p.

In general there will be more than one module (even up to isomorphism) that we can get from V through reduction mod p, as we can choose different $V_{\mathbf{Z}}$. But one can still show that they have the same composition factors. (In other words, the class of V_k in the Grothendieck group of G_k is uniquely determined by V.) This independence was proved in [Serre] generalising the corresponding statement for abstract finite groups due to Brauer. One can even generalise Brauer's lifting of idempotents if $\mathbf{Z}[G]$ is free. So every injective indecomposable G_k–module lifts to the p–adic completion of $\mathbf{Z}$. Furthermore, then Brauer's reciprocity law holds in this situation. These results were proved in [Green 1], and we follow Green's approach here.

In the case where $G_{\mathbf{Q}}$ is semi-simple, $V_{\mathbf{Z}}$ has usually been constructed as a $\mathrm{Dist}(G)$–stable lattice in V. Such a lattice is indeed a G–module (10.13). This is proved using a property of $\mathrm{Dist}(G)$ and $\mathbf{Z}[G]$ which holds (as proved by Bruhat) for any "smooth" G, cf. 10.11.

We can replace $\mathbf{Z}$ above by any Dedekind ring R and $\mathbf{F}_p$ by any residue field of R. Then the results will still hold and we do everything in this generality. (Therefore the term "mod p" occurs only in the title and the introduction of this chapter.)

10.1. (Restriction of Scalars) Let k' be a k–algebra and G a k–group functor. We observed in 2.7(6) that any G–module M leads in a natural way to a $G_{k'}$–module $M \otimes k'$: For any k'–algebra A' the group $G_{k'}(A') = G(A')$ acts as given on $M \otimes A' = (M \otimes k') \otimes_{k'} A'$.

There is a functor in the opposite direction: We can regard each $G_{k'}$–module V in a natural way as a G–module. For each k–algebra A the map $a \mapsto 1 \otimes a$ is a homomorphism of k–algebras $A \to k' \otimes A$, hence induces a group homomorphism $G(A) \to G(k' \otimes A) = G_{k'}(k' \otimes A)$ and thus an action of $G(A)$ on $V \otimes_{k'} (k' \otimes A) \simeq V \otimes A$. These actions are compatible with homomorphisms of k–algebras and lead therefore to a representation of G on V regarded as a k–module.

In the case of a group scheme we get the comodule map of V as a G–module (i.e., $V \to V \otimes k[G]$) from that as a $G_{k'}$–module (i.e., $V \to V \otimes_{k'} k'[G_{k'}]$) using the identification $V \otimes_{k'} k'[G_{k'}] = V \otimes_{k'} (k' \otimes k[G]) \simeq V \otimes k[G]$.

If M is a G–module, then the map $i_M : M \to M \otimes k'$, $m \mapsto m \otimes 1$ is a homomorphism of G–modules, if we regard the $G_{k'}$–module $M \otimes k'$ as a G–module as above. Indeed, the action of any $G(A)$ on $M \otimes k' \otimes A$ comes from the action of $G(k' \otimes A)$ on this module and the homomorphism $j_A : a \mapsto 1 \otimes a$ from A to $k' \otimes A$. We can regard $i_M \otimes \mathrm{id}_A : M \otimes A \to M \otimes k' \otimes A$ also as $\mathrm{id}_M \otimes j_A$; it is therefore compatible with the action.

The universal property of the tensor product implies that $\varphi \mapsto \varphi \circ i_M$ is a bijection $\mathrm{Hom}_{k'}(M \otimes k', V) \to \mathrm{Hom}_k(M, V)$ for any k–module M and any k'–module V. We claim that it yields a bijection

$$(1) \qquad \mathrm{Hom}_{G_{k'}}(M \otimes k', V) \xrightarrow{\sim} \mathrm{Hom}_G(M, V),$$

when M is a G–module and V a $G_{k'}$–module. As i_M is a homomorphism of G–modules, we have already checked one direction.

Suppose now that $\varphi \circ i_M$ is a homomorphism of G–modules and let us show that φ is a homomorphism of $G_{k'}$–modules. Consider any k'–algebra A' and the map $\varphi \otimes \mathrm{id}_{A'} : M \otimes k' \otimes_{k'} A' \to V \otimes_{k'} A'$. We have to show that it commutes with the action of $G_{k'}(A') = G(A')$. We can identify $M \otimes k' \otimes_{k'} A' \simeq M \otimes A'$ and then factor the map into first $(\varphi \circ i_M) \otimes \mathrm{id}_{A'} : M \otimes A' \to V \otimes A'$ and then the canonical map $V \otimes A' \to V \otimes_{k'} A'$. By assumption the first map is $G(A')$–equivariant, where we get the action of $G(A')$ on $V \otimes A'$ from that of $G_{k'}(k' \otimes A')$ on $V \otimes_{k'} (k' \otimes A') \simeq V \otimes A'$ via $A' \to k' \otimes A'$, $a \mapsto 1 \otimes a$. As $k' \otimes A' \to A'$, $b \otimes a \mapsto ba$ is a homomorphism of k'–algebras, the corresponding map $V \otimes A' \simeq V \otimes_{k'} (k' \otimes A') \to V \otimes_{k'} A'$ is compatible with $G_{k'}(k' \otimes A') \to G_{k'}(A')$, hence with the action of $G_{k'}(A')$. This is what we had to show.

10.2. Lemma: *Let k be an integral domain, G a flat k–group scheme, and M a G–module.*

a) *For all $a \in k$ the k–submodules*

$$\ker(a_{|M}) = \{\, m \in M \mid am = 0 \,\} \qquad \textit{and} \qquad aM = \{\, am \mid m \in M \,\}$$

are G–submodules of M.

b) *The torsion submodule*

$$M_{\mathrm{tor}} = \bigcup_{a \in k, a \neq 0} \ker(a_{|M})$$

is a G–submodule of M.

Proof: a) Multiplication by a is clearly an G–module endomorphism of M. Its kernel and image are G–submodules by 2.9(3). This yields the claim.

b) Let k' be the field of fractions of k. Then M_{tor} is the kernel of the homomorphism of G–modules $i_M : M \to M \otimes k'$ as in 10.1, hence a G–submodule, cf. 2.9(3).

10.3. Lemma: *Let k' be a k–algebra that is finitely presented as a k–module. Let G be a flat k–group scheme and let M, N be G–modules. If M is projective over k and if N is a direct summand of $E \otimes k[G]$ for some k–module E (regarded as a trivial G–module), then the canonical map*

$$(1) \qquad \mathrm{Hom}_G(M, N) \otimes k' \longrightarrow \mathrm{Hom}_{G_{k'}}(M \otimes k', N \otimes k')$$

is bijective.

Proof: If the result holds for M and N, then it also follows for direct summands. So we may assume that M is free over k and that $N = E \otimes k[G]$, hence that $N \otimes k' \simeq (E \otimes k') \otimes_{k'} k'[G_{k'}]$. Then 3.7(1) and Frobenius reciprocity yield isomorphisms

$$\mathrm{Hom}_G(M, N) \simeq \mathrm{Hom}_k(M, E)$$

and

$$\mathrm{Hom}_{G_{k'}}(M \otimes k', N \otimes k') \simeq \mathrm{Hom}_{k'}(M \otimes k', E \otimes k').$$

So we have to show that the canonical map $\mathrm{Hom}_k(M, E) \otimes k' \to \mathrm{Hom}_{k'}(M \otimes k', E \otimes k')$ is an isomorphism. This is clear for $M = k$ (both sides being isomorphic to $E \otimes k'$) whereas, in general, we have to use that tensoring with k' commutes with taking direct products, cf. [B2], chap. I, §2, exerc. 9a.

Remark: Let G be a k–group scheme such that $k[G]$ is a projective k–module. Then we can apply the lemma to $M = k[G] = N$ and get:

$$(\mathrm{End}_G\, k[G]) \otimes k' \simeq \mathrm{End}_{G_{k'}}(k'[G_{k'}]) \tag{2}$$

for each k–algebra k' that is finitely presented as a k–module.

10.4. (Lattices) Let R be a Dedekind ring and K its field of fractions. Recall that a *lattice* in a finite dimensional vector space V over K is a finitely generated R–submodule M of V such that the canonical map $M \otimes_R K \to V$ is an isomorphism. This map is always injective, so we can weaken the condition to "V is generated by M over K", cf. [B2], ch. VII, §4, n° 1, rem. 1. As R is a Dedekind ring, any such lattice is a projective R–module of rank equal to $\dim_K V$. If M is a lattice in V and V' a subspace of V, then $M \cap V'$ is a lattice in V' and $(M + V')/V'$ is one in V/V'. If $M_1 \subset V_1$ and $M_2 \subset V_2$ are lattices, then $M_1 \otimes_R M_2$ is one in $V_1 \otimes_K V_2$. (For more details, consult [B2], ch. VII, §4, n° 1.)

Lemma: *Let R be a Dedekind ring and G a flat group scheme over R. Let K denote the field of fractions of R. If V is a finite dimensional G_K–module, then there is a G–stable lattice in V.*

Proof: Let $v_1, v_2, \ldots, v_n$ be a basis of V. By 2.13(3) there is a G–submodule M of V containing all v_i that is finitely generated over R. As M generates V over K, it is a lattice.

10.5. Let us assume from now on (in this chapter) that R is a Dedekind ring that is not a field. Let K denote the field of fractions of R and let $\mathfrak{m}$ be a maximal ideal of R. Set $k = R/\mathfrak{m}$. Furthermore, let G be an R–group scheme such that $R[G]$ is a projective R–module.

Proposition: *Suppose that R is a complete discrete valuation ring. Then there is for each idempotent $e \in \mathrm{End}_G(k[G_k])$ an idempotent $\widetilde{e} \in \mathrm{End}_G(R[G])$ inducing e.*

Proof: The canonical map from

$$(\mathrm{End}_G\, R[G]) \otimes_R k \;\simeq\; \mathrm{End}_G\, R[G]/\mathfrak{m}\,\mathrm{End}_G\, R[G]$$

to $\mathrm{End}_{G_k} k[G_k]$ is an isomorphism by 10.3(2).

We want to apply Proposition 3.15 to the ring $\mathrm{End}_G R[G]$ and its chain of ideals $\mathfrak{m}_i = \mathfrak{m}^i \mathrm{End}_G R[G]$. If that is possible, then our claim follows immediately. So we have to prove that naturally

$$\mathrm{End}_G R[G] \xrightarrow{\sim} \varprojlim \mathrm{End}_G R[G]/\mathfrak{m}^i \mathrm{End}_G R[G]. \tag{1}$$

If M is a G–submodule of $R[G]$ that is finitely generated over R, then it is a free R–module and we have an isomorphism $\mathrm{Hom}_G(M, R[G]) \simeq M^*$ by Frobenius reciprocity. As R is complete, we get

$$\mathrm{Hom}_G(M, R[G]) \xrightarrow{\sim} \varprojlim \mathrm{Hom}_G(M, R[G])/\mathfrak{m}^i \mathrm{Hom}_G(M, R[G]).$$

This implies (1) as $R[G]$ is the direct limit of such M.

10.6. Corollary: *Let R be as in* 10.5. *For each indecomposable and injective G_k–module Q there is a direct summand $\widetilde{Q}$ of $R[G]$ with $Q \simeq \widetilde{Q} \otimes_R k$.*

Proof: We may assume that Q is a direct summand of $k[G_k]$. (Combine 3.16 and 3.10.) Therefore we can find $\varphi \in \mathrm{End}_{G_k}(k[G_k])$ idempotent with $Q \simeq \varphi(k[G_k])$. Let $\psi \in \mathrm{End}_G(R[G])$ be an idempotent inducing φ. Then $\psi(R[G])$ is a direct summand of $R[G]$ and

$$Q \simeq \varphi(k[G_k]) = (\psi \otimes \mathrm{id}_k)k[G_k] \simeq \psi(R[G]) \otimes k.$$

Remark: For $G = SL_2$ and $G = GL_2$ one can find explicit decompositions of $R[G]$ in [Winter] and [Sullivan 1].

10.7. (Reciprocity) For any finite dimensional G_K–module V we can find by 10.4 a G–stable lattice V_R in V and then form the G_k–module $V_k = V_R \otimes_R k$. We have obviously

$$\dim(V_k) = \mathrm{rk}_R(V_R) = \dim_K(V).$$

The choice of a G–stable lattice is not unique and different choices will lead in general to non-isomorphic G_k–modules. We claim, however, that the composition factors of V_k are uniquely determined by V.

Let E be a simple G_k–module and let Q_E be an injective hull of E, cf. 3.16/17. The multiplicity $[V_k : E]_{G_k}$ of E as a composition factor of V_k is then equal to (cf. 3.17(3))

$$[V_k : E]_{G_k} = \dim \mathrm{Hom}_{G_k}(V_k, E)/\dim \mathrm{End}_{G_k}(E). \tag{1}$$

Denote by $\widehat{R}$ the $\mathfrak{m}$–adic completion of R and by $\widehat{K}$ the field of fractions of $\widehat{R}$. We can identify k with the residue field of $\widehat{R}$. By 10.6 there is a direct summand $\widetilde{Q}_E$ of the $G_{\widehat{R}}$–module $\widehat{R}[G_{\widehat{R}}] = R[G] \otimes_R \widehat{R}$ with $\widetilde{Q}_E \otimes_{\widehat{R}} k \simeq Q_E$. Now 4.18(1) implies (as $V_k \simeq (V_R \otimes_R \widehat{R}) \otimes_{\widehat{R}} k$)

$$\mathrm{Hom}_{G_{\widehat{R}}}(V_R \otimes_R \widehat{R}, \widetilde{Q}_E) \otimes_{\widehat{R}} k \simeq \mathrm{Hom}_{G_k}(V_k, Q_E). \tag{2}$$

On the other hand, $(V_R \otimes_R \widehat{R}) \otimes_{\widehat{R}} \widehat{K} \simeq V \otimes_K \widehat{K}$ and $\widehat{K}$ is flat over $\widehat{R}$, so already 2.10(7) implies

$$\text{(3)} \qquad \operatorname{Hom}_{G_{\widehat{R}}}(V_R \otimes_R \widehat{R}, \widetilde{Q}_E) \otimes_{\widehat{R}} \widehat{K} \simeq \operatorname{Hom}_{G_{\widehat{K}}}(V \otimes_K \widehat{K}, \widetilde{Q}_E \otimes_{\widehat{R}} \widehat{K}).$$

We have by Frobenius reciprocity $\operatorname{Hom}_{G_{\widehat{R}}}(V_R \otimes_R \widehat{R}, \widehat{R}[G_{\widehat{R}}]) \simeq (V_R \otimes_R \widehat{R})^*$. Since $\widetilde{Q}_E$ is a direct summand of $\widehat{R}[G_{\widehat{R}}]$, this shows that $\operatorname{Hom}_{G_{\widehat{R}}}(V_R \otimes_R \widehat{R}, \widetilde{Q}_E)$ is finitely generated over $\widehat{R}$. It is therefore a lattice in the right hand side in (3); in particular, it is a free $\widehat{R}$–module of rank equal to the dimension over $\widehat{K}$ of the right hand side in (3). Now a look at (2) yields the "*Brauer reciprocity formula*"

$$\text{(4)} \qquad [V_k : E]_{G_k} = \dim_{\widehat{K}} \operatorname{Hom}_{G_{\widehat{K}}}(V \otimes_K \widehat{K}, \widetilde{Q}_E \otimes_{\widehat{R}} \widehat{K}) / \dim \operatorname{End}_{G_k}(E).$$

10.8. Assume in addition that each simple G_k–module E (resp. each simple G_K–module V) satisfies $\operatorname{End}_{G_k}(E) = k$ (resp. $\operatorname{End}_{G_K}(V) = K$) and that each G_K–module is semi-simple. (This is, e.g., satisfied for a split connected reductive group if $\operatorname{char}(K) = 0$, see Part II.) In order to simplify, let us assume that $R = \widehat{R}$.

Consider a simple G_K–module V and a simple G_k–module E. Let us construct $\widetilde{Q}_E$ and V_k as in 10.7. The semi-simplicity of $\widetilde{Q}_E \otimes_R K$ and $\operatorname{End}_{G_K}(V) = K$ imply that $\dim_K \operatorname{Hom}_{G_K}(V, \widetilde{Q}_E \otimes_R K)$ is equal to the multiplicity of V as a composition factor of $\widetilde{Q}_E \otimes_R K$. So 10.7(4) yields

$$\text{(1)} \qquad [V_k : E]_{G_k} = [\widetilde{Q}_E \otimes_R K : V]_{G_K}.$$

If G is the finite group scheme associated to an abstract finite group Γ by the procedure of 8.5.a (working over R instead of k), then we get here Brauer's original theorem.

10.9. (Grothendieck Groups) Return to the more general situation of 10.7. We can interpret the result as a statement about Grothendieck groups.

Recall that one can associate a Grothendieck group to each abelian category. One starts with the free abelian group generated by the objects of the category (let $[M]$ denote the generator corresponding to an object M) and one then divides out the subgroup generated by all $[M] - [M'] - [M'']$ for all short exact sequences $0 \to M' \to M \to M'' \to 0$ in the category.

Let us denote by $\mathcal{R}(G)$ the Grothendieck group of all those G–modules that are finitely generated over R. Define $\mathcal{R}(G_K)$ and $\mathcal{R}(G_k)$ by analogy. Then $\mathcal{R}(G_K)$ and $\mathcal{R}(G_k)$ are free abelian groups with the classes $[E']$ resp. $[E]$ of all simple G_K–modules E' resp. of all simple G_k–modules E as a basis. For any finite dimensional G_k–module M one has

$$[M] = \sum_E [M : E]_{G_k} [E]$$

where E runs through a system of representatives of isomorphism classes of simple G_k–modules. (Similarly for G_K.) In these cases (over a field) the Grothendieck groups have a natural ring structure induced by the tensor product, i.e., with $[M]\,[M'] = [M \otimes M']$.

We can now deduce from 10.7(4) that the class of $[V_k]$ of V_k is uniquely determined by V and does not depend on the choice of V_R. In this way one easily gets a homomorphism of rings $\mathcal{R}(G_K) \to \mathcal{R}(G_k)$ with $[V] \mapsto [V_k]$.

10.10. Let me mention some results about $\mathcal{R}(G)$ proved in [Serre]. The map $M \mapsto M \otimes_R K$ induces a homomorphism $\mathcal{R}(G) \to \mathcal{R}(G_K)$. Its kernel is equal to the subgroup $\mathcal{R}_{\text{tor}}(G)$ of $\mathcal{R}(G)$ generated by all $[M]$ such that M is a (finitely generated) torsion module (and a G–module). Lemma 10.4 implies that the map is surjective, i.e., that we get an exact sequence of the form

$$0 \to \mathcal{R}_{\text{tor}}(G) \longrightarrow \mathcal{R}(G) \longrightarrow \mathcal{R}(G_K) \to 0. \tag{1}$$

Consider the category of all G–modules that are finitely generated and projective over R and let $\mathcal{R}_{\text{pr}}(G)$ denote its Grothendieck group. The inclusion of categories induces a homomorphism from $\mathcal{R}_{\text{pr}}(G)$ to $\mathcal{R}(G)$ that turns out to be an isomorphism

$$\mathcal{R}_{\text{pr}}(G) \xrightarrow{\sim} \mathcal{R}(G). \tag{2}$$

The reduction modulo $\mathfrak{m}$ (i.e., $M \mapsto M \otimes_R k = M/\mathfrak{m}M$) yields a homomorphism $\mathcal{R}_{\text{pr}}(G) \to \mathcal{R}(G_k)$, hence by (2) also $\mathcal{R}(G) \to \mathcal{R}(G_k)$. One checks that $\mathcal{R}_{\text{tor}}(G)$ is mapped to 0 and gets $\mathcal{R}(G_K) \to \mathcal{R}(G_k)$ by (1). This is the same map as the one constructed using 10.7(4).

If R is a principal ideal domain, then $\mathcal{R}_{\text{tor}}(G) = 0$. (If M is a G–stable lattice in a finite dimensional G_K–module V, then $[M/aM] = 0$ in $\mathcal{R}(G)$ for all $a \in R$, $a \neq 0$, as $M \simeq aM$. One can show that $\mathcal{R}_{\text{tor}}(G)$ is generated by such $[M/aM]$.)

Let me point out that in [Serre] the R–module $R[G]$ is not assumed to be projective, only to be flat.

10.11. (Smooth Group Schemes) Set $\Pi(R)$ equal to the set of all maximal ideals of R.

Let us drop the assumption that $R[G]$ is projective. We shall assume instead that G is *smooth* (cf. [EGA], IV, §17 or [DG], I, §4, n° 4). Then $R[G]$ is a flat R–module and a finitely presented R–algebra. So the natural map from $R[G]$ to $K[G_K] = R[G] \otimes_R K$ is injective and we shall always identify $R[G]$ with its image. Furthermore, (cf. [B2], ch. II, §3, cor. 4 du th. 1)

$$R[G] = \bigcap_{\mathfrak{p} \in \Pi(R)} R[G]_{\mathfrak{p}}. \tag{1}$$

Note that we can regard $R[G]_{\mathfrak{p}}$ as $R_{\mathfrak{p}}[G_{\mathfrak{p}}]$.

Set $I = I_1 = \{f \in R[G] \mid f(1) = 0\}$. The smoothness of G implies also that all I^n/I^{n+1} are finitely generated and projective R–modules (cf. [EGA], 0_{IV}, 19.5.4). So G is infinitesimally flat and $\text{Dist}(G)$ resp. each $\text{Dist}_n(G)$ is naturally embedded into $\text{Dist}(G_K) \simeq \text{Dist}(G) \otimes_R K$ resp. $\text{Dist}_n(G_K) \simeq \text{Dist}_n(G) \otimes_R K$

By definition, the smoothness of G implies that G_K and all $G_{R/\mathfrak{p}}$ with $\mathfrak{p} \in \Pi(R)$ are smooth. If these schemes are in addition irreducible, then they are even integral (cf. [DG], II, §5, 2.1), so $K[G_K]$ and all $(R/\mathfrak{p})[G_{R/\mathfrak{p}}]$ are integral domains. Hence so is $R[G] \subset K[G_K]$. Therefore 7.14–7.16 can be applied to G in this case.

10.12. Proposition: *If G is smooth, if G_K and all $G_{R/\mathfrak{p}}$ with $\mathfrak{p} \in \Pi(R)$ are irreducible, then*

$$R[G] = \{f \in K[G_K] \mid \mu(f) \in R \textit{ for all } \mu \in \text{Dist}(G)\}. \tag{1}$$

Proof: Of course, we only have to prove one inclusion: "$\supset$". Because of 10.11(1) we can restrict ourselves to the corresponding result for all $R[G]_{\mathfrak{p}} = R_{\mathfrak{p}}[G_{\mathfrak{p}}]$. So we can assume that R is a discrete valuation ring with a unique maximal ideal $\mathfrak{p} = R\pi$. Let $\widehat{R}$ denote the $\mathfrak{p}$–adic completion of R.

For any $f \in K[G_K]$, $f \notin R[G]$ there is $n \in \mathbf{N}$ with $\pi^n f \notin R[G]$, but $\pi^{n+1} f \in R[G]$. So the claim will follow if we can prove:

$$\pi R[G] = \{f \in R[G] \mid \mu(f) \in R\pi \text{ for all } \mu \in \operatorname{Dist}(G)\}. \tag{2}$$

Set $I = \{f \in R[G] \mid f(1) = 0\}$ and $J = I + \pi R[G] = I \oplus \pi R1$. So $R[G]/J \simeq R/\mathfrak{p}$ and J is a maximal ideal of $R[G]$. It contains the ideal $\pi R[G]$ which is prime as $R[G]/\pi R[G] \simeq (R/\mathfrak{p})[G_{R/\mathfrak{p}}]$. If we localise at J, we get thus

$$\pi R[G] = R[G] \cap \pi R[G]_J. \tag{3}$$

Let A denote the J–adic completion of the local ring $R[G]_J$. Then A is a faithfully flat $R[G]_J$–module (cf. [B2], ch. III, §3, prop. 9), so [B2], chap. I, §3, prop. 9 together with (3) yields

$$\pi R[G] = R[G] \cap \pi A. \tag{4}$$

Because of (4) it will be enough to show

$$\{f \in R[G] \mid \mu(f) \in R\pi \text{ for all } \mu \in \operatorname{Dist}(G)\} \subset \pi A. \tag{5}$$

Then (2) will follow.

Choose $T_1, T_2, \dots, T_r \in JR[G]_J$ such that their images in $JR[G]_J/((JR[G]_J)^2 + \mathfrak{p}R[G]_J)$ form a basis of this vector space over $R/\mathfrak{p}$. As $R[G]_J = R + JR[G]_J$, we have

$$JR[G]_J = IR[G]_J + \mathfrak{p}R[G]_J = I + (JR[G]_J)^2 + \mathfrak{p}R[G]_J.$$

So we can (and shall) choose $T_1, T_2, \dots, T_r \in I$. Now the smoothness of G implies (cf. [DG], I, §4, 5.9)

$$\widehat{R}[\![T_1, T_2, \dots, T_r]\!] \xrightarrow{\sim} A. \tag{6}$$

For any $\alpha = (\alpha_1, \alpha_2, \dots \alpha_r) \in \mathbf{N}^r$ let $c_\alpha : A \to \widehat{R}$ be the map associating to each $f \in A$ the coefficient of $T^\alpha = \prod_{i=1}^r T_i^{\alpha_i}$ in the power series expansion of f arising from (6). Each $f \in I$ satisfies $c_0(f) = 0$ (as $I \subset JR[G]_J$), so c_α annihilates $I^{|\alpha|+1}$ where $|(\alpha_1, \alpha_2, \dots \alpha_r)| = \sum_i \alpha_i$. So (cf. [B2], ch. I, §2, prop. 10):

$$\begin{aligned}(c_\alpha)_{|R[G]} \in \operatorname{Hom}_R(R[G]/I^{|\alpha|+1}, \widehat{R}) &\simeq \operatorname{Hom}_R(R[G]/I^{|\alpha|+1}, R) \otimes_R \widehat{R} \\ &\simeq \operatorname{Dist}_{|\alpha|}(G) \otimes_R \widehat{R}.\end{aligned}$$

So, if $f \in R[G]$ with $\mu(f) \in R\pi$ for all $\mu \in \operatorname{Dist}(G)$, then $c_\alpha(f) \in \pi\widehat{R}$ for all $\alpha \in \mathbf{N}^r$, hence $f \in \pi A$. This implies (5) and the proposition.

Remark: This proposition was announced in [BrT], 3.5.3.1. The proof has been communicated to me by F. Bruhat.

10.13. Proposition: *Assume that G is smooth and that G_K and all $G_{R/\mathfrak{p}}$ with $\mathfrak{p} \in \Pi(R)$ are irreducible. Let V be a finite dimensional G_K–module and M an R–lattice in V. Then M is G–stable if and only if* $\operatorname{Dist}(G)M = M$.

Proof: As R is a Dedekind ring, the R–module M is projective of finite rank. We can add an R–module M' to M and $M' \otimes_R K$ to V (with trivial action of G_K). So we may assume that M is free over R.

Choose a basis $m_1, m_2, \dots, m_r$ of M over R. Then this is also a basis of V over K. There are (uniquely determined) $f_{ij} \in K[G_K]$ with

$$\Delta_V(m_i) = \sum_{j=1}^{r} m_j \otimes f_{ij}$$

for all i. Now M is G–stable if and only if $f_{ij} \in R[G]$ for all i, j. On the other hand, $\mu m_i = \sum_{j=1}^{r} \mu(f_{ij}) m_j$ for all $\mu \in \operatorname{Dist}(G)$, so $M = \operatorname{Dist}(G)M$ if and only if $\mu(f_{ij}) \in R$ for all i and j. Now the claim follows from 10.12(1).

Remark: This result generalises to all group schemes satisfying 10.12(1).

10.14. Return to the general situation from 10.1 and assume that $k' = k/\mathfrak{a}$ for some ideal $\mathfrak{a}$ in G. Let G be a flat k–group scheme.

We have by 10.1 adjoint functors $\mathcal{F}$ from $\{G_{k'}\text{–modules}\}$ to $\{G\text{–modules}\}$ (mapping any V to V considered as a G–module) and $\mathcal{G}$ from $\{G\text{–modules}\}$ to $\{G_{k'}\text{–modules}\}$ (mapping any M to $M \otimes_k k'$).

If V is a k'–module, then we have $V \otimes_k k' = V \otimes_k (k/\mathfrak{a}) \xrightarrow{\sim} V/\mathfrak{a}V = V$. If V is a $G_{k'}$–module, then this yields an isomorphism $\mathcal{G}(\mathcal{F}(V)) \xrightarrow{\sim} V$ of $G_{k'}$–modules: For each k'–algebra A' the induced isomorphism $\mathcal{G}(\mathcal{F}(V)) \otimes_{k'} A' = (V \otimes_k k') \otimes_{k'} A' \xrightarrow{\sim} V \otimes_{k'} A'$ is compatible with the action of $G_{k'}(A') = G(A') = G(k' \otimes_k A')$ since also $k' \otimes_k A' = (k/\mathfrak{a}) \otimes_k A' \xrightarrow{\sim} A'/\mathfrak{a}A' = A'$.

On the other hand, if M is a G–module with $\mathfrak{a}M = 0$, then $M \otimes_k k' = M \otimes_k (k/\mathfrak{a}) \xrightarrow{\sim} M/\mathfrak{a}M = M$. Then the homomorphism of G–modules $i_M : M \to M \otimes_k k' = \mathcal{F}(\mathcal{G}(M))$ in 10.1 identifies with the identity map on M and is an isomorphism.

We see thus that $\mathcal{F}$ and $\mathcal{G}$ induce equivalences of categories between $\{G_{k'}\text{–modules}\}$ and $\{G\text{–modules } M \text{ with } \mathfrak{a}M = 0\}$.

10.15. (Simple Modules) Let G again be a flat k–group scheme over arbitrary k. In 2.14 we considered simple G–modules in case k is a field. But the definition makes clearly sense over general k. If $\mathfrak{m}$ is a maximal ideal in k and if V is a simple $G_{k/\mathfrak{m}}$–module, then V considered as a G–module (as in 10.1) is simple: Any G–submodule is by 10.14 also a $G_{k/\mathfrak{m}}$–submodule. We claim that conversely:

Lemma: *If M is a simple G–module, then there exists a maximal ideal $\mathfrak{m}$ in k such that M is a simple $G_{k/\mathfrak{m}}$–module considered as a G–module.*

Proof: For each $a \in k$ the action of a on M is a homomorphism $M \to M$ of G–modules. So both kernel and image are G–submodules. The simplicity of G implies now that either $aM = 0$ or that $aM = M$ and $am \neq 0$ for all $m \in M$, $m \neq 0$. Set $\mathfrak{a} = \{a \in k \mid aM = 0\}$. This is clearly a prime ideal in k. We have $\mathfrak{a}M = 0$ and can thus (see 10.14) think of M as of a $G_{k/\mathfrak{a}}$–module. Then M is a simple

$G_{k/\mathfrak{a}}$–module. We have to show that $\mathfrak{a}$ is a maximal ideal in k, i.e., that $k/\mathfrak{a}$ is a field.

We may now replace k by $k/\mathfrak{a}$. So we may assume that k is an integral domain, that M is a simple G–module which is torsion free over k such that $aM = M$ for all $a \in k$, $a \neq 0$. So the actions of these a on M are bijective. This means that the k–module structure on M extends to the field of fractions K of k. It is therefore isomorphic as a k–module to a direct sum of copies of K. The simplicity of M implies by 2.13(3) that M is finitely generated over k. Therefore also K is finitely generated as a k–module. But this is possible only if $K = k$.

Part II

Representations of Reductive Groups

CHAPTER 1

Reductive Groups

The purpose of this first chapter is to introduce split reductive group schemes and their most important subgroups, to fix a lot of notations (which will be used throughout part II), and to mention (without proof) the main properties of these objects. Furthermore, the algebra of distributions on such a group scheme is described and the relationship between the representation theories of the group and its algebra of distributions is discussed.

In the case of an algebraically closed field the reader ought to be familiar with the notions and results described here from [Bo], [Hu2], or [Sp2]. (Except for the part on the algebra of distributions, of course.) So the reader is asked to believe that everything she or he knows (about these groups) extends nicely to the case of an arbitrary ground ring, which we assume to be an integral domain in order to simplify a few technical details.

The existence of these group schemes over $\mathbf{Z}$ (hence over any k) was first proved in [Chevalley]. They were then characterised by Demazure in [SGA3]. There one can find proofs of most of the results mentioned below and one can use Demazure's thesis [D] as a guide to where to find them. One can also find many results in [Borel 1].

The algebras of distributions were first determined in [Haboush 3]. At least in the semi-simple and simply connected case they had been used before as they coincide in this case (for $k = \mathbf{Z}$) with Kostant's $\mathbf{Z}$–form of the enveloping algebra of a complex semi-simple Lie algebra with the same root system. (See [St1], [Kostant 3], and [Borel 1] for more details.)

More or less the only result proved in this chapter is the isomorphism theorem: A connected reductive group (over an algebraically closed field) is determined uniquely (up to isomorphism) by its root datum. The proof given here (1.14) goes back to Takeuchi. It uses the algebra of distributions and avoids case-by-case considerations.

Throughout this chapter k denotes an integral domain.

1.1. (Split Reductive Groups) Let $G_{\mathbf{Z}}$ be a *split* and *connected reductive* algebraic $\mathbf{Z}$–group. Set $G_A = (G_{\mathbf{Z}})_A$ for any $\mathbf{Z}$–algebra A and $G = G_k$.

For an algebraically closed field K, the K–group G_K is reduced, connected, and reductive. The ring $\mathbf{Z}[G_{\mathbf{Z}}]$ is a free $\mathbf{Z}$–module, so $k[G]$ is free over k and G is flat. Furthermore, $\operatorname{Lie}(G_{\mathbf{Z}})$ is a free $\mathbf{Z}$–module of finite rank and we have

(1) $$\operatorname{Lie}(G) = \operatorname{Lie}(G_{\mathbf{Z}}) \otimes_{\mathbf{Z}} k.$$

Let $T_{\mathbf{Z}} \subset G_{\mathbf{Z}}$ be a *split maximal torus* of $G_{\mathbf{Z}}$; set $T_A = (T_{\mathbf{Z}})_A$ for any $\mathbf{Z}$–algebra A and $T = T_k$. Then $T_{\mathbf{Z}}$ is isomorphic to a direct product of, say, r copies of the multiplicative group over $\mathbf{Z}$. The integer r is uniquely determined as the rank of the free abelian group $X(T_{\mathbf{Z}})$; it is called the *rank* of G, denoted by $\operatorname{rk} G$.

For an algebraically closed field K, the K–group T_K is reduced; it is a maximal torus in G_K. The k–group T is isomorphic to $(G_m)^r$ and $X(T) \simeq X(T_{\mathbf{Z}})$ is isomorphic to $\mathbf{Z}^r$. (This uses the integrality of k, cf. I.2.5(1).) Any T–module M has a direct sum decomposition into weight spaces (cf. I.2.11(3))

(2) $$M = \bigoplus_{\lambda \in X(T)} M_\lambda.$$

The λ with $M_\lambda \neq 0$ are called the *weights* of M. If we apply this to the adjoint representation on Lie(G), then the corresponding decomposition has the form

(3) $$\operatorname{Lie} G \; = \; \operatorname{Lie} T \oplus \bigoplus_{\alpha \in R} (\operatorname{Lie} G)_\alpha .$$

Here R is the set of non-zero weights of Lie(G). So (3) amounts to

(4) $$(\operatorname{Lie} G)_0 \; = \; \operatorname{Lie} T.$$

The elements of R are called the *roots* of G with respect to T; the set R is called the *root system* of G with respect to T. For any α in R the "root subspace" $(\operatorname{Lie} G)_\alpha$ is a free k–module of rank 1.

Note that, using (1), R can be identified with the root system of $G_{\mathbf{Z}}$ with respect to $T_{\mathbf{Z}}$. One has for all $\alpha \in R$

(5) $$(\operatorname{Lie} G)_\alpha = (\operatorname{Lie} G_{\mathbf{Z}})_\alpha \otimes_{\mathbf{Z}} k.$$

One basic property of R is that

(6) $$R = -R.$$

1.2. (Root Subgroups) For each $\alpha \in R$ there is a *root homomorphism*

(1) $$x_\alpha : G_a \longrightarrow G,$$

with

(2) $$t\, x_\alpha(a)\, t^{-1} = x_\alpha(\alpha(t)\, a)$$

for any k–algebra A and all $t \in T(A)$, $a \in A$ such that the tangent map dx_α induces an isomorphism

(3) $$dx_\alpha : \operatorname{Lie}(G_a) \xrightarrow{\sim} (\operatorname{Lie} G)_\alpha .$$

Such a root homomorphism is unique up to multiplication by a unit in k (acting on G_a). We shall always assume that x_α arises (by base change) from a similar homomorphism over $\mathbf{Z}$ making it unique up to a sign change.

The functor $A \mapsto x_\alpha(G_a(A)) = x_\alpha(A)$ is a closed subgroup of G denoted by U_α. It is called the *root subgroup* of G corresponding to α. So x_α is an isomorphism $G_a \xrightarrow{\sim} U_\alpha$ and we have

$$\text{(4)} \qquad \operatorname{Lie}(U_\alpha) = (\operatorname{Lie} G)_\alpha.$$

We denote the corresponding group over $\mathbf{Z}$ by $U_{\alpha,\mathbf{Z}}$ and have then $U_\alpha = (U_{\alpha,\mathbf{Z}})_k$.

For two roots α, β with $\alpha + \beta \neq 0$, there are integers c_{ij} for all $i, j > 0$ with $i\alpha + j\beta \in R$ such that the commutator

$$(x_\alpha(a), x_\beta(b)) = x_\alpha(a)x_\beta(b)x_\alpha(-a)x_\beta(-b)$$

is given (for all A and all $a, b \in A$) by

$$\text{(5)} \qquad (x_\alpha(a), x_\beta(b)) = \prod_{i,j>0} x_{i\alpha+j\beta}(c_{ij}a^i b^j).$$

Here the product has to be taken in a fixed (but arbitrarily chosen) order. The c_{ij} will depend on that choice. (We have $c_{ij} \in \mathbf{Z}$ as we take the x_α over $\mathbf{Z}$.)

1.3. (Coroots and Rank-1-Subgroups) The set

$$\text{(1)} \qquad Y(T) = \operatorname{Hom}(G_m, T)$$

has a natural structure as an abelian group. We usually write the group law additively (as we do for $X(T) = \operatorname{Hom}(T, G_m)$). Now $T \simeq (G_m)^r$ implies $Y(T) \simeq \operatorname{End}(G_m)^r \simeq \mathbf{Z}^r$.

For any $\lambda \in X(T)$ and $\varphi \in Y(T)$ we have $\lambda \circ \varphi \in \operatorname{End}(G_m) \simeq \mathbf{Z}$, so there is a unique integer $\langle \lambda, \varphi \rangle$ such that $\lambda \circ \varphi$ is the map $a \mapsto a^{\langle \lambda, \varphi \rangle}$ on each $G_m(A) = A^\times$. This pairing $\langle\ ,\ \rangle$ on $X(T) \times Y(T)$ is easily seen to be bilinear and to induce an isomorphism $Y(T) \simeq \operatorname{Hom}_{\mathbf{Z}}(X(T), \mathbf{Z})$.

The definition of $Y(T)$ generalises obviously to any commutative group scheme (instead of T), similarly for the pairing. We have $Y(T_{\mathbf{Z}}) \simeq Y(T)$; this isomorphism is compatible with the pairings and with $X(T_{\mathbf{Z}}) \simeq X(T)$.

For any $\alpha \in R$ there is a homomorphism

$$\text{(2)} \qquad \varphi_\alpha : SL_2 \longrightarrow G$$

such that for a suitable normalisation of x_α and $x_{-\alpha}$

$$\text{(3)} \qquad \varphi_\alpha \begin{pmatrix} 1 & a \\ 0 & 1 \end{pmatrix} = x_\alpha(a) \qquad \text{and} \qquad \varphi_\alpha \begin{pmatrix} 1 & 0 \\ a & 1 \end{pmatrix} = x_{-\alpha}(a)$$

for any A and $a \in A$. If so, then

$$\text{(4)} \qquad n_\alpha(a) = x_\alpha(a)x_{-\alpha}(-a^{-1})x_\alpha(a) = \varphi_\alpha \begin{pmatrix} 0 & a \\ -a^{-1} & 0 \end{pmatrix} \in N_G(T)(A)$$

and

$$\text{(5)} \qquad \alpha^\vee(a) = n_\alpha(a)n_\alpha(1)^{-1} = \varphi_\alpha \begin{pmatrix} a & 0 \\ 0 & a^{-1} \end{pmatrix} \in T(A)$$

for any $a \in A^\times$ and any A. Obviously $\alpha^\vee \in Y(T)$. It is called the *coroot* or *dual root* corresponding to α. It is uniquely determined by α.

We shall usually choose $\varphi_{-\alpha}$ such that

(6)
$$\varphi_{-\alpha}\begin{pmatrix} a & b \\ c & d \end{pmatrix} = \varphi_\alpha \begin{pmatrix} d & c \\ b & a \end{pmatrix}$$

for all $\left(\begin{smallmatrix} a & b \\ c & d \end{smallmatrix}\right)$ in SL_2. Note that

$$\begin{pmatrix} d & c \\ b & a \end{pmatrix} = \begin{pmatrix} -1 & 0 \\ 0 & 1 \end{pmatrix} {}^t\!\begin{pmatrix} a & b \\ c & d \end{pmatrix}^{-1} \begin{pmatrix} -1 & 0 \\ 0 & 1 \end{pmatrix}^{-1}$$

for all matrices in SL_2. Note also that $(-\alpha)^\vee(a) = \alpha^\vee(a)^{-1}$ for all a.

If φ_α is not injective, then its kernel consists of all $\left(\begin{smallmatrix} a & 0 \\ 0 & a \end{smallmatrix}\right)$ with $a^2 = 1$. This occurs if and only if $\lambda(\alpha^\vee(a)) = 1$ for all such a and all $\lambda \in X(T)$. So:

(7)
$$\ker(\varphi_\alpha) \simeq \mu_{(2)} \iff \alpha^\vee \in 2Y(T).$$

The image of φ_α is normalised by T and intersects T in $\alpha^\vee(G_m)$ as any element in $\varphi_\alpha^{-1}(T)$ centralises all diagonal matrices in SL_2, hence is a diagonal matrix. So we have inside G the product subgroup

(8)
$$G_\alpha \;=\; \varphi_\alpha(SL_2)T \;\simeq\; (\varphi_\alpha(SL_2) \rtimes T)\,/\,\alpha^\vee(G_m).$$

Each G_α is a split and connected reductive group with maximal torus T and root system $\{\alpha, -\alpha\}$. One has $G_\alpha = G_{-\alpha}$ and

(9)
$$G_\alpha = C_G(\ker(\alpha)).$$

Let G' be a flat group scheme over k and let φ, φ' be homomorphisms $G \to G'$. Choose a basis S for the root system R (cf. 1.5 below). Then:

(10)
$$\varphi_{|G_\alpha} = (\varphi')_{|G_\alpha} \text{ for all } \alpha \in S \implies \varphi = \varphi'.$$

Indeed, let K be an algebraic closure of the field of fractions of k. As $k[G']$ is a flat k–module, the natural map $k[G'] \to K[G'_K]$ is injective. Therefore φ is uniquely determined by φ_K, hence also by $\varphi_{|G(K)}$ as G_K is reduced. The group $G(K)$ is generated by $T(K)$ and all $U_\alpha(K)$ with $\alpha \in S \cup (-S)$, hence by all $G_\alpha(K)$ with $\alpha \in S$. So φ is uniquely determined by all $\varphi_{|G_\alpha}$ with $\alpha \in S$.

1.4. (The Weyl Group) The set R together with the map $\alpha \mapsto \alpha^\vee$ is a root system (as in [B3], ch. VI, §1, n° 1) in the space generated by R in $X(T) \otimes_{\mathbf{Z}} \mathbf{R}$. The set $R^\vee = \{\alpha^\vee \mid \alpha \in R\}$ is called the *dual root sytem* of R. Let us denote by s_α for each α the corresponding reflection on $X(T)$

(1)
$$s_\alpha \lambda = \lambda - \langle \lambda, \alpha^\vee \rangle\, \alpha,$$

which we extend to the whole of $X(T) \otimes_{\mathbf{Z}} \mathbf{R}$ by extending $\alpha^\vee \in Y(T) \simeq X(T)^*$ to $X(T) \otimes_{\mathbf{Z}} \mathbf{R}$. Now

(2)
$$W = \langle s_\alpha \mid \alpha \in R \rangle$$

is the Weyl group of R.

Any $g \in N_G(T)(A)$ for some A acts through conjugation on T_A, hence also (linearly) on the $\mathbf{Z}$–modules $X(T_A)$ and $Y(T_A)$. If A is integral, then $X(T_A) \simeq X(T)$ and $Y(T_A) \simeq Y(T)$ and we get thus actions on $X(T)$ and $Y(T)$ for which the pairing $\langle\ ,\ \rangle$ on $X(T) \times Y(T)$ is invariant.

The action of any $n_\alpha(a)$ (with $\alpha \in R$, $a \in A$, A integral) on $X(T)$ is the same as that of s_α. We get for any integral A isomorphisms

$$W \simeq (N_G(T)/T)(A) \simeq N_G(T)(A)/T(A). \tag{3}$$

The last equality follows from the fact that each generator s_α of W has a representative $n_\alpha(1)$ in $N_G(T)(A)$, hence so has any $w \in W$. (More generally, $N_G(T)/T$ is isomorphic to the finite k–group associated to W as in I.8.5.a.)

Let us choose for any $w \in W$ a representative $\dot{w} \in N_G(T)(k)$. Then there is for any $\alpha \in R$ a constant $c_\alpha \in k^\times$ with

$$\dot{w}\, x_\alpha(a)\, \dot{w}^{-1} = x_{w\alpha}(c_\alpha\, a) \tag{4}$$

for all $a \in A$ and all A. (If we choose $\dot{w} \in N_{G_\mathbf{Z}}(T_\mathbf{Z})(\mathbf{Z})$, then $c_\alpha \in \{\pm 1\}$.) This implies

$$\dot{w}\, U_\alpha\, \dot{w}^{-1} = U_{w\alpha}. \tag{5}$$

1.5. (Simple Reflections) Choose a *positive system* $R^+ \subset R$ and denote by S the corresponding set of *simple* roots. Then $R^+ \subset \sum_{\alpha\in S} \mathbf{N}\alpha$ and $(R^+)^\vee \subset \sum_{\alpha\in S} \mathbf{N}\alpha^\vee$.

We define an order relation $\leq$ on $X(T)$ (and on $X(T) \otimes_\mathbf{Z} \mathbf{R}$) by

$$\lambda \leq \mu \iff \mu - \lambda \in \sum_{\alpha\in S} \mathbf{N}\alpha = \sum_{\beta\in R^+} \mathbf{N}\beta. \tag{1}$$

(So $R^+ = \{\alpha \in R \mid \alpha > 0\}$ and $-R^+ = \{\alpha \in R \mid \alpha < 0\}$.) Obviously $\lambda \leq \mu \iff -\mu \leq -\lambda$.

The Weyl group is generated already by the *simple reflections* with respect to R^+, i.e., by all s_α with $\alpha \in S$. The *length* $l(w)$ of any $w \in W$ is defined as the smallest m such that there exist $\beta_1, \beta_2, \dots, \beta_m \in S$ with $w = s_{\beta_1} s_{\beta_2} \dots s_{\beta_m}$. So $l(w) = 0$ if and only if $w = 1$, and $l(w) = 1$ if and only if $w = s_\alpha$ for some $\alpha \in S$. One has obviously $l(w) = l(w^{-1})$ for all $w \in W$.

One has for all $w \in W$ and $\alpha \in S$

$$l(ws_\alpha) = \begin{cases} l(w) + 1, & \text{if } w(\alpha) > 0, \\ l(w) - 1, & \text{if } w(\alpha) < 0, \end{cases} \tag{2}$$

and, symmetrically,

$$l(s_\alpha w) = \begin{cases} l(w) + 1, & \text{if } w^{-1}(\alpha) > 0, \\ l(w) - 1, & \text{if } w^{-1}(\alpha) < 0. \end{cases} \tag{3}$$

One can show

$$l(w) = |\{\alpha \in R^+ \mid w(\alpha) < 0\}| = |\{\alpha \in R^+ \mid w^{-1}(\alpha) < 0\}|. \tag{4}$$

As W permutes the positive systems simply transitively, there is a unique $w_0 \in W$ with $w_0(R^+) = -R^+$. Then $w_0^2(R^+) = R^+$, hence $w_0^2 = 1$. We have by (4)

(5) $$l(w_0) = |R^+|$$

and $l(ww_0) = l(w_0 w) = |R^+| - l(w)$ for all $w \in W$; if $w \neq w_0$, then $l(w) < |R^+|$. Obviously $\lambda \leq \mu \iff w_0\mu \leq w_0\lambda$ for all $\lambda, \mu \in X(T)$.

Set $\rho = \frac{1}{2}\sum_{\alpha \in R^+} \alpha \in X(T) \otimes_{\mathbf{Z}} \mathbf{Q}$. Then $2\rho \in \mathbf{Z}R \subset X(T)$ and

(6) $$\langle \rho, \beta^\vee \rangle = 1 \qquad \text{for all } \beta \in S.$$

This implies $s_\beta\rho - \rho \in \mathbf{Z}R$, hence $w\rho - \rho \in \mathbf{Z}R$ for all $w \in W$. Therefore the "dot action"

(7) $$w \bullet \lambda = w(\lambda + \rho) - \rho$$

of W on $X(T) \otimes_{\mathbf{Z}} \mathbf{R}$ maps $X(T)$ into itself.

1.6. (Semi-Simple Groups) The centre $Z(G)$ of G is equal to the intersection

(1) $$Z(G) = \bigcap_{\alpha \in R} \ker(\alpha) \subset T.$$

This group scheme is not necessarily reduced, i.e., it can happen (even if k is a field) that $k[Z(G)]$ contains nilpotent elements $\neq 0$. Obviously $Z(G)$ is isomorphic to the diagonalisable group scheme $\mathrm{Diag}(X(T)/\mathbf{Z}R)$, using the notation from I.2.5.

We call G *semi-simple* if $Z(G)$ is a finite group scheme. By the remark above we get

(2) $$G \text{ semi-simple} \iff (X(T) : \mathbf{Z}R) < \infty.$$

In the extreme case where $Z(G) = 1$ we call G *adjoint.* So we have

(3) $$G \text{ adjoint} \iff X(T) = \mathbf{Z}R.$$

If G is semi-simple, then S is a basis of the vector space $X(T) \otimes_{\mathbf{Z}} \mathbf{Q}$. Therefore $(\alpha^\vee)_{\alpha \in S}$ is a basis of $Y(T) \otimes_{\mathbf{Z}} \mathbf{Q} \simeq (X(T) \otimes_{\mathbf{Z}} \mathbf{Q})^*$. So there are ϖ_α ($\alpha \in S$) with $\langle \varpi_\alpha, \beta^\vee \rangle = \delta_{\alpha\beta}$ (the Kronecker delta) for all $\alpha, \beta \in S$. These ϖ_α are called the *fundamental weights.* They are, in general, only elements in $X(T) \otimes_{\mathbf{Z}} \mathbf{Q}$. We call (semi-simple) G *simply connected* if $\varpi_\alpha \in X(T)$ for all $\alpha \in S$. If so, then $(\varpi_\alpha)_{\alpha \in S}$ is a basis of $X(T)$ and $(\alpha^\vee)_{\alpha \in S}$ is the dual basis of $Y(T)$. So:

(4) $$G \text{ simply connected} \iff Y(T) = \mathbf{Z}R^\vee.$$

Observe that 1.5(6) implies

(5) $$\rho = \sum_{\alpha \in S} \varpi_\alpha \qquad \text{(for } G \text{ semisimple.)}$$

1.7. (Regular Subgroups) A subset $R' \subset R$ is called *closed* if $(\mathbf{N}\alpha+\mathbf{N}\beta)\cap R \subset R'$ for any $\alpha, \beta \in R'$. It is called *unipotent* (resp. *symmetric*) if $R' \cap (-R') = \emptyset$ (resp. $R' = -R'$).

For any $R' \subset R$ unipotent and closed we denote by $U(R')$ the closed subgroup generated by all U_α with $\alpha \in R'$. Using 1.2(5) one can show that the multiplication induces (for any ordering of R') an isomorphism of schemes over k

(1) $$\prod_{\alpha\in R'} U_\alpha \xrightarrow{\sim} U(R').$$

Obviously

(2) $$\operatorname{Lie} U(R') = \bigoplus_{\alpha\in R'} (\operatorname{Lie} G)_\alpha.$$

Each $U(R')$ is connected and unipotent. It is isomorphic to $\mathbf{A}^n$ with $n = |R'|$ as a scheme (hence reduced). It is normalised by T. Using (1), one can identify $k[U(R')]$ with a polynomial ring over k in indeterminates Y_α ($\alpha \in R'$) such that $Y_\alpha \in k[U(R')]_{-\alpha}$ with respect to the conjugation action of T (for all α).

If $R' \subset R$ is symmetric and closed, then let $G(R')$ be the closed subgroup generated by T and all U_α with $\alpha \in R'$. Then

(3) $$\operatorname{Lie} G(R') = \operatorname{Lie} T \oplus \bigoplus_{\alpha\in R'} (\operatorname{Lie} G)_\alpha.$$

The k–group $G(R')$ is reductive, split, and connected. It contains T as a maximal torus. Its root system is exactly R', and we can identify its Weyl group with $\langle s_\alpha \mid \alpha \in R'\rangle \subset W$.

We can take especially some $I \subset S$ and set $R_I = R\cap \mathbf{Z}I$. Then R_I is symmetric and closed. Set $L_I = G(R_I)$. Then L_I is split and reductive with Weyl group isomorphic to $W_I = \langle s_\alpha \mid \alpha \in I\rangle$.

1.8. (Borel Subgroups and Parabolics) Both R^+ and $-R^+$ are unipotent and closed subsets of R. Applying 1.7 to these sets, we get

(1) $$U^+ = U(R^+) \qquad\text{and}\qquad U = U(-R^+).$$

Then

(2) $$B^+ = U^+T = U^+ \rtimes T \qquad\text{and}\qquad B = UT = U \rtimes T$$

are *Borel subgroups* of G, with $B \cap B^+ = T$. Note that B corresponds to the **negative roots.**

If $\dot w_0 \in N_G(T)(k)$ is a representative for w_0, then $\dot w_0 U \dot w_0^{-1} = U^+$ and $\dot w_0 B \dot w_0^{-1} = B^+$.

For any $I \subset S$ the subsets $R^+ \setminus R_I$ and $(-R^+) \setminus R_I$ of R are closed and unipotent. So

(3) $$U_I^+ = U(R^+ \setminus R_I) \qquad\text{and}\qquad U_I = U((-R^+) \setminus R_I)$$

are closed unipotent subgroups of G. The commutator relations 1.2(5) imply that L_I normalises U_I^+ and U_I. One has $U_I^+ \cap L_I = 1 = U_I \cap L_I$, so we get semi-direct products inside G:

(4) $$P_I = U_I L_I = U_I \rtimes L_I \qquad \text{and} \qquad P_I^+ = U_I^+ L_I = U_I^+ \rtimes L_I$$

The P_I (resp. P_I^+) with $I \subset S$ are called the *standard parabolic subgroups* of G containing B (resp. B^+), and L_I is called the *standard Levi factor* of P_I (and of P_I^+) containing T. Furthermore, U_I resp. U_I^+ is the *unipotent radical* of P_I resp. of P_I^+.

Note that $P_\emptyset = B$ and $P_S = G$. For $I \subset J \subset S$ one has $P_I \subset P_J$; the quotient P_J/P_I is an irreducible and projective scheme over k. The inclusions $L_J \subset P_J \subset G$ induce an isomorphism and an embedding

(5) $$L_J/(L_J \cap P_I) \xrightarrow{\sim} P_J/P_I \hookrightarrow G/P_I.$$

If k is a field, then G/P_I is a projective and irreducible variety over k. This implies $k \simeq k[G/P_I] = \operatorname{ind}_{P_I}^G(k)$. Using I.3.5(3) we get for any k

(6) $$k[G/P_I] = k$$

since the k–module $k[G/P_I] \subset k[G]$ is torsion free with $k = k1$ as a direct summand.

1.9. (Big Cells and the Bruhat Decomposition) For any k–algebra k' that is a field, $G(k')$ is decomposed as the *disjoint* union (cf. [Bo], 14.12 for k' algebraically closed or [BoT], 2.11 for the general case):

(1) $$G(k') = \bigcup_{w \in W} B(k')\,\dot{w}\,B(k') = \bigcup_{w \in W} U(k')\,\dot{w}\,B(k').$$

(Note that $B(k')\dot{w}B(k') = U(k')\dot{w}B(k')$ as $B(k') = U(k') \rtimes T(k')$ and as $\dot{w}$ normalises $T(k')$.) This is the *Bruhat decomposition* of $G(k')$. As left multiplication with $\dot{w}_0$ is a bijection on $G(k')$ and as $\dot{w}_0 B(k')\dot{w}_0^{-1} = B^+(k')$, one also has (again disjointly)

(2) $$G(k') = \bigcup_{w \in W} B^+(k')\,\dot{w}\,B(k') = \bigcup_{w \in W} U^+(k')\,\dot{w}\,B(k').$$

Using the $|R^+|^{\text{th}}$ exterior power of the adjoint representation, one constructs (as a suitable matrix coefficient) a function $d \in k[G]$ with $d(u_1 t u_2) = (2\rho)(t)^{-1}$ and $d(u_1 \dot{w} t u_2) = 0$ for all $w \neq 1$, all $u_1 \in U^+(A)$, $t \in T(A)$, $u_2 \in U(A)$, and all A. So G_d is an open subscheme of G with $G_d(k') = U^+(k')B(k')$ for any k–algebra k' that is a field. One can show more precisely that G_d is dense in G and that the multiplication induces an isomorphism of $U^+ \times B$ onto G_d. Therefore we usually denote G_d by U^+B or U^+TU. Any subscheme of G like U^+B is called a *big cell* in G. (Another choice of T and B will lead to other big cells. For example, we get UTU^+ by keeping T and replacing B by B^+.)

Using 1.7(1), we get an isomorphism of schemes (given by multiplication, for any ordering of the roots):

(3) $$\prod_{\alpha \in R^+} U_\alpha \times T \times \prod_{\alpha \in R^+} U_{-\alpha} \xrightarrow{\sim} U^+B.$$

Therefore $k[U^+B]$ and its subalgebra $k[G]$ are integral domains. We get especially:

(4) *G is integral.*

If K is an extension field of the field of fractions of k, then

(5) $$k[G] = K[G_K] \cap k[U^+B]$$

(One reduces easily to the case where K is the field of fractions of k. One then has to show that the k–module $k[U^+B]/k[G]$ is torsion free. In the case $k = \mathbf{Z}$ one can use the argument in [Borel 1], Lemma 4.9. The general case then follows, as d comes from $\mathbf{Z}[G]$, hence $U^+B = (U^+_{\mathbf{Z}} B_{\mathbf{Z}})_k$.)

Let K be a k–algebra that is an algebraically closed field. Regard $G(K)$ as a variety over K. For any $w \in W$ one has

$$B^+(K)\,\dot{w}\,B(K) = U^+(K)\,\dot{w}\,B(K) = \dot{w}\,(\dot{w}^{-1}U^+(K)\,\dot{w})\,B(K).$$

Set $R' = \{\alpha \in R^+ \mid w^{-1}(\alpha) > 0\}$ and $R'' = \{\alpha \in R^+ \mid w^{-1}(\alpha) < 0\}$. Then R' and R'' are closed and unipotent subsets of R^+ with $R' \cup R'' = R^+$, $R' \cap R'' = \emptyset$. Therefore the multiplication induces an isomorphism $U(R') \times U(R'') \xrightarrow{\sim} U^+$. Obviously $\dot{w}^{-1}U(R')\dot{w} = U(w^{-1}R') \subset U^+$ and $\dot{w}^{-1}U(R'')\dot{w} = U(w^{-1}R'') \subset U$, hence

$$B^+(K)\,\dot{w}\,B(K) = \dot{w}\,U(w^{-1}R')(K)\,B(K) \subset \dot{w}\,U^+(K)\,B(K).$$

As $|R''| = l(w)$ and as the multiplication induces an isomorphism $\prod_{\alpha\in R'} U_\alpha \xrightarrow{\sim} U(R')$, we get:

(6) *Each $B^+(K)\dot{w}B(K)$ is a closed subvariety, isomorphic to $K^{|R^+|-l(w)} \times B(K)$, of $\dot{w}U^+(K)B(K)$.*

This implies, of course, that $B^+(K)\dot{w}B(K)$ is a locally closed subvariety of $G(K)$ of dimension equal to $|R^+|-l(w)+\dim B(K)$. Furthermore, the Bruhat decomposition implies:

(7) *The $\dot{w}U^+B$ with $w \in W$ form an open covering of G.*

The only $B^+(K)\dot{w}B(K)$ of dimension equal to $\dim G(K)$ is $B^+(K)B(K) = U^+(K)B(K)$, and the only ones of codimension 1 are the

$$B^+(K)\,\dot{s}_\alpha\,B(K) \subset \dot{s}_\alpha\,U^+(K)\,B(K)$$

with $\alpha \in S$ (cf. 1.5). So the complement of $U^+(K)B(K) \cup \bigcup_{\alpha\in S} \dot{s}_\alpha U^+(K)B(K)$ in $G(K)$ has codimension at least 2. As $G(K)$ is smooth, hence normal, this implies:

(8) *Any regular function on $U^+(K)B(K) \cup \bigcup_{\alpha\in S} \dot{s}_\alpha U^+(K)B(K)$ can be extended (uniquely) to a regular function on $G(K)$.*

1.10. (Local Triviality) Let $\pi : G \to G/B$ denote the canonical map. Each $\pi(\dot{w}U^+B)$ is open in G/B, cf. I.5.7(3). For any k–algebra K that is an algebraically closed field, $\pi(K)$ is surjective, so we get $(G/B)(K) = \bigcup_{w\in W} \pi(\dot{w}U^+B)(K)$ by 1.9(7), hence, cf. I.1.7:

(1) *The $\pi(\dot{w}U^+B)$ with $w \in W$ form an open covering of G/B.*

Because of 1.9(3) the map $(u,b) \mapsto \dot{w}ub$ is an isomorphism of schemes $U^+ \times B \xrightarrow{\sim} \dot{w}U^+B$. It is compatible with the right multiplication by B. Therefore $u \mapsto \dot{w}u$ is an isomorphism from U^+ onto the image functor of $\dot{w}U^+B$ in G/B. So this image functor is a faisceau and a scheme, hence equal to its associated faisceau $\pi(\dot{w}U^+B)$. Therefore $\pi_i : U^+ \to \pi(\dot{w}U^+B)$, $u \mapsto \pi(\dot{w}u)$ is an isomorphism of schemes, and $\sigma_i : \pi(\dot{w}U^+B) \to G$, $x \mapsto \dot{w}\pi_i^{-1}(x)$ is a local section, cf. I.5.16. Hence:

(2) *The canonical map $\pi : G \to G/B$ is locally trivial.*

We get also:

(3) *For any k–algebra k' that is a field, the canonical map $G(k') \to (G/B)(k')$ is surjective.*

For each $I \subset S$ there are similar results for G/P_I. To start with, the results in 1.9 generalise. For example, the multiplication induces an isomorphism of $U_I^+ \times P_I$ onto an open and dense subscheme of G denoted by $U_I^+P_I$ and containing U^+B. More precisely, we have (for any ordering of the roots) an isomorphism of schemes

(4) $$\prod_{\alpha \in R^+ \setminus R_I} U_\alpha \times L_I \times \prod_{\alpha \in R^+ \setminus R_I} U_{-\alpha} \xrightarrow{\sim} U_I^+ P_I.$$

As $U_I^+P_I \supset U^+B$, the $\dot{w}U_I^+P_I$ (resp. their images in G/P_I) with $w \in W$ form an open covering of G (resp. of G/P_I). Furthermore:

(5) *The canonical map $G \to G/P_I$ is locally trivial.*

Again $G(k')$ maps onto $(G/P_I)(k')$ for any k–algebra k' that is a field.

1.11. (The Lie Algebra) Set for any $\alpha \in R$

(1) $$X_\alpha = (dx_\alpha)(1) \in (\operatorname{Lie} G_{\mathbf{Z}})_\alpha$$

where we regard x_α as a homomorphism $G_{a,\mathbf{Z}} \to G_{\mathbf{Z}}$. Choose a basis $\varphi_1, \varphi_2, \dots, \varphi_r$ of $Y(T_{\mathbf{Z}}) \simeq Y(T)$ and set for each i $(1 \le i \le r)$

(2) $$H_i = (d\varphi_i)(1) \in \operatorname{Lie}(T_{\mathbf{Z}}).$$

Then $H_1, H_2, \dots, H_r$ is a basis of $\operatorname{Lie}(T_{\mathbf{Z}})$ and $(H_i)_{1 \le i \le r}$, $(X_\alpha)_{\alpha \in R}$ is a basis of $\operatorname{Lie}(G_{\mathbf{Z}})$. By 1.1(1) we get for any A that $(H_i \otimes 1)_i$, $(X_\alpha \otimes 1)_\alpha$ is a basis of $\operatorname{Lie}(G_A)$.

For any $\alpha \in R$ the element $H_\alpha = (d\alpha^\vee)(1) \in \operatorname{Lie}(T_{\mathbf{Z}})$ is equal to $[X_\alpha, X_{-\alpha}]$ and satisfies $(d\alpha)(H_\alpha) = 2$. (Compute in SL_2 as in 1.3.) This implies that $\operatorname{Lie}(T_{\mathbf{Z}})$ is an admissible lattice (in the sense of [B3], ch. VIII, §12, n° 6) in $\operatorname{Lie}(T_{\mathbf{C}}) = \operatorname{Lie}(T_{\mathbf{Z}}) \otimes_{\mathbf{Z}} \mathbf{C}$.

1.12. (The Algebra of Distributions) The element 1 in G is contained in the open subscheme U^+TU that is a product of copies of G_m and G_a. So I.7.4(2) and I.7.8 imply:

(1) *G is infinitesimally flat,*

and

(2) *The multiplication induces an isomorphism of k–modules*

$$\bigotimes_{\alpha\in R^+}\mathrm{Dist}(U_\alpha)\otimes\mathrm{Dist}(T)\otimes\bigotimes_{\alpha\in R^+}\mathrm{Dist}(U_{-\alpha})\xrightarrow{\sim}\mathrm{Dist}(G).$$

(We can take any ordering of the roots.)

Furthermore, we have for any k–algebra A

$$\mathrm{Dist}(G_A)\simeq\mathrm{Dist}(G)\otimes A. \tag{3}$$

Applying this to $(G_{\mathbf{Z}},\mathbf{Z})$ instead of (G,k) we get that $\mathrm{Dist}(G_A)\simeq\mathrm{Dist}(G_{\mathbf{Z}})\otimes_{\mathbf{Z}}A$ for all $\mathbf{Z}$–algebras A, in particular for $A=k$ and $A=\mathbf{C}$.

The descriptions of $\mathrm{Dist}(G_a)$ and $\mathrm{Dist}(G_m)$ in I.7.8 applied to each $U_{\alpha,\mathbf{Z}}\simeq G_{a,\mathbf{Z}}$ and to $T_{\mathbf{Z}}\simeq(G_{m,\mathbf{Z}})^r$ show that each $\mathrm{Dist}(U_{\alpha,\mathbf{Z}})$ and that $\mathrm{Dist}(T_{\mathbf{Z}})$ is a free $\mathbf{Z}$–module. Then (2) applied to $(G_{\mathbf{Z}},\mathbf{Z})$ shows that also $\mathrm{Dist}(G_{\mathbf{Z}})$ is free over $\mathbf{Z}$. We can therefore identify $\mathrm{Dist}(G_{\mathbf{Z}})$ with a $\mathbf{Z}$–submodule in $\mathrm{Dist}(G_{\mathbf{C}})\simeq\mathrm{Dist}(G_{\mathbf{Z}})\otimes_{\mathbf{Z}}\mathbf{C}$, hence with a $\mathbf{Z}$–submodule in $U(\mathrm{Lie}\,G_{\mathbf{C}})$.

Take X_α and H_i as in 1.11. The computations in I.7.8 show more precisely that the $X_\alpha^n/(n!)$ with $n\in\mathbf{N}$ form a basis of $\mathrm{Dist}(U_{\alpha,\mathbf{Z}})$ for each $\alpha\in R$ and that all

$$\binom{H_1}{m(1)}\binom{H_2}{m(2)}\cdots\binom{H_r}{m(r)}$$

form a basis of $\mathrm{Dist}(T_{\mathbf{Z}})$. Set $X_{\alpha,n}=X_\alpha^n/(n!)\otimes 1\in\mathrm{Dist}(U_\alpha)$ and $H_{i,m}=\binom{H_i}{m}\otimes 1\in\mathrm{Dist}(T)$. Then (2), (3), and the result over $\mathbf{Z}$ imply:

(4) *All* $\prod_{\alpha\in R^+}X_{\alpha,n(\alpha)}\prod_{i=1}^r H_{i,m(i)}\prod_{\alpha\in R^+}X_{-\alpha,n'(\alpha)}$ *with* $n(\alpha)$, $m(i)$, $n'(\alpha)\in\mathbf{N}$ *form a basis of* $\mathrm{Dist}(G)$.

Similarly, if $R'\subset R$ is closed and unipotent, then the algebra $\mathrm{Dist}\,U(R')$ has all $\prod_{\alpha\in R'}X_{\alpha,n(\alpha)}$ with $n(\alpha)\in\mathbf{N}$ as a basis. (Use 1.7(1).) If $I\subset S$ and if we take in (4) only $\alpha\in R_I$, then we get a basis for $\mathrm{Dist}(L_I)$. If we take in (4) all $\alpha\in R^+$ in the last product, but only $\alpha\in R^+\cap R_I$ in the first one, then we get a basis for $\mathrm{Dist}(P_I)$.

The description above shows that $\mathrm{Dist}(G_{\mathbf{Z}})$ is the $\mathbf{Z}$–algebra of $U(\mathrm{Lie}G_{\mathbf{C}})$ generated by all $X_\alpha^n/(n!)$ with $\alpha\in\mathbf{N}$ and $n\in\mathbf{N}$, and by all $\binom{H}{m}$ with $H\in\mathrm{Lie}(T_{\mathbf{Z}})$ and $m\in\mathbf{N}$. We have $\mathrm{Dist}(G_{\mathbf{Z}})\cap\mathrm{Lie}(G_{\mathbf{C}})=\mathrm{Lie}(G_{\mathbf{Z}})$, so th. 2 in [B3], ch. VIII, §12 implies that the $(X_\alpha)_{\alpha\in R}$ form a Chevalley system and that $\mathrm{Lie}(G_{\mathbf{Z}})$ is a Chevalley order (as defined there) in $\mathrm{Lie}(G_{\mathbf{C}})$. Furthermore, $\mathrm{Dist}(G_{\mathbf{Z}})$ is the corresponding biorder in $U(\mathrm{Lie}G_{\mathbf{C}})$. In the case where G is semi-simple and simply connected, this shows that $\mathrm{Dist}(G_{\mathbf{Z}})$ is Kostant's $\mathbf{Z}$–form of $U(\mathrm{Lie}G_{\mathbf{C}})$.

1.13. (Homomorphisms of Root Data) The *root datum* of G is the quadruple $(X(T),R,Y(T),R^\vee)$ together with the pairing of $X(T)$ and $Y(T)$ and the bijection $\alpha\mapsto\alpha^\vee$ from R to $R^\vee$, cf. [Sp2], 7.4.1.

Let G' be another connected and split reductive k–group, let $T'\subset G'$ be a split maximal torus, and let $(X(T'),R',Y(T'),R'^\vee)$ be the corresponding root datum. We denote by B', U'_α, … the objects for G' analogous to B, U_α, … .

A *homomorphism of root data* (from that of G to that of G') is a group homomorphism $f:X(T')\to X(T)$ that maps R' bijectively to R and such that the dual homomorphism $f^\vee:Y(T)\to Y(T')$ maps $f(\beta)^\vee$ to $\beta^\vee$ for each $\beta\in R'$. Then

$f \otimes \mathrm{id}_{\mathbf{Q}}$ maps $\mathbf{Z}R' \otimes_{\mathbf{Z}} \mathbf{Q}$ bijectively to $\mathbf{Z}R \otimes_{\mathbf{Z}} \mathbf{Q}$ and induces an isomorphism of root systems. There is an isomorphism of Weyl groups $W \xrightarrow{\sim} W'$, say $w \mapsto w'$, such that always $f \circ w' = w \circ f$ and $s_{f(\alpha)} \mapsto s_\alpha$ for all $\alpha \in R'$.

We say that a homomorphism $\psi : T \to T'$ is *compatible with the root data* if the induced map $\psi^* : X(T') \to X(T)$ is a homomorphism of root data. Of course, $\psi \mapsto \psi^*$ is a bijection from $\{$ homomorphisms $T \to T'$ compatible with the root data $\}$ to the set of all homomorphisms of root data, cf. I.2.5(2).

Fix a homomorphism $\psi : T \to T'$ compatible with the root data. If $\alpha(t) \neq 1$ for some $\alpha \in R$ and $t \in T(A)$, then $\beta(\psi(t)) = \psi^*(\beta)(t) \neq 1$ for the unique $\beta \in R'$ with $\psi^*(\beta) = \alpha$, hence $\psi(t) \neq 1$. This implies, cf. 1.6(1):

(1) $$\ker(\psi) \subset Z(G).$$

On the other hand, $\ker(\psi^*) \cap \mathbf{Z}R' = 0$ implies

(2) $$\psi(T)\, Z(G') = T'.$$

Consider again $\alpha \in R$ and $\beta \in R'$ with $\psi^*(\beta) = \alpha$. We get from 1.3(7)

$$\ker(\varphi_\alpha) \subset \ker(\varphi'_\beta).$$

Therefore we get a homomorphism $\psi_\alpha : \varphi_\alpha(SL_2) \to \varphi'_\beta(SL_2)$ with $\psi_\alpha \circ \varphi_\alpha = \varphi'_\beta$. It induces isomorphisms $U_\alpha \xrightarrow{\sim} U'_\beta$ and $U_{-\alpha} \xrightarrow{\sim} U'_{-\beta}$. We have $\psi_\alpha \circ \alpha^\vee = \beta^\vee = \psi \circ \alpha^\vee$. So ψ_α and ψ coincide on $\varphi_\alpha(SL_2) \cap T = \alpha^\vee(G_m)$. Therefore ψ_α can be extended to a homomorphism

(3) $$\psi_\alpha : G_\alpha \longrightarrow G'_\beta$$

with $(\psi_\alpha)_{|T} = \psi$ and $\psi_\alpha \circ \varphi_\alpha = \varphi'_\beta$.

1.14. Keep all the assumptions from 1.13.

Proposition: *Suppose that k is an algebraically closed field. There is a homomorphism $\varphi : G \to G'$ with $\varphi_{|T} = \psi$ and $\mathrm{im}(\varphi(k))\, Z(G')(k) = G'(k)$ and $\ker(\varphi) = \ker(\psi) \subset Z(G)$.*

Proof: We can assume that $\psi^*(S') = S$. Otherwise we can replace ψ by a suitable $\psi \circ \mathrm{Int}(\dot{w})$ with $w \in W$, similarly for φ.

On order to simplify notations we shall identify R and R' via ψ^*. So we have $\psi^*(\alpha) = \alpha$ and $\psi_*(\beta^\vee) = \beta^\vee$ for all α, β.

Set

$$H_0 = \{\, (t, \psi(t)) \mid t \in T(k) \,\} \subset G(k) \times G'(k)$$

and (for all $\alpha \in S$)

(1) $$H_\alpha = \{\, (g, \psi_\alpha(g)) \mid g \in G_\alpha(k) \,\} \subset G(k) \times G'(k).$$

These are closed subgroups of $G(k) \times G'(k)$ (with $H_0 \subset H_\alpha$ for all α), isomorphic to $T(k)$ resp. to $G_\alpha(k)$ via the first projection, hence irreducible. Therefore the subgroup H of $G(k) \times G'(k)$ generated by all H_α with $\alpha \in S$ is also closed and irreducible.

Let $\pi : H \to G(k)$ and $\pi' : H \to G'(k)$ denote the two projections. We want to show that π is an isomorphism of linear algebraic groups. Then $\pi' \circ \pi^{-1}$ is a homomorphism $G(k) \to G'(k)$ of algebraic groups inducing $\psi_\alpha(k)$ on each $G_\alpha(k)$ with $\alpha \in S$. As G is reduced, there is a unique homomorphism $\varphi : G \to G'$ of group schemes with $\varphi(k) = \pi' \circ \pi^{-1}$. As each G_α is reduced, $\varphi(k)_{|G_\alpha(k)} = \psi_\alpha(k)$ implies $\varphi_{|G_\alpha} = \psi_\alpha$ for all $\alpha \in S$, hence $\varphi_{|T} = (\psi_\alpha)_{|T} = \psi$.

We shall need several steps to reach our aim.

(2) π *is surjective.*

This is clear since $G(k)$ is generated by $T(k)$ and all $U_\alpha(k)$ with $\alpha \in S \cup (-S)$.

(3) $$G'(k) = \mathrm{im}(\pi')\, Z(G')(k).$$

The group $G'(k)$ is generated by $T'(k)$ and all $U'_\alpha(k)$ with $\alpha \in S\cup(-S)$. Both $U'_\alpha(k)$ and $U'_{-\alpha}(k)$ are (for any $\alpha \in S$) contained in the image of $\varphi'_\alpha(k) = \psi_\alpha(k) \circ \varphi_\alpha(k)$, hence in $\pi'(H_\alpha) \subset \pi'(H)$. Finally $T'(k) = \mathrm{im}(\psi(k))\, Z(G')(k)$ by 1.13(2).

(4) H *is reductive.*

The unipotent radical $R_u(H)$ of H is mapped under π resp. π' to a unipotent subgroup of $G(k)$ resp. $G'(k)$ which is normal by (2) resp. (3), hence equal to $\{1\}$. This implies $R_u(H) = \{1\}$.

We shall apply the notations $\mathrm{Lie}(H')$, $\mathrm{Dist}(H')$ to linear algebraic groups H' over k by regarding H' as a reduced group scheme.

(5) $d\pi : \mathrm{Lie}(H) \to \mathrm{Lie}(G)$ *is surjective.*

As the image of $d\pi$ contains $\mathrm{Lie}(G_\alpha) = \mathrm{Lie}(T) + (\mathrm{Lie}\, G)_\alpha + (\mathrm{Lie}\, G)_{-\alpha}$ for each $\alpha \in S$, this follows from

$$\mathrm{Lie}\, G = \mathrm{Lie}\, T + \sum_{\alpha\in S} \sum_{w\in W} \mathrm{Ad}(w)\, (\mathrm{Lie}\, G)_\alpha$$

and the compatibility of $d\pi$ with the adjoint action.

Set $V_\alpha = \{(g, \psi_\alpha(g)) \mid g \in U_\alpha(k)\}$ for each $\alpha \in S$ and define similarly $V_{-\alpha}$. The isomorphism $g \mapsto (g, \psi_\alpha(g))$ from $G_\alpha(k)$ to H_α maps the dense and open subset (a big cell) $U_\alpha(k)T(k)U_{-\alpha}(k)$ in $G_\alpha(k)$ to the dense and open subset $V_\alpha H_0 V_{-\alpha}$ in H_α. So:

(6) $$\mathrm{Dist}(H_\alpha) = \mathrm{Dist}(V_\alpha)\, \mathrm{Dist}(H_0)\, \mathrm{Dist}(V_{-\alpha}).$$

The algebra $\mathrm{Dist}(H)$ is generated by all $\mathrm{Dist}(H_\alpha)$ with $\alpha \in S$, cf. I.7.19. Let $\mathrm{Dist}(H)^+$ resp. $\mathrm{Dist}(H)^-$ denote the subalgebra generated by all $\mathrm{Dist}(V_\alpha)$ resp. $\mathrm{Dist}(V_{-\alpha})$ with $\alpha \in S$. If $\alpha, \beta \in S$ with $\alpha \neq \beta$, then V_α and $V_{-\beta}$ commute, hence so do $\mathrm{Dist}(V_\alpha)$ and $\mathrm{Dist}(V_{-\beta})$. Therefore we get using (6)

(7) $$\mathrm{Dist}(H) = \mathrm{Dist}(H)^+\, \mathrm{Dist}(H_0)\, \mathrm{Dist}(H)^-.$$

On the other hand, 1.12(2) applied to $G \times G'$ yields

(8) $$\mathrm{Dist}(G \times G') = \mathrm{Dist}(G \times G')^+ \otimes \mathrm{Dist}(T \times T') \otimes \mathrm{Dist}(G \times G')^-$$

where $\mathrm{Dist}(G \times G')^+$ resp. $\mathrm{Dist}(G \times G')^-$ is generated by all $\mathrm{Dist}(U_\beta \times 1)$ and $\mathrm{Dist}(1 \times U'_\beta)$ with $\beta \in R^+$ resp. $\beta \in -R^+$. Now $V_\alpha \subset U_\alpha \times U'_\alpha$ for all α implies $\mathrm{Dist}(H)^+ \subset \mathrm{Dist}(G \times G')^+$. One has similarly $\mathrm{Dist}(H)^- \subset \mathrm{Dist}(G \times G')^-$ and gets now from (7) and (8):

$$\mathrm{Dist}(H) \cap \mathrm{Dist}(T \times T') = \mathrm{Dist}(H_0). \tag{9}$$

As $H_0 \simeq T(k)$ is a torus, it is contained in a maximal torus H_1 of H. Necessarily $H_1 \subset C_{G(k)\times G'(k)}(H_0)$. As no root of $G \times G'$ with respect to $T \times T'$ (i.e., no $\alpha \times 1$ or $1 \times \alpha$ with $\alpha \in R$) vanishes on H_0, this centraliser is equal to $T(k) \times T'(k)$. Therefore $\mathrm{Dist}(H_1) \subset \mathrm{Dist}(T \times T') \cap \mathrm{Dist}(H) = \mathrm{Dist}(H_0)$ by (9), hence $H_1 \subset H_0$ as H_1 is irreducible, cf. I.7.17(7). This implies:

(10) *H_0 is a maximal torus of H.*

We already know H to be reductive by (4). Therefore each non-trivial normal subgroup of H intersects H_0 non-trivially. As π is injective on H_0, we get:

(11) *π is injective.*

Combining this with (2) and (5) we see that π is an isomorphism. Therefore we get $\varphi : G \to G'$ as mentioned above with $\varphi_{|G_\alpha} = \psi_\alpha$ for all $\alpha \in S$. This implies especially that $\varphi(\dot{s}_\alpha)$ is a representative for s_α in $N_{G'}(T')(k)$, cf. 1.3(4), hence for any $w \in W$ that $\varphi(\dot{w})$ is a representative in $N_{G'}(T')(k)$ for the corresponding element in W'. This implies that φ induces an isomorphism $U_\alpha \xrightarrow{\sim} U'_\alpha$ for all $\alpha \in R$. (It does so for all $\alpha \in S$ as $\varphi_{|U_\alpha} = (\psi_\alpha)_{|U_\alpha}$; we get it for arbitrary α by conjugation with suitable $\dot{w}$.) This implies $\ker(\varphi) \cap U^+B = \ker(\psi)$, hence $\mathrm{Dist}(\ker \varphi) = \mathrm{Dist}(\ker \psi)$. A look at the Bruhat decompositions of $G(k)$ and $G'(k)$ yields $\ker(\varphi(k)) = \ker(\psi(k))$. So $\ker(\varphi) = \ker(\psi)$ follows from I.7.17(8). Finally, (3) implies $G'(k) = \mathrm{im}(\varphi(k))\, Z(G')(k)$.

Remarks: 1) The homomorphisms φ as constructed in the proposition are called *central isogenies*.

There is a larger class of isogenies if $\mathrm{char}(k) \neq 0$. They can be constructed in the same way as above. For the necessary changes one may consult the original paper [Takeuchi]. (These more general isogenies no longer satisfy $\ker(\varphi) \subset T$.)

2) Any homomorphism $\varphi : G \to G'$ extending ψ satisfies $\ker(\varphi) = \ker(\psi)$ and $G'(k) = \mathrm{im}(\varphi(k))\, Z(G')(k)$. (Using $\varphi \circ \alpha^\vee = \psi \circ \alpha^\vee$ one can show that φ maps any U_α isomorphically to U'_α. Then one can argue as above.)

1.15. Keep all the assumptions from 1.13.

Proposition: *There is a homomorphism $\varphi : G \to G'$ with $\varphi_{|T} = \psi$ and* $\ker(\varphi) = \ker(\psi)$. *If ψ is an isomorphism, then so is φ.*

Proof: As in the proof in 1.14, we assume that $\psi^*(S') = S$ and identify R and R' via ψ^*. We similarly identify the two Weyl groups.

Let K be an algebraic closure of the field of fractions of k. By 1.14 we have a homomorphism $\varphi' : G_K \to G'_K$ restricting to ψ_K on T_K and to $(\psi_\alpha)_K$ on each $(G_\alpha)_K$. It therefore induces (for each $\alpha \in S$) an isomorphism $(U_\alpha)_K \xrightarrow{\sim} (U'_\alpha)_K$ with $\varphi'^* k[U'_\alpha] = k[U_\alpha]$. Furthermore, it maps any $\dot{s}_\alpha$ (which belongs to $N_G(T)(k)$) to a representative of s_α in $N_{G'}(T')(k)$, hence any $\dot{w} \in N_G(T)(k)$ to a representative of w again over k. Therefore φ' induces for all $\alpha \in R$ an isomorphism $(U_\alpha)_K \xrightarrow{\sim} (U'_\alpha)_K$ with $\varphi'^* k[U'_\alpha] = k[U_\alpha]$. As φ' induces ψ_K on T_K, we get that φ' maps $U_K^+ B_K$ to $U_K'^+ B'_K$ and satisfies $\varphi'^* k[U'^+ B'] \subset k[U^+ B]$. Now 1.9(5) implies $\varphi'^* k[G'] \subset k[G]$. Therefore there is a morphism $\varphi : G \to G'$ with $\varphi' = \varphi_K$.

The injectivity of the maps $k[G] \to K[G_K]$, $k[G_\alpha] \to K[(G_\alpha)_K]$, etc. implies that φ is a homomorphism with $\varphi_{|G_\alpha} = \psi_\alpha$ for all $\alpha \in S$, especially with $\varphi_{|T} = \psi$. By construction, φ induces an isomorphism $U_\alpha \xrightarrow{\sim} U'_\alpha$ for all $\alpha \in R$. Therefore the same argument as in 1.14 yields $\ker(\varphi) = \ker(\psi)$.

If ψ is an isomorphism, then so are all ψ_α with $\alpha \in S$. We can apply the same argument to ψ^{-1} and the ψ_α^{-1} as to ψ and the ψ_α. We get thus a homomorphism $\varphi_1 : G' \to G$ inducing ψ_α^{-1} on each G'_α. Then $\varphi_1 \circ \varphi$ is the identity on each G_α, hence the identity on G by 1.3(10). Similarly $\varphi \circ \varphi_1 = \mathrm{id}$. (One could also go back to the proof of 1.14 and show that π' is an isomorphism if π is so.)

1.16. Corollary: *There is an antiautomorphism τ of G with $\tau^2 = \mathrm{id}_G$ and $\tau_{|T} = \mathrm{id}_T$ and $\tau(U_\alpha) = U_{-\alpha}$ for all $\alpha \in R$.*

Proof: We want to apply 1.15 to $G' = G$, $T' = T$, and $\psi : T \to T$ with $\psi(t) = t^{-1}$ for any t. Then ψ is compatible with the root data as $\psi^*(\alpha) = -\alpha$ and $\psi_*(\alpha^\vee) = -\alpha^\vee$ for each $\alpha \in R$. The homomorphisms $\psi_\alpha : G_\alpha \to G_{-\alpha} = G_\alpha$ for all $\alpha \in S$ satisfy $\psi_\alpha \circ \varphi_\alpha = \varphi_{-\alpha}$, hence $\psi_\alpha^2 = \mathrm{id}_{G_\alpha}$. The extension $\varphi : G \to G$ with $\varphi_{|G_\alpha} = \psi_\alpha$ for all $\alpha \in S$ satisfies $\varphi(t) = t^{-1}$ for all $t \in T$ and $\varphi(U_\alpha) = U_{-\alpha}$ for all $\alpha \in R$ and $(\varphi^2)_{|G_\alpha} = \psi_\alpha^2 = \mathrm{id}$ for all $\alpha \in S$, hence $\varphi^2 = \mathrm{id}$ by 1.3(10). So $\tau : G \to G$ with $\tau(g) = \varphi(g^{-1})$ has the desired properties.

Remark: In the special case where $G = SL_n$ and $T = \{\text{diagonal matrices}\}$ we usually take $\tau(g) = {}^t g$ for all g.

1.17. (Quotients and Covering Groups) The proposition 1.15 implies that G is determined up to isomorphism by its root datum. (This fact is known as the *isomorphism theorem.*)

On the other hand, one has an *existence theorem*: To each "possible" root datum there corresponds a group. More precisely, if X and Y are free abelian groups of finite rank with a pairing inducing an isomorphism $Y \xrightarrow{\sim} \mathrm{Hom}_{\mathbf{Z}}(X, \mathbf{Z})$ and if $R' \subset X$, $R'^\vee \subset Y$ are finite subsets together with a bijection $\alpha \mapsto \alpha^\vee$ from R' to $R'^\vee$ such that R' and $R'^\vee$ induce a root system in $\mathbf{Z}R' \otimes_{\mathbf{Z}} \mathbf{Q}$ in the sense, say, of [B3], ch. VI, §1, n° 1, then there is a group (as in 1.1) having a root datum isomorphic to $(X, R', Y, R'^\vee)$.

In the case where $(X : \mathbf{Z}R') < \infty$ one can find a construction in [Borel 1]. This yields all semi-simple groups. In general one can take a quotient of a direct product of a semi-simple group with a torus. Let me describe in general what happens when we take a quotient of G by a group contained in $Z(G)$.

If X' is a subgroup of $X(T)$ with $\mathbf{Z}R \subset X'$, then $Z = \bigcap_{\lambda \in X'} \ker(\lambda)$ is a closed subgroup scheme of G contained in $Z(G)$. The quotient T/Z is a torus (isomorphic to $\mathrm{Diag}(X')$) and G/Z is a connected and split reductive group with split maximal torus T/Z. One has $k[G/Z] = \bigoplus_{\lambda \in X'} k[G]_\lambda$ where we take weight spaces with respect to the left (or right) regular representation. So $k[G/Z]$ is a direct summand of $k[G]$ as a k–module. The root system of G/Z is just $R \subset X' = X(T/Z)$ and the canonical map $G \to G/Z$ induces an isomorphism of U_α onto the corresponding group in G/Z (for any $\alpha \in R$). The coroots of G/Z are just the images of the coroots of G under the canonical map $Y(T) \to Y(T)/Y(Z) = Y(T/Z)$.

In the case where $(X(T) : X') < \infty$ (or, equivalently, where Z is a finite group scheme) we call G a *covering group* of G/Z.

On the other hand, consider a finitely generated subgroup X' of $X(T) \otimes_{\mathbf{Z}} \mathbf{Q}$ with $X(T) \subset X'$ and $\langle \lambda, \alpha^\vee \rangle \in \mathbf{Z}$ for all $\lambda \in X'$ and $\alpha \in R$. Then we can identify the dual lattice $Y' = \mathrm{Hom}_{\mathbf{Z}}(X', \mathbf{Z})$ with a subgroup of $Y(T)$ containing $R^\vee$, and $(X', R, Y', R^\vee)$ is a possible root datum. Let G' be a reductive group having this root datum. Then the inclusion of $X(T)$ into X' yields a homomorphism $G' \to G$ inducing an isomorphism $G'/(\bigcap_{\lambda \in X(T)} \ker \lambda) \xrightarrow{\sim} G$. So we can regard G' as a covering group of G. This construction leads to all covering groups of G.

For example, we can choose $\varpi'_\alpha \in X(T) \otimes_{\mathbf{Z}} \mathbf{Q}$ for each $\alpha \in S$ such that $\langle \varpi'_\alpha, \beta^\vee \rangle = \delta_{\alpha\beta}$ for all $\beta \in S$. If we now take $X' = X(T) + \sum_{\alpha \in S} \mathbf{Z}\varpi'_\alpha$, then we get a covering (G', T') of (G, T) such that for each $m \in \mathbf{N}$, $m > 0$ each $\lambda \in X(T)$ has a decomposition $\lambda = \lambda_0 + m\lambda_1$ with $\lambda_0, \lambda_1 \in X(T')$ and $0 \leq \langle \lambda, \alpha^\vee \rangle < m$ for all $\alpha \in S$.

1.18. (The Derived Group and Characters) Set

$$X_0(T) = \{\, \lambda \in X(T) \mid \langle \lambda, \alpha^\vee \rangle = 0 \text{ for all } \alpha \in R \}$$

and

$$X'_0(T) = X(T) \cap \sum_{\alpha \in R} \mathbf{Q}\alpha.$$

Then

(1) $$T_1 = \bigcap_{\lambda \in X_0(T)} \ker(\lambda) \qquad \text{and} \qquad T_2 = \bigcap_{\lambda \in X'_0(T)} \ker(\lambda)$$

are closed subgroups of T. In fact, as $X(T_1) \simeq X(T)/X_0(T)$ and $X(T_2) \simeq X(T)/X'_0(T)$ are torsion free, both T_1 and T_2 are tori. Obviously T_2 is contained in the centre $Z(G) = \bigcap_{\alpha \in R} \ker(\alpha)$. More precisely, it is the largest subtorus contained in the centre. (If k is a field, then T_2 is the reduced part of the connected component $Z(G)^0$ of the identity.)

The definitions yield

$$Y(T_1) \simeq Y(T) \cap \sum_{\alpha \in R} \mathbf{Q}\alpha^\vee$$

and

$$Y(T_2) \simeq \{ \varphi \in X(T) \mid \langle \alpha, \varphi \rangle = 0 \text{ for all } \alpha \in R \}.$$

This implies easily that $(X(T_1), \{\alpha_{|T_1} \mid \alpha \in R\}, Y(T_1), R^\vee)$ is a root datum. Let G_1 be a corresponding split, reductive, and connected group containing T_1 as a maximal torus. Obviously $(Y(T_1) : \mathbf{Z}R^\vee) < \infty$, hence also $(X(T_1) : \mathbf{Z}R) < \infty$. So G_1 is semi-simple. The inclusion of T_1 into T leads to a homomorphism of root data, hence to an injective homomorphism $G_1 \to G$. Its image contains all U_α with $\alpha \in R$. For any algebraically closed field K that is a k–algebra, $G_1(K)$ is equal to its derived group (being semi-simple). Its image in $G(K)$ contains the derived group of $G(K)$, which is generated by all $U_\alpha(K)$; so that image is equal to the derived group of $G(K)$. We shall call in general the image of G_1 in G the *derived group* of G and denote it by $\mathcal{D}G$. (In case k is a field, this is compatible with the general definition in [DG], II, §5, 4.8.)

As T_2 is central in G, we get a homomorphism

(2) $$\mathcal{D}G \times T_2 \longrightarrow G$$

induced by the multiplication. On the maximal tori the comorphism (on the character groups) of $T_1 \times T_2 \to T$ has kernel $X_0(T) \cap X_0'(T) = 0$. Therefore (2) is a quotient map, i.e.,

$$G \simeq (\mathcal{D}G \times T_2)/(T_1 \cap T_2)$$

with $T_1 \cap T_2$ embedded via $t \mapsto (t, t^{-1})$.

Any character $\mu \in X(G)$ has to be trivial on $\mathcal{D}G$. So it is determined by its restriction to $T_2 \subset X(G) \subset T$, and this restriction has to vanish on $T_1 \cap T_2$, or (equivalently) $\mu_{|T}$ has to vanish on T_1. Using the quotient map (2) we get also the converse, i.e.,

(3) $$X(G) \simeq X_0(T).$$

We can also apply this to each L_I with $I \subset S$. As any character of P_I has to vanish on each $U_\alpha \subset \mathcal{D}P_I$, hence on U_I, we get also

(4) $$X(P_I) \simeq X(L_I) \simeq \{\lambda \in X(T) \mid \langle \lambda, \alpha^\vee \rangle = 0 \text{ for all } \alpha \in I\}.$$

1.19. (Weight Spaces of G–modules) Any T–module M has a weight space decomposition $M = \bigoplus_{\lambda \in X(T)} M_\lambda$, cf. 1.1(2). The action of $\mathrm{Dist}(T)$ on M can be described as follows: Any $H_{i,m}$ as in 1.12 acts on M_λ as multiplication by $\binom{\langle \lambda, \varphi_i \rangle}{m}$. (Recall that $H_i = (d\varphi_i)(1)$ for some $\varphi_i \in Y(T)$ and apply I.7.13.) In fact, M_λ is the set of all $v \in M$ with $H_{i,m} v = \binom{\langle \lambda, \varphi_i \rangle}{m} v$ for all i and m. This follows because $\binom{a}{m} \equiv \binom{b}{m} \pmod p$ for all $m \in \mathbf{N}$ implies $a = b$ for all $a, b \in \mathbf{Z}$. So $\langle \lambda, \varphi_i \rangle$ is uniquely determined by the $\binom{\langle \lambda, \varphi_i \rangle}{m}$ modulo p.

We associated in I.2.11(6) to each T–module M, which is projective of finite rank over k, a formal character

$$\mathrm{ch}\, M \in \mathbf{Z}[X(T)].$$

(As k is an integral domain, a direct summand of a projective module of finite rank over k again has this property.) There are some elementary properties mentioned in I.2.11(7), (8) where we simply have to replace Λ by $X(T)$.

If M is a G–module and if $\dot{w} \in N_G(T)(k)$ is a representative of some $w \in W$, then an elementary calculation shows

(1) $$\dot{w}\, M_\lambda = M_{w(\lambda)} \qquad \text{for all } \lambda \in X(T).$$

So M_λ and $M_{w(\lambda)}$ have the same rank in case M is projective of finite rank over k. It follows that

(2) $$\mathrm{ch}\, M \in \mathbf{Z}[X(T)]^W$$

in this case. (Here W acts on $\mathbf{Z}[X(T)]$ via $w\, e(\lambda) = e(w\lambda)$.)

We can regard $\mathrm{Dist}(G)$ as a G–module under the adjoint action, cf. I.7.18, hence also as a T–module. Then we have

(3) $$X_{\alpha,n} \in \mathrm{Dist}(G)_{n\alpha} \qquad \text{for all } \alpha \in R \text{ and } n \in \mathbf{N}$$

and

(4) $$\mathrm{Dist}(T) \subset \mathrm{Dist}(G)_0.$$

Indeed, we can identify $k[U_\alpha]$ with a polynomial ring in one variable Y_α, which is a weight vector of weight $-\alpha$ for the adjoint action of T as $kY_\alpha \simeq (\mathrm{Lie}\, U_\alpha)^*$. Then Y_α^n has weight $-n\alpha$ and the "dual" vector $X_{\alpha,n}$ weight $n\alpha$. Of course, (4) is an immediate consequence of the commutativity of T.

If M is a $U_\alpha T$–module for some $\alpha \in R$, then (3) and I.7.18(1) imply for all λ and n

(5) $$X_{\alpha,n}\, M_\lambda \subset M_{\lambda+n\alpha}.$$

Recall from I.7.12 that the $X_{\alpha,n}$ determine the action of $U_\alpha \simeq G_a$ on any U_α–module M as follows: If $m \in M$ and $a \in A$ for some k–algebra A, then

(6) $$x_\alpha(a)\,(m \otimes 1) = \sum_{n \geq 0} (X_{\alpha,n} m) \otimes a^n.$$

Let M be a B^+–module (or even a G–module). Suppose that $\lambda \in X(T)$ is maximal among all weights of M (with respect to $\leq$ as in 1.5). So $\lambda + n\alpha$ is not a weight of M for any $\alpha \in R^+$ and $n > 0$, hence $X_{\alpha,n} M_\lambda = 0$ by (5). Therefore (6) implies:

(7) *If λ is maximal among the weights of M, then $M_\lambda \subset M^{U^+}$.*

We have in this case also that $\bigoplus_{\mu \neq \lambda} M_\mu$ is a B–submodule of M. Using this one gets:

(8) *If λ is maximal among the weights of M, then $\varphi \mapsto \varphi_{|M_\lambda}$ induces a bijection*

$$\mathrm{Hom}_B(M, k_\lambda) \xrightarrow{\sim} \mathrm{Hom}_k(M_\lambda, k).$$

(The inverse maps extends any $\psi : M_\lambda \to k$ by 0 on all M_μ with $\mu \neq \lambda$.)

1.20. (G–modules and Dist(G)–modules) The lemmata I.7.14–I.7.16 can be proved in a more direct way for our G (as in the case of G_a and G_m). For example, any $\mathrm{Dist}(G)$–submodule M' of a G–module M is also a submodule for $\mathrm{Dist}(T)$ and all $\mathrm{Dist}(U_\alpha)$, hence a T–submodule and a U_α–submodule for all α (using the cases G_a and G_m). In case k is a field, one can now use the fact that $G(K)$ is generated by $T(K)$ and the $U_\alpha(K)$ for an algebraic closure K of k. In general, one can use the density of U^+B in G: Assume as in I.7.15 that M/M' is projective over k. The inclusion $k[G] \subset k[U^+B]$ remains injective after tensoring with M/M'. As M' is stable under all of U^+B, the composed map

$$M' \longrightarrow M \otimes k[G] \longrightarrow (M/M') \otimes k[G] \longrightarrow (M/M') \otimes k[U^+B]$$

is zero. This implies I.7.15 for our G. For the other properties one can argue similarly.

Let M be a Dist(G)–module. Suppose that M has also a structure as a T–module such that:

(1) *The structures on M as a* Dist(T)*–module induced from the actions of* Dist(G) *and T coincide.*

The commutator formula

$$H_{i,m} X_{\alpha,n} = X_{\alpha,n} \binom{H_i + \langle n\alpha, \varphi_i \rangle}{m} = \sum_{r=0}^{m} \binom{\langle n\alpha, \varphi_i \rangle}{m-r} X_{\alpha,n} H_{i,r}$$

implies then $X_{\alpha,n} M_\lambda \subset M_{\lambda+n\alpha}$ for all $\alpha \in R$, $n \in \mathbf{N}$, and $\lambda \in X(T)$. We get more generally $\mathrm{Dist}(G)_\mu M_\lambda \subset M_{\lambda+\mu}$ for all $\lambda, \mu \in X(T)$ and hence

$$(2) \qquad t\,(\mu\,(t^{-1}\,(m \otimes 1))) = (\mathrm{Ad}(t)\,\mu)\,(m \otimes 1)$$

for all $m \in M$, $\mu \in \mathrm{Dist}(G)$, $t \in T(A)$, and any A. Let us suppose that this "Dist(G)–T–module" M is locally finite, i.e., that for each $m \in M$ there is a Dist(G)–T–submodule M' with $m \in M'$ and M' finitely generated over k. Then $M' = \bigoplus_{\lambda \in X(T)} M'_\lambda$ as a T–module; only finitely many M'_λ are non-zero, as M' is finitely generated. This implies that there exists for each $m \in M$ and $\alpha \in R$ an integer $n_\alpha(m)$ with $X_{\alpha,n} m = 0$ for all $n > n_\alpha(m)$. Then the action of $\mathrm{Dist}(U_\alpha)$ on M is induced by a structure as U_α–module. The comodule map sends m to $\sum_{n \geq 0} X_{\alpha,n} m \otimes X^n$ where $X \in k[U_\alpha]$ is defined by $X(x_\alpha(a)) = a$. So any $x_\alpha(a)$ acts on $m \otimes 1 \in M \otimes A$ via $x_\alpha(a)\,(m \otimes 1) = \sum_{n \geq 0} X_{\alpha,n} m \otimes a^n$.

There is a description of G in terms of generators and relations ([D], 4.4) that tells us when homomorphisms $T \to H$ and $U_\alpha \to H$ (for all $\alpha \in R$) into a group scheme H (or only a group faisceau H) over k can be glued together to a homomorphism $G \to H$. Consider especially $H = GL(M)$ and the representations above. Then the computations in [St1] show in the case of a ground field that these relations are satisfied. From this we can deduce the same for our integral domain k, e.g., for modules that are projective over k.

So for a projective k–module M there is a one-to-one correspondence between possible structures as a G–module and as a locally finite Dist(G)–T–module.

If k is a field and if G is semi-simple and simply connected, then any locally finite Dist(G)–module is in a natural (and unique) way also a Dist(G)–T–module. So in this case there is an equivalence of categories between { G–modules } and { locally finite Dist(G)–modules }. Probably because of some remark in [Humphreys 8] this result became known as Verma's conjecture although Verma must have been aware of a proof along the lines sketched above when writing [Verma]. These arguments were used in special cases in [Borel 1], §3 and [Wong 1], p. 46. When I needed the result for a larger class of modules, I sketched a proof in [Jantzen 3], p. 119. There is a completely different proof in [Sullivan 4] following earlier partial results in [Sullivan 2]. (These proofs work only for char(k) $\neq 0$, but for char(k) $= 0$ the result is classical.) In [Cline, Parshall, and Scott 6], 9.2/4 one can find another proof and a generalisation to B–modules.

1.21. (The Case GL_n) The general linear groups (cf. I.2.2) are (besides the tori) the simplest examples of reductive groups. Fix $n \in \mathbf{N}$, $n \geq 2$ and take $G = GL_n$. The conventions and notations introduced below will be used whenever we look at this example.

For all i, j $(1 \leq i, j \leq n)$ let E_{ij} be the $(n \times n)$–matrix with (i,j)–entry equal to 1 and all other entries equal to 0. The E_{ij} form a basis of the k–module $M_n(k)$ of all $(n \times n)$–matrices over k. Let us denote the dual basis of $M_n(k)^*$ by X_{ij} $(1 \leq i, j \leq n)$. So the X_{ij} are the matrix coefficients on $M_n(k)$; the k–algebra $k[G]$ is generated by the X_{ij} and by $\det(X_{ij})^{-1}$.

We choose $T \subset GL_n$ as the subfunctor such that $T(A)$ consists of all diagonal matrices in $GL_n(A)$ for all A, i.e., $T = V(\{X_{ij} \mid i \neq j\})$. Then T is isomorphic to a direct product of n copies of G_m. The $\varepsilon_i = (X_{ii})_{|T}$ with $1 \leq i \leq n$ are a basis of $X(T)$, the ε'_i $(1 \leq i \leq n)$ given by $\varepsilon'_i(a) = aE_{ii} + \sum_{j \neq i} E_{jj}$ form a basis of $Y(T)$. One has $\langle \varepsilon_i, \varepsilon'_j \rangle = \delta_{ij}$ (the Kronecker symbol).

The root system has the form

(1) $$R = \{\, \varepsilon_i - \varepsilon_j \mid 1 \leq i, j \leq n,\ i \neq j \,\}.$$

It is of type A_{n-1}; our notations will be consistent with [B3], ch. VI, planche I (for $l = n-1$). One has $(\operatorname{Lie} G)_{\varepsilon_i - \varepsilon_j} = kE_{ij}$ for $i \neq j$ and $(\operatorname{Lie} G)_0 = \operatorname{Lie} T = \sum_{i=1}^n kE_{ii}$. We take for $i \neq j$

(2) $$x_{\varepsilon_i - \varepsilon_j}(a) = 1 + aE_{ij}$$

and

(3) $$\varphi_{\varepsilon_i - \varepsilon_j}\begin{pmatrix} a & b \\ c & d \end{pmatrix} = aE_{ii} + bE_{ij} + cE_{ji} + dE_{jj} + \sum_{h \neq i,j} E_{hh}.$$

One has $(\varepsilon_i - \varepsilon_j)^\vee = \varepsilon'_i - \varepsilon'_j$ for all i, j. The Weyl group permutes $\{\varepsilon_1, \varepsilon_2, \dots, \varepsilon_n\}$ and can be identified with the symmetric group S_n: Map any $w \in W$ to the permutation σ with $w(\varepsilon_i) = \varepsilon_{\sigma(i)}$ for all i. Then $s_{\varepsilon_i - \varepsilon_j}$ is mapped to the transposition (i, j). The composition of this isomorphism with the canonical map $N_G(T)(k) \to W$ admits a section: Map any σ to the permutation matrix $\sum_{i=1}^n E_{\sigma(i),i}$.

We choose as set of positive roots

(4) $$R^+ = \{\, \varepsilon_i - \varepsilon_j \mid 1 \leq i < j \leq n \,\}$$

and get then

(5) $$S = \{\, \alpha_i = \varepsilon_i - \varepsilon_{i+1} \mid 1 \leq i < n \,\}.$$

The element w_0 corresponds to the permutation σ_0 with $\sigma_0(i) = n + 1 - i$ for all i. The Borel subgroup B (resp. B^+) is the functor associating to each A the group of lower (resp. upper) triangular matrices in $GL_n(A)$. Furthermore, $U(A)$ (resp. $U^+(A)$) consists of all matrices in $B(A)$ (resp. $B^+(A)$) such that all diagonal entries are equal to 1.

The centre of G is isomorphic to G_m:

(6) $$G_m \xrightarrow{\sim} Z(G), \qquad a \mapsto a \sum_{i=1}^n E_{ii}.$$

Let us identify GL_n and $GL(k^n)$ via the canonical basis $e_1, e_2, \ldots, e_n$ of k^n. (This is the *natural representation* of GL_n.) Set $V_i = \langle e_n, e_{n-1}, \ldots, e_{n+1-i} \rangle$ for $1 \leq i \leq n$. Then B is the stabiliser of the flag $(V_1 \subset V_2 \subset \cdots \subset V_{n-1})$, i.e., $B = \bigcap_{i=1}^{n-1} \mathrm{Stab}_G(V_i)$. The stabiliser of any partial flag $(V_{i_1} \subset V_{i_2} \subset \cdots \subset V_{i_r})$ with $i_1 < i_2 < \cdots < i_r$ is the parabolic subgroup P_I with $I = \{\alpha_i \mid 1 \leq i < n,\, i \neq n - i_h \text{ for } 1 \leq h \leq r\}$. One has especially

$$\mathrm{Stab}_G(V_i) = P_{S \setminus \{\alpha_{n-i}\}}. \tag{7}$$

The subgroup $X_0(T)$ of $X(T)$ is just $\mathbf{Z}(\varepsilon_1 + \varepsilon_2 + \cdots + \varepsilon_n)$. The homomorphism $\det : GL_n \to G_m$ restricts to $\varepsilon_1 + \varepsilon_2 + \cdots + \varepsilon_n$ on T. Hence:

$$X(G) = \{\, (\det)^r \mid r \in \mathbf{Z} \,\}. \tag{8}$$

The derived group $\mathcal{D}G$ of G is equal to $G' = SL_n$. Furthermore, $T' = T \cap G'$ is a maximal torus in G' with $X(T') = X(T)/\mathbf{Z}(\varepsilon_1 + \varepsilon_2 + \cdots + \varepsilon_n)$ and $Y(T') = \{\sum_{i=1}^n a_i \varepsilon_i' \mid \sum_{i=1}^n a_i = 0\}$. The canonical map $X(T) \to X(T')$ induces an isomorphism of root systems. The standard Borel and parabolic subgroups of G' are just $B \cap G'$ and $P_I \cap G'$. The homomorphisms x_α and φ_α from (2), (3) already take values in G' and yield thus the corresponding homomorphisms for G'.

CHAPTER 2

Simple G–Modules

We assume from now on in this book that G is a split reductive group scheme over a ground ring k as described in the previous chapter. We shall state at the beginning of each chapter what our assumptions on k are. For example, in this chapter we assume throughout that k is a field.

The first aim of this chapter is the classification of all simple G–modules. It turns out (2.7) that Cartan's classification for compact Lie groups carries over to this situation: The simple modules (up to isomorphism) correspond to the dominant weights of T. This was first proved by Chevalley (see [SC]) and there are other accounts of this fact in [Hu2], [St1], [St2], and [Borel 1].

Having established the classification we prove some elementary properties (2.8–2.14). Most of them do not require a reference to an original paper. See, however, the remark to 2.11.

The last subsections (2.15–2.18) contain some examples. In 2.15 the exterior powers of the natural representation of GL_n (or SL_n) are shown to be simple and the corresponding highest weights are determined. In 2.16 we look at some special induced representations of GL_n (or SL_n) and show that they are isomorphic to the symmetric powers of the natural representation (or rather of its dual). In the case $G = SL_2$ we get from this an explicit description of all simple G–modules. In 2.17/18 this example is generalised from GL_n to the symplectic groups and the special orthogonal groups. (In the last case we do not get the symmetric powers, but homomorphic images.)

2.1. As U and U^+ are unipotent, I.2.14(8) implies for any G–module $V \neq 0$:

(1) $$V^U \neq 0 \qquad \text{and} \qquad V^{U^+} \neq 0.$$

As T normalises U and U^+, these two subspaces are T–submodules of V, hence direct sums of their weight spaces. For any $\lambda \in X(T)$ the λ–weight space of V^U is the sum of all simple B–submodules of V isomorphic to k_λ (similarly for U^+ and B^+). We shall write λ instead of k_λ whenever no confusion is possible. So we can express (1) as follows:

(2) *There are* $\lambda, \lambda' \in X(T)$ *with* $\mathrm{Hom}_B(\lambda, V) \neq 0 \neq \mathrm{Hom}_{B^+}(\lambda', V)$.

If $\dim V < \infty$, then we can apply (2) to V^* and get:

(3) *If* $\dim V < \infty$, *then there are* $\lambda, \lambda' \in X(T)$ *with*

$$\mathrm{Hom}_B(V, \lambda) \neq 0 \neq \mathrm{Hom}_{B^+}(V, \lambda').$$

[We could also apply 1.19(8).] Using Frobenius reciprocity (I.3.4) this implies:

(4) *If* $\dim V < \infty$, *then there are* $\lambda, \lambda' \in X(T)$ *with*

$$\mathrm{Hom}_G(V, \mathrm{ind}_B^G \lambda) \neq 0 \neq \mathrm{Hom}_G(V, \mathrm{ind}_{B^+}^G \lambda').$$

In order to shorten many formulae we shall use the notation

(5) $$H^i(M) = R^i \mathrm{ind}_B^G M$$

for any B–module M and any $i \in \mathbf{N}$. (This notation is inspired by the isomorphism $R^i \mathrm{ind}_B^G M \simeq H^i(G/B, \mathcal{L}(M))$, cf. I.5.12.) In particular, we shall write

$$H^i(\lambda) = H^i(k_\lambda) = R^i \mathrm{ind}_B^G k_\lambda.$$

By I.5.12.c each $H^i(\lambda)$ is finite dimensional. In case $\lambda = 0$ our notation could lead to some confusion as $H^i(0)$ could also refer to the zero-module. So in that case we shall usually write $H^i(k)$. Note that

(6) $$H^0(k) = k$$

as $\mathrm{ind}_B^G k = k[G/B]$ and as G/B is projective and irreducible.

The tensor identity and (6) imply $H^0(M) \simeq M$ for any G–module M. The isomorphism is given by the evaluation map $f \mapsto f(1)$. For any $m \in M^B$ the subspace km is a trivial B–submodule of M, hence $\mathrm{ind}_B^G(km)$ a trivial G–submodule of $\mathrm{ind}_B^G M$. The isomorphism $\mathrm{ind}_B^G M \xrightarrow{\sim} M$ maps $\mathrm{ind}_B^G(km)$ to km. This implies (for any G–module M):

(7) $$M^B = M^G.$$

2.2. Proposition: *Let* $\lambda \in X(T)$ *with* $H^0(\lambda) \neq 0$.

a) *We have* $\dim H^0(\lambda)^{U^+} = 1$ *and* $H^0(\lambda)^{U^+} = H^0(\lambda)_\lambda$.

b) *Each weight* μ *of* $H^0(\lambda)$ *satisfies* $w_0\lambda \leq \mu \leq \lambda$.

Proof: Recall that

$$H^0(\lambda) = \{ f \in k[G] \mid f(gb) = \lambda(b)^{-1} f(g) \text{ for all } g \in G(A),\, b \in B(A), \text{ and all } A \}.$$

The action of G is given by left translation. So any $f \in H^0(\lambda)^{U^+}$ satisfies

$$f(u_1 t u_2) = \lambda(t)^{-1} f(1)$$

for all $u_1 \in U^+(A)$, $t \in T(A)$, $u_2 \in U(A)$, and all A. Thus $f(1)$ determines the restriction of f to U^+B, hence also f, as U^+B is dense in G (1.9). This implies $\dim H^0(\lambda)^{U^+} \leq 1$; equality follows now from 2.1(1) and our assumption that $H^0(\lambda) \neq 0$.

Furthermore, the evaluation map $\varepsilon : H^0(\lambda) \to \lambda$, $f \mapsto f(1)$ is a homomorphism of B–modules and is injective on $H^0(\lambda)^{U^+}$. This implies

$$H^0(\lambda)^{U^+} \subset H^0(\lambda)_\lambda.$$

If μ is a maximal weight of $H^0(\lambda)$ (or of any finite dimensional submodule, if we do not want to apply I.5.12.c), then $H^0(\lambda)_\mu \subset H^0(\lambda)^{U^+}$ by 1.19(7). Together with the inclusion $H^0(\lambda)^{U^+} \subset H^0(\lambda)_\lambda$ from above this implies $H^0(\lambda)^{U^+} = H^0(\lambda)_\lambda$ and $\mu \leq \lambda$ for each weight μ of $H^0(\lambda)$.

If μ is a weight of $H^0(\lambda)$, then so is $w_0\mu$ by 1.19(1), hence $w_0\mu \leq \lambda$ and $w_0\lambda \leq \mu$, cf. 1.5.

Remark: We can generalise part of the result as follows: Let $P = P_I \supset B$ be a standard parabolic subgroup of G (cf. 1.8) and set $\overline{P} = P_I^+ \supset B^+$. Denote the unipotent radical of $\overline{P}$ by $U(\overline{P}) = U_I^+$. Then $U(\overline{P})P$ is open and dense in G and the multiplication induces an isomorphism $U(\overline{P}) \times P \xrightarrow{\sim} U(\overline{P})P$ of schemes, cf. 1.10(4).

Let M be a P–module and let $\varepsilon_M : \operatorname{ind}_P^G M \to M$ denote the evaluation map $f \mapsto f(1)$. The proof of the proposition shows:

(1) *ε_M maps $(\operatorname{ind}_P^G M)^{U(\overline{P})}$ injectively into M and is a homomorphism of modules over $L_I = P \cap \overline{P}$.*

2.3. Corollary: *If $H^0(\lambda) \neq 0$, then $\operatorname{soc}_G H^0(\lambda)$ is simple.*

Proof: If L_1, L_2 are two distinct simple submodules of $H^0(\lambda)$, then $L_1 \oplus L_2 \subset H^0(\lambda)$, hence $L_1^{U^+} \oplus L_2^{U^+} \subset H^0(\lambda)^{U^+}$ and $\dim H^0(\lambda)^{U^+} \geq 2$ by 2.1(1), contradicting 2.2. Therefore $\operatorname{soc}_G H^0(\lambda)$ has to be simple.

2.4. Set

(1) $$L(\lambda) = \operatorname{soc}_G H^0(\lambda)$$

for any $\lambda \in X(T)$ with $H^0(\lambda) \neq 0$.

Proposition: a) *Any simple G–module is isomorphic to exactly one $L(\lambda)$ with $\lambda \in X(T)$ and $H^0(\lambda) \neq 0$.*

b) *Let $\lambda \in X(T)$ with $H^0(\lambda) \neq 0$. Then $L(\lambda)^{U^+} = L(\lambda)_\lambda$ and $\dim L(\lambda)^{U^+} = 1$. Any weight μ of $L(\lambda)$ satisfies $w_0\lambda \leq \mu \leq \lambda$. The multiplicity of $L(\lambda)$ as a composition factor of $H^0(\lambda)$ is equal to 1.*

Proof: b) As $L(\lambda)^{U^+} \neq 0$, the formulas $L(\lambda)^{U^+} = L(\lambda)_\lambda$ and $\dim L(\lambda)^{U^+} = 1$ follow immediately from 2.2. The same is true for $w_0\lambda \leq \mu \leq \lambda$. Finally, the multiplicity of $L(\lambda)$ in $H^0(\lambda)$ is at least 1 by construction, but cannot be larger as $\dim L(\lambda)_\lambda = 1 = \dim H^0(\lambda)_\lambda$ and as $V \mapsto V_\lambda$ is an exact functor.

a) The existence follows from 2.1(4), the uniqueness from the formula $L(\lambda)^{U^+} = L(\lambda)_\lambda$ in (b).

Remarks: 1) This proposition shows that λ is the largest weight of $L(\lambda)$ with respect to $\leq$. One usually calls λ the *highest weight* of $L(\lambda)$ and $L(\lambda)$ the simple G–module with highest weight λ.

2) Using $\dot{w}_0\, U\, \dot{w}_0^{-1} = U^+$ we see:

(2) $$\dim L(\lambda)^U = 1 \qquad \text{and} \qquad L(\lambda)^U = L(\lambda)_{w_0\lambda}.$$

3) Each $\mu \in X(T)$ with $\langle \mu, \alpha^\vee \rangle = 0$ for all $\alpha \in R$ extends to a character of G (see 1.18(3)), hence defines a one dimensional (hence simple) G–module k_μ. We have then $L(\mu) \simeq k_\mu$. Note that the tensor identity and 2.1(6) imply that $H^0(\mu) = H^0(k_\mu \otimes k) \simeq k_\mu \otimes H^0(k) \simeq k_\mu$, hence $L(\mu) = H^0(\mu)$.

2.5. Corollary: *Let $\lambda \in X(T)$ with $H^0(\lambda) \neq 0$. Then $L(\lambda)^* \simeq L(-w_0\lambda)$.*

Proof: If μ is a weight of $L(\lambda)^*$, then $-\mu$ is a weight of $L(\lambda)$, cf. I.2.11(11), hence satisfies $w_0\lambda \leq -\mu \leq \lambda$, hence $-\lambda \leq \mu \leq -w_0\lambda$. Furthermore $-w_0\lambda$ is a weight of $L(\lambda)^*$, because $w_0\lambda$ is one of $L(\lambda)$. As $L(\lambda)^*$ is simple, it has to be isomorphic to $L(-w_0\lambda)$.

2.6. It remains to determine the λ with $H^0(\lambda) \neq 0$. Set

$$\begin{aligned} X(T)_+ &= \{\, \lambda \in X(T) \mid \langle \lambda, \alpha^\vee \rangle \geq 0 \text{ for all } \alpha \in S \,\} \\ &= \{\, \lambda \in X(T) \mid \langle \lambda, \alpha^\vee \rangle \geq 0 \text{ for all } \alpha \in R^+ \,\}. \end{aligned}$$

The elements of $X(T)_+$ are called the *dominant weights* of T (with respect to R^+).

Proposition: *Let $\lambda \in X(T)$. The following are equivalent:*

(i) *λ is dominant.*

(ii) *$H^0(\lambda) \neq 0$.*

(iii) *There is a G–module V with $(V^{U^+})_\lambda \neq 0$.*

Proof: (iii) $\Rightarrow$ (ii) Using a composition series of a suitable finite dimensional submodule of V we reduce to the case where V is simple. Now (ii) follows from 2.4.

(ii) $\Rightarrow$ (iii) This is obvious by 2.2.

(ii) $\Rightarrow$ (i) Suppose $H^0(\lambda) \neq 0$. By 1.19(2) any $s_\alpha\lambda$ with $\alpha \in R^+$ is a weight of $H^0(\lambda)$, hence $s_\alpha\lambda \leq \lambda$ by 2.2. Now $s_\alpha\lambda = \lambda - \langle \lambda, \alpha^\vee \rangle \alpha$, so $\langle \lambda, \alpha^\vee \rangle \alpha \geq 0$, hence $\langle \lambda, \alpha^\vee \rangle \geq 0$.

(i) $\Rightarrow$ (ii) Because of I.3.5 we can assume that k is algebraically closed. We regard $G(k)$ as a variety over k and have to find a regular function $f : G(k) \to k$, $f \neq 0$ such that $f(gtu) = \lambda(t)^{-1} f(g)$ for all $g \in G(k)$, $t \in T(k)$, $u \in U(k)$.

On the open and dense subvariety $U^+(k)T(k)U(k)$ of $G(k)$ we consider the (regular) function f_λ given by $f_\lambda(u_1 t u_2) = \lambda(t)^{-1}$ for all $u_1 \in U^+(k)$, $t \in T(k)$, $u_2 \in U(k)$. Obviously $f_\lambda \neq 0$. If we can show that f_λ can be extended to a regular function on the whole of $G(k)$, then the extension of f_λ belongs to $H^0(\lambda)$ because the restriction map $k[G] \to k[U^+B]$ is an injective homomorphism of B–modules (for the right regular action). It then follows that $H^0(\lambda) \neq 0$.

It will be enough to show for each simple root α that f_λ can be extended to a regular function on $\dot{s}_\alpha U^+(k)B(k)$ (or rather that the restriction of f_λ to $\dot{s}_\alpha U^+(k)B(k) \cap U^+(k)B(k)$ can be extended). If so, then for all $\alpha, \beta \in S$ the extensions coincide on $\dot{s}_\alpha U^+(k)B(k) \cap \dot{s}_\beta U^+(k)B(k)$ since they coincide on $\dot{s}_\alpha U^+(k)B(k) \cap \dot{s}_\beta U^+(k)B(k) \cap U^+(k)B(k)$ and since the latter set is dense in $G(k)$. Therefore f_λ can be extended to a regular function on $U^+(k)B(k) \cup \bigcup_{\alpha \in S} \dot{s}_\alpha U^+(k)B(k)$, hence to all of $G(k)$ by 1.9(8).

Fix a simple root $\alpha \in S$. Set $U_1^+ = \langle U_\beta \mid \beta > 0, \beta \neq \alpha \rangle$. Then $U^+ = U_1^+ U_\alpha \simeq U_1^+ \times U_\alpha$ (as schemes) and $\dot{s}_\alpha$ normalises U_1^+. So

$$\dot{s}_\alpha U^+(k)B(k) = U_1^+(k)\dot{s}_\alpha U_\alpha(k)B(k).$$

We have $U_\alpha(k) = \{x_\alpha(a) \mid a \in k\}$. The map $(u_1, a, t, u) \mapsto u_1 \dot{s}_\alpha x_\alpha(a) t u$ is an isomorphism of varieties $U_1^+(k) \times k \times T(k) \times U(k) \xrightarrow{\sim} \dot{s}_\alpha U^+(k)B(k)$ that we use to identify both sides.

We may assume that $\dot{s}_\alpha = n_\alpha(1)$, cf. 1.3. Then

$$\dot{s}_\alpha \, x_\alpha(a) = x_\alpha(-a^{-1}) \, \alpha^\vee(-a^{-1}) \, x_{-\alpha}(a^{-1})$$

for all $a \neq 0$ by a calculation in SL_2. Then

$$u_1 \, \dot{s}_\alpha \, x_\alpha(a) \, t \, u = u_1 \, x_\alpha(-a^{-1}) \, \alpha^\vee(-a^{-1}) \, t \, x_{-\alpha}(\alpha(t) a^{-1}) \, u \in U^+(k)B(k)$$

for all u_1, t, u as above, hence

$$f_\lambda(u_1 \dot{s}_\alpha x_\alpha(a) t u) = \lambda(t)^{-1} \lambda(\alpha^\vee(-a^{-1}))^{-1} = \lambda(t)^{-1}(-a)^{\langle \lambda, \alpha^\vee \rangle}.$$

As $\langle \lambda, \alpha^\vee \rangle \geq 0$, this function on $U_1^+(k) \times (k \setminus \{0\}) \times T(k) \times U(k)$ can be extended (uniquely) to a regular function on $U_1^+(k) \times k \times T(k) \times U(k)$, hence f_λ to a regular function on $\dot{s}_\alpha U^+(k)B(k)$.

Remark: There are other proofs of "(i) $\Rightarrow$ (ii)", cf. [St1] and [Hu2], 31.4. One can also reduce by I.4.18.b to the case $k = \mathbf{C}$ and there apply a construction using the enveloping algebra of $\mathrm{Lie}(G_{\mathbf{C}})$. This construction can be imitated over any k using $\mathrm{Dist}(G)$. (There is for each $\lambda \in X(T)$ a simple $\mathrm{Dist}(G)$–module with highest weight λ, cf. [Haboush 3]. It has finite dimension if λ is dominant, cf. the proof of Lemma 1 in [Jantzen 3], and lifts to G for such λ, cf. 1.20.)

2.7. Corollary: *The $L(\lambda)$ with $\lambda \in X(T)_+$ are a system of representatives for the isomorphism classes of simple G–modules.*

Remark: We have by 2.4 and 2.6 for each $\lambda \in X(T)_+$

$$\text{(1)} \qquad \mathrm{ch}\, L(\lambda) = e(\lambda) + \sum_{\mu < \lambda} \dim(L(\lambda)_\mu)\, e(\mu).$$

This shows that the ch $L(\lambda)$ with $\lambda \in X(T)_+$ are linearly independent. As ch(?) is additive on short exact sequences, the map $V \mapsto \mathrm{ch}(V)$ induces a homomorphism from the Grothendieck group of { finite dimensional G–modules } to $\mathbf{Z}[X(T)]$ which is injective because a basis (the classes of the simple modules) is mapped to a linearly independent set. This implies that two finite dimensional G–modules have the same formal character if and only if the multiplicities of the simple G–modules as composition factors are the same in both cases. We usually denote these multiplicities by $[V : L(\lambda)]$ (for any finite dimensional G–module V and any $\lambda \in X(T)_+$).

2.8. Proposition: *Let $\lambda \in X(T)_+$. Then*

$$\mathrm{End}_G\, H^0(\lambda) = k = \mathrm{End}_G\, L(\lambda).$$

Proof: Using Frobenius reciprocity and 2.2 we get

$$\begin{aligned} \mathrm{End}_G\, H^0(\lambda) &\simeq \mathrm{Hom}_B(H^0(\lambda), \lambda) \\ &\subset \mathrm{Hom}_T(H^0(\lambda), \lambda) \simeq \mathrm{Hom}(H^0(\lambda)_\lambda, \lambda) \simeq k. \end{aligned}$$

On the other hand, the identity map is a non-zero element in $\mathrm{End}_G\, H^0(\lambda)$, so $\mathrm{End}_G\, H^0(\lambda) = k$.

Similarly,

$$\begin{aligned} \mathrm{End}_G\, L(\lambda) &\subset \mathrm{Hom}_G(L(\lambda), H^0(\lambda)) \simeq \mathrm{Hom}_B(L(\lambda), \lambda) \\ &\subset \mathrm{Hom}_T(L(\lambda), \lambda) \simeq \mathrm{Hom}(L(\lambda)_\lambda, \lambda) \simeq k. \end{aligned}$$

Again $\mathrm{End}_G\, L(\lambda) \neq 0$, hence $\mathrm{End}_G\, L(\lambda) = k$.

2.9. Corollary: *For any extension field k' of k the $G_{k'}$–module $L(\lambda)\otimes k'$ is the simple $G_{k'}$–module with highest weight λ.*

Proof: We know (by I.7.15/16) that $L(\lambda)$ is also a simple $\mathrm{Dist}(G)$–module and that $\mathrm{End}_{\mathrm{Dist}(G)} L(\lambda) = \mathrm{End}_G L(\lambda) = k$. Therefore Wedderburn's theorem implies that the canonical map from $\mathrm{Dist}(G)$ to $\mathrm{End}_k L(\lambda)$ is surjective. Hence so is that from $\mathrm{Dist}(G_{k'}) \simeq \mathrm{Dist}(G)\otimes k'$ to $\mathrm{End}_{k'}(L(\lambda)\otimes k') \simeq (\mathrm{End}_k L(\lambda))\otimes k'$. This implies the simplicity of $L(\lambda)\otimes k'$, hence the corollary. (Cf. also [B1], ch. VIII, §13, n° 5, cor. de la prop. 5.)

Remark: The corollary implies for all G–modules V:

$$\mathrm{soc}_{G_{k'}}(V\otimes k') = \mathrm{soc}_G(V)\otimes k'. \tag{1}$$

Here one inclusion ("$\supset$") is a trivial consequence of the corollary. In order to get "$\subset$" suppose that $\dim V < \infty$. The multiplicity of any $L(\lambda)$ in $\mathrm{soc}_G(V)$ is equal to $\dim \mathrm{Hom}_G(L(\lambda),V)$, that of $L(\lambda)\otimes k'$ in $\mathrm{soc}_{G_{k'}}(V\otimes k')$ equal to $\dim_{k'} \mathrm{Hom}_{G_{k'}}(L(\lambda)\otimes k', V\otimes k')$. Therefore equality follows from I.2.10(7).

Obviously (1) implies that V is a semi-simple G–module if and only if $V\otimes k'$ is a semi-simple $G_{k'}$–module.

Of course (1) generalises to the higher terms of the socle series, cf. I.2.14.

2.10. It follows from 2.8 for each $\lambda\in X(T)_+$ that the centre $Z(G)$ of G acts on $H^0(\lambda)$ through scalars, i.e., via a character. As $Z(G)\subset T$ and as $H^0(\lambda)_\lambda \neq 0$, this character has to be the restriction of λ to $Z(G)$.

This fact is a special case of the following statement:

(1) *For each $\lambda\in X(T)$ the group $Z(G)$ acts on each $H^i(\lambda)$ through the restriction of λ to $Z(G)$.*

Let ν denote the restriction of λ to $Z(G)$. Each $Z(G)$–module M is the direct sum of its weight spaces M_μ with $\mu\in X(Z(G))$ relative $Z(G)$. If M is a B–module, then each M_μ is a B–submodule because $Z(G)$ is central in B. If M is an injective B–module, then also each M_μ is an injective B–module.

Consider now an injective resolution $(Q^i)_{i\geq 0}$ of k_λ as a B–module. Then also $(Q^i_\nu)_{i\geq 0}$ is an injective resolution of k_λ. (Note that $(k_\lambda)_\nu = k_\lambda$.) Therefore we may assume that $Z(G)$ acts on each Q^i via ν. Now the definition of an induced representation (see I.3.3(2)) shows that $Z(G)$ acts also on each $\mathrm{ind}_B^G Q^i$ via ν. As $H^i(\lambda) = R^i \mathrm{ind}_B^G k_\lambda$ is the i–th cohomology group of the complex $\mathrm{ind}_B^G Q^\bullet$, the claim follows.

Let Z be a closed subgroup of $Z(G)$. For any $\lambda\in X(T)_+$ with $\lambda_{|Z} = 0$ the group Z acts trivially on $H^0(\lambda)$, hence also on $L(\lambda)$. Therefore $L(\lambda)$ is a G/Z–module and it is the simple module with highest weight $\lambda\in X(T/Z)\subset X(T)$. Recall that G/Z is a split reductive group with split maximal torus T/Z, cf. 1.17. Conversely, any simple G/Z–module is a simple G–module in a natural way, hence isomorphic to some $L(\lambda)$ with $\lambda_{|Z}=0$.

We can apply this argument to $(\mathcal{D}G\times T_2, G)$ instead of $(G, G/Z)$, using the notation from 1.18. As $T_2\subset Z(\mathcal{D}G\times T_2)$ acts through scalars on any simple $(\mathcal{D}G\times T_2)$–module, one gets:

(2) *Regarded as a $\mathcal{D}G$–module any $L(\lambda)$ with $\lambda \in X(T)_+$ is simple with highest weight $\lambda_{|T\cap\mathcal{D}G}$.*

If $G' \to G$ is a covering group, then each simple G–module is also a simple G'–module in an obvious way. If T' is the inverse image of T in G', then we have inclusions $X(T) \subset X(T')$ and $X(T)_+ \subset X(T')_+$. To each $\lambda \in X(T)_+$ we have a simple G–module $L(\lambda)$ with highest weight λ and a simple G'–module, say $L'(\lambda)$, with highest weight λ. Clearly $L'(\lambda)$ is isomorphic to $L(\lambda)$ considered as a G'–module. If M and N are G–modules, then one has for all $i \in \mathbf{N}$

(3) $$\operatorname{Ext}^i_{G'}(M,N) \simeq \operatorname{Ext}^i_G(M,N)$$

by I.6.8(2).

2.11. Recall the notations L_I, U_I^+, U_I from 1.8.

Proposition: *Let $I \subset S$ and $\lambda \in X(T)_+$. Then:*

a) $L(\lambda)^{U_I^+} = \bigoplus_{\nu\in\mathbf{Z}I} L(\lambda)_{\lambda-\nu}$.

b) *$\bigoplus_{\nu\in\mathbf{Z}I} L(\lambda)_{\lambda-\nu}$ is the simple L_I–module with highest weight λ.*

Proof: a) Set $M = L(\lambda)^{U_I^+}$. For any $\alpha \in R^+ \setminus R_I$, any $n > 0$, and any $\nu \in \mathbf{Z}I$ the element $\lambda - \nu + n\alpha$ is not a weight of $L(\lambda)$ as $\lambda - \nu + n\alpha \not\leq \lambda$. (There is a simple root $\beta \notin I$ occurring in α, hence in $n\alpha - \nu$ with a positive coefficient.) This implies $X_{\alpha,n}L(\lambda)_{\lambda-\nu} = 0$ by 1.19(5), hence $L(\lambda)_{\lambda-\nu} \subset M$ by 1.19(6).

As L_I normalises U_I^+, the subspace M is an L_I–submodule. It is especially the sum of its weight spaces and for each μ the sum $\bigoplus_{\nu\in\mathbf{Z}I} M_{\mu+\nu}$ is an L_I–submodule. (It is a $\operatorname{Dist}(L_I)$–submodule as $\operatorname{Dist}(L_I)$ is generated by $\operatorname{Dist}(T)$ and the $\operatorname{Dist}(U_\alpha)$ with $\alpha \in R_I$.) For each μ with $M_\mu \neq 0$, the L_I–submodule $\bigoplus_{\nu\in\mathbf{Z}I} M_{\mu+\nu}$ contains a non-zero vector invariant under the unipotent group $U^+ \cap L_I$. As U_I^+ fixes M and as $U^+ = U_I^+(U^+ \cap L_I)$, this vector is invariant under U^+, hence contained in $L(\lambda)^{U^+} = L(\lambda)_\lambda$. This implies $\mu - \lambda \in \mathbf{Z}I$, hence $M_\mu \subset \bigoplus_{\nu\in\mathbf{Z}I} L(\lambda)_{\lambda-\nu}$.

b) Keep the notation M from above. The proof of a) shows that $M_\lambda = L(\lambda)_\lambda$ is the space of all $(U^+ \cap L_I)$–invariants in M. This implies as in 2.3 that M has simple socle as an L_I–module and that this socle is the L_I–submodule generated by any $v \in L(\lambda)_\lambda$, $v \neq 0$, i.e.,

$$\operatorname{soc}_{L_I} M = \operatorname{Dist}(L_I)v.$$

On the other hand, the simplicity of $L(\lambda)$ implies $L(\lambda) = \operatorname{Dist}(G)v$. Using the basis of $\operatorname{Dist}(G)$ as in 1.12(4), but with R^+ replaced by $-R^+$, one gets

$$L(\lambda) = \operatorname{Dist}(U)v$$

and then (using 1.19(5))

$$\bigoplus_{\nu\in\mathbf{Z}I} L(\lambda)_{\lambda-\nu} = \operatorname{Dist}(L_I)v,$$

hence the simplicity of this direct sum.

Remarks: 1) The same proof as above shows

(1) $$H^0(\lambda)^{U_I^+} = \bigoplus_{\nu\in\mathbf{Z}I} H^0(\lambda)_{\lambda-\nu}.$$

2) This proposition appears for the first time in [Smith]. Part b) was certainly known before, cf., e.g., [Jantzen 1], Satz I.5, p. 15.

2.12. (Extensions) We can identify $\mathrm{Ext}^1_G(M_2, M_1)$ for all G–modules M_1, M_2 with the set of equivalence classes of all short exact sequences $0 \to M_1 \xrightarrow{i} M \xrightarrow{j} M_2 \to 0$ of G–modules.

Take, for example, $M_1 = M_2 = L(\lambda)$ for some $\lambda \in X(T)_+$ and consider an exact sequence as above. Choose $v \in L(\lambda)_\lambda$, $v \neq 0$, and $v' \in M$ with $j(v') = v$. As λ is the largest weight of M, we have $v' \in M^{U^+}$, cf. 1.19(7), hence

$$\mathrm{Dist}(G)v' = \mathrm{Dist}(U)v'$$

as in the proof of 2.11.b, and (using 1.19(5))

$$(\mathrm{Dist}(G)v')_\lambda = kv'.$$

This implies $i(v) \notin \mathrm{Dist}(G)v'$, hence $i(L(\lambda)) \cap \mathrm{Dist}(G)v' = 0$ by the simplicity of $L(\lambda)$. Therefore $M = i(L(\lambda)) \oplus \mathrm{Dist}(G)v'$ and the sequence splits. We have thus proved:

(1) $$\mathrm{Ext}^1_G(L(\lambda), L(\lambda)) = 0 \qquad \text{for all } \lambda \in X(T)_+.$$

In order to prove another elementary property of Ext^1_G let us use the antiautomorphism τ from 1.16. For any finite dimensional G–module M we define a new G–module ${}^\tau M$ as follows: Take ${}^\tau M = M^*$ as a vector space, but define the action of G on some $\varphi \in M^*$ via $g\varphi = \varphi \circ \tau(g)$ [instead of the usual $\varphi \circ g^{-1}$ as in M^*]. One has ch ${}^\tau M =$ ch M as $\tau_{|T} = \mathrm{id}$, and ${}^\tau({}^\tau M) \simeq M$ as $\tau^2 = 1$. If M is simple, then so is ${}^\tau M$. As any simple G–module is determined by its weights (2.4), we get

(2) $${}^\tau L(\lambda) \simeq L(\lambda) \qquad \text{for all } \lambda \in X(T)_+.$$

Any exact sequence $0 \to M_1 \to M \to M_2 \to 0$ of finite dimensional G–modules yields an exact sequence $0 \to {}^\tau M_2 \to {}^\tau M \to {}^\tau M_1 \to 0$ by taking the transposed maps of the original maps. This produces an isomorphism

(3) $$\mathrm{Ext}^1_G(M_2, M_1) \simeq \mathrm{Ext}^1_G({}^\tau M_1, {}^\tau M_2).$$

Applying this to simple modules we get by (2)

(4) $$\mathrm{Ext}^1_G(L(\lambda), L(\mu)) \simeq \mathrm{Ext}^1_G(L(\mu), L(\lambda)) \qquad \text{for all } \lambda, \mu \in X(T)_+.$$

2.13. We have $\dim H^0(\lambda) < \infty$ for all λ by I.5.12.c. (There is an elementary proof of this fact, cf. [Donkin 9], 1.8.) So we can define for each $\lambda \in X(T)_+$ a G–module via

(1) $$V(\lambda) = H^0(-w_0\lambda)^*.$$

The automorphism σ of G with $\sigma(g) = \tau(\dot{w}_0 g^{-1} \dot{w}_0^{-1})$ stabilises B and induces $-w_0$ on $X(T)$. So ${}^\sigma H^0(\lambda) \simeq H^0(-w_0\lambda)$ for all $\lambda \in X(T)_+$ by I.3.5(4). One has for all finite dimensional G–modules M that $({}^\sigma M)^* \simeq {}^\tau M$. (One gets at first that $({}^\sigma M)^*$ is isomorphic to ${}^\tau M$ twisted by the inner automorphism by $\dot{w}_0$. But twisting with any inner automorphism yields an isomorphic module.) It follows that

(2) $$V(\lambda) \simeq {}^\tau H^0(\lambda).$$

This shows that ch $V(\lambda) =$ ch $H^0(\lambda)$.

Lemma: *Let $\lambda \in X(T)_+$.*

a) *There are for each G–module M functorial isomorphisms*

$$\mathrm{Hom}_G(V(\lambda), M) \xrightarrow{\sim} \mathrm{Hom}_{B^+}(\lambda, M) \xrightarrow{\sim} (M^{U^+})_\lambda.$$

b) *The G–module $V(\lambda)$ is generated by a B^+–stable line of weight λ. Any G–module generated by a B^+–stable line of weight λ is a homomorphic image of $V(\lambda)$.*

Proof: a) Let $\varepsilon : H^0(-w_0\lambda) \to -w_0\lambda$ be the evaluation map as in I.3.4. Assume first that $\dim(M) < \infty$. We get canonical isomorphisms

$$\begin{aligned}\mathrm{Hom}_G(V(\lambda), M) &\xrightarrow{\sim} \mathrm{Hom}_G(M^*, H^0(-w_0\lambda)) \xrightarrow{\sim} \mathrm{Hom}_B(M^*, -w_0\lambda)\\ &\xrightarrow{\sim} \mathrm{Hom}_B(w_0\lambda, M) \xrightarrow{\sim} (M^U)_{w_0\lambda}\end{aligned}$$

mapping any φ first to φ^*, then to $\varepsilon \circ \varphi^*$, to $\varphi \circ \varepsilon^*$, and finally to $\varphi(\varepsilon^*(1))$. There are also natural isomorphisms $\mathrm{Hom}_{B^+}(\lambda, M) \xrightarrow{\sim} (M^{U^+})_\lambda$ given by $\psi \mapsto \psi(1)$ and $(M^U)_{w_0\lambda} \xrightarrow{\sim} (M^{U^+})_\lambda$ given by $m \mapsto \dot{w}_0\, m$. So $\gamma(\varphi) = \dot{w}_0\, \varphi(\varepsilon^*(1)) = \varphi(\dot{w}_0\, \varepsilon^*(1))$ defines a (functorial) isomorphism $\gamma : \mathrm{Hom}_G(V(\lambda), M) \xrightarrow{\sim} (M^{U^+})_\lambda$. For arbitrary M the claim follows by taking direct limits.

b) The element $v = \dot{w}_0\, \varepsilon^*(1)$ is obviously a B^+–eigenvector of weight λ. If $V(\lambda) \to M'$ is a homomorphism of G–modules with $\varphi(v) = 0$, then $\gamma(\varphi) = \varphi(v) = 0$, hence $\varphi = 0$. This shows that $V(\lambda)$ is generated by v. For any G–module M and any B^+–eigenvector $m \in M$ of weight λ we have a B^+–homomorphism $\lambda \to M$ with $1 \mapsto m$, hence a G–homomorphism $V(\lambda) \to M$ with $v \mapsto m$. If m generates M, then M is a homomorphic image of $V(\lambda)$.

Remarks: 1) Because of the universal property in b) we call $V(\lambda)$ the *universal highest weight module* of weight λ. (It is also often called a Weyl module, cf. 5.11 below.)

2) Suppose that λ is a *maximal* weight of a G–module M. Then $M_\lambda \subset M^{U^+}$ by 1.19(7) and we get $\mathrm{Hom}_G(V(\lambda), M) \xrightarrow{\sim} M_\lambda$. Furthermore, 1.19(8) implies in this case that $(M_\lambda)^* \xrightarrow{\sim} \mathrm{Hom}_G(M, H^0(\lambda))$ by Frobenius reciprocity.

2.14. We get from dualising 2.4(1):

(1) $$V(\lambda)/\operatorname{rad}_G V(\lambda) \simeq L(\lambda).$$

Proposition: *Let $\lambda, \mu \in X(T)_+$ with $\mu \not> \lambda$. Then*

$$\mathrm{Ext}^1_G(L(\lambda), L(\mu)) \simeq \mathrm{Hom}_G(\operatorname{rad}_G V(\lambda), L(\mu)).$$

Proof: We get from the short exact sequence (cf. (1))

$$0 \to \operatorname{rad}_G V(\lambda) \longrightarrow V(\lambda) \longrightarrow L(\lambda) \to 0$$

a long exact sequence

(2) $$\begin{aligned}0 &\to \mathrm{Hom}_G(L(\lambda), L(\mu)) \to \mathrm{Hom}_G(V(\lambda), L(\mu)) \to \mathrm{Hom}_G(\operatorname{rad}_G V(\lambda), L(\mu))\\ &\to \mathrm{Ext}^1_G(L(\lambda), L(\mu)) \to \mathrm{Ext}^1_G(V(\lambda), L(\mu)) \to \cdots.\end{aligned}$$

Any homomorphism from $V(\lambda)$ to a simple G–module has to annihilate $\text{rad}_G V(\lambda)$. Therefore the last map in the first row of (2) is the zero map. So the proposition will follow from (2) as soon as we prove $\text{Ext}^1_G(V(\lambda), L(\mu)) = 0$.

Consider an exact sequence of G–modules

$$0 \to L(\mu) \longrightarrow M \longrightarrow V(\lambda) \to 0. \tag{3}$$

Choose some $v \in M_\lambda$ that is mapped to a B^+–eigenvector generating $V(\lambda)$. By our assumption λ is a maximal weight of M, cf. 2.2. Therefore v is a B^+–eigenvector of weight λ, cf. 1.19(7). So the G–submodule M' of M generated by v is a homomorphic image of $V(\lambda)$ by 2.13. On the other hand, it is mapped onto $V(\lambda)$ in (3). So M' has to be mapped isomorphically onto $V(\lambda)$ in (3) and is therefore a complement to the kernel $L(\mu)$. So the exact sequence splits, hence $\text{Ext}^1_G(V(\lambda), L(\mu)) = 0$.

Remarks: 1) One has dually: If $\mu \not> \lambda$, then

$$\text{Ext}^1_G(L(\mu), L(\lambda)) \simeq \text{Hom}_G(L(\mu), H^0(\lambda)/\text{soc}_G H^0(\lambda)). \tag{4}$$

See also 4.14 below for a generalisation.

2) The proof of $\text{Ext}^1_G(V(\lambda), L(\mu)) = 0$ shows more generally: If V is a G–module with $\text{Ext}^1_G(V(\lambda), V) \neq 0$, then there exists a weight μ of V with $\mu > \lambda$. This implies, e.g.,

$$\text{Ext}^1_G(V(\lambda), V(\lambda)) = 0 \qquad \text{for all } \lambda \in X(T)_+. \tag{5}$$

2.15. Let $\lambda \in X(T)_+$. Suppose that M is a G–module with $\dim M_\lambda = 1$ and with all weights of the form $w\lambda$ with $w \in W$. By 1.19(2) all non-zero weight spaces of M have dimension 1. If $M' \subset M$ is a non-zero G–submodule, then there is a $\mu \in X(T)$ with $M'_\mu \neq 0$, hence with $M'_\mu = M_\mu$. Then $\mu \in W\lambda$ and by 1.19(1) all weight spaces are contained in M'. So M is simple, hence isomorphic to $L(\lambda)$. (Recall that $\lambda \in X(T)_+$ implies that $\lambda \geq w\lambda$ for all $w \in W$.)

Here is an example where this occurs. Suppose that $G = GL_n$ for some $n \geq 1$ and that T is the subgroup of diagonal matrices in G. Set $V = k^n$ and let $e_1, e_2, \dots, e_n$ denote the canonical basis of k^n. Then each e_i is an eigenvector for T; the corresponding weight will be denoted by ε_i (as in 1.21). Then $\varepsilon_1, \varepsilon_2, \dots, \varepsilon_n$ is a basis of $X(T)$.

Consider the G–module $\Lambda^m V$ for some m with $1 \leq m \leq n$. It has as a basis all $e_{i_1} \wedge e_{i_2} \wedge \cdots \wedge e_{i_m}$ with $1 \leq i_1 < i_2 < \cdots < i_m \leq n$. Such a basis vector is a weight vector of weight $\varepsilon_{i_1} + \varepsilon_{i_2} + \cdots + \varepsilon_{i_m}$. Since all these weights are distinct, all non-zero weight spaces have dimension 1. We can (as in 1.21) identify W with the full permutation group of $\{\varepsilon_1, \varepsilon_2, \dots, \varepsilon_n\}$ isomorphic to S_n. Therefore all weights of $\Lambda^m V$ are conjugate under W to $\varpi_m = \varepsilon_1 + \varepsilon_2 + \cdots + \varepsilon_m$. If we choose $R^+ = \{\varepsilon_i - \varepsilon_j \mid 1 \leq i < j \leq n\}$ as in 1.21, then $\varpi_m \in X(T)_+$. Therefore $\Lambda^m V$ is the simple G–module with highest weight ϖ_m.

We can replace G and T in this example by $G' = SL_n$ and $T' = T \cap SL_n$. Then one has $\varepsilon_1 + \varepsilon_2 + \cdots + \varepsilon_n = 0$ in $X(T')$ and the ε_i with $1 \leq i < n$ are a basis of $X(T')$. However, the weight spaces in $\Lambda^m V$ still have dimension 1 and so the same argument shows that $\Lambda^m V$ is the simple G'–module with highest weight ϖ_m. (Alternatively, we could refer to 2.10(2).)

In characteristic 0 for arbitrary G (say, semi-simple with R indecomposable) the situation described above occurs only for $\lambda = 0$ and for λ a *minuscule* fundamental weight, cf. [B3], ch. VI, §1, exerc. 24 and §4, exerc. 15. There are more cases in positive characteristic.

2.16. In this and the following subsections we want to describe $H^0(\lambda) = \operatorname{ind}_B^G \lambda$ in some special cases. Let us take first $G = GL_n$ with $n \geq 2$, let us use the notations introduced in 1.21 and set $V = k^n$. Let $P = P_{\{\alpha_1, \alpha_2, \dots, \alpha_{n-2}\}}$ be the stabiliser in G of the line $ke_n \subset k^n = V$. There is a character $\varpi \in X(P)$ such that

$$g(e_n \otimes 1) = e_n \otimes \varpi(g)^{-1}$$

for all $g \in P(A)$ and all A. By abuse of notation we write below ϖ also for the restrictions of ϖ to T and B.

We have for any $r \in \mathbf{N}$

$$\text{(1)} \quad H^0(r\varpi) \simeq \operatorname{ind}_B^G(r\varpi) \simeq \operatorname{ind}_P^G(\operatorname{ind}_B^P r\varpi) \simeq \operatorname{ind}_P^G(r\varpi \otimes \operatorname{ind}_B^P k) \simeq \operatorname{ind}_P^G(r\varpi),$$

using the transitivity of induction (I.3.5(2)), the tensor identity (I.3.6), and finally the equality $\operatorname{ind}_B^P k = k[P/B] = k$, cf. 1.8(5), (6).

Let $\overline{k}$ be an algebraically closed extension of k. As all groups concerned are reduced, we have

$$\text{(2)} \quad \operatorname{ind}_P^G(r\varpi) = \{f \in k[G] \mid f(gg_1) = \varpi(g_1)^{-r} f(g) \text{ for all } g \in G(\overline{k}),\ g_1 \in P(\overline{k})\}.$$

Set H equal to the kernel of ϖ in P. By (2) any $f \in \operatorname{ind}_P^G(r\varpi)$ is invariant under right translation by any element of $H(\overline{k})$. The map $g \mapsto ge_n$ induces an isomorphism of varieties $G(\overline{k})/H(\overline{k}) \xrightarrow{\sim} G(\overline{k})e_n = (V \otimes k) \setminus \{0\}$ as the tangent map is obviously surjective (cf. [Bo], 6.7). So $\overline{k}[G(\overline{k})/H(\overline{k})] = \overline{k}[(V \otimes \overline{k}) \setminus \{0\}]$. As $V \otimes \overline{k}$ is a smooth, hence normal variety of dimension at least 2, any regular function on $(V \otimes \overline{k}) \setminus \{0\}$ extends (uniquely, of course) to $V \otimes \overline{k}$. So $\overline{k}[(V \otimes \overline{k}) \setminus \{0\}] \simeq \overline{k}[V \otimes \overline{k}] \simeq S(V^*) \otimes \overline{k}$. (As usual, $S(?)$ denotes the symmetric algebra.) On the other hand, $\overline{k}[G(\overline{k})/H(\overline{k})] \simeq k[G/H] \otimes \overline{k}$ and there is a natural map $S(V^*) \to k[G/H]$. As it becomes an isomorphism after tensoring with $\overline{k}$, it was one already before. Hence

$$S(V^*) \simeq k[G/H].$$

Any $f \in S(V^*)$ is mapped to the function $g \mapsto f(ge_n)$. If $g_1 \in P(\overline{k})$, then $g_1 e_n = \varpi(g_1)^{-1} e_n$, hence $f(gg_1e_n) = f(\varpi(g_1)^{-1} ge_n)$. Therefore (2) implies

$$\text{(3)} \quad \operatorname{ind}_P^G(r\varpi) = \{f \in S(V^*) \mid f(av) = a^r f(v) \text{ for all } a \in \overline{k} \text{ and all } v \in V \otimes \overline{k}\}.$$

This space is obviously equal to $S^r(V^*)$, the homogeneous part of degree r in $S(V^*)$:

$$\text{(4)} \qquad H^0(r\varpi) \simeq \operatorname{ind}_P^G(r\varpi) \simeq S^r(V^*).$$

(It is left to the reader to realise also $S^r(V)$ as $H^0(\lambda)$ for suitable λ.)

If we denote the basis of V^* dual to $e_1, e_2, \dots, e_n$ by $X_1, X_2, \dots, X_n$, then we can identify $S(V^*)$ with the polynomial ring in $X_1, X_2, \dots, X_n$ and $S^r(V^*)$ with the subspace of all homogeneous polynomials of degree r. The weight of each e_i is ε_i, so each X_i has weight $-\varepsilon_i$ and any monomial $X_1^{a(1)}X_2^{a(2)}\dots X_n^{a(n)}$ has weight $-\sum_{i=1}^n a(i)\varepsilon_i$. Therefore distinct monomials in $S^r(V^*)$ have distinct weights, each weight space in $S^r(V^*)$ has dimension 1 and is spanned by a monomial $X_1^{a(1)}X_2^{a(2)}\dots X_n^{a(n)}$ with $\sum a(i) = r$. We have $\varpi = -\varepsilon_n$, so X_n^r is the unique monomial of weight $r\varpi$. It follows that X_n^r spans $H^0(r\varpi)^{U^+} = H^0(r\varpi)_{r\varpi}$, hence

(5) $$L(r\varpi) = \mathrm{Dist}(U)X_n^r.$$

The representation of G on $S(V^*)$ and on each $S^r(V^*)$ can be constructed via base change from a representation if $G_{\mathbf{Z}} = GL_{n,\mathbf{Z}}$ on $S(V_{\mathbf{Z}}^*)$ resp. on $S^r(V_{\mathbf{Z}}^*)$, where $V_{\mathbf{Z}}$ is the natural module of $G_{\mathbf{Z}}$. Each $G_{\mathbf{Z}}$–module $S^r(V_{\mathbf{Z}}^*)$ is a lattice in the corresponding $G_{\mathbf{Q}}$–module $S^r(V_{\mathbf{Q}}^*)$. So we can compute the action of any $X_{\alpha,m} \in \mathrm{Dist}(G)$ ($\alpha \in R$, $m \in \mathbf{N}$) on $S(V^*)$ from the action of the corresponding element in $\mathrm{Dist}(G_{\mathbf{Q}})$.

We identify $\mathrm{Lie}(G)$ as in 1.21 with $M_n(k)$ and may assume that $X_\alpha = E_{ij}$ for $\alpha = \varepsilon_i - \varepsilon_j$. Then we get $X_{\alpha,m} = (E_{ij})^m/m!$ in $\mathrm{Dist}(G_{\mathbf{Q}})$. As $E_{ij}e_l = \delta_{jl}e_i$ (with δ_{jl} the Kronecker symbol) we get $E_{ij}X_l = -\delta_{il}X_j$, hence

(6) $$E_{ij,m}X_1^{a(1)}X_2^{a(2)}\dots X_n^{a(n)} = (-1)^m\binom{a(i)}{m}\Big(\prod_{l\neq i,j} X_l^{a(l)}\Big)X_i^{a(i)-m}X_j^{a(j)+m}.$$

In the special case $n = 2$ this implies

(7) $$L(r\varpi) = \sum_{m\geq 0} kE_{21,m}X_2^r = \sum_{m=0}^{r} k\binom{r}{m}X_1^mX_2^{r-m}.$$

So we get for $n = 2$: if $\mathrm{char}(k) = p \neq 0$, then $L(r\varpi)$ is spanned by all $X_1^mX_2^{r-m}$ with $p \nmid \binom{r}{m}$; if $\mathrm{char}(k) = 0$, then $L(r\varpi) = S^r(V^*)$.

For all n one can determine the submodule structure of $H^0(r\varpi)$ for all r by elementary methods. This was first done for $n = 2$ in [Carter and Cline], see also [Cline 2] and [Deriziotis 2, 3]. For arbitrary n see [Doty 1].

For arbitrary λ it is much harder to determine the submodule structure of $H^0(\lambda)$, even in those cases where one knows all composition factors. There are partial results, especially for SL_3, cf. [Doty and Sullivan 1], [Irving 1], [Kühne-Hausmann]. The proofs there require more advanced techniques.

2.17. Consider $G = Sp_{2n}$ with $n \geq 1$, set $V = k^{2n}$ and choose $e \in V$, $e \neq 0$. Now V is a G–module in a natural way. Set P equal to the stabiliser in G of the line ke. Then P is a parabolic subgroup of G; we may assume that $P \supset B$. There is again $\varpi \in X(P)$ such that $g(e \otimes 1) = e \otimes \varpi(g)^{-1}$ for all $g \in P(A)$ and all A. Now the same arguments as in 2.16 yield isomorphisms

(1) $$H^0(r\varpi) \simeq \mathrm{ind}_P^G(r\varpi) \simeq S^r(V^*).$$

The symplectic form on V defines an isomorphism of G–modules $V \simeq V^*$, so the module in (1) is isomorphic also to $S^r(V)$.

2.18. Let us consider now a vector space V over k of dimension $2n+1 \geq 3$ or $2n \geq 4$ with a basis $(e_i)_{-n \leq i \leq n}$ dropping e_0 if dim $V = 2n$. Let q be the quadratic form $q(\sum_i x_i e_i) = \sum_{i=1}^{n} x_i x_{-i} + x_0^2$ dropping the term x_0^2 in case dim $V = 2n$. Let G be the corresponding special orthogonal group. (Its definition requires special care if $\mathrm{char}(k) = 2$, but I do not want to go into this.)

The diagonal matrices in G with respect to this basis form a split maximal torus T. If we order the basis elements as $e_1, e_2, \ldots, e_n, (e_0), e_{-n}, \ldots, e_{-2}, e_{-1}$, then the lower triangular matrices form a Borel subgroup B. The stabiliser of ke_{-1} in G is a parabolic subgroup $P \supset B$. Again let $\varpi \in X(P)$ denote the character such that P acts via $-\varpi$ on ke_{-1}. We get as before

$$H^0(r\varpi) \simeq \mathrm{ind}_P^G(r\varpi) \tag{1}$$

where we use the notation ϖ also for its restriction to T.

Let $\overline{k}$ be an algebraic closure of k and set $Y = \{v \in V \otimes \overline{k} \mid q(v) = 0\}$, using the obvious extension of q to $V \otimes \overline{k}$. The differential of q at any point $v \in V \otimes \overline{k}$, $v \neq 0$ is non-zero except for the case $\mathrm{char}(k) = 2$, dim $V = 2n+1$, and $v \in \overline{k}e_0$. As $\overline{k}e_0 \cap Y = 0$, we can always deduce from Serre's normality criterion ([M2], III, §8, prop. 2, p. 391) that Y is a normal variety and that q generates in $S(V^* \otimes \overline{k})$ the ideal of functions vanishing on Y. We get:

$$\overline{k}[Y \setminus \{0\}] = \overline{k}[Y] = S(V^* \otimes \overline{k})/qS(V^* \otimes \overline{k}). \tag{2}$$

If we denote by $H \subset P$ the kernel of ϖ, then $G(\overline{k})/H(\overline{k}) \simeq Y \setminus \{0\}$ as varieties (the tangent map being surjective), hence

$$k[G/H] \otimes \overline{k} \simeq \overline{k}[G(\overline{k})/H(\overline{k})] \simeq (S(V^*)/qS(V^*)) \otimes \overline{k}$$

and

$$k[G/H] \simeq S(V^*)/qS(V^*). \tag{3}$$

As in 2.14 we pick up in $\mathrm{ind}_P^G(r\varpi)$ the elements homogeneous of degree r in this ring, i.e., we get an exact sequence of G–modules

$$0 \to S^{r-2}(V^*) \longrightarrow S^r(V^*) \longrightarrow \mathrm{ind}_P^G(r\varpi) \to 0 \tag{4}$$

where the first map is multiplication with q.

Using the bilinear form associated to q we can replace V^* by V in (4), except for the case where both $\mathrm{char}(k) = 2$ and dim V is odd.

CHAPTER 3

Irreducible Representations of the Frobenius Kernels

In this chapter let p be a prime number and assume that k is a perfect field of characteristic p. (The assumption of perfectness is not really necessary and is only made for the sake of convenience as in Chapter I.9.)

In this chapter we want to describe the simple modules of the Frobenius kernels G_r of G. In the special case $r = 1$ we get thus the irreducible representations of $\mathrm{Lie}(G)$ as a p–Lie-algebra.

The classification of the simple G_r–modules parallels that of the simple G–modules. The latter ones have a one dimensional B^+–socle. If λ is the weight of T on this socle, then the module is isomorphic to the G–socle of $\mathrm{ind}_B^G \lambda$, hence uniquely determined by λ. Now the simple G_r–modules have a one dimensional B_r^+–socle; if λ is the weight of T_r on this socle, then the module is isomorphic to the G_r–socle of $\mathrm{ind}_{B_r}^{G_r} \lambda$, hence uniquely determined by λ. However, in contrast to the case for G, now all characters of T_r can occur as the weight of the B_r^+–socle of a simple G_r–module (3.10).

We now have a natural construction of our simple modules not only as socles, but also in the form $M/\mathrm{rad}_{G_r} M$ for a suitable G_r–module M, which can be chosen as coinduced from a one dimensional representation of B_r^+. Furthermore, all these modules, induced and coinduced from characters of B_r resp. B_r^+, have a natural interpretation when regarded as B_r^+ resp. B_r–modules: They yield then the indecomposable injective (= projective, cf. I.8) modules (3.8). This was first observed in [Humphreys 1] (for $r = 1$).

Now there is an intimate relationship between simple G–modules and simple G_r–modules. At this point let me describe it only for semi-simple and simply connected G. If we take a simple G–module $L(\lambda)$ such that its highest weight λ has coordinates (with respect to the fundamental weights) strictly less than p^r (for short: $\lambda \in X_r(T)$), then $L(\lambda)$ stays irreducible under restriction to G_r and we get thus each simple G_r–module exactly once (up to isomorphism). This was first proved in [Curtis 1] for $r = 1$. When the G_r for $r > 1$ were first looked at, it was obvious that Curtis's proof carried over to this more general situation, cf. [Humphreys 8].

Looking at arbitrary simple G–modules and at their structure as G_r–modules one gets an easy proof (due to [Cline, Parshall, and Scott 7]) of Steinberg's tensor product theorem (first proved in [Steinberg 2]): If $\lambda = \sum_{i=0}^{m} p^i \lambda_i$ with all $\lambda_i \in X_1(T)$, then (3.17)

$$L(\lambda) \simeq L(\lambda_0) \otimes L(\lambda_1)^{[1]} \otimes \cdots \otimes L(\lambda_m)^{[m]}$$

where an exponent $[i]$ denotes a twist of the representation with the ith power of the Frobenius endomorphism. This is a very important result that reduces the detailed description of all $L(\lambda)$ (e.g., the calculation of their formal characters) to the finite set of those $L(\lambda)$ with $\lambda \in X_1(T)$.

Finally, we introduce for each r the Steinberg module St_r (3.18). This is a simple G–module that is simple also as a G_r–module, but which is also isomorphic to a suitable $\operatorname{ind}_{B_r}^{G_r} \lambda$. We could even show now (but do so only in 11.8) that this is the only simple G_r–module with this property. Then we prove a formula about G–modules of the form $R^i \operatorname{ind}_B^G(M)^{[r]} \otimes St_r$ for some B–module M. This formula will play an important rôle in the next chapter and is due to [Andersen 5] and [Haboush 4] independently.

If $k = \mathbf{F}_p$ and if we regard St_r as a $G(\mathbf{F}_q)$–module where $q = p^r$, then we get the "original" Steinberg module from [Steinberg 1] that is responsible for this terminology.

3.1. (The Frobenius endomorphism on G) As G arises from $G_{\mathbf{Z}}$ through base change, we also get G from $G_{\mathbf{F}_p}$ through base change. Therefore (cf. I.9.4) any $G^{(r)}$ is isomorphic to G, and there is a Frobenius endomorphism $F = F_G : G \to G$ such that we get any F_G^r as $G \xrightarrow{F^r} G \xrightarrow{\sim} G^{(r)}$ using a suitable isomorphism. We get especially for all r

(1) $$G_r = \ker(F^r).$$

We also have $T = (T_{\mathbf{F}_p})_k$ and $U_\alpha = (U_{\alpha,\mathbf{F}_p})_k$ for all α. So F stabilises T and all U_α. We get any x_α from $x_{\alpha,\mathbf{Z}} : G_{a,\mathbf{Z}} \xrightarrow{\sim} U_{\alpha,\mathbf{Z}}$ and we have $X(T) = X(T_{\mathbf{Z}})$. Therefore the isomorphisms $G_a \xrightarrow{\sim} U_\alpha$ and $T \xrightarrow{\sim} (G_m)^s$ (where $s = \operatorname{rk} T$) are compatible with the usual Frobenius endomorphisms on G_a and G_m. This implies (for all A)

(2) $$F(t) = t^p \qquad \text{for all } t \in T(A)$$

and

(3) $$F(x_\alpha(a)) = x_\alpha(a^p) \qquad \text{for all } \alpha \in R \text{ and } a \in A.$$

All the groups introduced in Chapter 1 (e.g., B, B^+, P_I, P_I^+, $U(R')$, L_I) are F–stable and F restricts to a Frobenius endomorphism on them.

3.2. Lemma: *The multiplication induces isomorphisms of schemes (for any r)*

$$\prod_{\alpha \in R^+} U_{\alpha,r} \times T_r \times \prod_{\alpha \in R^+} U_{-\alpha,r} \xrightarrow{\sim} U_r^+ \times T_r \times U_r \xrightarrow{\sim} G_r.$$

Proof: As 1 is contained in the open subscheme U^+TU of G, its inverse image $G_r = F^{-r}(1)$ is contained in $F^{-r}(U^+TU) = U^+TU$, cf. I.9.3(5). The multiplication induces isomorphisms of schemes

$$\prod_{\alpha \in R^+} U_\alpha \times T \times \prod_{\alpha \in R^+} U_{-\alpha} \xrightarrow{\sim} U^+ \times T \times U \xrightarrow{\sim} U^+TU$$

where 1 corresponds to the family having all components equal to 1. Now F stabilises all factors and induces the Frobenius endomorphism on each of them. This implies the lemma.

3.3. Lemma: *The elements*

$$\prod_{\alpha\in R^+} X_{\alpha,n(\alpha)} \prod_i H_{i,m(i)} \prod_{\alpha\in R^+} U_{-\alpha,n'(\alpha)}$$

with $0 \le n(\alpha), m(i), n'(\alpha) < p^r$ *form a basis for* $\mathrm{Dist}(G_r)$.

Proof: We have by 3.2 (or using I.7.2(5)) an isomorphism of vector spaces given by multiplication

$$\bigotimes_{\alpha\in R^+} \mathrm{Dist}(U_{\alpha,r}) \otimes \mathrm{Dist}(T_r) \otimes \bigotimes_{\alpha\in R^+} \mathrm{Dist}(U_{-\alpha,r}) \xrightarrow{\sim} \mathrm{Dist}(G_r).$$

So the claim follows from the description of $\mathrm{Dist}(G_{a,r}) = M(G_{a,r})$ and $\mathrm{Dist}(G_{m,r}) = \mathrm{Dist}(\mu_{(q)}) = M(\mu_{(q)})$ where $q = p^r$ in I.8.9.

Remark: It is now also clear how to write down bases of $\mathrm{Dist}(U_r)$, $\mathrm{Dist}(U_r^+)$, $\mathrm{Dist}(T_r)$, $\mathrm{Dist}(B_r)$, $\mathrm{Dist}(B_r^+)$, ...

3.4. Recall that G (resp. B, B^+) acts on G_r (resp. B_r, B_r^+) by conjugation. This leads to a representation on $\mathrm{Dist}(G_r)$ (resp. $\mathrm{Dist}(B_r)$, $\mathrm{Dist}(B_r^+)$); then $\mathrm{Dist}(G_r)_l^{G_r}$ (resp. $\mathrm{Dist}(B_r)_l^{B_r}$, $\mathrm{Dist}(B_r^+)_l^{B_r^+}$) is a one dimensional submodule, cf. I.8.19, I.9.7.

Proposition: a) *The action of* G *on* $\mathrm{Dist}(G_r)_l^{G_r}$ *is trivial.*

b) *The action of* $B = TU$ *on* $\mathrm{Dist}(B_r)_l^{B_r}$ *is trivial on* U, *and is given by* $-2(p^r-1)\rho$ *on* T.

c) *The action of* $B^+ = TU^+$ *on* $\mathrm{Dist}(B_r^+)_l^{B_r^+}$ *is trivial on* U^+, *and is given by* $2(p^r-1)\rho$ *on* T.

Proof: a) The adjoint representation of G on $\mathrm{Lie}(G)$ factors through $G/Z(G)$, which is semi-simple and admits only the trivial character. Hence $\det \circ \mathrm{Ad} = 1$ and the claim follows from I.9.7.

b) We have $\mathrm{Lie}(B) = \mathrm{Lie}(T) \oplus \bigoplus_{\alpha\in R^+} (\mathrm{Lie}\, G)_{-\alpha}$, hence

$$\det(\mathrm{Ad}(t)) = (\mathrm{rk}(T)\, 0 + \sum_{\alpha\in R^+} (-\alpha))(t) = (-2\rho)(t)$$

for all $t \in T(A)$ and all A, so T acts by I.9.7 on $\mathrm{Dist}(B_r)_l^{B_r}$ via $(p^r-1)(-2\rho)$. On the other hand, the unipotent group U has only the trivial character, hence has to act trivially.

c) This is proved similarly.

3.5. Corollary: *Let* M *be a* B_r*–module and* M' *a* B_r^+*–module. Then*

(1) $$\mathrm{coind}_{B_r}^{G_r} M \simeq \mathrm{ind}_{B_r}^{G_r}(M \otimes 2(p^r-1)\rho)$$

and

(2) $$\mathrm{coind}_{B_r^+}^{G_r} M' \simeq \mathrm{ind}_{B_r^+}^{G_r}(M' \otimes (-2(p^r-1)\rho)).$$

In case $\dim M < \infty$ *resp.* $\dim M' < \infty$, *we have*

(3) $$(\mathrm{ind}_{B_r}^{G_r} M)^* \simeq \mathrm{ind}_{B_r}^{G_r}(M^* \otimes 2(p^r-1)\rho)$$

resp.

(4) $$(\mathrm{ind}_{B_r^+}^{G_r} M')^* \simeq \mathrm{ind}_{B_r^+}^{G_r}(M'^* \otimes (-2(p^r-1)\rho)).$$

Proof: This follows from I.8.17–19 and from 3.4.

3.6. (Induced and Coinduced Modules) The isomorphism of schemes $U_r^+ \times B_r \xrightarrow{\sim} G_r$ given by multiplication is compatible with the action of U_r^+ by left multiplication on U_r^+ and on G_r and with the action of B_r by right multiplication on B_r and on G_r. It is also compatible with the action of T_r by conjugation on U_r^+ and by left multiplication on B_r and on G_r. Therefore the isomorphisms of vector spaces $k[G_r] \xrightarrow{\sim} k[U_r^+] \otimes k[B_r]$ and $\mathrm{Dist}(U_r^+) \otimes \mathrm{Dist}(B_r) \xrightarrow{\sim} \mathrm{Dist}(G_r)$ are also compatible with the representations of U_r^+ (resp. B_r, T_r) induced by these actions.

We have for any B_r–module M

$$\mathrm{ind}_{B_r}^{G_r} M = (k[G_r] \otimes M)^{B_r} \simeq k[U_r^+] \otimes (k[B_r] \otimes M)^{B_r}$$

hence

(1) $$\mathrm{ind}_{B_r}^{G_r} M \simeq k[U_r^+] \otimes M.$$

This isomorphism is compatible with the representations of U_r^+ (acting via ρ_l on $k[U_r^+]$ and trivially on M) and of T_r (acting as given on M and via the conjugation action on $k[U_r^+]$). Similarly,

$$\mathrm{coind}_{B_r}^{G_r} M = \mathrm{Dist}(G_r) \otimes_{\mathrm{Dist}(B_r)} M \simeq \mathrm{Dist}(U_r^+) \otimes \mathrm{Dist}(B_r) \otimes_{\mathrm{Dist}(B_r)} M$$

hence

(2) $$\mathrm{coind}_{B_r}^{G_r} M \simeq \mathrm{Dist}(U_r^+) \otimes M.$$

Again this isomorphism of vector spaces is compatible with the representations of U_r^+ and T_r (constructed similarly as above).

By interchanging the rôles of U_r and U_r^+ we also get for each B_r^+–module M' isomorphisms

(3) $$\mathrm{ind}_{B_r^+}^{G_r} M' \simeq k[U_r] \otimes M'$$

and

(4) $$\mathrm{coind}_{B_r^+}^{G_r} M' \simeq \mathrm{Dist}(U_r) \otimes M'$$

of U_r–modules and of T_r–modules for the "obvious" representations on the right hand sides.

We have $\dim k[U_r] = p^{r \dim U} = p^{r|R^+|}$ by I.9.6(2) as U is reduced. This is also the dimension of $k[U_r^+]$, $\mathrm{Dist}(U_r)$, and $\mathrm{Dist}(U_r^+)$. So all induced or coinduced modules considered above have dimension equal to

(5) $$p^{r|R^+|} \dim(M).$$

3.7. Any $\lambda \in X(T)$ defines by restriction a character of T_r, which we usually also denote by λ (if no confusion is possible). We get from the restriction an exact sequence

(1) $$0 \to p^r X(T) \longrightarrow X(T) \longrightarrow X(T_r) \to 0$$

where the first map is the inclusion.

By extending its restriction to T_r trivially on U_r or U_r^+, any $\lambda \in X(T)$ defines a one dimensional module (usually denoted by λ) for B_r and B_r^+. We can induce and coinduce these modules to G_r. Let us introduce the following notations:

(2) $$Z_r(\lambda) = \operatorname{coind}_{B_r^+}^{G_r} \lambda,$$

(3) $$Z_r'(\lambda) = \operatorname{ind}_{B_r}^{G_r} \lambda.$$

Now 3.5 implies:

(4) $$Z_r(\lambda) \simeq \operatorname{ind}_{B_r^+}^{G_r} (\lambda - 2(p^r - 1)\rho),$$

(5) $$Z_r'(\lambda) \simeq \operatorname{coind}_{B_r}^{G_r} (\lambda - 2(p^r - 1)\rho),$$

(6) $$Z_r(\lambda)^* \simeq Z_r(2(p^r - 1)\rho - \lambda),$$

(7) $$Z_r'(\lambda)^* \simeq Z_r'(2(p^r - 1)\rho - \lambda).$$

Note that we have by 3.6(5)

(8) $$\dim Z_r(\lambda) = \dim Z_r'(\lambda) = p^{r|R^+|}$$

for all λ.

As $\lambda + p^r\nu$ and λ have the same restriction to T_r (for all $\lambda, \nu \in X(T)$), we get

(9) $$Z_r(\lambda + p^r\nu) = Z_r(\lambda),$$

(10) $$Z_r'(\lambda + p^r\nu) = Z_r'(\lambda).$$

It will become clear later on (Remark 3 in 9.6) that $Z_r(\lambda)$ and $Z_r'(\lambda)$ define the same element in the Grothendieck group of all finite dimensional G_r–modules. This is the reason for choosing the notation as above.

3.8. Proposition: *Let $\lambda \in X(T)$.*

a) *Considered as a B_r–module $Z_r(\lambda)$ is the projective cover of λ and the injective hull of $\lambda - 2(p^r - 1)\rho$.*

b) *Considered as a B_r^+–module $Z_r'(\lambda)$ is the injective hull of λ and the projective cover of $\lambda - 2(p^r - 1)\rho$.*

Proof: The injective hull of any simple module for groups like $B_r^+ = U_r^+ \rtimes T_r$ and $B_r = U_r \rtimes T_r$ has been determined in I.3.11 (cf. I.3.17). The injective hull, say, of λ for B_r^+ is $k[U_r^+] \otimes \lambda$ with U_r^+ acting only on the first factor (via ρ_l) and with T_r acting via the conjugation representation on $k[U_r^+]$ tensored with the restriction of λ. On the other hand, $Z_r'(\lambda)$ has exactly this form by 3.6(1). We get thus the first claim in b). The second claim in a) follows similarly using 3.7(4). For the claims concerning projective covers we now use 3.4 and I.8.13.

3.9. (Simple G_r–modules) It follows from 3.8 that all $Z_r(\lambda)^{U_r}$ and $Z'_r(\lambda)^{U_r^+}$ have dimension 1 as these spaces of fixed points are contained in the B_r– resp. the B_r^+–socle of the module. On the other hand, we have $M^{U_r} \neq 0 \neq M^{U_r^+}$ for any G_r–module M, as U_r and U_r^+ are unipotent. (Recall I.2.14(8).)

Arguing as in 2.3 we get:

(1) *Any $Z_r(\lambda)$ and any $Z'_r(\lambda)$ has a simple socle when considered as a G_r–module.*

Dualising (using 3.7(6), (7)) we get:

(2) *Any $Z_r(\lambda)/\operatorname{rad}_{G_r} Z_r(\lambda)$ and any $Z'_r(\lambda)/\operatorname{rad}_{G_r} Z'_r(\lambda)$ is a simple G_r–module.*

Arguing as in 2.1(4) we get:

(3) *For any simple G_r–module L there are $\lambda_1, \lambda_2, \lambda_3, \lambda_4 \in X(T)$ with*

$$L \simeq \operatorname{soc}_{G_r} Z'_r(\lambda_1) \simeq Z'_r(\lambda_3)/\operatorname{rad}_{G_r} Z'_r(\lambda_3)$$
$$\simeq \operatorname{soc}_{G_r} Z_r(\lambda_2) \simeq Z_r(\lambda_4)/\operatorname{rad}_{G_r} Z_r(\lambda_4).$$

Set for all $\lambda \in X(T)$

(4) $$L_r(\lambda) = \operatorname{soc}_{G_r} Z'_r(\lambda).$$

Obviously 3.7(10) implies (for all $\lambda, \nu \in X(T)$)

(5) $$L_r(\lambda + p^r\nu) \simeq L_r(\lambda).$$

3.10. Proposition: *We have for all $\lambda \in X(T)$*

(1) $$L_r(\lambda)^{U_r^+} \simeq \lambda,$$

(2) $$Z_r(\lambda)/\operatorname{rad}_{G_r} Z_r(\lambda) \simeq L_r(\lambda),$$

(3) $$\operatorname{End}_{G_r} L_r(\lambda) \simeq k.$$

If Λ is a set of representatives in $X(T)$ for $X(T)/p^rX(T)$, then each simple G_r–module is isomorphic to exactly one $L_r(\lambda)$ with $\lambda \in \Lambda$.

Proof: The first claim follows immediately from the definition of $L_r(\lambda)$ and from 3.8.b. This implies

$$\operatorname{Hom}_{B_r^+}(\lambda, L_r(\lambda)) \neq 0,$$

hence by I.8.14(3)

$$0 \neq \operatorname{Hom}_{G_r}(\operatorname{coind}_{B_r^+}^{G_r} \lambda, L_r(\lambda)) = \operatorname{Hom}_{G_r}(Z_r(\lambda), L_r(\lambda)).$$

As $L_r(\lambda)$ is simple, any non-zero homomorphism $Z_r(\lambda) \to L_r(\lambda)$ has to be surjective and to factor through $Z_r(\lambda)/\operatorname{rad}_{G_r} Z_r(\lambda)$. Now 3.9(2) implies (2). In order to get (3) observe that any $\varphi \in \operatorname{End}_{G_r} L_r(\lambda)$ stabilises the line $L_r(\lambda)^{U_r^+}$, hence has an eigenvalue in k. Therefore Schur's lemma implies (3).

It follows from (1) and 3.7(1), (10) that $L_r(\lambda)$, $L_r(\mu)$ are isomorphic if and only if $\lambda - \mu \in p^rX(T)$. This together with 3.9(3) gives the last claim.

Remark: As in 2.9, equation (3) implies that $L_r(\lambda)$ is absolutely irreducible, i.e., that for each (perfect) field $k' \supset k$ the module $L_r(\lambda) \otimes k'$ over $(G_r)_{k'} = (G_{k'})_r$ is the unique simple $(G_{k'})_r$–module such that $(T_{k'})_r$ acts via λ on the $(U_{k'}^+)_r$–fixed points. Furthermore, we have (as in 2.9(1))

(4) $$\operatorname{soc}_{(G_{k'})_r}(M \otimes k') = (\operatorname{soc}_{G_r} M) \otimes k'$$

for any G_r–module M.

3.11. As described in I.2.15, we can construct for each $g \in G(k)$ and each G_r–module M a new G_r–module ${}^gM$.

Proposition: *We have ${}^gL \simeq L$ for all simple G_r–modules L and all $g \in G(k)$.*

Proof: The "twist" $M \mapsto {}^gM$ commutes obviously with field extensions. Because of 3.10 and the remark to 3.10, we may assume that k is algebraically closed.

Suppose for the moment that $g \in B^+(k)$. Pick $v \in L^{U_r^+}$, $v \neq 0$. There exists $\lambda \in X(T)$ such that T_r acts via λ on v. Since B^+ normalises U^+ and U_r^+, we have also that $v \in ({}^gL)^{U_r^+}$. If $t \in T_r(A)$ for some A, then $g^{-1}tg = t(t^{-1}g^{-1}tg) \in B_r^+(A)$ and $t^{-1}g^{-1}tg \in U^+(A)$, hence $t^{-1}g^{-1}tg \in U_r^+(A)$. The action of t on ${}^gL \otimes A$ takes $v \otimes 1$ to $(g^{-1}tg)(v \otimes 1)$ in $L \otimes A$, hence to $t(v \otimes 1) = \lambda(t)(v \otimes 1)$. This shows that $({}^gL)^{U_r^+} \simeq \lambda$ and therefore ${}^gL \simeq L_r(\lambda) \simeq L$.

This proves the claim for all $g \in B^+(k)$, hence for all g that belong to a Borel subgroup of G. Since $G(k)$ is the union of its Borel subgroups, the claim follows.

3.12. Corollary: *We have*

$$\operatorname{soc}_{G_r} Z_r(\lambda) \simeq L_r(w_0 \bullet \lambda) \simeq Z_r'(\lambda)/\operatorname{rad}_{G_r} Z_r'(\lambda)$$

for all $\lambda \in X(T)$.

Proof: Recall that $w_0 \in W$ is the element with $w_0R^+ = -R^+$. It satisfies $w_0^2 = 1$ and $w_0\rho = -\rho$. So the dot action from 1.5(7) satisfies $w_0 \bullet \lambda = w_0\lambda - 2\rho$.

Choose $g \in N_G(T)(k)$ that is a representative for w_0. Then $gU^+g^{-1} = U$, hence $gU_r^+g^{-1} = U_r$. Identifying $L_r(\lambda)$ and ${}^gL_r(\lambda)$ as vector spaces we get

$$L_r(\lambda)^{U_r^+} = ({}^gL_r(\lambda))^{U_r}.$$

As T_r acts on the left hand side through λ, it acts on the right hand side through $w_0\lambda$. This implies using 3.11

$$0 \neq \operatorname{Hom}_{B_r}(w_0\lambda, {}^gL_r(\lambda)) \simeq \operatorname{Hom}_{B_r}(w_0\lambda, L_r(\lambda)),$$

hence (by 3.7(5) and 3.7(10))

$$\begin{aligned} 0 \neq \operatorname{Hom}_{G_r}(\operatorname{coind}_{B_r}^{G_r} w_0\lambda, L_r(\lambda)) &= \operatorname{Hom}_{G_r}(Z_r'(w_0\lambda + 2(p^r - 1)\rho), L_r(\lambda)) \\ &= \operatorname{Hom}_{G_r}(Z_r'(w_0 \bullet \lambda), L_r(\lambda)). \end{aligned}$$

Using $w_0^2 = 1$ we get also $\operatorname{Hom}_{G_r}(Z_r'(\lambda), L_r(w_0 \bullet \lambda)) \neq 0$ which yields (by 3.9(2)) the second isomorphism in the corollary.

We know by the above that $L_r(\lambda)$ is the unique simple G_r–module (up to isomorphism) with

$$L_r(\lambda)^{U_r} \simeq w_0\lambda.$$

Now 3.9(1) and 3.8.a yield the first isomorphism in the corollary.

3.13. Corollary: *We have for all $\lambda \in X(T)$*

$$L_r(\lambda)^* \simeq L_r(-w_0\lambda).$$

Proof: Obviously $L_r(\lambda)^*$ is again simple. We saw above $\operatorname{Hom}_{B_r}(w_0\lambda, L_r(\lambda)) \neq 0$ and get by dualising

$$0 \neq \operatorname{Hom}_{B_r}(L_r(\lambda)^*, -w_0\lambda) \simeq \operatorname{Hom}_{G_r}(L_r(\lambda)^*, Z_r'(-w_0\lambda))$$

which implies the corollary.

3.14. Lemma: *Let $\lambda \in X(T)_+$ and choose $v \in L(\lambda)_\lambda$, $v \neq 0$. Then* $\mathrm{Dist}(G_r)v$ *is a simple G_r–module isomorphic to $L_r(\lambda)$.*

Proof: We may assume that k is algebraically closed. As G is reduced, I.6.16 implies that $L(\lambda)$ is a semi-simple G_r–module, hence so is $\mathrm{Dist}(G_r)v$. We have $kv \simeq \lambda$ as a B_r^+–module, so we get a homomorphism

$$\varphi : Z_r(\lambda) = \mathrm{Dist}(G_r) \otimes_{\mathrm{Dist}(B_r^+)} \lambda \longrightarrow \mathrm{Dist}(G_r)v$$

mapping $1 \otimes 1$ to v. Obviously φ is surjective. Its image is semi-simple, hence by 3.10(2) even simple and isomorphic to $L_r(\lambda)$.

3.15. Set

$$X_r(T) = \{\, \lambda \in X(T) \mid 0 \leq \langle \lambda, \alpha^\vee \rangle < p^r \text{ for all } \alpha \in S \,\}. \tag{1}$$

Obviously

$$X_1(T) \subset X_2(T) \subset \cdots \subset X_r(T) \subset \cdots \subset X(T)_+.$$

Proposition: *For each $\lambda \in X_r(T)$ the simple G–module $L(\lambda)$ is also simple as a G_r–module and is isomorphic to $L_r(\lambda)$ for G_r.*

Proof: Choose v as in 3.14 and set $L = \mathrm{Dist}(G_r)v$. We have to show $L = L(\lambda)$. We can obviously assume that k is algebraically closed. We want to show that $gL = L$ for all $g \in G(k)$. Since G is reduced, this will imply that L is a G–submodule of the simple G–module $L(\lambda)$, cf. Remark I.2.8, hence that $L = L(\lambda)$ as claimed.

For each $g \in G$ the subspace $gL \simeq {}^gL$ is a simple G_r–submodule of $L(\lambda)$. So we have either $gL = L$ or $gL \cap L = 0$. If $g \in B^+(k)$, then $gv \in k^\times v$, hence $gL \cap L \neq 0$ and $gL = L$.

Consider now a simple root α and a representative of s_α in $N_G(T)(k)$, e.g., let us take $n_\alpha(1)$. Then $n_\alpha(1)v \in L(\lambda)_{s_\alpha\lambda}$ and $n_\alpha(1)v \neq 0$. The proof of 2.11.b shows

$$\bigoplus_{n \geq 0} L(\lambda)_{\lambda - n\alpha} = \mathrm{Dist}(U_{-\alpha})v = \bigoplus_{n \geq 0} kX_{-\alpha,n}v,$$

hence $L(\lambda)_{\lambda - n\alpha} = kX_{-\alpha,n}v$ for all $n \in \mathbf{N}$, and

$$L(\lambda)_{\lambda - n\alpha} \subset \mathrm{Dist}(G_r)v \qquad \text{for all } n < p^r.$$

Now $s_\alpha\lambda = \lambda - \langle \lambda, \alpha^\vee \rangle \alpha$ and $0 \leq \langle \lambda, \alpha^\vee \rangle < p^r$ since $\lambda \in X_r(T)$. This yields $n_\alpha(1)v \in \mathrm{Dist}(G_r)v = L$ and, therefore, $n_\alpha(1)L = L$.

As W is generated by the s_α with $\alpha \in S$, it follows that each $w \in W$ has a representative $\dot{w} \in N_G(T)(k)$ with $\dot{w}L = L$. Now the Bruhat decomposition $G(k) = \bigcup_{w \in W} B^+(k)\dot{w}B^+(k)$ shows that $gL = L$ for all $g \in G(k)$, hence the proposition.

Remarks: 1) There is an alternative approach to this result in [Cline, Parshall, and Scott 7] using ideas from [Cartier]. Given a simple G_r–module L, one gets from 3.11 a projective representation of G on L, i.e., a homomorphism $G \to PGL(L)$. This can be lifted to a representation $G' \to GL(L)$ of a certain covering group

G' of G. Here one uses the nontrivial result (due to Steinberg) that for semi-simple and simply connected groups any projective representation lifts to a linear representation. In case $L = L_r(\lambda)$ with $\lambda \in X_r(T)$ one then checks that the lifted G'–module is isomorphic to $L(\lambda)$.

There is still another proof in [Kempf 5], 4.2.

2) For G semi-simple and simply connected $X_r(T)$ is a system of representatives for $X(T)/p^r(T)$. So in this case any simple G_r–module can be lifted to G. In general one can replace G by a covering group such that $X_r(T)$ contains a system of representatives for $X(T)/p^r(T)$. One has, however, to observe for a covering $G' \to G$ that G_r is not necessarily a factor group of G'_r. For example, if $G = PGL(V)$ and $G' = SL(V)$ for some vector space V over k with $p \mid \dim V$, then $\text{Lie}(G')$ does not map onto $\text{Lie}(G)$, hence G'_1 not onto G_1.

3.16. If M is a G–module, then we can define another G–module structure on the vector space M by composing the given representation $G \to GL(M)$ with $F^r : G \to G$. We denote this new G–module by $M^{[r]}$. If the given representation of G on M is defined over $\mathbf{F}_p$, then $M^{[r]} \simeq M^{(r)}$, cf. I.9.10. Obviously G_r acts trivially on $M^{[r]}$.

On the other hand, consider a G–module V such that G_r acts trivially on V. Then the representation $G \to GL(V)$ factors through G/G_r. We know by I.9.5 that $G/G_r \simeq G$. More precisely, there is an isomorphism $\varphi : G/G_r \xrightarrow{\sim} G$ such that the composition of φ with the canonical map $G \to G/G_r$ is equal to $F^r : G \to G$. Therefore there exists a G–module M with $M^{[r]} = V$; the module M is uniquely determined by V and will be denoted by $V^{[-r]}$.

For example, we can apply this construction for any G–module V to the submodule V^{G_r} and, more generally, for any G–modules V, V' with $\dim(V') < \infty$ to $\text{Hom}_{G_r}(V', V)$. (This is a generalisation since $\text{Hom}_{G_r}(k, V) \simeq V^{G_r}$.)

Suppose that we have a system $X'_r(T)$ of representatives for $X(T)/p^r X(T)$ with $X'_r(T) \subset X_r(T)$. Then I.6.15(2) and 3.10(3) imply for any G–module M that $\text{soc}_{G_r} M$ is a G–submodule of M and that there is an isomorphism of G–modules

$$\text{(1)} \qquad \text{soc}_{G_r} M \simeq \bigoplus_{\lambda \in X'_r(T)} L(\lambda) \otimes \text{Hom}_{G_r}(L(\lambda), M).$$

If M is simple as a G–module, then we get $\text{soc}_{G_r} M = M$ (since the socle is non-zero), i.e., that M is semi-simple for G_r. This shows that any semi-simple G–module also is semi-simple for G_r, hence that

$$\text{(2)} \qquad \text{soc}_G M \subset \text{soc}_{G_r} M$$

for any G–module M. This follows also from I.6.16 using 3.10(4) and 2.9(1).

Proposition: *We have for all $\lambda \in X_r(T)$ and $\mu \in X(T)_+$*

$$L(\lambda + p^r \mu) \simeq L(\lambda) \otimes L(\mu)^{[r]}.$$

Proof: We can replace G by a covering group without changing either side of our claim. Thus we may assume that there exists $X'_r(T)$ as above and that $\lambda \in$

$X'_r(T)$. We can then apply (1) with $M = L(\lambda + p^r\mu)$. Since this module is simple, there is exactly one non-zero summand in (1). So there exists $\lambda' \in X'_r(T)$ with $M \simeq L(\lambda') \otimes \operatorname{Hom}_{G_r}(L(\lambda'), M)$. By the discussion above there exists a G–module V with $V^{[r]} \simeq \operatorname{Hom}_{G_r}(L(\lambda'), M)$. The simplicity of $M \simeq L(\lambda') \otimes V^{[r]}$ implies that V is simple; so there exists $\mu' \in X(T)_+$ with $V \simeq L(\mu')$. We have now $L(\lambda + p^r\mu) = M \simeq L(\lambda') \otimes L(\mu')^{[r]}$. A comparison of the highest weights yields $\lambda + p^r\mu = \lambda' + p^r\mu'$. Now $\lambda - \lambda' \in p^r X(T)$ and $\lambda, \lambda' \in X'_r(T)$ imply $\lambda = \lambda'$, hence $\mu = \mu'$ and the proposition.

Remark: Suppose there is $X'_r(T)$ as above. Let M be a G–module. The proposition shows for any $\lambda \in X_r(T)$ that $L(\lambda) \otimes \operatorname{soc}_G \operatorname{Hom}_{G_r}(L(\lambda), M)$ is a semi-simple G–module. On the other hand, each simple G–submodule of M is isomorphic to some $L(\lambda) \otimes L(\mu)^{[r]}$ with $\lambda \in X'_r(T)$ and $\mu \in X(T)_+$. Under the identification in (1) it corresponds to some $L(\lambda) \otimes V$ with $V \subset \operatorname{Hom}_{G_r}(L(\lambda), M)$ and $V \simeq L(\mu)^{[r]}$, hence $V \subset \operatorname{soc}_G \operatorname{Hom}_{G_r}(L(\lambda), M)$. This implies

(3) $$\operatorname{soc}_G M \simeq \bigoplus_{\lambda \in X'_r(T)} L(\lambda) \otimes \operatorname{soc}_G \operatorname{Hom}_{G_r}(L(\lambda), M).$$

3.17. We get from Proposition 3.16 by induction:

Corollary (Steinberg's Tensor Product Theorem): *Let* $\lambda = \sum_{i=0}^{m} p^i \lambda_i$ *with* $\lambda_0, \lambda_1, \dots, \lambda_m \in X_1(T)$. *Then*

$$L(\lambda) \simeq L(\lambda_0) \otimes L(\lambda_1)^{[1]} \otimes \cdots \otimes L(\lambda_m)^{[m]}.$$

3.18. (The Steinberg Module) We have for all $\lambda \in X(T)$ by 3.6(2)

(1) $$Z_r(\lambda) \simeq \operatorname{Dist}(U_r) \otimes \lambda.$$

So all elements (for some fixed ordering of R^+)

(2) $$\prod_{\alpha \in R^+} X_{-\alpha, n(\alpha)} \otimes 1$$

with $0 \le n(\alpha) < p^r$ form a basis of $Z_r(\lambda)$, cf. 3.3. Set $q = p^r$. We have

(3) $$Z_r(\lambda)^{U_r} = k \prod_{\alpha \in R^+} X_{-\alpha, q-1} \otimes 1$$

as the element on the right hand side is non-zero and obviously annihilated by all $X_{-\alpha,n}$ with $\alpha \in R^+$ and $0 < n < p^r$, hence invariant under U_r, and as $\dim Z_r(\lambda)^{U_r} = 1$ by 3.8.a.

Consider $\lambda \in X_r(T)$ and choose $v \in L(\lambda)_\lambda$, $v \neq 0$. We have then (cf. 3.15) a surjective homomorphism $Z_r(\lambda) \to L(\lambda)$ of G_r–modules mapping $1 \otimes 1$ to v, hence a basis element as in (2) to $\prod_{\alpha \in R^+} X_{-\alpha, n(\alpha)} v$. This element is a weight vector (for T) of weight $\lambda - \sum_{\alpha \in R^+} n(\alpha)\alpha$.

Suppose that $(p^r - 1)\rho \in X(T)$. (This holds automatically if $p \neq 2$ since $2\rho \in X(T)$; in case $p = 2$ we may at this point have to replace G by some covering.)

Obviously $\langle \rho, \alpha^\vee \rangle = 1$ for all $\alpha \in S$ implies $(p^r - 1)\rho \in X_r(T)$. We apply the preceding paragraph with $\lambda = (p^r - 1)\rho$. The lowest weight of $L(\lambda)$ is

$$w_0\lambda = w_0(p^r - 1)\rho = -(p^r - 1)\rho = \lambda - 2(p^r - 1)\rho = \lambda - \sum_{\alpha \in R^+} (p^r - 1)\alpha.$$

The elements $\prod_{\alpha \in R^+} X_{-\alpha, n(\alpha)} v$ with all $n(\alpha) < p^r$ generate $L(\lambda)$ over k. The only element among these generators that has weight $w_0\lambda$, is $\prod_{\alpha \in R^+} X_{-\alpha, q-1} v$ (where $q = p^r$ as above). So (3) implies that the epimorphism $Z_r(\lambda) \to L(\lambda)$ is injective on the U_r–socle of $Z_r(\lambda)$. Therefore the kernel has U_r–socle equal to 0, hence is 0 itself. This implies

$$L_r((p^r - 1)\rho) \simeq Z_r((p^r - 1)\rho) \simeq L((p^r - 1)\rho) \tag{4}$$

and by dimension comparison also

$$L_r((p^r - 1)\rho) = Z'_r((p^r - 1)\rho) = \operatorname{ind}_{B_r}^{G_r}(p^r - 1)\rho. \tag{5}$$

We denote the G–module $L((p^r - 1)\rho)$ by St_r and call it the rth *Steinberg module* for G (always supposing that $(p^r - 1)\rho \in X(T)$). Recall (2.4) that St_r is the socle of the induced module $H^0((q - 1)\rho) = \operatorname{ind}_B^G(q - 1)\rho$ where $q = p^r$. So we have

$$\begin{aligned} 0 \neq \operatorname{Hom}_G(St_r, \operatorname{ind}_B^G(q - 1)\rho) &\simeq \operatorname{Hom}_B(St_r, (q - 1)\rho) \\ &\simeq \operatorname{Hom}_{G_rB}(St_r, \operatorname{ind}_B^{G_rB}(q - 1)\rho). \end{aligned}$$

We know by I.6.13(1) that $\operatorname{ind}_B^{G_rB}(q - 1)\rho$ considered as a G_r–module is isomorphic to $\operatorname{ind}_{B_r}^{G_r}(q - 1)\rho$, hence to St_r by (4), (5). It therefore has the same dimension. As St_r is simple for G_r, hence also for G_rB, any non-zero homomorphism of G_rB–modules $St_r \to \operatorname{ind}_B^{G_rB}(q - 1)\rho$ has to be injective, hence bijective because of the equality of dimensions. So we get

$$St_r \simeq \operatorname{ind}_B^{G_rB}(p^r - 1)\rho \qquad \text{(as } G_rB\text{–modules).} \tag{6}$$

(We shall study modules of the form $\operatorname{ind}_{B_r}^{G_r} \lambda$ more systematically later on in Chapter 9.)

Remark: The results in (4)–(6) generalise from $\lambda = (p^r - 1)\rho$ to all $\lambda \in X(T)$ with $\langle \lambda, \alpha^\vee \rangle = p^r - 1$ for all $\alpha \in S$. (One argues as above using $w(\lambda - (p^r - 1)\rho) = \lambda - (p^r - 1)\rho$ for all $w \in W$, hence $w_0\lambda = \lambda - 2(p^r - 1)\rho$.) If G is semi-simple, then $\lambda = (p^r - 1)\rho$ is the only possible candidate, but for general reductive G there are others. For example, for $G = GL_n$ one can take $\lambda = (p^r - 1)\sum_{i=1}^n (n - i)\varepsilon_i$ with the ε_i as in 1.21. This choice has the advantage that $\lambda \in X(T)$ always, whereas $(p^r - 1)\rho \notin X(T)$ for $p = 2$ and (e.g.) $G = GL_3$. In the representation theory of the general linear groups it is more usual to call $L(\lambda)$ for this choice of λ a Steinberg module.

3.19. Proposition: *Suppose that $(p^r-1)\rho \in X(T)$. We have for each B–module M and each $i \in \mathbf{N}$ an isomorphism*

$$H^i((p^r-1)\rho \otimes M^{[r]}) \simeq St_r \otimes H^i(M)^{[r]}$$

of G–modules.

Proof: We can apply I.6.11 to $H = G_rB$ and $N = G_r$. We have $G_rB/G_r \simeq B/(B \cap G_r) = B/B_r$ and can thus regard $M^{[r]}$ as a G_rB/G_r–module. So I.6.11 yields isomorphisms

$$R^i \operatorname{ind}_{G_rB/G_r}^{G/G_r}(M^{[r]}) \simeq R^i \operatorname{ind}_{G_rB}^{G}(M^{[r]}). \tag{1}$$

The isomorphisms $G/G_r \simeq G$ and $G_rB/G_r \simeq B/B_r \simeq B$ induced by the Frobenius morphism are compatible with the inclusions $G_rB/G_r \subset G/G_r$ and $B \subset G$. So the left hand side in (1) is

$$R^i \operatorname{ind}_B^G(M) = H^i(M) \tag{2}$$

when regarded as a module over $G \simeq G/G_r$. If, as in (1), we take the structure as a G–module via $G \xrightarrow{\text{can}} G/G_r$, then we have to apply the Frobenius endomorphism again. So we get

$$H^i(M)^{[r]} \simeq R^i \operatorname{ind}_{G_rB}^{G}(M^{[r]}). \tag{3}$$

We can now tensor this isomorphism with St_r, use the generalised tensor identity (I.4.8) and the isomorphism from 3.18(6) to get

$$\begin{aligned} H^i(M)^{[r]} \otimes St_r &\simeq R^i \operatorname{ind}_{G_rB}^{G}(M^{[r]}) \otimes St_r \\ &\simeq R^i \operatorname{ind}_{G_rB}^{G}(M^{[r]} \otimes St_r) \\ &\simeq R^i \operatorname{ind}_{G_rB}^{G} \circ \operatorname{ind}_B^{G_rB}(M^{[r]} \otimes (p^r-1)\rho). \end{aligned}$$

As $G_rB/B \simeq G_r/B_r \simeq U_r^+$ is affine, the functor $\operatorname{ind}_B^{G_rB}$ is exact (I.5.13). Therefore the spectral sequence I.4.5.c degenerates in this case to give isomorphisms

$$R^i \operatorname{ind}_{G_rB}^{G} \circ \operatorname{ind}_B^{G_rB} \simeq R^i \operatorname{ind}_B^G.$$

Plugging this into the last formula we get the claim of the proposition.

Remark: Take $M = k$. As $H^0(k) \simeq k$ by 2.1(6), we get

$$H^0((p^r-1)\rho) \simeq St_r, \tag{4}$$

hence also $V((p^r-1)\rho) \simeq St_r$ by definition (2.13(1)) and 2.5.

CHAPTER 4

Kempf's Vanishing Theorem

Let k be a field throughout this chapter.

If $\lambda \in X(T)$ is dominant (equivalently, if $H^0(\lambda) \neq 0$), then all higher cohomology groups

$$H^i(\lambda) = H^i(G/B, \mathcal{L}(\lambda)) = R^i \operatorname{ind}_B^G \lambda \qquad \text{with } i > 0$$

are zero ("vanish"). This is Kempf's vanishing theorem, the main result in this chapter. In the case of $\operatorname{char}(k) = 0$ it had been known for a long time, but in the case of $\operatorname{char}(k) \neq 0$ it was only in 1976 that Kempf's proof of this theorem appeared. The proof given here (4.5) is due to [Andersen 5] and [Haboush 4] independently.

Before we enter the proof of this result, we discuss general properties of invertible sheaves, especially of ample ones, and describe the form Serre duality takes on G/B. This is done in 4.1–4.4. There is more known about line bundles on G/B than we discuss here. For more information see, e.g., [Kempf 3] or [Iversen].

After the proof of Kempf's vanishing we discuss some first applications. For example, we prove formulae comparing B– and G–cohomology (4.6/7). We deviate to prove results on B–cohomology (4.8–4.10), which are then applied to give some results on G–cohomology (4.11–4.13). For example, we describe the cohomology of the trivial module. Most of the results in 4.6–4.13 were first proved in [Cline, Parshall, Scott, and van der Kallen]. Corollary 4.12, however, was first proved by W. van der Kallen who has communicated his proof to me.

Many G–modules that occur "in nature" have an ascending filtration such that each successive factor is isomorphic to some $H^0(\lambda)$. We discuss such "good" filtrations in 4.16–4.24. There are some references mentioned in 4.20/21. One should add [Donkin 4] where the criterion in 4.16.b appears for the first time.

4.1. For any (reduced) parabolic subgroup $P \subset G$ the canonical map $\pi : G \to G/P$ is locally trivial by 1.10(5). This means that we can cover G/P by open subschemes Y for which there is a section $\sigma_Y : Y \to G$, i.e., a morphism with $\pi \circ \sigma_Y = \mathrm{id}_Y$.

Consider now a P–module M and the associated sheaf $\mathcal{L}(M) = \mathcal{L}_{G/P}(M)$ as in I.5.8. The local triviality of $G \to G/P$ implies by I.5.16(2):

(1) *The $\mathcal{O}_{G/P}$–module $\mathcal{L}(M)$ is locally free of rank* $\dim M$.

We have for all P–modules M, M' and for each open $Y \subset G/P$ a natural map

$$\mathcal{L}(M)(Y) \otimes_{\mathcal{O}_{G/P}(Y)} \mathcal{L}(M')(Y) \longrightarrow \mathcal{L}(M \otimes M')(Y)$$

mapping any $f_1 \otimes f_2$ to the function $g \mapsto f_1(g) \otimes f_2(g)$. Using the local triviality it is easy to show that we get thus an isomorphism

$$\mathcal{L}(M) \otimes_{\mathcal{O}_{G/P}} \mathcal{L}(M') \xrightarrow{\sim} \mathcal{L}(M \otimes M'). \tag{2}$$

Similarly one gets isomorphisms for each $r \in \mathbf{N}$

$$(3) \qquad \mathcal{L}(S^r M) \simeq S^r \mathcal{L}(M) \qquad \text{and} \qquad \mathcal{L}(\Lambda^r M) \simeq \Lambda^r \mathcal{L}(M)$$

where the symmetric and exterior powers on the right hand sides are taken within the category of $\mathcal{O}_{G/P}$–modules. Furthermore, one has in case $\dim M < \infty$

$$(4) \qquad \mathcal{L}(M^*) \simeq \mathcal{L}(M)^\vee$$

where $(\ )^\vee$ denotes the dual of a locally free sheaf, cf. [Ha], II, exerc. 5.1.

4.2. Let $I \subset S$ and set $P = P_I$. As G/P is projective, I.5.12.c implies:

(1) *If M is a finite dimensional P–module, then each $R^i \operatorname{ind}_P^G M$ is a finite dimensional G–module.*

The dimension $n(P) = \dim G/P$ is equal to the number of roots α with $U_\alpha \not\subset P_I = P$, hence:

$$(2) \qquad n(P) = |R^+ \setminus R_I|.$$

Recall from I.5.12.b that

$$(3) \qquad R^i \operatorname{ind}_P^G = 0 \qquad \text{for all } i > n(P).$$

For any P–module M there is an action of G on the associated bundle $G \times^P M_a$ (cf. I.5.14/15) via left multiplication on the first factor. This action is compatible with the canonical map $\pi_M : G \times^P M_a \to G/P$ and the obvious action on G/P. It is not difficult to show that any vector bundle over G/P with a G–action such that the projection to G/P is G–equivariant, has the form $G \times^P M_a$ for some P–module M. (Take M as the fibre over $1P \in (G/P)(k)$.) For example, in the case of the tangent bundle we have $M = \operatorname{Lie}(G)/\operatorname{Lie}(P)$, so we get using the comparison from I.5.15(1):

(4) *The tangent sheaf on G/P is $\mathcal{L}(\operatorname{Lie}(G)/\operatorname{Lie}(P))$.*

(Of course, this generalises to arbitrary G and P with P reduced.) The canonical sheaf of G/P (cf. [Ha], p. 180) is

$$\omega_{G/P} = \Lambda^{n(P)} \mathcal{L}(\operatorname{Lie}(G)/\operatorname{Lie}(P))^\vee \simeq \mathcal{L}(\Lambda^{n(P)}(\operatorname{Lie}(G)/\operatorname{Lie}(P))^*).$$

The weights of T on $\operatorname{Lie}(G)/\operatorname{Lie}(P)$ are the $\alpha \in R^+ \setminus R_I$. Therefore the exterior power $\Lambda^{n(P)}(\operatorname{Lie}(G)/\operatorname{Lie}(P))^*$ is the one dimensional P–module corresponding to the weight $-2\rho_P$ where

$$(5) \qquad \rho_P = \frac{1}{2} \sum_{\alpha \in R^+ \setminus R_I} \alpha \in X(T) \otimes_{\mathbf{Z}} \mathbf{Q}.$$

Hence

$$(6) \qquad \omega_{G/P} \simeq \mathcal{L}(-2\rho_P),$$

especially

$$\omega_{G/B} \simeq \mathcal{L}(-2\rho). \tag{7}$$

The correspondence between vector bundles and locally free sheaves (on G/P) maps bundles with G–actions as above to G–linearised sheaves as, e.g., in [MF], p. 30. On such a sheaf $\mathcal{L}$ the functorial property of sheaf cohomology induces a G–action on each cohomology group $H^i(G/P, \mathcal{L})$, cf. [MF], p. 32 for $i = 0$. For each P–module M and each i this action on $H^i(G/P, \mathcal{L}(M))$ coincides with that coming from $H^i(G/P, \mathcal{L}(M)) \simeq R^i \operatorname{ind}_P^G M$, i.e., from I.5.12.a. This can be checked directly for $i = 0$ and follows in general from the uniqueness of derived functors. This implies that Serre duality (cf. [Ha], III, 7.7) is compatible with the G–action. As G/P is smooth, the canonical sheaf is equal to the dualising sheaf (cf. [Ha], III, 7.12). So for any finite dimensional P–module M:

(8) *The G–module $R^i \operatorname{ind}_P^G M$ is dual to $R^{n(P)-i} \operatorname{ind}_P^G(M^* \otimes (-2\rho_P))$.*

We get (taking $P = B$) for all $\lambda \in X(T)$

$$H^i(\lambda) \simeq H^{n-i}(-(\lambda + 2\rho))^* \tag{9}$$

where $n = |R^+|$. We have, for example, (cf. 2.13(1))

$$V(\lambda) \simeq H^n(w_0\lambda - 2\rho) \simeq H^n(w_0 \bullet \lambda) \qquad \text{for all } \lambda \in X(T)_+. \tag{10}$$

4.3. (Ample Invertible Sheaves) Let us apply 4.1 to $G = GL(V)$ and $P = \operatorname{Stab}_G kv$ for some finite dimensional vector space V over k (with $\dim V \geq 2$) and some $v \in V$, $v \neq 0$.

We have $G/P \simeq \mathbf{P}(V) \simeq \mathbf{P}^{\dim(V)-1}$, the projective space associated to V. Let us denote the character of P that describes the action of P on kv, by $-\varpi$ as in 2.16. The arguments there show for any $Y \subset \mathbf{P}(V)$ open, that we can identify $\mathcal{L}(r\varpi)(Y)$ for each $r \in \mathbf{Z}$ with the space of all regular functions on $\pi^{-1}(Y) \cup \{0\} \subset V$ that are homogeneous of degree r. (Here π is the canonical map $V \setminus \{0\} \to \mathbf{P}(V)$.) So $\mathcal{L}(r\varpi)$ is the sheaf usually denoted by $\mathcal{O}_{\mathbf{P}(V)}(r)$, cf. [Ha], p. 117:

$$\mathcal{L}_{GL(V)/P}(r\varpi) = \mathcal{O}_{\mathbf{P}(V)}(r). \tag{1}$$

(I have to admit that the language above is not quite correct. We can argue like that only for k algebraically closed when working with varieties instead of schemes. The arguments carry over to arbitrary k and varieties with k–structures. As all our schemes are reduced, they correspond to such varieties.)

Recall that an invertible sheaf $\mathcal{L}$ on some algebraic scheme X over k is called *very ample* if there is an immersion $i : X \to \mathbf{P}(V)$ for some vector space V over k with $2 \leq \dim V < \infty$ such that $\mathcal{L} \simeq i^*\mathcal{O}_{\mathbf{P}(V)}(1)$, cf. [Ha], p. 120. It is called *ample* if there is an integer $m > 0$ such that $\mathcal{L}^m$ is very ample. The following three conditions on $\mathcal{L}$ are equivalent ([Ha], II.7.5):

(i) $\mathcal{L}$ is ample.
(ii) $\mathcal{L}^m$ is ample for some integer $m > 0$.
(iii) $\mathcal{L}^m$ is ample for all integers $m > 0$.

We shall have to use Serre's cohomological criterion for ampleness ([Ha], III.5.3):

(2) *An invertible sheaf $\mathcal{L}$ on an algebraic scheme X over k is ample if and only if there is, for each coherent sheaf $\mathcal{F}$ on X, an integer m_0 such that $H^i(X, \mathcal{F} \otimes \mathcal{L}^m) = 0$ for all $i > 0$ and all $m \geq m_0$.*

4.4. Proposition: *Let $\lambda \in X(T)$. Then $\mathcal{L}_{G/B}(\lambda)$ is ample if and only if $\langle \lambda, \alpha^\vee \rangle > 0$ for all $\alpha \in S$.*

Proof: Suppose at first that $\langle \lambda, \alpha^\vee \rangle > 0$ for all $\alpha \in S$. Then there is for each $\mu \in X(T)$ some integer $m > 0$ with $m\lambda + \mu \in X(T)_+$.

Denote by $\overline{k}$ some algebraic closure of k. We know (cf. [Bo], 5.1) that there is a representation $G \to GL(V)$ in some finite dimensional vector space V over k and a line $kv \subset V$ such that $B(\overline{k})$ is the stabiliser of $\overline{k}(v \otimes 1) \subset V \otimes \overline{k}$ in $G(\overline{k})$ and such that $\operatorname{Lie}(B)$ is the stabiliser of kv in $\operatorname{Lie}(G)$. Let $\mu \in X(T)$ denote the weight through which B acts on kv. Choose an integer $m > 0$ with $m\lambda + \mu \in X(T)_+$. we know from 2.6 (with B and B^+ interchanged) that there is a G–module V' and a line $kv' \subset V'$ such that $B(\overline{k}) \subset \operatorname{Stab}_{G(\overline{k})} \overline{k}(v' \otimes 1)$ and $\operatorname{Lie}(B) \subset \operatorname{Stab}_{\operatorname{Lie}(G)} kv'$ and such that B acts via $-(m\lambda + \mu)$ on kv'. Then an easy calculation shows that $B(\overline{k})$ is the stabiliser in $G(\overline{k})$ of $\overline{k}(v \otimes v' \otimes 1) \subset (V \otimes V') \otimes \overline{k}$ and that $\operatorname{Lie}(B)$ is the stabiliser in $\operatorname{Lie}(G)$ of $k(v \otimes v')$. Furthermore, B acts on $k(v \otimes v')$ via $-m\lambda$. The precise description of the stabilisers implies that $g \mapsto g(v \otimes v')$ defines an immersion $i : G/B \to \mathbf{P}(V \otimes V')$, cf. [Bo], 6.7. Now 4.3(1) and I.5.17(1) yield

$$i^* \mathcal{O}_{\mathbf{P}(V \otimes V')}(1) \simeq \mathcal{L}_{G/B}(m\lambda).$$

So $\mathcal{L}_{G/B}(m\lambda)$ is very ample. As the functor $\mathcal{L}_{G/B}$ commutes with tensor products, we have $\mathcal{L}_{G/B}(m\lambda) \simeq \mathcal{L}_{G/B}(\lambda)^m$; so $\mathcal{L}_{G/B}(\lambda)$ is ample.

Let us now prove the converse. Take $\alpha \in S$. There is a homomorphism $\varphi_\alpha : SL_2 \to G$ as in 1.3(2) mapping each $\left(\begin{smallmatrix} 1 & a \\ 0 & 1 \end{smallmatrix}\right)$ to $x_\alpha(a)$, each $\left(\begin{smallmatrix} 1 & 0 \\ a & 1 \end{smallmatrix}\right)$ to $x_{-\alpha}(a)$, and each $\left(\begin{smallmatrix} b & 0 \\ 0 & b^{-1} \end{smallmatrix}\right)$ to $\alpha^\vee(b)$. Let $B' \subset SL_2$ denote the Borel subgroup of lower triangular matrices. We can identify SL_2/B' with $\mathbf{P}^1$. If μ is the character $\left(\begin{smallmatrix} b & 0 \\ 0 & b^{-1} \end{smallmatrix}\right) \mapsto b^n$ for some $n \in \mathbf{Z}$, then $\mathcal{L}_{SL_2/B'}(\mu)$ is identified with $\mathcal{O}_{\mathbf{P}^1}(n)$, cf. 4.3(1). This implies

$$\overline{\varphi}_\alpha^* \mathcal{L}_{G/B}(\lambda) \simeq \mathcal{O}_{\mathbf{P}^1}(\langle \lambda, \alpha^\vee \rangle)$$

using I.5.17(1). Now $\overline{\varphi}_\alpha$ defines an isomorphism of $\mathbf{P}^1$ onto a closed subscheme of G/B, cf. 1.8(5). If $\mathcal{L}_{G/B}(\lambda)$ is ample, then so is $\overline{\varphi}_\alpha^* \mathcal{L}_{G/B}(\lambda)$, e.g., by [Ha], III, exerc. 5.7, hence $\langle \lambda, \alpha^\vee \rangle > 0$, e.g., by [Ha], II, exerc. 7.6.1.

Remarks: 1) Let $I \subset S$ and consider $P = P_I$. We know from 1.18(4) that any $\lambda \in X(T)$ with $\langle \lambda, \alpha^\vee \rangle = 0$ for all $\alpha \in I$ defines a one dimensional representation of P, hence an invertible sheaf $\mathcal{L}_{G/P}(\lambda)$ on G/P. Using more or less the same arguments as above one can show that $\mathcal{L}_{G/P}(\lambda)$ is ample if and only if $\langle \lambda, \alpha^\vee \rangle > 0$ for all $\alpha \in S \setminus I$.

2) For another approach to this result see [Kempf 3], 5.3.

3) We shall see in 8.5: If $\mathcal{L}_{G/B}(\lambda)$ is ample, then it is even very ample.

4.5. Proposition: *If $\lambda \in X(T)_+$, then $H^i(\lambda) = 0$ for all $i > 0$.*

Proof: Recall that $H^i(\lambda) = R^i \operatorname{ind}_B^G \lambda \simeq H^i(G/B, \mathcal{L}(\lambda))$. There is a covering $G_1 \times G_2 \to G$ with G_1 semi-simple and simply connected and with G_2 a torus. The inverse image of B (resp. of T) has the form $B_1 \times G_2$ (resp. $T_1 \times G_2$) with B_1 a Borel subgroup in G_1 and T_1 a split maximal torus in G_1 contained in B_1. Now λ defines a character of $T_1 \times G_2$, hence $\lambda_1 \in X(T_1)$ and $\lambda_2 \in X(G_2)$. One has $\langle \lambda, \alpha^\vee \rangle = \langle \lambda_1, \alpha^\vee \rangle$ for all $\alpha \in R$, hence $\lambda_1 \in X(T_1)_+$. We can now either apply I.6.11 and I.4.9(1) in order to show that it is enough to prove the proposition for G_1 and λ_1, or we can observe that $G/B \simeq G_1/B_1$ and that under this isomorphism $\mathcal{L}_{G/B}(\lambda) \simeq \mathcal{L}_{G_1/B_1}(\lambda_1)$. So we may assume that G semi-simple and simply connected.

We can therefore assume $\rho \in X(T)_+$. As $\langle \rho, \alpha^\vee \rangle > 0$ and $\langle \lambda, \alpha^\vee \rangle \geq 0$ for all $\alpha \in S$, we get $\langle \lambda + \rho, \alpha^\vee \rangle > 0$ for all $\alpha \in S$; so $\mathcal{L}(\lambda + \rho)$ is ample by 4.4. Assume now at first that $\mathrm{char}(k) = p \neq 0$. We apply 4.3(2) to $\mathcal{L} = \mathcal{L}(\lambda + \rho)$ and $\mathcal{F} = \mathcal{L}(-\rho)$. We get thus $r \in \mathbf{N}$ with

$$0 = H^i(G/B, \mathcal{L}(-\rho) \otimes \mathcal{L}(\lambda + \rho)^{p^r}) = H^i(p^r(\lambda + \rho) - \rho)$$

for all $i > 0$. We have by 3.19

$$H^i((p^r - 1)\rho + p^r\lambda) \simeq St_r \otimes H^i(\lambda)^{[r]}.$$

So the vanishing above implies $H^i(\lambda) = 0$ for all $i > 0$ as claimed.

As G and B arise from $G_\mathbf{Z}$ and $B_\mathbf{Z}$ via base change and as also $\lambda \in X(T) \simeq X(T_\mathbf{Z})$ comes from a $B_\mathbf{Z}$–module, we can apply the universal coefficient theorem I.4.18.b. The vanishing established by now for all $\mathbf{F}_p$ proves the vanishing of $R^i \operatorname{ind}_{B_\mathbf{Z}}^{G_\mathbf{Z}} \lambda$ for all $i > 0$; from this we get the result for all fields of characteristic 0.

Remarks: 1) In the case of characteristic 0 we can also apply Kodaira's vanishing theorem (cf. [Ha], III, 7.15) in order to prove the proposition. (One has to use that $\mathcal{L}(-2\rho)$ is the canonical sheaf of G/B, cf. 4.2(7).) There is another proof (for $\mathrm{char}(k) = 0$) in [Donkin 9], 12.3.

2) In the case of characteristic $p \neq 0$ the proposition was first proved in [Kempf 1]. We shall call it *Kempf's vanishing theorem* even in those cases where characteristic 0 is still included.

4.6. Proposition: *Let $I \subset S$ and set $P = P_I$.*

a) *We have*

$$R^i \operatorname{ind}_B^P k = \begin{cases} k, & \text{if } i = 0, \\ 0, & \text{if } i > 0. \end{cases}$$

b) *We have for any P–module M and any $i \geq 0$*

$$R^i \operatorname{ind}_B^G M \simeq R^i \operatorname{ind}_P^G M.$$

c) *We have for all $\lambda \in X(T)_+$ with $\langle \lambda, \alpha^\vee \rangle = 0$ for all $\alpha \in I$*

$$R^i \operatorname{ind}_P^G \lambda = 0 \qquad \text{for all } i > 0.$$

Proof: a) We apply I.6.11 to the normal subgroup U_I and we apply Kempf's vanishing theorem to the reductive group $L_I \simeq P/U_I$ and $B \cap L_I \simeq B/U_I$, which is a Borel subgroup of L_I. Then we get a) immediately from 2.1(6) for $i = 0$ and from 4.5 for $i > 0$.

b) We get from the generalised tensor identity (I.4.8) and from a)

$$R^i \operatorname{ind}_B^P M \simeq M \otimes R^i \operatorname{ind}_B^P k \simeq \begin{cases} M, & \text{if } i = 0, \\ 0, & \text{if } i > 0. \end{cases}$$

Now the spectral sequence from I.4.5.c directly implies b).

c) This follows from b) and from Kempf's vanishing theorem.

4.7. Corollary: *Let $I \subset S$ and set $P = P_I$. Let V, V' be G–modules and $i \in \mathbf{N}$.*

a) *For any $\lambda \in X(T)_+$ with $\langle \lambda, \alpha^\vee \rangle = 0$ for all $\alpha \in I$ there is an isomorphism*

$$\operatorname{Ext}_G^i(V', V \otimes H^0(\lambda)) \simeq \operatorname{Ext}_P^i(V', V \otimes \lambda).$$

b) *There is an isomorphism*

$$\operatorname{Ext}_G^i(V', V) \simeq \operatorname{Ext}_P^i(V', V).$$

c) *There is an isomorphism*

$$H^i(G, V) \simeq H^i(P, V).$$

Proof: a) The generalised tensor identity (I.4.8) and 4.6.b/c show

$$R^i \operatorname{ind}_P^G (V \otimes \lambda) \simeq V \otimes R^i \operatorname{ind}_P^G \lambda \simeq \begin{cases} V \otimes H^0(\lambda), & \text{if } i = 0, \\ 0, & \text{if } i > 0. \end{cases}$$

We apply now the spectral sequence I.4.5.a to $N = V'$ and $M = V \otimes \lambda$ and get a).

b) This is the special case $\lambda = 0$ of a).

c) This is the special case $V' = k$ of b).

Remark: Let M, M' be P–modules and let $\lambda \in X(T)$ with $\langle \lambda, \alpha^\vee \rangle \geq 0$ for all $\alpha \in I$. Then the same arguments as above yield isomorphisms

(1) $$\operatorname{Ext}_P^i(M', M \otimes \operatorname{ind}_B^P \lambda) \simeq \operatorname{Ext}_B^i(M', M \otimes \lambda)$$

and

(2) $$\operatorname{Ext}_P^i(M', M) \simeq \operatorname{Ext}_B^i(M', M)$$

and

(3) $$H^i(P, M) \simeq H^i(B, M).$$

4.8. We want to apply 4.7 to $P = B$ and to get information on G–cohomology using information on B–cohomology. We have to know more about the latter to do this.

Recall from I.3.11 (cf. I.3.17): For each $\lambda \in X(T)$ the injective hull Y_λ of $\lambda = k_\lambda$ as a B–module is

$$(1) \qquad Y_\lambda = k[U] \otimes \lambda$$

with T acting on $k[U]$ via the conjugation action and with U acting on $k[U]$ via ρ_l or ρ_r.

The isomorphism of varieties

$$\prod_{\alpha \in R^+} U_{-\alpha} \xrightarrow{\sim} U$$

given by the multiplication is compatible with the conjugation action of T. So $k[U]$ considered as a T–module can be identified with the symmetric algebra $S(\text{Lie}(U)^*)$. It is a polynomial ring in $|R^+|$ generators y_α ($\alpha \in R^+$), each y_α being a weight vector of weight α. This shows:

(2) *The weights of* $k[U]$ *belong to* $\mathbf{N}R^+ = \mathbf{N}S$.

(3) *Each weight space of* $k[U]$ *has finite dimension.*

(4) *The zero weight space of* $k[U]$ *is equal to* $k1$ *and has dimension* 1.

We define the *height* $\text{ht}(\lambda)$ of any $\lambda \in \mathbf{Z}S$ by $\text{ht}(\lambda) = \sum_{\alpha \in S} n_\alpha$ if $\lambda = \sum_{\alpha \in S} n_\alpha \alpha$. Combining (1) and (2)–(4) we get for any $\lambda \in X(T)$:

(5) *The weight spaces of* Y_λ *are finite dimensional. Each weight* μ *of* Y_λ *belongs to* $\lambda + \mathbf{N}R^+$ *and satisfies* $\text{ht}(\mu) \geq \text{ht}(\lambda)$. *If here* $\text{ht}(\mu) = \text{ht}(\lambda)$, *then* $\mu = \lambda$. *The* λ*–weight space of* Y_λ *has dimension* 1.

4.9. Lemma: *There is an injective resolution*

$$0 \longrightarrow k \longrightarrow I_0 \longrightarrow I_1 \longrightarrow I_2 \longrightarrow I_3 \longrightarrow \cdots$$

of k *as a* B*–module such that* I_0 *is the injective hull of* k, *such that all weight spaces of all* I_j *are finite dimensional, and such that all weights* μ *of* I_j *belong to* $\mathbf{N}R^+$ *and satisfy* $\text{ht}(\mu) \geq j$.

Proof: We construct the $(I_j)_j$ as the obvious minimal resolution: We start with I_0 equal to the injective hull of k. Consider some $j \geq 1$; suppose that we already have $I_0, I_1, \dots, I_{j-1}$. We then set I_j equal to the injective hull of $\text{soc}_B(I_{j-1}/\operatorname{im} I_{j-2})$. So $I_j = \bigoplus_{l \in J(j)} Y_{\lambda_{lj}}$ for some index set $J(j)$ such that $\text{soc}_B(I_{j-1}/\operatorname{im} I_{j-2}) = \bigoplus_{l \in J(j)} \lambda_{lj}$. The embedding of $\text{soc}_B(I_{j-1}/\operatorname{im} I_{j-2})$ into I_j can be extended to an embedding of $I_{j-1}/\operatorname{im} I_{j-2}$ and leads thus to an exact sequence $I_{j-2} \to I_{j-1} \to I_j$. (For $j = 1$ replace I_{-1} by k.)

Any λ_{lj} is a weight of I_{j-1}, so by induction $\lambda_{lj} \in \mathbf{N}R^+$. For any $\mu \in \mathbf{N}R^+$ there are only finitely many $\lambda \in \mathbf{N}R^+$ with $\mu \in \lambda + \mathbf{N}R^+$. As the λ–weight space of I_{j-1} is finite dimensional, there can be only finitely many $l \in J(j)$ with $\lambda = \lambda_{lj}$. So again by 4.8(5) the μ–weight space of I_j has finite dimension.

Any λ_{lj} is a weight of I_{j-1}, so there has to be a direct summand Y_λ of I_{j-1} such that λ_{lj} is a weight of Y_λ. By construction $\text{soc}_B(I_{j-1})$ is contained in the image of I_{j-2} (resp. of k for $j = 1$), so $\lambda \neq \lambda_{lj}$, hence $\text{ht}(\lambda_{lj}) > \text{ht}(\lambda)$. Now induction yields $\text{ht}(\mu) \geq j$ for any weight μ of I_j.

4.10. Proposition: *Let $P \supset B$ be a reduced parabolic subgroup of G and let M be a P–module.*

a) *If $\dim M < \infty$, then $\dim H^j(P, M) < \infty$ for all $j \in \mathbf{N}$.*

b) *If $H^j(P, M) \neq 0$ for some $j \in \mathbf{N}$, then there exists a weight λ of M with $-\lambda \in \mathbf{N}R^+$ and $\mathrm{ht}(-\lambda) \geq j$.*

Proof: Because of 4.7(3) we can assume that $P = B$. Tensoring the resolution $(I_\bullet)$ from 4.9 with M we get a resolution $(M \otimes I_\bullet)$ of M. So $H^\bullet(B, M)$ is the cohomology of the complex $(M \otimes I_\bullet)^B$. Therefore any $H^j(B, M)$ is a subquotient of

$$(M \otimes I_j)^T = \bigoplus_{\lambda \in X(T)} M_\lambda \otimes (I_j)_{-\lambda}.$$

Now 4.9 implies $\dim(M \otimes I_j)^T < \infty$ if $\dim M < \infty$, hence a). If $(M \otimes I_j)^T \neq 0$ then there has to be a weight λ of M such that $-\lambda$ is a weight of I_j. Applying 4.9 again yields b).

Remark: Using $\mathrm{Ext}^j_P(M, M') \simeq H^j(P, M^* \otimes M')$, cf. I.4.4, we get from a) that

$$\dim \mathrm{Ext}^j_P(M, M') < \infty \tag{1}$$

for all finite dimensional P–modules M and M' and all $j \in \mathbf{N}$.

4.11. Corollary: *We have for all $i \in \mathbf{N}$*

$$H^i(G, k) \simeq H^i(P, k) \simeq H^i(B, k) \simeq \begin{cases} k, & \text{if } i = 0, \\ 0, & \text{if } i > 0. \end{cases}$$

Proof: The claim for $i = 0$ is obvious. For $i > 0$ we can apply 4.10 observing that $\mathrm{ht}(0) = 0$.

4.12. Corollary: *Let $P \supset B$ be a reduced parabolic subgroup of G and let M be a finite dimensional P–module. If $\mathrm{char}(k) = p \neq 0$, then*

$$H^i(P, M) \simeq \varprojlim H^i(P_r, M) \qquad \text{for all } i \in \mathbf{N}.$$

Proof: Combine 4.10.a and 4.11 with I.9.9.

Remark: In the special case $P = B$ this result was proved in [Cline, Parshall, and Scott 6], in the special case $P = G$ and $p > h$ in [Friedlander and Parshall 3]. The general case is due to W. van der Kallen, cf. I.9.9.

The result for $P = B$ also yields immediately that

$$H^i(B, M) \simeq \varprojlim H^i(B_r T, M).$$

One can show more precisely, cf. [Cline, Parshall, and Scott 6], that there is some $r_0(M, i)$ such that $H^i(B, M) \simeq H^i(B_r T, M)$ for all $r > r_0(M, i)$. The $H^i(B_r, M)$, however, will not stabilise for large r (in general).

4.13. Recall the notation $V(\lambda) = H^0(-w_0\lambda)^*$ from 2.13.

Proposition: *Let $\lambda, \mu \in X(T)_+$. Then*

$$H^i(G, H^0(\lambda) \otimes H^0(\mu)) \simeq \begin{cases} k, & \text{if } i = 0 \text{ and } \lambda = -w_0\mu, \\ 0, & \text{otherwise,} \end{cases}$$

and

$$\mathrm{Ext}^i_G(V(\lambda), H^0(\mu)) \simeq \begin{cases} k, & \text{if } i = 0 \text{ and } \lambda = \mu, \\ 0, & \text{otherwise.} \end{cases}$$

Proof: We have by 4.7

$$H^i(G, H^0(\lambda) \otimes H^0(\mu)) \simeq H^i(B, H^0(\lambda) \otimes \mu) \simeq H^i(B, H^0(\mu) \otimes \lambda).$$

If this cohomology group is non-zero, then there have to be weights ν of $H^0(\lambda)$ and ν' of $H^0(\mu)$ with $\mu + \nu, \lambda + \nu' \in -\mathbf{N}R^+$ and $\mathrm{ht}(\mu + \nu), \mathrm{ht}(\lambda + \nu') \leq -i$. We have seen in 2.2 that $w_0\lambda$ (resp. $w_0\mu$) is the smallest weight of $H^0(\lambda)$ (resp. of $H^0(\mu)$); so we get $\mu + w_0\lambda, \lambda + w_0\mu \in -\mathbf{N}R^+$. Now $w_0^2 = 1$ implies $\lambda + w_0\mu = w_0(\mu + w_0\lambda)$, hence

$$\lambda + w_0\mu \in (-\mathbf{N}R^+) \cap w_0(-\mathbf{N}R^+) = (-\mathbf{N}R^+) \cap (\mathbf{N}R^+) = 0.$$

This yields $\lambda = -w_0\mu$. Furthermore,

$$0 = \mathrm{ht}(\lambda + w_0\mu) \leq \mathrm{ht}(\lambda + \nu') \leq -i$$

shows that $i = 0$. On the other hand

$$H^0(\lambda)^U \simeq w_0\lambda,$$

hence

$$(H^0(\lambda) \otimes (-w_0\lambda))^U \simeq k \simeq (H^0(\lambda) \otimes (-w_0\lambda))^B.$$

This implies that $H^0(G, H^0(\lambda) \otimes H^0(-w_0\lambda)) \simeq k$ and proves the first part of the proposition. The second one follows from the first one using I.4.4.

Remarks: 1) We have $H^0(0) = V(0) = k$. Therefore we get as a special case:

(1) *If $\mu \neq 0$, then $H^\bullet(G, H^0(\mu)) = 0$.*

2) Corollary 4.7.a implies for all $\lambda \in X(T)_+$, all $i \in \mathbf{N}$ and all finite dimensional G–modules V

$$\mathrm{Ext}^i_G(V, H^0(\lambda)) \simeq \mathrm{Ext}^i_B(V, \lambda) \simeq H^i(B, V^* \otimes \lambda).$$

Now 4.10 implies: *If $\mathrm{Ext}^i_G(V, H^0(\lambda)) \neq 0$, then V has a weight μ with $\mu \geq \lambda$ and* $\mathrm{ht}(\mu - \lambda) \geq i$. For example, this implies for all $\lambda \in X(T)_+$

(2) $$\mathrm{Ext}^i_G(L(\lambda), H^0(\lambda)) = 0 = \mathrm{Ext}^i_G(H^0(\lambda), H^0(\lambda)) \qquad \text{for all } i > 0.$$

We have for all finite dimensional G–modules M_1 and M_2 isomorphisms for each $i \in \mathbf{N}$

(3) $$\mathrm{Ext}^i_G(M_1, M_2) \simeq \mathrm{Ext}^i_G(M_2^*, M_1^*) \simeq \mathrm{Ext}^i_G({}^\tau M_2, {}^\tau M_1).$$

Here the first isomorphism comes from I.4.4(2). For the second one use that one gets ${}^\tau M$ from M^* by twisting with an automorphism of G and that this twisting is an equivalence of categories.

The statement on $\mathrm{Ext}^i_G(V, H^0(\lambda))$ above translates now into: *If* $\mathrm{Ext}^i_G(V(\lambda), V) \neq 0$, *then V has a weight μ with $\mu \geq \lambda$ and* $\mathrm{ht}(\mu - \lambda) \geq i$. This generalises the claim for $i = 1$ in Remark 2 in 2.14 that was proved by more elementary methods.

4.14. Proposition: *Let $\lambda, \mu \in X(T)_+$ with $\lambda \not> \mu$. Then*

$$\operatorname{Ext}^i_G(L(\lambda), L(\mu)) \simeq \operatorname{Ext}^{i-1}_G(L(\lambda), H^0(\mu)/\operatorname{soc}_G H^0(\mu))$$

for all $i \in \mathbf{N}$, $i > 0$.

Proof: We apply the functor $\operatorname{Hom}_G(L(\lambda), ?)$ to the short exact sequence

$$0 \to L(\mu) \longrightarrow H^0(\mu) \longrightarrow H^0(\mu)/\operatorname{soc}_G H^0(\mu) \to 0.$$

The long exact sequence for the Ext groups will yield the claim if we can show that

$$\operatorname{Ext}^i_G(L(\lambda), H^0(\mu)) = 0 \qquad \text{for all } i > 0.$$

(Note that always $\operatorname{Hom}_G(L(\lambda), L(\mu)) \simeq \operatorname{Hom}_G(L(\lambda), H^0(\mu))$ as $\operatorname{soc}_G H^0(\mu) = L(\mu)$.) This vanishing is equivalent to

$$0 = H^i(G, L(\lambda)^* \otimes H^0(\mu)) \simeq H^i(B, L(\lambda)^* \otimes \mu)$$

(using 4.7 and I.4.4). The smallest weight of $L(\lambda)^* \otimes \mu$ is $\mu - \lambda$. By our assumption $\mu - \lambda \not< 0$. In case $\lambda \neq \mu$ this implies $\mu - \lambda \notin -\mathbf{N}R^+$, hence $\nu \notin -\mathbf{N}R^+$ for each weight ν of $L(\lambda)^* \otimes \mu$. Now 4.10 yields the vanishing of $H^i(B, L(\lambda)^* \otimes \mu)$ in this case. In case $\lambda = \mu$ apply 4.13(2).

Remarks: 1) We get from the proposition by dualising (for all $i > 0$ and for λ, μ as above with $\lambda \not> \mu$)

$$\operatorname{Ext}^i_G(L(\mu), L(\lambda)) \simeq \operatorname{Ext}^{i-1}_G(\operatorname{rad}_G V(\mu), L(\lambda)).$$

2) Note that these results generalise 2.14 (proposition and first remark).

4.15. Lemma: *Let V be a G–module, let $\lambda \in X(T)_+$ with* $\operatorname{Hom}_G(L(\lambda), V) \neq 0$. *If* $\operatorname{Hom}_G(L(\mu), V) = 0 = \operatorname{Ext}^1_G(V(\mu), V)$ *for all $\mu \in X(T)_+$ with $\mu < \lambda$, then V contains a submodule isomorphic to $H^0(\lambda)$.*

Proof: The assumption implies $\operatorname{Ext}^1_G(L(\mu), V) = 0$ for all $\mu \in X(T)_+$ with $\mu < \lambda$, hence $\operatorname{Ext}^1_G(H^0(\lambda)/L(\lambda), V) = 0$. Therefore the given non-zero homomorphism $L(\lambda) \to V$ can be extended to $H^0(\lambda)$. As $\operatorname{soc}_G H^0(\lambda) = L(\lambda)$, this extension has to be injective.

4.16. An ascending chain

$$0 = V_0 \subset V_1 \subset V_2 \subset \cdots$$

of submodules of a G–module V is called a *good filtration* of V if $V = \bigcup_{i \geq 0} V_i$ and if each V_i/V_{i-1} is isomorphic to some $H^0(\lambda_i)$ with $\lambda_i \in X(T)_+$.

Proposition: *Let V be a G–module.*

a) *If V admits a good filtration, then the number of factors in the filtration isomorphic to $H^0(\lambda)$ is equal to* $\dim \mathrm{Hom}_G(V(\lambda), V)$ *for each $\lambda \in X(T)_+$.*

b) *Suppose that* $\dim \mathrm{Hom}_G(V(\lambda), V) < \infty$ *for all $\lambda \in X(T)_+$. Then the following three properties are equivalent:*
 (i) *V admits a good filtration.*
 (ii) $\mathrm{Ext}^i_G(V(\lambda), V) = 0$ *for all $\lambda \in X(T)_+$ and all $i > 0$.*
(iii) $\mathrm{Ext}^1_G(V(\lambda), V) = 0$ *for all $\lambda \in X(T)_+$.*

Proof: a) This follows immediately from 4.13. (Note that we admit here the possibility that $\mathrm{Hom}_G(V(\lambda), V)$ has countable infinite dimension; this corresponds to the case that infinitely many factors are isomorphic to $H^0(\lambda)$.)

b) The implication (i) $\Rightarrow$ (ii) is another consequence of 4.13, and the implication (ii) $\Rightarrow$ (iii) is trivial. Let us assume that (iii) is satisfied and prove (i).

Choose a total ordering $\lambda_0, \lambda_1, \dots$ of $X(T)_+$ such that $\lambda_i < \lambda_j$ implies $i < j$. We may assume $V \neq 0$ and get $\lambda \in X(T)_+$ with $\mathrm{Hom}_G(V(\lambda), V) \neq 0$. Let i be minimal with $\mathrm{Hom}_G(V(\lambda_i), V) \neq 0$. Each weight $\mu \in X(T)_+$ with $\mu < \lambda_i$ has the form λ_j with $j < i$. This implies $\mathrm{Hom}_G(V(\mu), V) = 0$, hence $\mathrm{Hom}_G(L(\mu), V) = 0$. Therefore 4.15 shows that V contains a submodule V_1 isomorphic to $H^0(\lambda)$. As V_1 satisfies (ii) by 4.13, we see that also V/V_1 satisfies (iii). Furthermore,

$$\mathrm{Hom}_G(V(\mu), V/V_1) \simeq \mathrm{Hom}_G(V(\mu), V) \qquad \text{for all } \mu \neq \lambda_i$$

whereas $\dim \mathrm{Hom}_G(V(\lambda_i), V/V_1) = \dim \mathrm{Hom}_G(V(\lambda_i), V) - 1$.

We now iterate this construction and get a chain $0 = V_0 \subset V_1 \subset V_2 \subset \cdots \subset V' = \bigcup_{i \geq 0} V_i \subset V$ of submodules of V such that each V_i/V_{i-1} is isomorphic to some $H^0(\mu_i)$ and such that $\dim \mathrm{Hom}_G(V(\lambda), V') = \dim \mathrm{Hom}_G(V(\lambda), V)$ for all $\lambda \in X(T)_+$. Then $\mathrm{Hom}_G(V(\lambda), V/V') = 0$ for all $\lambda \in X(T)_+$, hence $V/V' = 0$. So $V = V'$ has a good filtration.

Remarks: 1) We shall see in 5.6: If $\mathrm{char}(k) = 0$, then $H^0(\lambda) = L(\lambda)$ for all $\lambda \in X(T)_+$; furthermore, all G–modules are semi-simple. So in that case all G–modules of countable dimension have a good filtration. This notion is therefore of interest only if $\mathrm{char}(k) \neq 0$.

In [van der Kallen 5] a more general definition of a good filtration is used: The factors V_i/V_{i-1} no longer have to be isomorphic to some $H^0(\lambda_i)$ but to a (possibly infinite) direct sum of copies of $H^0(\lambda_i)$. This alternative definition has the advantage that one can remove the restriction "countable dimension" in the preceding paragraph and that one can drop the assumption "$\dim \mathrm{Hom}_G(V(\lambda), V) < \infty$" in the second part of the proposition.

2) Note that the notion of a good filtration does not depend on the choice of our Borel subgroup B: If B' is another Borel subgroup of G defined over k, then there exists $g \in G(k)$ with $B' = gBg^{-1}$. One has then $H^0(\lambda) = \mathrm{ind}_B^G \lambda \simeq \mathrm{ind}_{B'}^G(\lambda')$ with $\lambda'(b') = \lambda(g^{-1}b'g)$, cf. I.3.5(4). So the class of modules $\{H^0(\lambda) \mid \lambda \in X(T)_+\}$ is independent of the choice of B.

3) Let V be a G–module with $\dim \mathrm{Hom}_G(V(\lambda), V) < \infty$ for all $\lambda \in X(T)_+$ and let $k' \supset k$ be a field extension. Then V has a good filtration if and only if the

$G_{k'}$–module $V \otimes k'$ has a good filtration. This follows from the second part of the proposition together with I.4.13.a. (Note that $H^0(\lambda) \otimes k'$ is the analogue to $H^0(\lambda)$ for $G_{k'}$, cf. I.3.5(3); similarly for $V(\lambda) \otimes k'$.)

4) Let V be a finite dimensional G–module with a good filtration. Write ch $V = \sum_{i=1}^{r}$ ch $H^0(\lambda_i)$ with dominant weights $\lambda_1, \lambda_2, \dots, \lambda_r$ ordered such that $\lambda_i < \lambda_j$ implies $i < j$. Then there exists a good filtration $V_0 = 0 \subset V_1 \subset \cdots \subset V_r = V$ of V such that $V_i/V_{i-1} \simeq H^0(\lambda_i)$ for all $i > 0$.

This follows from the proof of "(iii) $\Rightarrow$ (i)" above, but can also be seen more directly as follows. Start with an arbitrary good filtration $V_0' = 0 \subset V_1' \subset \cdots \subset V_r' = V$ of V. Then there exists a permutation σ of $\{1, 2, \dots, r\}$ such that $V_i'/V_{i-1}' \simeq H^0(\lambda_{\sigma(i)})$ for all $i > 0$. If $\sigma(i) > \sigma(i+1)$ for some i, $0 < i < r$, then the extension

$$0 \to H^0(\lambda_{\sigma(i)}) \longrightarrow V_{i+1}'/V_{i-1}' \longrightarrow H^0(\lambda_{\sigma(i+1)}) \to 0$$

splits by Remark 2 in 4.13 since $\lambda_{\sigma(i+1)} \not> \lambda_{\sigma(i)}$. So $V_{i+1}'/V_{i-1}' \simeq H^0(\lambda_{\sigma(i)}) \oplus H^0(\lambda_{\sigma(i+1)})$ and we can replace V_i' by a submodule V_i'' between V_{i+1}' and V_{i-1}' such that $V_i''/V_{i-1}' \simeq H^0(\lambda_{\sigma(i+1)})$ and $V_{i+1}'/V_i'' \simeq H^0(\lambda_{\sigma(i)})$. Iterating we get finally the desired filtration.

4.17. Corollary: *Let* $0 \to V' \to V \to V'' \to 0$ *be an exact sequence of* G*–modules with* $\dim \operatorname{Hom}_G(V(\lambda), V'') < \infty$ *for all* $\lambda \in X(T)_+$. *If* V' *and* V *have good filtrations, then so does* V''.

This is now obvious.

Remark: Following [Friedlander and Parshall 2] one says that a G–module $V \neq 0$ has *good filtration dimension* n if $\operatorname{Ext}_G^j(V(\lambda), V) = 0$ for all $\lambda \in X(T)_+$ and all $j > n$ while there exists $\mu \in X(T)_+$ with $\operatorname{Ext}_G^n(V(\mu), V) \neq 0$. For more properties of this notion see [Friedlander and Parshall 2], 3.3–3.5.

4.18. Proposition: *Let* $\lambda \in X(T)_+$. *The injective hull of the* G*–module* $L(\lambda)$ *has a good filtration. Any* $H^0(\mu)$ *with* $\mu \in X(T)_+$ *occurs exactly* $[V(\mu) : L(\lambda)]$ *times as a factor in such a filtration.*

Proof: Denote that injective hull by Q. It obviously satisfies (ii) in 4.16.b. Furthermore, we have $\dim \operatorname{Hom}_G(V(\mu), Q) = [V(\mu) : L(\lambda)]$ by standard properties of injective hulls.

Remark: We can embed any G–module into an injective G–module using an embedding of its socle. Therefore the proposition implies: Any finite dimensional G–module can be embedded into a finite dimensional G–module admitting a good filtration.

4.19. An ascending chain

$$0 = V_0 \subset V_1 \subset V_2 \subset \cdots$$

of submodules of a G–module V is called a *Weyl filtration* of V if $V = \bigcup_{i \geq 0} V_i$ and if each V_i/V_{i-1} is isomorphic to some $V(\lambda_i)$ with $\lambda_i \in X(T)_+$.

One shows as in 4.16: If V is a G–module with a Weyl filtration, then the number of factors in the filtration isomorphic to $V(\lambda)$ is equal to $\dim \operatorname{Hom}_G(V, H^0(\lambda))$

for each $\lambda \in X(T)_+$. Furthermore one has then $\mathrm{Ext}^i_G(V, H^0(\lambda)) = 0$ for all $\lambda \in X(T)_+$ and all $i > 0$.

If V is a finite dimensional G–module, then V has a Weyl filtration if and only if V^* has a good filtration if and only if ${}^{\tau}V$ has a good filtration if and only if $\mathrm{Ext}^1_G(V, H^0(\lambda)) = 0$ for all $\lambda \in X(T)_+$. (This is obvious by 2.13(1), 2.13(2), or 4.13(3) and 4.16.)

Let V be a finite dimensional G–module with a Weyl filtration. Write ch $V = \sum_{i=1}^r \mathrm{ch}\, V(\lambda_i)$ with dominant weights $\lambda_1, \lambda_2, \dots, \lambda_r$ ordered such that $\lambda_i < \lambda_j$ implies $i > j$. Then there exists a Weyl filtration $V_0 = 0 \subset V_1 \subset \cdots \subset V_r = V$ of V such that $V_i/V_{i-1} \simeq V(\lambda_i)$ for all $i > 0$. (This follows by applying Remark 4 in 4.16 to ${}^{\tau}V$ or by carrying over the argument used there.)

4.20. Proposition: *Consider $k[G]$ as a $(G \times G)$–module via $\rho_l \times \rho_r$. Then $k[G]$ admits a good filtration. The factors are the $H^0(\lambda) \otimes H^0(-w_0\lambda)$ with $\lambda \in X(T)_+$ each occurring with multiplicity* 1.

Proof: Observe that $T \times T$ is a maximal torus in $G \times G$ and that $B \times B$ is a Borel subgroup. We can identify $X(T \times T)$ with $X(T) \times X(T)$ and get $H^0(\lambda, \mu) = H^0(\lambda) \otimes H^0(\mu)$ and $V(\lambda, \mu) = V(\lambda) \otimes V(\mu)$ with $(g_1, g_2) \in G \times G$ acting as $g_1 \otimes g_2$, cf. Lemma I.3.8.

As $1 \times G$ is normal in $G \times G$ and as we can identify $G \times 1$ with $(G \times G)/(1 \times G)$, we get from I.6.6(3) a spectral sequence

$$H^i(G \times 1, H^j(1 \times G, H^0(\lambda, \mu) \otimes k[G])) \Rightarrow H^{i+j}(G \times G, H^0(\lambda, \mu) \otimes k[G]).$$

As $k[G]$ is injective for G, this sequence degenerates and yields isomorphisms

$$H^i(G \times 1, H^0(\lambda) \otimes (H^0(\mu) \otimes k[G])^G) \simeq \mathrm{Ext}^i_{G \times G}(V(-w_0\lambda, -w_0\mu), k[G]).$$

We can identify $(H^0(\mu) \otimes k[G])^G \simeq \mathrm{ind}_G^G H^0(\mu) \simeq H^0(\mu)$. So the left hand side above is $H^i(G, H^0(\lambda) \otimes H^0(\mu))$, which we have computed in 4.13. Now the claim follows from 4.13 and 4.16.

Remark: Proposition 4.20 is due to Donkin (cf. [Donkin 13], 1.4(17)) and [Koppinen 3]. A special case had appeared earlier in [De Concini 2]. If $\mathrm{char}(k) = 0$, then this proposition together with the results in 5.6 below yields the classical result that $k[G]$ is isomorphic to the direct sum of all $L(\lambda) \otimes L(\lambda)^*$ with $\lambda \in X(T)_+$.

4.21. Proposition: *If V and V' are G–modules admitting a good filtration, then so is $V \otimes V'$.*

This was first proved in in type A and for large p in other types in [Wang 1]. Then Donkin proved it in most cases in [Donkin 9]. (He had to exclude the case where $\mathrm{char}(k) = 2$ and where G has a component of type E_7 or E_8.)

The first general proof was given in [Mathieu 3]. We shall discuss his proof in Chapter G. (The careful reader may check that we do not use Proposition 4.21 on our way.) Other accounts of Mathieu's proof can be found in [van der Kallen 5] and [Kaneda 8].

For proofs of the result in certain special cases see [Littelmann 2] and [Polo 4]. There are completely different proofs of the general result using "crystal" (or "canonical") bases, see [Paradowski] and [Kaneda 12].

Remark: If we embed G diagonally into $G \times G$, then the representation in Proposition 4.20 restricts to the conjugation action of G on $k[G]$. Now the present proposition implies that $k[G]$ has a good filtration as a G–module for this action.

4.22. Set $\mathfrak{g} = \mathrm{Lie}(G)$ and $I = \{f \in k[G] \mid f(1) = 0\}$. The isomorphism $\mathfrak{g}^* \simeq I/I^2$ is compatible with the coadjoint representation of G on $\mathfrak{g}^*$ and the conjugation action on I/I^2. Since G is smooth, we have also isomorphisms $S^j\mathfrak{g}^* \simeq I^j/I^{j+1}$ for all $j \in \mathbf{N}$. One can now show, cf. [Andersen and Jantzen], 4.4 and [Donkin 17], 2.1:

(1) *If G is semi-simple and simply connected or if $G = GL_n$ for some $n \geq 1$ and if* char(k) *is good for G, then $S\mathfrak{g}^*$ has a good filtration.*

Here char(k) is called *good* for G if either char$(k) = 0$ or is char$(k) = p > 0$ with $p > n_{\alpha\beta}$ for all $\alpha \in R^+$, $\beta \in S$ where $\alpha = \sum_{\beta \in S} n_{\alpha\beta}\beta$. So we exclude $p = 2$ (resp. 3, 5) if R has a component not of type A (resp. has a component of exceptional type, resp. of type E_8).

If char(k) is *not* good for G, then $S\mathfrak{g}^*$ usually does not admit a good filtration (according to S. Donkin and C. Musili, cf. [Donkin 15], p. 78).

4.23. Suppose that G is semi-simple and simply connected or that $G = GL_n$ for some $n \geq 1$. Let $A = k[G]$ or $A = S\mathfrak{g}^*$ with $\mathfrak{g} = \mathrm{Lie}(G)$; in the second case assume that char(k) is good for G. Consider A as a G–module under the conjugation resp. the coadjoint representation; so it has a good filtration by Remark 4.21 or by 4.22(1).

Set $C = A^G$; this is a subalgebra of A. Choose a total ordering $\lambda_1, \lambda_2, \lambda_3, \dots$ of $X(T)_+$ such that $\lambda_i < \lambda_j$ implies $i < j$. Then the main theorem (2.2) in [Donkin 17] says:

Proposition: *There exists a chain $A_0 = 0 \subset A_1 \subset A_2 \subset \cdots$ of (C, G)–submodules of A with $A = \bigcup_{i \geq 0} A_i$ and with isomorphisms of (C, G)–modules*

(1) $$A_i/A_{i+1} \simeq C \otimes H^0(\lambda_i)^{m(i)} \qquad \textit{where } m(i) = \dim H^0(\lambda)_0$$

for all $i > 0$.

Here "(C, G)–submodules" means that the A_i are both C–submodules and G–submodules of A. In (1) we consider $C \otimes H^0(\lambda_i)^{m(i)}$ as a G–module letting G act on the second factor only, and as a C–module letting C act on the first factor only; the isomorphism in (1) is then supposed to be compatible with both module structures.

If char$(k) = 0$, then A is isomorphic to the direct sum of all A_i/A_{i-1} (by 5.6 below). In this case the proposition is equivalent to a classical result from [Kostant 2] on $S\mathfrak{g}^*$ and a theorem from [Richardson] on $k[G]$.

4.24. When proving Proposition 4.21 (under minor restrictions) Donkin proved at the same time (under the same restrictions):

Proposition: *Let $I \subset S$. If V is a G–module with a good filtration, then V has also a good filtration when considered as an L_I–module.*

This was then proved in full generality in [Mathieu 3], see Chapter G. Note that the claim makes sense because L_I is again a reductive group of the same form as G. The results extends of course to all groups conjugate to some L_I, i.e., to all Levi factors of all reduced parabolic subgroups in G.

Let H be a closed, connected reductive subgroup in G. One calls (G, H) a *good pair* (in [Brundan 4]) or a *Donkin pair* (in [van der Kallen 6]) if for any G–module V with a good filtration also the H–module $\operatorname{res}_H^G V$ has a good filtration.

So the proposition says that (G, L_I) is a Donkin pair for each I. Proposition 4.21 can be interpreted as saying that $(G \times G, \Delta(G))$ is a Donkin pair where Δ is the diagonal embedding $g \mapsto (g, g)$. Other examples can be found in [Brundan 4], 3.3 and [van der Kallen 6]. In [Brundan 4], 4.4 it is conjectured that (G, H) is a Donkin pair in case H is the centraliser of a graph automorphism or an involution of G.

CHAPTER 5

The Borel-Bott-Weil Theorem and Weyl's Character Formula

Let k be a field throughout this chapter.

In characteristic 0 the Borel-Bott-Weil theorem gives complete information about all $H^i(\mu)$: If $\lambda \in X(T)_+$ and $w \in W$, then $H^i(w \bullet \lambda) = 0$ for $i \neq l(w)$ and $H^{l(w)}(w \bullet \lambda) \simeq H^0(\lambda)$. [Recall the dot action from 1.5(7).] Furthermore, $\operatorname{ch} H^0(\lambda)$ is given by Weyl's character formula. If $\mu \in X(T)$ with $\mu \notin W \bullet X(T)_+$, then $H^i(\mu) = 0$ for all i.

The result does not generalise to prime characteristic, not even in the weak version where we replace the isomorphism $H^{l(w)}(w \bullet \lambda) \simeq H^0(\lambda)$ from above by an identity of characters $\operatorname{ch} H^{l(w)}(w \bullet \lambda) = \operatorname{ch} H^0(\lambda)$. This weak version holds if and only if the root system has only irreducible components of rank 1 (assuming $\operatorname{char}(k) \neq 0$).

We start this chapter with an explicit description of the $H^i(\mu)$ in the case where G has semi-simple rank 1 (for short $\mathrm{rk}_{ss}G = 1$). More precisely, we take for arbitrary G a simple root $\alpha \in S$ and describe in 5.1–5.3 all $R^i \operatorname{ind}_B^{P(\alpha)}(\mu)$ where $P(\alpha) = P_{\{\alpha\}}$. We have $G = P(\alpha)$ if $\mathrm{rk}_{ss}G = 1$; so in this case we get the $H^i(\mu)$. On the other hand, the formula I.4.9(1) for $R^i \operatorname{ind}$ in the case of a semi-direct product reduces the computation of $R^i \operatorname{ind}_B^{P(\alpha)} M$ in general to the case where $\mathrm{rk}_{ss}G = 1$ and then to the well known computation of the cohomology of line bundles on the projective line $\mathbf{P}^1$. (We use the computations for SL_2 in 2.16 instead.) In this case ($\mathrm{rk}_{ss}G = 1$) we see that the weak version of the Borel-Bott-Weil theorem still holds, but a computation of the socle of $H^1(\mu)$ in 5.12 shows that we have, in general, no isomorphism $H^0(\lambda) \simeq H^1(s_\alpha \bullet \lambda)$ for $\lambda \in X(T)_+$ (and $\mathrm{rk}_{ss}G = 1$).

We then use the complete information about the $R^i \operatorname{ind}_B^{P(\alpha)}(\mu)$ to deduce results about the $H^i(\mu)$. For example, we prove the Borel-Bott-Weil theorem in characteristic 0 (5.5), while in prime characteristic we exhibit some $\lambda \in X(T)_+$ for which the $H^i(w \bullet \lambda)$ behave as in characteristic 0. We use here an approach to the Borel-Bott-Weil theorem due to [Demazure 5]; the part about prime characteristic was first proved in [Andersen 1].

Another application (of the description of all $R^i \operatorname{ind}_B^{P(\alpha)}(\mu)$) is a proof of Weyl's character formula for the $H^0(\lambda)$ with $\lambda \in X(T)_+$, more generally, for $\sum_{i \geq 0}(-1)^i \operatorname{ch} H^i(\mu)$ for all $\mu \in X(T)$. The approach used here is due to [Donkin 9].

The results on $R^\bullet \operatorname{ind}_B^{P(\alpha)}$ are used to determine all μ with $H^1(\mu) \neq 0$, the socles of these $H^1(\mu)$, and their largest weights. All this is due to [Andersen 2]. As a corollary we get a complete description of all $H^1(B, \mu)$ due to [Andersen 10].

5.1. Recall from 1.5 that the "dot action"

$$w \bullet \lambda = w(\lambda + \rho) - \rho \tag{1}$$

satisfies $w \bullet X(T) = X(T)$. In fact, we have $w \bullet \lambda - \lambda \in \mathbf{Z}R$ for all $w \in W$ and $\lambda \in X(T)$.

Let $\alpha \in S$. Set $P(\alpha) = P_{\{\alpha\}}$. Note that (1) says in particular

$$s_\alpha \bullet \lambda = s_\alpha \lambda - \alpha = \lambda - (\langle \lambda, \alpha^\vee \rangle + 1)\alpha \qquad \text{for all } \lambda \tag{2}$$

since $\langle \rho, \alpha^\vee \rangle = 1$.

5.2. Proposition: *Let $\alpha \in S$ and $\lambda \in X(T)$.*

a) *The unipotent radical of $P(\alpha)$ acts trivially on each $R^i \operatorname{ind}_B^{P(\alpha)} \lambda$.*

b) *If $\langle \lambda, \alpha^\vee \rangle = -1$, then $R^\bullet \operatorname{ind}_B^{P(\alpha)} \lambda = 0$.*

c) *If $\langle \lambda, \alpha^\vee \rangle = r \geq 0$, then $R^i \operatorname{ind}_B^{P(\alpha)} \lambda = 0$ for all $i \neq 0$ and $\operatorname{ind}_B^{P(\alpha)} \lambda$ has a basis $v_0, v_1, \dots, v_r$ such that for all i $(0 \leq i \leq r)$ and each k–algebra A:*

$$tv_i = (\lambda - i\alpha)(t) v_i \qquad \textit{for all } t \in T(A), \tag{1}$$

$$x_\alpha(a) v_i = \sum_{j=0}^{i} \binom{i}{j} a^{i-j} v_j \qquad \textit{for all } a \in A, \tag{2}$$

$$x_{-\alpha}(a) v_i = \sum_{j=i}^{r} \binom{r-i}{r-j} a^{j-i} v_j \qquad \textit{for all } a \in A. \tag{3}$$

d) *If $\langle \lambda, \alpha^\vee \rangle \leq -2$, then $R^i \operatorname{ind}_B^{P(\alpha)} \lambda = 0$ for all $i \neq 1$ and $R^1 \operatorname{ind}_B^{P(\alpha)} \lambda$ has a basis $v'_0, v'_1, \dots, v'_r$ where $r = -\langle \lambda, \alpha^\vee \rangle - 2$ such that for all i $(0 \leq i \leq r)$ and each k–algebra A:*

$$tv'_i = (s_\alpha \bullet \lambda - i\alpha)(t) v'_i \qquad \textit{for all } t \in T(A), \tag{1'}$$

$$x_\alpha(a) v'_i = \sum_{j=0}^{i} \binom{r-j}{r-i} a^{i-j} v'_j \qquad \textit{for all } a \in A, \tag{2'}$$

$$x_{-\alpha}(a) v'_i = \sum_{j=i}^{r} \binom{j}{i} a^{j-i} v'_j \qquad \textit{for all } a \in A. \tag{3'}$$

Proof: As the unipotent radical of $P(\alpha)$ is contained in U and as U acts trivially on $k_\lambda = \lambda$, the statement in a) is an immediate consequence of I.6.11. Furthermore, we may replace B and $P(\alpha)$ with their quotients by this radical. So we may assume that $G = P(\alpha)$ has semi-simple rank 1. Then G is a factor group of some group of the form $SL_2 \times T_2$ with T_2 a torus, cf. 1.18(2). This can be done in a way such that T is the image of $\{$ diagonal matrices in $SL_2 \} \times T_2$ and such that x_α and $x_{-\alpha}$ come from the "standard" root homomorphisms in SL_2. Using I.6.11 again, we may assume that $G = SL_2 \times T_2$.

Now I.4.9(1) implies that $R^i \operatorname{ind}_B^G \lambda$ is isomorphic to $R^i \operatorname{ind}_{B \cap SL_2}^{SL_2} \lambda$ as an SL_2–module. On the other hand, T_2 acts on each $R^i \operatorname{ind}_B^G \lambda$ through the restriction of λ

on T_2 by 2.10(1). This is compatible with our claims as α vanishes on T_2 and as $s_\alpha \bullet \lambda - \lambda \in \mathbf{Z}\alpha$. We may therefore assume that $G = SL_2$.

Now $\operatorname{ind}_B^G \lambda$ is described in 2.16. The character denoted by ϖ there maps (for SL_2) any $\left(\begin{smallmatrix} a & 0 \\ 0 & a^{-1} \end{smallmatrix}\right)$ to a. This implies $\langle \varpi, \alpha^\vee \rangle = 1$ and $X(T) = \mathbf{Z}\varpi$. So any $\lambda \in X(T)$ is equal to $\langle \lambda, \alpha^\vee \rangle \varpi$. For $\langle \lambda, \alpha^\vee \rangle = r \geq 0$ we get as $\operatorname{ind}_B^G \lambda$ the rth symmetric power of the dual of the natural representation of SL_2. Taking the basis consisting of all monomials (in two variables) and changing some signs, we get the action described in (1)–(3). If $\langle \lambda, \alpha^\vee \rangle < 0$, then $\operatorname{ind}_B^G \lambda = H^0(\lambda) = 0$ by 2.6.

Because of 4.2(3) we have $R^i \operatorname{ind}_B^G = 0$ for $i \neq 0, 1$. So it remains to look at $R^1 \operatorname{ind}_B^G$; here we can use Serre duality (4.2(9)):

$$H^1(\lambda) \simeq H^0(-(\lambda + 2\rho))^*.$$

If $\langle -(\lambda + 2\rho), \alpha^\vee \rangle < 0$, i.e., if $\langle \lambda, \alpha^\vee \rangle > -2$ this is zero, whereas for $\langle \lambda, \alpha^\vee \rangle \leq -2$ we get (1′)–(3′) using the dual basis up to sign changes.

5.3. Corollary: *Let $\alpha \in S$ and $\lambda \in X(T)$ with $\langle \lambda, \alpha^\vee \rangle \geq 0$.*

a) *If $\operatorname{char}(k) = 0$, then $\operatorname{ind}_B^{P(\alpha)} \lambda \simeq R^1 \operatorname{ind}_B^{P(\alpha)} (s_\alpha \bullet \lambda)$.*

b) *If $\operatorname{char}(k) = p \neq 0$ and if there are s, $m \in \mathbf{N}$ with $0 < s < p$ and $\langle \lambda, \alpha^\vee \rangle = sp^m - 1$, then $\operatorname{ind}_B^{P(\alpha)} \lambda \simeq R^1 \operatorname{ind}_B^{P(\alpha)} (s_\alpha \bullet \lambda)$.*

Proof: Let us use the notations from 5.2. Set $r = \langle \lambda, \alpha^\vee \rangle$. The map

$$R^1 \operatorname{ind}_B^{P(\alpha)} (s_\alpha \bullet \lambda) \longrightarrow \operatorname{ind}_B^{P(\alpha)} \lambda$$

taking each v_i' to $\binom{r}{i} v_i$ is a homomorphism of $P(\alpha)$–modules. This follows from 5.2 using elementary calculations. It is an isomorphism if each $\binom{r}{i}$ with $0 \leq i \leq r$ is non-zero in k. This is automatically satisfied if $\operatorname{char}(k) = 0$. If $\operatorname{char}(k) = p \neq 0$, then we have to determine the r for which no $\binom{r}{i}$ with $0 \leq i \leq r$ is divisible by p. This is done using the standard formula for binomial coefficients modulo p (cf. [Haboush 3], 5.1) and it immediately implies the claim.

Remark: Suppose that $G = P(\alpha)$. Then a comparison with 2.16(7) shows that the map $R^1 \operatorname{ind}_B^{P(\alpha)} (s_\alpha \bullet \lambda) \to \operatorname{ind}_B^{P(\alpha)} \lambda$ has image $L(\lambda)$.

5.4. Proposition: *Let $\alpha \in S$ and $\lambda \in X(T)$.*

a) *If $\langle \lambda, \alpha^\vee \rangle = -1$, then $H^\bullet(\lambda) = 0$.*

b) *If $\langle \lambda, \alpha^\vee \rangle \geq 0$, then (for all i)*

$$H^i(\lambda) \simeq R^i \operatorname{ind}_{P(\alpha)}^G (\operatorname{ind}_B^{P(\alpha)} \lambda) \simeq H^i(\operatorname{ind}_B^{P(\alpha)} \lambda).$$

c) *If $\langle \lambda, \alpha^\vee \rangle \leq -2$, then (for all i)*

$$H^i(\lambda) \simeq R^{i-1} \operatorname{ind}_{P(\alpha)}^G (R^1 \operatorname{ind}_B^{P(\alpha)} \lambda) \simeq H^{i-1}(R^1 \operatorname{ind}_B^{P(\alpha)} \lambda).$$

d) *Suppose that* $\langle\lambda,\alpha^\vee\rangle \geq 0$. *If* $\operatorname{char}(k) = 0$ *or if* $\operatorname{char}(k) = p \neq 0$ *and* $\langle\lambda,\alpha^\vee\rangle = sp^m - 1$ *for suitable* $s, m \in \mathbf{N}$ *with* $0 < s < p$, *then (for all* i*)*

$$H^i(\lambda) \simeq H^{i+1}(s_\alpha \bullet \lambda).$$

Proof: We use the spectral sequence

$$R^i \operatorname{ind}_{P(\alpha)}^G (R^j \operatorname{ind}_B^{P(\alpha)} \lambda) \Rightarrow R^{i+j} \operatorname{ind}_B^G \lambda = H^{i+j}(\lambda)$$

from I.4.5.c. If $\langle\lambda,\alpha^\vee\rangle = -1$, then all terms on the left hand side are zero by 5.2.b; this implies a). If $\langle\lambda,\alpha^\vee\rangle \neq -1$, then there is by 5.2 exactly one j with $R^j \operatorname{ind}_B^{P(\alpha)} \lambda$ is non-zero. So the spectral sequence degenerates and yields isomorphisms

$$R^i \operatorname{ind}_{P(\alpha)}^G (R^j \operatorname{ind}_B^{P(\alpha)} \lambda) \simeq R^{i+j} \operatorname{ind}_B^G \lambda$$

for all i (and that unique j). We get thus the first isomorphisms in b) and c). The second isomorphisms follow from 4.6.b. Finally d) is an immediate consequence of b), c), and 5.3.

5.5. Set

$$\overline{C}_{\mathbf{Z}} = \{\, \lambda \in X(T) \mid 0 \leq \langle\lambda+\rho,\beta^\vee\rangle \text{ for all } \beta \in R^+ \,\}$$

if $\operatorname{char}(k) = 0$, and

$$\overline{C}_{\mathbf{Z}} = \{\, \lambda \in X(T) \mid 0 \leq \langle\lambda+\rho,\beta^\vee\rangle \leq p \text{ for all } \beta \in R^+ \,\}$$

if $\operatorname{char}(k) = p > 0$. Recall the definition of $l(w)$ and its elementary properties from 1.5.

Corollary: a) *If* $\lambda \in \overline{C}_{\mathbf{Z}}$ *with* $\lambda \notin X(T)_+$, *then* $H^\bullet(w \bullet \lambda) = 0$ *for all* $w \in W$.

b) *If* $\lambda \in \overline{C}_{\mathbf{Z}} \cap X(T)_+$, *then we have for all* $w \in W$ *and* $i \in \mathbf{N}$

$$H^i(w \bullet \lambda) \simeq \begin{cases} H^0(\lambda), & \text{if } i = l(w), \\ 0, & \text{otherwise.} \end{cases}$$

Proof: If $\lambda \in \overline{C}_{\mathbf{Z}}$ with $\lambda \notin X(T)_+$, then there exists $\alpha \in S$ with $\langle\lambda,\alpha^\vee\rangle < 0$. As $0 \leq \langle\lambda+\rho,\alpha^\vee\rangle = \langle\lambda,\alpha^\vee\rangle + 1$, we get $\langle\lambda,\alpha^\vee\rangle = -1$. Therefore the claim in a) follows for $w = 1$ from 5.4.a.

On the other hand, the claim in b) follows for $w = 1$ from Kempf's vanishing theorem 4.5.

We now use induction on $l(w)$. We have just dealt with the case $l(w) = 0$ [equivalent to $w = 1$]. Suppose now $l(w) > 0$. Then there is some $\alpha \in S$ with $l(s_\alpha w) = l(w) - 1$, hence with $w^{-1}(\alpha) \in -R^+$, see 1.5(3). Set $\beta = -w^{-1}(\alpha) \in R^+$. We have

$$\begin{aligned}\langle s_\alpha w \bullet \lambda, \alpha^\vee\rangle &= \langle s_\alpha w(\lambda+\rho) - \rho, \alpha^\vee\rangle = -\langle w(\lambda+\rho),\alpha^\vee\rangle - 1 \\ &= \langle\lambda+\rho,\beta^\vee\rangle - 1 \geq -1.\end{aligned}$$

If $\operatorname{char}(k) = p > 0$, this calculation yields also that $\langle s_\alpha w \bullet \lambda,\alpha^\vee\rangle < p$.

Now 5.4.d (and 5.4.a) imply

$$H^i(w \bullet \lambda) \simeq H^{i-1}(s_\alpha w \bullet \lambda).$$

(Observe that $s_\alpha \bullet (w \bullet \lambda) = (s_\alpha w) \bullet \lambda$.) Now use induction.

Remarks: 1) If $\text{char}(k) = 0$, then each element of $X(T)$ has the form $w \bullet \lambda$ with $\lambda \in \overline{C}_{\mathbf{Z}}$ and $w \in W$. So in characteristic 0 the corollary yields all $H^i(\mu)$. It is known in this case as the *Borel-Bott-Weil theorem.*

2) We can avoid using Kempf's vanishing theorem by working with descending induction for all $H^i(w \bullet \lambda)$ with $i > l(w)$ using $H^i(\mu) = 0$ for $i > |R^+|$.

5.6. Corollary: *If* $\lambda \in X(T)_+ \cap \overline{C}_{\mathbf{Z}}$, *then* $L(\lambda) = H^0(\lambda)$.

Proof: We have $w_0(\rho) = -\rho$, hence $-w_0(\overline{C}_{\mathbf{Z}}) = \overline{C}_{\mathbf{Z}}$. Furthermore, we have

(1) $$w_0 \bullet (-w_0\lambda) = w_0(-w_0\lambda + \rho) - \rho = -\lambda - 2\rho$$

for any $\lambda \in X(T)$, hence by Serre duality (4.2(9))

(2) $$H^n(w_0 \bullet (-w_0\lambda)) \simeq H^0(\lambda)^*$$

where $n = |R^+| = l(w_0)$. So 5.5.b implies for each $\lambda \in X(T)_+ \cap \overline{C}_{\mathbf{Z}}$

(3) $$H^0(\lambda)^* \simeq H^0(-w_0\lambda).$$

We get from $L(-w_0\lambda) = \text{soc}_G\, H^0(-w_0\lambda)$ by dualising (cf. 2.5)

(4) $$H^0(\lambda)/\,\text{rad}_G\, H^0(\lambda) \simeq L(-w_0\lambda)^* \simeq L(\lambda) = \text{soc}_G\, H^0(\lambda).$$

As $L(\lambda)$ occurs with multiplicity 1 as a composition factor in $H^0(\lambda)$ by 2.4.b, this implies that $\text{rad}_G\, H^0(\lambda) = 0$ and $\text{soc}_G\, H^0(\lambda) = H^0(\lambda)$, hence $H^0(\lambda) = L(\lambda)$.

Remark: Combining the corollary with 2.12 and 2.14, we get

(5) $$\text{Ext}^1_G(L(\lambda), L(\mu)) = 0 \qquad \text{for all } \lambda, \mu \in \overline{C}_{\mathbf{Z}} \cap X(T)_+.$$

If $\text{char}(k) = 0$, then $X(T)_+ \subset \overline{C}_{\mathbf{Z}}$ and (5) implies $\text{Ext}^1_G(M, M') = 0$ for all finite dimensional G–modules M, M', hence (using the local finiteness):

(6) *If* $\text{char}(k) = 0$, *then all* G*–modules are semi-simple.*

If $\text{char}(k) = p > 0$, then one can show (see [Jantzen 16]) that all G–modules V with $\dim(V) \leq p$ are semi-simple. This had been conjectured in [Larsen], where a weaker result was proved. With additional work one can show (see [McNinch 1, 2]) that (e.g.) for almost simple G all V with $\dim(V) \leq \text{rk}(G) \cdot p$ are semi-simple.

5.7. We shall see in 5.18 that 5.5.b does not hold for all $\lambda \in X(T)_+$ if k has prime characteristic. There is, however, an Euler characteristic analogue that always holds.

Define for each finite dimensional B–module M the *Euler characteristic*

(1) $$\chi(M) = \sum_{i \geq 0} (-1)^i \,\text{ch}\, H^i(M).$$

This is a well defined element in $\mathbf{Z}[X(T)]^W$ as each $H^i(M)$ is finite dimensional (4.2(1)) and as $H^i(M) = 0$ for $i > |R^+|$ by 4.2(3). The long exact sequence for the derived functors of ind_B^G yields

(2) $$\chi(M) = \chi(M') + \chi(M'')$$

for any exact sequence $0 \to M' \to M \to M'' \to 0$ of finite dimensional B–modules.

The generalised tensor identity (I.4.8) implies

(3) $$\chi(V \otimes M) = \mathrm{ch}(V)\,\chi(M)$$

for each G–module V and each B–module M, both finite dimensional.

We shall use the notation $\chi(\lambda) = \chi(k_\lambda)$. Note that Kempf's vanishing theorem implies

(4) $$\chi(\lambda) = \mathrm{ch}\ H^0(\lambda) \qquad \text{for all } \lambda \in X(T)_+.$$

5.8. Lemma: a) *The* $\mathrm{ch}\, L(\lambda)$ *with* $\lambda \in X(T)_+$ *are a basis of* $\mathbf{Z}[X(T)]^W$.

b) *One has for all* $\lambda \in X(T)$ *and all* $\sum_\mu a(\mu)e(\mu) \in \mathbf{Z}[X(T)]^W$

$$\chi(\lambda) \sum_\mu a(\mu)e(\mu) = \sum_\mu a(\mu)\chi(\lambda + \mu).$$

Proof: a) Set

(1) $$\mathrm{sym}(\mu) = \sum_{\nu \in W\mu} e(\nu) \in \mathbf{Z}[X(T)]^W.$$

As each $\nu \in X(T)$ is conjugate under W to exactly one $\mu \in X(T)_+$, the $\mathrm{sym}(\mu)$ with $\mu \in X(T)_+$ form a basis of $\mathbf{Z}[X(T)]^W$.

If $\lambda \in X(T)_+$, then $\mathrm{ch}\, L(\lambda) \in \mathbf{Z}[X(T)]^W$ and $\dim L(\lambda)_\lambda = 1$, by 1.19(2) and 2.4, hence (with suitable $a_\mu \in \mathbf{N}$)

(2) $$\mathrm{ch}\, L(\lambda) = \mathrm{sym}(\lambda) + \sum_\mu a_\mu \mathrm{sym}(\mu)$$

where we sum over $\mu \in X(T)_+$ with $\mu \neq \lambda$. Again by 2.4 only $\mu < \lambda$ occur with non-zero a_μ. Therefore the transition matrix from the $(\mathrm{sym}(\lambda))_\lambda$ to the $(\mathrm{ch}\, L(\lambda))_\lambda$ is unipotent triangular, and also the $\mathrm{ch}\, L(\lambda)$ form a basis.

b) Because of a) we may assume $\sum_\mu a(\mu)e(\mu) = \mathrm{ch}\, V$ for some finite dimensional G–module V. Now the desired formula follows from 5.7(3), (2) as the B–module $V \otimes \lambda$ has a composition series with $\dim(V_\mu)$ factors $k_{\lambda+\mu}$ for each μ.

Remark: The same proof as for a) shows that also the $\mathrm{ch}\, H^0(\lambda) = \chi(\lambda)$ with $\lambda \in X(T)_+$ form a basis of $\mathbf{Z}[X(T)]^W$.

5.9. For any $\alpha \in S$ and $\lambda \in X(T)$ with $\langle \lambda + \rho, \alpha^\vee \rangle \geq 0$, Proposition 5.2 implies

$$\mathrm{ch}\ \mathrm{ind}_B^{P(\alpha)}(\lambda) = \mathrm{ch}\ R^1\, \mathrm{ind}_B^{P(\alpha)}(s_\alpha \bullet \lambda).$$

Plugging this into Proposition 5.4 we get

$$\chi(\lambda) = -\chi(s_\alpha \bullet \lambda).$$

This holds now for all $\lambda \in X(T)$; in case $\langle \lambda + \rho, \alpha^\vee \rangle = -1$ one uses 5.4.a. As $W = \langle s_\alpha \mid \alpha \in S \rangle$, this shows that

(1) $$\chi(w \bullet \lambda) = \det(w)\chi(\lambda) \qquad \text{for all } w \in W \text{ and } \lambda \in X(T).$$

We have by 2.1(6) and 5.7(4)

(2) $$\chi(0) = e(0) = 1.$$

Therefore 5.8.b implies for all $\sum_\mu a(\mu)e(\mu) \in \mathbf{Z}[X(T)]^W$

(3) $$\sum_\mu a(\mu)e(\mu) = \sum_\mu a(\mu)\chi(\mu).$$

Set for all $\lambda \in X(T) \otimes_{\mathbf{Z}} \mathbf{Q}$

(4) $$A(\lambda) = \sum_{w \in W} \det(w)e(w\lambda) \in \mathbf{Z}[X(T) \otimes_{\mathbf{Z}} \mathbf{Q}].$$

Then obviously $wA(\lambda) = \det(w)A(\lambda)$ for all w, hence $A(\lambda)A(\mu) \in \mathbf{Z}[X(T) \otimes_{\mathbf{Z}} \mathbf{Q}]^W$ for all $\lambda, \mu \in X(T) \otimes_{\mathbf{Z}} \mathbf{Q}$.

5.10. Proposition (Weyl's Character Formula): *We have for all* $\lambda \in X(T)$

$$\chi(\lambda) = A(\lambda + \rho)/A(\rho).$$

Proof: To start with, $A(\lambda + \rho)/A(\rho)$ is just an element in the fraction field of the integral domain $\mathbf{Z}[X(T) + \mathbf{Z}\rho]$, which is a localised polynomial ring. (We have $\mathbf{Z}[X(T) + \mathbf{Z}\rho] = \mathbf{Z}[e(\varpi_i), e(\varpi_i)^{-1} \mid i]$, if $(\varpi_i)_i$ is a basis for the free abelian group $X(T) + \mathbf{Z}\rho$.)

We get for any $\lambda \in X(T)$ using 5.9(3), (1)

$$\begin{aligned} A(\lambda+\rho)A(\rho) &= \sum_{w,w' \in W} \det(ww')e(w(\lambda+\rho) + w'\rho) \\ &= \sum_{w,w' \in W} \det(w')e(w(\lambda+\rho+w'\rho)) \\ &= \sum_{w,w' \in W} \det(w')\chi(w(\lambda+\rho+w'\rho)) \\ &= \sum_{w,w' \in W} \det(w')\chi(w \bullet (\lambda + w'\rho + w^{-1}\rho)) \\ &= \sum_{w,w' \in W} \det(ww')\chi(\lambda + w'\rho + w\rho). \end{aligned}$$

Since

$$A(\rho)^2 = \sum_{w,w' \in W} \det(ww')e(w\rho + w'\rho) \in \mathbf{Z}[X(T)]^W$$

we get using 5.8.b

$$A(\lambda+\rho)A(\rho) = A(\rho)^2\chi(\lambda).$$

As ρ has trivial stabiliser in W, cf. [Hu1], 13.2 and 13.3, we have $A(\rho) \neq 0$, hence we can cancel $A(\rho)$ in the integral domain $\mathbf{Z}[X(T) + \mathbf{Z}\rho]$ and get

$$A(\lambda+\rho) = A(\rho)\chi(\lambda),$$

hence the proposition.

5.11. Corollary: *Let* $\lambda \in X(T)_+$. *Then*

$$\operatorname{ch} V(\lambda) = \operatorname{ch} H^0(\lambda) = \chi(\lambda) = A(\lambda + \rho)/A(\rho).$$

Proof: The second equality is 5.7(4), the third one is Proposition 5.10. The first equality follows from 2.13(2) as observed there. Alternatively, we can argue as follows: Introduce an involution $x \mapsto x^*$ on $\mathbf{Z}[X(T) \otimes_{\mathbf{Z}} \mathbf{Q}]$ by

$$\Big(\sum_\mu a(\mu)e(\mu)\Big)^* = \sum_\mu a(\mu)e(-\mu).$$

We have then $\operatorname{ch}(V^*) = (\operatorname{ch} V)^*$ for each finite dimensional G–module V, cf. I.2.11(12). We get for all λ

$$A(\lambda)^* = \sum_{w \in W} \det(w)e(-w\lambda) = A(-\lambda) = \det(w_0)A(-w_0\lambda),$$

in particular $A(\rho)^* = \det(w_0)A(\rho)$. It follows that

$$\begin{aligned} \det(w_0)A(\rho)\operatorname{ch} V(\lambda) &= A(\rho)^* \operatorname{ch} H^0(-w_0\lambda)^* = (A(\rho)\chi(-w_0\lambda))^* \\ &= A(-w_0\lambda + \rho)^* = \det(w_0)A(\lambda+\rho) \end{aligned}$$

hence $\operatorname{ch} V(\lambda) = A(\lambda+\rho)/A(\rho)$ as claimed.

Remarks: 1) The modules $V(\lambda)$ are often called *Weyl modules.* This name was introduced in [Carter and Lusztig 1] where the case $G = GL_n$ is considered and where the $V(\lambda)$ are constructed in the "same" way as in Weyl's book on classical groups. Observe that by Serre duality

$$V(\lambda) \simeq H^0(-w_0\lambda)^* \simeq H^n(w_0\lambda - 2\rho) = H^n(w_0 \bullet \lambda)$$

where $n = |R^+|$.

2) Note that $\operatorname{ch} V(\lambda) = \operatorname{ch} H^0(\lambda)$ implies by Lemma 5.8.a that

$$[V(\lambda) : L(\mu)] = [H^0(\lambda) : L(\mu)] \qquad \text{for all } \lambda, \mu \in X(T)_+.$$

5.12. Assume in this subsection that G has semi-simple rank 1. Denote by α the only simple root. Let $\lambda \in X(T)$ with $\langle \lambda, \alpha^\vee \rangle \geq 0$ and consider the basis $(v_i')_i$ of $H^1(s_\alpha \bullet \lambda)$ as in 5.2.d. Recall from 5.3:

(1) *If* $\operatorname{char}(k) = 0$ *or if* $\operatorname{char}(k) = p > 0$ *and* $\langle \lambda, \alpha^\vee \rangle = ap^n - 1$ *for some* $a, n \in \mathbf{N}$ *with* $0 < a < p$, *then* $H^1(s_\alpha \bullet \lambda) \simeq L(\lambda)$.

We want to determine $\operatorname{soc}_G H^1(s_\alpha \bullet \lambda)$ in the remaining cases using explicit calculations. There is also a more conceptual approach, cf. 11.13.

Lemma: *Assume that* $\operatorname{char}(k) = p > 0$. *Write*

$$r = \langle \lambda, \alpha^\vee \rangle = \sum_{j=0}^n a_j p^j$$

with $0 \le a_j < p$ for all j and $a_n \neq 0$. Suppose that there is some $j < n$ with $a_j < p-1$ and set $m = \min\{j \mid a_j < p-1\}$.

a) *Then $H^1(s_\alpha \bullet \lambda)^{U^+} = \sum_{m'=m}^{n-1} kv'_{i(m')} + kv'_0$, where $i(m') = \sum_{j=0}^{m'} a_j p^j + 1$.*

b) $\operatorname{soc}_G H^1(s_\alpha \bullet \lambda) \simeq L(\lambda - (\sum_{j=0}^{n-1} a_j p^j + 1)\alpha) = L(s_\alpha \bullet \lambda + a_n p^n \alpha)$.

Proof: a) Obviously $H^1(s_\alpha \bullet \lambda)^{U^+}$ is spanned by its weight vectors, hence by the v'_i contained in it. Because of 5.2(2′) these are v'_0 and those v'_i with $i > 0$ such that p divides all $\binom{r-j}{r-i}$ with $0 \le j < i$.

Suppose that some $i > 0$ satisfies this condition. Consider the p–adic expansion $r - i = \sum_{j=0}^{n} b_j p^j$ with $0 \le b_j < p$ for all j. As $r - i < r$ we have $b_j < a_j$ for the largest j with $b_j \neq a_j$. Therefore there are j with $b_j < p-1$. Let s be minimal with $b_s < p - 1$. Using the standard formula for binomial coefficients modulo p (cf. [Haboush 3], 5.1) we see that p does not divide $\binom{r-i+p^s}{r-i}$, hence that $p^s > i$ and

$$r - i + p^s = \sum_{j=0}^{s-1} (p-1)p^j + (b_s + 1)p^s + \sum_{s+1}^{n} b_j p^j > r > r - i.$$

Comparing the p–adic expansions of $r - i + p^s$, r, and $r - i$ we get $b_j = a_j$ for $j > s$, and $b_s = a_s - 1$, and $a_j < p - 1$ for at least one $j < s$, hence $m < s$.

Furthermore,

$$r - i = \sum_{j=0}^{s-1} (p-1)p^j + (a_s - 1)p^s + \sum_{s+1}^{n} a_j p^j,$$

hence

$$i = p^s - \sum_{j=0}^{s-1} (p - 1 - a_j)p^j = \sum_{j=0}^{s-1} a_j p^j + 1 = i(s-1).$$

So at most the $v'_{i(m')}$ with $m \le m' < n$ occur in $H^1(s_\alpha \bullet \lambda)^{U^+}$ (besides v'_0).

On the other hand, consider m' with $m \le m' < n$. Let t be minimal for $t > m'$ and $a_t > 0$. Note that $i(m') < p^{m'+1} \le p^t$ since $a_m < p - 1$. We have

$$r - i(m') = \sum_{j=0}^{t-1} (p-1)p^j + (a_t - 1)p^t + \sum_{j=t+1}^{n} a_j p^j.$$

It follows that each $r - i(m') + j$ with $0 < j \le i(m')$ has a p–adic expansion $\sum_{l=0}^{n} c_l p^l$ where $c_l < p - 1$ for some $l < t$ since $i(m') < p^t$. Therefore p has to divide all $\binom{r-i(m')+j}{r-i(m')}$ by the standard formula for binomial coefficients. This yields the converse inclusion.

b) The socle of $H^1(s_\alpha \bullet \lambda)$ is generated by those weight vectors in $H^1(s_\alpha \bullet \lambda)^{U^+}$ (i.e., by the v'_i from a)) that generate a simple submodule. So we have to have $\dim(\operatorname{Dist}(G)v'_i)^{U^+} = 1$, hence $(\operatorname{Dist}(G)v'_i)^{U^+} = kv'_i$. Now 5.2(3′) shows that

$$\operatorname{Dist}(G)v'_0 = \sum_{j=0}^{r} kv'_j = H^1(s_\alpha \bullet \lambda),$$

so v'_0 cannot occur. If $m \le m' < n-1$ with, then p does not divide $\binom{i(n-1)}{i(m')}$ by the standard formula and by the definitions of $i(m')$ and $i(n-1)$. It follows that

$$v'_{i(n-1)} \in \operatorname{Dist}(G) v'_{i(m')}$$

by 5.2(3′), so $\operatorname{Dist}(G)v'_{i(m')}$ is not simple unless $i(m') = i(n-1)$. So only $v'_{i(n-1)}$ remains. Its weight is given by 5.2.d(1′) and is the one mentioned.

5.13. We want to apply 5.12 to get for arbitrary G all λ with $H^1(\lambda) \neq 0$ and then to determine the socles of these modules. If $\operatorname{char}(k) = 0$, then we can apply the Borel-Bott-Weil theorem (5.5) and get complete information. (So $H^1(\lambda) \neq 0$ if and only if there exists $\alpha \in S$ with $s_\alpha \bullet \lambda \in X(T)_+$; if so, then $H^1(\lambda) \simeq H^0(s_\alpha \bullet \lambda) = L(s_\alpha \bullet \lambda)$.)

Let us therefore assume from now on that $\operatorname{char}(k) = p \neq 0$.

If $\mu \in X(T)$ satisfies $\langle \mu + \rho, \alpha^\vee \rangle \ge 0$ for all $\alpha \in S$, then $H^1(\mu) = 0$ by Kempf's vanishing theorem or by 5.4.a. So, if $H^1(\mu) \neq 0$, then $\langle \mu + \rho, \alpha^\vee \rangle < 0$ for some $\alpha \in S$ and then $H^1(\mu) \simeq \operatorname{ind}_{P(\alpha)}^G (R^1 \operatorname{ind}_B^{P(\alpha)} \mu)$ by 5.4.c.

In other words, we get all non-zero $H^1(\mu)$'s by looking at all $\alpha \in S$ and $\lambda \in X(T)$ with $\langle \lambda + \rho, \alpha^\vee \rangle > 0$ and by taking

$$H^1(s_\alpha \bullet \lambda) \simeq \operatorname{ind}_{P(\alpha)}^G (R^1 \operatorname{ind}_B^{P(\alpha)} (s_\alpha \bullet \lambda)).$$

The unipotent radical of $P(\alpha)$ acts trivially on $R^1 \operatorname{ind}_B^{P(\alpha)} (s_\alpha \bullet \lambda)$ and the Levi factor $L_{\{\alpha\}}$ acts as on "the $H^1(s_\alpha \bullet \lambda)$ for $L_{\{\alpha\}}$", cf. the proof of 4.6.a. So 5.12 describes the socle of $R^1 \operatorname{ind}_B^{P(\alpha)} (s_\alpha \bullet \lambda)$ as an $L_{\{\alpha\}}$–module and it lists the U_α–invariant elements in this module. From this we get information on $H^1(s_\alpha \bullet \lambda)$ using a general result to be proved first.

5.14. Lemma: *Let $I \subset S$ and set $P = P_I$. Let M be a P–module.*

a) *Suppose that the socle of M as an L_I–module is simple with highest weight $\lambda \in X(T)$. If $\operatorname{ind}_P^G M \neq 0$, then $\lambda \in X(T)_+$ and $\operatorname{soc}_G \operatorname{ind}_P^G M \simeq L(\lambda)$.*

b) *Suppose that $\operatorname{ind}_P^G M \neq 0$ and that λ is a maximal weight of $\operatorname{ind}_P^G M$. Then $\lambda \in X(T)_+$ and $(M^{U^+ \cap L_I})_\lambda \neq 0$.*

c) *Let $\lambda \in X(T)_+$ with $(M^{U^+ \cap L_I})_\lambda \neq 0$. If U_I acts trivially on M, then λ is a weight of $\operatorname{ind}_P^G M$.*

Proof: a) The evaluation map $\varepsilon_M : \operatorname{ind}_P^G M \to M$ is a homomorphism of P–modules. According to the remark in 2.2 it is injective on $(\operatorname{ind}_P^G M)^{U(\overline{P})}$ where $U(\overline{P}) = U_I^+$. If $L(\mu)$ is a simple submodule of $\operatorname{ind}_P^G M$, then $L(\mu)^{U(\overline{P})}$ is the simple L_I–module with highest weight μ (by 2.11), hence so is $\varepsilon_M(L(\mu)^{U(\overline{P})})$. Our assumption yields $\lambda = \mu \in X(T)_+$ and $\operatorname{soc}_G(\operatorname{ind}_P^G M) \simeq L(\lambda)$.

b) If λ is a maximal weight of $\operatorname{ind}_P^G M$, then

$$0 \neq (\operatorname{ind}_P^G M)_\lambda \subset (\operatorname{ind}_P^G M)^{U^+} \subset (\operatorname{ind}_P^G M)^{U(\overline{P})}$$

by 1.19(7), hence $\lambda \in X(T)_+$ by 2.6 and

$$0 \neq \varepsilon_M((\operatorname{ind}_P^G M)_\lambda) \subset M_\lambda \cap M^{U^+ \cap L_I}.$$

c) Let $v \in M_\lambda$, $v \neq 0$ be $(U^+ \cap L_I)$–invariant. Using the left exactness of ind_P^G we may assume

$$\begin{aligned} M &= \operatorname{Dist}(P)v = \operatorname{Dist}(U)\operatorname{Dist}(T)\operatorname{Dist}(U^+ \cap L_I)v \\ &= \operatorname{Dist}(U)v = \operatorname{Dist}(U \cap L_I)v. \end{aligned}$$

Therefore $M' = \bigoplus_{\mu < \lambda} M_\mu$ is a B–submodule of M with $M = M' \oplus kv$ (as a vector space) and we have an exact sequence of B–modules

$$0 \to M' \longrightarrow M \longrightarrow \lambda \to 0,$$

hence an exact sequence of G–modules

$$\cdots \longrightarrow H^0(M) \longrightarrow H^0(\lambda) \longrightarrow H^1(M') \longrightarrow \cdots.$$

We know by 2.2 and 2.6 that $H^0(\lambda)_\lambda \neq 0$. If we can show that $H^1(M')_\lambda = 0$, then clearly $(\operatorname{ind}_P^G M)_\lambda \simeq H^0(M)_\lambda \neq 0$ (cf. 4.6.b) as claimed. Of course, it suffices to show that $H^1(\mu)_\lambda = 0$ for each weight μ of M'.

Suppose that $H^1(\mu)_\lambda \neq 0$ for such a μ. By 5.13 there exists $\alpha \in S$ with $\langle \mu + \rho, \alpha^\vee \rangle < 0$ and $H^1(\mu) \simeq \operatorname{ind}_{P(\alpha)}^G (R^1 \operatorname{ind}_B^{P(\alpha)} \mu)$. We know by 5.2 that $s_\alpha \bullet \mu$ is the largest weight of $R^1 \operatorname{ind}_B^{P(\alpha)} \mu$. As $H^1(\mu)_\lambda \neq 0$, there is a maximal weight λ' of $H^1(\mu)$ with $\lambda' \geq \lambda$. Now b) implies that λ' is a weight of $R^1 \operatorname{ind}_B^{P(\alpha)} \mu$, hence that $\lambda \leq \lambda' \leq s_\alpha \bullet \mu$. On the other hand, we have $\mu < \lambda$ by definition of M'. Now $\mu < \lambda \leq s_\alpha \bullet \mu = \mu - (\langle \mu, \alpha^\vee \rangle + 1)\alpha = s_\alpha \mu - \alpha$ implies $\lambda - \mu \in \mathbf{Z}\alpha$. All weights of $M = \operatorname{Dist}(L_I)v$ belong to $\lambda + \mathbf{Z}I$. So $\lambda - \mu \in \mathbf{Z}\alpha$ implies $\alpha \in I$. As M is an L_I–module, also $s_\alpha \mu$ is a weight of M. So we get $s_\alpha \mu \leq \lambda$ and $s_\alpha \bullet \mu < \lambda$ contradicting $\lambda \leq s_\alpha \bullet \mu$. Therefore $H^1(\mu)_\lambda = 0$ as claimed.

5.15. Proposition: *Suppose that* $\operatorname{char}(k) = p \neq 0$. *Let* $\alpha \in S$ *and* $\lambda \in X(T)$ *with* $\langle \lambda, \alpha^\vee \rangle \geq 0$.

a) *If* $\langle \lambda, \alpha^\vee \rangle = ap^n - 1$ *for some* $a, n \in \mathbf{N}$ *with* $0 < a < p$, *then*

$$H^1(s_\alpha \bullet \lambda) \neq 0 \iff \lambda \in X(T)_+$$

and

$$L(\lambda) \simeq \operatorname{soc}_G H^1(s_\alpha \bullet \lambda) \qquad \textit{in case } \lambda \in X(T)_+.$$

b) *Let* $\langle \lambda, \alpha^\vee \rangle = \sum_{j=0}^n a_j p^j$ *with* $0 \leq a_j < p$ *for all* j *and* $a_n \neq 0$. *Assume that there exists* $j < n$ *with* $a_j < p - 1$. *Then*

$$H^1(s_\alpha \bullet \lambda) \neq 0 \iff s_\alpha \bullet \lambda + a_n p^n \alpha \in X(T)_+$$

and if so, then

$$\operatorname{soc}_G H^1(s_\alpha \bullet \lambda) \simeq L(s_\alpha \bullet \lambda + a_n p^n \alpha).$$

If $\lambda \in X(T)_+$, then λ is the largest weight of $H^1(s_\alpha \bullet \lambda)$. If not, let m be minimal with $a_m < p-1$ and let $m' \geq m$ be minimal for $\mu = s_\alpha \bullet \lambda + \sum_{j=m'}^{n} a_j p^j \alpha \in X(T)_+$; then μ is the largest weight of $H^1(s_\alpha \bullet \lambda)$.

Proof: a) In this case $H^1(s_\alpha \bullet \lambda) \simeq H^0(\lambda)$ by 5.4.d. Therefore the claim follows from 2.6 and 2.4(1).

b) We apply 5.14 to $M = R^1 \operatorname{ind}_B^{P(\alpha)}(s_\alpha \bullet \lambda)$ and $P = P(\alpha)$. If $H^1(s_\alpha \bullet \lambda) \simeq \operatorname{ind}_{P(\alpha)}^G M \neq 0$ (cf. 5.13), then 5.14.a and 5.12 imply $s_\alpha \bullet \lambda + a_n p^n \alpha \in X(T)_+$ and $\operatorname{soc}_G H^1(s_\alpha \bullet \lambda) \simeq L(s_\alpha \bullet \lambda + a_n p^n \alpha)$. On the other hand, 5.12 and 5.14.c show that any dominant weight contained in $\{\lambda\} \cup \{\lambda - i(m')\alpha \mid m \leq m' < n\}$ is a weight of $\operatorname{ind}_{P(\alpha)}^G M$. So, if $s_\alpha \bullet \lambda + a_n p^n \alpha = \lambda - i(n-1)\alpha \in X(T)_+$, then $H^1(s_\alpha \bullet \lambda) \simeq \operatorname{ind}_{P(\alpha)}^G M \neq 0$.

If μ is a maximal weight of $H^1(s_\alpha \bullet \lambda)$, then μ is by 5.14.b also a weight of $R^1 \operatorname{ind}_B^{P(\alpha)}(s_\alpha \bullet \lambda)^{U_\alpha}$, hence by 5.12 either equal to λ or to one of the $\lambda - i(m')\alpha$. As

$$\lambda > \lambda - i(m)\alpha \geq \lambda - i(m+1)\alpha \geq \cdots \geq \lambda - i(n-1)\alpha$$

this leads to the claim on the largest weight of $H^1(s_\alpha \bullet \lambda)$.

Remark: If $H^1(s_\alpha \bullet \lambda) \neq 0$, then $\langle s_\alpha \bullet \lambda, \beta^\vee \rangle \geq 0$ for all $\beta \in S$, $\beta \neq \alpha$. Indeed, in the situation of b) we get

$$0 \leq \langle s_\alpha \bullet \lambda + a_n p^n \alpha, \beta^\vee \rangle \leq \langle s_\alpha \bullet \lambda, \beta^\vee \rangle$$

as $\langle \alpha, \beta^\vee \rangle \leq 0$. The argument in a) is similar.

5.16. Corollary: *Let $\mu \in X(T)$. If $H^1(\mu) \neq 0$, then $\operatorname{soc}_G H^1(\mu)$ is simple.*

Remark: This does not generalise to all H^i. For example, there is in [Jantzen 5] an example of a Weyl module $V(\lambda) \simeq H^n(w_0 \bullet \lambda)$ where $n = |R^+|$ (cf. Remark 1 in 5.11) with a socle that is not simple. In [Andersen 4], p. 58 there is even an example where some $H^i(\mu)$ is decomposable.

5.17. The description of all $H^1(\mu) \neq 0$ gets nicer if we look at it from the opposite point of view: We describe for any $\lambda \in X(T)_+$ all $\mu \in X(T)$ with $L(\lambda) \simeq \operatorname{soc}_G H^1(\mu)$.

Proposition: *Assume that* $\operatorname{char}(k) = p \neq 0$. *Let $\lambda \in X(T)_+$. Define for each $\alpha \in S$ and $n \in \mathbf{N}$ an integer $c_\lambda(n, \alpha)$ by*

$$(c_\lambda(n,\alpha) - 1)p^n < \langle \lambda + \rho, \alpha^\vee \rangle \leq c_\lambda(n,\alpha)p^n.$$

Then

$$\{\mu \in X(T) \mid L(\lambda) \simeq \operatorname{soc}_G H^1(\mu)\} = \{\lambda - c_\lambda(n,\alpha)p^n\alpha \,|\, \alpha \in S, n \in \mathbf{N}, c_\lambda(n,\alpha) < p\}.$$

Proof: Consider at first some μ with $L(\lambda) \simeq \operatorname{soc}_G H^1(\mu)$. By Remark 5.15 there is a unique $\alpha \in S$ with $\langle \mu + \rho, \alpha^\vee \rangle < 0$. Write $\langle s_\alpha \bullet \mu, \alpha^\vee \rangle = \sum_{j=0}^{n} a_j p^j$ with

$0 \leq a_j < p$ for all j and $a_n > 0$. If $a_j = p-1$ for all $j < n$, then we get from 5.15.a that $\lambda = s_\alpha \bullet \mu$, hence

$$\mu = s_\alpha \bullet \lambda = \lambda - \langle \lambda + \rho, \alpha^\vee \rangle \alpha = \lambda - (\sum_{j=0}^{n} a_j p^j + 1)\alpha = \lambda - (a_n + 1)p^n \alpha.$$

On the other hand, we have $\langle \lambda + \rho, \alpha^\vee \rangle = (a_n + 1)p^n$, hence $c_\lambda(n, \alpha) = a_n + 1 \leq p$ and $\mu = \lambda - c_\lambda(n, \alpha)p^n \alpha$. If $a_n = p - 1$, then $c_\lambda(n+1, \alpha) = 1$ and $\mu = \lambda - c_\lambda(n + 1, \alpha)p^{n+1}\alpha$.

Suppose now that $a_j < p - 1$ for some $j < n$. Then $\lambda = \mu + a_n p^n \alpha$ by 5.15.b. Furthermore,

$$\begin{aligned} \langle \lambda + \rho, \alpha^\vee \rangle &= 2a_n p^n + \langle \mu + \rho, \alpha^\vee \rangle = 2a_n p^n - \langle s_\alpha(\mu + \rho), \alpha^\vee \rangle \\ &= 2a_n p^n - \sum_{j=0}^{n} a_j p^j - 1 = (a_n - 1)p^n + \sum_{j=0}^{n-1} (p - 1 - a_j)p^j. \end{aligned}$$

As $p - 1 - a_j > 0$ for some $j < n$, we get $c_\lambda(n, \alpha) = a_n < p$, hence $\mu = \lambda - c_\lambda(n, \alpha)p^n \alpha$.

We have proved one inclusion so far. The converse requires more or less the same calculations and is left to the reader.

5.18. Corollary: *Assume that* $\operatorname{char}(k) = p \neq 0$. *Let* $\mu \in X(T)$. *The trivial* G*–module* $L(0) = k$ *is the socle of* $H^1(\mu)$ *if and only if there is a simple root* $\alpha \in S$ *and a natural number* $n \in \mathbf{N}$ *with* $\mu = -p^n \alpha$.

Proof: We have $\langle \rho, \alpha^\vee \rangle = 1$ for all $\alpha \in S$, hence $c_0(n, \alpha) = 1$ for all $n \in \mathbf{N}$. So the corollary follows immediately from 5.17.

Remark: Let $\alpha, \beta \in S$ with $\langle \alpha, \beta^\vee \rangle < 0$. If $n > 0$, then $s_\alpha \bullet (-p^n \alpha) = (p^n - 1)\alpha \notin X(T)_+$ as $\langle (p^n - 1)\alpha, \beta^\vee \rangle < 0$. So $\mu = -p^n \alpha$ is a weight with $H^1(\mu) \neq 0$, but $\mu \notin \bigcup_{l(w)=1} w \bullet X(T)_+$. This shows that the Borel-Bott-Weil theorem will not generalise to characteristic p — not even in its weak version, which says that if $\lambda \in X(T)_+$ and $H^i(w \bullet \lambda) \neq 0$, then $i = l(w)$ — as soon as R has an irreducible component of rank ≥ 2. This was first observed by Mumford for SL_3 and $p = 2$. Then Griffith determined for SL_3 and any p all μ and i with $H^i(\mu) \neq 0$. The results can be found in [Andersen 2]. (Note that we can restrict to $i = 0, 1$ in this case using Serre duality and $|R^+| = 3$. So the result follows from 2.6 and 5.15 and some combinatorial considerations.) Using additional ideas, all non-vanishing $H^i(\mu)$ for rank-2-groups were determined in [Andersen 9]. (There are some minor errors in the figure for type G_2 and the statements about the walls, cf. [Humphreys 21].)

5.19. We have by I.4.5.a for each $\lambda \in X(T)_+$ and $\mu \in X(T)$ a spectral sequence with

(1) $$E_2^{i,j} = \operatorname{Ext}_G^i(L(\lambda), H^j(\mu)) \Rightarrow \operatorname{Ext}_B^{i+j}(L(\lambda), \mu).$$

For $\mu \in X(T)_+$ we have used this sequence in 4.7.a (in a more general situation) to get isomorphisms

(2) $$\operatorname{Ext}_G^i(L(\lambda), H^0(\mu)) \simeq \operatorname{Ext}_B^i(L(\lambda), \mu) \qquad \text{for } \mu \in X(T)_+.$$

For $\mu \notin X(T)_+$ (and $\mathrm{char}(k) \neq 0$) the situation is more complicated. But we know at least that $H^0(\mu) = 0$ for such μ, hence $E_2^{i,0} = 0$. Therefore, the five term exact sequence corresponding to (1), cf. I.4.1(4), yields an isomorphism

$$\mathrm{Ext}^1_B(L(\lambda), \mu) \simeq \mathrm{Hom}_G(L(\lambda), H^1(\mu)) \qquad \text{for } \mu \notin X(T)_+. \tag{3}$$

The right hand side in (3) is completely known by 5.16/17. So we get under the assumption (and with the notations) from 5.17:

(4) *Let $\lambda \in X(T)_+$ and $\mu \in X(T)$, $\mu \notin X(T)_+$. If there are $\alpha \in S$ and $n \in \mathbf{N}$ with $\mu = \lambda - c_\lambda(n, \alpha)p^n\alpha$ and $c_\lambda(n, \alpha) < p$, then* $\mathrm{Ext}^1_B(L(\lambda), \mu) \simeq k$. *If not,then* $\mathrm{Ext}^1_B(L(\lambda), \mu) = 0$.

5.20. Proposition: *Assume that* $\mathrm{char}(k) = p \neq 0$.

a) *We have $H^1(B, -p^n\alpha) \simeq k$ for all $\alpha \in S$ and $n \in \mathbf{N}$.*

b) *For all other $\mu \in X(T)$ we have $H^1(B, \mu) = 0$.*

Proof: Part a) and Part b) for $\mu \notin X(T)_+$ are the special case $\lambda = 0$ of 5.19(4), cf. 5.18. It remains to show $H^1(B, \mu) = 0$ for all $\mu \in X(T)_+$. But $H^1(B, \mu) \neq 0$ implies by 4.10 that $\mu \in -\mathbf{N}R^+$ and $\mathrm{ht}(-\mu) \geq 1$. If we take a W–invariant scalar product $(\,,\,)$ on $X(T) \otimes_{\mathbf{Z}} \mathbf{R}$, then $\mu \in -\mathbf{N}R^+$ and $\mathrm{ht}(-\mu) \geq 1$ imply $(\mu, \rho) < 0$, whereas $\mu \in X(T)_+$ implies $(\mu, \rho) \geq 0$. So $H^1(B, \mu) = 0$ for $\mu \in X(T)_+$.

5.21. Let $I \subset S$ and set $P = P_I$. Write $H^i_I(\mu) = R^i \mathrm{ind}_B^P(\mu)$ for all i and μ. Regarded as an L_I–module, $H^i_I(\mu)$ is isomorphic to $R^i \mathrm{ind}_{B \cap L_I}^{L_I}(\mu)$, and U_I acts trivially on $H^i_I(\mu)$, cf. I.6.11. Set $L_I(\mu) = \mathrm{soc}_P H^0_I(\mu)$. So either $L_I(\mu)$ is 0 or a simple module with highest weight μ.

Let $\lambda \in X(T)_+$. We have $L_I(\lambda) \simeq \bigoplus_{\nu \in \mathbf{Z}I} L(\lambda)_{\lambda-\nu}$ by 2.11. Applying this in characteristic 0 (where $L(\lambda) = H^0(\lambda)$) we get $\dim H^0(\lambda)_{\lambda-\nu} = \dim H^0_I(\lambda)_{\lambda-\nu}$ for all $\nu \in \mathbf{Z}I$ using the independence of $\chi(\lambda) = \mathrm{ch}\, H^0(\lambda)$ of k. We have $H^0(\lambda) \simeq \mathrm{ind}_P^G H^0_I(\lambda)$ by the transitivity of induction. The evaluation map $\varepsilon : H^0(\lambda) \to H^0_I(\lambda)$ is (by 2.11(1) and 2.2(1)) injective on $\bigoplus_{\nu \in \mathbf{Z}I} H^0(\lambda)_{\lambda-\nu}$. So the equality of dimensions yields

$$H^0_I(\lambda) \simeq \bigoplus_{\nu \in \mathbf{Z}I} H^0(\lambda)_{\lambda-\nu}. \tag{1}$$

If we intersect a composition series of $H^0(\lambda)$ with $\bigoplus_{\nu \in \mathbf{Z}I} H^0(\lambda)_{\lambda-\nu}$, then we get (by (1) and 2.11) a composition series of $H^0_I(\lambda)$ with some factors equal to 0. This implies for all $\lambda, \mu \in X(T)_+$:

$$[H^0(\lambda) : L(\mu)] = [H^0_I(\lambda) : L_I(\mu)] \qquad \text{if } \lambda - \mu \in \mathbf{Z}I. \tag{2}$$

A result similar to (1) for the $V(\lambda)$ is proved in [Jantzen 1], p. 15. That it leads to results like (2) is clear and was mentioned in a similar situation in [J1], 1.18. It was applied in our situation in [Schaper], p. 65, and also appears in [Donkin 8].

CHAPTER 6

The Linkage Principle

In 6.1–6.11 we assume that p is a positive integer. From 6.12 on we assume that p is a prime and k a field of characteristic p.

Whereas in characteristic 0 each $H^0(\lambda)$ with $\lambda \in X(T)_+$ is simple, this is no longer so in prime characteristic (except in the trivial case $G = B = T$). We want to know more about the composition factors and their multiplicities in this case. As before we denote by $[V : L(\mu)]$ the multiplicity of any $L(\mu)$ with $\mu \in X(T)_+$ as a composition factor of any finite dimensional G–module V. So we ask for information about the matrix of all $[H^0(\lambda) : L(\mu)]$ with $\lambda, \mu \in X(T)_+$. The results so far show that $[H^0(\lambda) : L(\lambda)] = 1$, see 2.4, and that $[H^0(\lambda) : L(\mu)] \neq 0$ implies $\mu \leq \lambda$, see 2.2.b.

Assume that k a field of characteristic p. We want to show that we can replace "$\mu \leq \lambda$" in the last statement by "$\mu \uparrow \lambda$" where $\uparrow$ is an order relation refining $\leq$. Consider (affine) reflections of the form $x \mapsto x - (\langle x + \rho, \alpha^\vee \rangle - np)\alpha$ with $\alpha \in R$ and $n \in \mathbf{Z}$. Then by definition $\mu \uparrow \lambda$ holds (for arbitrary $\lambda, \mu \in X(T)$) if and only if either $\mu = \lambda$ or there are reflections $s_1, s_2, \dots, s_r$ of this form with

$$\lambda \geq s_1 \bullet \lambda \geq s_2 s_1 \bullet \lambda \geq \cdots \geq (s_r \dots s_2 s_1) \bullet \lambda = \mu.$$

So $\mu \uparrow \lambda$ implies $\mu \leq \lambda$, but the converse does not hold in general.

That $[H^0(\lambda) : L(\mu)] \neq 0$ should imply $\mu \uparrow \lambda$ was first conjectured in [Verma], then proved for type A_n and $p > n$ in [Jantzen 2], for $p \geq h$ in [Jantzen 3], and for arbitrary p in [Andersen 4]. There the statement is generalised from $H^0(\lambda)$ to all $H^i(w \bullet \lambda)$ with $i \in \mathbf{N}$ and $w \in W$. We follow here Andersen's approach (6.12–6.16). It also yields complete information about $[H^i(w \bullet \lambda) : L(\lambda)]$ for all w and i.

A careful analysis of Andersen's method also gives more detailed information about $[H^i(w \bullet \lambda) : L(\mu)]$ with μ not "too far away" from λ (in a sense made precise in 6.22). This was first observed in [Koppinen 1] where some special cases treated in [Jantzen 3] were generalised. Since writing the first version of this text, I have received copies of [Koppinen 6] and [Wong 3, 4] containing other generalisations of these results.

As a corollary to the above results we get the "linkage principle" (6.17): If $\mathrm{Ext}^1_G(L(\lambda), L(\mu)) \neq 0$ for some $\lambda, \mu \in X(T)_+$, then $\lambda \in W_p \bullet \mu$ where W_p is the group generated by all reflections as above. We call W_p the *affine Weyl group* of R. It is isomorphic to the affine Weyl group $W_a(R^\vee)$ as in [B3], ch. VI, §2. This chapter begins with a discussion of this group, looking at the system of alcoves and facets defined by it. Then we describe some properties of the order relation $\uparrow$, most of which are straightforward. (Proposition 6.8 and 6.11(5) were first proved in [Jantzen 3].)

6.1. Denote by $s_{\beta,r}$ for all $\beta \in R$ and $r \in \mathbf{Z}$ the affine reflection on $X(T)$ or $X(T) \otimes_{\mathbf{Z}} \mathbf{R}$ with

(1) $$s_{\beta,r}(\lambda) = \lambda - (\langle \lambda, \beta^\vee \rangle - r)\,\beta = s_\beta(\lambda) + r\,\beta$$

for all λ. Set W_p equal to the group generated by all $s_{\beta,np}$ with $\beta \in R$ and $n \in \mathbf{Z}$. We call W_p the *affine Weyl group* (associated to G and p). It is isomorphic to the affine Weyl group $W_a(R^\vee)$ as in [B3], ch. VI, §2, which is generated by all $s_{\beta,r}$ with $\beta \in R$ and $r \in \mathbf{Z}$. One easily shows (as in [B3], ch. VI, §2, prop. 1) that W_p is the semi-direct product of W and the group $p\mathbf{Z}R$ acting by translations on $X(T) \otimes_{\mathbf{Z}} \mathbf{R}$:

(2) $$W_p \;\simeq\; p\mathbf{Z}R \rtimes W.$$

(Use that $s_\beta s_{\beta,np}(\lambda) = \lambda - np\beta$ for all λ.)

We shall always consider the dot action $w \bullet \lambda = w(\lambda + \rho) - \rho$ of W_p on $X(T)$ and $X(T) \otimes_{\mathbf{Z}} \mathbf{R}$. So we regard $s_{\beta,np}$ as a reflection with respect to the hyperplane

(3) $$\{\, \lambda \in X(T) \otimes_{\mathbf{Z}} \mathbf{R} \mid \langle \lambda + \rho, \beta^\vee \rangle = np \,\}.$$

We can apply the general theory of reflection groups, as in [B3], ch. V, to the group W_p.

6.2. The reflection group W_p acting on $X(T) \otimes_{\mathbf{Z}} \mathbf{R}$ defines a system of *facets.* A facet (for W_p) is a non-empty subset of the form

(1) $$\begin{aligned} F = \{\, \lambda \in X(T) \otimes_{\mathbf{Z}} \mathbf{R} \mid \langle \lambda + \rho, \alpha^\vee \rangle &= n_\alpha p \text{ for all } \alpha \in R_0^+(F), \\ (n_\alpha - 1)p < \langle \lambda + \rho, \alpha^\vee \rangle &< n_\alpha p \text{ for all } \alpha \in R_1^+(F)\} \end{aligned}$$

for suitable integers $n_\alpha \in \mathbf{Z}$ and for a disjoint decomposition $R^+ = R_0^+(F) \dot\cup R_1^+(F)$. If so, then the closure $\overline{F}$ of F is equal to

(2) $$\begin{aligned} \overline{F} = \{\, \lambda \in X(T) \otimes_{\mathbf{Z}} \mathbf{R} \mid \langle \lambda + \rho, \alpha^\vee \rangle &= n_\alpha p \text{ for all } \alpha \in R_0^+(F), \\ (n_\alpha - 1)p \le \langle \lambda + \rho, \alpha^\vee \rangle &\le n_\alpha p \text{ for all } \alpha \in R_1^+(F)\}. \end{aligned}$$

We then call

(3) $$\begin{aligned} \widehat{F} = \{\, \lambda \in X(T) \otimes_{\mathbf{Z}} \mathbf{R} \mid \langle \lambda + \rho, \alpha^\vee \rangle &= n_\alpha p \text{ for all } \alpha \in R_0^+(F), \\ (n_\alpha - 1)p < \langle \lambda + \rho, \alpha^\vee \rangle &\le n_\alpha p \text{ for all } \alpha \in R_1^+(F)\} \end{aligned}$$

the *upper closure* of F. It is obvious that $\widehat{F} \subset \overline{F}$ and that both $\widehat{F}$ and $\overline{F}$ are unions of facets.

Any facet is an open subset in an affine subspace of $X(T) \otimes_{\mathbf{Z}} \mathbf{R}$, more precisely in $\{\lambda \mid \langle \lambda + \rho, \alpha^\vee \rangle = n_\alpha p \text{ for all } \alpha \in R_0^+(F)\}$ using the notations from above. The codimension of this subspace is equal to $\dim \sum_{\alpha \in R_0^+(F)} \mathbf{R}\alpha$.

A facet F is called an *alcove* if $R_0^+(F) = \emptyset$ (or, equivalently, if F is an open subset of $X(T) \otimes_{\mathbf{Z}} \mathbf{R}$). The alcoves (for W_p) are precisely the connected components of the complement in $X(T) \otimes_{\mathbf{Z}} \mathbf{R}$ to the union of all reflection hyperplanes, i.e., of

$$X(T) \otimes_{\mathbf{Z}} \mathbf{R} \setminus \bigcup_{\alpha \in R} \bigcup_{n \in \mathbf{Z}} \{\lambda \mid \langle \lambda + \rho, \alpha^\vee \rangle = np\}.$$

The union of the closures of the alcoves is all of $X(T)\otimes_{\mathbf{Z}}\mathbf{R}$. Any $\lambda\in X(T)\otimes_{\mathbf{Z}}\mathbf{R}$ and any facet belongs to the upper closure of exactly one alcove.

Let me quote from [B3], ch. V, §3, th. 2:

(4) *If F is an alcove for W_p, then its closure is a fundamental domain for W_p acting on $X(T)\otimes_{\mathbf{Z}}\mathbf{R}$. The group W_p permutes the alcoves simply transitively.*

As W_p stabilises $X(T)$, we get immediately:

(5) *If F is an alcove for W_p, then $\overline{F}\cap X(T)$ is a fundamental domain for W_p acting on $X(T)$.*

There is an alcove (our "standard alcove") to which we shall usually apply (5). Set

$$C=\{\,\lambda\in X(T)\otimes_{\mathbf{Z}}\mathbf{R}\mid 0<\langle\lambda+\rho,\alpha^\vee\rangle<p \text{ for all } \alpha\in R^+\,\}. \tag{6}$$

As $\{\alpha^\vee\mid\alpha\in S\}$ is linearly independent and as $\beta^\vee\in\sum_{\alpha\in S}\mathbf{N}\alpha^\vee$ for all $\beta\in R^+$, it is elementary to check that $C\neq\emptyset$; so C is indeed an alcove. Note that

$$\overline{C}\cap X(T)=\overline{C}_{\mathbf{Z}} \tag{7}$$

in the notation from 5.5 if $\operatorname{char}(k)=p$.

Suppose that $\lambda\in C\cap X(T)$. Then we have for all $\alpha\in S$

$$0<\langle\lambda+\rho,\alpha^\vee\rangle=\langle\lambda,\alpha^\vee\rangle+1$$

hence $\langle\lambda,\alpha^\vee\rangle\geq 0$ and $\lambda\in X(T)_+$. This implies $\langle\lambda,\beta^\vee\rangle\geq 0$ for all $\beta\in R^+$, hence

$$\langle\rho,\beta^\vee\rangle\leq\langle\lambda+\rho,\beta^\vee\rangle<p.$$

This shows:

$$C\cap X(T)\neq\emptyset \iff \langle\rho,\beta^\vee\rangle<p \text{ for all } \beta\in R^+ \iff 0\in C. \tag{8}$$

Set

$$h=\max\{\,\langle\rho,\beta^\vee\rangle+1\mid\beta\in R^+\,\}. \tag{9}$$

In case R is indecomposable, h is the *Coxeter number* of R, cf. [B3], ch. VI, §1, prop. 31. (Use that R and $R^\vee$ have the same Weyl group, hence the same Coxeter number.) In general, h is the maximum of the Coxeter numbers of the irreducible components of R. We can reformulate (8) as:

$$C\cap X(T)\neq\emptyset \iff p\geq h. \tag{10}$$

Note that we can replace here C by any alcove as these are all conjugate to C under W_p.

6.3. A facet F is called a *wall* if $|R_0^+(F)| = 1$, i.e., if there is a unique $\beta \in R^+$ with $\langle \lambda+\rho, \beta^\vee\rangle \in \mathbf{Z}p$ for all $\lambda \in F$. If so, then there is a unique reflection $s_F = s_{\beta,np}$ with $np = \langle \lambda+\rho, \beta^\vee\rangle$ for all $\lambda \in F$ in W_p which acts as the identity on F; one calls s_F the reflection with respect to F.

Let C' be an alcove for W_p. Denote by $\Sigma(C')$ the set of all reflections s_F where F is a wall (for W_p) with $F \subset \overline{C'}$. Then $\Sigma(C')$ generates W_p as a group, more precisely $(W_p, \Sigma(C'))$ is a Coxeter system (cf. [B3], ch. V, §3, th. 1). For any $\lambda \in \overline{C'}$ the stabiliser

$$W_p^0(\lambda) = \{\, w \in W_p \mid w \bullet \lambda = \lambda \,\}$$

is generated by

$$\Sigma^0(\lambda, C') = \{\, s \in \Sigma(C') \mid s \bullet \lambda = \lambda \,\}$$

and $(W_p^0(\lambda), \Sigma^0(\lambda, C'))$ is a Coxeter system, cf. [B3], ch. V, §3, prop. 1 and 2.

In the case of the standard alcove C from 6.2(6) we simply write $\Sigma = \Sigma(C)$ and $\Sigma^0(\lambda) = \Sigma^0(\lambda, C)$. The walls contained in $\overline{C}$ and hence also Σ can be described explicitly, cf. [B3], ch. VI, §2, prop. 5. We get that Σ consists of all s_α with $\alpha \in S$ and of all $s_{\beta,p}$ where β is the largest short root of an irreducible component of R.

Suppose for the moment that there are characters $\varpi'_\alpha \in X(T)$ for all $\alpha \in S$ with $\langle \varpi'_\alpha, \beta^\vee\rangle = \delta_{\alpha\beta}$ (the Kronecker delta) for all $\alpha, \beta \in S$. If G is semi-simple, then the ϖ'_α are just the fundamental weights (cf. 1.6), so the assumption amounts to G being simply connected in this case. In general, the restrictions of the ϖ'_α to $T \cap \mathcal{D}G$ are the fundamental weights of $\mathcal{D}G$, hence they exist if and only if $\mathcal{D}G$ is simply connected. We want to show:

(1) *Suppose that there exist ϖ'_α as above and that $p \geq h$. If R has a component of type E_8 (or F_4 or G_2) assume in addition that $p \neq 30$ (or $p \neq 12$ or $p \neq 6$ respectively). Then $F \cap X(T) \neq \emptyset$ for each wall F for W_p.*

As all alcoves are conjugate under W_p, we may assume that $F \subset \overline{C}$. Let us look first at a wall F given by $\langle \lambda+\rho, \alpha^\vee\rangle = 0$ for some $\alpha \in S$. Then $-\varpi'_\alpha \in F$ since $\langle -\varpi'_\alpha + \rho, \alpha^\vee\rangle = 0$ and $0 < \langle \rho - \varpi'_\alpha, \beta^\vee\rangle \leq \langle \rho, \beta^\vee\rangle < h \leq p$ for all $\beta \in R^+$, $\beta \neq \alpha$. (Note that $\langle \rho - \sum_{\gamma\in S} \varpi'_\gamma, \beta^\vee\rangle = 0$ for all $\beta \in R^+$ as $\langle \rho, \gamma^\vee\rangle = 1$ for all $\gamma \in S$. Note also that we can write $\beta^\vee = \sum_{\gamma\in S} m_\gamma \gamma^\vee$ with $m_\gamma \in \mathbf{N}$ and that $\beta \neq \alpha$ implies $m_\gamma > 0$ for some $\gamma \neq \alpha$.)

Now let β be the largest short root of some irreducible component of R. Write $\beta^\vee = \sum_{\alpha\in S} m_\alpha \alpha^\vee$. If $\beta' \in R^+$ belongs to the same component as β and if $\beta'^\vee = \sum_{\alpha\in S} m'_\alpha \alpha^\vee$, then $m'_\alpha \leq m_\alpha$ for all $\alpha \in S$. Suppose at first that there is some $\alpha \in S$ with $m_\alpha = 1$, i.e., with $\langle \varpi'_\alpha, \beta^\vee\rangle = 1$. (This is always satisfied if the component of β is not of type E_8, F_4, or G_2, cf. the tables in [B3], ch. VI.) Then $\lambda = (p - \langle \rho, \beta^\vee\rangle)\varpi'_\alpha$ will work as $\langle \lambda+\rho, \beta^\vee\rangle = p$, as $\langle \lambda+\rho, \gamma^\vee\rangle \geq \langle \rho, \gamma^\vee\rangle > 0$ for all $\gamma \in R^+$, as $\langle \lambda+\rho, \gamma^\vee\rangle = \langle \rho, \gamma^\vee\rangle < p$ for all $\gamma \in R^+$ not in the component of β, and $\langle \lambda+\rho, \gamma^\vee\rangle = \langle \lambda, \gamma^\vee\rangle + \langle \rho, \gamma^\vee\rangle < \langle \lambda, \gamma^\vee\rangle + \langle \rho, \beta^\vee\rangle \leq \langle \lambda+\rho, \beta^\vee\rangle = p$ for all $\gamma \in R^+$, $\gamma \neq \beta$ in the component of β.

If β belongs to a component of type E_8, F_4, or G_2, then $\langle \rho, \beta^\vee\rangle$ is equal to 29, 11, or 5 respectively, cf. the tables in [B3], ch. VI. There are $\alpha, \gamma \in S$ with $\langle \varpi'_\alpha, \beta^\vee\rangle = 2$ and $\langle \varpi'_\gamma, \beta^\vee\rangle = 3$. If p is odd, then now take $\lambda = \frac{1}{2}(p - \langle \rho, \beta^\vee\rangle)\varpi'_\alpha$, and if p is even, then we take $\lambda = \frac{1}{2}(p - \langle \rho, \beta^\vee\rangle - 3)\varpi'_\alpha + \varpi'_\gamma$. One argues as above to check that $\lambda \in F$.

6.4. Let us introduce an order relation $\uparrow$ on $X(T)$. We want $\lambda \uparrow \mu$ to hold if and only if there are $\mu_1, \mu_2, \dots, \mu_r \in X(T)$ and reflections $s_1, s_2, \dots, s_{r+1} \in W_p$ with

$$(1) \quad \lambda \le s_1 \bullet \lambda = \mu_1 \le s_2 \bullet \mu_1 = \mu_2 \le \cdots \le s_r \bullet \mu_{r-1} = \mu_r \le s_{r+1} \bullet \mu_r = \mu$$

or if $\mu = \lambda$. We have obviously

$$(2) \qquad \lambda \uparrow \mu \;\Rightarrow\; \lambda \le \mu \text{ and } \lambda \in W_p \bullet \mu.$$

If we write $s_i = s_{\beta_i, n_i p}$ with $\beta_i \in R^+$ and $n_i \in \mathbf{Z}$, then the condition $\mu_i \le s_{i+1} \bullet \mu_i$ in (1) amounts to $\langle \mu_i + \rho, \beta_{i+1}^\vee \rangle \le n_{i+1} p$ for $0 \le i \le r$ (setting $\mu_0 = \lambda$).

If $\lambda \in X(T)$ and $\alpha \in R^+$, then there are unique $n_\alpha, d_\alpha \in \mathbf{Z}$ with $\langle \lambda + \rho, \alpha^\vee \rangle = n_\alpha p + d_\alpha$ and $0 < d_\alpha \le p$. Then $s_{\alpha, n_\alpha p} \bullet \lambda = \lambda - d_\alpha \alpha \uparrow \lambda$. If $d_\alpha < p$, then $s_{\alpha,(n_\alpha - 1)p} \bullet (\lambda - d_\alpha \alpha) = \lambda - p\alpha \uparrow \lambda - d_\alpha \alpha$. We get in each case

$$(3) \qquad \lambda - p\alpha \uparrow \lambda \qquad \text{for all } \lambda \in X(T) \text{ and } \alpha \in R^+.$$

Another elementary property of $\uparrow$ is the following: We have for all $\lambda, \mu, \nu \in X(T)$

$$(4) \qquad \lambda \uparrow \mu \iff \lambda + p\nu \uparrow \mu + p\nu.$$

(If $\lambda \uparrow \mu$ and if we have $s_i = s_{\beta_i, n_i p}$ as in (1), then the $s_{\beta_i, (n_i + \langle \nu, \beta_i^\vee \rangle)p}$ will do for $\lambda + p\nu$ and $\mu + p\nu$.)

We have, furthermore,

(5) *If $\lambda \in X(T)$ with $\langle \lambda + \rho, \alpha^\vee \rangle \ge 0$ for all $\alpha \in R^+$, then $w \bullet \lambda \uparrow \lambda$ for all $w \in W$.*

This is proved using induction on $l(w)$. If $l(w) = 0$, then $w = 1$ and the claim is obvious. If $l(w) > 0$, then there exists $\alpha \in S$ with $l(s_\alpha w) = l(w) - 1$, i.e., with $w^{-1}\alpha < 0$. Then $\langle w(\lambda + \rho), \alpha^\vee \rangle = \langle \lambda + \rho, w^{-1}(\alpha)^\vee \rangle < 0$, hence $w \bullet \lambda \uparrow (s_\alpha w) \bullet \lambda$. Now use induction to get $(s_\alpha w) \bullet \lambda \uparrow \lambda$.

We have for all $w \in W$ and $\alpha \in R$ that $w s_{\alpha, rp} w^{-1} = s_{w\alpha, rp}$ for all r. Applying this to $w = w_0$ one gets for all $\lambda, \mu \in X(T)$

$$(6) \qquad \lambda \uparrow \mu \iff w_0 \bullet \mu \uparrow w_0 \bullet \lambda.$$

6.5. We can also define an order relation $\uparrow$ on the set of alcoves for W_p. Let C_1, C_2 be two alcoves. For any $\lambda_1 \in C_1 \cap X(T)$ there is a unique $\lambda_2 \in C_2 \cap W_p \bullet \lambda_1$ as C_2 is a fundamental domain for W_p. Then we want the following to hold:

$$(1) \qquad \lambda_1 \uparrow \lambda_2 \iff C_1 \uparrow C_2.$$

It is elementary to show that the left hand side in (1) does not depend on the choice of $\lambda_1 \in C_1 \cap X(T)$. So we could take (1) as the definition whenever $C_1 \cap X(T) \ne \emptyset$. As this is not true for small p by 6.2(10), we proceed differently.

If $\alpha \in R^+$ and $n \in \mathbf{Z}$, then either $\langle \lambda + \rho, \alpha^\vee \rangle < np$ for all $\lambda \in C_1$, or $\langle \lambda + \rho, \alpha^\vee \rangle > np$ for all $\lambda \in C_1$. In the first case set $C_1 \uparrow s_{\alpha,np} \bullet C_1$, and in the second one set $s_{\alpha,np} \bullet C_1 \uparrow C_1$. This is a definition of $C_1 \uparrow s \bullet C_1$ for any reflection $s \in W_p$ and any alcove C_1. Now we set $C_1 \uparrow C_2$ if and only if there are reflections $s_1, s_2, \dots, s_{r+1}$ in W_p with

$$C_1 \uparrow s_1 \bullet C_1 \uparrow s_2 s_1 \bullet C_1 \uparrow \cdots \uparrow s_{r+1} \dots s_2 s_1 \bullet C_1 = C_2$$

or if $C_1 = C_2$. One checks now easily that $\uparrow$ is an order relation and that (1) holds. (For the anti-symmetry consider $\lambda_1 \in C_1$ and $\lambda_2 \in C_2 \cap W_p \bullet \lambda_1$. If $C_1 \uparrow C_2$ and $C_1 \neq C_2$, then $\lambda_2 - \lambda_1 = \sum_{\alpha \in S} m_\alpha \alpha$ with $m_\alpha \in \mathbf{R}$, $m_\alpha \geq 0$ for all α, and $m_\alpha > 0$ for at least one α.)

For $\lambda_1 \in \overline{C_1} \cap X(T)$ only a weaker version of (1) holds: If λ_2 is the unique element in $\overline{C_2} \cap W_p \bullet \lambda_1$, then

(2) $$C_1 \uparrow C_2 \Rightarrow \lambda_1 \uparrow \lambda_2.$$

The same proofs as in 6.4 show for all alcoves C_1, C_2

(3) $$C_1 - p\alpha \uparrow C_1 \qquad \text{for all } \alpha \in R^+,$$

and for each $\nu \in X(T)$

(4) $$C_1 \uparrow C_2 \iff C_1 + p\nu \uparrow C_2 + p\nu.$$

Let us call an alcove C_1 *dominant* if $\langle \lambda + \rho, \alpha^\vee \rangle > 0$ for all $\alpha \in R^+$ and $\lambda \in C_1$. Then:

(5) *If C_1 is dominant, then $w \bullet C_1 \uparrow C_1$ for all $w \in W$.*

6.6. For any alcove C_1 and any $\alpha \in R^+$ there is a unique $n_\alpha \in \mathbf{Z}$ with $n_\alpha p < \langle \lambda + \rho, \alpha^\vee \rangle < (n_\alpha + 1)p$ for all $\lambda \in C_1$. If so, then set

(1) $$d(C_1) = \sum_{\alpha \in R^+} n_\alpha.$$

If C_1 is dominant (i.e., if $n_\alpha \geq 0$ for all $\alpha \in R^+$), then $d(C_1)$ is exactly the number of reflection hyperplanes for W_p such that C_1 and the standard alcove C lie on different sides of this hyperplane. (The equations of these hyperplanes are $\langle \lambda + \rho, \alpha^\vee \rangle = ip$ with $\alpha \in R^+$ and $1 \leq i \leq n_\alpha$.)

Lemma: *Let C_1 be an alcove for W_p and let s be a reflection in W_p. We have then $s \bullet C_1 \uparrow C_1 \iff d(s \bullet C_1) < d(C_1)$.*

Proof: Let n_α $(\alpha \in R^+)$ be as above and let m_α $(\alpha \in R^+)$ be the corresponding integers for $s \bullet C_1$. Pick $\lambda \in C_1$. For each $\alpha \in R^+$ there is $d_\alpha \in \mathbf{R}$ with $0 < d_\alpha < p$ and $\langle \lambda + \rho, \alpha^\vee \rangle = n_\alpha p + d_\alpha$. There are $\beta \in R^+$ and $r \in \mathbf{Z}$ with $s = s_{\beta, rp}$.

Suppose that $s \bullet C_1 \uparrow C_1$, i.e., that $r \leq n_\beta$. We have $s \bullet \lambda = \lambda - ((n_\beta - r)p + d_\beta)\beta$, hence $\langle s \bullet \lambda + \rho, \beta^\vee \rangle = (2r - n_\beta)p - d_\beta < n_\beta p$, hence $m_\beta < n_\beta$.

It is obvious for each $\alpha \in R^+$ that $m_\alpha \leq n_\alpha$ if $\langle \beta, \alpha^\vee \rangle \geq 0$, and $m_\alpha \geq n_\alpha$ if $\langle \beta, \alpha^\vee \rangle \leq 0$. Consider $\alpha \in R^+$ with $\langle \beta, \alpha^\vee \rangle < 0$. Then also $\langle \alpha, \beta^\vee \rangle < 0$, hence $\gamma = s_\beta(\alpha) = \alpha - \langle \alpha, \beta^\vee \rangle \beta \in R^+$ with $\gamma \neq \alpha$ and $\langle \beta, \gamma^\vee \rangle = -\langle \beta, \alpha^\vee \rangle > 0$. Furthermore, we have $\gamma^\vee = \alpha^\vee - \langle \beta, \alpha^\vee \rangle \beta^\vee$, hence

$$n_\gamma p + d_\gamma = \langle \lambda + \rho, \gamma^\vee \rangle = (n_\alpha - \langle \beta, \alpha^\vee \rangle n_\beta)p + d_\alpha - \langle \beta, \alpha^\vee \rangle d_\beta.$$

A simple calculation shows now

$$\langle s \bullet \lambda + \rho, \alpha^\vee \rangle = (n_\gamma + r\langle \beta, \alpha^\vee \rangle)p + d_\gamma$$

and

$$\langle s \bullet \lambda + \rho, \gamma^\vee \rangle = (n_\alpha - r\langle \beta, \alpha^\vee \rangle)p + d_\alpha,$$

hence $m_\alpha + m_\gamma = n_\alpha + n_\gamma$.

So the contributions to $d(C_1) - d(s \bullet C_1)$ from all $\alpha \in R^+$ with $\langle \beta, \alpha^\vee \rangle < 0$ cancel with the contributions from all $\gamma \in R^+$ with $\langle \beta, \gamma^\vee \rangle > 0$ and $s_\beta(\gamma) \in R^+$. As any other $\gamma \in R^+$ with $\langle \beta, \gamma^\vee \rangle \geq 0$ contributes something non-negative and as $\gamma = \beta$ contributes something positive, we get $d(C_1) > d(s \bullet C_1)$.

If $s \bullet C_1 \uparrow C_1$ does not hold, then $C_1 \uparrow s \bullet C_1$, hence $d(C_1) < d(s \bullet C_1)$ by the part already proved and by $C_1 = s \bullet (s \bullet C_1)$; the claim follows.

Remark: Let $\lambda \in X(T)$. Define $n_\alpha, d_\alpha \in \mathbf{Z}$ by $\langle \lambda + \rho, \alpha^\vee \rangle = n_\alpha p + d_\alpha$ and $0 < d_\alpha \leq p$ for all $\alpha \in R^+$. Set $d(\lambda) = \sum_{\alpha \in R^+} n_\alpha$. The same proof shows for any reflection $s \in W_p$ with $s \bullet \lambda \neq \lambda$:

(2) $$s \bullet \lambda \uparrow \lambda \iff d(s \bullet \lambda) < d(\lambda).$$

If C_1 is the unique alcove with $\lambda \in \widehat{C}_1$, then $d(\lambda) = d(C_1)$. One can also deduce (2) from the lemma by observing: If $s \bullet \lambda \uparrow \lambda$, then $d(s \bullet \lambda) \leq d(s \bullet C_1)$ as $s \bullet \lambda \in \overline{s \bullet C_1}$.

6.7. Lemma: *Let C_1 be an alcove for W_p and F a wall with $F \subset \overline{C}_1$ and $C_1 \uparrow s_F \bullet C_1$. Let $s_1, s_2, \dots, s_r$ be reflections in W_p and set $w_i = s_i s_{i-1} \dots s_1$ for $1 \leq i \leq r$. Suppose that $w_r \bullet C_1 \uparrow w_r s_F \bullet C_1$ and*

(1) $$w_r \bullet C_1 \uparrow w_{r-1} \bullet C_1 \uparrow \cdots \uparrow w_1 \bullet C_1 \uparrow C_1.$$

a) *We have*

(2) $$w_r s_F \bullet C_1 \uparrow w_{r-1} s_F \bullet C_1 \uparrow \cdots \uparrow w_1 s_F \bullet C_1 \uparrow s_F \bullet C_1$$

or there is some i $(1 \leq i \leq r)$ with

(3) $$w_r s_F \bullet C_1 \uparrow w_{r-1} s_F \bullet C_1 \uparrow \cdots \uparrow w_i s_F \bullet C_1 = w_{i-1} \bullet C_1.$$

b) *Suppose $d(w_j \bullet C_1) = d(C_1) - j$ for all j, $1 \leq j \leq r$. Then $d(w_j s_F \bullet C_1) = d(s_F \bullet C_1) - j$ for all j in case (2) holds, resp. for all $j \geq i$ in case (3) holds. If $w_j \bullet C_1$ is dominant for some j in case (2) holds, resp. for some $j \geq i$ in case (3) holds, then $w_j s_F \bullet C_1$ is dominant.*

Proof: We have $w_j s_F \bullet C_1 = w_j s_F w_j^{-1} \bullet (w_j \bullet C_1) = s_{F'} \bullet (w_j \bullet C_1)$ where $F' = w_j \bullet F$ is a wall contained in $\overline{w_j \bullet C_1}$. Therefore the two alcoves $w_j \bullet C_1$ and $w_j s_F \bullet C_1$ are separated only by one reflection hyperplane (with reflection $s_{F'}$). For any reflection $s \neq s_{F'}$ in W_p the alcoves $w_j \bullet C_1$ and $w_j s_F \bullet C_1$ lie on the same side of the reflection hyperplane; therefore $w_j \bullet C_1 \uparrow s \bullet (w_j \bullet C_1)$ holds if and only if $w_j s_F \bullet C_1 \uparrow s \bullet (w_j s_F \bullet C_1)$ holds.

a) If (2) does not hold, then choose $i \geq 1$ maximal such that $w_i s_F \bullet C_1 \uparrow w_{i-1} s_F \bullet C_1$ does not hold (where $w_0 = 1$ for the moment). As $w_{i-1} s_F \bullet C_1 = s_i w_i s_F \bullet C_1 = s_i \bullet (w_i s_F \bullet C_1)$, as $w_i \bullet C_1 \uparrow s_i \bullet (w_i \bullet C_1)$ holds and $w_i s_F \bullet C_1 \uparrow s_i \bullet (w_i s_F \bullet C_1)$ does not, the remarks above imply $s_i = w_i s_F w_i^{-1}$, hence $w_i s_F \bullet C_1 = s_i w_i \bullet C_1 = w_{i-1} \bullet C_1$. In this way we get (3).

b) Because of 6.6 we have in the case of (2)

$$d(C_1) - r = d(w_r \bullet C_1) < d(w_r s_F \bullet C_1) < d(w_{r-1} s_F \bullet C_1) < \cdots < d(s_F \bullet C_1)$$

and

$$d(s_F \bullet C_1) = d(C_1) + 1.$$

(This equality holds since C_1 and $s_F \bullet C_1$ are separated only by one reflection hyperplane.) This chain of inequalities can only be satisfied if $d(w_j s_F \bullet C_1) = d(C_1)+1-j = d(s_F \bullet C_1) - j$ for all j. The arguments in the case of (3) are similar.

It follows that $d(w_j \bullet C_1) = d(w_j s_F \bullet C_1) - 1$ for all j resp. for all $j \geq i$, hence

$$w_j \bullet C_1 \uparrow w_j s_F \bullet C_1 = (w_j s_F w_j^{-1}) \bullet (w_j \bullet C_1).$$

Suppose now that $w_j \bullet C_1$ is dominant. The hyperplane containing $w_j \bullet F$ cannot have an equation $\langle \lambda + \rho, \alpha^\vee \rangle = 0$ with $\alpha \in R$ as otherwise $w_j s_F w_j^{-1} = s_\alpha \in W$ and $w_j s_F \bullet C_1 \uparrow w_j \bullet C_1$ by 6.5(5). As this was the only reflection hyperplane separating $w_j \bullet C_1$ and $w_j s_F \bullet C_1$, these two alcoves lie on the same side of each hyperplane $\langle \lambda + \rho, \alpha^\vee \rangle = 0$ with $\alpha \in R$; therefore also $w_j s_F \bullet C_1$ is dominant.

6.8. Proposition: *Let C_1 be a dominant alcove for W_p. Let $\alpha \in R^+$ and $n \in \mathbf{N}$ with $0 < np < \langle \lambda + \rho, \alpha^\vee \rangle$ for all $\lambda \in C_1$. There is a unique $w \in W$ with $ws_{\alpha,np} \bullet C_1$ dominant, and there are dominant alcoves C_i' ($0 \leq i \leq r$ for some r) with $d(C_i') = d(C_1) - i$ and*

(1) $$ws_{\alpha,np} \bullet C_1 = C_r' \uparrow C_{r-1}' \uparrow \cdots \uparrow C_1' \uparrow C_0' = C_1.$$

Proof: The existence and uniqueness of w is clear as there is for each $\lambda \in C_1$ a unique $w \in W$ with $\langle ws_{\alpha,np} \bullet \lambda + \rho, \beta^\vee \rangle > 0$ for all $\beta \in S$.

Let us prove the existence of the C_i' using induction on $d(C_1)$. There has to be a wall $F \subset \overline{C}_1$ such that the supporting hyperplane of F separates C_1 and the standard alcove C. Write the equation of this hyperplane in the form $\langle \lambda + \rho, \beta^\vee \rangle = mp$ with $\beta \in R^+$ and $m \in \mathbf{Z}$. As C_1 is dominant, we have $m \geq 0$, as the hyperplane separates C_1 and C, we get $m > 0$ and $\langle \lambda + \rho, \beta^\vee \rangle > mp$ for all $\lambda \in C_1$. It follows that $s_F \bullet C_1 \uparrow C_1$ and $d(s_F \bullet C_1) = d(C_1) - 1$; furthermore $s_F \bullet C_1$ is dominant.

If $s_F = s_{\alpha,np}$, then $ws_{\alpha,np} \bullet C_1 = s_F \bullet C_1$ and (1) is satisfied with $r = 1$. So let us assume that $s_F \neq s_{\alpha,np}$. Then C_1 and $s_F \bullet C_1$ lie on the same side of the hyperplane $\langle \lambda + \rho, \alpha^\vee \rangle = np$. We can now apply induction to $s_F \bullet C_1$ and $s_{\alpha,np}$: There is $w' \in W$ with $w' s_{\alpha,np} \bullet (s_F \bullet C_1)$ dominant and there is a chain as in (1) from $w' s_{\alpha,np} s_F \bullet C_1$ to $s_F \bullet C_1$.

Suppose at first that $w' = w$. The alcoves $ws_{\alpha,np} \bullet C_1$ and $ws_{\alpha,np} s_F \bullet C_1$ are separated by one reflection hyperplane only. We have either $ws_{\alpha,np} \bullet C_1 \uparrow ws_{\alpha,np} s_F \bullet C_1$ and $d(ws_{\alpha,np} \bullet C_1) = d(ws_{\alpha,np} s_F \bullet C_1) - 1$, or $ws_{\alpha,np} s_F \bullet C_1 \uparrow ws_{\alpha,np} \bullet C_1$ and $d(ws_{\alpha,np} \bullet C_1) = d(ws_{\alpha,np} s_F \bullet C_1) + 1$. In the first case we prolong a chain as in (1) from $w' s_{\alpha,np} s_F \bullet C_1$ to $s_F \bullet C_1$ by adding $ws_{\alpha,np} \bullet C_1$ and C_1 at its two ends and get a chain as in (1). In the second case we apply Lemma 6.7 to $s_F \bullet C_1$ and $ws_{\alpha,np}$ instead of C_1 and w_r; this immediately yields (1).

Suppose now that $w' \neq w$. Then $w' s_{\alpha,np} \bullet C_1$ is not dominant, hence there is $\gamma \in S$ with $\langle w' s_{\alpha,np}(\lambda + \rho), \gamma^\vee \rangle < 0$ for all $\lambda \in C_1$. As $w' s_{\alpha,np} \bullet C_1$ and $w' s_{\alpha,np} s_F \bullet C_1$ lie on different sides of the hyperplane $\langle \lambda + \rho, \gamma^\vee \rangle = 0$, we get $s_\gamma w' s_{\alpha,np} = w' s_{\alpha,np} s_F$, hence $w = s_\gamma w'$ and $ws_{\alpha,np} \bullet C_1 = w' s_{\alpha,np} s_F \bullet C_1$. Therefore we get a chain from $ws_{\alpha,np} \bullet C_1$ to C_1 by adding C_1 to a chain from $w' s_{\alpha,np} s_F \bullet C_1$ to $s_F \bullet C_1$.

Remark: Note that $ws_{\alpha,np} \bullet C_1 \neq C_1$ as $s_{\alpha,np} \notin W$.

6.9. Corollary: *Let $\lambda \in X(T)$ with $\langle \lambda+\rho, \beta^\vee \rangle \geq 0$ for all $\beta \in S$. Let $\alpha \in R^+$ and $n \in \mathbf{N}$ with $0 < np < \langle \lambda+\rho, \alpha^\vee \rangle$. Let $w \in W$ with $\langle ws_{\alpha,np}(\lambda+\rho), \beta^\vee \rangle \geq 0$ for all $\beta \in S$. Then $ws_{\alpha,np} \bullet \lambda \uparrow \lambda$ and $ws_{\alpha,np} \bullet \lambda < \lambda$.*

Proof: There is a dominant alcove C_1 with $\lambda \in \overline{C_1}$. Then $np < \langle \mu+\rho, \alpha^\vee \rangle$ for all $\mu \in C_1$ and we can apply 6.8. If $w's_{\alpha,np} \bullet C_1$ is dominant for some $w' \in W$, then $\langle w's_{\alpha,np}(\lambda+\rho), \beta^\vee \rangle \geq 0$ for all $\beta \in S$, hence $w's_{\alpha,np} \bullet \lambda = ws_{\alpha,np} \bullet \lambda$. Now $w's_{\alpha,np} \bullet C_1 \uparrow C_1$ by 6.8, hence $w's_{\alpha,np} \bullet \lambda \uparrow \lambda$ by 6.5(2).

If $ws_{\alpha,np} \bullet \lambda = \lambda$, then

$$(\lambda+\rho, \lambda+\rho) = (w^{-1}(\lambda+\rho), w^{-1}(\lambda+\rho)) = (s_{\alpha,np}(\lambda+\rho), s_{\alpha,np}(\lambda+\rho)).$$

Now $s_{\alpha,np}(\lambda+\rho) = \lambda+\rho - m\alpha$ with $m = \langle \lambda+\rho, \alpha^\vee \rangle - np > 0$ yields

$$\begin{aligned}(s_{\alpha,np}(\lambda+\rho), s_{\alpha,np}(\lambda+\rho)) &= (\lambda+\rho-m\alpha, \lambda+\rho-m\alpha)\\ &= (\lambda+\rho, \lambda+\rho) - 2m(\lambda+\rho, \alpha) + m^2(\alpha,\alpha)\\ &= (\lambda+\rho, \lambda+\rho) - m(\alpha,\alpha)(\langle \lambda+\rho, \alpha^\vee \rangle - m)\\ &< (\lambda+\rho, \lambda+\rho),\end{aligned}$$

a contradiction. (We have used here that $\langle \lambda+\rho, \alpha^\vee \rangle = 2(\lambda+\rho,\alpha)/(\alpha,\alpha)$.) So $ws_{\alpha,np} \bullet \lambda \uparrow \lambda$ implies $ws_{\alpha,np} \bullet \lambda < \lambda$.

6.10. Corollary: *Let C', C'' be alcoves for W_p with $C'' \uparrow C'$. Then there are alcoves C_i $(0 \leq i \leq r = d(C') - d(C''))$ with $d(C_i) = d(C') - i$ for all i and*

$$C'' = C_r \uparrow C_{r-1} \uparrow \cdots \uparrow C_1 \uparrow C_0 = C'.$$

Proof: We may assume that $C'' = s \bullet C'$ for some reflection s in W_p. If we translate our alcoves by $p\nu$ for some $\nu \in X(T)$, then $d(C') - d(C'')$ and the relation $C'' \uparrow C'$ do not change. (We have $d(C'+p\nu) = d(C') + \sum_{\alpha \in R^+} \langle \nu, \alpha^\vee \rangle$.) We may therefore assume that C' and C'' are dominant. Then we get the chain from 6.8(1).

6.11. Let F be a facet. We shall use the notation from 6.2(1). Then $R_0(F) = R_0^+(F) \cup (-R_0^+(F))$ is a root system in its own right; its Weyl group can be identified with the group $W_p^0(F)$ generated by all $s_{\alpha,n_\alpha p}$ with $\alpha \in R_0^+(F)$. We have $W_p^0(F) = W_p^0(\lambda)$ for all $\lambda \in F$, cf. 6.3.

There is a unique alcove $C^- = C^-(F)$ with $F \subset \widehat{C^-}$. It is given by

(1) $$C^- = \{ \lambda \in X(T) \otimes_{\mathbf{Z}} \mathbf{R} \mid (n_\alpha - 1)p < \langle \lambda+\rho, \alpha^\vee \rangle < n_\alpha p \text{ for all } \alpha \in R^+. \}.$$

The group $W_p^0(F)$ permutes simply transitively the alcoves containing F in its closure:

(2) $$\{ C' \mid F \subset \overline{C'} \} = \{ w \bullet C^- \mid w \in W_p^0(F) \}.$$

At the other extreme, there is an alcove $C^+ = C^+(F)$ with $F \subset \overline{C^+}$ and $\langle \lambda+\rho, \alpha^\vee \rangle > n_\alpha p$ for all $\alpha \in R_0^+(F)$ and $\lambda \in C^+$. (It is the alcove containing F in its upper closure, if we work with $-R^+$ instead of R^+ as the positive system and if we forget the shift by ρ.) One proves by an induction argument similar to that in 6.4(5) using (2), cf. [Jantzen 3], Lemma 6:

(3) *If C' is an alcove with $F \subset \overline{C'}$, then $C^- \uparrow C' \uparrow C^+$.*

We shall use this to show:

(4) *Let $\lambda \in F \cap X(T)$ and $w \in W_p$. Then $\lambda \uparrow w \bullet \lambda \iff C^- \uparrow w \bullet C^-$.*

We know already one direction, cf. 6.5(2). Suppose now that $\lambda \uparrow w \bullet \lambda$. There are reflections $s_1, s_2, \dots, s_r$ in W_p with

$$w \bullet \lambda > s_1 w \bullet \lambda > s_2 s_1 w \bullet \lambda > \cdots > s_r \dots s_2 s_1 w \bullet \lambda = \lambda.$$

We get, as $w'w \bullet \lambda \in \overline{w'w \bullet C^-}$ for all $w' \in W_p$ and as $sw'w \bullet \lambda < w'w \bullet \lambda$ implies $sw'w \bullet C^- \uparrow w'w \bullet C^-$ for any reflection s in W_p,

$$s_r \dots s_2 s_1 w \bullet C^- \uparrow s_{r-1} \dots s_2 s_1 w \bullet C^- \uparrow \cdots \uparrow s_1 w \bullet C^- \uparrow w \bullet C^-.$$

Furthermore, $\lambda \in \overline{s_r \dots s_2 s_1 w \bullet C^-}$ implies $C^- \uparrow s_r \dots s_2 s_1 w \bullet C^-$ by (3), hence $C^- \uparrow w \bullet C^-$.

We conclude by proving the converse to (3), i.e.,

$$\{ C' \mid F \subset \overline{C'} \} = \{ C' \mid C^- \uparrow C' \uparrow C^+ \}. \tag{5}$$

For the purpose of the proof we change for the moment the definition of $\leq$ and say that $\lambda \leq \mu$ holds for $\lambda, \mu \in X(T) \otimes_{\mathbf{Z}} \mathbf{R}$ if $\mu - \lambda = \sum_{\alpha \in S} m_\alpha \alpha$ with all $m_\alpha \in \mathbf{R}$, $m_\alpha \geq 0$.

Suppose now that $C^- \uparrow C' \uparrow C^+$. Then there are $w_1 = 1, w_2, \dots, w_r \in W_p$ and reflections s_i in W_p such that $w_{i+1} = s_i w_i$ and (with a suitable j)

$$C^- = w_1 \bullet C^- \uparrow w_2 \bullet C^- \uparrow \cdots \uparrow w_j \bullet C^- = C' \uparrow \cdots \uparrow w_r \bullet C^- = C^+.$$

Note that $w_r \in W_p^0(F)$ since $F \subset \overline{C^+}$, see 6.11(2). So we get for any $\lambda \in F$

$$\lambda = w_1 \bullet \lambda \leq w_2 \bullet \lambda \leq \cdots \leq w_j \bullet \lambda \leq \cdots \leq w_r \bullet \lambda = \lambda,$$

hence $w_j \bullet \lambda = \lambda$ and $\lambda \in \overline{w_j \bullet C^-} = \overline{C'}$, hence $F \subset \overline{C'}$ by 6.11(2).

6.12. Lemma: *Let $\alpha \in S$ and $\lambda \in X(T)$ with $\langle \lambda + \rho, \alpha^\vee \rangle \geq 0$. Then there are finite dimensional B–modules $N_0^\alpha(\lambda)$, $N_1^\alpha(\lambda)$, $N_2^\alpha(\lambda)$ with*

$$\operatorname{ch} N_1^\alpha(\lambda) = \operatorname{ch} N_2^\alpha(\lambda) = \sum_n e(s_\alpha \bullet \lambda + np\alpha) \tag{1}$$

where the sum is over all $n \in \mathbf{N}$ with $0 < np < \langle \lambda + \rho, \alpha^\vee \rangle$, such that there are long exact sequences of G–modules

$$\cdots \to H^i(s_\alpha \bullet \lambda) \to H^{i-1}(\lambda) \to H^i(N_0^\alpha(\lambda)) \to H^{i+1}(s_\alpha \bullet \lambda) \to \cdots \tag{2}$$

and

$$(3) \qquad \cdots \to H^i(N_1^\alpha(\lambda)) \to H^i(N_0^\alpha(\lambda)) \to H^{i-1}(N_2^\alpha(\lambda)) \to H^{i+1}(N_1^\alpha(\lambda)) \to \cdots.$$

Proof: Set $r = \langle \lambda + \rho, \alpha^\vee \rangle$. In case $r \leq 1$ we set $N_i^\alpha(\lambda) = 0$ for $i = 0, 1, 2$. This is certainly compatible with (1) and (3). If $r = 0$, then $\langle \lambda, \alpha^\vee \rangle = -1$ and $s_\alpha \bullet \lambda = \lambda$, hence $H^\bullet(\lambda) = H^\bullet(s_\alpha \bullet \lambda) = 0$ by 5.4.a; so (2) is trivially exact. If $r = 1$, then we get from 5.4.d isomorphisms $H^i(\lambda) \simeq H^{i+1}(s_\alpha \bullet \lambda)$, hence (2). Therefore we assume from now on that $r \geq 2$.

Let us suppose for the time being that $\rho \in X(T)$. We shall write $H_\alpha^0(M) = \operatorname{ind}_B^{P(\alpha)} M$ for any B–module M (where $P(\alpha) = P_{\{\alpha\}}$ as in 5.1). We have described $H_\alpha^0(\lambda + \rho)$ in 5.2.c: There is a basis $(v_i)_{0 \leq i \leq r}$ such that T acts on v_i through $\lambda + \rho - i\alpha$, we know how U_α, $U_{-\alpha}$ act and that each $U_{-\beta}$ with $\beta \in R^+$, $\beta \neq \alpha$ acts trivially.

It is then clear that $H_\alpha^0(\lambda + \rho)^- = \sum_{i=1}^r k v_i$ and kv_r are B–submodules of $H_\alpha^0(\lambda + \rho)$ with $kv_r \simeq s_\alpha(\lambda + \rho)$. Set $H_\alpha^0(\lambda + \rho)^m = H_\alpha^0(\lambda + \rho)^- / kv_r$. We get thus some exact sequences of B–modules; after tensoring with $-\rho$ they have the form

$$(4) \qquad 0 \to H_\alpha^0(\lambda + \rho)^- \otimes (-\rho) \longrightarrow H_\alpha^0(\lambda + \rho) \otimes (-\rho) \longrightarrow \lambda \to 0$$

and

$$(5) \qquad 0 \to s_\alpha \bullet \lambda \longrightarrow H_\alpha^0(\lambda + \rho)^- \otimes (-\rho) \longrightarrow H_\alpha^0(\lambda + \rho)^m \otimes (-\rho) \to 0.$$

Consider also $H_\alpha^0(\lambda + \rho - \alpha)$. It has an analogous basis $\widetilde{v}_i$ $(0 \leq i \leq r - 2)$. The formulae in 5.2 show that the map $v_i \mapsto (r - i)\widetilde{v}_{i-1}$ induces a homomorphism of B–modules $H_\alpha^0(\lambda + \rho)^m \to H_\alpha^0(\lambda + \rho - \alpha)$, hence also $H_\alpha^0(\lambda + \rho)^m \otimes (-\rho) \to H_\alpha^0(\lambda + \rho - \alpha) \otimes (-\rho)$. Let us denote the kernel, cokernel, and image of this homomorphism by $N_1^\alpha(\lambda)$, $N_2^\alpha(\lambda)$, $N_3^\alpha(\lambda)$. We have, therefore, short exact sequences of B–modules

$$(6) \qquad 0 \to N_1^\alpha(\lambda) \longrightarrow H_\alpha^0(\lambda + \rho)^m \otimes (-\rho) \longrightarrow N_3^\alpha(\lambda) \to 0$$

and

$$(7) \qquad 0 \to N_3^\alpha(\lambda) \longrightarrow H_\alpha^0(\lambda + \rho - \alpha) \otimes (-\rho) \longrightarrow N_2^\alpha(\lambda) \to 0.$$

Furthermore, the explicit description of the map shows that $\operatorname{ch} N_1^\alpha(\lambda) = \operatorname{ch} N_2^\alpha(\lambda)$ are as stated in (1).

Since $\langle -\rho, \alpha^\vee \rangle = -1$, we have $R^\bullet \operatorname{ind}_B^{P(\alpha)}(-\rho) = 0$ by 5.2.b, hence

$$R^\bullet \operatorname{ind}_B^{P(\alpha)}(M \otimes (-\rho)) \simeq M \otimes R^\bullet \operatorname{ind}_B^{P(\alpha)}(-\rho) = 0$$

for any $P(\alpha)$–module M (using the generalised tensor identity I.4.8), hence finally $R^\bullet \operatorname{ind}_B^G(M \otimes (-\rho)) = 0$ (using the spectral sequence I.4.5.c). We can apply this to $M = H_\alpha^0(\lambda + \rho)$ and $M = H_\alpha^0(\lambda + \rho - \alpha)$. Therefore (4) and (7) yield isomorphisms for each i

$$(8) \qquad H^i(\lambda) \simeq H^{i+1}(H_\alpha^0(\lambda + \rho)^- \otimes (-\rho))$$

and

(9) $$H^i(N_2^\alpha(\lambda)) \simeq H^{i+1}(N_3^\alpha(\lambda)).$$

We now apply ind_B^G to (5) and (6) and get long exact sequences. They contain the right hand sides of (8) and (9) respectively and we use (8) and (9) to replace these terms with the left hand sides. We get then (2) and (3) with $N_0^\alpha(\lambda) = H_\alpha^0(\lambda+\rho)^m \otimes (-\rho)$.

This proves the lemma in case $\rho \in X(T)$. In general there is a central extension $G' \to G$ with a split maximal torus $T' \to T$ such that $\rho \in X(T') \supset X(T)$. Let $B' \subset G'$ be the inverse image of B. The kernel of the covering $G' \to G$ is a central subgroup scheme Z of G' contained in T'. We have then $T \simeq T'/Z$ and $B \simeq B'/Z$ and $G \simeq G'/Z$.

We can now carry out the constructions as above for G', B', and T'. The B'–modules $N_i^\alpha(\lambda)$ have all their weights in $X(T)$. Therefore Z acts trivially on them and they are B–modules in a natural way. Using $R^i \operatorname{ind}_B^G N \simeq R^i \operatorname{ind}_{B'}^{G'} N$ for any B–module N (cf. I.6.11) we see that the sequences (2), (3) for G' yield the desired sequences for G.

6.13. Proposition (The Strong Linkage Principle): *Let $\lambda \in X(T)$ with $\langle \lambda+\rho, \alpha^\vee\rangle \geq 0$ for all $\alpha \in R^+$. Let $\mu \in X(T)_+$. If $L(\mu)$ is a composition factor of some $H^i(w \bullet \lambda)$ with $w \in W$ and $i \in \mathbf{N}$, then $\mu \uparrow \lambda$.*

We shall prove this result via induction on λ and $\leq$ in the next subsections (6.14–6.16), always supposing that λ and μ are as above. We shall assume that 6.13 holds for all $\lambda' < \lambda$. We use the notations $N_i^\alpha(\lambda')$ as in 6.12.

6.14. Lemma: *Let $\alpha \in S$ and $w \in W$ with $\langle w(\lambda+\rho), \alpha^\vee\rangle \geq 0$. If $L(\mu)$ is a composition factor of some $H^i(N_0^\alpha(w \bullet \lambda))$, then $\mu \uparrow \lambda$ and $\mu < \lambda$.*

Proof: Because of 6.12(3) it is enough to prove the same result for $N_1^\alpha(w \bullet \lambda)$ and $N_2^\alpha(w \bullet \lambda)$, hence for each composition factor of these B–modules. By 6.12(1) we have to look at all $H^i(\lambda_1)$ with λ_1 of the form $\lambda_1 = s_\alpha w \bullet \lambda + np\alpha$ for some $n \in \mathbf{N}$ with $0 < np < \langle w(\lambda+\rho), \alpha^\vee\rangle = \langle \lambda+\rho, w^{-1}(\alpha)^\vee\rangle$. Our assumption on λ implies $\beta = w^{-1}(\alpha) \in R^+$. We have $\lambda_1 = s_\alpha w \bullet (\lambda - np\beta)$. There are $\lambda_2 \in X(T)$ and $w' \in W$ with $\lambda_1 = w' \bullet \lambda_2$ and $\langle \lambda_2+\rho, \gamma^\vee\rangle \geq 0$ for all $\gamma \in R^+$. As $\lambda_2 \in W \bullet (\lambda - np\beta)$ we get from 6.9 that $\lambda_2 \uparrow \lambda$ and $\lambda_2 < \lambda$. We can now apply the induction hypothesis to any composition factor $L(\mu)$ of some $H^i(w' \bullet \lambda_2)$ and get $\mu \uparrow \lambda_2$, hence $\mu \uparrow \lambda$ and $\mu \leq \lambda_2 < \lambda$.

6.15. Proposition: *Let $i \in \mathbf{N}$ and $w \in W$ with $l(w) \neq i$. If $L(\mu)$ is a composition factor of $H^i(w \bullet \lambda)$ with, then $\mu \uparrow \lambda$ and $\mu < \lambda$.*

Proof: Let us at first suppose that $i < l(w)$ and use induction on i. In case $i = 0$, we get $H^0(w \bullet \lambda) \neq 0$, hence $w \bullet \lambda \in X(T)_+$ by 2.6. As $\{\lambda' \in X(T) \otimes_{\mathbf{Z}} \mathbf{R} \mid \langle \lambda+\rho, \alpha^\vee\rangle \geq 0$ for all $\alpha \in R^+\}$ is a fundamental domain for the "dot action" of W on $X(T) \otimes_{\mathbf{Z}} \mathbf{R}$, we get $w \bullet \lambda = \lambda$ and $\lambda \in X(T)_+$. This implies that λ has trivial stabiliser in W (for the dot action). Therefore $w \bullet \lambda = \lambda$ implies $w = 1$ and $l(w) = 0 = i$ — a contradiction. This settles the case $i = 0$.

Suppose now that $i > 0$. If $\langle w(\lambda+\rho), \alpha^\vee\rangle \geq 0$ for all $\alpha \in S$, then $H^i(w \bullet \lambda) = 0$ by 4.5 and 5.4.a. Therefore we can find $\alpha \in S$ with $\langle w(\lambda + \rho), \alpha^\vee\rangle < 0$. We can now apply Lemma 6.12 to $s_\alpha w \bullet \lambda$ instead of λ. From 6.12(2) we get that $L(\mu)$ is a composition factor of $H^{i-1}(N_0^\alpha(s_\alpha w \bullet \lambda))$ or of $H^{i-1}(s_\alpha w \bullet \lambda)$. In the first case we apply Lemma 6.14, in the second one we use induction on i as $l(s_\alpha w) = l(w) - 1 > i - 1$. (Note that $\langle \lambda + \rho, w^{-1}(\alpha)^\vee\rangle < 0$, hence $w^{-1}(\alpha) < 0$, hence $l(s_\alpha w) = l(w) - 1$ by 1.5(3).)

This settles the case $i < l(w)$. The case $i > l(w)$ follows either using Serre duality, or by descending induction. (Then we have to use that $H^i(\lambda') = 0$ for $i > n = |R^+|$ and that $H^n(\lambda') \neq 0$ if and only if $\langle \lambda' + \rho, \alpha^\vee\rangle < 0$ for all $\alpha \in S$ by a special case of Serre duality.)

Remarks: 1) Observe that this settles the strong linkage principle for all λ as in 6.13 with $\lambda \notin X(T)_+$. For these we can find $\alpha \in S$ with $s_\alpha \bullet \lambda = \lambda$, hence $ws_\alpha \bullet \lambda = w \bullet \lambda$ for all w. Given $i \in \mathbf{N}$ we have $i \neq l(w)$ or $i \neq l(ws_\alpha)$ and can thus apply the proposition to $H^i(w \bullet \lambda) = H^i(ws_\alpha \bullet \lambda)$.

2) This proposition actually is in some aspect stronger than the strong linkage principle giving the additional information that $\mu < \lambda$ for $i \neq l(w)$. (For $\lambda \notin X(T)_+$ this is obvious, of course.)

6.16. Proposition: *Suppose $\lambda \in X(T)_+$. Then $L(\lambda)$ is a composition factor with multiplicity one of each $H^{l(w)}(w \bullet \lambda)$ with $w \in W$. Any composition factor $L(\mu)$ of $H^{l(w)}(w \bullet \lambda)$ satisfies $\mu \uparrow \lambda$.*

Proof: Suppose that either $\mu \uparrow \lambda$ does not hold or that $\mu = \lambda$.

For any $w \in W$ and $\alpha \in S$ with $l(s_\alpha w) = l(w) + 1$ we have by 6.12(2) an exact sequence

$$H^{l(w)}(N_0^\alpha(w \bullet \lambda)) \to H^{l(w)+1}(s_\alpha w \bullet \lambda) \to H^{l(w)}(w \bullet \lambda) \to H^{l(w)+1}(N_0^\alpha(w \bullet \lambda))$$

Lemma 6.14 implies that $L(\mu)$ is not a composition factor of any $H^i(N_0^\alpha(w \bullet \lambda))$. It is therefore not a composition factor of the kernel or of the cokernel of the homomorphism $H^{l(w)+1}(s_\alpha w \bullet \lambda) \to H^{l(w)}(w \bullet \lambda)$.

We can choose a sequence $w_0', w_1', \dots, w_n' \in W$ (where $n = |R^+|$) such that $l(w_i') = i$ for all i (hence $w_0' = 1$ and $w_n' = w_0$), such that there are simple roots $\alpha_i \in S$ with $w_i' = s_{\alpha_i} w_{i-1}'$ for all $i > 0$, and such that $w_{l(w)}' = w$. (We have $l(w_0 w^{-1}) = l(w_0) - l(w)$, cf. [B3], ch. IV, §1, exerc. 22a. By the definition of the length there are $\alpha_i \in S$ with $w = s_{\alpha_{l(w)}} \cdots s_{\alpha_2} s_{\alpha_1}$ and $w_0 w^{-1} = s_{\alpha_n} s_{\alpha_{n-1}} \cdots s_{\alpha_{l(w)+1}}$. Then take $w_i' = s_{\alpha_i} \cdots s_{\alpha_2} s_{\alpha_1}$ for all i. We have obviously $l(w_i') \leq i$ for all i; if $l(w_i') < i$ for some i, then the length of $w_0 = s_{\alpha_n} s_{\alpha_{n-1}} \cdots s_{\alpha_{i+1}} w_i'$ would be less than n — contradiction.)

We apply the argument above to each (w_i', α_i) instead of (w, α). We get thus a sequence of homomorphisms of G–modules

$$H^n(w_n' \bullet \lambda) \longrightarrow H^{n-1}(w_{n-1}' \bullet \lambda) \longrightarrow \cdots \longrightarrow H^1(w_1' \bullet \lambda) \longrightarrow H^0(w_0' \bullet \lambda).$$

As seen above, $L(\mu)$ does not occur in the kernel or cokernel of any of the maps $H^{i+1}(w_{i+1}' \bullet \lambda) \to H^i(w_i' \bullet \lambda)$. Therefore it occurs in each $H^i(w_i' \bullet \lambda)$ with the same multiplicity as in the image M of the composed map

$$H^n(w_0 \bullet \lambda) = H^n(w_n' \bullet \lambda) \longrightarrow H^0(w_0' \bullet \lambda) = H^0(\lambda).$$

As this observation applies in particular to $\mu = \lambda$ and as $L(\lambda)$ is a composition factor of $H^0(\lambda)$, it is also one of M. It follows that $M \neq 0$. Since $L(\lambda)$ occurs with multiplicity 1 in $H^0(\lambda)$, it does so in all $H^i(w_i' \bullet \lambda)$, in particular in $H^{l(w)}(w \bullet \lambda)$. (This proves the first claim of the proposition.)

We have by Serre duality

$$H^n(w_0 \bullet \lambda) = H^n(w_0\lambda - 2\rho) \simeq H^0(-w_0\lambda)^* = V(\lambda),$$

cf. 2.13. As $V(\lambda)/\operatorname{rad}_G V(\lambda) \simeq L(\lambda)$ by 2.14(1), also the non-zero homomorphic image M of $V(\lambda)$ satisfies

$$M/\operatorname{rad}_G M \simeq L(\lambda).$$

On the other hand $M \subset H^0(\lambda)$ implies

$$\operatorname{soc}_G M = L(\lambda).$$

As $L(\lambda)$ occurs with multiplicity 1 in $H^0(\lambda)$, this implies (as in 5.6) that

$$M = L(\lambda).$$

So, if μ does not satisfy $\mu \uparrow \lambda$, then it is not a composition factor of any $H^i(w_i' \bullet \lambda)$, in particular not of $H^{l(w)}(w \bullet \lambda)$.

Remark: The proof also shows that the composed map $H^n(w_0 \bullet \lambda) \to H^0(\lambda)$ is non-zero and has image $L(\lambda)$. In fact, any non-zero homomorphism from $H^n(w_0 \bullet \lambda)$ to $H^0(\lambda)$ has image $L(\lambda)$.

6.17. Corollary (The Linkage Principle): *Let $\lambda, \mu \in X(T)_+$ Then:*

$$\operatorname{Ext}^1_G(L(\lambda), L(\mu)) \neq 0 \Rightarrow \lambda \in W_p \bullet \mu.$$

Proof: Because of 2.12(4) we may assume $\mu \not> \lambda$. Therefore 2.14 implies:

$$\operatorname{Ext}^1_G(L(\lambda), L(\mu)) \neq 0 \Rightarrow [H^0(\lambda) : L(\mu)] = [V(\lambda) : L(\mu)] \neq 0,$$

hence $\mu \in W_p \bullet \lambda$ by the strong linkage principle.

6.18. Suppose that some $\lambda \in X(T)$ with $\langle \lambda + \rho, \beta^\vee \rangle \geq 0$ for all $\beta \in S$ has the property that there is no $\mu \in X(T)_+$ with $\mu \uparrow \lambda$ and $\mu \neq \lambda$. Then 6.16 and 6.15 imply for each $w \in W$ that $H^i(w \bullet \lambda) = 0$ for all $i \neq l(w)$ and $H^{l(w)}(w \bullet \lambda) \simeq L(\lambda)$ if $\lambda \in X(T)_+$, whereas $H^{l(w)}(w \bullet \lambda) = 0$ if $\lambda \notin X(T)_+$.

This argument can be applied especially in case $\lambda = 0$ as $(-\sum_{\alpha \in S} \mathbf{N}\alpha) \cap X(T)_+ = \{0\}$. We get thus

$$(1) \qquad H^i(w \bullet 0) \simeq \begin{cases} k, & \text{if } i = l(w), \\ 0, & \text{otherwise.} \end{cases}$$

We get furthermore:

(2) *If $\lambda < 0$ with $\langle \lambda + \rho, \beta^\vee \rangle \geq 0$ for all $\beta \in S$, then $H^\bullet(w \bullet \lambda) = 0$ for all $w \in W$.*

Proposition: *Let $n \in \mathbf{N}$. Then $H^i(\Lambda^n(\mathrm{Lie}\, G/\mathrm{Lie}\, B)^*) = 0$ for all $i \neq n$, and $H^n(\Lambda^n(\mathrm{Lie}\, G/\mathrm{Lie}\, B)^*)$ is the trivial G–module of dimension $|\{w \in W \mid l(w) = n\}|$.*

Proof: Let us abbreviate $M = (\mathrm{Lie}\, G/\mathrm{Lie}\, B)^*$. We shall show for each weight ν of $\Lambda^n M$ that $H^i(\nu) = 0$ for all $i \neq n$ while $H^n(\nu)$ is either 0 or the trivial G–module k; furthermore the number of weights ν with $H^n(\nu) \simeq k$ will turn out to be $r_n = |\{w \in W \mid l(w) = n\}|$. Then induction over a composition series of M as a B–module shows that $H^i(M) = 0$ for all $i \neq n$ while the G–module $H^n(M)$ has a composition series of length r_n with all factors isomorphic to k. As G is reductive (or by 2.12(1)), the trivial module has only trivial extensions with itself, so $H^n(M)$ is a trivial G–module.

The weights of $\mathrm{Lie}\, G/\mathrm{Lie}\, B$ are just the positive roots, hence those of M the negative roots (each with multiplicity 1). Set

$$\rho(J) = \sum_{\alpha \in J} \alpha$$

for each subset $J \subset R^+$. Then the weights of $\Lambda^n M$ are just the $-\rho(J)$ with $J \subset R^+$ and $|J| = n$.

One checks easily for each $w \in W$ that $w^{-1} \bullet 0 = w^{-1}(\rho) - \rho = -\rho(J_w)$ where $J_w = \{\alpha \in R^+ \mid w(\alpha) < 0\}$, hence $|J_w| = l(w) = l(w^{-1})$. So the $-\rho(J_w)$ with $l(w) = n$ are weights of $\Lambda^n M$ with $H^i(-\rho(J_w)) = 0$ for all $i \neq n$ and with $H^n(-\rho(J_w)) \simeq k$, see (1). Furthermore, we get thus r_n distinct weights: In fact, if $J \subset R^+$ with $-\rho(J) = w^{-1}(\rho) - \rho$, then $J = J_w$ according to [Kostant 1], 5.10.2.

Given an arbitrary $J \subset R^+$, there exists $w \in W$ with $\langle w(\rho - \rho(J)), \beta^\vee \rangle \geq 0$ for all $\beta \in S$. We have

$$\rho - \rho(J) = \frac{1}{2} \sum_{\beta \in J'} \beta - \frac{1}{2} \sum_{\alpha \in J} \alpha$$

where $J' = R^+ \setminus J$. Therefore $w(\rho - \rho(J))$ has the form $\rho - \rho(J_0)$ for some $J_0 \subset R^+$. This implies especially $w \bullet (-\rho(J)) = -\rho(J_0) \leq 0$. If $w \bullet (-\rho(J)) \neq 0$, then the argument above and (2) show that $H^\bullet(-\rho(J)) = 0$. If $w \bullet (-\rho(J)) = 0$, then $-\rho(J) = w^{-1}(\rho) - \rho$, hence $J = J_w$ by the result of Kostant quoted above. Now the claims in the first paragraph of the proof follow.

Remarks: 1) Let $I \subset S$. Then the weights of $\Lambda^n(\mathrm{Lie}\, G/\mathrm{Lie}\, P_I)^*$ are all $-\rho(J)$ with $J \subset R^+$ and $|J| = n$ and $J \cap \mathbf{Z}I = \emptyset$. The same arguments as above show that also $H^i(\Lambda^n(\mathrm{Lie}\, G/\mathrm{Lie}\, P_I)^*) = 0$ for all $i \neq n$ whereas $H^n(\Lambda^n(\mathrm{Lie}\, G/\mathrm{Lie}\, P_I)^*)$ is a trivial G–module; its dimension is now the number of all $w \in W$ with $l(w) = n$ and $w\alpha > 0$ for all $\alpha \in I$.

This proof works equally well in characteristic 0, so the result is true there also. A proof in that case appeared in [Marlin]. The present extension to arbitrary characteristic is due to Andersen, see his review (*Math. Reviews* **58** (1979), #22094) of [Marlin].

2) Note that we compute above the cohomology of the exterior powers of the cotangent bundle on G/B.

6.19. (Central Characters) By 2.8, the centre $Z(\mathrm{Dist}(G))$ of $\mathrm{Dist}(G)$ acts by scalars on each simple G–module, i.e., via some *central character* $Z(\mathrm{Dist}(G)) \to k$. Similarly, the subalgebra $U(\mathrm{Lie}G)^G$ of all G–invariant elements in the enveloping algebra $U(\mathrm{Lie}G)$ acts via some central character $U(\mathrm{Lie}G)^G \to k$. If there is a non-split extension of $L(\lambda)$ and $L(\mu)$ for some $\lambda, \mu \in X(T)_+$, then these two simple modules have to have the same central characters as otherwise the generalised eigenspaces of $Z(\mathrm{Dist}(G))$ or $U(\mathrm{Lie}G)^G$ would split the extension into a direct sum.

Using the Casimir operator in $U(\mathrm{Lie}G)$, the first results in the direction of 6.17 were proved in [Springer]. In [Humphreys 1] the linkage principle was proved (using $U(\mathrm{Lie}G)^G$) for all $p > h$ (the Coxeter number). Then $U(\mathrm{Lie}G)^G$ was determined for all p in [Kac and Weisfeiler]. Their result almost implied 6.17: One had to replace "$\lambda \in W_p \bullet \mu$" by "$\lambda \in W \bullet \mu + p\mathbf{Z}R$". This restriction made a difference only for those p dividing the order of $(X(T) \cap \mathbf{Q}R)/\mathbf{Z}R$ as $\mathrm{Ext}^1_G(L(\lambda), L(\mu)) \neq 0$ implies $\lambda - \mu \in \mathbf{Z}R$.

This exception was by then already known to be unnecessary for G of type A_n due to [Carter and Lusztig 1] where $Z(\mathrm{Dist}(G))$ was used. There are more results about this centre in [Haboush 3], but it is still not completely understood.

6.20. Recall the notation $d(\lambda)$ from the remark in 6.6.

Proposition: *Let $\lambda, \mu \in X(T)_+$ and $i \in \mathbf{N}$. If $\mathrm{Ext}^i_G(L(\lambda), H^0(\mu)) \neq 0$ or if $\mathrm{Ext}^i_G(H^0(\lambda), H^0(\mu)) \neq 0$, then $\mu \uparrow \lambda$ and $i \leq d(\lambda) - d(\mu)$.*

Proof: We first prove the claim concerning $L(\lambda)$ using induction on i. For $i = 0$ we have

$$\mathrm{Ext}^0_G(L(\lambda), H^0(\mu)) = \mathrm{Hom}_G(L(\lambda), H^0(\mu)) \simeq \begin{cases} k, & \text{if } \lambda = \mu, \\ 0, & \text{otherwise.} \end{cases}$$

Suppose now $i > 0$. If $\lambda = \mu$, then we have to show that $\mathrm{Ext}^i_G(L(\lambda), H^0(\lambda)) = 0$, which we proved already in 4.13(2).

So assume that $\lambda \neq \mu$. Then the short exact sequence $0 \to \mathrm{rad}_G V(\lambda) \to V(\lambda) \to L(\lambda) \to 0$ yields by 4.13 an isomorphism

(1) $$\mathrm{Ext}^i_G(L(\lambda), H^0(\mu)) \simeq \mathrm{Ext}^{i-1}_G(\mathrm{rad}_G V(\lambda), H^0(\mu)).$$

If this group in non-zero, then there is a composition factor $L(\lambda')$ of $\mathrm{rad}_G V(\lambda)$ with $\mathrm{Ext}^{i-1}_G(L(\lambda'), H^0(\mu)) \neq 0$. The strong linkage principle implies now $\lambda' \uparrow \lambda$; since $\lambda' \neq \lambda$ we get $d(\lambda') < d(\lambda)$ by the remark in 6.6. On the other hand, induction yields $\mu \uparrow \lambda'$ and $i - 1 \leq d(\lambda') - d(\mu)$, hence $\mu \uparrow \lambda$ and $i \leq d(\lambda) - d(\mu)$.

We now turn to the claim for $H^0(\lambda)$. If $\mathrm{Ext}^i_G(H^0(\lambda), H^0(\mu)) \neq 0$, then there exists a composition factor $L(\nu)$ of $H^0(\lambda)$ with $\mathrm{Ext}^i_G(L(\nu), H^0(\mu)) \neq 0$. The strong linkage principle implies $\nu \uparrow \lambda$, hence $d(\nu) \leq d(\lambda)$ by 6.6. Now apply the first part of the proof.

6.21. We have by 4.10(1)

(1) $$\dim \mathrm{Ext}^i_G(V, H^0(\mu)) < \infty$$

for each finite dimensional G–module V. Furthermore, 6.20 and induction on the length of V (or Remark 2 in 4.13) imply that there is for each such V an integer

$n(V)$ with $\operatorname{Ext}_G^i(V, H^0(\mu)) = 0$ for all $i > n(V)$ and all μ. We can therefore form the alternating sum

(2) $$c_\mu(V) = \sum_{i \geq 0} (-1)^i \dim \operatorname{Ext}_G^i(V, H^0(\mu)).$$

Like any Euler characteristic, the function c_μ is additive: Given finite dimensional G–modules V_j and integers a_j, then

$$\operatorname{ch}(V) = \sum_{j=1}^r a_j \operatorname{ch}(V_j) \Rightarrow c_\mu(V) = \sum_{j=1}^r a_j c_\mu(V_j).$$

There are unique integers b_μ with

(3) $$\operatorname{ch}(V) = \sum_{\mu \in X(T)_+} b_\mu \chi(\mu),$$

see Remark 5.8. We have $\chi(\mu) = \operatorname{ch} H^0(\mu) = \operatorname{ch} V(\mu)$, and 4.12 implies for all $\lambda, \mu \in X(T)_+$

(4) $$c_\mu(V(\lambda)) = \delta_{\lambda\mu}$$

(the Kronecker delta). Therefore (3) yields

(5) $$b_\mu = c_\mu(V).$$

In other words, we have for each finite dimensional G–module V

(6) $$\operatorname{ch}(V) = \sum_{\mu \in X(T)_+} \Big(\sum_{i \geq 0} (-1)^i \dim \operatorname{Ext}_G^i(V, H^0(\mu)) \Big) \chi(\mu).$$

6.22. We now return to the technique used in 6.14–6.16 to get information about composition factors $L(\mu)$ in $H^i(w \bullet \lambda)$ in cases where μ is not "to far" away from λ. More precisely, consider $\lambda, \mu \in X(T)_+$ with $\mu \uparrow \lambda$. We say that μ is *close to* λ if and only if:

(1) *There is no $\alpha \in R^+$ with $\mu \uparrow \lambda - p\alpha$.*

(2) *If $\mu \uparrow \lambda' \uparrow \lambda$ for some $\lambda' \in X(T)$, then $\lambda' \in X(T)_+$.*

Note that $\mu \uparrow \lambda' \uparrow \lambda$ and μ close to λ imply that λ' is close to λ (obviously) and that μ close to λ' (as $\lambda' - p\alpha \uparrow \lambda - p\alpha$ for all $\alpha \in R^+$).

Let $\lambda \in X(T)_+$ and let F_1 be the facet with $\lambda \in F_1$. Pick a facet F with $F \subset \overline{F}_1$. Recall the notation $W_p^0(F)$ from 6.11, and consider $\mu = w \bullet \lambda$ with $w \in W_p^0(F)$ and $\mu \uparrow \lambda$. Then 6.11(4), (5) imply that any λ' with $\mu \uparrow \lambda' \uparrow \lambda$ has the form $\lambda' = w' \bullet \lambda$ with $w' \in W_p^0(F)$. This shows especially that $\lambda' \neq \lambda - p\alpha$ for all $\alpha \in R^+$. If we know in addition that $\langle x + \rho, \beta^\vee \rangle > 0$ for all $\beta \in R^+$ and all $x \in F$, then we get in addition that $\lambda' \in X(T)_+$. So in this case μ is close to λ.

6.23. Proposition: *Let $\lambda,\mu\in X(T)_+$ with $\mu\uparrow\lambda$ and μ close to λ.*

a) *If $w\in W$ and $i\in\mathbf{N}$ with $i\neq l(w)$, then $L(\mu)$ is not a composition factor of $H^i(w\bullet\lambda)$.*

b) *For any $w\in W$ the multiplicity of $L(\mu)$ as a composition factor of $H^{l(w)}(w\bullet\lambda)$ is positive and equal to the multiplicity in $H^0(\lambda)$.*

Proof: We want to use induction on $\lambda-\mu$. The case $\lambda=\mu$ has been dealt with in 6.15/16. So let us suppose $\mu<\lambda$.

There are $n_\beta,d_\beta\in\mathbf{N}$ with $\langle\lambda+\rho,\beta^\vee\rangle=n_\beta p+d_\beta$ and $0<d_\beta\le p$ for all $\beta\in R^+$. Set $\lambda_\beta=\lambda-d_\beta\beta=s_{\beta,n_\beta p}\bullet\lambda$. The proof will show more precisely

$$(1)\qquad \max_{\beta>0}\,[H^0(\lambda_\beta):L(\mu)]\le[H^0(\lambda):L(\mu)]\le\sum_{\beta>0}[H^0(\lambda_\beta):L(\mu)]$$

using as before the notation $[?:L(\mu)]$ for the multiplicity of $L(\mu)$ as a composition factor.

Consider $w\in W$ and $\alpha\in S$ with $\beta=w^{-1}\alpha>0$. Let us look at the long exact sequences in 6.12 for $w\bullet\lambda$ instead of λ. If $L(\mu)$ is a composition factor of some $H^i(N_1^\alpha(w\bullet\lambda))$ or $H^i(N_2^\alpha(w\bullet\lambda))$, then there has to be a weight λ_1 of $N_1^\alpha(w\bullet\lambda)$ or of $N_2^\alpha(w\bullet\lambda)$ with $[H^i(\lambda_1):L(\mu)]\neq 0$. The possible λ_1 have the form $s_\alpha w\bullet\lambda+np\alpha$ with $0<np<\langle w(\lambda+\rho),\alpha^\vee\rangle=\langle\lambda+\rho,\beta^\vee\rangle$; we can rewrite this as $\lambda_1=ws_\beta\bullet\lambda+npw\beta=ws_{\beta,np}\bullet\lambda$. There is $w_1\in W$ such that $\lambda_2=w_1\bullet\lambda_1=w_1ws_{\beta,np}\bullet\lambda$ satisfies $\langle\lambda_2+\rho,\gamma^\vee\rangle\ge 0$ for all $\gamma\in S$. We have $\mu\uparrow\lambda_2\uparrow\lambda$ by 6.13 (as $[H^i(w_1^{-1}\bullet\lambda_2):L(\mu)]\neq 0$) and by 6.9. The closeness of μ to λ implies $\lambda_2\in X(T)_+$ and that μ is close to λ_2.

If $w_1\alpha<0$, then

$$\mu\uparrow\lambda_2=w_1s_\alpha w\bullet\lambda-np(-w_1\alpha)\uparrow w_1s_\alpha w\bullet\lambda\uparrow\lambda$$

(cf. 6.4(3), (5)), contradicting that μ is close to λ. Therefore $w_1\alpha>0$. As

$$\lambda_2=w_1s_\alpha w\bullet\lambda+npw_1\alpha=s_{w_1\alpha,np}s_{w_1\alpha}w_1s_\alpha w\bullet\lambda=s_{w_1\alpha,np}w_1w\bullet\lambda$$

and as

$$\langle w_1w(\lambda+\rho),w_1(\alpha)^\vee\rangle=\langle\lambda+\rho,\beta^\vee\rangle>np$$

we get $\mu\uparrow\lambda_2\uparrow w_1w\bullet\lambda\uparrow\lambda$. The closeness of μ to λ implies $w_1=w^{-1}$ and $\lambda_2=s_{\beta,np}\bullet\lambda$. Furthermore, we get $n_\beta=n>0$ and $\lambda_\beta=\lambda_2\in X(T)_+$. So, if $\lambda_\beta\notin X(T)_+$ (e.g., if $n_\beta=0$), then $[H^i(N_j^\alpha(w\bullet\lambda)):L(\mu)]=0$ for all i and for $j=1,2$, hence also for $j=0$.

Suppose on the other hand that $\lambda_\beta\in X(T)_+$, hence $n_\beta>0$. Then $w\bullet\lambda_\beta$ is a weight of $N_j^\alpha(w\bullet\lambda)$ (for $j=1,2$) with multiplicity 1, and it is the only weight λ_1 of $N_1^\alpha(w\bullet\lambda)$ or of $N_2^\alpha(w\bullet\lambda)$ with possibly $[H^i(\lambda_1):L(\mu)]\neq 0$. This implies

$$[H^i(N_j^\alpha(w\bullet\lambda)):L(\mu)]=[H^i(w\bullet\lambda_\beta):L(\mu)]$$

for $j=1,2$ and all i (if $n_\beta>0$). If this term is non-zero, then $\mu\uparrow\lambda_\beta\uparrow\lambda$ implies that μ is close to $\lambda_\beta<\lambda$. Now our induction on $\lambda-\mu$ yields (for $j=1,2$)

$$[H^i(N_j^\alpha(w\bullet\lambda)):L(\mu)]=\begin{cases}[H^0(\lambda_\beta):L(\mu)], & \text{for } i=l(w),\\ 0, & \text{otherwise.}\end{cases}$$

Therefore 6.12(3) yields

$$[H^i(N_0^\alpha(w \bullet \lambda)) : L(\mu)] = \begin{cases} [H^0(\lambda_\beta) : L(\mu)], & \text{for } i = l(w), l(w)+1, \\ 0, & \text{otherwise.} \end{cases}$$

We know in addition that $[H^0(\lambda_\beta) : L(\mu)] > 0$ in case $\mu \uparrow \lambda_\beta$.

We get from 6.12(2) — also if $\lambda_\beta \notin X(T)_+$ — that $L(\mu)$ is not a composition factor of the kernel of one of the maps $H^{i+1}(s_\alpha w \bullet \lambda) \to H^i(w \bullet \lambda)$ for $i < l(w)$, or of the cokernel if $i > l(w)$.

Furthermore, we get from 6.12(2) that the multiplicity of $L(\mu)$ both in the kernel and the cokernel of

$$H^{l(w)+1}(s_\alpha w \bullet \lambda) \to H^{l(w)}(w \bullet \lambda)$$

is equal to $[H^0(\lambda_\beta) : L(\mu)]$. (Note that $l(s_\alpha w) = l(w)+1$.) This implies

(2) $$[H^{l(w)+1}(s_\alpha w \bullet \lambda) : L(\mu)] = [H^{l(w)}(w \bullet \lambda) : L(\mu)] \geq [H^0(\lambda_\beta) : L(\mu)].$$

We can now prove the claim in a) for $i > l(w)$ using downward induction on $l(w)$. The claim is obvious for $w = w_0$ since all $H^i(\lambda')$ with $i > l(w_0) = \dim(G/B)$ are 0. If $w \neq w_0$, then we choose $\alpha \in S$ with $w^{-1}\alpha > 0$. Then $L(\mu)$ is by induction not a composition factor of $H^{i+1}(s_\alpha w \bullet \lambda)$ since $i+1 > l(s_\alpha w) = l(w)+1 > l(w)$. It is also not a composition factor of the cokernel of the map $H^{i+1}(s_\alpha w \bullet \lambda) \to H^i(w \bullet \lambda)$. This yields the claim.

We get similarly the claim in a) for $i < l(w)$ using normal induction on $l(w)$. Here the claim is trivial for $w = 1$. If $w \neq 1$, then we choose $\alpha \in S$ with $w^{-1}\alpha < 0$. Now apply the considerations above to $s_\alpha w$ and $i-1$ instead of w and i.

In order to prove b) we work with a sequence $w'_0, w'_1, \dots, w'_n$ as in the proof of 6.16. We get now from (2) that any $[H^{l(w)}(w \bullet \lambda) : L(\mu)]$ is equal to $[H^0(\lambda) : L(\mu)]$ and that the lower bound in (1) holds. [Recall that there exists for each $\beta \in R^+$ some $w \in W$ with $\alpha = w\beta \in S$.] For the upper bound in (1) use that $L(\mu)$ is not a composition factor of the image of $H^n(w_0 \bullet \lambda) \to H^0(\lambda)$.

Finally, $\mu \uparrow \lambda$ and $\mu < \lambda$ imply (by the definition of $\uparrow$) that there exists $\beta \in R^+$ with $\mu \uparrow \lambda_\beta$. Then we get above $[H^0(\lambda_\beta) : L(\mu)] > 0$, hence $[H^0(\lambda) : L(\mu)] > 0$.

6.24. Corollary: *Let $\lambda \in X(T)_+$. Suppose that $\mu \in X(T)$ is maximal for $\mu \uparrow \lambda$ and $\mu \neq \lambda$. If $\mu \in X(T)_+$ and if $\mu \neq \lambda - p\alpha$ for all $\alpha \in R^+$, then*

(1) $$[H^0(\lambda) : L(\mu)] = 1$$

and

(2) $$\operatorname{Hom}_G(H^0(\lambda), H^0(\mu)) \simeq k.$$

Proof: The assumption implies that μ is close to λ. Furthermore, μ has to be equal to λ_α for a unique $\alpha \in R^+$ (using the notations from the last proof). Therefore (1) follows from 6.23(1).

Denote the injective hull of the G–module $L(\mu)$ by Q_μ. As $\dim \operatorname{Hom}_G(V, Q_\mu) = [V : L(\mu)]$ for each finite dimensional G–module V, we get

$$\operatorname{Hom}_G(H^0(\lambda), Q_\mu) \simeq k$$

from (1). Let $\varphi : H^0(\lambda) \to Q_\mu$ be a non-zero homomorphism. We can regard $H^0(\mu)$ as a submodule of Q_μ because of $\mathrm{soc}_G H^0(\mu) = L(\mu)$. We want to prove that $\varphi(H^0(\lambda)) \subset H^0(\mu)$. Then (2) will follow immediately.

If $\varphi(H^0(\lambda)) \not\subset H^0(\mu)$, then $\varphi(H^0(\lambda))+H^0(\mu)$ contains $H^0(\mu)$ properly, so there is a submodule $V \subset \varphi(H^0(\lambda)) + H^0(\mu)$ with $V \supset H^0(\mu)$ and $V/H^0(\mu)$ simple, say $V/H^0(\mu) \simeq L(\mu')$ with $\mu' \in X(T)_+$. Then $L(\mu')$ is a composition factor of $H^0(\lambda)$ with $\mu' \neq \lambda$ since $L(\lambda) \subset \ker(\varphi)$. So we get $\mu' \uparrow \lambda$ and $\mu' < \lambda$. The extension $0 \to H^0(\mu) \to V \to L(\mu') \to 0$ cannot split as $V \subset Q_\mu$ has simple socle. It follows that $\mathrm{Ext}^1_G(L(\mu'), H^0(\mu)) \neq 0$, hence $\mu \uparrow \mu'$ by 6.20, and $\mu' \neq \mu$ by 4.13(2). This contradicts the maximality of μ for $\mu \uparrow \lambda$. Therefore there is no such V as above and we get $\varphi(H^0(\lambda)) \subset H^0(\mu)$.

Remark: Suppose that $\lambda \in C_1$ for some alcove C_1. Consider $\mu \in X(T)_+ \cap W_p \bullet \lambda$. Let C_2 denote the alcove with $\mu \in C_2$. Then μ satisfies the assumption of the corollary if and only if $C_2 \uparrow C_1$ and $d(C_2) = d(C_1) - 1$. This follows easily from 6.10.

6.25. The proof of 6.24(2) given here is taken from [Koppinen 6]. There, more generally, $\mathrm{Hom}_G(H^0(\lambda), H^0(\mu)) \neq 0$ is shown for all $\lambda, \mu \in X(T)_+$ with $\mu \uparrow \lambda$ and μ close to λ. (In fact, the situation there is still more general.) In special cases (type A, B), 6.24(2) had been proved in [Andersen 6] where 6.24(1) is also proved (without restriction).

Before the work of Andersen and Koppinen, homomorphisms had been constructed in a different way. Notice that one has by dualising, cf. 2.13(2):

$$\mathrm{Hom}_G(H^0(\lambda), H^0(\mu)) \simeq \mathrm{Hom}_G(V(\mu), V(\lambda)) \simeq V(\lambda)_\mu^{U^+}.$$

Take $v \in V(\lambda)_\lambda$, $v \neq 0$; then $V(\lambda) = \mathrm{Dist}(G)v$. In order to get a non-zero homomorphism $V(\mu) \to V(\lambda)$ we have to find $u \in \mathrm{Dist}(G)$ with $uv \neq 0$ and $uv \in V(\lambda)^{U^+}$ such that u is a weight vector of weight $\mu - \lambda$ for the adjoint action.

Take, for example, $\alpha \in S$ and $\mu = \lambda_\alpha = \lambda - d_\alpha \alpha$ (using the notations from 6.23). If $n_\alpha > 0$ and $d_\alpha < p$, then $u = X_{-\alpha,d_\alpha}$ will work. For $\alpha \notin S$ one has to replace $X_{-\alpha,d_\alpha}$ by more complicated elements in $\mathrm{Dist}(G)$. For R of type A this was done in [Carter and Lusztig 1] and (for more μ) in [Carter and Payne], for arbitrary R in [Franklin].

CHAPTER 7

The Translation Functors

We assume in this chapter that p is a prime and k a field of characteristic p.

We can decompose the category of all G–modules into a direct sum of subcategories as follows: Consider for each $\lambda \in X(T)$ the category $\mathcal{M}_\lambda$ of all G–modules having only composition factors of the form $L(\mu)$ with $\mu \in W_p \bullet \lambda$. The linkage principle easily implies (7.2/3) that $\{ G\text{–modules} \}$ is the direct product of all different $\mathcal{M}_\lambda$, i.e., of all $\mathcal{M}_\lambda$ with λ in a system of representatives for the W_p–orbits in $X(T)$. We usually take $\overline{C}_{\mathbf{Z}}$ (see 5.5) as our system of representatives.

We want to compare different $\mathcal{M}_\lambda$ and introduce "*translation functors*" T_λ^μ : $\mathcal{M}_\lambda \to \mathcal{M}_\mu$ for all $\lambda, \mu \in \overline{C}_{\mathbf{Z}}$, cf. 7.6. (For the sake of convenience we regard them as functors from $\{ G\text{–modules} \}$ to itself. In fact, we do not use the notation $\mathcal{M}_\lambda$ except in this introduction.) It turns out that T_λ^μ is an equivalence of categories if λ and μ belong to the same facet (7.9). This is sometimes called the *translation principle.*

Even in more general situations the T_λ^μ are useful. For example, if μ belongs to the closure of the facet of λ, then T_λ^μ behaves nicely on cohomology groups $H^i(w \bullet \lambda)$ and simple modules $L(w \bullet \lambda)$ with $w \in W_p$, cf. 7.11 and 7.15. One thus reduces the computation of the composition factors of the $H^i(w \bullet \mu)$, or of character formulae for the $L(w \bullet \mu)$, to the same problem for λ.

One gets also equalities for a priori unrelated multiplicities. For example, if $\lambda \in C \cap X(T)_+$ and $w \in W_p$ and $s \in \Sigma$ with $w \bullet \lambda \in X(T)_+$ and $w \bullet \lambda < ws \bullet \lambda$, then $[H^i(w_1 \bullet \mu) : L(w \bullet \lambda)] = [H^i(w_1 s \bullet \mu) : L(w \bullet \lambda)]$ for all $w_1 \in W_p$ and all i, see 7.18.

We have for any $\lambda \in \overline{C}_{\mathbf{Z}}$ and $w \in W_p$ with $w \bullet \lambda \in X(T)_+$

$$\chi(w \bullet \lambda) = \operatorname{ch} H^0(w \bullet \lambda) = \sum_{w'} b_{w,w'} \operatorname{ch} L(w' \bullet \lambda)$$

with $b_{w,w'} = [H^0(w \bullet \lambda) : L(w' \bullet \lambda)]$ where w' runs over representatives for W_p modulo the stabiliser of λ, such that $w' \bullet \lambda \in X(T)_+$ and $w' \bullet \lambda \uparrow w \bullet \lambda$. (This follows from 6.15). Since $b_{w,w} = 1$ by 2.4.b, we can invert the matrix of the $b_{w,w'}$ and get $a_{w,w'} \in \mathbf{Z}$ with

$$\operatorname{ch} L(w \bullet \lambda) = \sum_{w'} a_{w,w'} \chi(w' \bullet \lambda) \tag{1}$$

where we sum over w' as before. The results in this chapter show: If $p \geq h$ and if we know all $a_{w,w'}$ for one $\lambda \in C \cap X(T)$, then we know all $a_{w,w'}$ for all $\lambda \in \overline{C}_{\mathbf{Z}}$. For $\lambda \in C \cap X(T)$ there is a conjecture predicting the $a_{w,w'}$ in terms of Kazhdan-Lusztig polynomials as long as $w \bullet \lambda$ is "not too large". We shall state and discuss this conjecture (and the meaning of "not too large") in the next chapter (8.22).

The conjecture is related to the behaviour of the "wall crossing functors", certain compositions of two translation functors that we discuss briefly at the end of this chapter (7.21/22).

The main sources for this chapter are [Andersen 9], [Gabber and Joseph], [Jantzen 2, 3, 4, 5, 7], and [Koppinen 2].

Compared to the first edition the subsections 7.19 and 7.21 have been expanded and rearranged into the present 7.19/20. In particular, the old 7.19.b/c has been moved to 7.20, the old 7.21 to 7.19.b–d. The old 7.20 has been moved to 8.22.

7.1. (Blocks of Algebraic Groups) Let H be an algebraic group scheme over a field. Consider on the set of simple H–modules (or rather on the set of isomorphism classes of such modules) the smallest equivalence relation such that two simple H–modules L, L' are equivalent whenever $\operatorname{Ext}^1_H(L, L') \neq 0$. The equivalence classes are called the *blocks* of H. Let us denote the set of these blocks by $\mathcal{B}(H)$.

For any H–module M and any $b \in \mathcal{B}(H)$ let M_b be the sum of all submodules M' of M such that all composition factors of M' belong to b. Then M_b is the largest submodule with this property.

Lemma: *We have for all H–modules M, M'*

$$M = \bigoplus_{b \in \mathcal{B}(H)} M_b \tag{1}$$

and

$$\operatorname{Hom}_H(M, M') \simeq \prod_{b \in \mathcal{B}(H)} \operatorname{Hom}_H(M_b, M'_b). \tag{2}$$

Proof: It is obvious that the sum in (1) is direct and that $\operatorname{Hom}_H(M_b, M'_{b'}) = 0$ for two distinct blocks b, b'. It is therefore enough to show for any M that the submodule $N = \sum_{b \in \mathcal{B}(H)} M_b$ is all of M. We may assume that $\dim M < \infty$. Suppose that $N \neq M$. Then there is a submodule $N' \subset M$ with $N \subset N'$ and $N'/N = L$ simple. There is a block b with $L \in b$. Set $N'' = \bigoplus_{b' \neq b} M_{b'} \subset N \subset N'$. All composition factors of N'/N'' belong to b since they are either L or a composition factor of $N/N'' \simeq M_b$. By the definition of blocks the exact sequence $0 \to N'' \to N' \to N'/N'' \to 0$ splits. So there is a submodule E of N' with $N' = N'' \oplus E$. We have then $E \simeq N'/N''$, hence $E \subset M_b$. It follows that $N' \subset \bigoplus_{b'} M_{b'} = N$, a contradiction.

Remarks: 1) The lemma implies that $M \mapsto M_b$ is an exact functor for each b. Furthermore (1) generalises to

$$\operatorname{Ext}^i_H(M, M') \simeq \prod_{b \in \mathcal{B}(H)} \operatorname{Ext}^i_H(M_b, M'_b) \tag{3}$$

for all i.

2) For an indecomposable H–module $M \neq 0$ there is a unique block b such that $M = M_b$. Then b is called *the block of* M.

3) If H is a finite algebraic group (I.8), then the blocks of H correspond to the indecomposable two-sided ideals in $M(H)$. This is a general fact about representations of finite dimensional algebras.

4) If b is a block of H, then the L^* with L in b form again a block of H.

7.2. (Blocks of G) In the case of $H = G$ we can regard blocks also as subsets of $X(T)_+$ via $L(\lambda) \mapsto \lambda$. Denote for any $\lambda \in X(T)_+$ the block of λ (or of $L(\lambda)$) by $b(\lambda)$. Obviously, 6.17 implies

(1) $$b(\lambda) \subset W_p \bullet \lambda \cap X(T)_+.$$

We shall not need more precise information than this for our present purposes. Let me, however, mention without proof the exact result. Most of the time we have equality in (1):

(2) *Let $\lambda \in X(T)_+$. Suppose that there is some $\alpha \in R^+$ with $\langle \lambda + \rho, \alpha^\vee \rangle \notin \mathbf{Z}p$. Then $b(\lambda) = W_p \bullet \lambda \cap X(T)_+$.*

This is proved in [Donkin 5]. The case where $\langle \lambda + \rho, \alpha^\vee \rangle \notin \mathbf{Z}p$ for all $\alpha \in R^+$ is particularly easy and had been treated before, see [Humphreys and Jantzen].

For λ not satisfying the assumption in (2) we no longer have equality in (1). This was first observed in the case of SL_2 in [Winter], and then proved for all groups in [Haboush 3] for the $(p^r - 1)\rho$ with $r \in \mathbf{N}$, using central characters. In order to formulate the precise description of $b(\lambda)$ in general, let $W_p^{(r)}$ denote for any r the subgroup of W_p generated by W and all translations by $p^{r+1}\nu$ with $\nu \in \mathbf{Z}R$. So $W_p^{(r)} = W_{p^{r+1}}$ in the notations from 6.1; this group is isomorphic to W_p. Now:

(3) *Let $\lambda \in X(T)_+$. Let r be the largest integer such that p^r divides all $\langle \lambda+\rho, \alpha^\vee \rangle$ with $\alpha \in R$. Then $b(\lambda) = W_p^{(r)} \bullet \lambda \cap X(T)_+$.*

If G is semi-simple and simply connected, then we can write $\lambda = p^r(\lambda' + \rho) - \rho$ with $\lambda' \in X(T)_+$. Then $H^0(\lambda) \simeq H^0(\lambda')^{[r]} \otimes St_r$ by 3.19. So the composition factors of $H^0(\lambda)$ are the $L(p^r(\mu' + \rho) - \rho) \simeq L(\mu')^{[r]} \otimes St_r$ (cf. 3.16) with $L(\mu')$ a composition factor of $H^0(\lambda')$. Now at least $b(\lambda) \subset W_p^{(r)} \bullet \lambda$ follows easily, but the precise statement also requires not too much work, cf. 10.5 below or [Humphreys and Jantzen], 2.4.

One can now easily extend this result to G of the form $G_1 \times G_2$ with G_2 a torus and G_1 semi-simple and simply connected. In general G has a central extension $G' \to G$ with G' a direct product $G_1 \times G_2$ as before. If T' is the inverse image of T in G', then the blocks of G are the blocks of G' contained in $X(T) \subset X(T')$.

7.3. For any G–module V and any $\lambda \in X(T)$ set $\mathrm{pr}_\lambda V$ equal to the sum of all submodules in V such that all its composition factors have a highest weight in $W_p \bullet \lambda$. Then $\mathrm{pr}_\lambda V$ is the largest submodule with this property.

It is clear from 7.2(1) that $\mathrm{pr}_\lambda V$ is a direct sum of some V_b with $b \in \mathcal{B}(G)$. If Z is a system of representatives for the W_p–orbits in $X(T)$, then 7.1(1)–(3) yield

(1) $$V = \bigoplus_{\lambda \in Z} \mathrm{pr}_\lambda V$$

and

(2) $$\mathrm{Hom}_G(V, V') \simeq \prod_{\lambda \in Z} \mathrm{Hom}_G(\mathrm{pr}_\lambda V, \mathrm{pr}_\lambda V')$$

or, more generally, for all $i \in \mathbf{N}$

(3) $$\operatorname{Ext}^i_G(V,V') \simeq \prod_{\lambda\in Z} \operatorname{Ext}^i_G(\operatorname{pr}_\lambda V, \operatorname{pr}_\lambda V').$$

Each $\operatorname{pr}_\lambda$ is an exact functor.

Note that we have, by definition, for all $\mu \in X(T)_+$

(4) $$\operatorname{pr}_\lambda L(\mu) = \begin{cases} L(\mu), & \text{if } \mu \in W_p \bullet \lambda,\\ 0, & \text{otherwise.}\end{cases}$$

The strong linkage principle implies for all $\mu \in X(T)$ and $i \in \mathbf{N}$

(5) $$\operatorname{pr}_\lambda H^i(\mu) = \begin{cases} H^i(\mu), & \text{if } \mu \in W_p \bullet \lambda,\\ 0, & \text{otherwise.}\end{cases}$$

A similar result holds for the $V(\mu)$ with $\mu \in X(T)_+$ because $V(\mu)$ and $H^0(\mu)$ have the same composition factors.

Recall from 2.12 that ${}^\tau L \simeq L$ for all simple G–modules L. This implies: If V is a finite dimensional G–module with $V = \operatorname{pr}_\lambda V$ for some λ, then also ${}^\tau V = \operatorname{pr}_\lambda({}^\tau V)$. Using (1) one gets now for all finite dimensional G–modules V that

(6) $$\operatorname{pr}_\lambda({}^\tau V) \simeq {}^\tau(\operatorname{pr}_\lambda V) \qquad \text{for all } \lambda \in X(T).$$

One has also $\operatorname{pr}_{\lambda^*}(V^*) \simeq (\operatorname{pr}_\lambda V)^*$ where $\lambda^* \in Z$ is the representative for the W_p–orbit of $-w_0\lambda$. (Note that $-w_0$ permutes the orbits for the dot action of W_p.)

7.4. Let V be a finite dimensional G–module. The exactness of $\operatorname{pr}_\lambda$ implies: If $\operatorname{ch} V = \sum_{i=1}^r a_i \operatorname{ch} V_i$ for some $a_i \in \mathbf{Z}$ and some finite dimensional G–modules V_i, then $\operatorname{ch}(\operatorname{pr}_\lambda V) = \sum_{i=1}^r a_i \operatorname{ch}(\operatorname{pr}_\lambda V_i)$. For example, we can write

(1) $$\operatorname{ch} V = \sum_{\mu\in X(T)} a_\mu\, \chi(\mu)$$

with almost all $a_\mu = 0$ and all $a_\mu \in \mathbf{Z}$. (This expression is unique if we take only $\mu \in X(T)_+$.) Then

(2) $$\operatorname{ch}(\operatorname{pr}_\lambda V) = \sum_{\mu\in W_p\bullet\lambda} a_\mu\, \chi(\mu)$$

by 7.3(5) and 5.9(1). If M is another finite dimensional G–module, then

(3) $$\operatorname{ch}(M\otimes V) = \sum_{\mu\in X(T)} \sum_{\nu\in X(T)} a_\mu \dim(M_\nu)\, \chi(\mu+\nu)$$

by 5.8.b. (Regard $M\otimes ?$ as an exact functor!) Composing the functors $\operatorname{pr}_\lambda$ and $M\otimes ?$, we get now from (2) and (3)

(4) $$\operatorname{ch}(\operatorname{pr}_\lambda(M\otimes V)) = \sum_{\mu,\nu} a_\mu \dim(M_\nu)\, \chi(\mu+\nu)$$

where we sum over all pairs $(\mu,\nu) \in X(T)\times X(T)$ with $\mu+\nu \in W_p \bullet \lambda$.

7.5. Lemma: *Let $\lambda, \mu \in X(T)$ and let M be a finite dimensional G–module.*

a) *The functors $\mathrm{pr}_\mu \circ (M \otimes ?) \circ \mathrm{pr}_\lambda$ and $\mathrm{pr}_\lambda \circ (M^* \otimes ?) \circ \mathrm{pr}_\mu$ are adjoint to each other.*

b) *Let V be a finite dimensional G–module with $\mathrm{pr}_\lambda V = V$. Write*

$$\mathrm{ch}\, V = \sum_{w \in W_p} a_w\, \chi(w \bullet \lambda)$$

with $a_w \in \mathbf{Z}$, almost all $a_w = 0$. Then

$$\mathrm{ch}\,(\mathrm{pr}_\mu(M \otimes V)) = \sum_{w \in W_p} a_w \sum_{\nu} \dim(M_\nu)\, \chi(w \bullet (\lambda + \nu)) \tag{1}$$

where we sum over all $\nu \in X(T)$ with $\lambda + \nu \in W_p \bullet \mu$.

Proof: a) We have by 7.3(2) for all G–modules V, V' canonical isomorphisms (using I.4.4(1))

$$\begin{aligned}\mathrm{Hom}_G(\mathrm{pr}_\mu(M \otimes \mathrm{pr}_\lambda(V)), V') &\simeq \mathrm{Hom}_G(\mathrm{pr}_\mu(M \otimes \mathrm{pr}_\lambda(V)), \mathrm{pr}_\mu(V'))\\ &\simeq \mathrm{Hom}_G(M \otimes \mathrm{pr}_\lambda(V), \mathrm{pr}_\mu(V')) \simeq \mathrm{Hom}_G(\mathrm{pr}_\lambda(V), M^* \otimes \mathrm{pr}_\mu(V'))\\ &\simeq \mathrm{Hom}_G(\mathrm{pr}_\lambda(V), \mathrm{pr}_\lambda(M^* \otimes \mathrm{pr}_\mu(V'))) \simeq \mathrm{Hom}_G(V, \mathrm{pr}_\lambda(M^* \otimes \mathrm{pr}_\mu(V'))).\end{aligned}$$

This proves that $\mathrm{pr}_\mu \circ (M \otimes ?) \circ \mathrm{pr}_\lambda$ is left adjoint to $\mathrm{pr}_\lambda \circ (M^* \otimes ?) \circ \mathrm{pr}_\mu$. As the situation is symmetric in λ and μ and as $(M^*)^* \simeq M$, we also get that it is right adjoint.

b) We have (cf. 7.4(3))

$$\mathrm{ch}\,(M \otimes V) = \sum_{w \in W_p} a_w \sum_{\nu \in X(T)} \dim(M_\nu)\, \chi((w \bullet \lambda) + \nu).$$

As $(w \bullet \lambda) + \nu = w \bullet (\lambda + w_1 \nu)$ for a suitable $w_1 \in W$ and as $\mathrm{ch}\, M \in \mathbf{Z}[X(T)]^W$, we get

$$\mathrm{ch}\,(M \otimes V) = \sum_{w \in W_p} a_w \sum_{\nu \in X(T)} \dim(M_\nu)\, \chi(w \bullet (\lambda + \nu)).$$

From this (1) follows immediately (as in 7.4).

7.6. Consider $\lambda, \mu \in \overline{C}_{\mathbf{Z}}$, cf. 5.5, 6.2(7). There is a unique $\nu_1 \in X(T)_+ \cap W(\mu - \lambda)$. We define the *translation functor* T_λ^μ from λ to μ via

$$T_\lambda^\mu V = \mathrm{pr}_\mu\,(L(\nu_1) \otimes \mathrm{pr}_\lambda V) \tag{1}$$

for each G–module V. It is a functor from $\{\, G\text{–modules}\,\}$ to itself.

Lemma: *Let $\lambda, \mu \in \overline{C}_{\mathbf{Z}}$.*

a) *The functor T_λ^μ is exact.*

b) *The functors T_λ^μ and T_μ^λ are adjoint to each other.*

Proof: The first claim is obvious as T_λ^μ is a composition of exact functors. The second claim follows from 7.5.a as $-w_0\nu_1 \in X(T)_+ \cap W(\lambda - \mu)$ and as $L(\nu_1)^* \simeq L(-w_0\nu_1)$ by 2.5.

Remarks: 1) It will follow from 7.7 below that we can replace $L(\nu_1)$ in the definition of T_λ^μ by any finite dimensional G–module M with $\dim(M_{\nu_1}) = 1$ and such that all weights ν of M satisfy $\nu \leq \nu_1$. For example, $H^0(\nu_1)$ or $V(\nu_1)$ would do.

2) The adjointness of T_λ^μ and T_μ^λ yields for any G–module V an isomorphism of functors

$$\operatorname{Hom}_G(V, ?) \circ T_\mu^\lambda \simeq \operatorname{Hom}_G(T_\lambda^\mu V, ?),$$

hence also isomorphisms of derived functors (cf. I.4.1(3))

$$\operatorname{Ext}_G^i(V, ?) \circ T_\mu^\lambda \simeq \operatorname{Ext}_G^i(T_\lambda^\mu V, ?).$$

We have therefore for each G–module V' isomorphisms

$$\operatorname{Ext}_G^i(V, T_\mu^\lambda V') \simeq \operatorname{Ext}_G^i(T_\lambda^\mu V, V') \tag{2}$$

for all i.

3) Using 7.3(6) and ${}^\tau L(\nu_1) \simeq L(\nu_1)$ one shows for all λ and μ as above and all finite dimensional G–modules V that

$$ {}^\tau(T_\mu^\lambda V) \simeq T_\mu^\lambda({}^\tau V). \tag{3}$$

(Note that ${}^\tau(V_1 \otimes V_2) \simeq {}^\tau V_1 \otimes {}^\tau V_2$ for all finite dimensional G–modules V_1, V_2.) One can show similarly that $(T_\mu^\lambda V)^* \simeq T_{-w_0\mu}^{-w_0\lambda}(V^*)$.

7.7. Lemma: *Let $\lambda, \mu \in \overline{C}_{\mathbf{Z}}$ and let $\nu_1 \in X(T)_+ \cap W(\mu - \lambda)$.*

a) *We have $\lambda + w\nu \notin W_p \bullet \mu$ for all $w \in W$ and $\nu \in X(T)_+$ with $\nu < \nu_1$.*

b) *If $w \in W$ with $\lambda + w\nu_1 \in W_p \bullet \mu$, then there is some $w_1 \in W_p$ with $w_1 \bullet \lambda = \lambda$ and $w_1 \bullet \mu = \lambda + w\nu_1$.*

Proof: By Weyl's character formula the set of weights of $V(\nu_1)$ or $H^0(\nu_1)$ is independent of k and of char(k). Comparing with the situation over $\mathbf{C}$, cf. [Hu1], 21.3, we see that this set is equal to $\{\, w\nu \mid w \in W,\ \nu \in X(T)_+,\ \nu \leq \nu_1 \,\}$. So we can express the claim of the lemma as follows: If ν is a weight of $H^0(\nu_1)$ with $\lambda + \nu \in W_p \bullet \mu$, then $\nu \in W\nu_1$ and there exists $w_1 \in W_p$ with $w_1 \bullet \lambda = \lambda$ and $w_1 \bullet \mu = \lambda + \nu$.

Let us work with this reformulation. Suppose $\lambda + \nu \in W_p \bullet \mu$. There is some alcove C' with $\lambda + \nu \in \overline{C'}$. Let d be the number of reflection hyperplanes (for W_p) separating C' and our standard alcove C from 6.2(6). We use induction on d. If $d = 0$, then $C = C'$. Both μ and $\lambda + \nu$ belong to $\overline{C}$ and are conjugate under W_p, hence equal. So we can take $w_1 = 1$ and have $\nu = \mu - \lambda \in W\nu_1$.

Suppose now $d > 0$. There is a wall F of C' such that C and C' are on different sides of the reflection hyperplane containing F. There are $\alpha \in R^+$ and $r \in \mathbf{Z}$ such that this hyperplane is given by $\langle x + \rho, \alpha^\vee \rangle = rp$. Suppose that $\langle x + \rho, \alpha^\vee \rangle < rp$ for $x \in C'$ and $\langle x + \rho, \alpha^\vee \rangle > rp$ for $x \in C$. (The other case can be treated similarly.) Set $C'' = s_{\alpha,rp} \bullet C'$. Then C'' and C are separated by one reflection hyperplane less than C' and C, so we can apply induction in case $\lambda + \nu \in \overline{C''}$. Suppose therefore that $\lambda + \nu \notin \overline{C''}$, i.e., that $\langle \lambda + \nu + \rho, \alpha^\vee \rangle < rp$. As $\lambda \in \overline{C}$, we have $\langle \lambda + \rho, \alpha^\vee \rangle \geq rp$, hence $\langle \nu, \alpha^\vee \rangle < 0$. We can write $s_{\alpha,rp} \bullet (\lambda + \nu) \in W_p \bullet \mu$ in the form $\lambda + \nu'$ where

$$\nu' = s_\alpha(\nu) - (\langle \lambda + \rho, \alpha^\vee \rangle - rp)\alpha = \nu + (rp - \langle \lambda + \nu + \rho, \alpha^\vee \rangle)\alpha. \tag{1}$$

This shows that $\nu < \nu' \leq s_\alpha(\nu)$; therefore ν', too, has to be a weight of $H^0(\nu_1)$, cf. [Hu1], 21.3. As $\lambda + \nu' \in W_p \bullet \mu \cap \overline{C''}$, we can apply induction and get $\nu' \in W\nu_1$ and some $w_2 \in W_2$ with $w_2 \bullet \lambda = \lambda$ and $w_2 \bullet \mu = \lambda + \nu'$. As $\nu' \in W\nu_1$ is an extremal weight of $H^0(\nu_1)$, it is impossible that both $\nu' + \alpha$ and $\nu' - \alpha$ are weights of $H^0(\nu_1)$. So the inequalities $\nu < \nu' \leq s_\alpha(\nu)$ where all terms are weights of $H^0(\nu_1)$ yield $\nu' = s_\alpha(\nu)$, hence $\nu \in W\nu_1$. Furthermore, (1) implies that $\langle \lambda + \rho, \alpha^\vee \rangle = rp$, hence $\lambda = s_{\alpha,rp} \bullet \lambda = s_{\alpha,rp} w_2 \bullet \lambda$ whereas $s_{\alpha,rp} w_2 \bullet (\lambda + \nu) = s_{\alpha,rp} \bullet (\lambda + \nu') = \lambda + \nu$. So $w_1 = s_{\alpha,rp} w_2$ works.

Remark: Let M be a finite dimensional G–module.

a) If all composition factors of M have the form $L(\nu)$ with $\nu < \nu_1$, then the lemma and 7.5(1) imply $\mathrm{pr}_\mu(M \otimes V) = 0$ for all G–modules V with $\mathrm{pr}_\lambda V = V$. (One has to apply 7.5(1) to all finite dimensional submodules of V and then to take direct limits.)

b) If $[M : L(\nu_1)] = 1$ and if all other composition factors of M have the form $L(\nu)$ with $\nu < \nu_1$, then $\mathrm{pr}_\mu(M \otimes V) \simeq \mathrm{pr}_\mu(L(\nu_1) \otimes V)$ for all V as above. This is an immediate consequence of a) and the exactness of pr_μ. It proves the first remark in 7.6.

7.8. Proposition: *Let $\lambda, \mu \in \overline{C}_{\mathbf{Z}}$. Let V be a finite dimensional G–module with $\mathrm{pr}_\lambda V = V$. Write $\mathrm{ch}\, V = \sum_{w \in W_p} a_w \, \chi(w \bullet \lambda)$ with $a_w \in \mathbf{Z}$ and almost all $a_w = 0$. Then*

$$\mathrm{ch}\; T_\lambda^\mu V = \sum_{w \in W_p} a_w \sum_{w_1} \chi(w w_1 \bullet \mu) \tag{1}$$

where w_1 runs through a system of representatives for $\mathrm{Stab}_{W_p}(\lambda)/(\mathrm{Stab}_{W_p}(\lambda) \cap \mathrm{Stab}_{W_p}(\mu))$.

Proof: Take ν_1 as in 7.7 and apply 7.5(1) to $M = L(\nu_1)$. By 7.7.a all weights ν of $L(\nu_1)$ with $\lambda + \nu \in W_p \bullet \mu$ belong to $W\nu_1$, hence satisfy $\dim L(\nu_1)_\nu = 1$, cf. 1.19(2). Furthermore, by 7.7.b all such $\lambda + \nu$ have the form $w_1 \bullet \mu$ with $w_1 \in \mathrm{Stab}_{W_p}(\lambda)$.

On the other hand, any $w_1 \bullet \mu$ with $w_1 \in \mathrm{Stab}_{W_p}(\lambda)$ has the form $\lambda + \nu$ with $\nu \in W\nu_1$ as $\mu - \lambda \in W\nu_1$. Therefore the $\lambda + \nu$ occurring in 7.5(1) are exactly the $w_1 \bullet \mu$ with $w_1 \in \mathrm{Stab}_{W_p}(\lambda)$. Each $w_1 \bullet \mu$ occurs only once in 7.5(1), so we have to take a system of representatives as claimed.

7.9. Proposition: *Suppose $\lambda, \mu \in \overline{C}_{\mathbf{Z}}$ belong to the same facet. Then T_λ^μ induces an equivalence of categories from $\{ G$–modules V with $\mathrm{pr}_\lambda V = V \}$ to $\{ G$–modules V with $\mathrm{pr}_\mu V = V \}$. The functor $T_\mu^\lambda \circ T_\lambda^\mu$ is isomorphic to pr_λ.*

Proof: The adjointness of T_μ^λ and T_λ^μ (cf. 7.6.b) yields for each G–module V a canonical isomorphism

$$\mathrm{Hom}_G(V, T_\mu^\lambda T_\lambda^\mu V) \simeq \mathrm{Hom}_G(T_\lambda^\mu V, T_\lambda^\mu V). \tag{1}$$

Let $\varphi_V : V \to T_\mu^\lambda T_\lambda^\mu V$ correspond to the identity on $T_\lambda^\mu V$. Then $V \mapsto \varphi_V$ is a natural transformation from the functor $V \mapsto V$ to $T_\mu^\lambda \circ T_\lambda^\mu$.

Suppose for the moment that $\dim V < \infty$ and $\mathrm{pr}_\lambda V = V$. Since λ and μ belong to the same facet, they have the same stabiliser in W_p. Therefore 7.8 yields

$\operatorname{ch} T_\mu^\lambda T_\lambda^\mu V = \operatorname{ch} V$. If $V \neq 0$, this implies $T_\lambda^\mu V \neq 0$, hence $\varphi_V \neq 0$. If V is simple, then φ_V has to be an isomorphism. The same follows then for any V by induction on the length of V.

Let V be arbitrary again. If $\operatorname{pr}_\lambda V = V$, then the local finiteness implies now that φ_V is an isomorphism. In general, we have $T_\lambda^\mu(\operatorname{pr}_{\lambda'} V) = 0$ for all $\lambda' \notin W_p \bullet \lambda$, hence also $\varphi_V(\operatorname{pr}_{\lambda'} V) = 0$ for these λ'. So φ_V induces an isomorphism $\operatorname{pr}_\lambda V \xrightarrow{\sim} T_\mu^\lambda \circ T_\lambda^\mu V$. Therefore $\operatorname{pr}_\lambda$ is isomorphic to $T_\mu^\lambda \circ T_\lambda^\mu$.

This shows that the restriction of $T_\mu^\lambda \circ T_\lambda^\mu$ to $\{\, G$–modules V with $\operatorname{pr}_\lambda V = V\,\}$ is isomorphic to the identity. As the assumptions are symmetric in λ and μ, we get that T_λ^μ induces an equivalence of categories as claimed.

7.10. Let $\lambda, \mu \in \overline{C}_{\mathbf{Z}}$ and let $\nu_1 \in X(T)_+ \cap W(\mu - \lambda)$. We want to describe the effect of T_λ^μ on G–modules of the form $H^i(w \bullet \lambda) = R^i \operatorname{ind}_B^G(w \bullet \lambda)$ with $w \in W_p$ and $i \in \mathbf{N}$.

The generalised tensor identity (I.4.8) yields

$$(1) \qquad \begin{aligned} T_\lambda^\mu H^i(w \bullet \lambda) &= \operatorname{pr}_\mu (L(\nu_1) \otimes R^i \operatorname{ind}_B^G(w \bullet \lambda)) \\ &\simeq \operatorname{pr}_\mu R^i \operatorname{ind}_B^G(L(\nu_1) \otimes w \bullet \lambda). \end{aligned}$$

Let us take a composition series of $L(\nu_1)$ considered as a B–module and tensor it with $w \bullet \lambda$. We get a composition series of $L(\nu_1) \otimes w \bullet \lambda$ which we denote by

$$0 = M_0 \subset M_1 \subset M_2 \subset \cdots \subset M_r = L(\nu_1) \otimes w \bullet \lambda.$$

The factors M_j/M_{j-1} have dimension one. There is $\lambda_j \in X(T)$ with $M_j/M_{j-1} \simeq \lambda_j + w \bullet \lambda$. The λ_j are the weights of T on $L(\nu_1)$ counted with their multiplicities. We may and shall choose the composition series of $L(\nu_1)$ in such a way that $\lambda_j < \lambda_{j'}$ implies $j < j'$.

The short exact sequences $0 \to M_{j-1} \to M_j \to M_j/M_{j-1} \to 0$ with $1 \leq j \leq r$ yield long exact sequences (recall 2.1(5))

$$(2) \qquad \cdots \to H^i(M_{j-1}) \to H^i(M_j) \to H^i(\lambda_j + w \bullet \lambda) \to H^{i+1}(M_{j-1}) \to \cdots$$

hence (using the exactness of pr_μ)

(3)

$$\cdots \to \operatorname{pr}_\mu H^i(M_{j-1}) \to \operatorname{pr}_\mu H^i(M_j) \to \operatorname{pr}_\mu H^i(\lambda_j + w \bullet \lambda) \to \operatorname{pr}_\mu H^{i+1}(M_{j-1}) \to \cdots$$

where (by 7.3(5))

$$(4) \qquad \operatorname{pr}_\mu H^i(\lambda_j + w \bullet \lambda) = \begin{cases} H^i(\lambda_j + w \bullet \lambda), & \text{if } \lambda_j + w \bullet \lambda \in W_p \bullet \mu, \\ 0, & \text{otherwise.} \end{cases}$$

The term $\operatorname{pr}_\mu H^i(M_r)$ in (3) for $j = r$ is by (1) just $T_\lambda^\mu H^i(w \bullet \lambda)$.

7.11. There are a few cases where the long exact sequences from 7.10 collapse.

Proposition: *Let $\lambda, \mu \in \overline{C}_{\mathbf{Z}}$ and let F be the facet with $\lambda \in F$. If $\mu \in \overline{F}$, then*

$$T_\lambda^\mu H^i(w \bullet \lambda) \simeq H^i(w \bullet \mu)$$

for all $w \in W_p$ and $i \in \mathbf{N}$.

Proof: Since μ belongs to the closure of the facet of λ, the stabiliser of λ in W_p is contained in the stabiliser of μ. So by 7.7/8 there is exactly one l with $\lambda_l + w \bullet \lambda \in W_p \bullet \mu$ (in the notations of 7.10) and it satisfies $\lambda_l + w \bullet \lambda = w \bullet \mu$. Therefore 7.10(3) yields $\operatorname{pr}_\mu H^i(M_j) = 0$ for all $j < l$ and $\operatorname{pr}_\mu H^i(M_j) \simeq H^i(w \bullet \mu)$ for all $j \geq l$ (and all i). Taking $j = r$ we get the claim.

7.12. Recall the notation $\Sigma = \Sigma(C)$ from 6.3.

Proposition: *Let $\lambda, \mu \in \overline{C}_{\mathbf{Z}}$. Suppose that $\mu \in C$ and that there exists $s \in \Sigma$ with $\mathrm{Stab}_{W_p}(\lambda) = \{1, s\}$. Let $w \in W_p$ with $ws \bullet \mu < w \bullet \mu$. Then there is a long exact sequence of G–modules*

$$0 \to H^0(ws \bullet \mu) \to T_\lambda^\mu H^0(w \bullet \lambda) \to H^0(w \bullet \mu) \to H^1(ws \bullet \mu) \to \cdots$$
$$\cdots \to H^i(ws \bullet \mu) \to T_\lambda^\mu H^i(w \bullet \lambda) \to H^i(w \bullet \mu) \to H^{i+1}(ws \bullet \mu) \to \cdots .$$

Proof: By 7.7/8 there are exactly two l with $\lambda_l + w \bullet \lambda \in W_p \bullet \mu$. They satisfy $\lambda_l + w \bullet \lambda = w \bullet \mu$ resp. $= ws \bullet \mu$. As $ws \bullet \mu < w \bullet \mu$, the l with $\lambda_l + w \bullet \lambda = ws \bullet \mu$ is smaller than the one for $w \bullet \mu$. Now everything follows from 7.10.

7.13. Proposition: *Let $\lambda, \mu \in \overline{C}_{\mathbf{Z}}$ and $w \in W_p$ with $w \bullet \lambda \in X(T)_+$. Then $T_\lambda^\mu H^0(w \bullet \lambda)$ has a filtration such that the factors are the $H^0(ww_1 \bullet \mu)$ with $w_1 \in \mathrm{Stab}_{W_p}(\lambda)$ and $ww_1 \bullet \mu \in X(T)_+$. Each different $ww_1 \bullet \mu$ occurs exactly once.*

Proof: We know by 7.7/8 that the $\lambda_j + w \bullet \lambda$ (as in 7.10) in $W_p \bullet \mu$ are exactly the $ww_1 \bullet \mu$ with $w_1 \in \mathrm{Stab}_{W_p}(\lambda)$. As $w \bullet \lambda \in X(T)_+$ and as $w \bullet \lambda \in \overline{ww_1 \bullet C}$, we have $\langle x + \rho, \alpha^\vee \rangle > 0$ for all $x \in ww_1 \bullet C$ and $\alpha \in R^+$. Now $ww_1 \bullet \mu \in \overline{ww_1 \bullet C}$, hence $\langle ww_1 \bullet \mu + \rho, \alpha^\vee \rangle \geq 0$ for all $\alpha \in R^+$ and thus $H^i(ww_1 \bullet \mu) = 0$ for all $i > 0$ by Kempf's vanishing theorem and 5.4.a. Therefore 7.10(3) implies inductively $\mathrm{pr}_\mu H^i(M_j) = 0$ for all $i > 0$ and all j. Furthermore, we get for j with $\lambda_j + w \bullet \lambda \notin W_p \bullet \mu$ an isomorphism $\mathrm{pr}_\mu H^0(M_{j-1}) \simeq \mathrm{pr}_\mu H^0(M_j)$, and for j with $\lambda_j + w \bullet \lambda \in W_p \bullet \mu$ a short exact sequence

$$0 \to \mathrm{pr}_\mu H^0(M_{j-1}) \to \mathrm{pr}_\mu H^0(M_j) \to H^0(\lambda_j + w \bullet \lambda) \to 0.$$

This yields the claim.

Remark: It follows that T_λ^μ takes modules with a good filtration (as in 4.16) to modules with a good filtration. Using 7.6(3) one gets a similar result for modules with a Weyl filtration (as in 4.19).

7.14. Lemma: *Let $\lambda, \mu \in \overline{C}_{\mathbf{Z}}$ such that μ belongs to the closure of the facet containing λ. Let $w \in W_p$ with $w \bullet \lambda \in X(T)_+$. If $w \bullet \mu \notin X(T)_+$, then $T_\lambda^\mu L(w \bullet \lambda) = 0$. If $w \bullet \mu \in X(T)_+$, then either $T_\lambda^\mu L(w \bullet \lambda) \simeq L(w \bullet \mu)$ or $T_\lambda^\mu L(w \bullet \lambda) = 0$.*

Proof: There is a homomorphism (where $n = |R^+|$)

$$H^n(w_0 w \bullet \lambda) \simeq V(w \bullet \lambda) \longrightarrow H^0(w \bullet \lambda)$$

with image equal to $L(w \bullet \lambda)$, cf. 2.14(1), 4.2(10). The exactness of T_λ^μ together with 7.11 implies that $T_\lambda^\mu L(w \bullet \lambda)$ is the image of some homomorphism

$$H^n(w_0 w \bullet \mu) \longrightarrow H^0(w \bullet \mu). \tag{1}$$

If $w \bullet \mu \notin X(T)_+$, then $H^0(w \bullet \mu) = 0$, hence also $T_\lambda^\mu L(w \bullet \lambda) = 0$. If $w \bullet \mu \in X(T)_+$, then any non-zero homomorphism as in (1) has image $L(w \bullet \mu)$, cf. the remark in 6.16. This yields the claim.

7.15. Proposition: *Let $\lambda, \mu \in \overline{C}_{\mathbf{Z}}$ such that μ belongs to the closure of the facet containing λ. Let $w \in W_p$ with $w \bullet \lambda \in X(T)_+$ and denote by F the facet with $w \bullet \lambda \in F$. Then*

$$T_\lambda^\mu L(w \bullet \lambda) \simeq \begin{cases} L(w \bullet \mu), & \text{if } w \bullet \mu \in \widehat{F}, \\ 0, & \text{otherwise.} \end{cases}$$

Proof: By the definition (6.2(3)) of the upper closure, $w \bullet \mu \notin X(T)_+$ implies $w \bullet \mu \notin \widehat{F}$. So in this case the claim follows from 7.14. We assume from now on that $w \bullet \mu \in X(T)_+$.

We know that $T_\lambda^\mu H^0(w \bullet \lambda) \simeq H^0(w \bullet \mu)$ by 7.11, so there has to be a composition factor $L(w'w \bullet \lambda)$ with $w' \in W_p$ of $H^0(w \bullet \lambda)$ such that $L(w \bullet \mu) \subset H^0(w \bullet \mu)$ is a composition factor of $T_\lambda^\mu L(w'w \bullet \lambda)$. By 7.14 we get more precisely $T_\lambda^\mu L(w'w \bullet \lambda) \simeq L(w'w \bullet \mu)$, hence $w'w \bullet \mu = w \bullet \mu$ and $w' \in W_p^0(w \bullet \mu) = \mathrm{Stab}_{W_p}(w \bullet \mu)$. Furthermore, $L(w'w \bullet \lambda)$ is the only composition factor L of $H^0(w \bullet \lambda)$ with $T_\lambda^\mu L \simeq L(w \bullet \mu)$ as $L(w \bullet \mu)$ occurs with multiplicity one in $H^0(w \bullet \mu)$.

Suppose at first $w \bullet \mu \in \widehat{F}$. Then each reflection $s \in W_p^0(w \bullet \mu)$ satisfies $sw \bullet \lambda \geq w \bullet \lambda$. Now [B3], ch. VI, §1, prop. 18 implies $w'w \bullet \lambda \geq w \bullet \lambda$ for all $w' \in W_p^0(w \bullet \mu)$. Therefore $L(w \bullet \lambda)$ is the only composition factor of $H^0(w \bullet \lambda)$ having the form $L(w'w \bullet \lambda)$ with $w' \in W_p^0(w \bullet \mu)$. So the argument from above implies that $T_\lambda^\mu L(w \bullet \lambda) \simeq L(w \bullet \mu)$.

Suppose now $w \bullet \mu \notin \widehat{F}$. Then there is a reflection $s' \in W_p^0(w \bullet \mu)$ with $s'w \bullet \lambda < w \bullet \lambda$. We can find $w_1 \in W_p^0(w \bullet \mu)$ with $w_1 w \bullet \lambda \uparrow w \bullet \lambda$ and $sw_1 w \bullet \lambda \geq w_1 w \bullet \lambda$ for all reflections $s \in W_p^0(w \bullet \mu)$. (If $s'w \bullet \lambda$ does not work, then iterate.) We know by the first case that $T_\lambda^\mu L(w_1 w \bullet \lambda) \simeq L(w \bullet \mu)$. If we can show that $L(w_1 w \bullet \lambda)$ is a composition factor of $H^0(w \bullet \lambda)$, then the uniqueness statement from above implies $T_\lambda^\mu L(w \bullet \lambda) = 0$ as claimed. Because of 6.23.b it is enough to show that $w_1 w \bullet \lambda$ is close to $w \bullet \lambda$. This, however, follows from 6.22.

7.16. Let us denote by Q_λ the injective hull of the G–module $L(\lambda)$ (for any $\lambda \in X(T)_+$).

Corollary: *Let $\lambda, \mu \in \overline{C}_{\mathbf{Z}}$ and $w \in W_p$ with $w \bullet \mu \in X(T)_+$. Suppose that $w \bullet \mu$ belongs to the upper closure of the facet containing $w \bullet \lambda$. Then $T_\mu^\lambda Q_{w \bullet \mu} \simeq Q_{w \bullet \lambda}$.*

Proof: Note that $w \bullet \lambda \in X(T)_+$ because a weight in the upper closure of the facet containing $w \bullet \lambda$ is dominant; so the claim makes sense.

As $T_\mu^\lambda Q_{w \bullet \mu}$ is a direct summand of $Q_{w \bullet \mu} \otimes E$ for some G–module E, it is also an injective G–module (cf. I.3.10.c), hence a direct sum of some $Q_{w' \bullet \lambda}$ with $w' \in W_p$ and $w' \bullet \lambda \in X(T)_+$. Each such $Q_{w' \bullet \lambda}$ occurs exactly $\dim \mathrm{Hom}_G(L(w' \bullet \lambda), T_\mu^\lambda Q_{w \bullet \mu})$ times. The adjointness of T_μ^λ and T_λ^μ yields

$$\mathrm{Hom}_G(L(w' \bullet \lambda), T_\mu^\lambda Q_{w \bullet \mu}) \simeq \mathrm{Hom}_G(T_\lambda^\mu L(w' \bullet \lambda), Q_{w \bullet \mu}).$$

So the multiplicity of $Q_{w' \bullet \lambda}$ is either 0 or 1. It is 1 if and only if $w \bullet \mu = w' \bullet \mu$ belongs to the upper closure of the facet containing $w' \bullet \lambda$. This holds for $w' \bullet \lambda = w \bullet \lambda$ by assumption, but not for any other $w' \bullet \lambda$. (Let C' be the unique alcove with $w \bullet \mu \in \widehat{C'}$. If $w \bullet \mu$ belongs to the upper closure of the facet containing $w' \bullet \lambda$, then also $w' \bullet \lambda \in \widehat{C'}$. But $\overline{C'}$ intersects $W_p \bullet \lambda$ in one element only.)

7.17. Corollary: *Let $\lambda, \mu \in \overline{C}_{\mathbf{Z}}$ and $w \in W_p$ with $w \bullet \lambda \in X(T)_+$. Suppose that $w \bullet \mu$ belongs to the upper closure of the facet containing $w \bullet \lambda$.*

a) *We have for all $w_1 \in W_p$ and $i \in \mathbf{N}$*

$$[H^i(w_1 \bullet \lambda) : L(w \bullet \lambda)] = [H^i(w_1 \bullet \mu) : L(w \bullet \mu)].$$

b) *If $\operatorname{ch} L(w \bullet \lambda) = \sum_{w' \in W_p} a_{w,w'}\, \chi(w' \bullet \lambda)$ with almost all $a_{w,w'} = 0$, then*

$$\operatorname{ch} L(w \bullet \mu) = \sum_{w' \in W_p} a_{w,w'}\, \chi(w' \bullet \mu).$$

Because T_λ^μ is exact, these claims are immediate consequences of 7.11 and 7.15.

7.18. Proposition: *Let $\lambda \in C \cap X(T)$ and $w \in W_p$ with $w \bullet \lambda \in X(T)_+$. Let $s \in \Sigma$ with $w \bullet \lambda < ws \bullet \lambda$. Then*

$$[H^i(w_1 \bullet \lambda) : L(w \bullet \lambda)] = [H^i(w_1 s \bullet \lambda) : L(w \bullet \lambda)]$$

for all $w_1 \in W_p$ and $i \in \mathbf{N}$.

Proof: We may suppose that $\mathcal{D}G$ is simply connected as going to a central extension does not change the multiplicities we consider. The existence of $\lambda \in C \cap X(T)$ implies $p \geq h$ by 6.2(10). So there is by 6.3(1) some $\mu \in \overline{C}_{\mathbf{Z}}$ with $\Sigma^0(\mu) = \{s\}$ using a notation from 6.3. Now $w \bullet \lambda < ws \bullet \lambda$ implies that $w \bullet \mu$ is in the upper closure of $w \bullet C$. As $w_1 \bullet \mu = w_1 s \bullet \mu$, both sides in the proposition are (by 7.17.a) equal to $[H^i(w_1 \bullet \mu) : L(w \bullet \mu)]$.

7.19. Proposition: *Let $\lambda \in C \cap X(T)$ and $w \in W_p$ with $w \bullet \lambda \in X(T)_+$. Let $s \in \Sigma$ with $w \bullet \lambda < ws \bullet \lambda$ and let $\mu \in \overline{C}_{\mathbf{Z}}$ with $\Sigma^0(\mu) = \{s\}$.*

a) *There exists a short exact sequence of G–modules*

$$0 \to H^0(w \bullet \lambda) \longrightarrow T_\mu^\lambda H^0(w \bullet \mu) \longrightarrow H^0(ws \bullet \lambda) \to 0.$$

We have $\operatorname{soc}_G T_\mu^\lambda H^0(w \bullet \mu) \simeq L(w \bullet \lambda)$.

b) *If M is a G–module with $T_\lambda^\mu M = 0$, then*

$$\operatorname{Ext}_G^i(M, H^0(w \bullet \lambda)) \simeq \operatorname{Ext}_G^{i-1}(M, H^0(ws \bullet \lambda))$$

for all $i \in \mathbf{N}$.

c) *We have for all $i \in \mathbf{N}$*

$$\operatorname{Ext}_G^i(L(ws \bullet \lambda), H^0(w \bullet \lambda)) \simeq \begin{cases} k, & \text{if } i = 1, \\ 0, & \text{otherwise.} \end{cases}$$

d) *We have for all $i \in \mathbf{N}$*

$$\operatorname{Ext}_G^i(H^0(ws \bullet \lambda), H^0(w \bullet \lambda)) \simeq \begin{cases} k, & \text{if } i \in \{0,1\}, \\ 0, & \text{otherwise.} \end{cases}$$

Proof: a) The first claim is a special case of 7.13. (Note that $w \bullet \mu \in X(T)_+$ as $w \bullet \mu$ belongs to the upper closure of $w \bullet C$.) Let $w' \in W_p$ with $w' \bullet \lambda \in X(T)_+$. By the adjointness of T_λ^μ and T_μ^λ (7.6.b) and by 7.15 we get

$$\begin{aligned} &\operatorname{Hom}_G(L(w' \bullet \lambda), T_\mu^\lambda H^0(w \bullet \mu)) \\ &\qquad \simeq \operatorname{Hom}_G(T_\lambda^\mu L(w' \bullet \lambda), H^0(w \bullet \mu)) \simeq \begin{cases} k, & \text{if } w' = w, \\ 0, & \text{otherwise.} \end{cases} \end{aligned}$$

This yields the claim on $\operatorname{soc}_G T_\mu^\lambda H^0(w \bullet \mu)$.

b) Apply $\operatorname{Hom}_G(M, ?)$ to the short exact sequence in a). By the adjointness of T_λ^μ and T_μ^λ we get

$$\begin{aligned} \operatorname{Ext}_G^i(M, T_\mu^\lambda H^0(w \bullet \mu)) &\simeq \operatorname{Ext}_G^i(T_\lambda^\mu M, H^0(w \bullet \mu)) \\ &= \operatorname{Ext}_G^i(0, H^0(w \bullet \mu)) = 0, \end{aligned}$$

hence the claim.

c) We have $T_\lambda^\mu L(ws \bullet \lambda) = 0$ by 7.15. So c) follows from b) and 4.13(2).

d) We have

$$\begin{aligned} &\operatorname{Ext}_G^i(H^0(ws \bullet \lambda), T_\mu^\lambda H^0(w \bullet \mu)) \simeq \operatorname{Ext}_G^i(T_\lambda^\mu H^0(ws \bullet \lambda), H^0(w \bullet \mu)) \\ &\qquad \simeq \operatorname{Ext}_G^i(H^0(w \bullet \mu), H^0(w \bullet \mu)) \simeq \begin{cases} k, & \text{if } i = 0, \\ 0, & \text{if } i > 0 \end{cases} \end{aligned}$$

by the adjointness as in the proof of b), by 7.11, 2.8, and 4.13(2). Apply Hom_G $(H^0(ws \bullet \lambda), ?)$ to the short exact sequence in a). If $i > 1$, then we get using 4.13(2)

$$\operatorname{Ext}_G^i(H^0(ws \bullet \lambda), H^0(w \bullet \lambda)) \simeq \operatorname{Ext}_G^{i-1}(H^0(ws \bullet \lambda), H^0(ws \bullet \lambda)) = 0.$$

Furthermore, there is an exact sequence

$$\begin{aligned} 0 \to \operatorname{Hom}_G(H^0(ws \bullet \lambda), H^0(w \bullet \lambda)) &\to \operatorname{Hom}_G(H^0(ws \bullet \lambda), T_\mu^\lambda H^0(w \bullet \mu)) \\ \to \operatorname{Hom}_G(H^0(ws \bullet \lambda), H^0(ws \bullet \lambda)) &\to \operatorname{Ext}_G^1(H^0(ws \bullet \lambda), H^0(w \bullet \lambda)) \to 0. \end{aligned} \tag{1}$$

The two terms in the middle are isomorphic to k, by the computation above and by 2.8. Any homomorphism $\varphi : H^0(ws \bullet \lambda) \to T_\mu^\lambda H^0(w \bullet \mu)$ annihilates $L(ws \bullet \lambda) = \operatorname{soc}_G H^0(ws \bullet \lambda)$ as $\operatorname{soc}_G T_\mu^\lambda H^0(w \bullet \mu) \simeq L(w \bullet \lambda)$ by a). Therefore $L(ws \bullet \lambda)$ is not a composition factor of $\operatorname{im}(\psi \circ \varphi)$ where $\psi : T_\mu^\lambda H^0(w \bullet \mu) \to H^0(ws \bullet \lambda)$ is the map from the sequence in a). But $L(ws \bullet \lambda) = \operatorname{soc}_G H^0(ws \bullet \lambda)$, hence $\psi \circ \varphi = 0$. Therefore the middle map in (1) is zero, hence all terms isomorphic to k.

Remark: The fact that $\operatorname{Hom}_G(H^0(ws \bullet \lambda), H^0(w \bullet \lambda)) \simeq k$ is also a special case of 6.24(2).

7.20. Lemma: *Let $\lambda \in C \cap X(T)$ and $\mu \in \overline{C}_{\mathbf{Z}}$. Suppose that there is $s \in \Sigma$ with $\Sigma^0(\mu) = \{s\}$. Let $w \in W_p$ with $w \bullet \lambda \in X(T)_+$ and $w \bullet \lambda < ws \bullet \lambda$.*

a) *We have* $\operatorname{soc}_G T_\mu^\lambda L(w \bullet \mu) \simeq L(w \bullet \lambda) \simeq T_\mu^\lambda L(w \bullet \mu)/\operatorname{rad}_G T_\mu^\lambda L(w \bullet \mu)$.

b) *We have* $[T_\mu^\lambda L(w \bullet \mu) : L(w \bullet \lambda)] = 2$ *and* $[T_\mu^\lambda L(w \bullet \mu) : L(ws \bullet \lambda)] = 1$.

c) *Let $w' \in W_p$ with $w' \bullet \lambda \in X(T)_+$. If $[T_\mu^\lambda L(w \bullet \mu) : L(w' \bullet \lambda)] > 0$ and $w' \bullet \lambda \neq w \bullet \lambda$, then $w's \bullet \lambda < w' \bullet \lambda$. On the other hand, if $w's \bullet \lambda < w' \bullet \lambda$, then*

$$\begin{aligned}\operatorname{Ext}_G^1(L(w \bullet \lambda), L(w' \bullet \lambda)) &\simeq \operatorname{Hom}_G(\operatorname{rad}_G T_\mu^\lambda L(w \bullet \mu), L(w' \bullet \lambda))\\ &\simeq \operatorname{Hom}_G(L(w' \bullet \lambda), T_\mu^\lambda L(w \bullet \mu)/\operatorname{soc}_G T_\mu^\lambda L(w \bullet \mu)).\end{aligned}$$

Proof: a) The first claim is proved in the same way as the analogous claim in 7.19.a with $H^0(w \bullet \mu)$ replaced by $L(w \bullet \mu)$. For the second claim one argues similarly looking at all $\operatorname{Hom}_G(T_\mu^\lambda L(w \bullet \mu), L(w' \bullet \lambda))$ with $w' \bullet \lambda \in X(T)_+$.

b) We have a unique expression with $a_{w,w'} \in \mathbf{Z}$

$$\operatorname{ch} L(w \bullet \lambda) = \sum_{w'} a_{w,w'}\, \chi(w' \bullet \lambda) \tag{1}$$

summing over all $w' \in W_p$ with $w' \bullet \lambda \in X(T)_+$. We get now from 7.17.b

$$\operatorname{ch} L(w \bullet \mu) = \sum_{w'} a_{w,w'}\, \chi(w' \bullet \mu), \tag{2}$$

hence from 7.8(1)

$$\operatorname{ch} T_\mu^\lambda L(w \bullet \mu) = \sum_{w'} a_{w,w'} \Big(\chi(w' \bullet \lambda) + \chi(w's \bullet \lambda)\Big) \tag{3}$$

and

$$\begin{aligned}\operatorname{ch} T_\lambda^\mu T_\mu^\lambda L(w \bullet \mu) &= \sum_{w'} a_{w,w'} \Big(\chi(w' \bullet \mu) + \chi(w's \bullet \mu)\Big)\\ &= 2\sum_{w'} a_{w,w'}\, \chi(w' \bullet \mu) = 2 \operatorname{ch} L(w \bullet \mu).\end{aligned} \tag{4}$$

So $T_\lambda^\mu T_\mu^\lambda L(w \bullet \mu)$ is an extension of $L(w \bullet \mu)$ by itself, and 2.12(1) implies

$$T_\lambda^\mu T_\mu^\lambda L(w \bullet \mu) \simeq L(w \bullet \mu) \oplus L(w \bullet \mu). \tag{5}$$

We have by a) clearly $[T_\mu^\lambda L(w \bullet \mu) : L(w \bullet \lambda)] > 0$. If $[T_\mu^\lambda L(w \bullet \mu) : L(w \bullet \lambda)] = 1$, then a) implies $T_\mu^\lambda L(w \bullet \mu) \simeq L(w \bullet \lambda)$, hence $T_\lambda^\mu T_\mu^\lambda L(w \bullet \mu) \simeq L(w \bullet \mu)$ contradicting (5). If $[T_\mu^\lambda L(w \bullet \mu) : L(w \bullet \lambda)] \geq 3$, then $T_\lambda^\mu T_\mu^\lambda L(w \bullet \mu)$ has at least 3 composition factors isomorphic to $L(w \bullet \mu) \simeq T_\lambda^\mu L(w \bullet \lambda)$, another contradiction to (5). So $[T_\mu^\lambda L(w \bullet \mu) : L(w \bullet \lambda)] = 2$.

Since $L(w \bullet \mu) = \operatorname{soc}_G H^0(w \bullet \mu)$, we can regard $T_\mu^\lambda L(w \bullet \mu)$ as a submodule of $T_\mu^\lambda H^0(w \bullet \mu)$. By 7.19.a there is a submodule M of $T_\mu^\lambda H^0(w \bullet \mu)$ with $M \simeq H^0(w \bullet \lambda)$

and $T_\mu^\lambda H^0(w \bullet \mu)/M \simeq H^0(ws \bullet \lambda)$. We cannot have $T_\mu^\lambda L(w \bullet \mu) \subset M$ because $[M : L(w \bullet \lambda)] = 1$ and $[T_\mu^\lambda L(w \bullet \mu) : L(w \bullet \lambda)] = 2$. So $T_\mu^\lambda L(w \bullet \mu)$ has a non-zero homomorphic image in $H^0(ws \bullet \lambda)$. Since $L(ws \bullet \mu) = \mathrm{soc}_G H^0(ws \bullet \lambda)$, we get $[T_\mu^\lambda L(w \bullet \mu) : L(ws \bullet \lambda)] > 0$. On the other hand, we have by 7.19.a obviously $[T_\mu^\lambda H^0(w \bullet \mu) : L(ws \bullet \lambda)] = 1$, hence also $[T_\mu^\lambda L(w \bullet \mu) : L(ws \bullet \lambda)] = 1$.

c) Each composition factor $L(w' \bullet \lambda)$ of $T_\mu^\lambda L(w \bullet \mu)$ with $w' \bullet \lambda < w's \bullet \lambda$ satisfies $T_\lambda^\mu L(w' \bullet \lambda) \simeq L(w' \bullet \mu)$ by 7.15, hence $w' \bullet \lambda = w \bullet \lambda$ by (5). This yields the first claim.

If $w's \bullet \lambda < w' \bullet \lambda$, then $T_\lambda^\mu L(w' \bullet \lambda) = 0$, hence $\mathrm{Ext}_G^i(T_\mu^\lambda L(w \bullet \mu), L(w' \bullet \lambda)) = 0 = \mathrm{Ext}_G^i(L(w' \bullet \lambda), T_\mu^\lambda L(w \bullet \mu))$ for all i by the adjointness. Now we have by a) short exact sequences

(6) $$0 \to \mathrm{rad}_G T_\mu^\lambda L(w \bullet \mu) \longrightarrow T_\mu^\lambda L(w \bullet \mu) \longrightarrow L(w \bullet \lambda) \to 0$$

and

(7) $$0 \to L(w \bullet \lambda) \to T_\mu^\lambda L(w \bullet \mu) \to T_\mu^\lambda L(w \bullet \mu)/\mathrm{soc}_G T_\mu^\lambda L(w \bullet \mu) \to 0.$$

To get the final claim, apply $\mathrm{Hom}_G(?, L(w' \bullet \lambda))$ to (6) and $\mathrm{Hom}_G(L(w' \bullet \lambda), ?)$ to (7) and use 2.12(4).

Remark: One can show for all $w' \in W_p$ with $w' \bullet \lambda \in X(T)_+$ and $w' \neq ws$ that

(8) $$[T_\mu^\lambda L(w \bullet \mu) : L(w' \bullet \lambda)] \leq 2\, [H^0(w \bullet \lambda) : L(w' \bullet \lambda)],$$

cf. the argument in [Jantzen 2], p. 146, or in [J1], 2.18.c.

7.21. Suppose again that $\lambda \in C \cap X(T)$ and $\mu \in \overline{C}_{\mathbf{Z}}$ and $s \in \Sigma$ with $\Sigma^0(\mu) = \{s\}$. Set

(1) $$\Theta = T_\mu^\lambda \circ T_\lambda^\mu.$$

One often calls Θ a *wall crossing functor*.

Recall from 7.7 that T_μ^λ and T_λ^μ are adjoint to each other. We have for all G–modules M, N with $\mathrm{pr}_\lambda M = M$ and $\mathrm{pr}_\mu N = N$ isomorphisms given by adjunction

(2) $$\mathrm{adj}_1 : \mathrm{Hom}_G(N, T_\lambda^\mu M) \xrightarrow{\sim} \mathrm{Hom}_G(T_\mu^\lambda N, M)$$

and

(3) $$\mathrm{adj}_2 : \mathrm{Hom}_G(M, T_\mu^\lambda N) \xrightarrow{\sim} \mathrm{Hom}_G(T_\lambda^\mu M, N).$$

The functorial properties of these maps can be expressed by the following formulae where we write $T = T_\mu^\lambda$ and $T' = T_\lambda^\mu$. One has for all maps f, g such that the formulae make sense

$$\begin{aligned} \mathrm{adj}_1(f \circ g) &= \mathrm{adj}_1(f) \circ T(g), & \mathrm{adj}_2(f \circ g) &= \mathrm{adj}_2(f) \circ T'(g),\\ \mathrm{adj}_1(T'(f) \circ g) &= f \circ \mathrm{adj}_1(g), & \mathrm{adj}_2(T(f) \circ g) &= f \circ \mathrm{adj}_2(g),\\ \mathrm{adj}_1^{-1}(f \circ g) &= T'(f) \circ \mathrm{adj}_1^{-1}(g), & \mathrm{adj}_2^{-1}(f \circ g) &= T(f) \circ \mathrm{adj}_2^{-1}(g),\\ \mathrm{adj}_1^{-1}(f \circ T(g)) &= \mathrm{adj}_1^{-1}(f) \circ g, & \mathrm{adj}_2^{-1}(f \circ T'(g)) &= \mathrm{adj}_2^{-1}(f) \circ g. \end{aligned}$$

Taking $N = T_\lambda^\mu M$ in (2) and (3) we get homomorphisms

(4) $$j_M : \Theta M \longrightarrow M \qquad \text{and} \qquad i_M : M \longrightarrow \Theta M$$

via $j_M = \mathrm{adj}_1(\mathrm{id}_{T_\lambda^\mu M})$ and $i_M = \mathrm{adj}_2^{-1}(\mathrm{id}_{T_\lambda^\mu M})$. If M' is another G–module with $\mathrm{pr}_\lambda(M') = M'$, then the formulae imply for each homomorphism $f : M \to M'$ of G–modules

(5) $$j_{M'} \circ \Theta(f) = f \circ j_M \qquad \text{and} \qquad \Theta(f) \circ i_M = i_{M'} \circ f.$$

Note that

(6) $$j_M = 0 \iff T_\lambda^\mu M = 0 \iff i_M = 0$$

(and that $T_\lambda^\mu M = 0 \iff \Theta M = 0$).

If L is a simple module with $T_\lambda^\mu L \neq 0$, i.e., if $L \simeq L(w \bullet \lambda)$ for some $w \in W_p$ with $w{\bullet}\lambda < ws{\bullet}\lambda$ and $w{\bullet}\lambda \in X(T)_+$, then this shows (together with Lemma 7.20.a) that i_L induces an isomorphism $L \xrightarrow{\sim} \mathrm{soc}\,\Theta L$ and that j_L induces an isomorphism $\Theta L/\mathrm{rad}\,\Theta L \xrightarrow{\sim} L$. We have $\mathrm{soc}\,\Theta L \subset \mathrm{rad}\,\Theta L$ by 7.20.b. This implies:

(7) *If M is semi-simple, then $i_M(M) = \mathrm{soc}\,\Theta M$ and $\ker(j_M) = \mathrm{rad}\,\Theta M$ and $j_M \circ i_M = 0$.*

Let $w \in W_p$ with $w \bullet \lambda \in X(T)_+$; apply (2) and (3) to $M = H^0(w \bullet \lambda)$ and $N = H^0(w \bullet \mu) \simeq T_\lambda^\mu M$. Since $\mathrm{End}_G(N) \simeq k$ by 2.8, it follows that $\mathrm{Hom}_G(M, \Theta M)$ and $\mathrm{Hom}_G(\Theta M, M)$ are one dimensional in this case. As i_M and j_M are non-zero by (6), they have to be bases for these Hom spaces. This implies:

(8) *We may assume in Proposition 7.19.a that the non-zero maps in the short exact sequence are $i_{H^0(w\bullet\lambda)}$ and $j_{H^0(ws\bullet\lambda)}$.*

7.22. Keep the assumptions (on λ, μ, s) and notations from 7.21.

Lemma: *Let M, N be G–modules with $M = \mathrm{pr}_\lambda M$ and $N = \mathrm{pr}_\lambda N$. Then we have*

(1) $$\mathrm{adj}_1(\varphi) \circ i_M = j_N \circ \mathrm{adj}_2^{-1}(\varphi)$$

for all $\varphi \in \mathrm{Hom}_G(T_\lambda^\mu M, T_\lambda^\mu N)$, and

(2) $$T_\lambda^\mu(\psi) = \mathrm{adj}_2(i_N \circ \psi) = \mathrm{adj}_1^{-1}(\psi \circ j_M)$$

for all $\psi \in \mathrm{Hom}_G(M, N)$.

Proof: We use the abbreviations $T = T_\mu^\lambda$ and $T' = T_\lambda^\mu$ and get from the formulae following 7.21(3):

$$\mathrm{adj}_1(\varphi) = \mathrm{adj}_1(\mathrm{id}_{T'N} \circ \varphi) = \mathrm{adj}_1(\mathrm{id}_{T'N}) \circ T(\varphi) = j_N \circ T(\varphi)$$

and

$$\mathrm{adj}_2^{-1}(\varphi) = \mathrm{adj}_2^{-1}(\varphi \circ \mathrm{id}_{T'M}) = T(\varphi) \circ \mathrm{adj}_2^{-1}(\mathrm{id}_{T'M}) = T(\varphi) \circ i_M$$

for all φ as above. So both sides in (1) are equal to $j_N \circ T(\varphi) \circ i_M$.

We get similarly for all ψ as above

$$T'(\psi) = \mathrm{adj}_2(\mathrm{adj}_2^{-1}(\mathrm{id}_{T'N} \circ T'(\psi))) = \mathrm{adj}_2(\mathrm{adj}_2^{-1}(\mathrm{id}_{T'N}) \circ \psi) = \mathrm{adj}_2(i_N \circ \psi)$$

and

$$T'(\psi) = \mathrm{adj}_1^{-1}(\mathrm{adj}_1(T'(\psi) \circ \mathrm{id}_{T'M})) = \mathrm{adj}_1^{-1}(\psi \circ \mathrm{adj}_1(\mathrm{id}_{T'M})) = \mathrm{adj}_1^{-1}(\psi \circ j_M).$$

Remark: The lemma implies that we have a commutative diagram

$$\begin{array}{ccccc} \mathrm{Hom}_G(M,N) & \longrightarrow & \mathrm{Hom}_G(M,\Theta N) & \longrightarrow & \mathrm{Hom}_G(M,N) \\ \| & & \downarrow & & \| \\ \mathrm{Hom}_G(M,N) & \longrightarrow & \mathrm{Hom}_G(\Theta M,N) & \longrightarrow & \mathrm{Hom}_G(M,N) \end{array}$$

where the vertical map in the middle is equal to $\mathrm{adj}_1 \circ \mathrm{adj}_2$, where the maps in the upper row are $f \mapsto i_N \circ f$ and $g \mapsto j_N \circ g$, the maps in the lower row $f \mapsto f \circ j_M$ and $g \mapsto g \circ i_M$. We get now for each i an induced commutative diagram

$$\begin{array}{ccccc} \mathrm{Ext}^i_G(M,N) & \longrightarrow & \mathrm{Ext}^i_G(M,\Theta N) & \longrightarrow & \mathrm{Ext}^i_G(M,N) \\ \| & & \downarrow & & \| \\ \mathrm{Ext}^i_G(M,N) & \longrightarrow & \mathrm{Ext}^i_G(\Theta M,N) & \longrightarrow & \mathrm{Ext}^i_G(M,N) \end{array}$$

(Choose an injective resolution for N, apply the exact functor Θ to get an injective resolution for ΘN, and apply 7.21(5).)

CHAPTER 8

Filtrations of Weyl Modules

Assume that k is a field of characteristic $p \neq 0$. In this chapter we construct for any $\lambda \in X(T)_+$ a filtration $V(\lambda) = V(\lambda)^0 \supset V(\lambda)^1 \supset V(\lambda)^2 \supset \cdots$ of the Weyl module $V(\lambda)$, with $V(\lambda)^i = 0$ for large i and with $V(\lambda)/V(\lambda)^1 \simeq L(\lambda)$. Furthermore, we have an explicit "sum formula" for $\sum_{i>0} \operatorname{ch} V(\lambda)^i$, cf. 8.19. This formula combined with results in earlier chapters is, so far, the most efficient tool for computing all $\operatorname{ch} L(\mu)$ with $\mu \in X(T)_+$. For example, one gets them all for low rank groups (of type A_1, A_2, A_3, B_2, G_2). These examples were one of the motivating factors that led to Lusztig's conjecture on these characters; we discuss this conjecture in 8.22.

One can define a Weyl module $V(\lambda)_A$ for G_A over any ring A. The filtration mentioned above arises from a filtration of $V(\lambda)_{\mathbf{Z}}$ and the sum formula follows from a result over $\mathbf{Z}$. So we are concerned in this chapter most of the time with representations of $G_{\mathbf{Z}}$ or, more generally, of G_A for some Dedekind ring A of characteristic 0. (The assumption $\operatorname{char}(A) = 0$ can be dropped at the beginning.) Denote by K the field of fractions of A.

We start in 8.2 with a criterion for an A–lattice in a G_K–module to be G_A–stable. This is then applied to the $V(\lambda)_A$. We determine in 8.4 the annihilator in $\operatorname{Lie}(B_k)$ of a highest weight vector in $V(\lambda)$. This can be used to prove that ample line bundles on G_k/B_k are very ample.

We consider the induction functor from B_A to G_A and its derived functors (denoted by H^i_A, $i \geq 0$). We discuss in 8.7 for which λ and i the G_A–modules $H^i_A(\lambda)$ are torsion modules over A and determine the character of $H^i_A(\lambda)$ divided by its torsion submodule. We generalise Kempf's vanishing theorem to A (8.8) and identify $V(\lambda)_A$ in 8.9 with $H^n_A(w_0 \bullet \lambda)$ where $n = |R^+|$.

The filtrations are constructed using a certain homomorphism $\varphi : V(\lambda)_{\mathbf{Z}} \to H^0_{\mathbf{Z}}(\lambda)$: one sets $V(\lambda)^i_{\mathbf{Z}} = \{v \in V(\lambda)_{\mathbf{Z}} \mid \varphi(v) \in p^i H^0_{\mathbf{Z}}(\lambda)\}$ and defines $V(\lambda)^i$ as the image of $V(\lambda)^i_{\mathbf{Z}} \otimes_{\mathbf{Z}} k$ in $V(\lambda) = V(\lambda)_{\mathbf{Z}} \otimes_{\mathbf{Z}} k$. The sum formula follows from formulae for the determinant of φ on each weight space of $V(\lambda)_{\mathbf{Z}}$. In the case where G has semi-simple rank 1, one can write down the map explicitly and then read off the determinant (8.13/14). The general case is reduced to this special case (8.15/16).

The filtration can also be (and was, in fact, at first) constructed using a symmetric bilinear form (the "contravariant" form) on $V(\lambda)_{\mathbf{Z}}$. This alternative approach is discussed in 8.19. This bilinear form was first mentioned in [St1] and independently in [Wong 1]. Both authors prove: If we reduce the form modulo p, then its radical is the radical of the G–module $V(\lambda)$, so the factor module by this radical is $L(\lambda)$. For G of type A_n the determinants of this bilinear form (on the weight spaces) were computed in [Jantzen 1]; that paper contained also a conjecture for the general case. In [Jantzen 3] it was proved that the p–adic valuation of each determinant is equal to the conjectured one provided $p \geq h$ (the Coxeter number). That one can use the bilinear form to construct filtrations is one of the main ideas behind [Jantzen 3], but is made explicit only in [Jantzen 5]. The proof of the conjecture for all p together with the new construction via homomorphisms and reduction to the case of semi-simple rank 1 is due to [Andersen 12], where one can also find some results on arbitrary $H^{l(w)}(w \bullet \lambda)$ with $w \in W$. For all other non-trivial results one can find references within the text.

Compared to the first edition the remarks in 8.19 and the subsections 8.20–21 have been expanded and some material from 8.20 has been moved to 8.21. The new 8.22 contains expanded material from the old 7.20. Most of the old 8.23 has been integrated into the new Chapter D.

8.1. Let A be an integral domain and K its field of fractions.

We can identify $A[G_A]$ resp. $\mathrm{Dist}(G_A)$ with subalgebras of $K[G_K] \simeq A[G_A] \otimes_A K$ resp. $\mathrm{Dist}(G_K) \simeq \mathrm{Dist}(G_A) \otimes_A K$ as these A–algebras are free modules over A, cf. 1.1, 1.12. Similar statements hold with G_A replaced by the big cell $U_A^+ B_A$, cf. 1.9. Recall that $A[G_A]$ is a subalgebra of $A[U_A^+ B_A]$ and that $\mathrm{Dist}(G_A) = \mathrm{Dist}(U_A^+ B_A, 1)$, as $U_A^+ B_A$ is open in G_A. (We can also replace A by K in these statements.) Furthermore,

(1) $$A[G_A] = K[G_K] \cap A[U_A^+ B_A]$$

cf. 1.9(5). Using this we can show

(2) $$A[G_A] = \{\, f \in K[G_K] \mid \mu(f) \in A \text{ for all } \mu \in \mathrm{Dist}(G_A)\,\}.$$

By definition $A[G_A]$ is contained in the right hand side of (2). In order to get the other inclusion it is by (1) enough to show that the right hand side of (2) is contained in $A[U_A^+ B_A]$. So we want to prove that the scheme $X = U_A^+ B_A$ satisfies

(3) $$A[X] = \{\, f \in K[X_K] \mid \mu(f) \in A \text{ for all } \mu \in \mathrm{Dist}(X, 1)\,\}.$$

Now X is a direct product of schemes isomorphic to $G_{a,A}$ or $G_{m,A}$. Using the explicit description of $\mathrm{Dist}(G_a)$ and $\mathrm{Dist}(G_m)$ in I.7.3/8, an elementary computation (left to the reader) yields (3).

8.2. According to our general conventions from Chapter 1 the group G satisfies the assumptions of Proposition I.10.12 and we might have deduced 8.1(2) from that proposition. I have, however, preferred the above approach, as the smoothness of G is really proved by arguments like those leading to 1.9(5) and 8.1(2).

Arguing as in I.10.13 we get from 8.1(2):

(1) *Let V be a finite dimensional G_K–module and let M be an A–lattice in V. Then M is G_A–stable if and only if* $\mathrm{Dist}(G_A)\, M = M$.

8.3. Let $\lambda \in X(T)_+ = X(T_{\mathbf{Q}})_+$ and let $V_{\mathbf{Q}}$ be a simple $G_{\mathbf{Q}}$–module with highest weight λ. We know (by 5.6/11) that $V_{\mathbf{Q}} \simeq \mathrm{ind}_{B_{\mathbf{Q}}}^{G_{\mathbf{Q}}} \lambda$ and that $\mathrm{ch}(V_{\mathbf{Q}}) = \chi(\lambda)$ is given by Weyl's character formula. Choose $v \in V_{\mathbf{Q},\lambda}$, $v \neq 0$. Consider the basis of $\mathrm{Dist}(G_{\mathbf{Z}})$ as in 1.12(4), but with R^+ and $-R^+$ interchanged. As λ is maximal among the weights of V, we have $X_{\alpha,n}v = 0$ for all $\alpha > 0$ and $n > 0$ since $X_{\alpha,n}v \in V_{\mathbf{Q},\lambda+n\alpha}$. Furthermore, each $H_{i,m}$ acts as multiplication by some integer on v. Hence

(1) $$V_{\mathbf{Z}} := \operatorname{Dist}(G_{\mathbf{Z}})v = \sum \mathbf{Z} \prod_{\alpha>0} X_{-\alpha,n(\alpha)}v$$

summing over all R^+–tuples $(n(\alpha))_\alpha$ of natural numbers. Since $\prod_{\alpha>0} X_{-\alpha,n(\alpha)}v$ has weight $\lambda - \sum n(\alpha)\alpha$, only finitely many of these elements are non-zero. Therefore $V_{\mathbf{Z}}$ is a finitely generated $\mathbf{Z}$–module. As $V_{\mathbf{Q}}$ is simple, we have $V_{\mathbf{Q}} = \operatorname{Dist}(G_{\mathbf{Q}})v = \mathbf{Q}V_{\mathbf{Z}}$. Therefore $V_{\mathbf{Z}}$ is a lattice in $V_{\mathbf{Q}}$ and 8.2 implies:

(2) *$V_{\mathbf{Z}}$ is a $G_{\mathbf{Z}}$–stable lattice in $V_{\mathbf{Q}}$.*

We can therefore form for any ring A the G_A–module $V_A = V_{\mathbf{Z}} \otimes_{\mathbf{Z}} A$. If K is a field of characteristic 0, then $V_K \simeq V_{\mathbf{Q}} \otimes_{\mathbf{Q}} K$ and V_K is a simple G_K–module with highest weight λ, cf. 2.9. For an arbitrary field K, the G_K–module V_K is generated by the B_K^+–stable line $K(v \otimes 1)$ as $V_K = \operatorname{Dist}(G_K)(v \otimes 1)$, cf. I.7.15. Now 2.13 implies that V_K is a homomorphic image of $V(\lambda)_K = \operatorname{ind}_{B_K}^{G_K}(-w_0\lambda)^*$ (using obvious notations). Both $\operatorname{ch} V_K = \operatorname{ch} V_{\mathbf{Z}} = \operatorname{ch} V_{\mathbf{Q}}$ and $\operatorname{ch} V(\lambda)_K$ are given by Weyl's character formula, cf. 5.11, so we get an isomorphism

(3) $$V(\lambda)_K \simeq V_{\mathbf{Z}} \otimes_{\mathbf{Z}} K.$$

(This result is due to Humphreys, cf. [Jantzen 3], Satz 1. For G of type A_n it had been known before, cf. [Jantzen 2], Satz 12.) Because of (3) we denote $V_{\mathbf{Z}}$ henceforth by $V(\lambda)_{\mathbf{Z}}$ and set $V(\lambda)_A = V(\lambda)_{\mathbf{Z}} \otimes_{\mathbf{Z}} A$ for any ring A.

8.4. Keep the assumptions from 8.3. We have by 8.3(1)

(1) $$V_{\mathbf{Z},\lambda-n\alpha} = \mathbf{Z}X_{-\alpha,n}\,v$$

for all simple roots $\alpha \in S$ and all $n \in \mathbf{N}$. As $s_\alpha(\lambda) = \lambda - \langle \lambda, \alpha^\vee \rangle \alpha$ is a weight of $V_{\mathbf{Q}}$ and as $s_\alpha(\lambda) - r\alpha = s_\alpha(\lambda + r\alpha)$ with $r > 0$ is not (by 1.19(1)), we have $X_{-\alpha,\langle\lambda,\alpha^\vee\rangle}\,v \neq 0$ and $X_{-\alpha,n}\,v = 0$ for all $n > \langle \lambda, \alpha^\vee \rangle$.

Set

(2) $$I(\lambda) = \{\, \alpha \in S \mid \langle \lambda, \alpha^\vee \rangle = 0 \,\}.$$

Lemma: *Let $v \in V_{\mathbf{Z}}$ with $V_{\mathbf{Z},\lambda} = \mathbf{Z}v$. Let k be a field. Then $v \otimes 1 \in V_{\mathbf{Z}} \otimes_{\mathbf{Z}} k$ satisfies*
$$\{\, \alpha \in R^+ \mid X_{-\alpha}\,(v \otimes 1) = 0 \,\} = R^+ \cap R_{I(\lambda)}.$$

Proof: As λ is dominant, we have

(3) $$R_{I(\lambda)} = \{\, \alpha \in R \mid \langle \lambda, \alpha^\vee \rangle = 0 \,\}.$$

If $\alpha \in R^+ \cap R_{I(\lambda)}$, then $\lambda - \alpha = s_\alpha(\lambda + \alpha)$ is not a weight of $V_{\mathbf{Q}}$, hence $0 = X_{-\alpha}v \in V_{\mathbf{Q},\lambda-\alpha}$ and therefore also $X_{-\alpha}\,(v \otimes 1) = 0$.

Suppose now that $\alpha \notin R_{I(\lambda)}$. Using $\lambda([X_\alpha, X_{-\alpha}]) = \langle \lambda, \alpha^\vee \rangle \neq 0$ and $X_\alpha v = 0$, we get

$$X_\alpha X_{-\alpha} v = [X_\alpha, X_{-\alpha}]\, v = \langle \lambda, \alpha^\vee \rangle\, v \neq 0,$$

hence $X_{-\alpha} v \neq 0$. It is therefore enough to show for all $\alpha \in R^+$ that

$$\mathbf{Z} X_{-\alpha} v = \mathbf{Q} X_{-\alpha} v \cap V_{\mathbf{Z}}. \tag{4}$$

For $\alpha \in S$ this is a trivial consequence of (1). Suppose now that $\alpha \notin S$. Then there is some $\beta \in S$ with $\langle \alpha, \beta^\vee \rangle > 0$. Let us assume that α, β do not belong to a component of R of type G_2. Then $\alpha + \beta \notin R$, hence $[X_\beta, X_{-\alpha}] = \pm X_{-(\alpha-\beta)}$ as the $(X_\gamma)_{\gamma \in R}$ form a Chevalley system, cf. 1.12 and [B3], ch. VIII, §2, prop. 7.

Suppose (4) not to hold. Then there is an integer $m > 1$ with $m^{-1} X_{-\alpha} v \in V_{\mathbf{Z}}$, hence with

$$X_\beta(m^{-1} X_{-\alpha} v) = m^{-1} [X_\beta, X_{-\alpha}]\, v = \pm m^{-1} X_{-(\alpha-\beta)} v \in V_{\mathbf{Z}}.$$

If $X_{-(\alpha-\beta)} v \neq 0$, then this shows that (4) is false already for $\alpha - \beta$. In this case we use induction and get a contradiction.

Suppose now that $X_{-(\alpha-\beta)} v = 0$, i.e., that $\alpha - \beta \in R_{I(\lambda)}$. The $\mathbf{Z}$–module $V_{\mathbf{Z}, \lambda-\alpha}$ is spanned by all $\prod_{\gamma>0} X_{-\gamma, n(\gamma)} v$ with $\sum_{\gamma>0} n(\gamma)\gamma = \alpha$. We may assume that the order of the factors in these products has been chosen such that we first apply all $X_{-\gamma, n(\gamma)}$ with $\gamma \in R_{I(\lambda)}$ to v. Since $X_{-\gamma, n} v = 0$ for all $\gamma \in R_{I(\lambda)}$ and $n > 0$, we see that $V_{\mathbf{Z}, \lambda-\alpha}$ is spanned already by all $\prod_{\gamma>0} X_{-\gamma, n(\gamma)} v$ with $\sum_{\gamma>0} n(\gamma)\gamma = \alpha$ and with $n(\gamma) = 0$ for all $\gamma \in R_{I(\lambda)}$. Now $\alpha - \beta \in R_{I(\lambda)}$ and $\beta \in S$ imply: If $\alpha = \sum_{\gamma>0} n(\gamma)\gamma$, then there is at most one γ with $n(\gamma) > 0$ and $\gamma \notin R_{I(\lambda)}$. Since also $n(\gamma) = 0$ for all $\gamma \in R_{I(\lambda)}$, we get $n(\gamma) = 0$ for all but one γ, and $\alpha = n(\gamma)\gamma$ for that single γ. Then necessarily $\gamma = \alpha$ and $n(\gamma) = 1$. This shows that $V_{\mathbf{Z}, \lambda-\alpha} = \mathbf{Z} X_{-\alpha} v$, hence (4) in this case.

Instead of also going through the proof in case G_2 let me refer you to the original proof by V. V. Deodhar to be found in [Lakshmibai, Musili, and Seshadri 3], Lemma 5.8.

8.5. Lemma 8.4 implies that $\mathrm{Lie}(P^+_{I(\lambda)})$ is the stabiliser in $\mathrm{Lie}(G_k)$ of the line $k(v \otimes 1)$. It is easy to see that $P^+_{I(\lambda)}$ is the stabiliser in G_k of this line. Therefore $g \mapsto g\, k(v \otimes 1)$ defines an embedding of $G_k/P^+_{I(\lambda)}$ into the projective space $\mathbf{P}(V(\lambda)_k)$. This embedding gives rise to a very ample sheaf on $G_k/P^+_{I(\lambda)}$ (cf. [Ha], p. 120), which can be checked to be $\mathcal{L}_{G_k/P^+_{I(\lambda)}}(-\lambda)$.

Interchanging the rôles of R^+ and $-R^+$ we get especially:

(1) *For all $\lambda \in X(T)_+$ with $I(\lambda) = \emptyset$ the sheaf $\mathcal{L}_{G_k/B_k}(\lambda)$ is very ample.*

8.6. From now on (in this chapter) we assume that A is a Dedekind domain and that K is its field of fractions.

For any A–module M we denote its torsion submodule by M_{tor} and by $M_{\mathrm{fr}} = M/M_{\mathrm{tor}}$ its torsion free quotient. Note that M_{fr} is a projective A–module if M is finitely generated ([B2], ch. VII, §4, prop. 22).

If M is a G_A–module, then M_{tor} is a G_A–submodule (I.10.2). So also M_{fr} has a natural structure as a G_A–module. If M is finitely generated, then $\mathrm{ch}(M_{\mathrm{fr}})$ is defined and obviously equal to $\mathrm{ch}(M \otimes_A K)$.

We shall use (not only for our A, but for any ring) the notation

$$H_A^i(M) = R^i \operatorname{ind}_{B_A}^{G_A}(M) \simeq H^i(G_A/B_A, \mathcal{L}_{G_A/B_A}(M)) \tag{1}$$

for any B_A–module M and any $i \in \mathbf{Z}$. Of course, $H_A^i(M) = 0$ for $i < 0$.

Lemma: *Let M be a B_A–module such that M is a finitely generated torsion module over A.*

a) *Each $H_A^i(M)$ is a finitely generated torsion module over A.*

b) *The B_A–module M has finite length. Its composition factors have the form $(A/\mathfrak{p})_\lambda$ with $\lambda \in X(T)$ and a maximal ideal $\mathfrak{p}$ of A.*

Proof: a) Each H_A^i is a functor. It maps the B_A–endomorphism "multiplication with a on M" for any a to "multiplication with a on $H_A^i(M)$". Therefore $aM = 0$ implies $aH_A^i(M) = 0$. For the finite generation use I.5.12.c.

b) We have a direct sum decomposition $M = \bigoplus_{\lambda \in X(T)} M_\lambda$ with $M_\lambda \neq 0$ for at most finitely many λ. Any A–submodule of the form $\bigoplus_{\mu \leq \lambda} M_\mu$ is B_A–stable (cf. 1.19). So the M_μ can be regarded as the subquotients of some filtration of M, and we may restrict ourselves to the case $M = M_\lambda$. Then U_A acts trivially on M and T_A acts through scalars, hence any A–submodule is also a B_A–submodule. So we have to know that a finitely generated torsion module over A has finite length and that the simple A–modules have the form $A/\mathfrak{p}$ with $\mathfrak{p}$ a maximal ideal. The second part is trivial, of course, for any ring, for the first part compare [B2], ch. VII, §2, lemme 1.

8.7. Recall from I.4.18.b that we have for each $\mu \in X(T)$, each $i \in \mathbf{N}$, and each A–algebra A' an exact sequence of $G_{A'}$–modules

$$0 \to H_A^i(\mu) \otimes_A A' \longrightarrow H_{A'}^i(\mu) \longrightarrow \operatorname{Tor}_1^A(H_A^{i+1}(\mu), A') \to 0. \tag{1}$$

We can write any $\mu \in X(T)$ in the form $\mu = w \bullet \lambda$ with $w \in W$ and $\langle \lambda + \rho, \alpha^\vee \rangle \geq 0$ for all $\alpha \in R^+$. If $\operatorname{char}(K) = 0$, then we know by 5.5 that $H_K^i(\mu) = 0$ if $\lambda \notin X(T)_+$ or $i \neq l(w)$, whereas $\operatorname{ch} H_K^{l(w)}(w \bullet \lambda) = \chi(\lambda) = (-1)^{l(w)} \chi(w \bullet \lambda)$, see 5.11 and 5.9(1). Therefore (1) yields (using 6.15 for the claim on $H_A^i(w \bullet \lambda)_\lambda$):

(2) *Suppose that char$(K) = 0$. If $\lambda \notin X(T)_+$ or $i \neq l(w)$, then $H_A^i(w \bullet \lambda)$ is a torsion module over A with $H_A^i(w \bullet \lambda)_\lambda = 0$. We have*

$$\operatorname{ch} H_A^{l(w)}(w \bullet \lambda)_{\mathrm{fr}} = \chi(\lambda) = (-1)^{l(w)} \chi(w \bullet \lambda). \tag{3}$$

Of course $H_A^{l(w)}(w \bullet \lambda)_{\mathrm{fr}}$ is then a G_A–stable lattice on $H_K^{l(w)}(w \bullet \lambda)$. Furthermore $H_A^{l(w)}(w \bullet \lambda)_\lambda$ is a projective A–module of rank 1 (by 6.16), if $\lambda \in X(T)_+$. [It actually is free: argue as for 8.8(1) below.]

If we apply the above to $A = \mathbf{Z}$ and take an arbitrary field k in (1) for A', then we see that $H_{\mathbf{Z}}^{l(w)}(w \bullet \lambda)_{\mathrm{fr}} \otimes_{\mathbf{Z}} k$ is isomorphic to a subquotient of $H_k^{l(w)}(w \bullet \lambda)$. This implies for any $\nu \in X(T)$

$$\dim_k H_k^{l(w)}(w \bullet \lambda)_\nu \geq \dim_{\mathbf{Q}} H_{\mathbf{Q}}^{l(w)}(w \bullet \lambda)_\nu. \tag{4}$$

8.8. Let $\lambda \in X(T)$ with $\langle \lambda + \rho, \alpha^\vee \rangle \geq 0$ for all $\alpha \in R^+$.

Kempf's vanishing theorem and 5.4 imply $H^i_{\mathbf{Z}/(p)}(\lambda) = 0$ for all $i > 0$ and all prime numbers p. Furthermore ch $H^0_{\mathbf{Z}/(p)}(\lambda) = \chi(\lambda) =$ ch $H^0_{\mathbf{Z}}(\lambda)_{\text{fr}}$. Therefore 8.7(1) [applied to $A = \mathbf{Z}$ with at first $A' = \mathbf{Z}/(p)$ for all p and then with arbitrary A'] yields for any ring A'

(1) $H^0_{A'}(\lambda) \simeq H^0_{\mathbf{Z}}(\lambda) \otimes_{\mathbf{Z}} A'$ *is free over* A'

and

(2) $H^i_{A'}(\lambda) = 0$ *for all* $i > 0$.

We know by Serre duality (4.2(9)) that $H^i_{\mathbf{Z}/(p)}(w_0 \bullet \lambda) = 0$ for all $i < n = |R^+| = l(w_0)$ (and all prime numbers), hence that ch $H^n_{\mathbf{Z}/(p)}(w_0 \bullet \lambda) = \chi(\lambda) =$ ch $H^n_{\mathbf{Z}}(w_0 \bullet \lambda)_{\text{fr}}$. This implies as above:

(3) $H^i_{A'}(w_0 \bullet \lambda) = 0$ *for all* $i < n$

and

(4) $H^n_{A'}(w_0 \bullet \lambda) \simeq H^n_{\mathbf{Z}}(w_0 \bullet \lambda) \otimes_{\mathbf{Z}} A'$ *is free over* A'.

As $w_0 \bullet 0 = -2\rho$, we get especially that $H^n_{\mathbf{Z}}(-2\rho)$ is a free lattice in $H^n_{\mathbf{Q}}(-2\rho) \simeq L(0)_{\mathbf{Q}} \simeq \mathbf{Q}$, hence that $H^n_{\mathbf{Z}}(-2\rho) \simeq \mathbf{Z}$ (with trivial $G_{\mathbf{Z}}$–action). The cup product yields for any $\lambda \in X(T)_+$ a bilinear map $H^0_{\mathbf{Z}}(\lambda) \times H^n_{\mathbf{Z}}(-\lambda - 2\rho) \to H^n_{\mathbf{Z}}(-2\rho) \simeq \mathbf{Z}$. Reduction modulo p leads for each prime number p to the non-degenerate pairing $H^0_{\mathbf{Z}/(p)}(\lambda) \times H^n_{\mathbf{Z}/(p)}(-\lambda - 2\rho) \to \mathbf{Z}/(p)$ giving rise to Serre duality. Therefore the pairing over $\mathbf{Z}$ has to yield an isomorphism (first for $A' = \mathbf{Z}$ and then by (1) and (4) in general)

(5) $$H^n_{A'}(-\lambda - 2\rho) \simeq H^0_{A'}(\lambda)^*.$$

If $H^1_A(\mu) \neq 0$ for some $\mu \in X(T)$, then $\mu \notin X(T)_+$ by (2), hence $H^0_{A/\mathfrak{p}}(\mu) = 0$ for all $\mathfrak{p}$ by 2.6. Therefore 8.7(1) applied with $i = 0$ implies:

(6) $H^1_A(\mu)$ *is torsion free over* A *for all* $\mu \in X(T)$.

This last result is taken from [Andersen 14]. There one can also find estimates for the annihilator in $\mathbf{Z}$ of any $H^i_{\mathbf{Z}}(\mu)_{\text{tor}}$.

8.9. Lemma: *There is for each* $\lambda \in X(T)_+$ *and each ring* A' *an isomorphism*

$$H^n_{A'}(w_0 \bullet \lambda) \simeq V(\lambda)_{A'}$$

of $G_{A'}$*–modules.*

Proof: Using 8.8(4) we may restrict ourselves to the case $A' = \mathbf{Z}$. Each $H^n_{\mathbf{Z}}(w_0 \bullet \lambda)_\mu$ is a lattice in $H^n_{\mathbf{Q}}(w_0 \bullet \lambda)_\mu \simeq H^0_{\mathbf{Q}}(\lambda)_\mu$. Choose $v \in H^n_{\mathbf{Z}}(w_0 \bullet \lambda)_\lambda$ with $\mathbf{Z}v = H^n_{\mathbf{Z}}(w_0 \bullet \lambda)_\lambda$. We may assume that $V(\lambda)_{\mathbf{Z}} = \text{Dist}(G_{\mathbf{Z}})\, v$, cf. 8.3(1). Both $V(\lambda)_{\mathbf{Z}}$ and $H^n_{\mathbf{Z}}(w_0 \bullet \lambda)$ are $G_{\mathbf{Z}}$–stable lattices in $H^n_{\mathbf{Q}}(w_0 \bullet \lambda)$. We have $V(\lambda)_{\mathbf{Z}} \subset H^n_{\mathbf{Z}}(w_0 \bullet \lambda)$ since $v \in H^n_{\mathbf{Z}}(w_0 \bullet \lambda)$; we want to prove equality.

We have to show that $H^n_{\mathbf{Z}}(w_0 \bullet \lambda)/V(\lambda)_{\mathbf{Z}}$ has no p–torsion for any prime number p, i.e., that the induced map

(1) $$V(\lambda)_{\mathbf{Z}} \otimes_{\mathbf{Z}} \mathbf{F}_p \longrightarrow H^n_{\mathbf{Z}}(w_0 \bullet \lambda) \otimes_{\mathbf{Z}} \mathbf{F}_p \simeq H^n_{\mathbf{F}_p}(w_0 \bullet \lambda)$$

is injective. However, the $G_{\mathbf{F}_p}$–module $H^n_{\mathbf{F}_p}(w_0 \bullet \lambda)$ is isomorphic to $V(\lambda)_{\mathbf{F}_p}$, cf. 5.11, hence generated by its λ–weight space. Therefore $v \otimes 1$ generates the $G_{\mathbf{F}_p}$–module $H^n_{\mathbf{Z}}(w_0 \bullet \lambda) \otimes_{\mathbf{Z}} \mathbf{F}_p$. So the map in (1) is surjective, hence bijective, because both sides have the same dimension.

Remark: Combining (1) and 8.8(5) we get for all λ and A' as above

$$V(\lambda)_{A'} \simeq H^0_{A'}(-w_0\lambda)^*. \tag{2}$$

8.10. Denote the set of all maximal ideals in A by $\Pi(A)$. For each $\mathfrak{p} \in \Pi(A)$ let $\nu_{\mathfrak{p}}$ be the $\mathfrak{p}$–adic valuation of K. (If $a \in A$, $a \neq 0$, then $\nu_{\mathfrak{p}}(a) = r$ if and only if $a \in \mathfrak{p}^r$, $a \notin \mathfrak{p}^{r+1}$.) Denote by $D(A)$ the free abelian group in generators $[\mathfrak{p}]$, $\mathfrak{p} \in \Pi(A)$. The elements of $D(A)$ are called the *divisors* of A. For each $x \in K$, $x \neq 0$ denote by

$$\operatorname{div}(x) = \sum_{\mathfrak{p}\in\Pi(A)} \nu_{\mathfrak{p}}(x)\,[\mathfrak{p}] \tag{1}$$

the corresponding divisor.

To each finitely generated torsion module M over A one associates the divisor

$$\nu(M) = \sum_{\mathfrak{p}\in\Pi(A)} \nu_{\mathfrak{p}}(M)\,[\mathfrak{p}] \tag{2}$$

where $\nu_{\mathfrak{p}}(M)$ is the length of the $A_{\mathfrak{p}}$–module $M_{\mathfrak{p}} = M \otimes_A A_{\mathfrak{p}}$. One can find $\mathfrak{p}_1$, $\mathfrak{p}_2, \dots, \mathfrak{p}_s \in \Pi(A)$ and $n(i) \in \mathbf{N}$ with $M \simeq \bigoplus_{i=1}^s A/\mathfrak{p}_i^{n(i)}$, cf. [B2], ch. VII, §4, prop. 23. Then

$$\nu(M) = \sum_{i=1}^{s} n(i)\,[\mathfrak{p}_i]. \tag{3}$$

(The length of $(A/\mathfrak{p}^m)_{\mathfrak{p}}$ is m because $A_{\mathfrak{p}}$ is a principal ideal domain.)

One has for any $a \in A$, $a \neq 0$ (by [B2], ch. VII, §4, prop. 12)

$$\nu(A/(a)) = \operatorname{div}(a). \tag{4}$$

If $0 \to M_0 \to M_1 \to \cdots \to M_r = 0$ is an exact sequence of finitely generated torsion modules over A, then

$$\sum_{i=0}^{r} (-1)^i\,\nu(M_i) = 0. \tag{5}$$

(Cf. [B2], ch. VII, §4, cor. de la prop. 10.)

8.11. Any homomorphism $\varphi : M \to M'$ of A–modules maps M_{tor} to M'_{tor} and thus induces a homomorphism $M_{\mathrm{fr}} \to M'_{\mathrm{fr}}$ which we denote by φ_{fr}.

Consider now a homomorphism $\varphi : M \to M'$ as above with M and M' finitely generated over A such that $\varphi \otimes \mathrm{id}_K : M \otimes_A K \to M' \otimes_A K$ is bijective. We can identify $M \otimes_A K \simeq M_{\mathrm{fr}} \otimes_A K$ and $M' \otimes_A K \simeq M'_{\mathrm{fr}} \otimes_A K$. Then $\varphi \otimes \mathrm{id}_K$ corresponds to $\varphi_{\mathrm{fr}} \otimes \mathrm{id}_K$, so $\varphi_{\mathrm{fr}} \otimes \mathrm{id}_K$ is also bijective. Therefore φ_{fr} is injective (hence $\varphi^{-1}(M'_{\mathrm{tor}}) = M_{\mathrm{tor}}$) and $\operatorname{coker}(\varphi_{\mathrm{fr}})$ is a finitely generated torsion module. We have isomorphisms

$$\begin{aligned}\operatorname{coker}(\varphi_{\mathrm{fr}}) &= M'_{\mathrm{fr}}/\varphi_{\mathrm{fr}}(M_{\mathrm{fr}}) \simeq \big(M'/M'_{\mathrm{tor}}\big)/\big((\varphi(M) + M'_{\mathrm{tor}})/M'_{\mathrm{tor}}\big)\\ &\simeq M'/(\varphi(M) + M'_{\mathrm{tor}}) \simeq \big(M'/\varphi(M)\big)/\big((\varphi(M) + M'_{\mathrm{tor}})/\varphi(M)\big).\end{aligned}$$

Using $\varphi^{-1}(M'_{\text{tor}}) = M_{\text{tor}}$ we get

$$(\varphi(M) + M'_{\text{tor}})/\varphi(M) \simeq M'_{\text{tor}}/(\varphi(M) \cap M'_{\text{tor}}) = M'_{\text{tor}}/\varphi(M_{\text{tor}}),$$

hence an exact sequence

(1) $$0 \to \operatorname{coker}(\varphi_{\text{tor}}) \longrightarrow \operatorname{coker}(\varphi) \longrightarrow \operatorname{coker}(\varphi_{\text{fr}}) \to 0$$

where φ_{tor} is the restriction of φ to $M_{\text{tor}} \to M'_{\text{tor}}$.

For φ as above, set $\nu_{\mathfrak{p}}(\varphi) = \nu_{\mathfrak{p}}(\operatorname{coker}(\varphi_{\text{fr}}))$ for all $\mathfrak{p} \in \Pi(A)$ and

(2) $$\nu\,(\varphi) = \nu\,(\operatorname{coker}\varphi_{\text{fr}}))$$

If M_{fr} and M'_{fr} are free modules, then we have by [B2], ch. VII, §4, cor. de la prop. 14

(3) $$\nu\,(\varphi) = \operatorname{div}\,(\det(\varphi_{\text{fr}}))$$

where $\det(\varphi_{\text{fr}})$ is the determinant of φ_{fr} with respect to bases of M_{fr} and M'_{fr}. (Another choice of bases will multiply $\det(\varphi_{\text{fr}})$ by a unit in A and leave $\operatorname{div}\,(\det(\varphi_{\text{fr}}))$ unchanged.)

Let $\psi : M' \to M''$ be another homomorphism of finitely generated A–modules such that $\psi \otimes \mathrm{id}_K$ is bijective. Then $\psi \circ \varphi$ has the same property and we get

(4) $$\nu\,(\psi \circ \varphi) = \nu\,(\psi) + \nu\,(\varphi).$$

In fact, one easily checks (ψ_{fr} being injective) that there is an exact sequence

(5) $$0 \to \operatorname{coker}(\varphi_{\text{fr}}) \longrightarrow \operatorname{coker}(\psi \circ \varphi)_{\text{fr}} \longrightarrow \operatorname{coker}(\psi_{\text{fr}}) \to 0.$$

(Observe that $(\psi \circ \varphi)_{\text{fr}} = \psi_{\text{fr}} \circ \varphi_{\text{fr}}$.) One can also use (3): In order to get the coefficient of $[\mathfrak{p}]$ on both sides of (4) we may replace A by $A_{\mathfrak{p}}$, hence suppose the modules to be free. Then use $\det(\psi_{\text{fr}} \circ \varphi_{\text{fr}}) = \det(\psi_{\text{fr}}) \det(\varphi_{\text{fr}})$ and $\operatorname{div}(ab) = \operatorname{div}(a) + \operatorname{div}(b)$.

Lemma: *Let*

$$0 \to M_0 \xrightarrow{\varphi_0} M_1 \xrightarrow{\varphi_1} M_2 \xrightarrow{\varphi_2} \cdots \xrightarrow{\varphi_r} M_{r+1} \to 0$$

be an exact sequence of finitely generated A–modules. Suppose that there is some j $(0 \le j \le r)$ such that all M_i with $i \ne j, j+1$ are torsion modules. Then $\varphi_j \otimes \mathrm{id}_K$ is bijective and

(6) $$(-1)^j\,\nu(\varphi_j) = \sum_{i \ge 0} (-1)^i\,\nu(M_{i,\text{tor}}).$$

Proof: As tensoring with K is exact and as $M_i \otimes_A K = 0$ for $i \ne j, j+1$, we get that $\varphi_i \otimes \mathrm{id}_K = 0$ for $i \ne j$ and that $\varphi_j \otimes \mathrm{id}_K$ is bijective. The given exact sequence gives rise to exact sequences

$$0 \to M_0 \to M_1 \to \cdots \to M_{j,\text{tor}} \to M_{j+1,\text{tor}} \to \operatorname{coker}(\varphi_{j,\text{tor}}) \to 0$$

and (as $\operatorname{coker}(\varphi_j) \simeq M_{j+1}/\ker(\varphi_{j+1}) \simeq \varphi_{j+1}(M_{j+1})$)

$$0 \to \operatorname{coker}(\varphi_j) \to M_{j+2} \to \cdots \to M_{r+1} \to 0.$$

Now (6) follows from 8.10(5) and from (1).

8.12. If M is a T_A–module that is a finitely generated torsion module over A, then $M = \bigoplus_{\mu \in X(T)} M_\mu$ with all M_μ finitely generated torsion modules over A (and almost all $M_\mu = 0$). We can therefore set

$$\nu^c(M) = \sum_\mu \nu(M_\mu)\, e(\mu) \;\in\; D(A)[X(T)]. \tag{1}$$

Clearly 8.10(5) generalises to ν^c for T_A–modules. If M is a G_A–module, then obviously $\nu^c(M) \in D(A)[X(T)]^W$.

If $\varphi : M \to M'$ is a homomorphism of T_A–modules (finitely generated over A) such that $\varphi \otimes \mathrm{id}_K$ is bijective, then φ maps any M_μ to M'_μ and the induced map $M_\mu \otimes_A K \to M'_\mu \otimes_A K$ is bijective. Set

$$\nu^c(\varphi) = \sum_\mu \nu(\varphi_{|M_\mu})\, e(\mu) \;\in\; D(A)[X(T)]. \tag{2}$$

If φ is a homomorphism of G_A–modules, then $\nu^c(\varphi) \in D(A)[X(T)]^W$.

There are obvious generalisations of 8.11(3), (4), and (6) to the ν^c. We denote by $\nu^c_{\mathfrak p}(M)$ or $\nu^c_{\mathfrak p}(\varphi)$ for any $\mathfrak p \in \Pi(A)$ the coefficient of $[\mathfrak p]$ in $\nu^c(M)$ resp. $\nu^c(\varphi)$.

For any element $\sum_{\mu\in X(T)} r(\mu)e(\mu) \in D(A)[X(T)]$ set

$$\chi\Big(\sum_{\mu\in X(T)} r(\mu)e(\mu)\Big) = \sum_{\mu\in X(T)} r(\mu)\chi(\mu) \in D(A)[X(T)]^W. \tag{3}$$

Lemma: *Suppose that* $\mathrm{char}(K) = 0$. *Let M be a B_A–module that is a finitely generated torsion module over A. Then*

$$\sum_{i\ge 0} (-1)^i\, \nu^c(H^i_A(M)) = \chi\,(\nu^c(M)). \tag{4}$$

Proof: Both sides are additive on exact sequences. Therefore we can restrict ourselves to the case $M = (A/\mathfrak p)_\lambda$ for some $\mathfrak p \in \Pi(A)$ and $\lambda \in X(T)$, cf. 8.6.b. In this case our claim amounts to

$$\sum_{i\ge 0} (-1)^i\, \nu^c\, H^i_A((A/\mathfrak p)_\lambda) = \chi(\lambda)\,[\mathfrak p]. \tag{5}$$

We have $\nu^c\, H^i_A((A/\mathfrak p)_\lambda) \in \mathbf{Z}[X(T)][\mathfrak p]$ for each i because the proof of 8.6.a shows that $\mathfrak p\, H^i_A((A/\mathfrak p)_\lambda) = 0$. In order to compute the coefficient of $[\mathfrak p]$ we may replace A by $A_{\mathfrak p}$ as (by I.4.13.b)

$$H^i_A((A/\mathfrak p)_\lambda) \otimes_A A_{\mathfrak p} \simeq H^i_{A_{\mathfrak p}}((A/\mathfrak p)_\lambda \otimes_A A_{\mathfrak p}) \simeq H^i_{A_{\mathfrak p}}((A_{\mathfrak p}/\mathfrak p A_{\mathfrak p})_\lambda).$$

So we may assume that $\mathfrak p$ is a principal ideal $\mathfrak p = Aa$ for some $a \in A$. We have now a short exact sequence of B_A–modules

$$0 \to A_\lambda \xrightarrow{\varphi} A_\lambda \longrightarrow (A/\mathfrak p)_\lambda \to 0$$

where φ is multiplication by a. This leads to a long exact sequence

$$\cdots \to H^i_A(\lambda) \xrightarrow{\varphi_i} H^i_A(\lambda) \longrightarrow H^i_A((A/\mathfrak{p})_\lambda) \longrightarrow H^{i+1}_A(\lambda) \to \cdots$$

where φ_i again is multiplication by a. By 8.7(2) there is at most one j for which $H^j_A(\lambda)$ is not a torsion module, and this j satisfies $\operatorname{ch} H^j_A(\lambda)_{\mathrm{fr}} = (-1)^j \chi(\lambda)$ by 8.7(3). We can now apply Lemma 8.11 and get

$$(6) \qquad (-1)^j \, \nu^c(\varphi_j) = \sum_{i\geq 0} (-1)^i \, \nu^c(H^i_A((A/\mathfrak{p})_\lambda))$$

as the $\nu^c(H^i_A(\lambda))$ cancel.

We have $\operatorname{coker}(\varphi_{j,\mathrm{fr}}) \simeq H^j_A(\lambda)_{\mathrm{fr}}/aH^j_A(\lambda)_{\mathrm{fr}}$ because φ_j is multiplication with a on $H^j_A(\lambda)$. It follows that $\nu^c(\varphi_j) = \operatorname{ch}(H^j_A(\lambda)_{\mathrm{fr}}) \operatorname{div}(a) = (-1)^j \chi(\lambda)[\mathfrak{p}]$. Inserting this into (6) we get (5), hence the lemma.

8.13. Assume from now on (in this chapter) that $\operatorname{char}(K) = 0$.

Let $\alpha \in S$. For any B_A–module M and any $i \in \mathbf{N}$ set

$$H^i_{\alpha,A}(M) = R^i \operatorname{ind}_{B_A}^{P(\alpha)_A}(M) \simeq H^i(P(\alpha)_A/B_A, \mathcal{L}(M))$$

where $P(\alpha) = P_{\{\alpha\}}$ as in 5.1. There is obviously a result analogous to 8.7(1) for the $H^i_{\alpha,A}(\lambda)$ with $\lambda \in X(T)$. Arguing as in 8.8, we can generalise parts of Proposition 5.2 as follows:

(1) *If* $\langle \lambda + \rho, \alpha^\vee \rangle = 0$, *then* $H^\bullet_{\alpha,A}(\lambda) = 0$.

(2) *If* $\langle \lambda, \alpha^\vee \rangle \geq 0$, *then* $H^i_{\alpha,A}(\lambda) = 0$ *for all* $i \neq 0$ *and* $H^j_{\alpha,A}(s_\alpha \bullet \lambda) = 0$ *for all* $j \neq 1$.

Suppose $\langle \lambda, \alpha^\vee \rangle = r \geq 0$. Then $H^0_{\alpha,A}(\lambda)$ and $H^1_{\alpha,A}(s_\alpha \bullet \lambda)$ are lattices in $H^0_{\alpha,K}(\lambda)$ resp. $H^1_{\alpha,K}(s_\alpha \bullet \lambda)$ and they are free modules over A (as we can reduce to the case $A = \mathbf{Z}$). We can choose bases $(v_i)_{0\leq i\leq r}$ of $H^0_{\alpha,K}(\lambda)$ and $(v_i')_{0\leq i\leq r}$ of $H^1_{\alpha,K}(s_\alpha \bullet \lambda)$ as in 5.2.c/d (with $s_\alpha \bullet \lambda$ replacing λ in the second case). By multiplying all basis elements with the same scalar we may assume $Av_0 = H^0_{\alpha,A}(\lambda)_\lambda$ and $Av_0' = H^1_{\alpha,A}(s_\alpha \bullet \lambda)_\lambda$. The same argument as in 8.9 shows $H^1_{\alpha,A}(s_\alpha \bullet \lambda) = \operatorname{Dist}(U_{-\alpha,A})v_0'$, hence (using 5.2(3′))

$$(3) \qquad H^1_{\alpha,A}(s_\alpha \bullet \lambda) = \sum_{i=0}^{r} Av_i'.$$

We get now, arguing as in the proof of 8.8(5)

$$(4) \qquad H^0_{\alpha,A}(\lambda) = \sum_{i=0}^{r} Av_i.$$

The explicit description in 5.2 of the action of $P(\alpha)_K$ on our bases shows that the maps

$$(5) \qquad T_\alpha(s_\alpha \bullet \lambda) : H^1_{\alpha,A}(s_\alpha \bullet \lambda) \longrightarrow H^0_{\alpha,A}(\lambda), \qquad v_i' \mapsto \binom{r}{i} v_i$$

and

$$(6) \qquad T_\alpha(\lambda) : H^0_{\alpha,A}(\lambda) \longrightarrow H^1_{\alpha,A}(s_\alpha \bullet \lambda), \qquad v_i \mapsto (r-i)!\, i!\, v_i'$$

(both times for all i, $0 \leq i \leq r$) are homomorphisms of $P(\alpha)_A$–modules. Both $T_\alpha(s_\alpha \bullet \lambda) \circ T_\alpha(\lambda)$ and $T_\alpha(\lambda) \circ T_\alpha(s_\alpha \bullet \lambda)$ are multiplication with $r!$. (The map $T_\alpha(s_\alpha \bullet \lambda)$ has already been used in the proof of 5.3.)

8.14. Let $\alpha \in S$. By 8.13 all $H^i_{\alpha,A}(\mu)$ are free over A, so we can define $\chi_\alpha(\mu) = \sum_{i \geq 0}(-1)^i \operatorname{ch} H^i_{\alpha,A}(\mu)$ for all $\mu \in X(T)$. More precisely, we have $\chi_\alpha(\mu) = 0$ if $\langle \mu + \rho, \alpha^\vee \rangle = 0$; if $\langle \mu, \alpha^\vee \rangle = r \geq 0$, then

$$\chi_\alpha(\mu) = \sum_{i=0}^{r} e(\mu - i\alpha) = -\chi_\alpha(s_\alpha \bullet \mu). \tag{1}$$

Lemma: *Let $\lambda \in X(T)$ with $\langle \lambda, \alpha^\vee \rangle = r \geq 0$. Then*

$$\nu^c(T_\alpha(s_\alpha \bullet \lambda)) = -\sum_{j=1}^{r} \operatorname{div}(j)\, \chi_\alpha(\lambda - j\alpha) \tag{2}$$

and

$$\nu^c(T_\alpha(\lambda)) = \operatorname{div}(r!)\, \chi_\alpha(\lambda) + \sum_{j=1}^{r} \operatorname{div}(j)\, \chi_\alpha(\lambda - j\alpha). \tag{3}$$

Proof: By definition 8.13(3)–(5) imply

$$\nu^c(T_\alpha(s_\alpha \bullet \lambda)) = \sum_{i=0}^{r} \operatorname{div}(\binom{r}{i})\, e(\lambda - i\alpha).$$

We may drop the terms for $i = 0$ and $i = r$ as $\operatorname{div}(1) = 0$. Using the trivial equation

$$\operatorname{div}(\binom{r}{i}) = \sum_{j=1}^{i} (\operatorname{div}(r - j + 1) - \operatorname{div}(j))$$

the right hand side above is checked to be equal to

$$\sum_{j=1}^{r-1} \operatorname{div}(r - j + 1) \sum_{i=j}^{r-1} e(\lambda - i\alpha) - \sum_{j=2}^{r} \operatorname{div}(j) \sum_{i=j}^{r-1} e(\lambda - i\alpha)$$
$$= \sum_{j=2}^{r} \operatorname{div}(j) \Big(\sum_{i=r-j+1}^{r-1} e(\lambda - i\alpha) - \sum_{i=j}^{r-1} e(\lambda - i\alpha) \Big).$$

If $j < r - j + 1$, then the factor after $\operatorname{div}(j)$ in the last sum is equal to $-\sum_{i=j}^{r-j} e(\lambda - i\alpha) = -\sum_{i=0}^{r-2j} e((\lambda - j\alpha) - i\alpha)$. In this case $\langle \lambda - j\alpha, \alpha^\vee \rangle = r - 2j \geq 0$, so that the factor is equal to $-\chi_\alpha(\lambda - j\alpha)$. The other cases ($j = r - j + 1$, $j > r - j + 1$) can be treated similarly. We get thus (2).

As $T_\alpha(s_\alpha \bullet \lambda) \circ T_\alpha(\lambda)$ is multiplication with $r!$, we get $\nu^c(T_\alpha(s_\alpha \bullet \lambda) \circ T_\alpha(\lambda)) = \operatorname{div}(r!)\, \chi_\alpha(\lambda)$. Now (3) follows from (2) and 8.11(4).

8.15. For any $\alpha \in S$ the chain of subgroups $B_A \subset P(\alpha)_A \subset G_A$ gives rise to spectral sequences as in I.4.5.c. Using 8.13(2) we see that these spectral sequences induce (as in 5.4) for any $\mu \in X(T)$ with $\langle \mu, \alpha^\vee \rangle \geq 0$ isomorphisms

$$H^i_A(\mu) \simeq R^i \operatorname{ind}_{P(\alpha)_A}^{G_A} (H^0_{\alpha,A}(\mu)) \simeq H^i_A(H^0_{\alpha,A}(\mu)) \tag{1}$$

and

$$H^i_A(s_\alpha \bullet \mu) \simeq R^{i-1}\operatorname{ind}_{P(\alpha)_A}^{G_A}(H^1_{\alpha,A}(s_\alpha \bullet \mu)) \simeq H^{i-1}_A(H^1_{\alpha,A}(s_\alpha \bullet \mu)). \tag{2}$$

Suppose that there is some $j \in \mathbf{N}$ such that $H^j_A(\mu)$ is not a torsion module. Then j is unique by 8.7(2) and also $H^{j+1}(s_\alpha \bullet \mu)$ is not a torsion module. Applying H^i_A to the maps $T_\alpha(s_\alpha \bullet \mu)$ and $T_\alpha(\mu)$ gives rise to

$$\widetilde{T}_\alpha(s_\alpha \bullet \mu) : H^{j+1}(s_\alpha \bullet \mu) \longrightarrow H^j_A(\mu) \tag{3}$$

and

$$\widetilde{T}_\alpha(\mu) : H^j_A(\mu) \longrightarrow H^{j+1}(s_\alpha \bullet \mu) \tag{4}$$

such that both $\widetilde{T}_\alpha(s_\alpha \bullet \mu) \circ \widetilde{T}_\alpha(\mu)$ and $\widetilde{T}_\alpha(\mu) \circ \widetilde{T}_\alpha(s_\alpha \bullet \mu)$ are the multiplication with $\langle \mu, \alpha^\vee\rangle !$.

There is an exact sequence of B_A–modules

$$0 \to H^1_{\alpha,A}(s_\alpha \bullet \mu) \longrightarrow H^0_{\alpha,A}(\mu) \longrightarrow M \to 0$$

with $M = \operatorname{coker}(T_\alpha(s_\alpha \bullet \mu))$, hence $\nu^c(M) = \nu^c(T_\alpha(s_\alpha \bullet \mu))$. This gives rise to a long exact sequence of G_A–modules

$$0 \to H^1_A(s_\alpha \bullet \mu) \to H^0_A(\mu) \to H^0_A(M) \to H^2_A(s_\alpha \bullet \mu) \to H^1_A(\mu) \to H^1_A(M) \to \cdots$$

with all terms torsion modules except for $H^{j+1}_A(s_\alpha \bullet \mu)$ and $H^j_A(\mu)$. We can therefore apply Lemma 8.11 and get

$$\begin{aligned}(-1)^j \nu^c(\widetilde{T}_\alpha(s_\alpha \bullet \mu)) = \sum_{i \geq 0}(-1)^i \nu^c(H^i_A(M)) - \sum_{i\geq 0}(-1)^i \nu^c(H^i_A(\mu)_{\mathrm{tor}}) \\ - \sum_{i \geq 0}(-1)^i \nu^c(H^i_A(s_\alpha \bullet \mu)_{\mathrm{tor}}).\end{aligned} \tag{5}$$

(Note that $H^0_A(s_\alpha \bullet \mu) = 0$.)

We have by Lemma 8.12

$$\sum_{i\geq 0}(-1)^i \nu^c(H^i_A(M)) = \chi(\nu^c(M)).$$

The term $\nu^c(M) = \nu^c(T_\alpha(s_\alpha \bullet \mu))$ has been expressed in Lemma 8.14 in terms of the $\chi_\alpha(\mu')$. Using $\chi(\mu') + \chi(\mu' - (\langle \mu', \alpha^\vee\rangle + 1)\alpha) = \chi(\mu') + \chi(s_\alpha \bullet \mu') = 0$ one checks that $\chi(\chi_\alpha(\mu')) = \chi(\mu')$. Therefore (5) and Lemma 8.14 yield

$$\begin{aligned}(-1)^j \nu^c(\widetilde{T}_\alpha(s_\alpha \bullet \mu)) = - \sum_{i=1}^{\langle \mu, \alpha^\vee\rangle} \operatorname{div}(i)\chi(\mu - i\alpha) \\ - \sum_{i\geq 0}(-1)^i\big(\nu^c(H^i_A(\mu)_{\mathrm{tor}}) + \nu^c(H^i_A(s_\alpha \bullet \mu)_{\mathrm{tor}})\big).\end{aligned} \tag{6}$$

One proves similarly

$$(7)\qquad \begin{aligned}(-1)^j\nu^c(\widetilde{T}_\alpha(\mu)) = \operatorname{div}(\langle\mu,\alpha^\vee\rangle!)\chi(\mu) + \sum_{i=1}^{\langle\mu,\alpha^\vee\rangle}\operatorname{div}(i)\chi(\mu-i\alpha)\\ +\sum_{i\geq 0}(-1)^i\big(\nu^c(H^i_A(\mu)_{\mathrm{tor}})+\nu^c(H^i_A(s_\alpha\bullet\mu)_{\mathrm{tor}})\big).\end{aligned}$$

(One can also use that $\nu^c(\widetilde{T}_\alpha(s_\alpha\bullet\mu))+\nu^c(\widetilde{T}_\alpha(\mu)) = (-1)^j\operatorname{div}(\langle\mu,\alpha^\vee\rangle!)\chi(\mu)$.)

Suppose that above $\mu = w\bullet\lambda$ with $\lambda\in X(T)_+$ and $w\in W$. Then $j=l(w)$. We know that λ is not a weight of any $H^i_A(\mu)_{\mathrm{tor}}$ or $H^i_A(s_\alpha\bullet\mu)_{\mathrm{tor}}$ (see 8.7) or of any $\chi(\mu-i\alpha)=\chi(w\bullet(\lambda-iw^{-1}\alpha))$ with $0<i<\langle\mu+\rho,\alpha^\vee\rangle = \langle\lambda+\rho,w^{-1}\alpha^\vee\rangle$ (by 6.9). Therefore $\widetilde{T}_\alpha(s_\alpha\bullet\mu)$ maps $H^{j+1}_A(s_\alpha\bullet\mu)_\lambda$ bijectively to $H^j_A(\mu)_\lambda$.

8.16. Choose a reduced decomposition $w_0 = s_{\beta_n}s_{\beta_{n-1}}\dots s_{\beta_1}$ with all $\beta_i\in S$ and with $n=l(w_0)=|R^+|$. Given $\lambda\in X(T)_+$ we get for each i a homomorphism as in 8.15(3)

$$\widetilde{T}_{\beta_i}(s_{\beta_i}s_{\beta_{i-1}}\dots s_{\beta_1}\bullet\lambda) : H^i_A(s_{\beta_i}s_{\beta_{i-1}}\dots s_{\beta_1}\bullet\lambda)\to H^{i-1}_A(s_{\beta_{i-1}}\dots s_{\beta_1}\bullet\lambda).$$

(We have $\langle s_{\beta_{i-1}}\dots s_{\beta_1}(\lambda+\rho),\beta_i^\vee\rangle>0$ because of $s_{\beta_1}\dots s_{\beta_{i-1}}\beta_i>0$.) The composition of all these maps is a homomorphism

$$\widetilde{T}_{w_0}(w_0\bullet\lambda) : H^n_A(w_0\bullet\lambda)\longrightarrow H^0_A(\lambda).$$

Proposition: *We have for all $\lambda\in X(T)_+$*

$$(1)\qquad \nu^c(\widetilde{T}_{w_0}(w_0\bullet\lambda)) = -\sum_{\alpha\in R^+}\sum_{i=1}^{\langle\lambda+\rho,\alpha^\vee\rangle-1}\operatorname{div}(i)\,\chi(\lambda-i\alpha).$$

Proof: Set $\alpha_j = s_{\beta_1}\dots s_{\beta_{j-1}}\beta_j$ and $m(j)=\langle s_{\beta_{j-1}}\dots s_{\beta_1}(\lambda+\rho),\beta_i^\vee\rangle = \langle\lambda+\rho,\alpha_j^\vee\rangle$ for all j, $1\leq j\leq n$. Then R^+ is the set of all α_j; each $m(j)$ is positive. Now 8.11(4) and 8.15(6) yield

$$\begin{aligned}\nu^c(\widetilde{T}_{w_0}(w_0\bullet\lambda)) &= \sum_{j=1}^n \nu^c(\widetilde{T}_{\beta_j}(s_{\beta_j}s_{\beta_{j-1}}\dots s_{\beta_1}\bullet\lambda))\\
&= -\sum_{j=1}^n\sum_{i=1}^{m(j)-1}\operatorname{div}(i)\,(-1)^j\,\chi(s_{\beta_{j-1}}\dots s_{\beta_1}\bullet\lambda - i\beta_j)\\
&\quad -\sum_{j=1}^n(-1)^j\sum_{i\geq 0}(-1)^i(\nu^c\,H^i_A(s_{\beta_j}s_{\beta_{j-1}}\dots s_{\beta_1}\bullet\lambda)_{\mathrm{tor}}+\nu^c\,H^i_A(s_{\beta_{j-1}}\dots s_{\beta_1}\bullet\lambda)_{\mathrm{tor}})\\
&= -\sum_{j=1}^n\sum_{i=1}^{m(j)-1}\operatorname{div}(i)\,\chi(\lambda-i\alpha_j)+\sum_{i\geq 0}(-1)^i\,\nu^c(H^i_A(\lambda)_{\mathrm{tor}})\\
&\quad -\sum_{i\geq 0}(-1)^{i+n}\,\nu^c(H^i_A(w_0\bullet\lambda)_{\mathrm{tor}}).\end{aligned}$$

We have used here 5.9(1). As $H^i_A(\lambda)_{\mathrm{tor}}=0$ and $H^i_A(w_0\bullet\lambda)_{\mathrm{tor}}=0$ for all i by 8.8, we get (1).

Remarks: 1) The right hand side in (1) is independent of the choice of the reduced decomposition of w_0. This is clear for the left hand side: As $H^n_K(w_0\bullet\lambda)$ and $H^0_K(\lambda)$ are isomorphic, two homomorphisms differ only by a scalar in K. Any $\widetilde{T}_{w_0}(w_0\bullet\lambda)$ maps $H^n_A(w_0\bullet\lambda)_\lambda \simeq A$ bijectively to $H^0_A(\lambda)_\lambda \simeq A$. Therefore this scalar has to be a unit in A. But multiplying some φ with a unit does not change $\nu(\varphi)$. (As we can work with $A=\mathbf{Z}$, we can even choose $\widetilde{T}_{w_0}(w_0\bullet\lambda)$ uniquely up to a sign.)

2) We can similarly define for any $w\in W$ a homomorphism

$$\widetilde{T}_{w_0}(w\bullet\lambda): H_A^{l(w)}(w\bullet\lambda)\longrightarrow H_A^{l(w_0w)}(w_0w\bullet\lambda).$$

Arguing as above and using in addition 8.15(7) one proves

$$\begin{aligned}\nu^c(\widetilde{T}_{w_0}(w\bullet\lambda)) &= \sum_{\alpha\in R^+}\operatorname{sgn}(w\alpha)\sum_{i=1}^{\langle\lambda+\rho,\alpha^\vee\rangle-1}\operatorname{div}(i)\,\chi(\lambda-i\alpha)\\ &+\sum_{\substack{\alpha\in R^+\\ w\alpha>0}}\operatorname{div}(\langle\lambda+\rho,\alpha^\vee\rangle!)\chi(\lambda)-(-1)^{l(w)}\sum_{i\geq 0}(-1)^i\nu^c(H^i_A(w\bullet\lambda)_{\mathrm{tor}})\\ &+(-1)^{l(w)}\sum_{i\geq 0}(-1)^{i+n}\nu^c(H^i_A(w_0w\bullet\lambda)_{\mathrm{tor}}).\end{aligned}$$

(Here $\operatorname{sgn}(w\alpha)$ is $+1$ if $w\alpha>0$, and -1 otherwise.)

8.17. Let k be an integral domain and set $G=G_k$. Recall the anti-automorphism τ from 1.16 and the functor $M\mapsto {}^\tau M$ from 2.12. There k was a field and we defined ${}^\tau M$ for all finite dimensional G–modules. The definition, however, makes sense in general for G–modules that are finitely generated and projective over k. In particular, it can be applied to $V(\lambda)_k\simeq H^n_k(w_0\bullet\lambda)$ and to $H^0_k(\lambda)$ for any $\lambda\in X(X)^+$. Arguing as in 2.13 one gets (with σ as in 2.13)

$${}^\tau V(\lambda)_k\simeq {}^\tau H^n_k(w_0\lambda-2\rho)\simeq ({}^\sigma H^n_k(w_0\lambda-2\rho))^*\simeq (H^n_k(-\lambda-2\rho))^*\simeq H^0_k(\lambda)$$

using 8.8(5) for the last step.

Let M be a G–module that is finitely generated and projective over k. If $\varphi: M\to {}^\tau M$ is a homomorphism of G–modules, then we get a bilinear form on M setting

(1) $$(v,v')=\varphi(v)\,(v')\qquad\text{for all } v,v'\in M.$$

(Recall that ${}^\tau M$ is M^* with a twisted G–action.) This form satisfies

$$(g\,v,v')=\varphi(g\,v)\,(v')=(g\,\varphi(v))\,(v')=\varphi(v)\,(\tau(g)\,v'),$$

hence

(2) $$(g\,v,v')=(v,\tau(g)\,v')$$

for all $v,v'\in M$ and all $g\in G(A')$ for all k–algebras A'. (More rigorously, we should extend the bilinear form to $M\otimes_k A'$ and replace v and v' by $v\otimes 1$ and $v'\otimes 1$.)

Conversely, if we have on M a bilinear form $(,)$ satisfying (2), then we can define via (1) a map $M \to M^*$ that is a homomorphism $M \to {}^\tau M$ of G–modules.

Bilinear forms satisfying (2) were called *contravariant* by W. J. Wong (see [Wong 1]). Any contravariant form on a G–module M satisfies $(M_\lambda, M_\mu) = 0$ for all $\lambda, \nu \in X(T)$ with $\lambda \neq \nu$. (The corresponding homomorphism $M \to {}^\tau M$ maps M_λ to $({}^\tau M)_\lambda = (M^*)_{-\lambda}$; now use I.2.12(11).) If k is a field and if M a simple G–module, then 2.12(2) implies that there exists a non-degenerate contravariant form on M; any other contravariant form on M is then a multiple of the first form because of $\mathrm{Hom}_G(M, {}^\tau M) \simeq \mathrm{End}_G(M) \simeq k$ by 2.8. This form has to be symmetric: If $(,)$ is a contravariant form on M, then so is $(,)'$ with $(v, v')' = (v', v)$. (Recall that $\tau^2 = \mathrm{id}$.) So there is $a \in k$ with $(v', v) = a(v, v')$ for all v and v', hence $(v, v') = a^2(v, v')$, hence $a^2 = 1$. So the form is either symmetric or alternating. But the form has to be non-degenerate on each weight space, in particular on the top one, say M_λ. Since $\dim M_\lambda = 1$, this rules out the possibility that $(,)$ is alternating.

Now return to the situation of 8.16. We can compose $\widetilde{T}_{w_0}(w_0 \bullet \lambda)$ with an isomorphism $\psi : H^0_A(\lambda) \xrightarrow{\sim} {}^\tau V(\lambda)_A$ and get a homomorphism $\varphi : V(\lambda)_A \to {}^\tau V(\lambda)_A$ that induces an isomorphism of A–modules $V(\lambda)_{A,\lambda} \xrightarrow{\sim} ({}^\tau V(\lambda)_A)_\lambda$ on the top weight spaces.

Choose $v_1 \in V(\lambda)_A$ with $V(\lambda)_{A,\lambda} = Av_1$. Then $({}^\tau V(\lambda)_A)_\lambda = Af_1$ where $f_1 \in V(\lambda)^*_A$ is given by $f_1(v_1) = 1$ and $f_1(V(\lambda)_{A,\mu}) = 0$ for all $\mu \neq \lambda$. We have $\varphi(Av_1) = Af_1$. We can multiply ψ by a unit in A and may assume that $\varphi(v_1) = f_1$. So the associated contravariant form (as in (1)) on $V(\lambda)_A$ satisfies $(v_1, v_1) = 1$. Since $V(\lambda)_A$ is a lattice in $V(\lambda)_K$ and since $V(\lambda)_K$ is a simple G_K–module, the remarks above imply that $(,)$ is the unique contravariant form on $V(\lambda)_A$ such that $(v_1, v_1) = 1$. The form is symmetric and non-degenerate on $V(\lambda)_K$. Let $D_\lambda(\mu)$ denote the determinant of this form with respect to a basis of $V(\lambda)_{A,\mu}$ (for any $\mu \in X(T)$). Then $D_\lambda(\mu) \neq 0$ by the non degeneracy over K and we have

$$\sum_\mu \mathrm{div}(D_\lambda(\mu))\, e(\mu) = \nu^c(\varphi) = \nu^c(\widetilde{T}_{w_0}(w_0 \bullet \lambda)) \tag{3}$$

by 8.11(3) since ψ was an isomorphism over A, hence satisfies $\nu^c(\psi) = 0$.

Assume that $A = \mathbf{Z}$. Then (3) and 8.16(1) determine all $D_\lambda(\mu)$ up to sign. Now one can show that in this case $(,)$ is positive, hence that $D_\lambda(\mu) > 0$ is uniquely determined. This positivity was proved using a compact real form of G in [Jantzen 1], p. 35. For an algebraic proof, see [K], Thm. 11.7.

It should be mentioned that there is a direct way of constructing the contravariant form on $V(\lambda)_{\mathbf{Z}}$ that does not use the machinery developed so far in this chapter, see [St1], p. 228/9, [Wong 1], [Burgoyne], and [Jantzen 1], p. 7. This direct construction shows rather quickly for each $I \subset S$ that $\bigoplus_{\nu \in \mathbf{Z}I} V(\lambda)_{\mathbf{Z},\lambda-\nu}$ is isometric to the Weyl module with highest weight λ for the group $L_{I,\mathbf{Z}}$, see [Jantzen 1], Satz I.5.

If $\lambda = -w_0\lambda \in X(T)_+$, then there is a non-degenerate G_K–invariant bilinear form on $V(\lambda)_K$, unique up to a scalar, either symmetric or alternating, cf. [St1], p. 226/7, or [B3], ch. VIII, §7, prop. 12. The determinant of this invariant form differs from that of the contravariant form by a sign that can be computed. See [Jantzen 1], pp. 25–34 for the details.

8.18. Let $\mathfrak{p} \in \Pi(A)$ be a fixed maximal ideal in this subsection.

Consider a homomorphism $\varphi : M \to M'$ of finitely generated torsion free A–modules such that $\varphi \otimes \mathrm{id}_K$ is bijective. Set for all $i \in \mathbf{N}$

(1) $$M^i = \{ m \in M \mid \varphi(m) \in \mathfrak{p}^i M' \}.$$

Obviously $M = M^0 \supset M^1 \supset M^2 \supset \cdots$ is a chain of sublattices of M with $M^i \supset \mathfrak{p}^i M$ for all i. An elementary calculation shows that this construction commutes with localisation at $\mathfrak{p}$, i.e., if we define $(M_\mathfrak{p})^i$ with respect to $\varphi_\mathfrak{p} : M_\mathfrak{p} \to M'_\mathfrak{p}$, then $(M_\mathfrak{p})^i = (M^i)_\mathfrak{p}$ for all i.

Set $\overline{M} = M \otimes_A (A/\mathfrak{p})$ and denote the canonical image of $M^i \otimes_A (A/\mathfrak{p})$ in $\overline{M}$ by $\overline{M}^i$. Then $\overline{M} = \overline{M}^0 \supset \overline{M}^1 \supset \overline{M}^2 \supset \cdots$ is a chain of subspaces of $\overline{M}$. Obviously

(2) $$\overline{M}^1 = \ker(\varphi \otimes \mathrm{id}_{A/\mathfrak{p}}).$$

Furthermore, $\overline{M}$ and the $\overline{M}^i$ do not change if we replace A by $A_\mathfrak{p}$ and M by $M_\mathfrak{p}$ etc.

If A is a principal ideal domain, then we can find bases $(m_i)_{1 \le i \le r}$ of M and $(m'_i)_{1 \le i \le r}$ of M' and $a_i \in A$ such that $\varphi(m_i) = a_i m'_i$ for all i. Choose $p \in A$ with $\mathfrak{p} = Ap$. Then all m_i with $\nu_\mathfrak{p}(a_i) \ge j$ and all $p^{j-\nu_\mathfrak{p}(a_i)} m_i$ with $\nu_\mathfrak{p}(a_i) < j$ are a basis for M^j (for any j). Therefore all $m_i \otimes 1$ with $\nu_\mathfrak{p}(a_i) \ge j$ form a basis for $\overline{M}^j$. This shows in particular that $\overline{M}^j = 0$ for $j \gg 0$. As $\sum_{i=1}^r \nu_\mathfrak{p}(a_i) = \nu_\mathfrak{p}(\prod_{i=1}^r a_i) = \nu_\mathfrak{p}(\det(\varphi))$, we get

(3) $$\sum_{j>0} \dim_{A/\mathfrak{p}} \overline{M}^j = \nu_\mathfrak{p}(\det(\varphi)).$$

This equation will always hold for M, M' free, even if A is not a principal ideal domain, as both sides in (3) do not change when we replace A by $A_\mathfrak{p}$ and localise. The argument above, together with 8.10(4), also yields a proof of 8.11(3). Furthermore, we get (for any M, M', and A)

(4) $$\sum_{j>0} \dim_{A/\mathfrak{p}} \overline{M}^j = \nu_\mathfrak{p}(\varphi).$$

Suppose now that φ is a homomorphism of G_A–modules. Then all M^i are G_A–submodules of M and all $\overline{M}^i$ are $G_{A/\mathfrak{p}}$–submodules of $\overline{M}$. Applying (4) to all weight spaces we get

(5) $$\sum_{j>0} \mathrm{ch}(\overline{M}^j) = \nu_\mathfrak{p}^c(\varphi).$$

8.19. Let k be a field with $\mathrm{char}(k) = p \neq 0$. Set $G = G_k$ and write $\nu_p = \nu_{\mathbf{Z}p}$. We write $V(\mu)$ instead of $V(\mu)_k$ as in previous chapters. Keep these assumptions until 8.22.

Proposition: *For each $\lambda \in X(T)_+$ there is a filtration of G–modules*

$$V(\lambda) = V(\lambda)^0 \supset V(\lambda)^1 \supset V(\lambda)^2 \supset \cdots$$

such that

(1) $$\sum_{i>0} \operatorname{ch} V(\lambda)^i = \sum_{\alpha \in R^+} \sum_{0<mp<\langle \lambda+\rho, \alpha^\vee \rangle} \nu_p(mp) \chi(s_{\alpha,mp} \bullet \lambda)$$

and

(2) $$V(\lambda)/V(\lambda)^1 \simeq L(\lambda).$$

Proof: If we have a filtration of this type for $k = \mathbf{F}_p$, then we get it in general by tensoring over $\mathbf{F}_p$ with k. (Recall Corollary 2.9.) So we assume now that $k = \mathbf{F}_p$.

We construct then the filtration as in 8.18, working with $A = \mathbf{Z}$ and $\mathfrak{p} = \mathbf{Z}p$ and the map

$$\widetilde{T}_{w_0} = \widetilde{T}_{w_0}(w_0 \bullet \lambda) : V(\lambda)_{\mathbf{Z}} \simeq H^n_{\mathbf{Z}}(w_0 \bullet \lambda) \to H^0_{\mathbf{Z}}(\lambda).$$

Now 8.18(5) says that we get the left hand side in (1) from 8.16(1) by replacing each $\operatorname{div}(i)$ by $\nu_p(i)$. Obviously, we have to look only at those i of the form $i = mp$ with $m \in \mathbf{N}$. As

$$-\chi(\lambda - mp\alpha) = \chi(s_\alpha \bullet (\lambda - mp\alpha)) = \chi(s_\alpha \bullet \lambda + mp\alpha) = \chi(s_{\alpha,mp} \bullet \lambda)$$

we get (1).

We know that $\widetilde{T}_{w_0}$ maps $V(\lambda)_{\mathbf{Z},\lambda}$ bijectively to $H^0_{\mathbf{Z}}(\lambda)_\lambda$. Therefore $\widetilde{T}_{w_0} \otimes \mathrm{id}_k$ is a non-zero homomorphism from $V(\lambda) \simeq H^n(w_0 \bullet \lambda)$ to $H^0(\lambda)$. Now we get

$$V(\lambda)/V(\lambda)^1 \simeq \operatorname{im}(\widetilde{T}_{w_0} \otimes \mathrm{id}_k) \simeq L(\lambda)$$

by 8.18(2) and the remark to 6.16.

Remarks: 1) Let (as in 8.17) ψ be an isomorphism $H^0_{\mathbf{Z}}(\lambda) \xrightarrow{\sim} {}^\tau V(\lambda)_{\mathbf{Z}}$. Use $\varphi = \psi \circ \widetilde{T}_{w_0}$ to define a contravariant form on $V(\lambda)_{\mathbf{Z}}$ as in 8.17(1). We have then

$$V(\lambda)^i_{\mathbf{Z}} = \{ v \in V(\lambda)_{\mathbf{Z}} \mid (v, V(\lambda)_{\mathbf{Z}}) \subset \mathbf{Z}p^i \}.$$

Using this description of the filtration it is easy to show that we have an isomorphism

(3) $${}^\tau\big(V(\lambda)^i/V(\lambda)^{i+1}\big) \simeq V(\lambda)^i/V(\lambda)^{i+1}$$

for each i. As above, we may assume that $k = \mathbf{F}_p$. To start with, we get a symmetric contravariant form on $V(\lambda)^i_{\mathbf{Z}}$ setting $(v, v')_i = p^{-i}(v, v')$. If $v = pv_0 \in V(\lambda)^i_{\mathbf{Z}} \cap pV(\lambda)_{\mathbf{Z}}$, then we have $(v, v') = p(v_0, v') \in \mathbf{Z}p^{i+1}$ for all $v' \in V(\lambda)^i_{\mathbf{Z}}$, hence $(v, v')_i \in \mathbf{Z}p$. Therefore we get a symmetric contravariant form [again denoted by $(\ ,\ )_i$] on $V(\lambda)^i$ with $(v \otimes 1, v' \otimes 1)_i = (v, v')_i + \mathbf{Z}p$ for all $v, v' \in V(\lambda)^i_{\mathbf{Z}}$. It is then obvious that $V(\lambda)^{i+1}$ belongs to the radical of this form and that we get an induced form on $V(\lambda)^i/V(\lambda)^{i+1}$. Now (3) will follow if we can show that this induced form is non-degenerate. For this note that we can choose bases $(v_h)_{1\le h\le r}$ and $(v'_h)_{1\le h\le r}$ for $V(\lambda)_{\mathbf{Z}}$ such that $(v_h, v'_j) = a_h\delta_{hj}$ with suitable $a_h \in \mathbf{Z}$, $a_h \neq 0$. Then both the $v_h \otimes 1$ with $\nu_p(a_h) \geq i$ and the $v'_h \otimes 1$ with $\nu_p(a_h) \geq i$ are bases for $V(\lambda)^i$, and we have $(v_h \otimes 1, v'_j \otimes 1)_i = (p^{-i}a_h + \mathbf{Z}p)\delta_{h,j}$. This yields the required non-degeneracy.

2) One can construct filtrations for each $H^{l(w)}(w \bullet \lambda)$ with $w \in W$ using the maps from Remark 2 in 8.16, cf. [Andersen 12]. The filtrations there are shifted compared to our situation because there any $T_\alpha(\lambda)$ as in 8.13(6) is divided by the greatest common divisor of all $(r-i)!i!$.

3) Suppose that λ is p–regular, i.e., that $\langle\lambda + \rho, \alpha^\vee\rangle \notin \mathbf{Z}p$ for all $\alpha \in R^+$, or (equivalently) that the stabiliser of λ in W_p (for the dot action) is trivial. Then the summands in (1) are linearly independent: To start with, we have always $\chi(s_{\alpha,mp}\bullet\lambda) \neq 0$ since otherwise there exists a root β with $s_\beta\bullet(s_{\alpha,mp}\bullet\lambda) = s_{\alpha,mp}\bullet\lambda$ contradicting the assumption on $\mathrm{Stab}_{W_p}(\lambda)$. If, next, $\chi(s_{\alpha,mp}\bullet\lambda) = \pm\chi(s_{\beta,rp}\bullet\lambda)$ for some $\alpha, \beta \in R^+$ and $r, m \in \mathbf{Z}$, then there exists $w \in W$ with $s_{\alpha,mp}\bullet\lambda = (ws_{\beta,rp})\bullet\lambda$, cf. 5.9(1) and Remark 5.8. This implies $s_{\alpha,mp} = ws_{\beta,rp}$, hence $s_\alpha = ws_\beta$ and $mp\alpha = w(rp\beta) = ws_\beta(-rp\beta) = s_\alpha(-rp\beta)$. Applying $-s_\alpha$ to both sides we get $mp\alpha = rp\beta$, hence $\alpha = \beta$ and $m = r$.

8.20. In simple cases one can compute all composition factors of $V(\lambda)$ using 8.19(1) and the results of Chapters 6 and 7. If, for example, $\sum_{i>0} \mathrm{ch}\ V(\lambda)^i = 0$, then $V(\lambda)^1 = 0$ and $V(\lambda) \simeq L(\lambda)$. If, $\sum_{i>0} \mathrm{ch}\ V(\lambda)^i = \mathrm{ch}\, L(\mu)$ for some $\mu \in X(T)_+$, then necessarily $V(\lambda)^2 = 0$ and $V(\lambda)^1 \simeq L(\mu)$.

Let me next look at an explicit example. Consider $G = SL_4$. We write elements of $X(T)$ in the form $(r,s,t) = r\varpi_1 + s\varpi_2 + t\varpi_3$ where the ϖ_i are the fundamental weights in their usual numbering (as in the tables in [B3], ch. VI). Consider $\lambda_0 \in C \cap X(T)_+$ where C is our standard alcove. So we have $\lambda_0 + \rho = (r,s,t)$ with $r,s,t > 0$ and $r+s+t < p$. (So we assume that $p > 3$.) We have then $X_1(T) \cap W_p \bullet \lambda = \{\, \lambda_i \mid 0 \leq i \leq 5 \,\}$ where $\lambda_1 + \rho = (p-(s+t), s, p-(r+s))$ and

$$\lambda_2 + \rho = (p-s, s+t, p-(r+s+t)), \qquad \lambda_3 + \rho = (p-(r+s+t), r+s, p-s),$$
$$\lambda_4 + \rho = (p-(r+s), r+s+t, p-(s+t)), \qquad \lambda_5 + \rho = (p-r, r+s+t, p-t).$$

We want to determine all $\mathrm{ch}\, L(\lambda_i)$. Then we can use 7.17.b to find $\mathrm{ch}\, L(\mu)$ for all $\mu \in X_1(T)$. Afterwards Steinberg's tensor product theorem yields $\mathrm{ch}\, L(\mu)$ for all $\mu \in X(T)_+$. The calculation of the $\mathrm{ch}\, L(\lambda_i)$ will also involve λ'_0 and λ''_0 given by

$$\lambda'_0 + \rho = (p+s, t, p-(r+s+t)), \qquad \lambda''_0 + \rho = (p-(r+s+t), r, p+s).$$

These are elements in $X(T)_+ \cap W_p \bullet \lambda$ outside $X_1(T)$.

One starts by observing that $\sum_{i>0} \mathrm{ch}\ V(\lambda_0)^i = 0$ and hence $\mathrm{ch}\, L(\lambda_0) = \chi(\lambda_0)$. (Of course, we have known this since 5.6.) Next the sum formula 8.19(1) yields $\sum_{i>0} \mathrm{ch}\ V(\lambda_1)^i = \chi(\lambda_0) = \mathrm{ch}\, L(\lambda_0)$, hence $V(\lambda_1)^1 \simeq L(\lambda_0)$. It follows that

$$\chi(\lambda_1) = \mathrm{ch}\, L(\lambda_1) + \mathrm{ch}\, L(\lambda_0) \quad \text{and} \quad \mathrm{ch}\, L(\lambda_1) = \chi(\lambda_1) - \chi(\lambda_0).$$

Next the sum formula yields

$$\sum_{i>0} \mathrm{ch}\ V(\lambda_2)^i = \chi(\lambda_1) - \chi(\lambda_0) = \mathrm{ch}\, L(\lambda_1),$$

hence $V(\lambda_2)^1 \simeq L(\lambda_1)$. It follows that

$$\chi(\lambda_2) = \operatorname{ch} L(\lambda_2) + \operatorname{ch} L(\lambda_1) \quad \text{and} \quad \operatorname{ch} L(\lambda_2) = \chi(\lambda_2) - \chi(\lambda_1) + \chi(\lambda_0).$$

One gets similarly

$$\chi(\lambda_3) = \operatorname{ch} L(\lambda_3) + \operatorname{ch} L(\lambda_1) \quad \text{and} \quad \operatorname{ch} L(\lambda_3) = \chi(\lambda_3) - \chi(\lambda_1) + \chi(\lambda_0).$$

Then the sum formula yields

$$\sum_{i>0} \operatorname{ch} V(\lambda_0')^i = \chi(\lambda_2) - \chi(\lambda_1) + \chi(\lambda_0) = \operatorname{ch} L(\lambda_2),$$

hence $V(\lambda_0')^1 \simeq L(\lambda_2)$ and

$$\chi(\lambda_0') = \operatorname{ch} L(\lambda_0') + \operatorname{ch} L(\lambda_2), \qquad \operatorname{ch} L(\lambda_0') = \chi(\lambda_0') - \chi(\lambda_2) + \chi(\lambda_1) - \chi(\lambda_0).$$

One gets similarly (or by symmetry)

$$\chi(\lambda_0'') = \operatorname{ch} L(\lambda_0'') + \operatorname{ch} L(\lambda_3), \qquad \operatorname{ch} L(\lambda_0'') = \chi(\lambda_0'') - \chi(\lambda_3) + \chi(\lambda_1) - \chi(\lambda_0).$$

Next the sum formula yields

$$\sum_{i>0} \operatorname{ch} V(\lambda_4)^i = \chi(\lambda_3)+\chi(\lambda_2)+\chi(\lambda_0) = \operatorname{ch} L(\lambda_3)+\operatorname{ch} L(\lambda_2)+2\operatorname{ch} L(\lambda_1)+\operatorname{ch} L(\lambda_0).$$

This implies immediately for all $j \in \{0,3,2\}$ that $[V(\lambda_4) : L(\lambda_j)] = 1$ and that $L(\lambda_j)$ is a composition factor of $V(\lambda_4)^1/V(\lambda_4)^2$. However, the sum formula alone leaves us with two possibilities for $[V(\lambda_4) : L(\lambda_1)]$: This multiplicity could be equal to 1 or 2. However, Proposition 7.18 (applied with s equal to the reflection with respect to the third simple root) implies that $[V(\lambda_4) : L(\lambda_1)] = [V(\lambda_3) : L(\lambda_1)] = 1$. We get thus

$$\chi(\lambda_4) = \operatorname{ch} L(\lambda_4) + \operatorname{ch} L(\lambda_3) + \operatorname{ch} L(\lambda_2) + \operatorname{ch} L(\lambda_1) + \operatorname{ch} L(\lambda_0)$$

and

$$\operatorname{ch} L(\lambda_4) = \chi(\lambda_4) - \chi(\lambda_3) - \chi(\lambda_2) + \chi(\lambda_1) - 2\chi(\lambda_0).$$

Finally, the sum formula yields

$$\begin{aligned}\sum_{i>0} \operatorname{ch} V(\lambda_5)^i &= \chi(\lambda_4) + \chi(\lambda_0') + \chi(\lambda_0'') - \chi(\lambda_0) \\ &= \operatorname{ch} L(\lambda_4) + \operatorname{ch} L(\lambda_0') + \operatorname{ch} L(\lambda_0'') + 2\operatorname{ch} L(\lambda_3) + 2\operatorname{ch} L(\lambda_2) + \operatorname{ch} L(\lambda_1).\end{aligned}$$

Now Proposition 7.18 (applied with s equal to the reflection with respect to the second simple root) implies that $[V(\lambda_5) : L(\lambda_3)] = [V(\lambda_4) : L(\lambda_3)]$ and $[V(\lambda_5) : L(\lambda_2)] = [V(\lambda_4) : L(\lambda_2)]$. We get then

$$\chi(\lambda_5) = \operatorname{ch} L(\lambda_5)+\operatorname{ch} L(\lambda_4)+\operatorname{ch} L(\lambda_0')+\operatorname{ch} L(\lambda_0'')+\operatorname{ch} L(\lambda_3)+\operatorname{ch} L(\lambda_2)+\operatorname{ch} L(\lambda_1)$$

and

$$\operatorname{ch} L(\lambda_5) = \chi(\lambda_5) - \chi(\lambda_4) - \chi(\lambda_0') - \chi(\lambda_0'') + \chi(\lambda_3) + \chi(\lambda_2) - 2\chi(\lambda_1) + 3\chi(\lambda_0).$$

The character formulae above in type A_3 were first proved in [Jantzen 2]; there also the case $p \leq 3$ is treated. The same method works also for G of types A_1, A_2, B_2, and G_2 (assuming $p \neq 3, 5$ in the last case), see [Jantzen 3], p. 139/140. The result for G of type A_1 had been known for a long time, for G of type A_2 or B_2 there had been partial results by Braden (partly unpublished). For G of type G_2 the case $p = 3$ had already been treated in [Springer], and the case $p = 5$ was solved by J. Hagelskjær in his "speciale" from 1983 at Aarhus Universitet. (The results for $p = 5$ look as they do for $p \geq 7$ for the corresponding facets.)

In the "next" cases (G of type B_3, C_3, or A_4) the orbit $W_p \bullet \lambda$ of a p–regular $\lambda \in X(T)$ intersects $X_1(T)$ in 24 elements. The methods described above and some refinements (e.g., using wall crossing functors and Vogan's generalised τ–invariants, cf. [Jantzen 5], p. 299) lead to a computation of all but 1 or 2 multiplicities in these cases.

8.21. If we use the construction of the filtration of $V(\lambda)$ using the contravariant form as in Remark 1 in 8.19, then one gets $\dim L(\lambda)$ as the rank modulo p of the matrix of $(\, , \,)$ with respect to a basis (or just a generating system) of $V(\lambda)_{\mathbf{Z}}$. More precisely, one gets each $\dim L(\lambda)_\mu$ working with a generating system of $V(\lambda)_{\mathbf{Z},\mu}$. Computer programmes carrying out such calculations exist, see [Gilkey and Seitz] or `http://home.imf.au.dk/abuch/dynkin/index.html`. There are also computer programmes using other methods, such as embedding the simple modules into $\mathrm{Dist}_n(G)$, cf. [Koppinen 5], or working with modules of the form $Z_r(\lambda)$ as in 3.7, cf. `http://www.math.Virginia.EDU/~lls21/research_undergrad.html` and [Scott 5], p. 8.

Using such computer programmes or other direct computations one has a series of results (where all $\operatorname{ch} L(\mu)$ with $\mu \in X_1(T)$ are determined) in particular for $p = 2$ (see [Veldkamp 1], [Dowd and Sin], [Xu and Ye], [Ye and Zhou 1, 2], [Ye 5]) or $p = 3$ (see [Ye and Zhou 3, 4]).

Apart from that only special highest weights have been treated. The easiest case is that where all highest weights of $V(\lambda)$ belong to $W\lambda$. Then $V(\lambda) = L(\lambda)$ by 2.15. This happens if and only if λ is minuscule, i.e., if $\langle \lambda, \alpha^\vee \rangle \in \{0, 1\}$ for all $\alpha \in R^+$, cf. [B3], ch. VI, §1, exerc. 24 and §4, exerc. 15.

The next easiest case is that where λ is a root. If R is irreducible and G is semi-simple, then R contains a unique largest root $\widetilde{\alpha}$ and one has $V(\widetilde{\alpha})_{\mathbf{Q}} \simeq \mathrm{Lie}(G_{\mathbf{Q}})$ with the adjoint representation. If R has two root lengths (for R and G as before), then there is also $V(\alpha_0)$ with α_0 the largest short root. For classical groups the composition factors of $V(\widetilde{\alpha})$ and $V(\alpha_0)$ can be computed easily (and have been known for a long time). For exceptional groups they appear in [Dieudonné]. One can find the results in [Jantzen 1], p. 20/1, [Hogeweij], and [Hiß]. For the case of α_0 see also the generalisation in [Jones].

The next class of weights are the fundamental weights. We shall use the same notation for these weights as in [B3], ch. VI. Some of the fundamental weights are minuscule or a root and have already been discussed. For example, in type A_n all fundamental weights are minuscule. For R of type B_n or D_n, let $V_{\mathbf{Q}}$ denote the "natural" module for $G_{\mathbf{Q}}$, i.e., the module of dimension $2n + 1$ or $2n$ such that $G_{\mathbf{Q}}$

acts on $V_{\mathbf{Q}}$ as special orthogonal group $SO(V_{\mathbf{Q}})$. Then one has $V(\varpi_i)_{\mathbf{Q}} \simeq \Lambda^i V_{\mathbf{Q}}$ for $1 \le i \le n-1$ in type B_n, for $1 \le i \le n-2$ in type D_n. (One has $\Lambda^n V_{\mathbf{Q}} \simeq V(2\varpi_n)_{\mathbf{Q}}$ in type B_n while $\Lambda^{n-1} V_{\mathbf{Q}} \simeq V(\varpi_{n-1}+\varpi_n)_{\mathbf{Q}}$ and $\Lambda^n V_{\mathbf{Q}} \simeq V(2\varpi_{n-1})_{\mathbf{Q}} \oplus V(2\varpi_n)_{\mathbf{Q}}$ in type D_n.) Now all $V(\varpi_i)$ with i as above are simple if $\operatorname{char}(k) \neq 2$, and are not simple if $\operatorname{char}(k) = 2$ (except for the minuscule ϖ_1 in type D_n). This was first observed in [Wong], p. 65–67 using the contravariant form. (The basis used there for $V(\lambda)_{\mathbf{Z}}$ and the formula for the determinant was not quite correct and was corrected in [Jantzen 1], p. 38–44.) The same is actually also true for the $V(\mu)$ coming from the remaining exterior powers, cf. [McNinch 3], 3.4.

For R of type C_n the sum formula implies for $1 \le j \le n$

$$\text{(1)} \qquad \sum_{i>0} \operatorname{ch} V(\varpi_j)^i = \sum_{0<2h\le j} \big(\nu_p(n+1+h-j) - \nu_p(h)\big)\, \chi(\varpi_{j-2h})$$

where we use the convention that $\varpi_0 = 0$. It follows that $V(\varpi_j)^i$ is simple if and only if p does not divide any $\binom{n+1+h-j}{h}$ with $0 < h \le (j/2)$. This was first proved in [Premet and Suprunenko 1] just before the sum formula was proved also for the small p that are relevant here.

For R of exceptional type one can find in [Jantzen 15], 4.6 a list of all ϖ_j and p such that $V(\varpi_j)$ is not simple if $\operatorname{char}(K) = p$. In many cases the composition factors then can be found in [Gilkey and Seitz].

For R of type A_{n-1} there is in [Jantzen 1], p. 113 an explicit determination of all λ with $V(\lambda)$ simple. Let me quote the result without proof using the notation from 1.21, especially $R^+ = \{\varepsilon_i - \varepsilon_j \mid 1 \le i < j \le n\}$. Then $V(\lambda)$ is simple if and only if for each $\alpha \in R^+$ the following is satisfied: Write $\langle \lambda+\rho, \alpha^\vee\rangle = ap^s + bp^{s+1}$ with $a, b, s \in \mathbf{N}$ and $0 < a < p$. Then there have to be $\beta_0, \beta_1, \dots, \beta_b \in R^+$ with $\langle \lambda+\rho, \beta_i^\vee\rangle = p^{s+1}$ for $1 \le i \le b$ and $\langle \lambda+\rho, \beta_0^\vee\rangle = ap^s$, with $\alpha = \sum_{i=0}^b \beta_i$ and with $\alpha - \beta_0 \in R$. More explicitly, if $\alpha = \varepsilon_i - \varepsilon_j$ with $1 \le i < j \le n$, then there have to be integers $i = i_0 < i_1 < \cdots < i_b < i_{b+1} = j$ such that $\{\beta_i \mid 0 \le i \le b\} = \{\varepsilon_{i_m} - \varepsilon_{i_{m+1}} \mid 0 \le m \le b\}$ and $\beta_0 \in \{\varepsilon_i - \varepsilon_{i_1}, \varepsilon_{i_b} - \varepsilon_j\}$.

8.22. Let $\lambda \in X(T) \cap C$. Recall that the existence of λ implies $p \ge h$, see 6.2(10). Let W_p^+ denote the set of all $w \in W_p$ such that $\langle w(x+\rho), \alpha^\vee\rangle > 0$ for all $\alpha \in R^+$ and for all $x \in C$ (equivalently: for one $x \in C$). Then W_p^+ is also the set of all $w \in W_p$ with $w \bullet \lambda \in X(T)_+$. We have (unique) integers $a_{w,w'} \in \mathbf{Z}$ for all $w, w' \in W_p^+$ with

$$\text{(1)} \qquad \operatorname{ch} L(w \bullet \lambda) = \sum_{w' \in W_p^+} a_{w,w'}\, \chi(w' \bullet \lambda).$$

The $a_{w,w'}$ are by 7.17.b independent of the choice of λ. One gets the matrix of all $a_{w,w'}$ by inverting the matrix of all $[H^0(w \bullet \lambda) : L(w' \bullet \lambda)]$; therefore the strong linkage principle (6.13) tells us that $a_{w,w'} \neq 0$ implies $w' \bullet \lambda \uparrow w \bullet \lambda$.

If we know all $a_{w,w'}$ with $w \bullet \lambda \in X_1(T)$ (as in 3.15(1)), then we can apply 7.17.b and express all $\operatorname{ch} L(\mu)$ with $\mu \in X_1(T)$ as linear combinations of suitable $\chi(\mu)$. Using Steinberg's tensor product theorem (possibly for a covering group of G), we can then compute $\operatorname{ch} L(\mu)$ for all $\mu \in X(T)_+$.

We shall now state a conjecture of Lusztig that predicts the $a_{w,w'}$ in (1) within limits. One aspect of this conjecture is that the predicted $a_{w,w'}$ are in some sense independent of p. In order to make this independence clear we introduce some awkward notation that we intend to forget after the end of this subsection.

Recall from 6.1 that we have an isomorphism from the affine Weyl group $W_1 = W_a(R^\vee)$ to W_p mapping any $s_{\alpha,n}$ to $s_{\alpha,np}$. Let us denote this isomorphism by $w \mapsto w[p]$. The inverse image W_1^+ of W_p^+ is independent of p and can be defined analogously. Now the conjecture from [Lusztig 3] says:

Conjecture: *Let $w \in W_1^+$. If $\langle w[p](\lambda+\rho), \alpha^\vee\rangle \leq p(p-h+2)$ for all $\alpha \in R^+$, then*

$$a_{w[p],w'[p]} = \varepsilon(w)\,\varepsilon(w')\,P_{w_0 w', w_0 w}(1) \tag{2}$$

for all $w, w' \in W_1^+$.

Here $P_{x,y}$ denotes a Kazhdan-Lusztig polynomial for W_1 as introduced in [Kazhdan and Lusztig 1], see also [Hu3], 7.9. Furthermore $\varepsilon(w)$ is the sign of w: If w is the composition of some $w_1 \in W$ with an arbitrary translation, then $\varepsilon(w) = \det(w_1)$.

The definition of the Kazhdan-Lusztig polynomials involves the Bruhat-Chevalley ordering $\leq$ on W_1 (as in [Hu3], 5.9). If $P_{x,y} \neq 0$, then $x \leq y$. One can show for all $w, w' \in W_1^+$ that $w'[p] \bullet \lambda \uparrow w[p] \bullet \lambda$ holds if and only if $w' \leq w$ while $w_0 w'[p] \bullet \lambda \uparrow w_0 w[p] \bullet \lambda$ holds if and only if $w_0 w \leq w_0 w'$. The observation after (1) says that $a_{w[p],w'[p]} \neq 0$ implies $w'[p] \bullet \lambda \uparrow w[p] \bullet \lambda$, hence $w_0 w[p] \bullet \lambda \uparrow w_0 w'[p] \bullet \lambda$, cf. 6.4(6). So the non-zero terms in (2) satisfy $w' \leq w$ and $w_0 w' \leq w_0 w$.

As mentioned above, one aspect of this conjecture is the independence of p on the right hand side of (2). This independence makes it necessary to restrict the "domain of validity" of (2) somewhat — this is done in the conjecture by the bound involving $p(p-h+2)$ — because in general the $a_{w[p],w'[p]}$ will not be independent of p. This can be seen by looking closely at Steinberg's tensor product theorem.

Let me illustrate this by a simple example. Take $G = SL_3$. In this case ρ is equal to the largest root and $V(\rho)$ is equal to Lie(G) with the adjoint representation. It is now elementary to check that $V(\rho) = L(\rho)$ if $p \neq 3$, whereas $V(\rho)$ has composition factors $L(\rho)$ and $L(0)$, both with multiplicity 1, if $p = 3$. (The identity matrix spans for $p = 3$ a one dimensional submodule. One can also refer to 5.6 and 6.24(1).) It follows that

$$\operatorname{ch} L(\rho) = \sum_{w \in W} e(w\rho) + a\, e(0) \qquad \text{with } a = \begin{cases} 2, & \text{if } p \neq 3, \\ 1, & \text{if } p = 3. \end{cases} \tag{3}$$

Assume now that $p \geq 3$, so that $\lambda = 0 \in C$. Since $\rho \in R$ in this case, the translation by ρ belongs to W_1. Denote this element by x; it actually belongs to W_1^+ and we have $x[p] \bullet 0 = p\rho$. Steinberg's tensor product theorem (3.17) implies that $L(x[p] \bullet 0) = L(p\rho) \simeq L(\rho)^{[1]}$, hence by (3) that

$$\operatorname{ch} L(x[p] \bullet 0) = \sum_{w \in W} e(pw\rho) + a\, e(0) \tag{4}$$

with a as above. Note that $xw[p]\bullet 0 = p\rho + w\bullet 0 = (p-1)\rho + w\rho$ for all $w \in W$. One can use this to check that $xw \in W_1^+$ for all $w \in W$. If we now write $\operatorname{ch} L(x[p]\bullet 0)$ in the form as in (1), then we get

$$\text{(5)} \qquad \operatorname{ch} L(x[p]\bullet 0) = \sum_{w\in W} \det(w)\,\chi(xw[p]\bullet 0) + a\,\chi(0).$$

So $a_{x[p],w'[p]}$ depends on p for $w' = 1$. (In order to deduce (5) from (4) one has to show that $\sum_{w\in W} e(pw\rho) = \sum_{w\in W} \det(w)\chi((p-1)\rho + w\rho)$. For this one recalls the notation $A(\mu)$ from 5.9(4), multiplies both sides with $A(\rho)$, and applies Weyl's character formula.)

Return to the general set-up. Suppose for simplicity that G is semi-simple and simply connected. For an arbitrary $w \in W_1^+$ we can write $w[p]\bullet\lambda = p\mu_w + \nu$ with $\nu \in X_1(T)$ and $\mu_w \in X(T)_+$. Here μ_w depends only on w, not on p and λ (whereas ν does). Set then $\widehat{L}(w[p]\bullet\lambda) = H^0(\mu_w)^{[1]} \otimes L(\nu)$. Now Steinberg's tensor product theorem implies $\operatorname{ch}\widehat{L}(w[p]\bullet\lambda) = \operatorname{ch} L(w[p]\bullet\lambda)$ if and only if $H^0(\mu_w) = L(\mu_w)$. Suppose that we knew that the $a_{w[p],w'[p]}$ were independent of p for all w with $w[p]\bullet\lambda \in X_1(T)$, i.e., with $\mu_w = 0$. Then one can find for arbitrary $w \in W_1^+$ integers $a'_{w,w'}$ independent of p such that $\operatorname{ch}\widehat{L}(w[p]\bullet\lambda) = \sum_{w'\in W_1^+} a'_{w,w'}\,\chi(w'[p]\bullet\lambda)$. If $p \geq \langle \mu_w + \rho, \alpha^\vee\rangle$ for all $\alpha \in R^+$, then $H^0(\mu_w) = L(\mu_w)$ by 5.6, hence $a_{w[p],w'[p]} = a'_{w,w'}$. However, if $H^0(\mu_w) \neq L(\mu_w)$, then there has to be some w' with $a_{w[p],w'[p]} \neq a'_{w,w'}$. (Note that this is what happens for $p = 3$ in our example where $\mu_x = \rho$.) So we can expect the $a_{w[p],w'[p]}$ to be independent of p only when $H^0(\mu_w) = L(\mu_w)$. The condition "$\langle w[p](\lambda+\rho), \alpha^\vee\rangle \leq p(p-h+2)$ for all $\alpha \in R^+$" in Lusztig's conjecture has the purpose to make sure that $\langle \mu_w + \rho, \alpha^\vee\rangle \leq p$ for all $\alpha \in R^+$, hence that $H^0(\mu_w) = L(\mu_w)$.

The considerations sketched in the preceding paragraph show: If the $a_{w[p],w'[p]}$ are independent of p for all w with $w[p]\bullet\lambda \in X_1(T)$, then they are independent of p for all w satisfying the bound in the conjecture. One can actually show (see [Kato 2]): If (2) holds for all w with $w[p]\bullet\lambda \in X_1(T)$, then it holds for all w satisfying the bound in the conjecture. (Note that not all w with $w[p]\bullet\lambda \in X_1(T)$ satisfy that bound if $p < 2h-3$.)

When the conjecture first was formulated, it was supported by the results for low rank cases mentioned in 8.20, and by its analogy with the Kazhdan-Lusztig conjecture for simple highest weight modules for complex semi-simple Lie algebras (proved by Beilinson & Bernstein and by Brylinski & Kashiwara around 1980). It was then soon checked that Lusztig's conjecture is equivalent to certain other conjectures; we shall look at that point in Chapter C. In [Andersen, Jantzen, and Soergel] it was shown that there exists for each given root system an unknown bound $n_0 = n_0(R)$ such that the conjecture holds for groups with root system R if $p > n_0$. This involves a comparison with quantum groups at a root of unity, see H.12.

CHAPTER 9

Representations of G_rT and G_rB

Let p be a prime and k a field of characteristic p. We assume k to be perfect for the sake of convenience (as we did in I.9). Fix an integer $r \geq 1$. (We shall make some additional assumptions in 9.14 which are assumed to hold in 9.14–9.25.)

The representation theory of the group schemes G_rT and G_rB (and G_rB^+, of course) parallels to a large extent that of G and G_r. There is a classification of the simple modules by their highest weights (9.6). They can also be described as the (simple) socles of induced representations $\widehat{Z}'_r(\lambda) = \operatorname{ind}_B^{G_rB} \lambda$, which play the role of the G–modules $H^0(\lambda)$ and also of the $V(\lambda')$ as the class of all $\widehat{Z}'_r(\lambda)$ is stable under taking the dual (9.2). The formal characters of the simple G_rT– or G_rB–modules are linearly independent, so the formal character of any finite dimensional G_rT– or G_rB–module determines its composition factors.

There is a strong linkage principle for the $\widehat{Z}'_r(\lambda)$. We deduce it here (9.15) from the corresponding result for G. But it can also be proved directly; we do that here (9.11) only for $r = 1$ and refer in general for the direct proof to [Doty 5]. As a consequence we get (9.19) the linkage principle for $\operatorname{Ext}^1_{G_rT}$. We then prove it for $\operatorname{Ext}^1_{G_rB}$ using the spectral sequence relating $\operatorname{Ext}_{G_rB}$ to Ext_{G_r}, cf. 9.20/21. As a consequence many results about blocks and about the functors $\operatorname{pr}_\lambda$ and T_λ^μ carry over from G to G_rB and G_rT, sometimes in a simplified version (9.22).

Furthermore, this chapter contains some applications of the representation theory of G_rT and G_rB to that of G. Let me mention the most important one (9.17): Suppose all composition factors of some $\widehat{Z}'_r(\lambda)$ with $\lambda \in X(T)_+$ have a highest weight of the form $\lambda_0 + p^r\lambda_1$ with $\lambda_0 \in X_r(T)$ and $\lambda_1 \in \overline{C}_{\mathbf{Z}}$. (There are conditions that imply this, cf. 9.18.) Then $H^i(w\bullet\lambda) = 0$ for all $w \in W$ and $i \neq l(w)$. The composition factors of $H^0(\lambda)$ (and of any $H^{l(w)}(w \bullet \lambda)$) have the same highest weights as those of $\widehat{Z}'_r(\lambda)$; also the multiplicities coincide. As a consequence we get: If λ and $\lambda + p^r\nu$ for some $\nu \in X(T)$ both satisfy the assumption above, then we get the highest weights of the composition factors of $H^0(\lambda + p^r\nu)$ from those of $H^0(\lambda)$ by adding $p^r\nu$ because the analogous result for $\widehat{Z}'_r(\lambda + p^r\nu)$ and $\widehat{Z}'_r(\lambda)$ holds trivially. That there is such a behaviour of the composition factors was first proved in [Jantzen 4] using complicated calculations. The use of G_rT–modules simplified the proof considerably, cf. [Jantzen 7]. The experimental evidence available at the time led to the suggestion (cf. [Humphreys 10]) that a vanishing result as above ($H^i(w \bullet \lambda) = 0$ for $i \neq l(w)$) should hold. This was then proved in [Cline, Parshall, and Scott 10] whose approach we follow here.

Besides the papers already mentioned, the main sources for this chapter are [Andersen 15], [Andersen and Jantzen], [Cline, Parshall, and Scott 6, 9], [Humphreys 1], [Jantzen 6], and [O'Halloran 4].

Compared to the first edition the subsections 9.3 and 9.8–11 were inserted. This led to a shift of most old subsections: the subsections 9.3–6 were moved to 9.4–7, the old subsection 9.7 is now Remark 3 in 9.6, the subsections 9.8/9 were moved to 9.12/13, and the subsections 9.11-21 were moved to 9.14–24. The old subsection 9.10 is now the first part of 9.14. The new subsection 9.9 is based on parts of the old 11.1.

9.1. Set for all $\lambda \in X(T)$

(1) $$\widehat{Z}'_r(\lambda) = \operatorname{ind}_B^{G_rB} \lambda$$

and

(2) $$\widehat{Z}_r(\lambda) = \operatorname{coind}_{B^+}^{G_rB^+} \lambda.$$

(See I.8.20 for the definition of coind in this situation.) We have by construction resp. by I.6.13(1) isomorphisms of G_r–modules (cf. 3.7(2), (3))

(3) $$\widehat{Z}'_r(\lambda) \simeq Z'_r(\lambda) \qquad \text{and} \qquad \widehat{Z}_r(\lambda) \simeq Z_r(\lambda) \qquad (\text{over } G_r).$$

Similarly, one has isomorphisms of G_rT–modules

(4) $$\widehat{Z}'_r(\lambda) \simeq \operatorname{ind}_{B_rT}^{G_rT} \lambda \qquad (\text{over } G_rT)$$

and

(5) $$\widehat{Z}_r(\lambda) \simeq \operatorname{coind}_{B_r^+T}^{G_rT} \lambda \qquad (\text{over } G_rT).$$

As G_rT is not normal in G_rB, we cannot apply I.6.13(1) in order to prove (4). We have, however, $G_r \simeq U_r^+ \times B_r$, hence $G_rB \simeq U_r^+ \times B$ and $G_rT \simeq U_r^+ \times B_rT$ (as schemes, in all cases) so that both sides in (4) are isomorphic to

(6) $$k[U_r^+] \otimes \lambda$$

with T acting through the conjugation action on $k[U_r^+]$. Similarly, both sides in (5) are isomorphic to

(7) $$\operatorname{Dist}(U_r) \otimes \lambda$$

with the adjoint action of T on $\operatorname{Dist}(U_r)$, cf. 3.6.

Note that $\operatorname{ind}_B^{G_rB}$ and $\operatorname{coind}_B^{G_rB}$ (similarly for B^+ instead of B) are exact functors. (Use I.6.13(1) or the remark to I.8.20 together with I.8.16(1), (4).)

9.2. Lemma: *Let $\lambda \in X(T)$. Then:*

(1) $$\widehat{Z}_r(\lambda) \simeq \operatorname{ind}_{B^+}^{G_rB^+}(\lambda - 2(p^r-1)\rho) \simeq \widehat{Z}_r(2(p^r-1)\rho - \lambda)^*$$

(2) $$\widehat{Z}'_r(\lambda) \simeq \operatorname{coind}_B^{G_rB}(\lambda - 2(p^r-1)\rho) \simeq \widehat{Z}'_r(2(p^r-1)\rho - \lambda)^*$$

(3) $$\operatorname{ch} \widehat{Z}_r(\lambda) = \operatorname{ch} \widehat{Z}'_r(\lambda) = e(\lambda) \prod_{\alpha \in R^+} \frac{1 - e(-p^r\alpha)}{1 - e(-\alpha)}$$

(4) $$\widehat{Z}_r(\lambda + p^r\mu) \simeq \widehat{Z}_r(\lambda) \otimes p^r\mu \quad \textit{for all } \mu \in X(T)$$

(5) $$\widehat{Z}'_r(\lambda + p^r\mu) \simeq \widehat{Z}'_r(\lambda) \otimes p^r\mu \quad \textit{for all } \mu \in X(T).$$

Proof: (1) and (2) follow from 3.4 and I.8.20(2), (3). In order to get (3) we have to determine (by 9.1(6), (7)) the character of $k[U_r^+]$ and $\mathrm{Dist}(U_r)$. All $\prod_{\alpha\in R^+} X_{-\alpha,n(\alpha)}$ with $0 \le n(\alpha) < p^r$ for all α form a basis for $\mathrm{Dist}(U_r)$, cf. 3.3. As T acts through $-m\alpha$ on $X_{-\alpha,m}$, we get

$$\mathrm{ch}\,\mathrm{Dist}(U_r) = \prod_{\alpha\in R^+} (1+e(-\alpha)+\cdots+e(-(p^r-1)\alpha)) = \prod_{\alpha\in R^+} \frac{1-e(-p^r\alpha)}{1-e(-\alpha)}.$$

One computes $\mathrm{ch}\,k[U_r^+] = \mathrm{ch}(\mathrm{Dist}(U_r^+)^*)$ in a similar way. The claims in (4) and (5) follow from the tensor identity. We just have to observe that we can regard $p^r\mu$ as a G_rB–module (or a G_rB^+–module) with G_rU acting trivially and $G_rB/G_rU \simeq T/T_r$ via $p^r\mu$ (similarly for G_rB^+).

Remark: Note that (3) or its proof show that all weights μ of $\widehat{Z}_r(\lambda)$ satisfy

(6) $$\lambda - 2(p^r-1)\rho \le \mu \le \lambda$$

and that both λ and $\lambda - 2(p^r-1)\rho$ occur with multiplicity 1 as weights of $\widehat{Z}_r(\lambda)$.

9.3. Consider for all $w\in W$ and $\lambda \in X(T)$ the G_rT–module

(1) $$\widehat{Z}_r^w(\lambda) = \mathrm{coind}_{w(B_r^+T)}^{G_rT}\lambda$$

where $w(B_r^+T) = \dot{w}B_r^+T\dot{w}^{-1}$. (Recall that $\dot{w}$ is a representative of w in $N_G(T)(k)$. So we have also $w(B_r^+T) = (\dot{w}U^+\dot{w}^{-1})_rT$.) We get as special cases (by definition or by 9.2(2))

$$\widehat{Z}_r^1(\lambda) = \widehat{Z}_r(\lambda) \qquad \text{and} \qquad \widehat{Z}_r^{w_0}(\lambda) \simeq \widehat{Z}'_r(\lambda - 2(p^r-1)\rho).$$

We can alternatively construct the $\widehat{Z}_r^w(\lambda)$ as follows: Since $\dot{w}$ normalises T and the normal subgroup scheme G_r, it also normalises G_rT. We can therefore twist G_rT–modules with $\dot{w}$, cf. I.2.15. Now an analogue for coind to I.3.5(4) yields

(2) $$^{\dot{w}}\widehat{Z}_r(\lambda) \simeq \widehat{Z}_r^w(w\lambda)$$

for all $\lambda \in X(T)$ and $w\in W$.

We can identify $\widehat{Z}_r^w(\lambda)$ with $\mathrm{Dist}((\dot{w}U\dot{w}^{-1})_r)\otimes\lambda$ as a T–module. Calculations similar to those in 9.2 show that $\mathrm{ch}\,\mathrm{Dist}((\dot{w}U\dot{w}^{-1})_r)$ is equal to

$$\prod_{\alpha>0}\frac{1-e(-p^rw\alpha)}{1-e(-w\alpha)} = \prod_{\substack{\beta>0\\ w^{-1}\beta>0}}\frac{1-e(-p^r\beta)}{1-e(-\beta)}\prod_{\substack{\beta>0\\ w^{-1}\beta<0}}\frac{1-e(p^r\beta)}{1-e(\beta)}.$$

Rewriting the factors in the last product as

$$\frac{e(p^r\beta)\,(e(-p^r\beta)-1)}{e(\beta)\,(e(-\beta)-1)} = e((p^r-1)\beta)\,\frac{1-e(-p^r\beta)}{1-e(-\beta)}$$

and using $\sum_{\beta>0,\, w^{-1}\beta>0}\beta = \rho - w\rho$ one gets

$$\operatorname{ch}\operatorname{Dist}((\dot{w}U\dot{w}^{-1})_r) = e((p^r-1)(\rho - w\rho))\operatorname{ch}\operatorname{Dist}(U_r)$$

hence

$$\operatorname{ch}\widehat{Z}_r^w(\lambda) = e((p^r-1)(\rho - w\rho))\operatorname{ch}\widehat{Z}_r(\lambda). \tag{3}$$

One gets in particular for all $w \in W$

$$\operatorname{ch}\widehat{Z}_r^w(\lambda\langle w\rangle) = \operatorname{ch}\widehat{Z}_r(\lambda) \qquad \text{where } \lambda\langle w\rangle = \lambda + (p^r-1)(w\rho - \rho). \tag{4}$$

The anti-automorphism τ of G preserves G_rT. (We have $\tau(T) = T$ by construction of τ. Since τ commutes with the Frobenius automorphism, we get also $\tau(G_r) = G_r$.) We can therefore define for each finite dimensional G_rT–module M a G_rT–module ${}^\tau M$ proceeding as in 2.12. We have again $\operatorname{ch}({}^\tau M) = \operatorname{ch}(M)$ and ${}^\tau({}^\tau M) \simeq M$. The forgetful functor from $\{\,G\text{–modules}\,\}$ to $\{\,G_rT\text{–modules}\,\}$ commutes with the two functors $M \mapsto {}^\tau M$.

We claim that for all $\lambda \in X(T)$

$${}^\tau\widehat{Z}_r(\lambda) \simeq \widehat{Z}'_r(\lambda). \tag{5}$$

Indeed, the automorphism σ with $\sigma(g) = \tau(g^{-1})$ satisfies

$$\begin{aligned}{}^\tau\widehat{Z}_r(\lambda) &\simeq {}^\sigma(\widehat{Z}_r(\lambda)^*) \simeq {}^\sigma\widehat{Z}_r(2(p^r-1)\rho - \lambda) = {}^\sigma(\operatorname{ind}_{B_r^+T}^{G_rT}(-\lambda)) \\ &\simeq \operatorname{ind}_{\sigma(B_r^+T)}^{G_rT}(-\sigma\lambda)) \simeq \operatorname{ind}_{B_rT}^{G_rT}(\lambda) \simeq \widehat{Z}'_r(\lambda)\end{aligned}$$

since $\sigma(B^+) = B$ and since σ induces -1 on $X(T)$.

9.4. Lemma: *Let H be an finite subgroup scheme of G that is normalised by T. Let M be a finite dimensional HT–module. Then the following are equivalent:*

(i) *M is injective as an HT–module.*

(ii) *M is injective as an H–module.*

(iii) *M is projective as an H–module.*

(iv) *M is projective as an HT–module.*

Proof: We know by I.8.10 that (ii) and (iii) are equivalent. As H is normal in HT, we get (i) $\Rightarrow$ (ii) from I.6.5(2), I.4.12, and I.3.10.a. Lemma I.3.18.a yields the implication (iv) $\Rightarrow$ (iii).

We have for any HT–module N

$$\begin{aligned}\operatorname{Hom}_H(M,N) &= \bigoplus_\lambda \operatorname{Hom}_H(M,N)_\lambda = \bigoplus_\lambda \operatorname{Hom}_{TH}(M\otimes\lambda, N) \\ &= \bigoplus_\lambda \operatorname{Hom}_{TH}(M, N\otimes(-\lambda))\end{aligned}$$

where we sum over all $\lambda \in X(T)$ vanishing on $T \cap H$, cf. I.6.9(5). If $\operatorname{Hom}_H(M,?)$ is exact, then all $\operatorname{Hom}_{TH}(M\otimes\lambda,?)$ have to be exact, especially $\operatorname{Hom}_{TH}(M,?)$. This yields (iii) $\Rightarrow$ (iv). Interchanging M and N we get (ii) $\Rightarrow$ (i).

Remark: For each finite dimensional HT–module M the HT–module $\mathrm{ind}_T^{HT} M$ is injective (as all T–modules are injective) and finite dimensional (as $\mathrm{ind}_T^{HT} M \simeq M \otimes \mathrm{ind}_T^{HT} k$ and $\mathrm{ind}_T^{HT} k \simeq k[HT]^T \simeq k[HT/T] \simeq k[H/(H \cap T)]$). The obvious embedding of M into $M \otimes \mathrm{ind}_T^{HT} k \simeq \mathrm{ind}_T^{HT} M$ shows that each finite dimensional HT–module has a resolution by finite dimensional injective HT–modules. This implies (e.g.) for finite dimensional HT–modules M and N that $\mathrm{Ext}^i_{HT}(M, N)$ is also the i–th Ext group of M and N considered as objects in the full subcategory of all finite dimensional HT–modules. (A similar remark applies of course to H–modules.)

9.5. Lemma: *Let $\lambda \in X(T)$.*

a) *Considered as a B_rT–module, $\widehat{Z}_r(\lambda)$ is the projective cover of λ and the injective hull of $\lambda - 2(p^r - 1)\rho$.*

b) *Considered as a B_r^+T–module, $\widehat{Z}'_r(\lambda)$ is the injective hull of λ and the projective cover of $\lambda - 2(p^r - 1)\rho$.*

Proof: We have $B_r^+T \simeq U_r^+ \rtimes T$. The isomorphism $\widehat{Z}'_r(\lambda) \simeq k[U_r^+] \otimes \lambda$, cf. 9.1(6), is compatible with the action of B_r^+T. Therefore the description of injective hulls in I.3.11 yields the first claim in b).

We know by 9.4 that $\widehat{Z}'_r(\lambda)$ is also projective, and get from 9.2(2) that it maps onto the simple module $\lambda - 2(p^r - 1)\rho$. Since it is indecomposable, it has to be the projective cover of this module.

The proof of a) is similar.

Remark: This lemma implies that there are isomorphisms of T–modules

(1) $$\mathrm{soc}_{B_r^+} \widehat{Z}'_r(\lambda) \simeq \lambda \simeq \widehat{Z}_r(\lambda)/\mathrm{rad}_{B_r} \widehat{Z}_r(\lambda)$$

and

(2) $$\mathrm{soc}_{B_r} \widehat{Z}_r(\lambda) \simeq \lambda - 2(p^r - 1)\rho \simeq \widehat{Z}'_r(\lambda)/\mathrm{rad}_{B_r^+} \widehat{Z}'_r(\lambda).$$

9.6. Proposition: a) *For each $\lambda \in X(T)$ there is a simple G_rT–module $\widehat{L}_r(\lambda)$ with*

(1) $$\widehat{L}_r(\lambda)^{U_r^+} \simeq \lambda.$$

We have $\dim \widehat{L}_r(\lambda)_\lambda = 1$ *and* $\mathrm{End}_{G_rT} \widehat{L}_r(\lambda) \simeq k$. *Each weight μ of $\widehat{L}_r(\lambda)$ satisfies $\mu \le \lambda$.*

b) *Each simple G_rT–module is isomorphic to exactly one $\widehat{L}_r(\lambda)$ with $\lambda \in X(T)$.*

c) *Each $\widehat{L}_r(\lambda)$ with $\lambda \in X(T)$ can be uniquely extended to a G_rB–module and to a G_rB^+–module. Each simple G_rB–module (resp. simple G_rB^+–module) is isomorphic to exactly one $\widehat{L}_r(\lambda)$ thus extended.*

d) *We have isomorphisms of G_rB–modules (in (2), (3)) resp. of G_rB^+–modules (in (4), (5))*

(2) $$\widehat{L}_r(\lambda) \simeq \mathrm{soc}_{G_r} \widehat{Z}'_r(\lambda)$$

(3) $$\widehat{L}_r(2(p^r - 1)\rho - \lambda)^* \simeq \widehat{Z}'_r(\lambda)/\mathrm{rad}_{G_r} \widehat{Z}'_r(\lambda)$$

(4) $$\widehat{L}_r(\lambda) \simeq \widehat{Z}_r(\lambda)/\mathrm{rad}_{G_r} \widehat{Z}_r(\lambda)$$

(5) $$\widehat{L}_r(2(p^r - 1)\rho - \lambda)^* \simeq \mathrm{soc}_{G_r} \widehat{Z}_r(\lambda).$$

All these socles or radicals are also the socles or radicals as G_rT–modules and as G_rB–modules resp. G_rB^+–modules.

e) *Each $\widehat{L}_r(\lambda)$ is isomorphic to $L_r(\lambda)$ as a G_r–module.*

f) *We have for all $\lambda, \mu \in X(T)$ isomorphisms of G_rB–modules and of G_rB^+–modules*

$$\widehat{L}_r(\lambda + p^r\mu) \simeq \widehat{L}_r(\lambda) \otimes p^r\mu. \tag{6}$$

g) *If $\lambda \in X_r(T)$, then $\widehat{L}_r(\lambda) \simeq L(\lambda)$ as a G_rB– and as a G_rB^+–module.*

Proof: The same arguments as in 2.4 or as in 3.9/10 show that each simple G_rT–module (and each simple G_rB–module) can be embedded into some $\widehat{Z}'_r(\lambda)$, that each $\widehat{Z}'_r(\lambda)$ has a simple socle, denoted by $\widehat{L}_r(\lambda)$, when considered as a G_rT–module, and a simple socle, conceivably different, as a G_rB–module. Both socles satisfy (1), and we get thus bijections from $X(T)$ onto the set of isomorphism classes of simple G_rT– and of simple G_rB–modules. The statement in a) about weights is clear from the corresponding result for $\widehat{Z}'_r(\lambda)$, cf. 9.2(3). The claim in a) on the space of endomorphisms (and its analogue for G_rB) follows as in 3.10. It implies (as there) that the simple G_rT–modules (and the simple G_rB–modules) remain simple under field extension and that soc_{G_rT} (and soc_{G_rB}) commute with field extensions.

This reduces the proof of the next two claims to the case where k is algebraically closed: We have now for each G_rT–module M

$$\text{soc}_{G_rT} M \subset \text{soc}_{G_r} M \tag{7}$$

and for each G_rB–module M'

$$\text{soc}_{G_rB} M' \subset \text{soc}_{G_r} M'. \tag{8}$$

If k is algebraically closed, then we apply Remark I.6.16.

Now $\text{soc}_{G_r} \widehat{Z}'_r(\lambda)$ is simple, so (7) and (8) yield

$$\text{soc}_{G_rB} \widehat{Z}'_r(\lambda) = \text{soc}_{G_r} \widehat{Z}'_r(\lambda) = \text{soc}_{G_rT} \widehat{Z}'_r(\lambda). \tag{9}$$

This implies e) and c) for G_rB. By symmetry we get it also for G_rB^+. Furthermore, (2) follows from (9), and we get (3) by dualising using 9.2(2). One proves (4) and (5) similarly. Finally, f) is obvious, and g) follows from 3.15.

Remarks: 1) Choose a system X' of representatives for $X(T)/p^rX(T)$. We have for any G_rT–module M an isomorphism of G_rT–modules (cf. I.6.15(2))

$$\text{soc}_{G_r} M \simeq \bigoplus_{\lambda \in X'} \widehat{L}_r(\lambda) \otimes \text{Hom}_{G_r}(\widehat{L}_r(\lambda), M). \tag{10}$$

Each Hom space is a module over $G_rT/G_r \simeq T/T_r$, hence semi-simple. This implies, improving (7),

$$\text{soc}_{G_r} M = \text{soc}_{G_rT} M. \tag{11}$$

If M is even a G_rB–module, then (10) is an isomorphism of G_rB–modules. For any $\mu \in X(T)$ we have $\operatorname{Hom}_{G_rB}(\widehat{L}_r(\lambda + p^r\mu), M) = \operatorname{Hom}_{G_r}(\widehat{L}_r(\lambda), M)^U_{p^r\mu}$. Hence

$$\operatorname{soc}_{G_rB} M \simeq \bigoplus_{\lambda \in X'} \widehat{L}_r(\lambda) \otimes \operatorname{Hom}_{G_r}(\widehat{L}_r(\lambda), M)^U. \tag{12}$$

2) One gets from (2) and (4) that each non-zero homomorphism $\varphi : \widehat{Z}_r(\lambda) \to \widehat{Z}'_r(\lambda)$ of G_rT–modules has image $\widehat{L}_r(\lambda)$. One argues as in 6.16: The image of φ has both head and socle isomorphic to $\widehat{L}_r(\lambda)$. We have $[\widehat{Z}'_r(\lambda) : \widehat{L}_r(\lambda)] = 1$ because of $\dim \widehat{Z}'_r(\lambda)_\lambda = 1$. This implies that $\operatorname{im}(\varphi) \simeq \widehat{L}_r(\lambda)$. It follows also that $\dim \operatorname{Hom}_{G_rT}(\widehat{Z}_r(\lambda), \widehat{Z}'_r(\lambda)) = 1$. This can also be deduced from Frobenius reciprocity which implies $\operatorname{Hom}_{G_rT}(\widehat{Z}_r(\lambda), \widehat{Z}'_r(\lambda)) \simeq \operatorname{Hom}_{B_rT}(\widehat{Z}_r(\lambda), \lambda)$, and from $\dim \widehat{Z}_r(\lambda)_\lambda = 1$.

3) Obviously a) implies that the $\operatorname{ch} \widehat{L}_r(\lambda)$ with $\lambda \in X(T)$ are linearly independent. It follows (as in Remark 2.7) that the Grothendieck group of all finite dimensional G_rB–modules in embedded into $\mathbf{Z}[X(T)]$ via $V \mapsto \operatorname{ch} V$. Two G_rB–modules with the same formal character have the same composition factors with the same multiplicities. The same applies to G_rB^+ and G_rT.

For example, 9.2(3) implies that $\widehat{Z}_r(\lambda)$ and $\widehat{Z}'_r(\lambda)$ have the same composition factors with the same multiplicities when considered as G_rT–modules. The same holds then also for the G_r–modules $Z_r(\lambda)$ and $Z'_r(\lambda)$ by e) and by 9.1(3).

The functor $M \mapsto {}^\tau M$ (cf. 9.6) takes a simple G_rT–module to a simple G_rT–module with the same character. This implies

$${}^\tau\widehat{L}_r(\lambda) \simeq \widehat{L}_r(\lambda) \qquad \text{for all } \lambda \in X(T). \tag{13}$$

9.7. We want to prove analogues for G_rT to the results for G_r in 3.12/13 that decrlbe the dual of a simple module and the socles of modules of the form $Z_r(\mu)$. The proofs in the general case require going to some covering group. So we start with some remarks on the behaviour of G_rT under this transition.

Consider a covering group $\widetilde{G}$ of G with maximal torus $\widetilde{T}$ covering T. Let Z denote the kernel of $\widetilde{G} \to G$; we have then $\widetilde{G}/Z \xrightarrow{\sim} G$ and $\widetilde{T}/Z \xrightarrow{\sim} T$. In general, G_r is not a factor group of $\widetilde{G}_r$, see Remark 2 in 3.15. However, G_rT is a factor group of $\widetilde{G}_r\widetilde{T}$: The multiplication induces isomorphisms of schemes $U_r \times T \times U_r^+ \xrightarrow{\sim} G_rT$ and $\widetilde{U}_r \times \widetilde{T} \times \widetilde{U}_r^+ \xrightarrow{\sim} \widetilde{G}_r\widetilde{T}$ (using obvious notations). The covering $\widetilde{G} \to G$ restricts to isomorphisms $\widetilde{U} \xrightarrow{\sim} U$ and $\widetilde{U}^+ \xrightarrow{\sim} U^+$, hence also $\widetilde{U}_r \xrightarrow{\sim} U_r$ and $\widetilde{U}_r^+ \xrightarrow{\sim} U_r^+$. This implies that $\widetilde{G}_r\widetilde{T}/Z \xrightarrow{\sim} G_rT$. It follows that we can identify $\{\, G_rT\text{–modules} \,\}$ with the full subcategory of all $\widetilde{G}_r\widetilde{T}$–modules M such that all weights of M belong to $X(T) \subset X(\widetilde{T})$.

One gets similarly isomorphisms $\widetilde{G}_r\widetilde{B}/Z \xrightarrow{\sim} G_rB$ and $\widetilde{G}_r\widetilde{B}^+/Z \xrightarrow{\sim} G_rB^+$ (with obvious notations).

Let $\lambda \in X(T)$. There are $\widetilde{G}$ and $\widetilde{T}$ as above such that λ decomposes $\lambda = \lambda_0 + p^r\lambda_1$ with $\lambda_0 \in X_r(\widetilde{T})$ and $\lambda_1 \in X(\widetilde{T})$. We get from 9.6.f/g isomorphisms of $\widetilde{G}_r\widetilde{B}$– and $\widetilde{G}_r\widetilde{B}^+$–modules $\widehat{L}_r(\lambda) \simeq L(\lambda_0) \otimes p^r\lambda_1$, hence by 2.5

$$\widehat{L}_r(\lambda)^* \simeq L(-w_0\lambda_0) \otimes (-p^r\lambda_1) \simeq \widehat{L}_r(-w_0(\lambda_0 + p^r\lambda_1)) \otimes p^r(w_0\lambda_1 - \lambda_1)$$

hence

(1) $$\widehat{L}_r(\lambda)^* \simeq \widehat{L}_r(-w_0\lambda) \otimes p^r(w_0\lambda_1 - \lambda_1).$$

This is now an isomorphism of G_rB– and G_rB^+–modules as $w_0\lambda_1 - \lambda_1 \in \mathbf{Z}R \subset X(T)$. (Note that $w_0\lambda_1 - \lambda_1 \in \mathbf{Z}R \subset X(T)$ is independent of the decomposition $\lambda = \lambda_0 + p^r\lambda_1$: If $\lambda = \lambda_0' + p^r\lambda_1'$ is another decomposition with $\lambda_0' \in X_r(\widetilde{T})$ and $\lambda_1' \in X(\widetilde{T})$, then $\langle \lambda_0' - \lambda_0, \alpha^\vee \rangle = 0$ and $\langle \lambda_1' - \lambda_1, \alpha^\vee \rangle = 0$ for all $\alpha \in S$, hence $w(\lambda_1' - \lambda_1) = \lambda_1' - \lambda_1$ for all $w \in W$.) Furthermore,

$$\widehat{L}_r(\lambda)^{U_r} \simeq L(\lambda_0)^{U_r} \otimes p^r\lambda_1 \simeq w_0\lambda_0 + p^r\lambda_1$$

hence

(2) $$\widehat{L}_r(\lambda)^{U_r} \simeq w_0\lambda + p^r(\lambda_1 - w_0\lambda_1).$$

We can also decompose (in a suitable $\widetilde{T}$)

$$\lambda = p^r\lambda_1' + (\lambda_0' - \rho)$$

with $\lambda_0' \in X_r(\widetilde{T})$ and $\lambda_1' \in X(\widetilde{T})$. Then (1) and 9.6(5) yield

(3) $$\operatorname{soc}_{G_r} \widehat{Z}_r(\lambda) \simeq \widehat{L}_r(p^r\lambda_1' + w_0(\lambda_0' + \rho)) \simeq \widehat{L}_r(w_0 \bullet \lambda) \otimes p^r(\lambda_1' - w_0\lambda_1')$$

as $2(p^r - 1)\rho - \lambda = (p^r - 1)\rho - \lambda_0' + p^r(\rho - \lambda_1')$ and as $(p^r - 1)\rho - \lambda_0' \in X_r(\widetilde{T})$.

It follows from 9.3(2) that $\operatorname{soc}_{G_r} \widehat{Z}_r^w(\mu) \simeq {}^{\dot{w}}(\operatorname{soc}_{G_r} \widehat{Z}_r(w^{-1}\mu))$ for all $w \in W$ and $\mu \in X(T)$. Together with (3) this makes it possible to determine the socles of all $\widehat{Z}_r^w(\mu)$ provided we know ${}^{\dot{w}}L$ for all simple G_rT–modules L. We have ${}^{\dot{w}}V \simeq V$ for each G–module V. Therefore we get for each $\lambda \in X(T)$ (writing $\lambda = \lambda_0 + p^r\lambda_1$ as above, going to a covering if necessary)

$${}^{\dot{w}}(\widehat{L}_r^w(\lambda)) \simeq {}^{\dot{w}}L(\lambda_0) \otimes {}^{\dot{w}}(p^r\lambda_1) \simeq L(\lambda_0) \otimes p^r w\lambda_1,$$

hence

(4) $${}^{\dot{w}}(\widehat{L}_r^w(\lambda)) \simeq \widehat{L}_r(\lambda) \otimes p^r(w\lambda_1 - \lambda_1).$$

This final result makes then [as in (1)] sense for our original G_rT as $w\lambda_1 - \lambda_1 \in \mathbf{Z}R$.

9.8. Let $\lambda \in X(T)$. Since $\operatorname{ind}_{B_rT}^{G_rT}$ is exact, we get from I.4.6.a for each G_rT–module V and each i an isomorphism

(1) $$\operatorname{Ext}^i_{G_rT}(V, \widehat{Z}_r'(\lambda)) \simeq \operatorname{Ext}^i_{B_rT}(V, \lambda).$$

Proceeding as in 4.9 one can construct an injective resolution

$$0 \longrightarrow \lambda \longrightarrow I_0 \longrightarrow I_1 \longrightarrow I_1 \longrightarrow \cdots$$

of λ as a B_rT–module with $I_0 = \widehat{Z}_r(\lambda + 2(p^r - 1)\rho)$, cf. 9.5.a, such that all weights μ of I_j (for all j) satisfy $\mu \geq \lambda$ and $\text{ht}(\mu - \lambda) \geq j$. One gets then (cf. 4.10) for any B_rT–module M: If $\text{Ext}^i_{B_rT}(M, \lambda) \neq 0$, then M has a weight μ with $\mu \geq \lambda$ and $\text{ht}(\mu - \lambda) \geq i$. Now (1) implies:

(2) *If V is a G_rT–module with $\text{Ext}^i_{G_rT}(V, \widehat{Z}'_r(\lambda)) \neq 0$ for some $i \in \mathbf{N}$, then V has a weight μ with $\mu \geq \lambda$ and $\text{ht}(\mu - \lambda) \geq i$.*

We get in particular:

(3) $$\text{Ext}^i_{G_rT}(\widehat{L}_r(\lambda), \widehat{Z}'_r(\lambda)) = 0 = \text{Ext}^i_{G_rT}(\widehat{Z}'_r(\lambda), \widehat{Z}'_r(\lambda)) \qquad \text{for all } i > 0.$$

9.9. Lemma: *Let $\lambda, \mu \in X(T)$ and $i \in \mathbf{N}$. Then*

$$\text{Ext}^i_{G_rT}(\widehat{Z}_r(\lambda), \widehat{Z}'_r(\mu)) \simeq \text{Ext}^i_{G_rT}(\widehat{Z}'_r(\mu), \widehat{Z}_r(\lambda)) \simeq \begin{cases} k, & \textit{if } \lambda = \mu \textit{ and } i = 0, \\ 0, & \textit{otherwise.} \end{cases}$$

Proof: We have $\text{Ext}^i_{G_rT}(\widehat{Z}_r(\lambda), \widehat{Z}'_r(\mu)) \simeq \text{Ext}^i_{B_rT}(\widehat{Z}_r(\lambda), \mu)$, cf. 9.8(1). As $\widehat{Z}_r(\lambda)$ is a projective B_rT–module (9.5), these Ext-groups vanish for $i > 0$. As $\widehat{Z}_r(\lambda)$ is the projective cover of the B_rT–module λ, we get $\dim \text{Hom}_{B_rT}(\widehat{Z}_r(\lambda), \mu) = \delta_{\lambda\mu}$.

The claim for $\text{Ext}^i_{G_rT}(\widehat{Z}'_r(\mu), \widehat{Z}_r(\lambda))$ follows analogously, replacing B_rT by B^+_rT.

Remarks: 1) The same proof as above (working with 3.8 instead of 9.5) shows also

(1) $$\text{Ext}^i_{G_r}(Z_r(\lambda), Z'_r(\mu)) \simeq \text{Ext}^i_{G_r}(Z'_r(\mu), Z_r(\lambda)) \simeq \begin{cases} k, & \text{if } \lambda - \mu \in p^r X(T) \\ & \text{and } i = 0, \\ 0, & \text{otherwise.} \end{cases}$$

Alternatively, this follows from the lemma using I.6.9(5).

2) We can now prove the following analogue to 6.21(6): If M is a finite dimensional G_rT–module, then

(2) $$\text{ch}(M) = \sum_{\mu \in X(T)} \Big(\sum_{i \geq 0} (-1)^i \dim \text{Ext}^i_{G_rT}(M, \widehat{Z}'_r(\mu))\Big) \text{ch}\, \widehat{Z}_r(\mu).$$

To start with, denote the right hand side in (2) by $c_r(M)$ and set $c_r(\lambda) = c_r(\widehat{L}_r(\lambda))$ for all $\lambda \in X(T)$, i.e.,

(3) $$c_r(\lambda) = \sum_{\mu \in X(T)} \Big(\sum_{i \geq 0} (-1)^i \dim \text{Ext}^i_{G_rT}(\widehat{L}_r(\lambda), \widehat{Z}'_r(\mu))\Big) \text{ch}\, \widehat{Z}_r(\mu).$$

By 9.8(2) we know that $\text{Ext}^i_{G_rT}(\widehat{L}_r(\lambda), \widehat{Z}'_r(\mu)) \neq 0$ implies $\lambda \geq \mu$ and $\text{ht}(\lambda - \mu) \geq i$. So each sum over i in (3) [and, similarly, in (2)] has at most finitely many non-zero terms. Since each $\nu \in X(T)$ is a weight of only finitely many $\widehat{Z}_r(\mu)$, the coefficient of $e(\nu)$ in any $c_r(\lambda)$ [and in any $c_r(M)$] is a well defined integer. However, conceivably there might be infinitely many ν such that the coefficient of $e(\nu)$ is non-zero. So our $c_r(M)$ may not belong to $\mathbf{Z}[X(T)]$, but to some suitable completion. We know at least that all $e(\nu)$ that occur in $c_r(\lambda)$ satisfy $\nu \leq \lambda$. Furthermore, $e(\lambda)$ occurs by

9.8(3) with coefficient 1 in $c_r(\lambda)$. The additivity of the Euler characteristic implies for all M that

$$c_r(M) = \sum_{\lambda\in X(T)} [M : \widehat{L}_r(\lambda)]c_r(\lambda). \tag{4}$$

(So it is enough to prove (2) for simple M.) Now the lemma shows that

$$c_r(\widehat{Z}_r(\nu)) = \operatorname{ch}\widehat{Z}_r(\nu)$$

for all $\nu \in X(T)$, hence $\sum_{\lambda\in X(T)}[\widehat{Z}_r(\nu) : \widehat{L}_r(\lambda)]\, c_r(\lambda) = \operatorname{ch}\widehat{Z}_r(\nu)$ and

$$\sum_{\lambda\leq\nu}[\widehat{Z}_r(\nu) : \widehat{L}_r(\lambda)]\,(c_r(\lambda) - \operatorname{ch}\widehat{L}_r(\lambda)) = 0. \tag{5}$$

We now want to use induction on m to show that the coefficient of $e(\mu)$ in $c_r(\nu)$ is equal to the coefficient of $e(\mu)$ in $\operatorname{ch}\widehat{L}_r(\nu)$ whenever $\mu \leq \nu$ and $\operatorname{ht}(\nu-\mu) \leq m$. (This will prove that $c_r(\nu) = \operatorname{ch}\widehat{L}_r(\nu)$ for all ν, because the coefficient of $e(\mu)$ in $c_r(\nu)$ or $\operatorname{ch}\widehat{L}_r(\nu)$ can be non-zero only if $\mu \leq \nu$.) For $m = 0$ we have $\mu = \nu$ and we have seen above that the coefficient of $e(\nu)$ in $c_r(\nu)$ is equal to 1, hence equal to the coefficient in $\operatorname{ch}\widehat{L}_r(\nu)$. In general, look at (5): For each $\lambda \neq \nu$ in that sum, we have either $\mu \not\leq \lambda$ or $\mu \leq \lambda < \nu$, hence $\operatorname{ht}(\lambda-\mu) < \operatorname{ht}(\nu-\mu)$. In both cases we know that the coefficient of $e(\mu)$ in $c_r(\lambda)$ is equal to the coefficient of $e(\mu)$ in $\operatorname{ch}\widehat{L}_r(\lambda)$. Since $[\widehat{Z}_r(\nu) : \widehat{L}_r(\nu)] = 1$, we get now the same for ν instead of λ, hence our claim.

9.10. Let $\alpha \in S$ and set $P^+(\alpha) = P^+_{\{\alpha\}}$. Then $P^+(\alpha)_rT$ contains both B_r^+T and $s_\alpha(B_r^+T) = \dot{s}_\alpha B_r^+T\dot{s}_\alpha^{-1}$. For all $\lambda,\mu \in X(T)$ let (for the moment) $M(\lambda)$ denote the $P^+(\alpha)_rT$–module coinduced from the B_r^+T–module λ, and $N(\mu)$ the $P^+(\alpha)_rT$–module coinduced from the $s_\alpha(B_r^+T)$–module μ. We have then $M(\lambda) \simeq \operatorname{Dist}((U_{-\alpha})_r) \otimes \lambda$ and $N(\mu) \simeq \operatorname{Dist}((U_\alpha)_r) \otimes \mu$ as T–modules. There is a basis $(v_i)_{0\leq i<q}$ for $M(\lambda)$ where $q = p^r$, such that T acts on v_i via $\lambda - i\alpha$ and such that for all i,j with $0 \leq i,j < q$

$$X_{-\alpha,j}v_i = \binom{i+j}{j}v_{i+j}, \qquad X_{\alpha,j}v_i = \binom{\langle\lambda,\alpha^\vee\rangle - i + j}{j}v_{i-j} \tag{1}$$

where we use the convention that $v_m = 0$ for $m < 0$ or $m \geq q$. Similarly, $N(\mu)$ has a basis $(v_i')_{0\leq i<q}$ such that T acts on v_i via $\mu + i\alpha$ and such that for all i,j with $0 \leq i,j < q$

$$X_{\alpha,j}v_i' = \binom{i+j}{j}v_{i+j}', \qquad X_{-\alpha,j}v_i' = (-1)^j\binom{\langle\mu,\alpha^\vee\rangle - j + i}{j}v_{i-j}'. \tag{2}$$

All $X_{\beta,j}$ with $\beta \in R^+ \setminus \{\alpha\}$ and $j > 0$ act as 0 on $M(\lambda)$ and $N(\mu)$.

Using the universal property of the coinduced module $M(\lambda)$ or an elementary calculation one checks that there is a homomorphism

$$\varphi : M(\lambda) \longrightarrow N(\lambda - (q-1)\alpha) \tag{3}$$

of $P^+(\alpha)_rT$–modules with

$$(4) \qquad \varphi(v_i) = (-1)^i \binom{\langle \lambda, \alpha^\vee \rangle}{i} v'_{q-1-i} \qquad \text{for all } i,\ 0 \leq i < q.$$

The kernel of φ is spanned by all v_i such that p divides $\binom{\langle \lambda, \alpha^\vee \rangle}{i}$. This can be evaluated using the standard formula for the reduction modulo p of a binomial coefficient. For example, one checks that φ is an isomorphism if and only if $\langle \lambda, \alpha^\vee \rangle$ is congruent to $p^r - 1$ modulo p^r (equivalently: if $p^r \mid \langle \lambda + \rho, \alpha^\vee \rangle = \langle \lambda, \alpha^\vee \rangle + 1$).

The description of $\ker(\varphi)$ is particularly simple for $r = 1$: Write $\langle \lambda + \rho, \alpha^\vee \rangle = mp + d$ with $d, m \in \mathbf{Z}$ and $0 < d \leq p$. Then $\ker(\varphi)$ is spanned by all v_i with $i \geq d$ (for $r = 1$). If $d = p$, then φ is an isomorphism. If $d < p$, then we claim that we have an exact sequence (for $r = 1$)

$$(5) \qquad \cdots \to M(\lambda - (p+d)\alpha) \to M(\lambda - p\alpha) \to M(\lambda - d\alpha) \to \ker(\varphi) \to 0.$$

Indeed, let $(\widetilde{v}_i)_{0 \leq i < q}$ be the basis for $M(\lambda - d\alpha)$ analogous to the basis of all v_i for $M(\lambda)$. Then there is (by the universal property of the coinduced module or by an elementary calculation) a homomorphism $\psi : M(\lambda - d\alpha) \to M(\lambda)$ mapping each $\widetilde{v}_i = X_{-\alpha,i}\widetilde{v}_0$ to $X_{-\alpha,i}v_d = \binom{i+d}{i} v_{d+i}$. Then $\ker(\varphi) = \operatorname{im}(\psi)$. The kernel of ψ is spanned by all $\widetilde{v}_i$ with $i \geq p - d$. Since $\langle \lambda - d\alpha + \rho, \alpha^\vee \rangle = (m-1)p + (p-d)$, we see that $\ker(\psi)$ is also the kernel of the homomorphism $M(\lambda - d\alpha) \to N(\lambda - d\alpha - (p-1)\alpha)$ analogous to φ. So $\ker(\psi)$ is the image of some homomorphism from $M(\lambda - d\alpha - (p-d)\alpha) = M(\lambda - p\alpha)$ to $M(\lambda - d\alpha)$. Now iterate to get the exact sequence above.

We apply (for arbitrary r) coinduction from $P^+(\alpha)_rT$ to G_rT to the modules and homomorphisms above. It takes $M(\lambda)$ to $\widehat{Z}_r(\lambda)$ and $N(\mu)$ to $\widehat{Z}_r^{s_\alpha}(\mu)$. The homomorphism φ above coinduces a homomorphism of G_rT–modules

$$(6) \qquad \widetilde{\varphi} : \widehat{Z}_r(\lambda) \to \widehat{Z}_r^{s_\alpha}(\lambda - (q-1)\alpha).$$

Since coinduction is exact and takes non-zero modules to non-zero modules, we see that $\widetilde{\varphi}$ is an isomorphism if and only if φ is an isomorphism, i.e., if and only if $p^r \mid \langle \lambda + \rho, \alpha^\vee \rangle$. If $r = 1$ and $\langle \lambda + \rho, \alpha^\vee \rangle = mp + d$ with $d, m \in \mathbf{Z}$ and $0 < d < p$, then we have a long exact sequence

$$(7) \qquad \cdots \to \widehat{Z}_1(\lambda - (p+d)\alpha) \to \widehat{Z}_1(\lambda - p\alpha) \to \widehat{Z}_1(\lambda - d\alpha) \to \ker(\widetilde{\varphi}) \to 0.$$

9.11. If M is a G_rT–module and $w \in W$, then we denote by ${}^wM$ the G_rT–module we get from twisting M by the automorphism $g \mapsto \dot{w} g \dot{w}^{-1}$ of G_rT, cf. 2.15. (We ignore the dependence on the choice of the representative $\dot{w}$ of w.) A result for coind analogous to I.3.5(4) yields then that ${}^w\widehat{Z}_r(\lambda) \simeq \widehat{Z}_r^w(w\lambda)$ for all λ, and, more generally, ${}^w\widehat{Z}_r^x(\lambda) \simeq \widehat{Z}_r^{wx}(w\lambda)$ for all $x \in W$. (One could also combine I.3.5(4) with I.8.17.)

Twisting the homomorphism $\widetilde{\varphi}$ from 9.10(6) with w we get a homomorphism $\widehat{Z}_r^w(w\lambda) \to \widehat{Z}_r^{ws_\alpha}(w\lambda - (q-1)w\alpha)$, hence for all μ a homomorphism $\widehat{Z}_r^w(\mu) \to \widehat{Z}_r^{ws_\alpha}(\mu - (q-1)w\alpha)$. Applying this to $\mu = \lambda\langle w \rangle = \lambda + (q-1)(w\rho - \rho)$ as in 9.3(4), we get a homomorphism

$$(1) \qquad \psi : \widehat{Z}_r^w(\lambda\langle w \rangle) \to \widehat{Z}_r^{ws_\alpha}(\lambda\langle ws_\alpha \rangle).$$

(Note that $ws_\alpha\rho - \rho = (w\rho - \rho) - w\alpha$.) Now ψ is an isomorphism if and only if $p^r \mid \langle\lambda+\rho, w\alpha^\vee\rangle$. If $r = 1$ and $\langle\lambda+\rho, w\alpha^\vee\rangle = mp+d$ with $d, m \in \mathbf{Z}$ and $0 < d < p$, then we have a long exact sequence

$$\cdots \to \widehat{Z}_1^w((\lambda - p\alpha)\langle w\rangle) \to \widehat{Z}_1^w((\lambda - d\alpha)\langle w\rangle) \to \ker(\psi) \to 0. \tag{2}$$

These claims follow from those at the end of 9.10 by elementary calculations.

Choose now a reduced decomposition $w_0 = s_1 s_2 \dots s_n$ where $s_i = s_{\alpha_i}$ with $\alpha_i \in S$ and $n = |R^+|$. Set $x_i = s_1 s_2 \dots s_{i-1}$ and $\beta_i = x_i\alpha_i$ for $1 \le i \le n$. Then $R^+ = \{\beta_i \mid 1 \le i \le n\}$. We get for all i a homomorphism

$$\varphi_i : \widehat{Z}_r^{x_i}(\lambda\langle x_i\rangle) \to \widehat{Z}_r^{x_{i+1}}(\lambda\langle x_{i+1}\rangle) \tag{3}$$

as a special case of (1). The composition $\varphi = \varphi_n \circ \dots \circ \varphi_2 \circ \varphi_1$ of these maps is a homomorphism

$$\varphi : \widehat{Z}_r(\lambda) \longrightarrow \widehat{Z}_r^{w_0}(\lambda\langle w_0\rangle) \simeq \widehat{Z}'_r(\lambda). \tag{4}$$

Now φ is an isomorphism if and only if all φ_i are isomorphisms if and only if $p^r \mid \langle\lambda + \rho, \beta^\vee\rangle$ for all $\beta \in R^+$. By Remark 2 in 9.6 this is also equivalent to $\widehat{Z}_r(\lambda) \simeq \widehat{L}_r(\lambda)$. (Here one direction can also be deduced from 3.18.)

One gets easily from 9.3(4) that $\dim \widehat{Z}_r(\lambda)_{\lambda\langle w\rangle} = 1$ for all $w \in W$. More precisely, if $\widehat{Z}_r(\lambda)_\lambda = kv_\lambda$, then $\widehat{Z}_r(\lambda)_{\lambda\langle w\rangle} = \prod_\beta X_{-\beta,q-1}v_\lambda$ where the product is over all $\beta \in R^+$ with $w^{-1}\beta < 0$. This description can be used to show that $\widetilde{\varphi}$ as in 9.10(6) is bijective on the $\lambda\langle w\rangle$–weight spaces with $w^{-1}\alpha > 0$. Then one can deduce that the φ_i in (3) and hence also φ in (4) are bijective on the λ–weight spaces. In particular, one gets $\varphi \neq 0$, hence

$$\operatorname{im}(\varphi) \simeq \widehat{L}_r(\lambda).$$

Suppose now that $r = 1$. Let us reprove the last statement in a more detailed way. Write $\langle\lambda + \rho, \beta_i^\vee\rangle = m_i p + d_i$ with $m_i, d_i \in \mathbf{Z}$ and $0 < d_i \le p$. If $d_i = p$, then φ_i is an isomorphism. If $d_i < p$, then $\ker(\varphi_i)$ is a homomorphic image of $\widehat{Z}_1^{x_i}((\lambda - d_i\beta_i)\langle x_i\rangle)$. Now $\operatorname{ch} \widehat{Z}_1^{x_i}((\lambda - d_i\beta_i)\langle x_i\rangle) = \operatorname{ch} \widehat{Z}_1(\lambda - d_i\beta_i)$, see 9.3(4), shows that λ is not a weight of $\ker(\varphi_i)$. This holds therefore for all i. We get now as above that $\varphi \neq 0$ and $\operatorname{im}(\varphi) \simeq \widehat{L}_1(\lambda)$.

We get now in addition: If $\widehat{L}_1(\mu)$ is a composition factor of $\widehat{Z}_1(\lambda)$ with $\mu \neq \lambda$, then $\widehat{L}_1(\mu)$ is a composition factor of $\ker(\varphi)$, hence of $\ker(\varphi_i)$ for some i, hence of $\widehat{Z}_1(\lambda - d_i\beta_i)$ for some i with $d_i < p$. Since $\lambda - d_i\beta_i = s_{\beta_i, m_i p} \bullet \lambda \uparrow \lambda$, we get now using induction on $\lambda - \mu$ a strong linkage principle for the $\widehat{Z}_1(\lambda)$: If $[\widehat{Z}_1(\lambda) : \widehat{L}_1(\mu)] \neq 0$, then $\mu \uparrow \lambda$.

This approach to the strong linkage principle for the $\widehat{Z}_1(\lambda)$ appears first in [Doty 4] where one works with $\widehat{Z}'_1(\lambda)$ instead of $\widehat{Z}_1(\lambda)$. Before that this principle (or rather its extension to $\widehat{Z}_r(\lambda)$ for all r) had been deduced from the strong linkage principle (6.13) for G–modules; we give that earlier proof below in 9.15. If we had worked a little harder on the description of the kernel of φ as in 9.10(3) for higher r, then we could also have used the present approach to get the strong linkage principle for $\widehat{Z}_r(\lambda)$ for all r. This was carried out in [Doty 5] where this principle for G_rT–modules is then used to deduce the strong linkage principle for G–modules of the form $V(\lambda)$ or $H^0(\lambda)$.

One can use φ and the φ_i to construct a filtration

$$\widehat{Z}_1(\lambda) = \widehat{Z}_1(\lambda)^0 \supset \widehat{Z}_1(\lambda)^1 \supset \widehat{Z}_1(\lambda)^2 \supset \cdots \tag{5}$$

such that $\widehat{Z}_1(\lambda)/\widehat{Z}_1(\lambda)^1 \simeq \widehat{L}_1(\lambda)$ and (using the notation from above)

$$\sum_{i>0} \operatorname{ch} \widehat{Z}_1(\lambda)^i = \sum_{j,\, d_j<p} \Big(\sum_{l\geq 0} \operatorname{ch} \widehat{Z}_1(\lambda - (pl + d_j)\beta_j) - \sum_{l>0} \operatorname{ch} \widehat{Z}_1(\lambda - pl\beta_j)\Big). \tag{6}$$

(Note that any $e(\mu)$ occurs in only finitely many summands on the right hand side with a non-zero coefficient; so this sum makes sense.) If $n(\lambda)$ is the number of j with $d_j < p$, then $\widehat{Z}_1(\lambda)^{n(\lambda)+1} = 0$ while $\widehat{Z}_1(\lambda)^{n(\lambda)}$ is the simple socle of $\widehat{Z}_1(\lambda)$. One has

$$^{\tau}(\widehat{Z}_1(\lambda)^i/\widehat{Z}_1(\lambda)^{i+1}) \simeq \widehat{Z}_1(\lambda)^i/\widehat{Z}_1(\lambda)^{i+1} \tag{7}$$

for all i. This can be found in [Andersen, Jantzen, and Soergel], 6.6. The construction of the filtration involves the study of a suitable category of representations of Lie(G) over the polynomial ring over k and is then similar to the construction in Chapter 8.

9.12. Lemma: *There is for each $\lambda \in X(T)$ and each $i \in \mathbf{N}$ an isomorphism of G–modules*

$$H^i(\lambda) \simeq R^i \operatorname{ind}_{G_rB}^G \widehat{Z}'_r(\lambda).$$

Proof: We want to apply the spectral sequence from I.4.5.c to the chain of subgroups $B \subset G_rB \subset G$. As $G_rB/B \simeq G_r/B_r \simeq U_r^+$ is affine, induction from B to G_rB is exact (I.5.13). Therefore the spectral sequence degenerates, and we get for each B–module M and each $i \in \mathbf{N}$ an isomorphism

$$H^i(M) \simeq R^i \operatorname{ind}_{G_rB}^G(\operatorname{ind}_B^{G_rB} M). \tag{1}$$

Taking $M = \lambda$ yields the lemma.

Remark: Let M be a G_rB–module. We have $\widehat{Z}'_r(\lambda) \otimes M \simeq \operatorname{ind}_B^{G_rB}(\lambda \otimes M)$. Therefore the same proof as above shows

$$H^i(\lambda \otimes M) \simeq R^i \operatorname{ind}_{G_rB}^G(\widehat{Z}'_r(\lambda) \otimes M). \tag{2}$$

9.13. Let H be a reduced and F–stable subgroup of G. Consider an H–module M. Since F^r maps G_rH to H, we get on $M^{[r]}$ not only a structure as an H–module, but even as a G_rH–module (with G_r acting trivially, of course). In the case where $H = B$ and $M = \mu$ for some $\mu \in X(T)$, this is just the structure as a G_rB–module on $p^r\mu$ that we have been using since 9.2.

We have by I.6.11 isomorphisms of G–modules

$$R^i \operatorname{ind}_{G_rH}^G(M^{[r]}) \simeq R^i \operatorname{ind}_{G_rH/G_r}^{G/G_r}(M^{[r]}) \tag{1}$$

where the right hand side is considered as a G–module via the canonical map $G \to G/G_r$. As F^r induces isomorphisms $G/G_r \xrightarrow{\sim} G$ and $H/H_r \xrightarrow{\sim} H$ compatible with the inclusions $H \hookrightarrow G$ and $H/H_r \simeq G_rH/G_r \hookrightarrow G/G_r$, we can identify $R^i \operatorname{ind}_{G_rH/G_r}^{G/G_r}$ with $R^i \operatorname{ind}_H^G$. The structure as an H–module on $M^{[r]}$ we now have to take, is that of M (as the identification $H/H_r \xrightarrow{\sim} H$ is given by F^r). Similarly, the canonical map $G \to G/G_r$ has now to be replaced by $F^r : G \to G$. So (1) takes the form

$$(2) \qquad R^i \operatorname{ind}_{G_rH}^G(M^{[r]}) \simeq (R^i \operatorname{ind}_H^G M)^{[r]}.$$

Therefore the generalised tensor identity (I.4.8) implies for each G–module V

$$(3) \qquad R^i \operatorname{ind}_{G_rH}^G(V \otimes M^{[r]}) \simeq V \otimes (R^i \operatorname{ind}_H^G M)^{[r]}.$$

9.14. It will be convenient to assume from now on (until 9.22) that each $\lambda \in X(T)$ can be written in the form $\lambda = \lambda_0 + p^r\lambda_1$ with $\lambda_0 \in X_r(T)$ and $\lambda_1 \in X(T)$. This is automatically satisfied if $\mathcal{D}G$ is simply connected. For arbitrary G there is always a covering group satisfying this assumption. Results like those in 9.15, 9.19/20, and 9.22 also hold for arbitrary G and can be proved by at first going to a suitable covering. The same remark applies to the results in 9.16 except for 9.16(3). See also Remark 5 below.

A decomposition $\lambda = \lambda_0 + p^r\lambda_1$ as above is unique for G semi-simple. This is no longer so in general, as we can take any $\nu \in X(T)$ with $\langle \nu, \alpha^\vee \rangle = 0$ for all $\alpha \in S$ (i.e., any $\nu \in X_0(T)$ in the notation from 1.18) and get then with $\lambda = (\lambda_0 - p^r\nu) + p^r(\lambda_1 + \nu)$ another decomposition of this type. It will be convenient to remove this ambiguity by choosing a system of representatives $X_r'(T) \subset X_r(T)$ for $X(T)/p^rX(T)$. Then any $\lambda \in X(T)$ has a unique decomposition $\lambda = \lambda_0 + p^r\lambda_1$ with $\lambda_0 \in X_r'(T)$ and $\lambda_1 \in X(T)$. Recall that then $\widehat{L}_r(\lambda) \simeq L(\lambda_0) \otimes p^r\lambda_1$.

Proposition: *Let $\lambda \in X(T)_+$. Suppose each composition factor of $\widehat{Z}_r'(\lambda)$ has the form $\widehat{L}_r(\mu_0 + p^r\mu_1)$ with $\mu_0 \in X_r'(T)$ and $\mu_1 \in X(T)$ such that*

$$\langle \mu_1 + \rho, \beta^\vee \rangle \geq 0 \qquad \textit{for all } \beta \in S.$$

Then $H^0(\lambda)$ has a filtration with factors of the form $L(\mu_0) \otimes H^0(\mu_1)^{[r]}$ with $\mu_0 \in X_r'(T)$ and $\mu_1 \in X(T)_+$. Each such tensor product occurs $[\widehat{Z}_r'(\lambda) : \widehat{L}_r(\mu_0 + p^r\mu_1)]$ times as a factor in the filtration.

Proof: Consider a composition factor $\widehat{L}_r(\mu_0 + p^r\mu_1)$ as above. We have $\widehat{L}_r(\mu) \simeq L(\mu_0) \otimes p^r\mu_1$, hence

$$R^i \operatorname{ind}_{G_rB}^G \widehat{L}_r(\mu_0 + p^r\mu_1) \simeq L(\mu_0) \otimes H^i(\mu_1)^{[r]}$$

by 9.13(3). By Kempf's vanishing theorem (4.5) and by 5.4.a we have $H^i(\mu_1) = 0$ for all $i > 0$, hence $R^i \operatorname{ind}_{G_rB}^G \widehat{L}_r(\mu) = 0$ for each composition factor $\widehat{L}_r(\mu)$ of $\widehat{Z}_r'(\lambda)$ (and all $i > 0$). Therefore, if $0 \subset M_1 \subset M_2 \subset \cdots \subset M_s = \widehat{Z}_r'(\lambda)$ is a composition series, then

$$0 \subset \operatorname{ind}_{G_rB}^G M_1 \subset \operatorname{ind}_{G_rB}^G M_2 \subset \cdots \subset \operatorname{ind}_{G_rB}^G M_s = \operatorname{ind}_{G_rB}^G \widehat{Z}_r'(\lambda)$$

is a filtration with factors isomorphic to $\operatorname{ind}_{G_rB}^G(M_i/M_{i-1})$. These factors have the desired form and satisfy the claim about the multiplicities.

Remarks: 1) By duality $V(\lambda)$ has (under the same assumption as in the proposition) a filtration with factors of the form $L(\mu_0) \otimes V(\mu_1)^{[r]}$ satisfying the same rule for the multiplicities as above. If $v \in V(\lambda)_\lambda$, $v \neq 0$, then $\mathrm{Dist}(G_r)v$ is a G_rB^+–submodule of $V(\lambda)$ isomorphic to $\widehat{Z}_r(\lambda)$ for λ as above, cf. [Jantzen 7], 6.1. If we take a composition series of this G_rB^+–submodule and consider the G–modules generated by its terms, then we get the filtration mentioned on $V(\lambda)$. For more details, see [Jantzen 7], 3.8.

2) There is for each $\lambda \in X(T)$ an Euler characteristic identity

$$\chi(\lambda) = \sum_{\mu_0 \in X'_r(T)} \sum_{\mu_1 \in X(T)} [\widehat{Z}'_r(\lambda) : \widehat{L}_r(\mu_0 + p^r\mu_1)] \operatorname{ch} L(\mu_0)\chi(\mu_1)^{[r]}. \tag{1}$$

This is an immediate consequence of 9.12 and the additivity of Euler characteristics. For a direct proof, using Weyl's character formula, see [Jantzen 7], 3.1.

3) We shall describe in 9.18 conditions that imply that λ satisfies the assumption in the proposition.

4) Any homomorphism $\varphi : \widehat{Z}'_r(\lambda) \to \widehat{Z}'_r(\mu)$ of G_rB–modules (for any $\lambda, \mu \in X(T)$) induces a homomorphism

$$\psi = \operatorname{ind}_{G_rB}^G(\varphi) : H^0(\lambda) \to H^0(\mu).$$

For λ as in the proposition we get from the proof above that $R^1 \operatorname{ind}_{G_rB}^G \ker(\varphi) = 0$, hence

$$\operatorname{im}(\psi) = \operatorname{ind}_{G_rB}^G(\operatorname{im}\varphi). \tag{2}$$

Take $\alpha \in S$ and write $\langle \lambda + \rho, \alpha^\vee \rangle = m_\alpha p^r + d_\alpha$ with $0 \le d_\alpha < p$ and $m_\alpha \in \mathbf{Z}$. Suppose $d_\alpha \neq 0$. Then there is an obvious homomorphism $\widehat{Z}'_r(\lambda) \to \widehat{Z}'_r(\lambda - d_\alpha \alpha)$. (Its construction is dual to that of $\widehat{Z}_1(\lambda - d_\alpha\alpha) \to \widehat{Z}_1(\lambda)$ in 9.10(7).) One can describe the image explicitly and gets from (2) a character formula for the image of the induced homomorphism $H^0(\lambda) \to H^0(\lambda - d_\alpha\alpha)$. See [Jantzen 7], 3.10–3.12 for details (where the dual situation is considered) and see [Andersen 6], 4.3 for another approach that leads to different conditions for the formula to hold.

5) As mentioned above, for arbitrary G there will not exist a system of representatives $X'_r(T) \subset X_r(T)$ for $X(T)/p^rX(T)$. However, we can always choose a subset $X'_r(T) \subset X_r(T)$ such that no two elements in $X'_r(T)$ are congruent to each other modulo $p^rX(T)$. Now the proposition extends to G. A similar remark applies to Proposition 9.17.

9.15. Corollary: *Let $\lambda, \mu \in X(T)$.*

a) *If $\widehat{L}_r(\mu)$ is a composition factor of $\widehat{Z}_r(\lambda)$ (or, equivalently, of $\widehat{Z}'_r(\lambda)$), then $\mu \uparrow \lambda$.*

b) *If $L_r(\mu)$ is a composition factor of $Z_r(\lambda)$ (or, equivalently, of $Z'_r(\lambda)$), then $\mu \in W_p \bullet \lambda + p^rX(T)$.*

Proof: a) As $\dim \widehat{Z}_r(\lambda) < \infty$, there are only finitely many $\mu'_0 \in X'_r(T)$ and $\mu'_1 \in X(T)$ such that $\widehat{L}_r(\mu'_0 + p^r\mu'_1)$ is a composition factor of $\widehat{Z}_r(\lambda)$. We can therefore find $\nu \in X(T)$ with $\nu+\mu_1' \in X(T)_+$ for all μ_1' occurring. We get the highest weights of the composition factors of $\widehat{Z}_r(\lambda+p^r\nu) \simeq \widehat{Z}_r(\lambda) \otimes p^r\nu$ by adding $p^r\nu$ to those for $\widehat{Z}_r(\lambda)$. Therefore $\widehat{Z}_r(\lambda+p^r\nu)$ satisfies the assumptions of 9.14.

Now suppose that $\widehat{L}_r(\mu)$ is a composition factor of $\widehat{Z}_r(\lambda)$. Decompose $\mu = \mu_0 + p^r\mu_1$ with $\mu_0 \in X_r'(T)$ and $\mu_1 \in X(T)$. Now 9.14 implies that $L(\mu_0) \otimes H^0(\mu_1+\nu)^{[r]}$ is a subquotient of $H^0(\lambda+p^r\nu)$. This subquotient has highest weight $\mu_0 + p^r(\mu_1+\nu) = \mu + p^r\nu$, so $L(\mu+p^r\nu)$ is a composition factor of $H^0(\lambda+p^r\nu)$. The strong linkage principle (6.13) implies $\mu+p^r\nu \uparrow \lambda+p^r\nu$, hence $\mu \uparrow \lambda$ by 6.4(4).

b) If $L_r(\mu)$ is a composition factor of $Z_r(\lambda)$ as a G_r–module, then there is some $\nu \in X(T)$ such that $\widehat{L}_r(\mu+p^r\nu)$ is a composition factor of $\widehat{Z}_r(\lambda)$ as a G_rT–module. So b) follows from a).

9.16. Assume for the moment that $\rho \in X(T)$. We can express 3.18(6) in the form $\widehat{Z}_r'((p^r-1)\rho) \simeq St_r$. Therefore $\widehat{Z}_r'((p^r-1)\rho)$ can be extended to a G–module which implies

(1) $$\operatorname{ch} \widehat{Z}_r((p^r-1)\rho) \in \mathbf{Z}[X(T)]^W.$$

This can also be proved directly using 9.2(3) which yields

$$\operatorname{ch} \widehat{Z}_r((p^r-1)\rho) = \prod_{\alpha \in R^+} \frac{e(\frac{p^r\alpha}{2}) - e(-\frac{p^r\alpha}{2})}{e(\frac{\alpha}{2}) - e(-\frac{\alpha}{2})}$$

(in $\mathbf{Z}[X(T) \otimes_{\mathbf{Z}} \mathbf{Q}]$).

For an arbitrary $\lambda \in X(T)$ we have $\operatorname{ch} \widehat{Z}_r(\lambda) = e(\lambda-(p^r-1)\rho) \operatorname{ch} \widehat{Z}_r((p^r-1)\rho)$ by 9.2(3), and get from (1) for all $w \in W$

$$\operatorname{ch} \widehat{Z}_r(p^r\rho + w \bullet \lambda) = e(w(\lambda+\rho)) \operatorname{ch} \widehat{Z}_r((p^r-1)\rho) = w\big(e(\lambda+\rho) \operatorname{ch} \widehat{Z}_r((p^r-1)\rho)\big)$$

hence

(2) $$\operatorname{ch} \widehat{Z}_r(p^r\rho + w \bullet \lambda) = w \operatorname{ch} \widehat{Z}_r(p^r\rho + \lambda).$$

We have for each finite dimensional G_rT–module M

(3) $$\operatorname{ch} M = \sum_{\mu_0 \in X_r'(T)} \sum_{\mu_1 \in X(T)} [M : \widehat{L}_r(\mu_0 + p^r\mu_1)]\, e(p^r\mu_1) \operatorname{ch} L(\mu_0).$$

Take $M = \widehat{Z}_r(p^r\rho + \lambda)$ and apply any $w \in W$ to the equation. We get

$$\operatorname{ch} \widehat{Z}_r(p^r\rho+w\bullet\lambda) = \sum_{\mu_0 \in X_r'(T)} \sum_{\mu_1 \in X(T)} [\widehat{Z}_r(p^r\rho+\lambda) : \widehat{L}_r(\mu_0+p^r\mu_1)] e(p^r w\mu_1) \operatorname{ch} L(\mu_0)$$

hence, by comparing coefficients with (3) for $M = \widehat{Z}_r(p^r\rho + w \bullet \lambda)$

$$[\widehat{Z}_r(p^r\rho+\lambda) : \widehat{L}_r(\mu_0 + p^r\mu_1)] = [\widehat{Z}_r(p^r\rho + w \bullet \lambda) : \widehat{L}_r(\mu_0 + p^r w\mu_1)].$$

As, generally, $[\widehat{Z}_r(p^r\nu + \lambda) : \widehat{L}_r(\mu)] = [\widehat{Z}_r(\lambda) : \widehat{L}_r(\mu - p^r\nu)]$, we get

(4) $$[\widehat{Z}_r(\lambda) : \widehat{L}_r(\mu_0 + p^r\mu_1)] = [\widehat{Z}_r(w \bullet \lambda) : \widehat{L}_r(\mu_0 + p^r w \bullet \mu_1)].$$

This equation holds also when $\rho \notin X(T)$. (One proves it first for a suitable covering group of G.)

Lemma: *Let $\mu_0 \in X'_r(T)$ and $\mu_1, \lambda \in X(T)$.*

a) *If $(p^r-1)\rho \in X(T)$, then*

$$[\widehat{Z}_r(\lambda + (p^r-1)\rho) : \widehat{L}_r(\mu_0)] = [\widehat{Z}_r(w\lambda + (p^r-1)\rho) : \widehat{L}_r(\mu_0)] \tag{5}$$

for all $w \in W$.

b) *If $[\widehat{Z}_r(\lambda) : \widehat{L}_r(\mu_0 + p^r\mu_1)] \neq 0$, then*

$$p^r\mu_1 + \mu_0 \uparrow \lambda \uparrow p^r(\mu_1 + 2\rho) + w_0 \bullet \mu_0. \tag{6}$$

Proof: a) Shifting $p^r\mu_1$ and $p^r w \bullet \mu_1$ to the other side in (4), we get

$$[\widehat{Z}_r(\lambda - p^r\mu_1) : \widehat{L}_r(\mu_0)] = [\widehat{Z}_r(w \bullet \lambda - p^r w \bullet \mu_1) : \widehat{L}_r(\mu_0)].$$

Substituting $\lambda + p^r\mu_1 + (p^r-1)\rho$ for λ, we get (5).

b) The assumption implies that $[\widehat{Z}_r(w_0 \bullet \lambda) : \widehat{L}_r(\mu_0 + p^r w_0 \bullet \mu_1)] \neq 0$ by (4) , hence $\mu_0 + p^r w_0(\mu_1 + 2\rho) \uparrow w_0 \bullet \lambda$. As w_0 reverses $\uparrow$, we get

$$\lambda \uparrow w_0 \bullet (\mu_0 + p^r w_0(\mu_1 + 2\rho)) = p^r(\mu_1 + 2\rho) + w_0 \bullet \mu_0.$$

Remark: Note that (5) applied to $\lambda = (p^r-1)\rho + w_0\mu_0$ and $w = w_0$ yields

$$[\widehat{Z}_r(w_0 \bullet \mu_0 + 2p^r\rho) : \widehat{L}_r(\mu_0)] = [\widehat{Z}_r(\mu_0) : \widehat{L}_r(\mu_0)] = 1.$$

In general (5) implies: If we know all $[\widehat{Z}_r(\lambda + (p^r-1)\rho) : \widehat{L}_r(\mu_0)]$ with $\mu_0 \in X_r(T)$ and $\lambda \in X(T)_+$, then we know all $[\widehat{Z}_r(\lambda') : \widehat{L}_r(\mu')]$.

For $r = 1$ and $p \geq 2h-2$ one has a converse to (6): Let $\mu_0 \in X_1(T)$. For each $\lambda \in X(T)_+$ with $\lambda - \rho \uparrow w_0 \bullet \mu_0 + p\rho$ (equivalently: with $\lambda + (p-1)\rho \uparrow w_0 \bullet \mu_0 + 2p\rho$) one has $[\widehat{Z}_1(\lambda + (p-1)\rho) : \widehat{L}_1(\mu_0)] \neq 0$. This was proved in [Ye 1] for p–regular μ_0, and in [Doty and Sullivan 5], 2.7 in general. The proof uses results that are here discussed in Chapter 11 below, especially in 11.13. More precisely, one shows for $\lambda, \lambda' \in X(T)_+$ with $\lambda \uparrow \lambda'$ and $\lambda - \rho, \lambda' - \rho \uparrow w_0 \bullet \mu_0 + p\rho$ that

$$[\widehat{Z}_1(\lambda' + (p-1)\rho) : \widehat{L}_1(\mu_0)] \leq [\widehat{Z}_1(\lambda + (p-1)\rho) : \widehat{L}_1(\mu_0)]. \tag{7}$$

Now apply this with $\lambda' = (p-1)\rho + w_0\mu_0$.

9.17. Proposition: *Let $\lambda \in X(T)$. Suppose that all composition factors of $\widehat{Z}_r(\lambda)$ have the form $\widehat{L}_r(\mu_0 + p^r\mu_1)$ with $\mu_0 \in X'_r(T)$ and $\mu_1 \in \overline{C}_{\mathbf{Z}}$. Let $w \in W$. Then $H^i(w \bullet \lambda) = 0$ for all $i \neq l(w)$, and each composition factor of $H^{l(w)}(w \bullet \lambda)$ has the form $L(\mu_0 + p^r\mu_1)$ with $\mu_0 \in X'_r(T)$ and $\mu_1 \in X(T)_+ \cap \overline{C}_{\mathbf{Z}}$. Its multiplicity is equal to $[\widehat{Z}_r(\lambda) : \widehat{L}_r(\mu_0 + p^r\mu_1)]$.*

Proof: By 9.16(4) the composition factors of $\widehat{Z}'_r(w \bullet \lambda)$ have the form $\widehat{L}_r(\mu_0 + p^r w \bullet \mu_1) \simeq L(\mu_0) \otimes p^r w \bullet \mu_1$ with μ_0 and μ_1 as above, each of them occurring $[\widehat{Z}_r(\lambda) : \widehat{L}_r(\mu_0 + p^r\mu_1)]$ times. Now 5.5 implies $H^i(w \bullet \mu_1) = 0$ if $i \neq l(w)$ or $\mu_1 \notin X(T)_+$, whereas $H^{l(w)}(w \bullet \mu_1) \simeq L(\mu_1)$ if $\mu_1 \in X(T)_+$, cf. 5.6. Therefore 9.13(3) yields $R^i \operatorname{ind}_{G_rB}^G \widehat{L}_r(\mu_0 + p^r w \bullet \mu_1) = 0$ if $i \neq l(w)$ or $\mu_1 \notin X(T)_+$ whereas

$$R^{l(w)} \operatorname{ind}_{G_rB}^G \widehat{L}_r(\mu_0 + p^r w \bullet \mu_1) \simeq L(\mu_0) \otimes L(\mu_1)^{[r]} \simeq L(\mu_0 + p^r\mu_1)$$

if $\mu_1 \in X(T)_+$. It follows that $R^i \operatorname{ind}_{G_rB}^G \widehat{Z}'_r(w \bullet \lambda) = 0$ for all $i \neq l(w)$, whereas $R^{l(w)} \operatorname{ind}_{G_rB}^G$ is exact on the subquotients of $\widehat{Z}'_r(w \bullet \lambda)$ and maps a composition series (as a G_rB–module) to a composition series (as a G–module) with some factors zero, perhaps. Now the proposition follows from Lemma 9.12.

Remarks: 1) Suppose that we have not just $\mu \in \overline{C}_{\mathbf{Z}}$, but even $\mu \in \overline{C}_{\mathbf{Z}} \cap X(T)_+$ in the assumption of the proposition. Let $\nu \in X(T)_+$ such that $\lambda + p^r\nu$ satisfies the same assumption as λ. Then we get the composition factors of $H^0(\lambda + p^r\nu)$ from those of $H^0(\lambda)$ by adding $p^r\nu$ to the highest weights. This follows from the proposition as the same statement holds for $\widehat{Z}_r(\lambda + p^r\nu)$ and $\widehat{Z}_r(\lambda)$. So the "pattern" of the highest weights of the composition factors of $H^0(\lambda + p^r\nu)$ and $H^0(\lambda)$ coincide. For $r = 1$ this is the "generic decomposition behaviour" from [Jantzen 4].

2) Write $w \bullet \lambda = \mu_0 + p^r w \bullet \mu_1$ with $\mu_0 \in X_r'(T)$ and $\mu_1 \in X(T)$. As $\widehat{L}_r(w \bullet \lambda)$ is a composition factor of $\widehat{Z}_r(w \bullet \lambda)$ we get $\mu_1 \in \overline{C}_{\mathbf{Z}}$ for λ as in the proposition. Suppose that even $\mu_1 \in \overline{C}_{\mathbf{Z}} \cap X(T)_+$. Then the fact that $\widehat{L}_r(w \bullet \lambda)$ is a submodule of $\widehat{Z}_r'(w \bullet \lambda)$ by 9.6(2) implies (arguing as in the proof of the proposition) that $L(\mu_0 + p^r\mu_1)$ is a submodule of $H^{l(w)}(w \bullet \lambda)$.

9.18. We want to describe some conditions on λ that imply the assumptions in 9.14 or 9.17. It will be convenient to assume $\rho \in X(T)$ and to write $\lambda + \rho = p^r\lambda_1 + \lambda_0$ with $\lambda_0 \in X_r'(T)$ and $\lambda_1 \in X(T)$.

For each $\alpha \in R$ there is a unique $\beta \in W\alpha \cap X(T)_+$. Set then $h_\alpha = \langle \rho, \beta^\vee \rangle + 1$. (So h as in 6.2(9) is the maximum of all h_α.)

Lemma: *Let $\widehat{L}_r(\mu)$ be a composition factor of $\widehat{Z}_r(\lambda)$. Decompose $\mu = p^r\mu_1 + \mu_0$ with $\mu_0 \in X_r'(T)$ and $\mu_1 \in X(T)$. Then:*

a) *If $\langle \lambda_1, \alpha^\vee \rangle \geq h_\alpha - 2$ for all $\alpha \in R^+$, then $\langle \mu_1 + \rho, \beta^\vee \rangle \geq 0$ for all $\beta \in R^+$.*

b) *If $\langle \lambda_1, \alpha^\vee \rangle \geq h_\alpha - 1$ for all $\alpha \in R^+$, then $\mu_1 \in X(T)_+$.*

c) *If $\langle \lambda_1, \alpha^\vee \rangle \geq h_\alpha - 2$ for all $\alpha \in R^+$ and $\langle \lambda, \beta^\vee \rangle \leq p^r(p - h_\beta + 1)$ for all $\beta \in R^+ \cap X(T)_+$, then $\mu_1 \in \overline{C}_{\mathbf{Z}}$.*

Proof: There are $w \in W$ and $\nu \in X(T)$ with $\nu + \rho \in X(T)_+$ such that $p^r(\lambda_1 - \mu_1 - \rho) + \lambda_0 = w(\nu + \rho)$, hence $\lambda = p^r(\mu_1 + \rho) + w \bullet \nu$.

We have by assumption

$$\begin{aligned} 0 \neq [\widehat{Z}_r(\lambda) : \widehat{L}_r(\mu)] &= [\widehat{Z}_r(p^r(\mu_1 + \rho) + w \bullet \nu) : \widehat{L}_r(p^r\mu_1 + \mu_0)] \\ &= [\widehat{Z}_r((p^r - 1)\rho + w(\nu + \rho)) : \widehat{L}_r(\mu_0)] = [\widehat{Z}_r(p^r\rho + \nu) : \widehat{L}_r(\mu_0)] \end{aligned}$$

using 9.16(5) for the last equality. Now 9.16(6) implies

$$p^r\rho + \nu \uparrow 2p^r\rho + w_0 \bullet \mu_0$$

hence

$$\nu \uparrow (p^r - 2)\rho + w_0\mu_0.$$

This yields for all $\alpha \in R^+$ (with $\beta \in W\alpha \cap X(T)_+$)

$$\begin{aligned} \langle w(\nu + \rho), \alpha^\vee \rangle &= \langle \nu + \rho, w^{-1}\alpha^\vee \rangle \leq \langle \nu + \rho, \beta^\vee \rangle \\ &\leq \langle (p^r - 1)\rho + w_0\mu_0, \beta^\vee \rangle \leq (p^r - 1)(h_\alpha - 1) \end{aligned}$$

hence

$$\begin{aligned} p^r\langle \lambda_1, \alpha^\vee \rangle &\leq \langle p^r\lambda_1 + \lambda_0, \alpha^\vee \rangle = \langle \lambda + \rho, \alpha^\vee \rangle \\ &= \langle p^r(\mu_1 + \rho) + w(\nu + \rho), \alpha^\vee \rangle < p^r(\langle \mu_1 + \rho, \alpha^\vee \rangle + h_\alpha - 1) \end{aligned}$$

and

$$\langle \mu_1+\rho,\alpha^\vee\rangle > \langle \lambda_1,\alpha^\vee\rangle - (h_\alpha - 1).$$

Now a) and b) follow immediately.

In order to prove c) is is now enough to show that $\langle \mu_1+\rho,\beta^\vee\rangle \le p$ for all $\beta\in R^+\cap X(T)_+$. For such β obviously

$$\begin{aligned}\langle \mu_1+\rho,\beta^\vee\rangle &\le p^{-r}\langle p^r(\mu_1+\rho)+\mu_0,\beta^\vee\rangle = p^{-r}\langle \mu,\beta^\vee\rangle + \langle\rho,\beta^\vee\rangle\\ &\le p^{-r}\langle\lambda,\beta^\vee\rangle + h_\beta - 1\end{aligned}$$

hence the claim.

Remark: Obviously $\lambda_0\in X_r(T)$ implies $\langle\lambda_0-\rho,\alpha^\vee\rangle < p^r\langle\rho,\alpha^\vee\rangle$ for all $\alpha\in R^+$. We could replace the condition "$\langle\lambda,\beta^\vee\rangle\le p^r(p-h_\beta+1)$" by "$\langle\lambda_1,\beta^\vee\rangle\le p-2(h_\beta-1)$", and the condition "$\langle\lambda_1,\alpha^\vee\rangle\ge h_\alpha-1$" by "$\langle\lambda,\alpha^\vee\rangle\ge 2p^r(h_\alpha-1)$".

9.19. We have $\operatorname{Ext}^i_{G_rT}(M_1,M_2)\simeq \operatorname{Ext}^i_{G_rT}(M_2^*,M_1^*)\simeq \operatorname{Ext}^i_{G_rT}({}^\tau M_2,{}^\tau M_1)$ for all finite dimensional G_rT–modules M_1 and M_2. (Argue as for 4.13(3).) Therefore 9.6(13) implies for all $\lambda,\mu\in X(T)$

(1)
$$\operatorname{Ext}^i_{G_rT}(\widehat{L}_r(\lambda),\widehat{L}_r(\mu))\simeq \operatorname{Ext}^i_{G_rT}(\widehat{L}_r(\mu),\widehat{L}_r(\lambda))$$

for all i. One gets similarly

(2)
$$\operatorname{Ext}^i_{G_r}(L_r(\lambda),L_r(\mu))\simeq \operatorname{Ext}^i_{G_r}(L_r(\mu),L_r(\lambda)).$$

One can also show that $\operatorname{Ext}^i_{G_rB}(\widehat{L}_r(\lambda),\widehat{L}_r(\mu))\simeq \operatorname{Ext}^i_{G_rB^+}(\widehat{L}_r(\mu),\widehat{L}_r(\lambda))$. Note that the groups above are related by I.6.9(4), (5): We have

(3)
$$\operatorname{Ext}^i_{G_r}(L_r(\lambda),L_r(\mu))\simeq \bigoplus_{\nu\in X(T)} \operatorname{Ext}^i_{G_rT}(\widehat{L}_r(\lambda+p^r\nu),L_r(\mu))$$

for all $i\in\mathbf{N}$.

Lemma: *Let $\lambda,\mu\in X(T)$.*

a) *If $\operatorname{Ext}^1_{G_rT}(\widehat{L}_r(\lambda),\widehat{L}_r(\mu))\ne 0$, then $\mu\in W_p\bullet\lambda$.*

b) *If $\operatorname{Ext}^1_{G_r}(L_r(\lambda),L_r(\mu))\ne 0$, then $\mu\in W_p\bullet\lambda+p^rX(T)$.*

Proof: a) Using (1) we may assume $\lambda\not<\mu$. Each $\widehat{Z}_r(\lambda)$ considered either as a G_rB^+–module or as a G_rT–module has by I.8.20(1) a universal property similar to that of a Weyl module $V(\lambda')$, cf. 2.13.a. Therefore one can argue as in 2.14 and gets

(4)
$$\operatorname{Ext}^1_{G_rT}(\widehat{L}_r(\lambda),\widehat{L}_r(\mu))\simeq \operatorname{Hom}_{G_rT}(\operatorname{rad}_{G_rT}\widehat{Z}_r(\lambda),\widehat{L}_r(\mu)).$$

(Similarly with G_rT replaced by G_rB^+.) Now the claim follows from 9.15.

b) This is a consequence of a) and of (3).

Remarks: 1) One gets also from (4) or from the arguments used for 2.12(1)

(5)
$$\operatorname{Ext}^1_{G_rT}(\widehat{L}_r(\lambda),\widehat{L}_r(\lambda))=0 \qquad \text{for all } \lambda\in X(T).$$

The same holds for G_rB and G_rB^+, but *not* for G_r. There are, however, only few exceptions, see 12.9 below.

2) Using block theory one can extend the lemma to all Ext^i with $i\ge 0$, cf. 7.1(3), similarly for the next result.

9.20. Proposition: *Let $\lambda, \mu \in X(T)$. Then* $\operatorname{Ext}^1_{G_rB}(\widehat{L}_r(\lambda), \widehat{L}_r(\mu)) \neq 0$ *implies* $\mu \in W_p \bullet \lambda$.

Proof: Let us use the spectral sequence I.6.6(1) for the normal subgroup G_r of G_rB. If $\operatorname{Ext}^1_{G_rB}(\widehat{L}_r(\lambda), \widehat{L}_r(\mu)) \neq 0$, then the five term exact sequence, cf. I.4.1(4), implies that $\operatorname{Ext}^1_{G_r}(\widehat{L}_r(\lambda), \widehat{L}_r(\mu))^B \neq 0$ or that $H^1(B, \operatorname{Hom}_{G_r}(\widehat{L}_r(\lambda), \widehat{L}_r(\mu))^{[-r]}) \neq 0$. In the first case one also has $0 \neq \operatorname{Ext}^1_{G_r}(\widehat{L}_r(\lambda), \widehat{L}_r(\mu))^T \simeq \operatorname{Ext}^1_{G_rT}(\widehat{L}_r(\lambda), \widehat{L}_r(\mu))$ and gets $\mu \in W_p \bullet \lambda$ from 9.19.a. In the second case there has to be some $\nu \in X(T)$ with $\mu = \lambda + p^r\nu$ and $H^1(B, \nu) \neq 0$. Then $\nu \in \mathbf{Z}R$ by 4.10, so again $\mu \in W_p \bullet \lambda$.

9.21. Here is a more detailed result:

Proposition: *Let $\lambda, \mu \in X(T)$. Write $\lambda = \lambda_0 + p^r\lambda_1$ and $\mu = \mu_0 + p^r\mu_1$ with $\lambda_0, \mu_0 \in X'_r(T)$ and $\lambda_1, \mu_1 \in X(T)$.*

a) *If $\mu_1 - \lambda_1 \in X(T)_+$, then*

$$\operatorname{Ext}^1_{G_rB}(\widehat{L}_r(\lambda), \widehat{L}_r(\mu)) \simeq \operatorname{Ext}^1_G(L(\lambda_0), L(\mu_0) \otimes H^0(\mu_1 - \lambda_1)^{[r]}).$$

b) *Suppose $\mu_1 - \lambda_1 \notin X(T)_+$. If $\lambda_0 = \mu_0$ and $\mu_1 - \lambda_1 = -p^i\alpha$ for some $\alpha \in S$ and $i \in \mathbf{N}$, then $\operatorname{Ext}^1_{G_rB}(\widehat{L}_r(\lambda), \widehat{L}_r(\mu)) \simeq k$. Otherwise $\operatorname{Ext}^1_{G_rB}(\widehat{L}_r(\lambda), \widehat{L}_r(\mu)) = 0$.*

Proof: Observe that by I.4.4

$$\operatorname{Ext}^1_{G_rB}(\widehat{L}_r(\lambda), \widehat{L}_r(\mu)) \simeq \operatorname{Ext}^1_{G_rB}(L(\lambda_0), L(\mu_0) \otimes p^r(\mu_1 - \lambda_1)).$$

Consider the spectral sequence from I.4.5.a for $H = G_rB$, $N = L(\lambda_0)$, and $M = L(\mu_0) \otimes p^r(\mu_1 - \lambda_1)$. Recall from 9.13(3) that

$$R^i \operatorname{ind}^G_{G_rB}(L(\mu_0) \otimes p^r(\mu_1 - \lambda_1)) \simeq L(\mu_0) \otimes H^i(\mu_1 - \lambda_1)^{[r]}.$$

If $\mu_1 - \lambda_1 \in X(T)_+$, then these terms vanish for $i > 0$ and we get isomorphisms

$$(1) \qquad \operatorname{Ext}^i_{G_rB}(\widehat{L}_r(\lambda), \widehat{L}_r(\mu)) \simeq \operatorname{Ext}^i_G(L(\lambda_0), L(\mu_0) \otimes H^0(\mu_1 - \lambda_1)^{[r]}).$$

Suppose $\mu_1 - \lambda_1 \notin X(T)_+$. Then $H^0(\mu_1 - \lambda_1) = 0$, so all $E_2^{i,0}$–terms in the spectral sequence are zero. The five term exact sequence I.4.1(4) yields an isomorphism $E^1 \simeq E_2^{0,1}$, i.e.,

$$(2) \qquad \operatorname{Ext}^1_{G_rB}(\widehat{L}_r(\lambda), \widehat{L}_r(\mu)) \simeq \operatorname{Hom}_G(L(\lambda_0), L(\mu_0) \otimes H^1(\mu_1 - \lambda_1)^{[r]}).$$

Considered as a G_r–module $L(\mu_0) \otimes H^1(\mu_1 - \lambda_1)^{[r]}$ is a direct sum of copies of the simple module $L(\mu_0)$. If the space in (2) is non-zero, then $L(\lambda_0)$, $L(\mu_0)$ have to be isomorphic G_r–modules, hence $\lambda_0 = \mu_0$. If so, then $\operatorname{Hom}_{G_r}(L(\lambda_0), L(\mu_0) \otimes H^1(\mu_1 - \lambda_1)^{[r]})$ can be identified with $H^1(\mu_1 - \lambda_1)^{[r]}$, hence the right hand side in (2) with $H^1(\mu_1 - \lambda_1)^G \simeq \operatorname{Hom}_G(k, H^1(\mu_1 - \lambda_1))$. Now b) follows from 5.18.

Remark: There is, of course, a similar result for G_rB^+.

9.22. Similarly as in 7.2, we can regard the blocks of G_rB, G_rB^+, G_rT, or G_r as a subset of $X(T)$ via $\lambda \mapsto \widehat{L}_r(\lambda)$. (In the case of G_r, we ought to take $X(T)/p^rX(T)$ and $\lambda \mapsto L_r(\lambda)$.) By 9.21 or 9.19.a the block of λ for G_rB, G_rB^+, or G_rT is contained in $W_p \bullet \lambda$. The block of λ for G_r is contained in $W_p \bullet \lambda + p^rX(T)$ by 9.19.b. One can show more precisely (cf. [Jantzen 6], 5.5):

(1) *Let $\lambda \in X(T)$. If $\langle \lambda + \rho, \alpha^\vee \rangle \in \mathbf{Z}p^r$ for all $\alpha \in R^+$, then the block of λ for G_r is equal to $\lambda + p^rX(T)$. Otherwise let m be the smallest integer such that there is $\alpha \in R^+$ with $\langle \lambda + \rho, \alpha^\vee \rangle \notin \mathbf{Z}p^m$. Then the block of λ for G_r is equal to $W \bullet \lambda + p^m\mathbf{Z}R + p^rX(T)$.*

For $r = 1$ (where we get $W \bullet \lambda + pX(T)$) this result follows easily from 9.16(4) and was already proved in [Humphreys 1].

As in 7.3 we can define functors pr_λ on the categories of G_rB–, G_rB^+–, and G_rT–modules. Obviously 7.3(1)–(3) generalise. There are statements analogous to 7.3(4), (5) with $L(\mu)$ replaced by $\widehat{L}_r(\mu)$ (and arbitrary $\mu \in X(T)$) and with $H^i(\mu)$ replaced by $\widehat{Z}'_r(\mu)$ or $\widehat{Z}_r(\mu)$. The functors pr_λ on G_rB–modules and pr_λ on G_rT–modules commute (almost by definition) with the forgetful functor from $\{ G_rB$–modules$\}$ to $\{ G_rT$–modules$\}$. Consider $\mu \in X(T)_+$ and decompose $\mu = \mu_0 + p^r\mu_1$ with $\mu_0 \in X_r(T)$ and $\mu_1 \in X(T)_+$. Then $L(\mu) \simeq L(\mu_0) \otimes L(\mu_1)^{[r]}$, so the G_rB–composition factors of $L(\mu)$ are precisely all $\widehat{L}_r(\mu_0 + p^r(\mu_1 - \nu)) \simeq L(\mu_0) \otimes p^r(\mu_1 - \nu)$ with $\nu \in \mathbf{Z}R$ and $\mu_1 - \nu$ a weight of $L(\mu_1)$. Then $\mu_0 + p^r(\mu_1 - \nu) = \mu - p^r\nu \in W_p \bullet \mu$. Therefore a functor pr_λ on G_rB–modules applied to $L(\mu)$ yields $L(\mu)$ if $\mu \in W_p \bullet \lambda$, and 0 otherwise. This shows that the pr_λ for G and for G_rB commute with the forgetful functor from $\{ G$–modules$\}$ to $\{ G_rB$–modules$\}$. — This is the reason for using the same notation for all these functors.

Furthermore, we can use the same definition as in 7.6 in order to define translation functors T_λ^μ for all $\lambda, \mu \in \overline{C}_\mathbf{Z}$ on G_rB–, G_rB^+–, and G_rT–modules. Like the pr_λ, the T_λ^μ also commute with the forgetful functors $\{ G$–modules$\} \to \{ G_rB$–modules$\} \to \{ G_rT$–modules$\}$. Lemma 7.6 and Proposition 7.9 generalise to the new situation.

Consider a finite dimensional G_rB–module E and $\lambda \in X(T)$. Then $E \otimes \widehat{Z}'_r(\lambda)$ has a filtration with factors of the form $\widehat{Z}'_r(\lambda + \nu)$ each occurring $\dim(E_\nu)$ times. Indeed, the tensor identity implies

$$E \otimes \widehat{Z}'_r(\lambda) \simeq E \otimes \mathrm{ind}_B^{G_rB} \lambda \simeq \mathrm{ind}_B^{G_rB}(E \otimes \lambda).$$

The B–module $E \otimes \lambda$ has a composition series with each $\lambda + \nu$ occurring $\dim(E_\nu)$ times. Now use the exactness of $\mathrm{ind}_B^{G_rB}$, cf. 9.1. (If we assume E to be a G_rT–module only, then we get similarly a filtration of $E \otimes \widehat{Z}'_r(\lambda)$ as a G_rT–module.)

It follows that

$$\mathrm{ch}(E \otimes \widehat{Z}'_r(\lambda)) = \sum_{\nu \in X(T)} \dim(E_\nu) \, \mathrm{ch}\, \widehat{Z}'_r(\lambda + \nu).$$

Using this one generalises 7.5(1) to G_rB–modules (or G_rT–modules) V such that $\mathrm{ch}(V)$ is a finite linear combination of $\mathrm{ch}\,\widehat{Z}'_r(w \bullet \lambda)$ with $w \in W_p$, replacing all $\chi(?)$ by $\mathrm{ch}\,\widehat{Z}'_r(?)$. Furthermore, it is obvious how to generalise 7.13 to G_rB or

G_rT, replacing all H^0 by $\widehat{Z}'_r$ and dropping the assumption $w \bullet \lambda \in X(T)_+$ and the condition $ww_1 \bullet \mu \in X(T)_+$. The special cases 7.11 and 7.19.a yield:

(2) *Let $\lambda, \mu \in \overline{C}_{\mathbf{Z}}$ and let F be the facet with $\lambda \in F$. If $\mu \in \overline{F}$, then $T_\lambda^\mu \widehat{Z}'_r(w \bullet \lambda) \simeq \widehat{Z}'_r(w \bullet \mu)$ for all $w \in W_p$.*

(3) *Let $\lambda \in C \cap X(T)$ and $\mu \in \overline{C}_{\mathbf{Z}}$. Suppose that there is $s \in \Sigma$ with $\Sigma^0(\mu) = \{s\}$. Let $w \in W_p$ with $w \bullet \lambda < ws \bullet \lambda$. Then there is an exact sequence of G_rB–modules*

$$0 \to \widehat{Z}'_r(w \bullet \lambda) \longrightarrow T_\mu^\lambda \widehat{Z}'_r(w \bullet \mu) \longrightarrow \widehat{Z}'_r(ws \bullet \lambda) \to 0.$$

There are analogous results for the $\widehat{Z}_r(\lambda)$ that can be proved similarly using 9.2(1) or using a tensor identity for coind. One has to observe that $w \bullet \lambda$ and $ws \bullet \lambda$ in (3) have to be switched when dealing with the $\widehat{Z}_r$.

Using 7.15 one shows:

(4) *Let $\lambda, \mu \in \overline{C}_{\mathbf{Z}}$ such that μ belongs to the closure of the facet containing λ. Let $w \in W_p$ and denote by F the facet with $w \bullet \lambda \in F$. Then*

$$T_\lambda^\mu \widehat{L}_r(w \bullet \lambda) \simeq \begin{cases} \widehat{L}_r(w \bullet \mu), & \text{if } w \bullet \mu \in \widehat{F}, \\ 0, & \text{otherwise.} \end{cases}$$

One has to decompose $w \bullet \lambda = \lambda_0 + p^r \lambda_1$ and $w \bullet \mu = \mu_0 + p^r \lambda_1$ with $\lambda_0 \in X_r(T)$ and $\lambda_1 \in X(T)$. Then there are $\lambda' \in \overline{C}_{\mathbf{Z}}$ and $w_1 \in W_p$ with $\lambda_0 = w_1 \bullet \lambda'$. It then follows that $\mu' = w_1^{-1} \bullet \mu_0 \in \overline{C}_{\mathbf{Z}}$ and that

$$T_\lambda^\mu \widehat{L}_r(w \bullet \lambda) \simeq T_\lambda^\mu (L(\lambda_0) \otimes p^r \lambda_1) \simeq (T_{\lambda'}^{\mu'} L(\lambda_0)) \otimes p^r \lambda_1.$$

Now apply 7.15 to $T_{\lambda'}^{\mu'} L(\lambda_0)$ and note that $\mu_0 \in X_r(T)$ in case $w \bullet \mu \in \widehat{F}$.

Also 7.16–7.20 lead to analogous results for G_rT–modules. We get in particular for all $w \in W_p$ and $s \in \Sigma$ with $w \bullet \lambda < ws \bullet \lambda$ that

(5) $\operatorname{Hom}_{G_1T}(\widehat{Z}'_r(ws \bullet \lambda), \widehat{Z}'_r(w \bullet \lambda)) \simeq k \simeq \operatorname{Hom}_{G_1T}(\widehat{Z}_r(w \bullet \lambda), \widehat{Z}_r(ws \bullet \lambda))$

proceeding as for 7.19.d. (Replace the references to 4.18(2) by references to 9.8(3).)

9.23. Let me conclude this chapter with some results relating the representation theory of G to that of G_r rather than G_rT or G_rB.

For any G–modules V, V' the spectral sequence I.6.6(1) for $N = G_r$ and $M = k$ leads to the following five term exact sequence

$$\begin{aligned} 0 &\to H^1(G/G_r, \operatorname{Hom}_{G_r}(V, V')) \to \operatorname{Ext}^1_G(V, V') \to \operatorname{Ext}^1_{G_r}(V, V')^{G/G_r} \\ &\to H^2(G/G_r, \operatorname{Hom}_{G_r}(V, V')) \to \operatorname{Ext}^2_G(V, V'). \end{aligned}$$

As $G/G_r \simeq G$ and as $H^i(G, k) = 0$ for all $i > 0$ (by 4.11), we get:

(1) *If $\operatorname{Hom}_{G_r}(V, V') = \operatorname{Hom}_G(V, V')$, then $\operatorname{Ext}^1_G(V, V') \xrightarrow{\sim} \operatorname{Ext}^1_{G_r}(V, V')^{G/G_r}$,*

especially

(2) *If $V^G = V^{G_r}$, then $H^1(G, V) \simeq H^1(G_r, V)^{G/G_r}$.*

Of course, this generalises from (G, G_r) to any pair (H, N) with N normal in H and $H^1(H/N, k) = 0 = H^2(H/N, k)$. For example, we can take $(H, N) = (B, B_r)$, cf. 4.11.

9.24. Proposition: *Let V, V' be finite dimensional G–modules such that each composition factor of V and of V' has its highest weight in $X_r(T)$. Then:*

a) $\mathrm{Hom}_{G_r}(V, V') = \mathrm{Hom}_G(V, V')$.

b) *Any G_r–submodule of V is also a G–submodule.*

Proof: a) We use induction on the length of V and V'. If both are simple, then the claim follows from 3.15 and 3.10(3). Suppose next that V_1 is a submodule of V' such that the claim is known to hold for V'/V_1 and for V_1 instead of V'. Now apply the functors $\mathcal{F} = \mathrm{Hom}_G(V, ?)$ and $\mathcal{F}' = \mathrm{Hom}_{G_r}(V, ?)$ to the short exact sequence $0 \to V_1 \to V' \to V'/V_1 \to 0$. We get a commutative diagram with exact rows

$$\begin{array}{ccccccccc} 0 & \to & \mathcal{F}V_1 & \to & \mathcal{F}V' & \to & \mathcal{F}(V'/V_1) & \to & \mathrm{Ext}^1_G(V, V_1) \\ & & \downarrow & & \downarrow & & \downarrow & & \downarrow \\ 0 & \to & \mathcal{F}'V_1 & \to & \mathcal{F}'V' & \to & \mathcal{F}'(V'/V_1) & \to & \mathrm{Ext}^1_{G_r}(V, V_1) \end{array}$$

The first and third vertical maps are isomorphisms by our inductive assumption, and the last one is injective by 9.23(1). Now some diagram chasing shows that also the second vertical map is an isomorphism.

b) Let us use induction on $\dim V$. Let M be a non-zero G_r–submodule of V. There is $\lambda \in X_r(T)$ and an injective homomorphism of G_r–modules $\varphi : L(\lambda) \to M \subset V$. By a) this map is also a homomorphism of G–modules. By induction $M/\varphi(L(\lambda))$ is a G–submodule of $V/\varphi(L(\lambda))$, hence M one of V.

Remark: Under our assumption any $G(\mathbf{F}_q)$–submodule (where $q = p^r$) of V is also a G–submodule, cf. [Cline, Parshall, Scott, and van der Kallen], 7.5.

CHAPTER 10

Geometric Reductivity and Other Applications of the Steinberg Modules

Let p be a prime and k a field of characteristic p. For the sake of simplicity let us assume that k is perfect. Furthermore, we shall assume that $(p-1)\rho \in X(T)$, hence that $(p^r-1)\rho \in X(T)$ for all $r \in \mathbf{N}$.

The Steinberg modules St_r have already played an important role in the proof of Kempf's vanishing theorem. We give some more applications in this chapter. Over and over again we shall have to use that St_r is both simple and injective as a G_r–module (10.1/2).

We prove in 10.7 that G is geometrically reductive. This had been conjectured by Mumford and was proved in [Haboush 1]. It was later generalised in [Seshadri 1] to arbitrary ground rings, cf. 10.8 below for a slightly weaker result. This is not the place to discuss the importance of this result in algebraic geometry; for that let me refer you to [MF].

The proof of Mumford's conjecture given here follows [Humphreys 8] and uses only the injectivity of St_r. The original approach also required the fact that certain injective B–modules (resp. G–modules) are direct limits of modules of the form $St_r \otimes V_r$ with suitable B–modules (resp. G–modules) V_r. Such results are proved here in 10.12/13. They imply that any finite dimensional B–module (resp. G–module) can be mapped injectively into some $St_r \otimes V_r$. For simple M such maps can be constructed explicitly, cf. 10.15 for the case of G. Using these embeddings and I.6.10, one gets for all B–modules (resp. G–modules) V that the natural maps $H^i(B,V) \to H^i(B,V^{[r]})$ resp. $H^i(G,V) \to H^i(G,V^{[r]})$ induced by the Frobenius endomorphism are injective. One also gets more complicated results (cf. 10.14, 10.16) reducing the computation of Ext-groups for G to that for G_1 (10.17).

Besides the papers already mentioned above, the main references for this chapter are [Andersen 10], [Andersen and Jantzen], [Cline, Parshall, and Scott 8], [Cline, Parshall, Scott, and van der Kallen], [Donkin 5, 7], [Humphreys 7], [Humphreys and Jantzen], [O'Halloran 2], [Pfautsch 2], and [Sullivan 3].

10.1. Recall from 3.18 the definition of the rth Steinberg module St_r for any $r \in \mathbf{N}$, $r > 0$:

(1) $$St_r = L((p^r-1)\rho).$$

We know by 3.19(4) that (as G–modules)

(2) $$St_r = H^0((p^r-1)\rho) \simeq V((p^r-1)\rho) \simeq St_r^*$$

and by 3.18(6) that (as G_rB–modules)

(3) $$St_r \simeq \widehat{Z}'_r((p^r-1)\rho).$$

Similarly one has (as G_rB^+–modules)

(4) $$St_r \simeq \widehat{Z}_r((p^r-1)\rho).$$

10.2. Proposition: *The Steinberg module St_r is (for all $r > 0$) injective and projective both as a G_rT–module and as a G_r–module.*

Proof: By Lemma 9.4 it suffices to prove that St_r is projective as a G_rT–module. It is enough to show that $\operatorname{Ext}^1_{G_rT}(St_r, \widehat{L}_r(\mu)) = 0$ for all $\mu \in X(T)$. In case $(p^r-1)\rho \not< \mu$ we apply 9.19(4) and get, since $\operatorname{rad}_{G_rT}\widehat{Z}_r((p^r-1)\rho) = 0$ by 10.1(4)

$$\operatorname{Ext}^1_{G_rT}(St_r, \widehat{L}_r(\mu)) \simeq \operatorname{Hom}_{G_rT}(0, \widehat{L}_r(\mu)) = 0.$$

If $(p^r-1)\rho < \mu$, then we get from 9.19(1), (4)

$$\operatorname{Ext}^1_{G_rT}(St_r, \widehat{L}_r(\mu)) \simeq \operatorname{Ext}^1_{G_rT}(\widehat{L}_r(\mu), St_r) \simeq \operatorname{Hom}_{G_rT}(\operatorname{rad}_{G_rT}\widehat{Z}_r(\mu), St_r).$$

As St_r is simple, any non-zero homomorphism to St_r has to be surjective. As

$$\dim \operatorname{rad}_{G_rT}\widehat{Z}_r(\mu) < \dim \widehat{Z}_r(\mu) = \dim St_r$$

we get $\operatorname{Hom}_{G_rT}(\operatorname{rad}_{G_rT}\widehat{Z}_r(\mu), St_r) = 0$, hence the claim.

10.3. As the Steinberg module St_r is both simple and injective as a G_r–module, it belongs to a block of its own. The representation theory of G_r is equivalent to that of $\operatorname{Dist}(G_r)$. So there is a central idempotent $e_r \in \operatorname{Dist}(G_r)$ corresponding to St_r. Any G_r–module M decomposes $M = e_rM \oplus (1-e_r)M$ where e_rM is isomorphic to a direct sum of copies of St_r, and where $(1-e_r)M$ has no composition factor isomorphic to St_r.

We claim that (for the adjoint action of G)

(1) $$e_r \in \operatorname{Dist}(G_r)^G.$$

We may (by I.2.10(3)) assume that k is algebraically closed. Then $G(k)$ permutes the central idempotents in $\operatorname{Dist}(G_r)$ under the adjoint action. As the number of these idempotents is finite, all are fixed by the connected algebraic group $G(k)$, hence by G as $G(k)$ is dense in G. So we get (1).

10.4. Proposition: *Let H be a closed subgroup scheme of G with $G_r \subset H$ for some $r > 0$.*

a) *For any H–module V the subspaces e_rV and $(1-e_r)V$ are H–submodules of V with $V = e_rV \oplus (1-e_r)V$. The map $\varphi \otimes x \mapsto \varphi(x)$ induces an isomorphism of H–modules*

$$\operatorname{Hom}_{G_r}(St_r, V) \otimes St_r \xrightarrow{\sim} e_rV.$$

b) *The functor* $M \mapsto M \otimes St_r$ *is an equivalence of categories*

$$\{\,H/G_r\text{–modules}\,\} \xrightarrow{\sim} \{\,H\text{–modules } V \text{ with } e_r V = V\,\}.$$

Proof: a) The first part follows easily from 10.3(1), the second part is a special case of I.6.15(2) as $\mathrm{End}_{G_r}(St_r) \simeq k$, cf. 3.10(3).

b) For any H–module V with $e_r V = V$ we have by a) a natural isomorphism $\mathrm{Hom}_{G_r}(St_r, V) \otimes St_r \xrightarrow{\sim} V$. Note that G_r acts trivially on $\mathrm{Hom}_{G_r}(St_r, V)$ so that we can regard this space as an H/G_r–module.

For any H/G_r–module M the map $\psi : M \to \mathrm{Hom}_{G_r}(St_r, M \otimes St_r)$ with $\psi(m)(x) = m \otimes x$ for all $m \in M$ and $x \in St_r$ is obviously an injective homomorphism of H/G_r–modules. It follows from a) that ψ is bijective.

Therefore $M \mapsto M \otimes St_r$ and $V \mapsto \mathrm{Hom}_{G_r}(St_r, V)$ are inverse equivalences of categories.

10.5. Let us apply 10.4 to $H = G$. As $G/G_r \simeq G$ via F_G^r, we can formulate 10.4.b in this case as follows:

(1) *The functors* $M \mapsto M^{[r]} \otimes St_r$ *is an equivalence of categories from* $\{\,G$–*modules*$\,\}$ *to* $\{\,G$–*modules* V *with* $e_r V = V\,\}$.

We know (by Steinberg's tensor product theorem) that this functor maps any simple module $L(\lambda)$ with $\lambda \in X(T)_+$ to the simple module $L(p^r\lambda + (p^r - 1)\rho)$. This implies a corresponding statement for the injective hulls: Denote the injective hull of $L(\lambda)$ as a G–module by Q_λ. Then

$$Q_{p^r\lambda + (p^r-1)\rho} \simeq Q_\lambda^{[r]} \otimes St_r \tag{2}$$

for all $\lambda \in X(T)_+$.

We can apply (1) to get results on blocks (cf. 7.2) for G. Because of the direct sum decomposition $V = e_r V \oplus (1 - e_r)V$ in 10.4.a there are two types of blocks. A block of the first type will contain only modules V with $e_r V = 0$, and a block of the second type will contain only modules V with $e_r V = V$. The equivalence of categories (1) maps any block to a block of the second type.

Let us identify blocks with subsets of $X(T)_+$. If $b \subset X(T)_+$ is a block, then $\{p^r\mu + (p^r - 1)\rho \mid \mu \in b\}$ is by (1) also a block. If G is semi-simple and simply connected, then this reduces the determination of all blocks to that of all blocks b with $b + \rho \not\subset pX(T)$. (Given $\lambda \in X(T)_+$ let $s \in \mathbf{N}$ be minimal for $\langle \lambda + \rho, \alpha^\vee \rangle \in \mathbf{Z}p^s$ for all $\alpha \in R^+$. Then there exists $\mu \in X(T)_+$ with $\lambda = p^s\mu + (p^s - 1)\rho$. Then the block of λ is equal to $p^s b + (p^s - 1)\rho$ where b is the block of μ. We have $\mu + \rho \notin pX(T)$, hence $b + \rho \not\subset pX(T)$.) Note that this shows that 7.2(3) follows from 7.2(2) for G semi-simple and simply connected.

It is left to the reader to formulate and prove analogous results on injective modules and blocks for groups like G_{r+s}, $G_{r+s}T$, $G_{r+s}B$. For example, if we take $H = G_{r+s}$, then H/G_r identifies with G_s, cf. I.9.5. The equivalence of categories in 10.4.b maps then a block b of G_s to a block b' of G_{r+s}. We can identify b with a two-sided ideal in $\mathrm{Dist}(G_s)$ and b' with a two-sided ideal in $\mathrm{Dist}(G_{r+s})$. Considered as a left module we have $b \simeq \bigoplus_E Q_E^{\dim E}$ where Q_E is the injective hull of E and where the sum runs over a system of representatives for the isomorphism classes of simple G_s–modules in b. Set $m = \dim St_r$. Then the simple modules in b' are the $E^{[r]} \otimes St_r$ with E as above, their injective hulls are $Q_E^{[r]} \otimes St_r$, so b' is as a left module isomorphic to

$$b' \simeq \bigoplus_E (E^{[r]} \otimes St_r)^{m \dim E} \simeq (b^{[r]} \otimes St_r)^m.$$

This leads to algebra isomorphisms

$$b' \simeq \operatorname{End}_{G_{r+s}}(b')^{\mathrm{op}} \simeq M_m(\operatorname{End}_{G_{r+s}}(b^{[r]} \otimes St_r))^{\mathrm{op}}.$$

Now the equivalence of categories implies

$$b' \simeq M_m(b) \tag{3}$$

as algebras. (Cf. [Pfautsch 2] for the first proof of this result.)

10.6. Lemma: *Let M, M' be finite dimensional G–modules with $\langle \mu, \alpha^\vee \rangle < p^r$ for all $\alpha \in R^+$ and all $\mu \in X(T)$ with $M_\mu \neq 0$ or $(M')_\mu \neq 0$. Then:*

a) $\operatorname{Ext}^1_G(e_r(M \otimes St_r), e_r(M' \otimes St_r)) = 0$.

b) *The G–module $e_r(M \otimes St_r)$ is semi-simple.*

c) $\operatorname{Ext}^1_G(M, \operatorname{End}(St_r)) = 0$.

Proof: Let L be a composition factor of $e_r(M \otimes St_r)$, and L' one of $e_r(M' \otimes St_r)$. We want to show that $\operatorname{Ext}^1_G(L, L') = 0$. Then a) and b) [setting $M' = M$] will follow.

As $e_r L = L$, the highest weight of L has the form $p^r\nu + (p^r - 1)\rho$ with $\nu \in X(T)_+$. Similarly, there exists $\nu' \in X(T)_+$ with $L' \simeq L(p^r\nu' + (p^r - 1)\rho)$. There have to be weights μ of M and μ' of M' with $p^r\nu \leq \mu$ and $p^r\nu' \leq \mu'$. Now our assumption implies that $p^r\langle \nu, \alpha^\vee \rangle < p^r$ and $p^r\langle \nu', \alpha^\vee \rangle < p^r$ for all $\alpha \in R^+ \cap X(T)_+$, hence $\langle \nu, \alpha^\vee \rangle \leq 0$ and $\langle \nu', \alpha^\vee \rangle \leq 0$. On the other hand $\nu, \nu' \in X(T)_+$, so we get $\langle \nu, \beta^\vee \rangle = 0$ and $\langle \nu', \beta^\vee \rangle = 0$ for all $\beta \in S$. So $L(\nu) = H^0(\nu)$ and $L(\nu') = H^0(\nu')$ are one dimensional G–modules that cannot extend each other. We get $\operatorname{Ext}^1_G(L(\nu), L(\nu')) = 0$, hence $\operatorname{Ext}^1_G(L, L') = 0$ by 10.5(1).

For c) we use that $St_r = e_r St_r$, hence

$$\begin{aligned}\operatorname{Ext}^1_G(M, \operatorname{End}(St_r)) &\simeq \operatorname{Ext}^1_G(M, St_r^* \otimes St_r) \simeq \operatorname{Ext}^1_G(M \otimes St_r, St_r)\\ &= \operatorname{Ext}^1_G(M \otimes St_r, e_r St_r) \simeq \operatorname{Ext}^1_G(e_r(M \otimes St_r), e_r St_r)\end{aligned}$$

and apply a) with $M' = k$.

Remark: We can replace our assumption on μ by $\langle \mu, \alpha^\vee \rangle < p^r(p - \langle \rho, \alpha^\vee \rangle)$ for all $\alpha \in R^+ \cap X(T)_+$. Then we get above $\nu, \nu' \in \overline{C}_{\mathbf{Z}}$, hence $\operatorname{Ext}^1_G(L(\nu), L(\nu')) = 0$ by 5.6(5). So we get the same conclusion as before. This leads to a strengthening of the lemma in case $p > \langle \rho, \alpha^\vee \rangle$ for all α, i.e., for $p \geq h$, cf. 6.2(9).

10.7. Proposition: *Let M be a finite dimensional G–module. There is for any $m \in M^G$, $m \neq 0$ some $n \in \mathbf{N}$ and some $f \in S^{p^n}(M^*)^G$ with $f(m) \neq 0$.*

Proof: Choose $r \in \mathbf{N}$ with $p^r > \langle \mu, \alpha^\vee \rangle$ for all weights μ of M and all $\alpha \in R^+$. For any $m \in M^G$, $m \neq 0$ the subspace $km \otimes St_r$ of $M \otimes St_r$ is a G–submodule isomorphic to St_r, hence contained in $e_r(M \otimes St_r)$. By 10.6.b and 10.4.a this submodule is a direct summand of $M \otimes St_r$. Therefore we can find a homomorphism $\varphi : M \otimes St_r \to St_r$ with $\varphi(m \otimes x) = x$ for all $x \in St_r$.

We have natural isomorphisms

$$\mathrm{Hom}_G(M \otimes St_r, St_r) \xrightarrow{\sim} \mathrm{Hom}_G(M, St_r^* \otimes St_r) \xrightarrow{\sim} \mathrm{Hom}_G(M, \mathrm{End}(St_r)).$$

Under these bijections φ is mapped to a homomorphism $\psi : M \to \mathrm{End}(St_r)$ of G–modules with $\psi(m) = \mathrm{id}$. Then $f = \det \circ \psi \in S^n(M^*)^G$ with $n = \dim(St_r) = p^{r|R^+|}$ satisfies $f(m) = 1$.

Remarks: 1) This proposition is usually stated in the form: *G is geometrically reductive.*

2) Note that the proposition also holds when G does not satisfy the standard assumption in this chapter that $(p-1)\rho \in X(T)$: We can replace G by a suitable covering group without changing the assumption or the conclusion in the proposition. Also the assumption that k should be perfect can be removed: The result over the algebraic closure of k implies the result over k.

3) A similar (and simpler) result holds for fields of characteristic 0: There the complete reducibility of M (see 5.5(6)) yields a $\varphi \in (M^*)^G$ with $\varphi(m) \neq 0$.

10.8. Proposition: *Let A be a Dedekind ring and let M be a G_A–module which is projective of finite rank over A. Let A' be an A–algebra which is a field. If $m \in M \otimes_A A'$, $m \neq 0$ is a fixed point for $G_{A'}$, then there exist $n \in \mathbf{N}$ and $f \in S^n(M^*)^G$ with $f(m) \neq 0$.*

Proof: In order to simplify our notations let us assume that k is the image of A in A' under the structural morphism. (The necessary modifications when $\mathrm{char}(A') = 0$ or when A is mapped injectively into A', are easy and are left to the reader.)

Choose $r \in \mathbf{N}$ such that $p^r > \langle \mu, \alpha^\vee \rangle$ for each weight μ of the G–module $\bar{M} = M \otimes_A k$ and each $\alpha \in R^+$. We get as in the proof of 10.7 a homomorphism of G–modules

$$\varphi : \bar{M} \longrightarrow \bar{M}^G \otimes \mathrm{End}(St_r)$$

with $\varphi(m') = m' \otimes \mathrm{id}$ for all $m' \in \bar{M}^G$. We have $m \in \bar{M}^G \otimes A'$ by I.2.10(3), so there is $\varphi_1 \in (\bar{M}^G)^*$ with $\varphi_1 \otimes \mathrm{id}_{A'}(m) \neq 0$. Then $\psi = (\varphi_1 \bar{\otimes} \mathrm{id}) \circ \varphi$ is a homomorphism of G–modules $\psi : \bar{M} \to \mathrm{End}(St_r)$ with $\psi \otimes \mathrm{id}_{A'}(m) = a' \,\mathrm{id}$ for some $a' \in A'$, $a' \neq 0$.

Set $V = H^0_A((p^r - 1)\rho)$, cf. 8.6. This is a G_A–module, free of finite rank over A with $V \otimes_A k \simeq St_r$, cf. 8.8(1). We have then also $\mathrm{End}_A(V) \otimes_A k \simeq \mathrm{End}(St_r)$. Suppose that we can find a homomorphism $\psi_1 : M \to \mathrm{End}_A(V)$ of G_A–modules with $\psi = \psi_1 \otimes \mathrm{id}_k$. Then $f = \det \circ \psi_1 \in S^n(M^*)^G$ with $n = \dim St_r$ will satisfy $f(m) = (a')^n \neq 0$.

So our claim will follow if we can show that the natural map is an isomorphism

$$\mathrm{Hom}_{G_A}(M, \mathrm{End}_A(V)) \otimes_A k \xrightarrow{\sim} \mathrm{Hom}_G(\bar{M}, \mathrm{End}(St_r)).$$

Because of I.4.18.a this will follow from

$$\mathrm{Tor}_1^A(\mathrm{Ext}^1_{G_A}(M, \mathrm{End}_A(V)), k) = 0,$$

hence (using I.4.18.a again) from $\mathrm{Ext}^1_G(\bar{M}, \mathrm{End}(St_r)) = 0$. But this holds by 10.6.c.

10.9. (In this section k may be any field.) Using 10.7 (or Remark 3 in 10.7) one can show for any finite dimensional G–module M that $S(M)^G$ is finitely generated as a k–algebra. This was proved by Nagata. You can find a proof in [Sp1], 2.4.

More generally, for any affine algebraic scheme X over k on which G acts the k–algebra $k[X]^G$ is finitely generated. Then $X' = Sp_k(k[X]^G)$ is an affine algebraic scheme over k with a morphism $\pi : X \to X'$ induced by the inclusion $k[X]^G \hookrightarrow k[X]$. Obviously π is constant on G–orbits. One can show that (X', π) is a quotient scheme (in the sense of I.5) and has nice geometric properties, cf. [MF], Thm. A.1.1.

Using Proposition 10.8 one can generalise these results to arbitrary ground rings, see Part II of [Seshadri 1].

For an arbitrary algebraic group scheme H over k and a finite dimensional H–module V the k–algebra $S(V)^H$ need not be finitely generated. The first counter-example was discovered by Nagata. There are, however, important cases where one has finite generation even for non-reductive H. For example:

(1) *If V is a finite dimensional G–module and if H is the unipotent radical of a reduced parabolic subgroup of G, then $S(V)^H$ is finitely generated.*

This was proved by Hochschild and Mostow in case $\mathrm{char}(k) = 0$ and by Grosshans in general, see [Grosshans].

10.10. It has always been an important problem in invariant theory to determine generators for algebras of the form $S(V)^G$ with V a finite dimensional G–module. Most of the classical work concerned groups over $\mathbf{C}$, but there has also been some investigations in prime characteristics.

During the first decades of the 20th century there were many computations of invariants for the finite groups $G(q) = G(\mathbf{F}_q)$, mainly by L. E. Dickson and his collaborators. See [Dic], vol. 3, ch. 19 and [Ru] for surveys, and [B3], ch. V, §5, exerc. 6 for an example. For any finite dimensional G–module V and any $n \in \mathbf{N}$ one has $S^n(V)^G = S^n(V)^{G(p^r)}$ for r large enough, so the computations for $G(p^r)$ also yield results for G.

In [De Concini and Procesi 1] generators and relations are computed for $S(M)^G$ where

$$M = V \oplus V \oplus \cdots \oplus V \oplus V^* \oplus V^* \oplus \cdots \oplus V^* = V^r \oplus (V^*)^s$$

($r, s \in \mathbf{N}$) and where G is one of $GL(V)$, $SL(V)$, $Sp(V, f)$, $SO(V, f')$ for some finite dimensional vector space V. (Here f resp. f' is a non-degenerate alternating resp. symmetric bilinear form on V. For $\mathrm{char}(k) = 2$ the definition of $SO(V, f')$ requires extra care.)

The results in that paper show (among other things) that $S(M)^G$ has the "same form" as over the complex numbers. In particular the dimension of each $S^i(M)^G$ is the same. This can be deduced a priori from the fact that all these $S(M)$ have a good filtration (as in 4.16), cf. [Andersen and Jantzen], 4.9.

10.11. We get from 10.1(1), (4) and 3.8.a, 2.4(2) for all $r > 0$

(1) $$(St_r)^U \simeq -(p^r-1)\rho \simeq (St_r)^{U_r}.$$

Define a B–module St'_r via

(2) $$St'_r = St_r \otimes (p^r-1)\rho.$$

Then

(3) $$(St'_r)^B \simeq k \simeq (St'_r)^{B_r}.$$

Lemma: *Let $\mu \in X(T)$.*

a) *Considered as a B_rT–module $St'_r \otimes \mu$ is the injective hull of μ (for each $r > 0$).*

b) *One has for each $r > 0$*

$$\mathrm{Hom}_B(St'_r \otimes \mu, St'_{r+1} \otimes \mu) = \mathrm{Hom}_{B_{r+1}T}(St'_r \otimes \mu, St'_{r+1} \otimes \mu) \simeq k.$$

Each non-zero homomorphism $\gamma_r : St'_r \otimes \mu \to St'_{r+1} \otimes \mu$ of B–modules is injective.

Proof: a) As St_r is the injective hull of the B_rT–module $-(p^r-1)\rho$ by 9.5.a, the tensor product $St_r \otimes ((p^r-1)\rho + \mu)$ is the injective hull of μ.

b) By a) and by standard properties of injective hulls, $\mathrm{Hom}_{B_{r+1}T}(St'_r \otimes \mu, St'_{r+1} \otimes \mu)$ has dimension equal to the multiplicity of μ as a $B_{r+1}T$–composition factor of $St'_r \otimes \mu$, hence equal to 1 as μ is the lowest weight of $St'_r \otimes \mu$ occurring with multiplicity 1 as a weight.

Steinberg's tensor product theorem says $St_{r+1} \simeq St_r \otimes St_1^{[r]}$, hence $St'_{r+1} \otimes \mu \simeq (St'_r \otimes \mu) \otimes {St'_1}^{[r]}$. By (3) there is an embedding of k into ${St'_1}^{[r]}$ as a B–module, hence an injective homomorphism $St'_r \otimes \mu \to St'_{r+1} \otimes \mu$ of B–modules. It is also a homomorphism of $B_{r+1}T$–modules and therefore it is unique up to a multiple. The claim follows.

10.12. Choose γ_r as in 10.11.b and take the direct limit below with respect to the γ_r.

Proposition: *Let $\mu \in X(T)$. Then $\varinjlim(St'_r \otimes \mu)$ is an injective hull of μ as a B–module.*

Proof: We know that the injective hull of μ is isomorphic to $k[U] \otimes \mu$, cf. 4.8(1). Because of $\mathrm{soc}_B(St'_r \otimes \mu) = \mu$, we can embed $St'_r \otimes \mu$ into $k[U] \otimes \mu$ as a B–module. This embedding is unique (up to a scalar) by the same argument as in 10.11.b. So we get an ascending chain

(1) $$St'_1 \otimes \mu \subset St'_2 \otimes \mu \subset St'_3 \otimes \mu \subset \cdots \subset k[U] \otimes \mu.$$

We have to show that

(2) $$k[U] \otimes \mu = \bigcup_{r>0} St'_r \otimes \mu.$$

This will certainly follow if we can prove for all $\lambda \in X(T)$ that there is $r \in \mathbf{N}$ with

(3) $$\dim k[U]_\lambda = \dim(St'_r)_\lambda = \dim k[U_r]_\lambda.$$

(Compare 9.1(4) for the last equality.)

We have described $k[U]$ as a T–module in 4.8: It is isomorphic to a polynomial ring in generators y_α ($\alpha \in R^+$) with y_α of weight α. So $\dim k[U]_\lambda$ is equal to the number of R^+–tuples $(n_\alpha)_\alpha$ of non-negative integers with $\lambda = \sum_\alpha n_\alpha \alpha$. (This number is usually called Kostant's partition function of λ.) On the other hand, $\dim k[U_r]_\lambda$ is equal to the number of such tuples with $n_\alpha < p^r$ for all α, cf. 9.2(3). So we get equality for large r.

Remarks: We have by Steinberg's tensor product theorem $St_r \otimes (St_m)^{[r]} \simeq St_{r+m}$. This implies $St'_r \otimes (St'_m)^{[r]} \simeq St'_{r+m}$. Now obviously

$$k[U] \simeq \varinjlim St'_m \simeq St'_r \otimes (\varinjlim St'_m)^{[r]}$$

hence

(4) $$k[U] \simeq St'_r \otimes k[U]^{[r]}.$$

Note that we can use $St'_r \otimes (St'_1)^{[r]} \simeq St'_{r+1}$ together with the embedding of k into St'_1 as a B–submodule in order to get the embeddings $St'_r \to St'_{r+1}$ (and then also $St'_r \otimes \mu \to St'_{r+1} \otimes \mu$ from 10.11.b).

10.13. Let us write $k[G]_\mu$ for the μ–weight space of $k[G]$ with respect to the *right regular* representation ρ_r.

Corollary: *One has for all* $\lambda \in X(T)$

(1) $$k[G]_\lambda \simeq \varinjlim St_r \otimes H^0((p^r-1)\rho - \lambda)$$

Proof: Obviously

$$k[G]_\lambda \simeq \operatorname{ind}_T^G(-\lambda) \simeq \operatorname{ind}_B^G(\operatorname{ind}_T^B(-\lambda)).$$

On the other hand, $\operatorname{ind}_T^B(-\lambda)$ is by I.3.11(1) the injective hull of $-\lambda$ as a B–module. So 10.12 implies

$$k[G]_\lambda \simeq \operatorname{ind}_B^G \varinjlim (St_r \otimes ((p^r-1)\rho - \lambda))$$

hence (1), cf. I.3.3.b.

Remark: The special case $\lambda = 0$ yields

(2) $$k[G]_0 = k[G]^T \simeq \varinjlim St_r \otimes St_r.$$

10.14. Proposition: *Let* $H \in \{U, B\}$ *and let* M_1, M_2 *be* H*–modules. The natural maps (induced by the Frobenius endomorphism)*

(1) $$\operatorname{Ext}^i_H(M_1, M_2) \longrightarrow \operatorname{Ext}^i_H(M_1^{[r]}, M_2^{[r]})$$

are injective for all $i \in \mathbf{N}$. There is an exact sequence

$$0 \longrightarrow \mathrm{Ext}^1_H(M_1, M_2) \longrightarrow \mathrm{Ext}^1_H(M_1^{[r]}, M_2^{[r]}) \longrightarrow \mathrm{Hom}_H(M_1, H^1(H_r, k)^{[-r]} \otimes M_2) \longrightarrow 0. \tag{2}$$

Proof: Because of 10.11(3) we can apply I.6.10 to the pair (H, H_r) and to $E = k$ using $Q = St'_r$. We thus get (1) and (2) using the identification $H/H_r \xrightarrow{\sim} H$ via F^r. The map in (1) is the base map of the Hochschild-Serre spectral sequence for the normal subgroup H_r of H. If we realise Ext-groups as equivalence classes of finite exact sequences, then the base maps arise by regarding H/H_r–modules as H–modules via the canonical map $H \to H/H_r$. Having identified H/H_r with H this means that we twist the H–modules with F^r. So the map in (1) is given by the Frobenius endomorphism.

Remarks: 1) Taking $M_1 = k$ we get for any H–module M an exact sequence

$$0 \to H^1(H, M) \longrightarrow H^1(H, M^{[r]}) \longrightarrow (H^1(H_r, k)^{[-r]} \otimes M)^H \to 0. \tag{3}$$

Furthermore, all maps (induced by the Frobenius endomorphism)

$$H^i(H, M) \longrightarrow H^i(H, M^{[r]}) \tag{4}$$

are injective. (If we use the Hochschild complex $C^\bullet(H, M)$ to compute the cohomology, then the map in (4) is induced by $(F^r)^* : k[H] \to k[H]$.)

2) We can apply I.6.10 also to any pair (H_{r+s}, H_r) with $r, s > 0$ taking E, Q as before. We can identify $H_{r+s}/H_r \simeq H_s$ via F^r. Let us denote by $M^{[r]}$ any H_s–module M regarded as an H_{r+s}–module via $F^r : H_{r+s} \to H_s$. Now I.6.10 implies, for example, that the Frobenius endomorphism induces injective maps

$$H^i(H_s, M) \longrightarrow H^i(H_{r+s}, M^{[r]}). \tag{5}$$

3) Applying 4.7.c to $P = B$, we see that the Frobenius endomorphism induces for any G–module M injective maps

$$H^i(G, M) \longrightarrow H^i(G, M^{[r]}). \tag{6}$$

We shall give a direct proof below (10.16). For $\dim M < \infty$ one can show that the sequence of injective maps

$$H^i(G, M) \longrightarrow H^i(G, M^{[1]}) \longrightarrow H^i(G, M^{[2]}) \longrightarrow H^i(G, M^{[3]}) \longrightarrow \cdots \tag{7}$$

has to stabilise. The limit is called the *generic cohomology* of M. Consult [Cline, Parshall, Scott, and van der Kallen] for proofs and more precise information.

10.15. Lemma: *Let $\lambda \in X_r(T)$ and $r \in \mathbf{N}$, $r > 0$. Let V be one of the modules $L((p^r-1)\rho + w_0\lambda)$, $V((p^r-1)\rho + w_0\lambda)$, $H^0((p^r-1)\rho + w_0\lambda)$. Then*

$$\mathrm{Hom}_G(L(\lambda), St_r \otimes V) = \mathrm{Hom}_{G_r}(L(\lambda), St_r \otimes V) \simeq k.$$

Proof: We have for any $H \in \{G, G_r\}$

$$\mathrm{Hom}_H(L(\lambda), St_r \otimes V) \simeq \mathrm{Hom}_H(L(\lambda) \otimes V^*, St_r) \simeq \mathrm{Hom}_H(e_r(L(\lambda) \otimes V^*), St_r).$$

Now V^* is one of the modules

$$L((p^r-1)\rho - \lambda), V((p^r-1)\rho - \lambda), H^0((p^r-1)\rho - \lambda)$$

cf. 2.5, 2.13. So $(p^r-1)\rho$ is the highest weight of $L(\lambda) \otimes V^*$; it occurs with multiplicity 1. Therefore St_r is a composition factor of $L(\lambda) \otimes V^*$ (as a G–module) with multiplicity 1. All other composition factors L of $L(\lambda) \otimes V^*$ have a highest weight strictly less that $(p^r-1)\rho$, hence satisfy $e_r L = 0$. So $e_r(L(\lambda) \otimes V^*) \simeq St_r$ and the claim follows from 3.10(3).

10.16. Proposition: *Let $\lambda \in X_r(T)$ and $r \in \mathbf{N}$, $r > 0$. For all G–modules M_1, M_2 the natural maps (induced by the Frobenius endomorphism)*

(1) $$\operatorname{Ext}^i_G(M_1, M_2) \longrightarrow \operatorname{Ext}^i_G(L(\lambda) \otimes M_1^{[r]}, L(\lambda) \otimes M_2^{[r]})$$

are injective for all $i \in \mathbf{N}$. There is an exact sequence

(2) $$\begin{aligned} 0 \longrightarrow \operatorname{Ext}^1_G(M_1, M_2) &\longrightarrow \operatorname{Ext}^1_G(L(\lambda) \otimes M_1^{[r]}, L(\lambda) \otimes M_2^{[r]}) \\ &\longrightarrow \operatorname{Hom}_G(M_1, \operatorname{Ext}^1_{G_r}(L(\lambda), L(\lambda))^{[-r]} \otimes M_2) \longrightarrow 0. \end{aligned}$$

Proof: Because of 10.15 we can apply I.6.10 for the pair (G, G_r) to $E = L(\lambda)$ and $Q = St_r \otimes L((p^r - 1)\rho + w_0\lambda)$. So we get the claim arguing as in 10.14.

Remark: As before we get for any G–module M an exact sequence

(3) $$0 \to H^1(G, M) \longrightarrow H^1(G, M^{[r]}) \longrightarrow (H^1(G_r, k)^{[-r]} \otimes M)^G \to 0.$$

10.17. For all $\lambda \in X_r(T)$ and $\mu, \mu' \in X(T)_+$ there is by 10.16(2) an exact sequence

(1) $$\begin{aligned} 0 \longrightarrow \operatorname{Ext}^1_G(L(\mu), L(\mu')) &\longrightarrow \operatorname{Ext}^1_G(L(\lambda + p^r\mu), L(\lambda + p^r\mu')) \\ &\longrightarrow \operatorname{Hom}_G(L(\mu), \operatorname{Ext}^1_{G_r}(L(\lambda), L(\lambda))^{[-r]} \otimes L(\mu')) \longrightarrow 0. \end{aligned}$$

This implies:

(2) $\operatorname{Ext}^1_{G_r}(L(\lambda), L(\lambda)) = 0 \Rightarrow \operatorname{Ext}^1_G(L(\mu), L(\mu')) \simeq \operatorname{Ext}^1_G(L(\lambda + p^r\mu), L(\lambda + p^r\mu'))$.

We shall see later on (12.9) that the assumption in (2) is always satisfied if $p \neq 2$, and most of the time if $p = 2$.

If $\lambda' \in X_r(T)$ with $\lambda \neq \lambda'$, then $\operatorname{Hom}_{G_r}(L(\lambda), L(\lambda') \otimes L(\mu')^{[r]}) = 0$. So in the spectral sequence I.6.6(1) computing $\operatorname{Ext}^\bullet_G(L(\lambda) \otimes L(\mu)^{[r]}, L(\lambda') \otimes L(\mu')^{[r]})$ all $E_2^{i,0}$–terms are zero. Therefore the five term exact sequence yields an isomorphism

(3) $\operatorname{Ext}^1_G(L(\lambda + p^r\mu), L(\lambda' + p^r\mu')) \simeq \operatorname{Hom}_G(L(\mu), \operatorname{Ext}^1_{G_r}(L(\lambda), L(\lambda'))^{[-r]} \otimes L(\mu'))$.

One should regard these results as part of a programme to compute inductively Ext^1_G for two simple modules (assuming $\operatorname{Ext}^1_{G_1}$ to be known). See [Donkin 7] or [Andersen 10] for more details. For some explicit calculations of Ext-groups in low rank cases, see [Cline 1], [Ye 3, 4], [Liu and Ye], and [Dowd and Sin]. The case $G = SL_3$ was treated by S. El-B. Yehia in a Ph. D. thesis from 1982 at the University of Warwick.

10.18. If H is a subgroup scheme of G with containing G_r, then we can apply I.6.10 to (H, G_r) and to the same E and Q as in 10.16. For example, we can take $H = G_rT$ or $H = G_rB$ or $H = G_{r+s}$ with $s > 0$.

Let me mention explicitly results for G_rB that are analogous to 10.17(1), (3). Let $\mu, \mu' \in X(T)$ and $\lambda, \lambda' \in X_r(T)$ with $\lambda \neq \lambda'$. We get an exact sequence

(1) $$\begin{aligned} 0 \longrightarrow \operatorname{Ext}^1_B(\mu, \mu') &\longrightarrow \operatorname{Ext}^1_{G_rB}(\widehat{L}_r(\lambda + p^r\mu), \widehat{L}_r(\lambda + p^r\mu')) \\ &\longrightarrow (\operatorname{Ext}^1_{G_r}(L(\lambda), L(\lambda))^{[-r]} \otimes (\mu' - \mu))^B \longrightarrow 0 \end{aligned}$$

and an isomorphism

(2) $\operatorname{Ext}^1_{G_rB}(\widehat{L}_r(\lambda + p^r\mu), \widehat{L}_r(\lambda' + p^r\mu')) \simeq (\operatorname{Ext}^1_{G_r}(L(\lambda), L(\lambda'))^{[-r]} \otimes (\mu' - \mu))^B$.

One may compare these results to 9.21.

If we take $H = G_{r+s}$ with $s > 0$, then we get as in 10.14(5) injective maps (induced by the Frobenius endomorphism) for each G_s–module M

(3) $$H^i(G_s, M) \longrightarrow H^i(G_{r+s}, M^{[r]}).$$

CHAPTER 11

Injective G_r–Modules

Let p, k, r be as in Chapter 9.

Assume for the moment that G is semi-simple and simply connected. We know since Chapter 3 that simple G_r–modules lift to G. In [Jeyakumar 1] it was shown that this holds also for the injective hulls of these modules in case $G = SL_2$ and $r = 1$. It was then conjectured in [Humphreys and Verma] and proved for large p in [Ballard 3] that this statement is true for any G. For small p the conjecture is still open except for G of type A_2 or for some "special" simple modules. For large p two such G–structures on these injective hulls are equivalent.

The original motive for trying to construct a G–structure seems to have been that these G–modules (in those cases where they are known to exist) are also injective for the finite group $G(p^r)$ of points of G over the field of p^r–elements. In fact, they are in those cases quite close to the principal indecomposable $kG(p^r)$–modules, cf. [Humphreys 7], [Chastkofsky 2], and [Jantzen 8].

There are, however, by now applications also to the representation theory of G. Take $\lambda \in X_r(T)$ and suppose that the injective hull $Q_r(\lambda)$ of $L(\lambda) \simeq L_r(\lambda)$ as a G_r–module lifts to G. Then $\operatorname{soc}_{G_r} Q_r(\lambda) \simeq L(\lambda)$ implies $\operatorname{soc}_G Q_r(\lambda) \simeq L(\lambda)$ by 3.16(2), hence $\operatorname{soc}_G M \simeq L(\lambda)$ for any non-zero G–submodule $M \subset Q_r(\lambda)$. One can show that certain G–modules of the form $H^i(\mu)$ can be embedded into certain $Q_r(\lambda)$ and get in this way information about the submodule structure of these $H^i(\mu)$.

If all $Q_r(\lambda)$ with $r > 0$ and $\lambda \in X_r(T)$ lift to G, then one can describe the injective indecomposable G–modules as direct limits of such lifted G_r–modules.

The main results in this chapter go back to [Humphreys 7], [Ballard 3], [Jantzen 6, 7], [Donkin 2, 6], and [Cline, Parshall, and Scott 10].

Compared to the first edition Part b) of Lemma 11.1 and 11.1(2) were moved to 9.9. There are a few added remarks and some revisions.

11.1. Lemma: *The G_rT–module $\widehat{Z}_r(\lambda) \otimes \widehat{Z}'_r(\mu)$ is injective for all $\lambda, \mu \in X(T)$.*

Proof: Considered as a B_rT–module, $\widehat{Z}_r(\lambda)$ is injective (9.5), hence so are $\widehat{Z}_r(\lambda) \otimes \mu$ and (as a G_rT–module)

$$\operatorname{ind}_{B_rT}^{G_rT}(\widehat{Z}_r(\lambda) \otimes \mu) \simeq \widehat{Z}_r(\lambda) \otimes \operatorname{ind}_{B_rT}^{G_rT} \mu \simeq \widehat{Z}_r(\lambda) \otimes \widehat{Z}'_r(\mu)$$

using the tensor identity. (Cf. I.3.10.c and I.3.9.a.)

Remarks: 1) The same proof as above (working with 3.8 instead of 9.5) shows also

(1) $\qquad Z_r(\lambda) \otimes Z'_r(\mu)$ *is an injective G_r–module.*

2) One can get the lemma also as follows: We have as B_r^+T–modules $\operatorname{ind}_T^{B_r^+T}\mu \simeq \widehat{Z}'_r(\mu)$ for all $\mu \in X(T)$, cf. 9.5.b and I.3.11(1). Each $\lambda \in X(T)$ extends to B_r^+T, so the tensor identity implies $\operatorname{ind}_T^{B_r^+T}(\lambda+\mu) \simeq \lambda \otimes \widehat{Z}'_r(\mu)$, hence (using the tensor identity again and transitivity of induction)

$$\text{(2)} \qquad \operatorname{ind}_T^{G_rT}(\lambda+\mu) \simeq \widehat{Z}'_r(\mu) \otimes \operatorname{ind}_{B_r^+T}^{G_rT}(\lambda) \simeq \widehat{Z}'_r(\mu) \otimes \widehat{Z}_r(\lambda + 2(p^r-1)\rho).$$

Now use that any T–module is injective, hence so is any $\operatorname{ind}_T^{G_rT}\nu$.

11.2. Proposition: *Let M be a finite dimensional G_rT–module.*

a) *M is injective as a B_r–module if and only if M has a filtration (as a G_rT–module) with factors of the form $\widehat{Z}_r(\lambda)$, $\lambda \in X(T)$.*

b) *Let $\lambda \in X(T)$. Suppose M has a filtration as in* a). *Then the number of factors isomorphic to $\widehat{Z}_r(\lambda)$ is equal to*

$$\dim \operatorname{Hom}_{G_rT}(\widehat{Z}'_r(\lambda), M) = \dim \operatorname{Hom}_{G_rT}(M, \widehat{Z}'_r(\lambda)).$$

Proof: As any $\widehat{Z}_r(\lambda)$ is injective as a B_r–module (see 3.8), any filtration as in a) splits for B_r, and M is a direct sum of injective B_r–modules, hence itself injective. This is one direction in a). Furthermore b) is obvious from 9.9.

Suppose now that M is injective (and projective) for B_r, hence also for B_rT, cf. 9.4. Assume $M \neq 0$. Choose $\lambda \in X(T)$ maximal among the weights of M and take $m \in M_\lambda$, $m \neq 0$. Then there are homomorphisms $\varphi : \lambda \to M$ of B_r^+T–modules and $\psi : M \to \lambda$ of B_rT–modules with $m \in \operatorname{im}(\varphi)$ and $\psi(m) \neq 0$. By the universal property of $\widehat{Z}_r(\lambda) \simeq \operatorname{coind}_{B_r^+T}^{G_rT}\lambda$ there is a homomorphism $\varphi' : \widehat{Z}_r(\lambda) \to M$ of G_rT–modules with $m \in \operatorname{im}(\varphi')$. As M is projective for B_rT, there exists a homomorphism $\psi' : M \to \widehat{Z}_r(\lambda)$ of B_rT–modules such that ψ is the composition of ψ with the obvious map $\widehat{Z}_r(\lambda) \to \lambda$. Then $\psi \circ \varphi' : \widehat{Z}_r(\lambda) \to \lambda$ is surjective and factors over $\psi' \circ \varphi' \in \operatorname{End}_{B_rT} \widehat{Z}_r(\lambda)$. As $\widehat{Z}_r(\lambda)$ is the projective cover of λ, the map $\psi' \circ \varphi'$ has to be surjective, hence bijective. So φ' is injective and $\varphi'(\widehat{Z}_r(\lambda))$ is a G_rT–submodule of M isomorphic to $\widehat{Z}_r(\lambda)$. This submodule is injective for B_rT, hence a direct summand. Therefore the G_rT–module $M/\varphi'(\widehat{Z}_r(\lambda))$ is also injective for B_rT. Now apply induction on $\dim M$ to this factor module.

Remarks: 1) The same result holds, when we replace B by B^+ and interchange $\widehat{Z}_r$ and $\widehat{Z}'_r$. More generally, we can take any $w \in W$; if M is injective for $\dot{w}B_r\dot{w}^{-1}$, then it has a filtration with factors of the form $\widehat{Z}_r^w(\lambda)$. One can similarly generalise the next remarks.

2) If M is a G_rB^+–module, then we can can choose the filtration in the proposition as a filtration consisting of G_rB^+–modules.

3) If a G_r–module M admits a filtration with all factors of the form $Z_r(\lambda')$ with $\lambda' \in X(T)$, then the number of factors isomorphic to a given $Z_r(\lambda)$ is equal to $\dim \operatorname{Hom}_{G_r}(Z'_r(\lambda), M) = \dim \operatorname{Hom}_{G_r}(M, Z'_r(\lambda))$.

4) Let M be a G_rT–module injective for B_r. We can write $\operatorname{ch}(M) = \sum_{i=1}^s \operatorname{ch} \widehat{Z}_r(\lambda_i)$ with suitable $\lambda_i \in X(T)$. We may assume that the numbering is chosen such that $\lambda_i > \lambda_j$ implies $i < j$. Then the proof of the proposition shows: We can find a filtration $0 = M_0 \subset M_1 \subset \cdots \subset M_s = M$ with $M_i/M_{i-1} \simeq \widehat{Z}_r(\lambda_i)$ for all i.

11.3. Let us denote by $\widehat{Q}_r(\lambda)$ the injective hull of $\widehat{L}_r(\lambda)$ as a G_rT–module, and by $Q_r(\lambda)$ the injective hull of $L_r(\lambda)$ as a G_r–module (for any $\lambda \in X(T)$). One has for all $\lambda, \mu \in X(T)$

(1) $$Q_r(\lambda) \simeq Q_r(\lambda + p^r\mu)$$

(trivially) and by 9.6(6) and I.3.10.c

(2) $$\widehat{Q}_r(\lambda) \otimes p^r\mu \simeq \widehat{Q}_r(\lambda + p^r\mu).$$

By 9.4 each $\widehat{Q}_r(\lambda)$ is injective also as a G_r–module. We have for any $\mu \in X(T)$

$$\begin{aligned}\operatorname{Hom}_{G_r}(L_r(\mu), \widehat{Q}_r(\lambda)) &\simeq \bigoplus_{\nu \in X(T)} \operatorname{Hom}_{G_rT}(\widehat{L}_r(\mu) \otimes p^r\nu, \widehat{Q}_r(\lambda))\\ &\simeq \bigoplus_{\nu \in X(T)} \operatorname{Hom}_{G_rT}(\widehat{L}_r(\mu + p^r\nu), \widehat{Q}_r(\lambda))\\ &\simeq \begin{cases} k, & \text{if } \lambda - \mu \in p^rX(T),\\ 0, & \text{otherwise.}\end{cases}\end{aligned}$$

This implies:

(3) *Regarded as a G_r–module, $\widehat{Q}_r(\lambda)$ is isomorphic to $Q_r(\lambda)$.*

11.4. Proposition: *Let $\lambda \in X(T)$. The G_rT–module $\widehat{Q}_r(\lambda)$ admits filtrations $0 = M_0 \subset M_1 \subset \cdots \subset M_s = \widehat{Q}_r(\lambda)$ and $0 = M'_0 \subset M'_1 \subset \cdots \subset M'_s = \widehat{Q}_r(\lambda)$ such that each factor has the form $M_i/M_{i-1} \simeq \widehat{Z}_r(\lambda_i)$ resp. $M'_i/M'_{i-1} \simeq \widehat{Z}'_r(\lambda'_i)$ with $\lambda_i, \lambda'_i \in X(T)$. For each $\mu \in X(T)$ the number of i $(1 \le i \le s)$ with $\lambda_i = \mu$ resp. with $\lambda'_i = \mu$ is equal to $[\widehat{Z}_r(\mu) : \widehat{L}_r(\lambda)] = [\widehat{Z}'_r(\mu) : L_r(\lambda)]$.*

Proof: It follows from I.5.13 and I.4.12 that $\widehat{Q}_r(\lambda)$ is injective for B_r and B_r^+. Furthermore, 11.3(3) implies that $\dim \widehat{Q}_r(\lambda) < \infty$ as $Q_r(\lambda)$ is a direct summand of the finite dimensional algebra $k[G_r]$. Now 11.2 yields the existence of desired filtrations and implies that the numbers counted at the end are equal to $\dim \operatorname{Hom}_{G_rT}(\widehat{Z}_r(\mu), \widehat{Q}_r(\lambda))$ resp. to $\dim \operatorname{Hom}_{G_rT}(\widehat{Z}'_r(\mu), \widehat{Q}_r(\lambda))$. By standard properties of injective hulls (cf. I.3.17(3)) these dimensions are equal to $[\widehat{Z}_r(\mu) : \widehat{L}_r(\lambda)]$ resp. $[\widehat{Z}'_r(\mu) : \widehat{L}_r(\lambda)]$. These two numbers coincide by Remark 3 in 9.6.

Remarks: 1) The length of the filtrations is clearly $s = \dim \widehat{Q}_r(\lambda)/p^{r|R^+|}$.

2) There is a more direct proof for the fact that $\widehat{Q}_r(\lambda)$ is injective for B_r and B_r^+ and that $\dim \widehat{Q}_r(\lambda) < \infty$: The embeddings $\widehat{L}_r(\lambda) \hookrightarrow \widehat{Z}'_r(\lambda)$ and $k = \widehat{L}_r(0) \hookrightarrow \widehat{Z}_r(0)^*$ yield an inclusion

$$\widehat{L}_r(\lambda) \simeq \widehat{L}_r(\lambda) \otimes k \hookrightarrow \widehat{Z}'_r(\lambda) \otimes \widehat{Z}_r(0)^*.$$

By 11.1.a the right hand side is injective for G_rT, hence contains $\widehat{Q}_r(\lambda)$ as a direct summand. It is injective for B_r and B_r^+ by 3.8, hence so is $\widehat{Q}_r(\lambda)$. Furthermore, we do not only get $\dim \widehat{Q}_r(\lambda) < \infty$, but also $s \le p^{r|R^+|}$.

3) Let M be a finite dimensional G_rT–module. The proof of the proposition or the preceding remark yields one direction of:

$$\text{(1)} \qquad M \text{ injective for } G_rT \iff M \text{ injective for } B_r \text{ and } B_r^+.$$

In order to prove the other direction, suppose M to be injective for B_r and B_r^+. Then M has a filtration with factors of the form $\widehat{Z}_r(\lambda)$ and a filtration with factors of the form $\widehat{Z}'_r(\lambda')$. Therefore $M \otimes M^* \simeq \operatorname{End}(M)$ has a filtration with factors of the form $\widehat{Z}_r(\lambda) \otimes \widehat{Z}'_r(\mu)$, hence is injective as a G_rT–module by 11.1.a. By I.3.10.c also $M \otimes \operatorname{End}(M)$ is injective. Now identify M with the submodule $M \otimes \mathrm{id}_M$ of $M \otimes \operatorname{End}(M)$. This submodule is a direct summand (hence injective) because the map $M \otimes \operatorname{End}(M) \to M$ with $m \otimes \varphi \mapsto \varphi(m)$ splits the inclusion.

4) Given $w \in W$, each $\widehat{Q}_r(\lambda)$ is also injective for $\dot{w}B_r\dot{w}^{-1}$ and has a filtration where all factors have the form $\widehat{Z}_r^w(\mu)$.

11.5. Let $\lambda \in X(T)$. If $[\widehat{Z}_r(\mu) : \widehat{L}_r(\lambda)] \neq 0$, then $\lambda \leq \mu$. Therefore Remark 4 in 11.2 together with 11.4 shows that $\widehat{Z}_r(\lambda)$ occurs at the top of a filtration as in 11.4. So we have an epimorphism

$$\text{(1)} \qquad \widehat{Q}_r(\lambda) \longrightarrow \widehat{Z}_r(\lambda) \longrightarrow 0.$$

Similarly, we have an embedding

$$\text{(2)} \qquad 0 \longrightarrow \widehat{Z}'_r(\lambda) \longrightarrow \widehat{Q}_r(\lambda).$$

As $\widehat{L}_r(\lambda) \simeq \widehat{Z}_r(\lambda)/\operatorname{rad}_{G_rT} \widehat{Z}_r(\lambda)$, we also get an epimorphism

$$\widehat{Q}_r(\lambda) \longrightarrow \widehat{L}_r(\lambda) \longrightarrow 0.$$

We know by 9.4 that $\widehat{Q}_r(\lambda)$ is also a projective G_rT–module. Being indecomposable, it has to be the projective cover of a simple G_rT–module. Therefore;

(3) *$\widehat{Q}_r(\lambda)$ is the projective cover of $\widehat{L}_r(\lambda)$ as a G_rT–module.*

We get thus $\widehat{Q}_r(\lambda)/\operatorname{rad}_{G_rT} \widehat{Q}_r(\lambda) \simeq \widehat{L}_r(\lambda)$, and $\widehat{Z}_r(\lambda)$ is the only module of the form $\widehat{Z}_r(\mu)$ that is a homomorphic image of $\widehat{Q}_r(\lambda)$.

Now 11.3(3) implies:

(4) *$Q_r(\lambda)$ is the projective cover of $L_r(\lambda)$ as a G_r–module.*

This is, however, clear already by I.8.13 and 3.4.a.

The functor $M \mapsto {}^\tau M$ is a self-duality on the category of finite dimensional G_rT–modules. It follows that ${}^\tau\widehat{Q}_r(\lambda)$ is injective and projective in this subcategory, hence (see Remark 9.4) as a G_rT–module. The socle of ${}^\tau\widehat{Q}_r(\lambda)$ is isomorphic to ${}^\tau(\widehat{Q}_r(\lambda)/\operatorname{rad}\widehat{Q}_r(\lambda)) \simeq {}^\tau\widehat{L}_r(\lambda) \simeq \widehat{L}_r(\lambda)$, cf. 9.6(13). This implies that

$$\text{(5)} \qquad {}^\tau\widehat{Q}_r(\lambda) \simeq \widehat{Q}_r(\lambda).$$

11.6. Lemma: a) *Let $\lambda \in X(T)$. Suppose $\lambda = \lambda_0 + p^r\lambda_1$ with $\lambda_0 \in X_r(T)$ and $\lambda_1 \in X(T)$. If μ is a weight of $\widehat{Q}_r(\lambda)$, then*

$$\lambda - 2(p^r-1)\rho \leq \mu \leq w_0\lambda_0 + p^r\lambda_1 + 2(p^r-1)\rho.$$

Both $\lambda - 2(p^r-1)\rho$ and $w_0\lambda_0 + p^r\lambda_1 + 2(p^r-1)\rho$ occur with multiplicity 1 as weights of $\widehat{Q}_r(\lambda)$.

b) *The $\operatorname{ch}\widehat{Q}_r(\lambda)$ with $\lambda \in X(T)$ are linearly independent.*

Proof: a) By 9.2(6) the weights ν of $\widehat{Z}_r(\mu)$ satisfy $\mu - 2(p^r-1)\rho \leq \nu \leq \mu$; both $\mu - 2(p^r-1)\rho$ and μ occur with multiplicity 1 as weights of $\widehat{Z}_r(\mu)$. Therefore the claim in a) will follow from 11.4 if we can show that the μ with $[\widehat{Z}_r(\mu) : \widehat{L}_r(\lambda)] \neq 0$ satisfy $\lambda \leq \mu \leq w_0\lambda_0 + p^r\lambda_1 + 2(p^r-1)\rho$ and that this multiplicity is 1 for the two extreme weights.

The lower bound ($\lambda \leq \mu$) and the equality $[\widehat{Z}_r(\lambda) : \widehat{L}_r(\lambda)] = 1$ are obvious. If $\widehat{L}_r(\lambda)$ is a composition factor of $\widehat{Z}_r(\mu)$, then the lowest weight $w_0\lambda_0 + p^r\lambda_1$ of $\widehat{L}_r(\lambda) \simeq L(\lambda_0) \otimes p^r\lambda_1$ has to be greater than or equal to the lowest weight $\mu - 2(p^r-1)\rho$ of $\widehat{Z}_r(\mu)$, hence $\mu \leq w_0\lambda_0 + p^r\lambda_1 + 2(p^r-1)\rho$. For μ equal to this weight, we have

$$[\widehat{Z}_r(\mu) : \widehat{L}_r(\lambda)] = [\widehat{Z}_r(w_0(\lambda_0 - (p^r-1)\rho) + (p^r-1)\rho) : \widehat{L}_r(\lambda_0)] = 1$$

by 9.16(5). (We could also use the fact that $\widehat{L}_r(\lambda)$ is the only simple G_rT–module with lowest weight $w_0\lambda_0 + p^r\lambda_1$.)

b) This is an immediate consequence of a) using (if necessary) a suitable covering of G where we can decompose each $\lambda = \lambda_0 + p^r\lambda_1$ as in a).

Remark: Take λ as in a) and set $\lambda' = w_0\lambda_0 + p^r\lambda_1 + 2(p^r-1)\rho = w_0\bullet\lambda_0 + p^r(\lambda_1 + 2\rho)$. The lemma and Remark 4 in 11.2 show that there is an injective homomorphism $\iota : \widehat{Z}_r(\lambda') \to \widehat{Q}_r(\lambda)$ of G_rT–modules. The analogue to that remark for B_r^+ instead of B_r implies that there is a surjective homomorphism $\pi : \widehat{Q}_r(\lambda) \to \widehat{Z}'_r(\lambda')$. Both maps induce isomorphisms on the λ'–weight spaces. Therefore the composition $\pi \circ \iota : \widehat{Z}_r(\lambda') \to \widehat{Z}'_r(\lambda')$ is non-zero. Now Remark 2 in 9.6 implies that $\pi \circ \iota$ has image $\widehat{L}_r(\lambda')$.

11.7. Recall from I.2.15 that we can define for each $g \in G(k)$ and each G_r–module M a new G_r–module ${}^gM$. Similarly, we can define for each $g \in N_G(T)(k)$ and each G_rT–module V a new G_rT–module ${}^gV$.

Lemma: *Let $\lambda \in X(T)$ and $g \in G(k)$.*

a) *The G_r–modules $Q_r(\lambda)$ and ${}^gQ_r(\lambda)$ are isomorphic.*

b) *If $\lambda \in X_r(T)$ and $g \in N_G(T)(k)$, then the G_rT–modules $\widehat{Q}_r(\lambda)$ and ${}^g\widehat{Q}_r(\lambda)$ are isomorphic.*

c) *We have $\operatorname{ch}\widehat{Q}_r(\lambda) \in \mathbf{Z}[X(T)]^W$ for each $\lambda \in X_r(T)$.*

Proof: a) The functor $M \mapsto {}^gM$ is an equivalence of categories. It maps the injective hull of $L_r(\lambda)$ to the injective hull of ${}^gL_r(\lambda)$. Now apply Proposition 3.11.

b) Here $\widehat{L}_r(\lambda)$ is just $L(\lambda)$ regarded as a G_rT–module. The action of g yields an isomorphism $L(\lambda) \xrightarrow{\sim} {}^gL(\lambda)$, hence $\widehat{L}_r(\lambda) \xrightarrow{\sim} {}^g\widehat{L}_r(\lambda)$. Now argue as in a).

c) If $g \in N_G(T)(k)$ is a representative for $w \in W \simeq N_G(T)/T$, then $\mathrm{ch}({}^gV) = w\,\mathrm{ch}(V)$ for any finite dimensional G_rT–module V, cf. I.2.15(2). As any $w \in W$ has a representative in $N_G(T)(k)$, this shows that b) implies c).

Remark: One can also prove c) using 11.4 and 9.16(5).

11.8. Lemma: *Let $\lambda \in X(T)$. The following are equivalent:*

(i) $\widehat{Z}_r(\lambda)$ *is simple.*

(ii) $\widehat{Q}_r(\lambda) \simeq \widehat{Z}_r(\lambda)$.

(iii) $\langle \lambda + \rho, \alpha^\vee \rangle \in \mathbf{Z}p^r$ *for all* $\alpha \in S$.

Proof: (i) $\Rightarrow$ (ii): If $\widehat{Z}_r(\lambda) \simeq \widehat{L}_r(\lambda)$, then dimension considerations imply $[\widehat{Z}_r(\mu) : \widehat{L}_r(\lambda)] = 0$ for all $\mu \neq \lambda$. So (ii) follows from 11.4.

(ii) $\Rightarrow$ (i): If $\widehat{Q}_r(\lambda) \simeq \widehat{Z}_r(\lambda)$, then $\widehat{L}_r(\lambda)$ is isomorphic to the socle and the head of $\widehat{Z}_r(\lambda)$. Since $[\widehat{Z}_r(\lambda) : \widehat{L}_r(\lambda)] = 1$, this implies $\widehat{Z}_r(\lambda) \simeq \widehat{L}_r(\lambda)$.

(i) $\Leftrightarrow$ (iii): The homomorphism $\varphi : \widehat{Z}_r(\lambda) \to \widehat{Z}'_r(\lambda)$ from 9.11(4) has image $\widehat{L}_r(\lambda)$. There (i) holds if and only if φ is an isomorphism if and only if all φ_i in 9.11(3) are isomorphisms. By 9.10/11 this holds if and only if $\langle \lambda + \rho, \beta^\vee \rangle \in \mathbf{Z}p^r$ for all $\beta \in R^+$. Since $\beta^\vee \in \sum_{\alpha \in S} \mathbf{Z}\alpha^\vee$, this condition is equivalent to the one in (iii).

Remark: We have $L(\lambda) \simeq \widehat{L}_r(\lambda) \subset \widehat{Z}'_r(\lambda)$ for all $\lambda \in X_r(T)$, hence $\dim L(\lambda) \leq p^{r|R^+|}$. The lemma implies for all $\lambda \in X_r(T)$:

(1)
$$\dim L(\lambda) = p^{r|R^+|} \iff \langle \lambda, \alpha^\vee \rangle = p^r - 1 \text{ for all } \alpha \in S.$$

11.9. Suppose from now on $(p^r - 1)\rho \in X(T)$ and set $St_r = L((p^r - 1)\rho)$ as before. We get from 11.8 another approach to the result proved in 10.2, that St_r is injective as a G_rT–module and as a G_r–module. More precisely, we have (as G_rT–modules)

(1)
$$St_r \simeq \widehat{Q}_r((p^r - 1)\rho).$$

Consider $\lambda \in X_r(T)$ and let V be one of the G–modules $L((p^r - 1)\rho + w_0\lambda)$, $H^0((p^r - 1)\rho + w_0\lambda)$, $V((p^r - 1)\rho + w_0\lambda)$. We have seen in 10.15 that

(2)
$$\mathrm{Hom}_G(L(\lambda), St_r \otimes V) \simeq \mathrm{Hom}_{G_r}(L(\lambda), St_r \otimes V) \simeq k.$$

As $St_r \otimes V$ is injective as a G_rT–module, this implies:

(3) *In a direct sum decomposition of $St_r \otimes V$ into indecomposable modules for G_rT (resp. for G_r) there is a unique summand Q with $Q \simeq \widehat{Q}_r(\lambda)$ (resp. with $Q \simeq Q_r(\lambda)$). Furthermore, $\mathrm{soc}_{G_r} Q$ is a G–submodule of $St_r \otimes V$ isomorphic to $L(\lambda)$.*

We would very much like to be able to choose Q as a G–submodule. We shall see that this is possible if λ is "large" or if p is big.

Let us observe at once (for any $\lambda \in X_r(T)$): If there is a G–structure on $\widehat{Q}_r(\lambda)$, then

$$\text{soc}_G \widehat{Q}_r(\lambda) \simeq L(\lambda) \tag{4}$$

as $\text{soc}_G \widehat{Q}_r(\lambda)$ is semi-simple for G_r, hence contained in the G_r–socle which already is simple for G_r.

11.10. Let $\lambda, \mu \in \overline{C}_{\mathbf{Z}}$ and $w \in W_p$. If $w \bullet \mu$ is in the upper closure of the facet containing $w \bullet \lambda$, then the same proof as in 7.16 (using 9.22(4) instead of 7.15) shows

$$T_\mu^\lambda \widehat{Q}_r(w \bullet \mu) \simeq \widehat{Q}_r(w \bullet \lambda). \tag{1}$$

We want to apply this to $w \bullet \mu = (p^r - 1)\rho$. This weight belongs to the upper closure of the alcove $w_0 \bullet C + p^r\rho$, the "top alcove for p^r". Let $w_1 \in W_p$ be the element with $w_1 \bullet C = w_0 \bullet C + p^r\rho$. (In case $\rho \in \mathbf{Z}R$ one can take the composition of w_0 with the translation by $p^r\rho$.)

Consider $\lambda' \in X(T)_+ \cap w_1 \bullet \overline{C}$, set $\lambda = w_1^{-1} \bullet \lambda'$ and $\mu = w_1^{-1} \bullet (p^r - 1)\rho$. If $\langle \lambda' + \rho, \alpha^\vee \rangle > p^r \langle \rho, \alpha^\vee \rangle - p$ for all $\alpha \in R^+$, then $(p^r - 1)\rho$ is in the upper closure of the facet containing λ', hence $\widehat{Q}_r(\lambda') \simeq T_\mu^\lambda St_r$. As St_r is a G–module, in this case also $\widehat{Q}_r(\lambda')$ has a G–module structure as T_μ^λ commutes with the forgetful functor from $\{\, G\text{–modules} \,\}$ to $\{\, G_rT\text{–modules} \,\}$.

We claim that $\widehat{Q}_r(\lambda') \simeq T_\mu^\lambda St_r$ holds for *all* $\lambda' \in X(T)_+ \cap w_1 \bullet \overline{C}$. As we cannot apply (1) in general, we need a different proof. There exists $\nu \in X(T)_+$ with $\lambda' = w_0\nu + (p^r - 1)\rho$ since $\lambda' \in X_r(T)$. We have $T_\mu^\lambda St_r = \text{pr}_\lambda(St_r \otimes L(\nu))$. So 7.7/8 yield (using 5.8.b)

$$\begin{aligned} \text{ch}\, T_\mu^\lambda St_r &= \sum_w \chi(w\nu + (p^r - 1)\rho) = \sum_w e(w\nu) \cdot \text{ch}\, St_r \\ &= \sum_w \text{ch}\, \widehat{Z}_r((p^r - 1)\rho + w\nu) \end{aligned} \tag{2}$$

summing over a system of representatives for $W/\text{Stab}_W \nu$. We know by 11.9(3) that $\widehat{Q}_r(\lambda')$ is a direct summand of $St_r \otimes L(\nu)$, hence that $\widehat{Q}_r(\lambda') \subset T_\mu^\lambda St_r$. It is therefore enough to show $\dim \widehat{Q}_r(\lambda') \geq \dim T_\mu^\lambda St_r$.

We have by 11.4

$$\text{ch}\, \widehat{Q}_r(\lambda') = \sum_{\mu' \in X(T)} [\widehat{Z}_r(\mu') : \widehat{L}_r(\lambda')]\, \text{ch}\, \widehat{Z}_r(\mu')$$

and by 9.16(5) for all $w \in W$

$$\begin{aligned} [\widehat{Z}_r(w\nu + (p^r - 1)\rho) : \widehat{L}_r(\lambda')] &= [\widehat{Z}_r(w_0\nu + (p^r - 1)\rho) : \widehat{L}_r(\lambda')] \\ &= [\widehat{Z}_r(\lambda') : \widehat{L}_r(\lambda')] = 1. \end{aligned}$$

Comparing this to (2) we get $\dim \widehat{Q}_r(\lambda') \geq \dim T_\mu^\lambda St_r$, hence $\widehat{Q}_r(\lambda') \simeq T_\mu^\lambda St_r$. So we have proved:

Lemma: *Let $\lambda' \in X(T)_+ \cap (w_0 \bullet \overline{C} + p\rho)$. Then the representation of G_rT on $\widehat{Q}_r(\lambda')$ can be extended to a representation of G on $\widehat{Q}_r(\lambda')$ such that $\widehat{Q}_r(\lambda') = \mathrm{pr}_{\lambda'}(St_r \otimes E)$ for some finite dimensional G–module E. One has*

$$\mathrm{ch}\,\widehat{Q}_r(\lambda') = \sum_w \mathrm{ch}\,\widehat{Z}_r(w(\lambda' - (p^r - 1)\rho) + (p^r - 1)\rho) \tag{3}$$

summing over representatives for $W/\mathrm{Stab}_W(\lambda' - (p^r - 1)\rho)$.

Remark: Comparing with 11.4 we see that $[\widehat{Z}_r(w(\lambda' - (p^r - 1)\rho) + (p^r - 1)\rho) : \widehat{L}_r(\lambda')] = 1$ for all $w \in W$ and we get for all $\mu' \notin W(\lambda' - (p^r - 1)\rho) + (p^r - 1)\rho$ that $[\widehat{Z}_r(\mu') : \widehat{L}_r(\lambda')] = 0$. (This could also have been deduced from Lemma 9.16.) Replacing λ' by $\lambda' - p^r\rho$, we get (using 9.2(4) and 9.6(6) and working, if necessary over a covering) the following, nicer formulation of this fact:

(4) *Let $\lambda' \in X(T) \cap w_0 \bullet \overline{C}$ such that $\langle \lambda', \alpha^\vee \rangle \geq -p^r$ for all simple roots α. Then we have $[\widehat{Z}_r(w \bullet \lambda') : \widehat{L}_r(\lambda')] = 1$ for all $w \in W$ and $[\widehat{Z}_r(\mu') : \widehat{L}_r(\lambda')] = 0$ for all $\mu' \notin W \bullet \lambda'$.*

Note that the example $\lambda' = -(p+1)\rho$ for $G = \mathrm{SL}(2)$ and $r = 1$ shows that the condition on $\langle \lambda', \alpha^\vee \rangle$ is needed.

11.11. For large p not only those $\widehat{Q}_r(\lambda')$ mentioned in the last lemma have a G–structure, but all $\widehat{Q}_r(\lambda')$ with $\lambda' \in X_r(T)$. Furthermore two such G–structures are equivalent. I want to give only a sketch of the proof of this result and to refer for details (of this approach) to [Jantzen 4], section 4.

For two G–modules V_1, V_2 the space $\mathrm{Hom}_{G_r}(V_1, V_2)$ is a G–submodule of $\mathrm{Hom}(V_1, V_2)$ on which G_r acts trivially. Therefore there is a G–module M with $\mathrm{Hom}_{G_r}(V_1, V_2) \simeq M^{[r]}$. We have $M = \bigoplus_{\nu \in \overline{C}_{\mathbf{Z}}} \mathrm{pr}_\nu(M)$ by 7.3(1), hence

$$\mathrm{Hom}_{G_r}(V_1, V_2) = \bigoplus_{\nu \in \overline{C}_{\mathbf{Z}}} \mathrm{Hom}^\nu_{G_r}(V_1, V_2) \tag{1}$$

where $\mathrm{Hom}^\nu_{G_r}(V_1, V_2) = (\mathrm{pr}_\nu\, M)^{[r]}$. Let us suppose $p \geq h$ so that $0 \in C$. Obviously

$$\mathrm{Hom}_G(V_1, V_2) = (\mathrm{Hom}_{G_r}(V_1, V_2))^G \subset \mathrm{Hom}^0_{G_r}(V_1, V_2). \tag{2}$$

Consider now some $\lambda \in X_r(T)$ and look at $M = St_r \otimes V$ with V as in 11.9(2), (3). We identify $\widehat{Q}_r(\lambda)$ with a direct summand of M as a G_rT–module, and we identify $L(\lambda)$ with a G–submodule of M; we have then $L(\lambda) = \mathrm{soc}_{G_r} \widehat{Q}_r(\lambda)$. Let Q_λ denote the injective hull of $L(\lambda)$ as a G–module. The inclusion of $L(\lambda)$ into M can be extended to a homomorphism of G–modules $\varphi : M \to Q_\lambda$. Then φ is injective on $\widehat{Q}_r(\lambda)$ as it is so on its socle. So $\varphi(\widehat{Q}_r(\lambda))$ is isomorphic to $\widehat{Q}_r(\lambda)$, hence a direct summand of $\varphi(M)$ as a G_rT–module since it is injective. So, if $\varphi(M)$ is an indecomposable G_rT–module, then $\varphi(M) = \varphi(\widehat{Q}_r(\lambda))$ and $\widehat{Q}_r(\lambda) \simeq \varphi(M)$ has a structure as a G–module.

If we could prove that $\mathrm{soc}_{G_r} \varphi(M) \simeq L(\lambda)$, then we could deduce that $\varphi(M)$ is indecomposable for G_rT and we could conclude as above. Now $\varphi(M) \subset Q_\lambda$

implies $\mathrm{soc}_G\,\varphi(M) \simeq L(\lambda)$. The isotypic components of $\mathrm{soc}_{G_r}\,\varphi(M)$ are by 3.11 G–submodules of $\varphi(M)$. So $\mathrm{soc}_{G_r}\,\varphi(M)$ has to be isotypic of type $L(\lambda)$, hence

$$\mathrm{soc}_{G_r}\,\varphi(M) \simeq \mathrm{Hom}_{G_r}(L(\lambda),\varphi(M)) \otimes L(\lambda) \simeq \bigoplus_{\nu\in \overline{C}_{\mathbf{Z}}} \mathrm{Hom}^{\nu}_{G_r}(L(\lambda),\varphi(M)) \otimes L(\lambda).$$

The indecomposability of $\varphi(M)$ for G implies that at most one summand here is non-zero; as $k \simeq \mathrm{Hom}_G(L(\lambda),\varphi(M)) \subset \mathrm{Hom}^0_{G_r}(L(\lambda),\varphi(M))$, we get

$$\mathrm{soc}_{G_r}\,\varphi(M) \simeq \mathrm{Hom}^0_{G_r}(L(\lambda),\varphi(M)) \otimes L(\lambda).$$

So it is enough to prove $\mathrm{Hom}^0_{G_r}(L(\lambda),\varphi(M)) = \mathrm{Hom}_G(L(\lambda),\varphi(M))$. If not, then there is some $\nu \in X(T)_+ \cap W_p \bullet 0$, $\nu \neq 0$ such that $L(p^r\nu) \simeq L(\nu)^{[r]}$ is a composition factor of $\mathrm{Hom}^0_{G_r}(L(\lambda),\varphi(M))$. Then $\lambda + p^r\nu$ has to be a weight of M, hence $\lambda + p^r\nu \leq w_0\lambda + 2(p^r-1)\rho$. One can check that the smallest possible ν have the form $s_{\alpha,p}\bullet 0$ with $\alpha \in R \cap X(T)_+$ such that α is a short root in its component of R. If $p \geq 2(h-1)$, elementary estimates show that all these ν cannot satisfy an inequality as above. So we get a G–structure on $\widehat{Q}_r(\lambda)$.

Let us say that a G–module N is *p^r–bounded* if each weight μ of N satisfies $\langle \mu, \alpha^\vee\rangle < 2p^r\langle \rho, \alpha^\vee\rangle$ for all (short) $\alpha \in R \cap X(T)_+$. For $p \geq 2(h-1)$ the module M above is p^r–bounded, but $\lambda + p^r\nu$ cannot be the weight of a p^r–bounded module. More or less the same argument also proves (for $p \geq 2(h-1)$) that $\mathrm{Hom}^0_{G_r}(N,\widehat{Q}_r(\lambda)) = \mathrm{Hom}_G(N,\widehat{Q}_r(\lambda))$ for any p^r–bounded G–module N (where we now regard $\widehat{Q}_r(\lambda)$ as a G–module). This implies that $\mathrm{Hom}_G(?,\widehat{Q}_r(\lambda))$ is exact on the full subcategory of all p^r–bounded G–modules. Therefore $\widehat{Q}_r(\lambda)$ can be characterised (for $p \geq 2(h-1)$) as the injective hull of $L(\lambda)$ in this subcategory. This implies the uniqueness of the G–structure on $\widehat{Q}_r(\lambda)$ up to equivalence.

Unfortunately, this characterisation as an injective hull will not carry over to smaller p, see [Jantzen 7], 4.6.

Remark: The indecomposable injective G_r–modules are the indecomposable direct summands of $k[G_r]$ under the left (or right) regular representation. Now this representation itself can easily be lifted to G (as is pointed out in [Koppinen 4]): We have (cf. I.3.7(2) and I.3.5(2))

$$k[G_r] \simeq \mathrm{ind}_1^{G_r} k \simeq \mathrm{ind}_{U_r}^{G_r} k[U_r].$$

Now $k[U_r]$ is isomorphic to each $\widehat{Z}_r(\lambda)$ as a U_r–module, cf. 9.1(6), hence in particular to St_r. So the tensor identity yields

$$k[G_r] \simeq St_r \otimes \mathrm{ind}_{U_r}^{G_r} k.$$

For any G_r–module M and any submodule M' of M, the injectivity of $St_r \otimes M'$ yields a direct sum decomposition

$$St_r \otimes M \simeq St_r \otimes M' \,\oplus\, St_r \otimes (M/M').$$

It follows that there are simple G_r–modules L_i ($1 \leq i \leq s$ for some s) with

$$k[G_r] \simeq \bigoplus_{i=1}^{s} St_r \otimes L_i.$$

As each simple G_r–module lifts to some covering of G, so does $k[G_r]$. Unfortunately, this does not help us to lift the indecomposable summands. Furthermore, there may exist non-equivalent liftings of $k[G_r]$ to G, cf. [Koppinen 4], section 8.

11.12. For any group scheme H and any H–module M set $\mathrm{hd}_H M = M/\operatorname{rad}_H M$, the *head* of M. If M is finite dimensional, then $\mathrm{hd}_H M$ is the largest semi-simple homomorphic image of M. Any surjection $M \to M'$ of finite dimensional H–modules leads to a split epimorphism $\mathrm{hd}_H M \to \mathrm{hd}_H M'$.

As any simple G–module is semi-simple for G_r, we have for each G–module M a surjection $\mathrm{hd}_{G_r} M \to \mathrm{hd}_G M$. Note that 9.6(3), (4) describe $\mathrm{hd}_{G_r} \widehat{Z}'_r(\lambda)$ and $\mathrm{hd}_{G_r} \widehat{Z}_r(\lambda)$. By definition the projective covers $\widehat{Q}_r(\lambda)$ satisfy $\mathrm{hd}_{G_r} \widehat{Q}_r(\lambda) \simeq \widehat{L}_r(\lambda)$. We shall use this fact to determine $\mathrm{hd}_G H^0(\mu)$ in certain cases.

Return to the situation from 11.10. Consider $\lambda \in X(T)_+$ with $\lambda + (p^r - 1)\rho \in p^r\rho + \overline{C}$. Then $w_0\lambda + (p^r - 1)\rho \in p^r\rho + w_0 \bullet \overline{C}$ and $\widehat{Q}_r(w_0\lambda + (p^r - 1)\rho)$ is of the form $T_{\mu'}^{\lambda'} St_r$ for suitable λ', μ'. By 7.13 this translated module has a filtration with factors of the form $H^0(w\lambda + (p^r - 1)\rho)$ with $w \in W$, such that $H^0(w_0\lambda + (p^r - 1)\rho)$ occurs at the bottom and $H^0(\lambda + (p^r - 1)\rho)$ at the top of the filtration.

The surjection $\widehat{Q}_r(w_0\lambda+(p^r-1)\rho) \to H^0(\lambda+(p^r-1)\rho)$ leads by the observations above to surjective maps

$$\widehat{L}_r(w_0\lambda + (p^r - 1)\rho) \xrightarrow{\sim} \mathrm{hd}_{G_r} \widehat{Q}_r(w_0\lambda + (p^r - 1)\rho) \longrightarrow \mathrm{hd}_{G_r} H^0(\lambda + (p^r - 1)\rho) \longrightarrow \mathrm{hd}_G H^0(\lambda + (p^r - 1)\rho).$$

The simplicity of $\widehat{L}_r(w_0\lambda + (p^r - 1)\rho)$ implies that these maps are isomorphisms; we get:

$$\text{(1)} \qquad \mathrm{hd}_{G_r} H^0(\lambda + (p^r - 1)\rho) = \mathrm{hd}_G H^0(\lambda + (p^r - 1)\rho) \simeq L(w_0\lambda + (p^r - 1)\rho).$$

11.13. Suppose in this subsection that $p \geq 2h-2$. So we know by 11.11 that each $\widehat{Q}_r(\lambda_0)$ with $\lambda_0 \in X_r(T)$ has a structure as a G–module extending the given representation of G_rT. We may suppose that $\widehat{Q}_r(\lambda_0)$ is a direct summand of $St_r \otimes H^0(\lambda')$ for a suitable λ'. Therefore 4.21 implies that $\widehat{Q}_r(\lambda_0)$ has a good filtration. (Alternatively, one can show that one gets $\widehat{Q}_r(\lambda_0)$ by applying a sequence of translation functors to St_r and then taking a direct summand. In this way one gets a proof of the existence of a good filtration of $\widehat{Q}_r(\lambda_0)$ that does not require 4.21, see [Jantzen 7], 5.6.) This filtration satisfies (for $p \geq 2h-2$) a reciprocity law similar to that in 11.4, cf. [Jantzen 7], 5.9.) Lemma 11.6.a implies that $H^0(w_0\lambda_0 + 2(p^r - 1)\rho)$ occurs as the top factor in this filtration. Now one can extend the argument used for 11.12(1).

More generally, one gets a good filtration for all $H^0(\lambda_1)^{[r]} \otimes \widehat{Q}_r(\lambda_0)$ with $\lambda_1 \in X(T)_+$ and λ_0 as above, using the fact that $H^0(\lambda_1)^{[r]} \otimes St_r \simeq H^0(p^r\lambda_1 + (p^r - 1)\rho)$ by 3.19. The largest weight of $H^0(\lambda_1)^{[r]} \otimes \widehat{Q}_r(\lambda_0)$ is $p^r\lambda_1 + w_0\lambda_0 + 2(p^r - 1)\rho)$. So the corresponding H^0 occurs as the top factor in the good filtration.

Lemma: *Suppose that $p \geq 2h - 2$. Let $\lambda_0 \in X_r(T)$ and $\lambda_1 \in X(T)_+$. Then*

$$\mathrm{hd}_G H^0(p^r\lambda_1 + \lambda_0 + (p^r - 1)\rho) \simeq (\mathrm{hd}_G H^0(\lambda_1))^{[r]} \otimes L(w_0\lambda_0 + (p^r - 1)\rho).$$

Proof: Applying the consideration above to λ_0 replaced by $w_0\lambda_0 + (p^r - 1)\rho$ we get that there is an epimorphism of G–modules

$$\text{(1)} \qquad H^0(\lambda_1)^{[r]} \otimes \widehat{Q}_r(w_0\lambda_0 + (p^r - 1)\rho) \longrightarrow H^0(p^r\lambda_1 + \lambda_0 + (p^r - 1)\rho).$$

Set $\lambda = p^r\lambda_1 + \lambda_0$ and $\lambda' = w_0\lambda_0 + (p^r-1)\rho$. Now $\operatorname{hd}_{G_r}\widehat{Q}_r(\lambda') \simeq L(\lambda')$ implies, as G_r acts trivially on $H^0(\lambda_1)^{[r]}$

(2) $$\operatorname{hd}_{G_r}\big(H^0(\lambda_1)^{[r]} \otimes \widehat{Q}_r(\lambda')\big) \simeq H^0(\lambda_1)^{[r]} \otimes L(\lambda').$$

This yields

(3) $$\operatorname{hd}_{G_r} H^0(\lambda + (p^r-1)\rho) \simeq M^{[r]} \otimes L(\lambda')$$

where $M = \operatorname{Hom}_{G_r}(H^0(\lambda + (p^r-1)\rho), L(\lambda'))^{*[-r]}$ is isomorphic to a non-zero homomorphic image of

$$\operatorname{Hom}_{G_r}(H^0(\lambda_1)^{[r]} \otimes \widehat{Q}_r(\lambda'), L(\lambda'))^{*[-r]} \simeq H^0(\lambda_1)^{[r]}.$$

We want to show that $M \simeq H^0(\lambda_1)$ so that

(4) $$\operatorname{hd}_{G_r} H^0(\lambda + (p^r-1)\rho) \simeq H^0(\lambda_1)^{[r]} \otimes L(\lambda').$$

Because of $L(\lambda_1) = \operatorname{soc}_G H^0(\lambda_1)$ and $[H^0(\lambda_1) : L(\lambda_1)] = 1$, it suffices to show $[M : L(\lambda_1)] \neq 0$.

We have by 9.6(3) an epimorphism of G_rB–modules

$$\widehat{Z}'_r(\lambda_0 + (p^r-1)\rho) \longrightarrow \widehat{L}_r((p^r-1)\rho - \lambda_0)^*.$$

The right hand side is isomorphic to

$$L((p^r-1)\rho - \lambda_0)^* \simeq L(w_0\lambda_0 + (p^r-1)\rho) = L(\lambda').$$

We get therefore also an epimorphism

$$\varphi : \widehat{Z}'_r(\lambda + (p^r-1)\rho) \longrightarrow L(\lambda') \otimes p^r\lambda_1.$$

Applying $\operatorname{ind}_{G_rB}^G$ yields a homomorphism of G–modules (cf. 9.12, 9.13(3))

$$\operatorname{ind}(\varphi) : H^0(\lambda + (p^r-1)\rho) \longrightarrow L(\lambda') \otimes H^0(\lambda_1)^{[r]}.$$

This map induces a homomorphism $H^0(\lambda_1)^{[r]*} \to \operatorname{Hom}_{G_r}(H^0(\lambda + (p^r-1)\rho), L(\lambda'))$ of G–modules, hence a homomorphism

$$M = \operatorname{Hom}_{G_r}(H^0(\lambda + (p^r-1)\rho), L(\lambda'))^{*[-r]} \longrightarrow H^0(\lambda_1).$$

If $\operatorname{ind}(\varphi) \neq 0$, then also the two subsequent maps are non-zero; it follows that $[M : L(\lambda_1)] \neq 0$, hence (4).

As φ and $\operatorname{ind}(\varphi)$ are compatible with the evaluation maps, and as $\varphi \neq 0$, we get $\operatorname{ind}(\varphi) \neq 0$ and then (4), if we can show:

(5) *The evaluation map $H^0(\lambda + (p^r-1)\rho) \to \widehat{Z}'_r(\lambda + (p^r-1)\rho)$ is surjective.*

Since $\lambda + (p^r-1)\rho$ is the highest weight of $L(\lambda) \otimes St_r$ and since it occurs with multiplicity 1, we get

$$\begin{aligned} k &\simeq \operatorname{Hom}_B(L(\lambda) \otimes St_r, \lambda + (p^r-1)\rho) \simeq \operatorname{Hom}_G(L(\lambda) \otimes St_r, H^0(\lambda + (p^r-1)\rho)) \\ &\simeq \operatorname{Hom}_{G_rB}(L(\lambda) \otimes St_r, \widehat{Z}'_r(\lambda + (p^r-1)\rho)). \end{aligned}$$

This leads to a commutative diagram

$$\begin{array}{ccc} L(\lambda)\otimes St_r & \xrightarrow{\ \psi\ } & H^0(\lambda+(p^r-1)\rho) \\ & \underset{\psi'}{\searrow} & \big\downarrow \varepsilon \\ & & \widehat{Z}'_r(\lambda+(p^r-1)\rho) \end{array}$$

where ε is the evaluation map and where ψ, ψ' are non-zero on the $(\lambda+(p^r-1)\rho)$–weight spaces. Because of $St_r \simeq \widehat{Z}'_r((p^r-1)\rho)$ the tensor product $L(\lambda)\otimes St_r$ is injective as a B_r^+–module, hence has by Remark 2 in 11.2 (applied with B and B^+ switched) a filtration as a G_rB–module with factors of the form $\widehat{Z}'_r(\mu)$. By weight considerations the factor $\widehat{Z}'_r(\lambda+(p^r-1)\rho)$ occurs as the top quotient. Since the Hom space has dimension 1, we get that ψ' is surjective. Hence so is ε; now (5) and (4) follow.

Now (4) implies easily the claim in the lemma since in general $\mathrm{hd}_G V = \mathrm{hd}_G(\mathrm{hd}_{G_r} V)$ for any G–module V.

Remarks: 1) Consider for example R of type A_1. Assume that $\rho \in X(T)$ and suppose that $\lambda \in X(T)$ satisfies $\lambda+\rho = m\rho$ with $m=\sum_{i=0}^n a_i p^i$ with integers a_i such that $0 \le a_i < p$ for all i and $a_n \neq 0$. Then

$$\lambda-(p^n-1)\rho = (a_n-1)p^n\rho + \sum_{i=0}^{n-1} a_i p^i \rho \in X(T)_+.$$

So the lemma implies

$$\begin{aligned} \mathrm{hd}_G H^0(\lambda) &\simeq L((a_n-1)\rho)^{[n]} \otimes L\big((p^n-1)\rho - \sum_{i=0}^{n-1} a_i p^i \rho\big) \\ &\simeq L\big((a_n p^n - \sum_{i=0}^{n-1} a_i p^i - 1)\rho\big). \end{aligned}$$

Using Serre duality this gives another approach to 5.12.b.

2) One has for any finite dimensional module M over a group scheme H that $\mathrm{soc}_H(M^*) \simeq (\mathrm{hd}_H M)^*$. Using this one can translate the lemma and 11.12(1) into statements on socles of Weyl modules, cf. [Jantzen 7], section 6.

11.14. Assume again that $p \ge 2h-2$. Suppose we can decompose each $\lambda \in X(T)$ as $\lambda = \lambda_0 + p^r\lambda_1$ with $\lambda_0 \in X_r(T)$ and $\lambda_1 \in X(T)$. Recall the numbers h_α from 9.18. Fix $w \in W$. Let us say that $\lambda \in X(T)$ is (w,r)*–generic* if each $\mu \in X(T)$ with $[\widehat{Q}_r(\lambda) : \widehat{L}_r(\mu)] \neq 0$ has the form $\mu = p^r w \bullet \mu_1 + \mu_0$ with $\mu_1 \in \overline{C}_{\mathbf{Z}}$ and $\mu_0 \in X_r(T)$. Under these assumptions:

Lemma: a) *If $\lambda_1 \in X(T)$ satisfies $2(h_\beta - 1) \le \langle \lambda_1+\rho, \beta^\vee\rangle \le p-2(h_\beta-1)$ for all $\beta \in R^+$, then $\lambda = p^r w \bullet \lambda_1 + \lambda_0$ is (w,r)–generic for all $\lambda_0 \in X_r(T)$.*

b) *Let $\lambda_1 \in X(T)_+$ and $\lambda_0 \in X_r(T)$ such that $\lambda = p^r w \bullet \lambda_1 + \lambda_0$ is (w,r)–generic. Then*

$$\mathrm{soc}_G\, H^{l(w)}(\lambda) = \mathrm{soc}_{G_r}\, H^{l(w)}(\lambda) \simeq L(\lambda_0 + p^r\lambda_1).$$

Proof: a) Consider the order relation $\le_{\mathbf{Q}}$ on $X(T)$ with $\lambda' \le_{\mathbf{Q}} \mu'$ if and only if $\mu' - \lambda' \in \sum_{\alpha\in S}(\mathbf{Q}_{\ge 0})\alpha$. Then each weight of $\widehat{Q}_r(\lambda_0)$ is $\le_{\mathbf{Q}} 2(p^r-1)\rho$. So any G–composition factor has the form $L(p^r\mu_1+\mu_0)$ with $\mu_0 \in X_r(T)$ and $\mu_1 \in X(T)_+$ and $p^r\mu_1+\mu_0 \le_{\mathbf{Q}} 2(p^r-1)\rho$, and any G_rT–composition factor the form $\widehat{L}_r(p^r w'\mu_1'+\mu_0)$ with $\mu_0\in X_r(T)$, $\mu_1'\in X(T)_+$, $w'\in W$, and $p^r\mu_1'+\mu_0 \le_{\mathbf{Q}} 2(p^r-1)\rho$; this implies $\mu_1' \le_{\mathbf{Q}} 2\rho$. So any G_rT–composition factor of $\widehat{Q}_r(\lambda)\simeq \widehat{Q}_r(\lambda_0)\otimes p^r w\bullet\lambda_1$ has the form $\widehat{L}_r(p^r(w\bullet\lambda_1 + w'\mu_1')+\mu_0)$ with $w'\in W$ and μ_1' as above. Now $\mu_1' \le_{\mathbf{Q}} 2\rho$ yields $|\langle w'\mu_1', \beta^\vee\rangle| \le 2(h_\beta - 1)$ for all $\beta\in R$. So, if λ_1 satisfies the inequalities in the lemma, then clearly $\lambda_1 + w'\mu_1' \in \overline{C}_{\mathbf{Z}}$ in all cases.

b) We have $\widehat{Q}_r(\lambda)\simeq\widehat{Q}_r(\lambda_0)\otimes p^r w\bullet\lambda_1$ as a G_rT–module. We can extend $\widehat{Q}_r(\lambda_0)$ to a G–module and regard $p^r w\bullet\lambda_1$ as a G_rB–module. Thus we get a G_rB–module structure on $\widehat{Q}_r(\lambda)$. We have then by 9.13(3) for all i

$$R^i\operatorname{ind}_{G_rB}^G \widehat{Q}_r(\lambda)\simeq \widehat{Q}_r(\lambda_0)\otimes H^i(w\bullet\lambda_1)^{[r]}.$$

By assumption any $\mu\in X(T)$ with $[\widehat{Q}_r(\lambda):\widehat{L}_r(\mu)]$ has the form $\mu = p^r w\bullet\mu_1+\mu_0$ with $\mu_1\in\overline{C}_{\mathbf{Z}}$ and $\mu_0\in X_r(T)$. Therefore $R^i\operatorname{ind}_{G_rB}^G\widehat{L}_r(\mu)\simeq L(\mu_0)\otimes H^i(w\bullet\mu_1)^{[r]}$ is 0 for $i\ne l(w)$, and isomorphic to $L(\mu_0)\otimes L(\mu_1)^{[r]} = L(\mu_0+p^r\mu_1)$ for $i = l(w)$. [In case $\mu_1\notin X(T)_+$ interpret $L(\mu_1)$ etc. as 0.] The argument from 9.17 shows then that $R^{l(w)}\operatorname{ind}_{G_rB}^G$ is exact on the submodules of $\widehat{Q}_r(\lambda)$. The embedding of $\widehat{Z}_r'(\lambda)$ into $\widehat{Q}_r(\lambda)$ leads therefore to an embedding of $H^{l(w)}(\lambda)\simeq R^{l(w)}\operatorname{ind}_{G_rB}^G\widehat{Z}_r'(\lambda)$ into $R^{l(w)}\operatorname{ind}_{G_rB}^G\widehat{Q}_r(\lambda)\simeq\widehat{Q}_r(\lambda_0)\otimes H^{l(w)}(\lambda_1)^{[r]}$. This implies for the G_r–socles

$$\operatorname{soc}_{G_r} H^{l(w)}(\lambda)\hookrightarrow L(\lambda_0)\otimes H^{l(w)}(w\bullet\lambda_1)^{[r]}\simeq L(\lambda_0+p^r\lambda_1),$$

hence the claim. (Note that $\lambda_1\in\overline{C}_{\mathbf{Z}}$ since $[\widehat{Q}_r(\lambda):\widehat{L}_r(\lambda)]\ne 0$.)

Remark: There is a dual statement for heads that follows from b) using Serre duality.

11.15. Steinberg's tensor product theorem yields an isomorphism of G–modules

$$St_r\otimes St_1^{[r]}\simeq St_{r+1}$$

as $(p^r-1)\rho + p^r(p-1)\rho = (p^{r+1}-1)\rho$, hence an equality of formal characters

$$\operatorname{ch}\widehat{Z}_r((p^r-1)\rho)\ \operatorname{ch}\widehat{Z}_1((p-1)\rho)^{[r]} = \operatorname{ch}\widehat{Z}_{r+1}((p^{r+1}-1)\rho).$$

Using $\operatorname{ch}\widehat{Z}_r(\lambda) = e(\lambda-(p^r-1)\rho)\operatorname{ch}\widehat{Z}_r((p^r-1)\rho)$, cf. 9.2(3), we get for all $\lambda,\lambda'\in X(T)$

$$\operatorname{ch}\widehat{Z}_{r+1}(\lambda+p^r\lambda') = \operatorname{ch}\widehat{Z}_r(\lambda)\ \operatorname{ch}\widehat{Z}_1(\lambda')^{[r]}. \tag{1}$$

(This can also be deduced directly from 9.2(3).)

We have $G_rT = F^{-r}(T)$, hence $F^{r'}(G_rT)\subset G_{r-r'}T$ for any $r'<r$. Therefore we can form for any $G_{r-r'}T$–module M a G_rT–module $M^{[r']}$ by composing the given representation $G_{r-r'}T\to GL(M)$ with $F^{r'}$. In the case where M is a G–module this is compatible with the usual definition. Also regarded as a T–module (via $T\subset G_rT$) we get the usual structure, especially $\operatorname{ch}(M^{[r']}) = \operatorname{ch}(M)^{[r']}$ in case $\dim M<\infty$.

Lemma: *Let $\lambda_0 \in X_r(T)$ and $\lambda_1 \in X(T)$. Then there is an isomorphism of G_rT–modules*

$$\widehat{Q}_{r+1}(\lambda_0 + p^r\lambda_1) \simeq \widehat{Q}_r(\lambda_0) \otimes \widehat{Q}_1(\lambda_1)^{[r]}. \tag{2}$$

Proof: Replacing G by a covering group, we may assume that there exists a system of representatives $X_r'(T)$ for $X(T)/p^rX(T)$ with $X_r'(T) \subset X_r(T)$, cf. 9.13.

Set $\lambda = \lambda_0 + p^r\lambda_1$. As any $\widehat{L}_r(\mu)$ with $\mu \in X_r'(T)$ can be extended to a G–module, we have an isomorphism of $G_{r+1}T$–modules, cf. I.6.15(2),

$$\operatorname{soc}_{G_r} \widehat{Q}_{r+1}(\lambda) \simeq \bigoplus_{\mu \in X_r'(T)} \widehat{L}_r(\mu) \otimes \operatorname{Hom}_{G_r}(\widehat{L}_r(\mu), \widehat{Q}_{r+1}(\lambda)).$$

Any non-zero term on the right hand side makes a non-trivial contribution to $\operatorname{soc}_{G_{r+1}T} \widehat{Q}_{r+1}(\lambda) \simeq \widehat{L}_{r+1}(\lambda) \simeq \widehat{L}_r(\lambda_0) \otimes \widehat{L}_1(\lambda_1)^{[r]}$. So

$$\operatorname{soc}_{G_r} \widehat{Q}_{r+1}(\lambda) \simeq \widehat{L}_r(\lambda_0) \otimes \operatorname{Hom}_{G_r}(\widehat{L}_r(\lambda_0), \widehat{Q}_{r+1}(\lambda)) \tag{3}$$

and

$$\operatorname{soc}_{G_{r+1}T} \operatorname{Hom}_{G_r}(\widehat{L}_r(\lambda_0), \widehat{Q}_{r+1}(\lambda)) \simeq \widehat{L}_1(\lambda_1)^{[r]}. \tag{4}$$

For any $G_{r+1}T/G_r$–module N the spectral sequence (cf. I.6.6)

$$\operatorname{Ext}^i_{G_{r+1}T/G_r}(N, \operatorname{Ext}^j_{G_r}(\widehat{L}_r(\lambda_0), \widehat{Q}_{r+1}(\lambda))) \Rightarrow \operatorname{Ext}^{i+j}_{G_{r+1}T}(N \otimes \widehat{L}_r(\lambda_0), \widehat{Q}_{r+1}(\lambda))$$

degenerates as $\widehat{Q}_{r+1}(\lambda)$ is injective not only for $G_{r+1}T$, but also for G_rT (and G_r) because $G_{r+1}T/G_rT \simeq G_{r+1}/G_r \simeq G_1$ is affine, cf. I.5.13 and I.4.12. Therefore

$$\operatorname{Ext}^i_{G_{r+1}T/G_r}(N, \operatorname{Hom}_{G_r}(\widehat{L}_r(\lambda_0), \widehat{Q}_{r+1}(\lambda))) = 0$$

for all $i > 0$. So the Hom-space is an injective $G_{r+1}T/G_r$–module. Using the identification $G_{r+1}T/G_r \xrightarrow{\sim} G_1T$ via F^r we can rewrite the module in the form $M^{[r]}$ for some G_1T–module M which has to be injective. The description of the socle in (4) implies now

$$\operatorname{Hom}_{G_r}(\widehat{L}_r(\lambda_0), \widehat{Q}_{r+1}(\lambda_0 + p^r\lambda_1)) \simeq \widehat{Q}_1(\lambda_1)^{[r]}. \tag{5}$$

On the other hand, there are $\mu_1, \mu_2, \dots, \mu_m \in X(T)$ such that $\widehat{Q}_{r+1}(\lambda) \simeq \bigoplus_{i=1}^m \widehat{Q}_r(\mu_i)$ as a G_rT–module (being injective). Because of (3) there have to $\nu_i \in X(T)$ with $\mu_i = \lambda_0 + p^r\nu_i$, hence

$$\widehat{Q}_{r+1}(\lambda) \simeq \bigoplus_{i=1}^m \widehat{Q}_r(\lambda_0) \otimes p^r\nu_i.$$

We must have $\sum_{i=1}^m e(\nu_i) = \operatorname{ch} \widehat{Q}_1(\lambda_1)$ by (3). So $\widehat{Q}_1(\lambda_1)^{[r]} \simeq \bigoplus_{i=1}^m (p^r\nu_i)$ as G_rT–modules (with G_r acting trivially), hence (2).

Remark: We get from (2) an equality of formal characters. It yields by induction for all $\lambda_0, \lambda_1, \dots, \lambda_{r-1} \in X_1(T)$ and $\lambda = \sum_{i=0}^{r-1} p^i \lambda_i$:

$$\text{(6)} \qquad \operatorname{ch} \widehat{Q}_r(\lambda) = \operatorname{ch} \widehat{Q}_1(\lambda_0) \operatorname{ch} \widehat{Q}_1(\lambda_1)^{[1]} \operatorname{ch} \widehat{Q}_1(\lambda_2)^{[2]} \dots \operatorname{ch} \widehat{Q}_1(\lambda_{r-1})^{[r-1]}.$$

11.16. Let us denote the injective hull of the G–module $L(\lambda)$ by Q_λ (for any $\lambda \in X(T)_+$).

Proposition: *Let $\lambda \in X_r(T)$. Suppose that the G_rT–module structure on $\widehat{Q}_r(\lambda)$ can be lifted to G.*

a) *There is for each $\lambda' \in X(T)_+$ an isomorphism of G–modules*

$$\widehat{Q}_r(\lambda) \otimes Q_{\lambda'}^{[r]} \simeq Q_{\lambda + p^r \lambda'}.$$

b) *There is for each $\lambda' \in X(T)$ and each $r' > r$ an isomorphism of $G_{r'}T$–modules*

$$\widehat{Q}_r(\lambda) \otimes \widehat{Q}_{r'-r}(\lambda')^{[r]} \simeq \widehat{Q}_{r'}(\lambda + p^r \lambda').$$

Proof: Suppose for the moment that G is a group scheme over some field k with a normal subgroup N. Let V_1 be a G–module which is injective as an N–module, and let V_2 be an injective G/N–module. We claim that $V_1 \otimes V_2$ is an injective G–module: Let $0 \to M' \to M \to M'' \to 0$ be a short exact sequence of G–modules. Then the injectivity of V_1 over N shows that the natural sequence

$$0 \to \operatorname{Hom}_N(M'', V_1) \longrightarrow \operatorname{Hom}_N(M, V_1) \longrightarrow \operatorname{Hom}_N(M', V_1) \to 0$$

is an exact sequence of G/N–modules. So is then

$$0 \to \operatorname{Hom}_N(M'', V_1) \otimes V_2 \longrightarrow \operatorname{Hom}_N(M, V_1) \otimes V_2 \longrightarrow \operatorname{Hom}_N(M', V_1) \otimes V_2 \to 0.$$

As V_2 is injective for G/N, so are all terms in this sequence by I.3.10.c. Therefore the sequence splits and remains exact when we take fixed points under G/N. Using the natural identifications

$$(\operatorname{Hom}_N(M, V_1) \otimes V_2)^{G/N} \simeq (\operatorname{Hom}_N(M, V_1 \otimes V_2))^{G/N} \simeq \operatorname{Hom}_G(M, V_1 \otimes V_2),$$

similarly for M' and M'', we get an exact sequence of vector spaces

$$0 \to \operatorname{Hom}_G(M'', V_1 \otimes V_2) \longrightarrow \operatorname{Hom}_G(M, V_1 \otimes V_2) \longrightarrow \operatorname{Hom}_G(M', V_1 \otimes V_2) \to 0.$$

One can check that we get here the natural maps. Therefore $V_1 \otimes V_2$ is injective for G.

In our case this general consideration shows that the left hand side in a) is injective for G, the one in b) for $G_{r'}T$. (Recall that $G/G_r \simeq G$ and $G_{r'}T/G_r \simeq G_{r'-r}T$, with both isomorphisms induced by F^r.)

It remains to determine the socles of these tensor products. They are contained in the G_r–socles, hence equal to

$$L(\lambda) \otimes \operatorname{soc}_G Q_{\lambda'}^{[r]} \simeq L(\lambda) \otimes (\operatorname{soc}_G Q_{\lambda'})^{[r]} \simeq L(\lambda) \otimes L(\lambda')^{[r]} \simeq L(\lambda + p^r \lambda')$$

resp. to

$$\begin{aligned} L(\lambda) \otimes \operatorname{soc}_{G_{r'}T}(Q_{r'-r}(\lambda')^{[r]}) &\simeq L(\lambda) \otimes (\operatorname{soc}_{G_{r'-r}T} Q_{r'-r}(\lambda'))^{[r]} \\ &\simeq L(\lambda) \otimes \widehat{L}_{r'-r}(\lambda')^{[r]} \simeq \widehat{L}_{r'}(\lambda + p^r \lambda'). \end{aligned}$$

This implies the proposition.

Remarks: 1) In b) it suffices to require that $\widehat{Q}_r(\lambda)$ lifts to $G_{r'}T$.

2) Assume that $X(T) = X_1(T) + pX(T)$. Suppose that each $\widehat{Q}_1(\mu)$ with $\mu \in X_1(T)$ lifts to a G–module. Each $\lambda \in X_r(T)$ can be expanded $\lambda = \sum_{i=0}^{r-1} p^i\lambda_i$ with all $\lambda_i \in X_1(T)$. Then

$$\widehat{Q}_1(\lambda_0) \otimes \widehat{Q}_1(\lambda_1)^{[1]} \otimes \widehat{Q}_1(\lambda_2)^{[2]} \otimes \cdots \otimes \widehat{Q}_1(\lambda_{r-1})^{[r-1]} \tag{2}$$

has a structure as a G–module. The proposition yields by induction that this tensor product is isomorphic to $\widehat{Q}_r(\lambda)$ as a G_rT–module.

This shows: If each $\widehat{Q}_1(\mu)$ with $\mu \in X_1(T)$ lifts to a G–module, then so does each $\widehat{Q}_r(\lambda)$ with $\lambda \in X_r(T)$ for any r. If these G–structures are unique (e.g., for $p \geq 2h-2$), then we get an isomorphism of G–modules between $\widehat{Q}_r(\lambda)$ and the tensor product in (2).

For example, for R of type A_1 we have $X_1(T) \subset \overline{C} = w_0 \bullet \overline{C} + p\rho$, hence get a G–structure on each $\widehat{Q}_1(\lambda)$ with $\lambda \in X_1(T)$ from 11.10. For R of type A_2 each $\lambda \in X_1(T)$, not in $w_0 \bullet \overline{C} + p\rho$, lies in C. So we have $p \geq h$ by 6.2(10) and there is by 6.3(1) some $\mu \in X_1(T)$ lying on the common wall of C and $w_0 \bullet \overline{C} + p\rho$. Then we have a G–structure on $\widehat{Q}_1(\mu)$ by 11.10 and get one on $\widehat{Q}_1(\lambda)$ as $\widehat{Q}_1(\lambda) \simeq T_\mu^\lambda \widehat{Q}_1(\mu)$ by 11.10(1). (Recall that translation functors commute with the forgetful functor from $\{ G\text{–modules} \}$ to $\{ G_1T\text{–modules} \}$, see 9.22.) So in these two cases (types A_1 and A_2) we get G–structures on all $\widehat{Q}_r(\lambda)$ for all p.

11.17. Let $\lambda \in X(T)_+$. There exists r with $\lambda \in X_r(T)$. Suppose that $\widehat{Q}_r(\lambda)$ and $\widehat{Q}_1(0)$ lift to G–modules. Using 11.16 we get for each $s \geq r$ a G–structure on $\widehat{Q}_s(\lambda)$ such that (as G–modules)

$$\widehat{Q}_{s+1}(\lambda) \simeq \widehat{Q}_s(\lambda) \otimes \widehat{Q}_1(0)^{[s]}. \tag{1}$$

The embedding $k = L(0) = \operatorname{soc}_G \widehat{Q}_1(0) \hookrightarrow \widehat{Q}_1(0)$ yields an embedding $\widehat{Q}_s(\lambda) \hookrightarrow \widehat{Q}_{s+1}(\lambda)$ for each s, hence a directed system of injective homomorphisms of G–modules

$$\widehat{Q}_r(\lambda) \hookrightarrow \widehat{Q}_{r+1}(\lambda) \hookrightarrow \widehat{Q}_{r+2}(\lambda) \hookrightarrow \cdots. \tag{2}$$

Proposition: *The direct limit of the $\widehat{Q}_s(\lambda)$ with $s \geq r$ is an injective hull of $L(\lambda)$ as a G–module.*

Proof: Denote this limit by Q. We have $\operatorname{soc}_G \widehat{Q}_s(\lambda) \simeq L(\lambda)$ for all s; this implies that also $\operatorname{soc}_G Q \simeq L(\lambda)$. So it suffices to show that Q is an injective G–module.

Since each G–module is locally finite, we have only to check that $\operatorname{Hom}_G(?, Q)$ is exact on finite dimensional G–modules.

Consider a short exact sequence $0 \to M' \to M \to M'' \to 0$ of finite dimensional G–modules. There exists $s' \geq r$ such that all composition factors of M have a highest weight in $X_{s'}(T)$. It is enough to show that

$$0 \to \operatorname{Hom}_G(M'', \widehat{Q}_s(\lambda)) \to \operatorname{Hom}_G(M, \widehat{Q}_s(\lambda)) \to \operatorname{Hom}_G(M', \widehat{Q}_s(\lambda)) \to 0 \tag{3}$$

is exact for all $s \geq s'$.

As $\widehat{Q}_s(\lambda)$ is injective for G_s, we know that

$$(4) \qquad 0 \to \operatorname{Hom}_{G_s}(M'', \widehat{Q}_s(\lambda)) \to \operatorname{Hom}_{G_s}(M, \widehat{Q}_s(\lambda)) \to \operatorname{Hom}_{G_s}(M', \widehat{Q}_s(\lambda)) \to 0$$

is an exact sequence of G–modules. From this we get (3) by taking G–fixed points as soon as we can show that $\operatorname{Ext}^1_G(k, \operatorname{Hom}_{G_s}(M'', \widehat{Q}_s(\lambda))) = 0$. The weights of $\operatorname{Hom}_{G_s}(M'', \widehat{Q}_s(\lambda))$ are the $p^s\nu$ with $\nu \in X(T)$ such that $\widehat{L}_s(\lambda - p^s\nu)$ is a G_sT–composition factor of M'' (as $\operatorname{Hom}_{G_s}(M'', \widehat{Q}_s(\lambda))_\mu \simeq \operatorname{Hom}_{G_sT}(M'' \otimes \mu, \widehat{Q}_s(\lambda))$). This implies $\langle \nu, \alpha^\vee \rangle = 0$ for all $\alpha \in R$ as all composition factors of M'' have a highest weight in $X_s(T)$. Therefore all G–composition factors of $\operatorname{Hom}_{G_s}(M'', \widehat{Q}_s(\lambda))$ have dimension 1 and do not extend k. So the Ext^1–group vanishes and the proposition follows.

Remark: In case $p \geq 2h-2$ we get the injectivity of Q also from the fact that $\widehat{Q}_s(\lambda)$ is injective in the subcategory of p^s–bounded G–modules (cf. 11.11) and that each finite dimensional G–module is p^s–bounded for large s.

CHAPTER 12

Cohomology of the Frobenius Kernels

Let p, k, r be as in Chapter 9.

It would be nice to compute all $\mathrm{Ext}^j_{G_r}(L(\mu), H^0(\lambda))$ with $\mu \in X_r(T)$ and $\lambda \in X(T)_+$ and then to apply the spectral sequence

$$\mathrm{Ext}^i_G(H^0(\nu)^*, \mathrm{Ext}^j_{G_r}(L(\mu), H^0(\lambda))^{[-r]}) \Rightarrow \mathrm{Ext}^{i+j}_G(H^0(\nu)^{*[r]} \otimes L(\mu), H^0(\lambda)).$$

It would be even nicer if all $\mathrm{Ext}^j_{G_r}(L(\mu), H^0(\lambda))$ had a good filtration (as in 4.16) as then the spectral sequence would degenerate and yield isomorphisms

$$\mathrm{Hom}_G(H^0(\nu)^*, \mathrm{Ext}^j_{G_r}(L(\mu), H^0(\lambda))^{[-r]}) \simeq \mathrm{Ext}^j_G(H^0(\nu)^{*[r]} \otimes L(\mu), H^0(\lambda)).$$

If $r = 1$ and if λ, μ are p–regular, then the dimension on the right hand is expected to be a coefficient of a Kazhdan-Lusztig polynomial, cf. 7.22. So one may hope to get these coefficients also on the left hand side.

The only case where these speculations work so far, is $\mu = 0$ and $r = 1$ under the additional assumption that $p > h$ (insuring 0 to be p–regular), see 12.15. For $\lambda = 0$ we get here $H^1(G_1, k)$; one interesting aspect of this case is that $H^\bullet(G_1, k)$ is isomorphic (for $p > h$) to the algebra of regular functions on the nilpotent cone in $\mathrm{Lie}(G)$.

The other important result in this chapter is that

$$\mathrm{Ext}^1_{G_r}(L(\lambda), H^0(\lambda)) = 0 = \mathrm{Ext}^1_{G_r}(L(\lambda), L(\lambda))$$

for all $\lambda \in X_r(T)$ if $p \neq 2$ (and most of the time also if $p = 2$).

The main results of this chapter are due to [Andersen 10], [Friedlander and Parshall 1, 2, 3], and [Andersen and Jantzen].

We shall use in this chapter the notations $\mathfrak{u} = \mathrm{Lie}\, U$, $\mathfrak{b} = \mathrm{Lie}\, B$, $\mathfrak{g} = \mathrm{Lie}\, G$.

Compared to the first edition a few statements look now differently thanks to the introduction of the notation $\mathcal{M}_r$ in 12.4. In Lemma 12.4 itself the old Part b) was merged into Part a), and the old Part c) is now Part b).

12.1. In I.9.14 we have seen a spectral sequence converging to $H^\bullet(H_r, M)$ for each reduced algebraic group H, each H_r–module M, and each r. The only non-zero $E_1^{i,j}$–terms with $i + j = 1$ were all

(1) $$E_1^{p^s, 1-p^s} = M \otimes ((\mathrm{Lie} H)^*)^{(s)} \qquad (0 \leq s \leq r-1).$$

Lemma: *If $p \neq 2$ or if R does not have a component of type C_n (for each $n \geq 1$), then $H^1(B_r, k) = 0$ for all r.*

Proof: We have $H^i(B_r, k) = H^i(U_r, k)^{T_r}$ for all i by I.6.9(3). We can compute these cohomology groups using the spectral sequence from I.9.14 for $H = U$ and taking fixed points under T_r. The terms contributing to $H^1(B_r, k)$ are

$$(E_1^{p^s, 1-p^s})^{T_r} = (\mathfrak{u}^{*[s]})^{T_r}$$

with $0 \leq s \leq r-1$ where $\mathfrak{u} = \operatorname{Lie} U$. (We have been able to replace the "(s)" from (1) by "$[s]$" because the adjoint representations of G, B, T, U are defined over $\mathbf{F}_p$.) The weights of $\mathfrak{u}^*$ are just the positive roots, so the weights of $(\mathfrak{u}^{*[s]})^{T_r}$ are the $p^s\alpha$ with $\alpha \in R^+$ and $p^s\alpha \in p^r X(T)$. For any such α all $\langle \alpha, \gamma^\vee \rangle$ with $\gamma \in R$ have to be divisible by p^{r-s}, hence by p as $s < r$. However, the classification of root systems shows that given $\alpha \in R$ there exists $\gamma \in R$ with $\langle \alpha, \gamma^\vee \rangle = 1$, except in the case where α belongs to a component of type C_n (for some $n \geq 1$) and is a long root in that component (for $n \geq 2$ only). In this special case the set of all non-zero $\langle \alpha, \gamma^\vee \rangle$ is equal to $\{2, -2\}$. This implies the lemma.

Remark: Suppose $p = 2$. Set V equal to the sum of all $\mathfrak{u}_\alpha$ with $\alpha \in 2X(T)$. The proof above shows that all $(E_1^{i,1-i})^{T_r}$ are zero except for $(E_1^{p^{r-1}, 1-p^{r-1}})^{T_r} \simeq V^{[r-1]}$.

So $H^1(B_r, k)$ has to be a submodule of $V^{[r-1]}$. For $r = 1$ one can compute the differentials in the complex from I.9.14 and gets $H^1(B_1, k) \simeq V$, cf. [Andersen and Jantzen], 6.19. Now we can use the injectivity of the maps in 10.14(5) to get

(2) $$H^1(B_r, k) \simeq V^{[r-1]}$$

for all r.

Note that $V = 0$ for G semi-simple and adjoint, even if R has a component of type C_n for some n.

12.2. We want to apply I.6.12 to $(G, H, N) = (G, B, G_r)$. Then $H \cap N = B_r$ and $G/N = G/G_r \simeq G$ and $H/(H \cap N) = B/B_r \simeq B$ (via F^r). So we get for each B–module M two spectral sequences (with $E_2^{n,m}$–terms $H^n(G_r, R^m \operatorname{ind}_B^G M)$ or $(R^n \operatorname{ind}_B^G (H^m(B_r, M)^{[-r]}))^{[r]}$) converging to the same abutment. If $R^m \operatorname{ind}_B^G M = 0$ for all $m > 0$, then the first spectral sequence degenerates and its abutment is just $H^n(G_r, \operatorname{ind}_B^G M)$:

(1) *If $R^m \operatorname{ind}_B^G M = 0$ for all $m > 0$, then there is a spectral sequence with*
$$E_2^{n,m} = R^n \operatorname{ind}_B^G (H^m(B_r, M)^{[-r]}) \Rightarrow H^{n+m}(G_r, \operatorname{ind}_B^G M)^{[-r]}.$$

So Kempf's vanishing theorem implies for all $\lambda \in X(T)_+$ that there is a spectral sequence

(2) $$E_2^{n,m} = R^n \operatorname{ind}_B^G (H^m(B_r, \lambda)^{[-r]}) \Rightarrow H^{n+m}(G_r, H^0(\lambda))^{[-r]}.$$

If $\lambda \notin p^r X(T)$, then $H^0(B_r, \lambda) = 0$; so each $E_2^{n,0}$–term vanishes. If $\lambda = p^r \mu$ with $\mu \in X(T)$, then $\mu \in X(T)_+$; now Kempf's vanishing theorem applied to $\mu \simeq H^0(B_r, \lambda)^{[-r]}$ shows that all $E_2^{n,0}$ with $n > 0$ vanish. Therefore the five term exact sequence yields (in either case) an isomorphism:

(3) $$H^1(G_r, H^0(\lambda))^{[-r]} \simeq \operatorname{ind}_B^G (H^1(B_r, \lambda)^{[-r]}) \qquad \text{for each } \lambda \in X(T)_+.$$

Taking $\lambda = 0$ shows by 12.1:

Proposition: *If $p \neq 2$ or if R does not have a component of type C_n (for each $n \geq 1$), then $H^1(G_r, k) = 0$ for all r.*

Remark: Suppose $p = 2$ and R of type C_n for some $n \geq 1$. Suppose that G is semi-simple and simply connected. Then $H^1(G_r, k)^{[-r]} \simeq H^0(\varpi_1)$ where ϖ_1 is the first fundamental weight as in [B3], ch. VI, planche III. (So $H^0(\varpi_1)$ is the natural representation of G as a symplectic group.) This follows easily from (3) and the remark in 12.1, cf. the argument in [Andersen and Jantzen], 6.19 for $r = 1$. One can also go back to [Cline, Parshall, and Scott 6], 6.10 for an earlier proof.

12.3. We want to generalise Proposition 12.2 and show that $\operatorname{Ext}^1_{G_r}(L(\lambda), L(\lambda)) = 0$ for all $\lambda \in X_r(T)$ under the same assumptions as in 12.2 (which is the special case $\lambda = 0$). As above, we have to prove at first a result on B_r–cohomology, namely $\operatorname{Ext}^1_{B_r}(L(\lambda), \lambda) = 0$.

Suppose that this is not true. Then the B–module $\operatorname{Ext}^1_{B_r}(L(\lambda), \lambda)^{[-r]}$ has a non-trivial socle, so there is $\nu \in X(T)$ with

(1) $$\operatorname{Hom}_{B/B_r}(p^r\nu, \operatorname{Ext}^1_{B_r}(L(\lambda), \lambda)) \neq 0.$$

This is just the $E_2^{0,1}$–term in the spectral sequence (from I.6.6(1))

(2) $$\operatorname{Ext}^n_{B/B_r}(p^r\nu, \operatorname{Ext}^m_{B_r}(L(\lambda), \lambda)) \Rightarrow \operatorname{Ext}^{n+m}_B(L(\lambda) \otimes p^r\nu, \lambda).$$

We have by 9.6(4) an exact sequence of G_rT–modules (with suitable M)

(3) $$0 \to M \longrightarrow \widehat{Z}_r(\lambda) \longrightarrow L(\lambda) \to 0$$

since $\lambda \in X_r(T)$. Considered as a B_r–module, $\widehat{Z}_r(\lambda)$ is isomorphic to $Z_r(\lambda)$, hence the projective cover of λ (3.8.a). So we have an isomorphism of B–modules

(4) $$\operatorname{Hom}_{B_r}(\widehat{Z}_r(\lambda), \lambda) \simeq k.$$

Therefore we get (using I.4.2(1), I.4.4) that a part of the five term exact sequence corresponding to the spectral sequence (2) has the form

(5) $$\begin{aligned} 0 &\to H^1(B, -\nu) \to \operatorname{Ext}^1_B(L(\lambda), \lambda - p^r\nu) \\ &\to \operatorname{Hom}_B(\nu, \operatorname{Ext}^1_{B_r}(L(\lambda), \lambda)^{[-r]}) \to H^2(B, -\nu). \end{aligned}$$

We want to show that $E_2^{1,0} \simeq E^1$ and that $E_2^{2,0} = 0$. Then also $E_2^{0,1} = 0$ contradicting (1).

12.4. Set $\mathcal{M}_r = \{\sum_{\alpha \in R^+} m_\alpha \alpha \mid m_\alpha \in \mathbf{Z}, 0 \leq m_\alpha < p^r \text{ for all } \alpha\}$. By 9.2(3) the set of weights of $\widehat{Z}_r(\lambda)$ is (for any $\lambda \in X(T)$) equal to $\{\lambda - \mu \mid \mu \in \mathcal{M}_r\}$.

Lemma: *Let $\lambda \in X_r(T)$. If ν is a weight of $\operatorname{Ext}^1_{B_r}(L(\lambda), \lambda)^{[-r]}$, then:*

a) *$p^r\nu > 0$, $p^r\nu \in \mathcal{M}_r$, and $\lambda - p^r\nu \notin X(T)_+$.*

b) *If $p \neq 2$ or if R does not have a component of type C_n (for each $n \geq 1$), then $\dim H^1(B, -\nu) = \dim \operatorname{Ext}^1_B(L(\lambda), \lambda - p^r\nu)$.*

Proof: a) As $\widehat{Z}_r(\lambda)$ is projective for B_r, we get from 12.3(3) an isomorphism of B–modules

$$\operatorname{Ext}^1_{B_r}(L(\lambda), \lambda) \simeq \operatorname{Hom}_{B_r}(M, \lambda).$$

Therefore $\lambda - p^r\nu$ is a weight of M, hence a weight of $\widehat{Z}_r(\lambda)$ different from λ. This implies $\lambda - p^r\nu < \lambda$ (hence $p^r\nu > 0$) and $p^r\nu \in \mathcal{M}_r$. If $\lambda - p^r\nu \in X(T)_+$, then $\lambda \in X_r(T)$ implies $-\nu \in X(T)_+$, hence $-p^r\nu \in X(T)_+$, contradicting $p^r\nu > 0$.

b) We may assume that $\operatorname{Ext}^1_B(L(\lambda), \lambda - p^r\nu) \neq 0$ since otherwise the claim is obvious by 12.3(5). Now 5.19(4) implies by a) that $\operatorname{Ext}^1_B(L(\lambda), \lambda - p^r\nu) \simeq k$ and that there are $\alpha \in S$ and $n \in \mathbf{N}$ with $\lambda - p^r\nu = \lambda - c_\lambda(n,\alpha)p^n\alpha$ and $0 < c_\lambda(n,\alpha) < p$ where $c_\lambda(n,\alpha)$ is determined by

$$(c_\lambda(n,\alpha) - 1)p^n < \langle \lambda + \rho, \alpha^\vee \rangle \leq c_\lambda(n,\alpha)p^n.$$

If $p \neq 2$ or if R has no component of type C, then the same argument as in 12.1 shows that $p^r\nu = c_\lambda(n,\alpha)p^n\alpha$ with $\nu \in X(T)$ implies $n \geq r$. This yields $c_\lambda(n,\alpha) = 1$ since $\lambda \in X_r(T)$. So we get $\nu = p^{n-r}\alpha$, hence $H^1(B, -\nu) \simeq k$ by 5.20. The claim follows.

12.5. Lemma: *Let $\mu \in X(T)$ with $H^2(B,\mu) \neq 0$.*

a) *There are $\alpha, \beta \in S$ and $i, m \in \mathbf{N}$ with $i > 0$ and $i + p^m\langle \beta, \alpha^\vee \rangle > 0$ such that $\mu = -(i\alpha + p^m\beta)$.*

b) *We have $-p^r\mu \notin \mathcal{M}_r$ for all $r \in \mathbf{N}$.*

Proof: a) We have $\operatorname{ht}(-\mu) \geq 2$ by 4.10, hence $\mu \notin X(T)_+$. So we can find $\alpha \in S$ with $\langle \mu, \alpha^\vee \rangle < 0$.

We have by I.4.5.b for each B–module M a spectral sequence

$$H^n(P(\alpha), R^m \operatorname{ind}_B^{P(\alpha)} M) \Rightarrow H^{n+m}(B, M)$$

where $P(\alpha) = P_{\{\alpha\}}$. Furthermore, 4.7(3) implies

$$H^n(P(\alpha), R^m \operatorname{ind}_B^{P(\alpha)} M) \simeq H^n(B, R^m \operatorname{ind}_B^{P(\alpha)} M).$$

If $\langle \mu, \alpha^\vee \rangle = -1$, then $R^\bullet \operatorname{ind}_B^{P(\alpha)} \mu = 0$ by 5.2.b, hence $H^\bullet(B,\mu) = 0$. So we must have $\langle \mu, \alpha^\vee \rangle < -1$. Then the spectral sequence and 5.2.d yield isomorphisms

$$H^{n-1}(B, R^1 \operatorname{ind}_B^{P(\alpha)} \mu) \simeq H^n(B, \mu).$$

The weights of $R^1 \operatorname{ind}_B^{P(\alpha)} \mu$ are the $\mu + j\alpha$ with $0 < j < \langle -\mu, \alpha^\vee \rangle$ (see 5.2.d), so $H^2(B,\mu) \neq 0$ implies that there exists i with $0 < i < \langle -\mu, \alpha^\vee \rangle$ such that $H^1(B, \mu + i\alpha) \neq 0$. By 5.20 there are $m \in \mathbf{N}$ and $\beta \in S$ with $\mu + i\alpha = -p^m\beta$, hence with $\mu = -(i\alpha + p^m\beta)$. Furthermore

$$i < \langle -\mu, \alpha^\vee \rangle = 2i + p^m \langle \beta, \alpha^\vee \rangle,$$

i.e., $i + p^m\langle \beta, \alpha^\vee \rangle > 0$.

b) Suppose that $-p^r\mu \in \mathcal{M}_r$. Write $-p^r\mu = \sum_{\gamma \in R^+} m_\gamma \gamma$ with $0 \leq m_\gamma < p^r$ for all γ. Since s_α permutes $R^+ \setminus \{\alpha\}$ (because α is simple), we get

$$-s_\alpha(p^r\mu) = \sum_{\gamma \neq \alpha} m_\gamma s_\alpha(\gamma) - m_\alpha \alpha \geq -(p^r - 1)\alpha,$$

hence $\alpha \leq p^r(\alpha - s_\alpha\mu)$. On the other hand, $\mu = -(i\alpha + p^m\beta)$ implies

$$\begin{aligned} \alpha - s_\alpha\mu &= \alpha - i\alpha + p^m\beta - p^m\langle \beta, \alpha^\vee \rangle \alpha \\ &= p^m\beta - (i + p^m\langle \beta, \alpha^\vee \rangle - 1)\alpha \leq p^m\beta, \end{aligned}$$

hence $\alpha \leq p^{r+m}\beta$. As both α and β are simple, this yields $\beta = \alpha$. We get now $\alpha - s_\alpha\mu = -(i + p^m - 1)\alpha$, hence $\alpha \leq -p^r(i + p^m - 1)\alpha$ — contradiction!

Remark: For more information on $H^2(B,\mu)$ consult [O'Halloran 5], 1.4 and [Andersen 10], 3.7.

12.6. It is now trivial to combine 12.4/5 with the remarks in 12.3 and to conclude:

Proposition: *If $p \neq 2$ or if R does not have a component of type C_n (for each $n \geq 1$), then $\operatorname{Ext}^1_{B_r}(L(\lambda),\lambda) = 0$ for all $\lambda \in X_r(T)$.*

12.7. Let E be a finite dimensional G–module and M a B–module. We can apply the spectral sequences from 12.2 to $E^* \otimes M$. Using the generalised tensor identity (i.e., $R^n \operatorname{ind}_B^G(E^* \otimes M) \simeq E^* \otimes R^n \operatorname{ind}_B^G M$) and the relationship between Ext-groups and cohomology (e.g., $H^m(B_r, E^* \otimes M) \simeq \operatorname{Ext}^m_{B_r}(E,M)$) we can write the $E_2^{n,m}$–terms as

$$\operatorname{Ext}^n_{G_r}(E, R^m \operatorname{ind}_B^G M)^{[-r]} \qquad \text{resp.} \qquad R^n \operatorname{ind}_B^G(\operatorname{Ext}^m_{B_r}(E,M)^{[-r]}).$$

So we get as in 12.2:

(1) *If $R^m \operatorname{ind}_B^G M = 0$ for all $m > 0$, then there is a spectral sequence with*

$$E_2^{n,m} = R^n \operatorname{ind}_B^G(\operatorname{Ext}^m_{B_r}(E,M)^{[-r]}) \Rightarrow \operatorname{Ext}^{n+m}_{G_r}(E, \operatorname{ind}_B^G M)^{[-r]}.$$

In the special case $M = k_\lambda$ with $\lambda \in X(T)_+$ we get

(2) $$E_2^{n,m} = R^n \operatorname{ind}_B^G(\operatorname{Ext}^m_{B_r}(E,\lambda)^{[-r]}) \Rightarrow \operatorname{Ext}^{n+m}_{G_r}(E, H^0(\lambda))^{[-r]}.$$

12.8. Lemma: *There is for all $\lambda, \mu \in X_r(T)$ a natural isomorphism*

$$\operatorname{Ext}^1_{G_r}(L(\mu), H^0(\lambda))^{[-r]} \simeq \operatorname{ind}_B^G(\operatorname{Ext}^1_{B_r}(L(\mu),\lambda)^{[-r]}).$$

Proof. Consider the spectral sequence from 12.7(2) for $E = L(\mu)$. It suffices to show that all $E_2^{n,0}$–terms with $n > 0$ are zero; then the claim follows from the five term exact sequence.

In case $\mu = \lambda$ we get $\operatorname{Hom}_{B_r}(L(\lambda),\lambda) \simeq k$ from 12.3(4), hence

$$R^n \operatorname{ind}_B^G(\operatorname{Hom}_{B_r}(L(\lambda),\lambda)) = 0 \qquad \text{for all } n > 0$$

by Kempf's vanishing theorem. If $\mu \neq \lambda$, then 12.3(3) yields $\operatorname{Hom}_{B_r}(L(\mu),\lambda) \hookrightarrow \operatorname{Hom}_{B_r}(Z_r(\mu),\lambda) = 0$ (cf. 3.8.a), so again $R^n \operatorname{ind}_B^G(\operatorname{Hom}_{B_r}(L(\mu),\lambda)) = 0$.

Remark: We can extend the lemma to all $\lambda, \mu \in X(T)_+$ that have a decomposition $\lambda = \lambda_1 + p^r\lambda_2$ and $\mu = \mu_1 + p^r\mu_2$ with $\lambda_1, \mu_1 \in X_r(T)$ and $\lambda_2, \mu_2 \in X(T)_+$. Then 3.16 implies

$$\operatorname{Hom}_{B_r}(L(\mu),\lambda)^{[-r]} \simeq L(\mu_2)^* \otimes \lambda_2 \otimes \operatorname{Hom}_{B_r}(L(\mu_1),\lambda_1)^{[-r]}.$$

In case $\mu_1 \neq \lambda_1$ this space is zero. In case $\mu_1 = \lambda_1$ we can apply the generalised tensor identity and Kempf's vanishing theorem to get

$$R^n \operatorname{ind}_B^G(L(\mu_2)^* \otimes \lambda_2) \simeq L(\mu_2)^* \otimes R^n \operatorname{ind}_B^G(\lambda_2) = 0$$

for $n > 0$.

12.9. Proposition: *If $p \neq 2$ or if R does not have a component of type C_n (for each $n \geq 1$), then*

$$\mathrm{Ext}^1_{G_r}(L(\lambda), L(\lambda)) = 0 = \mathrm{Ext}^1_{G_r}(L(\lambda), H^0(\lambda))$$

for all $\lambda \in X_r(T)$.

Proof: The claim about $H^0(\lambda)$ is an immediate consequence of 12.8 and 12.6. Applying $\mathrm{Hom}_{G_r}(L(\lambda), ?)$ to the exact sequence $0 \to L(\lambda) \to H^0(\lambda) \to H^0(\lambda)/L(\lambda) \to 0$ yields a surjection

$$\mathrm{Hom}_{G_r}(L(\lambda), H^0(\lambda)/L(\lambda)) \longrightarrow \mathrm{Ext}^1_{G_r}(L(\lambda), L(\lambda)).$$

Now $\mathrm{Hom}_{G_r}(L(\lambda), H^0(\lambda)/L(\lambda))$ is a G–module of the form $M^{[r]}$ for some other G–module M. If $M = 0$, then the proposition follows. If not, then choose a dominant weight ν of M. Now $p^r\nu$ is a weight of $\mathrm{Hom}_{G_r}(L(\lambda), H^0(\lambda)/L(\lambda))$, hence $\lambda + p^r\nu$ a weight of $H^0(\lambda)/L(\lambda)$, cf. 3.16(1). This implies $\lambda + p^r\nu < \lambda$, hence $p^r\nu < 0$; this is impossible for dominant ν.

12.10. Set $\mathfrak{g} = \mathrm{Lie}\, G$.

Lemma: *Suppose $p > h$. If λ is a weight of $\Lambda\mathfrak{g}^*$ vanishing on T_1, then $\lambda = 0$.*

Proof: As λ is a sum of pairwise distinct roots, we have $\lambda \leq 2\rho$ and λ vanishes on $Z(G)$. It is therefore enough to show that λ vanishes on $T \cap \mathcal{D}G$. Now $\lambda(T_1) = 1$ implies $\lambda((T \cap \mathcal{D}G)_1) = \lambda(T_1 \cap \mathcal{D}G) = 1$. So we may assume G to be semi-simple. Note that $\lambda(T_1) = 1$ is equivalent to $\lambda \in pX(T)$. When we replace G by a covering group, then we enlarge $X(T)$, hence also $pX(T)$. Therefore we may assume G to be simply connected. Now we can easily restrict ourselves to the case where R is indecomposable, which we want to assume from now on.

Let α_0 be the largest short root in R. Write $\lambda = p\mu$. We may assume λ to be dominant since $\Lambda\mathfrak{g}^*$ is a G–module, hence also $\mu \in X(T)_+$. Then

$$0 \leq p\langle \mu, \alpha_0^\vee \rangle \leq \langle 2\rho, \alpha_0^\vee \rangle = 2(h-1) < 2p,$$

hence $\langle \mu, \alpha_0^\vee \rangle \in \{0, 1\}$. If $\langle \mu, \alpha_0^\vee \rangle = 0$, then $\mu = 0$ and $\lambda = 0$ as desired. If $\langle \mu, \alpha_0^\vee \rangle = 1$, then μ is a minuscule fundamental weight. The minuscule weights are a system of representatives for the non-zero classes in $X(T)/\mathbf{Z}R$, cf. [B3], ch. VI, §1, exerc. 24c and §2, exerc. 5a. As $p > h$ does not divide the index of connection $(X(T) : \mathbf{Z}R)$, we get also $p\mu = \lambda \notin \mathbf{Z}R$, a contradiction.

12.11. Let us write (as in I.4.26) $S'(V)$ for the symmetric algebra $S(V)$ over a vector space V when we grade this algebra such that each $S^i(V)$ gets degree $2i$. So the homogeneous components of odd degree in $S'(V)$ are zero.

Proposition: *If $p > h$, then we have an isomorphism*

$$H^\bullet(B_1, k) \;\simeq\; S'(\mathfrak{u}^*)^{[1]}$$

of B–modules and of k–algebras.

Proof: As in 12.1 we take the spectral sequence from I.9.14 for the group U (and for $r = 1$), take fixed points under T_1, and get a spectral sequence converging to $H^\bullet(B_1, k)$. We use the form of the sequence in I.9.16, hence

$$(E_0^{i,j})^{T_1} = S^i(\mathfrak{u}^*)^{[1]} \otimes (\Lambda^{j-i}\mathfrak{u}^*)^{T_1}.$$

Because of 12.10 we may replace T_1 by T on the right hand side. No weight of some $\Lambda^l\mathfrak{u}^*$ with $l > 0$ is zero as it is a sum of l positive roots. So we get $(E_0^{i,j})^{T_1} = 0$ for $i \neq j$ whereas $(E_0^{i,i})^{T_1} = S^i(\mathfrak{u}^*)^{[i]}$. As each differential d_m has bidegree $(m, 1-m)$, the spectral sequence degenerates and yields isomorphisms $(E_0^{i,i})^{T_1} \simeq H^{2i}(B_1, k)$ whereas each $H^{2i+1}(B_1, k)$ is zero. We get thus an isomorphism of vector spaces as in our claim. The map is compatible with the B–action and the cup-product as the spectral sequence is so, cf. I.9.13.

Remarks: a) For $p \leq h$ the cohomology looks different, cf. [Andersen and Jantzen], section 6. For $r > 1$ the algebra $H^\bullet(B_r, k)$ gets more and more complicated, cf. [Andersen and Jantzen], 2.4.

b) If we did not care about the precise structure as a B–module and as a k–algebra, then we could have proved the proposition by applying the Lyndon-Hochschild-Serre spectral sequence several times to a T–stable central series of U with factors isomorphic to the different root subgroups U_α, cf. the approach in [O'Halloran 3].

12.12. Let V be a vector space and $V' \subset V$ a subspace. We have then for each $n \in \mathbf{N}$ an exact sequence (the *Koszul resolution*)

$$(1) \qquad \begin{aligned} \cdots \longrightarrow S^{n-i}V \otimes \Lambda^i V' \xrightarrow{\varphi_i} S^{n-i+1}V \otimes \Lambda^{i-1}V' \longrightarrow \cdots \\ \longrightarrow S^{n-1}V \otimes V' \longrightarrow S^n V \longrightarrow S^n(V/V') \longrightarrow 0 \end{aligned}$$

where φ_i is given by

$$\varphi_i(x \otimes (v_1' \wedge \cdots \wedge v_i')) = \sum_{j=1}^{i} (-1)^j x v_j' \otimes (v_1' \wedge \cdots \wedge v_{j-1}' \wedge v_{j+1}' \wedge \cdots \wedge v_i')$$

for all $v_l' \in V'$, $x \in S^{n-i}V$. The map $S^n V \to S^n(V/V')$ is induced by the canonical map $V \to V/V'$. If V is an H–module for some group scheme H and if V' is an H–submodule, then (1) is an exact sequence of H–modules.

Lemma: *Let $n \in \mathbf{N}$.*

a) *We have $R^i \operatorname{ind}_B^G S^n\mathfrak{u}^* = 0 = R^i \operatorname{ind}_B^G S^n\mathfrak{b}^*$ for all $i > 0$.*

b) *If p is good for R, then $\operatorname{ind}_B^G S^n\mathfrak{u}^*$ and $\operatorname{ind}_B^G S^n\mathfrak{b}^*$ have good filtrations.*

Proof: We prove first the statements involving $\mathfrak{b}^*$. Take the exact sequence (1) with $\mathfrak{g}^* = V$ and $(\mathfrak{g}/\mathfrak{b})^* \simeq \mathfrak{b}^\perp = V' \subset V$. It yields many short exact sequences of B–modules

$$(2) \qquad 0 \to L_{j+1} \longrightarrow S^{n-j}\mathfrak{g}^* \otimes \Lambda^j(\mathfrak{g}/\mathfrak{b})^* \longrightarrow L_j \to 0$$

where $L_0 \simeq S^n\mathfrak{b}^*$ and $L_j = 0$ for $j > \dim \mathfrak{g}/\mathfrak{b}$. The generalised tensor identity yields

$$R^i \operatorname{ind}_B^G(S^{n-j}\mathfrak{g}^* \otimes \Lambda^j(\mathfrak{g}/\mathfrak{b})^*) \simeq S^{n-j}\mathfrak{g}^* \otimes R^i \operatorname{ind}_B^G(\Lambda^j(\mathfrak{g}/\mathfrak{b})^*).$$

By 6.18 this term vanishes for $i \neq j$. Now induction on j from above yields $R^i \operatorname{ind}_B^G(L_j) = 0$ for $i \neq j$, in particular $R^i \operatorname{ind}_B^G(S^n\mathfrak{b}^*) = 0$ for $i \neq 0$. We know by 6.18 also that G acts trivially on each $R^i \operatorname{ind}_B^G(\Lambda^j(\mathfrak{g}/\mathfrak{b})^*)$, so 4.22 implies (for good

p) that each $S^{n-j}\mathfrak{g}^* \otimes R^i \operatorname{ind}_B^G(\Lambda^j(\mathfrak{g}/\mathfrak{b})^*)$ has a good filtration. Again induction on j from above shows that each $R^i \operatorname{ind}_B^G(L_j)$ has a good filtration, hence so has $\operatorname{ind}_B^G S^n\mathfrak{b}^*$ (taking $j = 0$).

Let us now turn to $\mathfrak{u}^*$. Considered as a B–module under the adjoint action it is a homomorphic image of $\mathfrak{b}^*$ with kernel $\mathfrak{u}^\perp \simeq \mathfrak{h}^*$. So B acts trivially on this kernel. Consider (1) with $V = \mathfrak{b}^*$ and $V/V' = \mathfrak{u}^*$. Any $S^{n-j}V \otimes \Lambda^j V'$ is a direct sum of copies of $S^{n-j}\mathfrak{b}^*$ as a B–module. We get now the claims on $\mathfrak{u}^*$ from those on $\mathfrak{b}^*$ using short exact sequences as in (2) and induction.

12.13. Proposition: *Suppose $p > h$. Then*

$$H^i(G_1, k)^{[-1]} \simeq \begin{cases} \operatorname{ind}_B^G(S^{i/2}\mathfrak{u}^*), & \text{for } i \text{ even,} \\ 0, & \text{for } i \text{ odd.} \end{cases}$$

Each $H^i(G_1, k)^{[-1]}$ has a good filtration. If $i > 0$, then $H^0(0) \simeq k$ does not occur in a good filtration of $H^i(G_1, k)^{[-1]}$.

Proof: Except for the last statement everything follows immediately from 12.11 using the spectral sequence 12.2(2) and the results in 12.12. The multiplicity of $H^0(0)$ is a good filtration of $H^i(G_1, k)^{[-1]}$ is equal to

$$\dim \operatorname{Hom}_G(k, \operatorname{ind}_B^G S^{i/2}\mathfrak{u}^*) = \dim \operatorname{Hom}_B(k, S^{i/2}\mathfrak{u}^*).$$

This is 0 for $i > 0$ as 0 is not a weight of $S^{i/2}\mathfrak{u}^*$ in that case.

Remarks: 1) The direct sum $\bigoplus_{i\geq 0} \operatorname{ind}_B^G(S^i\mathfrak{u}^*)$ has a natural structure as a graded k–algebra with the multiplication induced by the one in $S(\mathfrak{u}^*)$. The isomorphism in the proposition is compatible with this multiplication and the cup-product on $H^\bullet(G_1, k)$ as already the one in 12.11 was so.

2) The situation for $p \leq h$ is different, similarly for the $H^\bullet(G_r, k)$ with $r > 1$, cf. [Andersen and Jantzen], 3.10 and section 6.

12.14. Let us assume in this subsection that k is algebraically closed. The restriction of functions $S(\mathfrak{g}^*) \to S(\mathfrak{u}^*)$ induces a homomorphism of G–modules and k–algebras $\varphi : S(\mathfrak{g}^*) \to \operatorname{ind}_B^G S(\mathfrak{u}^*)$ mapping any $f \in S(\mathfrak{g}^*)$ to the function $g \mapsto (g^{-1}f)_{|\mathfrak{u}}$. Obviously

$$\ker(\varphi) = \{\, f \in S(\mathfrak{g}^*) \mid f(G\,\mathfrak{u}) = 0 \,\}.$$

It is well known that $G\mathfrak{u}$ is the closed subvariety $\mathcal{N}$ of all nilpotent elements in $\mathfrak{g}$, cf. [Bo], 14.26. So we get an injective homomorphism of k–algebras

$$k[\mathcal{N}] \longrightarrow H^\bullet(G_1, k). \tag{1}$$

One can check that (1) is an isomorphism by comparing dimensions of homogeneous parts, cf. [Friedlander and Parshall 2], 2.6 or [Andersen and Jantzen], 3.9.

12.15. Suppose that $p > h$ and (in order to simplify) that G is semi-simple and simply connected. The description of the blocks of G_r (cf. 9.22) implies that the any $\mu \in X(T)_+$ with $H^\bullet(G_1, H^0(\mu)) \neq 0$ has the form $\mu = w \bullet 0 + p\lambda$ with $w \in W$ and $\lambda \in X(T)$. For such μ one can show:

$$(1) \qquad H^i(G_1, H^0(\mu))^{[-1]} \simeq \begin{cases} \operatorname{ind}_B^G(S^{(i-l(w))/2}\mathfrak{u}^* \otimes \lambda), & \text{if } i - l(w) \text{ even,} \\ 0, & \text{if } i - l(w) \text{ odd.} \end{cases}$$

This was proved in most cases in [Andersen and Jantzen] (if R does not have a component of type E or F or if $\langle \lambda, \beta^\vee \rangle \geq h - 1$ for all $\beta \in S$), in the general case in [Kumar, Lauritzen, and Thomsen], Thm. 8. One can also show that $H^i(G_1, H^0(\mu))^{[-1]}$ has a good filtration for each i if $p > h$, and one can compute the factors in the good filtration using suitable partition functions, cf. [Andersen and Jantzen], 4.5, 3.8. The numbers one gets can be identified (using [Kato 2]) with certain coefficients of Kazhdan-Lusztig polynomials.

It had been expected for some time (see the conjecture in [Donkin 15], p. 79) that for a G–module V with a good filtration all $H^i(G_r, V)^{[-r]}$ have a good filtration. The result on the $H^i(G_1, H^0(\mu))^{[-1]}$ mentioned above was evidence supporting this conjecture. However, the conjecture has turned out to be wrong in this generality, see [van der Kallen 4].

CHAPTER 13

Schubert Schemes

Throughout this chapter let k be a Dedekind ring (including the case of a field).

For the purpose of this introduction let us, however, assume that k is an algebraically closed field. It is then well known (cf. [Bo], 14.12) that $G(k)$ is the disjoint union of all $B(k)\dot{w}B(k)$ with $w \in W$. This leads then to a disjoint decomposition of $(G/B)(k)$ into all $B(k)\dot{w}B(k)/B(k)$, which turn out to be locally closed subvarieties isomorphic to affine spaces. They are called the *Bruhat cells* in $(G/B)(k)$ and their closures are called the *Schubert varieties* in $(G/B)(k)$. (Over more general rings one gets schemes. That is the reason for the title of this chapter.) These constructions can also be generalised from G/B to G/P for any parabolic subgroup of G.

The flag variety $(G/B)(k)$ itself is a special case of a Schubert variety, equal to the closure of $B(k)\dot{w}_0B(k)/B(k)$ where w_0 is the longest element in W, with $w_0(S) = -S$. A major part of the representation theory of G is the study of cohomology groups of line bundles on G/B. We shall see in the next chapter (14) that some methods used for G/B can also be used to study the cohomology of line bundles on each Schubert variety. In order to prepare for this we discuss in the present chapter elementary properties (together with an example in SL_n, cf. 13.9), and we describe the Bott-Samelson schemes. These are desingularisations of the Schubert schemes which will play an important role in the next chapter. They were first described in [Demazure 3] and, independently, by H. C. Hansen.

13.1. For each $w \in W$ set

(1) $$R_1(w) = \{\, \alpha \in R^+ \mid w^{-1}(\alpha) < 0 \,\}$$

and

(2) $$R_2(w) = \{\, \alpha \in R^+ \mid w^{-1}(\alpha) > 0 \,\}.$$

Each $R' \in \{R_1(w), R_2(w), -R_1(w), -R_2(w)\}$ is obviously a closed and unipotent subset (cf. 1.7) of R. So the multiplication induces an isomorphism of schemes from $\prod_{\alpha\in R'} U_\alpha$ (for any ordering of R') onto a closed subgroup scheme $U(R')$ of G. As a scheme $U(R')$ is isomorphic to $\mathbf{A}^{|R'|}$. Observe that $|R_1(w)| = l(w)$ and $|R_2(w)| = n - l(w)$ where $n = |R^+|$. Set

(3) $$U_1(w) = U(-R_1(w)), \qquad U_2(w) = U(-R_2(w))$$

and

(4) $$U_1^+(w) = U(R_1(w)), \qquad U_2^+(w) = U(R_2(w)).$$

One has obviously

$$(5)\qquad \dot w^{-1}U_1(w)\dot w = U_1^+(w^{-1}),\qquad \dot w^{-1}U_2(w)\dot w = U_2(w^{-1})$$

and

$$(6)\qquad \dot w^{-1}U_1^+(w)\dot w = U_1(w^{-1}),\qquad \dot w^{-1}U_2^+(w)\dot w = U_2^+(w^{-1}),$$

where (as usual) $\dot w \in N_G(T)(k)$ is a representative for w. The multiplication induces (again by 1.7) isomorphisms of schemes

$$(7)\qquad U_1(w)\times U_2(w) \xrightarrow{\sim} U$$

and

$$(8)\qquad U_1^+(w)\times U_2^+(w) \xrightarrow{\sim} U^+.$$

Obviously (5) implies $U_1(w) \subset \dot wU^+\dot w^{-1}$ and $U_2(w)\cap \dot wU^+\dot w^{-1} \subset \dot wU\dot w^{-1}\cap \dot wU^+\dot w^{-1} = \dot w(U\cap U^+)\dot w^{-1} = 1$. So we get from (7)

$$(9)\qquad U_1(w) = U\cap \dot wU^+\dot w^{-1}.$$

Similarly one proves

$$(10)\qquad U_2(w) = U\cap \dot wU\dot w^{-1}.$$

Let $w_1, w_2 \in W$ with $l(w_1)+l(w_2) = l(w_1w_2)$. Then $R_1(w_1w_2)$ is the disjoint union of $R_1(w_1)$ and of $w_1R_1(w_2)$, cf. [B3], ch. VI, §1, cor. 2 de la prop. 17. Therefore the multiplication induces an isomorphism of schemes

$$(11)\qquad U_1(w_1)\times \dot w_1U_1(w_2)\dot w_1^{-1} \xrightarrow{\sim} U_1(w_1w_2).$$

Suppose that $w = s_{\alpha_1}s_{\alpha_2}\dots s_{\alpha_r}$ with all $\alpha_i\in S$ is a reduced decomposition, i.e., with $r = l(w)$. Using induction and (11) we get an isomorphism of schemes

$$(12)\qquad U_{-\alpha_1}\times U_{-\alpha_2}\times\dots\times U_{-\alpha_r} \xrightarrow{\sim} U_1(w),$$

$(u_1,u_2,\dots,u_r)\mapsto u_1\dot s_{\alpha_1}u_2\dot s_{\alpha_2}\dots u_r\dot s_{\alpha_r}\dot w^{-1}$ if we choose $\dot w = \dot s_{\alpha_1}\dot s_{\alpha_2}\dots\dot s_{\alpha_r}$.

13.2. Consider for any $w\in W$ the map $m_w : G\times G\to G$ with $(g,g')\mapsto g\dot wg'$ for all g,g'. We shall always choose $\dot w = 1$ if $w = 1$. So m_1 is just the usual multiplication in G.

We want to describe $m_w(B\times B)$. Any $b\in B(A)$ has a unique decomposition $b = u_1u_2t$ with $t\in T(A)$ and $u_i\in U_i(w)(A)$ for $i = 1,2$ (for any k–algebra A). Then

$$m_w(b,b') = u_1u_2t\dot wb' = m_w(u_1,(\dot w^{-1}u_2\dot w)(\dot w^{-1}t\dot w)b')$$

for all $b'\in B(A)$. Now $\dot w^{-1}u_2\dot w\in U(A)$ and $\dot w^{-1}t\dot w\in T(A)$, hence

$$\begin{aligned} m_w(B\times B) &= m_w(U_1(w)\times B) = \dot w\, m_1(\dot w^{-1}U_1(w)\dot w\times B)\\ &= \dot w\, m_1(U_1^+(w^{-1})\times B)\subset \dot w\, m_1(U^+\times B).\end{aligned}$$

Recall that $(u, b) \mapsto \dot{w}ub$ induces an isomorphism of schemes from $U^+ \times B$ onto an open subscheme (a big cell) of G, cf. 1.9. Therefore the closed subscheme $U_1^+(w^{-1}) \times B$ of $U^+ \times B$ is mapped isomorphically onto a closed subscheme of this big cell. As conjugating with $\dot{w}$ is again an isomorphism, we get:

(1) *The map m_w induces an isomorphism of schemes from $U_1(w) \times B$ onto a closed subscheme of a big cell in G.*

We denote this subscheme of G by BwB. It is obviously equal to the image functor $m_w(B \times B)$ and independent of the special choice of the representative $\dot{w}$. We have an isomorphism $BwB \simeq U_1(w) \times B \simeq \mathbf{A}^{l(w)} \times B$, hence:

(2) *Each BwB is an integral and affine scheme.*

The big cell in (1) is an open subscheme in G of the form G_d for some $d \in k[G]$, $d \neq 0$, see 1.9. There exists a prime ideal I in $k[G_d] = k[G]_d$ with $BwB = V(I)$. Then $J = k[G] \cap I$ is a prime ideal in $k[G]$ such that $V(J)$ is the closure $\overline{BwB}$ of BwB in G. We have then $I = J_d$ and $V(I) = V(J) \cap G_d$, i.e., $BwB = \overline{BwB} \cap G_d$. This shows:

(3) *Each $\overline{BwB}$ is an integral and affine scheme. It contains BwB as an open subscheme.*

Furthermore:

(4) *Each $\overline{BwB}$ is stable under left and right multiplication by B.*

Indeed, the multiplication map $G \times G \to G$ is continuous and the closure of $BwB \times B$ (resp. of $B \times BwB$) is $\overline{BwB} \times B$ (resp. $B \times \overline{BwB}$).

As G is integral and as $Bw_0B = \dot{w}_0\, m_1(U^+ \times B)$ is open and non-empty, we get

(5) $$\overline{Bw_0B} = G.$$

Recall that we have associated to each subset $I \subset S$ a subgroup W_I of W (see 1.7) and a parabolic subgroup P_I of G with $P_I \supset B$ (see 1.8). One has $\dot{s}_\alpha \in P_I$ for all $\alpha \in I$, cf. 1.3(4), hence $\dot{w} \in P_I$ for all $w \in W_I$. This implies $BwB \subset P_I$ and (since P_I is closed in G) $\overline{BwB} \subset P_I$ for all $w \in W_I$.

More precisely: If $w \in W_I$, then $R_1(w) \subset \mathbf{Z}I$, hence $U_1(w) \subset B_I = B \cap L_I$. The isomorphism of schemes $L_I \times U_I \xrightarrow{\sim} P_I$ (given by multiplication, see 1.8(4)) induces isomorphisms $B_I \times U_I \xrightarrow{\sim} B$ and $B_IwB_I \times U_I = m_w(U_1(w) \times B_I) \times U_I \xrightarrow{\sim} m_w(U_1(w) \times B) = BwB$, hence $\overline{B_IwB_I} \times U_I \xrightarrow{\sim} \overline{BwB}$.

There is a unique element $w_I \in W_I$ with $w_I(\alpha) < 0$ for all $\alpha \in I$. Then (5) applied to L_I yields $\overline{B_IwB_I} = L_I$, so the discussion above shows that

(6) $$\overline{Bw_IB} = P_I.$$

Note that $\overline{Bw_IB}$ is smooth as P_I is so. In general $\overline{BwB}$ will *not* be smooth.

For any k–algebra k' that is a field we have the Bruhat decomposition of $G(k')$; this means that $G(k')$ is the disjoint union of all $B(k')\dot{w}B(k')$ with $w \in W$, cf. [Bo], 14.12 for the algebraically closed case, [BoT], 2.11 for the general case. This implies for any k–algebra A that the $B(A)\dot{w}B(A)$ are pairwise disjoint. If we work with "schemes" as in I.1.11, then we can express this result as follows:

(7) $$|G| = \bigcup_{w\in W} |BwB|.$$

13.3. Let us denote the canonical map $G \to G/B$ by π. We know that π is locally trivial, cf. 1.10(2). More precisely, each $\dot{w}U^+$ with $w \in W$ is mapped isomorphically onto an open subscheme of G/B. Since $U_1(w)\dot{w} = \dot{w}U_1^+(w^{-1}) \subset \dot{w}U^+$, see 13.1(5), we see that $g \mapsto \pi(g\dot{w})$ is an isomorphism from $U_1(w)$ onto a closed subscheme of that open subscheme of G/B. The image of this map is equal to $\pi(BwB) = \pi \circ m_w(U_1(w) \times B)$ and it is obviously a quotient scheme of $BwB \simeq U_1(w) \times B$ by B. So we denote it by BwB/B. We call the subschemes BwB/B of G/B the *Bruhat cells* in G/B. The isomorphism above

(1) $$U_1(w) \xrightarrow{\sim} BwB/B, \qquad g \mapsto \pi(g\dot{w})$$

induces an isomorphism

$$\mathbf{A}^{l(w)} \simeq BwB/B.$$

One has as in 13.2(7) a disjoint decomposition

(2) $$|G/B| = \bigcup_{w\in W} |BwB/B|.$$

We denote the closure of BwB/B in G/B by $X(w)$ and call the $X(w)$ the *Schubert schemes* in G/B. Note that I.5.21(2) implies $X(w) = \overline{BwB}/B$. As $\overline{BwB}$ is stable under left multiplication by B, so is $\overline{BwB}/B = X(w)$. As G/B is projective and as $BwB/B \simeq \mathbf{A}^{l(w)}$, we get:

(3) *Each $X(w)$ is an integral and projective scheme. We have* $\dim X(w) = l(w)$ *if k is a field.*

If $I \subset S$ and if w_I is as in 13.2(6), then obviously:

(4) $$X(w_I) = P_I/B \simeq L_I/(L_I \cap B),$$

especially

(5) $$X(w_0) = G/B.$$

In these special cases the Schubert schemes are smooth. (More precisely, P_I/B has an open covering by all $w(U^+ \cap L_I)B/B \simeq \mathbf{A}^{m(I)}$ with $w \in W_I$ where $m(I) = |R^+ \cap \mathbf{Z}I|$.) In general, the $X(w)$ will *not* be smooth.

Let us write $P(\alpha) = P_{\{\alpha\}}$ for all $\alpha \in S$. Then (4) together with the known result for SL_2 or PGL_2 yields

(6) $$X(s_\alpha) = P(\alpha)/B \simeq \mathbf{P}^1.$$

13.4. Let $w_1, w_2, \dots, w_r \in W$. Then

$$\prod_{i=1}^r \overline{Bw_iB} = \overline{Bw_1B} \times \overline{Bw_2B} \times \cdots \times \overline{Bw_rB}$$

is a closed subscheme of $G^r = G\times G\times\cdots\times G$, and $\prod_{i=1}^r Bw_iB$ is an open subscheme of $\prod_{i=1}^r \overline{Bw_iB}$. The map

$$\sigma_r : (g_1, g_2, \dots, g_r) \mapsto (g_1, g_1g_2, \dots, g_1g_2\cdots g_r)$$

is an automorphism of G^r as a scheme. Taking images under σ_r we see therefore (setting $g_0 = 1$ and using some abuse of notation)

(1) $\quad V(w_1, w_2, \dots, w_r) = \{\,(g_1, g_2, \dots, g_r) \in G^r \mid g_{i-1}^{-1}g_i \in \overline{Bw_iB} \text{ for all } i\,\}$

is a closed subscheme of G^r containing

(1′) $\quad V'(w_1, w_2, \dots, w_r) = \{\,(g_1, g_2, \dots, g_r) \in G^r \mid g_{i-1}^{-1}g_i \in Bw_iB \text{ for all } i\,\}$

as an open subscheme.

There is on G^r a (free) right action of B^r given by

$$(g_1, g_2, \dots, g_r)(b_1, b_2, \dots, b_r) = (g_1b_1, g_2b_2, \dots, g_rb_r)$$

and a (free) left action of B given by

$$b(g_1, g_2, \dots, g_r) = (bg_1, bg_2, \dots, bg_r).$$

Both actions stabilise $V(w_1, w_2, \dots, w_r)$ and $V'(w_1, w_2, \dots, w_r)$. We can identify the quotients by B^r with subschemes of $(G/B)^r$ on which B operates from the left, i.e., we set (with $g_0 = 1$ and again some abuse of notation)

(2) $\quad X(w_1, w_2, \dots, w_r) = V(w_1, w_2, \dots, w_r)/B^r$

$$= \{\,(g_1B, g_2B, \dots, g_rB) \in (G/B)^r \mid g_{i-1}^{-1}g_i \in \overline{Bw_iB} \text{ for all } i\,\}$$

and

(2′) $\quad X'(w_1, w_2, \dots, w_r) = V'(w_1, w_2, \dots, w_r)/B^r$

$$= \{\,(g_1B, g_2B, \dots, g_rB) \in (G/B)^r \mid g_{i-1}^{-1}g_i \in Bw_iB \text{ for all } i\,\}.$$

Note that this notation is compatible with the one in 13.3 in case $r = 1$.

We have by I.5.21(1):

(3) $\quad X(w_1, w_2, \dots, w_r)$ *is a closed subscheme of* $(G/B)^r$.

As G/B is projective, we get in particular:

(4) $\quad X(w_1, w_2, \dots, w_r)$ *is a projective scheme.*

Furthermore, I.5.21(2), (3) imply:

(5) $\quad X'(w_1, w_2, \dots, w_r)$ *is open in* $X(w_1, w_2, \dots, w_r)$ *and its closure is equal to* $X(w_1, w_2, \dots, w_r)$.

Since the canonical map $G \to G/B$ is locally trivial, so is $G^r \to (G/B)^r$. Therefore:

(6) $\quad$ *The canonical maps* $V(w_1, \dots, w_r) \to X(w_1, \dots, w_r)$ *and* $V'(w_1, \dots, w_r) \to X'(w_1, \dots, w_r)$ *are locally trivial.*

Remark: If $f : \widetilde{G} \to G$ is a covering group, then f induces an isomorphism of schemes $\overline{f} : \widetilde{G}/\widetilde{B} \xrightarrow{\sim} G/B$ where $\widetilde{B} = f^{-1}(B)$. Then $\overline{f}$ induces an isomorphism $\widetilde{B}w\widetilde{B}/\widetilde{B} \xrightarrow{\sim} BwB/B$ for each $w \in W$ as well as an isomorphism between the closures of these subschemes. Furthermore, we get isomorphisms between the analogues of any $X(w_1, w_2, \dots, w_r)$ or $X'(w_1, w_2, \dots, w_r)$ in $(\widetilde{G}/\widetilde{B})^r$ and $X(w_1, w_2, \dots, w_r)$ or $X'(w_1, w_2, \dots, w_r)$. Similar remarks apply to G/P and all BwP/P and $X(w)_P$ to be considered below in 13.8.

13.5. Let $w'_1, w'_2, \dots, w'_s, w_1, w_2, \dots, w_r \in W$. The map

$$(g'_1, g'_2, \dots, g'_s, g_1, g_2, \dots, g_r) \mapsto (g'_1, g'_2, \dots, g'_s, g'_s g_1, g'_s g_2, \dots, g'_s g_r)$$

induces isomorphism of schemes

$$(1) \qquad V(w'_1, w'_2, \dots, w'_s) \times V(w_1, w_2, \dots, w_r) \xrightarrow{\sim} V(w'_1, \dots, w'_s, w_1, \dots, w_r)$$

and

$$(1') \quad V'(w'_1, w'_2, \dots, w'_s) \times V'(w_1, w_2, \dots, w_r) \xrightarrow{\sim} V'(w'_1, \dots, w'_s, w_1, \dots, w_r).$$

The right action of $B^{r+s} = B^s \times B^r$ (as in 13.4) on the right hand sides in (1) and (1′) yields the following action on the left hand sides:

$$\begin{aligned}(g'_1, \dots, g'_s, g_1, \dots, g_r)&(b'_1, \dots, b'_s, b_1, \dots, b_r)\\ &= (g'_1 b'_1, \dots, g'_s b'_s, b'^{-1}_s g_1 b_1, \dots, b'^{-1}_s g_r b_r).\end{aligned}$$

The left action of B on the right hand sides yields the following action on the left hand sides:

$$b(g'_1, \dots, g'_s, g_1, \dots, g_r) = (bg'_1, \dots, bg'_s, g_1, \dots, g_r).$$

If we divide by the right action of B^{r+s} in (1) and (1′), then we get isomorphism of schemes (with the notation from I.5.14)

$$(2) \qquad V(w'_1, w'_2, \dots, w'_s) \times^{B^s} X(w_1, w_2, \dots, w_r) \xrightarrow{\sim} X(w'_1, \dots, w'_s, w_1, \dots, w_r)$$

and

$$(2') \quad V'(w'_1, w'_2, \dots, w'_s) \times^{B^s} X'(w_1, w_2, \dots, w_r) \xrightarrow{\sim} X'(w'_1, \dots, w'_s, w_1, \dots, w_r).$$

These isomorphisms are compatible with the left actions of B (which acts on the left hand sides only on $V(w'_1, w'_2, \dots, w'_s)$ resp. on $V'(w'_1, w'_2, \dots, w'_s)$).

Let $w \in W$. We get from (2′) an isomorphism $BwB \times^B X'(w_1, w_2, \dots, w_r) \xrightarrow{\sim} X'(w, w_1, w_2, \dots, w_r)$. We have by 13.2(1) an isomorphism $U_1(w) \times B \xrightarrow{\sim} BwB$ compatible with the right action of B. We get therefore an isomorphism

$$U_1(w) \times X'(w_1, w_2, \dots, w_r) \xrightarrow{\sim} X'(w, w_1, w_2, \dots, w_r).$$

Induction shows now that we have an isomorphism

$$(3) \qquad U_1(w_1) \times U_1(w_2) \times \dots \times U_1(w_r) \xrightarrow{\sim} X'(w_1, w_2, \dots, w_r)$$

given by

$$(u_1, u_2, \dots, u_r) \mapsto (u_1 \dot{w}_1 B, u_1 \dot{w}_1 u_2 \dot{w}_2 B, \dots, u_1 \dot{w}_1 u_2 \dot{w}_2 \dots u_r \dot{w}_r B).$$

As each $U_1(w)$ is isomorphic to the affine space $\mathbf{A}^{l(w)}$, we get

$$(4) \qquad X'(w_1, w_2, \dots, w_r) \simeq \mathbf{A}^{\sum_{i=1}^r l(w_i)}.$$

So this is a smooth and integral affine scheme. Now 13.4(5) implies:

(5) *Each $X(w_1, w_2, \ldots, w_r)$ is an integral scheme.*

Since the canonical map $G \to G/B$ is locally trivial, so is its restriction to any $\overline{BwB} \to X(w)$. This implies for any scheme Z with a left action of B: If $X(w)$ and Z are smooth, then so is $\overline{BwB} \times^B Z$. (It has an open covering by subschemes isomorphic to suitable $X_i \times Z$ with $X_i \subset X(w)$ open.) Applying (2) repeatedly in the case $s = 1$ we get therefore:

(6) *If each $X(w_i)$ with $1 \leq i \leq r$ is smooth, then so is $X(w_1, w_2, \ldots, w_r)$.*

Set $w = w_1 w_2 \ldots w_r$. Obviously (3) and 13.1(11) imply:

(7) *If $l(w) = \sum_{i=1}^r l(w_i)$, then the projection $(G/B)^r \to G/B$ onto the last factor induces an isomorphism of schemes $X'(w_1, w_2, \ldots, w_r) \xrightarrow{\sim} X'(w) = BwB/B$.*

The inverse image of $X(w)$ under this morphism is a closed subfunctor of $(G/B)^r$ containing $X'(w_1, w_2, \ldots, w_r)$, hence also its closure $X(w_1, w_2, \ldots, w_r)$. So:

(8) *If $l(w) = \sum_{i=1}^r l(w_i)$, then we get a morphism $\varphi : X(w_1, w_2, \ldots, w_r) \to X(w)$ which induces an isomorphism of suitable open and dense subschemes.*

Suppose for the moment that $1 \in \overline{BwB}(k)$ for all $w \in W$. (We shall see in 13.6 that this holds.) Then we have $(1, 1, \ldots, 1) \in V(w_1, w_2, \ldots, w_r)$ for all $w_1, w_2, \ldots, w_r$. So the isomorphism in (1) gives rise to a closed embedding

$$\psi : V(w'_1, w'_2, \ldots, w'_s) \longrightarrow V(w'_1, w'_2, \ldots, w'_s, w_1, w_2, \ldots, w_r)$$

given by $(g'_1, g'_2, \ldots, g'_s) \mapsto (g'_1, g'_2, \ldots, g'_s, g'_s, g'_s, \ldots, g'_s)$. It commutes with the left action of B (i.e., we have $\psi(bx) = b\psi(x)$ for all b in B and x in $V(w'_1, w'_2, \ldots, w'_s)$) and it is compatible with the right actions of B^s and B^{r+s} in the following way:

$$\psi(x\,(b_1, b_2, \ldots, b_s)) = \psi(x)\,(b_1, b_2, \ldots, b_s, b_s, b_s, \ldots, b_s)$$

for all x in $V(w'_1, w'_2, \ldots, w'_s)$ and b_i in B. Therefore ψ induces a morphism (also denoted by ψ) of the quotient schemes

$$X(w'_1, w'_2, \ldots, w'_s) \longrightarrow X(w'_1, w'_2, \ldots, w'_s, w_1, w_2, \ldots, w_r)$$

which obviously commutes with the left action of B. It is a closed embedding, i.e., an isomorphism from the first scheme onto a closed subscheme of the second scheme. In fact, the image is just $X(w'_1, w'_2, \ldots, w'_s, 1, 1, \ldots, 1)$. The explicit expression for ψ above shows that we get a commutative diagram

$$\begin{array}{ccc} X(w'_1, w'_2, \ldots, w'_s) & \xrightarrow{\psi} & X(w'_1, w'_2, \ldots, w'_s, w_1, w_2, \ldots, w_r) \\ & \searrow \quad \swarrow & \\ & G/B & \end{array} \tag{9}$$

where the maps to G/B are (as in (7), (8)) given by the projection from $(G/B)^s$ or $(G/B)^{r+s}$ onto the last factor.

13.6. Let $s_1, s_2, \dots, s_r$ be simple reflections. So there are $\alpha_i \in S$ with $s_i = s_{\alpha_i}$ for $1 \le i \le r$. Combining 13.5(6) and 13.3(6) we get:

(1) *$X(s_1, s_2, \dots, s_r)$ is smooth.*

Let us be more precise. We have

$$X(s_1, s_2, \dots, s_r) \simeq \overline{Bs_1B} \times^B X(s_2, \dots, s_r) = P(\alpha_1) \times^B X(s_2, \dots, s_r).$$

We can cover $P(\alpha_1)/B \simeq \mathbf{P}^1$ with two (open) affine lines; their inverse images in $X(s_1, s_2, \dots, s_r)$ are isomorphic to $\mathbf{A}^1 \times X(s_2, \dots, s_r)$. Iterating we get:

(2) *$X(s_1, s_2, \dots, s_r)$ has an open covering by subschemes isomorphic to $\mathbf{A}^r$.*

Set $w = s_1 s_2 \dots s_r$ and suppose that this is a *reduced decomposition* of w, i.e., that $r = l(w)$. Then we have by 13.5(7), (8) an isomorphism of schemes

(3) $$X'(s_1, s_2, \dots, s_r) \xrightarrow{\sim} BwB/B$$

induced by a morphism

(4) $$\varphi : X(s_1, s_2, \dots, s_r) \longrightarrow X(w).$$

This morphism is by (1) a desingularisation of $X(w)$. One can easily check that the complement of $X'(s_1, s_2, \dots, s_r)$ in $X(s_1, s_2, \dots, s_r)$ is a divisor with normal crossings, but we shall not use this fact. At least in the case $w = w_0$ one usually calls $X(s_1, s_2, \dots, s_r)$ a *Bott-Samelson* scheme, cf. [Demazure 3].

As $1 \in P(\alpha_i) = \overline{Bs_iB}$ for all i, it is clear for any sequence $1 \le i_1 < i_2 < \dots < i_m \le r$ that $Bs_{i_1}s_{i_2} \dots s_{i_m}B/B$ is contained in the image of φ, hence that

(5) $$X(s_{i_1}s_{i_2} \dots s_{i_m}) \subset X(w)$$

and that

(6) $$B(k)\dot{s}_{i_1}\dot{s}_{i_2} \dots \dot{s}_{i_m}B(k) \subset \overline{BwB}(k).$$

One gets especially $1 \in \overline{BwB}(k)$ for all w, as needed for the construction of 13.5(9).

13.7. It is obvious that we can define an order relation on W setting

(1) $$w_1 \le w_2 \iff X(w_1) \subset X(w_2).$$

This order is called the *Bruhat* or *Bruhat-Chevalley order.* As $X(w_1)$ is the closure of Bw_1B/B and as $X(w_2)$ is stable under left multiplication by B, we get:

(2) $$w_1 \le w_2 \iff Bw_1B/B \subset X(w_2) \iff \dot{w}_1B \in X(w_2)(k).$$

Proposition: *Let $w, w' \in W$. Suppose that k is an algebraically closed field. Let $w = s_1 s_2 \dots s_r$ be a reduced decomposition of $w \in W$ where $r = l(w)$ and $s_i = s_{\alpha_i}$*

with $\alpha_i \in S$. *Then* $w' \le w$ *if and only if there exist* m *and* $1 \le i_1 < i_2 < \cdots < i_m \le r$ *with* $w' = s_{i_1} s_{i_2} \dots s_{i_m}$.

Proof: We may regard the map $\varphi : X(s_1, s_2, \dots, s_r) \to X(w)$ from 13.6(4) as a morphism of projective varieties over k. The image is closed and contains the open and dense subset $(BwB/B)(k)$. This implies

$$X(w)(k) = \varphi(X(s_1, s_2, \dots, s_r)(k)) = P(\alpha_1)(k)P(\alpha_2)(k) \dots P(\alpha_r)(k)/B(k).$$

The Bruhat decomposition of each $P(\alpha_i)(k)$ (or of its Levi factor) yields $P(\alpha_i)(k) = B(k) \cup B(k)\dot{s}_i B(k)$, so $X(w)(k)$ is the union of all

$$B(k)\dot{s}_{i_1} B(k)\dot{s}_{i_2} B(k) \dots B(k)\dot{s}_{i_m} B(k)/B(k).$$

One has for all $\alpha \in S$ and $w_1 \in W$

$$B(k)\dot{s}_\alpha B(k)\dot{w}_1 B(k) \subset B(k)\dot{s}_\alpha \dot{w}_1 B(k) \cup B(k)\dot{w}_1 B(k),$$

cf. [Bo], (v) on p. 195. This easily implies that

$$X(w)(k) = \bigcup B(k)\dot{s}_{i_1} \dot{s}_{i_2} \dots \dot{s}_{i_m} B(k)/B(k).$$

The proposition follows.

Remark: One can show using Proposition 14.16 that this result extends to arbitrary k.

13.8. Let $I \subset S$ and set $P = P_I$. The canonical map $\pi' : G \to G/P$ is locally trivial, cf. 1.10(5). Therefore $\overline{\pi} : G/B \to G/P$ is also locally trivial, with fibres isomorphic to P/B.

For any $w' \in W_I$ one has $\dot{w}' \in P(k)$. Therefore the image of $B \times P$ under $m_w : (g, g') \mapsto g\dot{w}g'$ and under $\pi' \circ m_w$ depend only on the coset wW_I. There is a special set of coset representatives (cf. [B3], ch. IV, §1, exerc. 3):

$$(1)\qquad \begin{aligned} W^I &= \{w \in W \mid w(\alpha) > 0 \text{ for all } \alpha \in I\} = \{w \in W \mid R_1(w^{-1}) \subset R^+ \setminus \mathbf{Z}I\} \\ &= \{w \in W \mid U_1^+(w^{-1}) \subset U_I^+\}, \end{aligned}$$

cf. 1.8. Then $l(ww') = l(w) + l(w')$ for all $w \in W^I$ and $w' \in W_I$.

Arguing as in 13.2 one proves for any $w \in W^I$ that m_w induces an isomorphism from $U_1(w) \times P$ onto a closed subscheme (denoted by BwP) of an open subscheme of G, and that $\pi' \circ m_w$ induces an isomorphism from $U_1(w)$ onto a closed subscheme (denoted by BwP/P) of an open subscheme of G/P. These schemes are integral, hence so are their closures $\overline{BwP} \subset G$ and $X(w)_P = \overline{BwP}/P \subset G/P$. The $BwP/P \simeq \mathbf{A}^{l(w)}$ with $w \in W^I$ are called *generalised Bruhat cells* and the $X(w)_P$ are called *generalised Schubert schemes.*

Let $w \in W^I$ and $w' \in W_I$. As $l(ww') = l(w) + l(w')$, we have a commutative diagram of morphisms:

$$\begin{array}{ccccccc}
 & & & & U_1(w) \times U_1(w') & \longrightarrow & U_1(w) \\
 & & & & \downarrow & & \downarrow \\
BwB \times^B X'(w) & \xrightarrow{\sim} & X'(w, w') & \xrightarrow{\sim} & Bww'B/B & \xrightarrow{\overline{\pi}} & BwP/P \\
\downarrow & & \downarrow & & \downarrow & & \downarrow \\
\overline{BwB} \times^B X(w') & \xrightarrow{\sim} & X(w, w') & \longrightarrow & X(ww') & \xrightarrow{\overline{\pi}} & X(w)_P
\end{array}$$

The first horizontal map is the projection onto the first factor. The two vertical maps in the upper half of the diagram are the isomorphism $U_1(w) \times U_1(w') \xrightarrow{\sim} Bww'B/B$ with $(g_1, g_2) \mapsto \pi(g_1 \dot{w} g_2 \dot{w}')$, cf. 13.1(11) and 13.3(1), and the isomorphism $U_1(w) \xrightarrow{\sim} BwP/P$ with $g \mapsto \pi'(g\dot{w})$ from above. The vertical maps in the lower half of the diagram are natural inclusions. The isomorphisms at the beginning of rows 2 and 3 come from 13.5(2′), (2) — note that $BwB = V'(w)$ and $\overline{BwB} = V(w)$ — and the second maps in these rows come from 13.5(7), (8).

We can take especially $w' = w_I$ where $X(w_I) = P$, cf. 13.2(6). The diagram shows $X(ww_I) \subset \overline{\pi}^{-1}X(w)_P$ and (looking at the first two maps in the last row, restricted to $BwB \times^B X(w') = BwB \times^B P$) that $BwP/B \subset X(ww_I)$, hence also $\overline{BwP}/B = \overline{BwP/B} \subset X(ww_I)$. On the other hand, $X(w)_P = \overline{BwP}/P$, hence $\overline{\pi}^{-1}X(w)_P = \overline{BwP}/B$. This implies

(2) $$X(ww_I) = \overline{\pi}^{-1}X(w)_P.$$

Furthermore, the restriction of $\overline{\pi}$ to $X(ww_I) \to X(w)_P$ is locally trivial with fibres P/B. This implies as $k[P/B] = k$:

(3) $$\overline{\pi}_* \mathcal{O}_{X(ww_I)} = \mathcal{O}_{X(w)_P}.$$

If $w \in W^I$ and if $w = s_1 s_2 \dots s_r$ is a reduced decomposition, then we get as in 13.6(4) a morphism $X(s_1, s_2, \dots, s_r) \to X(w)_P$ which induces an isomorphism between suitable open and dense subschemes. Arguing as in 13.7 we can deduce for all $w, w' \in W^I$ (if k is an algebraically closed field)

(4) $$X(w)_P \subset X(w')_P \iff X(w) \subset X(w') \iff w \leq w'.$$

(Using Remark 14.16 this can be extended to any k.)

13.9. Let us illustrate the Bruhat decomposition by an example. Take $G = SL_m$ or $G = GL_m$ for some $m \in \mathbf{N}$ and suppose (as in 1.21) that B and T are the subgroup schemes of all lower triangular resp. all diagonal matrices in G.

Consider any k–algebra A. For all i, j with $1 \leq i, j \leq m$ and any $(m \times m)$–matrix g over A denote by $c_{ij}(g)$ the matrix consisting of the entries in the first j rows and the last $m - i + 1$ columns of g. So, if $g = (a_{rs})_{1 \leq r,s \leq m}$, then $c_{ij}(g) = (a_{rs})_{1 \leq r \leq j, i \leq s \leq m}$. For all r with $1 \leq r \leq \min(m - i + 1, j)$ denote by $\mathfrak{d}_{ij}^r(g)$ the ideal in A generated by all $(r \times r)$–minors of the matrix $c_{ij}(g)$.

We claim now, for all g, i, j, r as above:

(1) $$\mathfrak{d}_{ij}^r(bgb') = \mathfrak{d}_{ij}^r(g) \qquad \text{for all } b, b' \in B(A).$$

Indeed, if we multiply g with an element of $T(A)$ on the left or on the right, then we multiply each row resp. each column of g with a unit in A. Then also each minor considered above is multiplied with a unit in A. So $\mathfrak{d}_{ij}^r(g)$ does not change.

If we multiply g (on the left or on the right) with an elementary unipotent matrix $b \in B(A)$, i.e., with a matrix with all diagonal entries equal to 1 and with only one non-zero entry off the diagonal, then we add a multiple of one row of g to another row of g, or we add a multiple of one column of g to another column of g. Furthermore, as b is lower triangular, we get $c_{ij}(bg)$ resp. $c_{ij}(gb)$ from $c_{ij}(g)$ by the same elementary operation (or possibly do not change $c_{ij}(g)$ at all). Obviously these elementary operations on $c_{ij}(g)$ do not change the ideal $\mathfrak{d}^r_{ij}(g)$.

Any matrix in $B(A)$ is a product of matrices as above, so (1) follows.

Denote the canonical basis of k^m or of A^m by $e_1, e_2, \dots, e_m$. We can identify W with the symmetric group S_m such that any representative $\dot\sigma \in N_G(T)(A)$ of some $\sigma \in S_m \simeq W$ is given by $\dot\sigma\, e_i = a_i\, e_{\sigma(i)}$ for some $a_i \in A^\times$ and all i. Obviously, any minor of $\dot\sigma$ is either a product of some a_i (hence in $A^\times$) or 0. This implies for all i, j, r as before:

(2) *Either* $\mathfrak{d}^r_{ij}(\dot\sigma) = A$ *or* $\mathfrak{d}^r_{ij}(\dot\sigma) = 0$.

We claim:

$$B(A)\dot\sigma B(A) = \{\, g \in G(A) \mid \mathfrak{d}^r_{ij}(g) = \mathfrak{d}^r_{ij}(\dot\sigma) \text{ for all } i,\, j,\, r\,\}. \tag{3}$$

By (1) one inclusion is trivial. Consider, on the other hand, $g \in G(A)$ with $\mathfrak{d}^r_{ij}(g) = \mathfrak{d}^r_{ij}(\dot\sigma)$ for all i, j, r. Suppose that there is some $j \geq 1$ such that the first $j-1$ rows of g and $\dot\sigma$ are equal. We want to find some $g' \in B(A)gB(A)$ such that the first j rows of g and $\dot\sigma$ are equal. Then (3) follows by induction on j from above.

Let $l_1 > l_2 > \cdots > l_j$ be the numbers $\sigma^{-1}(1), \sigma^{-1}(2), \dots, \sigma^{-1}(j)$ ordered by size. Then one easily checks (for all $r \leq j$) that $\mathfrak{d}^r_{ij}(\dot\sigma) = 0$ for all $i > l_r$, and $\mathfrak{d}^r_{ij}(\dot\sigma) = A$ for all $i \leq l_r$. (This shows, incidentally, that σ is determined by the family of all $\mathfrak{d}^r_{ij}(\dot\sigma)$, hence that the $B(A)\dot\sigma B(A)$ are pairwise disjoint.) Write $l_s = \sigma^{-1}(j)$. We can subtract from the jth row of g suitable multiples of the first $j-1$ rows to get zero entries in all positions (j,i) with $i > l_s$. (This is obvious for i of the form l_r and follows in general from the vanishing of suitable $\mathfrak{d}^r_{ij}(g) = \mathfrak{d}^r_{ij}(\dot\sigma)$.) Now $\mathfrak{d}^s_{l_s j}(g) = A$ implies that the (j, l_s)-entry in our new g is a unit in A. We can subtract multiples of the l_s–column from the other columns in order to get zero entries in all positions (j,i) with $i < l_s$.

So we have replaced g by some $g' \in U(A)gU(A)$ such that the first $j-1$ rows of g' and $\dot\sigma$ coincide (as before) and such that g' has only one non-zero entry in the jth row. It occurs in the same column as in $\dot\sigma$ and is a unit in A. In case $G = SL_m$ and $j = m$, we deduce from $\det(g) = 1 = \det(\dot\sigma)$ that $g = \dot\sigma$. In all other cases, we can multiply g' by some $t \in T(A)$ without changing the first $j-1$ rows so that we get the same entry in the jth row as in $\dot\sigma$. This completes the inductive step, hence the proof of (3).

Note that any condition of the form $\mathfrak{d}^r_{ij}(g) = 0$ (resp. $\mathfrak{d}^r_{ij}(g) = A$) defines a closed (resp. open) subfunctor of G, cf. I.1.4/5. So we have shown here by elementary methods, that each $B\sigma B$ is the intersection of a closed and an open subscheme of G. In fact, we can write down explicitly ideals I_1 and I_2 in $k[G]$ with $B\sigma B = V(I_1) \cap D(I_2)$.

13.10. We want to also look at some $B\sigma P$ for some $P \neq B$ in the example from 13.9. Keep all notations introduced there. We want to take $P = P_I$ with $I = S \setminus \{\alpha\}$ for some $\alpha \in S$. There exists $d \in \mathbf{N}$, $1 \leq d < m$ such that P is the stabiliser of $\sum_{i=m-d+1}^{m} k e_i$ and G/P can be identified with the Grassmannian $\mathcal{G}_{d,m}$, cf. I.5.6. Assume k to be an algebraically closed field. (Using 14.16 one can generalise to any k.)

Identifying $W \simeq S_m$ as before, we get

$$W^I = \{\sigma \in S_m \mid \sigma(1) < \sigma(2) < \cdots < \sigma(m-d),\\ \sigma(m-d+1) < \sigma(m-d+2) < \cdots < \sigma(m)\}.$$

So we get a bijection from W^I onto the set of all sequences $(j_1, j_2, \dots, j_d)$ with $1 \le j_1 < j_2 < \cdots < j_d \le m$: Map σ to the sequence with $j_r = \sigma(m-d+r)$ for all r. Let us write $B(j_1, j_2, \dots, j_d)P$ instead of $B\sigma P$ when σ corresponds to $(j_1, j_2, \dots, j_d)$. Using 13.7 and 13.8(4) one can show for all sequences as above:

(1) $$B(j_1, j_2, \dots, j_d)P \subset \overline{B(j_1', j_2', \dots, j_d')P} \iff j_r' \le j_r \text{ for all } r.$$

Not all $\mathfrak{d}_{i,j}^r$ will be invariant under multiplication by $P(A)$ on the right, but the $\mathfrak{d}_{m-d+1,j}^r$ are easily checked to be so. Arguing as in 13.9 one gets for each sequence $(j_1, j_2, \dots, j_d)$ as above:

(2) $(B(j_1, j_2, \dots, j_d)P)\,(A)$ is the set of all $g \in G(A)$ with $\mathfrak{d}_{m-d+1,j}^r(g) = 0$ for all $j < j_r$, and $\mathfrak{d}_{m-d+1,j}^r(g) = A$ for all $j \ge j_r$ and for all r, $1 \le r \le d$.

For each r $(1 \le r \le d)$ there is some $\sigma_r \in W^I$ corresponding to the sequence $(1, 2, \dots, r, m-d+1, m-d+2, \dots, m-r)$. By (2) any $g \in \overline{B\sigma_r P}(A)$ satisfies $\mathfrak{d}_{m-d+1,m-d}^{r+1}(g) = 0$, whereas any $g' \in B\sigma P(A) \not\subset \overline{B\sigma_r P}(A)$ satisfies $\mathfrak{d}_{m-d+1,m-d}^{r+1}(g') = A$. If A is a field, then we get

(3) $$\overline{B\sigma_r P}(A) = \{\, g \in G(A) \mid \operatorname{rk} c_{m-d+1,m-d}(g) \le r \,\}.$$

In general $\overline{B\sigma_r P}$ is the unique integral and closed subscheme of G such that (3) holds over any field, cf. 13.8.

Let $M_{m-d,d}$ be the scheme that associates to any A the affine space of all $(m-d) \times d$ matrices over A. We can identify $M_{m-d,d}$ with U_I^+ such that any matrix h is mapped to the unique $g \in U_I^+$ with $c_{m-d+1,m-d}(g) = h$. We can identify U_I^+, hence $M_{m-d,d}$ with an open subscheme of G/P, cf. 1.10. We get from (3):

(4) *$M_{m-d,d} \cap X(\sigma_r)_P$ is the unique integral and closed subscheme of $M_{m-d,d}$ such that its points over any field A are just all $h \in M_{m-d,d}(A)$ with* $\operatorname{rk}(h) \le r$.

These subschemes are usually called *determinantal varieties* (at least in the case of a ground field).

One can also define determinantal varieties of symmetric or alternating matrices. They can then be identified with intersections of suitable generalised Schubert schemes with the big cell in an appropriate flag variety for an orthogonal or symplectic group. See [Lakshmibai and Seshadri 1], sections 4/5 for details.

CHAPTER 14

Line Bundles on Schubert Schemes

Let k be (as in Chapter 13) a Dedekind ring, possibly a field.

Each line bundle $\mathcal{L}(\lambda)$ with $\lambda \in X(T)$ restricts to a line bundle (also denoted by $\mathcal{L}(\lambda)$) on each Schubert scheme $X(w)$. One main result of this chapter will be (14.15) that for $\lambda \in X(T)_+$ all $H^i(X(w), \mathcal{L}(\lambda))$ with $i > 0$ vanish and the natural restriction map from $H^0(G/B, \mathcal{L}(\lambda))$ to $H^0(X(w), \mathcal{L}(\lambda))$ is surjective. Each $H^0(X(w), \mathcal{L}(\lambda))$ has a natural structure as a T–module and there is a character formula (14.18) for this T–module (if $\lambda \in X(T)_+$). One main ingredient of the proof of these results is a comparison with the desingularisation of $X(w)$ described in Chapter 13. Via this comparison one also gets that each $X(w)$ is a normal scheme (14.15). Furthermore, one can extend all this to generalised Schubert schemes.

These results were announced in [Demazure 3] in case k is a field of characteristic 0. His proof, however, contains a gap that was not discovered until 1983. Then there was found (for all k) a proof of the vanishing part of the theorem in [Mehta and Ramanathan 1] for ample $\mathcal{L}(\lambda)$ which then led to proof of normality in [Seshadri 4], [Ramanan and Ramanathan], and [Andersen 13]. We follow here Andersen's approach which is closest to the proof of Kempf's vanishing theorem as in Chapter 4.

It should be mentioned that some of the main techniques (14.13/14) of the proof are due to Kempf, but appear in print only in [Demazure 3], 5.1.

On our way to the proof of the main theorem we show (14.8) that a B–module (projective of finite rank over k) extends to a G–module if and only if it extends to all $P(\alpha)$ with $\alpha \in S$. This was first proved in [Cline, Parshall, and Scott 4].

One can identify the closures of the G–orbits in $G/B \times G/B \simeq G \times^B G/B$ with the $G \times^B X(w)$ with $w \in W$. These closures have properties similar to those of the $X(w)$: Given $\lambda, \mu \in X(T)_+$, one has $H^i(G \times^B X(w), \mathcal{L}(\lambda, \mu)) = 0$ for all $i > 0$ and the restriction map from $H^0(G/B \times G/B, \mathcal{L}(\lambda, \mu)) \simeq H^0(\lambda) \otimes H^0(\mu)$ to $H^0(G \times^B X(w), \mathcal{L}(\lambda, \mu))$ is surjective. In the first edition of this book I asked whether this was true, and these results were proved while the book was in the press, see [Kumar 1] and [Mehta and Ramanathan 2].

In the special case $w = 1$, one gets that $G \times^B X(1)$ is just G/B diagonally embedded into $G/B \times G/B$. Any $\mathcal{L}(\lambda, \mu)$ restricts to $\mathcal{L}(\lambda+\mu)$ on $G \times^B X(1) \simeq G/B$, so the vanishing part holds by Kempf's vanishing theorem. (If $\lambda, \mu \in X(T)_+$, then also $\lambda + \mu \in X(T)_+$.) The surjectivity amounts to the cup product inducing a surjective homomorphism $H^0(\lambda) \otimes H^0(\mu) \to H^0(\lambda + \mu)$. This follows from 4.21, but also has a direct proof due to [Ramanan and Ramanathan] that preceded the general result and that we give here in 14.20. It implies for very ample $\mathcal{L}(\lambda)$ that each $X(w)$ is projectively normal with respect to the embedding defined by $\mathcal{L}(\lambda)$.

We do not go through the proof in the general case here. It can easily deduced from the compatible Frobenius splitting of the $G\times^B X(w)$ in $G/B\times G/B$, see F.23.

Compared to the first edition Subsection 14.22 was moved to 14.23 and a new Subsection 14.22 was added.

14.1. Any BwB is isomorphic to some $\mathbf{A}^m\times G_m^l$, hence $k[BwB]$ a free k–module. It contains $k[\overline{BwB}]$ as a submodule which is hence torsion free. Since we assume that k is a Dedekind ring, this implies that $k[\overline{BwB}]$ is a flat k–module. (If M is a torsion free k–module, then its localisation at each maximal ideal of k is torsion free, hence flat as that localisation is a principal ideal domain. Now apply [B2], ch. II, §3, cor. de la prop. 15 to get that M itself is flat.) So $\overline{BwB}$ is a flat scheme. Similarly (or using I.5.7(2)) we see that $\overline{BwB/B}=\overline{BwB}/B=X(w)$ is flat, and more generally:

(1) *All* $V(w_1,w_2,\dots,w_r)$ *and* $X(w_1,w_2,\dots,w_r)$ *are flat.*

Fix $w_1,\ w_2,\dots,w_r\in W$ and abbreviate the two schemes in (1) by V and X. Regard any B–module as a B^r–module by making the first $r-1$ copies of B act trivially. We can then associate to M a sheaf $\mathcal{L}_X(M)$ on $V\simeq V/B^r$ as in I.5.8. As the quotient map $V\to X$ is locally trivial by 13.4(6), we may conclude from I.5.16(2):

(2) *If M is a projective k–module of rank m, then $\mathcal{L}_X(M)$ is a locally free sheaf of $\mathcal{O}_X$–modules of rank m.*

The left actions of B on V and X, which are compatible with the canonical map $V\to X$, make any $\mathcal{L}_X(M)$ into a B–linearised sheaf. So any cohomology group $H^j(X,\mathcal{L}_X(M))$ is a B–module in a natural way. The action on $H^0(X,\mathcal{L}_X(M))=\operatorname{Mor}(V,M_a)^B=(k[V]\otimes M)^B$ is just the one induced by the left action on V, hence on $k[V]$. As we assume k to be a Dedekind domain, we can conclude that $H^0(X,\mathcal{L}_X(M))$ is flat if M is so, as it is a submodule of the torsion free module $k[V]\otimes M$. This will no longer be true for higher H^j. Recall that (by I.5.11) the $H^j(X,\mathcal{L}_X(?))$ are the derived functors of $H^0(X,\mathcal{L}_X(?))$.

For $0<i<r$ set $X_i=X(w_1,w_2,\dots,w_i)$ and $X_i'=X(w_{i+1},w_{i+2},\dots,w_r)$. Define similarly X_r and X_0', set $X_0=X_r'=X(1)$. We have by 13.5(2) isomorphisms of schemes

$$V(w_1,w_2,\dots,w_i)\times^{B^i}X_i'\xrightarrow{\sim}X \tag{3}$$

for $0\le i\le r$. (For $i=0$ replace $V(w_1,w_2,\dots,w_i)$ by the point $\{1\}$, similarly below. For $i=r$ replace $V(w_{i+1},w_{i+2},\dots,w_r)$ below by $\{1\}$.) The isomorphism (3) gives rise to a morphism $\pi_i : X\to V(w_1,w_2,\dots,w_i)/B^i=X_i$ which is locally trivial with fibres X_i', cf. I.5.16, and which is compatible with the left action of B. As $X_i'=V(w_{i+1},w_{i+2},\dots,w_r)/B^{r-i}$ and as $V(w_{i+1},w_{i+2},\dots,w_r)$ is affine, I.5.19(1) yields isomorphisms of B–linearised sheaves

$$R^j(\pi_i)_*\mathcal{L}_X(M)\simeq\mathcal{L}_{X_i}(H^j(X_i',\mathcal{L}_{X_i'}(M)))\qquad\text{(for all } j\ge 0). \tag{4}$$

Recall from 13.5(9) that we have a B–equivariant closed embedding $\psi_i : X_i\to X$. Its construction shows that we can apply I.5.20 using $\alpha' : B^i\to B^r$ with

$(b_1, b_2, \dots, b_i) \mapsto (b_1, b_2, \dots, b_i, b_i, b_i, \dots, b_i)$. As we regard M as a B^i–module resp. as a B^r–module via the action of the last component, we get from I.5.20 an isomorphism of B–linearised sheaves

$$\psi_i^* \mathcal{L}_X(M) \simeq \mathcal{L}_{X_i}(M). \tag{5}$$

14.2. Let $\alpha \in S$, let M be a B–module and M_1 a $P(\alpha)$–module that both are flat over k.

If we combine I.5.12.b and I.4.18.b, then $P(\alpha)/B \simeq \mathbf{P}^1$ implies

$$H^i(P(\alpha)/B, \mathcal{L}_{P(\alpha)/B}(M)) = 0 \qquad \text{for all } i \geq 2. \tag{1}$$

The generalised tensor identity (I.4.8) yields

$$H^i(P(\alpha)/B, \mathcal{L}_{P(\alpha)/B}(M_1)) \simeq M_1 \otimes H^i(P(\alpha)/B, \mathcal{L}_{P(\alpha)/B}(k))$$

for all i, so we get from 5.2 (or from the well-known description of $H^i(\mathbf{P}^1, \mathcal{O}(0))$):

$$H^0(P(\alpha)/B, \mathcal{L}_{P(\alpha)/B}(M_1)) \simeq M_1, \tag{2}$$

$$H^1(P(\alpha)/B, \mathcal{L}_{P(\alpha)/B}(M_1)) = 0. \tag{3}$$

Lemma: *If M is a homomorphic image of M_1 as a B–module, then $H^1(P(\alpha)/B, \mathcal{L}_{P(\alpha)/B}(M)) = 0$.*

Proof: Suppose $M \simeq M_1/M'$ where M' is a B–submodule of M_1. As k is a Dedekind domain, also M' is flat. So (1) and (3) imply $H^2(P(\alpha)/B, \mathcal{L}_{P(\alpha)/B}(M')) = 0 = H^1(P(\alpha)/B, \mathcal{L}_{P(\alpha)/B}(M_1))$. Now $H^1(P(\alpha)/B, \mathcal{L}_{P(\alpha)/B}(M))$ occurs in a long exact sequence just between these two other groups, hence also vanishes.

14.3. Let $s_1, s_2, \dots, s_r$ be simple reflections. Set $X_i = X(s_1, s_2, \dots, s_i)$ for $0 < i \leq r$ and $X'_i = X(s_{i+1}, s_{i+2}, \dots, s_r)$ for $0 \leq i < r$. Set $X_0 = X'_r = X(1)$ and $X = X_r = X'_0$. There are $\alpha_i \in S$ with $s_i = s_{\alpha_i}$, hence $X(s_i) = P(\alpha_i)/B$.

Lemma: *Let M be a B–module that is flat over k.*

a) *If M can be extended to a $P(\alpha_i)$–module for all i $(1 \leq i \leq r)$, then $M \simeq H^0(X, \mathcal{L}_X(M))$ as a B–module.*

b) *If each B–module $H^0(X'_i, \mathcal{L}_{X'_i}(M))$ with $1 \leq i \leq r$ is a homomorphic image of a G–module that is flat over k, then $H^j(X, \mathcal{L}_X(M)) = 0$ for all $j > 0$.*

Proof: a) We have $X'_{r-1} = X(s_r) = P(\alpha_r)/B$, hence $H^0(X'_{r-1}, \mathcal{L}_{X'_{r-1}}(M)) \simeq M$ by 14.2(2). If we apply 14.1(4) with $j = 0$ and $i = r-1$ and take global sections, then we get

$$H^0(X, \mathcal{L}_X(M)) \simeq H^0(X_{r-1}, \mathcal{L}_{X_{r-1}}(M)).$$

Now we can use induction on r.

b) We get from 14.1 for $0 \leq i < j \leq r$ morphisms $\pi_{ij} : X_j \to X_i$ with $\pi_{i_1 i_2} \circ \pi_{i_2 i_3} = \pi_{i_1 i_3}$ for $0 \leq i_1 < i_2 < i_3 \leq r$. Set $M'_i = H^0(X'_i, \mathcal{L}_{X'_i}(M))$ for all i. This is a B–module that is flat over k, cf. 14.1. So our assumption implies by Lemma 14.2:

$$H^j(P(\alpha_i)/B, \mathcal{L}_{P(\alpha_i)/B}(M'_i)) = 0 \qquad \text{for all } j > 0.$$

Therefore the isomorphism

$$R^j(\pi_{i-1,i})_*\mathcal{L}_{X_i}(M_i') \simeq \mathcal{L}_{X_{i-1}}(H^j(P(\alpha_i)/B, \mathcal{L}_{P(\alpha_i)/B}(M)))$$

from 14.1(4) yields $R^j(\pi_{i-1,i})_*\mathcal{L}_{X_i}(M_i') = 0$ for all $j > 0$. Now the Leray spectral sequence

$$H^l(X_{i-1}, R^j(\pi_{i-1,i})_*\mathcal{L}_{X_i}(M_i')) \Rightarrow H^{l+j}(X_i, \mathcal{L}_{X_i}(M_i'))$$

degenerates and yields isomorphisms

$$H^j(X_{i-1}, (\pi_{i-1,i})_*\mathcal{L}_{X_i}(M_i')) \simeq H^j(X_i, \mathcal{L}_{X_i}(M_i')).$$

As $\mathcal{L}_{X_i}(M_i') \simeq (\pi_{ir})_*\mathcal{L}_X(M)$ (again by 14.1(4)) we can write this isomorphism in the form

$$H^j(X_{i-1}, (\pi_{i-1,r})_*\mathcal{L}_X(M)) \simeq H^j(X_i, (\pi_{i,r})_*\mathcal{L}_X(M)).$$

Iterating we get

$$H^j(X_0, (\pi_{0,r})_*\mathcal{L}_X(M)) \simeq H^j(X, \mathcal{L}_X(M)).$$

As $X_0 = X(1)$ is a point, these groups vanish for $j > 0$.

14.4. We have $P = (P_{\mathbf{Z}})_k$ for any standard parabolic subgroup P of G. As $X(s_\alpha) = P(\alpha)/B$ for all $\alpha \in S$, this implies $X(s_\alpha) = (X(s_\alpha)_{\mathbf{Z}})_k$ where we use (here and below) the index $\mathbf{Z}$ to indicate the corresponding construction over $\mathbf{Z}$. Now I.5.5(4) yields for all simple reflections $s_1, s_2, \dots, s_r$

(1) $$X(s_1, s_2, \dots, s_r) \simeq (X(s_1, s_2, \dots, s_r)_{\mathbf{Z}})_k.$$

Let M be a $B_{\mathbf{Z}}$–module that is free and finitely generated over $\mathbf{Z}$ and let $\mathcal{L}(M)_{\mathbf{Z}}$ be the associated sheaf on $X_{\mathbf{Z}} = X(s_1, s_2, \dots, s_r)_{\mathbf{Z}}$. Then $\mathcal{L}(M)_k = (\mathcal{L}(M)_{\mathbf{Z}})_k$ is the associated sheaf on $X = X(s_1, s_2, \dots, s_r)$ corresponding to the B–module M_k. According to [M1], p. 46 there is a complex $(C^i)_{i\geq 0}$ of free $\mathbf{Z}$–modules of finite rank such that the complex $(C^i \otimes_{\mathbf{Z}} A)_{i\geq 0}$ computes the cohomology $H^\bullet(X_A, \mathcal{L}(M)_A)$, for each $\mathbf{Z}$–algebra A. This yields as in I.4.18 exact sequences (for all i and A)

$$0 \longrightarrow H^i(X_{\mathbf{Z}}, \mathcal{L}(M)_{\mathbf{Z}}) \otimes_{\mathbf{Z}} A \longrightarrow H^i(X_A, \mathcal{L}(M)_A)$$
$$\longrightarrow \operatorname{Tor}_1^{\mathbf{Z}}(H^{i+1}(X_{\mathbf{Z}}, \mathcal{L}(M)_{\mathbf{Z}}), A) \longrightarrow 0.$$

This implies:

(2) *If there is for each prime number p a field $K(p)$ of characteristic p with $H^i(X_{K(p)}, \mathcal{L}(M)_{K(p)}) = 0$ for all $i > 0$, then $H^i(X_A, \mathcal{L}(M)_A) = 0$ for all $\mathbf{Z}$–algebras A and all $i > 0$.*

14.5. Let X be an integral scheme. Then for any open and affine $X' \subset X$, $X' \neq \emptyset$ the ring $k[X']$ is an integral domain. Its ring of fractions is independent of X' and is denoted by $k(X)$. (If $X'' \subset X' \subset X$ both are open and affine, $X'' \neq \emptyset$, then the restriction of functions $k[X'] \to k[X'']$ induces an isomorphism of fields of fractions.) The field $k(X)$ is called the *function field* of X, cf. [Ha], II, exerc. 3.6.

Let $\varphi : Y \to X$ be a morphism of integral schemes. Suppose that φ is dominant, i.e., that $\varphi(Y)$ is dense in X, cf. [Ha], II, exerc. 3.7. There are $X' \subset X$ and $Y' \subset Y$ open and affine, $Y' \neq \emptyset$ with $\varphi(Y') \subset X'$. Then the comorphism $\varphi^* : k[X'] \to k[Y']$ is injective (by the dominance of φ), hence induces a homomorphism $\varphi^* : k(X) \to k(Y)$ of the fields of fractions. It is easily checked to be independent of the choice of X' and Y'.

Recall that X is called *normal* if each $k[X']$ with $X' \subset X$ open and affine is integrally closed. It is enough to check this for an open covering of X.

Lemma: *Let $\varphi : Y \to X$ be a dominant and projective morphism of noetherian and integral schemes such that φ induces an isomorphism $k(X) \xrightarrow{\sim} k(Y)$ of function fields.*

a) *If X is normal, then $\varphi_* \mathcal{O}_Y = \mathcal{O}_X$.*

b) *If Y is normal and if $\varphi_* \mathcal{O}_Y = \mathcal{O}_X$, then X is normal.*

Proof: As the statements are local, we may assume that X is affine.

a) The comorphism φ^* induces an inclusion from $k[X]$ into $\varphi_* \mathcal{O}_Y(X) = \mathcal{O}_Y(Y)$ compatible with the isomorphism $k(X) \xrightarrow{\sim} k(Y)$. So we can regard $\varphi_* \mathcal{O}_Y(X)$ as a subalgebra of $k(X)$ containing $k[X]$.

As φ is projective, the $\mathcal{O}_X$–module $\varphi_* \mathcal{O}_Y$ is coherent (cf. [Ha], III, 8.8). So $\varphi_* \mathcal{O}_Y(X)$ is a finitely generated $k[X]$–module, hence a subalgebra of $k(X)$ that is integral over $k[X]$. As $k[X]$ is integrally closed, we get $\varphi_* \mathcal{O}_Y(X) = k[X] = \mathcal{O}_X(X)$, hence $\varphi_* \mathcal{O}_Y = \mathcal{O}_X$.

b) As Y is normal and as φ is dominant, there is a factorisation

$$\varphi : Y \xrightarrow{\alpha} \widetilde{X} \xrightarrow{\beta} X$$

where $\beta : \widetilde{X} \to X$ is the normalisation of X, cf. [Ha], II, exerc. 3.8. So we have in a natural way

$$\mathcal{O}_X \subset \beta_* \mathcal{O}_{\widetilde{X}} \subset \varphi_* \mathcal{O}_Y .$$

Now our assumption yields $\mathcal{O}_X = \beta_* \mathcal{O}_{\widetilde{X}}$, and $k[X] = k[\widetilde{X}]$ is integrally closed.

14.6. We collect here some properties of $\mathcal{O}_X$–modules that will be used later on. Let $\varphi : Y \to X$ be a morphism of schemes and let $\mathcal{M}$ be a locally free $\mathcal{O}_X$–module of finite rank.

We have for all $i \geq 0$ and all $\mathcal{O}_Y$–modules $\mathcal{F}$ the projection formula (cf. [Ha], exerc. 8.3):

(1) $$R^i \varphi_* (\mathcal{F} \otimes \varphi^* \mathcal{M}) \simeq (R^i \varphi_* \mathcal{F}) \otimes \mathcal{M}.$$

Taking $\mathcal{F} = \mathcal{O}_Y$, we get

(2) $$R^i \varphi_* (\varphi^* \mathcal{M}) \simeq (R^i \varphi_* \mathcal{O}_Y) \otimes \mathcal{M}.$$

There is for each $\mathcal{O}_Y$–module $\mathcal{F}$ the Leray spectral sequence $H^j(X, R^i \varphi_* \mathcal{F}) \Rightarrow H^{i+j}(Y, \mathcal{F})$. If $R^i \varphi_* \mathcal{F} = 0$ for all $i > 0$, then this spectral sequence degenerates and yields isomorphisms $H^j(X, \varphi_* \mathcal{F}) \simeq H^j(Y, \mathcal{F})$ for all j. So (2) yields:

(3) *If $R^i \varphi_* \mathcal{O}_Y = 0$ for all $i > 0$, then $H^j(Y, \varphi^* \mathcal{M}) \simeq H^j(X, (\varphi_* \mathcal{O}_Y) \otimes \mathcal{M})$ for all $j \in \mathbf{N}$. If in addition $\varphi_* \mathcal{O}_Y = \mathcal{O}_X$, then even $H^j(Y, \varphi^* \mathcal{M}) \simeq H^j(X, \mathcal{M})$ for all j.*

Suppose now that X is a noetherian and projective scheme. Let $\mathcal{L}$ be an ample line bundle on X. By the cohomological criterion for ampleness (cf. [Ha], III, 5.3) there is for any coherent $\mathcal{O}_X$–module $\mathcal{F}$ an integer $n(\mathcal{F}) \in \mathbf{N}$ such that $H^i(X, \mathcal{F} \otimes \mathcal{L}^n) = 0$ for all $n \geq n(\mathcal{F})$ and all $i > 0$.

Consider a homomorphism $\alpha : \mathcal{F} \to \mathcal{F}'$ of coherent $\mathcal{O}_X$–modules. We can choose n_0 such that $H^1(X, \ker(\alpha) \otimes \mathcal{L}^n) = 0 = H^1(X, \mathrm{im}(\alpha) \otimes \mathcal{L}^n)$ for all $n \geq n_0$. The short exact sequences $0 \to \ker(\alpha) \to \mathcal{F} \to \mathrm{im}(\alpha) \to 0$ and $0 \to \mathrm{im}(\alpha) \to \mathcal{F}' \to \mathrm{coker}(\alpha) \to 0$ give rise to long exact sequences (after tensoring with $\mathcal{L}^n$) which for $n \geq n_0$ yield an exact sequence

$$H^0(X, \mathcal{F} \otimes \mathcal{L}^n) \to H^0(X, \mathcal{F}' \otimes \mathcal{L}^n) \to H^0(X, \mathrm{coker}(\alpha) \otimes \mathcal{L}^n) \to 0.$$

We may choose n_0 even so large that $\mathrm{coker}(\alpha) \otimes \mathcal{L}^n$ is generated by its global sections (for all $n \geq n_0$). Therefore we get:

(4) *The homomorphism α is surjective if and only if there is $n_0 \in \mathbf{N}$ such that $H^0(X, \mathcal{F} \otimes \mathcal{L}^n) \to H^0(X, \mathcal{F}' \otimes \mathcal{L}^n)$ is surjective for all $n \geq n_0$.*

Let X' be a closed subscheme of X and denote the inclusion by $i : X' \hookrightarrow X$. It gives rise to a short exact sequence of $\mathcal{O}_X$–modules

$$0 \to \mathcal{I}_{X'} \longrightarrow \mathcal{O}_X \longrightarrow i_*\mathcal{O}_{X'} \to 0$$

where $\mathcal{I}_{X'}$ is the sheaf of ideals defining X', cf. [Ha], p. 115. This sequence remains exact when we tensor with $\mathcal{M} \otimes \mathcal{L}^n$ for any $n \in \mathbf{N}$ (with $\mathcal{M}$ as above). There is $n_0 \in \mathbf{N}$ with $H^1(X, \mathcal{I}_{X'} \otimes \mathcal{M} \otimes \mathcal{L}^n) = 0$ for all $n \geq n_0$. So we get an exact sequence (for $n \geq n_0$)

$$\begin{aligned} 0 \longrightarrow H^0(X, \mathcal{I}_{X'} \otimes \mathcal{M} \otimes \mathcal{L}^n) &\longrightarrow H^0(X, \mathcal{M} \otimes \mathcal{L}^n) \\ \longrightarrow H^0(X, (i_*\mathcal{O}_{X'}) \otimes \mathcal{M} \otimes \mathcal{L}^n) &\longrightarrow 0. \end{aligned}$$

By (2) we can identify the last term with

$$H^0(X, i_*i^*(\mathcal{M} \otimes \mathcal{L}^n)) \simeq H^0(X', i^*(\mathcal{M} \otimes \mathcal{L}^n)).$$

The last map is simply given by restriction of sections from X to X'. So we see:

(5) *There does not exist $n_0 \in \mathbf{N}$ such that for all $n \geq n_0$ the restriction map $H^0(X, \mathcal{M} \otimes \mathcal{L}^n) \to H^0(X', i^*(\mathcal{M} \otimes \mathcal{L}^n))$ is surjective.*

14.7. Let $w_0 = s_1 s_2 \dots s_n$ be a reduced decomposition (with $s_i = s_{\alpha_i}$ and $\alpha_i \in S$) of the longest element $w_0 \in W$, cf. 1.5. So $n = l(w_0) = |R^+|$ and $X(w_0) = G/B$.

Set $Z = X(s_1, s_2, \dots, s_n)$. The projection onto the last factor induces a morphism $\varphi' : Z \to G/B$ which maps the open subscheme $X'(s_1, s_2, \dots, s_n)$ isomorphically to the open subscheme Bw_0B/B of G/B, cf. 13.6(3), (4). It induces therefore an isomorphism of the function fields $k(G/B) \xrightarrow{\sim} k(Z)$. Both G/B and Z admit open coverings by affine spaces, hence are normal. So Lemma 14.5.a yields

$$\varphi'_*\mathcal{O}_Z = \mathcal{O}_{G/B}. \tag{1}$$

Let $P = P_I$ with $I \subset S$ be a standard parabolic subgroup of G and denote by $\pi' : G/B \to G/P$ the canonical map. Set $\varphi = \pi' \circ \varphi' : Z \to G/P$. Applying 13.8(3) to $ww_I = w_0$, then we get $(\pi')_*\mathcal{O}_{G/B} = \mathcal{O}_{G/P}$, hence

$$\varphi_*\mathcal{O}_Z = \mathcal{O}_{G/P}. \tag{2}$$

We have by I.5.17(1) for any P–module M an isomorphism

$$\varphi^*\mathcal{L}_{G/P}(M) \simeq \mathcal{L}_Z(M).$$

Note that $\mathcal{L}_Z(M)$ depends only on the B–module structure on M.

Suppose now that M is projective of finite rank over k. Then $\mathcal{L}_{G/P}(M)$ is locally free of finite rank, see I.5.16(2). We can therefore apply the projection formula 14.6(1), (2) and get for all $i \geq 0$ using (3)

$$R^i\varphi_*\mathcal{L}_Z(M) \simeq \mathcal{L}_{G/P}(M) \otimes R^i\varphi_*\mathcal{O}_Z, \tag{4}$$

hence by (2)

$$\varphi_*\mathcal{L}_Z(M) \simeq \mathcal{L}_{G/P}(M). \tag{5}$$

Taking global sections yields

$$H^0(Z, \mathcal{L}_Z(M)) \simeq H^0(G/P, \mathcal{L}_{G/P}(M)). \tag{6}$$

14.8. Keep the assumptions and notations from 14.7.

Proposition: *Let M be a B–module that is projective of finite rank over k.*

a) *The B–module structure on $H^0(Z, \mathcal{L}_Z(M))$ can be extended to a G–module structure. Considered as a k–module, $H^0(Z, \mathcal{L}_Z(M))$ is projective of finite rank.*

b) *If the B–module structure on M can be extended to each $P(\alpha)$ with $\alpha \in S$, then it can be extended to G.*

Proof: We apply 14.7 in the case $P = B$. As $\varphi : Z \to G/B$ commutes with the left action of B, so does the isomorphism in 14.7(6). As the right hand side there extends to G, so does the left hand side. This proves the first part in a). We know that $H^0(Z, \mathcal{L}_Z(M))$ is finitely generated over k, see [Ha], III, 5.2(a), and that it is flat over k, see 14.1. As we assume k to be a Dedekind domain, we get now that $H^0(Z, \mathcal{L}_Z(M))$ is projective from [B2], ch. VII, §4, prop. 22.

For M as in b) we have $M \simeq H^0(Z, \mathcal{L}_Z(M))$ by Lemma 14.3.a. So b) follows from a).

Remark: The G–module structures in a) and b) are unique. Indeed, in case k is a field this follows from the general fact that $\mathrm{Hom}_G(M, M') = \mathrm{Hom}_B(M, M')$ for all G–modules M and M', see 4.7.b. The proof given there generalises to k provided the modules are flat over k. (We have to use that 4.6.a extends by 8.8.)

14.9. Let $w \in W$. The results in 14.7 generalise from G/B to $X(w)$, if $X(w)$ is normal. To be more precise, let $w = s_1 s_2 \dots s_r$ be a reduced decomposition of w. Set $X = X(s_1, s_2, \dots, s_r)$. Choose $I \subset S$ with $w \in W^I$. (Note that we can always take $I = \emptyset$.) Set $P = P_I$.

We have by 13.8 a morphism $\varphi : X \to X(w)_P$ that induces an isomorphism between the open and dense subschemes $X'(s_1, s_2, \dots, s_r)$ and BwP/P. (It is also the composite of $X \to X(w)$ as in 13.6(4) with $X(w) \to X(w)_P$ induced by the canonical map $G/B \to G/P$.) So Lemma 14.5 implies:

$$X(w)_P \text{ normal} \iff \varphi_*\mathcal{O}_X = \mathcal{O}_{X(w)_P}. \tag{1}$$

As in 14.7 we get for any P–module M

(2) $$\varphi^* \mathcal{L}_{X(w)_P}(M) \simeq \mathcal{L}_X(M).$$

If $i : X(w)_P \to G/P$ is the inclusion induced by the inclusion $\overline{BwP} \to G$, then (also by I.5.17(1))

(3) $$i^* \mathcal{L}_{G/P}(M) \simeq \mathcal{L}_{X(w)_P}(M).$$

Suppose now that M is projective of finite rank over k. Then we get as in 14.7:

(4) *Suppose that $X(w)_P$ is normal. Then $R^j\varphi_* \mathcal{L}_X(M) \simeq \mathcal{L}_{X(w)_P}(M) \otimes R^j \varphi_* \mathcal{O}_X$ for all $j \geq 0$ and $\varphi_* \mathcal{L}_X(M) \simeq \mathcal{L}_{X(w)_P}(M)$. The restriction map*

$$H^0(X(w)_P, \mathcal{L}_{X(w)_P}(M)) \longrightarrow H^0(X, \mathcal{L}_X(M))$$

is an isomorphism of B–modules.

Let w_I be (as before) the longest element in W_I. Recall that the canonical map $G/B \to G/P$ induces a locally trivial map $X(ww_I) \to X(w)_P$ with fibres P/B. As P/B has an open covering by affine spaces, we can deduce, cf. [B2], ch. V, §1, cor. 1 de la prop. 13:

(5) *If $X(w)_P$ is normal, then so is $X(ww_I)$.*

14.10. Let p be a prime number. Assume until 14.12 that k is an algebraically closed field of characteristic p and that $\rho \in X(T)$.

As G is defined and split over $\mathbf{F}_p$, we have a Frobenius endomorphism F on G that stabilises B and T and induces multiplication by p on $X(T)$. We can choose all $\dot{w} \in G(\mathbf{F}_p)$, so all $V'(w_1, \ldots, w_m)$ and $V(w_1, \ldots, w_m)$ are also F–stable. The action of F on $V(w_1, \ldots, w_m)$ is compatible with that on B and B^m and the left resp. right actions of these groups on $V(w_1, \ldots, w_m)$, i.e., we have $F(bxb') = F(b)F(x)F(b')$ for all $b \in B$, $x \in V(w_1, \ldots, w_m)$, $b' \in B^m$. Therefore we also get an action of F on $X(w_1, \ldots, w_m)$ that is still compatible with the left action of B.

Let $w_0 = s_1 s_2 \ldots s_n$ be a reduced decomposition of w_0 and set $Z = X(s_1, s_2, \ldots, s_n)$ as in 14.7. Fix some i $(1 \leq i \leq n)$ and set $X = X(s_1, s_2, \ldots, s_i)$. Let M, M' be B–modules and consider $\mathcal{L}(M) = \mathcal{L}_X(M)$. (We shall drop the index to $\mathcal{L}$ also for other schemes.) If $f \in H^0(X, \mathcal{L}(M))$, then $(F^r)^* f = f \circ F^r \in H^0(X, \mathcal{L}(M^{[r]}))$ for all $r \in \mathbf{N}$, $r > 0$, and $f \mapsto (F^r)^* f$ is easily checked to be a (functorial) homomorphism of B–modules

$$H^0(X, \mathcal{L}(M))^{[r]} \longrightarrow H^0(X, \mathcal{L}(M^{[r]})).$$

On the other hand, we have also a (functorial) homomorphism

$$H^0(X, \mathcal{L}(M)) \otimes H^0(X, \mathcal{L}(M')) \longrightarrow H^0(X, \mathcal{L}(M \otimes M'))$$

where any $f \otimes f'$ on the left hand side is sent to the map $v \mapsto f(v) \otimes f'(v)$ on $V(s_1, s_2, \ldots, s_i)$. (This is just the cup product in degree zero.) Combining such maps we get a (functorial) homomorphism

(1) $$H^0(X, \mathcal{L}(M))^{[r]} \otimes H^0(X, \mathcal{L}(M')) \longrightarrow H^0(X, \mathcal{L}(M^{[r]} \otimes M')).$$

We shall look at this map in the case where M' is the one dimensional module given by $(p^r - 1)\rho$.

For any standard parabolic subgroup $P \supset B$ of G we have by 3.19 (applied to a Levi factor of P) an isomorphism

$$\begin{aligned}(2)\qquad & H^0(P/B, \mathcal{L}(M))^{[r]} \otimes H^0(P/B, \mathcal{L}((p^r-1)\rho)) \\ & \qquad \xrightarrow{\sim} H^0(P/B, \mathcal{L}(M^{[r]} \otimes (p^r-1)\rho)).\end{aligned}$$

One can check that these isomorphisms arise in the same manner as the map in (1).

We can especially take $P = G$. Now (2) and 14.7(6) imply that (1) is an isomorphism in the case $X = Z$ and $M' = (p^r - 1)\rho$:

$$(3)\qquad H^0(Z, \mathcal{L}(M))^{[r]} \otimes H^0(Z, \mathcal{L}((p^r-1)\rho)) \xrightarrow{\sim} H^0(Z, \mathcal{L}(M^{[r]} \otimes (p^r-1)\rho)).$$

14.11. Keep the assumptions and notations from 14.10. Fix a B–module M and denote the homomorphism from 14.10(1) [with $M' = (p^r - 1)\rho$] by γ:

$$(1)\qquad H^0(X, \mathcal{L}(M))^{[r]} \otimes H^0(X, \mathcal{L}((p^r-1)\rho)) \xrightarrow{\gamma} H^0(X, \mathcal{L}(M^{[r]} \otimes (p^r-1)\rho)).$$

Let $\varepsilon : H^0(X, \mathcal{L}((p^r-1)\rho)) \to k$ be the evaluation map $f \mapsto f(1, 1, \ldots, 1)$. This is a homomorphism of B–modules.

Lemma: *There is a (functorial) homomorphism of B–modules*

$$\gamma' : H^0(X, \mathcal{L}(M^{[r]} \otimes (p^r-1)\rho)) \longrightarrow H^0(X, \mathcal{L}(M))^{[r]} \otimes (p^r-1)\rho$$

such that $\gamma' \circ \gamma = \mathrm{id} \otimes \varepsilon$.

Proof. Recall that $X = X(s_1, s_2, \ldots, s_i)$. We want to prove the lemma using induction on i. If $i = 1$, then $X = X(s_1) \simeq P(\alpha)/B$ for some $\alpha \in S$. In this case γ is an isomorphism by 14.10(2), and we can take $\gamma' = (\mathrm{id} \otimes \varepsilon) \circ \gamma^{-1}$.

Suppose now $i > 1$. Set $X = X'(s_1, s_2, \ldots, s_{i-1})$ and $P = P(\alpha)$ where $\alpha \in S$ is the simple root with $s_i = s_\alpha$. We get from 14.1(4) an isomorphism of B–modules

$$(2)\qquad H^0(X, \mathcal{L}(M^{[r]} \otimes (p^r-1)\rho)) \simeq H^0(X', \mathcal{L}(H^0(P/B, \mathcal{L}(M^{[r]} \otimes (p^r-1)\rho)))).$$

We have already treated the case of P/B, hence get by the functoriality of $H^0(X', \mathcal{L}(?))$ a homomorphism from the right hand side in (2) to

$$H^0(X', \mathcal{L}(H^0(P/B, \mathcal{L}(M)^{[r]} \otimes (p^r-1)\rho)),$$

then (by applying the induction to X') a map to

$$H^0(X', \mathcal{L}(H^0(P/B, \mathcal{L}(M)))^{[r]} \otimes (p^r-1)\rho$$

which we can identify [by 14.1(4)] with

$$H^0(X, \mathcal{L}(M))^{[r]} \otimes (p^r-1)\rho.$$

So we have constructed a (functorial) map γ'. It is left to the reader to check that, indeed, $\gamma' \circ \gamma = \mathrm{id} \otimes \varepsilon$.

14.12. Keep the assumptions and notations of the last two subsections. Recall that the morphism $\psi : X \to Z$ as in 13.5(9) satisfies $\psi^*\mathcal{L}_Z(M) \simeq \mathcal{L}_X(M)$ for any B–module M, hence induces a restriction map $H^0(Z,\mathcal{L}(M)) \to H^0(X,\mathcal{L}(M))$.

We have by 13.5(9) a commutative diagram of morphisms

$$\begin{array}{ccc} X & \xrightarrow{\psi} & Z \\ \downarrow & & \downarrow \\ X(w) & \longrightarrow & G/B \end{array} \tag{1}$$

where the vertical maps are as in 14.1(4). We get therefore a commutative diagram of restriction maps

$$\begin{array}{ccc} H^0(G/B,\mathcal{L}(M)) & \longrightarrow & H^0(X(w),\mathcal{L}(M)) \\ \downarrow & & \downarrow \\ H^0(Z,\mathcal{L}(M)) & \longrightarrow & H^0(X,\mathcal{L}(M)). \end{array} \tag{2}$$

(So far we did not use our assumption on k.)

Proposition: *Let $\lambda \in X(T)_+$. If $X(w)$ is normal, then the restriction map $H^0(Z,\mathcal{L}(\lambda)) \to H^0(X,\mathcal{L}(\lambda))$ is surjective.*

Proof: The line bundle $\mathcal{L}(\lambda+\rho)$ on G/B is ample, cf. 4.4, and the sheaf $\mathcal{L}(-\rho)$ is coherent. So we can find by 14.6(5) and 14.9(3) some $r \in \mathbf{N}$ so that the restriction map

$$H^0(G/B,\mathcal{L}(p^r(\lambda+\rho)-\rho)) \longrightarrow H^0(X(w),\mathcal{L}(p^r(\lambda+\rho)-\rho))$$

is surjective.

Consider the diagram (2) for $M = p^r(\lambda+\rho)-\rho$. The vertical maps are isomorphisms by 14.7(6) resp. 14.9(4). (This is where we use the normality of $X(w)$.) The top horizontal map is surjective by our choice of r, hence so is the lower horizontal map

$$H^0(Z,\mathcal{L}(p^r(\lambda+\rho)-\rho)) \longrightarrow H^0(X,\mathcal{L}(p^r(\lambda+\rho)-\rho)).$$

On the other hand, there is by 14.11 a commutative diagram

$$\begin{array}{ccc} H^0(Z,\mathcal{L}(\lambda))^{[r]} \otimes H^0(Z,\mathcal{L}((p^r-1)\rho)) & \xrightarrow{\sim} & H^0(Z,\mathcal{L}(p^r(\lambda+\rho)-\rho)) \\ \beta\downarrow & & \downarrow\beta' \\ H^0(X,\mathcal{L}(\lambda))^{[r]} \otimes H^0(Z,\mathcal{L}((p^r-1)\rho)) & \xrightarrow{\gamma} & H^0(X,\mathcal{L}(p^r(\lambda+\rho)-\rho)) \\ \mathrm{id}\otimes\varepsilon \searrow & & \swarrow \gamma' \\ & H^0(X,\mathcal{L}(\lambda))^{[r]} \otimes (p^r-1)\rho & \end{array}$$

The evaluation map

$$St_r \simeq H^0(Z,\mathcal{L}((p^r-1)\rho)) \longrightarrow H^0(X,\mathcal{L}((p^r-1)\rho)) \longrightarrow (p^r-1)\rho$$

is surjective (cf. 2.2(1)), hence so are ε and $\mathrm{id}\otimes\varepsilon = \gamma' \circ \gamma$, hence also γ'. We have seen above that β' is surjective, hence so are $\gamma' \circ \beta'$ and $(\mathrm{id}\otimes\varepsilon)\circ\beta$. Then, necessarily, the restriction map $H^0(Z,\mathcal{L}(\lambda)) \to H^0(X,\mathcal{L}(\lambda))$ is surjective.

Remark: Recall that we assume $\rho \in X(T)$ in 14.10–12. Note however, that Proposition 14.12 holds without this assumption since we can replace G by a suitable covering group. This replacement does not change X and Z or the cohomology groups, see Remark 13.4.

14.13. Proposition: *Let $\varphi : Y \to X$ be a morphism of noetherian and projective schemes. Let $\mathcal{L}$ be an ample line bundle on X. If there is an integer $m_0 \in \mathbf{N}$ such that $H^i(Y, \varphi^*\mathcal{L}^m) = 0$ for all $i > 0$ and all $m \geq m_0$, then $R^i\varphi_*\mathcal{O}_Y = 0$ for all $i > 0$.*

Proof: As each $R^i\varphi_*\mathcal{O}_Y$ is ample (cf. [Ha], III, 8.8) and as $\mathcal{L}$ is ample, there is for each i some $m(i) \in \mathbf{N}$ with

$$H^j(X, R^i\varphi_*(\varphi^*\mathcal{L}^m)) \simeq H^j(X, (R^i\varphi_*\mathcal{O}_Y) \otimes \mathcal{L}^m) = 0$$

for all $m \geq m(i)$ and all $j > 0$, cf. also 14.6(2). In the Leray spectral sequence

$$H^j(X, R^i\varphi_*(\varphi^*\mathcal{L}^m)) \Rightarrow H^{j+i}(Y, \varphi^*\mathcal{L}^m)$$

any given $H^r(Y, \varphi^*\mathcal{L}^m)$ will involve only finitely many $R^i\varphi_*(\varphi^*\mathcal{L}^m)$. We can therefore choose $m_i \in \mathbf{N}$ such that for all $m > m_i$

$$H^i(Y, \varphi^*\mathcal{L}^m) \simeq H^0(X, R^i\varphi_*(\varphi^*\mathcal{L}^m)).$$

So our assumption implies for all i and all $m \geq m_i$

$$H^0(X, (R^i\varphi_*\mathcal{O}_Y) \otimes \mathcal{L}^m) = 0. \tag{1}$$

On the other hand, as $\mathcal{L}$ is ample, each $(R^i\varphi_*\mathcal{O}_Y) \otimes \mathcal{L}^m$ is generated by its global sections for m large enough. So (1) implies $(R^i\varphi_*\mathcal{O}_Y)\otimes\mathcal{L}^m = 0$, hence $R^i\varphi_*\mathcal{O}_Y = 0$.

14.14. Proposition: *Let $\varphi : Y \to X$ be a dominant morphism of noetherian and projective schemes. Let $\mathcal{L}$ be an ample line bundle on X. Let $Y' \subset Y$ and $X' \subset X$ be closed subschemes such that φ induces a dominant morphism $\varphi' : Y' \to X'$. If there is an integer $m_0 \in \mathbf{N}$ such that the restriction map $H^0(Y, \varphi^*\mathcal{L}^m) \to H^0(Y', \varphi'^*\mathcal{L}^m)$ is surjective for all $m \geq m_0$ and if $\varphi_*\mathcal{O}_Y = \mathcal{O}_X$, then $\varphi'_*\mathcal{O}_{Y'} = \mathcal{O}_{X'}$.*

Proof: Let us denote the inclusions by $i : X' \hookrightarrow X$ and $j : Y' \hookrightarrow Y$. We have a commutative diagram for each $m \in \mathbf{N}$

$$\begin{array}{ccc} H^0(X, \varphi_*\mathcal{O}_Y \otimes \mathcal{L}^m) & \xrightarrow{\sim} & H^0(Y, \varphi^*\mathcal{L}^m) \\ \downarrow & & \downarrow \\ H^0(X', \varphi'_*\mathcal{O}_{Y'} \otimes i^*\mathcal{L}^m) & \xrightarrow{\sim} & H^0(Y', j^*\varphi^*\mathcal{L}^m). \end{array}$$

The horizontal maps are induced by φ resp. φ' (using $j^*\varphi^* = \varphi'^*i^*$). They are isomorphisms by the projection formula, see 14.6(2). The vertical maps arise from restriction. The right one is surjective by assumption for $m \geq m_0$, hence so is the left one.

As $\varphi_*\mathcal{O}_Y = \mathcal{O}_X$ and as $i^*\mathcal{O}_X = \mathcal{O}_{X'}$, we can also express this as saying that

$$H^0(X', \mathcal{O}_{X'} \otimes i^*\mathcal{L}^m) \longrightarrow H^0(X', \varphi'_*\mathcal{O}_{Y'} \otimes i^*\mathcal{L}^m)$$

is surjective for all $m \geq m_0$. This map is induced by the natural map $\mathcal{O}_{X'} \to \varphi'_*\mathcal{O}_{Y'}$. As these sheaves are coherent and as $i^*\mathcal{L}$ is ample (cf. [Ha], III, exerc. 5.7.a), we get from 14.6(4) that $\mathcal{O}_{X'} \to \varphi'_*\mathcal{O}_{Y'}$ is surjective. As φ' is dominant, this map on $\mathcal{O}_{X'}$ is also injective, hence an isomorphism.

14.15. Let $w \in W$. Choose a reduced decomposition $w = s_1 s_2 \dots s_r$ of w and extend it to a reduced decomposition $w_0 = s_1 s_2 \dots s_r s_{r+1} \dots s_n$ of w_0. Set $X = X(s_1, s_2, \dots, s_r)$ and $Z = X(s_1, s_2, \dots, s_n)$. Denote by $\psi : X \to Z$ the closed embedding as in 13.5(9). We have commutative diagrams as in 14.12(1), (2).

Let $I \subset S$ and set $P = P_I$. There is a unique decomposition $w = w_1 w_2$ with $w_1 \in W^I$ and $w_2 \in W_I$. Set $X(w)_P = X(w_1)_P$. The canonical map $\overline{\pi} : G/B \to G/P$ maps BwB/B to Bw_1P/B, hence the closure $X(w)$ to $\overline{Bw_1P/P} = X(w_1)_P = X(w)_P$. Let us denote the composite map $X \to X(w) \to X(w)_P$ by φ.

Proposition: a) *$X(w)_P$ is normal.*

b) *$\varphi_* \mathcal{O}_X = \mathcal{O}_{X(w)_P}$ and $R^j \varphi_* \mathcal{O}_X = 0$ for all $j > 0$.*

c) *One has for each locally free sheaf $\mathcal{M}$ of finite rank on $X(w)_P$ natural isomorophisms (for all $j \in \mathbf{N}$)*

$$H^j(X(w)_P, \mathcal{M}) \simeq H^j(X, \varphi^* \mathcal{M}).$$

d) *For all $\lambda \in X(T)_+$ the restriction map $H^0(Z, \mathcal{L}(\lambda)) \to H^0(X, \mathcal{L}(\lambda))$ is surjective. One has $H^j(X, \mathcal{L}(\lambda)) = 0$ for all $j > 0$.*

e) *For all $\lambda \in X(T)_+$ with $\langle \lambda, \alpha^\vee \rangle = 0$ for all $\alpha \in I$ the restriction map*

$$H^0(G/P, \mathcal{L}(\lambda)) \to H^0(X(w)_P, \mathcal{L}(\lambda))$$

is surjective. One has $H^j(X(w)_P, \mathcal{L}(\lambda)) = 0$ for all $j > 0$.

Proof: Assume at first that k is an algebraically closed field of prime characteristic.

We use induction on $l(w)$ and assume the result for all w' with $l(w') < l(w)$ and for all possible P.

We look first at the case $P = B$. If $w = 1$, then $X(w)$ is a point, so a) is obvious. If $w \neq 1$, then there is some $\alpha \in S$ with $l(ws_\alpha) < l(w)$. Then $ws_\alpha \in W^{\{\alpha\}}$ and we know by induction that $X(ws_\alpha)_{P(\alpha)}$ is normal. Now the normality of $X(w)$ follows from 14.9(5) since $s_\alpha = w_{\{\alpha\}}$. As $\varphi : X \to X(w)$ induces an isomorphism of open and dense subsets, Lemma 14.5.a implies $\varphi_* \mathcal{O}_X = \mathcal{O}_{X(w)}$.

We can apply the induction hypothesis to each $s_{i+1} s_{i+2} \dots s_r$ with $1 \leq i < r$. Setting $X'_i = X(s_{i+1}, s_{i+2}, \dots, s_r)$ we get especially for all $\lambda \in X(T)_+$ that the restriction map $H^0(Z, \mathcal{L}(\lambda)) \to H^0(X'_i, \mathcal{L}(\lambda))$ is surjective. Now Proposition 14.8 and Lemma 14.3.b show that $H^j(X, \mathcal{L}(\lambda)) = 0$ for all $j > 0$.

Choose $\mu \in X(T)_+$ with $\langle \mu, \alpha^\vee \rangle > 0$ for all $\alpha \in S$. Then $\mathcal{L}(\mu)$ is an ample line bundle on G/B, hence also on $X(w)$, cf. 4.4 and [Ha], III, exerc. 5.7.a. We have just proved $H^j(X, \mathcal{L}(m\mu)) = 0$ for all $j > 0$ and all $m \in \mathbf{N}$. So Proposition 14.13 implies $R^j \varphi_* \mathcal{O}_X = 0$ for all $j > 0$.

So far we have proved [always in case $P = B$] a), b), and the second part of d). We get c) from 14.6(3) and 14.9(2). The surjectivity of the restriction map in d) follows from 14.12. [This is the only part of the proof where our assumption on k is used.] Finally, c) together with d) imply e).

Now we have to look at an arbitrary P. Choose $\mu \in X(T)_+$ with $\langle \mu, \alpha^\vee \rangle = 0$ for all $\alpha \in I$ and $\langle \mu, \alpha^\vee \rangle > 0$ for all $\alpha \in S \setminus I$. Then $\mathcal{L}_{G/P}(\mu)$ is ample, cf. Remark 1 in 4.4. There is a commutative diagram

$$\begin{array}{ccc} X & \xrightarrow{\psi} & Z \\ \varphi\downarrow & & \downarrow\varphi' \\ X(w)_P & \longrightarrow & G/P \end{array}$$

with φ, φ' dominant. We have $\varphi'^*\mathcal{L}_{G/P}(m\mu) = \mathcal{L}_Z(m\mu)$ and $\varphi^*\mathcal{L}_{X(w)_P}(m\mu) = \mathcal{L}_X(m\mu)$ for all $m \in \mathbf{N}$. [Here we have added for clarity the indices to $\mathcal{L}$ that we suppress otherwise.] By d) the restriction map $H^0(Z,\mathcal{L}(m\mu)) \to H^0(X,\mathcal{L}(m\mu))$ is surjective. As we know $\varphi'_*\mathcal{O}_X = \mathcal{O}_{G/P}$ by 14.7(2), Proposition 14.14 implies $\varphi_*\mathcal{O}_X = \mathcal{O}_{X(w)_P}$. The vanishing of $H^j(X,\mathcal{L}(m\mu))$ for all $m \in \mathbf{N}$ and $j > 0$ implies by Proposition 14.13 that $R^j\varphi_*\mathcal{O}_X = 0$ for all $j > 0$. This proves b).

If $w \notin W^I$, then $X(w)_P = X(w_1)_P$ for some w_1 with $l(w_1) < l(w)$ and we get the normality of $X(w)_P$ by induction. If $w \in W^I$, then a) follows from 14.9(1). As 14.9(2) also generalises to the case where $w \notin W^I$, we get c) from 14.6(3), and we get e) from c) and d).

Let us look at arbitrary k. It is clear that we only have to prove the surjectivity of $H^0(Z,\mathcal{L}(\lambda)) \to H^0(X,\mathcal{L}(\lambda))$ for all $\lambda \in X(T)_+$. Then the arguments from above go through. We want to apply 14.4. The vanishing in d) over (enough) fields of prime characteristic implies $H^j(X_{\mathbf{Z}},\mathcal{L}(\lambda)_{\mathbf{Z}}) = 0 = H^j(Z_{\mathbf{Z}},\mathcal{L}(\lambda)_{\mathbf{Z}})$ for all $j > 0$ by 14.4(2). The universal coefficient formulae in 14.4 yield therefore $H^0(X,\mathcal{L}(\lambda)) \simeq H^0(X_{\mathbf{Z}},\mathcal{L}(\lambda)_{\mathbf{Z}}) \otimes_{\mathbf{Z}} k$ and $H^0(Z,\mathcal{L}(\lambda)) \simeq H^0(Z_{\mathbf{Z}},\mathcal{L}(\lambda)_{\mathbf{Z}}) \otimes_{\mathbf{Z}} k$; also the restriction map over k is induced by the restriction map over $\mathbf{Z}$. It is therefore enough to prove the surjectivity over $\mathbf{Z}$. This in turn follows from the surjectivity over fields of prime characteristic (one for each prime) where we know the result already.

Remark: The last part of the proof shows that $H^0(X(s_1,s_2,\dots,s_r)_{\mathbf{Z}},\mathcal{L}(\lambda)_{\mathbf{Z}})$ is torsion free for all $\lambda \in X(T)_+$, hence free of finite rank over $\mathbf{Z}$. The same holds then by c) for $H^0(X(w)_{\mathbf{Z}},\mathcal{L}(\lambda)_{\mathbf{Z}})$ and (when defined) for $H^0(X(w)_{P,\mathbf{Z}},\mathcal{L}(\lambda)_{\mathbf{Z}})$

14.16. If k is a field of characteristic 0, then the last proposition implies (using general theorems) that each $X(w)_P$ has rational singularities and is Cohen-Macaulay, cf. [Demazure 3], 5.4, cor. 2. One can prove the Cohen-Macaulay property over any field, cf. [Ramanathan 1].

Let $w \in W$ and keep the notations from 14.15. We can identify BwB and $(B_{\mathbf{Z}}wB_{\mathbf{Z}})_k$, hence also BwB/B and $(B_{\mathbf{Z}}wB_{\mathbf{Z}}/B_{\mathbf{Z}})_k$, cf. I.5.5(3). So $(X(w)_{\mathbf{Z}})_k$ is a closed subscheme of $G/B \simeq (G_{\mathbf{Z}}/B_{\mathbf{Z}})_k$ containing BwB/B, hence containing $X(w)$ as a closed subscheme. Denote the inclusion by $i : X(w) \hookrightarrow (X(w)_{\mathbf{Z}})_k$. We want to prove equality:

Proposition: *We have $X(w) = (X(w)_{\mathbf{Z}})_k$ for all $w \in W$.*

Proof: Let us abbreviate $Y = (X(w)_{\mathbf{Z}})_k$. The closed embedding i leads to a short exact sequence

(1) $$0 \to \mathcal{I} \longrightarrow \mathcal{O}_Y \longrightarrow i_*\mathcal{O}_{X(w)} \to 0$$

with a suitable ideal sheaf $\mathcal{I}$. For all $\lambda \in X(T)_+$ with $\langle\lambda,\alpha^\vee\rangle > 0$ for all $\alpha \in S$ the line bundle $\mathcal{L}_{G/B}(\lambda)$ is ample, hence so is $\mathcal{L}_Y(\lambda)$. After replacing λ by some $m\lambda$ with

$m > 0$, we may assume that $H^1(Y, \mathcal{I} \otimes \mathcal{L}(\lambda)) = 0$ and that $\mathcal{I} \otimes \mathcal{L}(\lambda)$ is generated by its global sections. As $i^*\mathcal{L}_Y(\lambda) \simeq \mathcal{L}_{X(w)}(\lambda)$ and as $(i_*\mathcal{O}_{X(w)}) \otimes \mathcal{L}_Y(\lambda) \simeq i_*i^*\mathcal{L}_Y(\lambda)$, cf. 14.6(2), we get an exact sequence of k–modules

$$0 \to H^0(Y, \mathcal{I} \otimes \mathcal{L}_Y(\lambda)) \longrightarrow H^0(Y, \mathcal{L}_Y(\lambda)) \longrightarrow H^0(X(w), \mathcal{L}_{X(w)}(\lambda)) \to 0.$$

We have an isomorphism $H^0(X(w), \mathcal{L}_{X(w)}(\lambda)) \xrightarrow{\sim} H^0(X, \mathcal{L}_X(\lambda))$ that is induced by $\varphi : X \to X(w)$, cf. 14.15.c. So we get an exact sequence

$$(2) \qquad 0 \to H^0(Y, \mathcal{I} \otimes \mathcal{L}_Y(\lambda)) \longrightarrow H^0(Y, \mathcal{L}_Y(\lambda)) \longrightarrow H^0(X, \mathcal{L}_X(\lambda)) \to 0$$

where the last map is induced by $i \circ \varphi$.

We observed already in 14.4 that $X = (X_{\mathbf{Z}})_k$. Obviously $i \circ \varphi = (\varphi_{\mathbf{Z}})_k$ where $\varphi_{\mathbf{Z}} : X_{\mathbf{Z}} \to X(w)_{\mathbf{Z}}$ is the analogue to φ over $\mathbf{Z}$. We know by 14.15 (applied to $k = \mathbf{Z}$) that $\varphi_{\mathbf{Z}}$ induces an isomorphism

$$(3) \qquad H^0(X(w)_{\mathbf{Z}}, \mathcal{L}(\lambda)_{\mathbf{Z}}) \xrightarrow{\sim} H^0(X_{\mathbf{Z}}, \mathcal{L}(\lambda)_{\mathbf{Z}})$$

dropping the index to $\mathcal{L}$. The vanishing of $H^j(X_{\mathbf{Z}}, \mathcal{L}(\lambda)_{\mathbf{Z}})$ for all $j > 0$ yields $H^0(X, \mathcal{L}_X(\lambda)) \simeq H^0(X_{\mathbf{Z}}, \mathcal{L}(\lambda)_{\mathbf{Z}}) \otimes_{\mathbf{Z}} k$ as already used in the proof in 14.15. Similarly, the vanishing of $H^j(X(w)_{\mathbf{Z}}, \mathcal{L}(\lambda)_{\mathbf{Z}})$ for all $j > 0$ yields $H^0(Y, \mathcal{L}_Y(\lambda)) \simeq H^0(X(w)_{\mathbf{Z}}, \mathcal{L}(\lambda)_{\mathbf{Z}}) \otimes_{\mathbf{Z}} k$. Therefore (3) shows that $i \circ \varphi$ induces an isomorphism $H^0(Y, \mathcal{L}_Y(\lambda)) \xrightarrow{\sim} H^0(X, \mathcal{L}_X(\lambda))$. Now (2) implies that $H^0(Y, \mathcal{I} \otimes \mathcal{L}_Y(\lambda)) = 0$. As $\mathcal{I} \otimes \mathcal{L}_Y(\lambda)$ is generated by its global sections, thie implies $\mathcal{I} = 0$, hence the proposition.

Remark: One can prove similarly for all $P \supset B$ that $(X(w)_{P,\mathbf{Z}})_k \simeq X(w)_P$.

14.17. Let $\alpha \in S$. Assume for the moment that k is a field. Set for any finite dimensional B–module M

$$(1) \qquad \begin{aligned} \chi_\alpha(M) &= \sum_{i \geq 0} (-1)^i \operatorname{ch} H^i(P(\alpha)/B, \mathcal{L}(M)) \\ &= \operatorname{ch} H^0(P(\alpha)/B, \mathcal{L}(M)) - \operatorname{ch} H^1(P(\alpha)/B, \mathcal{L}(M)). \end{aligned}$$

If $M = k_\lambda$ for some $\lambda \in X(T)$, then $\chi_\alpha(M) = \chi_\alpha(\lambda)$ can be computed using 5.2; we get

$$(2) \quad \chi_\alpha(\lambda) = \begin{cases} e(\lambda) + e(\lambda - \alpha) + \cdots + e(s_\alpha \lambda), & \text{if } \langle \lambda, \alpha^\vee \rangle \geq 0, \\ 0, & \text{if } \langle \lambda, \alpha^\vee \rangle = -1, \\ -(e(s_\alpha\lambda - \alpha) + e(s_\alpha\lambda - 2\alpha) + \cdots + e(\lambda + \alpha)), & \text{if } \langle \lambda, \alpha^\vee \rangle \leq -2. \end{cases}$$

This shows that

$$(3) \qquad (1 - e(-\alpha))\, \chi_\alpha(\lambda) = e(\lambda) - e(-\alpha) e(s_\alpha \lambda)$$

in all cases. Therefore any $\chi - e(-\alpha)\chi$ with $\chi \in \mathbf{Z}[X(T)]$ is divisible within the integral domain $\mathbf{Z}[X(T)]$ by $1 - e(-\alpha)$. The corresponding statement holds for any $\beta \in R$ as β is conjugate to some simple root under the Weyl group. We can therefore define

$$\sigma_\beta(\chi) = \frac{\chi - e(-\beta)s_\beta(\chi)}{1 - e(-\beta)} \tag{4}$$

for any $\beta \in R$ and $\chi \in \mathbf{Z}[X(T)]$. Obviously σ_β is an endomorphism of $\mathbf{Z}[X(T)]$ considered as an additive group; it satisfies $s_\beta\,\sigma_\beta(\chi) = \sigma_\beta(\chi)$ for all $\chi \in \mathbf{Z}[X(T)]$ and $\sigma_\beta^2 = \sigma_\beta$. The additivity of the Euler characteristic implies for any M as above

$$\chi_\alpha(M) = \sigma_\alpha(\mathrm{ch}(M)). \tag{5}$$

14.18. Proposition: *Let $w \in W$ and let $w = s_1 s_2 \dots s_r$ be a reduced decomposition of w. Set $\sigma_i = \sigma_{\alpha_i}$ where $\alpha_i \in S$ with $s_i = s_{\alpha_i}$. Let $I \subset S$ and set $P = P_I$.*

a) *If k is a field, then*

$$\sum_{i \geq 0} \mathrm{ch}\, H^i(X(w)_P, \mathcal{L}(M)) = \sigma_1\sigma_2 \dots \sigma_r(\mathrm{ch}(M))$$

for any finite dimensional P–module M.

b) *If $\lambda \in X(T)_+$ with $\langle \lambda, \alpha^\vee \rangle = 0$ for all $\alpha \in I$, then*

$$\mathrm{ch}\, H^0(X(w)_P, \mathcal{L}(\lambda)) = \sigma_1\sigma_2 \dots \sigma_r(e(\lambda)).$$

Proof: a) We know by 14.15.c that we may as well compute

$$\sum_{i \geq 0} \mathrm{ch}\, H^i(X(s_1, s_2, \dots, s_r), \mathcal{L}(M)).$$

There is a Leray spectral sequence corresponding to $\pi_1 : X(s_1, s_2, \dots, s_r) \to P(\alpha_1)/B$:

$$\begin{aligned} H^i(P(\alpha_1)/B, R^j(\pi_1)_*\mathcal{L}(M) &= H^i(P(\alpha_1)/B, \mathcal{L}(H^j(X(s_2, \dots, s_r), \mathcal{L}(M)))) \\ &\Rightarrow H^{i+j}(X(s_1, s_2, \dots, s_r), \mathcal{L}(M)) \end{aligned}$$

cf. 14.1(4). Therefore our alternating sum is equal to

$$\sum_{j \geq 0} (-1)^j\, \chi_{\alpha_1}(\mathrm{ch}\, H^j(X(s_2, \dots, s_r), \mathcal{L}(M))).$$

Now use induction and 14.17(5).

b) In the case of a field use a) and the vanishing in 14.15.e. In general, use the isomorphism

$$H^0(X(s_1, s_2, \dots, s_r), \mathcal{L}(\lambda)) \simeq H^0(X(s_1, s_2, \dots, s_r)_{\mathbf{Z}}, \mathcal{L}(\lambda)_{\mathbf{Z}}) \otimes_{\mathbf{Z}} k$$

as in the proof of 14.15.

14.19. Let $\lambda \in X(T)_+$ and $w \in W$. By 14.15.e the restriction

$$H^0(\lambda) = H^0(G/B, \mathcal{L}(\lambda)) \longrightarrow H^0(X(w), \mathcal{L}(\lambda))$$

is surjective. We want to describe its kernel.

Set $I = \{\alpha \in S \mid \langle \lambda, \alpha^\vee \rangle = 0\}$ and $P = P_I$. We can then regard $\mathcal{L}(\lambda)$ as a sheaf on G/P and on $X(w)_P$. The cohomology groups do not change (e.g., by 14.15.c), so we may as well compute the kernel of

$$H^0(\lambda) \simeq H^0(G/P, \mathcal{L}(\lambda)) \longrightarrow H^0(X(w)_P, \mathcal{L}(\lambda)).$$

We may assume $w \in W^I$. As $X(w)_P$ is the closure of BwP/P, we get the same kernel by looking at

(1) $$H^0(\lambda) \simeq H^0(G/P, \mathcal{L}(\lambda)) \longrightarrow H^0(BwP/P, \mathcal{L}(\lambda)).$$

Choose $v \in V = V(-w_0\lambda) \simeq H^0(\lambda)^*$ with $V_{-\lambda} = kv$. Then $g \mapsto gv$ gives rise to a closed embedding of G/P into the projective space $\mathbf{P}(V)$, cf. 8.5. Identify any $f \in V^*$ with the function $g \mapsto f(gv)$ in $H^0(\lambda) \subset k[G]$. Then the kernel in (1) consists exactly of all $f \in V^*$ with $(f \otimes 1)(b\dot{w}(v \otimes 1)) = 0$ for all $b \in B(A)$ and all k–algebras A. As T stabilises kv and as $\dot{w}$ normalises T, we may replace $B(A)$ by $U(A)$ in this condition. Certainly any f with $f(\mathrm{Dist}(U)\dot{w}v) = 0$ will do as $u\dot{w}(v \otimes 1) \in \mathrm{Dist}(U_A)\dot{w}(v \otimes 1)$ for all $u \in U(A)$ and all A and as $\mathrm{Dist}(U_A)\dot{w}(v \otimes 1)$ is the canonical image of $\mathrm{Dist}(U)\dot{w}v \otimes A$ in $V \otimes A$. On the other hand, working with the field of fractions of k (or rather with an extension of that if k is finite) we see that any f in the kernel has to annihilate $\mathrm{Dist}(U)\dot{w}v$. This implies

(2) $$H^0(X(w), \mathcal{L}(\lambda)) \simeq H^0(\lambda)/\{f \in H^0(\lambda) \mid f(\mathrm{Dist}(U)\dot{w}v) = 0\}.$$

The fact that $H^0(X(w), \mathcal{L}(\lambda)) \simeq H^0(X(w)_\mathbf{Z}, \mathcal{L}(\lambda)_\mathbf{Z}) \otimes_\mathbf{Z} k$ for all k implies easily that (at first for $k = \mathbf{Z}$ and then in general) that $\mathrm{Dist}(U)\dot{w}v$ is a direct summand of $V = V(-w_0\lambda)$ as a k–module, hence

(3) $$H^0(X(w), \mathcal{L}(\lambda)) \simeq (\mathrm{Dist}(U)\dot{w}v)^*.$$

14.20. Assume in this subsection that k is a field with $\mathrm{char}(k) = p \neq 0$.

For any B–module M the Frobenius endomorphism induces a homomorphism of G–modules $H^0(G/B, \mathcal{L}(M))^{[r]} \to H^0(G/B, \mathcal{L}(M^{[r]}))$. If M' is another B–module, then the cup product induces a homomorphism of G–modules

$$H^0(G/B, \mathcal{L}(M^{[r]})) \otimes H^0(G/B, \mathcal{L}(M')) \longrightarrow H^0(G/B, \mathcal{L}(M^{[r]} \otimes M')).$$

In case $M' = (p^r - 1)\rho$ we get as the composed map just the isomorphism

$$H^0(G/B, \mathcal{L}(M))^{[r]} \otimes St_r \xrightarrow{\sim} H^0(G/B, \mathcal{L}(M^{[r]} \otimes (p^r - 1)\rho))$$

from 3.19.

Let us use the notation $H^0(\lambda) = H^0(G/B, \mathcal{L}(\lambda))$ as before. By composing maps induced by the Frobenius endomorphism and the cup product (as above) we get a commutative diagram of homomorphisms of G–modules for all $\lambda, \mu \in X(T)_+$ if $(p^r - 1)\rho \in X(T)$:

$$
\begin{array}{ccccc}
 & & H^0(p^r(\lambda+\mu+\rho)-\rho) & & \\
 & \overset{\varphi_1}{\nearrow} & \uparrow{\scriptstyle \varphi_2} & \overset{\sim}{\nwarrow} & \\
H^0(p^r\lambda)\otimes H^0(p^r(\mu+\rho)-\rho) & & H^0(p^r(\lambda+\rho)-\rho)\otimes H^0(\mu)^{[r]} & & H^0(\lambda+\mu)^{[r]}\otimes St_r \\
\sim\uparrow & \nearrow & & \overset{\sim}{\nwarrow} & \uparrow{\scriptstyle \varphi_3} \\
H^0(p^r\lambda)\otimes H^0(\mu)^{[r]}\otimes St_r & & & & H^0(\lambda)^{[r]}\otimes H^0(\mu)^{[r]}\otimes St_r
\end{array}
$$

Here three maps (indicated by $\sim$) are isomorphisms by 3.19. We can read off from the diagram: If φ_1 is surjective, then so is φ_2, hence also φ_3 and finally also $H^0(\lambda) \otimes H^0(\mu) \to H^0(\lambda + \mu)$.

Proposition: *The cup product induces for all $\lambda, \mu \in X(T)_+$ a surjective homomorphism of G–modules $H^0(\lambda) \otimes H^0(\mu) \to H^0(\lambda + \mu)$.*

Proof: We may assume $\rho \in X(T)$ by going to a covering group if necessary. Suppose at first that $\langle \lambda, \alpha^\vee \rangle > 0$ for all $\alpha \in S$, i.e., suppose that $\mathcal{L}(\lambda)$ is ample. Then $\mathcal{L}(\lambda, \mu + \rho)$ is ample on $G/B \times G/B$. Denote the diagonal embedding of G/B into $G/B \times G/B$ by i. By 14.6(5) we can find some r such that the restriction map

$$H^0(G/B \times G/B, \mathcal{L}(p^r\lambda, p^r(\mu+\rho) - \rho)) \to H^0(G/B, i^*\mathcal{L}(p^r\lambda, p^r(\mu+\rho) - \rho))$$

is surjective. Using the Künneth formula we can identify the left hand side with $H^0(p^r\lambda) \otimes H^0(p^r(\mu+\rho) - \rho)$ whereas the right hand side is just $H^0(p^r\lambda + p^r(\mu + \rho) - \rho)$. It is left to the reader to check that the restriction map is just the cup product. So above φ_1 is surjective, hence also $H^0(\lambda) \otimes H^0(\mu) \to H^0(\lambda + \mu)$.

Now let λ be arbitrary in $X(T)_+$. As $\langle p^r(\mu+\rho) - \rho, \alpha^\vee \rangle > 0$ for all $\alpha \in S$, we can apply the first case to $(p^r(\mu+\rho) - \rho, p^r\lambda)$ instead of (λ, μ). This yields (for any r) the surjectivity of φ_1 in the diagram as the order of the two factors does not matter for the cup product on the H^0 level. We conclude as before.

Remark: This proposition is a special case of 4.21. Note that the result extends trivially to fields of characteristic 0 since the map is non-zero and since in that case any $H^0(\nu)$ is simple.

14.21. Let $\lambda, \mu \in X(T)_+$. The surjectivity of $H^0(\lambda) \otimes H^0(\mu) \to H^0(\lambda+\mu)$ implies by 14.15.e for all $w \in W$ that the cup product

$$(1) \qquad H^0(X(w), \mathcal{L}(\lambda)) \otimes H^0(X(w), \mathcal{L}(\mu)) \longrightarrow H^0(X(w), \mathcal{L}(\lambda + \mu))$$

is also surjective. We can replace here $X(w)$ by $X(w)_P$ with $P = P_I$ if $\langle \lambda, \alpha^\vee \rangle = \langle \mu, \alpha^\vee \rangle = 0$ for all $\alpha \in I$.

Fix $\lambda \in X(T)_+$, set $I = \{\alpha \in S \mid \langle \lambda, \alpha^\vee \rangle = 0\}$ and $P = P_I$. Then the very ample line bundle $\mathcal{L}_{G/P}(\lambda)$ defines an embedding of G/P, hence of each $X(w)_P$ into $\mathbf{P}(V)$ where $V = V(-w_0\lambda) = H^0(\lambda)^*$, cf. 14.19. We get from (1), by induction, that the cup product induces surjective maps

$$(2) \qquad \bigotimes{}^m H^0(\lambda) \longrightarrow \bigotimes{}^m H^0(X(w)_P, \mathcal{L}(\lambda)) \longrightarrow H^0(X(w)_P, \mathcal{L}(m\lambda)).$$

As the multiplication inside $\bigoplus_{m\geq 0} H^0(X(w)_P, \mathcal{L}(m\lambda))$ is commutative, we also get surjective maps

$$(3) \qquad S^m H^0(\lambda) \longrightarrow S^m H^0(X(w)_P, \mathcal{L}(\lambda)) \longrightarrow H^0(X(w)_P, \mathcal{L}(m\lambda)).$$

Now $S^m H^0(\lambda)$ is just $H^0(\mathbf{P}(V), \mathcal{O}_{\mathbf{P}(V)}(m))$. So the surjectivity in (3) means that $X(w)_P$ is projectively normal with respect to the embedding given by $\mathcal{L}(\lambda)$, cf. [Ha], II, exerc. 5.14.d, i.e., that the homogeneous coordinate ring of $X(w)_P$ is integrally closed.

14.22. Suppose that k is algebraically closed. Let $\lambda \in X(T)_+$. Consider

$$A = \bigoplus_{n\geq 0} H^0(n\lambda).$$

This is a graded and commutative algebra under the cup product. Elements in $G(k)$ act on A via algebra automorphisms. We denote the graded parts also by A_n. We know by 14.20 that A is generated over $k = A_0$ by A_1.

Proposition: a) *The ring A is an integrally closed integral domain.*

b) *Suppose that $\lambda \neq 0$. Choose $v \in V(-w_0\lambda)_{-\lambda}$, $v \neq 0$. Then $\overline{G(k)v} = G(k)v \cup \{0\}$ and there is an algebra isomorphism $k[\overline{G(k)v}] \xrightarrow{\sim} A$.*

Proof: If $\dim H^0(\lambda) = 1$, then also $\dim H^0(n\lambda) = 1$ for all $n \geq 0$; if $H^0(\lambda) = kv$, then $H^0(n\lambda) = kv^n$ and A identifies thus with a polynomial ring in one variable. This settles in particular a) in case $\lambda = 0$.

Suppose now $\lambda \neq 0$. It follows that we can regard A as a subalgebra of $k[G]$: Each $H^0(n\lambda)$ is in a natural way a subspace of $k[G]$ and their sum in $k[G]$ is direct because they belong to distinct weight spaces for the right regular action of T. And the multiplication in $k[G]$ is the same as the one coming from the cup product. This shows in particular that A is an integral domain.

Let us abbreviate $V = V(-w_0\lambda) = H^0(\lambda)^*$. Since $\lambda \neq 0$ we have $k^\times v \subset G(k)v$. The stabiliser of the line kv in $G(k)$ is a parabolic subgroup $P(k)$. Therefore the projective variety $G(k)/P(k)$ maps onto the $G(k)$ orbit of the line kv in the projective space of V. This implies that this orbit is closed, hence that $G(k)v \cup \{0\}$ is closed in V. As $0 \in \overline{k^\times v} \subset \overline{G(k)v}$ this shows that $\overline{G(k)v} = G(k)v \cup \{0\}$.

The closed subset $\overline{G(k)v} \subset V$ is stable under scalar multiplication. Therefore the algebra $k[\overline{G(k)v}]$ of regular functions on $\overline{G(k)v}$ is graded and the surjective restriction homomorphism from $S(V^*) = S(H^0(\lambda))$ onto $k[\overline{G(k)v}]$ preserves the grading. On the other hand, each function f on $\overline{G(k)v}$ defines via $f_1(g) = f(gv)$ a regular function on $G(k)$, i.e., an element of $k[G]$. If f is homogeneous of degree n, then $f_1 \in H^0(n\lambda)$. As $G(k)v$ is dense in $\overline{G(k)v}$, we get thus an injective algebra homomorphism $k[\overline{G(k)v}] \to A$ that preserves the grading.

Both maps $S(V^*) \to k[\overline{G(k)v}]$ and $k[\overline{G(k)v}] \to A$ commute with the obvious actions of $G(k)$. Hence so does their composition $S(V^*) \to A$. This composition restricts in degree 1 to a non-zero map from $V^* = H^0(\lambda)$ to $A_1 = H^0(\lambda)$. Since $\operatorname{End}_G H^0(\lambda) = k$, this restriction is bijective. As A is generated by A_1, this implies that the composed map is surjective. Then also $k[\overline{G(k)v}] \to A$ is surjective, hence bijective.

By construction this isomorphism above factors as $k[\overline{G(k)v}] \to k[G(k)v] \to A$ and here the second map is again injective. Since the total map is bijective, both separate maps are also bijective. The $G(k)$–orbit $G(k)v$ is smooth and irreducible. Therefore the local ring at each point is integrally closed, hence so is $k[G(k)v]$ which is their intersection taken in the fraction field of $k[G(k)v]$. This implies that A is integrally closed. (We get also that $\overline{G(k)v}$ is a normal variety.)

14.23. Let us assume for the sake of simplicity that G is semi-simple and simply connected. Let $\varpi_1, \varpi_2, \dots, \varpi_l$ be the fundamental weights. Consider $\lambda \in X(T)_+$. There are unique integers $n(i) \geq 0$ with $\lambda = \sum_{i=1}^{l} n(i)\varpi_i$. Then for all $w \in W$ we have by 14.20 and 14.15.e a surjection

$$\bigotimes^{n(1)} H^0(\varpi_1) \otimes \bigotimes^{n(2)} H^0(\varpi_2) \otimes \cdots \otimes \bigotimes^{n(l)} H^0(\varpi_l) \longrightarrow H^0(X(w), \mathcal{L}(w)).$$

So, if $(f_{ij})_{1 \leq j \leq d(i)}$ is a basis for $H^0(\varpi_i) \subset k[G]$ for each i, then one can choose monomials in these f_{ij} (with suitable homogeneity conditions) such that their restrictions to $\overline{BwB}$ form a basis of $H^0(X(w), \mathcal{L}(\lambda))$.

One would like to choose the bases $(f_{ij})_j$ such that there is a nice rule for which monomials to choose. This is the aim of the *standard monomial theory* which has its origin in work by Hodge (for Grassmannians over fields in characteristic 0). The strongest result in this direction can be found in [Lakshmibai, Musili, and Seshadri 3] where the classical groups are treated. There is a survey of the theory in [Musili and Seshadri 1].

CHAPTER A

Truncated Categories and Schur Algebras

Throughout this chapter we assume that k is a field.

For each subset π of $X(T)_+$ let $\mathcal{C}(\pi)$ denote the category of all G–modules M such that all composition factors of M have a highest weight in π. These $\mathcal{C}(\pi)$ are the truncated categories in the title of this chapter.

We call π saturated if $\mu \in \pi$ implies $\nu \in \pi$ for all dominant ν with $\nu < \mu$. We assume in this introduction from now on that π is saturated.

For $G = GL_n$ and any integer $d \geq 0$ the category of all polynomial representations of G that are homogeneous of degree d (i.e., such that all matrix coefficients are homogeneous polynomials of degree d on GL_n) is a special case of such a category $\mathcal{C}(\pi)$ with saturated and finite π, see A.3.

The first main result on these truncated categories is the fact (see A.10) that we get for two modules in $\mathcal{C}(\pi)$ the same Ext^i groups when we work in $\mathcal{C}(\pi)$ or in the category of all G–modules. If π is in addition finite, then there is a bound depending only on π such that all Ext^i groups between two modules in $\mathcal{C}(\pi)$ vanish for i greater than this bound, see A.11.

A main tool in proving these results is the functor O_π that associates to any G–module M the largest submodule $O_\pi(M)$ of M belonging to $\mathcal{C}(\pi)$. This functor is then in particular applied to $M = k[G]$ considered as a G–module via the right regular representation. Now $O_\pi(k[G])$ turns out to be a subcoalgebra of $k[G]$, see A.14.

Assume that π is finite. Then $O_\pi(k[G])$ is finite dimensional. So the dual space $S_G(\pi)$ of this coalgebra has a natural structure as a (finite dimensional) algebra. The category $\mathcal{C}(\pi)$ is then equivalent to the category of all $S_G(\pi)$–modules, see A.17.

One calls $S_G(\pi)$ a generalised Schur algebra. In the special case of GL_n mentioned above, one gets here the classical Schur algebras as in [Green 2]. Here the general theory leads to close connections between the representation theories of general linear and symmetric groups that we discuss in A.21–A.23.

In the last subsections of this chapter we look at the relationship between polynomial representations of GL_n and polynomial functors (see A.26). This relation is a main tool in the work by Friedlander and Suslin on the cohomology of finite group schemes.

The main sources for this chapter are [Donkin 12, 14], [Friedlander and Suslin], [Green 2].

A.1. For each subset $\pi \subset X(T)_+$ let $\mathcal{C}(\pi)$ denote the category of all G–modules M such that all composition factors of M (i.e., all composition factors of all finite dimensional submodules of M) have the form $L(\mu)$ with $\mu \in \pi$. We regard $\mathcal{C}(\pi)$

as a full subcategory of the category of all G–modules and call it the *truncated category* associated to π.

For example, $\mathcal{C}(X(T)_+)$ is the category of all G–modules. On the other hand, $\mathcal{C}(\{0\})$ is the category of all trivial G–modules. More generally, if $\mu \in X(T)_+$, then $\mathcal{C}(\{\mu\})$ is the category of all G–modules that are direct sums of copies of $L(\mu)$. [This requires a short argument: Let M belong to $\mathcal{C}(\{\mu\})$. Each finite dimensional submodule N of M has a composition series with all factors isomorphic to $L(\mu)$. Since $\mathrm{Ext}^1_G(L(\mu), L(\mu)) = 0$ by 2.12(1), we get that N is a direct sum of copies of $L(\mu)$. So is then M by the local finiteness of G–modules.]

Let $\pi \subset X(T)_+$. The full subcategory $\mathcal{C}(\pi)$ of the category of all G–modules is clearly closed under taking subquotients and extensions. If $(M_i)_{i\in I}$ is a family of submodules in a G–module M such that each M_i belongs to $\mathcal{C}(\pi)$, then also $\sum_{i\in I} M_i$ belongs to $\mathcal{C}(\pi)$. (Using local finiteness one reduces easily to the case where $|I| < \infty$ and where $\dim(M_i) < \infty$ for all i.) Denote by $O_\pi(M)$ the sum of all submodule in M that belong to $\mathcal{C}(\pi)$:

(1) $$O_\pi(M) = \sum_{M'\subset M,\, M'\in\mathcal{C}(\pi)} M'.$$

Then also $O_\pi(M)$ belongs to $\mathcal{C}(\pi)$ and is the largest submodule of M with this property. We call O_π the *truncation functor* associated to π.

The description of $\mathcal{C}(\{0\})$ above shows that $O_{\{0\}}(M) = M^G$ for all M. As in this special case one gets in general:

(2) *Each O_π is a left exact functor.*

Clearly each O_π commutes with intersections of submodules and arbitrary direct sums. It is right adjoint to the inclusion of $\mathcal{C}(\pi)$ into the category of all G–modules; in other words, the inclusion of $O_\pi(M)$ into M induces a bijection

(3) $$\mathrm{Hom}_{\mathcal{C}(\pi)}(V, O_\pi(M)) \xrightarrow{\sim} \mathrm{Hom}_G(V, M) \qquad \text{for all } V \text{ in } \mathcal{C}(\pi).$$

This implies (as in I.3.9.a):

(4) *Each O_π maps injective G–modules to injective objects in $\mathcal{C}(\pi)$.*

If π and σ are two subsets of $X(T)_+$, then we have clearly

(5) $$O_{\pi\cap\sigma}(M) = O_\sigma(O_\pi(M)) = O_\pi(O_\sigma(M))$$

for all G–modules M. In particular, if $\sigma \subset \pi$, then $O_\sigma(M) \subset O_\pi(M)$. If $\pi_1 \subset \pi_2 \subset \pi_3 \subset \cdots$ is an ascending chain of subsets of $X(T)_+$, then

(6) $$O_\pi(M) = \bigcup_{i\geq 1} O_{\pi_i}(M)$$

for $\pi = \bigcup_{i\geq 1} \pi_i$. Indeed, the inclusion "$\supset$" follows from the preceding observation. On the other hand, if $v \in O_\pi(M)$, then the G–submodule V of M generated by v is contained in $O_\pi(M)$. We have $\dim(V) < \infty$, so there exists an $m \in \mathbf{N}$ such that all composition factors of V have the form $L(\mu)$ with $\mu \in \pi_m$. It follows that $V \subset O_{\pi_m}(M)$, hence $v \in \bigcup_{i\geq 1} O_{\pi_i}(M)$.

A.2. A subset π of $X(T)_+$ is called *saturated* if for all $\lambda \in \pi$ and $\mu \in X(T)_+$ with $\mu \leq \lambda$ also $\mu \in \pi$. For example, for any $\nu \in X(T)_+$ the set $\{\mu \in X(T)_+ \mid \mu \leq \nu\}$ is saturated. For saturated π the category $\mathcal{C}(\pi)$ can be alternatively described as follows:

(1) *If $\pi \subset X(T)_+$ is saturated, then a G–module M belongs to $\mathcal{C}(\pi)$ if and only if each dominant weight of M belongs to π if and only if each weight of M belongs to $W\pi$.*

(The equivalence of the last two conditions follows from 1.19(2) as $W\pi \cap X(T)_+ = \pi$. If all dominant weights of a G–module M belong to π, then so does the highest weight of each composition factor of M and M is in $\mathcal{C}(\pi)$. Conversely, if M belongs to $\mathcal{C}(\pi)$ and if ν is a dominant weight of M, then there exist a composition factor $L(\mu)$ of M with $L(\mu)_\nu \neq 0$. Then we get $\nu \leq \mu$ and $\mu \in \pi$. Now the saturation of π implies $\nu \in \pi$.)

We have of course by definition $O_\pi(L(\mu)) = L(\mu)$ if $\mu \in \pi$ while $O_\pi(L(\mu)) = 0$ if $\mu \notin \pi$ (for arbitrary π). For saturated π we get:

(2) *If $\pi \subset X(T)_+$ is saturated, then we have for all $\mu \in X(T)_+$*

$$O_\pi(H^0(\mu)) = \begin{cases} H^0(\mu), & \text{if } \mu \in \pi, \\ 0, & \text{if } \mu \notin \pi. \end{cases}$$

Indeed, all dominant weights λ of $H^0(\mu)$ satisfy $\lambda \leq \mu$; if $\mu \in \pi$, then we get now $\lambda \in \pi$ by the saturation of π, hence $H^0(\mu)$ in $\mathcal{C}(\pi)$. On the other hand, if $O_\pi(H^0(\mu)) \neq 0$, then $O_\pi(H^0(\mu))$ has to contain the simple socle $L(\mu)$ of $H^0(\mu)$; this forces $\mu \in \pi$.

A.3. Here is an important example. Consider $G = GL_n$ with $n \geq 2$ and use the notations from 1.21. Note that $X(T)_+$ consists of all $\lambda = \sum_{i=1}^n a_i\varepsilon_i$ with all $a_i \in \mathbf{Z}$ and $a_1 \geq a_2 \geq \cdots \geq a_n$ (as $\langle \lambda, \alpha_i^\vee \rangle = a_i - a_{i+1}$). For each $d \in \mathbf{N}$ set now

$$(1) \qquad \pi(n,d) = \{\sum_{i=1}^n a_i\varepsilon_i \in X(T) \mid a_1 \geq a_2 \geq \cdots \geq a_n \geq 0, \sum_{i=1}^n a_i = d\}.$$

This is a subset of $X(T)_+$. It is clearly in bijection with the set of all partitions of d into at most n parts. As $W \simeq S_n$ acts by permuting the ε_i, we get

$$(2) \qquad W\pi(n,d) = \{\sum_{i=1}^n a_i\varepsilon_i \in X(T) \mid \text{all } a_i \geq 0, \sum_{i=1}^n a_i = d\}.$$

Note that $\pi(n,d)$ is saturated: Consider $\lambda = \sum_{i=1}^n a_i\varepsilon_i \in \pi(n,d)$ and $\mu = \sum_{i=1}^n b_i\varepsilon_i \in X(T)_+$ with $\mu \leq \lambda$. Then there are integers $c_i \geq 0$ with $\lambda - \mu = \sum_{i=1}^{n-1} c_i(\varepsilon_i - \varepsilon_{i+1})$. It follows that $\sum_{i=1}^n b_i = \sum_{i=1}^n a_i = d$ and $b_n = a_n + c_{n-1} \geq 0$. As μ is dominant, we have also $b_1 \geq b_2 \geq \cdots \geq b_n$, hence $\mu \in \pi(n,d)$.

Now A.2(1) says that $\mathcal{C}(\pi(n,d))$ is the category of all G–modules M such that all weights of M belong to $W\pi(n,d)$. There is, however, another and more natural description of this category.

Let $M_{n,k}$ denote the functor associating to each k–algebra A the set $M_n(A)$ of all $(n\times n)$–matrices over A. Then $M_{n,k}$ is an affine scheme isomorphic to $\mathbf{A}^{n^2}$ and $k[M_{n,k}]$ is the polynomial ring over k in the matrix coefficients X_{ij}, $1\le i,j\le n$ (cf. 1.21). Now GL_n is the open subfunctor of $M_{n,k}$ determined by det (equal to $D(\det)$ in the notation of I.1.5) and we have $k[GL_n]=k[M_{n,k}][\det^{-1}]$. Since $k[M_{n,k}]$ is an integral domain, we can regard $k[M_{n,k}]$ as a subalgebra of $k[GL_n]$ that we call the subalgebra of *polynomial functions* on GL_n.

Given a GL_n–module V, one associates to each $v\in V$ and each $\varphi\in V^*$ a "matrix coefficient" $c_{\varphi,v}\in k[GL_n]$ such that $c_{\varphi,v}(g)=(\varphi\otimes \mathrm{id}_A)(g(v\otimes 1))$ for all $g\in GL_n(A)$ and all A. If the comodule map Δ_V sends v to $\sum_s v_s\otimes f_s$ with all $v_s\in V$ and $f_s\in k[GL_n]$, then $c_{\varphi,v}=\sum_h \varphi(v_s)f_s$. [This construction works of course for any algebraic group, not just for GL_n, cf. A.12.]

A GL_n–module V is now called a *polynomial* GL_n*–module* and the corresponding representation $GL_n\to GL(V)$ a *polynomial representation*, if all $c_{\varphi,v}$ with $v\in V$ and $\varphi\in V^*$ belong to the subalgebra $k[M_{n,k}]$ of $k[GL_n]$. Furthermore a polynomial GL_n–module V is called *homogeneous of degree* d if all $c_{\varphi,v}$ are homogeneous polynomials of degree d in the X_{ij}.

Equivalently, V is polynomial if and only if Δ_V maps V to the subspace $V\otimes k[M_{n,k}]$ of $V\otimes k[GL_n]$. And V is then homogeneous of degree d if Δ_V maps V to $V\otimes k[M_{n,k};d]$ where $k[M_{n,k};d]$ is the space of homogeneous polynomials of degree d in the X_{ij}.

Proposition: *A* GL_n*–module belongs to* $\mathcal{C}(\pi(n,d))$ *if and only if it is polynomial and homogeneous of degree* d.

Here one direction is straightforward: Let V be a GL_n–module that is polynomial and homogeneous of degree d and let μ be a weight of V. Write $\mu=\sum_{i=1}^n a_i\varepsilon_i$ with all $a_i\in\mathbf{Z}$. Pick $v\in V_\mu$, $v\neq 0$ and choose $\varphi(v)=1$. We have for all $t\in T(A)$ and all A

$$c_{\varphi,v}(t)=(\varphi\otimes\mathrm{id}_A)(t(v\otimes 1))=(\varphi\otimes\mathrm{id}_A)(\mu(t)(v\otimes 1))=\mu(t)=\prod_{i=1}^n\varepsilon_i(t)^{a_i}.$$

This shows that the restriction of $c_{\varphi,v}$ to T is equal to $\prod_{i=1}^n X_{ii}^{a_i}$. On the other hand, $c_{\varphi,v}$ is a polynomial in the X_{ij} homogeneous of degree d. So the restriction of $c_{\varphi,v}$ to T is a polynomial in the X_{ii} homogeneous of degree d. It follows that $a_i\ge 0$ for all i and $\sum_{i=1}^n a_i=d$, hence that $\mu\in W\pi(n,d)$.

The proof of the other direction will be given in the next subsection. We shall here follow the approach in [Friedlander and Suslin]. (The techniques used there are on the whole unrelated to the main topic of this chapter.) The proposition was first proved in [Donkin 14], 1.2. We shall make some comments on that proof in A.18. Another approach to this result can be found in the appendix to [Nakano 1].

A.4. *Proof of Proposition A.3*: It remains to show: Let V be a GL_n–module in $\mathcal{C}(\pi(n,d))$. Then each matrix coefficient $c_{\varphi,v}$ with $v\in V$ and $\varphi\in V^*$ is a polynomial in the X_{ij} homogeneous of degree d.

We can replace V by the GL_n–submodule generated by v without changing $c_{\varphi,v}$. So we may assume that V is finite dimensional. Choose a basis $(v_s)_{1\le s\le u}$ for V over k such that each v_s is a weight vector relative to T. There are $\psi_{ts}\in k[GL_n]$

such that the comodule map Δ_V is given by $\Delta_V(v_s) = \sum_{t=1}^{u} v_t \otimes \psi_{ts}$ for all s. So for any A each $g \in G(A)$ acting on $V \otimes A$ has matrix $(\psi_{ts}(g))_{s,t}$ with respect to the basis of all $v_s \otimes 1$.

Now each matrix coefficient $c_{\varphi,v}$ is a linear combination of the ψ_{ts}: If $v = \sum_{s=1}^{u} a_s v_s$, then $c_{\varphi,v} = \sum_{s=1}^{u} a_s \varphi(v_t)\psi_{ts}$. On the other hand, each ψ_{ts} is a special matrix coefficient: take $v = v_s$ and φ such that $\varphi(v_l) = \delta_{lt}$. So we have to show that each ψ_{ts} is a polynomial in the X_{ij} homogeneous of degree d.

Our assumption says (since we have chosen the v_s as weight vectors for T) that each ψ_{ts} with $t \neq s$ restricts to 0 on T while the restriction of any ψ_{tt} to T has the form $\prod_{i=1}^{n} X_{ii}^{a(i,t)}$ with integers $a(i,t) \geq 0$ and $\sum_{i=1}^{n} a(i,t) = d$.

As $k[GL_n] = k[M_{n,k}][\det^{-1}]$ we can certainly find $m \in \mathbf{N}$ such that $\det^m \psi_{ts} \in k[M_{n,k}]$ for all s and t. Choose m minimal with this property. We want to show that $m = 0$; then we know at least that the ψ_{ts} are polynomials in the X_{ij}.

Suppose that $m > 0$. Set $\psi'_{ts} = \det^m \psi_{ts}$ for all s and t; so we have $\psi'_{ts} \in k[M_{n,k}]$. We are going to show that det divides each ψ'_{ts} in $k[M_{n,k}]$. This leads to a contradiction since then already $\det^{m-1} \psi_{ts} = \det^{-1} \psi'_{ts} \in k[M_{n,k}]$ for all s and t violating the minimality of m.

The irreducibility of det implies by the Hilbert Nullstellensatz that it suffices to show that

$$\text{(1)} \qquad \psi'_{ts}(h) = 0 \qquad \text{for all } h \in M_n(\overline{k}) \text{ with } \det(h) = 0$$

for some algebraic closure $\overline{k}$ of k.

This is easy if h belongs to the set D of all diagonal matrices in $M_n(\overline{k})$: The description of $(\psi_{ts})_{|T}$ above shows that $(\psi'_{ts})_{|T} = 0$ for $s \neq t$ and $(\psi'_{tt})_{|T} = (\det^m \prod_{i=1}^{n} X_{ii}^{a(i,t)})_{|T}$. As $T(\overline{k})$ is dense in D, we get also for all $h \in D$ that $\psi'_{ts}(h) = 0$ for $s \neq t$ and $\psi'_{tt}(h) = \det(h)^m \prod_{i=1}^{n} X_{ii}^{a(i,t)}(h)$. So $\det(h) = 0$ implies $\psi'_{ts}(h) = 0$ for all s and t.

Let now $\gamma' : M_n(\overline{k}) \to M_u(\overline{k})$ denote the map that takes any $h \in M_n(\overline{k})$ to the matrix of all $\psi'_{ts}(h)$, and let $\gamma : GL_n(\overline{k}) \to GL_u(\overline{k})$ denote the map that takes any $g \in GL_n(\overline{k})$ to the matrix of all $\psi_{ts}(g)$. As $\psi'_{ts} \in k[M_{n,k}]$ and $\psi_{ts} \in k[GL_n]$, these are morphisms of varieties over $\overline{k}$. We have $\gamma'(g) = \det(g)^m \gamma(g)$ for all $g \in GL_n(\overline{k})$ by the definition of the ψ'_{ts}. We have $\gamma(g_1 g_2) = \gamma(g_1)\gamma(g_2)$ for all $g_1, g_2 \in GL_n(\overline{k})$ since $\gamma(g)$ is just the matrix of g acting on $V \otimes \overline{k}$. Now $\gamma'(g) = \det(g)^m \gamma(g)$ implies that also $\gamma'(g_1 g_2) = \gamma'(g_1)\gamma'(g_2)$ for all $g_1, g_2 \in GL_n(\overline{k})$. And as $GL_n(\overline{k})$ is dense in $M_n(\overline{k})$ we get also $\gamma'(h_1 h_2) = \gamma'(h_1)\gamma'(h_2)$ for all $h_1, h_2 \in M_n(\overline{k})$.

Consider now $h \in M_n(\overline{k})$ with $\det(h) = 0$. There exist $g_1, g_2 \in GL_n(\overline{k})$ such that $g_1 h g_2$ is diagonal. Then also $\det(g_1 h g_2) = 0$, hence $\psi'_{ts}(g_1 h g_2) = 0$ for all s and t by the discussion of the "diagonal case". This means $0 = \gamma'(g_1 h g_2) = \gamma'(g_1)\gamma'(h)\gamma'(g_2)$. As $\gamma'(g_i) \in GL_u(\overline{k})$ for both i, we get $\gamma'(h) = 0$, i.e., $\psi'_{ts}(h) = 0$ for all s and t as desired.

We know by now that each ψ_{ts} is a polynomial in the X_{ij}. The description of the restrictions of the ψ_{ts} to T shows that any scalar matrix $a1 \in T(\overline{k})$ with $a \in \overline{k}^{\times}$ acts as multiplication by a^d on $V \otimes \overline{k}$. This implies for any $g \in GL_n(\overline{k})$ that $\gamma(ag) = a^d \gamma(g)$, hence that $\psi_{ts}(ag) = a^d \psi_{ts}(g)$ for all a. Therefore the polynomial ψ_{ts} in the X_{ij} has to be homogeneous of degree d. The claim follows.

A.5. We return to general G. If M is a G–module with a good filtration (see 4.16), then we denote by $(M : H^0(\lambda))$ the number of factors isomorphic to $H^0(\lambda)$ in such a filtration (for any $\lambda \in X(T)_+$). Recall that this number is equal to $\dim \mathrm{Hom}_G(V(\lambda), M)$, see 4.16.a. For the sake of convenience we restrict ourselves here to modules where all $(M : H^0(\lambda))$ are finite.

Lemma: *Let $\pi \subset X(T)_+$ be saturated. If M is a G–module with a good filtration such that all $(M : H^0(\lambda))$ are finite, then $O_\pi(M)$ has a good filtration with*

$$(O_\pi(M) : H^0(\lambda)) = \begin{cases} (M : H^0(\lambda)), & \text{if } \lambda \in \pi, \\ 0, & \text{otherwise} \end{cases}$$

for all $\lambda \in X(T)_+$.

Proof: Suppose at first that π is finite. We can choose a numbering $\lambda_1, \lambda_2, \lambda_3, \dots$ of $X(T)_+$ such that $\lambda_i < \lambda_j$ implies $i < j$ and such that $\pi = \{\lambda_1, \lambda_2, \dots, \lambda_r\}$ for some r. One has then (see 4.16) a chain $0 = M_0 \subset M_1 \subset M_2 \subset \cdots$ of submodules of M with $M = \bigcup_i M_i$ and with $M_i/M_{i-1} \simeq H^0(\lambda_i)^{m(i)}$ where $m(i) = (M : H^0(\lambda_i))$ for all $i > 0$.

Now each weight of M_r is the weight of some $H^0(\lambda_i)$ with $i \le r$, hence belongs to $W\pi$. So we get $M_r \subset O_\pi(M)$. We claim that we have here equality; this will certainly imply the claim (for finite π). Well, if $M_r \neq O_\pi(M)$, then there exists a simple submodule $L \subset O_\pi(M)/M_r$. We have $L \simeq L(\lambda)$ for some $\lambda \in X(T)_+$, and since L is a composition factor of $O_\pi(M)$ we have $\lambda \in \pi$. On the other hand, L has a non-zero image in some M_i/M_{i-1} with $i > r$. As $H^0(\lambda_i)^{m(i)}$ has socle $L(\lambda_i)^{m(i)}$, this implies $\lambda = \lambda_i$ with $i > r$, hence $\lambda_i \notin \pi$ — contradiction!

Now suppose that π is infinite. Then we can choose a numbering $\mu_1, \mu_2, \mu_3, \dots$ of π such that $\mu_i < \mu_j$ implies $i < j$. Set $\pi_i = \{\mu_1, \mu_2, \dots, \mu_i\}$ for all $i > 0$. Then each π_i is saturated. The inclusion $\pi_i \subset \pi_{i+1}$ implies $O_{\pi_i}(M) \subset O_{\pi_{i+1}}(M)$ for all i. The proof in the finite case shows that $O_{\pi_{i+1}}(M)/O_{\pi_i}(M) \simeq H^0(\mu_{i+1})^{m(i+1)}$ for all $i \ge 0$ with $m(i)$ as above. We have $O_\pi(M) = \bigcup_{i \ge 1} O_{\pi_i}(M)$ by A.1(6). This implies the claim.

A.6. For each $\lambda \in X(T)_+$ let Q_λ denote the injective hull of $L(\lambda)$ as a G–module. So Q_λ is indecomposable and has socle $L(\lambda)$. This implies that $O_\pi(Q_\lambda) = 0$ for any π with $\lambda \notin \pi$.

On the other hand, if $\lambda \in \pi$, then $L(\lambda)$ is also the socle of $O_\pi(Q_\lambda)$. Since $O_\pi(Q_\lambda)$ is injective in $\mathcal{C}(\pi)$ by A.1(4), this implies that $O_\pi(Q_\lambda)$ is an injective hull of $L(\lambda)$ in $\mathcal{C}(\pi)$.

If π is saturated, then A.5 and 4.18 show that each $O_\pi(Q_\lambda)$ has a good filtration with factors of the form $H^0(\mu)$ with $\mu \in \pi$ and that the multiplicity of any $H^0(\mu)$ with $\mu \in \pi$ as a factor in such a filtration is equal to $[V(\mu) : L(\lambda)]$.

This shows in particular: If π is finite (and saturated), then each $O_\pi(Q_\lambda)$ is finite dimensional. (One can check that one does not need to assume π to be saturated for this last statement.)

If π is finite and saturated and if μ is maximal in π, then Proposition 4.18 and Lemma A.5 imply that $O_\pi(I(\mu)) \simeq H^0(\mu)$; so $H^0(\mu)$ is an injective hull of $L(\mu)$ in $\mathcal{C}(\pi)$. On the other hand, the argument used for proving the splitting of 2.14(3) shows that $V(\mu)$ is projective in $\mathcal{C}(\pi)$, in fact a projective cover for $L(\mu)$.

In general ${}^\tau O_\pi(Q_\lambda)$ is (for π as above) a projective cover for $L(\lambda)$. For this note first that $M \mapsto {}^\tau M$ preserves $\mathcal{C}(\pi)$. In order to prove projectivity one has to show that each surjection $M \twoheadrightarrow {}^\tau O_\pi(Q_\lambda)$ with M in $\mathcal{C}(\pi)$ splits. For finite dimensional M this follows from the splitting of $O_\pi(Q_\lambda) \hookrightarrow {}^\tau M$. In general use that there exists a finite dimensional submodule M' of M that maps onto ${}^\tau O_\pi(Q_\lambda)$.

A.7. For each $\pi \subset X(T)_+$ the left exact functor O_π has right derived functors $R^l O_\pi$, $l \geq 0$ such that $R^0 O_\pi = O_\pi$.

Lemma: *All $R^l O_\pi$ commute with direct limits.*

Proof: We first look at $l = 0$. Suppose that we have a pre-ordered set I, for all $i \in I$ a G–module M_i, and for all $i, j \in I$ with $i \leq j$ a G–module homomorphism $f_{ji} : M_i \to M_j$ such that each f_{ii} is the identity map and such that $f_{ji} \circ f_{ih} = f_{jh}$ whenever $h \leq i \leq j$ in I. We also assume that I is directed: for all $i, j \in I$ one can find $l \in I$ with $i \leq l$ and $j \leq l$.

Let M denote the direct limit of the M_i with respect to the f_{ij}. So this is a G–module with homomorphisms $f_i : M_i \to M$ such that $f_j \circ f_{ji} = f_i$ whenever $i \leq j$ in I. Furthermore M has a suitable universal property. Since I is directed, one has $M = \bigcup_{i \in I} f_i(M_i)$; if $x \in M_i$, then $f_i(x) = 0$ if and only if there exists $j \in I$, $j \geq i$ with $f_{ji}(x) = 0$.

Each f_{ji} restricts to a morphism $O_\pi(f_{ji}) : O_\pi(M_i) \to O_\pi(M_j)$. These $O_\pi(f_{ji})$ satisfy the same assumptions as the f_{ji}. We can therefore form the direct limit, say N, of the $O_\pi(M_i)$ with respect to the $O_\pi(f_{ji})$. Let $f'_i : O_\pi(M_i) \to N$ denote the maps analogous to the f_i. Since N is the union of all $f'_i(O_\pi(M_i))$, it is clear that N belongs to $\mathcal{C}(\pi)$.

We have $(f_j)_{|O_\pi(M_j)} \circ O_\pi(f_{ji}) = (f_i)_{|O_\pi(M_i)}$ for all $i \leq j$ in I. So the universal property of N yields a homomorphism $\varphi : N \to M$ with $\varphi \circ f'_i = (f_i)_{|O_\pi(M_i)}$ for all i. As N belongs to $\mathcal{C}(\pi)$, it is clear that $\varphi(N) \subset O_\pi(M)$. We have to show that φ is an isomorphism from N onto $O_\pi(M)$. [This is what "O_π commutes with direct limits" means.]

Let $x \in \ker(\varphi)$. There exist i and $x' \in O_\pi(M_i)$ with $x = f'_i(x')$. Now $0 = \varphi(x) = f_i(x')$ means that there exists $j \geq i$ with $f_{ji}(x') = 0$. But then also $O_\pi(f_{ji})(x') = 0$, hence $x = f'_i(x') = 0$. So φ is injective.

Let $v \in O_\pi(M)$. Let V be the G–module generated by v in $O_\pi(M)$. As V is finite dimensional, there exists $i \in I$ with $V \subset f_i(M_i)$. Now we can find a finite dimensional submodule $\widetilde{V} \subset M_i$ with $V = f_i(\widetilde{V})$. [Pick a basis for V and choose for each basis vector an inverse image in M_i. Then let $\widetilde{V}$ be the G–submodule generated by these inverse images.] Let $x_1, x_2, \dots, x_s$ be a basis for $\ker(f_i) \cap \widetilde{V}$. There exists $j \geq i$ with $f_{ji}(x_l) = 0$ for all l. Then f_j induces an isomorphism $f_{ji}(\widetilde{V}) \xrightarrow{\sim} V$. Now $V \subset O_\pi(M)$ implies $f_{ji}(\widetilde{V}) \subset O_\pi(M_j)$. So V is contained in the image of $(f_j)_{|O_\pi(M_j)}$, hence in that of φ. In particular, we have $v \in \varphi(N)$. The claim for O_π follows.

And now for the $R^l O_\pi$ with $l > 0$: Given a directed system as above, we construct a new system with $M'_i = M_i \otimes k[G]$ for all i and $f'_{ji} = f_{ji} \otimes \mathrm{id}_{k[G]}$ for all $i \leq j$. As tensoring with $k[G]$ is exact, we get $\varinjlim M'_i \simeq (\varinjlim M_i) \otimes k[G]$. All M'_i and their limit are injective G–modules, see I.3.10, hence annihilated by all $R^l O_\pi$ with $l > 0$. We have for each i an embedding of G–modules $u_i : M_i \to M_i \otimes k[G]$

given by $x \mapsto x \otimes 1$. These u_i define a monomorphism from $(M_i)_i$ to $(M'_i)_i$ in the abelian category of all directed systems of G–modules, indexed by I. Now argue as in [Ha], III, Proof of Prop. 2.9.

A.8. Lemma: *Let $\pi \subset X(T)_+$ be saturated. If M is a G–module with a good filtration, then $R^i O_\pi(M) = 0$ for all $i > 0$.*

Proof: We use induction on i. Fix $i > 0$ and suppose that $R^j O_\pi(V) = 0$ for all G–modules V with a good filtration and all j with $0 < j < i$.

It will be enough to show $R^i H^0(\lambda) = 0$ for all $\lambda \in X(T)_+$. If we have that, then we get $R^i M = 0$ for all finite dimensional M with a good filtration using induction on the length of a good filtration and the long exact derived sequence associated to a short exact sequence. Finally, in the general case M is the direct limit of finite dimensional submodules with a good filtration; now $R^i O_\pi(M) = 0$ follows from the finite dimensional case using Lemma A.7.

So let $\lambda \in X(T)_+$. There is by 4.18 a short exact sequence

$$0 \to H^0(\lambda) \longrightarrow Q_\lambda \longrightarrow Q \to 0$$

with Q_λ as in A.6 and where Q has a good filtration. Applying O_π to this sequence, we get a long exact sequence that breaks into small pieces because Q_λ is injective. We get thus an exact sequence

$$(1) \qquad 0 \to O_\pi(H^0(\lambda)) \longrightarrow O_\pi(Q_\lambda) \longrightarrow O_\pi(Q) \longrightarrow R^1 O_\pi(H^0(\lambda)) \to 0$$

and for all $j \geq 1$ an isomorphism $R^j O_\pi(Q) \xrightarrow{\sim} R^{j+1} O_\pi(H^0(\lambda))$. If $i > 1$, then we get now immediately $R^i O_\pi(H^0(\lambda)) \simeq R^{i-1} O_\pi(Q) = 0$ from our induction. If $i = 1$, then we have to observe that in (1) the map $O_\pi(Q_\lambda) \to O_\pi(Q)$ is surjective. If $\lambda \notin \pi$, then this follows from the fact that $O_\pi(Q) = 0$ because Q has a good filtration with factors $H^0(\nu)$ such that $\nu > \lambda$, hence $\nu \notin \pi$ and $O_\pi(H^0(\nu)) = 0$. If $\lambda \in \pi$, then $O_\pi(H^0(\lambda)) = H^0(\lambda)$ and the surjectivity follows from a general observation: If V' is a submodule of a G–module V such that V' belongs to $\mathcal{C}(\pi)$, then $O_\pi(V/V') = O_\pi(V)/V'$.

If M is finite dimensional with a good filtration, then we get $R^i O_\pi(M) = 0$ by induction on the length of a good filtration. In general, M is the direct limit of finite dimensional submodules with a good filtration; now $R^i O_\pi(M) = 0$ follows from the finite dimensional case using Lemma A.7.

Remark: One can show (see [Donkin 12], (2.1d)): If V is a finite dimensional G–module with $R^i O_\pi(V) = 0$ for all $i > 0$ and all saturated $\pi \subset X(T)_+$, then V has a good filtration.

A.9. Lemma: *Let $\pi \subset X(T)_+$ be saturated. If V is a G–module in $\mathcal{C}(\pi)$, then $R^i O_\pi(V) = 0$ for all $i > 0$.*

Proof: We use induction on i. Fix $i > 0$ and suppose that $R^j O_\pi(V') = 0$ for all G–modules V' in $\mathcal{C}(\pi)$ and all j with $0 < j < i$.

It will be enough to show $R^i O_\pi L(\lambda) = 0$ for all $\lambda \in \pi$. If we have that, then we get $R^i O_\pi(V) = 0$ for all finite dimensional V in $\mathcal{C}(\pi)$ using induction on the length of V and the long exact derived sequence associated to a short exact sequence. Finally, in the general case V is the direct limit of its finite dimensional submodules; now $R^iO_\pi(V) = 0$ follows from the finite dimensional case using Lemma A.7.

So let $\lambda \in \pi$. There is a short exact sequence

$$0 \to L(\lambda) \longrightarrow H^0(\lambda) \longrightarrow M \to 0$$

with M in $\mathcal{C}(\pi)$. Applying O_π to this sequence, we get a long exact sequence that breaks into small pieces because $R^jO_\pi H^0(\lambda) = 0$ for all $j > 0$ by A.8. We get thus an exact sequence

$$(1) \qquad 0 \to L(\lambda) \longrightarrow H^0(\lambda) \longrightarrow M \longrightarrow R^1O_\pi L(\lambda) \to 0$$

and for all $j \geq 1$ an isomorphism $R^jO_\pi(M) \xrightarrow{\sim} R^{j+1}O_\pi L(\lambda)$. Here we have already used that O_π takes all terms in our short exact sequence to itself. We get in particular from (1) that $R^1O_\pi L(\lambda) = 0$. If $i > 1$, then we use that $R^iO_\pi L(\lambda) \simeq R^{i-1}O_\pi(M) = 0$ by induction.

A.10. Let $\pi \subset X(T)_+$. We have for all G–modules M and V with V in $\mathcal{C}(\pi)$ a spectral sequence with

$$(1) \qquad E_2^{i,j} = \operatorname{Ext}^i_{\mathcal{C}(\pi)}(V, R^jO_\pi(M)) \Rightarrow \operatorname{Ext}^{i+j}_G(V, M).$$

This is a special case of Grothendieck's spectral sequence (see I.4.1) since O_π takes injective G–modules to injective objects in $\mathcal{C}(\pi)$, see A.1(4), and since we have a natural isomorphism of functors $\operatorname{Hom}_{\mathcal{C}(\pi)}(V, ?) \circ O_\pi \simeq \operatorname{Hom}_G(V, ?)$ by A.1(3).

Proposition: *Let $\pi \subset X(T)_+$ be saturated. Let V and V' be G–modules in $\mathcal{C}(\pi)$. Then there are isomorphisms*

$$\operatorname{Ext}^i_{\mathcal{C}(\pi)}(V, V') \simeq \operatorname{Ext}^i_G(V, V')$$

for all $i \geq 0$.

Proof: Apply (1) with $M = V'$. As $R^jO_\pi(V') = 0$ for all $j > 0$ by A.9, the spectral sequence degenerates and yields the isomorphisms as claimed.

A.11. Proposition: *Let $\pi \subset X(T)_+$ be saturated and finite. Then we have $\operatorname{Ext}^i_{\mathcal{C}(\pi)}(V, V') = 0$ for all finite dimensional G–modules V and V' in $\mathcal{C}(\pi)$ and all $i > 2(|\pi| - 1)$.*

Proof: We use induction on $|\pi|$. If $|\pi| = 1$, say $\pi = \{\lambda\}$, then each module in $\mathcal{C}(\pi)$ is a direct sum of copies of $L(\lambda)$, hence semi-simple. So indeed all $\operatorname{Ext}^i_{\mathcal{C}(\pi)}(V, V')$ with $i > 0$ vanish.

Suppose now that $|\pi| > 1$. Choose a maximal element $\lambda \in \pi$ and set $\sigma = \pi \backslash \{\lambda\}$. Then also σ is saturated and we can apply induction to σ since $|\sigma| = |\pi| - 1$.

It is clearly enough to show that $\operatorname{Ext}^i_{\mathcal{C}(\pi)}(L(\mu), L(\nu)) = 0$ for all $\mu, \nu \in \pi$ and all $i > 2(|\pi| - 1)$. The general case follows then by induction on the lengths of V and V' using the usual long exact sequences. If both μ and ν belong to σ, then this follows from the induction assumption and the fact that $\operatorname{Ext}^i_{\mathcal{C}(\sigma)}(V, V') = \operatorname{Ext}^i_G(V, V') = \operatorname{Ext}^i_{\mathcal{C}(\pi)}(V, V')$ for all V and V' in $\mathcal{C}(\sigma)$.

The maximality of λ in π implies that $H^0(\lambda)$ is injective in $\mathcal{C}(\pi)$, cf A.6. We have a short exact sequence $0 \to L(\lambda) \to H^0(\lambda) \to M \to 0$ with a suitable G–module M that belongs to $\mathcal{C}(\sigma)$. The injectivity of $H^0(\lambda)$ implies that we have for all $j > 0$ and each V in $\mathcal{C}(\pi)$ an isomorphism $\mathrm{Ext}^j_{\mathcal{C}(\pi)}(V, M) \xrightarrow{\sim} \mathrm{Ext}^{j+1}_{\mathcal{C}(\pi)}(V, L(\lambda))$. We have by induction $\mathrm{Ext}^j_{\mathcal{C}(\pi)}(L(\mu), M) = 0$ for all $\mu \in \sigma$ and all $j > 2|\sigma| - 2$ and get now $\mathrm{Ext}^i_{\mathcal{C}(\pi)}(L(\mu), L(\lambda)) = 0$ for all $i > 2|\sigma| - 1 = 2|\pi| - 3$.

The maximality of λ in π implies also that $V(\lambda)$ is projective in $\mathcal{C}(\pi)$. We have a short exact sequence $0 \to M' \to V(\lambda) \to L(\lambda) \to 0$ with a suitable G–module M' that belongs to $\mathcal{C}(\sigma)$. Now the projectivity of $V(\lambda)$ shows that we have for all $j > 0$ and each V in $\mathcal{C}(\pi)$ an isomorphism $\mathrm{Ext}^j_{\mathcal{C}(\pi)}(M', V) \xrightarrow{\sim} \mathrm{Ext}^{j+1}_{\mathcal{C}(\pi)}(L(\lambda), V)$. This implies as above $\mathrm{Ext}^i_{\mathcal{C}(\pi)}(L(\lambda), L(\mu)) = 0$ for all $\mu \in \sigma$ and all $i > 2|\pi| - 3$.

The result in the preceding paragraph yields (using induction on the length of M') that $\mathrm{Ext}^i_{\mathcal{C}(\pi)}(L(\lambda), M') = 0$ for all $i > 2|\pi| - 3$, hence taking $V = L(\lambda)$, that $\mathrm{Ext}^i_{\mathcal{C}(\pi)}(L(\lambda), L(\lambda)) = 0$ for all $i > 2|\pi| - 2$.

Remark: It is possible to extend the result to infinite dimensional V and V'; this is left to the interested reader. The proposition says in particular that the category $\mathcal{C}(\pi)$ has *finite global dimension* and shows that this dimension (the largest m such that there exist V and V' with $\mathrm{Ext}^m_{\mathcal{C}(\pi)}(V, V') \neq 0$) is bounded above by $2|\pi| - 2$. A closer look at the proof (see the original proof in [Donkin 12], (2.2f)) shows that one can replace this bound by $2s$ where s is the largest number such that there exist chains $\lambda_0 < \lambda_1 < \cdots < \lambda_s$ with all $\lambda_i \in \pi$. Working with blocks one can sharpen this result and replace "$<$" by "$\uparrow$ and $\neq$", see the remarks at the end of the first section of [Totaro]. For some calculations of the exact value of this global dimension for certain cases where $G = GL_n$ and $\pi = \pi(n, d)$ as in A.3, see [Totaro] or [Parker 1, 2]. For a quantum analogue, see [Do].

A.12. As mentioned in A.3, the construction of the matrix coefficients there works more generally: Given a G–module V, one associates to each $v \in V$ and each $\varphi \in V^*$ the function $c_{\varphi,v} \in k[G]$ such that $c_{\varphi,v}(g) = (\varphi \otimes \mathrm{id}_A)(g(v \otimes 1))$ for all $g \in G(A)$ and all A. If the comodule map Δ_V sends v to $\sum_s v_s \otimes f_s$ with all $v_s \in V$ and $f_s \in k[G]$, then $c_{\varphi,v} = \sum_h \varphi(v_s) f_s$.

Let us call the subspace of $k[G]$ spanned by all $c_{\varphi,v}$ the *coefficient space* of V and denote it by $\mathrm{cf}(V)$. If we write $\Delta_V(v) = \sum_s v_s \otimes f_s$ as above and assume the v_s to be linearly independent, then we get $f_s \in \mathrm{cf}(V)$ for each s choosing $\varphi = \varphi_s$ with $\varphi_s(v_t) = \delta_{st}$. This shows that

(1) $$\Delta_V(V) \subset V \otimes \mathrm{cf}(V).$$

Let us write in the following subsections $k[G]_r$ for $k[G]$ considered a G–module under the right regular representation ρ_r, and $k[G]_l$ for $k[G]$ under the left regular representation ρ_l.

Note next:

(2) *For each $\varphi \in V^*$ the map $c_\varphi : V \to k[G]_r$ with $c_\varphi(v) = c_{\varphi,v}$ is a homomorphism of G–modules.*

It is clear that c_φ is linear. The compatibility with G follows from I.2.8(2), (5) or from I.3.7(5). Alternatively one may observe for any k–algebra A that $c_\varphi \otimes \mathrm{id}_A$

maps any $x \in V \otimes A$ to the function on G_A with

$$(c_\varphi \otimes \mathrm{id}_A)(x)(h) = (c_\varphi \otimes \mathrm{id}_{A'})(h(x \otimes 1))$$

for all $h \in G(A')$ and any A–algebra A'. Then we get for all $g \in G(A)$

$$(c_\varphi \otimes \mathrm{id}_A)(gx)(h) = (c_\varphi \otimes \mathrm{id}_{A'})(hg(x \otimes 1)) = (c_\varphi \otimes \mathrm{id}_A)(x)(hg)$$

hence $(c_\varphi \otimes \mathrm{id}_A)(gx) = \rho_r(g)((c_\varphi \otimes \mathrm{id}_A)(x))$.

One can show similarly:

(3) *If $\dim(V) < \infty$, then for each $v \in V$ the map $V^* \to k[G]_l$ with $\varphi \mapsto c_{\varphi,v}$ is a homomorphism of G–modules.*

Alternatively, this can be deduced from (2) as follows: We identify V and $(V^*)^*$ as usual. We get then for each $v \in V$ and $\varphi \in V^*$ a matrix coefficient $c_{v,\varphi}$ for V^*. We have for all $g \in G(A)$ and all A

$$\begin{aligned} c_{v,\varphi}(g) &= (v \otimes \mathrm{id}_A)(g(\varphi \otimes 1)) = (g(\varphi \otimes \mathrm{id}_A))(v \otimes 1) \\ &= (\varphi \otimes \mathrm{id}_A)(g^{-1}(v \otimes 1)) = c_{\varphi,v}(g^{-1}) \end{aligned}$$

hence $c_{v,\varphi} = \sigma_G(c_{\varphi,v})$ with σ_G as in I.2.3. As $\varphi \mapsto c_{v,\varphi}$ is a homomorphism $V^* \to k[G]_r$ and as σ_G is an isomorphism $k[G]_r \to k[G]_l$, see I.2.7, with $\sigma_G^2 = \mathrm{id}$, we get the claim. This argument shows also that

(4) $$\mathrm{cf}(V^*) = \sigma_G(\mathrm{cf}(V)).$$

Note that everything in this subsection works with G replaced by any group scheme over a field. (For more general ground rings one has to impose conditions on V for (1) to hold or for (3) and (4) to make sense.)

A.13. Let $\pi \subset X(T)_+$. Set $O_\pi^r(k[G]) = O_\pi(k[G]_r)$ and $O_\pi^l(k[G]) = O_\pi(k[G]_l)$.

Lemma: *Let $\pi \subset X(T)_+$ and let M be a G–module. Then the following are equivalent:*

(i) *M belongs to $\mathcal{C}(\pi)$.*

(ii) $\mathrm{cf}(M) \subset O_\pi^r(k[G])$.

(iii) $\Delta_M(M) \subset M \otimes O_\pi^r(k[G])$.

Proof: The implication (i) ⇒ (ii) follows from A.12(2) and (ii) ⇒ (iii) follows from A.12(1). Furthermore, (iii) ⇒ (ii) is clear from the description of $c_{\varphi,v}$ in terms of $\Delta_M(v)$ at the beginning of A.12.

Let us prove (ii) ⇒ (i). Pick $v \in M$ and set V equal to the G–submodule of M generated by v. Then $\dim(V) < \infty$; choose a basis $\varphi_1, \varphi_2, \dots, \varphi_m$ for V^*. Consider the map

$$V \longrightarrow O_\pi^r(k[G])^m, \qquad x \mapsto (c_{\varphi_1,x}, c_{\varphi_2,x}, \dots, c_{\varphi_m,x}).$$

This is by A.12(2) a homomorphism of G–modules and it is injective: If $x \in V$, $x \neq 0$, then there exists i with $\varphi_i(x) \neq 0$, hence with $c_{\varphi_i,x}(1) = \varphi_i(1x) = \varphi_i(x) \neq 0$.

So V is isomorphic to a submodule of $O_\pi^r(k[G])^m$. As $O_\pi^r(k[G])^m$ belongs to $\mathcal{C}(\pi)$, so does V and we get $v \in O_\pi(M)$.

Remark: We get actually a more precise result for all $v \in M$:

(1) $$v \in O_\pi(M) \iff c_{\varphi,v} \in O^r_\pi(k[G]) \text{ for all } \varphi \in M^*.$$

Here "$\Rightarrow$" is more or less obvious: If $v \in O_\pi(M)$ and if ψ is the restriction of φ to $O_\pi(M)$, then one has $c_{\varphi,v} = c_{\psi,v} \in O^r_\pi(k[G])$. On the other hand, each c_φ as in A.12(2) maps the submodule V of M generated by v to the submodule of $k[G]$ generated by $c_{\varphi,v}$ with respect to ρ_r. So, if $c_{\varphi,v} \in O^r_\pi(k[G])$ for all φ, then also $c_{\varphi,x} \in O^r_\pi(k[G])$ for all $x \in V$. We can now argue as in the proof above and get $V \in \mathcal{C}(\pi)$ and hence $v \in O_\pi(M)$. (Alternatively, we can apply the lemma to V.)

It now follows that

(2) $$O_\pi(M) = \Delta_M^{-1}(M \otimes O^r_\pi(k[G])).$$

Here "$\subset$" is obvious by the lemma and "$\supset$" follows from (1): If $\Delta_M(v) \in M \otimes O^r_\pi(k[G])$, then $c_{\varphi,v} \in O^r_\pi(k[G])$ for all $\varphi \in M^*$.

A.14. Let $\pi \subset X(T)_+$. We have

(1) $$\sigma_G(O^r_\pi(k[G])) = O^l_\pi(k[G])$$

as σ_G is an isomorphism of G–modules $k[G]_r \xrightarrow{\sim} k[G]_l$.

Let us show that

(2) $$O^r_\pi(k[G]) = \sum_V \mathrm{cf}(V)$$

where we sum over all finite dimensional G–modules V in $\mathcal{C}(\pi)$. Here the inclusion "$\supset$" is clear by Lemma A.13. On the other hand, consider $f \in O^r_\pi(k[G])$ and let V be the submodule of $k[G]$ with respect to ρ_r generated by f. Then V is finite dimensional and contained in $O^r_\pi(k[G])$, hence belongs to $\mathcal{C}(\pi)$. Let φ denote the restriction of the counit ε_G to V. Then $c_{\varphi,f} = f$ by I.2.3(2), hence $f \in \mathrm{cf}(V)$.

Set

(3) $$\pi^* = \{ -w_0\lambda \mid \lambda \in \pi \}.$$

As $L(\lambda)^* \simeq L(-w_0\lambda)$ for all $\lambda \in X(T)_+$, we see that a finite dimensional G–module V belongs to $\mathcal{C}(\pi)$ if and only if V^* belongs to $\mathcal{C}(\pi^*)$. Now (2) and A.12(4) imply that

(4) $$\sigma_G(O^r_\pi(k[G])) = O^r_{\pi^*}(k[G])).$$

Combining this with (1) we get

(5) $$O^r_\pi(k[G]) = O^l_{\pi^*}(k[G]).$$

This shows in particular that $O^r_\pi(k[G])$ is a submodule of $k[G]$ also for ρ_l. This can alternatively be seen as follows: Since $\Delta_{\rho_r} = \Delta_G$ by I.2.8(5), we get from Lemma A.13 applied to $M = O^r_\pi(k[G])$ that

(6) $$\Delta_G(O^r_\pi(k[G])) \subset O^r_\pi(k[G]) \otimes O^r_\pi(k[G]).$$

This implies that $O^r_\pi(k[G])$ is a subcoalgebra of $k[G]$. (We shall return to this fact.) It also shows using the description of Δ_{ρ_l} in I.2.8(6) that

$$\Delta_{\rho_l}(O^r_\pi(k[G])) \subset O^r_\pi(k[G]) \otimes \sigma_G(O^r_\pi(k[G])).$$

This implies by I.2.9 that $O^r_\pi(k[G])$ is a submodule for ρ_l.

A.15. We can consider $k[G]$ as a $(G\times G)$–module via $\rho_l \times \rho_r$. By the results in A.14 each $O_\pi^r(k[G])$ is a $(G\times G)$–submodule of $k[G]$.

Lemma: *If $\pi \subset X(T)_+$ is finite and saturated, then the $(G\times G)$–module $O_\pi^r(k[G])$ admits a good filtration. The factors are the $H^0(-w_0\lambda)\otimes H^0(\lambda)$ with $\lambda\in\pi$ each occurring with multiplicity* 1. *We have*

$$\dim O_\pi^r(k[G]) = \sum_{\lambda\in\pi} (\dim V(\lambda))^2. \tag{1}$$

Proof: We can choose a numbering $\lambda_1, \lambda_2, \lambda_3, \dots$ of $X(T)_+$ such that $\lambda_i < \lambda_j$ implies $i<j$ and such that $\pi = \{\lambda_1, \lambda_2, \dots, \lambda_m\}$ for some m. There is by Proposition 4.20 a good filtration $0 = V_0 \subset V_1 \subset V_2 \subset \cdots$ of the $(G\times G)$–module $k[G]$ with $V_i/V_{i-1} \simeq H^0(-w_0\lambda_i)\otimes H^0(\lambda_i)$. Considered as a G–module under ρ_r, any V_i/V_{i-1} is isomorphic to a direct sum of $\dim V(\lambda_i)$ copies of $H^0(\lambda_i)$. [Here we use that $V(\lambda)$, $H^0(-w_0\lambda)$ and $H^0(\lambda)$ all have the same dimension, given by Weyl's dimension formula.] Now we get $O_\pi^r(k[G]) = V_m$ as in the proof of Lemma A.5. This then implies the claims.

Remark: Except for (1), the results extend to infinite saturated subsets, cf. A.5.

A.16. Let $\pi\subset X(T)_+$ be finite. Then $O_\pi^r(k[G])$ is finite dimensional. (For saturated π this is clear by Lemma A.15; in general, we can use that π can be embedded into a finite and saturated subset of $X(T)_+$, for example the set of all $\mu\in X(T)_+$ such that there exists $\lambda\in\pi$ with $\mu\le\lambda$.)

Set $S_G(\pi) = O_\pi^r(k[G])^*$. This vector space has a natural structure as an associative k–algebra because $O_\pi^r(k[G])$ is a subcoalgebra of $k[G]$ by A.14(6): For any $\gamma_1,\gamma_2\in S_G(\pi)$ their product is defined by $(\gamma_1\gamma_2)(f) = (\gamma_1\otimes\gamma_2)(\Delta_G f)$ for all $f\in O_\pi^r(k[G])$. The associativity of this product follows from I.2.3(1). The restriction of the counit ε_G to $O_\pi^r(k[G])$ is by I.2.3(2) a 1 in this algebra.

We call $S_G(\pi)$ the *generalised Schur algebra* associated to G and π.

Proposition: *Let $\pi\subset X(T)_+$ be finite. Then there is a surjective homomorphism of associative k–algebras from* Dist(G) *onto $S_G(\pi)$ that maps any $\mu\in$* Dist(G) *to its restriction on $O_\pi^r(k[G])$. The kernel of this homomorphism is the two-sided ideal $I_G(\pi)$ of all $\mu\in$* Dist(G) *with $\mu V = 0$ for all G–modules V in $\mathcal{C}(\pi)$.*

Proof: Let I_1 denote the ideal of all $f\in k[G]$ with $f(1) = 0$. Recall from I.7 that Dist(G) is the set of all $\mu\in k[G]^*$ with $\mu(I_1^{n+1}) = 0$ for some $n\in\mathbf{N}$. So the restriction to $O_\pi^r(k[G])$ is a well defined linear map from Dist(G) to $S_G(\pi)$. A comparison of the definition of the multiplication on Dist(G) in I.7.7(1) with the definition above shows that this map is a homomorphism of associative k–algebras; it maps the 1 in Dist(G) (equal to ε_G) to the 1 in $S_G(\pi)$.

As G is integral, we have $\bigcap_{n\ge 0} I_1^{n+1} = 0$. Since $\dim O_\pi^r(k[G]) < \infty$ we can therefore find $m\in\mathbf{N}$ with $O_\pi^r(k[G])\cap I_1^{m+1} = 0$. Therefore each element in $O_\pi^r(k[G])$ can be extended to an element in $k[G]^*$ with $\mu(I_1^{m+1}) = 0$. This shows that the homomorphism Dist$(G)\to S_G(\pi)$ is surjective.

Recall the construction of the structure as a Dist(G)–module on a G–module M from I.7.11(2). If $\mu\in$ Dist(G) is in the kernel of our homomorphism, i.e., if $\mu(O_\pi^r(k[G])) = 0$, then we get $\mu M = 0$ for all M in $\mathcal{C}(\pi)$ from (iii) in A.13.

Suppose conversely that $\mu \in \mathrm{Dist}(G)$ satisfies $\mu V = 0$ for all V in $\mathcal{C}(\pi)$. We get then $\mu f = 0$ for all $f \in O^r_\pi(k[G])$ for the $\mathrm{Dist}(G)$–module structure coming from ρ_r. This means, if $\Delta_G(f) = \sum_{i=1}^s f_i \otimes f'_i$, then $0 = \sum_{i=1}^s \mu(f'_i) f_i$. We may assume the f_i to be linearly independent and get then $\mu(f'_i) = 0$ for all i; as $f = \sum_{i=1}^s f_i(1) f'_i$ by I.2.3(2), we get $\mu(f) = 0$. So μ is in the kernel of our homomorphism.

Remark: As all G–modules are locally finite, we can describe $I_G(\pi)$ also as the set of all $\mu \in \mathrm{Dist}(G)$ with $\mu V = 0$ for all finite dimensional G–modules V in $\mathcal{C}(\pi)$.

A.17. Let again $\pi \subset X(T)_+$ be finite. The same construction that associates to a $k[G]$–comodule (i.e., a G–module) a module over $\mathrm{Dist}(G)$, see I.7.11, associates to each comodule over $O^r_\pi(k[G])$ — a subcoalgebra of $k[G]$ by A.14(6) — a module over $S_G(\pi)$. Since $O^r_\pi(k[G])$ and $S_G(\pi) \simeq O^r_\pi(k[G])^*$ are finite dimensional, we have for each vector space V over k a natural isomorphism

$$\mathrm{Hom}(S_G(\pi) \otimes V, V) \xrightarrow{\sim} \mathrm{Hom}(V, V \otimes O^r_\pi(k[G])).$$

Using this isomorphism one can check that conversely each $S_G(\pi)$–module structure on V comes from a comodule structure over $O^r_\pi(k[G])$, cf. the analogous situation in I.8.6. We get therefore an equivalence of categories

(1) $$\{\, O^r_\pi(k[G])\text{–comodules} \,\} \xrightarrow{\sim} \{\, S_G(\pi)\text{–modules} \,\}.$$

On the other hand, by Lemma A.13 a G–module in $\mathcal{C}(\pi)$ is “the same” as an $O^r_\pi(k[G])$–comodule. So we get thus:

(2) *The category $\mathcal{C}(\pi)$ is equivalent to the category of all $S_G(\pi)$–modules.*

Furthermore, this equivalence is compatible with the usual functor from G–modules to $\mathrm{Dist}(G)$–modules and the isomorphism $\mathrm{Dist}(G)/I_G(\pi) \xrightarrow{\sim} S_G(\pi)$ from Lemma A.16. In particular, $\mathcal{C}(\pi)$ is also equivalent to the category of all $\mathrm{Dist}(G)$–modules annihilated by $I_G(\pi)$.

If π is in addition saturated, then the description of the $(G\times G)$–module structure in Lemma A.15 implies (together with the isomorphism $V(\lambda) \simeq H^0(-w_0\lambda)^*$) that $S_G(\pi)$ as a $(G \times 1)$–module has a filtration (a “Weyl filtration”) with factors $V(\lambda)$ with $\lambda \in \pi$, each $V(\lambda)$ occurring $\dim V(\lambda)$ times. The corresponding $\mathrm{Dist}(G)$–module structure is the one by left multiplication on $\mathrm{Dist}(G)/I_G(\pi) \simeq S_G(\pi)$, cf. the discussion of the analogous module structure in I.8.6.

A.18. Consider the example $G = GL_n$ and $\pi = \pi(n,d)$ as in A.3. We claim that in this case

(1) $$O^r_{\pi(n,d)}(k[GL_n]) = k[M_{n,k}; d]$$

where $k[M_{n,k}; d]$ denotes (as in A.3) the space of homogeneous polynomials of degree d in the matrix coefficients X_{ij}.

Note that $\Delta_G(X_{ij}) = \sum_{l=1}^n X_{il} \otimes X_{lj}$, cf. I.2.4, implies

(2) $$\Delta_G(k[M_{n,k}; d]) \subset k[M_{n,k}; d] \otimes k[M_{n,k}; d].$$

By Proposition A.3 a G–module V belongs to $\mathcal{C}(\pi(n,d))$ if and only if $\Delta_V(V) \subset V \otimes k[M_{n,k};d]$. Applying this to $V = k[M_{n,k};d]$ with $\Delta_{\rho_r} = \Delta_G$, we get the inclusion "$\supset$" in (1). The inclusion "$\subset$" follows from A.14(2).

In Donkin's original proof of Proposition A.3 one uses the "easy part" of that result to get the inclusion "$\supset$" in (1). Then he compares dimensions using A.15(1) to get equality. Now Proposition A.3 follows from (1) and Lemma A.13.

Set

(3) $$S(n,d) = S_{GL_n}(\pi(n,d)).$$

These $S(n,d)$ are the original Schur algebras from [Green 2], 2.3.

Set $E = k^n$ equal to the natural module for GL_n. Then the d–th tensor power $\bigotimes^d E$ is a polynomial GL_n–module homogeneous of degree d, hence an $S(n,d)$–module. It turns out (see [Green 2], 2.6.c) that $\bigotimes^d E$ is a faithful $S(n,d)$–module and that this action of $S(n,d)$ on $\bigotimes^d E$ yields an isomorphism

(4) $$S(n,d) \simeq \operatorname{End}_{kS_d}(\bigotimes{}^d E)$$

where S_d is the symmetric group on d letters acting on $\bigotimes^d E$ by permuting the factors in this tensor product.

A.19. Return to general G. If M is a $(G \times G)$–module, then we denote weight spaces with respect to $T \times T$ by $M_{(\lambda,\mu)}$. We write then

(1) $$M_{(\bullet,\mu)} = \bigoplus_{\lambda \in X(T)} M_{(\lambda,\mu)} \qquad \text{and} \qquad M_{(\lambda,\bullet)} = \bigoplus_{\mu \in X(T)} M_{(\lambda,\mu)}.$$

If we consider $k[G]$ as a $(G \times G)$–module via $\rho_l \times \rho_r$, then $k[G]_{(\lambda,\mu)}$ consists of all $f \in k[G]$ with $f(t_1^{-1} g t_2) = \lambda(t_1)\mu(t_2) f(g)$ for all $t_1, t_2 \in T(A)$, $g \in G(A)$ and all A. Note that $\lambda \neq -\mu$ implies $f(1) = 0$ for all $f \in k[G]_{(\lambda,\mu)}$ as $f(t) = \lambda(t)^{-1} f(1) = \mu(t) f(1)$ for all $t \in T(A)$ and all A.

Let $\pi \in X(T)_+$ be a finite and saturated subset. For each $\lambda \in W\pi$ define $\xi_\lambda \in S_G(\pi) = O_\pi^r(k[G])^*$ as the composition

(2) $$\xi_\lambda : O_\pi^r(k[G]) \longrightarrow O_\pi^r(k[G])_{(\bullet,\lambda)} \longrightarrow k$$

where the first map is the projection with respect to the weight space decomposition of $O_\pi^r(k[G])$ (with all $O_\pi^r(k[G])_{(\mu,\nu)}$ with $\nu \neq \lambda$ in the kernel) and where the second map is the restriction of ε_G. Since $\varepsilon_G(f) = f(1)$, the remark at the end of the preceding paragraph shows that ξ_λ is also the composition

(3) $$\xi_\lambda : O_\pi^r(k[G]) \longrightarrow O_\pi^r(k[G])_{(-\lambda,\lambda)} \longrightarrow k$$

where again the first map is the projection with respect to the weight space decomposition. This description shows that $\xi_\lambda \in S_G(\pi)_{(\lambda,-\lambda)}$.

Lemma: *Let V be an $S_G(\pi)$–module. Then ξ_λ acts as the identity on V_λ and annihilates all V_μ with $\mu \neq \lambda$.*

Proof: Regard V as a G–module in $\mathcal{C}(\pi)$. The G–module homomorphism property, see I.3.7(5), of the comodule map $\Delta_V : V \to V \otimes O_\pi^r(k[G])$ implies for any $\mu \in X(T)$

$$\Delta_V(V_\mu) \subset V \otimes O_\pi^r(k[G])_{(\bullet,\mu)}.$$

Given $v \in V_\mu$ we can write $\Delta_V(v) = \sum_{i=1}^s v_i \otimes f_i$ with $v_i \in V$ and $f_i \in O_\pi^r(k[G])_{(\nu_i,\mu)}$ for some $\nu_i \in X(T)$. We get then $\xi_\lambda v = \sum_{i=1}^s \xi_\lambda(f_i) v_i$. Now $\mu \neq \lambda$ yields $\xi_\lambda v = 0$ whereas we get $\xi_\lambda v = \sum_{i=1}^s f_i(1) v_i = 1\, v = v$ if $\mu = \lambda$.

A.20. Keep the notations and assumptions from A.19. Since the left action of G on $S_G(\pi)$ corresponds to the $S_G(\pi)$–module structure on itself by left multiplication, the lemma shows (using $\xi_\lambda \in S_G(\pi)_{(\lambda,-\lambda)}$) that $\xi_\lambda^2 = \xi_\lambda$ for all λ and $\xi_\lambda\xi_\mu = 0$ for all $\mu \neq \lambda$. We also get $1 = \sum_{\lambda\in W\pi} \xi_\lambda$.

We get for any $x \in S_G(\pi)_{(\mu,\nu)}$ and all λ that $\xi_\lambda x = x$ if $\mu = \lambda$, and $\xi_\lambda x = 0$ if $\mu \neq \lambda$. A simple calculation, similar to the one in the proof, shows that $x\xi_\lambda = x$ if $\nu = -\lambda$, and $x\xi_\lambda = 0$ if $\nu \neq -\lambda$. This shows that

$$\xi_\lambda S_G(\pi)\xi_\lambda = S_G(\pi)_{(\lambda,-\lambda)}. \tag{1}$$

If e is an idempotent element in a ring A, then eAe is a ring under the restrictions of the compositions from A with e as the neutral element for the multiplication. We can apply this observation to the idempotent element ξ_λ in $S_G(\pi)$. Now the general theory on the relations between modules over A and over eAe as developed in [Green 2], 6.2 (see also [Jantzen and Seitz], Section 2) says:

Proposition: *The maps $L \mapsto \xi_\lambda L$ induces a bijection from the set of isomorphism classes of simple $S_G(\pi)$–modules L with $\xi_\lambda L \neq 0$ onto the set of isomorphism classes of simple $\xi_\lambda S_G(\pi)\xi_\lambda$–modules.*

Since the simple $S_G(\pi)$–modules are the $L(\mu)$ with $\mu \in \pi$, we get more explicitly: Each simple $\xi_\lambda S_G(\pi)\xi_\lambda$–module is isomorphic to exactly one $\xi_\lambda L(\mu) = L(\mu)_\lambda$ with $\mu \in \pi$ and $L(\mu)_\lambda \neq 0$.

A.21. Keep the notations and assumptions from A.19 and A.20, but suppose now that $G = GL_n$ and $\pi = \pi(n, d)$ for some $d \leq n$. Denote as in 1.21 by $e_1, e_2, \dots, e_n$ the standard basis of the natural G–module k^n such that each e_i has weight ε_i. The matrix coefficient X_{ij} as in A.3 is given by $g(e_j) = \sum_{i=1}^n X_{ij}(g)e_i$ and has weight $(-\varepsilon_i, \varepsilon_j)$. The monomials $X_{i_1j_1}X_{i_2j_2}\dots X_{i_dj_d}$ with $1 \leq j_1 \leq j_2 \leq \dots \leq j_d \leq n$ and arbitrary $i_1, i_2, \dots, i_d$ in $\{1, 2, \dots, n\}$ are a basis for $O_\pi^r(k[G]) = k[M_{n,k}; d]$, cf. A.18(1). Such a monomial is a weight vector of weight $(-\sum_{l=1}^n \varepsilon_{i_l}, \sum_{l=1}^n \varepsilon_{j_l})$.

Set $\varpi = \varpi_d = \sum_{i=1}^d \varepsilon_i$. For each $\sigma \in S_d$ (the symmetric group) let g_σ denote the element in $GL_n(k)$ with $g_\sigma(e_i) = e_{\sigma(i)}$ for all $i \leq d$ and $g_\sigma(e_i) = e_i$ for all $i > d$. One has then $g_{\sigma_1}g_{\sigma_2} = g_{\sigma_1\sigma_2}$ for all $\sigma_1, \sigma_2 \in S_d$. Each g_σ normalises the standard maximal torus T of all diagonal matrices in G. This induces an action of g_σ on $X(T)$ that fixes ϖ; it follows that $g_\sigma V_\varpi = V_\varpi$ for any G–module V. So these V_ϖ get a kS_d–module structure that we shall call the standard module structure.

Lemma: *There exists an algebra isomorphism*

$$kS_d \xrightarrow{\sim} \xi_\varpi S(n, d)\xi_\varpi \tag{1}$$

such that for each G–module V in $\mathcal{C}(\pi(n, d))$ the kS_d–module structure derived from (1) on $\xi_\varpi V = V_\varpi$ coincides with the standard structure.

Proof: The discussion of the basis of $k[M_{n,k}; d] = O_\pi^r(k[G])$ shows that the monomials $X_{\sigma(1)1}X_{\sigma(2)2}\dots X_{\sigma(d)d}$ with $\sigma \in S_d$ form a basis for $k[M_{n,k}; d]_{(-\varpi,\varpi)}$. Let $(\eta_\sigma)_{\sigma\in S_d}$ denote the dual basis for $S(n, d)_{(\varpi,-\varpi)} \simeq (k[M_{n,k}; d]_{(-\varpi,\varpi)})^*$ (extended by 0 to the other weight spaces of $k[M_{n,k}; d]$). Then there is obviously an isomorphism of vector spaces $kS_d \xrightarrow{\sim} \xi_\varpi S(n, d)\xi_\varpi$ mapping any σ to η_σ.

An elementary calculation shows that $\eta_\sigma(f) = f(g_\sigma)$ for all $\sigma \in S_d$ and $f \in k[M_{n,k}; d]_{(\bullet,\varpi)}$. (It suffices to consider monomials for f.) This implies for each G–module V in $\mathcal{C}(\pi(n,d))$ that the action of $\eta_\sigma \in S(n,d)$ on V_ϖ is equal to the action of g_σ. Applying this to the left action on $S(d,n)_{(\varpi,\bullet)}$ one gets $\eta_{\sigma_1}\eta_{\sigma_2} = \eta_{\sigma_1\sigma_2}$ for all $\sigma_1, \sigma_2 \in S_d$. (Apply both sides to $\xi_\varpi \in S(d,n)_{(\varpi,\bullet)}$.) This implies that our isomorphism of vector spaces $kS_d \xrightarrow{\sim} \xi_\varpi S(n,d)\xi_\varpi$ is an algebra isomorphism. The lemma follows.

A.22. Keep the notations and assumptions from A.21. Combining Lemma A.21 and Proposition A.20 we see that the simple kS_d–modules are the $L(\lambda)_\varpi$ with $\lambda \in \pi(n,d)$ and $L(\lambda)_\varpi \neq 0$. We see also that these simple modules are absolutely simple because $L(\lambda)$ has this property. (So if k' is an extension field of k, then the simple $k'S_d$–modules have the form $(L(\lambda) \otimes k')_\varpi \simeq L(\lambda)_\varpi \otimes k'$.)

Look first at the case where $\mathrm{char}(k) = 0$. Here we have $L(\lambda) = V(\lambda)$ and it is known that $V(\lambda)_\mu \neq 0$ for all $\lambda, \mu \in X(T)_+$ with $\mu \leq \lambda$. It is easy to check that $\varpi \leq \lambda$ for all $\lambda \in \pi(n,d)$. So we regain here the classical one-to-one correspondence between partitions of d (in bijection with $\pi(n,d)$ as $d \leq n$) and the isomorphism classes of irreducible representations of the symmetric group S_d over a field of characteristic 0. (This correspondence has been normalised such that the trivial one dimensional representation corresponds to the partition (d) and the sign representation to the partition $(1 \geq 1 \geq \cdots \geq 1)$.)

Suppose next that $\mathrm{char}(k) = p > 0$. We can decompose any $\lambda \in \pi(n,d)$ as $\lambda = \lambda_1 + \lambda_2$ such that $\lambda_1 \in X_1(T) \cap \pi(n,d_1)$ and $\lambda_2 \in \pi(n,d_2)$ for suitable $d_1, d_2 \in \mathbf{N}$; we have then necessarily $d = d_1 + pd_2$. Any weight of $L(\lambda_i)$ is a sum of d_i not necessarily distinct ε_j's. Therefore any weight of $L(\lambda) \simeq L(\lambda_1) \otimes L(\lambda_2)^{[1]}$ is a linear combination of at most $d_1 + d_2$ distinct ε_j's. If $d_2 > 0$ (equivalently: if $\lambda \notin X_1(T)$), then $d_1 + d_2 < d_1 + pd_2 = d$, so ϖ cannot be a weight of $L(\lambda)$.

Set $\pi_1(n,d) = \pi(n,d) \cap X_1(T)$. The main theorem in [Suprunenko 1] says that $L(\lambda)$ and $V(\lambda)$ have the same weights if $\lambda \in X_1(T)$. We get now for any $\lambda \in \pi(n,d)$ that $L(\lambda)_\varpi \neq 0$ if and only if $\lambda \in \pi_1(n,d)$. (One could here replace the reference to [Suprunenko 1] by a counting argument using Brauer's theorem on the number of irreducible modular characters of a finite group.)

Note that $\lambda = \sum_{i=1}^n m_i\varepsilon_i \in \pi(n,d)$ belongs to $\pi_1(n,d)$ if and only if $m_i - m_{i+1} < p$ for all i, $1 \leq i < n$. If one thinks of elements in $\pi(n,d)$ as of partitions of d, then those belonging to $\pi_1(n,d)$ are called the *column p–regular partitions of d*.

Assume still that $\mathrm{char}(k) = p > 0$. The kS_d–module $V(\mu)_\varpi$ is a reduction modulo p of the simple $\mathbf{C}S_d$–module $V(\mu)_{\mathbf{C},\varpi}$ because we consider in both cases the "standard" module structure, cf. A.21. Taking weight spaces is exact. So for all $\mu \in \pi(n,d)$ and $\lambda \in \pi_1(n,d)$ the GL_n–multiplicity $[V(\mu) : L(\lambda)]$ is equal to the S_d –multiplicity $[V(\mu)_\varpi : L(\lambda)_\varpi]$. In this way the decomposition matrix for S_d is a submatrix of the decomposition matrix for GL_n (for any $n \geq d$). This was first proved in [James 1].

Conversely, according to [Erdmann 4] each GL_n–multiplicity $[V(\mu) : L(\lambda)]$ with $\lambda, \mu \in X(T)_+$ is also an entry in the decomposition matrix for S_r for suitable r.

The restriction of the contravariant form on $V(\mu)_{\mathbf{Q}}$ as in 8.17(1) to $V(\mu)_{\mathbf{Q},\varpi}$ yields (for any $\mu \in \pi(n,d)$) an S_d–invariant form. (We may assume that $\tau(g) = {}^tg$ for all g in GL_n and we have ${}^tg_\sigma = g_{\sigma^{-1}}$ for all $\sigma \in S_d$. Use 8.17(2).) This form can be constructed within the representation theory of S_d (this is due to A. Young) and its determinant on $V(\mu)_{\mathbf{Z},\varpi}$ was computed in [James and Murphy]. Their formula can also be deduced from 8.19 and one can actually use the form to define filtrations and prove sum formulas as in 8.18/19, see [Schaper] or [Benson 2].

A.23. Keep the notations and assumptions from A.21. The sum of weight spaces $S(n,d)_{(\bullet,-\varpi)} = S(n,d)\xi_\varpi$ is a left module for $S(n,d)$ (hence a GL_n–module in $\mathcal{C}(\pi(n,d))$ and a right module for $\xi_\varpi S(n,d)\xi_\varpi$ (hence a right module for kS_d). Denote the natural GL_n–module by $E = k^n$. We claim that there is an isomorphism

$$\bigotimes^d E \xrightarrow{\sim} S(n,d)_{(\bullet,-\varpi)} = S(n,d)\xi_\varpi \tag{1}$$

that is compatible with the actions both of GL_n and of S_d. Here the action of GL_n on the left hand side comes from the natural action on k^n while S_d acts by permuting the factors in the tensor product.

Note first that for each j the subspace $V_j = \sum_{i=1}^m kX_{ij} \subset k[M_{n,k};1]$ is a GL_n–submodule for ρ_l isomorphic to E^*. Next the multiplication is an isomorphism of GL_n–modules

$$V_1 \otimes V_2 \otimes \cdots \otimes V_d \xrightarrow{\sim} k[M_{n,k};d]_{(\bullet,\varpi)} = O_\pi^r(k[GL_n])_{(\bullet,\varpi)}. \tag{2}$$

(Note that the right hand side has basis all $X_{i_1 1}X_{i_2 2}\dots X_{i_d d}$ with arbitrary i_l in $\{1,2,\dots,n\}$.) Taking duals we get a map as in (1) that is at least an isomorphism of GL_n–modules. It sends the basis of all $e_{i_1}\otimes e_{i_2}\otimes\cdots\otimes e_{i_d}$ to the basis dual to that of the monomials. Now an elementary calculation (left to the reader) shows that the map is also compatible with the action of any $\sigma \in S_d$ (acting on the right hand side via right multiplication by η_σ).

Since ξ_ϖ is idempotent, the map $\varphi \mapsto \varphi(\xi_\varpi)$ is an isomorphism

$$\operatorname{Hom}_{S(n,d)}(S(n,d)\xi_\varpi, M) \xrightarrow{\sim} \xi_\varpi M \tag{3}$$

for each $S(n,d)$–module M. We get in particular that $\operatorname{End}_{S(n,d)}(S(n,d)\xi_\varpi) \simeq \xi_\varpi S(n,d)\xi_\varpi$; this is actually an isomorphism of algebras if we replace $\xi_\varpi S(n,d)\xi_\varpi$ by its opposite ring $(\xi_\varpi S(n,d)\xi_\varpi)^{\mathrm{op}}$. In view of (1) and Lemma A.21 this can be reformulated as

$$\operatorname{End}_{GL_n}(\bigotimes^d E) \simeq kS_d. \tag{4}$$

(Recall that we assume $n \geq d$ here.) This is in characteristic 0 a classical result that goes back to Schur. For arbitrary k is was proved in [Carter and Lusztig 1], section 3.1.

In general, (3) translates into: We have for each GL_n– module M in $\mathcal{C}(\pi(n,d))$ an isomorphism

$$\operatorname{Hom}_{GL_n}(\bigotimes^d E, M) \xrightarrow{\sim} M_\varpi. \tag{5}$$

This implies, for example, that

$$\bigotimes^d E/\operatorname{rad}\bigotimes^d E \simeq \bigoplus_{\lambda\in\pi_1(n,d)} L(\lambda)^{\dim L(\lambda)_\varpi} \tag{6}$$

where we set $\pi_1(n,d) = \pi(n,d)$ in case $\operatorname{char}(k) = 0$. In that case ($\operatorname{char}(k) = 0$) the radical is of course 0 and we see that there the dimension of a simple kS_d–module has not only the interpretation as the dimension of some $L(\lambda)_\varpi$, but also as the multiplicity of $L(\lambda)$ in $\bigotimes^d E$. For an interpretation of (6) in prime characteristic, see E.17.

A.24. Return to general G. Let $\pi \subset X(T)_+$ be finite and saturated. In this case we can use A.2(1) as a definition of $\mathcal{C}(\pi)$ that works over general ground rings. (Recall that we assume in this chapter that k is a field.)

Write $I_G(\pi)_{\mathbf{Q}}$ (resp. $S_G(\pi)_{\mathbf{Q}}$) for $I_G(\pi)$ (resp. $S_G(\pi)$) in case $k = \mathbf{Q}$. We know that $\mathrm{Dist}(G_{\mathbf{Z}})$ is a subring of $\mathrm{Dist}(G_{\mathbf{Q}})$, see 1.12; set

(1) $$I_G(\pi)_{\mathbf{Z}} = \mathrm{Dist}(G_{\mathbf{Z}}) \cap I_G(\pi)_{\mathbf{Q}}.$$

Since $G_{\mathbf{Q}}$–modules are semi-simple, any $G_{\mathbf{Q}}$–module in $\mathcal{C}(\pi)$ is a direct sum of some $V(\lambda)_{\mathbf{Q}}$ with $\lambda \in \pi$. It follows that an element in $\mathrm{Dist}(G_{\mathbf{Z}})$ belongs to $I_G(\pi)_{\mathbf{Z}}$ if and only if it annihilates all $V(\lambda)_{\mathbf{Q}}$ with $\lambda \in \pi$ if and only if it annihilates all $V(\lambda)_{\mathbf{Z}}$ with $\lambda \in \pi$. This shows (using the extension of the definition of $\mathcal{C}(\pi)$ to rings):

(2) *$I_G(\pi)_{\mathbf{Z}}$ is the set of all elements in* $\mathrm{Dist}(G_{\mathbf{Z}})$ *that annihilate all $G_{\mathbf{Z}}$–modules that are free of finite rank over* $\mathbf{Z}$ *and belong to* $\mathcal{C}(\pi)$.

(Note: If V is such a $G_{\mathbf{Z}}$–module, then $V \otimes_{\mathbf{Z}} \mathbf{Q}$ is a $G_{\mathbf{Q}}$–module in $\mathcal{C}(\pi)$, hence annihilated by $I_G(\pi)$. And V is mapped injectively to $V \otimes_{\mathbf{Z}} \mathbf{Q}$.)

Set

(3) $$S_G(\pi)_{\mathbf{Z}} = \mathrm{Dist}(G_{\mathbf{Z}})/I_G(\pi)_{\mathbf{Z}}.$$

The action of $\mathrm{Dist}(G_{\mathbf{Z}})$ on the $V(\lambda)_{\mathbf{Z}}$ defines a ring homomorphism

$$\mathrm{Dist}(G_{\mathbf{Z}}) \longrightarrow \prod_{\lambda \in \pi} \mathrm{End}_{\mathbf{Z}}(V(\lambda)_{\mathbf{Z}})$$

with kernel $I_G(\pi)_{\mathbf{Z}}$. This implies:

(4) *$S_G(\pi)_{\mathbf{Z}}$ is a free* $\mathbf{Z}$*–module of finite rank.*

By (1) and Proposition A.16 we can identify $S_G(\pi)_{\mathbf{Z}}$ with a subring of $S_G(\pi)_{\mathbf{Q}}$. Now $\mathrm{Dist}(G_{\mathbf{Q}}) = \mathrm{Dist}(G_{\mathbf{Z}}) \otimes_{\mathbf{Z}} \mathbf{Q}$ shows that $S_G(\pi)_{\mathbf{Z}}$ generates $S_G(\pi)_{\mathbf{Q}}$ over $\mathbf{Q}$. Now (4) shows that $S_G(\pi)_{\mathbf{Z}}$ is (modulo the identification) a lattice in $S_G(\pi)_{\mathbf{Q}}$; we have an isomorphism

(5) $$S_G(\pi)_{\mathbf{Z}} \otimes_{\mathbf{Z}} \mathbf{Q} \simeq S_G(\pi)_{\mathbf{Q}}.$$

This extends to arbitrary fields:

Lemma: *There is an isomorphism $S_G(\pi)_{\mathbf{Z}} \otimes_{\mathbf{Z}} k \simeq S_G(\pi)$.*

Proof: Set $d(\pi) = \sum_{\lambda \in \pi} (\dim V(\lambda))^2$. This is by A.15(1) the dimension over k of $O_\pi^r(k[G])$ and $S_G(\pi)$; it is also the dimension over $\mathbf{Q}$ of $O_\pi^r(\mathbf{Q}[G_{\mathbf{Q}}])$ and $S_G(\pi)_{\mathbf{Q}}$ and [by (5)] the rank over $\mathbf{Z}$ of $S_G(\pi)_{\mathbf{Z}}$, hence also the dimension over k of $S_G(\pi)_{\mathbf{Z}} \otimes_{\mathbf{Z}} k$.

Recall from 1.1 that $\mathbf{Z}[G_{\mathbf{Z}}]$ is a free $\mathbf{Z}$–module with $\mathbf{Z}[G_{\mathbf{Z}}] \otimes_{\mathbf{Z}} k = k[G]$. We first apply this to $k = \mathbf{Q}$; set

$$O_\pi^r(\mathbf{Z}[G_{\mathbf{Z}}]) = \mathbf{Z}[G_{\mathbf{Z}}] \cap O_\pi^r(\mathbf{Q}[G_{\mathbf{Q}}]).$$

It follows that $O_\pi^r(\mathbf{Z}[G_{\mathbf{Z}}])$ is a lattice in $O_\pi^r(\mathbf{Q}[G_{\mathbf{Q}}])$. It is free over $\mathbf{Z}$ of finite rank equal to $d(\pi)$.

Since $\mathbf{Z}[G_{\mathbf{Z}}]/O_\pi^r(\mathbf{Z}[G_{\mathbf{Z}}])$ is torsion free by the definition of $O_\pi^r(\mathbf{Z}[G_{\mathbf{Z}}])$, we can identify $O_\pi^r(\mathbf{Z}[G_{\mathbf{Z}}]) \otimes_{\mathbf{Z}} k$ with a subspace of $k[G]$. We claim that then

$$(6) \qquad O_\pi^r(\mathbf{Z}[G_{\mathbf{Z}}]) \otimes_{\mathbf{Z}} k = O_\pi^r(k[G]).$$

Since both sides above have dimension $d(\pi)$, it suffices to prove the inclusion "$\subset$". However, $O_\pi^r(\mathbf{Z}[G_{\mathbf{Z}}])$ is a $G_{\mathbf{Z}}$–submodule of $\mathbf{Z}[G_{\mathbf{Z}}]$ with respect to ρ_r such that all weights belong to $W\pi$ (since $O_\pi^r(\mathbf{Q}[G_{\mathbf{Q}}])$ has this property). Therefore $O_\pi^r(\mathbf{Z}[G_{\mathbf{Z}}]) \otimes_{\mathbf{Z}} k$ shares this property which implies $O_\pi^r(\mathbf{Z}[G_{\mathbf{Z}}]) \otimes_{\mathbf{Z}} k \subset O_\pi^r(k[G])$ by the definition of O_π.

We can now use (6) to show that the isomorphism $\mathrm{Dist}(G_{\mathbf{Z}}) \otimes_{\mathbf{Z}} k \xrightarrow{\sim} \mathrm{Dist}(G)$ as in 1.12 maps $I_G(\pi)_{\mathbf{Z}} \otimes_{\mathbf{Z}} k$ to $I_G(\pi)$. Any $\mu \in I_G(\pi)_{\mathbf{Z}}$ satisfies $\mu(O_\pi^r(\mathbf{Z}[G_{\mathbf{Z}}])) = 0$ by definition, so (6) implies $(\mu \otimes 1)(O_\pi^r(k[G])) = 0$, hence $\mu \otimes 1 \in I_G(\pi)$.

Now it is clear that the isomorphism $\mathrm{Dist}(G_{\mathbf{Z}}) \otimes_{\mathbf{Z}} k \xrightarrow{\sim} \mathrm{Dist}(G)$ induces a surjective homomorphism from $S_G(\pi)_{\mathbf{Z}} \otimes_{\mathbf{Z}} k$ onto $S_G(\pi)$. As both sides have the same dimension, we are done.

Remark: We can now define for each $\mathbf{Z}$–algebra A a generalised Schur algebra via $S_G(\pi)_A = S_G(\pi)_{\mathbf{Z}} \otimes_{\mathbf{Z}} A$. If A is a Dedekind domain, then one can construct $S_G(\pi)_A$ also in a way similar to (3) working with the field of fractions of A, see [Donkin 25], Lemma 1.2a (or [Donkin 12], 3.2b for the case of a principal ideal domain).

A.25. (Skew Modules) Let H be a group scheme over k. Let V be a $(G \times H)$–module. Each $h \in H(k)$ acts as a G–endomorphism on V, hence preserves each $O_\pi(V)$ with $\pi \subset X(T)_+$. If $H(k)$ is dense in H, then this implies that each $O_\pi(V)$ is a $(G \times H)$–submodule of V. Now this holds also without the density assumption: The comodule map $\Delta_V : V \to V \otimes k[G]$ of V considered as a G–module only is a homomorphism of H–modules if we let H act trivially on $k[G]$, see the proof of Lemma I.3.2. Therefore $O_\pi(V) = \Delta_V^{-1}(V \otimes O_\pi^r(k[G]))$, see A.13(2), is also an H–submodule of V, hence a $G \times H$–submodule.

Let $\mu \in X(T)_+$. Set $\pi(\mu) = \{\nu \in X(T)_+ \mid \nu \leq \mu\}$ and $\pi(\mu)' = \pi(\mu) \setminus \{\mu\}$. These are two finite and saturated subsets of $X(T)_+$. Both $O_{\pi(\mu)}(V)$ and $O_{\pi(\mu)'}(V)$ are $G \times H$–modules for any $G \times H$–module V. We have clearly $O_{\pi(\mu)'}(V) \subset O_{\pi(\mu)}(V)$. So also

$$\mathcal{F}_\mu(V) = O_{\pi(\mu)}(V)/O_{\pi(\mu)'}(V)$$

is a $G \times H$–module and

$$F_\mu(V) = \mathrm{Hom}_G(H^0(\mu), \mathcal{F}_\mu(V)) \simeq (H^0(\mu)^* \otimes \mathcal{F}_\mu(V))^G$$

is an H–module, cf. I.3.2.

This construction (taken from [Donkin 13]) may look strange at first sight. It makes more sense if we assume that V considered as a G–module has a good filtration with $(V : H^0(\lambda)) < \infty$ for all $\lambda \in X(T)_+$. Then Lemma A.5 shows that $\mathcal{F}_\mu(V)$ as a G–module is isomorphic to $H^0(\mu)^m$ with $m = (V : H^0(\mu))$. As $\mathrm{End}_G(H^0(\mu)) \simeq k$, it follows that $v \otimes \varphi \mapsto \varphi(v)$ defines an isomorphism

$$H^0(\mu) \otimes F_\mu(V) = H^0(\mu) \otimes \mathrm{Hom}_G(H^0(\mu), \mathcal{F}_\mu(V)) \xrightarrow{\sim} \mathcal{F}_\mu(V)$$

of vector spaces. This is actually an isomorphism of $(G \times H)$–modules when we regard $H^0(\mu) \otimes F_\mu(V)$ as a $G \times H$–module with G acting on the first factor and H

on the second one. It now follows (using a suitable total ordering of $X(T)_+$) that V has a filtration as a $(G \times H)$–module with factors $H^0(\mu) \otimes F_\mu(V)$ with $\mu \in X(T)_+$.

Consider now the case that $G = GL_n$ and $H = GL_m$ for some integers n, $m > 1$. In this case the construction above leads to certain modules constructed before differently. Regard $G \times H$ as a subgroup of $\widehat{G} = GL_{n+m}$ via the obvious action on $k^{n+m} = k^n \times k^m$. Each $\lambda \in \pi(n+m, d)$ defines an induced $\widehat{G}$–module $\widehat{H}^0(\lambda)$ that restricts to a $(G \times H)$–module. So we can define for any $\mu \in \pi(n, e)$ a GL_m–module $F_\mu(\widehat{H}^0(\lambda))$. These modules are then according to [Donkin 13], 2.3 isomorphic to the skew Schur modules introduced in [Akin, Buchsbaum, and Weyman]. Furthermore, each $\widehat{H}^0(\lambda)$ has a good filtration as a G–module (in fact, even as a $(G \times H)$–module) so that the remarks in the preceding paragraph apply.

In [Donkin 13] one calls in the general case $F_\mu(V)$ a *skew module* for H, if H is a connected and reductive group, if $G \times H$ is identified with a closed subgroup of a connected and reductive group $\widehat{G}$, and if V is the restriction to $G \times H$ of a $\widehat{G}$–module induced from a one dimensional representation of a Borel subgroup of $\widehat{G}$ such that V has a good filtration as a $(G \times H)$–module.

A.26. (Polynomial Functors) If k is an infinite field, then a polynomial functor is defined as a functor F from the category of all finite dimensional vector spaces over k to itself such that for all finite dimensional vector spaces V and V' over k the map

$$F_{V,V'} : \operatorname{Hom}(V, V') \longrightarrow \operatorname{Hom}(F(V), F(V')) \tag{1}$$

is a polynomial map. If so, then F is called homogeneous of degree d if $F_{V,V'}$ is homogeneous polynomial of degree d. (As all maps between finite dimensional vector spaces over finite fields are polynomial, the definition has to be modified if k is finite. This leads to the notion of a strict polynomial functor, see [Friedlander and Suslin], Def. 2.1.)

For the time being continue to assume that k is infinite. Examples for polynomial functors (all homogeneous of degree d) are the tensor power $V \mapsto \bigotimes^d V$, the exterior power $V \mapsto \Lambda^d V$, the symmetric power $V \mapsto S^d V$, and the "d–th divided power" $V \mapsto S^d(V^*)^*$. The last objects will play an important role in a moment, so let me add a few words on them. Each $f \in S^d(V^*)$ can be regarded as a homogeneous polynomial function $V \to k$ of degree d. So each $v \in V$ defines an evaluation map $ev_v \in S^d(V^*)^*$ via $ev_v(f) = f(v)$. (This is the restriction to $S^d(V^*)$ of the more frequently considered evaluation map $S(V^*) \to k$.) Each non-zero $f \in S^d(V^*)$ defines a non-zero function $V \to k$ as k is infinite. This implies that the intersection of the kernels of all ev_v is 0, hence that $S^d(V^*)^*$ is spanned as a vector space by all ev_v with $v \in V$.

We want to establish a connection between homogeneous polynomial functors of degree d and $S(n, d)$–modules. Fix for the moment n and denote by E the standard module k^n for GL_n. The subspace of $k[GL_n]$ generated by the matrix coefficients X_{ij} identifies with $\operatorname{End}(E)^*$, hence the space of $k[M_{n,k}; d]$ of homogeneous polynomials of degree d in the X_{ij} with $S^d(\operatorname{End}(E)^*)$, and the dual space $S(n, d)$ with $S^d(\operatorname{End}(E)^*)^*$. Now $S(n, d)$ is spanned by all ev_φ with $\varphi \in \operatorname{End}(E)$ and one checks now that the multiplication is given by

$$ev_\varphi \, ev_\psi = ev_{\varphi \circ \psi} \qquad \text{for all } \varphi, \psi \in \operatorname{End}(E). \tag{2}$$

Let F be a homogeneous polynomial functor of degree d. A homogeneous polynomial map of degree d from a finite dimensional vector space M_1 to a finite dimensional vector space M_2 corresponds to an element in $S^d(M_1^*) \otimes M_2$. Using this and the isomorphism $M_1 \otimes \operatorname{Hom}(M_2, M_3) \simeq \operatorname{Hom}(M_1^* \otimes M_2, M_3)$, we see that the polynomial map in (1) corresponds to a linear map (again denoted by $F_{V,V'}$)

$$F_{V,V'} : S^d(\operatorname{Hom}(V,V')^*)^* \otimes F(V) \longrightarrow F(V'). \tag{3}$$

The functor F maps then any $\varphi : V \to V'$ to $F(\varphi) : F(V) \to F(V')$ with

$$F(\varphi)\,(x) = F_{V,V'}(ev_\varphi \otimes x) \qquad \text{for all } x \in F(V). \tag{4}$$

Taking $V = V' = E$, we get in particular a linear map

$$F_{E,E} : S(n,d) \otimes F(E) = S^d(\operatorname{End}(E)^*)^* \otimes F(E) \longrightarrow F(E). \tag{5}$$

This turns $F(E)$ into an $S(n,d)$–module such that $u\,x = F_{E,E}(u \otimes x)$ for all $u \in S(n,d)$ and $x \in F(E)$: We get that

$$ev_\varphi\, x = F(\varphi)(x) \quad \text{for all } \varphi \in \operatorname{End}(E) \tag{6}$$

hence

$$ev_\varphi\,(ev_\psi\, x) = F(\varphi)(F(\psi)(x)) = F(\varphi\,\psi)(x) = ev_{\varphi\,\psi}\, x = (ev_\varphi\, ev_\psi)\, x.$$

And the 1 in $S(n,d)$ (equal to ev_{id}) acts as the identity because of $F(\mathrm{id}_E) = \mathrm{id}_{F(E)}$.

A natural transformation $F_1 \to F_2$ of homogeneous polynomial functors of degree d induces a linear map $F_1(E) \to F_2(E)$. The definition of a natural transformation says then that we have for all $\varphi \in \operatorname{End}(E)$ a commutative diagram

$$\begin{array}{ccc} F_1(E) & \longrightarrow & F_2(E) \\ {\scriptstyle F_1(\varphi)}\downarrow & & \downarrow{\scriptstyle F_2(\varphi)} \\ F_1(E) & \longrightarrow & F_2(E) \end{array}$$

which means that $F_1(E) \to F_2(E)$ commutes with the action of each ev_φ, hence is a homomorphism of $S(n,d)$–modules. We have thus constructed a functor $F \mapsto F(E)$ from the category of all homogeneous polynomial functors of degree d to the category of all finite dimensional $S(n,d)$–modules.

We next construct a functor in the opposite direction. For each finite dimensional vector space V over k the space $S^d(\operatorname{Hom}(E,V)^*) \simeq S^d(E \otimes V^*)$ is a polynomial GL_n–module, homogeneous of degree d, hence an $S(n,d)$–module. The dual space $S^d(\operatorname{Hom}(E,V)^*)^*$ is therefore a right module over $S(n,d)$. This right module structure is given by

$$ev_\varphi \cdot ev_\psi = ev_{\varphi\circ\psi} \qquad \text{for all } \varphi \in \operatorname{Hom}(E,V),\ \psi \in \operatorname{End}(E). \tag{7}$$

(The action of ev_ψ on any $f \in S^d(\operatorname{Hom}(E,V)^*)$ is given by $(ev_\psi \cdot f)(\varphi) = f(\varphi\circ\psi)$ for all $\varphi \in \operatorname{Hom}(E,V)$ extending the action of $GL_n = GL(E)$ via $(g\cdot f)(\varphi) = f(\varphi\circ g)$.)

Now each finite dimensional $S(n,d)$–module M defines a functor Φ_M from finite dimensional vector spaces over k to itself by

$$\Phi_M(V) = S^d(\mathrm{Hom}(E,V)^*)^* \otimes_{S(n,d)} M. \tag{8}$$

Indeed, any linear map $\gamma : V \to V'$ of finite dimensional vector spaces over k induces a linear map $\varphi \mapsto \gamma \circ \varphi$ from $\mathrm{Hom}(E,V)$ to $\mathrm{Hom}(E,V')$, hence a degree preserving algebra homomorphism $\gamma^* : S(\mathrm{Hom}(E,V')^*) \to S(\mathrm{Hom}(E,V)^*)$ such that $\gamma^*(f)(\varphi) = f(\gamma \circ \varphi)$, hence a linear map $\gamma_* : S^d(\mathrm{Hom}(E,V)^*)^* \to S^d(\mathrm{Hom}(E,V')^*)^*$ with

$$\gamma_*(ev_\varphi)(f) = ev_\varphi(\gamma^*(f)) = \gamma^*(f)(\varphi) = f(\gamma \circ \varphi) = ev_{\gamma\circ\varphi}(f),$$

i.e., with $\gamma_*(ev_\varphi) = ev_{\gamma\circ\varphi}$. This description shows that γ_* is a homomorphism of right modules over $S(n,d)$. Therefore γ_* induces a linear map $\Phi_M(\gamma) : \Phi_M(V) \to \Phi_M(V')$. In this way we turn Φ_M into a functor. One checks easily that $\gamma \mapsto \gamma_*$ is a homogeneous polynomial of degree d, so Φ_M is a homogeneous polynomial functor of degree d.

If we apply Φ_M to E, then we get a natural isomorphism

$$\Phi_M(E) = S^d(\mathrm{End}(E)^*)^* \otimes_{S(n,d)} M \xrightarrow{\sim} M \tag{9}$$

since $S^d(\mathrm{End}(E)^*)^* = S(n,d)$ and since a comparison of (7) and (2) shows that the right action of $S(n,d)$ on $S^d(\mathrm{Hom}(E,V)^*)^*$ is equal to right multiplication in case $V = E$.

This shows that the composition of first the functor $M \mapsto \Phi_M$ and then the functor $F \mapsto F(E)$ is isomorphic to the identity functor on the category of all finite dimensional $S(n,d)$–modules.

The composition the other way round will in general not be isomorphic to the identity functor: If $n = \dim(E) < d$, then the functor F with $F(V) = \Lambda^d V$ satisfies $F(E) = 0$, but $F \neq 0$.

There is, however, a natural transformation from this other composition to the identity functor: To start with, we have for any homogeneous polynomial functor F of degree d for each V a map

$$F_{E,V} : S^d(\mathrm{Hom}(E,V)^*)^* \otimes F(E) \longrightarrow F(V)$$

with $F_{E,V}(ev_\varphi \otimes x) = F(\varphi)(x)$, see (3). So $S^d(\mathrm{Hom}(E,V)^*)^*$ is a right module over $S(n,d)$, and $F(E)$ is a left module over $S(n,d)$. We claim that $F_{E,V}$ factors over

$$\overline{F}_{E,V} : \Phi_{F(E)} = S^d(\mathrm{Hom}(E,V)^*)^* \otimes_{S(n,d)} F(E) \longrightarrow F(V). \tag{10}$$

We have to show that $F_{E,V}(ev_\varphi\, u \otimes x - ev_\varphi \otimes u\, x) = 0$ for all $u \in S(n,d)$, $x \in F(E)$, and $\varphi \in \mathrm{Hom}(E,V)$. It suffices to consider $u = ev_\psi$ with $\psi \in \mathrm{End}(E)$. Then

$$\begin{aligned} F_{E,V}(ev_\varphi ev_\psi \otimes x) &= F_{E,V}(ev_{\varphi\circ\psi} \otimes x) = F(\varphi \circ \psi)(x) = F(\varphi)(F(\psi)(x)) \\ &= F(\varphi)(ev_\psi\, x) = F_{E,V}(ev_\varphi \otimes ev_\psi\, x) \end{aligned}$$

and the claim follows.

Now (10) yields the promised natural transformation. As the example with the exterior powers shows, the map in (10) need not be an isomorphism. However, it is so in the very special case where $F(V) = S^d(\mathrm{Hom}(E,V)^*)^*$ for all V: In this case $F(E)$ is just $S(n,d)$ with its usual module structure by left multiplication and (10) is bijective. Now it is shown in [Friedlander and Suslin], Thm. 2.10 that each homogeneous polynomial functor of degree d is a homomorphic image of the functor $V \mapsto S^d(\mathrm{Hom}(E,V)^*)^*$ provided that $\dim(E) = n \geq d$. Using this one gets then (see [Friedlander and Suslin], Thm. 3.2):

Proposition: *If $n = \dim(E) \geq d$, then $F \mapsto F(E)$ and $M \mapsto \Phi_M$ are quasi-inverse equivalences of categories.*

A.27. Keep the notations from A.26. We say that a polynomial functor over k has finite degree if it is a finite direct sum of homogeneous polynomial functors over k, cf. [Friedlander and Suslin], 2.6. Let $\mathcal{P}$ denote the category of all polynomial functors of finite degree over k. This is an abelian category.

Assume for the sake of simplicity that k is an algebraically closed field of characteristic $p > 0$. (It requires only minor modifications to extend the following to all fields of characteristic p, see [Friedlander and Suslin].) Recall from I.9.10: For any vector space V over k and any $r \in \mathbf{N}$ we denote by $V^{(r)}$ the vector space that is equal to V as an abelian group and where any $a \in k$ acts as $a^{p^{-r}}$ does on V.

For each r the map $V \mapsto V^{(r)}$ defines a functor $I^{(r)}$ from the category of all finite dimensional vector spaces over k to itself: To any linear map $f : V \to V'$ associate $f^{(r)} : V^{(r)} \to V'^{(r)}$ with $f^{(r)}(v) = f(v)$ for all v. If $(v_i)_{i \in I}$ is a basis for V over k, and if $(v'_j)_{j \in J}$ is a basis for V' over k, then $(v_i)_{i \in I}$ is also a basis for $V^{(r)}$ over k, and $(v'_j)_{j \in J}$ is also a basis for $V'^{(r)}$ over k. If now f has matrix (a_{ji}) with respect to these bases, then $f^{(r)}$ has matrix $(a_{ji}^{p^r})$. This shows that $I^{(r)}$ is a homogeneous polynomial functor of degree p^r over k.

Assume now that $n = \dim(E) \geq p^r$. The equivalence of categories in Proposition A.26 maps $I^{(r)}$ to $E^{(r)}$. The equivalence between $S(n, p^r)$–modules and the category $\mathcal{C}(\pi(n, p^r))$ takes it then to the $GL_(E)$–module $E^{(r)} \simeq E^{[r]}$. Using Propositions A.10 and A.26 one gets now isomorphisms of Ext-algebras

$$\operatorname{Ext}^\bullet_{GL(E)}(E^{(r)}, E^{(r)}) \simeq \operatorname{Ext}^\bullet_{S(n,p^r)}(E^{(r)}, E^{(r)}) \simeq \operatorname{Ext}^\bullet_{\mathcal{P}}(I^{(r)}, I^{(r)}). \tag{1}$$

The main result in Section 4 of [Friedlander and Suslin] is the calculation of $\operatorname{Ext}^\bullet_{\mathcal{P}}(I^{(r)}, I^{(r)})$. One gets (see [Friedlander and Suslin], Thm. 4.5)

$$\dim \operatorname{Ext}^i_{\mathcal{P}}(I^{(r)}, I^{(r)}) = \begin{cases} 1, & \text{if } i \text{ even, } 0 \leq i < 2p^r, \\ 0, & \text{otherwise.} \end{cases} \tag{2}$$

The multiplication on this algebra can be described as follows: Choose for all s, $1 \leq s \leq r$ a basis element $e_{r,s}$ for $\operatorname{Ext}^{2p^{s-1}}_{\mathcal{P}}(I^{(r)}, I^{(r)})$. Then $e_{r,s}^p = 0$ for all s; all

$$e_{r,1}^{m(1)} e_{r,2}^{m(2)} \dots e_{r,r}^{m(r)} \qquad \text{with } 0 \leq m(i) < p \text{ for all } i$$

form a basis for $\operatorname{Ext}^\bullet_{\mathcal{P}}(I^{(r)}, I^{(r)})$, cf. [Friedlander and Suslin], 4.10.

Twisting with the s–th power of the Frobenius endomorphism one gets for each $s \leq r$ a homomorphism

$$\operatorname{Ext}^\bullet_{\mathcal{P}}(I^{(r-s)}, I^{(r-s)}) \longrightarrow \operatorname{Ext}^\bullet_{\mathcal{P}}(I^{(r)}, I^{(r)})$$

that we denote by $e \mapsto e^{(s)}$. It corresponds to the map $\operatorname{Ext}^\bullet_{GL(E)}(E^{(r-s)}, E^{(r-s)}) \to \operatorname{Ext}^\bullet_{GL(E)}(E^{(r)}, E^{(r)})$ induced by the s–th power of the Frobenius endomorphism; this map is injective by 10.16. One can choose the basis elements above such that $e_{s,s}^{(r-s)} = e_{r,s}$.

Here the extension $e_{1,1} \in \operatorname{Ext}^2_{\mathcal{P}}(I^{(1)}, I^{(1)}) \simeq \operatorname{Ext}^2_{GL(E)}(E^{(r)}, E^{(r)})$ has the following elementary interpretation: It corresponds to the extension

$$0 \to E^{(r)} \longrightarrow S^p(E) \longrightarrow S^p(E^*)^* \longrightarrow E^{(r)} \to 0$$

where $E^{(r)} \to S^p(E)$ is given by $v \mapsto v^p$, and where the middle map takes any $v_1 v_2 \dots v_p$ (all $v_i \in E$) to the linear function on $S^p(E^*)$ that maps any $f_1 f_2 \dots f_p$ (all $f_i \in E^*$) to

$$\sum_{\sigma \in S_p} f_1(v_{\sigma(1)}) f_2(v_{\sigma(2)}) \dots f_p(v_{\sigma(p)}).$$

Note that

$$\operatorname{Ext}^\bullet_{GL(E)}(E^{(r)}, E^{(r)}) \simeq H^\bullet(GL(E), E^{*(r)} \otimes E^{(r)})$$

and that $E^* \otimes E \simeq \mathfrak{gl}(E) = \operatorname{Lie}(GL(E))$. So we have for $n \geq p^r$ an isomorphism

$$H^\bullet(GL(E), \mathfrak{gl}(E)^{(r)}) \simeq \operatorname{Ext}^\bullet_{\mathcal{P}}(I^{(r)}, I^{(r)})$$

and we can regard the e_{rs} as elements in $H^\bullet(GL(E), \mathfrak{gl}(E)^{(r)})$. We have now a restriction map from the cohomology of $GL(E)$ to that of its Frobenius kernel $GL(E)_1$. The main result (Thm. 6.2) of Section 6 in [Friedlander and Suslin] says now that the restriction of e_{rr} to $H^{2p^{r-1}}(GL(E)_1, \mathfrak{gl}(E)^{(r)})$ is non-zero. This is then the crucial tool in [Friedlander and Suslin] for proving finiteness results for the cohomology of finite group schemes over k, see I.9.21.

CHAPTER B

Results over the Integers

In this chapter we extend certain results proved earlier in the case of a ground field to arbitrary principal ideal domains k.

For example, we show that all $H^j(B, M)$ or $H^j(G, M)$ are finitely generated over k if M is a B– or G–module that is finitely generated over k, see B.2 and B.5. This generalises Proposition 4.10. We shall see that the description in 4.13 of the Ext groups between Weyl modules and their duals generalises to k, see B.4. In B.9 we get a generalisation of the characterisation in 4.16 of modules with a good filtration. Here the proof, communicated to me by S. Donkin, uses the truncation functors O_π^r from the preceding chapter.

We assume throughout that k is an integral domain. Most results will require more restrictive conditions on k.

B.1. We now extend to k some results proved in Chapter 4 in the case of a ground field. Recall the definition of $\text{ht}(\lambda)$ from 4.8.

Lemma: *Let M be a B–module. If $H^j(B, M) \neq 0$ for some $j \in \mathbf{N}$, then there exists a weight λ of M with $-\lambda \in \mathbf{N}R^+$ and $\text{ht}(-\lambda) \geq j$.*

Proof: We have $H^j(B, M) \simeq H^j(U, M)^T$ by I.6.9(3). We calculate $H^\bullet(U, M)$ using the normalised Hochschild complex $C_0^\bullet(U, M)$ as in Remark 2 in I.4.16. The action of T on this complex (via the given action on M and the conjugation action on $I_1 = \ker(\varepsilon_U) \subset k[U]$) is compatible with the action of T on $H^\bullet(U, M)$ used above, cf. I.6.7.

Now $H^j(B, M) \neq 0$ implies $0 \neq C_0^j(U, M)^T = (M \otimes \bigotimes^j I_1)^T$ as the T–fixed point functor is exact. So there have to be weights λ of M and $\nu_1, \nu_2, \dots, \nu_j$ of I_1 with $\lambda + \sum_{i=1}^j \nu_i = 0$. Now the claim follows because all ν_i belong to $\mathbf{N}R^+$ and satisfy $\text{ht}(\nu_i) > 0$, cf. 4.8(2), (4). (The results quoted from 4.8 extend to all k since the isomorphisms as in 1.7(1) exist over all k.)

B.2. Lemma: *Let M be a B–module that is finitely generated over k. If k is noetherian, then each $H^j(B, M)$ is a finitely generated module over k.*

Proof: We saw in the proof in B.1 that $H^j(B, M)$ is a subquotient of $C_0^j(U, M)^T$. Therefore it suffices to show that $C_0^j(U, M)^T$ is finitely generated over k. We have

$$C_0^j(U, M)^T = \bigoplus M_\mu \otimes (I_1)_{\nu_1} \otimes (I_1)_{\nu_2} \otimes \cdots \otimes (I_1)_{\nu_j}$$

where we sum over all $\mu, \nu_1, \nu_2, \dots, \nu_j \in X(T)$ with $\mu + \sum_{i=1}^j \nu_i = 0$. As M is finitely generated over k, there are only finitely many μ with $M_\mu \neq 0$, and all M_μ are finitely generated over k. All ν with $(I_1)_\nu \neq 0$ belong to $\mathbf{N}R^+$ and satisfy

$\mathrm{ht}(\nu) > 0$. Therefore there are for each μ only finitely many j–tuples $(\nu_1, \nu_2, \dots, \nu_j)$ with $\sum_{i=1}^{j} \nu_i = -\mu$ and all $(I_i)_{\nu_i} \neq 0$. Now the claim follows, because all $(I_1)_\nu$ are finitely generated over k, by the same arguments as for 4.8(3).

B.3. We shall use the notation $H^i(\mu) = R^i \operatorname{ind}_B^G \mu$ from 2.1 also over k. (If we consider more than one ground ring, then we write $H^i_k(\mu)$ as in 8.6.) We know by 8.8(2) that $H^i(\lambda) = 0$ for all $i > 0$ in case λ is dominant. We get therefore as in 4.7:

Lemma: *Let M and N be G–modules such that N is flat over k. Then we have*

(1) $$\operatorname{Ext}^i_G(M, N \otimes H^0(\lambda)) \simeq \operatorname{Ext}^i_B(M, N \otimes \lambda)$$

for all $\lambda \in X(T)_+$ and all $i \in \mathbf{N}$.

Recall that the proof uses the spectral sequence I.4.5.a and the generalised tensor identity I.4.8. We need N to be flat in order to be able to apply I.4.8.

Note that we get as special cases (for $M = k$)

(2) $$H^i(G, N \otimes H^0(\lambda)) \simeq H^i(B, N \otimes \lambda)$$

and (taking $\lambda = 0$)

(3) $$H^i(G, N) \simeq H^i(B, N).$$

(As in the lemma N is flat over k and λ dominant.)

If M is a G–module that is finitely generated and projective over k, then we have
$$\operatorname{Ext}^j_G(M, H^0(\lambda)) \simeq H^j(G, M^* \otimes H^0(\lambda)) \simeq H^j(B, M^* \otimes \lambda)$$
for all $\lambda \in X(T)_+$ and all j. So we get from B.1:

(4) *Let M be a G–module that is finitely generated and projective over k and let $\lambda \in X(T)_+$. If $\operatorname{Ext}^j_G(M, H^0(\lambda)) \neq 0$ for some $j \in \mathbf{N}$, then there exists a weight μ of M such that $\mu \geq \lambda$ and $\mathrm{ht}(\mu - \lambda) \geq j$.*

As a special case we get for all $\lambda \in X(T)_+$

(5) $$\operatorname{Ext}^i_G(H^0(\lambda), H^0(\lambda)) = 0 \qquad \text{for all } i > 0$$

cf. Remark 2 in 4.17.

B.4. We shall write $V(\lambda)$ instead of $V(\lambda)_k$ when it is clear which k we consider. Recall that $V(\lambda) \simeq H^0(-w_0\lambda)^*$, see 8.9(2).

Proposition: *Let $\lambda, \mu \in X(T)_+$. Then*
$$H^i(G, H^0(\lambda) \otimes H^0(\mu)) \simeq \begin{cases} k, & \text{if } i = 0 \text{ and } \lambda = -w_0\mu, \\ 0, & \text{otherwise,} \end{cases}$$
and
$$\operatorname{Ext}^i_G(V(\lambda), H^0(\mu)) \simeq \begin{cases} k, & \text{if } i = 0 \text{ and } \lambda = \mu, \\ 0, & \text{otherwise.} \end{cases}$$

We can basically copy the proof in 4.17 using B.1 and B.3 instead of 4.10 and 4.7. By 8.8(1) the sets of weights of any $H^0(\mu)$ is independent of μ and $H^0(\mu)$ is free (hence flat) over k. Since any $V(\lambda)$ is free of finite rank over k, we can use I.4.4 to deduce the second part of the proposition from the first one.

Remark: Let $\lambda \in X(T)_+$. If M is a G–module that is finitely generated and projective over k, then we have by I.4.4

$$\operatorname{Ext}^j_G(V(\lambda), M) \simeq \operatorname{Ext}^j_G(M^*, H^0(-w_0\lambda)).$$

So B.3(4) implies: If $\operatorname{Ext}^j_G(V(\lambda), M) \neq 0$ for some $j \in \mathbf{N}$, then there exists a weight μ of M such that $\mu \geq \lambda$ and $\operatorname{ht}(\mu - \lambda) \geq j$.

B.5. Lemma: *Let M be a G–module that is finitely generated over k. If k is principal ideal domain, then each $H^j(G, M)$ is a finitely generated module over k.*

Proof: If M is torsion free, then this follows from B.3(3) and Lemma B.2. In general there is a short exact sequence $0 \to M' \to M \to M'' \to 0$ of G–modules such that M' is a torsion module over k and M'' torsion free, cf. I.10.2. Since the claim holds for M'', it suffices to prove it for M'.

So we may assume that M is a torsion module and that k is not a field. Then M has finite length already as a k–module, hence also as a G–module. Using induction, it is enough to treat the case where M is simple.

If M is simple, then there exists a prime element $q \in k$ such that M is a simple $G_{k/(q)}$–module considered as a G–module, cf. I.10.15. There exists $\lambda \in X(T)_+$ such that M is the socle $L_{k/(q)}(\lambda)$ of $H^0_{k/(q)}(\lambda)$. We shall use induction on λ to show that each $H^j(G, L_{k/(q)}(\lambda))$ is a finitely generated module over k.

We get from 8.8(1) that $H^0_{k/(q)}(\lambda) \simeq H^0_k(\lambda)/qH^0_k(\lambda)$. So we have a short exact sequence of G–modules

$$0 \to H^0_k(\lambda) \longrightarrow H^0_k(\lambda) \longrightarrow H^0_{k/(q)}(\lambda) \to 0 \tag{1}$$

where the first non-zero map is multiplication by q. All $H^j(G, H^0_k(\lambda))$ are finitely generated over k by the claim in the torsion free case. So the long exact cohomology sequence associated to (1) yields this finiteness also for all $H^j(G, H^0_{k/(q)}(\lambda))$.

Now use the short exact sequence of G–modules

$$0 \to L_{k/(q)}(\lambda) \longrightarrow H^0_{k/(q)}(\lambda) \longrightarrow H^0_{k/(q)}(\lambda)/\operatorname{soc} H^0_{k/(q)}(\lambda) \to 0 \tag{2}$$

and the fact that all composition factors of $H^0_{k/(q)}(\lambda)/\operatorname{soc} H^0_{k/(q)}(\lambda)$ have the form $L_{k/(q)}(\mu)$ with $\mu < \lambda$ to deduce the claim for $L_{k/(q)}(\lambda)$.

B.6. Proposition: *Let M and N be G–modules that are finitely generated over k. If k is principal ideal domain, then each $\operatorname{Ext}^j_G(M, N)$ is a finitely generated module over k.*

Proof: If M is torsion free (hence free), then $\operatorname{Ext}^j_G(M, N) \simeq H^j(G, M^* \otimes N)$ and $M^* \otimes N$ is finitely generated over k. So in this case the claim follows from Lemma B.5.

In the general case, one proceeds as in the proof of B.5: One gets the claim first for $M = H^0_{k/(q)}(\lambda)$ with $\lambda \in X(T)_+$ and q a prime element in k, then for $M = L_{k/(q)}(\lambda)$ and the same λ and q, then for all M that are torsion modules over k, and finally for arbitrary M.

B.7. Let π be a finite and saturated (see A.2) subset of $X(T)_+$. Recall from the proof of Lemma A.24 the definition of $O_\pi^r(\mathbf{Z}[G_\mathbf{Z}]) = \mathbf{Z}[G_\mathbf{Z}] \cap O_\pi^r(\mathbf{Q}[G_\mathbf{Q}])$. This is a $(G_\mathbf{Z} \times G_\mathbf{Z})$–submodule of $\mathbf{Z}[G_\mathbf{Z}]$ that is free of finite rank over $\mathbf{Z}$. Since $\mathbf{Z}[G_\mathbf{Z}]/O_\pi^r(\mathbf{Z}[G_\mathbf{Z}])$ is torsion free, we can identify $O_\pi^r(\mathbf{Z}[G_\mathbf{Z}]) \otimes_\mathbf{Z} k$ with a $(G \times G)$–submodule of $k[G]$ that is free of finite rank over k. We shall use the notation

(1) $$O_\pi^r(k[G]) = O_\pi^r(\mathbf{Z}[G_\mathbf{Z}]) \otimes_\mathbf{Z} k.$$

This is by A.24(1) compatible with the notation from A.13 in case k is a field. We get for any k–algebra A that

(2) $$O_\pi^r(k[G]) \otimes A \simeq O_\pi^r(A[G_A]).$$

Lemma: *Suppose that k is a principal ideal domain. Then we have*

(3) $$\operatorname{Ext}_G^i(M, O_\pi^r(k[G])) = 0 \qquad \textit{for all } i > 0$$

for each G–module M that is finitely generated and projective over k and such that all weights of M belong to $W\pi$.

Proof: We get from I.4.4 and I.4.18 for each k–algebra A an exact sequence

(4) $$\begin{aligned} 0 \to \operatorname{Ext}_G^i(M, O_\pi^r(k[G])) \otimes A &\to \operatorname{Ext}_{G_A}^i(M \otimes A, O_\pi^r(A[G_A])) \\ &\to \operatorname{Tor}_1^k(\operatorname{Ext}_G^{i+1}(M, O_\pi^r(k[G])), A) \to 0. \end{aligned}$$

If A is a field, then M belongs to the category $\mathcal{C}(\pi)$ for G_A. Since $A[G_A]$ is an injective G_A–module in this case, we get from A.1(4) that $O_\pi^r(A[G_A])$ is injective in $\mathcal{C}(\pi)$. Now Proposition A.10 implies that the middle term in (4) is 0 if $i > 0$.

It follows (for $i > 0$) that $\operatorname{Ext}_G^i(M, O_\pi^r(k[G])) \otimes A = 0$ for each k–algebra A that is a field. This implies $\operatorname{Ext}_G^i(M, O_\pi^r(k[G])) = 0$ because this Ext group is finitely generated over k by Proposition B.6.

B.8. Keep the assumptions and notations from B.7. For each prime number p the $(G_{\mathbf{F}_p} \times G_{\mathbf{F}_p})$–module $O_\pi^r(\mathbf{Z}[G_\mathbf{Z}]) \otimes \mathbf{F}_p \simeq O_\pi^r(\mathbf{F}_p[G_{\mathbf{F}_p}])$ has a good filtration by Lemma A.15. Therefore Lemma B.9 implies that also $O_\pi^r(\mathbf{Z}[G_\mathbf{Z}])$ has a good filtration. The factors are determined by the formal character; so one has as over a field that the factors are all $H_\mathbf{Z}^0(-w_0\lambda) \otimes H_\mathbf{Z}^0(\lambda)$ with $\lambda \in \pi$, each occurring once.

Extending scalars from $\mathbf{Z}$ to k we get for all k that $O_\pi^r(k[G])$ has a good filtration as a $(G \times G)$–module with factors all $H^0(-w_0\lambda) \otimes H^0(\lambda)$ with $\lambda \in \pi$, each occurring once.

For each $\nu \in X(T)$ denote as in A.19 by $O_\pi^r(k[G])_{(\nu,\bullet)}$ the ν–weight space of $O_\pi^r(k[G])$ with respect to ρ_l, equal to the direct sum of all $O_\pi^r(k[G])_{(\nu,\mu)}$ with $\mu \in X(T)$. Then $O_\pi^r(k[G])_{(\nu,\bullet)}$ is a G–submodule of $O_\pi^r(k[G])$ with respect to ρ_r and $O_\pi^r(k[G])$ is the direct sum of all $O_\pi^r(k[G])_{(\nu,\bullet)}$. Taking a weight space is an exact functor. This shows that each $O_\pi^r(k[G])_{(\nu,\bullet)}$ has a good filtration (as a G–module under ρ_r); the factors have the form $H^0(\lambda)$ with $\lambda \in \pi$, each occurring $\dim(H^0(-w_0\lambda)_\nu)$ times. This shows in particular that we have for each $\lambda \in \pi$ a short exact sequence of G–modules

(1) $$0 \to H^0(\lambda) \longrightarrow O_\pi^r(k[G])_{(-w_0\lambda,\bullet)} \longrightarrow Q \to 0$$

where Q has a good filtration with factors of the form $H^0(\mu)$ with $\mu \in \pi$ and $\mu > \lambda$. (Note that $H^0(-w_0\mu)_{-w_0\lambda} \neq 0$ implies $\mu \geq \lambda$.)

Proposition: *Suppose that k is a principal ideal domain. Let M be a G–module that is free of finite rank over k. If $\text{Ext}^1_G(M, H^0(\lambda)) = 0$ for all $\lambda \in X(T)_+$, then also $\text{Ext}^i_G(M, H^0(\lambda)) = 0$ for all $\lambda \in X(T)_+$ and all $i > 0$.*

Proof: We can find a finite and saturated subset π of $X(T)_+$ such that all dominant weights of M belong to π. If $\lambda \in X(T)_+$ with $\lambda \notin \pi$, then we have $\text{Ext}^i_G(M, H^0(\lambda)) = 0$ for all $i \geq 0$ by B.3(4).

We now use induction on i to prove the claim for all $\lambda \in \pi$. Lemma B.7 implies that $\text{Ext}^i_G(M, O^r_\pi(k[G])) = 0$ for all $i > 0$. It follows that $\text{Ext}^i_G(M, O^r_\pi(k[G])_{(\nu,\bullet)}) = 0$ for each ν and all $i > 0$ because $O^r_\pi(k[G])_{(\nu,\bullet)})$ is a direct summand of $O^r_\pi(k[G])$. Given $\lambda \in \pi$ we now consider the short exact sequence (1) and get isomorphisms $\text{Ext}^i_G(M, Q) \xrightarrow{\sim} \text{Ext}^{i+1}_G(M, H^0(\lambda))$ for all $i > 0$. Using the good filtration for Q and the induction hypothesis for i, we get the claim for $i + 1$.

Remark: One gets under the same assumptions on k and M: If $\text{Ext}^1_G(V(\lambda), M) = 0$ for all $\lambda \in X(T)_+$, then also $\text{Ext}^i_G(V(\lambda), M) = 0$ for all $\lambda \in X(T)_+$ and all $i > 0$. (Use I.4.4.)

B.9. We can define good filtrations of G–modules also for arbitrary k. If a G–module M has a good filtration, then M is free over k (since all $H^0(\mu)$ are so) and for each k–algebra k' the $G_{k'}$–module $M \otimes k'$ has a good filtration (since $H^0(\mu) \otimes k' \simeq H^0_{k'}(\mu)$, cf. 8.8(1)). This proves directly the implication "(i) $\Rightarrow$ (iv)" in the following result:

Lemma: *Suppose that k is a principal ideal domain. Let M be a G–module that is free of finite rank over k. Then the following properties are equivalent:*

(i) *M has a good filtration.*

(ii) $\text{Ext}^i_G(V(\lambda), M) = 0$ *for all* $\lambda \in X(T)_+$ *and all* $i > 0$.

(iii) $\text{Ext}^1_G(V(\lambda), M) = 0$ *for all* $\lambda \in X(T)_+$.

(iv) *For each maximal ideal $\mathfrak{m}$ in k the $G_{k/\mathfrak{m}}$–module $M \otimes k/\mathfrak{m}$ has a good filtration.*

Proof: Here "(i) $\Rightarrow$ (ii)" follows from B.4. The implication "(ii) $\Rightarrow$ (iii)" is trivial and "(iii) $\Rightarrow$ (ii)" follows from B.8.

Using $\text{Ext}^i_G(V(\lambda), M) \simeq H^i(G, V(\lambda)^* \otimes M)$, cf. I.4.4, and the analogous result over $G_{k/\mathfrak{m}}$, one gets from I.4.18.a that (ii) implies $\text{Ext}^1_{G_{k/\mathfrak{m}}}(V(\lambda)_{k/\mathfrak{m}}, M \otimes k/\mathfrak{m}) = 0$ for all $\lambda \in X(T)_+$. Now apply 4.16.b to get (iv). (Recall that $V(\lambda) \otimes k/\mathfrak{m} = V(\lambda)_{k/\mathfrak{m}}$ by 8.3.)

It remains to show "(iv) $\Rightarrow$ (i)". We use induction on the rank of M. We may assume that $M \neq 0$. As M is finitely generated, we can find a maximal weight μ of M. Then M_μ is free of finite rank over k. Denote the rank by s. There exists for each $f \in (M_\mu)^*$ a G–module homomorphism $M \to H^0(\mu)$ inducing f on $M_\mu \to H^0(\mu)_\mu$. Applying this to all elements in a basis for $(M_\mu)^*$, we get a G–module homomorphism $\varphi : M \to H^0(\mu)^s$ such that the map is bijective on the μ–weight spaces: $M_\mu \xrightarrow{\sim} H^0(\mu)^s_\mu$.

We want to show that φ is surjective. Since we are dealing with free modules of finite rank over a principal ideal domain, it suffices to show for each maximal ideal $\mathfrak{m}$ in k that the induced map over $k/\mathfrak{m}$ is surjective. Fix $\mathfrak{m}$ and set $k' = k/\mathfrak{m}$. By assumption $M_{k'} = M \otimes k'$ has a good filtration. Since μ is a maximal weight of $M_{k'}$ we can arrange the filtration such that the factors isomorphic to $H^0_{k'}(\mu)$

occur at the top of the filtration. So there exists a $G_{k'}$–submodule $N \subset M_{k'}$ such that N has a good filtration with factors of the form $H^0_{k'}(\nu)$ with $\nu \not\geq \mu$ and such that $M_{k'}/N$ is isomorphic to a direct sum of copies of $H^0_{k'}(\mu)$. The number of these summands is equal to the dimension of the μ–weight space of $H^0_{k'}(\mu)$, hence equal to s. We have $\operatorname{Hom}_{G_{k'}}(N, H^0_{k'}(\mu)) = 0$ since each non-zero submodule of $H^0_{k'}(\mu)$ has weight μ while μ is not a weight of N. Therefore $\varphi_{k'} : M_{k'} \to H^0_{k'}(\mu)^s$ maps N to 0 and induces a homomorphism $\psi : H^0_{k'}(\mu)^s \to H^0_{k'}(\mu)^s$. As φ is bijective on the μ–weight spaces, so are $\varphi_{k'}$ and ψ. Therefore ψ is injective on the socle of $H^0_{k'}(\mu)^s$, hence injective altogether. It follows that ψ is also surjective. This then implies the same for $\varphi_{k'}$ and yields also that N is the kernel of $\varphi_{k'}$.

We have now shown that $\varphi : M \to H^0(\mu)^s$ is surjective. Let M' denote the kernel of φ. This is then a G–module that is free of finite rank over k. Its rank is less than that of M. Furthermore $M' \otimes k/\mathfrak{m}$ identifies (for each $\mathfrak{m}$) with the kernel of $\varphi_{k/\mathfrak{m}}$; so the preceding paragraph shows that $M' \otimes k/\mathfrak{m}$ has a good filtration. Now induction implies that M' has a good filtration. It follows that so does M as $M/M' \simeq H^0(\mu)^s$.

B.10. We can similarly define Weyl filtrations over k. If a G–module M has a Weyl filtration, then M is free over k (since all $V(\mu)$ are so) and for each k–algebra k' the $G_{k'}$–module $M \otimes k'$ has a Weyl filtration (since $V(\mu) \otimes k' \simeq H(\mu)_{k'}$, cf. 8.3).

If M is a G–module that is free of finite rank over k, then 8.9(2) shows that M has a Weyl filtration if and only if M^* has a good filtration. Furthermore, I.4.4 shows in this case that $\operatorname{Ext}^i_G(M, H^0(\lambda)) \simeq \operatorname{Ext}^i_G(V(-w_0\lambda), M^*)$ for all $\lambda \in X(T)_+$ and $i \in \mathbf{N}$. Since $M^* \otimes k/\mathfrak{m}$ is isomorphic to the dual of $M \otimes k/\mathfrak{m}$ for each maximal ideal $\mathfrak{m}$ in k, Lemma B.9 translates immediately to the following result (that can also be proved directly):

Lemma: *Suppose that k is a principal ideal domain. Let M be a G–module that is free of finite rank over k. Then the following properties are equivalent:*

(i) *M has a Weyl filtration.*

(ii) *$\operatorname{Ext}^i_G(M, H^0(\lambda)) = 0$ for all $\lambda \in X(T)_+$ and all $i > 0$.*

(iii) *$\operatorname{Ext}^1_G(M, H^0(\lambda)) = 0$ for all $\lambda \in X(T)_+$.*

(iv) *For each maximal ideal $\mathfrak{m}$ in k the $G_{k/\mathfrak{m}}$–module $M \otimes k/\mathfrak{m}$ has a Weyl filtration.*

B.11. Let K denote the field of fractions of k. Suppose that $\operatorname{char}(K) = 0$. Let $\lambda \in X(T)_+$. Identify $H^0(\lambda)$ with a G–submodule of $H^0_K(\lambda) \simeq H^0(\lambda) \otimes K$ via $v \mapsto v \otimes 1$. Fix $v_\lambda \in H^0(\lambda)_\lambda$ with $H^0(\lambda)_\lambda = k v_\lambda$.

Identify also $V(\lambda)$ with a G–submodule of $V(\lambda)_K \simeq V(\lambda) \otimes K$. Fix $v'_\lambda \in V(\lambda)_\lambda$ with $V(\lambda)_\lambda = k v'_\lambda$. We have then $V(\lambda) = \operatorname{Dist}(G) v'_\lambda$.

As $\operatorname{char}(K) = 0$, we can identify $V(\lambda)_K$ with $H^0_K(\lambda)$ such that v'_λ is mapped to v_λ. We have now

$$V(\lambda) = \operatorname{Dist}(G) v_\lambda \subset H^0(\lambda) \subset H^0_K(\lambda) = V(\lambda)_K. \tag{1}$$

We want to show (within this set-up):

Lemma: *Suppose that k is a principal ideal domain. Let M be a non-zero G–submodule of $H^0_K(\lambda)$ such that M is finitely generated over k. Then there exists $a \in K$, $a \neq 0$ with*

$$V(\lambda) \subset aM \subset H^0(\lambda). \tag{2}$$

Proof: We have $M = \bigoplus_\mu M_\mu$ and each M_μ is a free k–module of finite rank because k is a principal ideal domain. As we assume $\text{char}(K) = 0$, the G_K–module $H^0_K(\lambda)$ is simple and therefore M generates $H^0_K(\lambda)$ over K. Since M is free k–module of finite rank, this means that the natural map $M \otimes K \to H^0_K(\lambda)$ is an isomorphism and induces isomorphisms $M_\mu \otimes K \to H^0_K(\lambda)_\mu$. In particular, M_λ is a free k–module of rank 1. So there exists $a \in K$, $a \neq 0$ with $M_\lambda = k\, a v_\lambda$. Replacing M by $a^{-1}M$ we may assume that $M_\lambda = k\, v_\lambda$.

We now have to show that $V(\lambda) \subset M \subset H^0(\lambda)$. Since M is a G–submodule of $H^0_K(\lambda)$, it is stable under $\text{Dist}(G)$. This implies that $V(\lambda) = \text{Dist}(G) v_\lambda \subset M$.

Set $M' = M + H^0(\lambda)$. This is another G–submodule of $H^0_K(\lambda)$ that is finitely generated over k; it satisfies $M'_\lambda = k v_\lambda$. We have $H^0(\lambda) \subset M'$ and want to prove equality: $H^0(\lambda) = M'$. If so, then clearly $M \subset H^0(\lambda)$ as claimed.

For each prime element $q \in k$ the inclusion $H^0(\lambda) \subset M'$ induces a homomorphism

$$\psi_q : H^0_{k/(q)}(\lambda) \simeq H^0(\lambda)/qH^0(\lambda) \longrightarrow M'/qM'$$

of $G_{k/(q)}$–modules. It is an isomorphism on the λ–weight space since $H^0(\lambda)_\lambda = k v_\lambda = M'_\lambda$. Therefore ψ_q is non-zero on the simple socle of $H^0_{k/(q)}(\lambda)$, hence is injective. As $H^0(\lambda)/qH^0(\lambda)$ and M'/qM' have the same dimension over $k/(q)$ (equal to the dimension of $H^0_K(\lambda)$ over K), this implies that ψ_q is also surjective. It follows that $(M'/H^0(\lambda)) \otimes k/(q) = 0$. This holds for all prime elements q in k. We get therefore $M'/H^0(\lambda) = 0$ and thus $H^0(\lambda) = M'$ as desired.

CHAPTER C

Lusztig's Conjecture and Some Consequences

We assume in this chapter that p is a prime and k a field of characteristic p.

The Lusztig conjecture [short: (LC)] as stated in 8.22(1) predicts the characters of the simple G–modules within certain limits. We shall describe in this chapter some results that follow from the truth of (LC). It turns out that several of these results in turn imply (LC).

One of these results is the interpretation of the coefficients of the Kazhdan-Lusztig polynomials as dimensions of certain Ext groups between simple modules and induced modules of the form $H^0(\mu)$. Other results involve the semi-simplicity of certain modules that arise when applying wall crossing functors to simple modules, the non-vanishing of the Ext^1–group between certain simple modules, and some compatibility of the filtrations from Chapter 8 with certain homomorphisms between certain Weyl modules. (See C.2 and C.9 for more precise statements.) In fact, if (LC) holds, then one gets all Ext–groups between simple modules in its region of validity. (This is discussed in Subsection C.10. There, and only there, the reader is supposed to have some familiarity with derived categories.)

In these results as well as in (LC) we look at modules with a p–regular highest weight. Whereas (LC) immediately leads to a character formula also for simple modules with non-regular highest weights (for p big enough), the other results do not carry over that easily to the non-regular case. (Some of those results would not make sense in the non-regular case.)

If p is large enough, then (LC) predicts all $\mathrm{ch}\, L(w \bullet \lambda)$ with $\lambda \in C \cap X(T)$ and $w \bullet \lambda \in X_1(T)$, hence all $\mathrm{ch}\, \widehat{L}_1(\mu) = \mathrm{ch}\, L(\mu)$ with $\mu \in X_1(T)$ using translation functors, hence $\mathrm{ch}\, \widehat{L}_1(\mu)$ for all $\mu \in X(T)$ using 9.6(6).

If (LC) holds, then one can prove results for G_1T–modules analogous to those for G–modules referred to above. In some cases the results for G_1T can be deduced directly from the results for G, but it will be more natural and convenient to work within the framework of G_1T–modules. That is done in the subsections from C.11 on. In many cases the argument will be similar to that for G and we often simply refer to the proofs in that case.

Other applications of (LC) (on the radical series of certain standard modules) will be discussed in the next chapter.

We assume for the sake of simplicity that $\mathcal{D}G$ is simply connected. That allows us to apply 6.3(1). Most final results can be proved for general G by going to a suitable covering. We assume throughout that $C \cap X(T) \neq \emptyset$, hence that $p \geq h$. When dealing with G_1T we use the stronger assumption that $p > h$ so that we can apply 12.13. We then also assume as before that k is perfect.

The main sources for this chapter are [Andersen 11, 12], [Cline, Parshall, and Scott 16, 17], [Kaneda 5]. Much of the work presented here was inspired by work on the Kazhdan-Lusztig conjecture in characteristic 0 by Vogan and by Gabber and Joseph.

C.1. It will be convenient to use a different normalisation of the Kazhdan-Lusztig polynomials than in 8.22. Let us first carry over the ordering $\uparrow$ to W_p.

Since $x \mapsto x \bullet C$ is a bijection from W_p to the set of all alcoves for W_p, we get an order relation on W_p by setting $x \uparrow y$ if and only if $x \bullet C \uparrow y \bullet C$. (One can check that this order relation and the Bruhat-Chevalley ordering on W_p restrict to the same order relation on the subset W_p^+ of all $w \in W_p$ with $w \bullet \lambda \in X(T)_+$ for all $\lambda \in C \cap X(T)$.) Set $d(w) = d(w \bullet C)$ for all $w \in W_p$. Then $x \uparrow y$ implies $d(x) \leq d(y)$ with equality if and only if $x = y$. For all $w \in W_p$ and $s \in \Sigma$ we have $w \uparrow ws$ if and only if $d(w) < d(ws)$, cf. 6.6.

We shall normalise the Kazhdan-Lusztig polynomials as in Section 3 of [Soergel 4]. Let v be an indeterminate over $\mathbf{Z}$. One has then for all $x, y \in W_p^+$ polynomials $m_{y,x} \in \mathbf{Z}[v]$. They satisfy $m_{x,x} = 1$ for all x; if $y \neq x$, then $m_{y,x} \in v\mathbf{Z}[v]$. Furthermore $m_{y,x} \neq 0$ implies $y \uparrow x$. If v^i occurs with a non-zero coefficient in $m_{y,x}$, then $d(x) - d(y) \equiv i \pmod 2$. (We shall refer to this fact as the *parity property* of $m_{y,x}$.)

We have in particular $m_{1,1} = 1$ and $m_{y,1} = 0$ for $y \neq 1$. This is the starting point for an inductive computation of all $m_{y,x}$. For $x \in W_p^+$ and $s \in \Sigma$ with $x \uparrow xs$, one defines $m_{y,x}^s \in \mathbf{Z}[v]$ via

$$(1) \qquad m_{y,x}^s = \begin{cases} m_{ys,x} + v^{-1} m_{y,x}, & \text{if } ys \uparrow y \text{ and } ys \in W_p^+, \\ (v + v^{-1})\, m_{y,x}, & \text{if } ys \uparrow y \text{ and } ys \notin W_p^+, \\ m_{ys,x} + v\, m_{y,x}, & \text{if } y \uparrow ys. \end{cases}$$

Then one has $m_{xs,x}^s = 1$ and

$$(2) \qquad m_{y,xs} = m_{y,x}^s - \sum_{z \uparrow x,\, z \neq x} m_{z,x}^s(0)\, m_{y,z}.$$

(One knows that $m_{z,x}^s(0) \neq 0$ implies that $z = xs$ or $z \uparrow x$, $z \neq x$.) For any $w \in W_p^+$ with $w \neq 1$, there exists $s \in \Sigma$ with $ws \uparrow w$ and $ws \in W_p^+$. We now apply (2) with $x = ws$; we may assume by induction that all terms on the right hand side of (2) are known and compute thus $m_{y,w}$. (For all this, compare Section 3 in [Soergel 4].)

C.2. Since we assume that $p \geq h$ we can choose $\lambda \in C \cap X(T)$ and for each $s \in \Sigma$ an element $\mu(s) \in \overline{C}_{\mathbf{Z}}$ with $\Sigma^0(\mu(s)) = \{s\}$. Set

$$\Theta_s = T_{\mu(s)}^{\lambda} \circ T_{\lambda}^{\mu(s)}.$$

For each $w \in W_p^+$ with $w \bullet \lambda < ws \bullet \lambda$ we have $\Theta_s L(w \bullet \lambda) \simeq T_{\mu(s)}^{\lambda} L(w \bullet \mu(s))$. Therefore Lemma 7.20.a implies that we can set

$$(1) \qquad \beta_s L(w \bullet \lambda) = \operatorname{rad}_G \Theta_s L(w \bullet \lambda) / \operatorname{soc}_G \Theta_s L(w \bullet \lambda).$$

Suppose that $\mathcal{W}$ is a subset of W_p^+ such that for all $w \in \mathcal{W}$ and $w' \in W_p^+$ with $w' \uparrow w$ also $w' \in \mathcal{W}$. (For example, we could take for $\mathcal{W}$ the set of all $w \in W_p^+$ with $\langle w(\lambda + \rho), \alpha^\vee \rangle \leq p(p - h + 2)$ for all $\alpha \in R^+$.)

Proposition: a) *Let $\lambda \in C \cap X(T)$. The following are equivalent:*

(i) *We have* $\operatorname{ch} L(x \bullet \lambda) = \sum_{y \in W_p^+} m_{y,x}(-1)\, \chi(y \bullet \lambda)$ *for all* $x \in \mathcal{W}$.

(ii) *The G–module $\beta_s L(w \bullet \lambda)$ is semi-simple for all $w \in \mathcal{W}$ and $s \in \Sigma$ with $w \uparrow ws$ and $ws \in \mathcal{W}$.*

(iii) *We have* $\operatorname{Ext}^1_G(L(w \bullet \lambda), L(ws \bullet \lambda)) \neq 0$ *for all $w \in \mathcal{W}$ and $s \in \Sigma$ with $ws \in \mathcal{W}$.*

b) *If these equivalent conditions hold, then we get for all $x \in \mathcal{W}$ and $y \in W^+$ that*

$$m_{y,x} = \sum_{i \geq 0} \dim \operatorname{Ext}^i_G(L(x \bullet \lambda), H^0(y \bullet \lambda))\, v^i$$

and for all i that $\operatorname{Ext}^i_G(L(x \bullet \lambda), H^0(y \bullet \lambda)) \neq 0$ *implies* $d(x) - d(y) \equiv i$ *mod* 2.

The formula in (i) is equivalent to the one in Lusztig's conjecture, i.e., to 8.22(2). This requires a comparison of the $m_{y,x}$ with the original Kazhdan-Lusztig polynomials, cf. [Soergel 4], 3.4.

In the representation theory of semi-simple complex Lie algebras Vogan conjectured in a paper from 1979 that modules analogous to the $\beta_s L(w \bullet \lambda)$ should be semi-simple; he then showed in another paper from 1979 that this conjecture was (in the case of highest weight modules) equivalent to the Kazhdan-Lusztig conjecture. His results were then extended to the situation considered here in [Andersen 11]. There the equivalence of (i) and (ii) in a) as well as b) was proved. The equivalence of (iii) with the other conditions was shown in [Cline, Parshall, and Scott 17]. We shall discuss the proofs in the next subsections.

C.3. We keep $\lambda \in C \cap X(T)$. For each $y \in W_p^+$ and each finite dimensional G–module V with $\operatorname{pr}_\lambda V = V$ set

$$P[V, y] = \sum_{i \geq 0} \dim \operatorname{Ext}^i_G(V, H^0(y \bullet \lambda))\, v^i \in \mathbf{Z}[v]. \tag{1}$$

Recall from 6.21 that the dimensions in (1) are finite, that only finitely many of them are non-zero, and that by 6.21(6)

$$\operatorname{ch}(V) = \sum_{y \in W_p^+} P[V, y](-1)\, \chi(y \bullet \lambda). \tag{2}$$

Proposition 4.13 yields for all $w, y \in W_p^+$

$$P[V(w \bullet \lambda), y] = \begin{cases} 1, & \text{if } y = w, \\ 0, & \text{otherwise.} \end{cases} \tag{3}$$

Note: If $V = \bigoplus_{w \in W_p^+} L(w \bullet \lambda)^{m(w)}$ with $m(w) \in \mathbf{N}$ is a semi-simple finite dimensional G–module, then any $m(w) = \dim \operatorname{Hom}_G(V, H^0(w \bullet \lambda))$ is the coefficient $P[V, w](0)$ of v^0 in $P[V, w]$.

We introduce some abbreviations for the conditions in Proposition C.2. Let $x \in W_p^+$. We say that (A, x) holds if all $a_{x,y}$ as in 8.22(1) are equal to $m_{y,x}(-1)$, i.e.,

$$(A, x) \ :\Longleftrightarrow \ \operatorname{ch} L(x \bullet \lambda) = \sum_{y \in W_p^+} m_{y,x}(-1)\, \chi(y \bullet \lambda).$$

Two other conditions that may or may not hold, are

$$(B, x) \ :\Longleftrightarrow \ P[L(x \bullet \lambda), y] = m_{y,x} \text{ for all } y \in W_p^+$$

and

$$(C, x) :\Longleftrightarrow \ \forall i, y\text{: If } \operatorname{Ext}^i_G(L(x \bullet \lambda), H^0(y \bullet \lambda)) \neq 0, \text{ then } d(x) - d(y) \equiv i \bmod 2.$$

Furthermore, we set for any $s \in \Sigma$ with $x \uparrow xs$

$$(D, x, s) \ :\Longleftrightarrow \ \text{The } G\text{–module } \beta_s L(x \bullet \lambda) \text{ is semi-simple}$$

and

$$(E, x, s) \ :\Longleftrightarrow \ \operatorname{Ext}^1_G(L(x \bullet \lambda), L(xs \bullet \lambda)) \neq 0.$$

We start by observing some obvious connections between these conditions. Note first, however:

(4) *Condition* $(B, 1)$ *holds.*

This follows from (3) because of $L(\lambda) = V(\lambda)$. Observe next for any $x \in W_p^+$:

(5) $$(B, x) \implies (A, x) \text{ and } (C, x).$$

Here (A, x) follows immediately from (2). For (C, x) one uses the parity properties of the $m_{y,x}$ mentioned in C.1.

Lemma: *Let* $w, x \in W_p^+$ *with* $x \uparrow w$. *If* (C, w) *holds and* $d(w) - d(x)$ *is even, then* $\operatorname{Ext}^1_G(L(w \bullet \lambda), L(x \bullet \lambda)) = 0$.

Proof: If $x \neq w$, then $\operatorname{Hom}_G(\operatorname{rad}_G V(w \bullet \lambda), H^0(x \bullet \lambda)) \simeq \operatorname{Ext}^1_G(L(w \bullet \lambda), H^0(x \bullet \lambda))$ by 6.20(1). We have $\operatorname{Ext}^1_G(L(w \bullet \lambda), L(x \bullet \lambda)) \simeq \operatorname{Hom}_G(\operatorname{rad}_G V(w \bullet \lambda), L(x \bullet \lambda))$ by 2.14(1). Combining this with the obvious inclusion of $\operatorname{Hom}_G(\operatorname{rad}_G V(w \bullet \lambda), L(x \bullet \lambda))$ into $\operatorname{Hom}_G(\operatorname{rad}_G V(w \bullet \lambda), H^0(x \bullet \lambda))$ we get now an exact sequence

$$0 \longrightarrow \operatorname{Ext}^1_G(L(w \bullet \lambda), L(x \bullet \lambda)) \longrightarrow \operatorname{Ext}^1_G(L(w \bullet \lambda), H^0(x \bullet \lambda)).$$

This holds trivially also for $x = w$ where both terms are 0, cf. 2.12(1) and 4.13(2). So far we have just used that $x \uparrow w$. Now (C, w) yields the claim.

C.4. Keep the assumptions and notations from C.2/3. Consider $w \in W_p^+$ and $s \in \Sigma$ with $w \uparrow ws$. We can apply Lemma 7.20.c with $\mu = \mu(s)$. It implies that all composition factors of $\beta_s L(w \bullet \lambda)$ have the form $L(w' \bullet \lambda)$ with $w's \bullet \lambda < w' \bullet \lambda$, hence with $T_\lambda^\mu L(w' \bullet \lambda) = 0$. This implies

(1) $$T_\lambda^{\mu(s)} \beta_s L(w \bullet \lambda) = 0.$$

Lemma: *Condition (E, w, s) holds if and only if $\beta_s L(w \bullet \lambda)$ contains a submodule isomorphic to $L(ws \bullet \lambda)$. If so, then $\mathrm{Ext}_G^1(L(w \bullet \lambda), L(ws \bullet \lambda)) \simeq k$ and $L(ws \bullet \lambda)$ is isomorphic to a direct summand of $\beta_s L(w \bullet \lambda)$.*

Proof: If $\beta_s L(w \bullet \lambda)$ contains a submodule isomorphic to $L(ws \bullet \lambda)$, then we get from Lemma 7.20.b/c that $\mathrm{Ext}_G^1(L(w \bullet \lambda), L(ws \bullet \lambda)) \simeq k$. This yields then (E, w, s).

Conversely, if (E, w, s) holds, then $M = \beta_s L(w \bullet \lambda)$ contains by Lemma 7.20.c submodules M_1 and M_2 with $M_1 \simeq L(ws \bullet \lambda) \simeq M/M_2$; now Lemma 7.20.b implies that $M_1 \cap M_2 = 0$, hence $M = M_1 \oplus M_2$. So in this case M actually has a direct summand isomorphic to $L(ws \bullet \lambda)$.

C.5. Lemma: *Let $w \in W_p^+$ and $s \in \Sigma$ with $w \uparrow ws$. If (C, w) holds, then*

$$P[\beta_s L(w \bullet \lambda), y] = \begin{cases} P[L(w \bullet \lambda), ys] + v^{-1}\, P[L(w \bullet \lambda), y] & \textit{if } ys \uparrow y \textit{ and } ys \in W_p^+, \\ (v + v^{-1})\, P[L(w \bullet \lambda), y] & \textit{if } ys \uparrow y \textit{ and } ys \notin W_p^+, \\ P[L(w \bullet \lambda), ys] + v\, P[L(w \bullet \lambda), y] & \textit{if } y \uparrow ys. \end{cases}$$

Proof: Set
$$Q_s(w \bullet \lambda) = \Theta_s L(w \bullet \lambda) / \operatorname{soc} \Theta_s L(w \bullet \lambda).$$

We get from Lemma 7.20.a short exact sequences

(1) $$0 \to L(w \bullet \lambda) \to \Theta_s L(w \bullet \lambda) \to Q_s(w \bullet \lambda) \to 0$$

and

(2) $$0 \to \beta_s L(w \bullet \lambda) \to Q_s(w \bullet \lambda) \to L(w \bullet \lambda) \to 0.$$

Let $y \in W_p^+$ with $ys \uparrow y$. We want to show first that

(3) $$P[Q_s(w \bullet \lambda), y] = \begin{cases} P[L(w \bullet \lambda), ys] & \text{if } ys \in W_p^+, \\ vP[L(w \bullet \lambda), y] & \text{otherwise.} \end{cases}$$

Set $\mu = \mu(s)$. The second case in (3) is easy: If $ys \notin W_p^+$, then $y \bullet \mu \notin X(T)_+$, hence $T_\lambda^\mu H^0(y \bullet \lambda) = 0$ by 7.11, and $\mathrm{Ext}_G^\bullet(T_\mu^\lambda L(w \bullet \mu), H^0(y \bullet \lambda)) = 0$ by adjointness. Now recall that $\Theta_s L(w \bullet \lambda) = T_\mu^\lambda L(w \bullet \mu)$ and apply $\mathrm{Hom}_G(?, H^0(y \bullet \lambda))$ to (1). The corresponding long exact sequence then yields the claim in this case.

If $ys \in W_p^+$, then we have a short exact sequence

(4) $$0 \to H^0(ys \bullet \lambda) \to \Theta_s H^0(y \bullet \lambda) \to H^0(y \bullet \lambda) \to 0$$

as in 7.19.a. Now apply $\mathrm{Hom}_G(?, H^0(y \bullet \lambda))$ to (1) and $\mathrm{Hom}_G(L(w \bullet \lambda), ?)$ to (4). One gets two long exact sequences

$$\to \mathrm{Ext}^i_G(L_w, H^0_{ys}) \to \mathrm{Ext}^i_G(L_w, \Theta_s H^0_y) \to \mathrm{Ext}^i_G(L_w, H^0_y) \to \mathrm{Ext}^{i+1}_G(L_w, H^0_{ys}) \to$$

and

$$\to \mathrm{Ext}^i_G(Q_s, H^0_y) \to \mathrm{Ext}^i_G(\Theta_s L_w, H^0_y) \to \mathrm{Ext}^i_G(L_w, H^0_y) \to \mathrm{Ext}^{i+1}_G(Q_s, H^0_y) \to$$

where we use obvious (I hope) abbreviations like $L_w = L(w \bullet \lambda)$ and $H^0_y = H^0(y \bullet \lambda)$. These long exact sequences are related by commutative diagrams of the form

$$\begin{array}{ccc} \mathrm{Ext}^i_G(L_w, \Theta_s H^0_x) & \xrightarrow{\varphi_i} & \mathrm{Ext}^i_G(L_w, H^0_x) \\ \downarrow & & \| \\ \mathrm{Ext}^i_G(\Theta_s L_w, H^0_x) & \xrightarrow{\varphi'_i} & \mathrm{Ext}^i_G(L_w, H^0_x) \end{array}$$

where φ_i and φ'_i come from the long exact sequences and where the vertical map on the left is a composition of two adjunction isomorphisms as in 7.7. The commutativity of these diagrams follows from Remark 7.22 because we can choose the maps $L(w \bullet \lambda) \to \Theta_s L(w \bullet \lambda)$ in (1) and $\Theta_s H^0(y \bullet \lambda) \to H^0(y \bullet \lambda)$ in (4) equal to $i_{L(w \bullet \lambda)}$ and $j_{H^0(y \bullet \lambda)}$, cf. 7.21(8). It follows that $\ker(\varphi_i)$ and $\ker(\varphi'_i)$ have the same dimension, similarly for the cokernels. The dimension of $\mathrm{Ext}^i_G(L(w \bullet \lambda), H^0(ys \bullet \lambda))$ is equal to the sum of the dimensions of $\ker(\varphi_i)$ and $\mathrm{coker}(\varphi_{i-1})$, the dimension of $\mathrm{Ext}^i_G(Q_s(w \bullet \lambda), H^0(y \bullet \lambda))$ equal to the sum of the dimensions of $\ker(\varphi'_i)$ and $\mathrm{coker}(\varphi'_{i-1})$. Therefore these dimensions are equal to each other.

We have now proved (3). So far we did not use (C, w). It will be used for the next step. We apply $\mathrm{Hom}_G(?, H^0(y \bullet \lambda))$ for $ys \uparrow y$ to (2). Using (C, w) and (3) we see that many terms in the long exact sequence vanish and that the remaining terms yield short exact sequences from which we can read off the claim in the lemma in the first two cases.

In case $y \uparrow ys$, we use that $P[\beta_s L(w \bullet \lambda), y] = v\, P[\beta_s L(w \bullet \lambda), ys]$ by C.4(1) and Proposition 7.19.b. And we know $P[\beta_s L(w \bullet \lambda), ys]$ from the first case of the lemma.

C.6. Lemma: *Let again $w \in W_p^+$ and $s \in \Sigma$ with $w \uparrow ws$.*

a) *If (D, w, s) holds and if (B, x) holds for all $x \in W_p^+$ with $x \uparrow w$, then (B, ws) holds.*

b) *(C, w) and (E, w, s) $\Longrightarrow$ (C, ws).*

Proof: a) We can apply Lemma C.5 because in particular (B, w) holds, hence also (C, w). Comparing with C.1(1) we get $m^s_{y,w} = P[\beta_s L(w \bullet \lambda), y]$ for all $y \in W_p^+$. Now (D, w, s) yields for all y

$$m^s_{y,w} = P[\beta_s L(w \bullet \lambda), y] = P[L(ws \bullet \lambda), y] + \sum_z m(z) P[L(z \bullet \lambda), y]$$

where the sum is over $z \in W_p^+$ with $z \uparrow w$, $z \neq w$ and where $m(z)$ is the multiplicity of $L(z \bullet \lambda)$ in $\beta_s L(w \bullet \lambda)$, hence

$$m(z) = P[\beta_s L(w \bullet \lambda), z](0) = m_{z,w}^s(0).$$

Now a comparison with C.1(2) yields the claim.

b) Recall from Lemma C.4: If (E, w, s) holds, then $L(ws \bullet \lambda)$ is isomorphic to a direct summand in $\beta_s L(w \bullet \lambda)$. Then $\text{Ext}_G^i(L(ws \bullet \lambda), H^0(y \bullet \lambda)) \neq 0$ implies $\text{Ext}_G^i(\beta_s L(w \bullet \lambda), H^0(y \bullet \lambda)) \neq 0$, hence v^i occurs with a non-zero coefficient in $P[\beta_s L(w \bullet \lambda), y]$. Now apply Lemma C.5.

C.7. (*Proof of Proposition C.2.*) The proof of the implication "(ii) $\Rightarrow$ (i)" in C.2.a is now easy: We know by C.3(4) that $(B, 1)$ holds. If (D, w, s) always holds, then induction using Lemma C.6.a yields (B, x) for all $x \in \mathcal{W}$, hence (A, x) and (C, x). This proves at the same time that the claims in C.2.b follow from (ii).

Consider now the implication "(i) $\Rightarrow$ (ii)". Suppose that (A, x) holds for all $x \in \mathcal{W}$ and that there exist $w \in \mathcal{W}$ and $s \in \Sigma$ with $w \uparrow ws$ and $ws \in \mathcal{W}$ such that (D, w, s) does not hold. Pick w and s with this property such that $d(w)$ is minimal. Then we know by Lemma C.6.a and induction that (B, x) holds for all $x \in W_p^+$ with $x \uparrow w$, hence also by Lemma C.5 that $m_{y,w}^s = P[\beta_s L(w \bullet \lambda), y]$ for all $y \in W_p^+$. We get now

$$\begin{aligned}
\text{ch}\, \beta_s L(w \bullet \lambda) &= \sum_{y \in W_p^+} m_{y,w}^s(-1)\, \chi(y \bullet \lambda) \\
&= \sum_{z \in \mathcal{W}} m_{z,w}^s(0) \sum_{y \in W_p^+} m_{y,z}(-1)\, \chi(y \bullet \lambda) \\
&= \sum_{z \in \mathcal{W}} m_{z,w}^s(0)\, \text{ch}\, L(z \bullet \lambda)
\end{aligned}$$

where we use C.3(2), C.1(2), and [for the last step] the assumption that (A, z) holds for all $z \in \mathcal{W}$. It follows that

$$[\beta_s L(w \bullet \lambda) : L(z \bullet \lambda)] = m_{z,w}^s(0) = \dim \text{Hom}_G(\beta_s L(w \bullet \lambda), H^0(z \bullet \lambda))$$

for all z. (Recall the definition of $P[V, x]$.) One shows using 7.20.a and 7.6(3) that ${}^\tau\beta_s L(w \bullet \lambda) \simeq \beta_s L(w \bullet \lambda)$ and gets now a contradiction to the choice of w and s from the following fact: *Let V be a finite dimensional G–module with ${}^\tau V \simeq V$ and with $[V : L(\mu)] = \dim \text{Hom}_G(V, H^0(\mu))$ for all $\mu \in X(T)_+$. Then V is semi-simple.* We postpone a proof of this fact to the next subsection.

We now turn to the equivalence of (iii) with the other conditions. Note that we may restrict in (iii) to x and s with $x \uparrow xs$; this follows from 2.12(4). Lemma 7.20.b/c shows that (D, w, s) implies (E, w, s). This yields the implication "(ii) $\Rightarrow$ (iii)" in Proposition C.2. In order to get the implication "(iii) $\Rightarrow$ (ii)" one uses induction and reduces the claim to the following result:

(1) *If (E, w, s) holds and if (C, x) holds for all $x \in W_p^+$ with $x \uparrow w$, then (D, w, s) holds.*

Proof: If $\mathrm{Hom}_G(\beta_s L(w{\bullet}\lambda), H^0(y{\bullet}\lambda)) \neq 0$, then v^0 occurs with a non-zero coefficient in $P[\beta_s L(w \bullet \lambda), y]$. Now (C, w) and Lemma C.5 imply that in this case $d(y) - d(w)$ is odd and either $y = ws$ or $y \uparrow x$. In both cases we know that (C, y) holds, either by Lemma C.6.b or by our assumption. The definition of (C, y) implies now for all $y_1, y_2 \in W_p^+$ with $\mathrm{Hom}_G(\beta_s L(w \bullet \lambda), H^0(y_i \bullet \lambda)) \neq 0$ for both i, that $\mathrm{Ext}^1_G(L(y_1 \bullet \lambda), H^0(y_2 \bullet \lambda)) = 0$. Now the semi-simplicity of $\beta_s L(w \bullet \lambda)$ follows from the following fact (also to be proved in the next subsection): *Let V be a finite dimensional G–module with ${}^\tau V \simeq V$. If $\mathrm{Ext}^1_G(L(\mu), H^0(\nu)) = 0$ for all $\mu, \nu \in X(T)^+$ with $\mathrm{Hom}_G(V, H^0(\mu)) \neq 0$ and $\mathrm{Hom}_G(V, H^0(\nu)) \neq 0$, then V is semi-simple.*

C.8. If V is a finite dimensional G–module then we have for all $\mu \in X(T)$ isomorphisms

$$\mathrm{Hom}_G(V(\mu), V) \simeq \mathrm{Hom}_G({}^\tau V, {}^\tau V(\mu)) \simeq \mathrm{Hom}_G({}^\tau V, H^0(\mu)) \tag{1}$$

by 2.13(2). We get in particular $\mathrm{Hom}_G(V(\mu), V) \simeq \mathrm{Hom}_G(V, H^0(\mu))$ if ${}^\tau V \simeq V$. Therefore the two semi-simplicity claims in C.7 follow from the following, more symmetric statements:

Lemma: *Let V be a finite dimensional G–module.*

a) *If*

$$[V : L(\mu)] = \dim \mathrm{Hom}_G(V, H^0(\mu)) = \dim \mathrm{Hom}_G(V(\mu), V)$$

for all $\mu \in X(T)_+$, then V is semi-simple.

b) *Suppose that $\mathrm{Ext}^1_G(L(\nu), H^0(\mu)) = 0$ for all $\mu, \nu \in X(T)_+$ with $\mathrm{Hom}_G(V(\mu), V) \neq 0$ and $\mathrm{Hom}_G(V(\nu), V) \neq 0$. Suppose also that $\mathrm{Ext}^1_G(L(\nu), H^0(\mu)) = 0$ for all $\mu, \nu \in X(T)_+$ with $\mathrm{Hom}_G(V, H^0(\mu)) \neq 0$ and $\mathrm{Hom}_G(V, H^0(\nu)) \neq 0$. Then V is semi-simple.*

Proof: In both cases we use induction on $\dim V$. We may assume that $V \neq 0$. We shall then find submodules V_1 and V_2 with $V = V_1 \oplus V_2$ such that V_1 is semi-simple and $V_1 \neq 0$. Then also V_2 satisfies the assumption on V (in the first case one has to use the semi-simplicity of V_1 to see this) and one can apply induction to V_2.

a) Let $\mu \in X(T)_+$ be minimal for $m = [V : L(\mu)] > 0$. Set $V_1 \subset V$ equal to the sum of all submodules of V isomorphic to $L(\mu)$. Each non-zero homomorphism $\varphi : V(\mu) \to V$ satisfies $\varphi(V(\mu)) \simeq L(\mu)$ because otherwise $\varphi(V(\mu))$ and hence V had a composition factor $L(\nu)$ with $\nu < \mu$. We get thus $\varphi(V(\mu)) \subset V_1$. It follows that $\mathrm{Hom}_G(V(\mu), V_1) \simeq \mathrm{Hom}_G(V(\mu), V)$ has dimension m, hence $V_1 \simeq L(\mu)^m$.

We have $\mathrm{ch}({}^\tau V) = \mathrm{ch}(V)$, hence $[{}^\tau V : L(\nu)] = [V : L(\nu)]$ for all $\nu \in X(T)_+$. By (1) also ${}^\tau V$ satisfies the assumption in a). So there is a submodule M of ${}^\tau V$ with $M \simeq L(\mu)^m$. Recall that ${}^\tau V$ is just V^* with a twisted G–action. Therefore

$$V_2 = \{\, v \in V \mid f(v) = 0 \text{ for all } f \in M \,\}$$

is a submodule of V. One gets that ${}^\tau(V/V_2) \simeq N \simeq L(\mu)^m$, hence $V/V_2 \simeq ({}^\tau L(\mu))^m \simeq L(\mu)^m$. Now $m = [V : L(\mu)]$ implies that $[V_2 : L(\mu)] = 0$, hence $V_1 \cap V_2 = 0$. Comparing dimensions we see that $V = V_1 \oplus V_2$. Now we apply induction to V_2 as described above.

b) Let now μ be maximal for $m = [V : L(\mu)] > 0$. Set $V_1 \subset V$ equal to the sum of all submodules of V isomorphic to $L(\mu)$. We claim now that $(V/V_1)_\mu = 0$. Suppose that this is not the case. Then there exists $v \in V_\mu$ with $v \notin V_1$. The maximality of μ implies that μ is also maximal for $V_\mu \neq 0$. It follows that $v \in V^{U^+}$. So 2.13.a yields a homomorphism $\varphi : V(\mu) \to V$ with $v \in \operatorname{im}(\varphi)$, hence with $\operatorname{im}(\varphi) \not\subset V_1$. It follows that $\operatorname{im}(\varphi)$ is not isomorphic to $L(\mu)$ and that $\varphi(\operatorname{rad}_G V(\mu)) \neq 0$.

Let ν be a maximal weight of $\varphi(\operatorname{rad}_G V(\mu))$. Then there exist (by 2.13.a) a non-zero homomorphism $V(\nu) \to \varphi(\operatorname{rad}_G V(\mu)) \subset V$. Both $\operatorname{Hom}_G(V(\nu), V)$ and $\operatorname{Hom}_G(V(\mu), V)$ are now non-zero. Therefore $\operatorname{Ext}^1_G(L(\mu), H^0(\nu)) = 0$ holds by our assumption. We have also $\operatorname{Hom}_G(V(\mu), H^0(\nu)) = 0$ because of $\nu < \mu$. Applying $\operatorname{Hom}_G(?, H^0(\nu))$ to the short exact sequence $0 \to \operatorname{rad}_G V(\mu) \to V(\mu) \to L(\mu) \to 0$, we get now that $\operatorname{Hom}_G(\operatorname{rad}_G V(\mu), H^0(\nu)) = 0$.

On the other hand, because ν is a maximal weight of $\varphi(\operatorname{rad}_G V(\mu))$ we have $\operatorname{Hom}_B(\varphi(\operatorname{rad}_G V(\mu)), \nu) \neq 0$, hence $\operatorname{Hom}_G(\varphi(\operatorname{rad}_G V(\mu)), H^0(\nu)) \neq 0$ by Frobenius reciprocity, hence $\operatorname{Hom}_G(\operatorname{rad}_G V(\mu), H^0(\nu)) \neq 0$ — a contradiction!

So we get $(V/V_1)_\mu = 0$ as claimed, hence $[V/V_1 : L(\mu)] = 0$ and $V_1 \simeq L(\mu)^m$. By (1) also ${}^\tau V$ satisfies the assumption in b). We find now V_2 with $V = V_1 \oplus V_2$ arguing as in a).

C.9. Let $w \in W_p^+$ and $s \in \Sigma$ with $w \bullet \lambda < ws \bullet \lambda$. There exists a unique root $\beta \in R^+$ and a unique integer $r > 0$ with $w \bullet \lambda = s_{\beta,rp} \bullet (ws \bullet \lambda)$. Let us assume that

(1) $\qquad p$ does not divide r.

Then the sum formula 8.19(1) implies

(2) $$\sum_{i>0} [V(ws \bullet \lambda)^i : L(w \bullet \lambda)] = 1.$$

Indeed: The sum formula contains the summand $\nu_p(rp)\chi(s_{\beta,rp} \bullet (ws \bullet \lambda))$; it contributes 1 to the sum in (2). The remaining summands are integer multiples of some $\chi(\lambda')$ with $\lambda' \in X(T)_+$ and $\lambda' \uparrow ws \bullet \lambda$, $\lambda' \neq ws \bullet \lambda$ and (by Remark 3 in 8.19) $\lambda' \neq w \bullet \lambda$. Since $w \bullet \lambda$ is maximal for $w \bullet \lambda \uparrow ws \bullet \lambda$ (by Corollary 6.10), we get $[V(\lambda') : L(w \bullet \lambda)] = 0$ for all the other summands, hence (2).

Now (2) implies that $L(w \bullet \lambda)$ is a composition factor with multiplicity 1 of $V(ws \bullet \lambda)^1 / V(ws \bullet \lambda)^2$, and not a composition factor of $V(ws \bullet \lambda)^2$.

Dualising 6.24(2) or 7.19.c and twisting by τ we get a non-zero homomorphism of G–modules $\varphi : V(w \bullet \lambda) \to V(ws \bullet \lambda)$. This homomorphism is unique up to a scalar. Its image $\varphi(V(w \bullet \lambda))$ is clearly contained in $\operatorname{rad}_G V(ws \bullet \lambda) = V(ws \bullet \lambda)^1$. Since $L(w \bullet \lambda)$ is a composition factor of $\varphi(V(w \bullet \lambda))$, but not of $V(ws \bullet \lambda)^2$, we see that $\varphi(V(w \bullet \lambda))$ cannot be contained in $V(ws \bullet \lambda)^2$. So φ induces a non-zero homomorphism $\overline{\varphi} : V(w \bullet \lambda) \to V(ws \bullet \lambda)^1 / V(ws \bullet \lambda)^2$.

Consider the condition (F, w, s) defined by

$$(F, w, s) \; :\Longleftrightarrow \; \varphi(V(w \bullet \lambda)^1) \subset V(ws \bullet \lambda)^2.$$

If (F, w, s) holds, then $V(w \bullet \lambda)^1 = \operatorname{rad}_G V(w \bullet \lambda)$ is contained in the kernel of $\overline{\varphi}$ and the image of $\overline{\varphi}$ is isomorphic to $L(w \bullet \lambda)$. This proves one direction of

(3) $$(F, w, s) \iff \operatorname{Hom}_G(L(w \bullet \lambda), V(ws \bullet \lambda)^1 / V(ws \bullet \lambda)^2) \neq 0.$$

Conversely, if this Hom space is non-zero, then $M = V(ws\bullet\lambda)^1/V(ws\bullet\lambda)^2$ contains a submodule N isomorphic to $L(w\bullet\lambda)$. Now $[M : L(w\bullet\lambda)] = 1 = [\mathrm{im}(\overline{\varphi}) : L(w\bullet\lambda)]$ implies $[M/\mathrm{im}(\overline{\varphi}) : L(w\bullet\lambda)] = 0]$, hence $N \subset \mathrm{im}(\overline{\varphi})$. The unique maximal submodule $\overline{\varphi}(\mathrm{rad}_G V(w\bullet\lambda))$ of $\mathrm{im}(\overline{\varphi})$ cannot contain a submodule isomorphic to $L(w\bullet\lambda)$. Therefore $N = \mathrm{im}(\overline{\varphi})$ and $V(w\bullet\lambda)^1 \subset \ker(\overline{\varphi})$, hence (F, w, s).

Dualising and twisting by τ, we get using 8.19(3) and 2.12(2)

$$\mathrm{Hom}_G(L(w\bullet\lambda), V(ws\bullet\lambda)^1/V(ws\bullet\lambda)^2) \simeq \mathrm{Hom}_G(V(ws\bullet\lambda)^1/V(ws\bullet\lambda)^2, L(w\bullet\lambda)).$$

Since $[V(ws\bullet\lambda)^2 : L(w\bullet\lambda)] = 0$, any homomorphism from $V(ws\bullet\lambda)^1$ to $L(w\bullet\lambda)$ maps $V(ws\bullet\lambda)^2$ to 0. This shows that

$$\mathrm{Hom}_G(V(ws\bullet\lambda)^1/V(ws\bullet\lambda)^2, L(w\bullet\lambda)) \simeq \mathrm{Hom}_G(\mathrm{rad}_G V(ws\bullet\lambda), L(w\bullet\lambda)).$$

Combining Proposition 2.14 with these three results, we get now (in the terminology of C.3):

$$(4) \qquad (F, w, s) \iff (E, w, s)$$

assuming (1). It follows that C.7(1) remains correct if we replace (E, w, s) by (F, w, s).

Consider a subset $\mathcal{W}$ of W_p^+ as in Proposition C.2. such that $\langle w(\lambda+\rho), \alpha^\vee\rangle < p^2$ for all $w \in \mathcal{W}$ and $\alpha \in R^+$. Then we can add to the equivalent conditions in C.2.a another one:

(iv) *For all $w \in \mathcal{W}$ and $s \in \Sigma$ with $w \uparrow ws$ and $ws \in \mathcal{W}$ any non-zero homomorphism $\varphi : V(w\bullet\lambda) \to V(ws\bullet\lambda)$ satisfies $\varphi(V(w\bullet\lambda)^1) \subset V(ws\bullet\lambda)^2$.*

This was first observed in [Cline, Parshall, and Scott 17]. Previously it had been shown in [Andersen 12] (inspired by work on highest weight modules over **C** in [Gabber and Joseph]) that this would follow from the stronger condition $(F, w, s)^+$ defined as

$$(F, w, s)^+ \;:\iff\; \varphi(V(w\bullet\lambda)^i) \subset V(ws\bullet\lambda)^{i+1} \text{ for all } i$$

for all w and s as above. This stronger condition would then even imply that all $V(w\bullet\lambda)^i/V(w\bullet\lambda)^{i+1}$ with $w \in \mathcal{W}$ are semisimple, see [Andersen 12], 6.12, and that the polynomials $\sum_{i\geq 0}[V(w\bullet\lambda)^i/V(w\bullet\lambda)^{i+1} : L(x\bullet\lambda)]v^i$ with $w, x \in \mathcal{W}$ are "inverse Kazhdan-Lusztig polynomials" (up to normalisation), see [Andersen 16], 2.4. For highest weight modules over **C** the analogue of $(F, w, s)^+$ was proved in [Beilinson and Bernstein] relating the filtrations here to geometrically constructed filtrations.

C.10. Let $\mathcal{W}$ be a subset of W_p^+ as in C.2.

Proposition: *If $\mathcal{W}$ satisfies the equivalent conditions* (i)–(iii) *in Proposition C.2, then any $\mathrm{Ext}^i_G(L(x\bullet\lambda), L(y\bullet\lambda))$ with $x, y \in \mathcal{W}$ and $i \in \mathbf{N}$ has dimension equal to*

$$(1) \qquad \sum_{w\in\mathcal{W}}\sum_{j=0}^{i} \dim \mathrm{Ext}^j_G(L(x\bullet\lambda), H^0(w\bullet\lambda)) \dim \mathrm{Ext}^{i-j}_G(L(y\bullet\lambda), H^0(w\bullet\lambda)).$$

Proof: Consider the full subcategory $\mathcal{C}$ of all G–modules M such that all composition factors of M have the form $L(w\bullet\lambda)$ with $w\in\mathcal{W}$. Then $\mathcal{C}$ is a block of a category of the form $\mathcal{C}(\pi)$ for some saturated $\pi\subset X(T)_+$ as in Chapter A. Therefore Proposition A.10 and 7.3(3) imply for all M, N in $\mathcal{C}$ and all i that

$$(2)\qquad \operatorname{Ext}^i_{\mathcal{C}}(M,N)\;\simeq\;\operatorname{Ext}^i_G(M,N).$$

We shall have to work in the bounded derived category $D^b(\mathcal{C})$. Each object M in $\mathcal{C}$ defines an object, again denoted by M, in $D^b(\mathcal{C})$; this object is represented by the complex with M in degree 0 and with 0 in all other degrees. One has for two G–modules M and N in $\mathcal{C}$ regarded as objects in $D^b(\mathcal{C})$

$$(3)\qquad \operatorname{Hom}^n_D(M,N)\;\simeq\;\operatorname{Ext}^n_{\mathcal{C}}(M,N)$$

where Hom_D denotes morphisms in $D^b(\mathcal{C})$ and $\operatorname{Hom}^n_D(X,Y)=\operatorname{Hom}^n(X[n],Y)$ for arbitrary X and Y in $D^b(\mathcal{C})$. Here $X[n]$ denotes "X shifted by n to the *right*"; so we have $H^i(X)\simeq H^{i-n}(X)$ for all i. (In many books one finds an opposite convention; if you follow that one, then replace below each $[n]$ by $[-n]$.)

Define now two subcategories $\mathcal{E}^L$ and $\mathcal{E}^R$ of $D^b(\mathcal{C})$. An object X in $D^b(\mathcal{C})$ belongs to $\mathcal{E}^L$ if and only if for all $x\in\mathcal{W}$ and all $n\in\mathbf{Z}$:

$$(\mathcal{E}^L)\qquad \operatorname{Hom}^n_D(X,H^0(x\bullet\lambda))\neq 0\implies n\equiv d(x)\pmod 2.$$

And X in $D^b(\mathcal{C})$ belongs to $\mathcal{E}^R$ if and only if for all $x\in\mathcal{W}$ and all $n\in\mathbf{Z}$:

$$(\mathcal{E}^R)\qquad \operatorname{Hom}^n_D(V(x\bullet\lambda),X)\neq 0\implies n\equiv d(x)\pmod 2.$$

For example, any $L(w\bullet\lambda)$ with $w\in\mathcal{W}$ and $d(w)\equiv 0\pmod 2$ belongs both to $\mathcal{E}^L$ and $\mathcal{E}^R$. As (C,w) holds by our assumptions, we get the condition for $\mathcal{E}^L$ from (2) and (3). For $\mathcal{E}^R$ one uses that

$$\operatorname{Ext}^n_G(V(x\bullet\lambda),L(w\bullet\lambda))\simeq\operatorname{Ext}^n_G({}^{\tau}L(w\bullet\lambda),{}^{\tau}V(x\bullet\lambda))\simeq\operatorname{Ext}^n_G(L(w\bullet\lambda),H^0(x\bullet\lambda)).$$

(Recall 4.13(3).) Similarly, if $w\in\mathcal{W}$ and $d(w)\equiv 1\pmod 2$, then $L(w\bullet\lambda)[1]$ belongs to $\mathcal{E}^L\cap\mathcal{E}^R$.

Thanks to these observations the dimension formula in (1) follows easily from the following more general result: Let X and Y be objects in $D^b(\mathcal{C})$ such that all $H^r(X)$ and $H^s(Y)$ are finite dimensional. If X belongs to $\mathcal{E}^L$ and if Y belongs to $\mathcal{E}^R$, then one has for all i

(4)
$$\dim\operatorname{Hom}^i_D(X,Y)=\sum_{w\in\mathcal{W}}\sum_{j\in\mathbf{Z}}\dim\operatorname{Hom}^j_D(X,H^0(w\bullet\lambda))\,\dim\operatorname{Hom}^{i-j}_D(V(w\bullet\lambda),Y).$$

Note: If (4) holds for X and Y, then $\operatorname{Hom}^i_D(X,Y)=0$ for all *odd* i because the j–th summand on the right hand side can be non-zero only if $j\equiv d(w)\equiv i-j\pmod 2$ for some w.

Note that (4) obviously holds if $X = V(x \bullet \lambda)[n]$ with $n \equiv d(x) \pmod 2$ since then $\dim \mathrm{Hom}_D^j(X, H^0(w \bullet \lambda)) = \dim \mathrm{Ext}_G^{j-n}(V(x \bullet \lambda), H^0(w \bullet \lambda)) = \delta_{jn}\delta_{xw}$ by 4.13.

Note also: If $X'' \to X \to X' \to$ is a distinguished triangle in $D^b(\mathcal{C})$ such that all terms belong to $\mathcal{E}^L$ and if (4) holds for X' and X'', then (4) also holds for X. Indeed, if we apply the functor $\mathrm{Hom}_D(?, Y)$ to the distinguished triangle, then we get a long exact sequence. Since (4) holds for X' and X'', we get for all odd i that $\mathrm{Hom}_D^i(X'', Y) = \mathrm{Hom}_D^i(X', Y) = 0$. It follows that $\dim \mathrm{Hom}_D^i(X, Y) = \dim \mathrm{Hom}_D^i(X'', Y) + \mathrm{Hom}_D^i(X', Y)$ for all i. One gets the same additivity for each $\dim \mathrm{Hom}_D^i(X, H^0(w \bullet \lambda))$, hence the claim.

Let now $x \in \mathcal{W}$ such that $w \bullet \lambda$ is maximal for $\sum_r [H^r(X) : L(x \bullet \lambda)] \neq 0$. We want to prove (4) using induction on $w \bullet \lambda$ and (for fixed w) on $\sum_r [H^r(X) : L(x \bullet \lambda)]$.

Set $\mathcal{W}'$ equal to the set of all $w \in \mathcal{W}$ with $w \bullet \lambda \not> x \bullet \lambda$. Let $\mathcal{C}'$ be the analogue to $\mathcal{C}$ constructed with $\mathcal{W}'$ instead of $\mathcal{W}$, i.e., $\mathcal{C}'$ is the full subcategory of all G–modules M such that all composition factors of M have the form $L(w \bullet \lambda)$ with $w \in \mathcal{W}'$. By our choice of x, all $H^r(X)$ belong to $\mathcal{C}'$. By construction $x \bullet \lambda$ is maximal among all $w \bullet \lambda$ with $w \in \mathcal{W}'$. This implies that $V(x \bullet \lambda)$ is projective in $\mathcal{C}'$ and $H^0(x \bullet \lambda)$ is injective in $\mathcal{C}'$, cf. A.6.

We have a natural inclusion of $\mathcal{C}'$ into $\mathcal{C}$; it induces a functor of derived categories $D^b(\mathcal{C}') \to D^b(\mathcal{C})$. This functor turns out to be a full embedding; an object Z in $D^b(\mathcal{C})$ is isomorphic to an object in the image of $D^b(\mathcal{C}')$ if and only if all $H^r(Z)$ belong to $\mathcal{C}'$, see [Cline, Parshall, and Scott 13], 3.7.

It follows that we may assume that X belongs to the image of $D^b(\mathcal{C}')$. Now the projectivity of $V(x \bullet \lambda)$ in $\mathcal{C}'$ implies for all n that we have natural isomorphism $\mathrm{Hom}_D(V(x \bullet \lambda)[n], X) \simeq \mathrm{Hom}_G(V(x \bullet \lambda), H^n(X))$, whereas the injectivity of $H^0(x \bullet \lambda)$ yields isomorphisms $\mathrm{Hom}_D(X, H^0(x \bullet \lambda)[n]) \simeq \mathrm{Hom}_G(X, H^0(x \bullet \lambda))$.

By our choice of X there exists some m such that $[H^m(X) : L(x \bullet \lambda)] \neq 0$. The maximality of $x \bullet \lambda$ implies that

$$\mathrm{Hom}_G(V(x \bullet \lambda), H^m(X)) \neq 0 \neq \mathrm{Hom}_G(H^m(X), H^0(x \bullet \lambda)).$$

The second inequality implies in particular that $\mathrm{Hom}_D^m(X, H^0(x \bullet \lambda) \neq 0$, hence $m \equiv d(x) \pmod 2$. This implies that $V(x \bullet \lambda)[m]$ belongs to $\mathcal{E}^L$.

Choose a non-zero homomorphism $f : V(x \bullet \lambda) \to H^m(X)$. Denote by $\widehat{f} : V(x \bullet \lambda)[m] \to X$ the corresponding morphism in $D^b(\mathcal{C}')$. Set X' equal to the mapping cone of $\widehat{f}$. We have then in $D^b(\mathcal{C}')$ a distinguished triangle

$$V(x \bullet \lambda)[m] \longrightarrow X \longrightarrow X' \longrightarrow$$

that we regard as a distinguished triangle in $D^b(\mathcal{C})$. We now apply any functor $\mathrm{Hom}_D(?, H^0(w \bullet \lambda))$ to this distinguished triangle and get a long exact sequence that shows immediately that X' belongs to $\mathcal{E}^L$. (Use that X belongs to $\mathcal{E}^L$ and that $\dim \mathrm{Hom}_D^i(V(x \bullet \lambda)[m], H^0(w \bullet \lambda)) = \delta_{xw}\delta_{mi}$.)

One has by construction $H^r(X') = H^r(X)$ for all $r \neq m, m-1$; furthermore $H^m(X')$ is isomorphic to the cokernel of f and there is an exact sequence

$$0 \to H^{m-1}(X) \longrightarrow H^{m-1}(X') \longrightarrow \ker(f) \to 0.$$

Comparing $H^\bullet(X')$ to $H^\bullet(X)$, we see that we may pick up some extra composition factors in $H^{m-1}(X')$, but that they all have a highest weight $< x \bullet \lambda$. Furthermore, the multiplicity of $L(x \bullet \lambda)$ in degree m goes down. It follows that we can apply induction and may assume that X' satisfies (4). Then it follows also for X.

C.11. We are now going to look at the G_1T–version of the results in the earlier subsections. We assume from now on that $p > h$.

Working with G_1T–modules we have to replace the Kazhdan-Lusztig polynomials as in C.1 by the "generic Kazhdan-Lusztig polynomials" $q_{y,x}$ normalised as in [Soergel 4], 6.1. (Actually, Soergel uses alcoves as indices to these polynomials. We use the bijection $w \mapsto w \bullet C$ between W_p and the set of alcoves to get elements in W_p as indices.)

For each $\nu \in X(T)$ the translation by $p\nu$ permutes the reflection hyperplanes for W_p (given by equations of the form $\langle \mu + \rho, \alpha^\vee \rangle = mp$ with $\alpha \in R$ and $m \in \mathbf{Z}$), hence the alcoves for W_p (the connected components of the complement of the union of those hyperplanes). So there is for each $w \in W_p$ a unique ${}^\nu w \in W_p$ with ${}^\nu w \bullet C = p\nu + w \bullet C$. If $\nu \in \mathbf{Z}R$, then we have ${}^\nu w = t_{p\nu} \circ w$ where $t_{p\nu} \in W_p$ is the translation by $p\nu$.

The polynomials $q_{y,x}$ are related to the polynomials $m_{y,x}$ by the property that

$$q_{y,x} = m_{{}^\mu y, {}^\mu x} \qquad \text{for all sufficiently dominant } \mu \tag{1}$$

cf. [Soergel 4], Thm. 6.1. Here "sufficiently dominant" means that there exists $r \in \mathbf{Z}$, depending on x and y, such that (1) holds for all $\mu \in r\rho + X(T)_+$. For any $\nu \in X(T)$ and any $r_1, r_2 \in \mathbf{Z}$ the intersection of $\nu + r_1\rho + X(T)_+$ and $r_2\rho + X(T)_+$ is not empty. This implies

$$q_{y,x} = q_{{}^\nu y, {}^\nu x} \qquad \text{for all } x, y \in W_p \text{ and } \nu \in X(T). \tag{2}$$

One checks easily that $d(p\nu + C') = d(C') + \sum_{\alpha \in R^+} \langle \nu, \alpha^\vee \rangle$ for all $\nu \in X(T)$ and all alcoves C'. This implies that $d({}^\nu x) = d(x) + \sum_{\alpha \in R^+} \langle \nu, \alpha^\vee \rangle$, hence $d({}^\nu x) - d({}^\nu y) = d(x) - d(y)$ for all $x, y \in W_p$. Now (1) shows that the polynomials $q_{y,x}$ inherit the parity properties of the $m_{y,x}$ mentioned in C.1.

In analogy with C.1(1) we set for all $w, y \in W_p$ and $s \in \Sigma$ with $w \uparrow ws$

$$q^s_{y,w} = \begin{cases} q_{ys,w} + v^{-1}\, q_{y,w}, & \text{if } ys \uparrow y, \\ q_{ys,w} + v\, q_{y,w}, & \text{if } y \uparrow ys. \end{cases} \tag{3}$$

One has then as in C.1(2)

$$q^s_{y,w} = q_{y,ws} + \sum_{z \uparrow w,\, z \neq w} q^s_{z,w}(0)\, q_{y,z}. \tag{4}$$

In fact, this follows from C.1(2) using (1).

C.12. We introduce a G_1T–analogue to the notation $P[V, y]$ from C.3. Set for each finite dimensional G_1T–module V with $\mathrm{pr}_\lambda(V) = V$ and for each $y \in W_p$

$$\widehat{P}\,[V, y] = \sum_{i \geq 0} \dim \mathrm{Ext}^i_{G_1T}(V, \widehat{Z}'_1(y \bullet \lambda))\, v^i \in \mathbf{Z}[v]. \tag{1}$$

We have then by 9.9(3)

$$\mathrm{ch}(V) = \sum_{y \in W_p} \widehat{P}\,[V, y](-1)\, \mathrm{ch}\, \widehat{Z}_1(y \bullet \lambda). \tag{2}$$

Set in particular

$$\widehat{P}\,[x, y] = \widehat{P}\,[\widehat{L}_1(x \bullet \lambda), y] \qquad \text{for all } x \in W_p. \tag{3}$$

By the translation principle, $\widehat{P}\,[x, y]$ is independent of the choice of λ.

Lemma: a) *We have* $\widehat{P}\,[x,y] = \widehat{P}\,[{}^\nu x, {}^\nu y]$ *for all* $x, y \in W_p$ *and* $\nu \in X(T)$.

b) *Let* $w \in W_p$ *and* $s \in \Sigma$ *with* $w \uparrow ws$. *Then we have*

$$\widehat{P}\,[V,w] = v\,\widehat{P}\,[V,ws] \tag{4}$$

for all finite dimensional G_1T*–modules* V *with* $T_\lambda^{\mu(s)} V = 0$.

c) *Let* $\nu \in \mathbf{Z}R$. *Then*

$$\widehat{P}\,[1, t_{p\nu}] = \sum_{i \geq 0} \dim(S^i(\mathfrak{u}^*)_{-\nu})\, v^{2i} \qquad \textit{and} \qquad \widehat{P}\,[1, t_{p\nu} w] = v^{l(w)}\,\widehat{P}\,[1, t_{p\nu}]$$

for all $w \in W$.

Proof: a) The set $p\nu + W_p \bullet \lambda$ is again a W_p–orbit in $X(T)$, consisting of p–regular elements. So there is some $\lambda_\nu \in C \cap X(T)$ with $p\nu + W_p \bullet \lambda = W_p \bullet \lambda_\nu$. It then follows that ${}^\nu w \bullet \lambda_\nu = p\nu + w \bullet \lambda$. Since obviously

$$\begin{aligned} \operatorname{Ext}^i_{G_1T}(\widehat{L}_1(x \bullet \lambda), \widehat{Z}'_1(y \bullet \lambda)) &\simeq \operatorname{Ext}^i_{G_1T}(\widehat{L}_1(x \bullet \lambda + p\nu), \widehat{Z}'_1(y \bullet \lambda + p\nu)) \\ &= \operatorname{Ext}^i_{G_1T}(\widehat{L}_1({}^\nu x \bullet \lambda_\nu), \widehat{Z}'_1({}^\nu y \bullet \lambda_\nu)) \end{aligned}$$

for all i, the claim follows because our polynomials are independent of λ.

b) We have by 9.22(3) an exact sequence

$$0 \to \widehat{Z}'_1(w \bullet \lambda) \longrightarrow T^\lambda_{\mu(s)} \widehat{Z}'_1(w \bullet \mu(s)) \longrightarrow \widehat{Z}'_1(ws \bullet \lambda) \to 0. \tag{5}$$

Now $T_\lambda^{\mu(s)} V = 0$ implies by adjunction

$$\operatorname{Ext}^i_{G_rT}(V, T^\lambda_{\mu(s)} \widehat{Z}'_1(w \bullet \mu(s))) \simeq \operatorname{Ext}^i_{G_rT}(T_\lambda^{\mu(s)} V, \widehat{Z}'_1(w \bullet \mu(s))) = 0$$

for all i. Therefore (5) induces isomorphisms

$$\operatorname{Ext}^i_{G_rT}(V, \widehat{Z}'_1(ws \bullet \lambda)) \xrightarrow{\sim} \operatorname{Ext}^{i+1}_{G_rT}(V, \widehat{Z}'_1(w \bullet \lambda)).$$

The claim follows.

c) We can apply b) to $V = \widehat{L}_1(\lambda)$ with $s = s_\alpha$ for all $\alpha \in S$. Given $w \in W$, we have $(t_{p\nu} w) \uparrow (t_{p\nu} w) s_\alpha$ if and only if $w \uparrow w s_\alpha$. Now the second claim follows by induction on $l(w)$.

We may suppose that $\lambda = 0$, hence $\widehat{L}_1(\lambda) = k$. Therefore the coefficient of v^i in $\widehat{P}\,[1, t_{p\nu}]$ is equal to the dimension of $H^i(G_1T, \widehat{Z}'_1(p\nu)) \simeq H^i(G_1, k)_{-p\nu}$, cf. I.6.9(5). So the first claim follows from 12.13. (Recall that we assume now $p > h$.)

C.13. We introduce now and in the next subsection some conditions analogous to those in C.3. Given $x \in W_p$, we set

$$(\widehat{A}, x) \ :\Longleftrightarrow \ \operatorname{ch} \widehat{L}_1(x \bullet \lambda) = \sum_{y \in W_p} q_{y,x}(-1) \operatorname{ch} \widehat{Z}_1'(y \bullet \lambda)$$

and

$$(\widehat{B}, x) \ :\Longleftrightarrow \ \widehat{P}[x, y] = q_{y,x} \text{ for all } y \in W_p$$

and

$$(\widehat{C}, x) :\Longleftrightarrow \forall i, y\colon \operatorname{Ext}^i_{G_1T}(\widehat{L}_1(x \bullet \lambda), \widehat{Z}_1'(y \bullet \lambda)) \neq 0 \Rightarrow d(x) - d(y) \equiv i \bmod 2.$$

We have now:

(1) *Condition* $(\widehat{B}, 1)$ *holds.*

This follows from Lemma C.12.c and Kato's description of the $q_{y,x}$, cf. [Soergel 4], Thm. 6.3 combined with the beginning of the proof of Prop. 4.16 there.

Observe next for any $x \in W_p$:

$$(2) \qquad (\widehat{B}, x) \implies (\widehat{A}, x) \text{ and } (\widehat{C}, x).$$

Here $(\widehat{A}, x)$ follows immediately from C.12(2). For $(\widehat{C}, x)$ one uses that the $q_{y,x}$ inherit the parity properties of the $m_{y,x}$.

A comparison of C.11(2) with Lemma C.12.a implies:

(3) *If* $(\widehat{B}, x)$ *holds for some* $x \in W_p$, *then* $(\widehat{B}, {}^\nu x)$ *holds for all* $\nu \in X(T)$.

Similarly, Lemma C.12.a and the formula $d({}^\nu x) - d({}^\nu y) = d(x) - d(y)$ from C.11 yield:

(4) *If* $(\widehat{C}, x)$ *holds for some* $x \in W_p$, *then* $(\widehat{C}, {}^\nu x)$ *holds for all* $\nu \in X(T)$.

Exercise: Let $w, x \in W_p$ with $x \uparrow w$. Show: *If* $(\widehat{C}, w)$ *holds and* $d(w) - d(x)$ *is even, then* $\operatorname{Ext}^1_{G_1T}(\widehat{L}_1(w \bullet \lambda), \widehat{L}_1(x \bullet \lambda)) = 0$. (Hints: Lemma C.3, Lemma 9.9, and 9.19(4).)

C.14. Let $w \in W_p$ and $s \in \Sigma$ with $w \uparrow ws$. The same arguments as in 7.20 show

$$\begin{aligned}\widehat{L}_1(w \bullet \mu(s)) &\simeq \operatorname{soc}_{G_rT} T^\lambda_{\mu(s)} \widehat{L}_1(w \bullet \mu(s)) \\ &\simeq T^\lambda_{\mu(s)} \widehat{L}_1(w \bullet \mu(s)) / \operatorname{rad}_{G_1T} T^\lambda_{\mu(s)} \widehat{L}_1(w \bullet \mu(s)).\end{aligned}$$

Set again

$$\beta_s \widehat{L}_1(w \bullet \lambda) = \operatorname{rad}_{G_1T} T^\lambda_{\mu(s)} \widehat{L}_1(w \bullet \mu(s)) / \operatorname{soc}_{G_rT} T^\lambda_{\mu(s)} \widehat{L}_1(w \bullet \mu(s)).$$

Consider in analogy to C.3 the conditions

$$(\widehat{D}, w, s) \ :\Longleftrightarrow \ \text{The } G_1T\text{–module } \beta_s \widehat{L}_1(w \bullet \lambda) \text{ is semi-simple}$$

and

$$(\widehat{E}, w, s) \ :\Longleftrightarrow \ \mathrm{Ext}^1_{G_1T}(\widehat{L}_1(w \bullet \lambda), \widehat{L}_1(ws \bullet \lambda)) \neq 0.$$

We have as in C.4 that $(\widehat{E}, w, s)$ holds if and only if $\widehat{L}_1(ws \bullet \lambda)$ is a direct summand of $\beta_s \widehat{L}_1(w \bullet \lambda)$ and get that

(1) $$(\widehat{D}, w, s) \implies (\widehat{E}, w, s)$$

by an analogue to Lemma 7.20.b.

There is by 9.22(5) a non-zero homomorphism $\varphi : \widehat{Z}_1(w \bullet \lambda) \to \widehat{Z}_1(ws \bullet \lambda)$, unique up to multiplication by a scalar. Now consider analogously to C.9 the condition

$$(\widehat{F}, w, s) \ :\Longleftrightarrow \ \varphi(\widehat{Z}_1(w \bullet \lambda)^1) \subset \widehat{Z}_1(ws \bullet \lambda)^2$$

involving the filtration from 9.11(5).

Since $w \bullet \lambda$ is maximal for $w \bullet \lambda \uparrow ws \bullet \lambda$ and $w \bullet \lambda \neq ws \bullet \lambda$, the sum formula 9.11(6) implies that $\widehat{L}_1(w \bullet \lambda)$ is a composition factor (with multiplicity 1) of $\widehat{Z}_1(ws \bullet \lambda)^1$, but not of $\widehat{Z}_1(ws \bullet \lambda)^2$. Therefore the same argument as in C.9 shows that

(2) $$(\widehat{E}, w, s) \iff (\widehat{F}, w, s).$$

The arguments used to prove Lemma C.5 yield now for all $x \in W_p$

(3) $$(\widehat{C}, w) \Rightarrow \widehat{P}[\beta_s \widehat{L}_1(w \bullet \lambda), x] = \begin{cases} \widehat{P}[w, xs] + v^{-1}\widehat{P}[w, x], & \text{if } xs \uparrow x, \\ \widehat{P}[w, xs] + v\widehat{P}[w, x], & \text{if } x \uparrow xs. \end{cases}$$

Here the second line follows from the first one: Use $T_\lambda^{\mu(s)} \beta_s \widehat{L}_1(w \bullet \lambda) = 0$ and C.11(4). One has as in C.6:

(4) *If $(\widehat{D}, w, s)$ holds, then*

$$\widehat{P}\,[\beta_s \widehat{L}_1(w \bullet \lambda), y] = \widehat{P}\,[ws, y] + \sum_{z \in W_p,\, z \uparrow w} \widehat{P}\,[\beta_s \widehat{L}_1(w \bullet \lambda), z](0)\, \widehat{P}\,[z, y].$$

One gets using (3)

(5) $$(\widehat{C}, w) \text{ and } (\widehat{E}, w, s) \Rightarrow (\widehat{C}, ws)$$

and

(6) $$(\widehat{B}, w) \implies \widehat{P}\,[\beta_s \widehat{L}_1(w \bullet \lambda), y] = q^s_{y,w} \text{ for all } y \in W_p.$$

C.15. Set

$$\mathcal{W}_1 = \{\, w \in W_p \mid w \bullet \lambda \in X_1(T) \,\}.$$

This is a finite subset of W_p contained in W_p^+. Since $\mathcal{D}G$ is simply connected, there is for each $\mu \in X(T)$ some $\mu_1 \in X_1(T)$ such that $\mu - \mu_1 \in pX(T)$. This implies for each $w \in W_p$ that there exist $w_1 \in \mathcal{W}_1$ and $\nu \in X(T)$ with $w \bullet C = w_1 \bullet C + p\nu$, i.e., with $w = {}^{\nu}w_1$.

Lemma: *If $(\widehat{E}, w, s)$ holds for all $w \in \mathcal{W}_1$ and $s \in \Sigma$ with $w \uparrow ws$ and $ws \in \mathcal{W}_1$, then $(\widehat{D}, w, s)$ holds for all these w and s and $(\widehat{C}, x)$ holds for all $x \in W_p$.*

Proof: We look first at $(\widehat{C}, x)$. Since x can be written $x = {}^{\nu}x_1$ with $\nu \in X(T)$ and $x_1 \in \mathcal{W}_1$, we get from C.13(4) that it suffices to prove $(\widehat{C}, x)$ for $x \in \mathcal{W}_1$. Here we now use induction on $d(x)$. If $x = 1$, then we are done by C.12(1). If $x \neq 1$, then there exists a wall H of $x \bullet C$ separating $x \bullet C$ from C. As $x \bullet C$ is dominant, we have then $s_H x \bullet C \uparrow x \bullet C$ and $s_H x \in W_p^+$, in fact $s_H x \in \mathcal{W}_1$. There is $s \in \Sigma$ with $s_H x = xs$; it follows that $xs \uparrow x$ and $xs \in \mathcal{W}_1$. Then $d(xs) < d(x)$, so $(\widehat{C}, xs)$ holds by induction. Now apply C.14(5) to $w = xs$.

The claim on $(\widehat{D}, w, s)$ follows by an argument as for C.7(1). This requires a G_1T–version of Lemma C.8 where one replaces all $H^0(\mu)$ by $\widehat{Z}_1'(\mu)$, and all $V(\mu)$ by $\widehat{Z}_1(\mu)$. That is easily done.

C.16. Lemma: *Let $w \in \mathcal{W}_1$ and $s \in \Sigma$ with $w \uparrow ws$ and $ws \in \mathcal{W}_1$. If $(\widehat{D}, w, s)$ holds and if $(\widehat{B}, x)$ holds for all $x \in \mathcal{W}_1$ with $d(x) \leq d(w)$, then $(\widehat{B}, ws)$ holds.*

Proof: We know by assumption that in particular $(\widehat{B}, w)$ holds and get for all $y \in W_p$ that $\widehat{P}\,[\beta_s \widehat{L}_1(w \bullet \lambda), y] = q^s_{y,w}$, see C.14(6). Since $(\widehat{D}, w, s)$ holds, we get from C.14(4) that

$$q^s_{y,w} = \widehat{P}\,[ws, y] + \sum_{z \uparrow w,\, z \neq w} q^s_{z,w}(0)\, \widehat{P}\,[z, y]$$

for all y. A comparison with C.11(4) shows: If we know that $(\widehat{B}, z)$ holds for all composition factors $\widehat{L}_1(z \bullet \lambda)$ of $\beta_s \widehat{L}_1(w \bullet \lambda)$ with $z \neq ws$, then $(\widehat{B}, ws)$ follows.

Now $w \in \mathcal{W}_1$ implies that $\widehat{L}_1(w \bullet \lambda)$ is just $L(w \bullet \lambda)$ considered as a G_1T–module. A comparison of the definitions shows that $\beta_s \widehat{L}_1(w \bullet \lambda)$ is $\beta_s L(w \bullet \lambda)$ considered as a G_1T–module. Now the G–composition factors of $\beta_s L(w \bullet \lambda)$ different from $L(ws \bullet \lambda) \simeq \widehat{L}_1(ws \bullet \lambda)$ have the form $L(v \bullet \lambda)$ with $v \in W_p^+$ and $v \uparrow w$, $v \neq w$, hence $d(v) < d(w)$. Fix such v and write $v = {}^{\nu}v_1$ with $\nu \in X(T)_+$ and $v_1 \in \mathcal{W}_1$. We have then $d(v_1) \leq d(v) < d(v)$, so $(\widehat{B}, v_1)$ holds by assumption.

Using the notation λ_ν as in the proof of Lemma C.12.a, we have $v \bullet \lambda = p\nu + v_1 \bullet \lambda_{-\nu}$, hence $L(v \bullet \lambda) \simeq L(v_1 \bullet \lambda_{-\nu}) \otimes L(\nu)^{[1]}$. Therefore $L(v \bullet \lambda)$ as a G_1T–module is the direct sum over all $\widehat{L}_1(p\mu + v_1 \bullet \lambda_{-\nu}) = \widehat{L}_1({}^{\mu}v_1 \bullet \lambda)$ with μ a weight of $L(\nu)$ (taken with its multiplicity). Since $(\widehat{B}, v_1)$ holds, so do all $(\widehat{B}, {}^{\mu}v_1)$ by C.13(3). Now the claim follows.

C.17. Proposition: a) *Let $\lambda \in C \cap X(T)$. The following are equivalent:*

(i) *We have* $\operatorname{ch} \widehat{L}_1(x \bullet \lambda) = \sum_{y \in W_p} q_{y,x}(-1) \operatorname{ch} \widehat{Z}_1'(y \bullet \lambda)$ *for all* $x \in W_p$.

(ii) *The G_1T–module $\beta_s \widehat{L}_1(w \bullet \lambda)$ is semi-simple for all $w \in \mathcal{W}_1$ and $s \in \Sigma$ with $w \uparrow ws$ and $ws \in \mathcal{W}_1$.*

(iii) *We have* $\operatorname{Ext}^1_{G_1T}(\widehat{L}_1(w \bullet \lambda), \widehat{L}_1(ws \bullet \lambda)) \neq 0$ *for all $w \in \mathcal{W}_1$ and $s \in \Sigma$ with $ws \in \mathcal{W}_1$.*

b) *If these equivalent conditions hold, then*

$$q_{y,x} = \sum_{i \geq 0} \dim \operatorname{Ext}^i_{G_1T}(\widehat{L}_1(x \bullet \lambda), \widehat{Z}'_1(y \bullet \lambda))\, v^i$$

for all $x, y \in W_p$.

Proof: The equivalence of (ii) and (iii) follows from C.14(1) and Lemma C.15.

Assume that (ii) holds. Then we get the claim in b) [and hence (i), cf. C.13(2)] as follows: By C.13(3) it suffices to look at $x \in \mathcal{W}_1$. We use induction on $d(x)$ for $x \in \mathcal{W}_1$. If $x = 1$, then we can refer to C.13(1). If $x \neq 1$ we choose $s \in \Sigma$ with $xs \uparrow x$ and $xs \in \mathcal{W}_1$. Now we apply induction and Lemma C.16.

For the implication "(i) $\Rightarrow$ (ii)" one argues as in C.7. Details are left to the reader. (One needs a G_1T–analogue to Lemma C.8.a.)

Remarks: 1) Let us show that (ii) implies $(\widehat{D}, w, s)$ for all $w \in W_p$ and $s \in \Sigma$ with $w \uparrow ws$. (In some cases this follows from Lemma C.15. However, that lemma would not yield the result when $w \in \mathcal{W}_1$ and $ws \notin \mathcal{W}_1$.) To start with, we have

$$\begin{aligned}
\operatorname{ch} \beta_s \widehat{L}_1(w \bullet \lambda) &= \sum_{y \in W_p} q^s_{y,w}(-1) \operatorname{ch} \widehat{Z}'_1(y \bullet \lambda) \\
&= \sum_{z \in W_p} q^s_{z,w}(0) \sum_{y \in W_p} q_{y,z}(-1) \operatorname{ch} \widehat{Z}'_1(y \bullet \lambda) \\
&= \sum_{z \in W_p} q^s_{z,w}(0) \operatorname{ch} \widehat{L}_1(z \bullet \lambda)
\end{aligned}$$

where we use first $(\widehat{A}, w)$ and a simple calculation, then C.11(4), and finally $(\widehat{A}, z)$.

The parity properties of the $q_{y,x}$ imply: If $q^s_{z,w}(0) \neq 0$, then $d(z) \equiv d(ws) \pmod 2$. So the calculation above shows: If $\widehat{L}_1(z_1 \bullet \lambda)$ and $\widehat{L}_1(z_2 \bullet \lambda)$ are composition factors of $\beta_s \widehat{L}_1(w \bullet \lambda)$, then $d(z_1) \equiv d(z_2) \pmod 2$. Now the exercise in C.13 yields that $\operatorname{Ext}^1_{G_1T}(\widehat{L}_1(z_1 \bullet \lambda), \widehat{L}_1(z_2 \bullet \lambda)) = 0$. This implies the semi-simplicity of $\beta_s \widehat{L}_1(w \bullet \lambda)$.

2) If the root system R has no component of type E_8, F_4, or G_2, then the assumption in the proposition can be replaced by "If $(\widehat{E}, w, s_\alpha)$ holds for all $w \in W$ and $\alpha \in S$ with $w \uparrow ws_\alpha$", see [Cline, Parshall, and Scott 16], 5.10(b′).

C.18. We conclude this chapter by comparing the conditions in the G– and G_1T–cases.

Let $w \in \mathcal{W}_1$ and $s \in \Sigma$ with $w \uparrow ws$ and $ws \in \mathcal{W}_1$. Then $\beta_s \widehat{L}_1(w \bullet \lambda)$ is $\beta_s L(w \bullet \lambda)$ considered as a G_1T–module, as in the proof of Lemma C.16. It is therefore clear that (D, w, s) implies $(\widehat{D}, w, s)$, and that (E, w, s) implies $(\widehat{E}, w, s)$. So Proposition C.2 implies:

If (A, x) holds for all $x \in W_p^+$ such that there exists $y \in \mathcal{W}_1$ with $x \uparrow y$, then the equivalent conditions in Proposition C.17 are satisfied. In particular, $(\widehat{A}, x)$, $(\widehat{B}, x)$, and $(\widehat{C}, x)$ hold then for all $x \in W_p$.

Note that, conversely, $(\widehat{E}, w, s)$ implies (E, w, s) (for $w \in \mathcal{W}_1$ and $s \in \Sigma$ with $w \uparrow ws$ and $ws \in \mathcal{W}_1$). Indeed, $(\widehat{E}, w, s)$ implies that $\mathrm{Hom}_{G_1}(L(ws\bullet\lambda), \beta_s L(w\bullet\lambda)) \neq 0$. This Hom-space is a G–module. If ν is a dominant weight of this module, then we have $\mathrm{Hom}_{G_1T}(\widehat{L}_1(ws\bullet\lambda+p\nu), \beta_s L(w\bullet\lambda)) \neq 0$ and $ws\bullet\lambda+p\nu$ is (by the simplicity of $\widehat{L}_1(ws\bullet\lambda+p\nu)$ a weight of $\beta_s L(w\bullet\lambda)$. On the other hand, all weights of $\beta_s L(w\bullet\lambda)$ are $\leq ws \bullet \lambda$. Now $ws \bullet \lambda + p\nu \leq ws \bullet \lambda$ and $\nu \in X(T)_+$ imply $\nu = 0$. Therefore 0 is the only weight of the G–module $\mathrm{Hom}_{G_1}(L(ws \bullet \lambda), \beta_s L(w \bullet \lambda))$. Therefore this G–module is trivial, hence equal to $\mathrm{Hom}_G(L(ws \bullet \lambda), \beta_s L(w \bullet \lambda))$. And since the module is non-zero, we get (E, w, s).

CHAPTER D

Radical Filtrations and Kazhdan-Lusztig Polynomials

We assume in this chapter that p is a prime number with $p > h$ and that k is a perfect field of characteristic p.

In this chapter we look at some consequences of Lusztig's conjecture concerning the structure of certain "standard" modules. More precisely, we shall look at the radical filtrations of G_1T–modules of the form $\widehat{Z}_1(\mu)$ and $\widehat{Q}_1(\mu)$ for p–regular μ. There are also similar results of a more limited scope on Weyl modules that we do not get into here, see [Andersen 16].

Any module M over any algebraic group over k has an ascending socle series (see I.2.14) and a descending radical series. If M is finite dimensional, then both series contain the same number of terms (not counting repetitions of M or 0); this number is called the Loewy length of M. (This is discussed in more detail in D.1.)

Lusztig's conjecture gets involved with Loewy length because the condition $(\widehat{D}, w, s)$ in C.14 can be restated as saying that a certain module, $\Theta_s \widehat{L}_1(w \bullet \lambda)$, has Loewy length 3. The first result of this chapter shows then: If $(\widehat{D}, w, s)$ holds for all possible w, then one gets an upper bound for the Loewy length of all modules of the form $\Theta_s M$, see D.2.

Using this fact one gets quickly (see D.4) that all $\widehat{Z}_1(\mu)$ with p–regular μ have the same Loewy length; similarly for the corresponding $\widehat{Q}_1(\mu)$. A little more effort shows then (in D.8) that the first length is $|R^+|+1$, the second one $2|R^+|+1$. The proof uses the functors Θ_s, homomorphisms between modules of the form $\widehat{Z}_1^w(\mu\langle w\rangle)$ and the filtrations of the $\widehat{Q}_1(\mu)$ with factors of the form $\widehat{Z}_1(\mu')$.

The same tools are also involved in the following subsections where we prove some inductive formulae that allow (see D.13) the calculation of the multiplicity of each simple module in each factor $\operatorname{rad}^i \widehat{Z}_1(\mu)/\operatorname{rad}^{i+1} \widehat{Z}_1(\mu)$ of the radical filtration. These multiplicities turn out to be coefficients of Kazhdan-Lusztig polynomials. On the way we see (in D.12) that the $\widehat{Z}_1(\mu)$ are rigid; this means that their radical and socle filtrations coincide up to numbering (cf. D.9).

The chapter concludes with similar results for the radical series of the $\widehat{Q}_1(\mu)$; here I refer for most of the proof to the literature.

We assume for the sake of simplicity that $\mathcal{D}G$ is simply connected. Final results usually generalise to arbitrary G by going to a suitable covering.

The main results in this chapter are due to [Andersen and Kaneda 1] and [Irving 2].

D.1. Let H be an algebraic group scheme over a field k. Recall the definition of the Loewy (or socle) series of all $\operatorname{soc}_i M$ with $i \geq 0$ of an H–module M in I.2.14.

(We set $\operatorname{soc}_0 M = 0$.) We have also the radical series of all $\operatorname{rad}^i M = \operatorname{rad}^i_H M$ with $i \geq 0$ defined by $\operatorname{rad}^0 M = M$ and $\operatorname{rad}^{i+1} M = \operatorname{rad}_H(\operatorname{rad}^i M)$ for all $i \geq 0$. If $\varphi : M \to M'$ is a homomorphism of H–modules, then

(1) $$\varphi(\operatorname{soc}_i M) \subset \operatorname{soc}_i M' \qquad \text{and} \qquad \varphi(\operatorname{rad}^i M) \subset \operatorname{rad}^i M'$$

for all i. We have

(2) $$\varphi \text{ injective} \Rightarrow \operatorname{soc}_i M = \varphi^{-1}(\operatorname{soc}_i M')$$

and

(3) $$\varphi \text{ surjective} \Rightarrow \operatorname{rad}^i M' = \varphi(\operatorname{rad}^i M)$$

for all i. This says in particular for a submodule N of M that

(4) $$\operatorname{soc}_i N = N \cap \operatorname{soc}_i M \qquad \text{and} \qquad \operatorname{rad}^i(M/N) = (\operatorname{rad}^i M + N)/N$$

for all i.

If M is finite dimensional, then the *Loewy length* of M is defined as the smallest integer $r \geq 0$ with $\operatorname{soc}_r M = M$; we denote the Loewy length of M by $\ell\ell(M)$. Since $M/\operatorname{soc}_{\ell\ell(M)-1} M$ is semi-simple, we get $\operatorname{rad} M \subset \operatorname{soc}_{\ell\ell(M)-1} M$. Now $\operatorname{soc}_{\ell\ell(M)-1} M$ has Loewy length $\ell\ell(M) - 1$. Therefore we get using induction

(5) $$\operatorname{rad}^i M \subset \operatorname{soc}_{\ell\ell(M)-i} M \qquad \text{for all } i,\ 0 \leq i \leq \ell\ell(M).$$

This shows in particular that $\operatorname{rad}^{\ell\ell(M)} M = 0$. Using an inductive argument in the opposite direction one shows now that the Loewy length of M is also the smallest integer $r \geq 0$ with $\operatorname{rad}^r M = 0$. Note that $\ell\ell(M) = 0$ if and only if $M = 0$, and that $\ell\ell(M) = 1$ if and only if M is semi-simple and non-zero.

Continue to assume $\dim(M) < \infty$. Set $V^\perp = \{f \in V^* \mid f(V) = 0\}$ for any subspace V of M. The restriction of functions induces an isomorphism $M^*/V^\perp \xrightarrow{\sim} V^*$ and we have $V \xrightarrow{\sim} (V^\perp)^\perp$ under the natural isomorphism $M \xrightarrow{\sim} (M^*)^*$. For subspaces $V_1 \subset V_2$ of M the restriction induces an isomorphism $V_1^\perp/V_2^\perp \xrightarrow{\sim} (V_2/V_1)^*$. This is an isomorphism of H–modules if V_1 and V_2 are H–submodules of M.

We get in particular for all $i > 0$ that

$$(\operatorname{soc}_{i-1} M)^\perp/(\operatorname{soc}_i M)^\perp \simeq (\operatorname{soc}_i M/\operatorname{soc}_{i-1} M)^*$$

is semi-simple, hence by induction that $\ell\ell(M^*/(\operatorname{soc}_i M)^\perp) \leq i$ and thus

$$\operatorname{rad}^i(M^*) \subset (\operatorname{soc}_i M)^\perp.$$

Similarly, we have isomorphisms

$$(\operatorname{rad}^i M)^\perp/(\operatorname{rad}^{i-1} M)^\perp \simeq (\operatorname{rad}^{i-1} M/\operatorname{rad}^i M)^*,$$

hence $\ell\ell((\operatorname{rad}^i M)^\perp) \leq i$ and $(\operatorname{rad}^i M)^\perp \subset \operatorname{soc}_i(M^*)$. Applying this to M^* instead of M and identifying $(M^*)^*$ with M, we get $(\operatorname{rad}^i(M^*))^\perp \subset \operatorname{soc}_i(M)$, hence

$$\operatorname{rad}^i(M^*) = ((\operatorname{rad}^i(M^*))^\perp)^\perp \supset \operatorname{soc}_i(M)^\perp.$$

Since we have already the reversed inclusion, we get equality:

$$\text{(6)} \qquad \operatorname{rad}^i(M^*) = \operatorname{soc}_i(M)^\perp.$$

Applying this to M^* and taking orthogonal spaces we get also

$$\text{(7)} \qquad (\operatorname{rad}^i M)^\perp = \operatorname{soc}_i(M^*).$$

This shows in particular that

$$\text{(8)} \qquad \ell\ell(M^*) = \ell\ell(M).$$

D.2. Now return to our general set-up. Suppose that we have chosen $\lambda \in C \cap X(T)$ and $\mu(s)$ for all $s \in \Sigma$ as in C.2. Set $\Theta_s = T^\lambda_{\mu(s)} \circ T^{\mu(s)}_\lambda$ for all s.

Fix s for the moment and write $\mu = \mu(s)$. We need some G_1T–analogues of the results in 7.21. We have again for each $w \in W_p$ either $ws \uparrow w$ (which implies $T^\mu_\lambda \widehat{L}_1(w \bullet \lambda) = 0$, hence $\Theta_s \widehat{L}_1(w \bullet \lambda) = 0$) or $w \uparrow ws$. In the second case we have $T^\mu_\lambda \widehat{L}_1(w \bullet \lambda) \simeq \widehat{L}_1(w \bullet \mu)$, hence $\Theta_s \widehat{L}_1(w \bullet \lambda) \simeq T^\lambda_\mu \widehat{L}_1(w \bullet \mu)$. Then $\Theta_s \widehat{L}_1(w \bullet \lambda)$ has head and socle isomorphic to $\widehat{L}_1(w \bullet \lambda)$, cf. C.14; the quotient of the radical by the socle was called $\beta_s \widehat{L}_1(w \bullet \lambda)$ in C.14. Each composition factor L of $\beta_s \widehat{L}_1(w \bullet \lambda)$ satisfies $\Theta_s L = 0$. We have $[\beta_s \widehat{L}_1(w \bullet \lambda) : \widehat{L}_1(ws \bullet \lambda)] = 1$; if $[\beta_s \widehat{L}_1(w \bullet \lambda) : \widehat{L}_1(x \bullet \lambda)] \neq 0$ and $x \neq ws$, then $x \uparrow w$.

One gets that $\beta_s \widehat{L}_1(w \bullet \lambda)$ is semi-simple if and only if $\ell\ell(\Theta_s \widehat{L}_1(w \bullet \lambda)) = 3$. It follows that the condition $(\widehat{D}, w, s)$ from C.14 holds for all w (and fixed s) if and only if $\ell\ell(\Theta_s L) \leq 3$ for all simple G_1T–modules L with $\operatorname{pr}_\lambda L = L$.

We want to show following [Irving 2]:

Proposition: *If $\ell\ell(\Theta_s L) \leq 3$ for all simple G_1T–modules L with $\operatorname{pr}_\lambda L = L$, then $\ell\ell(\Theta_s M) \leq \ell\ell(M) + 2$ for all G_1T–modules M with $\operatorname{pr}_\lambda M = M$.*

Proof: Set $\mu = \mu(s)$. We have for all G_1T–modules M, N with $\operatorname{pr}_\lambda M = M$ and $\operatorname{pr}_\mu N = N$ isomorphisms given by adjunction

$$\operatorname{adj}_1 : \operatorname{Hom}_{G_1T}(N, T^\mu_\lambda M) \xrightarrow{\sim} \operatorname{Hom}_{G_1T}(T^\lambda_\mu N, M)$$

and

$$\operatorname{adj}_2 : \operatorname{Hom}_{G_1T}(M, T^\lambda_\mu N) \xrightarrow{\sim} \operatorname{Hom}_{G_1T}(T^\mu_\lambda M, N).$$

Taking $N = T^\mu_\lambda M$ we get thus homomorphisms

$$\text{(1)} \qquad j_M : \Theta_s M \to M \qquad \text{and} \qquad i_M : M \to \Theta_s M$$

via $j_M = \operatorname{adj}_1(\operatorname{id}_{T^\mu_\lambda M})$ and $i_M = \operatorname{adj}_2^{-1}(\operatorname{id}_{T^\mu_\lambda M})$. Note that

$$j_M = 0 \iff T^\mu_\lambda M = 0 \iff i_M = 0$$

(and that $T^\mu_\lambda M = 0 \iff \Theta_s M = 0$).

If L is a simple module with $T^\mu_\lambda L \neq 0$, then this shows (together with the description of the socle and of the head of $\Theta_s L$ in C.14) that i_L induces an isomorphism $L \xrightarrow{\sim} \operatorname{soc} \Theta_s L$ and that j_L induces an isomorphism $\Theta_s L / \operatorname{rad} \Theta_s L \xrightarrow{\sim} L$. This implies:

(2) *If M is semi-simple, then $i_M(M) = \operatorname{soc}\Theta_s M$ and $\ker(j_M) = \operatorname{rad}\Theta_s M$ and $j_M \circ i_M = 0$.*

Fix for the moment M, set $M_1 = \operatorname{rad} M$ and $M_2 = M/\operatorname{rad} M$. Let $\iota : M_1 \to M$ and $\pi : M \to M_2$ denote the inclusion or the canonical map. The G_1T–analogue of 7.21(5) shows that we have a commutative diagram

$$\begin{array}{ccccccccc}
0 \to & \Theta_s M_1 & \overset{\Theta_s(\iota)}{\longrightarrow} & \Theta_s M & \overset{\Theta_s(\pi)}{\longrightarrow} & \Theta_s M_2 & \to 0 \\
& {\scriptstyle j_{M_1}}\big\downarrow & & {\scriptstyle j_M}\big\downarrow & & \big\downarrow{\scriptstyle j_{M_2}} & \\
0 \to & M_1 & \overset{\iota}{\longrightarrow} & M & \overset{\pi}{\longrightarrow} & M_2 & \to 0 \\
& {\scriptstyle i_{M_1}}\big\downarrow & & {\scriptstyle i_M}\big\downarrow & & \big\downarrow{\scriptstyle i_{M_2}} & \\
0 \to & \Theta_s M_1 & \overset{\Theta_s(\iota)}{\longrightarrow} & \Theta_s M & \overset{\Theta_s(\pi)}{\longrightarrow} & \Theta_s M_2 & \to 0
\end{array}$$

The middle row is exact by construction, the other rows are so by the exactness of the translation functors.

The semi-simplicity of M_2 implies by (2) that

$$\operatorname{soc}\Theta_s M_2 = i_{M_2}(M_2) = \Theta_s(\pi) \circ i_M(M).$$

We have $i_M(M) \subset \operatorname{soc}_{\ell\ell(M)}\Theta_s M$ by D.1(2), hence

(3) $$\operatorname{soc}\Theta_s M_2 \subset \Theta_s(\pi)(\operatorname{soc}_{\ell\ell(M)}\Theta_s M).$$

Now consider

$$\gamma_M : \Theta_s M \longrightarrow M_2, \qquad \gamma_M = \pi \circ j_M = j_{M_2} \circ \Theta_s(\pi).$$

We want to show that

(4) $$\ker(\gamma_M) = \operatorname{rad}(\Theta_s M) + i_M(M) + \Theta_s(\iota)(\Theta_s M_1).$$

Here the inclusion "$\supset$" is easy: The radical of $\Theta_s M$ is contained in the kernel of γ_M because the image of γ_M is contained in the semi-simple module M_2, hence semi-simple. We have $\gamma_M \circ \Theta_s(\iota) = j_{M_2} \circ \Theta_s(\pi \circ \iota) = 0$ and $\gamma_M \circ i_M = j_{M_2} \circ \Theta_s(\pi) \circ i_M = j_{M_2} \circ i_{M_2} \circ \pi = 0$ by (2).

We now turn to the reverse inclusion "$\subset$". Set $M' = i_M(M) + \Theta_s(\iota)(M_1)$. We have $\Theta_s(\pi)(M') = \Theta_s(\pi) \circ i_M(M) = i_{M_2}(M_2) = \operatorname{soc}\Theta_s M_2$ using (2) for the last equality. Therefore $\Theta_s(\pi)$ induces an isomorphism

$$\overline{\Theta}_s(\pi) : \Theta_s M/M' \xrightarrow{\sim} \Theta_s M_2/\operatorname{soc}\Theta_s M_2.$$

Since $M' \subset \ker(\gamma_M)$, we get an induced map $\overline{\gamma}_M : \Theta_s M/M' \to M_2$. The under $\overline{\Theta}_s(\pi)$ corresponding map $\Theta_s M_2/\operatorname{soc}\Theta_s M_2 \to M_2$ is induced by j_{M_2}, hence has kernel

$$\operatorname{rad}\Theta_s M_2 \,/\, \operatorname{soc}\Theta_s M_2 = \operatorname{rad}(\Theta_s M_2/\operatorname{soc}\Theta_s M_2)$$

by (2). Therefore $\overline{\gamma}_M$ has kernel $\operatorname{rad}(\Theta_s M/M') = (\operatorname{rad}(\Theta_s M) + M')/M'$. This implies (4).

We now start with the proof of the proposition in earnest. We use induction on $\ell\ell(M)$. By assumption the claim holds if $\ell\ell(M) = 1$. So suppose that $\ell\ell(M) = r > 1$ and that the claim holds for all modules of Loewy length $< r$. We want to show first (with notations as above) that

$$\Theta_s(\iota)^{-1}(\operatorname{rad}^2 \Theta_s M) \subset \operatorname{soc}_r \Theta_s M_1. \tag{5}$$

For this, consider the commutative diagram

$$\begin{array}{ccccc} \Theta_s M_1 & \xrightarrow{\Theta_s(\iota)} & \Theta_s M & \xrightarrow{a} & \Theta_s M/\operatorname{rad}^2 \Theta_s M \\ {\scriptstyle j_{M_1}}\downarrow & & {\scriptstyle j_M}\downarrow & & \downarrow{\scriptstyle \overline{\jmath}} \\ M_1 & \xrightarrow{\iota} & M & \xrightarrow{b} & M/\operatorname{rad}^2 M \end{array}$$

where a and b are the canonical maps and where we get a map $\overline{\jmath}$ making the diagram commutative because $j_M(\operatorname{rad}^2 \Theta_s M) \subset \operatorname{rad}^2 M$. Our claim says that $\ker(a\circ\Theta(\iota)) \subset \operatorname{soc}_r(\Theta_s M_1)$. It suffices to show that the kernel of $\overline{\jmath}\circ a\circ\Theta(\iota) = b\circ\iota\circ j_{M_1}$ is contained in $\operatorname{soc}_r(\Theta_s M_1)$. Now $b\circ\iota$ identifies with the canonical map $M_1 \to M_1/\operatorname{rad} M_1$ since $M_1 = \operatorname{rad} M$. Therefore $b\circ\iota\circ j_{M_1}$ is just the map γ_{M_1} that we get applying the constructions above to M_1 instead of M. Setting $M_1' = \operatorname{rad} M_1$ and denoting the inclusion $M_1' \hookrightarrow M_1$ by ι', we have therefore by (4)

$$\ker(b\circ\iota\circ j_{M_1}) = \operatorname{rad}(\Theta_s M_1) + i_{M_1}(M_1) + \Theta_s(\iota')(\Theta_s M_1').$$

Now $\ell\ell(M) = r$ implies $\ell\ell(M_1) = r-1$ and $\ell\ell(M_1') = r-2$. So we have $i_{M_1}(M_1) \subset \operatorname{soc}_{r-1}\Theta_s M_1$. Induction implies $\ell\ell(\Theta_s M_1') \le r$, hence $\Theta_s(\iota')(\Theta_s M_1') \subset \operatorname{soc}_r \Theta_s M_1$, and it implies $\ell\ell(\Theta_s M_1) \le r+1$, hence $\operatorname{rad}(\Theta_s M_1) \subset \operatorname{soc}_r \Theta_s M_1$. So all three summands adding up to $\ker(b\circ\iota\circ j_{M_1})$ are contained in $\operatorname{soc}_r \Theta_s M_1$. This yields (5).

Set $P = \Theta_s M/\operatorname{soc}_r \Theta_s M$. We have to show that $\operatorname{soc}_{r+2}(\Theta_s M) = \Theta_s M$. That will follow, if we can show that $\ell\ell(P) \le 2$, or equivalently, that $\operatorname{rad}^2 P = 0$. We have a short exact sequence $0 \to P_1 \to P \to P_2 \to 0$ where

$$P_1 = \Theta_s M_1/\operatorname{soc}_r \Theta_s M_1 \qquad \text{and} \qquad P_2 = \Theta_s M_2/\Theta_s(\pi)(\operatorname{soc}_r \Theta_s M).$$

(Note that $\Theta_s(\iota)^{-1}\operatorname{soc}_r \Theta_s M = \operatorname{soc}_r \Theta_s M_1$ by D.1(4).) Since M_2 is semi-simple, our general assumption and (3) imply that $\ell\ell(P_2) \le 2$, hence $\operatorname{rad}^2 P_2 = 0$. So $\operatorname{rad}^2 P$ is by D.1(4) contained in the image of P_1. The inverse image of $\operatorname{rad}^2 P$ in P_1 is equal to

$$(\Theta_s(\iota)^{-1}(\operatorname{rad}^2 \Theta_s M + \operatorname{soc}_r \Theta_s M) + \operatorname{soc}_r \Theta_s M_1)/\operatorname{soc}_r \Theta_s M_1.$$

Now (5) and D.1(4) show that this inverse image is 0. This implies $\operatorname{rad}^2 P = 0$, hence the proposition.

D.3. Let $w \in W_p$ and $s \in \Sigma$ with $ws \uparrow w$. We have then

$$T_\lambda^{\mu(s)} \widehat{Z}_1(w\bullet\lambda) \simeq \widehat{Z}_1(w\bullet\mu(s)) \simeq T_\lambda^{\mu(s)} \widehat{Z}_1(w\bullet\lambda) \tag{1}$$

and there is a short exact sequence

$$(2) \qquad 0 \to \widehat{Z}_1(w \bullet \lambda) \xrightarrow{\iota} \Theta_s \widehat{Z}_1(w \bullet \lambda) \xrightarrow{\pi} \widehat{Z}_1(ws \bullet \lambda) \to 0.$$

We have $T_\lambda^{\mu(s)} \widehat{L}_1(w \bullet \lambda) = 0$ by 9.22(4) and $\widehat{Z}_1(w \bullet \lambda)/\operatorname{rad} \widehat{Z}_1(w \bullet \lambda) \simeq \widehat{L}_1(w \bullet \lambda)$ by 9.6(4). Therefore we get an isomorphism induced by the inclusion

$$(3) \qquad \Theta_s \operatorname{rad} \widehat{Z}_1(w \bullet \lambda) \xrightarrow{\sim} \Theta_s \widehat{Z}_1(w \bullet \lambda).$$

We get $\operatorname{Hom}_{G_1T}(\Theta_s \widehat{Z}_1(w \bullet \lambda), \widehat{L}_1(w \bullet \lambda)) \simeq \operatorname{Hom}_{G_1T}(\widehat{Z}_1(w \bullet \lambda), \Theta_s \widehat{L}_1(w \bullet \lambda)) = 0$ by the self-adjointness of Θ_s. Therefore $\widehat{L}_1(w \bullet \lambda)$ is not a composition factor of $\Theta_s \widehat{Z}_1(w \bullet \lambda)/\operatorname{rad} \Theta_s \widehat{Z}_1(w \bullet \lambda)$. So $\widehat{Z}_1(w \bullet \lambda)/\operatorname{rad} \widehat{Z}_1(w \bullet \lambda) \simeq \widehat{L}_1(w \bullet \lambda)$ implies that

$$(4) \qquad \iota(\widehat{Z}_1(w \bullet \lambda)) \subset \operatorname{rad} \Theta_s \widehat{Z}_1(w \bullet \lambda)$$

and π induces an isomorphism

$$(5) \qquad \Theta_s \widehat{Z}_1(w \bullet \lambda)/\operatorname{rad} \Theta_s \widehat{Z}_1(w \bullet \lambda) \xrightarrow{\sim} \widehat{Z}_1(ws \bullet \lambda)/\operatorname{rad} \widehat{Z}_1(ws \bullet \lambda).$$

Now $ws \uparrow w$ implies $w_0 w \uparrow w_0 ws$, hence $\Theta_s \widehat{L}_1(w_0 ws \bullet \lambda) = 0$. The translation functors commute with any $? \otimes p\nu$ with $\nu \in \mathbf{Z}R$. Therefore 9.7(3) yields

$$(6) \qquad \Theta_s \operatorname{soc} \widehat{Z}_1(ws \bullet \lambda) = 0.$$

As Θ_s is self-adjoint, this implies $\operatorname{Hom}_{G_1T}(\operatorname{soc} \widehat{Z}_1(ws \bullet \lambda), \Theta_s \widehat{Z}_1(w \bullet \lambda)) = 0$, hence

$$(7) \qquad \pi(\operatorname{soc} \Theta_s \widehat{Z}_1(w \bullet \lambda)) = 0.$$

So ι induces an isomorphism

$$(8) \qquad \operatorname{soc} \widehat{Z}_1(w \bullet \lambda) \xrightarrow{\sim} \operatorname{soc} \Theta_s \widehat{Z}_1(w \bullet \lambda).$$

D.4. Let us say that condition $(\widehat{D})$ holds if $(\widehat{D}, w, s)$ holds for all $w \in W_p$ and $s \in \Sigma$ with $w \uparrow ws$. By Remark 1 in C.17 this is equivalent to the truth of the assumption (ii) in Proposition C.17, hence to the truth of the G_1T–version of the Lusztig conjecture.

Lemma: *Suppose that* $(\widehat{D})$ *holds. Then:*

a) *All* $\widehat{Q}_1(w \bullet \lambda)$ *with* $w \in W_p$ *have the same Loewy length.*

b) *All* $\widehat{Z}_1(w \bullet \lambda)$ *with* $w \in W_p$ *have the same Loewy length.*

Proof: We have for any finite dimensional G_1T–module M and any $\nu \in X(T)$

$$(1) \qquad \ell\ell(M \otimes p\nu) = \ell\ell(M)$$

because M and $M \otimes p\nu$ have isomorphic submodule lattices.

a) Let us observe first for all $w \in W_p$ and $s \in \Sigma$: *If* $w \uparrow ws$, *then* $\widehat{Q}_1(w \bullet \lambda)$ *is isomorphic to a direct summand of* $\Theta_s \widehat{Q}_1(ws \bullet \lambda)$. Since $\Theta_s \widehat{Q}_1(ws \bullet \lambda)$ is projective, it suffices to show that

$$0 \neq \operatorname{Hom}_{G_1T}(\Theta_s \widehat{Q}_1(ws \bullet \lambda), \widehat{L}_1(w \bullet \lambda)) \simeq \operatorname{Hom}_{G_1T}(\widehat{Q}_1(ws \bullet \lambda), \Theta_s \widehat{L}_1(w \bullet \lambda)).$$

The second Hom space has dimension $[\Theta_s \widehat{L}_1(w \bullet \lambda) : \widehat{L}_1(ws \bullet \lambda)]$, hence is non-zero by the G_1T–analogue to 7.20.b.

Let us now use this observation to prove (for all $w \in W_p$ and $s \in \Sigma$)

$$w \uparrow ws \implies \ell\ell(\widehat{Q}_1(w \bullet \lambda)) \leq \ell\ell(\widehat{Q}_1(ws \bullet \lambda)). \tag{2}$$

Well, $w \uparrow ws$ implies $\Theta_s \widehat{L}_1(ws \bullet \lambda) = 0$. So Θ_s annihilates both $\operatorname{soc} \widehat{Q}_1(ws \bullet \lambda)$ and $\widehat{Q}_1(ws \bullet \lambda)/\operatorname{rad} \widehat{Q}_1(ws \bullet \lambda)$. It follows that

$$\Theta_s \widehat{Q}_1(ws \bullet \lambda) \;\simeq\; \Theta_s(\operatorname{rad} \widehat{Q}_1(ws \bullet \lambda)/\operatorname{soc} \widehat{Q}_1(ws \bullet \lambda)).$$

Now $\ell\ell((\operatorname{rad} \widehat{Q}_1(ws \bullet \lambda)/\operatorname{soc} \widehat{Q}_1(ws \bullet \lambda)) = \ell\ell(\widehat{Q}_1(ws \bullet \lambda)) - 2$ implies by D.2 (and our assumption) that

$$\ell\ell(\Theta_s \widehat{Q}_1(ws \bullet \lambda)) \;\leq\; \ell\ell(\widehat{Q}_1(ws \bullet \lambda)).$$

As $\widehat{Q}_1(w \bullet \lambda)$ is isomorphic to a direct summand of $\Theta_s \widehat{Q}_1(ws \bullet \lambda)$, this implies (2).

For each $w \in W$ (the ordinary Weyl group) with $w \neq 1$ we have $\alpha \in S$ with $l(ws_\alpha) < l(w)$, hence with $w \uparrow ws_\alpha$. Then (2) implies $\ell\ell(\widehat{Q}_1(w \bullet \lambda)) \leq \ell\ell(\widehat{Q}_1(ws_\alpha \bullet \lambda))$. Now induction on $l(w)$ yields

$$\ell\ell(\widehat{Q}_1(w \bullet \lambda)) \leq \ell\ell(\widehat{Q}_1(\lambda)) \qquad \text{for all } w \in W. \tag{3}$$

For each $w \in W_p^+$ (i.e., with $w \bullet \lambda \in X(T)_+$) with $w \neq 1$ there exists $s \in \Sigma$ with $ws \uparrow w$ and $ws \in W_p^+$. Then (2) implies $\ell\ell(\widehat{Q}_1(ws \bullet \lambda)) \leq \ell\ell(\widehat{Q}_1(w \bullet \lambda))$. Now induction on $d(w)$ yields

$$\ell\ell(\widehat{Q}_1(\lambda)) \leq \ell\ell(\widehat{Q}_1(w \bullet \lambda)) \qquad \text{for all } w \in W_p^+. \tag{4}$$

For each $w \in W$ we can find $\mu \in \mathbf{Z}R$ with $w \bullet \lambda + p\mu \in X(T)_+$. Then the composition of w with the translation by $p\mu$ belongs to W_p^+. Now (4), (1), and (3) imply

$$\ell\ell(\widehat{Q}_1(\lambda)) \leq \ell\ell(\widehat{Q}_1(w \bullet \lambda + p\mu)) = \ell\ell(\widehat{Q}_1(w \bullet \lambda)) \leq \ell\ell(\widehat{Q}_1(\lambda)).$$

(We also used 11.3(2).) It follows that $\ell\ell(\widehat{Q}_1(w \bullet \lambda)) = \ell\ell(\widehat{Q}_1(\lambda))$ for all $w \in W$. Another application of (1) extends this result to all $w \in W_p$.

b) We start by showing for all $w \in W_p$ and $s \in \Sigma$ that

$$\ell\ell(\Theta_s \widehat{Z}_1(w \bullet \lambda)) \;\leq\; \ell\ell(\widehat{Z}_1(w \bullet \lambda)) + 1. \tag{5}$$

If $ws \uparrow w$, then $\Theta_s \widehat{Z}_1(w \bullet \lambda) \simeq \Theta_s(\operatorname{rad} \widehat{Z}_1(w \bullet \lambda))$ by D.3(3). Now use that $\ell\ell(\operatorname{rad} \widehat{Z}_1(w \bullet \lambda)) = \ell\ell(\widehat{Z}_1(w \bullet \lambda)) - 1$ and apply D.2.

If $w \uparrow ws$, then D.3(6), applied to ws instead of w, implies that $\Theta_s \widehat{Z}_1(w \bullet \lambda) \simeq \Theta_s(\widehat{Z}_1(w \bullet \lambda)/\operatorname{soc} \widehat{Z}_1(w \bullet \lambda))$. Now we can argue as above.

We show next for all $w \in W_p$ and $s \in \Sigma$ that

$$\ell\ell(\widehat{Z}_1(w \bullet \lambda)) \leq \ell\ell(\Theta_s \widehat{Z}_1(w \bullet \lambda)) - 1. \tag{6}$$

If $ws \uparrow w$, then this follows from D.3(4) because ι is injective and because in general $\ell\ell(\operatorname{rad} M) = \ell\ell(M) - 1$. In this case we get also that $\ell\ell(\widehat{Z}_1(ws \bullet \lambda)) \leq \ell\ell(\Theta_s \widehat{Z}_1(w \bullet \lambda)) - 1$ using D.3(7), the surjectivity of π, and the general fact that $\ell\ell(M/\operatorname{soc} M) = \ell\ell(M) - 1$.

The last statement (applied to ws instead of w) yields (6) in case $w \uparrow ws$.

Combining (5) and (6) we get

$$\ell\ell(\Theta_s \widehat{Z}_1(w \bullet \lambda)) = \ell\ell(\widehat{Z}_1(w \bullet \lambda)) + 1 \tag{7}$$

for all $w \in W_p$ and $s \in \Sigma$. As $\Theta_s \widehat{Z}_1(w \bullet \lambda) \simeq \Theta_s \widehat{Z}_1(ws \bullet \lambda)$ for all w and s, this implies

$$\ell\ell(\widehat{Z}_1(ws \bullet \lambda)) = \ell\ell(\widehat{Z}_1(w \bullet \lambda)) \tag{8}$$

for all $w \in W_p$ and $s \in \Sigma$. Since W_p is generated by the $s \in \Sigma$, we get now $\ell\ell(\widehat{Z}_1(w \bullet \lambda)) = \ell\ell(\widehat{Z}_1(\lambda))$ for all $w \in W_p$, i.e., the claim in the lemma.

D.5. Lemma: *We have* $\ell\ell(\widehat{Q}_1(w_0 \bullet \lambda)) \geq 2\ell\ell(\widehat{Z}_1(\lambda)) - 1$.

Proof: Lemma 11.6 implies that λ is the largest weight of $\widehat{Q}_1(w_0 \bullet \lambda)$ and that $\dim(\widehat{Q}_1(w_0 \bullet \lambda)_\lambda) = 1$. Let m be maximal for $(\operatorname{rad}^m \widehat{Q}_1(w_0 \bullet \lambda))_\lambda \neq 0$. We have then $\widehat{Q}_1(w_0 \bullet \lambda)_\lambda \subset \operatorname{rad}^m \widehat{Q}_1(w_0 \bullet \lambda)$.

By Remark 11.6 there exist an injection $\iota : \widehat{Z}_1(\lambda) \to \widehat{Q}_1(w_0 \bullet \lambda)$ and a surjection $\pi : \widehat{Q}_1(w_0 \bullet \lambda) \to \widehat{Z}'_1(\lambda)$. Both maps induce isomorphisms of the λ–weight spaces.

Since $\widehat{L}_1(\lambda) = \operatorname{soc} \widehat{Z}'_1(\lambda)$ and since λ is not a weight of $\operatorname{rad}^{m+1} \widehat{Q}_1(w_0 \bullet \lambda)$, we get $\pi(\operatorname{rad}^{m+1} \widehat{Q}_1(w_0 \bullet \lambda)) = 0$. So $\widehat{Q}_1(w_0 \bullet \lambda)/\operatorname{rad}^{m+1} \widehat{Q}_1(w_0 \bullet \lambda)$ maps onto $\widehat{Z}'_1(\lambda)$; this implies

$$\ell\ell(\widehat{Z}'_1(\lambda)) \leq \ell\ell(\widehat{Q}_1(w_0 \bullet \lambda)/\operatorname{rad}^{m+1} \widehat{Q}_1(w_0 \bullet \lambda)) = m + 1.$$

The map ι takes the standard generator (of weight λ) of $\widehat{Z}_1(\lambda)$ into $\operatorname{rad}^m \widehat{Q}_1(w_0 \bullet \lambda)$. This implies $\iota(\widehat{Z}_1(\lambda)) \subset \operatorname{rad}^m \widehat{Q}_1(w_0 \bullet \lambda)$, hence by the injectivity of ι

$$\ell\ell(\widehat{Z}_1(\lambda)) \leq \ell\ell(\operatorname{rad}^m \widehat{Q}_1(w_0 \bullet \lambda)) = \ell\ell(\widehat{Q}_1(w_0 \bullet \lambda)) - m.$$

It follows that

$$\ell\ell(\widehat{Q}_1(w_0 \bullet \lambda)) \geq \ell\ell(\widehat{Z}_1(\lambda)) + \ell\ell(\widehat{Z}'_1(\lambda)) - 1.$$

Since $\widehat{Z}'_1(\lambda) \simeq {}^\tau\widehat{Z}_1(\lambda)$, see 9.3(5), both modules have the same Loewy length. The claim follows.

Remark: Given $w \in W_p$ there exists by Lemma 11.6 an element $w' \in W_p$ such that $w' \bullet \lambda$ is the largest weight of $\widehat{Q}_1(w \bullet \lambda)$. Then the same argument as above shows that $\ell\ell(\widehat{Q}_1(w \bullet \lambda)) \geq 2\ell\ell(\widehat{Z}_1(w' \bullet \lambda)) - 1$.

D.6. Lemma: *We have* $\ell\ell(\widehat{Z}_1(\mu)) \geq |R^+| + 1$ *for all* $\mu \in W_p \bullet \lambda$.

Proof: Consider homomorphisms $\varphi_i : \widehat{Z}_1^{x_i}(\mu\langle x_i\rangle) \to \widehat{Z}_1^{x_{i+1}}(\mu\langle x_{i+1}\rangle)$ as in 9.11(3). Let M_i denote the image of $\widehat{Z}_1(\mu)$ under $\varphi_{i-1} \circ \cdots \circ \varphi_1$. So we have $M_1 = \widehat{Z}_1(\mu)$ and (see 9.11) $M_{n+1} = \widehat{L}_1(\mu)$ where $n = |R^+|$. Each $\widehat{Z}_1^{x_i}(\mu\langle x_i\rangle)$ has a simple socle and φ_i (for $i \leq n$) annihilates this socle because φ_i is not injective. Each M_i is a non-zero (as $M_{n+1} \neq 0$) submodule of $\widehat{Z}_1^{x_i}(\mu\langle x_i\rangle)$, hence has the same socle as this module. It follows that $\varphi_i(\operatorname{soc} M_i) = 0$, hence that $M_{i+1} = \varphi_i(M_i)$ is a homomorphic image of $M_i/\operatorname{soc} M_i$. This implies that $\ell\ell(M_{i+1}) \leq \ell\ell(M_i) - 1$. Now induction yields $\ell\ell(M_{i+1}) \leq \ell\ell(M_1) - i$, in particular

$$1 = \ell\ell(\widehat{L}_1(\mu)) = \ell\ell(M_{n+1}) \leq \ell\ell(M_1) - n = \ell\ell(\widehat{Z}_1(\mu)) - n,$$

i.e., the claim.

D.7. Lemma: *Let* $w_0 = s_1 s_2 \dots s_n$ *be a reduced decomposition in* W *with* $s_i = s_{\alpha_i}$ *for suitable* $\alpha_i \in S$. *Then* $\widehat{Q}_1(w_0 \bullet \lambda)$ *is isomorphic to a direct summand of* $\Theta_{s_1}\Theta_{s_2}\dots\Theta_{s_n}\widehat{L}_1(w_0 \bullet \lambda)$.

Proof: Let us note first for all $\alpha \in S$ and $w \in W_p$ with $w \neq w_0$ that

(1) $$[\Theta_{s_\alpha}\widehat{L}_1(w \bullet \lambda) : \widehat{L}_1(w_0 \bullet \lambda)] = 0.$$

Indeed, any composition factor L of $\Theta_{s_\alpha}\widehat{L}_1(w \bullet \lambda)$ different from $\widehat{L}_1(w \bullet \lambda)$ satisfies $\Theta_{s_\alpha} L = 0$, see D.2. But $w_0 \uparrow w_0 s_\alpha$ for all $\alpha \in S$ implies $\Theta_{s_\alpha}\widehat{L}_1(w_0 \bullet \lambda) \neq 0$.

The exactness of the Θ_{s_α} implies now using induction that

(2) $$[\Theta_{s_n}\dots\Theta_{s_2}\Theta_{s_1}\widehat{L}_1(w \bullet \lambda) : \widehat{L}_1(w_0 \bullet \lambda)] = 0$$

for all $w \neq w_0$.

Set $M = \Theta_{s_1}\Theta_{s_2}\dots\Theta_{s_n}\widehat{L}_1(w_0 \bullet \lambda)$. We get from (2) using the adjointness properties of the Θ_s for all $w \neq w_0$

$$\operatorname{Hom}_{G_1T}(\widehat{L}_1(w \bullet \lambda), M) \simeq \operatorname{Hom}_{G_1T}(\Theta_{s_n}\dots\Theta_{s_2}\Theta_{s_1}\widehat{L}_1(w \bullet \lambda), \widehat{L}_1(w_0 \bullet \lambda)) = 0$$

and

$$\operatorname{Hom}_{G_1T}(M, \widehat{L}_1(w \bullet \lambda)) \simeq \operatorname{Hom}_{G_1T}(\widehat{L}_1(w_0 \bullet \lambda), \Theta_{s_n}\dots\Theta_{s_2}\Theta_{s_1}\widehat{L}_1(w \bullet \lambda)) = 0.$$

This implies:

(3) *Both* $\operatorname{soc} M$ *and* $M/\operatorname{rad} M$ *are isotypic of type* $\widehat{L}_1(w_0 \bullet \lambda)$.

The description of $\Theta_s\widehat{Z}_1(w \bullet \lambda)$ shows that $M' = \Theta_{s_1}\Theta_{s_2}\dots\Theta_{s_n}\widehat{Z}_1(w_0 \bullet \lambda)$ has a filtration where all factors have the form $\widehat{Z}_1(x \bullet \lambda)$ with $x \in W$. Since M is a homomorphic image of M', this implies that all weights of M are $\leq \lambda$. On the other hand, one checks inductively that $\widehat{L}_1(w_0 s_n \dots s_i \bullet \lambda)$ is a composition factor of $\Theta_{s_i}\dots\Theta_{s_n}\widehat{L}_1(w_0 \bullet \lambda)$, hence $\widehat{L}_1(\lambda)$ one of M. Therefore λ is the largest weight of M. It follows that there exists a non-zero homomorphism of G_1T–modules $\iota_M : \widehat{Z}_1(\lambda) \to M$. Now (3) implies that $\operatorname{soc}\iota_M(\widehat{Z}_1(\lambda))$ is isotypic of type $\widehat{L}_1(w_0 \bullet \lambda)$. On the other hand, we have $\operatorname{soc} \widehat{Z}_1(\lambda) \simeq \widehat{L}_1(w_0 \bullet \lambda)$ by 9.7(3) and $[\widehat{Z}_1(\lambda) : \widehat{L}_1(w_0 \bullet \lambda)] = [\widehat{Z}_1(w_0 \bullet \lambda) : \widehat{L}_1(w_0 \bullet \lambda)] = 1$ by 9.16(4). This shows that ι_M is injective. (Otherwise ι_M annihilates $\operatorname{soc} \widehat{Z}_1(\lambda)$ and the image of ι_M has no composition factor $\widehat{L}_1(w_0 \bullet \lambda)$.)

Recall from the proof of Lemma D.5 (or from Remark 11.6) the injection $\iota : \widehat{Z}_1(\lambda) \to \widehat{Q}_1(w_0 \bullet \lambda)$ and the surjection $\pi : \widehat{Q}_1(w_0 \bullet \lambda) \to \widehat{Z}'_1(\lambda)$. Because $\widehat{Q}_1(w_0 \bullet \lambda)$ is an injective G_1T–module and because ι_M is injective, there is a homomorphism of G_1T–modules $\varphi : M \to \widehat{Q}_1(w_0 \bullet \lambda)$ with $\varphi \circ \iota_M = \iota$.

We have $\pi \circ \varphi \circ \iota_M = \pi \circ \iota \neq 0$, hence $\pi \circ \varphi \neq 0$. By (3) the head of $\pi \circ \varphi(M)$ is isotypic of type $\widehat{L}_1(w_0 \bullet \lambda)$. Now $\widehat{Z}'_1(\lambda) \simeq {}^{\tau}\widehat{Z}_1(\lambda)$ implies that $\widehat{Z}'_1(\lambda)/\operatorname{rad} \widehat{Z}'_1(\lambda) \simeq \widehat{L}_1(w_0 \bullet \lambda)$ and $[\widehat{Z}'_1(\lambda) : \widehat{L}_1(w_0 \bullet \lambda)] = 1$. This implies that $\pi \circ \varphi(M) \not\subset \operatorname{rad} \widehat{Z}'_1(\lambda)$, hence that $\pi \circ \varphi(M) = \widehat{Z}'_1(\lambda)$.

Let $\pi' : \widehat{Q}_1(w_0 \bullet \lambda) \to \widehat{L}_1(w_0 \bullet \lambda)$ be the composition of π with the canonical map from $\widehat{Z}'_1(\lambda)$ onto $\widehat{Z}'_1(\lambda)/\operatorname{rad} \widehat{Z}'_1(\lambda) \simeq \widehat{L}_1(w_0 \bullet \lambda)$. This map has kernel $\operatorname{rad} \widehat{Q}_1(w_0 \bullet \lambda)$. The surjectivity of $\pi \circ \varphi$ implies that of $\pi' \circ \varphi$. So $\varphi(M)$ is not contained in $\operatorname{rad} \widehat{Q}_1(w_0 \bullet \lambda)$, hence equal to $\widehat{Q}_1(w_0 \bullet \lambda)$. Since $\widehat{Q}_1(w_0 \bullet \lambda)$ is a projective G_1T–module, the surjection $\varphi : M \to \widehat{Q}_1(w_0 \bullet \lambda)$ splits. The claim follows.

D.8. Proposition: *Suppose that* $(\widehat{D})$ *holds. Then*

$$\ell\ell(\widehat{Q}_1(\mu)) = 2|R^+| + 1 \tag{1}$$

and

$$\ell\ell(\widehat{Z}_1(\mu)) = |R^+| + 1 \tag{2}$$

for all $\mu \in W_p \bullet \lambda$.

Proof: Lemma D.7, Proposition D.2, Lemma D.5, and Lemma D.6 imply

$$2|R^+| + 1 \geq \ell\ell(\widehat{Q}_1(\mu)) \geq 2\ell\ell(\widehat{Z}_1(\mu)) - 1 \geq 2|R^+| + 1.$$

So all inequalities are equalities. Now the claim follows from Lemma D.4.

Remarks: 1) We get now from D.4(7) assuming $(\widehat{D})$ that

$$\ell\ell(\Theta_s \widehat{Z}_1(\mu)) = |R^+| + 2 \tag{3}$$

for all $\mu \in W_p \bullet \lambda$ and $s \in \Sigma$. One can show conversely: If (3) holds for all μ and s, then $(\widehat{D})$ holds, see [Andersen and Kaneda 1], 5.8.

2) We have

$$\widehat{Z}_1^w(\mu\langle w\rangle) \simeq {}^{\dot{w}}\widehat{Z}_1(w^{-1} \bullet \mu + p\,(\rho - w^{-1}\rho)) \tag{4}$$

by 9.3(2) and a simple calculation. Since $\rho - w^{-1}\rho \in \mathbf{Z}R$, we have $w^{-1} \bullet \mu + p(\rho - w^{-1}\rho \in W_p \bullet \mu$. As twisting with $\dot{w}$ does not change the Loewy length, we get assuming $(\widehat{D})$

$$\ell\ell\,(\widehat{Z}_1^w(\mu\langle w\rangle)) = |R^+| + 1 \qquad \text{for all } \mu \in W_p \bullet \lambda \text{ and } w \in W. \tag{5}$$

D.9. In the general situation from D.1 we call a finite dimensional H–module M *rigid* if $\operatorname{rad}^i M = \operatorname{soc}_{\ell\ell(M)-i} M$ for all i, $0 \le i \le \ell\ell(M)$.

If L is a simple H–module, then we say that L is *rigidly placed* in M if

$$(1)\qquad [\operatorname{rad}^i M : L] = [\operatorname{soc}_{\ell\ell(M)-i} M : L] \qquad \text{for all } i,\ 0 \le i \le \ell\ell(M).$$

Since we always have the inclusion $\operatorname{rad}^i M \subset \operatorname{soc}_{\ell\ell(M)-i} M$, cf. D.1(5), we see that M is rigid if and only if all simple H–modules are rigidly placed in M.

For M and L as above we introduce the polynomial

$$(2)\qquad RM(M,L) = \sum_{i\ge 0} [\operatorname{rad}^i M / \operatorname{rad}^{i+1} M : L]\, v^i \in \mathbf{Z}[v]$$

in the indeterminate v. (R = radical, M = multiplicity)

Now return to our standard set-up over G_1T with a fixed $\lambda \in C \cap X(T)$.

Lemma: *Let $\mu \in W_p \bullet \lambda$ and $w \in W$. If $(\widehat{D})$ holds, then $\widehat{L}_1(\mu)$ is rigidly placed in $\widehat{Z}_1^w(\mu\langle w\rangle)$ and we have*

$$(3)\qquad RM(\widehat{Z}_1^w(\mu\langle w\rangle), \widehat{L}_1(\mu)) = v^{l(w)}.$$

Proof: Set $n = |R^+|$. We have $[\widehat{Z}_1^w(\mu\langle w\rangle) : \widehat{L}_1(\mu)] = 1$ as $\operatorname{ch} \widehat{Z}_1^w(\mu\langle w\rangle) = \operatorname{ch} \widehat{Z}_1(\mu)$, cf. 9.3(4). Since $\ell\ell(\widehat{Z}_1^w(\mu\langle w\rangle)) = n+1$ by D.8(5), our claim is equivalent to

$$(4)\quad [\operatorname{rad}^{l(w)} \widehat{Z}_1^w(\mu\langle w\rangle) : \widehat{L}_1(\mu)] > 0 \qquad \text{and} \qquad [\operatorname{soc}_{n-l(w)} \widehat{Z}_1^w(\mu\langle w\rangle) : \widehat{L}_1(\mu)] = 0.$$

(Recall D.1(5).)

We can find a reduced decomposition $w_0 = s_1 s_2 \dots s_n$ (where $s_i = s_{\alpha_i}$ for some $\alpha_i \in S$) such that $w = s_1 s_2 \dots s_j$ for some j, $0 \le j \le n$. One gets then $j = l(w)$. Adopt the notation from the proof of D.6. We have seen there that $\ell\ell(M_{i+1}) \le \ell\ell(M_1) - i$ for all i. If we have here strict inequality for some i, then we get also a strict inequality for all larger i, in particular $\ell\ell(M_{n+1}) < \ell\ell(M_1) - n$. This is a contradiction as $M_{n+1} \simeq \widehat{L}_1(\mu)$ has Loewy length 1, while $M_1 \simeq \widehat{Z}_1(\mu)$ has Loewy length $n+1$. It follows that $\ell\ell(M_{i+1}) = \ell\ell(M_1) - i = n - i$ for all i, hence

$$(5)\qquad M_{i+1} \not\subset \operatorname{soc}_{n-i-1} \widehat{Z}^{x_{i+1}}(\mu\langle x_{i+1}\rangle).$$

Each $\widehat{Z}^{x_i}(\mu\langle x_i\rangle)$ has a simple head and a simple socle. Since φ_i is not surjective, we have therefore $\varphi_i(\widehat{Z}^{x_i}(\mu\langle x_i\rangle)) \subset \operatorname{rad} \widehat{Z}^{x_{i+1}}(\mu\langle x_{i+1}\rangle)$. This implies inductively that

$$(6)\qquad M_{i+1} \subset \operatorname{rad}^i \widehat{Z}^{x_{i+1}}(\mu\langle x_{i+1}\rangle).$$

Note that $\widehat{L}_1(\mu) \simeq M_{i+1}/\operatorname{rad} M_{i+1}$. If N is a submodule of $\widehat{Z}^{x_{i+1}}(\mu\langle x_{i+1}\rangle)$, then $M_{i+1} \subset N$ implies $[N : \widehat{L}_1(\mu)] > 0$. On the other hand, if $M_{i+1} \not\subset N$, then $M_{i+1} \cap N \subset \operatorname{rad} M_{i+1}$ and $\widehat{L}_1(\mu)$ is a composition factor of $\widehat{Z}^{x_{i+1}}(\mu\langle x_{i+1}\rangle)/N$. This shows that $[N : \widehat{L}_1(\mu)] > 0$ if and only if $M_{i+1} \subset N$.

Now (4) follows from (5) and (6) taking $i = j = l(w)$.

Remark: We have ${}^{\dot w}\widehat{L}_1(\mu) \simeq \widehat{L}_1(\mu + p(w^{-1}\mu_1 - \mu_1))$ by 9.7(4) when we decompose $\mu = \mu_0 + p^r\mu_1$ with $\mu_0 \in X_1(T)$ and $\mu_1 \in X(T)$ (maybe over some covering). Using D.8(4) the lemma implies that also $\widehat{L}_1(\mu + p(w^{-1}\mu_1 - \mu_1))$ is rigidly placed in $\widehat{Z}_1(w^{-1} \bullet \mu + p(\rho - w^{-1}\rho)$ with corresponding RM–polynomial $v^{l(w)}$. Tensoring with $p(\mu_1 - w^{-1}\mu_1)$ we get that $\widehat{L}_1(\mu)$ is rigidly placed in $\widehat{Z}_1(w^{-1} \bullet \mu - p(w^{-1} \bullet \mu_1 - \mu_1)$ with corresponding RM–polynomial $v^{l(w)}$.

Consider the special case $\mu = w_0 \bullet \lambda$. Here we have $\mu_1 = -\rho$, hence $w^{-1} \bullet \mu_1 = \mu_1$. We get that $\widehat{L}_1(w_0 \bullet \lambda)$ is rigidly placed in $\widehat{Z}_1(w^{-1}w_0 \bullet \lambda)$ with corresponding RM–polynomial $v^{l(w)}$. Substituting w for $w^{-1}w_0$ we get:

(7) *If $(\widehat{D})$ holds, then $\widehat{L}_1(w_0 \bullet \lambda)$ is rigidly placed in $\widehat{Z}_1(w \bullet \lambda)$ and we have*

$$RM(\widehat{Z}_1(w \bullet \lambda), \widehat{L}_1(w_0 \bullet \lambda)) = v^{|R^+| - l(w)} \qquad \text{for all } w \in W.$$

D.10. Lemma: *Suppose that $(\widehat{D})$ holds. Let $w \in W_p$ and $s \in \Sigma$ with $ws \uparrow w$. If a simple G_1T–module L is rigidly placed in $\widehat{Z}_1(w \bullet \lambda)$ and in $\widehat{Z}_1(ws \bullet \lambda)$, then L is also rigidly placed in $\Theta_s\widehat{Z}_1(w \bullet \lambda)$ and we have*

(1) $$RM(\Theta_s\widehat{Z}_1(w \bullet \lambda), L) = RM(\widehat{Z}_1(ws \bullet \lambda), L) + v\, RM(\widehat{Z}_1(w \bullet \lambda), L).$$

Proof: Consider the short exact sequence

$$0 \to \widehat{Z}_1(w \bullet \lambda) \xrightarrow{\iota} \Theta_s\widehat{Z}_1(w \bullet \lambda) \xrightarrow{\pi} \widehat{Z}_1(ws \bullet \lambda) \to 0$$

from D.3(2). Set $n = |R^+|$. We know from D.3 that $\pi(\operatorname{soc}\Theta_s\widehat{Z}_1(w \bullet \lambda)) = 0$ and $\iota(\widehat{Z}_1(w \bullet \lambda)) \subset \operatorname{rad}\Theta_s\widehat{Z}_1(w \bullet \lambda)$. This implies (for all i) that

$$\pi(\operatorname{soc}_{n+2-i}\Theta_s\widehat{Z}_1(w \bullet \lambda)) \subset \operatorname{soc}_{n+1-i}\widehat{Z}_1(ws \bullet \lambda)$$

and

$$\iota(\operatorname{rad}^{i-1}\widehat{Z}_1(w \bullet \lambda)) \subset \operatorname{rad}^i\Theta_s\widehat{Z}_1(w \bullet \lambda).$$

On the other hand, we have by D.1(2), (3)

$$\iota^{-1}(\operatorname{soc}_{n+2-i}\Theta_s\widehat{Z}_1(w \bullet \lambda)) = \operatorname{soc}_{n+2-i}\widehat{Z}_1(w \bullet \lambda)$$

and

$$\pi(\operatorname{rad}^i\Theta_s\widehat{Z}_1(w \bullet \lambda)) = \operatorname{rad}^i\widehat{Z}_1(ws \bullet \lambda).$$

It follows that

$$[\operatorname{soc}_{n+2-i}\Theta_s\widehat{Z}_1(w \bullet \lambda) : L] \le [\operatorname{soc}_{n+2-i}\widehat{Z}_1(w \bullet \lambda) : L] + [\operatorname{soc}_{n+1-i}\widehat{Z}_1(ws \bullet \lambda) : L]$$

and

$$[\operatorname{rad}^i\Theta_s\widehat{Z}_1(w \bullet \lambda) : L] \ge [\operatorname{rad}^{i-1}\widehat{Z}_1(w \bullet \lambda) : L] + [\operatorname{rad}^i\widehat{Z}_1(ws \bullet \lambda) : L].$$

By our assumption on L and by D.8(2) the right hand sides of these two inequalities are equal. So we get

$$[\operatorname{rad}^i\Theta_s\widehat{Z}_1(w \bullet \lambda) : L] \ge [\operatorname{soc}_{n+2-i}\Theta_s\widehat{Z}_1(w \bullet \lambda) : L],$$

hence equality by D.1(5), since we know from D.8(3) that $\Theta_s\widehat{Z}_1(w \bullet \lambda)$ has Loewy length $n+2$. Therefore L is rigidly placed in $\Theta_s\widehat{Z}_1(w \bullet \lambda)$. Furthermore, also the earlier inequalities have to have been equalities. We get in particular

$$[\operatorname{rad}^i\Theta_s\widehat{Z}_1(w \bullet \lambda) : L] = [\operatorname{rad}^{i-1}\widehat{Z}_1(w \bullet \lambda) : L] + [\operatorname{rad}^i\widehat{Z}_1(ws \bullet \lambda) : L].$$

This implies (1) taking the difference of this equation for i and for $i-1$.

D.11. Set $RM(w,x) = RM(\widehat{Z}_1(w\bullet\lambda), \widehat{L}_1(x\bullet\lambda))$ for all $w, x \in W_p$.

Lemma: *Suppose that $(\widehat{D})$ holds. Let $w, x \in W_p$ and $s \in \Sigma$ with $ws \uparrow w$ and $xs \uparrow x$. If $\widehat{L}_1(x\bullet\lambda)$ is rigidly placed in $\Theta_s\widehat{Z}_1(w\bullet\lambda)$, then $\widehat{L}_1(xs\bullet\lambda)$ is rigidly placed in $\widehat{Z}_1(w\bullet\lambda)$ and in $\widehat{Z}_1(ws\bullet\lambda)$ and $RM(\Theta_s\widehat{Z}_1(w\bullet\lambda), \widehat{L}_1(x\bullet\lambda))$ is equal to*

$$RM(w,xs) + \sum_{y;\, x\uparrow y} [\Theta_s\widehat{L}_1(y\bullet\lambda) : \widehat{L}_1(x\bullet\lambda)]\, RM(w,y) \tag{1}$$

and to

$$v\, RM(ws,xs) + v \sum_{y;\, x\uparrow y} [\Theta_s\widehat{L}_1(y\bullet\lambda) : \widehat{L}_1(x\bullet\lambda)]\, RM(ws,y). \tag{1'}$$

Proof: Set $L = \widehat{L}_1(x\bullet\lambda)$. The assumption $xs \uparrow x$ implies that $\Theta_s L = 0$. This implies for each G_1T–module V that $\operatorname{Hom}_{G_1T}(\Theta_s V, L) \simeq \operatorname{Hom}_{G_1T}(V, \Theta_s L) = 0$ and $\operatorname{Hom}_{G_1T}(L, \Theta_s V) \simeq \operatorname{Hom}_{G_1T}(\Theta_s L, V) = 0$, hence

$$[\Theta_s V : L] = [\operatorname{rad}\Theta_s V : L] \qquad \text{and} \qquad [\operatorname{soc}\Theta_s V : L] = 0. \tag{2}$$

Set $M = \widehat{Z}_1(w\bullet\lambda)$. Let us identify $\Theta_s M_1$ for each submodule M_1 of M with a submodule of $\Theta_s M$. Our assumption $(\widehat{D})$ implies that $\ell\ell(\Theta_s \operatorname{soc}_{n+1-i} M) \le n+3-i$, hence

$$\Theta_s \operatorname{soc}_{n+1-i} M \subset \operatorname{soc}_{n+3-i} \Theta_s M$$

and $\operatorname{rad}\Theta_s \operatorname{soc}_{n+1-i} M \subset \operatorname{soc}_{n+2-i}\Theta_s M$. Now (2) yields

$$[\Theta_s \operatorname{soc}_{n+1-i} M : L] \le [\operatorname{soc}_{n+2-i}\Theta_s M : L]. \tag{3}$$

Let N_i be the submodule of $\Theta_s M$ with $N_i/\Theta_s \operatorname{rad}^i M = \operatorname{soc}(\Theta_s M/\Theta_s \operatorname{rad}^i M)$. We get from $(\widehat{D})$ that

$$\ell\ell(\Theta_s M/\Theta_s \operatorname{rad}^i M) = \ell\ell(\Theta_s \operatorname{rad} M/\Theta_s \operatorname{rad}^i M) = \ell\ell(\Theta_s(\operatorname{rad} M/\operatorname{rad}^i M)) \le i+1.$$

(Recall that $\Theta_s M = \Theta_s \operatorname{rad} M$ by D.3(3).) This implies that $\operatorname{rad}^i \Theta_s M \subset N_i$. We know by (2) that $[N_i/\Theta_s \operatorname{rad}^i M : L] = 0$ and get now

$$[\operatorname{rad}^i \Theta_s M : L] \le [\Theta_s \operatorname{rad}^i M : L]. \tag{4}$$

Our assumption that L is rigidly placed in $\Theta_s M$ says that $[\operatorname{soc}_{n+2-i}\Theta_s M : L] = [\operatorname{rad}^i \Theta_s M : L]$. So we get from (3) and (4)

$$[\Theta_s \operatorname{soc}_{n+1-i} M : L] \le [\Theta_s \operatorname{rad}^i M : L].$$

Now the inclusion $\operatorname{rad}^i M \subset \operatorname{soc}_{n+1-i} M$ yields the reverse inequality, hence equality. It follows that

$$[\Theta_s(\operatorname{soc}_{n+1-i} M/\operatorname{rad}^i M) : L] = 0 \tag{5}$$

(by the exactness of Θ_s) and

$$[\operatorname{rad}^i \Theta_s M : L] = [\Theta_s \operatorname{rad}^i M : L]. \tag{6}$$

Now (5) implies (using the exactness of Θ_s once more) that

$$0 = \sum_{y \in W_p} [\operatorname{soc}_{n+1-i} M/\operatorname{rad}^i M : \widehat{L}_1(y \bullet \lambda)]\,[\Theta_s \widehat{L}_1(y \bullet \lambda) : L]$$

hence $[\operatorname{soc}_{n+1-i} M/\operatorname{rad}^i M : \widehat{L}_1(y \bullet \lambda)] = 0$ for all y with $[\Theta_s \widehat{L}_1(y \bullet \lambda) : L] \neq 0$. Since $[\Theta_s \widehat{L}_1(xs \bullet \lambda) : L] = 1$, this implies in particular that (for all i)

$$[\operatorname{soc}_{n+1-i} M/\operatorname{rad}^i M : \widehat{L}_1(xs \bullet \lambda)] = 0.$$

This shows that $\widehat{L}_1(xs \bullet \lambda)$ is rigidly placed in $M = \widehat{Z}_1(w \bullet \lambda)$. Furthermore (6) and the exactness of Θ_s imply

$$[\operatorname{rad}^i \Theta_s M/\operatorname{rad}^{i+1} \Theta_s M : L] = [\Theta_s(\operatorname{rad}^i M/\operatorname{rad}^{i+1} M) : L].$$

This yields (1).

Set $M' = \widehat{Z}_1(ws \bullet \lambda)$. We proceed more or less as above. We identify $\Theta_s M'$ with $\Theta_s M$ via D.3(1). We have $\Theta_s \operatorname{soc} M' = 0$ by D.3(6). It follows that $\ell\ell(\Theta_s \operatorname{soc}_{n+1-i} M') \leq n+2-i$ for all i, hence

$$\Theta_s \operatorname{soc}_{n+1-i} M' \subset \operatorname{soc}_{n+2-i} \Theta_s M'$$

and $\operatorname{rad} \Theta_s \operatorname{soc}_{n+1-i} M' \subset \operatorname{soc}_{n+1-i} \Theta_s M'$. So (2) implies

(7) $$[\Theta_s \operatorname{soc}_{n+1-i} M' : L] \leq [\operatorname{soc}_{n+1-i} \Theta_s M' : L].$$

Let N_i' be the submodule of $\Theta_s M'$ with

$$N_i'/\Theta_s \operatorname{rad}^i M' = \operatorname{soc}(\Theta_s M'/\Theta_s \operatorname{rad}^i M') \simeq \operatorname{soc} \Theta_s(M'/\operatorname{rad}^i M').$$

Now $\ell\ell(\Theta_s(M'/\operatorname{rad}^i M')) \leq i+2$ implies $\operatorname{rad}^{i+1} \Theta_s M' \subset N_i'$, hence using (2)

(8) $$[\operatorname{rad}^{i+1} \Theta_s M' : L] \leq [\Theta_s \operatorname{rad}^i M' : L].$$

Now the same arguments as above show that $\widehat{L}_1(xs \bullet \lambda)$ is rigidly placed in $M' = \widehat{Z}_1(ws \bullet \lambda)$. Furthermore, we get equality in (8) and can then deduce (1′).

Remark: Note that $RM(w,x) \neq 0$ implies $[\widehat{Z}_1(w \bullet \lambda) : \widehat{L}_1(x \bullet \lambda)] \neq 0$, hence $x \uparrow w$ and $d(x) \leq d(w)$. We can now show using induction on $d(w) - d(xs)$: *Suppose that $(\widehat{D})$ holds and that all $\widehat{Z}_1(\mu)$ with $\mu \in W_p \bullet \lambda$ are rigid. Let $w, x \in W_p$ and $s \in \Sigma$ with $ws \uparrow w$ and $xs \uparrow x$. Then*

(9) $$RM(w, xs) = v\, RM(ws, xs).$$

The assumptions here imply using Lemma D.10 that the sums in (1) and in (1′) are equal. Now use that all y occurring in that sum satisfy $\Theta_s \widehat{L}_1(y \bullet \lambda) \neq 0$, hence $y \uparrow ys$, and $d(y) > d(x) > d(xs)$, hence by induction $RM(w,y) = v\, RM(ws,y)$.

D.12. Proposition: *Suppose that $(\widehat{D})$ holds. Then all $\widehat{Z}_1(\mu)$ with $\mu \in W_p \bullet \lambda$ are rigid.*

Proof: We want to show for each $x \in W_p$ that $\widehat{L}_1(x \bullet \lambda)$ is rigidly placed in all $\widehat{Z}_1(\mu)$ with $\mu \in W_p \bullet \lambda$. That will imply the proposition, cf. D.9.

If our claim holds for some $\widehat{L}_1(x \bullet \lambda)$, then it also holds for all $\widehat{L}_1(x \bullet \lambda + p\nu) \simeq \widehat{L}_1(x \bullet \lambda) \otimes p\nu$ with $\nu \in \mathbf{Z}R$, as $\widehat{Z}_1(\mu + p\nu) \simeq \widehat{Z}_1(\mu) \otimes p\nu$ and as tensoring with $p\nu$ does not change the socle or the radical series.

Furthermore, if the claim holds for some $\widehat{L}_1(x \bullet \lambda)$, then it also holds for all $\widehat{L}_1(xs \bullet \lambda)$ with $s \in \Sigma$ and $xs \uparrow x$: This follows from Lemma D.10 (which implies that $\widehat{L}_1(x \bullet \lambda)$ is rigidly placed in all $\Theta_s \widehat{Z}_1(\mu)$) and from Lemma D.11.

Now $\widehat{L}_1(w_0 \bullet \lambda)$ is rigidly placed in all $\widehat{Z}_1(\mu)$: If $\mu \in W \bullet \lambda$, then this holds by D.9(7). If $\mu \in W_p \bullet \lambda \setminus W \bullet \lambda$, then $[\widehat{Z}_1(\mu) : \widehat{L}_1(w_0 \bullet \lambda)] = 0$, cf. 9.16(b), and the claim holds trivially.

There exists $x_1 \in W_p$ with $x_1 \bullet \lambda = w_0 \bullet \lambda + 2p\rho$. The observation above shows that $\widehat{L}_1(x_1 \bullet \lambda)$ is rigidly placed in all $\widehat{Z}_1(\mu)$. Note that $x_1 \in W_p^+$, i.e., that $x_1 \bullet \lambda \in X(T)_+$.

Suppose that our claim holds for some $x \in W_p^+$ with $x \neq 1$. There exists $s \in \Sigma$ with $xs \uparrow x$ and $xs \in W_p^+$. By the remark above the claim holds then also for xs. Starting with $x = x_1$ we get now inductively that $\widehat{L}_1(\lambda)$ is rigidly placed in all $\widehat{Z}_1(\mu)$. Then an induction on $l(w)$ yields the same for all $\widehat{L}_1(w \bullet \lambda)$ with $w \in W$. Since $W_p \bullet \lambda = W \bullet \lambda + p\mathbf{Z}R$ we get now the claim in general.

D.13. Assume that $(\widehat{D})$ holds. I want to show in this subsection that the results above allow the calculation of all $RM(w,x)$ with $w, x \in W_p$.

We noted before that $RM(w,x) \neq 0$ implies $x \uparrow w$, hence $d(x) \leq d(w)$. So we can use induction on $d(w) - d(x)$. If $d(w) - d(x) = 0$ and $x \uparrow w$, then $x = w$. In this case $[\widehat{Z}_1(\mu) : \widehat{L}_1(\mu)] = 1$ and $\widehat{Z}_1(\mu)/\operatorname{rad}\widehat{Z}_1(\mu) \simeq \widehat{L}_1(\mu)$ imply $RM(w,w) = 1$ for all $w \in W_p$.

Fix now some integer $d > 0$ and suppose that we know all $RM(w,x)$ with $d(w) - d(x) < d$.

Set $W_p^- = w_0 W_p^+$. (So W_p^- consists of all $w \in W_p$ such that $-w \bullet \lambda$ is dominant.) We shall first determine all $RM(w,x)$ with $d(w) - d(x) = d$ first in case $x \in W_p^-$ using downward induction with respect to $\uparrow$.

If $x = w_0$, then we know all $RM(w,x)$ by D.9(7) since $RM(w,x) = 0$ for $w \notin W$ (as in the proof of D.12). Let now $x \in W_p^-$ with $x \neq w_0$. Then there exists $s \in \Sigma$ with $x \uparrow xs$ and $xs \in W_p^-$. Let $w \in W_p$ with $d(w) - d(x) = d$. If $ws \uparrow w$, then we know $RM(ws,x)$ by induction and get $RM(w,x) = vRM(ws,x)$ from D.11(9). So assume that $w \uparrow ws$. We have now $d(ws) = d(w) + 1$ and $d(xs) = d(x) + 1$, hence $d(ws) - d(xs) = d > d(w) - d(xs)$. So we know $RM(ws,xs)$ and $RM(w,xs)$ by our inductions. Now D.10(1) gives us $f = RM(\Theta_s \widehat{Z}_1(w \bullet \lambda), \widehat{L}_1(xs \bullet \lambda))$. We get from D.11(1′) that

$$f = v\,RM(w,x) + v \sum_{y;\, xs \uparrow y} [\beta_s \widehat{L}_1(y \bullet \lambda) : \widehat{L}_1(xs \bullet \lambda)] RM(w,y)$$

with $\beta_s \widehat{L}_1(y \bullet \lambda)$ as in C.14. (Recall that any simple G_1T–module not isomorphic to $\widehat{L}_1(y \bullet \lambda)$ has the same multiplicity in $\beta_s \widehat{L}_1(y \bullet \lambda)$ as in $\Theta_s \widehat{L}_1(y \bullet \lambda)$.) The $RM(w,y)$

occurring in this sum are known since $d(y) > d(xs)$, hence $d(w) - d(y) < d$. Finally the $[\beta_s \widehat{L}_1(y \bullet \lambda) : \widehat{L}_1(xs \bullet \lambda)]$ are known: Either we apply Proposition C.17 and get them as $\widehat{P}[\beta_s \widehat{L}_1(y \bullet \lambda), xs](0)$, or we observe that this multiplicity is equal to (cf. 9.19(4))

$$\begin{aligned}\dim \operatorname{Ext}^1_{G_1T}(\widehat{L}_1(y \bullet \lambda), \widehat{L}_1(xs \bullet \lambda)) &= \dim \operatorname{Hom}_{G_1T}(\operatorname{rad} \widehat{Z}_1(y \bullet \lambda), \widehat{L}_1(xs \bullet \lambda)) \\ &= [\operatorname{rad} \widehat{Z}_1(y \bullet \lambda)/ \operatorname{rad}^2 \widehat{Z}_1(y \bullet \lambda) : \widehat{L}_1(xs \bullet \lambda)]\end{aligned}$$

i.e., equal to the coefficient of v in $RM(y, xs)$. (And this polynomial is known since $y \uparrow w$ implies $d(y) - d(xs) \leq d(w) - d(xs) < d$.) So we get also $RM(w, x)$.

Once we have all $RM(w, x)$ with $d(w) - d(x) = d$ and $x \in W_p^-$, then we get them in general tensoring with $p\nu$ for suitable $\nu \in \mathbf{Z}R$.

One can now show (comparing inductive formulae) that the $RM(w, x)$ are suitably normalised Kazhdan-Lusztig polynomials, see [Andersen and Kaneda 1], Thm. 6.3. In the set-up of [Soergel 4] one gets here certain periodic polynomials as in [Soergel 4], 4.4. If one indexes the polynomials there by elements of the affine Weyl group instead of by alcoves (identifying the two sets via $w \mapsto w \bullet C$), then one gets

(1) $$RM(w, x) = p_{w_0 w, w_0 x}.$$

The starting data for our induction (i.e., $RM(w, w) = 1$ and D.9(7)) and the inductive formulae show (tracing the arguments):

(2) *If v^i occurs with a non-zero coefficient in $RM(w, x)$, then $i \equiv d(w) - d(x) \pmod 2$.*

This follows also from the interpretation as Kazhdan-Lusztig polynomials, see [Andersen and Kaneda 1], 6.5(ii).

D.14. Proposition: *Suppose that $(\widehat{D})$ holds. Then all $\widehat{Q}_1(\mu)$ with $\mu \in W_p \bullet \lambda$ are rigid. One has for all $w, x \in W_p$*

(1) $$RM(\widehat{Q}_1(w \bullet \lambda), \widehat{L}_1(x \bullet \lambda)) = \sum_{y \in W_p} RM(y, w) RM(y, x).$$

At this point I want to describe only the main step of the proofs and refer for crucial details to [Andersen and Kaneda 1].

Let me write here f_i for the coefficient of v^i in any $f \in \mathbf{Z}[v]$. We have then $\sum_i f_i = f(1)$, in particular $\sum_i RM(y, w)_i = [\widehat{Z}_1(y \bullet \lambda) : \widehat{L}_1(w \bullet \lambda)]$ for all $y, w \in \mathbf{Z}$.

Recall from 11.4 that $\widehat{Q}_1(w \bullet \lambda)$ has a filtration with factors $\widehat{Z}_1(y \bullet \lambda)$, each $\widehat{Z}_1(y \bullet \lambda)$ occurring $[\widehat{Z}_1(y \bullet \lambda) : \widehat{L}_1(w \bullet \lambda)]$ times. This implies for all $x \in W_p$

$$\begin{aligned}[\widehat{Q}_1(w \bullet \lambda) : \widehat{L}_1(x \bullet \lambda)] &= \sum_{y \in W_p} [\widehat{Z}_1(y \bullet \lambda) : \widehat{L}_1(w \bullet \lambda)] \, [\widehat{Z}_1(y \bullet \lambda) : \widehat{L}_1(x \bullet \lambda)] \\ &= \sum_{y \in W_p} \sum_{i,j} RM(y, w)_i \, RM(y, x)_j,\end{aligned}$$

hence

$$[\widehat{Q}_1(w \bullet \lambda) : \widehat{L}_1(x \bullet \lambda)] = \sum_i \Big(\sum_{y \in W_p} RM(y,w)RM(y,x) \Big)_i . \tag{2}$$

Now we need two crucial formulae: One has for all i

$$[\widehat{Q}_1(w \bullet \lambda)/\operatorname{rad}^i \widehat{Q}_1(w \bullet \lambda) : \widehat{L}_1(x \bullet \lambda)] \le \sum_{j<i} \Big(\sum_{y \in W_p} RM(y,w)RM(y,x) \Big)_j \tag{3}$$

and

$$\Big(\sum_{y \in W_p} RM(y,w)RM(y,x) \Big)_i = \Big(\sum_{y \in W_p} RM(y,w)RM(y,x) \Big)_{2n-i} \tag{4}$$

where $n = |R^+|$.

We have by D.1(7) for each finite dimensional G_1T–module M isomorphisms ${}^{\tau}(M/\operatorname{rad}^i M) \simeq \operatorname{soc}_i({}^{\tau}M)$. As ${}^{\tau}(\widehat{Q}_1(w \bullet \lambda)) \simeq \widehat{Q}_1(w \bullet \lambda)$, see 11.5(5), this implies for all simple G_1T–modules L that $[\widehat{Q}_1(w \bullet \lambda)/\operatorname{rad}^i \widehat{Q}_1(w \bullet \lambda) : L] = [\operatorname{soc}_i \widehat{Q}_1(w \bullet \lambda) : L]$. Since $\ell\ell(\widehat{Q}_1(w \bullet \lambda)) = 2n+1$ by Proposition D.8, we get now for all i

$$\begin{aligned}
&[\operatorname{rad}^i \widehat{Q}_1(w \bullet \lambda) : \widehat{L}_1(x \bullet \lambda)] \\
&\quad \le [\operatorname{soc}_{2n+1-i} \widehat{Q}_1(w \bullet \lambda) : \widehat{L}_1(x \bullet \lambda)] \\
&\quad = [\widehat{Q}_1(w \bullet \lambda)/\operatorname{rad}^{2n+i-1} \widehat{Q}_1(w \bullet \lambda) : \widehat{L}_1(x \bullet \lambda)] \\
&\quad \le \sum_{j>i} \Big(\sum_{y \in W_p} RM(y,w)\, RM(y,x) \Big)_{2n+1-j} \\
&\quad = \sum_{j>i} \Big(\sum_{y \in W_p} RM(y,w)\, RM(y,x) \Big)_{j-1} \\
&\quad = \sum_{j} \Big(\sum_{y \in W_p} RM(y,w)\, RM(y,x) \Big)_j - \sum_{j<i} \Big(\sum_{y \in W_p} RM(y,w)\, RM(y,x) \Big)_j \\
&\quad \le [\widehat{Q}_1(w \bullet \lambda) : \widehat{L}_1(x \bullet \lambda)] - [\widehat{Q}_1(w \bullet \lambda)/\operatorname{rad}^i \widehat{Q}_1(w \bullet \lambda) : \widehat{L}_1(x \bullet \lambda)] \\
&\quad = [\operatorname{rad}^i \widehat{Q}_1(w \bullet \lambda) : \widehat{L}_1(x \bullet \lambda)]
\end{aligned}$$

using D.1(5) for the first inequality, and later on (3), (4), (2), and again (3). It follows that all inequalities are equalities. This shows that $\widehat{L}_1(x \bullet \lambda)$ is rigidly placed in $\widehat{Q}_1(w \bullet \lambda)$, hence the rigidity of $\widehat{Q}_1(w \bullet \lambda)$ as x is arbitrary. We also get equality in (3) which yields (1) by taking differences.

So we have reduced the proof of the proposition to that of (3) and (4). In [Andersen and Kaneda 1], 7.1, one gets (4) from the identification of the $RM(w,x)$ with Kazhdan-Lusztig polynomials, while (3) is proved in [Andersen and Kaneda 1], Prop. 3.7 by analysing the filtration of $\widehat{Q}_1(w \bullet \lambda)$ with factors of the form $\widehat{Z}_1(y \bullet \lambda)$.

CHAPTER E

Tilting Modules

Throughout this chapter let k be a field of characteristic $p > 0$. All G–modules will be assumed to be finite dimensional over k.

A G–module is called a tilting module if it has both a good filtration and a Weyl filtration (see 4.16/19).

It is easy to classify all tilting modules: There is for each dominant weight λ an indecomposable tilting module $T(\lambda)$ with highest weight λ; this module satisfies $\dim T(\lambda)_\lambda = 1$ and is unique up to isomorphism. An arbitrary tilting module is then a direct sum of such $T(\lambda)$s, see E.6.

It is not so easy to describe the $T(\lambda)$ more precisely, in particular their characters. One has $T(\lambda) = H^0(\lambda)$ if $H^0(\lambda)$ is simple (E.1), but that happens only for a small portion of all dominant weights. For $p \geq 2h-2$ we have seen in 11.11 that the indecomposable projective G_rT–modules $\widehat{Q}_r(\lambda)$ with $\lambda \in X_r(T)$ lift to G. One gets thus a tilting module $\widehat{Q}_r(\lambda) \simeq T(2(p^r-1)\rho + w_0\lambda)$, see E.9. In this case one gets from 11.4 a character formula in terms of the composition factor multiplicities of the G_rT–modules $\widehat{Z}_r(\mu)$.

In general, there is a conjectured character formula for $T(\lambda)$ in case λ is not "too large", see E.10. Here the meaning of "too large" is similar to that in the case of Lusztig's conjecture (in 8.22). In fact, the conjecture on the tilting modules would imply Lusztig's conjecture.

This conjecture involves the value at 1 of a Kazhdan-Lusztig polynomial. As in other cases, one expects also here that the individual coefficients of these polynomials have a representation theoretic interpretation: They should describe the factors in a certain filtration associated to the tilting modules, cf. E.24. The construction of this filtration requires the description of tilting modules for groups over complete discrete valuation rings that we achieve in E.22.

One reason for the interest in these tilting modules comes from the representation theories of symmetric and general linear groups. The tensor powers of the natural module of GL_n are tilting modules; the multiplicity of some $T(\lambda)$ as an indecomposable summand in the d–th tensor power is equal to the dimension of some irreducible representation of the symmetric group S_d, see E.17. This fact has been used by several people to compute such dimensions effectively.

Another interesting feature of the category of tilting modules is the fact that it is closed under tensor products, see E.7. This has led to the description of some remarkable "fusion rings", cf. E.13.

The term tilting module comes from the representation theory of finite dimensional algebras where they appear as modules inducing the *tilting functors* of Brenner and Butler. These functors generalise functors in the representation theory of quivers which relate the representations of two quivers in case one quiver arises from the other by reversing the orientation of certain arrows.

The connection of that theory with our tilting G–modules arises from the truncated categories $\mathcal{C}(\pi)$ as in Chapter A. If π is a saturated and finite set of dominant weights, then $\mathcal{C}(\pi)$ is equivalent to the category of modules over a generalised Schur algebra $S_G(\pi)$. Then the direct sum of all indecomposable tilting G–modules that belong to $\mathcal{C}(\pi)$ is a tilting module for $S_G(\pi)$ in the sense of the representation theory of finite dimensional algebras.

If we were to follow the conventions in the representation theory of finite dimensional algebras, the tilting modules defined above should be called *partial* tilting modules. However, we stick to the different convention that has become standard in the representation theory of algebraic groups.

The adaptation of the general concept to the present set-up was done in [Ringel] and [Donkin 23]. The other main sources for this chapter are [Andersen 26, 27, 28, 29], [Donkin 24], and [Mathieu 5].

E.1. A finite dimensional G–module V is called a *tilting module* if V has both a good filtration (as in 4.16) and a Weyl filtration (as in 4.19). This is equivalent to that both V and V^* have a good filtration.

For example, if $H^0(\lambda)$ is simple, for some $\lambda \in X(T)_+$, then $H^0(\lambda) \simeq V(\lambda)$ is a tilting module for G. This holds in particular for all $\lambda \in X(T)_+$ with $\langle \lambda+\rho, \alpha^\vee\rangle \leq p$ for all $\alpha \in R^+$. (This shows that in characteristic 0 each finite dimensional G–module is tilting; so this notion is not interesting in that case and therefore we restrict here to prime characteristic.)

Let μ be a dominant weight in the upper closure of the first alcove C such that $\langle \mu+\rho, \beta^\vee\rangle = p$ for exactly one positive root β. Let λ be a dominant weight in the interior of C. Then the G–module $T_\mu^\lambda H^0(\mu)$ has a filtration with factors $H^0(\lambda)$, $H^0(s \bullet \lambda)$ where $s = s_{\beta,p}$, and a filtration with factors $V(s \bullet \lambda)$, $V(\lambda)$. So $T_\mu^\lambda H^0(\mu)$ is a tilting module for G. (In the future we shall usually drop "for G" in case it is clear which group we consider.)

Remark 7.13 shows more generally that translation functors take tilting modules to tilting modules.

If k' is an extension field of k and if M is a tilting module for G, then $M \otimes k'$ is a tilting module for $G_{k'}$ since $H^0(\mu) \otimes k' \simeq H^0_{k'}(\mu)$ and $V(\mu) \otimes k' \simeq V(\mu)_{k'}$ for all $\mu \in X(T)_+$.

Proposition: *Let V be a finite dimensional G–module. Then the following three properties are equivalent:*

(i) *V is a tilting module.*

(ii) $\operatorname{Ext}^i_G(V(\lambda), V) = 0 = \operatorname{Ext}^i_G(V, H^0(\lambda))$ *for all* $\lambda \in X(T)_+$ *and all* $i > 0$.

(iii) $\operatorname{Ext}^1_G(V(\lambda), V) = 0 = \operatorname{Ext}^1_G(V, H^0(\lambda))$ *for all* $\lambda \in X(T)_+$.

Here "(i) $\Rightarrow$ (ii)" follows from 4.13, and "(ii) $\Rightarrow$ (iii)" is trivial. Finally, we get "(iii) $\Rightarrow$ (i)" from 4.16.b and $\operatorname{Ext}^1_G(V(\lambda), V^*) \simeq \operatorname{Ext}^1_G(V, H^0(-w_0\lambda))$.

E.2. Corollary: *Let V_1 and V_2 be finite dimensional G–modules.*

a) *The direct sum $V_1 \oplus V_2$ is a tilting module if and only if both V_1 and V_2 are tilting modules.*

b) *If* V_1 *and* V_2 *are tilting modules, then* $\mathrm{Ext}^i_G(V_1, V_2) = 0$ *for all* $i > 0$.

This is now clear. (In b) it suffices to assume that V_1 has a Weyl filtration and V_2 a good filtration.)

E.3. Lemma: *For each dominant* $\lambda \in X(T)$ *there exists a tilting module* M *with* $\dim M_\lambda = 1$ *such that* $M_\mu \neq 0$ *implies* $\mu \leq \lambda$.

Proof: Let $\mathcal{C}(\leq \lambda)$ denote the full subcategory of all G–modules M such that $M_\mu \neq 0$ implies $\mu \leq \lambda$. Choose a numbering $\lambda_1, \lambda_2, \dots, \lambda_r$ of the set of all dominant weights ν with $\nu \leq \lambda$ such that $\lambda_i < \lambda_j$ implies $i < j$. So we have in particular $\lambda = \lambda_r$.

We want to construct G–modules M_i, $1 \leq i \leq r$ using downward induction on i such that each M_i belongs to $\mathcal{C}(\leq \lambda)$, has a Weyl filtration, satisfies $\dim(M_i)_\lambda = 1$ and $\mathrm{Ext}^1_G(V(\lambda_j), M_i) = 0$ for all $j \geq i$. We get from Remark 2 in 4.13 that $\mathrm{Ext}^1_G(V(\mu), M_i) = 0$ for all $\mu \in X(T)_+$ with $\mu \not\leq \lambda$ since M_i does not have a weight $\mu' > \mu$. It follows then that M_1 satisfies (iii) in Proposition E.1, hence the conditions in the present lemma.

We start with $M_r = V(\lambda) = V(\lambda_r)$. It satisfies the conditions above by 2.2 and Remark 2 in 4.13.

Let now $i < r$ and suppose that we already have constructed M_{i+1}. Set $d = \dim \mathrm{Ext}^1_G(V(\lambda_i), M_{i+1})$. We have $d < \infty$ by 4.10(1). If $d = 0$, then we can set $M_i = M_{i+1}$. If not, then choose a basis $\xi_1, \xi_2, \dots, \xi_d$ for $\mathrm{Ext}^1_G(V(\lambda_i), M_{i+1})$ over k. Denote by $\xi \in \mathrm{Ext}^1_G(V(\lambda_i)^d, M_{i+1})$ the element with $\iota_j^*(\xi) = \xi_j$ for all j, $1 \leq j \leq d$. Here ι_j denotes the embedding of $V(\lambda_i)$ into $V(\lambda_i)^d$ as the j–th summand, and ι_j^* denotes the induced map $\mathrm{Ext}^1_G(V(\lambda_i)^d, M_{i+1}) \to \mathrm{Ext}^1_G(V(\lambda_i), M_{i+1})$. We get an isomorphism $\mathrm{Ext}^1_G(V(\lambda_i)^d, M_{i+1}) \xrightarrow{\sim} \mathrm{Ext}^1_G(V(\lambda_i), M_{i+1})^d$ defined by the d–tuple $(\iota_1^*, \iota_2^*, \dots, \iota_d^*)$.

Let

$$0 \to M_{i+1} \longrightarrow M_i \xrightarrow{\pi} V(\lambda_i)^d \to 0 \tag{1}$$

be a representative for the extension ξ. Applying $\mathrm{Hom}_G(V(\lambda_i), ?)$ we get a long exact sequence part of which has the form

$$\begin{aligned} \mathrm{Hom}_G(V(\lambda_i), M_i) \to \mathrm{Hom}_G(V(\lambda_i), V(\lambda_i)^d) \to \mathrm{Ext}^1_G(V(\lambda_i), M_{i+1}) \\ \to \mathrm{Ext}^1_G(V(\lambda_i), M_i) \to \mathrm{Ext}^1_G(V(\lambda_i), V(\lambda_i)^d). \end{aligned} \tag{2}$$

If we construct the long sequence with $\mathrm{Hom}_G(V(\lambda_i)^d, ?)$ instead, then the map $\mathrm{Hom}_G(V(\lambda_i)^d, V(\lambda_i)^d) \to \mathrm{Ext}^1_G(V(\lambda_i)^d, M_{i+1})$ takes the identity to ξ. It follows by functoriality, that the map $\mathrm{Hom}_G(V(\lambda_i), V(\lambda_i)^d) \to \mathrm{Ext}^1_G(V(\lambda_i), M_{i+1})$ takes each $\iota_j = \iota_j^*(\mathrm{id})$ to $\xi_j = \iota_j^*(\xi)$. Therefore $\mathrm{Hom}_G(V(\lambda_i), V(\lambda_i)^d) \to \mathrm{Ext}^1_G(V(\lambda_i), M_{i+1})$ takes a basis to a basis, hence is bijective.

Since $\mathrm{Ext}^1_G(V(\lambda_i), V(\lambda_i)^d) = 0$, we get now that also $\mathrm{Ext}^1_G(V(\lambda_i), M_i) = 0$. We have for all $j > i$ that $\lambda_i \not\geq \lambda_j$, hence $\mathrm{Ext}^1_G(V(\lambda_j), V(\lambda_i)) = 0$; therefore $\mathrm{Ext}^1_G(V(\lambda_j), M_{i+1}) = 0$ implies $\mathrm{Ext}^1_G(V(\lambda_j), M_i) = 0$.

Remarks: 1) The module M constructed in the proof is indecomposable. In fact, all M_i that occur above are indecomposable. In order to see this, note first that

$$\varphi(M_{i+1}) \subset M_{i+1} \qquad \text{for all } \varphi \in \mathrm{End}_G(M_i). \tag{3}$$

This holds since $\mathrm{Hom}_G(V(\lambda_j), V(\lambda_i)) = 0$ for all $j > i$ (as $\lambda_j \not\leq \lambda_i$) and since M_{i+1} has a Weyl filtration where all factors have the form $V(\lambda_j)$ with $j > i$; this implies $\mathrm{Hom}_G(M_{i+1}, V(\lambda_i)) = 0$ and thus $\mathrm{Hom}_G(M_{i+1}, M_i/M_{i+1}) = 0$.

Let us show next by downward induction on i for all $\varphi \in \mathrm{End}_G(M_i)$

$$\varphi((M_i)_\lambda) = 0 \implies \varphi \text{ is nilpotent.}$$

To start with, as $(M_i)_\lambda$ has dimension 1, we have $(M_i)_\lambda = (M_{i+1})_\lambda$. We know by (3) that $\varphi(M_{i+1}) \subset M_{i+1}$ and get now by induction that the restriction of φ to M_{i+1} is nilpotent. We now replace φ by some power and may assume that $\varphi(M_{i+1}) = 0$. So φ factors over a homomorphism $\overline{\varphi}$ from $V(\lambda_i)^d \simeq M_i/M_{i+1}$ to M_i. We saw above in the proof that the second map in (2) is an isomorphism. Therefore the first map in (2) is 0. This map takes any $\psi : V(\lambda_i) \to M_i$ to $\pi \circ \psi : V(\lambda_i) \to V(\lambda_i)^d$. So we get $\pi \circ \psi = 0$ for all these ψ. It follows that $\pi \circ \overline{\varphi} = 0$, hence $\overline{\varphi}(V(\lambda_i)^d) = \varphi(M_i) \subset M_{i+1}$ and hence $\varphi^2 = 0$.

The set J_i of all $\varphi \in \mathrm{End}_G(M_i)$ with $\varphi((M_i)_\lambda) = 0$ is clearly a two-sided ideal in $\mathrm{End}_G(M_i)$ with $\mathrm{End}_G(M_i) = k\,\mathrm{id} + J_i$ since $\dim(M_i)_\lambda = 1$. As all elements in J_i are nilpotent, all elements in $\mathrm{End}_G(M_i)$ not in J_i are invertible. This shows that $\mathrm{End}_G(M_i)$ is a local ring with maximal ideal J_i and hence that M_i is indecomposable.

2) If $V(\mu)$ occurs in a Weyl filtration of M, then $\mu \uparrow \lambda$. Indeed, by construction μ has to be one of the λ_i. We use downward induction on i. The case $i = r$ is trivial. Suppose now that $i < r$ and that $\mu = \lambda_i$. If $V(\mu)$ occurs in the Weyl filtration of M, then $\mathrm{Ext}^1_G(V(\lambda_i), M_{i+1}) \neq 0$. Then there exists $j > i$ such that $V(\lambda_j)$ occurs in a Weyl filtration of M_{i+1} and $\mathrm{Ext}^1_G(V(\lambda_i), V(\lambda_j)) \neq 0$. Now induction yields $\lambda_j \uparrow \lambda$. On the other hand, $0 \neq \mathrm{Ext}^1_G(V(\lambda_i), V(\lambda_j)) \simeq \mathrm{Ext}^1_G(H^0(\lambda_j), H^0(\lambda_i))$, cf. 2.12(3), implies by 6.20 that $\lambda_i \uparrow \lambda_j$. The claim follows.

E.4. For each $\lambda \in X(T)^+$ let $T(\lambda)$ denote an indecomposable tilting module with $\dim T(\lambda)_\lambda = 1$ such that $T(\lambda)_\mu \neq 0$ implies $\mu \leq \lambda$. Lemma E.3 implies that such a module exists since the module M constructed there has an indecomposable direct summand with a one dimensional λ-weight space and this summand is again a tilting module by Corollary E.2. In fact, Remark 1 in E.3 shows that we can take as $T(\lambda)$ the module constructed in the proof of Lemma E.3. (We shall soon see that $T(\lambda)$ is determined uniquely up to isomorphism.)

As λ is the largest weight of $T(\lambda)$, we can choose the good filtration of $T(\lambda)$ such that $H^0(\lambda)$ occurs as its top factor (cf. Remark 4 in 4.16). So we have a surjective homomorphism $\pi_\lambda : T(\lambda) \to H^0(\lambda)$ of G–modules such that also $\ker(\pi_\lambda)$ has a good filtration. Since $\dim T(\lambda)_\lambda = 1$, the restriction of π_λ to the λ–weight spaces is an isomorphism $H^0(\lambda)_\lambda \xrightarrow{\sim} H^0(\lambda)_\lambda$.

Applying this argument to $T(\lambda)^*$ and dualising, we get an injective homomorphism $\iota_\lambda : V(\lambda) \to T(\lambda)$ of G–modules such that $T(\lambda)/\operatorname{im}(\iota_\lambda)$ has a Weyl filtration. Now ι_λ restricts to an isomorphism $V(\lambda)_\lambda \xrightarrow{\sim} T(\lambda)_\lambda$.

Lemma: *Let V be a G–module.*

a) *If V has a good filtration, then the map* $\mathrm{Hom}_G(T(\lambda), V) \to \mathrm{Hom}_G(V(\lambda), V)$ *with* $\varphi \mapsto \varphi \circ \iota_\lambda$ *is surjective.*

b) *If V has a Weyl filtration, then the map* $\mathrm{Hom}_G(V, T(\lambda)) \to \mathrm{Hom}_G(V, H^0(\lambda))$ *with* $\varphi \mapsto \pi_\lambda \circ \varphi$ *is surjective.*

Proof: A look at a suitable long exact sequence shows that the claims follow once we know that $\mathrm{Ext}^1_G(T(\lambda)/\mathrm{im}(\iota_\lambda), V) = 0$ in a) and $\mathrm{Ext}^1_G(V, \ker(\pi_\lambda)) = 0$ in b). Both Ext groups have the form $\mathrm{Ext}^1_G(M_1, M_2)$ where M_1 has a Weyl filtration and M_2 a good filtration. So the claim follows from 4.13.

E.5. Keep the assumptions and notations from E.4. The ring $\mathrm{End}_G(T(\lambda))$ is local since $T(\lambda)$ is indecomposable. Each $\varphi \in \mathrm{End}_G(T(\lambda))$ with $\varphi(T(\lambda)_\lambda) = 0$ is not bijective, hence belongs to the radical of $\mathrm{End}_G(T(\lambda))$. As $T(\lambda)_\lambda$ has dimension 1, any $\varphi \in \mathrm{End}_G(T(\lambda))$ can be written as $\varphi = a\,\mathrm{id}_{T(\lambda)} + \varphi_1$ with $a \in k$ and $\varphi_1(T(\lambda)_\lambda) = 0$. It follows that

$$\mathrm{End}_G(T(\lambda)) = k\,\mathrm{id}_{T(\lambda)} \oplus \mathrm{rad}\,\mathrm{End}_G(T(\lambda)) \tag{1}$$

(as vector spaces) and

$$\mathrm{rad}\,\mathrm{End}_G(T(\lambda)) = \{\,\varphi \in \mathrm{End}_G(T(\lambda)) \mid \varphi(T(\lambda)_\lambda) = 0\,\}. \tag{2}$$

If k' is an extension field of k, then $\mathrm{End}_G(T(\lambda)) \otimes k' \simeq \mathrm{End}_{G_{k'}}(T(\lambda) \otimes k')$, cf. I.2.10(7). Now (1) implies easily that $\mathrm{End}_{G_{k'}}(T(\lambda) \otimes k')$ is a local ring with radical equal to $(\mathrm{rad}\,\mathrm{End}_G(T(\lambda))) \otimes k'$. So $T(\lambda) \otimes k'$ is an indecomposable $G_{k'}$–module. (The same argument shows also that each M_i in E.3 is absolutely indecomposable.)

Lemma: *Let Q be a tilting module with $Q_\lambda \neq 0$ such that λ is maximal among the weights of Q. Then there exist submodules Q_1, Q_2 of Q with $Q = Q_1 \oplus Q_2$ and $Q_1 \simeq T(\lambda)$.*

Proof: Choose $v \in Q_\lambda$, $v \neq 0$. We have then homomorphisms of G–modules $\varphi : V(\lambda) \to Q$ with $v \in \mathrm{im}(\varphi)$, and $\psi : Q \to H^0(\lambda)$ with $\psi(v) \neq 0$, cf. Remark 2 in 2.13. Then $\psi \circ \varphi$ restricts to an isomorphism $V(\lambda)_\lambda \xrightarrow{\sim} H^0(\lambda)_\lambda$.

Lemma E.4 implies that there exists a homomorphism of G–modules $g : Q \to T(\lambda)$ with $\pi_\lambda \circ g = \psi$ and a homomorphism of G–modules $f : T(\lambda) \to Q$ with $f \circ \iota_\lambda = \varphi$.

Now $\psi \circ \varphi = \pi_\lambda \circ g \circ f \circ \iota_\lambda$ implies that $g \circ f$ is non-zero on $T(\lambda)_\lambda$, hence bijective on $T(\lambda)_\lambda$ as $\dim T(\lambda)_\lambda = 1$. This implies that the $g \circ f$ is not in the radical of the local ring $\mathrm{End}_G(T(\lambda))$. Therefore $g \circ f$ is bijective. It follows that $f(T(\lambda)) \simeq T(\lambda)$ and that $Q = f(T(\lambda)) \oplus \ker(g)$, hence the claim.

Remark: The proof shows more generally: Let Q be a tilting module. If there exist homomorphisms $\varphi : V(\lambda) \to Q$ and $\psi : Q \to H^0(\lambda)$ with $\psi \circ \varphi \neq 0$, then Q has a direct summand isomorphic to $T(\lambda)$. (Note that $\psi \circ \varphi \neq 0$ implies that $\psi \circ \varphi$ restricts to an isomorphism $V(\lambda)_\lambda \xrightarrow{\sim} H^0(\lambda)_\lambda$.)

E.6. Proposition: *There is for each $\lambda \in X(T)_+$ an indecomposable tilting module $T(\lambda)$ with $\dim T(\lambda)_\lambda = 1$ such that $T(\lambda)_\mu \neq 0$ implies $\mu \leq \lambda$. These properties determine $T(\lambda)$ uniquely up to isomorphism. If Q is an arbitrary tilting module, then there exist unique integers $n(\nu) \geq 0$, almost all 0 with*

$$Q \simeq \bigoplus_{\nu \in X(T)_+} T(\nu)^{n(\nu)}. \tag{1}$$

Proof: The existence of $T(\lambda)$ follows from Lemma E.3 (as observed at the beginning of E.4), the uniqueness from Lemma E.5. If Q is an arbitrary tilting module, then we use induction on $\dim(Q)$ in order to get a decomposition as in (1): If $Q \neq 0$, then we choose a maximal weight λ of Q and get by Lemma E.5 a decomposition $Q \simeq T(\lambda) \oplus Q'$. Then Q' is a tilting module of smaller dimension; so we can apply induction to Q'. The uniqueness in (1) follows of course from the Krull-Schmidt theorem; one can also use that the formal characters $\operatorname{ch} T(\nu)$ are obviously linearly independent.

Remark: The uniqueness implies for all $\lambda \in X(T)_+$ that

$$T(\lambda)^* \simeq T(-w_0\lambda) \qquad \text{and} \qquad {}^\tau(T(\lambda)) \simeq T(\lambda). \tag{2}$$

(Note that V^* and ${}^\tau V$ are tilting if V is so.) For each extension field k' of k the $G_{k'}$–module $T(\lambda) \otimes k'$ is the analogue to $T(\lambda)$, because it is indecomposable (see E.5), tilting, and has the same formal character as $T(\lambda)$.

E.7. The following properties of the $T(\lambda)$ are easily checked, cf. [Donkin 23], 1.2 and 1.4:

If we have direct product decompositions $G = G_1 \times G_2$ and $T = T_1 \times T_2$ with $T_i \subset G_i$, then we get $T(\lambda) \simeq T_1(\lambda_1) \otimes T_2(\lambda_2)$ where $T_i(\mu)$ denotes an indecomposable tilting module for G_i and where λ_i is the restriction of λ to T_i.

If $\widetilde{G} \to G$ is a covering group, then any $T(\lambda)$ is also an indecomposable tilting module for $\widetilde{G}$.

Restricted to the derived group $\mathcal{D}G$ of G any $T(\lambda)$ yields the indecomposable tilting module for $\mathcal{D}G$ with highest weight the restriction of λ to $T \cap \mathcal{D}G$.

Proposition: *The tensor product of two tilting modules is tilting.*

This is an immediate consequence of Proposition 4.21.

E.8. Lemma: *Let $\lambda \in X(T)_+$ and $r \in \mathbf{N}$, $r > 0$. Then $T(\lambda)$ is projective as a G_rT–module if and only if $\langle \lambda, \alpha^\vee \rangle \geq p^r - 1$ for all simple roots α.*

Proof: We may assume that $(p^r - 1)\rho \in X(T)$: If not, replace G by a suitable covering group. This does not change $T(\lambda)$. We get then that $T(\lambda)$ is projective or not for some covering of G_rT. Now observe that the discussion at the beginning of 9.7 shows that a G_rT–module is projective for such a covering if and only if it is projective for G_rT.

If now $\langle \lambda, \alpha^\vee \rangle \geq p^r - 1$ for all simple roots α, then $\lambda - (p^r - 1)\rho \in X(T)_+$. We have $T((p^r - 1)\rho) = St_r = H^0((p^r - 1)\rho)$ as $H^0((p^r - 1)\rho)$ is simple, see 3.19(4). Now $St_r \otimes T(\lambda - (p^r - 1)\rho)$ has highest weight λ and is tilting (by E.7). Therefore

$T(\lambda)$ is isomorphic to a direct summand of $St_r \otimes T(\lambda - (p^r - 1)\rho)$. As St_r is projective for G_rT, see 10.2, so is $St_r \otimes T(\lambda - (p^r - 1)\rho)$, hence also $T(\lambda)$.

Suppose conversely that $T(\lambda)$ is projective as a G_rT–module. Fix a simple root α and consider the subgroup G_α of G as in 1.3(8), generated by U_α, $U_{-\alpha}$, and T. Since $T(\lambda)$ is projective for G_rT, it is also projective for $(G_\alpha)_rT$, cf. 9.4, I.4.12, and I.5.13. Therefore also $M = \bigoplus_{m\in\mathbf{Z}} T(\lambda)_{\lambda-m\alpha}$ is projective for $(G_\alpha)_rT$ as it is a direct summand of $T(\lambda)$ as $(G_\alpha)_rT$–module. It therefore has a filtration with factors of the form $\widehat{Z}_r^\alpha(\mu)$, the analogues to $\widehat{Z}_r(\mu)$ for $(G_\alpha)_rT$, see 11.4. Since λ is the largest weight of M, one of these factors has to be $\widehat{Z}_r^\alpha(\lambda)$. Now $\lambda - (p^r - 1)\alpha$ is a weight of $\widehat{Z}_r^\alpha(\lambda)$, hence of M and of $T(\lambda)$. Then also

$$s_\alpha(\lambda - (p^r - 1)\alpha) = \lambda + (p^r - 1 - \langle \lambda, \alpha^\vee \rangle)\alpha$$

is a weight of $T(\lambda)$. Now the maximality of λ implies $p^r - 1 \le \langle \lambda, \alpha^\vee \rangle$ as claimed.

E.9. Suppose in this subsection that $(p^r - 1)\rho \in X(T)$ for all $r \in \mathbf{N}$. If $p \ge 2h - 2$, then each projective indecomposable G_rT–module $\widehat{Q}_r(\lambda)$ with $\lambda \in X_r(T)$ extends to a G–module, see 11.11. The lifted module has a good filtration, see 11.13. Since ${}^\tau\widehat{Q}_r(\lambda) \simeq \widehat{Q}_r(\lambda)$ [first by 11.5(5) over G_rT, but then also as a G–module by the uniqueness up to isomorphism of the G–module structure], we get that $\widehat{Q}_r(\lambda)$ also has a Weyl filtration, hence is tilting. Since it is indecomposable also for G, it has to be one of the $T(\mu)$. Now Lemma 11.6 gives us the highest weight of $\widehat{Q}_r(\lambda)$. We get thus:

(1) $$p \ge 2h - 2 \implies \widehat{Q}_r(\lambda) \simeq T(2(p^r - 1)\rho + w_0\lambda) \text{ for all } \lambda \in X_r(T).$$

A conjecture of Donkin ([Donkin 23], 2.2) says that this should hold without restriction on p. By Lemma E.8 the problem is to show that $T(2(p^r - 1)\rho + w_0\lambda)$ is still indecomposable as a G_r–module.

The G–module structure on $\widehat{Q}_r(\lambda)$ for $p \ge 2h - 2$ is constructed by realising $\widehat{Q}_r(\lambda)$ as a direct summand of $St_r \otimes L((p^r - 1)\rho + w_0\lambda)$. A theorem of Pillen (see [Pillen] or [Donkin 23], 2.5) shows that $T(2(p^r - 1)\rho + w_0\lambda)$ is for any p a direct summand of $St_r \otimes L((p^r - 1)\rho + w_0\lambda)$.

Note that $\{ 2(p^r - 1)\rho + w_0\lambda \mid \lambda \in X_r(T) \} = (p^r - 1)\rho + X_r(T)$.

Lemma: *Let $\mu \in (p^r - 1)\rho + X_r(T)$ and $\nu \in X(T)_+$. Then $T(\mu) \otimes T(\nu)^{[r]}$ is a tilting module. If $T(\mu)$ is indecomposable as a G_r–module, then $T(\mu) \otimes T(\nu)^{[r]} \simeq T(\mu + p^r\nu)$.*

Proof: We have $St_r \otimes H^0(\lambda)^{[r]} \simeq H^0((p^r - 1)\rho + p^r\lambda)$ for all $\lambda \in X(T)_+$ by Proposition 3.19. This implies that $St_r \otimes T(\nu)^{[r]}$ has a good filtration. Now ${}^\tau(St_r \otimes T(\nu)^{[r]}) \simeq {}^\tau St_r \otimes ({}^\tau T(\nu))^{[r]} \simeq St_r \otimes T(\nu)^{[r]}$ shows that $St_r \otimes T(\nu)^{[r]}$ also has a Weyl filtration, hence is tilting.

Now $T(\mu)$ is isomorphic to a direct summand of $St_r \otimes T(\mu - (p^r - 1)\rho)$, hence $T(\mu) \otimes T(\nu)^{[r]}$ to one of $(St_r \otimes T(\nu)^{[r]}) \otimes T(\mu - (p^r - 1)\rho)$. Therefore it is tilting by E.7.

If $T(\mu)$ is indecomposable as a G_r–module, then it is actually absolutely indecomposable because it is isomorphic to some $\widehat{Q}_r(\mu')$ and the $\widehat{Q}_r(\mu')$ are absolutely indecomposable. Since also the $T(\lambda)$ are absolutely indecomposable (as

G–modules), the final claim in the lemma follows from the following fact: If M and N are finite dimensional G–modules such that M is absolutely indecomposable as a G_r–module and N absolutely indecomposable as a G–module, then $M \otimes N^{[r]}$ is absolutely indecomposable as a G–module, see [Donkin 3], the lemma in Section 2.

Remark: Consider the case $G = SL_2$ and $r = 1$. Here the assumption that $p \geq 2h - 2$ in (1) is unnecessary since all $\widehat{Q}_1(\lambda)$ are covered by Lemma 11.10. So in this case $T(\mu)$ in the lemma is always indecomposable for G_1. The lemma yields thus an inductive formula for the computation of all $T(m\varpi)$, where $\varpi = \rho$ is the unique fundamental weight, once we know the $T(m\varpi)$ for $0 \leq m \leq 2(p-1)$. But these $T(m\varpi)$ are covered by the easy examples in E.1 (or for $p = 2$ and $m = 2$ by 11.10). This was first observed by Donkin, see [Donkin 23], Example 2 on p. 47.

E.10. The $\chi(\mu)$ with $\mu \in X(T)_+$ are a basis for $\mathbf{Z}[X(T)]^W$ over $\mathbf{Z}$. We denote by $(\chi : \chi(\mu))$ the coefficients of any $\chi \in \mathbf{Z}[X(T)]^W$ with respect to this basis:

(1)
$$\chi = \sum_{\mu \in X(T)_+} (\chi : \chi(\mu))\, \chi(\mu).$$

If $\chi = \mathrm{ch}(M)$ for some G–module M, then we write $(M : \chi(\mu)) = (\mathrm{ch}(M) : \chi(\mu))$. If M has a good filtration (resp. a Weyl filtration), then $(M : \chi(\mu))$ is the number of factors isomorphic to $H^0(\mu)$ (resp. to $V(\mu)$) in such a filtration.

In order to understand all tilting modules one should know all $(T(\lambda) : \chi(\mu))$. They are not known in general. Remark 2 in E.3 shows that $(T(\lambda) : \chi(\mu)) \neq 0$ implies $\mu \uparrow \lambda$. In the simple cases mentioned in E.1 all $(T(\lambda) : \chi(\mu))$ can be read off from the description of $T(\lambda)$.

If $p \geq 2h - 2$, then E.9(1) together with Proposition 11.4 gives us a method to compute all $(T(\lambda) : \chi(\mu))$ with $\lambda \in (p^r - 1) + X_r(T)$ provided we know the characters of all simple G–modules. In fact, one can use E.9(1) to translate [Jantzen 7], 5.9 and gets for $p \geq 2h - 2$ and $\lambda \in X_r(T)$ that

(2)
$$(T(2(p^r - 1)\rho + w_0\lambda) : \chi(\mu)) = [V(\mu) : L(\lambda)]$$

for all $\mu \in X(T)_+$ with $\mu \uparrow 2(p^r - 1)\rho + w_0\lambda$.

Lemma E.9 shows that one can calculate all $(T(\mu + p^r\nu) : \chi(\lambda))$ for μ and ν as in that lemma if one knows all $(T(\mu) : \chi(\lambda'))$ and $(T(\nu) : \chi(\lambda''))$. For $G = SL_2$ one gets thus from Donkin's formula a method to calculate all $(T(\lambda) : \chi(\mu))$.

Suppose that $\lambda_0 \in X(T)_+ \cap C$ with C as in 6.2(6). (Recall that λ_0 exists only if $p \geq h$.) Given $w \in W_p$ with $w \bullet \lambda_0 \in X(T)_+$ any $\mu \in X(T)_+$ with $(T(w \bullet \lambda_0) : \chi(\mu)) \neq 0$ has the form $\mu = x \bullet \lambda_0$ with $x \in W_p$ and $x \uparrow w$. One expects now (see [Andersen 26], Remark (i) in 3.6)

(3) *Conjecture*: $(T(w \bullet \lambda_0) : \chi(x \bullet \lambda_0)) = n_{x,w}(1)$ if $\langle w(\lambda_0 + \rho), \alpha^\vee \rangle < p^2$ for all $\alpha \in R^+$

where $n_{x,w}$ is a certain "parabolic" Kazhdan-Lusztig polynomial normalised as in [Soergel 4], 3.3. This conjecture is analogous to the conjecture in [Soergel 4], 7.2 for quantum analogues of our tilting modules. (There one does not need the bound involving p^2 and the conjecture has been proved in many cases, see [Soergel 5], Section 5.)

The truth of this conjecture would imply for $p \geq 2h-2$ the truth of the Lusztig conjecture as in 8.22(2): It would determine via E.9(1) the formal characters of all $\widehat{Q}_1(\mu)$ with $\mu \in X_1(T) \cap W_p \bullet \lambda_0$. Using 11.3(2) and Proposition 11.4 one gets then all $[\widehat{Z}_1(w \bullet \lambda_0) : \widehat{L}_1(x \bullet \lambda_0)]$ with $w, x \in W_p$, hence all $\operatorname{ch} \widehat{L}_1(x \bullet \lambda_0)$. One can check that one would get the same character formulae as predicted by Lusztig.

In case $G = GL_n$ there is a formula due to Donkin that expresses $(T(\lambda) : \chi(\mu))$ in terms of composition multiplicities. Tensoring with powers of the determinant, one restricts to the case where $\lambda \in \pi(n,d)$ for some d. (Cf. A.3(1) for the notation.) We may then also assume $\mu \in \pi(n,d)$ since we should have $\mu \leq \lambda$. We can now regard λ and μ as partitions of d into at most n parts. Then we can form their transposed partitions that we shall denote by ${}^t\lambda$ and ${}^t\mu$. These are again partitions of d, though not necessarily into at most n parts. (If $\lambda = \sum_{i=1}^{n} m_i \varepsilon_1$, then ${}^t\lambda$ has m_1 parts.) Fix now some $m > 1$ such that ${}^t\lambda$ and ${}^t\mu$ have at most m parts. We can then regard ${}^t\lambda$ and ${}^t\mu$ as elements in $\pi(m,d)$. Now [Donkin 23], 3.10 says that

$$(4) \qquad (T(\lambda; GL_n) : \chi(\mu; GL_n)) = [V({}^t\mu; GL_m) : L({}^t\lambda; GL_m)].$$

Here we have added "GL_n" and "GL_m" so to make clear for which groups we consider the modules. (Note that it does not matter which m we choose as long as the right hand side in (4) is defined; this independence follows from 2.11.)

E.11. For the tilting modules covered by E.9(1) we have by 11.10(1) some information on their behaviour under translation functors. That result generalises to all tilting modules. Before we state the general result, let us consider a special case.

Let $\lambda, \mu \in \overline{C}_{\mathbf{Z}}$. Suppose that $\lambda \in C$ and that there exists $s \in \Sigma$ with $\operatorname{Stab}_{W_p}(\mu) = \{1, s\}$. Let $w \in W_p$ with $w \bullet \mu \in X(T)_+$. We have then also $w \bullet \lambda \in X(T)_+$ and $ws \bullet \lambda \in X(T)_+$. Replacing w by ws if necessary, we may assume that $ws \bullet \lambda < w \bullet \lambda$. We claim that now:

$$(1) \quad T_\mu^\lambda T(w \bullet \mu) \simeq T(w \bullet \lambda) \qquad \text{and} \qquad T_\lambda^\mu T(w \bullet \lambda) \simeq T(w \bullet \mu) \oplus T(w \bullet \mu).$$

We know for all $x \in W_p$ with $x \bullet \mu \in X(T)_+$ that $T_\mu^\lambda V(x \bullet \mu)$ has a filtration with factors $V(x \bullet \lambda)$ and $V(xs \bullet \lambda)$ while $T_\lambda^\mu V(x \bullet \lambda)$ and $T_\lambda^\mu V(xs \bullet \lambda)$ both are isomorphic to $V(x \bullet \mu)$, cf. 7.13. This implies for any finite dimensional G–module M with $\operatorname{pr}_\mu M = M$ that $\operatorname{ch}(T_\lambda^\mu T_\mu^\lambda M) = 2 \operatorname{ch}(M)$. Since the translation functors take tilting modules to tilting modules and since tilting modules are determined by their formal character, this implies

$$(2) \qquad T_\lambda^\mu T_\mu^\lambda T(w \bullet \mu) \simeq T(w \bullet \mu) \oplus T(w \bullet \mu).$$

If $V(x \bullet \mu)$ with $x \in W_p$ and $x \bullet \mu \in X(T)_+$ occurs in a Weyl filtration of $T(w \bullet \mu)$, it has to satisfy $x \bullet \mu \uparrow w \bullet \mu$; this then implies $x \bullet \lambda \uparrow w \bullet \lambda$. It follows that $w \bullet \lambda$ is the largest weight of $T_\mu^\lambda T(w \bullet \mu)$, hence that $T(w \bullet \lambda)$ is a direct summand of $T_\mu^\lambda T(w \bullet \mu)$.

Now (2) shows that $T_\lambda^\mu T(w \bullet \lambda)$ is isomorphic to either $T(w \bullet \mu) \oplus T(w \bullet \mu)$ or to $T(w \bullet \mu)$. We have by 7.19.d a non-zero homomorphism $H^0(w \bullet \lambda) \to H^0(ws \bullet \lambda)$. Combing this with the surjective homomorphism $\pi_{w \bullet \lambda} : T(w \bullet \lambda) \to H^0(w \bullet \lambda)$ as in E.4 we get $\operatorname{Hom}_G(T(w \bullet \lambda), H^0(ws \bullet \lambda)) \neq 0$. This implies that $V(ws \bullet \lambda)$ occurs at least once in a Weyl filtration of $T(w \bullet \lambda)$. Since also $V(w \bullet \lambda)$ occurs, it follows that a Weyl filtration of $T_\lambda^\mu T(w \bullet \lambda)$ has at least two factors isomorphic to $V(w \bullet \mu)$. Therefore $T_\lambda^\mu T(w \bullet \lambda)$ cannot be isomorphic to $T(w \bullet \mu)$. This yields the second claim in (1).

We get furthermore, that $T_\mu^\lambda T(w \bullet \mu)$ is the direct sum of $T(w \bullet \lambda)$ and a tilting module Q with $T_\lambda^\mu Q = 0$. This means that all factors $V(z \bullet \lambda)$ in a Weyl filtration of Q satisfy $T_\lambda^\mu V(z \bullet \lambda) = 0$. But any such $V(z \bullet \lambda)$ occurs also in a Weyl filtration of $T_\mu^\lambda T(w \bullet \mu)$, hence in a Weyl filtration of some $T_\mu^\lambda V(x \bullet \mu)$ with $x \bullet \mu \in X(T)_+$. But both factors in $T_\mu^\lambda V(x \bullet \mu)$, i.e., both $V(x \bullet \lambda)$ and $V(xs \bullet \lambda)$, are taken to $V(x \bullet \mu)$ under T_λ^μ, not to 0. This shows that $Q = 0$ as claimed.

Note that (1) implies for all $\lambda \in C \cap X(T)_+$ and all $w \in W_p$ with $w \bullet \lambda \in X(T)_+$, $ws \bullet \lambda \in X(T)_+$, and $ws \bullet \lambda < w \bullet \lambda$ that

$$(3) \qquad (T(w \bullet \lambda) : \chi(x \bullet \lambda)) = \begin{cases} (T(w \bullet \lambda) : \chi(xs \bullet \lambda)), & \text{if } xs \bullet \lambda \in X(T)_+ \\ 0, & \text{otherwise} \end{cases}$$

for all $x \in W_p$ with $x \bullet \lambda \in X(T)_+$.

Proposition: *Let $\lambda, \mu \in \overline{C}_{\mathbf{Z}}$ such that μ belongs to the closure of the facet of λ. Let $w \in W_p$ such that $w \bullet \mu \in X(T)_+$ and such that $w \bullet \lambda$ is maximal among all $wx \bullet \lambda$ with $x \in \mathrm{Stab}_{W_p}(\mu)$. Then $T_\mu^\lambda T(w \bullet \mu) \simeq T(w \bullet \lambda)$, and $T_\lambda^\mu T(w \bullet \lambda)$ is isomorphic to the direct sum of $(\mathrm{Stab}_{W_p}(\mu) : \mathrm{Stab}_{W_p}(\lambda))$ copies of $T(w \bullet \mu)$.*

This is proved in [Andersen 28], 5.2. One proceeds more or less as in the special case (1) above, but at one point a new argument is needed: One has to know that each $V(wx \bullet \lambda)$ with $x \in \mathrm{Stab}_{W_p}(\mu)$ occurs as a factor in a Weyl filtration of $T(w \bullet \lambda)$. If $\lambda \in C$, then one can deduce this from (3) because $\mathrm{Stab}_{W_p}(\mu)$ is generated by the $s \in \Sigma^0(\mu)$ and since $ws \bullet \lambda < w \bullet \lambda$ for all these s. In general, one uses results in [Koppinen 6] mentioned in 6.25 that imply $\mathrm{Hom}_G(V(wx \bullet \lambda), V(w \bullet \lambda)) \neq 0$ for all these x.

Note that this result reduces for $p \geq h$ the calculation of all $\mathrm{ch}\, T(\mu)$ with $\mu \in X(T)_+$ to that for all $\mu \in X(T)_+ \cap W_p \bullet 0$.

E.12. The $\mathrm{ch}\, T(\lambda)$ with $\lambda \in X(T)_+$ are a basis of $\mathbf{Z}[X(T)]^W$, cf. 5.8. So there are integers $a_{\lambda,\mu}^\nu$ for all $\lambda, \mu, \nu \in X(T)_+$ with

$$(1) \qquad \mathrm{ch}\, T(\lambda)\, \mathrm{ch}\, T(\mu) = \sum_{\nu \in X(T)_+} a_{\lambda,\mu}^\nu\, \mathrm{ch}\, T(\nu).$$

As a tensor product of tilting modules again is tilting, the coefficient $a_{\lambda,\mu}^\nu$ is equal to the multiplicity of $T(\nu)$ as a direct summand of $T(\lambda) \otimes T(\mu)$.

As long as we do not know the $\mathrm{ch}\, T(\lambda)$ there is little hope to determine the $a_{\lambda,\mu}^\nu$ in general. However, we know that $T(\lambda) = V(\lambda)$ (hence $\mathrm{ch}\, T(\lambda) = \chi(\lambda)$) for all $\lambda \in C \cap X(T)_+$. This leads to the following partial result (where W_p^+ denotes the set of all $w \in W_p$ with $w \bullet C$ dominant):

Proposition: a) *Let $Q \simeq \bigoplus_{\nu \in X(T)_+} T(\nu)^{n(\nu)}$ be a tilting module for G. Then we have*

$$(2) \qquad n(\lambda) = \sum_{w \in W_p^+} (-1)^{l(w)} (Q : \chi(w \bullet \lambda))$$

for all $\lambda \in C \cap X(T)_+$.

b) *Let* $\lambda, \mu, \nu \in C \cap X(T)_+$. *Then the multiplicity of* $V(\lambda)$ *as a direct summand in* $V(\mu) \otimes V(\nu)$ *is equal to*

(3)
$$\sum_{w \in W_p} (-1)^{l(w)} \dim V(\mu)_{w \bullet \lambda - \nu}.$$

Proof: a) It suffices to prove this for $Q = T(\mu)$ with $\mu \in X(T)_+$. We have to show that the sum in (2) is equal to $\delta_{\lambda\mu}$. Now $\operatorname{ch} T(\mu)$ is a linear combination of $\chi(\mu')$ with $\mu' \in W_p \bullet \mu$. This shows for $\mu \notin W_p \bullet \lambda$ that the sum is equal to 0. Suppose next that $Q = T(z \bullet \lambda)$ for some $z \in W_p^+$. If $z = 1$, then $\operatorname{ch} T(\lambda) = \chi(\lambda)$ shows that the sum in (2) is equal to 1. If $z \neq 1$, then there exists $s \in \Sigma$ with $zs \bullet \lambda < z \bullet \lambda$ and $zs \bullet \lambda \in X(T)_+$. In this case E.11(3) yields $(Q : \chi(w \bullet \lambda)) = 0$ if $ws \bullet \lambda \notin X(T)_+$, while $(Q : \chi(w \bullet \lambda)) = (Q : \chi(ws \bullet \lambda))$ if $ws \bullet \lambda \in X(T)_+$. In the second case we get

$$(-1)^{l(w)}(Q : \chi(w \bullet \lambda)) + (-1)^{l(ws)}(Q : \chi(ws \bullet \lambda)) = 0.$$

This shows that the sum in (2) is 0.

b) Let us denote the wanted multiplicity by a. We want to apply (2) to $Q = T(\mu) \otimes T(\nu) = V(\mu) \otimes V(\nu)$. We have

$$\operatorname{ch}(Q) = \chi(\mu)\,\chi(\nu) = \sum_{\omega \in X(T)} \dim V(\mu)_\omega\, \chi(\nu + \omega)$$

cf. 5.8.b. So

$$a = \sum_{w \in W_p^+} (-1)^{l(w)} \sum_{\omega \in X(T)} \dim V(\mu)_\omega\, (\chi(\nu + \omega) : \chi(w \bullet \lambda)).$$

Now 5.9(1) shows: If $\nu + \omega = x \bullet (w \bullet \lambda)$ for some $x \in W$, then $(\chi(\nu+\omega) : \chi(w \bullet \lambda)) = (-1)^{l(x)}$; if there is no such x, then $(\chi(\nu + \omega) : \chi(w \bullet \lambda)) = 0$. We get

$$a = \sum_{w \in W_p^+} (-1)^{l(w)} \sum_{x \in W} (-1)^{l(x)} \dim V(\mu)_{xw \bullet \lambda - \nu}.$$

Now use that each $z \in W_p$ can be written uniquely $z = xw$ with $x \in W$ and $w \in W_p^+$ and recall that always $l(xw) \equiv l(x) + l(w) \pmod 2$.

E.13. As noted above the $\operatorname{ch} T(\lambda)$ with $\lambda \in X(T)_+$ are a basis for $\mathbf{Z}[X(T)]^W$. Let J denote the span of all $\operatorname{ch} T(\lambda)$ with $\lambda \in X(T)_+$, $\lambda \notin C$. Then:

(1) *J is an ideal in* $\mathbf{Z}[X(T)]^W$.

This claim is equivalent to:

(2) *Let* $\lambda, \mu, \nu \in X(T)_+$ *such that* $T(\lambda)$ *is a direct summand of* $T(\mu) \otimes T(\nu)$. *If* $\mu \notin C$, *then* $\lambda \notin C$.

We may assume $p \geq h$. (If $p < h$, then $C \cap X(T)_+ = \emptyset$ and the claim holds trivially.) There are now different ways to proceed. Using E.11(3) one can check that J is spanned by all $\chi(\mu)$ with $\mu \in X(T)_+ \setminus W_p \bullet C$ together with all $\chi(w \bullet \lambda) + \chi(ws \bullet \lambda)$ with $\lambda \in C \cap X(T)_+$, $w \in W_p^+$, and $s \in \Sigma$ such that also $ws \in W_p^+$. Then an elementary character calculation yields the claim, see [Andersen and Paradowski], Lemma 2.9.

One can also use E.11(3) and Weyl's character formula to show that $\dim T(\nu)$ is divisible by p if and only if $\nu \notin C$ (assuming $p \geq h$). Now one has to apply the following lemma (see [Georgiev and Mathieu 1], 2.7, or [Mathieu 5], 8.5): If M is an indecomposable G–module of dimension divisible by p and if N is an arbitrary finite dimensional G–module, then all direct summands of $M \otimes N$ have a dimension divisible by p.

One can also use [Jantzen 14], Lemma 2.3 together with E.11(1) and show that the support variety of any $T(\nu)$ with $\nu \notin C$ (considered as a module over Lie(G)) is a proper subset of the variety $\mathcal{N}$ of all nilpotent elements in Lie(G), whereas any $T(\nu)$ with $\nu \in C$ has support variety equal to $\mathcal{N}$ because p does not divide $\dim T(\nu)$. Then one uses standard properties of support varieties.

One gets now a ring structure on $\mathbf{Z}[X(T)]^W/J$. The classes $[\chi(\mu)]$ of the $\mu \in C \cap X(T)_+$ are a basis over $\mathbf{Z}$ for this ring. This ring is often called a *fusion ring* (see, e.g., [Andersen and Paradowski] or [Georgiev and Mathieu 1]). The multiplication on this ring is given by E.12(3). This product formula is called the *modular Verlinde formula* in [Mathieu 5], 9.5. One can also define a *fusion category* whose objects are the finite direct sums of G–modules of the form $V(\mu)$ with $\mu \in C \cap X(T)_+$ (with the usual morphisms); in the notation of A.1 this is the full subcategory of all finite dimensional modules in $\mathcal{C}(C \cap X(T)_+)$. The point is now that one defines a tensor product $\underline{\otimes}$ in this category: Given V_1, V_2 in this category, we can decompose $V_1 \otimes V_2 = M \oplus N$ such that M belongs again to the category and such that N is a direct sum of $T(\lambda)$ with $\lambda \notin C$; then set $V_1 \underline{\otimes} V_2 = M$. Then the usual properties of a tensor category are satisfied. For the first construction of this fusion category see [Gelfand and Kazhdan].

E.14. Fix $\lambda \in X(T)_+$ and $n \in \mathbf{N}$, $n > 0$ and set $Q = T(\lambda)^n$. Then $\mathrm{End}_G(Q)$ is in a natural way isomorphic to the matrix ring $M_n(E)$ where (for the moment) $E = \mathrm{End}_G(T(\lambda))$. Now each $M_n(E)$–module is isomorphic to a module of the form V^n where V is an E–module and where matrices in $M_n(E)$ acts in the obvious way on V^n. Since E is a local ring, it has exactly one simple module (up to isomorphism). By E.5(1) this module has dimension 1; any $a\,\mathrm{id}_{T(\lambda)}$ with $a \in k$ acts as multiplication by a while elements in $\mathrm{rad}\,E$ acts as 0. It follows that $\mathrm{End}_G(Q)$ has exactly one simple module (up to isomorphism) and that this simple module has dimension n. So each $\mathrm{End}_G(Q)$–module of dimension n has to be simple.

This simple module can be described as follows: The space $\mathrm{Hom}_G(V(\lambda), Q)$ has a natural structure as a module over $\mathrm{End}_G(Q)$ via $\varphi \cdot f = \varphi \circ f$ for all $\varphi \in \mathrm{End}_G(Q)$ and $f \in \mathrm{Hom}_G(V(\lambda), Q)$. Pick $v_\lambda \in V(\lambda)_\lambda$, $v_\lambda \neq 0$. Since λ is the largest weight of $Q = T(\lambda)$, the map $f \mapsto f(v_\lambda)$ is an isomorphism of vector spaces

$$\mathrm{Hom}_G(V(\lambda), Q) \xrightarrow{\sim} Q_\lambda^{U^+} = Q_\lambda \tag{1}$$

cf. Remark 2 in 2.13. As $\dim T(\lambda)_\lambda = 1$, we get that $\dim \mathrm{Hom}_G(V(\lambda), Q) = n$. So $\mathrm{Hom}_G(V(\lambda), Q)$ has to be a simple module over $\mathrm{End}_G(Q)$. (Alternatively, we could consider Q_λ as a module over $\mathrm{End}_G(Q)$.)

E.15. We want to describe the simple modules over $\mathrm{End}_G(Q)$ for an arbitrary tilting module Q for G. Set for all $\lambda \in X(T)_+$

$$(1) \qquad F_\lambda(Q) = \mathrm{Hom}_G(V(\lambda), Q) \qquad \text{and} \qquad F'_\lambda(Q) = \mathrm{Hom}_G(Q, H^0(\lambda)).$$

Fix a basis η_λ for $\mathrm{Hom}_G(V(\lambda), H^0(\lambda))$. For any $\varphi \in F_\lambda(Q)$ and $\psi \in F'_\lambda(Q)$ there exists $a(\varphi, \psi) \in k$ with $\psi \circ \varphi = a(\varphi, \psi)\, \eta_\lambda$. Consider now the linear map

$$(2) \qquad \Phi : F_\lambda(Q) \longrightarrow F'_\lambda(Q)^*, \qquad \Phi(\varphi)(\psi) = a(\varphi, \psi).$$

Both $F_\lambda(Q)$ and $F'_\lambda(Q)^*$ are natural modules over $\mathrm{End}_G(Q)$ (via $\gamma \cdot \varphi = \gamma \circ \varphi$ and $(\gamma \cdot f)(\psi) = f(\psi \circ \gamma)$ for all $\gamma \in \mathrm{End}_G(Q)$ and $f \in F'_\lambda(Q)^*$) and Φ is obviously a homomorphism of $\mathrm{End}_G(Q)$–modules. Therefore

$$(3) \qquad F^\vee_\lambda(Q) = \Phi(F_\lambda(Q))$$

is an $\mathrm{End}_G(Q)$–submodule of $F'_\lambda(Q)^*$.

If Q is a direct sum, $Q = \bigoplus_{i=1}^m Q_i$, then $F_\lambda(Q) \simeq \bigoplus_{i=1}^m F_\lambda(Q_i)$ and $F'_\lambda(Q) \simeq \bigoplus_{i=1}^m F'_\lambda(Q_i)$. Then Φ identifies with the direct sum of all $F_\lambda(Q_i) \to F'_\lambda(Q_i)^*$ and we get

$$(4) \qquad F^\vee_\lambda(Q) \simeq \bigoplus_{i=1}^m F^\vee_\lambda(Q_i).$$

This is an isomorphism compatible with the obvious action of the subalgebra $\prod_{i=1}^m \mathrm{End}_G(Q_i)$ of $\mathrm{End}_G(Q)$.

If $Q = T(\lambda)$, then $F_\lambda(Q) = k\iota_\lambda$ and $F'_\lambda(Q) = k\pi_\lambda$ in the notation of E.4. We have $\pi_\lambda \circ \iota_\lambda \neq 0$; so in this case Φ is a bijection $F_\lambda(Q) \xrightarrow{\sim} F'_\lambda(Q)^*$.

If $Q = T(\mu)$ with $\mu \neq \lambda$, then Remark E.5 shows that $\psi \circ \varphi = 0$ for all $\varphi \in F_\lambda(Q)$ and $\psi \in F'_\lambda(Q)$. So we get here $F^\vee_\lambda(Q) = 0$.

In general, fix a decomposition

$$(5) \qquad Q = \bigoplus_{\mu \in X(T)_+} Q^\mu \qquad \text{with } Q^\mu \simeq T(\mu)^{a(\mu)} \text{ for all } \mu \in X(T)_+$$

with almost all $Q^\mu = 0$ (equivalently: with almost all $a(\mu) = 0$). Then the above discussion shows that Φ induces an isomorphism of vector spaces

$$(6) \qquad F_\lambda(Q^\lambda) \xrightarrow{\sim} F^\vee_\lambda(Q).$$

E.16. Keep the notations from E.15 and fix a decomposition as in E.15(5).

Proposition: *Each $F^\vee_\lambda(Q)$ with $Q^\lambda \neq 0$ is a simple module over $\mathrm{End}_G(Q)$ of dimension $a(\lambda)$. Each simple $\mathrm{End}_G(Q)$–module is isomorphic to exactly one $F^\vee_\lambda(Q)$ with $Q^\lambda \neq 0$.*

Proof: For each μ let ε_μ denote the endomorphism of Q that is the identity on Q^μ and acts as 0 on all Q^ν with $\nu \neq \mu$. Then ε_μ is idempotent, we have $\varepsilon_\mu \varepsilon_\nu = 0$ whenever $\nu \neq \mu$ and we have $1 = \sum_\mu \varepsilon_\mu$. (This sum is finite since $\varepsilon_\mu = 0$ if $Q^\mu = 0$.)

Fix $\lambda \in X(T)_+$. We have $\dim F_\lambda^\vee(Q) = \dim F_\lambda(Q^\lambda) = a(\lambda)$ by E.15(6) and the argument in E.14.

We can identify the algebra $\varepsilon_\lambda \operatorname{End}_G(Q)\varepsilon_\lambda$ with the algebra $\operatorname{End}_G(Q^\lambda)$. We know from E.14 that $F_\lambda(Q^\lambda)$ is a simple module over $\operatorname{End}_G(Q^\lambda) \simeq \varepsilon_\lambda \operatorname{End}_G(Q)\varepsilon_\lambda$. The restriction of Φ to $F_\lambda(Q^\lambda)$ is an isomorphism of $\varepsilon_\lambda \operatorname{End}_G(Q)\varepsilon_\lambda$–modules. Therefore $F_\lambda^\vee(Q)$ is simple for $\varepsilon_\lambda \operatorname{End}_G(Q)\varepsilon_\lambda$, hence also for $\operatorname{End}_G(Q)$.

Any ε_μ acts as the identity on each $F_\lambda(Q^\mu) \subset F_\lambda(Q)$ and as 0 on each $F_\lambda(Q^\nu) \subset F_\lambda(Q)$ with $\nu \neq \mu$. Since $F_\lambda^\vee(Q) = \Phi(F_\lambda(Q^\lambda))$, this shows that ε_μ acts as the identity on $F_\mu^\vee(Q)$ and as 0 on each $F_\lambda^\vee(Q)$ with $\lambda \neq \mu$. Therefore the $F_\lambda^\vee(Q)$ with $Q^\lambda \neq 0$ are pairwise non-isomorphic over $\operatorname{End}_G(Q)$.

By [Green 2], Thm. 6.2.g the map $M \mapsto \varepsilon_\lambda M$ induces a bijection between the isomorphism classes of simple modules M over $\operatorname{End}_G(Q)$ with $\varepsilon_\lambda M \neq 0$ and the isomorphism classes of simple modules over $\varepsilon_\lambda \operatorname{End}_G(Q)\varepsilon_\lambda$. As $\operatorname{End}_G(Q^\lambda)$ has only one simple module (up to isomorphism), see E.14, so does $\varepsilon_\lambda \operatorname{End}_G(Q)\varepsilon_\lambda$. It follows that $F_\lambda^\vee(Q)$ is (up to isomorphism) the only simple module over $\operatorname{End}_G(Q)$ not annihilated by ε_λ.

If M is an arbitrary simple module over $\operatorname{End}_G(Q)$, then $1 = \sum_\lambda \varepsilon_\lambda$ shows that there exists λ with $\varepsilon_\lambda M \neq 0$. The proposition follows.

Remark: Suppose that $Q^\lambda \neq 0$ and choose a decomposition $Q^\lambda = T(\lambda) \oplus M$ with suitable M. We can thus regard ι_λ as in E.4 as an element in $F_\lambda(Q^\lambda) \subset F_\lambda(Q)$. Suppose that $\varphi \in F_\lambda(Q^\mu)$ for some $\mu \neq \lambda$. We get from E.4.a a homomorphism $f : T(\lambda) \to Q^\mu$ with $f \circ \iota_\lambda = \varphi$. Composing f with the projection $Q \to T(\lambda)$ with kernel equal to the direct sum of M and all Q^ν with $\nu \neq \lambda$ we get $\gamma \in \operatorname{End}_G(Q)$ with $\gamma \circ \iota_\lambda = \varphi$ and $\gamma(Q^\nu) = 0$ for all $\nu \neq \lambda$. It follows that $\gamma \in \varepsilon_\mu \operatorname{End}_G(Q)\varepsilon_\lambda$. Now $\mu \neq \lambda$ implies that γ annihilates all simple modules over $\operatorname{End}_G(Q)$, hence that γ belongs to the radical of $\operatorname{End}_G(Q)$. The definition of the $\operatorname{End}_G(Q)$–module structure on $F_\lambda(Q)$ shows that $\gamma \cdot \iota_\lambda = \gamma \circ \iota_\lambda = \varphi$, hence $\varphi \in \operatorname{rad} \operatorname{End}_G(Q) \cdot F_\lambda(Q) = \operatorname{rad} F_\lambda(Q)$. Since the kernel of Φ is the sum of all $F_\lambda(Q^\mu)$ with $\mu \neq \lambda$ this shows that this kernel is equal to the radical of $F_\lambda(Q)$, hence also the only maximal submodule in $F_\lambda(Q)$.

E.17. Consider as an example the group $G = GL_n$ for some $n \in \mathbf{N}$, $n \geq 2$. Let V denote the natural G–module $V = k^n$. We have $V = H^0(\varepsilon_1) = V(\varepsilon_1)$ in the notations from 1.21. So V is a tilting module for G, hence so is each tensor power $\bigotimes^d V$ by Proposition 4.21. So there exists a direct sum decomposition

(1) $$\bigotimes{}^d V \simeq \bigoplus_{\lambda \in X(T)_+} T(\lambda)^{a(n,d,\lambda)}.$$

Any λ occurring here has the form $\lambda = \sum_{i=1}^n a_i \varepsilon_i$ with $\sum_{i=1}^n a_i = d$ and all $a_i \geq 0$ (because λ is a weight of $\bigotimes^d V$) and with $a_1 \geq a_2 \geq \cdots \geq a_n$ (because λ is dominant). In other words, λ belongs to the set $\pi(n,d)$ from A.3(1).

The symmetric group S_d acts on $\bigotimes^d V$ by permuting the factors in the tensor product. This action commutes with that of G and we get thus a homomorphism

(2) $$kS_d \longrightarrow \operatorname{End}_{GL_n}(\bigotimes{}^d V)$$

where kS_d denotes the group algebra of S_d over k. According to [Carter and Lusztig 1], 3.1 this map is an isomorphism in case $n \geq d$, cf. A.23(5). By [De Concini and Procesi 1], Thm. 4.1 the map in (2) is surjective for arbitrary d and n. (An alternative approach to this result is sketched at the top of page 399 in [Donkin 21].) Therefore Proposition E.16 implies that the non-zero $a(n,d,\lambda)$ in (1) are the dimensions of some irreducible representations of S_d over k; in case $n \geq d$ they are the dimensions of all irreducible representations of S_d over k.

One approach to the representation theory of S_d over an arbitrary field K proceeds as follows (cf. [JK], 7.1): One associates to each partition μ of d a *Specht module* S_K^μ. If $\mathrm{char}(K) = 0$, then each S_K^μ is a simple KS_d–module and each simple KS_d–module is isomorphic to exactly one S_K^μ. (This parametrisation of the simple modules is made such that the trivial one dimensional representation corresponds to the partition (d) and the sign representation to the partition $(1 \geq 1 \geq \cdots \geq 1)$.) If $\mathrm{char}(K) = p > 0$, then the S_K^μ are usually not simple. A partition $\mu = (\mu_1 \geq \mu_2 \geq \cdots \geq \mu_r > 0)$ is called *p–singular* if there exists i with $\mu_{i+1} = \mu_{i+2} = \cdots = \mu_{i+p} > 0$; otherwise μ is called *p–regular.* Now one shows: If μ is p–regular, then S_K^μ has a unique maximal submodule, so $D_K^\mu = S_K^\mu / \mathrm{rad}\, S_K^\mu$ is simple; each simple KS_d–module is isomorphic to exactly one D_K^μ with p–regular μ.

In order to connect this with the construction above, identify $\lambda = \sum_{i=1}^n m_i\varepsilon_i \in \pi(n,d)$ with the partition $(m_1 \geq m_2 \geq \cdots \geq m_n \geq 0)$ of d into at most n parts, cf. A.3. Now the kS_d–module $F_\lambda(\bigotimes^d V) = \mathrm{Hom}_{GL_n}(V(\lambda), \bigotimes^d V) \simeq (\bigotimes^d V)_\lambda^{U^+}$ is isomorphic to the Specht module S_k^λ, see [JK], 8.4.2. One has $a(n,d,\lambda) \neq 0$ if and only if λ is p–regular, see [Erdmann 2], Prop. 4.2. (In case $d \leq n$ we shall see this at the end of the next subsection. In the general case one can prove the claim looking at the restriction from GL_n to GL_{n-1}.) This implies:

(3) *The simple quotient* $F_\lambda^\vee((\bigotimes^d V)^\lambda)$ *of* $F_\lambda(\bigotimes^d V)$ *is isomorphic to* D_k^λ.

We get in particular that $\dim D_k^\lambda = a(n,d,\lambda)$.

In some cases this connection has been used to compute the dimensions of simple kS_d–module. Donkin's inductive formula for the tilting modules for SL_2 (see Remark E.9) was used in [Erdmann 4] to calculate $\dim D_k^\lambda$ in case λ is a partition with two parts. (The results in that case could also be deduced from work of James.) In [Mathieu 4] this dimension is determined when λ belongs to the interior of the first alcove C as in 6.2(6). (If $\lambda = \sum_{i=1}^n m_i\varepsilon_i$, then one has $\lambda \in C$ if and only if $m_1 - m_n < p - n + 1$.) Work in [Rasmussen] leads to a calculation of this dimension when $m_1 - m_{n-1} < p - n + 2$ or $m_2 - m_n < p - n + 2$ (using the same notation).

E.18. Keep the assumptions from E.17, but assume now that $n \geq d$. We have seen in A.22 a description of the simple kS_d–modules that is different from the one above. We now want to make a comparison.

Recall from A.23 that $\bigotimes^d V$ is isomorphic to $S(n,d)\xi_\varpi$ for a suitable idempotent $\xi_\varpi \in S(n,d)$. This implies in particular that $\bigotimes^d V$ is a projective $S(n,d)$–module. Therefore each $T(\mu)$ with $a(n,d,\mu) > 0$ is an indecomposable projective $S(n,d)$–module, hence the projective cover of some simple $S(n,d)$–module $L(\lambda)$. Then $T(\mu)/\mathrm{rad}\, T(\mu) \simeq L(\lambda)$ shows that $L(\lambda)$ is a homomorphic image of $\bigotimes^d V$, hence by A.23(6) that $\lambda \in \pi_1(n,d) = \pi(n,d) \cap X_1(T)$. Furthermore, all these λ occur as homomorphic image of $\bigotimes^d V$. This shows: There exists for each $\lambda \in \pi_1(n,d)$ some $\lambda' \in \pi(n,d)$ such that $T(\lambda')$ is the projective cover of $L(\lambda)$ as an $S(n,d)$–module.

The idempotent $\varepsilon_{\lambda'} \in kS_d$ as in the proof of Proposition E.16 acts as the identity in $D_k^{\lambda'} = F_{\lambda'}^{\vee}(\bigotimes^d V)$ and on $L(\lambda)_\varpi$. (In the second case use that the bijection in A.23(5) is kS_d–linear.) It follows that

$$D_k^{\lambda'} \simeq L(\lambda)_\varpi.$$

Since $\operatorname{End}_{S(n,d)}(L(\lambda)) = k$, we get then for each finite dimensional $S(n,d)$–module M that

$$\dim \operatorname{Hom}_{S(n,d)}(T(\lambda'), M) = [M : L(\lambda)]. \tag{1}$$

This shows in particular for all $\mu \in \pi(n,d)$

$$(T(\lambda') : \chi(\mu)) = \dim \operatorname{Hom}_{S(n,d)}(T(\lambda'), H^0(\mu)) = [H^0(\mu) : L(\lambda)]. \tag{2}$$

It follows that

$$\lambda' = \max\{\, \mu \in \pi(n,d) \mid [H^0(\mu) : L(\lambda)] \neq 0 \,\}. \tag{3}$$

Recall from A.22 that the simple kS_d–modules are the $L(\lambda)_\varpi$ with $\lambda \in \pi_1(n,d)$ where $\varpi = \sum_{i=1}^d \varepsilon_i$. Tensoring with the sign representation permutes the simple kS_d–modules. So there exists for each $\lambda \in \pi_1(n,d)$ some $\lambda^o \in \pi_1(n,d)$ with

$$L(\lambda)_\varpi \otimes \operatorname{sgn} \simeq L(\lambda^o)_\varpi.$$

We have clearly $(\lambda^o)^o = \lambda$. The map $\lambda \mapsto \lambda^o$ (sometimes called the Mullineux map) depends on k, or rather on $\operatorname{char}(k)$. In characteristic 0 the corresponding map is given by taking the transposed ${}^t\mu$ of a partition μ, i.e., we have $V(\mu)_{\mathbf{Q},\varpi} \otimes \operatorname{sgn} \simeq V({}^t\mu)_{\mathbf{Q},\varpi}$ for all $\mu \in \pi(n,d)$. (Note that $d \leq n$ implies that also ${}^t\mu \in \pi(n,d)$.) Reduction modulo p shows then that $V({}^t\mu)_\varpi$ has the same composition factors (with multiplicities) as $V(\mu)_\varpi \otimes \operatorname{sgn}$. This implies for all $\mu \in \pi(n,d)$ and $\lambda \in \pi_1(n,d)$

$$\begin{aligned}[H^0(\mu) : L(\lambda)] &= [H^0(\mu)_\varpi : L(\lambda)_\varpi] = [H^0(\mu)_\varpi \otimes \operatorname{sgn} : L(\lambda)_\varpi \otimes \operatorname{sgn}]\\ &= [H^0({}^t\mu)_\varpi : L(\lambda^o)_\varpi] = [H^0({}^t\mu) : L(\lambda^o)].\end{aligned} \tag{4}$$

One can check that $\mu \mapsto {}^t\mu$ reverses the order relation $\leq$ on $\pi(n,d)$. We want to use this to show that

$$\lambda' = {}^t(\lambda^o). \tag{5}$$

Indeed, we have $[H^0(\lambda') : L(\lambda)] \neq 0$, hence $[H^0({}^t\lambda') : L(\lambda^o)] \neq 0$; this implies $\lambda^o \leq {}^t\lambda'$ and thus ${}^t(\lambda^o) \geq \lambda'$. But (4) implies also that $[H^0({}^t(\lambda^o)) : L(\lambda)] \neq 0$, so the maximality of λ' in (3) yields (5).

If λ runs through $\pi_1(n,d)$, then so does λ^o and therefore λ' runs over $\{{}^t\mu \mid \mu \in \pi_1(n,d)\}$. The latter set is just the set of all p–regular partitions of d. So we see under our the present assumption $d \leq n$ that the μ with $a(n,d,\mu) > 0$ are indeed the p–regular partitions of d. We get also from (3)–(5) that

$$(T(\lambda') : \chi(\mu)) = [V(\mu) : L(({}^t\lambda')^o)] = [V({}^t\mu) : L({}^t\lambda')]$$

i.e., the special case of E.10(4) where λ' is p–regular and $d \leq n$.

E.19. We can define tilting modules for G_A for arbitrary A since we can define good filtrations and Weyl filtrations in this generality, see B.9 and B.10. By definition any tilting module for G_A is free of finite rank over A.

Lemma: *If A is a principal ideal domain, then there exists for each $\lambda \in X(T)_+$ a tilting module M for G_A such that M_λ is free of rank 1 over A and such that $M_\mu \neq 0$ implies $\mu \leq \lambda$.*

Proof: One proceeds as for the proof of Lemma E.3: One chooses a suitable numbering for the dominant weights $\leq \lambda$ and constructs, beginning with $M_r = V(\lambda)_A$ a series of G_A–modules M_i. Each M_i is free of finite rank over A, has a Weyl filtration, and satisfies the same conditions on the weights as M. The only point to be observed is that the $\mathrm{Ext}^1_{G_A}(V(\lambda_i)_A, M_{i+1})$ need not be free over A. But it is still finitely generated over A (by B.6) and we now choose $\xi_1, \xi_2, \dots, \xi_d$ as generators of that Ext group over A. Since they are not necessarily a basis, the map $\mathrm{Hom}_{G_A}(V(\lambda_i)^d_A, V(\lambda_i)_A) \to \mathrm{Ext}^1_{G_A}(V(\lambda_i)_A, M_{i+1})$ as in E.3(2) is possibly no longer bijective, but it is still surjective. And that suffices to make $\mathrm{Ext}^1_{G_A}(V(\lambda_i)_A, M_i)$ disappear.

At the end of this construction one has to use Lemma B.9 to make sure that $M = M_1$ has a good filtration.

E.20. Let A be a principal ideal domain. We can then decompose any G_A–module that is finitely generated over A into a direct sum of indecomposables. Such decompositions will, in general, have no uniqueness properties.

Suppose from now on that A is a complete discrete valuation ring. Let $\mathfrak{m}$ denote the maximal ideal in A.

If M is a G_A–module that is free of finite rank over A, then M is indecomposable as a G_A–module if and only if it is indecomposable as a $\mathrm{Dist}(G_A)$–module and we have $\mathrm{End}_{G_A}(M) = \mathrm{End}_{\mathrm{Dist}(G_A)}(M)$, cf. I.7.15/16 or 1.20. We can replace here $\mathrm{Dist}(G_A)$ by its image in $\mathrm{End}_A(M)$, i.e., by an A–algebra that is finitely generated as an A–module. Therefore the general theory shows (for the present A) that M is indecomposable as a G_A–module if and only if $\mathrm{End}_{G_A}(M)$ is a local ring. Furthermore, the decomposition of M into indecomposable summands satisfies the Krull-Schmidt theorem.

Fix now for each $\lambda \in X(T)_+$ an indecomposable tilting module $T_A(\lambda)$ such that $T_A(\lambda)_\lambda$ has rank 1 and such that $T_A(\lambda)_\mu \neq 0$ implies $\mu \leq \lambda$. For each $\varphi \in \mathrm{End}_{G_A}(T_A(\lambda))$ there exists $a \in A$ such that φ acts as multiplication by a on $T_A(\lambda)_\lambda$. Then $a \mapsto a + \mathfrak{m}$ is a ring homomorphism $\mathrm{End}_{G_A}(T_A(\lambda)) \to A/\mathfrak{m}$. This homomorphism is obviously surjective (already on the scalar multiples of the identity map), so its kernel has to be the unique maximal ideal in $\mathrm{End}_{G_A}(T_A(\lambda))$:

(1) $$\mathrm{rad}\,\mathrm{End}_{G_A}(T_A(\lambda)) = \{\varphi \in \mathrm{End}_{G_A}(T_A(\lambda)) \mid \varphi(T_A(\lambda)_\lambda) \subset \mathfrak{m}T_A(\lambda)_\lambda\}.$$

We see also that

(2) $$\mathrm{End}_{G_A}(T_A(\lambda)) = A\,\mathrm{id}_{T_A(\lambda)} + \mathrm{rad}\,\mathrm{End}_{G_A}(T_A(\lambda)).$$

The existence of the two filtrations on $T_A(\lambda)$ implies that $\mathrm{Ext}^i_{G_A}(T_A(\lambda), T_A(\lambda)) = 0$ for all $i > 0$, see B.4. Now I.4.18.a implies that

(3) $$\mathrm{End}_{G_{A/\mathfrak{m}}}(T_A(\lambda) \otimes_A A/\mathfrak{m}) \simeq \mathrm{End}_{G_A}(T_A(\lambda)) \otimes_A A/\mathfrak{m}.$$

Using this one sees easily that $\mathrm{End}_{G_{A/\mathfrak{m}}}(T_A(\lambda) \otimes_A A/\mathfrak{m})$ is a local ring, hence that $T_A(\lambda) \otimes_A A/\mathfrak{m}$ is indecomposable. Since it is tilting and has the same formal character as $T_A(\lambda)$, it follows that $T_A(\lambda) \otimes_A A/\mathfrak{m}$ is the indecomposable tilting module for $G_{A/\mathfrak{m}}$ with highest weight λ.

E.21. Keep the assumptions from E.20. In particular A is again a complete discrete valuation ring.

The filtrations on $T_A(\lambda)$ show that there exist homomorphisms of G_A–modules $\pi_\lambda : T_A(\lambda) \to H^0_A(\lambda)$ and $\iota_\lambda : V(\lambda)_A \to T_A(\lambda)$ that induce bijections on the λ–weight spaces such that $\ker(\pi_\lambda)$ has a good filtration and such that $T_A(\lambda)/\operatorname{im}(\iota_\lambda)$ has a Weyl filtration.

One gets then as in Lemma E.4 for any G_A–module V: If V has a good filtration, then $\mathrm{Hom}_{G_A}(T_A(\lambda), V) \to \mathrm{Hom}_{G_A}(V(\lambda)_A, V)$ with $\varphi \mapsto \varphi \circ \iota_\lambda$ is surjective. If V has a Weyl filtration, then $\mathrm{Hom}_{G_A}(V, T_A(\lambda)) \to \mathrm{Hom}_{G_A}(V, H^0_A(\lambda))$ with $\varphi \mapsto \pi_\lambda \circ \varphi$ is surjective.

Lemma: *Let Q be a tilting module for G_A. Then Q has a direct summand isomorphic to $T_A(\lambda)$ if and only if there exist homomorphisms of G_A–modules $f_1 : Q \to H^0_A(\lambda)$ and $f_2 : V(\lambda)_A \to Q$ with $f_1 \circ f_2 = \pi_\lambda \circ \iota_\lambda$.*

Proof: Suppose that we have f_1 and f_2 with this property. The observation preceding the lemma shows that there exist homomorphisms of G_A–modules $\varphi_1 : Q \to T_A(\lambda)$ with $\pi_\lambda \circ \varphi_1 = f_1$ and $\varphi_2 : T_A(\lambda) \to Q$ with $\varphi_2 \circ \iota_\lambda = f_2$. Then $\varphi_1 \circ \varphi_2$ is an endomorphism of $T_A(\lambda)$ that is bijective on $T_A(\lambda)_\lambda$ because of $\pi_\lambda \circ \varphi_1 \circ \varphi_2 \circ \iota_\lambda = f_1 \circ f_2 = \pi_\lambda \circ \iota_\lambda$. Now E.20(1) implies that $\varphi_1 \circ \varphi_2$ is bijective, hence that $\varphi_2(T_A(\lambda))$ is a direct summand of Q isomorphic to $T_A(\lambda)$. This proves one direction of the claim; the other one is obvious.

E.22. One can now immediately generalise Lemma E.5 and Proposition E.6 to our present situation. (In the proof of E.5 choose now v as part of a basis for Q_λ over A.) We get:

Proposition: *Suppose that A is a complete discrete valuation ring. There is for each $\lambda \in X(T)_+$ an indecomposable tilting module $T_A(\lambda)$ with $T_A(\lambda)_\lambda = 1$ free of rank 1 over A such that $T_A(\lambda)_\mu \neq 0$ implies $\mu \leq \lambda$. These properties determine $T_A(\lambda)$ uniquely up to isomorphism. If Q is an arbitrary tilting module, then there exist unique integers $n(\nu) \geq 0$, almost all 0 with*

$$Q \simeq \bigoplus_{\nu \in X(T)_+} T_A(\nu)^{n(\nu)}. \tag{1}$$

If A is a $\mathbf{Q}$–algebra, then one has $T_A(\lambda) = V(\lambda)_A = H^0_A(\lambda)$ for all λ. If A is an $\mathbf{F}_p$–algebra, then $T_A(\lambda) \simeq T_{\mathbf{F}_p}(\lambda) \otimes_{\mathbf{F}_p} A$ for all λ. The really interesting case is the one, where A has characteristic 0 and the residue field of A has characteristic $p > 0$. In this case A is an algebra over the ring $\mathbf{Z}_p$ of all p–adic integers. One gets then each $T_A(\lambda)$ by extension of scalars from $T_{\mathbf{Z}_p}(\lambda)$. Therefore we do not lose anything by restricting us to the case $A = \mathbf{Z}_p$.

E.23. Assume that A is the ring of all p–adic integers. Keep the notations $T_A(\lambda)$, ι_λ, π_λ as in E.21. Since k has characteristic p, we can regard is as an A–algebra via $A \to A/(p) \simeq \mathbf{F}_p \subset k$. We have $T_A(\lambda) \otimes_A \mathbf{F}_p \simeq T_{\mathbf{F}_p}(\lambda)$ by the observation at the end of E.20, hence $T_A(\lambda) \otimes_A k \simeq T(\lambda)$, cf. Remark E.6.

Recall that $\mathrm{Hom}_{G_A}(V(\lambda)_A, H^0_A(\lambda))$ is free of rank 1 over A (e.g., from B.4) and note that $\pi_\lambda \circ \iota_\lambda$ is a basis of $\mathrm{Hom}_{G_A}(V(\lambda)_A, H^0_A(\lambda))$ over A (because it is bijective on the λ–weight spaces).

We now generalise E.15(1) and set for each tilting module Q for G_A and each $\lambda \in X(T)_+$

$$(1) \qquad F_\lambda(Q) = \mathrm{Hom}_{G_A}(V(\lambda)_A, Q) \qquad \text{and} \qquad F'_\lambda(Q) = \mathrm{Hom}_{G_A}(Q, H^0_A(\lambda)).$$

Because Q has a good filtration, $F_\lambda(Q)$ is free over A of rank equal to the number of factors isomorphic to $H^0_A(\lambda)$ in a good filtration of Q; because Q has a Weyl filtration, $F'_\lambda(Q)$ is free over A of rank equal to the number of factors isomorphic to $V(\lambda)_A$ in a Weyl filtration of Q. Both these ranks are equal $(\mathrm{ch}\, Q : \chi(\lambda))$ in the notation of E.10.

Set $\overline{Q} = Q \otimes_A k$; so this is a tilting module for G. We have $\mathrm{Ext}^1_{G_A}(V(\lambda)_A, Q) = 0$ and $\mathrm{Ext}^1_{G_A}(Q, H^0_A(\lambda)) = 0$ and can therefore identify (using I.4.18)

$$(2) \qquad F_\lambda(\overline{Q}) \simeq F_\lambda(Q) \otimes_A k \qquad \text{and} \qquad F'_\lambda(\overline{Q}) \simeq F'_\lambda(Q) \otimes_A k.$$

For any $\varphi \in F_\lambda(Q)$ and $\psi \in F'_\lambda(Q)$ we have $\psi \circ \varphi \in \mathrm{Hom}_{G_A}(V(\lambda)_A, H^0_A(\lambda)) = A\, \pi_\lambda \circ \iota_\lambda$. So there exists $a(\varphi, \psi) \in A$ with $\psi \circ \varphi = a(\varphi, \psi)\, \pi_\lambda \circ \iota_\lambda$. We get now an A–linear map

$$(3) \qquad \Phi : F_\lambda(Q) \longrightarrow F'_\lambda(Q)^*, \qquad \Phi(\varphi)(\psi) = a(\varphi, \psi).$$

Claim: *The map Φ is injective.*

Proof: Let K denote the field of fractions for A. Any non-zero $\varphi \in F_\lambda(Q)$ induces an embedding $\varphi \otimes \mathrm{id}_K : V(\lambda)_K \to Q \otimes_A K$ of G_K–modules. Since $\mathrm{char}(K) = 0$, all G_K–modules are semi-simple. So there exists a homomorphism $\psi_K : Q \otimes_A K \to V(\lambda)_K$ of G_K–modules with $\psi_K \circ (\varphi \otimes \mathrm{id}_K)$ equal to the identity on $V(\lambda)_K$. Since $H^0_A(\lambda)$ is a lattice in $H^0_K(\lambda) = V(\lambda)_K$ and since Q is a lattice in $Q \otimes_A K$, we can find a p–power p^s with $p^s \psi_K(Q) \subset H^0_A(\lambda)$. Then the restriction of $p^s \psi_K$ to Q defines an element $\psi \in F'_\lambda(Q)$ with $\psi \circ \varphi \neq 0$. It follows that $0 \neq a(\varphi, \psi) = \Phi(\varphi)(\psi)$, hence the claim.

E.24. Keep the assumptions from E.23. We can imitate the construction from 8.18/19 and construct some filtrations. Set for all $j \geq 0$

$$(1) \qquad F_\lambda(Q)^j = \{\, \varphi \in F_\lambda(Q) \mid \Phi(\varphi) \in p^j F'_\lambda(Q)\, \}.$$

So the $F_\lambda(Q)^j$ form a descending chain of A–submodules of $F_\lambda(Q)$ with $F_\lambda(Q)^0 = F_\lambda(Q)$. If we do not want to mention Φ, then we can write

$$F_\lambda(Q)^j = \{\varphi \in F_\lambda(Q) \mid \psi \circ \varphi \in p^j\, \mathrm{Hom}_{G_A}(V(\lambda)_A, H^0_A(\lambda))\ \forall\, \psi \in F'_\lambda(Q)\}.$$

This leads now to a filtration on $F_\lambda(\overline{Q})$ setting $F_\lambda(\overline{Q})^j$ equal to the image of $F_\lambda(Q)^j \otimes_A k$ under the map

$$F_\lambda(Q)^j \otimes_A k \longrightarrow F_\lambda(Q) \otimes_A k \simeq F_\lambda(\overline{Q})$$

induced by the inclusion of $F_\lambda(Q)^j$ into $F_\lambda(Q)$.

Since Φ is injective (and A a principal ideal domain) there exist bases $\varphi_1, \varphi_2, \dots, \varphi_s$ of $F_\lambda(Q)$ over A and $\psi_1, \psi_2, \dots, \psi_s$ of $F'_\lambda(Q)$ over A such that $\Phi(\varphi_i) = p^{m(i)}\psi_i$ with non-negative integers $m(1) \leq m(2) \leq \cdots \leq m(s)$. Then each $F_\lambda(Q)^j$ has basis the φ_i with $m(i) \geq j$ and the $p^{j-m(i)}\varphi_i$ with $m(i) < j$. It follows that $F_\lambda(\overline{Q})^j$ has basis the $\varphi_i \otimes 1$ with $m(i) \geq j$. We get in particular $F_\lambda(\overline{Q})^j = 0$ for $j \gg 0$. We can therefore define a polynomial $f_\lambda[Q]$ in an indeterminate v setting

$$f_\lambda[Q] = \sum_{j\geq 0} \dim(F_\lambda(\overline{Q})^j / F_\lambda(\overline{Q})^{j+1})\, v^j \in \mathbf{Z}[v]. \tag{2}$$

In the set-up above the coefficient of v^j is the number of all i with $m(i) = j$.

Note that

$$f_\lambda[Q](1) = \dim(F_\lambda(\overline{Q})) = (Q : \chi(\lambda)) \tag{3}$$

is the number of factors isomorphic to $V(\lambda)$ in a Weyl filtration of $\overline{Q}$.

For example, we have $F_\lambda(T_A(\lambda)) = A\iota_\lambda$ and $F'_\lambda(T_A(\lambda)) = A\pi_\lambda$ and by definition $\Phi(\iota_\lambda)(\pi_\lambda) = 1$. This implies that

$$f_\lambda[T_A(\lambda)] = 1. \tag{4}$$

On the other hand, we claim that

$$f_\lambda[T_A(\mu)] \in v\mathbf{Z}[v] \qquad \text{if } \mu \neq \lambda. \tag{5}$$

Indeed, if this is wrong, then we get in the notation above $m(1) = 0$. This means, we get $\varphi \in F_\lambda(T_A(\mu))$ and $\psi \in F_\lambda(T_A(\mu))$ with $\psi \circ \varphi = \pi_\lambda \circ \iota_\lambda$. But then Lemma E.21 implies that $T_A(\mu)$ has a direct summand isomorphic to $T_A(\lambda)$, a contradiction as $T_A(\mu)$ is indecomposable.

If Q is a direct sum, $Q = \bigoplus_{i=1}^m Q_i$, then also $\overline{Q} = \bigoplus_{i=1}^m \overline{Q}_i$, and all our constructions are compatible with this direct sum decomposition, from $F_\lambda(Q) \simeq \bigoplus_{i=1}^m F_\lambda(Q_i)$ to $F_\lambda(\overline{Q})^j \simeq \bigoplus_{i=1}^m F_\lambda(\overline{Q}_i)^j$. It follows that

$$f_\lambda[\bigoplus_{i=1}^m Q_i] = \sum_{i=1}^m f_\lambda[Q_i]. \tag{6}$$

Applying this to a direct decomposition $Q = \bigoplus_{\nu \in X(T)_+} T_A(\nu)^{n(\nu)}$ as in E.22(1) one gets using (4) and (5)

$$f_\lambda[Q](0) = n(\lambda). \tag{7}$$

In [Andersen 28], Thm. 2.2 Andersen proves a sum formula for this filtration assuming $p \geq h$. (This restriction is expected to be unnecessary.) We state here the result without proof:

$$\sum_{j>0} \dim F_\lambda(\overline{Q})^j = -\sum_{\alpha \in R^+} \sum_{m \notin I(\lambda,\alpha)} \nu_p(m)(Q : \chi(\lambda - m\alpha)). \tag{8}$$

Here $I(\lambda, \alpha)$ denotes the set of all integers m with $0 \leq m \leq \langle \lambda+\rho, \alpha^\vee \rangle$. Furthermore we use here an extension of the notation $(\chi : \chi(\mu))$ as in E.10 to all $\mu \in X(T)$: If $\chi(\mu) = 0$, then set $(\chi : \chi(\mu)) = 0$. If $\chi(\mu) \neq 0$, then there exists a unique $w \in W$ with $w \bullet \mu \in X(T)_+$ and we set $(\chi : \chi(\mu)) = \det(w)(\chi : \chi(w \bullet \mu))$, cf. 5.9. For an application of this sum formula, see [Andersen 28], 2.13.

Let $\lambda_0 \in C \cap X(T)_+$. Recall the conjecture on $(T(w \bullet \lambda_0) : \chi(x \bullet \lambda_0))$ from E.10(3). It is actually a corollary [by (3)] of the following conjecture by Andersen (cf. [Andersen 27], 3.4(iv)):

(9) *Conjecture:* $f_{x \bullet \lambda_0}[T(w \bullet \lambda_0)] = n_{x,w}$ if $\langle w(\lambda_0 + \rho), \alpha^\vee \rangle < p^2$ for all $\alpha \in R^+$.

In Section 3 of [Andersen 26] it is shown that (9) would follow from some nice behaviour of the filtrations under translation functors.

CHAPTER F

Frobenius Splitting

The technique of "Frobenius splitting" is a powerful method in the algebraic geometry of varieties in prime characteristic. It can also lead to important results on varieties in characteristic 0 that are defined over the integers.

This technique was developed in [Mehta and Ramanathan 1] at about the same time when the first edition of this book was written. It led soon to alternative proofs of the results described here in Chapter 14. In fact, several of those results were first proved using Frobenius splitting.

If X is a variety over a finite prime field $\mathbf{F}_p$, then we have a Frobenius endomorphism $F_X : X \to X$, cf. Chapter I.9. It induces a homomorphism F_X^* from the structure sheaf $\mathcal{O}_X$ to the direct image sheaf $(F_X)_*\mathcal{O}_X$. One then says that X is Frobenius split if F_X^* is injective and if its image is a direct summand of $(F_X)_*\mathcal{O}_X$ as an $\mathcal{O}_X$–module (for the module structure given by F_X^*). A Frobenius splitting is now a projection map $(F_X)_*\mathcal{O}_X \to \mathcal{O}_X$.

The definition of "Frobenius split" for varieties over arbitrary ground fields of characteristic $p > 0$ (see F.8) requires the notion of the "Frobenius twisted" varieties $X^{(1)}$ as in I.9.2. The first subsections of this chapter (until F.7) repeat this construction and go through its properties at greater length than in I.9.2. However, we restrict as then to perfect ground fields for the sake of simplicity.

Having stated the definition in F.8, we then prove some properties of Frobenius split varieties. For example, if $\mathcal{L}$ is an ample line bundle on a Frobenius split projective variety X, then all $H^i(X, \mathcal{L})$ with $i > 0$ vanish, see F.10. If X is in addition smooth and irreducible, then one can replace $\mathcal{L}$ in this statement by its tensor product with the canonical sheaf on X.

The discussion of these general properties occupies the subsections until F.16. We then return to our reductive group G and prove (in F.18) that flag varieties are Frobenius split. In fact, one can choose the Frobenius splitting such that it induces a Frobenius splitting of each Schubert variety, see F.22. One gets then immediately one of the main results in Chapter 14: For all $w \in W$ and all dominant λ the restriction map $H^0(G/B, \mathcal{L}(\lambda)) \to H^0(X(w), \mathcal{L}(\lambda))$ is surjective and all $H^i(X(w), \mathcal{L}(\lambda))$ with $i > 0$ vanish, cf. 14.15.e. We do not reprove the normality of the Schubert varieties; for a proof in the context of the present techniques see [Mehta and Srinivas 1].

We also show that $G/B \times G/B$ has a Frobenius splitting that induces a Frobenius splitting of the closure of each G–orbit on $G/B \times G/B$ with respect to the diagonal action, see F.23. This yields then another proof of the surjectivity of the cup product $H^0(\lambda) \otimes H^0(\mu) \to H^0(\lambda + \mu)$, cf. 14.20.

In the next chapter we shall use Frobenius splitting to prove the main theorems on good filtrations from Chapter 4.

Besides the foundational paper [Mehta and Ramanathan 1], the main sources for this chapter are [van der Kallen 5], [Kaneda 7, 9], [Lauritzen and Thomsen], [Mehta and Ramanathan 2], [Ramanan and Ramanathan], [Ramanathan 1, 2].

Assume throughout that k is a perfect field of characteristic $p > 0$. From F.17 on we assume k to be algebraically closed. From F.3 on all schemes over k are assumed to be algebraic schemes over k.

F.1. We start by repeating some definitions and constructions from Chapter I.9. For each vector space V over k and each each $m \in \mathbf{Z}$ we define $V^{(m)}$ as the vector space over k that coincides with V as an additive group but where each $b \in k$ acts as $b^{p^{-m}}$ does on V. If V is a k–algebra, then each $V^{(m)}$ with the given multiplication is again a k–algebra.

We have associated in I.9.2 to each k–functor X new k–functors $X^{(r)}$. In this chapter we shall deal only with the case $r = 1$ and adopt (for this chapter only) the notation $X' = X^{(1)}$. So X' is by definition given by

$$X'(A) = X(A^{(-1)}) \qquad \text{for all } k\text{–algebras } A.$$

This construction is functorial: To each morphism $\varphi : X \to Y$ of k–functors one associates the morphism $\varphi' : X' \to Y'$ such that $\varphi'(A)$ from $X'(A) = X(A^{(-1)})$ to $Y'(A) = Y(A^{(-1)})$ is just $\varphi(A^{(-1)})$.

It is then clear (and was mentioned in I.9.2) that the functor $X \mapsto X'$ maps subfunctors to subfunctors, commutes with taking intersections and inverse images of subfunctors and with taking direct and fibre products. It takes local functors (see I.1.8) to local functors.

If X is an affine scheme, $X = Sp_k R$ for some k–algebra R, then we identify

$$(Sp_k R)' \simeq Sp_k(R^{(1)})$$

using that $\mathrm{Hom}_{k-\mathrm{alg}}(R, A^{(-1)}) = \mathrm{Hom}_{k-\mathrm{alg}}(R^{(1)}, A)$ for each k–algebra A. If now I is an ideal in R, then one gets $V(I)' = V(I^{(1)})$ and $D(I)' = D(I^{(1)})$. This implies now for arbitrary X:

(1) *If $U \subset X$ is an open subfunctor, then $U' \subset X'$ is open; if $Z \subset X$ is a closed subfunctor, then $Z' \subset X'$ is closed.*

(If you want to check this using the definitions in I.1.7 and I.1.12, note that a morphism $Sp_k R \to X'$ for some k–algebra R corresponds to an element in $X'(R) = X(R^{(-1)})$, hence to a morphism $Sp_k(R^{(-1)}) \to X$.)

If A is a field, then so is $A^{(-1)}$. Therefore the definition in I.1.7 shows immediately: *If $(Y_i)_i$ is an open covering of X, then $(Y_i')_i$ is an open covering of X'.* We get finally:

(2) *If X is a scheme (resp. an algebraic scheme) over k, then so is X'.*

(Note for the case of an algebraic scheme: If R is a finitely generated k–algebra, then so is $R^{(1)}$.)

F.2. Let $F_X : X \to X'$ denote the morphism such that each $F_X(A) : X(A) \to X'(A) = X(A^{(-1)})$ is given by $X(\gamma_A)$ where $\gamma_A : A \to A^{(-1)}$ satisfies $\gamma_A(a) = a^p$ for all $a \in A$. In I.9.2 this morphism was denoted by F_X^1. We again call F_X the (relative) *Frobenius morphism* on X. The map $X \mapsto F_X$ is a natural transformation from the identity functor to the functor $X \mapsto X'$: We have $F_Y \circ \varphi = \varphi' \circ F_X$ for any morphism $\varphi : X \to Y$ as above.

If $X = Sp_k R$ is an affine scheme and if we identify X' with $Sp_k R^{(1)}$, then the comorphism $F_X^* : R^{(1)} \to R$ maps any $f \in R$ to f^p. (Recall that we get the comorphism by applying F_X to $\mathrm{id}_R \in X(R)$; this yields γ_R that we have to interpret as a homomorphism $R^{(1)} \to R$.)

If A is an algebraically closed field (or more generally, a perfect field), then γ_A is bijective; hence so is then $F_X(A)$. Recall from the final remark in I.1.6 that an open subfunctor is determined by its values on all A that are algebraically closed fields. It follows for each open subfunctor $U \subset X$ that $U = (F_X)^{-1}(U')$, and for each open subfunctor $V \subset X'$ that $V = (F_X^{-1}V)'$. So:

(1) *The map $U \mapsto U'$ is a bijection from the set of all open subfunctors of X to the set of all open subfunctors of X'.*

If $\varphi : X \to Y$ is a morphism of schemes over k and if $\varphi' : X' \to Y'$ is the corresponding morphism, then one gets for each open U in Y

$$(\varphi')^{-1}(U') = \varphi^{-1}(U)'. \tag{2}$$

Again it suffices to check equality for points over an algebraically closed field. There one uses that $\varphi' \circ F_X = F_Y \circ \varphi$ and that $F_X(A)$ and $F_Y(A)$ are bijective.

Lemma: *If X is an algebraic scheme over k, then $F_X : X \to X'$ is a finite morphism.*

Proof: Pick open and affine subschemes $U_1, U_2, \dots, U_r$ of X that cover X. Then the U_i' form an open and affine covering of X' with $U_i = F_X^{-1}(U_i')$ for all i. This shows that F_X is an affine morphism. Each $k[U_i]$ is a finitely generated k–algebra, say $k[f_1, f_2, \dots, f_r]$. Now $k[U_i']$ identifies with $k[U_i]^{(1)}$; its image under F_X^* is the subalgebra $k[f_1^p, f_2^p, \dots, f_r^p]$ of $k[U_i]$. Now $k[U_i]$ is finitely generated as a module over this subalgebra by all monomials $f_1^{m(1)} f_2^{m(2)} \dots f_r^{m(r)}$ with $m(i) < p$ for all i. The claim follows.

Remark: Recall from I.1.11 that there is an equivalence of categories $X \mapsto |X|$ from our functorial schemes to the schemes in standard textbooks on algebraic geometry, e.g., in [Ha]. It induces a bijection between the set of all open subfunctors of X and the set of all open subsets of $|X|$. For an affine scheme $X = Sp_k R$ one gets that $|Sp_k R|$ is the prime spectrum of R with the usual Zariski topology.

In this affine case the formula $F_X^*(f) = f^p$ for the comorphism F_X^* implies easily $(F_X^*)^{-1}(P) = P^{(1)}$ for any prime ideal P in R. It follows that the morphism $|X| \to |X'|$ corresponding to F_X is a homeomorphism of topological spaces. Using open affine coverings one extends this result to general X. This fact is often used to identify $|X'|$ and $|X|$ as topological spaces so that F_X becomes the identity map.

Similarly, if k is algebraically closed and if one only considers varieties over k, then one usually replaces the functor X by $X(k)$ with a suitable topology and a sheaf of functions. Again one can identify $X(k)$ and $X'(k)$ via F_X as topological spaces.

We shall not make such identifications here. They would not work very well with our functorial approach. Furthermore, in many cases X will be defined over $\mathbf{F}_p$; then X' has a natural identification with X for which F_X usually is not the identity map. In fact, if we take $k = \mathbf{F}_p$, then $A^{(m)} = A$ for all $m \in \mathbf{Z}$ and all A, hence $X' = X$, but $F_X(A) = X(\gamma_A)$ is, in general, not the identity.

F.3. Let X be a scheme over k. (Recall that we from now on assume that our schemes are algebraic over k. This is done so that we can apply Lemma F.2. However many results still would work in greater generality.)

Let us use the following convention: If $\mathcal{M}$ is a sheaf of k–vector spaces on X, then $\mathcal{M}^{(1)}$ is the sheaf of k–vector spaces on X' such that $\mathcal{M}^{(1)}(U') = \mathcal{M}(U)^{(1)}$ for all open subschemes $U \subset X$. (This makes sense thanks to F.2(2).)

Lemma: *There exists a natural isomorphism*

(1) $$\mathcal{O}_X^{(1)} \xrightarrow{\sim} \mathcal{O}_{X'}$$

of sheaves of k–algebras on X'. If we use it to identify both sides, then the comorphism $F_X^* : \mathcal{O}_{X'} \to (F_X)_*\mathcal{O}_X$ is given by $f \mapsto f^p$ for each $f \in \mathcal{O}_{X'}(U') = \mathcal{O}_X(U)^{(1)}$ and each open $U \subset X$.

Proof: Any $f \in \mathcal{O}_X(X) = \mathrm{Mor}(X, \mathbf{A}^1)$ defines a morphism $\overline{f} : X' \to \mathbf{A}^1$, i.e., an element in $\mathcal{O}_{X'}(X')$, such that $\overline{f}(A)$ is equal to

$$\overline{f}(A) : X'(A) = X(A^{(-1)}) \xrightarrow{f(A^{(-1)})} A^{(-1)} = A.$$

One checks immediately that $\overline{f}$ indeed is a morphism, i.e., a natural transformation of functors. Both $\mathcal{O}_X(X)$ and $\mathcal{O}_{X'}(X')$ are k–algebras under pointwise addition, multiplication and scalar multiplication. The map $f \mapsto \overline{f}$ is easily seen to be compatible with addition and multiplication whereas one gets $\overline{cf} = c^p\,\overline{f}$ for all $c \in k$ due to the final identification of $A^{(-1)}$ and A in the definition of $\overline{f}$. This means that $f \mapsto \overline{f}$ is a k–algebra homomorphism $\mathcal{O}_X(X)^{(1)} \to \mathcal{O}_{X'}(X')$.

We get similarly for each open subscheme $U \subset X$ a k–algebra homomorphism $\mathcal{O}_X(U)^{(1)} \to \mathcal{O}_{X'}(U')$. These maps are compatible with restrictions and yield a homomorphism $\mathcal{O}_X^{(1)} \to \mathcal{O}_{X'}$ of sheaves of k–algebras on X'. If $U = Sp_k R$ is affine, hence $U' \simeq Sp_k(R^{(1)})$, then the map from $\mathcal{O}_X(U)^{(1)} = R^{(1)}$ to $\mathcal{O}_{X'}(U') = R^{(1)}$ is the identity. It follows that our homomorphism of sheaves is an isomorphism $\mathcal{O}_X^{(1)} \xrightarrow{\sim} \mathcal{O}_{X'}$ as claimed.

We have noticed in F.2 that the formula for F_X^* is correct in the affine case. Using affine coverings it follows in general.

Remark: We shall usually identify $\mathcal{O}_{X'}$ with $\mathcal{O}_X^{(1)}$ using the isomorphism in the lemma. Note that this identification is compatible with morphisms:

Let $\varphi : X \to Y$ be a morphism of schemes over k and let $\varphi' : X' \to Y'$ be the corresponding morphism. We get (under these identifications both on X' and on Y') for each open $U \subset Y$ that $\varphi'^* : \mathcal{O}_{Y'}(U') \to \mathcal{O}_{X'}(\varphi'^{-1}U')$ is identified with $\varphi^* : \mathcal{O}_Y(U)^{(1)} \to \mathcal{O}_X(\varphi^{-1}U)^{(1)}$. Indeed: Recall that $\varphi'^{-1}U' = (\varphi^{-1}U)'$ by F.2(2). And we have in the notation of the proof

$$\varphi'^*(\overline{f})(A) = \overline{f}(A) \circ \varphi'(A) = f(A^{-1}) \circ \varphi(A^{-1}) = \varphi^*(f)(A^{-1}) = \overline{\varphi^*(f)}(A)$$

for any $f \in \mathcal{O}_Y(U)$.

F.4. Let X again be a scheme over k. Let us write F instead of F_X when it is obvious which X we mean.

If a sheaf $\mathcal{M}$ on X is an $\mathcal{O}_X$–module, then the sheaf $\mathcal{M}^{(1)}$ on X is a module over $\mathcal{O}_{X'} = \mathcal{O}_X^{(1)}$ in a natural way. If $\mathcal{M}$ is coherent (resp. locally free of rank m), then $\mathcal{M}^{(1)}$ has the same property. The identity map yields an isomorphism of vector spaces

$$\text{(1)} \qquad \operatorname{Hom}_{\mathcal{O}_X}(\mathcal{M},\mathcal{N})^{(1)} \xrightarrow{\sim} \operatorname{Hom}_{\mathcal{O}_{X'}}(\mathcal{M}^{(1)},\mathcal{N}^{(1)})$$

for any $\mathcal{O}_X$–module $\mathcal{N}$.

On the other hand, also the direct image sheaf $F_*\mathcal{M}$ is a module over $\mathcal{O}_{X'}$. It satisfies $F_*\mathcal{M}(U') = \mathcal{M}(U)$ for all open $U \subset X$; now $\mathcal{M}(U)$ keeps its vector space structure over k whereas the $\mathcal{O}_{X'}$–module is changed: Any $f \in \mathcal{O}_{X'}(U') = \mathcal{O}_X(U)^{(1)}$ acts on $\mathcal{M}(U)$ as $F^*(f) = f^p \in \mathcal{O}_X(U)$ does.

One has

$$\text{(2)} \qquad H^i(X',\mathcal{M}^{(1)}) \simeq H^i(X,\mathcal{M})^{(1)}$$

for all $i \in \mathbf{N}$ and all $\mathcal{O}_X$–modules $\mathcal{M}$ because the cohomology groups do not depend on the $\mathcal{O}_{X'}$–module structure, whereas the k–structure on the cohomology groups does.

If $\varphi : X \to Y$ is a morphism of schemes, then the associated morphism $\varphi' : X' \to Y'$ satisfies $\varphi'_*(\mathcal{M}^{(1)}) = (\varphi_*\mathcal{M})^{(1)}$ for any $\mathcal{O}_X$–module $\mathcal{M}$ and $\varphi'^*(\mathcal{N}^{(1)}) = (\varphi^*\mathcal{N})^{(1)}$ for any $\mathcal{O}_Y$–module $\mathcal{N}$.

If $Z \subset X$ is a closed subscheme, then Z' is a closed subscheme in X' as observed before. If $\mathcal{I}_Z$ is the ideal sheaf of Z in $\mathcal{O}_X$, and if $\mathcal{I}_{Z'}$ is the ideal sheaf of Z' in $\mathcal{O}_{X'}$, then we have

$$\text{(3)} \qquad \mathcal{I}_{Z'} = \mathcal{I}_Z^{(1)}$$

under the identification from Lemma F.3. (The isomorphism in F.3(1) clearly maps $\mathcal{I}_Z^{(1)}$ into $\mathcal{I}_{Z'}$. One gets equality looking at affine open subschemes where one uses the formula $V(I)' = V(I^{(1)})$ mentioned in F.1.)

One gets for the sheaves of differentials $\Omega_{X'/k} = \Omega_{X/k}^{(1)}$ and for the canonical sheaves if X and X' are smooth

$$\text{(4)} \qquad \omega_{X'} = \omega_X^{(1)}.$$

F.5. In the second half of this chapter we shall apply the general theory form the first half to varieties with actions of G or B, such as flag varieties and Schubert varieties. We get then also involved with line bundles that are G– or B–linearised in the sense of [MF]. However, it will be conceptually easier to deal with sheaves with an action of $G(k)$ or $B(k)$ considered as an abstract group. We now go through the definitions needed in that context.

Let H be an abstract group acting on a scheme X over k. We have then for each open $U \subset X$ a k–algebra isomorphism $h^* : \mathcal{O}_X(U) \xrightarrow{\sim} \mathcal{O}_X(h^{-1}U)$. These maps are compatible with restrictions from open subschemes to smaller ones. One has $h_1^* \circ h_2^* = (h_2h_1)^*$ for all $h_1, h_2 \in X$. If $U \subset X$ is open and H–stable, then

$\mathcal{O}_X(U)$ is a kH–module via $h\,f = (h^{-1})^*(f)$ for all $h \in H$ and $f \in \mathcal{O}_X(U)$; here H acts on $\mathcal{O}_X(U)$ via algebra automorphisms.

An H–equivariant $\mathcal{O}_X$–module is an $\mathcal{O}_X$–module $\mathcal{M}$ together with isomorphisms of abelian groups $h^* : \mathcal{M}(U) \xrightarrow{\sim} \mathcal{M}(h^{-1}U)$ for all $U \subset X$ open and $h \in H$ such that $h^*(f\,v) = h^*(f)\,h^*(v)$ for all $v \in \mathcal{M}(U)$ and $f \in \mathcal{O}_X(U)$; furthermore these maps have to be compatible with restrictions from open subschemes to smaller ones and to satisfy $h_1^* \circ h_2^* = (h_2h_1)^*$ for all $h_1, h_2 \in X$. An obvious example is $\mathcal{O}_X$ itself.

Tensor products of H–equivariant $\mathcal{O}_X$–modules are again H–equivariant, similarly for symmetric and exterior powers.

If $\mathcal{M}$ is an H–equivariant $\mathcal{O}_X$–module and if $U \subset X$ is open and H–stable, then $\mathcal{M}(U)$ is a kH–module via $h\,f = (h^{-1})^*(f)$ for all $h \in H$ and $f \in \mathcal{M}(U)$.

If $\mathcal{M}$ and $\mathcal{N}$ are H–equivariant $\mathcal{O}_X$–modules, then $\mathrm{Hom}_{\mathcal{O}_X}(\mathcal{M},\mathcal{N})$ is a kH–module via $h \cdot \gamma = (h^{-1})^* \circ \gamma \circ h^*$ for all $h \in H$ and $\gamma \in \mathrm{Hom}_{\mathcal{O}_X}(\mathcal{M},\mathcal{N})$. In more detail $h \cdot \gamma$ maps any $\mathcal{M}(U)$ according to

$$\mathcal{M}(U) \xrightarrow{h^*} \mathcal{M}(h^{-1}U) \xrightarrow{\gamma} \mathcal{N}(h^{-1}U) \xrightarrow{(h^{-1})^*} \mathcal{N}(U).$$

We call $\gamma \in \mathrm{Hom}_{\mathcal{O}_X}(\mathcal{M},\mathcal{N})$ a *homomorphism of H–equivariant $\mathcal{O}_X$–modules* if $h \cdot \gamma = \gamma$ for all $h \in H$. If also $\mathcal{L}$ is an H–equivariant $\mathcal{O}_X$–module, then

$$\mathrm{Hom}_{\mathcal{O}_X}(\mathcal{L},\mathcal{M}) \otimes \mathrm{Hom}_{\mathcal{O}_X}(\mathcal{M},\mathcal{N}) \to \mathrm{Hom}_{\mathcal{O}_X}(\mathcal{L},\mathcal{N}), \qquad \delta \otimes \gamma \mapsto \gamma \circ \delta$$

is a homomorphism of kH–modules.

If $U \subset X$ is open and H–stable and if $\mathcal{M}$ and $\mathcal{N}$ are as above, then the kH–module structures on $\mathcal{M}(U)$ and $\mathcal{N}(U)$ make $\mathrm{Hom}(\mathcal{M}(U),\mathcal{N}(U))$ into a kH–module (via $h \cdot \varphi = h \circ \varphi \circ h^{-1}$). It contains $\mathrm{Hom}_{\mathcal{O}_X(U)}(\mathcal{M}(U),\mathcal{N}(U))$ as a submodule; the natural map $\mathrm{Hom}_{\mathcal{O}_X}(\mathcal{M},\mathcal{N}) \to \mathrm{Hom}_{\mathcal{O}_X(U)}(\mathcal{M}(U),\mathcal{N}(U))$ is a homomorphism of kH–modules.

F.6. Let H and X be as in F.5. The functoriality of $X \to X'$ leads to an action of H on X'; the map $\varphi \mapsto \varphi'$ as in F.1 takes the action of any $h \in H$ on X to the action of h on X'. We have $h^{-1}(U') = (h^{-1}U)'$ for all $U \subset X$ open and $h \in H$, see F.1(1). The comorphism $h^* : \mathcal{O}_{X'}(U') \to \mathcal{O}_{X'}(h^{-1}(U'))$ is equal to $h^* : \mathcal{O}_X(U)^{(1)} \xrightarrow{\sim} \mathcal{O}_X(h^{-1}U)^{(1)}$ under the identification $\mathcal{O}_X^{(1)} \xrightarrow{\sim} \mathcal{O}_{X'}$ from F.3(1), see Remark F.3.

If $\mathcal{M}$ is an H–equivariant $\mathcal{O}_X$–module, then $\mathcal{M}^{(1)}$ and $F_*\mathcal{M}$ are H–equivariant $\mathcal{O}_{X'}$–modules: One defines h^* via

$$\mathcal{M}^{(1)}(U') = \mathcal{M}(U)^{(1)} \xrightarrow{h^*} \mathcal{M}(h^{-1}U)^{(1)} = \mathcal{M}^{(1)}(h^{-1}(U'))$$

and

$$F_*\mathcal{M}(U') = \mathcal{M}(U) \xrightarrow{h^*} \mathcal{M}(h^{-1}U) = F_*\mathcal{M}(h^{-1}(U'))$$

and checks easily the defining properties of an H–equivariant $\mathcal{O}_{X'}$–module. We write in these cases occasionally $h^{*(1)}$ or $F_*(h^*)$ in order to distinguish them from the original h^*.

If $\mathcal{M}$ and $\mathcal{N}$ are H–equivariant $\mathcal{O}_X$–modules, then the isomorphism

$$\mathrm{Hom}_{\mathcal{O}_X}(\mathcal{M},\mathcal{N})^{(1)} \xrightarrow{\sim} \mathrm{Hom}_{\mathcal{O}_{X'}}(\mathcal{M}^{(1)},\mathcal{N}^{(1)})$$

is an isomorphism of kH–modules and the natural map $\gamma \mapsto F_*(\gamma)$ is a homomorphism

$$\mathrm{Hom}_{\mathcal{O}_X}(\mathcal{M},\mathcal{N}) \to \mathrm{Hom}_{\mathcal{O}_{X'}}(F_*\mathcal{M}, F_*\mathcal{N})$$

of kH–modules.

F.7. Lemma: *If $\mathcal{L}$ is a line bundle on X, then there is an isomorphism*

$$F^*(\mathcal{L}^{(1)}) \xrightarrow{\sim} \mathcal{L}^p \tag{1}$$

of $\mathcal{O}_X$–modules and an isomorphism

$$\mathcal{L}^{(1)} \otimes_{\mathcal{O}_{X'}} F_*\mathcal{O}_X \xrightarrow{\sim} F_*\mathcal{L}^p \tag{2}$$

of $\mathcal{O}_{X'}$–modules.

Proof: Interpreting $\mathcal{L}^p$ as the symmetric power $S^p(\mathcal{L})$, we have a natural map $f \otimes a \mapsto f\,a^p$ from $\mathcal{O}_X(U) \otimes_{\mathcal{O}_X(U)^{(1)}} \mathcal{L}(U)^{(1)}$ to $S^p(\mathcal{L}(U))$ for any open U in X. This yields the map in (1); one checks locally that it is an isomorphism.

We have an isomorphism $\mathcal{L}^{(1)} \otimes_{\mathcal{O}_{X'}} F_*\mathcal{O}_X \xrightarrow{\sim} F_*(F^*(\mathcal{L}^{(1)}) \otimes_{\mathcal{O}_X} \mathcal{O}_X)$ coming from the projection formula. Combining this with (1) we get the second claim. The map in (2) takes $a \otimes f$ to $a^p f$ for any $a \in \mathcal{L}^{(1)}(U') = \mathcal{L}(U)^{(1)}$ and $f \in (F_*\mathcal{O}_X)(U') = \mathcal{O}_X(U)$.

Remarks: 1) If we tensor $F^* : \mathcal{O}_{X'} \to F_*\mathcal{O}_X$ with the identity on $\mathcal{L}^{(1)}$, then we get a homomorphism of $\mathcal{O}_{X'}$–modules

$$F_{\mathcal{L}} : \mathcal{L}^{(1)} \longrightarrow F_*\mathcal{L}^p \tag{3}$$

that takes $a \in \mathcal{L}^{(1)}(U') = \mathcal{L}(U)^{(1)}$ to $a^p \in (F_*\mathcal{L}^p)(U') = \mathcal{L}^p(U)$. This map corresponds to the one in (1) under the adjunction map $\mathrm{Hom}_{\mathcal{O}_{X'}}(\mathcal{L}^{(1)}, F_*\mathcal{L}^p) \xrightarrow{\sim} \mathrm{Hom}_{\mathcal{O}_X}(F^*\mathcal{L}^{(1)}, \mathcal{L}^p)$.

2) If H is a group acting on X and if $\mathcal{L}$ is an H–equivariant line bundle on X, then the maps in (1)–(3) are homomorphisms of H–equivariant $\mathcal{O}_X$– or $\mathcal{O}_{X'}$–modules.

F.8. Each k–algebra A is a module over $A^{(1)}$ via the homomorphism $\gamma_A : A^{(1)} \to A$ with $a \mapsto a^p$. A *Frobenius splitting* for A is a homomorphism $\sigma : A \to A^{(1)}$ of $A^{(1)}$–modules such that $\sigma(a^p) = a$ for all $a \in A$, i.e., such that $\sigma \circ \gamma_A$ is the identity on $A^{(1)}$. Note that the $A^{(1)}$–linearity means that $\sigma(a_1 + a_2) = \sigma(a_1) + \sigma(a_2)$ and $\sigma(a_1^p a_2) = a_1 \sigma(a_2)$ for all $a_1, a_2 \in A$.

An arbitrary $A^{(1)}$–linear map $\sigma : A \to A^{(1)}$ satisfies $\sigma(a^p) = a\,\sigma(1)$ for all $a \in A$; so σ is a Frobenius splitting for A if and only if $\sigma(1) = 1$.

For example, the polynomial ring $A = k[T_1, T_2, \dots, T_n]$ has a Frobenius splitting: We can take for σ the additive map with

$$cT_1^{m(1)}T_2^{m(2)}\dots T_n^{m(n)} \mapsto \begin{cases} c^{p^{-1}} T_1^{m(1)/p} T_2^{m(2)/p} \dots T_n^{m(n)/p}, & \text{if } p \mid m(i) \text{ for all } i, \\ 0, & \text{otherwise,} \end{cases}$$

for all $c \in k$ and all $m(i) \in \mathbf{N}$.

We say that a scheme X over k is *Frobenius split* if there exists a homomorphism $\sigma : F_*\mathcal{O}_X \to \mathcal{O}_{X'}$ of $\mathcal{O}_{X'}$–modules such that $\sigma \circ F^*$ is the identity on $\mathcal{O}_{X'}$. Any σ with this property is called a *splitting* (or a splitting for X).

If σ is such a splitting, then each $\sigma(U')$ with U open in X is a Frobenius splitting for $\mathcal{O}_X(U)$: It is a homomorphism of modules over $\mathcal{O}_{X'}(U') = \mathcal{O}_X(U)^{(1)}$ from $(F_*\mathcal{O}_X)(U') = \mathcal{O}_X(U)$ to $\mathcal{O}_{X'}(U') = \mathcal{O}_X(U)^{(1)}$ that maps any $F^*(f) = f^p$ to f. And the $\mathcal{O}_X(U)^{(1)}$–module structure on $\mathcal{O}_X(U)$ is given via $f \mapsto F^*(f) = f^p$.

We get as in the algebra case:

(1) *An arbitrary homomorphism $\sigma : F_*\mathcal{O}_X \to \mathcal{O}_{X'}$ of $\mathcal{O}_{X'}$–modules is a splitting if and only if $\sigma(1) = 1$.*

If $X = Sp_k R$ is an affine scheme, then a splitting for X is the same as a Frobenius splitting for R. So the example above shows that each affine space $\mathbf{A}^n$ is Frobenius split. More generally, any smooth affine scheme is Frobenius split, see [Hochster and Roberts], Props. 5.9 and 5.14.

It is clear that any open subscheme of a Frobenius split scheme X is again Frobenius split. Note that:

(2) *Each Frobenius split scheme is reduced.*

Indeed, if σ is a splitting, then we have $\sigma(f^p) = f$ for all $f \in \mathcal{O}_{X'}(U') = \mathcal{O}_X(U)^{(1)}$ and any U. So $f^p = 0$ implies $f = 0$. It follows that $\mathcal{O}_X(U)$ cannot contain any nilpotent element other than 0.

F.9. Proposition: *Let X be a smooth and irreducible projective variety over k. Then there is a natural isomorphism* $\mathrm{Hom}_{\mathcal{O}_{X'}}(F_*\mathcal{O}_X, \mathcal{O}_{X'}) \simeq H^0(X, \omega_X^{1-p})$.

Proof: Set $n = \dim(X)$. We get isomorphisms

$$\begin{aligned}\mathrm{Hom}_{\mathcal{O}_{X'}}(F_*\mathcal{O}_X, \mathcal{O}_{X'}) &\simeq \mathrm{Hom}_{\mathcal{O}_{X'}}(F_*\mathcal{O}_X \otimes_{\mathcal{O}_{X'}} \omega_{X'}, \omega_{X'}) \\ &\simeq H^n(X', F_*\mathcal{O}_X \otimes_{\mathcal{O}_{X'}} \omega_{X'})^* \\ &\simeq H^n(X', F_*(\mathcal{O}_X \otimes_{\mathcal{O}_X} F^*\omega_{X'}))^* \simeq H^n(X, F^*\omega_{X'})^* \\ &\simeq H^n(X, \omega_X^p)^* \simeq H^0(X, \omega_X^{1-p})\end{aligned}$$

where we use that tensoring with an invertible sheaf does not change the Hom space, Serre duality for X' (which again is smooth, irreducible, and projective of dimension n), the projection formula, the fact that F is affine, the isomorphisms $F^*\omega_{X'} \simeq F^*(\omega_X^{(1)}) \simeq \omega_X^p$ from F.4(4) and F.7(1), and finally Serre duality for X.

Remarks: 1) If H is a group acting on X, then ω_X is an H–equivariant $\mathcal{O}_X$–module. Then the isomorphism in the proposition is H–equivariant as all isomorphisms used above were natural.

2) One can actually show that the sheaf $\mathcal{H}om_{\mathcal{O}_{X'}}(F_*\mathcal{O}_X, \mathcal{O}_{X'})$ is isomorphic to $F_*(\omega_X^{1-p})$,

F.10. Let $\mathcal{L}$ be a line bundle on a scheme X over k. If $\sigma : F_*\mathcal{O}_X \to \mathcal{O}_{X'}$ is an $\mathcal{O}_{X'}$–module homomorphism, then we get tensoring with the identity on $\mathcal{L}^{(1)}$ an $\mathcal{O}_{X'}$–module homomorphism

(1) $$\sigma_{\mathcal{L}} : F_*\mathcal{L}^p \longrightarrow \mathcal{L}^{(1)}$$

where we use the isomorphism from F.7(2). If σ is a splitting, then the composition $\sigma_{\mathcal{L}} \circ F_{\mathcal{L}}$ with $F_{\mathcal{L}}$ as in F.7(3) is the identity on $\mathcal{L}^{(1)}$ and $\mathcal{L}^{(1)}$ is isomorphic to a direct summand of $F_*\mathcal{L}^p$. For $\mathcal{L} = \mathcal{O}_X$ one gets $\sigma_{\mathcal{O}_X} = \sigma$.

Proposition: *Let X be a projective scheme over k and let $\mathcal{L}$ be an ample line bundle on X. If X is Frobenius split, then $H^i(X,\mathcal{L}) = 0$ for all $i > 0$. If X is smooth, irreducible, and Frobenius split, then $H^i(X, \omega_X \otimes \mathcal{L}) = 0$ for all $i > 0$.*

Proof: Since X is Frobenius split, $\mathcal{L}^{(1)}$ is a direct summand of $F_*\mathcal{L}^p$. It follows that $H^i(X', \mathcal{L}^{(1)})$ is a direct summand of $H^i(X', F_*\mathcal{L}^p)$ for all i. The latter cohomology group is isomorphic to $H^i(X, \mathcal{L}^p)$ because F is affine, see Lemma F.2. So F.4(1) shows that $H^i(X,\mathcal{L})^{(1)}$ is a direct summand of $H^i(X,\mathcal{L}^p)$. Iterating we get that $H^i(X,\mathcal{L})^{(r)}$ is a direct summand of $H^i(X,\mathcal{L}^{p^r})$ for all $r \in \mathbf{N}$. If now $\mathcal{L}$ is ample, then we can choose $r \geq 0$ such that $H^i(X,\mathcal{L}^{p^r}) = 0$ for all $i > 0$. The claim follows.

Assume now in addition that X is smooth and irreducible of dimension n. The same argument as above shows that each $H^j(X,\mathcal{L}^{-1})^{(r)}$ is a direct summand of $H^j(X,\mathcal{L}^{-p^r})$. Serre duality says that $H^j(X,\mathcal{L}^{-p^r}) \simeq H^{n-j}(X, \omega_X \otimes \mathcal{L}^{p^r})^*$ and the ampleness of $\mathcal{L}$ implies that the latter group is 0 for $r \gg 0$ and $n-j > 0$. It follows that $H^j(X,\mathcal{L}^{-1}) = 0$ for all $j < n$. Now use that $H^i(X,\omega_X \otimes \mathcal{L}) \simeq H^{n-i}(X,\mathcal{L}^{-1})$ by Serre duality.

Remark: If H is a group acting on X and if $\mathcal{L}$ is an H–equivariant line bundle on X, then the map $\sigma \mapsto \mathrm{id} \otimes \sigma$ is easily seen to be a homomorphism

$$\mathrm{Hom}_{\mathcal{O}_{X'}}(F_*\mathcal{O}_X, \mathcal{O}_{X'}) \to \mathrm{Hom}_{\mathcal{O}_{X'}}(\mathcal{L}^{(1)} \otimes_{\mathcal{O}_{X'}} F_*\mathcal{O}_X, \mathcal{L}^{(1)} \otimes_{\mathcal{O}_{X'}} \mathcal{O}_{X'})$$

of kH–modules. Since the isomorphism $\mathcal{L}^{(1)} \otimes_{\mathcal{O}_{X'}} F_*\mathcal{O}_X \xrightarrow{\sim} F_*(\mathcal{L}^p)$ from F.7(2) and the obvious isomorphism $\mathcal{L}^{(1)} \otimes_{\mathcal{O}_{X'}} \mathcal{O}_{X'} \xrightarrow{\sim} \mathcal{L}^{(1)}$ both are isomorphisms of H–equivariant $\mathcal{O}_{X'}$–modules, it follows that

(2) $$\mathrm{Hom}_{\mathcal{O}_{X'}}(F_*\mathcal{O}_X, \mathcal{O}_{X'}) \to \mathrm{Hom}_{\mathcal{O}_{X'}}(F_*(\mathcal{L}^p), \mathcal{L}^{(1)}), \qquad \sigma \mapsto \sigma_{\mathcal{L}}$$

is a homomorphism of kH–modules.

F.11. Let $A = \bigoplus_{n \geq 0} A_n$ be a graded k–algebra. We call a Frobenius splitting $\sigma : A \to A^{(1)}$ a *graded Frobenius splitting* for A if $\sigma(A_{mp}) \subset A_m^{(1)}$ for all m and if $\sigma(A_n) = 0$ for all n with $p \nmid n$.

Let $\mathcal{L}$ be a line bundle on a scheme X over k. For any open $U \subset X$ set

(1) $$\Gamma_*(U,\mathcal{L}) = \bigoplus_{n \geq 0} H^0(U, \mathcal{L}^n).$$

This is a graded k–algebra under the cup-product. If H is a group acting on X and if $\mathcal{L}$ is H–equivariant, then so are all $\mathcal{L}^m$; then H acts on $\Gamma_*(U,\mathcal{L})$ via graded algebra automorphisms.

Let $\sigma \in \mathrm{Hom}_{\mathcal{O}_{X'}}(F_*\mathcal{O}_X, \mathcal{O}_{X'})$ and let

$$\widetilde{\sigma}_{\mathcal{L},U} : \Gamma_*(U,\mathcal{L}) \longrightarrow \Gamma_*(U,\mathcal{L})^{(1)}$$

denote the linear map with $\widetilde{\sigma}_{\mathcal{L},U}(\Gamma_*(U,\mathcal{L})_n) = 0$ if $p \nmid n$, whereas $\widetilde{\sigma}_{\mathcal{L},U}$ restricts to $\sigma_{\mathcal{L}^m} : \mathcal{L}^{mp}(U) \to \mathcal{L}^m(U)^{(1)}$ on the remaining terms.

Lemma: *The map $\widetilde{\sigma}_{\mathcal{L},U}$ is a homomorphism of $\Gamma_*(U,\mathcal{L})^{(1)}$–modules. If σ is a splitting for X, then $\widetilde{\sigma}_{\mathcal{L},U}$ is a graded Frobenius splitting for $\Gamma_*(U,\mathcal{L})$.*

Proof: The second claim follows from the first one since the condition on the grading holds by construction and since

$$\widetilde{\sigma}_{\mathcal{L},U}(1) = \sigma_{\mathcal{L}^0}(1) = \sigma_{\mathcal{O}_X}(1) = \sigma(1) = 1.$$

As $\widetilde{\sigma}_{\mathcal{L},U}$ is additive by construction, it suffices to show that

$$\widetilde{\sigma}_{\mathcal{L},U}(a^p b) = a\,\widetilde{\sigma}_{\mathcal{L},U}(b) \qquad \text{for all } a \in \Gamma_*(U,\mathcal{L})^{(1)},\ b \in \Gamma_*(U,\mathcal{L}). \tag{3}$$

Because $a \mapsto a^p$ is additive, it suffices to take a and b homogeneous in (3), say $a \in \mathcal{L}^n(U)^{(1)}$ and $b \in \mathcal{L}^{mp}(U)$. Now the products in (3) arise from identifying $\mathcal{L}^n \otimes \mathcal{L}^m \simeq \mathcal{L}^{n+m}$ and $\mathcal{L}^{np} \otimes \mathcal{L}^{mp} \simeq \mathcal{L}^{(n+m)p}$. Without these identifications (3) amounts to

$$\sigma_{\mathcal{L}^n \otimes \mathcal{L}^m}(a^p \otimes b) = a \otimes \sigma_{\mathcal{L}^m}(b) \tag{4}$$

for all a and b.

Let us show more generally for any line bundles $\mathcal{L}_1$ and $\mathcal{L}_2$ on X that

$$\sigma_{\mathcal{L}_1 \otimes \mathcal{L}_2}(a^p \otimes b) = a \otimes \sigma_{\mathcal{L}_2}(b) \qquad \text{for all } a \in \mathcal{L}_1(U)^{(1)},\ b \in \mathcal{L}_2^p(U). \tag{5}$$

Then we get (4) as a special case.

Note that we can identify $\mathcal{L}_1^{(1)} \otimes_{\mathcal{O}_{X'}} \mathcal{L}_2^{(1)}$ and $(\mathcal{L}_1 \otimes_{\mathcal{O}_X} \mathcal{L}_2)^{(1)}$: Considered as sheaves of abelian groups they actually are equal. It follows that we have a commutative diagram

$$\begin{array}{ccc}
\mathcal{L}_1^{(1)} \otimes_{\mathcal{O}_{X'}} \mathcal{L}_2^{(1)} \otimes_{\mathcal{O}_{X'}} F_*\mathcal{O}_X & \xrightarrow{\sim} & \mathcal{L}_1^{(1)} \otimes_{\mathcal{O}_{X'}} F_*\mathcal{L}_2^p \\
\sim\big\downarrow & & \big\downarrow\gamma \\
(\mathcal{L}_1 \otimes_{\mathcal{O}_X} \mathcal{L}_2)^{(1)} \otimes_{\mathcal{O}_{X'}} F_*\mathcal{O}_X & \xrightarrow{\sim} & F_*((\mathcal{L}_1 \otimes_{\mathcal{O}_X} \mathcal{L}_2)^p)
\end{array}$$

where the vertical arrow on the left arises from an identification as above and the horizontal arrows come from F.7(2). The map γ is defined by this diagram. Going back to the description of the map in F.7(2) we see that $\gamma(a \otimes b) = a^p \otimes b$ for all a and b as in (5).

We get from σ another commutative diagram

$$\begin{array}{ccccc}
\mathcal{L}_1^{(1)} \otimes_{\mathcal{O}_{X'}} \mathcal{L}_2^{(1)} \otimes_{\mathcal{O}_{X'}} F_*\mathcal{O}_X & \to & \mathcal{L}_1^{(1)} \otimes_{\mathcal{O}_{X'}} \mathcal{L}_2^{(1)} \otimes_{\mathcal{O}_{X'}} \mathcal{O}_{X'} & \to & \mathcal{L}_1^{(1)} \otimes_{\mathcal{O}_{X'}} \mathcal{L}_2^{(1)} \\
\sim\big\downarrow & & \sim\big\downarrow & & \big\downarrow\sim \\
(\mathcal{L}_1 \otimes_{\mathcal{O}_X} \mathcal{L}_2)^{(1)} \otimes_{\mathcal{O}_{X'}} F_*\mathcal{O}_X & \to & (\mathcal{L}_1 \otimes_{\mathcal{O}_X} \mathcal{L}_2)^{(1)} \otimes_{\mathcal{O}_{X'}} \mathcal{O}_{X'} & \to & (\mathcal{L}_1 \otimes_{\mathcal{O}_X} \mathcal{L}_2)^{(1)}
\end{array}$$

where the vertical arrows come from identifications as above, where the two horizontal arrows on the left arise from tensoring σ with the identity on the other factors and where the two horizontal arrows on the right are obvious identifications. Combining the two diagrams and recalling the definition of $\sigma_{\mathcal{L}}$ in F.10(1), we get (5).

Remark: If $V \subset U$ is another open subscheme of X, then $\widetilde{\sigma}_{\mathcal{L},U}$ and $\widetilde{\sigma}_{\mathcal{L},V}$ commute by construction with the restriction maps $\Gamma_*(U,\mathcal{L}) \to \Gamma_*(V,\mathcal{L})$ and $\Gamma_*(U,\mathcal{L})^{(1)} \to \Gamma_*(V,\mathcal{L})^{(1)}$.

F.12. Let X be a scheme over k and $Z \subset X$ a closed subscheme. Suppose that X is Frobenius split. Then one says that Z is *compatibly split* in X if there exists a splitting $\sigma : F_*\mathcal{O}_X \to \mathcal{O}_{X'} = \mathcal{O}_X^{(1)}$ for X such that

$$\sigma(F_*\mathcal{I}_Z) \subset \mathcal{I}_{Z'} = \mathcal{I}_Z^{(1)} \tag{1}$$

where $\mathcal{I}_Z$ denotes as in F.4 the ideal sheaf in $\mathcal{O}_X$ corresponding to Z, similarly for $\mathcal{I}_{Z'} \subset \mathcal{O}_{X'}$. One says in this case more precisely that Z is *compatibly σ–split.*

For any $U \subset X$ open and any $f \in \mathcal{I}_{Z'}(U') = \mathcal{I}_Z(U)^{(1)}$ we have $F^*(f) = f^p \in (F_*\mathcal{I}_Z)(U')$, hence $f = \sigma(F^*(f)) \in \sigma((F_*\mathcal{I}_Z)(U'))$. In other words, we have for any splitting σ of X that $\mathcal{I}_{Z'} \subset \sigma(F_*\mathcal{I}_Z)$, hence that

$$\sigma(F_*\mathcal{I}_Z) \subset \mathcal{I}_{Z'} \iff \sigma(F_*\mathcal{I}_Z) = \mathcal{I}_{Z'}. \tag{2}$$

If these equivalent conditions hold, i.e., if Z is compatibly σ–split, then σ induces a homomorphism of $\mathcal{O}_{X'}$–modules

$$F_*(\mathcal{O}_X/\mathcal{I}_Z) \simeq F_*\mathcal{O}_X/F_*\mathcal{I}_Z \longrightarrow \mathcal{O}_{X'}/\mathcal{I}_{Z'}$$

where we use that F is affine, hence F_* exact. Pulling this map back to Z', we get a homomorphism of $\mathcal{O}_{Z'}$–modules $F_*\mathcal{O}_Z \to \mathcal{O}_{Z'}$ that maps 1 to 1, i.e., a splitting for Z. We see in particular that a compatibly split closed subscheme is itself Frobenius split and reduced.

If $U \subset X$ is an open subscheme and if Z is compatibly σ–split, then $Z \cap U$ is obviously compatibly $\sigma_{|U}$–split.

If $X = Sp_k R$ is affine, then a splitting σ of X is given by a Frobenius splitting $\sigma_R : R \to R^{(1)}$ for R, see F.8. If now $Z = V(I)$ for an ideal I in R, then Z is compatibly σ-split if and only if $\sigma_R(I) \subset I^{(1)}$.

Lemma: *Let X be a Frobenius split scheme over k with splitting σ. Let Z be an integrally closed subscheme of X and U an open subscheme of X with $|U| \cap |Z| \neq \emptyset$. If the closed subscheme $U \cap Z$ of U is compatibly $\sigma_{|U}$–split, then Z is compatibly σ–split.*

Proof: We have to show that $\sigma(F_*\mathcal{I}_Z) \subset \mathcal{I}_{Z'}$. It is enough to check this on an open covering. We may therefore assume that X is an affine scheme, $X = Sp_k R$. Then $Z = V(I)$ for some prime ideal I in R. We can replace U by a smaller open subscheme and may assume that $U = D(f)$ for some $f \in R$, hence that $U = Sp_K R_f$. The assumption that $|U| \cap |Z| \neq \emptyset$ means that $f \notin I$. We have $U \cap Z = V(I_f) \subset U$.

Now σ is given by $\sigma_R : R \to R^{(1)}$, hence $\sigma_{|U}$ by the localisation $\sigma_f : R_f \to R_f^{(1)}$ with $\sigma_f(gf^{-pr}) = \sigma_R(g)f^{-r}$ for all $g \in R$ and $r \in \mathbf{N}$. The assumption that $U \cap Z$ of U is compatibly $\sigma_{|U}$–split means that $\sigma_f(I_f) \subset I_f^{(1)}$. This implies that $\sigma_R(I) \subset I^{(1)}$ and hence the claim because I is equal the inverse image of I_f under the natural map $R \to R_f$. (More generally, if S is a multiplicative subset in R and P a prime ideal in R with $S \cap P = \emptyset$, then P is the inverse image of $S^{-1}P$ under the natural map $R \to S^{-1}R$.)

F.13. If Z is a closed subscheme of a scheme X over k, then $|Z|$ is a noetherian topological space since our schemes are assumed to be algebraic. So $|Z|$ has a decomposition $|Z| = V_1 \cup V_2 \cup \cdots \cup V_r$ into irreducible components. For each i there exists then a unique reduced closed subscheme Z_i of X with $|Z_i| = V_i$, cf. [DG], I, §2, 4.11. We call then $Z_1, Z_2, \ldots, Z_r$ the *reduced irreducible components* of Z.

Lemma: *Let X be a Frobenius split scheme over k with splitting σ.*

a) *If Z_1 and Z_2 are compatibly σ–split closed subschemes of X, then also $Z_1 \cap Z_2$ is compatibly σ–split.*

b) *Let Z be a closed subscheme of X. Let $Z_1, Z_2, \ldots, Z_r$ denote the reduced irreducible components of Z. If Z is compatibly σ–split, then so are all Z_i.*

Proof: a) The ideal sheaf of $Z_1 \cap Z_2$ is equal to $\mathcal{I}_{Z_1} + \mathcal{I}_{Z_2}$. (Note that we work with schemes, not with varieties.) If now $\sigma(F_*\mathcal{I}_{Z_1}) \subset \mathcal{I}_{Z_1}^{(1)}$ and $\sigma(F_*\mathcal{I}_{Z_2}) \subset \mathcal{I}_{Z_2}^{(1)}$, then clearly

$$\sigma(F_*(\mathcal{I}_{Z_1} + \mathcal{I}_{Z_2})) = \sigma(F_*(\mathcal{I}_{Z_1}) + F_*(\mathcal{I}_{Z_2})) \subset \mathcal{I}_{Z_1}^{(1)} + \mathcal{I}_{Z_2}^{(1)} = (\mathcal{I}_{Z_1} + \mathcal{I}_{Z_2})^{(1)}$$

hence the claim.

b) Let $U \subset X$ be the open subscheme such that $|U|$ is the complement of $\bigcup_{i \geq 2} |Z_i|$ in $|X|$, cf. [DG], I, §1, 4.12. Then $|U| \cap |Z| = |U| \cap |Z_1| \neq \emptyset$. Since Z is Frobenius split, it is reduced. Therefore both $U \cap Z$ and $U \cap Z_1$ are reduced closed subschemes of U with $|U \cap Z| = |U \cap Z_1|$. It follows that $U \cap Z = U \cap Z_1$. Since Z is compatibly σ–split, the intersection $U \cap Z$ is compatibly $\sigma_{|U}$–split. Now Lemma F.12 applied to Z_1 instead of Z implies that Z_1 is compatibly σ–split. The claim follows.

Remark: Note that a) implies for Z_1 and Z_2 as in that claim that the scheme theoretic intersection of Z_1 and Z_2 is reduced.

F.14. Proposition: *Let X be a projective scheme over k and Z a closed subscheme of X. Suppose that X is Frobenius split and that Z is compatibly σ–split for some splitting σ of X. Then the restriction map $H^0(X, \mathcal{L}) \to H^0(Z, \mathcal{L})$ is surjective for each ample line bundle $\mathcal{L}$ on X.*

Proof: We have a commutative diagram

$$\begin{array}{ccccc} \mathcal{O}_{X'} & \longrightarrow & F_*\mathcal{O}_X & \longrightarrow & \mathcal{O}_{X'} \\ \downarrow & & \downarrow & & \downarrow \\ \mathcal{O}_{Z'} & \longrightarrow & F_*\mathcal{O}_Z & \longrightarrow & \mathcal{O}_{Z'} \end{array}$$

where the vertical arrows are given by restriction, the horizontal arrows on the left by F^*, the horizontal arrows on the right by σ. The compositions of the horizontal maps are the identity.

Tensoring with $\mathcal{L}^{(1)}$ and taking global sections we get (cf. F.10) a commutative diagram

$$\begin{array}{ccccc} H^0(X, \mathcal{L})^{(1)} & \longrightarrow & H^0(X, \mathcal{L}^p) & \longrightarrow & H^0(X, \mathcal{L})^{(1)} \\ \downarrow & & \downarrow & & \downarrow \\ H^0(Z, \mathcal{L})^{(1)} & \longrightarrow & H^0(Z, \mathcal{L}^p) & \longrightarrow & H^0(Z, \mathcal{L})^{(1)} \end{array}$$

where the compositions of the horizontal maps still are the identity. This shows: If the restriction map $H^0(X, \mathcal{L}^p) \to H^0(Z, \mathcal{L}^p)$ is surjective, then also the restriction map $H^0(X, \mathcal{L}) \to H^0(Z, \mathcal{L})$ is surjective.

Now use that for ample $\mathcal{L}$ the restriction map $H^0(X, \mathcal{L}^m) \to H^0(Z, \mathcal{L}^m)$ is surjective for $m \gg 0$, cf. 14.6(5).

F.15. Let $\varphi : Y \to X$ be a morphism of schemes over k. If $W \subset Y$ and $Z \subset X$ are closed subschemes, then the image faisceau $\varphi(W)$ is contained in Z if and only if the comorphism $\varphi^* : \mathcal{O}_X \to \varphi_* \mathcal{O}_Y$ maps $\mathcal{I}_Z$ to $\varphi_* \mathcal{I}_W$.

Suppose now that $\varphi^* : \mathcal{O}_X \to \varphi_* \mathcal{O}_Y$ is an isomorphism. (This condition is usually expressed as: "Suppose that $\mathcal{O}_X = \varphi_* \mathcal{O}_Y$".) If $W \subset Y$ is a closed subscheme, then $\varphi_* \mathcal{I}_W$ identifies now with a quasi-coherent ideal sheaf in $\mathcal{O}_X$. So there is a closed subscheme $Z \subset X$ with $\mathcal{I}_Z = \varphi_* \mathcal{I}_W$ under our identification. Then Z is the smallest closed subscheme in X containing $\varphi(W)$, i.e., we have $Z = \overline{\varphi(W)}$, cf. I.1.12.

Proposition: *Let $\varphi : Y \to X$ be a morphism of schemes over k with $\mathcal{O}_X = \varphi_* \mathcal{O}_Y$.*

a) *If Y is Frobenius split, then so is X.*

b) *If Y is Frobenius split and if $W \subset Y$ is a compatibly split closed subscheme, then $\overline{\varphi(W)}$ is compatibly split in X.*

Proof: a) By assumption φ^* induces an isomorphism $\mathcal{O}_X \xrightarrow{\sim} \varphi_* \mathcal{O}_Y$. It is clear that also the corresponding morphism $\varphi' : Y' \to X'$ with $\varphi' \circ F_Y = F_X \circ \varphi$ leads to an isomorphism $\varphi'^* : \mathcal{O}_{X'} \xrightarrow{\sim} \varphi'_* \mathcal{O}_{Y'}$. Given a splitting $\sigma : (F_Y)_* \mathcal{O}_Y \to \mathcal{O}_{Y'}$ one gets now a splitting $\sigma_X : (F_X)_* \mathcal{O}_X \to \mathcal{O}_{X'}$ by setting

$$\sigma_X = (\varphi'^*)^{-1} \circ \varphi'_*(\sigma) \circ (F_X)_*(\varphi^*).$$

One checks that σ_X is $\mathcal{O}_{X'}$–linear and maps 1 to 1.

b) Suppose that W is compatibly σ–split for σ as above, i.e., that $\sigma((F_Y)_* \mathcal{I}_W) \subset \mathcal{I}_W^{(1)}$. Set $Z = \overline{\varphi(W)}$; so we have $\mathcal{I}_Z = \varphi_* \mathcal{I}_W$, hence $\mathcal{I}_Z^{(1)} = \varphi'_* \mathcal{I}_W^{(1)}$. We claim that Z is compatibly σ_X–split. So we have to show that $\sigma_X((F_X)_* \mathcal{I}_Z) \subset \mathcal{I}_Z^{(1)}$. We have

$$(F_X)_*(\varphi^*)\,((F_X)_* \mathcal{I}_Z) = (F_X)_* \, \varphi^*(\mathcal{I}_Z) = (F_X)_* \, \varphi_* \, \mathcal{I}_W = \varphi'_* \, (F_Y)_* \, \mathcal{I}_W$$

hence

$$\varphi'_*(\sigma) \circ ((F_X)_*(\varphi^*)\,((F_X)_* \mathcal{I}_Z)) = \varphi'_* \, \sigma((F_Y)_* \mathcal{I}_W) \subset \varphi'_* \, \mathcal{I}_W^{(1)}$$

which yields

$$\sigma_X((F_X)_* \mathcal{I}_Z) \subset (\varphi'^*)^{-1}(\varphi'_* \, \mathcal{I}_W^{(1)}) = \mathcal{I}_Z^{(1)}$$

and the claim.

Remark: If φ is an automorphism of a scheme X over k, then clearly $\varphi_* \mathcal{O}_X = \mathcal{O}_X$. In this case one gets: If σ is a splitting for X, then also $\varphi \cdot \sigma = (\varphi'^*)^{-1} \circ \varphi'_*(\sigma) \circ F_*(\varphi^*)$ is a splitting where φ' is the automorphism of X' with $\varphi' \circ F_X = F_X \circ \varphi$. If now Z is a compatibly σ–split closed subscheme of X, then $\varphi(Z)$ is compatibly $\varphi \cdot \sigma$–split.

F.16. Let H be a group acting on a scheme X over k and let $\mathcal{L}$ be an H–equivariant line bundle on X. We have a homomorphism of kH–modules

$$\varphi_1 : \mathcal{L}(X) \xrightarrow{\sim} \mathrm{Hom}_{\mathcal{O}_X}(\mathcal{O}_X, \mathcal{L}) \xrightarrow{F_*} \mathrm{Hom}_{\mathcal{O}_{X'}}(F_*\mathcal{O}_X, F_*\mathcal{L}) \tag{1}$$

where the first map takes any $v \in \mathcal{L}(X)$ to the morphism $\mathcal{O}_X \to \mathcal{L}$ with $f \mapsto fv$. This is map is easily checked to be kH–linear; so is the second map in (1) as observed in F.6.

Now $\mathcal{L}(X) \otimes \mathcal{O}_{X'}$ is the $\mathcal{O}_{X'}$–module that associates $\mathcal{L}(X) \otimes \mathcal{O}_{X'}(U')$ to any open $U' \subset X'$. Here $\mathcal{O}_{X'}$ acts on the second factor only; so $\mathcal{L}(X)\otimes\mathcal{O}_{X'}$ is isomorphic as an $\mathcal{O}_{X'}$–module to a direct sum of $\dim \mathcal{L}(X)$ copies of $\mathcal{O}_{X'}$. We have a natural structure as an H–equivariant $\mathcal{O}_{X'}$–module on $\mathcal{L}(X)\otimes\mathcal{O}_{X'}$: Define h^* from $\mathcal{L}(X)\otimes \mathcal{O}_{X'}(U')$ to $\mathcal{L}(X) \otimes \mathcal{O}_{X'}(h^{-1}U')$ as the tensor product of h^{-1} acting on $\mathcal{L}(X)$ and of h^* on $\mathcal{O}_{X'}(h^{-1}U')$.

With this H-structure it is easy to check that the map

$$\varphi_2 : \mathcal{L}(X)^* \longrightarrow \mathrm{Hom}_{\mathcal{O}_{X'}}(\mathcal{L}(X) \otimes \mathcal{O}_{X'}, \mathcal{O}_{X'}) \tag{2}$$

that maps any $\eta \in \mathcal{L}(X)^*$ to the morphism $\mathcal{L}(X)\otimes\mathcal{O}_{X'} \to \mathcal{O}_{X'}$ with $v\otimes f \mapsto \eta(v)f$ for all $v \in \mathcal{L}(X)$ and $f \in \mathcal{O}_{X'}(U')$ is a homomorphism of kH–modules.

We have a homomorphism of $\mathcal{O}_{X'}$–modules

$$\theta : \mathcal{L}(X) \otimes \mathcal{O}_{X'} \longrightarrow F_*\mathcal{L} \tag{3}$$

that maps any $v \otimes f$ with $v \in \mathcal{L}(X)$ and $f \in \mathcal{O}_{X'}(U') = \mathcal{O}_X(U)^{(1)}$ to $f^p v \in \mathcal{L}(U) = F_*\mathcal{L}(U')$. This is easily checked to be a homomorphism of H–equivariant $\mathcal{O}_{X'}$–modules.

In this set-up we get now:

Lemma: *Suppose that θ is an isomorphism. Then*

$$\psi : \mathcal{L}(X)^* \otimes \mathcal{L}(X) \to \mathrm{Hom}_{\mathcal{O}_{X'}}(F_*\mathcal{O}_X, \mathcal{O}_{X'}), \qquad \eta \otimes v \mapsto \varphi_2(\eta) \circ \theta^{-1} \circ \varphi_1(v)$$

is a homomorphism of kH–modules. Furthermore $\psi(\eta \otimes v)$ is a splitting for X if and only if $\eta(v) = 1$.

Proof: The kH–linearity follows from the earlier statements. For the splitting, note that $\varphi_1(v)(1) = v$ and $\theta(v\otimes 1) = v$ and $\varphi_2(\eta)(v\otimes 1) = \eta(v)$, hence $\psi(\eta\otimes v)(1) = \eta(v)$.

Remark: One can weaken the condition that θ is an isomorphism. It suffices to assume that there exists a morphism $\theta' : F_*\mathcal{L} \to \mathcal{L}(X)\otimes\mathcal{O}_{X'}$ of H–equivariant $\mathcal{O}_{X'}$–modules such that $\theta' \circ \theta$ is the identity. Then replace θ^{-1} by θ' in the definition of ψ.

F.17. We now return to the usual set-up in Part II. However, we shall assume that k is algebraically closed. It will be easier to work with actions of $G(k)$ and of its subgroups on several schemes than with actions of the corresponding algebraic group. The main results in the end will usually easily extend to arbitrary k.

We shall assume until Subsection F.24 that $\rho \in X(T)$. However, the main results (Propositions F.18 and F.22/23) hold without this assumption: One can replace G by a covering group satisfying the assumption without changing the varieties considered. The assumption implies $(p-1)\rho \in X(T)$. So we can consider the first Steinberg module $St_1 = L((p-1)\rho)$ that we shall now denote by St.

When dealing with G, B, or G/B we may always assume $k = \mathbf{F}_p$. If we do so, then $G' = G$, $B' = B$, and $(G/B)' = G/B$. If $\pi : G \to G/B$ denotes the canonical map, then $\pi \circ F_G = F_{G/B} \circ \pi$. We know by I.9.5 that F_G induces isomorphisms $G/G_1 \xrightarrow{\sim} G$ and $B/B_1 \xrightarrow{\sim} B$. This implies by I.6.2(4) that $G_1B = F_G^{-1}(B)$. Therefore $F_{G/B} \circ \pi$ induces for each A an embedding of $G(A)/(G_1B)(A)$ into $(G/B)(A)$, hence by I.5.5(4) an embedding of G/G_1B into G/B. Using once more that F_G induces $G/G_1 \xrightarrow{\sim} G$, one gets that this embedding is an isomorphism. Extending the field from $\mathbf{F}_p$ to arbitrary k, we see that there is an isomorphism

$$\pi_1 : G/G_1B \xrightarrow{\sim} (G/B)' \tag{1}$$

such that $F_{G/B} \circ \pi$ is the composition of the canonical map $G \to G/G_1B$ and π_1. If we denote by $\pi' : G/B \to G/G_1B$ the canonical map, then $\pi_1 \circ \pi' = F_{G/B}$.

If M is a B–module, then we get from I.5.12(4) a natural isomorphism

$$\pi'_* \,\mathcal{L}_{G/B}(M) \simeq \mathcal{L}_{G/G_1B}(\operatorname{ind}_B^{G_1B} M). \tag{2}$$

Applying $(\pi_1)_*$ we get an isomorphism of $G(k)$–equivariant $\mathcal{O}_{(G/B)'}$–modules

$$(F_{G/B})_* \,\mathcal{L}_{G/B}(M) \simeq (\pi_1)_* \,\mathcal{L}_{G/G_1B}(\operatorname{ind}_B^{G_1B} M). \tag{3}$$

Lemma: *Set $X = G/B$. There is an isomorphism*

$$\theta : St \otimes \mathcal{O}_{X'} \xrightarrow{\sim} F_*\mathcal{L}((p-1)\rho) \tag{4}$$

of $\mathcal{O}_{X'}$–modules with $\theta(v \otimes f) = f^p v$ for all $v \in St$ and $f \in \mathcal{O}_{X'}(U')$ and all U open in X.

Proof: Apply (3) to $M = (p-1)\rho$. The left hand side in (3) is equal to $F_*\mathcal{L}((p-1)\rho)$ in our usual notation where we drop indices G/B. The right hand side in (3) involves $\operatorname{ind}_B^{G_1B}((p-1)\rho)$; this is just St considered as a G_1B–module, cf. 3.18(6). Since St is a G–module, I.5.12(3) implies that $\mathcal{L}_{G/G_1B}(St) \simeq St \otimes \mathcal{L}_{G/G_1B}(k) = St \otimes \mathcal{O}_{G/G_1B}$. Applying the isomorphism π_1 we get $St \otimes \mathcal{O}_{X'} \simeq (\pi_1)_*\mathcal{L}_{G/G_1B}(St)$. So (3) shows that there exists an isomorphism $\theta : St \otimes \mathcal{O}_{X'} \xrightarrow{\sim} F_*\mathcal{L}((p-1)\rho)$ of G–equivariant $\mathcal{O}_{X'}$–modules.

It remains to show that we can choose θ as described in the lemma. Since θ is $\mathcal{O}_{X'}$–linear, we have $\theta(v \otimes f) = f \cdot \theta(v \otimes 1) = f^p\, \theta(v \otimes 1)$ for f and v as above because the action of $f \in \mathcal{O}_{X'}(U')$ on the direct image sheaf is given by multiplication by f^p, cf. F.4. In order to find $\theta(v \otimes 1)$ it suffices to look at the map on the global sections

$$\theta(X') : St = St \otimes \mathcal{O}_{X'}(X') \xrightarrow{\sim} \mathcal{L}((p-1)\rho)(X) = H^0((p-1)\rho) = St$$

where we identify St with $St \otimes \mathcal{O}_{X'}(X')$ via $v \mapsto v \otimes 1$. This map is a homomorphism of G–modules and bijective. Since St is simple, $\theta(X')$ has to be a multiple of the identity map. Multiplying with a scalar, we may therefore assume that $\theta(v \otimes 1) = v$ for all $v \in St$. The claim follows.

F.18. Proposition: *The variety G/B is Frobenius split.*

Proof: By Lemma F.17 we can apply Lemma F.16 with $\mathcal{L} = \mathcal{L}((p-1)\rho)$.

Remarks: 1) Let $P \supset B$ be a reduced parabolic subgroup of G. The canonical map $\varphi : G/B \to G/P$ satisfies

$$\varphi_* \mathcal{O}_{G/B} = \varphi_* \mathcal{L}_{G/B}(k) \simeq \mathcal{L}_{G/P}(\operatorname{ind}_B^P k) \simeq \mathcal{O}_{G/P} \tag{1}$$

by I.5.12(4) and 4.6.a. Therefore Proposition F.15.a implies now that also G/P is Frobenius split.

2) Recall from 4.2(7) that $\omega_{G/B} \simeq \mathcal{L}(-2\rho)$. Using this the proposition together with Proposition F.10 implies $H^i(\lambda) = 0$ for all $i > 0$ for any $\lambda \in X(T)$ such that $\mathcal{L}(\lambda + 2\rho)$ is ample. We know by Proposition 4.4 that $\mathcal{L}(\lambda)$ is ample if and only if $\langle \lambda, \alpha^\vee \rangle > 0$ for all simple roots α. So the vanishing we get here includes Kempf's vanishing theorem (4.5) as well as Proposition 5.4.a except for the vanishing of H^0 in the latter case.

For dominant λ we can prove this vanishing result also as follows: Observe that the splittings we get from our construction factor over $F_*\mathcal{L} = F_*\mathcal{L}((p-1)\rho)$. So the identity on $\mathcal{O}_{X'}$ factors as $\mathcal{O}_{X'} \to F_*\mathcal{L}((p-1)\rho) \to \mathcal{O}_{X'}$. One has for all $\mu \in X(T)$ isomorphisms

$$\begin{aligned} F_*(\mathcal{L}((p-1)\rho)) \otimes_{\mathcal{O}_{X'}} \mathcal{L}(\mu)^{(1)} &\xrightarrow{\sim} F_*(\mathcal{L}((p-1)\rho) \otimes_{\mathcal{O}_X} F^*\mathcal{L}(\mu)^{(1)}) \\ &\xrightarrow{\sim} F_*(\mathcal{L}((p-1)\rho) \otimes_{\mathcal{O}_X} \mathcal{L}(p\mu)) \\ &\xrightarrow{\sim} F_*\,\mathcal{L}(p(\mu+\rho)-\rho) \end{aligned} \tag{2}$$

coming from the projection formula, from F.7(1), and from 4.1(2). Arguing as in F.10 one gets that any $H^i(\lambda)^{(1)}$ is a direct summand of $H^i(p(\lambda+\rho)-\rho)$, hence by induction $H^i(\lambda)^{(r)}$ a direct summand of $H^i(p^r(\lambda+\rho)-\rho)$. Now one can argue as in the second paragraph of the proof of Proposition 4.5 where we used the more precise result that $H^i(p^r(\lambda+\rho)-\rho)$ is isomorphic to the tensor product $St_r \otimes H^i(\lambda)^{(r)}$. — Note that we did not use the smoothness of G/B in this proof. Therefore this technique works also for (e.g.) Schubert varieties that are not smooth.

F.19. We have $\omega_X \simeq \mathcal{L}(-2\rho)$ for $X = G/B$, see 4.2(7). Therefore Proposition F.9 implies that $\operatorname{Hom}_{\mathcal{O}_{X'}}(F_*\mathcal{O}_X, \mathcal{O}_{X'})$ is isomorphic to $H^0(2(p-1)\rho)$ as a $kG(k)$–module. This shows in particular that $\operatorname{Hom}_{\mathcal{O}_{X'}}(F_*\mathcal{O}_X, \mathcal{O}_{X'})$ actually is a G–module. The map ψ from Lemma F.16 with $X = G/B$ and $\mathcal{L} = \mathcal{L}((p-1)\rho)$ is a homomorphism of G–modules

$$\psi : St^* \otimes St \longrightarrow \operatorname{Hom}_{\mathcal{O}_{X'}}(F_*\mathcal{O}_X, \mathcal{O}_{X'}) \simeq H^0(2(p-1)\rho). \tag{1}$$

Denote the image of $\eta \otimes v$ by $\sigma_{\eta,v}$. Recall that $\sigma_{\eta,v}$ is a splitting if and only if $\eta(v) = 1$.

We have $-w_0\rho = \rho$, so $St = L((p-1)\rho)$ is isomorphic to its dual module. There exists a non-degenerate G–invariant bilinear form $(\,,\,)$ on St. Then $u \mapsto (u, ?)$ is an isomorphism of G–modules $St \xrightarrow{\sim} St^*$. If η is the image of u under this isomorphism, then we write $\sigma_{u,v}$ instead of $\sigma_{\eta,v}$. We get now a homomorphism of G–modules that we again call ψ

$$\psi : St \otimes St \longrightarrow \operatorname{Hom}_{\mathcal{O}_{X'}}(F_*\mathcal{O}_X, \mathcal{O}_{X'}), \qquad u \otimes v \mapsto \sigma_{u,v}. \tag{2}$$

Now $\psi(u \otimes v)$ is a splitting for G/B if and only if $(u, v) = 1$.

We have $\mathrm{Hom}_G(St \otimes St, H^0(2(p-1)\rho)) \simeq \mathrm{Hom}_B(St \otimes St, 2(p-1)\rho)$ by Frobenius reciprocity. If $v_1 \in St$ is a non-zero weight vector of weight $(p-1)\rho$, then $v_1 \otimes v_1$ spans the $2(p-1)\rho$ weight space in $St \otimes St$. It follows that this Hom space has at most dimension 1 and that any homomorphism from $St \otimes St$ to $H^0(2(p-1)\rho)$ is completely determined by the image of $v_1 \otimes v_1$. This implies that the map ψ in (2) is uniquely determined up to a scalar factor by the fact that it is a homomorphism of G–modules. This implies in particular that the map ψ' with $\psi'(b \otimes a) = \psi(a \otimes b)$ is a scalar multiple of ψ. And since $\psi'(v_1 \otimes v_1) = \psi(v_1 \otimes v_1)$ determines the map, we get $\psi' = \psi$. In other words, we have

$$\sigma_{u,v} = \sigma_{v,u} \qquad \text{for all } u, v \in St. \tag{3}$$

F.20. If s is a global section of a line bundle $\mathcal{L}$ on a scheme X, then s defines a homomorphism of $\mathcal{O}_X$–modules $s^\vee : \mathcal{L}^{-1} \to \mathcal{O}_X$. The image of $s^\vee$ is a coherent ideal sheaf in $\mathcal{O}_X$. So there exists a closed subscheme $Z(s)$ of X with $\mathcal{I}_{Z(s)} = s^\vee(\mathcal{L}^{-1})$. One calls $Z(s)$ the *zero scheme* of s.

We want to apply this to $X = G/B$ and $\mathcal{L} = \mathcal{L}(\rho)$. So any $s \in H^0(\rho)$ defines a homomorphism $s^\vee : \mathcal{L}(-\rho) \to \mathcal{O}_X$, mapping a local section $f \in \mathcal{L}(-\rho)(U)$ to $sf \in \mathcal{O}_X(U)$, and a zero scheme $Z(s)$. Regarding s as a function on G, we have $s^{p-1} \in H^0((p-1)\rho) = St$. If $s \neq 0$, then $s^{p-1} \neq 0$ because $k[G]$ is an integral domain. So we can then find $\eta \in St^*$ with $\eta(s^{p-1}) = 1$.

Proposition: *Let $s \in H^0(\rho)$, $s \neq 0$ and let $\eta \in St^*$ with $\eta(s^{p-1}) = 1$. Then the zero scheme $Z(s)$ is compatibly $\sigma_{\eta, s^{p-1}}$–split in G/B.*

Proof: We have to show that $\sigma = \varphi_2(\eta) \circ \theta^{-1} \circ \varphi_1(s^{p-1})$ maps $F_*(s^\vee \mathcal{L}(-\rho))$ to $(s^\vee \mathcal{L}(-\rho))^{(1)} \subset \mathcal{O}_X^{(1)} = \mathcal{O}_{X'}$. Since $F_*(s^\vee \mathcal{L}(-\rho))$ is the image of $F_*(\mathcal{L}(-\rho))$ under $F_*(s^\vee)$, we have to show that $\sigma \circ F_*(s^\vee)$ maps $F_*(\mathcal{L}(-\rho))$ to $(s^\vee \mathcal{L}(-\rho))^{(1)}$.

By definition $\varphi_1(s^{p-1})$ is the map $F_*\mathcal{O}_X \to F_*\mathcal{L}((p-1)\rho)$ that maps any $f \in (F_*\mathcal{O}_X)(U') = \mathcal{O}_X(U)$ to $s^{p-1}f$. Therefore $\varphi_1(s^{p-1}) \circ F_*(s^\vee)$ is the map $F_*\mathcal{L}(-\rho) \to F_*\mathcal{L}((p-1)\rho)$ with $f \mapsto s^p f$.

We have by F.18(2) applied to $\mu = -\rho$ an isomorphism $F_*(\mathcal{L}((p-1)\rho)) \otimes_{\mathcal{O}_{X'}} \mathcal{L}(-\rho)^{(1)} \xrightarrow{\sim} F_* \mathcal{L}(-\rho)$. One checks that it maps $f_1 \otimes f_2$ to $f_1 f_2^p$. Denote the inverse of this isomorphism by ψ_1. It follows that we have a commutative diagram

$$\begin{array}{ccc}
F_*\mathcal{L}(-\rho) & \xrightarrow{F_*(s^\vee)} & F_*\mathcal{O}_X \\
\psi_1 \big\downarrow & & \big\downarrow \varphi_1(s^{p-1}) \\
F_*(\mathcal{L}((p-1)\rho)) \otimes_{\mathcal{O}_{X'}} \mathcal{L}(-\rho)^{(1)} & \xrightarrow{\psi_2} & F_*\mathcal{L}((p-1)\rho)
\end{array}$$

where ψ_2 maps $f_1 \otimes f_2$ to $f_1 f_2^p s^p$.

A look at the definition of θ shows that we have a commutative diagram

$$\begin{array}{ccc}
(St \otimes \mathcal{O}_{X'}) \otimes_{\mathcal{O}_{X'}} \mathcal{L}(-\rho)^{(1)} & \xrightarrow{\psi_3} & St \otimes \mathcal{O}_{X'} \\
\theta\otimes\mathrm{id} \big\downarrow & & \big\downarrow \theta \\
F_*(\mathcal{L}((p-1)\rho)) \otimes_{\mathcal{O}_{X'}} \mathcal{L}(-\rho)^{(1)} & \xrightarrow{\psi_2} & F_*\mathcal{L}((p-1)\rho)
\end{array}$$

where ψ_3 maps any $v \otimes f_1 \otimes f_2$ to $v \otimes f_1 f_2 s$. Identify $(St \otimes \mathcal{O}_{X'}) \otimes_{\mathcal{O}_{X'}} \mathcal{L}(-\rho)^{(1)}$ with $St \otimes \mathcal{L}(-\rho)^{(1)}$ and combine the two commutative diagrams to a new one:

$$\begin{array}{ccc} F_*\mathcal{L}(-\rho) & \xrightarrow{F_*(s^\vee)} & F_*\mathcal{O}_X \\ \psi_4 \big\downarrow & & \big\downarrow \theta^{-1}\circ\varphi_1(s^{p-1}) \\ St \otimes \mathcal{L}(-\rho)^{(1)} & \xrightarrow{\psi_5} & St \otimes \mathcal{O}_{X'} \end{array}$$

where $\psi_5(v \otimes f) = v \otimes sf$.

Now $\varphi_2(\eta) \circ \psi_5$ maps $v \otimes f$ to $\eta(v) fs$. It therefore factors $\varphi_2(\eta) \circ \psi_5 = \psi_7 \circ \psi_6$ where $\psi_6 : St \otimes \mathcal{L}(-\rho)^{(1)} \to \mathcal{L}(-\rho)^{(1)}$ maps $v \otimes f$ to $\eta(v) f$ and where $\psi_7 : \mathcal{L}(-\rho)^{(1)} \to \mathcal{O}_{X'}$ maps f to sf. We get altogether that $\sigma \circ F_*(s^\vee) = \psi_7 \circ \psi_6 \circ \psi_4$. Now the claim follows since ψ_7 takes values in $(s^\vee \mathcal{L}(-\rho))^{(1)}$.

F.21. Lemma: a) *Let $s \in H^0(\rho)$ be a non-zero root vector of weight $-\rho$. Then the zero set $V(s)(k)$ of s regarded as a regular function on $G(k)$ is the union of all $\overline{Bs_\alpha w_0 B}(k)$ with $\alpha \in S$.*

b) *For each $w \in W$ with $l(w) \le |R^+| - 2$ there exist $w_1, w_2 \in W$ such that $l(w_1) = l(w_2) = l(w)+1$ and such that $X(w)$ is a reduced irreducible component of $X(w_1) \cap X(w_2)$.*

Proof: a) We have $s(b_1 g b_2) = \rho(b_1)\rho(b_2)^{-1} s(g)$ for all $g \in G(k)$ and $b_1, b_2 \in B(k)$. (For b_2 use the definition of $H^0(\rho)$. For b_1 note that $-\rho = w_0\rho$ is the lowest weight of $H^0(\rho)$; so ks is a B–submodule of $H^0(\rho)$.) This shows in particular that $V(s)(k)$ is stable under left and right multiplication by $B(k)$. We get also that

$$\rho(t)s(\dot{w}) = s(t\dot{w}) = s(\dot{w}(\dot{w}^{-1}t\dot{w})) = \rho(\dot{w}^{-1}t\dot{w})^{-1}s(\dot{w}) = (w\rho)(t)^{-1}s(\dot{w})$$

for all $w \in W$ and $t \in T(k)$. If $s(\dot{w}) \neq 0$, then this implies $\rho(t) = (-w\rho)(t)$ for all $t \in T(k)$, hence $w\rho = -\rho$ and $w = w_0$. It follows that $s(\dot{w}) = 0$ for all $w \neq w_0$, hence $B(k)\dot{w}B(k) \subset V(s)(k)$. On the other hand, we have $s(\dot{w}_0) \neq 0$ as otherwise s would vanish on $G(k) = \bigcup_{w \in W} B(k)\dot{w}B(k)$, hence $s = 0$. By 13.6(6) any $B(k)\dot{w}B(k)$ with $w \neq w_0$ is contained in some $\overline{Bs_\alpha w_0 B}(k)$ with $\alpha \in S$. It follows that $V(s)(k) = \bigcup_{\alpha \in S} \overline{Bs_\alpha w_0 B}(k)$ as claimed.

b) Since $l(w) \le |R^+| - 2$, one can find simple roots α and β with $l(s_\alpha w s_\beta) = l(w) + 2$. We have then $l(s_\alpha w) = l(w) + 1 = l(ws_\beta)$ and $s_\alpha w \neq ws_\beta$. This implies $X(s_\alpha w) \neq X(ws_\beta)$. We get $X(w) \subset X(s_\alpha w) \cap X(ws_\beta)$ from 13.6(5). Using that each $X(w')$ has dimension $l(w')$ we deduce first that $|X(w)|$ has to be an irreducible component of $|X(s_\alpha w)| \cap |X(ws_\beta)|$ and then using 13.3(3) that $X(w)$ has to be a reduced irreducible component of $X(s_\alpha w) \cap X(ws_\beta)$.

F.22. Proposition: *There exists a splitting σ for G/B such that all Schubert schemes $X(w)$ with $w \in W$ are compatibly σ–split.*

Proof: Let $s \in H^0(\rho)$ be a non-zero root vector of weight $-\rho = w_0\rho$. Then Proposition F.20 yields a splitting σ of G/B such that $Z(s)$ is compatibly σ–split. We want to show that all Schubert schemes $X(w)$ are compatibly σ–split for this σ. We shall use downward induction on $l(w)$; if $l(w) = |R^+|$, then $w = w_0$; this case is trivial since $X(w_0) = G/B$.

If $l(w) = |R^+| - 1$, then there exists a simple root α with $w = s_\alpha w_0$. We want to show that $X(w)$ in this case is a reduced irreducible component of $Z(s)$. Then Lemma F.13.b yields the claim for w.

Let $\pi : G \to G/B$ denote the canonical map. Then $\pi^{-1}(Z(s)(k))$ is equal to the zero set $V(s)(k)$ of s regarded as a function on $G(k)$. Therefore Lemma F.21.a implies that $Z(s)(k)$ is the union of all $X(s_\alpha w_0)(k)$ with $\alpha \in S$. This is the decomposition of $Z(s)(k)$ into indecomposable components (as a topological space) as all $X(s_\alpha w_0)(k)$ with $\alpha \in S$ have the same dimension. In a variety such as $X = G/B$ over an algebraically closed field the map $Y \mapsto Y(k)$ is a bijection between the reduced closed subschemes of X and the closed subsets of $X(k)$. As the $X(w)$ are integral by 13.3(3), they are reduced. Therefore the $X(s_\alpha w_0)(k)$ with $\alpha \in S$ are the reduced irreducible components of $Z(s)$.

We have now seen that all $X(w)$ with $l(w) = |R^+| - 1$ are compatibly σ–split. Consider now some w with $l(w) \leq |R^+| - 2$ and assume inductively that all $X(w')$ with $l(w') > l(w)$ are compatibly σ–split. By Lemma F.21.b there exist $w_1, w_2 \in W$ such that $l(w_1) = l(w_2) = l(w) + 1$ and such that $X(w)$ is a reduced irreducible component of $X(w_1) \cap X(w_2)$. By induction $X(w_1)$ and $X(w_2)$ are compatibly σ–split. Now Lemma F.13.a implies that so is $X(w)$.

Remarks: 1) If $P \supset B$ is a reduced parabolic subgroup of G, then Proposition F.15 yields now that G/P has a splitting σ such that all $X(w)_P$ are compatibly σ–split.

2) By construction our splitting factors over $F_*\mathcal{L}((p-1)\rho)$. Arguing as in the final paragraph of Remark 2 in F.18 one can now show for all $w \in W$ and all $\lambda \in X(T)_+$ that $H^i(X(w), \mathcal{L}(\lambda)) = 0$ for all $i > 0$ and that the restriction map $H^0(G/B, \mathcal{L}(\lambda)) \to H^0(X(w), \mathcal{L}(\lambda))$ is surjective. We have thus achieved a proof of Proposition 14.15.e that avoids the use of the Bott-Samelson schemes. One can now also give such a proof of the normality of all $X(w)_P$, see [Mehta and Srinivas 1].

F.23. The group $B \times B$ acts on $G \times G$ from the right via $(g_1, g_2)(b_1, b_2) = (g_1 b_1, b_1^{-1} g_2 b_2)$. The quotient for this action identifies with $G \times^B G/B$, cf. I.5.18(1). On the other hand, the automorphism $\widetilde{\varphi}$ of $G \times G$ as a scheme with $\widetilde{\varphi}(g_1, g_2) \mapsto (g_1, g_1 g_2)$ transforms the action into the usual one by right multiplication. It therefore induces an isomorphism of the quotients:

$$(1) \qquad \varphi : G \times^B G/B \xrightarrow{\sim} G/B \times G/B, \qquad (g_1, g_2 B)B \mapsto (g_1 B, g_1 g_2 B).$$

For each $w \in W$ the subschemes $G \times BwB$ and $G \times \overline{BwB}$ of $G \times G$ are stable under the action of $B \times B$. Their quotients identify with the subschemes $G \times^B BwB/B$ and $G \times^B X(w)$ of $G \times^B G/B$. We get from I.5.21(2) that $G \times^B X(w)$ is the closure of $G \times^B BwB/B$. Set $\mathbf{O}(w) = \varphi(G \times^B BwB/B)$; then $\overline{\mathbf{O}(w)} = \varphi(G \times^B X(w))$.

For $w = 1$ we get here that $\mathbf{O}(1) = \overline{\mathbf{O}(1)}$ is the diagonal in $G/B \times G/B$: It is the faisceau associated to the functor that associates to each A all (gB, gB) with $g \in G(A)$. The surjectivity of $G(k) \to (G/B)(k)$ shows for any w that $\mathbf{O}(w)(k)$ consists of all $(gB, g\dot{w}B)$ with $g \in G(k)$. In other words, $\mathbf{O}(w)(k)$ is the orbit of $(1B, \dot{w}B)$ under the diagonal action of $G(k)$ on $(G/B \times G/B)(k)$. The Bruhat decomposition of $G(k)$ implies that each orbit of $G(k)$ on $(G/B \times G/B)(k)$ is equal to $\mathbf{O}(w)(k)$ for exactly one $w \in W$.

The local triviality of the canonical map $G \to G/B$ implies the local triviality of the map $G \times^B G/B \to G/B$ with $(g_1, g_2B)B \mapsto g_1B$, cf. I.5.16. We get thus an open covering $(U_i)_i$ of $G \times^B G/B \to G/B$ with isomorphisms $U_i \xrightarrow{\sim} \mathbf{A}^n \times G/B$ where $n = |R^+|$. These isomorphisms map any $U_i \cap (G \times^B BwB/B)$ isomorphically to $\mathbf{A}^n \times BwB/B$, any $U_i \cap (G \times^B X(w))$ isomorphically to $\mathbf{A}^n \times X(w)$. It follows that the $G \times^B BwB/B$ and $G \times^B X(w)$ are integral schemes, cf. 13.3(3); hence so are the $\mathbf{O}(w)$ and the $\overline{\mathbf{O}(w)}$. The openness of $Bw_0B = \dot{w}_0U^+B/B$ in G/B implies that $\mathbf{O}(w_0)$ is open in $G/B \times G/B$.

Denote for the moment by Z the reduced closed subscheme of $G/B \times G/B$ such that $Z(k)$ is the complement to $\mathbf{O}(w_0)(k)$ in $(G/B \times G/B)(k)$. The arguments used in the proof of Proposition F.22 show that the $\overline{\mathbf{O}(s_\alpha w_0)}$ with $\alpha \in S$ are the reduced irreducible components of Z. (One gets this first for the intersections with all U_i of the inverse images under φ.) One gets also for each $w \in W$ with $l(w) \leq n-2$ that there exist $w_1, w_2 \in W$ with $l(w_1) = l(w) + 1 = l(w_2)$ such that $\overline{\mathbf{O}(w)}$ is a reduced irreducible component of $\overline{\mathbf{O}(w_1)} \cap \overline{\mathbf{O}(w_2)}$. This implies (arguing as in F.22): If σ is a splitting for $G/B \times G/B$ such that Z is compatibly σ–split, then all $\overline{\mathbf{O}(w)}$ are compatibly σ–split.

Proposition: *There exists a splitting σ for $G/B \times G/B$ such that all $\overline{\mathbf{O}(w)}$ with $w \in W$ are compatibly σ–split.*

Proof: By the preceding remarks and by Proposition F.20 (applied to $G \times G$ instead of G) it suffices to find $s \in H^0(\rho, \rho) \simeq H^0(\rho) \otimes H^0(\rho)$ such that $Z(s)(k)$ is the complement to $\mathbf{O}(w_0)(k)$ in $(G/B \times G/B)(k)$.

We have $H^0(\rho) \otimes H^0(\rho) \simeq V(\rho)^* \otimes H^0(\rho) \simeq \mathrm{Hom}(V(\rho), H^0(\rho))$. Since $\mathrm{Hom}_G(V(\rho), H^0(\rho))$ has dimension 1, so has $(H^0(\rho) \otimes H^0(\rho))^G$. Pick $s \in (H^0(\rho) \otimes H^0(\rho))^G$, $s \neq 0$. We want to show that $\mathbf{O}(w)(k) \subset Z(s)(k)$ for all $w \neq w_0$ and that $\mathbf{O}(w_0)(k) \cap Z(s)(k) = \emptyset$. This will yield the result by our discussion of Z preceding the proposition.

The function $s \in k[G \times G]$ satisfies $s(gg_1b_1, gg_2b_2) = \rho(b_1b_2)^{-1}s(g_1, g_2)$ for all $g, g_1, g_2 \in G(k)$ and $b_1, b_2 \in B(k)$. Using this one checks that the function $t \in k[G]$ with $t(g) = s(1, g)$ satisfies $t(b_1gb_2) = \rho(b_1b_2^{-1})t(g)$. Now the proof of Lemma F.21.a shows that $t(\dot{w}) = 0$ for all $w \in W$, $w \neq w_0$. If also $t(\dot{w}_0) = 0$, then $t = 0$, hence $s(g_1, g_2) = s(1, g_1^{-1}g_2) = 0$ for all g_1, g_2, hence $s = 0$. Since we chose $s \neq 0$, this means that $t(\dot{w}_0) \neq 0$.

Recall that $\widetilde{\varphi}(g_1, g_2) = (g_1, g_1g_2)$, hence that $s \circ \widetilde{\varphi}(g_1, g_2) = s(g_1, g_1g_2) = s(1, g_2)$. It follows that $s \circ \widetilde{\varphi}(G \times BwB)(k) = 0$ for all $w \neq w_0$, hence that $\mathbf{O}(w)(k) = \varphi(G \times^B BwB)(k) \subset Z(s)(k)$ for these w. On the other hand $s \circ \widetilde{\varphi}$ has no zero on $G(k) \times (Bw_0B)(k)$. This implies that $Z(s)(k) \cap \mathbf{O}(w_0)(k) = \emptyset$.

Remark: The splitting constructed above factors over $F_*\mathcal{L}((p-1)\rho, (p-1)\rho)$. As before this implies for all $\lambda, \mu \in X(T)_+$ and all $w \in W$ that $H^i(\overline{\mathbf{O}(w)}, \mathcal{L}(\lambda, \mu)) = 0$ for all $i > 0$; furthermore the restriction map from $H^0(G/B \times G/B, \mathcal{L}(\lambda, \mu)) \simeq H^0(\lambda) \otimes H^0(\mu)$ to $H^0(\overline{\mathbf{O}(w)}, \mathcal{L}(\lambda, \mu))$ is surjective. If we identify G/B with $\mathbf{O}(1)$ via the diagonal embedding, then the restriction of $\mathcal{L}(\lambda, \mu)$ to $\mathbf{O}(1)$ identifies with $\mathcal{L}(\lambda + \mu)$. So we get here Proposition 14.20 as a special case.

F.24. Denote by v^+ and v^- weight vectors in St of weight $(p-1)\rho$ and $-(p-1)\rho$ such that $(v^+, v^-) = 1$. Here $(\, , \,)$ is a non-degenerate G–invariant symmetric bilinear form on St as in F.19. One can find v^+ and v^- as above because the

G–invariance implies $(St_\mu, St_\nu) = 0$ whenever $\mu + \nu \neq 0$; therefore $(\,,\,)$ induces a perfect pairing $St_\mu \times St_{-\mu} \to k$ for all μ.

Lemma: a) *One can choose a splitting σ for $X = G/B$ in Proposition F.22 and for $X = G/B \times G/B$ in Proposition F.23 such that there exists a homomorphism of G–modules*

$$\psi : St \otimes St \longrightarrow \operatorname{Hom}_{\mathcal{O}_{X'}}(F_*\mathcal{O}_X, \mathcal{O}_{X'})$$

with $\sigma = \psi(v^+ \otimes v^-)$.

b) *If σ is a splitting as in* a), *then $h \cdot \sigma = \sigma$ for all $h \in T(k)$ and there are for each $\alpha \in -S$ elements $\sigma_{i,\alpha} \in \operatorname{Hom}_{\mathcal{O}_{X'}}(F_*\mathcal{O}_X, \mathcal{O}_{X'})$ for $0 \leq i < p$ such that*

$$x_\alpha(t) \cdot \sigma = \sum_{i=0}^{p-1} t^i \sigma_{i,\alpha}$$

for all $t \in k$.

Proof: a) For $X = G/B$ we take ψ as in F.19(2). Then $\psi(v^+ \otimes v^-)$ is a splitting for X. For any $s \in H^0(\rho)_{-\rho}$, $s \neq 0$ its power s^{p-1} is a non-zero multiple of v^-. If $s^{p-1} = av^-$ with $a \neq 0$, then $\psi(v^+ \otimes v^-) = \psi(a^{-1}v^+ \otimes av^-)$ shows that $\psi(v^+ \otimes v^-)$ satisfies Proposition F.22.

Suppose now that $X = G/B \times G/B$ with the diagonal action of G. Apply F.19 to $G \times G$ instead of G. The analogue to St for $G \times G$ is $St \otimes St$. We get a non-degenerate $(G \times G)$–invariant bilinear form on $St \otimes St$ from our old form on St by setting $(u \otimes v, u' \otimes v') = (u, u')(v, v')$. There is a homomorphism

$$\psi_1 : St \otimes St \otimes St \otimes St \longrightarrow \operatorname{Hom}_{\mathcal{O}_{X'}}(F_*\mathcal{O}_X, \mathcal{O}_{X'})$$

of $G \times G$–modules; for $x = \sum_i x_i \otimes x'_i$ and $y = \sum_j y_j \otimes y'_j$ in $St \otimes St$ the image $\psi_1(x \otimes y)$ is a splitting for X if and only if $1 = (x, y) = \sum_{i,j}(x_i, y_j)(x'_i, y'_j)$.

Choose a pair $(u_i)_i$ and $(v_i)_i$ of dual bases for St. So both families are bases and we have $(u_i, v_j) = \delta_{ij}$ for all i and j. We may assume that $u_1 = v^-$ and $v_1 = v^+$ with v^+ and v^- as before. The element $\sum_i u_i \otimes v_i$ in $St \otimes St$ is G–invariant because the form is G–invariant. In fact, it spans $(St \otimes St)^G$ because $(St \otimes St)^G \simeq \operatorname{End}_G(St)$ has dimension 1. Now the map

$$\psi_2 : St \otimes St \to St \otimes St \otimes St \otimes St, \qquad x \mapsto x \otimes \sum_i u_i \otimes v_i$$

is a G–module homomorphism. So is then $\psi = \psi_1 \circ \psi_2$. And $\psi(v^+ \otimes v^-)$ is a splitting for X as $\sum_i (v^+, u_i)(v^-, v_i) = 1$. In Proposition F.23 we construct the splitting using some $s \in (H^0(\rho) \otimes H^0(\rho))^G$, $s \neq 0$. Then s^{p-1} is a non-zero multiple of $\sum_i u_i \otimes v_i$. We conclude as in the case of G/B that $\psi(v^+ \otimes v^-)$ satisfies Proposition F.23.

b) The element $v^+ \otimes v^-$ has weight 0 in $St \otimes St$. So we have $h(v^+ \otimes v^-) = v^+ \otimes v^-$ for all $h \in T(k)$, hence $h \cdot \sigma = \sigma$.

Let $\alpha \in -S$. Since $-(p-1)\rho$ is the lowest weight in St, we have $x_\alpha(t)v^- = v^-$ for all $t \in k$. Recall from 1.19(6) that the actions of U_α and $\mathrm{Dist}(U_\alpha)$ on a U_α–module are related by $x_\alpha(t)\,v = \sum_{i\geq 0} t^i X_{\alpha,i} v$ for all v in the module. We get now

$$x_\alpha(t)\cdot\sigma = \psi\,(x_\alpha(t)v^+ \otimes v^-) = \sum_{i\geq 0} t^i\,\psi\,(X_{\alpha,i}v^+ \otimes v^-)$$

i.e., $x_\alpha(t)\cdot\sigma = \sum_{i\geq 0} t^i\,\sigma_{i,\alpha}$ with $\sigma_{i,\alpha} = \psi\,(X_{\alpha,i}v^+ \otimes v^-)$. We have to show that $\sigma_{i,\alpha} = 0$ for $i \geq p$.

Now $\alpha \in -S$ implies $s_\alpha((p-1)\rho) = (p-1)(\rho+\alpha)$. Therefore any $(p-1)\rho + i\alpha$ with $i \geq p$ is not a weight of St since $s_\alpha((p-1)\rho + i\alpha) = (p-1)\rho - (i-p+1)\alpha > (p-1)\rho$. It follows that $X_{\alpha,i}v^+ = 0$ for $i \geq p$, hence $\sigma_{i,\alpha} = 0$ for these i.

CHAPTER G

Frobenius Splitting and Good Filtrations

In this chapter k is an algebraically closed field of characteristic $p > 0$.

The aim of this chapter is the proof (in G.15 and G.17) of the main results on good filtrations from Chapter 4 (Propositions 4.21 and 4.24). These results were first proved in full generality by Mathieu and we follow here his approach using Frobenius splitting.

The proofs of both results have many features in common; in this introduction I shall concentrate on Proposition 4.21. It basically amounts to showing that each G–module of the form $H^0(\lambda) \otimes H^0(\mu)$ with $\lambda, \mu \in X(T)_+$ has a good filtration. There is a line bundle $\mathcal{L}(\lambda, \mu)$ on $G/B \times G/B$ such that $A'_1 := H^0(\lambda) \otimes H^0(\mu)$ identifies with $H^0(G/B \times G/B, \mathcal{L}(\lambda, \mu))$. This G–module is naturally embedded into the B–module $A_1 := H^0(G/B \times Bw_0B/B, \mathcal{L}(\lambda, \mu))$ because $G/B \times Bw_0B/B$ is open and dense in $G/B \times G/B$.

Now the idea is to consider not only the pair $A'_1 \subset A_1$, but more generally the direct sum A' of all $A'_n = H^0(n\lambda) \otimes H^0(n\mu)$ with $n \in \mathbf{N}$ as a graded G–subalgebra of the graded B–algebra $A = \bigoplus_{n \geq 0} A_n$ with $A_n = H^0(G/B \times G/B, \mathcal{L}(n\lambda, n\mu))$. (The terms G–subalgebra and B–algebra indicate that $G(k)$ and $B(k)$ act via algebra automorphisms.)

The pair $A' \subset A$ has the following properties: Each A_n is an injective B–module. (This follows easily from earlier results in I.3.11, see G.15.) The Frobenius splitting for $(G \times G)/(B \times B)$ established in the preceding chapter yields a graded Frobenius splitting $\sigma : A \to A$ with $\sigma(A') \subset A'$. The last lemma (F.24) in that chapter shows that the Frobenius splitting has an additional property (it is "B–canonical", see G.1) which implies that σ maps B–submodules to B–submodules and preserves weight spaces (see G.2).

Now a general proposition (G.3) shows for any pair $A' \subset A$ as in the preceding paragraph that the G–module A'_1 has a good filtration.

The proof of this proposition occupies the greatest part of this chapter, from G.4 to G.14. It is carried out in two stages: First (G.4–G.9) one proves the result when there exists some $\nu \in X(T)$ such that T acts on each $(A_n)^U$ via the weight $n\nu$. In this case one can show that each A_n has a largest G–submodule (which then has to contain A'_n) and that this largest G–submodule is a direct sum of copies of $H^0(nw_0\nu)$, see G.8. This allows one (with some work) to reduce to the study of G–subalgebras of G–algebras of the form $\bigoplus_{n \geq 0} H^0(n\mu)$ and to apply earlier results (from 14.22) on such G–algebras.

The second part of the proof (in G.10–G.14) is then a reduction to the first case. This involves certain truncation functors O^B_μ for B–modules that are analogous to the truncation functors O_π for G–modules considered in Chapter A. In fact, there is for each μ some $\pi = \pi(\mu)$ such that $O^B_\mu M = O_\pi M$ for any G–module M. Applying such truncation functors one gets compatible filtrations for A' and A such that factors of subsequent terms satisfy the assumption in the first case.

The main sources for this chapter are [Mathieu 3, 5]. For alternative proofs see the references in 4.21.

G.1. Let X be a variety over k on which $B(k)$ acts. Then a splitting σ for X is called *B–canonical* if $h \cdot \sigma = \sigma$ for all $h \in T(k)$ and if there exists for each $\alpha \in -S$ elements $\sigma_{i,\alpha} \in \mathrm{Hom}_{\mathcal{O}_{X'}}(F_*\mathcal{O}_X, \mathcal{O}_{X'})$ for $0 \leq i < p$ such that $x_\alpha(t) \cdot \sigma = \sum_{i=0}^{p-1} t^i \sigma_{i,\alpha}$ for all $t \in k$.

So Lemma F.24 (applied to a suitable covering group) says that the splittings in Propositions F.22 and F.23 can be chosen to be B–canonical. In order to understand the meaning of this notion let us look first at a slightly different situation.

Let α be a root and let M and N be U_α–modules. Then also $N^{(1)}$ and $\mathrm{Hom}(M, N^{(1)})$ are U_α–modules. We write the action of any $g \in U_\alpha(k)$ on any $\sigma \in \mathrm{Hom}(M, N^{(1)})$ as $g \cdot \sigma = g \circ \sigma \circ g^{-1}$.

In order to distinguish between the two k–structures on N and $N^{(1)}$ we write $t \cdot v$ for the action on $N^{(1)}$ and $t\, v$ for the action on N, for all $t \in k$ and $v \in N$. So we have $t^p \cdot v = t\, v$. Note that the action of $x_\alpha(t)$ on some $z \in N^{(1)}$ is the same as on z regarded as an element in N; the transition from N to $N^{(1)}$ changes only the k–structure, not the actions of elements in $U_\alpha(k)$.

Lemma: *Let $\sigma \in \mathrm{Hom}(M, N^{(1)})$. Then the following are equivalent:*

(i) *There exist $\sigma_i \in \mathrm{Hom}(M, N^{(1)})$ for $0 \leq i < p$ such that $x_\alpha(t) \cdot \sigma = \sum_{i=0}^{p-1} t^i \cdot \sigma_i(v)$ for all $t \in k$.*

(ii) *We have $X_{\alpha,n} \circ \sigma = \sigma \circ X_{\alpha,np}$ for all $n \in \mathbf{N}$.*

Proof: (i) $\Rightarrow$ (ii): We suppose that we have the σ_i as in (i). Plugging $t = 0$ into the assumption yields $\sigma_0 = \sigma$. We have on the one hand for all $t \in k$ and all $v \in M$

$$(x_\alpha(t^p) \cdot \sigma)(v) = \sum_{i=0}^{p-1} t^{pi} \cdot \sigma_i(v) = \sum_{i=0}^{p-1} t^i\, \sigma_i(v).$$

On the other hand, we get from the definition using that $(-1)^i = (-1)^{ip}$ in k

$$\begin{aligned}(x_\alpha(t^p) \cdot \sigma)(v) &= x_\alpha(t^p)\, \sigma(x_\alpha(-t^p)\, v)) \\ &= \sum_{j \geq 0} t^{pj} X_{\alpha,j}\, \sigma(\sum_{i \geq 0} (-t)^{pi} X_{\alpha,i}\, v) \\ &= \sum_{i,j \geq 0} (-1)^i t^{i+pj} X_{\alpha,j}\, \sigma(X_{\alpha,i}\, v).\end{aligned}$$

Comparing the powers of t with an exponent divisible by p yields

$$\sigma(v) = \sum_{i,j \geq 0} (-1)^{ip} t^{pi+pj} X_{\alpha,j}\, \sigma(X_{\alpha,pi}\, v) = x_\alpha(t^p) \sum_{i \geq 0} (-t)^{ip}\, \sigma(X_{\alpha,pi}\, v),$$

hence

$$\sum_{i \geq 0} (-t)^{pi} X_{\alpha,i}\, \sigma(v) = x_\alpha(-t^p) \sigma(v) = \sum_{i \geq 0} (-t)^{ip}\, \sigma(X_{\alpha,pi}\, v).$$

It follows that $X_{\alpha,i}\,\sigma(v) = \sigma(X_{\alpha,pi}\,v)$ for all i as claimed: compare the terms for different powers of t.

(ii) $\Rightarrow$ (i): The assumption now implies for all $v \in M$ and $t \in k$

$$x_\alpha(t)\,\sigma(v) = \sum_{n\geq 0} t^n X_{\alpha,n}\,\sigma(v) = \sigma(\sum_{n\geq 0} t^{np} X_{\alpha,np} v).$$

This yields that

$$\begin{aligned}(x_\alpha(t)\cdot\sigma)(v) &= \sigma(\sum_{n\geq 0} t^{np} X_{\alpha,np} x_\alpha(-t)\,v) = \sigma(\sum_{n,m\geq 0}(-1)^m t^{np+m} X_{\alpha,np}X_{\alpha,m} v)\\ &= \sigma(\sum_{m=0}^{p-1}(-1)^m t^m X_{\alpha,m} v)\end{aligned}$$

where we use that $X_{\alpha,r}X_{\alpha,s} = \binom{r+s}{s} X_{\alpha,r+s}$ and $\sum_{n,m\geq 0}(-1)^m\binom{np+m}{m}t^{np+m} = \sum_{m=0}^{p-1}(-1)^m t^m$ (left as an exercise). We get now (i) with $\sigma_i(v) = (-1)^i\sigma(X_{\alpha,i}v)$.

G.2. A B–*algebra* is a k–algebra A that is a B–module such that the multiplication induces a homomorphism $A\otimes A \to A$ of B–modules. Then $B(k)$ acts acts via algebra automorphisms on A. (One defines similarly G–algebras.)

A Frobenius splitting $\sigma : A \to A^{(1)}$ for a B–algebra A is called B–canonical if $h\cdot\sigma = \sigma$ for all $h \in T(k)$ and if there exists for each $\alpha \in -S$ elements $\sigma_{i,\alpha} \in \mathrm{Hom}_{\mathcal{O}_{X'}}(F_*\mathcal{O}_X, \mathcal{O}_{X'})$ for $0 \leq i < p$ such that $x_\alpha(t)\cdot\sigma = \sum_{i=0}^{p-1} t^i\sigma_{i,\alpha}$ for all $t \in k$.

For example, let X be a variety over k on which $B(k)$ acts. Let $\mathcal{L}$ be a $B(k)$–equivariant line bundle on X and let $Y \subset X$ be an open $B(k)$–stable subscheme of X. Then also each $\mathcal{L}^m$ is a $B(k)$–equivariant line bundle on X and each $H^0(Y,\mathcal{L}^m)$ is a $kB(k)$–module. Suppose that this module structure extends to a B–module structure. Then the algebra $\Gamma_*(Y,\mathcal{L}) = \bigoplus_{m\geq 0} H^0(Y,\mathcal{L}^m)$ as in F.11(1) is a B–algebra.

Suppose now that we have a B–canonical splitting σ for X. Then the graded Frobenius splitting $\widetilde{\sigma}_{\mathcal{L},Y}$ for $\Gamma_*(Y,\mathcal{L})$ as in Lemma F.11 is B–canonical. Indeed, since any $\sigma \mapsto \sigma_{\mathcal{L}}$ is a homomorphism of $kB(k)$–modules (cf. F.10(2)), we get $h\cdot\sigma_{\mathcal{L}^m} = (h\cdot\sigma)_{\mathcal{L}^m} = \sigma_{\mathcal{L}^m}$ for all $h \in T(k)$ and all m. Observe next that the identity $x_\alpha(t)\cdot\sigma = \sum_{i=0}^{p-1} t^i\sigma_{i,\alpha}$ for all $t\in k$ implies $x_\alpha(t)\cdot\sigma_{\mathcal{L}^m} = \sum_{i=0}^{p-1} t^i\sigma_{i,\alpha,\mathcal{L}^m}$ with suitable $\sigma_{i,\alpha,\mathcal{L}^m} \in \mathrm{Hom}_{\mathcal{O}_{X'}}(F_*\mathcal{L}^{pm}, \mathcal{L}^{m(1)})$. Restricting to Y we get the claim for $\widetilde{\sigma}_{\mathcal{L},Y}$.

In our applications X will be G/B or $G/B\times G/B$ and $\mathcal{L}$ will have the form $\mathcal{L}(\mu)$ or $\mathcal{L}(\mu,\nu)$. In this case any $H^0(Y,\mathcal{L}^m)$ as above will be automatically a B–module.

Lemma: *Let A be a B–algebra with a B–canonical Frobenius splitting σ. Then σ preserves weight spaces and maps B–submodules to B–submodules.*

Proof: The first claim is obvious because of $h\cdot\sigma = \sigma$ for all $h\in T(k)$. Let $M \subset A$ be a B–submodule. Then $\sigma(M)$ is clearly stable under $T(k)$, hence under T. So it suffices to show that $\sigma(M)$ is stable under U. This holds if and only if $\sigma(M)$ is stable under all $x_\alpha(t)$ with $\alpha \in -R^+$ and $t \in k$. This is equivalent to $X_{\alpha,n}\,\sigma(M) \subset \sigma(M)$

for all $n \in \mathbf{N}$. Now σ satisfies (i) in Lemma G.1. So we get from (ii) in that lemma that

$$X_{\alpha,n}\,\sigma(M) = \sigma(X_{\alpha,np}\,M) \subset \sigma(M) \qquad \text{for all } \alpha \in -S$$

(and all $n \in \mathbf{N}$) using the submodule property of M. The $X_{\alpha,n}$ with $\alpha \in -S$ generate the algebra $\operatorname{Dist}(U)$, see [Jantzen 1], Satz I.7. The claim follows since $\operatorname{Dist}(U)$ contains all $X_{\alpha,n}$ with $\alpha \in -R^+$.

G.3. Let V be a B–module. A G–submodule of V is a B–submodule M of V such that the B–module structure on M extends to a G–module structure. Such an extension is then unique, see Remark 14.8.

A G–subalgebra of a B–algebra A is a subalgebra A' that is also a G–submodule. It is then also a G–algebra, i.e., the multiplication map $A' \otimes A' \to A'$ is a homomorphism of G–modules. This follows from 4.7.b because the map is already a homomorphism of B–modules.

The aim in the remainder of this chapter is to prove Propositions 4.21 and 4.24. The main tool for proving the existence of good filtrations will be:

Proposition: *Let $A = \bigoplus_{n\geq 0} A_n$ be a graded B–algebra such that each A_n is an injective B–module. Let $A' = \bigoplus_{n\geq 0} A'_n$ be a graded G–subalgebra of A with $\dim A'_1 < \infty$. If A has a graded and B–canonical Frobenius splitting σ with $\sigma(A') \subset A'$, then the G–module A'_1 has a good filtration.*

The proof of this proposition will occupy the next subsections and will be concluded in G.14. We shall then apply it to a B–algebra of the form $\Gamma_*(Y, \mathcal{L})$ and a G–subalgebra of the form $\Gamma_*(X, \mathcal{L})$ where Y is a B–stable open subscheme in a G–scheme X.

G.4. Let $\lambda \in X(T)_+$. Consider

$$A = \Gamma_*(G/B, \mathcal{L}(\lambda)) = \bigoplus_{n\geq 0} H^0(n\lambda).$$

This is a graded G–algebra. We denote the graded parts usually by A_n.

Lemma: *Let $A' \subset A$ be a graded G–subalgebra with $A'_1 \neq 0$. Then there exists for all $f \in A_1$ an integer $m_0 \in \mathbf{N}$ such that $f^{p^m} \in A'$ for all $m \geq m_0$.*

Proof: If $\dim H^0(\lambda) = 1$, then A is generated by $(A_1)_\lambda = A_1$ and we have $A' = A$. So the claim is trivial in this case.

Assume that $\dim H^0(\lambda) > 1$. Set $M = A'_1$. This is a non-zero G–submodule of $A_1 = H^0(\lambda)$, hence contains the socle $L(\lambda)$ of $H^0(\lambda)$. The subalgebra of A generated by M is contained in A', graded and G–stable. It clearly suffices to prove our claim for this subalgebra instead of A'. So assume from now on that A' is generated by M as a k–algebra.

Pick a non-zero weight vector $v \in V = V(-w_0\lambda) = H^0(\lambda)^*$ of weight $-\lambda$. We know by 14.22 that we can identify A with the ring of regular functions on $\overline{G(k)v} = \{0\} \cup G(k)v$ and that A is integrally closed. The isomorphism $A \xrightarrow{\sim} k[\overline{G(k)v}]$ maps any $f \in A_1 = H^0(\lambda) = V^*$ to its restriction to $\overline{G(k)v} \subset V$.

Choose $f_0 \in H^0(\lambda)_\lambda$ with $f_0(v) = 1$. We have $kf_0 = H^0(\lambda)_\lambda \subset L(\lambda) \subset M$. Let I denote the ideal in A generated by M. This ideal is $G(k)$–stable. Therefore also the zero-set Z of I in $\overline{G(k)v}$ is $G(k)$–stable. As I is contained in $\bigoplus_{n>0} A_n$, the point 0 belongs to Z. On the other hand, $f_0(v) = 1$ and $f_0 \in M$ imply $v \notin Z$. Now the stability of Z under $G(k)$ yields $Z = \{0\}$. Therefore the Hilbert Nullstellensatz says that the radical of I in A is equal to the ideal of $\{0\}$ in A, in other words, that

$$\sqrt{I} = \bigoplus_{n>0} A_n. \tag{1}$$

It follows that A is finitely generated as an A'–module: if $f_1, f_2, \dots, f_r$ is a basis for A_1 over k, then A is spanned over k by all monomials in the f_i. By (1) there exists an integer $s \geq 0$ such that $f_i^s \in I = MA$ for all i. Then one checks using induction on the total degree that A is spanned over A' by all monomials $f_1^{n_1} f_2^{n_2} \dots f_r^{n_r}$ with all $n_i < s$. And the number of these monomials is finite.

The inclusion of A' into A induces a morphism $\varphi : \operatorname{Max}(A) \to \operatorname{Max}(A')$ of their maximal spectra. This morphism is surjective because A is a finitely generated A'–module.

The surjection $S(H^0(\lambda)) \to A$ identifies $\operatorname{Max}(A)$ with $G(k)v \cup \{0\} \subset V = H^0(\lambda)^*$. The surjection $S(M) \to A'$ identifies $\operatorname{Max}(A')$ with a closed subset of M^*. The morphism φ is the restriction of the natural map $H^0(\lambda)^* \to M^*$. This map commutes with the action of $G(k)$. We get thus that $\operatorname{Max}(A') = G(k)v' \cup \{0\}$ where $v' \in M^*$ is the image of $v \in H^0(\lambda)^*$. Both kv and kv' have stabiliser $P_I(k)$ in $G(k)$ where I is the set of simple roots α with $\langle \lambda, \alpha^\vee \rangle = 0$. (If $\langle \lambda, \alpha^\vee \rangle \neq 0$, then $\dot{s}_\alpha v$ and $\dot{s}_\alpha v'$ have weight $s_\alpha(-\lambda) \neq -\lambda$.) The stabiliser of v and v' is then the kernel of λ on $P_I(k)$. It follows that φ is a bijection from $\operatorname{Max}(A)$ to $\operatorname{Max}(A')$.

Denote by K (resp. $K' \subset K$) a fraction field of A (resp. of A'). Let A'' denote the integral closure of A' in K'. Then A'' has a natural grading extending that of A', see [B2], chap. V, §1, prop. 21. Since A is integrally closed, we have $A'' \subset A$. It follows that also A'' is finitely generated as an A'–module, and A' as an A''–module. Furthermore φ factors

$$\varphi : \operatorname{Max}(A) \xrightarrow{\varphi_1} \operatorname{Max}(A'') \xrightarrow{\varphi_2} \operatorname{Max}(A').$$

Here φ_2 and φ_1 are surjective. Since φ is bijective, also φ_2 and φ_1 have to be so.

The separable degree $[K : K']_s$ is equal to the cardinality of a general fibre of φ_1, hence equal to 1. Therefore $K \supset K'$ is a finite purely inseparable field extension. Its degree is a power p^s of p. We have $a^{p^s} \in K'$ for all $a \in K$. By integrality we get $a^{p^s} \in A''$ for all $a \in A$.

The bijectivity of φ_2 implies (as we shall see in a moment) that $A''_n = A'_n$ for all $n \gg 0$. Increasing s we may assume that this holds for $n = p^s$ and get that $a^{p^s} \in A''$ for all $a \in A_1$.

We have to check that $A''_n = A'_n$ for all $n \gg 0$. There exist homogeneous $a_1, a_2, \dots, a_m \in A''$ with $A'' = \sum_{i=1}^m A'a_i$. There exists $s \in A'$, $s \neq 0$ with $sa_i \in A'$ for all i, hence with $sA'' \subset A'$. Therefore the localisations $(A')_s$ and $(A'')_s$ coincide. This means that φ_2 is an isomorphism on the open subset $D(s) \subset \operatorname{Max}(A')$. Since φ_2 is $G(k)$–equivariant, it is an isomorphism on each $gD(s)$ with $g \in G(k)$, hence on the union of all these $gD(s)$. This union has to contain $G(k)v'$ since otherwise it would be equal to $\{0\}$ and not open. This implies that φ_2 is an isomorphism on each $D(f)$ with $f \in A'_1$, $f \neq 0$. It follows that $(A')_f = (A'')_f$ for all these f. Choose now a basis $h_1, h_2, \dots, h_r$ for A'_1 over k. It then follows that we can find an integer $n \geq 0$ with $h_i^n a_j \in A'$ for all i and j. Now $A'' = \sum_{i=1}^m A' a_i$ is generated over k by all $h_1^{c_1} h_2^{c_2} \dots h_r^{c_r} a_j$. If $c_i \geq n$ for some i, then this expression belongs to A'. Therefore almost all $h_1^{c_1} h_2^{c_2} \dots h_r^{c_r} a_j$ belong to A', and since these elements are homogeneous, we get $A''_u = A'_u$ for all large u as claimed.

G.5. Let $\lambda \in X(T)_+$. Consider a graded G–algebra $A = \bigoplus_{n \geq 0} A_n$ such that each A_n is isomorphic as a G–module to a (possibly infinite) direct sum of copies of $H^0(n\lambda)$. The algebra considered in G.4 is an example of such an algebra.

We have for each n an isomorphism of G–modules

$$\text{(1)} \qquad \operatorname{Hom}_G(H^0(n\lambda), A_n) \otimes H^0(n\lambda) \xrightarrow{\sim} A_n, \qquad \varphi \otimes v \mapsto \varphi(v)$$

because of $\operatorname{End}_G(H^0(n\lambda)) = k \operatorname{id}$, cf. 2.8. Choose for each $n \in \mathbf{N}$ a (non-zero) highest weight vector $v_n \in H^0(n\lambda)_{n\lambda}$. Set $C_n = (A_n)_{n\lambda}$ for all n. Then we have for each n an isomorphism of vector spaces

$$\text{(2)} \qquad \operatorname{Hom}_G(H^0(n\lambda), A_n) \xrightarrow{\sim} C_n, \qquad \varphi \mapsto \varphi(v_n).$$

The direct sum $C = \bigoplus_{n \geq 0} C_n$ is a graded subalgebra of A. We can use the isomorphisms in (2) to transport the multiplication to $\bigoplus_{n \geq 0} \operatorname{Hom}_G(H^0(n\lambda), A_n)$. We get then for all homomorphisms of G–modules $\varphi_1 : H^0(n\lambda) \to A_n$ and $\varphi_2 : H^0(m\lambda) \to A_m$ a homomorphism of G–modules $\varphi_1 \varphi_2 : H^0((n+m)\lambda) \to A_{n+m}$ determined by the condition that $(\varphi_1\varphi_2)(v_{n+m}) = \varphi_1(v_n)\, \varphi_2(v_m)$. Note that the multiplication depends on the choice of the v_r.

We want to describe $\varphi_1\varphi_2$ differently. There is a homomorphism of G–modules $\gamma_{n,m} : H^0(n\lambda) \otimes H^0(m\lambda) \to H^0((n+m)\lambda)$ with $\gamma_{n,m}(v_n \otimes v_m) = v_{n+m}$. We claim that

$$\text{(3)} \quad \varphi_1(v)\varphi_2(v') = (\varphi_1\varphi_2) \circ \gamma_{n,m}(v \otimes v') \qquad \text{for all } v \in H^0(n\lambda),\ v' \in H^0(m\lambda).$$

All weights of $\ker(\gamma_{n,m})$ are strictly less than $(n+m)\lambda$. This implies clearly that $\operatorname{Hom}_G(\ker(\gamma_{n,m}), H^0((n+m)\lambda)) = 0$, hence also $\operatorname{Hom}_G(\ker(\gamma_{n,m}), A_{n+m}) = 0$. So $\ker(\gamma_{n,m})$ is contained in the kernel of the homomorphism $\psi : H^0(n\lambda) \otimes H^0(m\lambda) \to A_{n+m}$ with $v \otimes v' \mapsto \varphi_1(v)\varphi_2(v')$. This implies, since $\gamma_{n,m}$ is surjective by 14.20, that there exists $\psi' \in \operatorname{Hom}_G(H^0((n+m)\lambda), A_{n+m})$ with $\psi = \psi' \circ \gamma_{n,m}$. Looking at the image of v_{n+m} we get that $\psi' = \varphi_1\varphi_2$, hence (3).

Consider now a graded subalgebra $D = \bigoplus_{n \geq 0} D_n$ of C. For each $n \geq 0$ let $\widetilde{D}_n = \{\varphi \in \operatorname{Hom}_G(H^0(n\lambda), A_n) \mid \varphi(v_n) \in D_n\}$ denote the inverse image of D_n under the isomorphism in (2). Set $j(D)_n$ equal to the image of $\widetilde{D}_n \otimes H^0(n\lambda)$ under the isomorphism in (1) and set $j(D) = \bigoplus_{n \geq 0} j(D)_n \subset A$.

Lemma: *In this setup $j(D)$ is a graded G–subalgebra of A. Each $j(D)_n$ is isomorphic as a G–module to a direct sum of copies of $H^0(n\lambda)$. We have $(j(D)_n)_{n\lambda} = D_n$. If M is a G–submodule of some A_n with $M_{n\lambda} \subset D_n$, then $M \subset j(D)_n$.*

Proof: It is clear by construction that $j(D)_n$ is isomorphic as a G–module to a direct sum of copies of $H^0(n\lambda)$ and that $\widetilde{D}_n = \operatorname{Hom}_G(H^0(n\lambda), j(D)_n)$. This implies that $(j(D)_n)_{n\lambda} = D_n$.

We have to check that $j(D)_n\, j(D)_m \subset j(D)_{n+m}$. It suffices to show that

$$\varphi_1(H^0(n\lambda))\,\varphi_2(H^0(m\lambda)) \subset j(D)_{n+m} \tag{4}$$

for all $\varphi_1 \in \widetilde{D}_n$, $\varphi_2 \in \widetilde{D}_m$. However the left hand side is by (3) contained in $(\varphi_1\varphi_2)(H^0((n+m)\lambda))$ and we have $\varphi_1\varphi_2 \in \widetilde{D}_{n+m}$ since $(\varphi_1\varphi_2)(v_{n+m}) = \varphi_1(v_n)\,\varphi_2(v_m) \in D_n\, D_m \subset D_{n+m}$.

Let now M be a G–submodule of some A_n with $M_{n\lambda} \subset D_n$. Choosing a complement to $\widetilde{D}_n$ in $\operatorname{Hom}_G(H^0(n\lambda), A_n)$ and applying (1) we get a G–submodule $N \subset A_n$ with $A_n = j(D)_n \oplus N$. Let $\pi : A_n \to N$ denote the projection with kernel $j(D)_n$. We have $M_{n\lambda} \subset D_n \subset j(D_n)$, hence $0 = \pi(M_{n\lambda}) = \pi(M)_{n\lambda}$. On the other hand, we have $(A_n)^U = (A_n)_{n\lambda}$ since $H^0(n\lambda)^U = H^0(n\lambda)_{n\lambda}$. It follows that $\pi(M)^U = 0$, hence $\pi(M) = 0$ and thus $M \subset j(D)_n$.

G.6. Lemma: *Let $\lambda \in X(T)_+$. Let $A = \bigoplus_{n\geq 0} A_n$ be a graded G–algebra such that each A_n is isomorphic as a G–module to a direct sum of copies of $H^0(n\lambda)$. Let A' be a graded G–subalgebra of A. If A'_1 is not isomorphic to a direct sum of copies of $H^0(\lambda)$, then there exists $f \in A$ with $f \notin A'$ and $f^p \in A'$.*

Proof: We apply Lemma G.5 to $D = \bigoplus_{n\geq 0}(A'_n)_{n\lambda}$ and get that $j(D)$ is a G–subalgebra of A that contains A' and such that $(j(D)_n)_{n\lambda} = (A'_n)_{n\lambda}$. It is clearly enough to prove the lemma for $j(D)$ instead of A. So we replace A by $j(D)$, hence assume that $(A'_n)_{n\lambda} = (A_n)_{n\lambda}$ for all n. The assumption that A'_1 is not isomorphic to a direct sum of copies of $H^0(\lambda)$ implies that $A'_1 \neq A_1$.

If $A'_1 \neq A_1$, then there exists a submodule N of A_1 with $N \simeq H^0(\lambda)$ and $N \not\subset A'_1$. Let v be a basis vector for N_λ. Note that $v \in A'_1$ since $(A'_1)_\lambda = (A_1)_\lambda$. We now apply Lemma G.5 to $D = \bigoplus_{n\geq 0} kv^n$. We have then $j(D)_1 = N$.

If there exists $n \geq 0$ with $v^n = 0$, then we have also $v^m = 0$ and hence $j(D)_m = 0$ for all $m \geq n$. This implies $f_1^m = 0$ for all $f_1 \in j(D)_1 = N$, hence $f_1^m \in A'$. Choose $s > 0$ with $p^s \geq m$. Take $f_1 \in N$ with $f_1 \notin A'_1$. Then $f_1^{p^s} \in A'$ implies that there exists $r \geq 0$ with $f_1^{p^r} \notin A'$ and $f_1^{p^{r+1}} \in A'$. Now $f = f_1^{p^r}$ satisfies our claim.

If $v^n \neq 0$ for all $n \geq 0$, then $j(D)_n \simeq H^0(n\lambda)$ for all n and $j(D)$ is isomorphic to the algebra in Lemma G.4. We apply that lemma to $j(D)$ and its subalgebra $A' \cap j(D)$. This is possible since $v \in A'_1$, hence $(A' \cap j(D))_1 \neq 0$. It follows that $f_1^{p^s} \in A'$ for all $f_1 \in j(D)_1$ and suitable s. Now argue as in the preceding paragraph.

G.7. For each $\nu \in X(T)$ denote by $I_B(\nu)$ the injective hull of k_ν as a B–module. Note that $\operatorname{soc}_B I_B(\nu) = k_\nu$ implies that $I_B(\nu)^U = k_\nu$. Recall from 4.8 (and I.3.11) that $I_B(\nu) \simeq \operatorname{ind}_T^B k_\nu$. All weights of $I_B(\nu)$ are $\geq \nu$ by 4.8(5).

Lemma: *If $\nu \in X(T)$ with $w_0\nu \in X(T)_+$, then $I_B(\nu)$ contains a unique G–submodule isomorphic to $H^0(w_0\nu)$.*

Proof: We have $\operatorname{soc}_B H^0(w_0\nu) = H^0(w_0\nu)^U = H^0(w_0\nu)_\nu \simeq k_\nu$. This follows from 2.2.a using $\dot{w}_0 U \dot{w}_0^{-1} = U^+$, cf. 2.4(2). Therefore $H^0(w_0\nu)$ can be embedded as a B–submodule into $I_B(\nu)$. It is then by the definition in G.3 a G–submodule of $I_B(\nu)$. The uniqueness follows from $\dim \operatorname{Hom}_B(H^0(w_0\nu), I_B(\nu)) = \dim(H^0(w_0\nu)_\nu) = 1$, cf. I.3.17(3).

Remark: We can describe the embedding more explicitly: Map any $f \in H^0(w_0\nu) = \operatorname{ind}_B^G k_{w_0\nu}$ to $f' \in I_B(\nu) \simeq \operatorname{ind}_T^B k_\nu$ with $f'(b) = f(b\dot{w}_0)$, cf. G.15 below.

G.8. Let $\nu \in X(T)$ and let Q be an injective B–module with $Q^U = Q_\nu$. Then each indecomposable summand of Q is isomorphic to $I_B(\nu)$, cf. I.3.16. Fix a decomposition $Q = \bigoplus_{i\in I} Q_i$ with $Q_i \simeq I_B(\nu)$ for all i. If $w_0\nu \in X(T)_+$ denote by $M_i \subset Q_i$ the unique G–submodule isomorphic to $H^0(w_0\nu)$; if $w_0\nu \notin X(T)_+$, then set $M_i = 0$ for all i. Set $m_G(Q) = \bigoplus_{i\in I} M_i$. So this is a G–submodule of Q and we claim in this situation:

Lemma: *Each G–submodule of Q is contained in $m_G(Q)$.*

Proof: Let N be a G–submodule of Q. Suppose first that $w_0\nu \notin X(T)_+$. Now $Q^U = Q_\nu$ implies $N^U = N_\nu$. It follows from 2.4(2) for any G–module V that the weights of T on V^U belong to $w_0X(T)_+$. Therefore $N^U = N_\nu$ implies $N = 0 \subset m_G(Q)$.

Suppose now that $w_0\nu \in X(T)_+$. Consider first the case where I is finite, say $|I| = r$. We have now $\dim \operatorname{Hom}_B(N, I_B(\nu)) = \dim N_\nu$, hence $\dim \operatorname{Hom}_B(N, Q) = r \dim N_\nu$. Note that this dimension is finite since $N_\nu = N^U \subset Q^U$ and since $\dim Q^U = r \dim I_B(\nu)^U = r$.

On the other hand, since all weights of Q are $\geq \nu$, we get from 1.19(2) that weights of the G–module $N \subset Q$ are $\leq w_0\nu$. This implies that

$$\dim \operatorname{Hom}_G(N, H^0(w_0\nu)) = \dim \operatorname{Hom}_B(N, k_{w_0\nu}) = \dim N_{w_0\nu} = \dim N_\nu$$

by 1.19(8) and 1.19(2), hence $\dim \operatorname{Hom}_G(N, m_G(Q)) = r \dim \operatorname{Hom}_G(N, H^0(w_0\nu))$ $= r \dim N_\nu = \dim \operatorname{Hom}_B(N, Q)$. Since the inclusion of $m_G(Q)$ into Q induces an inclusion of $\operatorname{Hom}_G(N, m_G(Q))$ into $\operatorname{Hom}_B(N, Q)$, the dimension comparison yields $\operatorname{Hom}_B(N, Q) = \operatorname{Hom}_G(N, m_G(Q))$. And since the inclusion of N into Q belongs to $\operatorname{Hom}_B(N, Q)$, we get $N \subset m_G(Q)$ as claimed.

Let now I be arbitrary. For each $x \in N$ there exists a finite dimensional G–submodule $N' \subset N$ with $x \in N'$. There exists a finite subset $J \subset I$ with $N' \subset Q' = \bigoplus_{j\in J} Q_j$. The preceding paragraph yields now that $N' \subset m_G(Q') \subset m_G(Q)$, hence $x \in m_G(Q)$. The claim follows.

G.9. Proposition: *Let $\lambda \in X(T)_+$. Let $A = \bigoplus_{n\geq 0} A_n$ be a graded B–algebra such that each A_n is an injective B–module with $(A_n)^U = (A_n)_{nw_0\lambda}$. Let $A' = \bigoplus_{n\geq 0} A'_n$ be a graded G–subalgebra of A. If A has a graded Frobenius splitting σ with $\sigma(A') \subset A'$, then the G–module A'_1 has a good filtration.*

Proof: We can apply the construction of G.8 to each A_n with $\nu = nw_0\lambda$. Set $A''_n = m_G(A_n)$. Then A''_n is a G–submodule of A_n that is isomorphic to a direct sum of copies of $H^0(n\lambda)$. Lemma G.8 implies that $A'_n \subset A''_n$ for all n.

Set $A'' = \bigoplus_{n\geq 0} A''_n$. We claim that A'' is a subalgebra of A. We have to show for all n and m: If M is a G–submodule of A_n isomorphic to $H^0(n\lambda)$ and if M' is a G–submodule of A_m isomorphic to $H^0(m\lambda)$, then $M \cdot M' \subset m_G(A_{n+m})$. Equivalently: If $\varphi_1 : H^0(n\lambda) \to A_n$ and $\varphi_2 : H^0(m\lambda) \to A_m$ are homomorphisms of B–modules, then

$$\psi : H^0(n\lambda) \otimes H^0(m\lambda) \to A_{n+m}, \qquad v_1 \otimes v_2 \mapsto \varphi_1(v_1)\,\varphi_2(v_2)$$

takes values in $m_G(A_{n+m})$. Now recall that all weights of the kernel N of

$$\gamma_{n,m} : H^0(n\lambda) \otimes H^0(m\lambda) \to H^0((n+m)\lambda)$$

as in G.5 are $< (m+n)\lambda$. Since N is a G–module, we get that $(m+n)w_0\lambda$ is not a weight of N, hence that $\mathrm{Hom}_B(N, I_B((m+n)w_0\lambda)) = 0$ and finally that $\mathrm{Hom}_B(N, A_{n+m}) = 0$. Therefore ψ factors over $\gamma_{n,m}$ and we get a homomorphism of B–modules $\psi' : H^0((n+m)\lambda) \to A_{n+m}$ with $\psi' \circ \gamma_{n,m} = \psi$. So our claim follows from $\psi'(H^0((n+m)\lambda)) \subset m_G(A_{n+m})$: If $\psi' = 0$, then this is trivial; if not, then ψ' is injective and $\psi'(H^0((n+m)\lambda))$ is a G–submodule of A_{n+m}.

We have now shown that A'' is a subalgebra of A. It follows that A'' is a G–algebra satisfying the assumptions of Lemma G.6 containing A' as a G–subalgebra. Any $f \in A''$ with $f^p \in A'$ satisfies $f = \sigma(f^p) \in \sigma(A') \subset A'$. So Lemma G.6 implies that A'_1 is isomorphic to a direct sum of copies of $H^0(\lambda)$, hence is goofy.

G.10. Proposition G.9 is a special case of Proposition G.3. In order to reduce the general case to the special case we now introduce some truncation functors on B–modules that are analogous to the O_π on G–modules in Chapter A. In fact, when applied to G–modules the new truncation functors will yield some of the old ones.

Choose a total ordering $\preccurlyeq$ of $X(T)$ that refines the usual ordering and such that $\mu \preccurlyeq \nu$ implies $\mu + \lambda \preccurlyeq \nu + \lambda$ for all $\lambda, \mu, \nu \in X(T)$. (Such an ordering can be constructed as follows: Choose an injective $\mathbf{Q}$–linear map $f : X(T) \otimes_{\mathbf{Z}} \mathbf{Q} \to \mathbf{R}$ such that $f(\alpha) > 0$ for all $\alpha \in S$. This is possible, because we can find a basis for $X(T) \otimes_{\mathbf{Z}} \mathbf{Q}$ containing S. Now set $\mu \preccurlyeq \nu$ if and only if $f(\mu) \le f(\nu)$.)

For each B–module M and each $\mu \in X(T)$ set $O^B_\mu(M)$ equal to the sum of all B–submodules M' of M such that all weights λ of M' satisfy $\lambda \succcurlyeq \mu$.

Then $O^B_\mu(M)$ is clearly itself a B–submodule of M, in fact the largest one having only weights $\lambda \succcurlyeq \mu$. It is then clear that we have for any submodule N of M

$$O^B_\mu(N) = N \cap O^B_\mu(M) \tag{1}$$

and for all $\mu, \nu \in X(T)$

$$\mu \preccurlyeq \nu \implies O^B_\mu(M) \supset O^B_\nu(M). \tag{2}$$

Each $x \in M$ belongs to some finite dimensional B–submodule M_1 of M. Now M_1 has only finitely many weights. So we can find $\mu, \nu \in X(T)$ such that $\mu \succ$ all weights of M_1 and $\nu \preccurlyeq$ all weights of M_1. Then we get $M_1 \cap O^B_\mu(M) = 0$ and $M_1 \subset O^B_\nu(M)$. This implies

$$\bigcap_{\mu \in X(T)} O^B_\mu(M) = 0 \qquad \text{and} \qquad \bigcup_{\mu \in X(T)} O^B_\mu(M) = M. \tag{3}$$

All weights of the injective hull $I_B(\nu)$ of k_ν as a B–module are $\geq \nu$ in the usual ordering. Any non-zero submodule M' of $I_B(\nu)$ contains $I_B(\nu)^U \simeq k_\nu$. This implies

$$O^B_\mu(I_B(\nu)) = \begin{cases} I_B(\nu), & \text{if } \nu \succcurlyeq \mu, \\ 0, & \text{if } \nu \prec \mu. \end{cases} \tag{4}$$

One gets similarly for all $\lambda \in X(T)_+$

(5) $$O_\mu^B(H^0(\lambda)) = \begin{cases} H^0(\lambda), & \text{if } w_0\lambda \succcurlyeq \mu, \\ 0, & \text{if } w_0\lambda \prec \mu. \end{cases}$$

G.11. For all $\mu \in X(T)$ and all B–modules M set $O_{\succ\mu}^B(M)$ equal to the sum of all B–submodules M' of M such that all weights λ of M' satisfy $\lambda \succ \mu$. Then $O_{\succ\mu}^B(M)$ is the largest B–submodule of M having this property. We have clearly $O_\nu^B(M) \subset O_{\succ\mu}^B(M)$ for all $\nu \succ \mu$. Any finite dimensional B–submodule of $O_{\succ\mu}^B(M)$ is contained in some $O_\nu^B(M)$ with $\nu \succ \mu$. Now the local finiteness of M implies that

(1) $$O_{\succ\mu}^B(M) = \bigcup_{\nu\succ\mu} O_\nu^B(M) \subset O_\mu^B(M).$$

If N is a B–submodule of M, then

(2) $$O_{\succ\mu}^B(N) = N \cap O_{\succ\mu}^B(M).$$

If $\varphi : M \to M'$ is a homomorphism of B–modules, then we have clearly

(3) $$\varphi\, O_\mu^B(M) \subset O_\mu^B(M') \qquad \text{and} \qquad \varphi\, O_{\succ\mu}^B(M) \subset O_{\succ\mu}^B(M').$$

In fact, we can weaken the assumption here and get:

(4) *Let M and M' be B–modules and let $\varphi : M \to M'$ be a homomorphism of T–modules such that φ maps B–submodules to B–submodules. Then $\varphi\, O_\mu^B(M) \subset O_\mu^B(M')$ and $\varphi\, O_{\succ\mu}^B(M) \subset O_{\succ\mu}^B(M')$ for all $\mu \in X(T)$.*

(Recall from G.2 that B–canonical Frobenius splittings satisfy this strange assumption.)

We have clearly for all B–modules M and N

(5) $$O_\mu^B(M \oplus N) = O_\mu^B(M) \oplus O_\mu^B(N) \quad \text{and} \quad O_{\succ\mu}^B(M \oplus N) = O_{\succ\mu}^B(M) \oplus O_{\succ\mu}^B(N).$$

The compatibility of $\preccurlyeq$ with addition implies for the tensor product of M and N

(6) $$O_\mu^B(M) \otimes O_\nu^B(N) \subset O_{\mu+\nu}^B(M \otimes N)$$

and

(7) $$O_{\succ\mu}^B(M) \otimes O_\nu^B(N) + O_\mu^B(M) \otimes O_{\succ\nu}^B(N) \subset O_{\succ\mu+\nu}^B(M \otimes N).$$

If M is a B–algebra, then one gets from (3) and (6) or (7)

(8) $$O_\mu^B(M)\, O_\nu^B(M) \subset O_{\mu+\nu}^B(M)$$

and

(9) $$O_{\succ\mu}^B(M)\, O_\nu^B(M) + O_\mu^B(M)\, O_{\succ\nu}^B(M) \subset O_{\succ\mu+\nu}^B(M).$$

Lemma: a) *We have*

$$(O_\mu^B(M)/O_{\succ\mu}^B(M))^U = (O_\mu^B(M)/O_{\succ\mu}^B(M))_\mu$$

for each B–module M and each $\mu \in X(T)$.

b) *If M is an injective B–module, then each $O_\mu^B(M)/O_{\succ\mu}^B(M)$ is a direct sum of copies of $I_B(\mu)$.*

Proof: a) The inclusion "$\supset$" is obvious because $O_\mu^B(M)/O_{\succ\mu}^B(M)$ has no weight less than μ. On the other hand, consider some ν and some $v \in O_\mu^B(M)_\nu$ such that $v + O_{\succ\mu}^B(M)$ is fixed by U and non-zero. Then the B–submodule kBv generated by v in M is contained in $kv + O_{\succ\mu}^B(M)$. If $\nu \succ \mu$, then this submodule is contained in $O_{\succ\mu}^B(M)$ contradicting $v + O_{\succ\mu}^B(M) \neq 0$. Since $v \in O_\mu^B(M)$ it follows that $\nu = \mu$ as claimed.

b) We have a direct sum decomposition $M = \bigoplus_{i\in I} M_i$ with $M_i \simeq I_B(\mu_i)$ for a suitable $\mu_i \in X(T)$. Then $O_\mu^B(M)$ is the direct sum of the M_i with $\mu_i \succcurlyeq \mu$, and $O_{\succ\mu}^B(M)$ is the direct sum of the M_i with $\mu_i \succ \mu$. So $O_\mu^B(M)/O_{\succ\mu}^B(M)$ is isomorphic to the direct sum of all M_i with $\mu_i = \mu$.

G.12. Lemma: *If M is a G–module, then all $O_\mu^B(M)$ and $O_{\succ\mu}^B(M)$ are G–submodules of M.*

Proof: It suffices to show that $O_\mu^B(M)$ and $O_{\succ\mu}^B(M)$ are stable under $\mathrm{Dist}(G)$.

As $O_\mu^B(M)$ is a B–submodule, it is stable under $\mathrm{Dist}(B)$. This implies that $\mathrm{Dist}(G)\, O_\mu^B(M) = \mathrm{Dist}(U^+)\, O_\mu^B(M)$. It ν is a weight of $\mathrm{Dist}(U^+)\, O_\mu^B(M)$, then there exists a weight ν' of $O_\mu^B(M)$ with $\nu \geq \nu'$. It follows that $\nu \succcurlyeq \nu' \succcurlyeq \mu$. This implies that $\mathrm{Dist}(G) O_\mu^B(M) \subset O_\mu^B(M)$. The claim for $O_{\succ\mu}^B(M)$ follows then from the definition.

Remark: If M is a G–module, then $O_\mu^B(M) = O_{\succ\mu}^B(M)$ for all μ with $w_0\mu \notin X(T)_+$. This follows from Lemma G.11.a because the U–fixed points in a G–module have weights in $w_0 X(T)_+$.

G.13. Fix $\mu \in X(T)$. Set $\widetilde{\pi} = \{\lambda \in X(T) \mid w_0\lambda \succcurlyeq \mu\}$ and $\pi = \widetilde{\pi} \cap X(T)_+$ and $\pi' = \pi \setminus \{w_0\mu\}$. We claim now using the notation from A.1:

Lemma: *We have*

$$O_\mu^B(M) = O_\pi(M) \qquad \textit{and} \qquad O_{\succ\mu}^B(M) = O_{\pi'}(M)$$

for each G–module M.

Proof: The compatibility of $\succcurlyeq$ with $\geq$ implies that π is a saturated subset of $X(T)_+$ in the terminology from A.2. If ν is a weight of $O_\mu^B(M)$, then also $w_0\nu$ is such a weight, so we get $w_0\nu \succcurlyeq \mu$ and $\nu \in \widetilde{\pi}$. So we get from A.2(1) that $O_\mu^B(M) \subset O_\pi(M)$. On the other hand, suppose that ν is a weight of $O_\pi(M)$. Let ν^+ denote the unique element in $W\nu \cap X(T)_+$. Then also ν^+ is a weight of $O_\pi(M)$, hence $\nu^+ \in \pi$. It follows that $\nu \geq w_0\nu^+ \succcurlyeq \mu$, hence the first claim.

If $w_0\mu \notin X(T)_+$, then $\pi' = \pi$ and $O_{\succ\mu}^B(M) = O_\mu^B(M)$ by Remark G.12. So in this case the second claim follows immediately.

Suppose now that $w_0\mu \in X(T)_+$. Since μ is not a weight of $O^B_{\succ\mu}(M)$, neither is $w_0\mu$. Therefore $O^B_{\succ\mu}(M) \subset O^B_\mu(M) = O_\pi(M)$ implies $O^B_{\succ\mu}(M) \subset O_{\pi'}(M)$. On the other hand, since $w_0\mu$ is not a weight of $O_{\pi'}(M)$, neither is μ. Therefore all weights of $O_{\pi'}(M) \subset O_\pi(M) = O^B_\mu(M)$ satisfy $\succ \mu$. This yields the other inclusion.

Remark: Now Lemma A.5 implies: Suppose that M has a good filtration such that all $(M : H^0(\lambda))$ are finite. Then also $O^B_\mu(M)$ and $O^B_{\succ\mu}(M)$ have this property. So does then $O^B_\mu(M)/O^B_{\succ\mu}(M)$ by 4.17. In fact, Lemma A.5 yields then that $O^B_\mu(M)/O^B_{\succ\mu}(M)$ is isomorphic to a direct sum of copies of $H^0(w_0\mu)$.

Conversely: If each $O^B_\mu(M)/O^B_{\succ\mu}(M)$ with $w_0\mu \in X(T)_+$ is isomorphic to a direct sum of copies of $H^0(w_0\mu)$, then M is goofy.

G.14. (*Proof of Proposition G.3*) Recall that we consider in Proposition G.3 a graded B–algebra $A = \bigoplus_{n\geq 0} A_n$ and a graded G–subalgebra A'. For each $\mu \in X(T)$ set

$$A^\mu = \bigoplus_{n\geq 0} O^B_{n\mu}(A_n) \qquad \text{and} \qquad IA^\mu = \bigoplus_{n\geq 0} O^B_{\succ n\mu}(A_n)$$

and similarly

$$A'^\mu = \bigoplus_{n\geq 0} O^B_{n\mu}(A'_n) \qquad \text{and} \qquad IA'^\mu = \bigoplus_{n\geq 0} O^B_{\succ n\mu}(A'_n).$$

Then A^μ is by G.11(8), (9) a graded B–subalgebra of A containing IA^μ as a graded B–stable ideal. So

$$C^\mu = A^\mu/IA^\mu = \bigoplus_{n\geq 0} O^B_{n\mu}(A_n)/O^B_{\succ n\mu}(A_n)$$

has a natural structure as a graded B–algebra.

We have $A'^\mu = A' \cap A^\mu$ and $IA'^\mu = A' \cap IA^\mu$. So

$$C'^\mu = A'^\mu/IA'^\mu$$

is in a natural way a subalgebra of C^μ. It is in fact a G–subalgebra because Lemma G.12 implies that the B–module structures on all $O^B_{n\mu}(A'_n)$ and $O^B_{\succ n\mu}(A'_n)$ extend to G–module structures. We have $C'^\mu = 0$ if $w_0\mu \notin X(T)_+$ by Remark G.12.

By assumption in Proposition G.3 each A_n is an injective B–module. Therefore Lemma G.11 implies that each $(C^\mu)_n$ is isomorphic to a direct sum of copies of $I_B(n\mu)$.

Let now σ be a graded and B–canonical Frobenius splitting for A such that $\sigma(A') \subset A'$. Then G.11(4) implies that $\sigma\, O^B_{np\mu}(A_{np}) \subset O^B_{np\mu}(A_n^{(1)}) = O^B_{n\mu}(A_n)^{(1)}$ for all $\mu \in X(T)$ and similarly $\sigma\, O^B_{\succ np\mu}(A'_{np}) \subset O^B_{\succ n\mu}(A'_n)^{(1)}$. It follows that σ induces a graded Frobenius splitting $C^\mu \to C^{\mu(1)}$ that maps C'^μ to $(C'^\mu)^{(1)}$.

Now Proposition G.9 implies that $C_1'^\mu$ has a good filtration if $w_0\mu \in X(T)^+$. Since $C'^\mu = 0$ if $w_0\mu \notin X(T)^+$, this shows that the $O^B_\mu(A'_1)$ with $\mu \in X(T)$ form a filtration of A'_1 such that each subsequent factor has a good filtration. Therefore A'_1 itself is goofy.

G.15. In order to be able to apply Proposition G.3 we have to know that certain B–modules are injective. In our applications this will follow from:

Lemma: *Let $\mu \in X(T)$. Then $H^0(Bw_0B/B, \mathcal{L}(\mu))$ is isomorphic to $I_B(w_0\mu)$ as a B–module.*

Proof: By definition $H^0(Bw_0B/B, \mathcal{L}(\mu))$ is equal to

$$\{ f \in k[B\dot{w}_0B] \mid f(gb) = \mu(b)^{-1} f(g) \text{ for all } f \in G(k),\ b \in B(k) \}$$

The map $(b, u) \mapsto b\dot{w}_0u$ is an isomorphism of schemes $B \times U \xrightarrow{\sim} B\dot{w}_0B$. We get an injective map from $H^0(B\dot{w}_0B/B, \mathcal{L}(\mu))$ to $k[B]$ sending any f to the function f' with $f'(b) = f(b\dot{w}_0)$; we recover f via $f(b\dot{w}_0u) = f'(b)$. This map is clearly a B–module homomorphism for the module structure arising from the left actions of B. The image of our map is

$$\{f \in k[B] \mid f(bt) = w_0(\mu)(t)^{-1} f(b) \text{ for all } b \text{ in } B(k) \text{ and } t \text{ in } T(k)\} = \operatorname{ind}_T^B k_{w_0\mu}.$$

Now the claim follows from 4.8, cf. I.3.11.

G.15. (*Proof of Proposition 4.21*) Let $\lambda, \mu \in X(T)_+$. We want to show that $H^0(\lambda) \otimes H^0(\mu)$ has a good filtration.

Choose a B–canonical splitting σ for $X = G/B \times G/B$, cf. G.1. Consider the open B–stable subscheme $Y = G/B \times Bw_0B/B$ of X. Set $A = \Gamma_*(Y, \mathcal{L}(\lambda, \mu))$ and $A' = \Gamma_*(X, \mathcal{L}(\lambda, \mu))$. Then A is a graded B–algebra with a graded and B–canonical Frobenius splitting $\widetilde{\sigma}$, see G.2. And A' identifies with a G–subalgebra of A. The Frobenius splitting $\widetilde{\sigma}$ restricts to a Frobenius splitting for A', see Remark F.11.

We have $A'_1 = H^0(X, \mathcal{L}(\lambda, \mu)) \simeq H^0(\lambda) \otimes H^0(\mu)$. So our claim follows from Proposition G.3 if we can show that all A_n are injective B–modules. Well, we have

$$\begin{aligned} A_n = H^0(Y, \mathcal{L}(n\lambda, n\mu)) &\simeq H^0(G/B, \mathcal{L}(n\lambda)) \otimes H^0(Bw_0B/B, \mathcal{L}(n\mu)) \\ &\simeq H^0(n\lambda) \otimes I_B(nw_0\mu) \end{aligned}$$

where we used Lemma G.15 for the last step. Now injectivity follows from I.3.10.c.

G.17. (*Proof of Proposition 4.24*) Let $\lambda \in X(T)_+$ and $I \subset S$. We want to show that $H^0(\lambda)$ has a good filtration when considered as a module for the Levi factor L_I.

Set $G' = L_I$, $B' = B \cap L_I$, and $T' = T \subset L_I$. Then the triple (G', B', T') satisfies the same assumptions as the triple (G, B, T). We shall apply Proposition G.3 to (G', B', T') instead of (G, B, T).

Set $A = \Gamma_*(Bw_0B/B, \mathcal{L}(\lambda))$ and $A' = \Gamma_*(G/B, \mathcal{L}(\lambda))$. Then A is a graded B–algebra with a graded and B–canonical Frobenius splitting $\widetilde{\sigma}$ such that A contains A' as a graded G–subalgebra of A with $\widetilde{\sigma}(A') \subset A'$, cf. the argument in G.15.

Applying a forgetful functor we now consider A as a graded B'–algebra and A' as a graded G'–subalgebra. The graded Frobenius splitting $\widetilde{\sigma}$ for A is also B'–canonical as $T' = T$ and as the simple roots for G' belong to the simple roots for G. Since $A'_1 = H^0(\lambda)$ our claim will follow from Proposition G.3 as soon as we show that all A_n are injective B'–modules.

We have $A_n = H^0(Bw_0B/B, \mathcal{L}(n\lambda))$. So Lemma G.15 says that A_n is an injective B–module. We know that multiplication induces an isomorphism of schemes $U_I \times (L_I \cap U) \xrightarrow{\sim} U$ with U_I as in 1.8(3), hence an isomorphism of schemes $U_I \times (L_I \cap B) \xrightarrow{\sim} B$. It follows that $B/B' \simeq U_I$ is affine. Therefore 5.13.b and Remark 4.12 imply that A_n is injective also as a B'–module, and we are done.

CHAPTER H

Representations of Quantum Groups

We assume in this chapter that G is semi-simple and simply connected. Let p be a fixed prime number.

Quantum groups are mathematical objects introduced around 1985 by Drinfel′d and Jimbo independently. The construction of a quantum group involves the choice of a root system, of a ground field, and of a parameter q in that ground field. If one chooses the root system R of G, a ground field of characteristic 0, and a primitive p–th root of unity as q, then the representation theory of the quantum group (in characteristic 0) has many features in common with the representation theory of G in characteristic p. Furthermore, using techniques of reduction modulo p, one can use results for the quantum group to prove results for G.

The purpose of this chapter is to describe these similarities and connections. This is done in the form of a survey without proofs: Anything else would require writing another book.

Quantum groups come in two forms: as quantised enveloping algebras and as deformations of the algebra of regular functions on an algebraic group. We consider here only the first kind. Furthermore, we restrict ourselves to the case where the root system is finite and do not look at quantum analogues of general Kac-Moody algebras.

When working at a root of unity, there are two possible choices available for the quantised enveloping algebra: One can use a construction by Lusztig involving divided powers, or one can, following De Concini and Kac, avoid divided powers. Here we only look at Lusztig's version. For a survey of the representation theory of the other version, see [De Concini and Procesi 5].

The first subsections of this chapter describe the construction of the quantum group and then look in H.7 at some special features of the root-of-unity case. We then turn to the representation theory. For q not a root of unity one gets (H.9) a theory similar to that of G in characteristic 0. If q is a root of unity, then we get many results that generalise theorems from the representation theory of G in characteristic p such as Steinberg's tensor product theorem, Kempf's vanishing theorem, Lusztig's conjecture, see H.10–H.12. One can imitate the theory of G_1– and of G_1T–modules (H.13) and gets results on the cohomology generalising those from Chapter 12, see H.16. Also the theory of tilting modules can be carried over to the quantum case, see H.15.

We could, but shall not, discuss truncated categories for quantum groups generalising results in Chapter A. This would lead in a natural way to the theory of q–Schur algebras. Instead, we refer for this topic to the monograph [Do].

H.1. (Gaussian Binomial Coefficients) Let v be an indeterminate over $\mathbf{Q}$. Consider the fraction field $\mathbf{Q}(v)$ of the polynomial ring $\mathbf{Q}[v]$. Set for all $a \in \mathbf{Z}$

(1) $$[a] = \frac{v^a - v^{-a}}{v - v^{-1}}$$

and for all $a, n \in \mathbf{Z}$

(2) $$\begin{bmatrix} a \\ n \end{bmatrix} = \frac{[a][a-1]\dots[a-n+1]}{[1][2]\dots[n]}$$

if $n > 0$ while $\begin{bmatrix} a \\ 0 \end{bmatrix} = 1$. Note that (2) makes sense since $[m] \neq 0$ for $m \neq 0$. Set $[0]^! = 1$ and $[n]^! = [1][2]\dots[n]$ for all $n \in \mathbf{Z}$ with $n > 0$. One can check that all elements defined above actually belong to the subring $\mathbf{Z}[v, v^{-1}]$ of $\mathbf{Q}(v)$, cf. [J2], 0.1.

H.2. (Quantised Enveloping Algebras) We can choose a W–invariant scalar product $(\,,\,)$ on $X(T) \otimes_{\mathbf{Z}} \mathbf{R}$ such that $(\alpha, \alpha) = 2$ for each $\alpha \in R$ that is a short root in its irreducible component of R. (If all roots in such a component have the same length, then they all are short.) If α is a long root in an irreducible component of R of type B_n, C_n, or F_4, then $(\alpha, \alpha) = 4$; if α is a long root in an irreducible component of R of type G_2, then $(\alpha, \alpha) = 6$. (See [J2], 4.1.) Set for each $\alpha \in R$

(1) $$d_\alpha = \frac{(\alpha, \alpha)}{2} \in \{1, 2, 3\}.$$

We have then for all $\lambda \in X(T)$ and $\alpha \in R$

(2) $$(\lambda, \alpha) = \langle \lambda, \alpha^\vee \rangle \, \frac{(\alpha, \alpha)}{2} = d_\alpha \, \langle \lambda, \alpha^\vee \rangle \in \mathbf{Z}.$$

There is for each $\alpha \in R$ an endomorphism of $\mathbf{Q}(v)$ that maps v to v^{d_α}. We denote by $[a]_\alpha$ (or $[n]^!_\alpha$ or $\begin{bmatrix} a \\ n \end{bmatrix}_\alpha$) the image of $[a]$ (or $[n]^!$ or $\begin{bmatrix} a \\ n \end{bmatrix}$) under this endomorphism. So we have (e.g.) $[a]_\alpha = (v^{ad_\alpha} - v^{-ad_\alpha})/(v^{d_\alpha} - v^{-d_\alpha})$.

Let U_v denote the algebra over $\mathbf{Q}(v)$ with generators E_α, F_α, K_α, and K_α^{-1} (for all $\alpha \in S$) and relations (for all $\alpha, \beta \in S$)

(R1) $$K_\alpha K_\alpha^{-1} = 1 = K_\alpha^{-1} K_\alpha, \qquad K_\alpha K_\beta = K_\beta K_\alpha,$$

(R2) $$K_\alpha E_\beta K_\alpha^{-1} = v^{(\alpha,\beta)} E_\beta,$$

(R3) $$K_\alpha F_\beta K_\alpha^{-1} = v^{-(\alpha,\beta)} F_\beta,$$

(R4) $$E_\alpha F_\beta - F_\beta E_\alpha = \delta_{\alpha\beta} \frac{K_\alpha - K_\alpha^{-1}}{v^{d_\alpha} - v^{-d_\alpha}}$$

where $\delta_{\alpha\beta}$ is the Kronecker delta, and (for $\alpha \neq \beta$)

(R5) $$\sum_{s=0}^{1-a_{\alpha\beta}} (-1)^s \begin{bmatrix} 1 - a_{\alpha\beta} \\ s \end{bmatrix}_\alpha E_\alpha^{1-a_{\alpha\beta}-s} E_\beta E_\alpha^s = 0,$$

(R6) $$\sum_{s=0}^{1-a_{\alpha\beta}} (-1)^s \begin{bmatrix} 1 - a_{\alpha\beta} \\ s \end{bmatrix}_\alpha F_\alpha^{1-a_{\alpha\beta}-s} F_\beta F_\alpha^s = 0,$$

where $a_{\alpha\beta} = \langle \beta, \alpha^\vee \rangle$ for all α, β.

We call U_v a *quantised enveloping algebra.* One can now show (cf. [J2], 4.11) that U_v has a unique Hopf algebra structure with comultiplication Δ, counit ε, and antipode σ such that for all $\alpha \in S$

$$\begin{aligned} \Delta(E_\alpha) &= E_\alpha \otimes 1 + K_\alpha \otimes E_\alpha, & \varepsilon(E_\alpha) &= 0, & \sigma(E_\alpha) &= -K_\alpha^{-1} E_\alpha, \\ \Delta(F_\alpha) &= F_\alpha \otimes K_\alpha^{-1} + 1 \otimes F_\alpha, & \varepsilon(F_\alpha) &= 0, & \sigma(F_\alpha) &= -F_\alpha K_\alpha, \\ \Delta(K_\alpha) &= K_\alpha \otimes K_\alpha, & \varepsilon(K_\alpha) &= 1, & \sigma(K_\alpha) &= K_\alpha^{-1}. \end{aligned}$$

H.3. (Triangular Decomposition) Denote by U_v^+ (resp. U_v^-) the subalgebra of U_v generated by all E_α (resp. F_α) with $\alpha \in S$. Set U_v^0 equal to the subalgebra of U_v generated by all K_α and K_α^{-1} with $\alpha \in S$. It is clear by (R1) that U_v^0 is a commutative algebra. We can define for each λ in the root lattice $\mathbf{Z}R$ an element K_λ in U_v^0 by

$$K_\lambda = \prod_{\beta \in S} K_\beta^{m_\beta} \qquad \text{if} \quad \lambda = \sum_{\beta \in S} m_\beta \beta. \tag{1}$$

We have then obviously $K_\lambda K_\mu = K_{\lambda+\mu}$ for all $\lambda, \mu \in \mathbf{Z}\Phi$. One can show that the K_λ are a basis of U_v^0 over $\mathbf{Q}(v)$, cf. [J2], 4.21.d. Furthermore, the multiplication map

$$U_v^- \otimes U_v^0 \otimes U_v^+ \longrightarrow U_v, \qquad u_1 \otimes u_2 \otimes u_3 \mapsto u_1 u_2 u_3 \tag{2}$$

is an isomorphism of vector spaces, cf. [J2], 4.21.a.

H.4. (Braid Group Action) In order to get bases for U_v^+ and U_v^- one uses a certain braid group action on U_v. There is for each $\alpha \in S$ an automorphism T_α of U_v (cf. [J2], 8.14) such that

$$T_\alpha(K_\mu) = K_{s_\alpha \mu} = T_\alpha^{-1}(K_\mu) \qquad \text{for all } \mu \in \mathbf{Z}R, \tag{1}$$

and

$$\begin{aligned} T_\alpha(E_\alpha) &= -F_\alpha K_\alpha, & T_\alpha^{-1}(E_\alpha) &= -K_\alpha^{-1} F_\alpha, \\ T_\alpha(F_\alpha) &= -K_\alpha^{-1} E_\alpha, & T_\alpha^{-1}(F_\alpha) &= -E_\alpha K_\alpha \end{aligned} \tag{2}$$

and for all $\beta \in S$, $\beta \neq \alpha$ (with $r = -\langle \beta, \alpha^\vee \rangle$)

$$T_\alpha(E_\beta) = \sum_{i=0}^{r} (-1)^i v^{-id_\alpha} E_\alpha^{(r-i)} E_\beta E_\alpha^{(i)}, \tag{3}$$

$$T_\alpha(F_\beta) = \sum_{i=0}^{r} (-1)^i v^{id_\alpha} F_\alpha^{(i)} F_\beta F_\alpha^{(r-i)}, \tag{4}$$

where we use the notation

$$E_\gamma^{(m)} = \frac{E_\gamma^m}{[m]_\gamma^!} \qquad \text{and} \qquad F_\gamma^{(m)} = \frac{F_\gamma^m}{[m]_\gamma^!} \tag{5}$$

for all $m \geq 0$ and for all simple roots γ.

The automorphisms T_α turn out to satisfy the *braid relations.* This means for all simple roots α and β with $\alpha \neq \beta$: If $s_\alpha s_\beta \in W$ has order m, then

$$T_\alpha T_\beta \ldots = T_\beta T_\alpha \ldots$$

where we have m factors on both sides. This implies (cf. [J2], 8.18) that we can define for each $w \in W$ an automorphism T_w of U_v as follows: For $w = 1$ set $T_1 = 1$ (the identity). For $w \neq 1$ choose a reduced expression $w = s_{\alpha_1} s_{\alpha_2} \ldots s_{\alpha_t}$ and set

$$T_w = T_{\alpha_1} T_{\alpha_2} \ldots T_{\alpha_t}. \tag{6}$$

(The braid relations yield that the right hand side in (6) is independent of the reduced expression.)

One shows next (see [J2], 8.20) for all $\alpha \in S$ and $w \in W$: If $w\alpha > 0$, then $T_w(E_\alpha) \in U_v^+$; if $w\alpha \in S$, then $T_w(E_\alpha) = E_{w\alpha}$. Finally one shows (cf. [J2], 8.24)

Proposition: *Let w_0 be the longest element in W and let $w_0 = s_{\alpha_1} s_{\alpha_2} \ldots s_{\alpha_t}$ be a reduced expression. Then all products*

$$T_{\alpha_1} T_{\alpha_2} \ldots T_{\alpha_{t-1}}(E_{\alpha_t}^{a_t}) \;\ldots\; T_{\alpha_1} T_{\alpha_2}(E_{\alpha_3}^{a_3})\, T_{\alpha_1}(E_{\alpha_2}^{a_2})\, E_{\alpha_1}^{a_1} \tag{7}$$

with all $a_i \in \mathbf{Z}, a_i \geq 0$ are a basis of U_v^+.

One can show similarly that all products

$$T_{\alpha_1} T_{\alpha_2} \ldots T_{\alpha_{t-1}}(F_{\alpha_t}^{a_t}) \;\ldots\; T_{\alpha_1} T_{\alpha_2}(F_{\alpha_3}^{a_3})\, T_{\alpha_1}(F_{\alpha_2}^{a_2})\, F_{\alpha_1}^{a_1} \tag{8}$$

are a basis of U_v^-. In both cases one gets another basis when one arranges the factors in the products in the reversed order.

H.5. (Integral Forms) Set $\mathcal{A} = \mathbf{Z}[v, v^{-1}]$. Denote by $U_\mathcal{A}$ the $\mathcal{A}$–subalgebra of U_v generated by all $E_\alpha^{(n)}$, $F_\alpha^{(n)}$, and $K_\alpha^{\pm 1}$ with $\alpha \in S$ and $n \in \mathbf{Z}, n \geq 0$. Set $U_\mathcal{A}^0 = U_v^0 \cap U_\mathcal{A}$; denote by $U_\mathcal{A}^+$ (resp. $U_\mathcal{A}^-$) the $\mathcal{A}$–subalgebra of U_v generated by all $E_\alpha^{(n)}$ (resp. all $F_\alpha^{(n)}$).

One can now show that the muliplication map induces an isomorphism

$$U_\mathcal{A}^- \otimes U_\mathcal{A}^0 \otimes U_\mathcal{A}^+ \xrightarrow{\sim} U_\mathcal{A}, \tag{1}$$

cf. [J2], 11.1(8). This implies in particular that $U_\mathcal{A}^+ = U_v^+ \cap U_\mathcal{A}$ and $U_\mathcal{A}^- = U_v^- \cap U_\mathcal{A}$. One also can show for all $\alpha \in S$ and all $a, n \in \mathbf{Z}$ with $n \geq 0$ that

$$\begin{bmatrix} K_\alpha; a \\ n \end{bmatrix} = \prod_{i=1}^{n} \frac{K_\alpha v^{(a-i+1)d_\alpha} - K_\alpha^{-1} v^{-(a-i+1)d_\alpha}}{v^{id_\alpha} - v^{-id_\alpha}} \tag{2}$$

belongs to $U_\mathcal{A}^0$. In fact, $U_\mathcal{A}^0$ is generated as an $\mathcal{A}$–algebra by all $K_\alpha^{\pm 1}$ and all $\left[\begin{smallmatrix} K_\alpha; a \\ n \end{smallmatrix}\right]$ with $\alpha \in S$ and $a, n \in \mathbf{Z}$, $n \geq 0$.

It turns out that the automorphisms T_α with $\alpha \in S$ induce automorphisms of $U_\mathcal{A}$, see [Lusztig 8], Thm. 6.6. If $w_0 = s_{\alpha_1} s_{\alpha_2} \dots s_{\alpha_t}$ is a reduced expression of the longest element in $w_0 \in W$, then all products

$$T_{\alpha_1} T_{\alpha_2} \dots T_{\alpha_{t-1}}(E_{\alpha_t}^{(a_t)}) \dots T_{\alpha_1} T_{\alpha_2}(E_{\alpha_3}^{(a_3)})\, T_{\alpha_1}(E_{\alpha_2}^{(a_2)})\, E_{\alpha_1}^{(a_1)} \tag{3}$$

with all $a_i \in \mathbf{Z}, a_i \geq 0$, are a basis of $U_\mathcal{A}^+$ over $\mathcal{A}$, and all products

$$T_{\alpha_1} T_{\alpha_2} \dots T_{\alpha_{t-1}}(F_{\alpha_t}^{(a_t)}) \dots T_{\alpha_1} T_{\alpha_2}(F_{\alpha_3}^{(a_3)})\, T_{\alpha_1}(F_{\alpha_2}^{(a_2)})\, F_{\alpha_1}^{(a_1)} \tag{4}$$

are a basis of $U_\mathcal{A}^-$ over $\mathcal{A}$, cf. [Lusztig 8], Thm. 6.7 and Appendix. Finally, all

$$\prod_{\alpha \in S} (K_\alpha^{\delta_\alpha} \begin{bmatrix} K_\alpha; 0 \\ c_\alpha \end{bmatrix}) \tag{5}$$

with $\delta_\alpha \in \{0, 1\}$ and $c_\alpha \in \mathbf{Z}$, $c_\alpha \geq 0$, are a basis of $U_\mathcal{A}^0$ over $\mathcal{A}$.

H.6. (Specialisation) For any commutative ring k and any invertible element $q \in k$ there is a unique ring homomorphism $e_q : \mathcal{A} = \mathbf{Z}[v, v^{-1}] \to k$ that maps v to q. We set then

$$U_{q,k} = U_\mathcal{A} \otimes_\mathcal{A} k \tag{1}$$

regarding k as an $\mathcal{A}$–module via e_q. We define analogously $U_{q,k}^+$, $U_{q,k}^-$, and $U_{q,k}^0$. Using H.5(1) we see that we can identify these three algebras with subalgebras of $U_{q,k}$ and that they then satisfy an analogue to H.5(1) over k. Each T_α induces an automorphism of $U_{q,k}$. We shall usually write $E_\alpha^{(n)}$ instead of $E_\alpha^{(n)} \otimes 1$ when it is clear that we are working inside $U_{q,k}$, similarly for other elements in $U_\mathcal{A}$. With this convention the elements in H.5(3)–(5) form bases over k for $U_{q,k}^+$, $U_{q,k}^-$, and $U_{q,k}^0$ respectively.

The algebras $U_\mathcal{A}$, $U_\mathcal{A}^+$, $U_\mathcal{A}^-$, and $U_\mathcal{A}^0$ can be described by generators and relations, see [Lusztig 8], Thm. 6.6. This yields then descriptions of all $U_{q,k}$, $U_{q,k}^+$, $U_{q,k}^-$, and $U_{q,k}^0$ by generators and relations.

For example, if k is a field and if q is not a root of unity, then $U_{q,k}$ is generated by all E_α, F_α, K_α, and K_α^{-1} with $\alpha \in S$ — here we adopt the convention mentioned above — with relations (R1)–(R6) as in H.2, only with v replaced by q.

Consider next the case where k is a field of characteristic 0 and $q = 1$. We have $e_1([a]_\alpha) = a$ and $e_1(\left[\begin{smallmatrix} a \\ n \end{smallmatrix}\right]_\alpha) = \binom{a}{n}$ for all a and n. It follows that $E_\alpha^{(r)} = E_\alpha^r / r!$ in $U_{1,k}$, similarly for $F_\alpha^{(r)}$. So also $U_{1,k}$ is generated by all E_α, F_α, K_α, and K_α^{-1} with $\alpha \in S$. Using this and Serre's descriptions of the enveloping algebra $U(\mathfrak{g}_k)$ of $\mathfrak{g}_k = \operatorname{Lie}(G_k)$ by generators and relations, one can check that there is a surjective homomorphism of k–algebras

$$U_{1,k} \longrightarrow U(\mathfrak{g}_k) \tag{2}$$

that maps each E_α (resp. F_α) with $\alpha \in S$ to X_α (resp. to $X_{-\alpha}$), that maps each K_α to 1 and each $\left[\begin{smallmatrix} K_\alpha;0 \\ n \end{smallmatrix}\right]$ to $\binom{H_\alpha}{n}$. (See 1.11 for the introduction of X_α and H_α.)

It then follows that any $E_\alpha^{(n)}$ (resp. $F_\alpha^{(n)}$) is mapped to $X_\alpha^n/n!$ (resp. to $X_{-\alpha}^n/n!$). One can also show for all $\alpha \in S$ and $w \in W$ that $T_w(E_\alpha)$ is mapped to $\pm X_{w\alpha}$, and $T_w(F_\alpha)$ to $\pm X_{-w\alpha}$.

A comparison of the Poincaré-Birkhoff-Witt basis for $U(\mathfrak{g}_k)$ with the basis for $U_{1,k}$ coming from H.5(3)–(5) shows then that the kernel of the homomorphism in (2) is the ideal in $U_{1,k}$ generated by all $K_\alpha - 1$ with $\alpha \in S$. Note that the relations (R1)–(R3) imply that each K_α is central in $U_{1,k}$. Furthermore, we have

$$K_\alpha - K_\alpha^{-1} = (v^{d_\alpha} - v^{-d_\alpha}) \begin{bmatrix} K_\alpha; 0 \\ 1 \end{bmatrix}$$

in $U_\mathcal{A}$, hence $K_\alpha = K_\alpha^{-1}$ and thus $K_\alpha^2 = 1$ in $U_{1,k}$. It follows that the K_α generate a subalgebra of dimension $2^{|S|}$ of the centre of $U_{1,k}$; all $\prod_{\alpha \in S'} K_\alpha$ with $S' \subset S$ are a basis of this subalgebra. The algebra $U_\mathcal{A}$ is a free module over this subalgebra; one gets a basis from the elements in H.5(3)–(5) by taking only those products in H.5(5) where $\delta_\alpha = 0$ for all α.

The preceding remarks apply in particular to $k = \mathbf{Q}$. We can identify $U_{1,\mathbf{Z}}$ with a subring of $U_{1,\mathbf{Q}}$. (The obvious map is injective because it takes a basis to a basis.) The restriction of the homomorphism $U_{1,\mathbf{Q}} \to U(\mathfrak{g}_\mathbf{Q})$ from (1) maps $U_{1,\mathbf{Z}}$ onto $\mathrm{Dist}(G_\mathbf{Z})$, cf. 1.12(4). (Recall that we assume that G is semi-simple and simply connected.)

We have $U_{1,A} \simeq U_{1,\mathbf{Z}} \otimes_\mathbf{Z} A$ for any commutative ring A. It follows (cf. 1.12(3)) that we have for any A a surjective homomorphism

$$U_{1,A} \longrightarrow \mathrm{Dist}(G_A) \tag{3}$$

with kernel generated by all $K_\alpha - 1$. As before, the K_α are central in $U_{1,A}$ and satisfy $K_\alpha^2 = 1$.

H.7. (Roots of Unity) Consider next the case where k is a field of characteristic 0 and where q is a primitive l–th root of unity for some integer $l > 1$. Assume that l is odd and that l is prime to 3 if R has a component of type G_2. This assumption implies that also q^{d_α} is a primitive l–th root of unity for all $\alpha \in S$; it also simplifies several formulae in case $d_\alpha = 1$ for all α.

One checks now for all $a, n \in \mathbf{N}$: If $a = a' + la''$ and $n = n' + ln''$ with $a'', n'' \in \mathbf{N}$ and $0 \le a', n' < l$, then

$$e_q\Big(\begin{bmatrix} a \\ n \end{bmatrix}_\alpha\Big) = \binom{a''}{n''}\, e_q\Big(\begin{bmatrix} a' \\ n' \end{bmatrix}_\alpha\Big) \tag{1}$$

for all $\alpha \in S$, see [Lusztig 5], Prop. 3.2. Furthermore, one has $e_q([a]^!_\alpha) \neq 0$ if and only if $a'' = 0$. Using the obvious general equation in $U_\mathcal{A}$

$$E_\alpha^{(r)}\, E_\alpha^{(s)} = \begin{bmatrix} r+s \\ r \end{bmatrix}_\alpha E_\alpha^{(r+s)}$$

(in particular for $r = 1$ and $r = l$) one gets for all $a = a' + la''$ as above and for all $\alpha \in S$

$$E_\alpha^{(a)} = \frac{E_\alpha^{a'}\, (E_\alpha^{(l)})^{a''}}{e_q([a']^!_\alpha)\, (a'')!}\,.$$

There is a similar formula for $F_\alpha^{(a)}$. This implies:

(2) *The algebra* $U_{q,k}$ *is generated over* k *by all* E_α, $E_\alpha^{(l)}$, F_α, $F_\alpha^{(l)}$, *and* $K_\alpha^{\pm l}$ *with* $\alpha \in S$.

One has $E_\alpha^l = F_\alpha^l = 0$ in $U_{q,k}$; this follows easily from $e_q([l]^!_\alpha) = 0$. The defining relations for U_v imply easily that each K_α^l is central in $U_{q,k}$. An argument similar to the one used in H.6 for $l = 1$ yields that $K_\alpha^{2l} = 1$ for all $\alpha \in S$.

Let $u_{q,k}$ denote the subalgebra of $U_{q,k}$ generated by all E_α, F_α, and $K_\alpha^{\pm l}$ with $\alpha \in S$. This algebra is finite dimensional: one has

$$\dim u_{q,k} = 2^{|S|}\, l^{\dim(G)} \tag{3}$$

cf. [Lusztig 8], Thm. 8.3. In more detail: The subalgebra $u_{q,k}^+ = u_{q,k} \cap U_{q,k}^+$ is generated by all E_α; it has a basis like that in H.5(3), but with the extra condition that $0 \le a_i < l$ for all i. Similarly, the subalgebra $u_{q,k}^- = u_{q,k} \cap U_{q,k}^-$ is generated by all F_α; it has a basis like that in H.5(4), but with the extra condition that $0 \le a_i < l$ for all i. All $\prod_{\alpha\in S} K_\alpha^{m_\alpha}$ with $0 \le m_\alpha < 2l$ form a basis for $u_{q,k}^0 = u_{q,k} \cap U_{q,k}^0$. The multiplication map is an isomorphism

$$u_{q,k}^- \otimes u_{q,k}^0 \otimes u_{q,k}^+ \xrightarrow{\sim} u_{q,k}$$

of vector spaces.

Using the description of $U_{q,k}$ with generators and relations one can show (see [Lusztig 8], Thm. 8.10) that there exists a (unique) homomorphism

$$\Phi : U_{q,k} \longrightarrow U_{1,k} \tag{4}$$

of algebras over k such that $\Phi(K_\alpha^{\pm 1}) = K_\alpha^{\pm 1}$ and (for all $m \in \mathbf{N}$)

$$\Phi(E_\alpha^{(m)}) = \begin{cases} E_\alpha^{(m/l)}, & \text{if } l \mid m, \\ 0, & \text{otherwise}, \end{cases} \qquad \Phi(F_\alpha^{(m)}) = \begin{cases} F_\alpha^{(m/l)}, & \text{if } l \mid m, \\ 0, & \text{otherwise}, \end{cases}$$

for all $\alpha \in S$. One thinks of Φ as a *lifting of the Frobenius morphism to characteristic* 0, cf. [Lusztig 8], 8.15. It is clear that Φ is surjective.

H.8. (Weights) For each group homomorphism $\omega : \mathbf{Z}R \to \{\pm 1\}$ and each $\lambda \in X(T)$ we have an algebra homomorphism

$$\varepsilon_{\omega,\lambda} : U_v^0 \longrightarrow \mathbf{Q}(v) \qquad \text{with } \varepsilon_{\omega,\lambda}(K_\mu) = \omega(\mu)\, v^{(\lambda,\mu)} \text{ for all } \mu \in \mathbf{Z}R. \tag{1}$$

We use the notation $\mathbf{1}$ for the homomorphism ω with $\omega(\mu) = 1$ for all $\mu \in \mathbf{Z}R$.

A simple calculation shows for arbitrary ω that

$$\varepsilon_{\omega,\lambda}\left(\begin{bmatrix} K_\alpha; a \\ n \end{bmatrix}\right) = \omega(\alpha)^n \begin{bmatrix} \langle \lambda, \alpha^\vee\rangle + a \\ n \end{bmatrix}_\alpha \in \mathcal{A} \tag{2}$$

for all $\alpha \in S$ and $a, n \in \mathbf{Z}$, $n > 0$. This implies that $\varepsilon_{\omega,\lambda}(U_{\mathcal{A}}^0) \subset \mathcal{A}$, cf. H.5. It follows that $\varepsilon_{\omega,\lambda}$ induces for all k and q a homomorphism $U_{q,k}^0 \to k$ which we again denote by $\varepsilon_{\omega,\lambda}$.

If M is a $U^0_{q,k}$–module, then we set

(3) $$M_{\omega,\lambda} = \{\, m \in M \mid um = \varepsilon_{\omega,\lambda}(u)m \text{ for all } u \in U^0_{q,k}\,\}$$

for all ω and λ. We shall often write M_λ instead of $M_{\mathbf{1},\lambda}$.

For each ω as above the homomorphism $\varepsilon_{\omega,0}$ extends to an algebra homomorphism $U_v \to \mathbf{Q}(v)$ taking all E_α and F_α to 0, cf. [J2], 5.3(2). It maps $U_{\mathcal{A}}$ to $\mathcal{A}$ and thus induces for all k and q a homomorphism $\varepsilon_\omega : U_{q,k} \to k$ that maps all $E_\alpha^{(r)}$ and $F_\alpha^{(r)}$ with $r > 0$ to 0.

The extension of $\varepsilon_{\mathbf{1},0}$ to U_v is just the counit ε of the Hopf algebra U_v. So we have just observed that $\varepsilon(U_{\mathcal{A}}) \subset \mathcal{A}$. One can also show that the comultiplication Δ satisfies $\Delta(U_{\mathcal{A}}) \subset U_{\mathcal{A}} \otimes U_{\mathcal{A}}$ and that the antipode σ satisfies $\sigma(U_{\mathcal{A}}) = U_{\mathcal{A}}$, cf. [J2], 4.9. It follows that we get a Hopf algebra structure on any $U_{q,k}$; we use also there the notations Δ, ε, and σ. In the situation considered in H.6 the subalgebra $u_{q,k}$ is clearly a Hopf subalgebra.

For arbitrary q and k the subalgebra $U^0_{q,k}$ of $U_{q,k}$ is a Hopf subalgebra. It follows that we can define a product $\chi\chi'$ of algebra homomorphisms $\chi, \chi' : U^0_{q,k} \to k$ via $\chi\chi' = (\chi \overline{\otimes} \chi') \circ \Delta$, cf. the definition in I.7.7(1). One gets then $\varepsilon_{\omega,\lambda}\varepsilon_{\omega',\lambda'} = \varepsilon_{\omega\omega',\lambda+\lambda'}$ for all ω, ω', λ, λ' as above, where $\omega\omega'$ is defined by $(\omega\omega')(\mu) = \omega(\mu)\omega'(\mu)$.

H.9. (Modules I) Suppose that k is a field with $\mathrm{char}(k) \neq 2$ and that q is not a root of unity. Then the $\varepsilon_{\omega,\lambda}$ as in H.8 induce obviously pairwise distinct homomorphisms $U^0_{q,k} \to k$. Any finite dimensional $U_{q,k}$–module M is the direct sum of all $M_{\omega,\lambda}$ cf. [J2], 5.1.

For each $\lambda \in X(T)_+$ and each ω there is a simple and finite dimensional $U_{q,k}$–module $V_q(\omega,\lambda)$ generated by a vector $v \in V_q(\omega,\lambda)_{\omega,\lambda}$, $v \neq 0$ such that $E_\alpha^{(n)} v = 0$ for all $\alpha \in S$ and all $n > 0$. Each simple finite dimensional $U_{q,k}$–module is isomorphic to exactly one $V_q(\omega,\lambda)$. Each finite dimensional $U_{q,k}$–module is a direct sum of simple submodules, cf. [J2], 5.10, 5.17, 6.26.

In the case where $\omega(\mu) = 1$ for all μ we write simply $V_q(\lambda)$ instead of $V_q(\omega,\lambda)$. One has then for arbitrary ω

(1) $$V_q(\omega,\lambda) \simeq V_q(\lambda) \otimes k_\omega$$

where k_ω is k regarded as a $U_{q,k}$–module via the homomorphism $\varepsilon_\omega : U_{q,k} \to k$, cf. [J2], 5.2/3.

Each $V_q(\omega,\lambda)$ is the direct sum of all $V_q(\omega,\lambda)_{\omega,\lambda-\mu}$ with $\mu \in \mathbf{Z}R$, $\mu \geq 0$. One can define the formal character of $V_q(\omega,\lambda)$ as $\sum_{\nu\in X(T)} \dim V_q(\omega,\lambda)_{\omega,\nu} e(\nu)$ where the $e(\nu)$ are as in I.2.11 the standard basis for $\mathbf{Z}[X(T)]$. One gets now for all $\lambda \in X(T)_+$ and all ω

(2) $$\mathrm{ch}\, V_q(\omega,\lambda) = \chi(\lambda)$$

with $\chi(\lambda)$ as in 5.7, cf. [J2], 5.15, 6.26.

H.10. (Modules II) Assume now that k is a field of characteristic 0 and that q is a primitive l–th root of unity for some odd integer $l > 1$ such that l is prime to 3 if R has a component of type G_2.

Also in this case the $\varepsilon_{\omega,\lambda}$ as in H.8 induce pairwise distinct homomorphisms $U^0_{q,k} \to k$, see [Lusztig 5], Cor. 3.3, and [Andersen, Polo, and Wen 1], Lemma 9.1. Note that $\varepsilon_{\omega,\lambda}(K^l_\alpha) = \varepsilon_\omega(K^l_\alpha) = \omega(\alpha)^l = \omega(\alpha)$ for all $\alpha \in S$ and all ω, λ.

Recall that $K^{2l}_\alpha = 1$ in $U_{q,k}$. Therefore K^l_α acts as an involution on any $U_{q,k}$–module M. So M is the direct sum of the $+1$ and -1 eigenspaces of K^l_α. Since the K_α with $\alpha \in S$ commute with each other, it follows that M is the direct sum of all

$$M^\omega = \{m \in M \mid K^l_\alpha m = \omega(\alpha)m \text{ for all } \alpha \in S\}.$$

Each M^ω is a $U_{q,k}$–submodule of M because each K^l_α is central in $U_{q,k}$. If M is simple, then $M = M^\omega$ for some ω.

We say that a $U_{q,k}$–module M has type ω if $M = M^\omega$. The argument above shows that each simple $U_{q,k}$–module is of type ω for some ω. Furthermore, one checks easily that $M \mapsto M \otimes k_\omega$ is an equivalence of categories between $\{U_q(q,k)$–modules of type $\mathbf{1}\}$ and $\{U_q(q,k)$–modules of type $\omega\}$, where k_ω is (as in H.9(2)) k regarded as a $U_{q,k}$–module via the homomorphism $\varepsilon_\omega : U_{q,k} \to k$.

Because of these facts we restrict ourselves from now on to $U_{q,k}$–modules of type $\mathbf{1}$. One gets now for each $\lambda \in X(T)_+$ a simple and finite dimensional $U_{q,k}$–module $L_q(\lambda)$ of type $\mathbf{1}$ generated by a vector $v \in L_q(\lambda)_\lambda$, $v \neq 0$ such that $E^{(n)}_\alpha v = 0$ for all $\alpha \in S$ and all $n > 0$. Each simple finite dimensional $U_{q,k}$–module of type $\mathbf{1}$ is isomorphic to exactly one $L_q(\lambda)$, see [Lusztig 5], Prop. 6.4, [Andersen, Polo, and Wen 1], Cor. 6.2. There exist finite dimensional $U_{q,k}$–modules that are not semi-simple. However, each finite dimensional $U_{q,k}$–module M is the direct sum of all M_ν with $\nu \in X(T)$, see [Andersen, Polo, and Wen 1], Thm. 9.12.

Set $X_l = \{\lambda \in X(T) \mid 0 \leq \langle \lambda, \alpha^\vee \rangle < l \text{ for all } \alpha \in S\}$. Each $\lambda \in X(T)$ can be decomposed uniquely $\lambda = \lambda' + l\lambda''$ with $\lambda' \in X_l$ and $\lambda'' \in X(T)$; we have then $\lambda \in X(T)_+$ if and only if $\lambda'' \in X(T)_+$. Now Lusztig has proved the following analogue to *Steinberg's tensor product theorem* (see [Lusztig 5], Thm. 7.4):

$$L_q(\lambda) \simeq L_q(\lambda') \otimes L_q(l\lambda''). \tag{1}$$

Furthermore the second factor in (1) has a description as a *Frobenius twist*: we can regard any G_k–module V as a module over $\mathrm{Dist}(G_k)$, hence using the surjective homomorphism in H.6(3) as a module over $U_{1,k}$. Finally we use the surjective homomorphism Φ from H.7(4) to make V into a $U_{q,k}$–module that we denote by $V^{[1]}$. If V is simple, then so is $V^{[1]}$ by the surjectivity of the maps involved. Since we assume $\mathrm{char}(k) = 0$, the Weyl module $V(\lambda)_k$ is simple for all $\lambda \in X(T)_+$. Looking closely at the action of $U^0_{q,k}$ one gets now:

$$V(\lambda)^{[1]}_k \simeq L_q(l\lambda) \qquad \text{for all } \lambda \in X(T)_+. \tag{2}$$

H.11. (Induction) Keep the assumptions on k, q, and l from H.10. In [Lusztig 5] the modules $L_q(\lambda)$ were constructed as quotients of $U_{q,k}$ considered as a module over itself. A different approach is used in [Andersen, Polo, and Wen 1]: Denote by $U^{\leq 0}_{q,k}$ the subalgebra of $U_{q,k}$ generated by $U^-_{q,k}$ and $U^0_{q,k}$; then the multiplication map is a bijection $U_{q,k}^{-}\otimes U_{q,k}^{0}\to U_{q,k}^{\leq 0}$. Let us call a $U_{q,k}$–module M (resp. a $U_{q,k}^{\leq 0}$–module M) *integrable* if M is the direct sum of all $M_\lambda = M_{\mathbf{1},\lambda}$ and if there exists for each $m\in M$ an integer $a=a(m)$ such that $E_\alpha^{(r)}m = 0 = F_\alpha^{(r)}m$ (resp. $F_\alpha^{(r)}m=0$) for all $\alpha\in S$ and all $r>a$.

Given a $U_{q,k}^{\leq 0}$–module M set

$$\widetilde{M} = \{\, f\in \operatorname{Hom}_k(U_{q,k}, M) \mid f(ux) = uf(x) \text{ for all } u\in U_{q,k}^{\leq 0},\ x\in U_{q,k}\,\}.$$

This is a $U_{q,k}$–module via $(xf)(y) = f(yx)$ for all $x,y\in U_{q,k}$. Set now

$$H_q^0(M) = \{\, v\in \bigoplus_{\lambda\in X(T)} \widetilde{M}_\lambda \mid E_\alpha^{(r)}v = 0 = F_\alpha^{(r)}v \text{ for all } \alpha\in S \text{ and all } r\gg 0\,\}.$$

This subspace turns out to be a $U_{q,k}$–submodule of $\widetilde{M}$, cf. [Andersen, Polo, and Wen 1], 1.5, in fact the largest integrable $U_{q,k}$–submodule of $\widetilde{M}$; we call $H_q^0(M)$ the $U_{q,k}$–module induced by M.

Each $\lambda\in X(T)$ defines a one-dimensional $U_{q,k}^{\leq 0}$–module k_λ where each $F_\alpha^{(m)}$ with $m>0$ acts as 0. Write $H_q^0(\lambda) = H_q^0(k_\lambda)$. One gets that $H_q^0(\lambda)\neq 0$ if and only if $\lambda\in X(T)_+$. If so, then $H_q^0(\lambda)$ is finite dimensional, its character is equal to $\chi(\lambda)$, and $H_q^0(\lambda)$ contains a unique simple submodule: $L_q(\lambda)$. (See [Andersen, Polo, and Wen 1], 6.1/2, 5.7, 5.12.)

Regard H_q^0 as a functor H_q^0 to the category of all integrable $U_{q,k}^{\leq 0}$–modules to the category of all integrable $U_{q,k}$–modules. Then H_q^0 has right derived functors that we denote by H_q^i.

Now H_q^0 and the H_q^i have many properties similar to those of the induction functors and their derived functors that are studied in this book. There are analogues to general results (such as Frobenius reciprocity, the tensor identity) and to special results for reductive groups (such as Kempf's vanishing theorem). One has in particular with $n=|R^+|$

$$H_q^n(w_0\bullet\lambda)\simeq H_q^0(-w_0\lambda)^* \qquad \text{for all } \lambda\in X(T)_+, \tag{1}$$

see [Andersen, Polo, and Wen 1], 7.3, [Andersen and Wen], 3.2. Furthermore, one can prove linkage principles and gets translation functors with their usual properties, one can construct filtrations similar to those in Chapter 8 and gets a sum formula analogous to the one in 8.19.

Most of these results were proved in [Andersen, Polo, and Wen 1]; one can find a good survey in [Andersen 20]. For type A see also [Parshall and Wang 1].

The case where one admits above k of prime characteristic is investigated in [Andersen and Wen], [Thams 1, 2]. For the case where one allows also even l, see [Andersen 25].

H.12. (Lusztig's Conjecture) Keep the assumptions from H.10. Among the results proved in [Andersen, Polo, and Wen 1] is a linkage principle. In one version it states for all $\lambda,\mu\in X(T)_+$: If $L_q(\mu)$ is a composition factor of $H_q^0(\lambda)$, then $\mu\in W_l\bullet\lambda$. (Recall that W_l is defined in 6.1 since p in that subsection can be any positive integer.) It follows as in the introduction to Chapter 7 for all $\lambda\in X(T)_+$

that the character $\operatorname{ch} L_q(\lambda)$ is a linear combination of all $\operatorname{ch} H_q^0(\mu) = \chi(\mu)$ with $\mu \in X(T)_+ \cap W_l \bullet \lambda$, $\mu \leq \lambda$.

Each dominant weight can be written in the form $w \bullet \lambda$ with $w \in W_l$ and $\lambda \in X(T)$ with $0 \leq \langle \lambda + \rho, \alpha^\vee \rangle \leq l$ for all $\alpha \in R^+$ such that $w \bullet \lambda$ belongs to the upper closure of the alcove $w \bullet C$. Here C is the standard alcove as in 6.2(6), with p replaced by l. In [Lusztig 5], 8.2, Lusztig stated the following conjecture analogous to his conjecture for algebraic groups as in 8.22(2): One should have

$$\operatorname{ch} L_q(w \bullet \lambda) = \sum_{w' \in W_l, w' \bullet \lambda \in X(T)_+} \varepsilon(w)\varepsilon(w') P_{w_0 w', w_0 w}(1)\, \chi(w' \bullet \lambda). \tag{1}$$

As before $\varepsilon(w)$ denotes the sign of w and any $P_{x,y}$ is a Kazhdan-Lusztig polynomial.

This conjecture has by now been proved in most cases following a programme outlined by Lusztig (at least in the case where all roots have the same length):

- Relate $U_{q,k}$–modules to representations of affine Lie algebras. This was carried out in [Kazhdan and Lusztig 3–6]. (For the case of two root lengths see [Lusztig 9], 8.4, Errata.)
- Relate representations of affine Lie algebras to the cohomology of certain simple perverse sheaves on an affine flag manifold. This was done in [Kashiwara and Tanisaki 1, 3].
- Relate this cohomology to Kazhdan-Lusztig polynomials. The necessary information for the case where all roots have the same length was already available from [Kazhdan and Lusztig 2]. The other cases were then treated in [Lusztig 9].

The only step requiring some restrictions is the first one, see the restriction on p' in [Lusztig 9], 8.1. In any case (1) holds if $l > h$, the Coxeter number. This follows from the fact that (1) holds by the results mentioned above for almost all l and from Cor. 16.24 in [Andersen, Jantzen, and Soergel] together with the tensor product theorem H.10(1).

Assume now for the moment that $l = p$ is a prime number and that $k = \mathbf{Q}(q)$ is the p–th cyclotomic field. Then $\mathbf{Z}[q]$ is the ring of all algebraic integers in k and $1 - q$ generates the unique maximal ideal in $\mathbf{Z}[q]$ containing p. Let A denote the localisation of $\mathbf{Z}[q]$ at $(1 - q)$. Then A is a discrete valuation ring with residue field $\mathbf{F}_p$. We regard $U_{q,A}$ as a subring of $U_{q,k}$; this is possible because $U_{\mathcal{A}}$ is free over $\mathcal{A}$. Take an arbitrary field $\mathbf{F}$ of characteristic $p > 0$ and regard it as an A–algebra via the embedding of the residue field of A into $\mathbf{F}$. We can then identify $U_{q,A} \otimes_A \mathbf{F} = (U_{\mathcal{A}} \otimes_{\mathcal{A}} A) \otimes_A \mathbf{F}$ with $U_{1,\mathbf{F}}$ since the image of q in $\mathbf{F}$ is equal to 1.

Let $\mu \in X(T)_+$. We can find in $L_q(\mu)$ an A–lattice $L_q(\mu)_A$ stable under $U_{q,A}$. For example, we can take $L_q(\mu)_A = U_{q,A} v$ for some $v \in L_q(\mu)_\mu$, $v \neq 0$. We have then $L_q(\mu)_A = \bigoplus_{\nu \in X(T)} L_q(\mu)_{A,\nu}$ where $L_q(\mu)_{A,\nu} = L_q(\mu)_A \cap L_q(\mu)_\nu$. Now $L_q(\mu)_{\mathbf{F}} = L_q(\mu)_A \otimes_A \mathbf{F}$ has a natural structure as a module over $U_{q,A} \otimes_A \mathbf{F} \simeq U_{1,\mathbf{F}}$. Any K_α with $\alpha \in S$ acts on each $L_q(\mu)_{A,\nu}$ as multiplication by $q^{(\nu,\alpha)}$, hence on $L_q(\mu)_{\mathbf{F}}$ as the identity. It follows (see H.6) that the action of $U_{1,\mathbf{F}}$ factors through an action of $\operatorname{Dist}(G_{\mathbf{F}})$. By 1.20 this action comes from a structure as a $G_{\mathbf{F}}$–module on $L_q(\mu)_{\mathbf{F}}$. Checking the action of $U_{q,A}^0$ and $U_{1,\mathbf{F}}^0$ one gets that $L_q(\mu)_{A,\nu} \otimes_A \mathbf{F}$ is the ν–weight space of $L_q(\mu)_{\mathbf{F}}$. So $L_q(\mu)_{\mathbf{F}}$ and $L_q(\mu)$ have the same formal character. In particular, μ is the largest weight and it occurs with multiplicity 1. Therefore $L(\mu)$ is a composition factor with multiplicity 1 of $L_q(\mu)_{\mathbf{F}}$.

If now $p > h$ and if $\langle \mu + \rho, \alpha^\vee \rangle \le p(p-h+2)$ for all $\alpha \in R^+$, then Lusztig's conjecture 8.22(2) combined with (1) predicts that $L(\mu)$ has the same formal character as $L_q(\mu)_{\mathbf{F}}$, hence that $L_q(\mu)_{\mathbf{F}}$ is irreducible. In fact, for given $p > h$ that old conjecture is equivalent to the irreducibility of all $L_q(\mu)_{\mathbf{F}}$ with μ satisfying those inequalities. The quantum result implies at least that Lusztig's conjecture 8.22(2) yields an upper bound for the dimensions of the weight spaces of $L_q(\mu)_{\mathbf{F}}$.

Recall from 8.22 that some bound on μ in that conjecture is necessary because of Steinberg's tensor product theorem. No such bound is needed in the quantum case since the tensor product theorem has here a somewhat different form, see H.10(1), (2). One might hope that $L_q(\mu)_{\mathbf{F}}$ is simple for all $\mu \in X_l$ as they do not involve the tensor product theorem. This, however, will not be true for small values of p: See the example in type A_{p+2} in [Andersen 20], 7.9.

H.13. (Infinitesimal Theory) Drop the special assumptions on k and l from the end of the preceding subsection and return to the more general situation from H.10. The finite dimensional subalgebra $u_{q,k}$ of $U_{q,k}$ described in H.7 plays a role similar to that of the first Frobenius kernel of a reductive group in prime characteristic: Each $L_q(\lambda)$ with $\lambda \in X_l$ is simple as a $u_{q,k}$–module. Each simple $u_{q,k}$–module is isomorphic to exactly one $L_q(\lambda) \otimes k_\omega$ with $\lambda \in X_l$ and ω a group homomorphism $\mathbf{Z}R \to \{\pm 1\}$, cf. [Lusztig 7], Prop. 5.11, [Andersen, Polo, and Wen 2], 0.9.

One can also imitate the theory of G_1T–modules by considering $u_{q,k}U^0_{q,k}$–modules. Here the simple modules "of type $\mathbf{1}$" (i.e., where each K^l_α acts as 1) are parametrised by elements in $X(T)$. If $\lambda \in X(T)$ is written as $\lambda = \lambda' + l\lambda''$ with $\lambda' \in X_l$ and $\lambda'' \in X(T)$, then the corresponding simple module is $\widetilde{L}_q(\lambda) \simeq L_q(\lambda') \otimes k_{l\lambda''}$ where $k_{l\lambda''}$ is a one dimensional module annihilated by all E_α and F_α while $U^0_{q,k}$ acts via $\varepsilon_{\mathbf{1},l\lambda''}$. (One can also look at $u_{q,k}U^{\le 0}_{q,k}$–modules and thus generalise the theory of G_1B–modules.)

There are induction functors Z from $\{\, u^{\le 0}_{q,k}\text{–modules}\,\}$ to $\{\, u_{q,k}\text{–modules}\,\}$, and $\widetilde{Z}$ from $\{\, u^-_{q,k}U^0_{q,k}\text{–modules}\,\}$ to $\{\, u_{q,k}U^0_{q,k}\text{–modules}\,\}$ analogous to the induction functors from B_1 to G_1 or from B_1T to G_1T. Then any $L_q(\lambda)$ with $\lambda \in X_l$ is (considered as a $u_{q,k}$–module) isomorphic to the socle of $Z(k_\lambda)$, each simple $u_{q,k}U^0_{q,k}$–module $\widetilde{L}_q(\lambda)$ with $\lambda \in X(T)$ is isomorphic to the socle of $\widetilde{Z}(k_\lambda)$.

For any $\lambda \in X_l$ let $Q_q(\lambda)$ denote the injective hull of $L_q(\lambda)$ as a $u_{q,k}$–module. As pointed out in Chapter 11, one of the main problems in the representation theory of Frobenius kernels (Do projective indecomposable modules lift to the group G?) has been open for about 25 years. The analogous problem in the quantum case has an easy solution: For each $\lambda \in X(T)_+$ the simple $U_{q,k}$–module $L_q(\lambda)$ has an injective hull $I_q(\lambda)$ in the category of integrable $U_{q,k}$–modules. One gets now:

(1) *If $\lambda \in X_l$, then $I_q(\lambda)$ is isomorphic to $Q_q(\lambda)$ as a $u_{q,k}$–module.*

Furthermore, if one decomposes an arbitrary $\lambda \in X(T)_+$ as $\lambda = \lambda' + l\lambda''$ with $\lambda' \in X_l$ and $\lambda'' \in X(T)_+$, then

(2) $$I_q(\lambda) \simeq I_q(\lambda') \otimes L_q(l\lambda'').$$

Each $\widetilde{L}_q(\lambda)$ with $\lambda \in X(T)$ has an injective hull $\widetilde{Q}_q(\lambda)$ in the category of all integrable $u_{q,k}U^0_{q,k}$–modules. (A $u_{q,k}U^0_{q,k}$–module M is called integrable if it is the direct sum of all $M_{1,\nu}$ with $\nu \in X(T)$.) One can show: If $\lambda = \lambda' + l\lambda''$ with $\lambda' \in X_l$ and $\lambda'' \in X(T)$, then

$$(3) \qquad \widetilde{Q}_q(\lambda) \simeq I_q(\lambda') \otimes k_{l\lambda''}.$$

On the other hand, $\widetilde{Q}_q(\lambda)$ has a filtration with factors of the form $\widetilde{Z}(\mu)$ with $\mu \in X(T)$; the number of factors isomorphic to a fixed $\widetilde{Z}(\mu)$ is equal to the multiplicity of $\widetilde{L}_q(\lambda)$ as a composition factor of $\widetilde{Z}(\mu)$. These results, as well as those mentioned above in this subsection can be found in [Andersen, Polo, and Wen 2].

H.14. (Comparison with Prime Characteristics) Return to the more restrictive situation at the end of H.12 where $l = p$ is a prime number and $k = \mathbf{Q}(q)$. For $p \geq 2h - 2$ Lusztig's conjecture 8.22(2) together with H.12(1) predicts that $\operatorname{ch} L(\mu) = \operatorname{ch} L_q(\mu)$ for all $\mu \in X(T)_+$ with $\langle \mu + \rho, \alpha^\vee \rangle < p$ for all simple roots $\alpha \in S$. This is equivalent to $\operatorname{ch} \widetilde{L}_1(\mu) = \operatorname{ch} \widetilde{L}_q(\mu)$ for all $\mu \in X(T)$ where $\widetilde{L}_1(\mu)$ is as in Chapter 9, hence to $[\widetilde{Z}_1(\mu) : \widetilde{L}_1(\nu)] = [\widetilde{Z}_q(\mu) : \widetilde{L}_q(\nu)]$ for all $\mu, \nu \in X(T)$, hence to $\operatorname{ch} \widetilde{Q}_1(\lambda) = \operatorname{ch} \widetilde{Q}_q(\lambda)$ for all $\lambda \in X(T)$. Comparing H.13(3) and 11.3(2) one sees that it is enough to have $\operatorname{ch} \widetilde{Q}_1(\lambda) = \operatorname{ch} \widetilde{Q}_q(\lambda)$ for all $\lambda \in X(T)_+$ with $\langle \lambda + \rho, \alpha^\vee \rangle < p$ for all $\alpha \in S$.

In [Andersen, Jantzen, and Soergel] it is shown that this equality holds for $p \gg 0$. More precisely, there is an unknown bound $n(R)$ depending on the root system R such that equality holds for $p > n(R)$. By H.12 this then implies Lusztig's conjecture for these p.

H.15. (Tilting Modules) Suppose that k has characteristic 0 and that q is a primitive l-th root of unity with $l > h$ odd (and prime to 3 if R has a component of type G_2). For each $\lambda \in X(T)_+$ set $V_q(\lambda) = H_q^n(w_0 \bullet \lambda)$ where $n = |R^+|$; we call $V_q(\lambda)$ the Weyl module for $U_{q,k}$ with highest weight λ.

One defines as in 4.16/19 good filtrations of $U_{q,k}$-modules (i.e., filtrations with factors of the form $H_q^0(\lambda)$) and Weyl filtrations (with factors of the form $V_q(\lambda)$). There is then an analogue to Proposition 4.16 thanks to the fact that the Ext groups in the category of all integrable $U_{q,k}$-modules satisfy for all $\lambda, \mu \in X(T)_+$

$$(1) \qquad \operatorname{Ext}^i(V_q(\lambda), H_q^0(\mu)) \simeq \begin{cases} k, & \text{if } i = 0 \text{ and } \lambda = \mu, \\ 0, & \text{otherwise,} \end{cases}$$

see [Andersen 21], 1.4.

The tensor product of two modules with a good filtration (resp. a Weyl filtration) has again a good filtration (resp. a Weyl filtration), see [Andersen, Polo, and Wen 1], 5.14, [Andersen 21], 2.3, [Paradowski], 3.3.

Call a finite dimensional $U_{q,k}$-module V a tilting module if V has both a good filtration and a Weyl filtration (or, equivalently by H.11(1), if both V and V^* are goofy). One can then show that there exists for each dominant λ an indecomposable tilting module $T_q(\lambda)$, unique up to isomorphism, such that $\dim T_q(\lambda)_\lambda = 1$ and such that $T_q(\lambda)_\mu \neq 0$ implies $\mu \leq \lambda$, see [Andersen 21], 2.5, [Paradowski], 5.2. An arbitrary tilting module for $U_{q,k}$ is then a direct sum of suitable $T_q(\lambda)$.

One has as in E.1: If $H_q^0(\lambda)$ is simple, then it is a tilting module and hence isomorphic to $T_q(\lambda)$.

The linkage principle implies that the factors in a good filtration of any $T_q(\lambda)$ have the form $H^0_q(w \bullet \lambda)$ with $w \in W_l$ and $w \bullet \lambda \in X(T)_+$. For $l \geq h$ (the Coxeter number of R) the multiplicities in these filtrations were conjectured in [Soergel 4], 7.1/2; the conjecture was proved in many cases in [Soergel 5], Section 5. If λ has trivial stabiliser in W_l, then we can write λ uniquely $\lambda = w \bullet \lambda_0$ with $w \in W_l$ and with $0 < \langle \lambda_0 + \rho, \alpha^\vee \rangle < l$ for all $\alpha \in R^+$. Then the multiplicity of any $H^0_q(x \bullet \lambda_0)$ with $x \in W_l$ and $x \bullet \lambda_0 \in X(T)_+$ in a good filtration of $T_q(w \bullet \lambda_0)$ should be equal to the value at one $n_{x,w}(1)$ of a certain Kazhdan-Lusztig polynomial. The multiplicities for weights "on walls" would then follow by translation arguments, see also [Andersen 28], 5.3(i).

As in the "classical case" in E.24(9), the coefficients of the polynomials $n_{x,w}$ are expected to describe layers in a suitable filtration, see Section 3 in [Andersen 28]. This conjecture is supported by a sum formula, see [Andersen 28], 3.6.

The injective hull $I_q(\lambda)$ of any simple $U_{q,k}$–module $L_q(\lambda)$ with $\lambda \in X(T)_+$ is an indecomposable tilting module: If $\lambda = \lambda' + l\lambda''$ with $\lambda' \in X_l$ and $\lambda'' \in X(T)_+$, then

(2) $$I_q(\lambda) \simeq T_q(2(l-1)\rho + w_0\lambda' + l\lambda''),$$

see [Andersen 21], 5.8.

The tensor product of two tilting modules is again a tilting module; this follows from the analogous result for modules with a good filtration.

Let us call a subset $J \subset X(T)_+$ a *tensor ideal* in $X(T)_+$ if for all $\mu \in J$ and all $\lambda \in X(T)_+$ each indecomposable summand of $T_q(\mu) \otimes T_q(\lambda)$ is isomorphic to some $T_q(\nu)$ with $\nu \in J$.

According to [Andersen 21], 3.8/5, the set of all $\mu \in X(T)_+$ with μ *not* in the interior of the first dominant alcove relative to W_l (i.e., with $\langle \mu + \rho, \alpha^\vee \rangle \geq l$ for at least one $\alpha \in R^+$) is a tensor ideal. This leads then (as in the situation described in E.13) to the construction of a tensor category and of invariants of 3–manifolds, see [Andersen 21], Section 4, [Andersen 22], [Andersen and Paradowski].

Further tensor ideals are described in [Ostrik 1]; that work uses Soergel's results on the good filtration multiplicities for the $T_q(\mu)$ and is inspired by [Humphreys 27].

H.16. (Cohomology) In [Ginzburg and Kumar] a quantum analogue of 12.13/14 is proved: Consider the set-up as in H.7. The elements K_α^l with $\alpha \in S$ are central in $U_{q,k}$. Divide the finite dimensional subalgebra $u_{q,k}$ by the ideal generated by all $K_\alpha^l - 1$ and set

(1) $$\overline{u}_{q,k} = u_{q,k}/(K_\alpha^l - 1 \mid \alpha \in S).$$

This algebra has then dimension $l^{\dim G}$.

The main result in [Ginzburg and Kumar] is then the calculation of the cohomology of $\overline{u}_{q,k}$ with coefficients in the trivial module k assuming that $l > h$ (and satisfies the other assumptions in H.7). One gets:

Proposition: *All $H^{2i+1}(\overline{u}_{q,k})$ are 0. The direct sum of all $H^{2i}(\overline{u}_{q,k})$ is isomorphic as a graded algebra to the ring of regular functions on the nilpotent cone in the Lie algebra $\mathfrak{g}_k$.*

This result can be used to define support varieties for $\overline{u}_{q,k}$–modules, see [Parshall and Wang 3] and [Ostrik 2].

Another interesting explicit result on the cohomology of $\overline{u}_{q,k}$–modules can be found in [Ostrik 4].

References

The following list of references consists of two parts. Part A contains textbooks and long articles of a similar nature whereas Part B contains ordinary papers published in journals or proceedings volumes. At the end of Part A we have listed some conference proceedings and similar collections containing more than one paper from Part B in order to give the full bibliographical data only once. We refer to an item in Part A by a code like [B1] or [Bo], to an item in Part B by giving the full name of the author(s) together with a number (if necessary).

Part A

[A] S. Anantharaman: Schémas en groupes, espaces homogènes et espaces algébriques sur une base de dimension 1, *Bull. Soc. math. France, Mémoire* **33** (1973), 5–79

[Bo] A. Borel: *Linear Algebraic Groups*, 2nd ed. (Graduate Texts in Math. **126**), New York etc. 1991 (Springer)

[BoT] A. Borel, J. Tits: Groupes réductifs, *Publ. Math. Inst. Hautes Études Sci.* **27** (1965), 55–151

[B1] N. Bourbaki: *Algèbre*, Paris 1958 (ch. I), 1962 (ch. II), 1971 (ch. III, 2nd ed.), 1959 (ch. IV/V), 1964 (ch. VI/VII, 2nd ed.), 1958 (ch. VIII), 1959 (ch. IX), 1980 (ch. X) (Hermann: ch. I–IX, Masson: ch. X)

[B2] N. Bourbaki: *Algèbre commutative*, Paris 1961 (ch. I/II), 1962 (ch. III/IV), 1964 (ch. V/VI), 1965 (ch. VII) (Hermann)

[B3] N. Bourbaki: *Groupes et algèbres de Lie*, Paris 1971 (ch. I), 1972 (ch. II/III), 1968 (ch. IV–VI), 1975 (ch. VII/VIII) (Hermann)

[BrT] F. Bruhat, J. Tits: Groupes réductifs sur un corps local II: Schémas en groupes, Existence d'une donnée radicielle valuée, *Publ. Math. Inst. Hautes Études Sci.* **60** (1984), 197–376

[Ca1] R. W. Carter: *Simple Groups of Lie Type* (Pure and Applied Math. **28**), London etc. 1972 (Wiley)

[Ca2] R. W. Carter: *Finite Groups of Lie Type: Conjugacy Classes and Complex Characters*, Chichester etc. 1985 (Wiley)

[Ch] C. Chevalley: *Théorie des groupes de Lie, tome II: Groupes algébriques* (Actualités Sci. Ind. **1152**), Paris 1951 (Hermann)

[CR] C. W. Curtis, I. Reiner: *Representation Theory of Finite Groups and Associative Algebras* (Pure and Applied Math. **11**), New York etc. 1962 (Interscience)

[D] M. Demazure: Schémas en groupes réductifs, *Bull. Soc. math. France* **93** (1965), 369–413

[DG] M. Demazure, P. Gabriel: *Groupes algébriques, tome I*, Paris / Amsterdam 1970 (Masson / North-Holland)

[DG′] M. Demazure, P. Gabriel: *Introduction to Algebraic Geometry and Algebraic Groups* (North-Holland Math. Studies **39**), Amsterdam etc. 1980 (North-Holland)

[SGA3] M. Demazure, A. Grothendieck (dirig.): *Schémas en groupes, Séminaire de Géométrie Algébrique du Bois Marie 1962/64 (SGA 3)* (Lecture Notes in Math. **151–153**), Berlin etc. 1970 (Springer)

[Dic] L. E. Dickson: *History of the Theory of Numbers, vol. III: Quadratic and Higher Forms*, Washington 1923 (Carnegie Institution)

[Dix] J. Dixmier: *Algèbres enveloppantes*, Paris etc. 1974 (Gauthier-Villars)

[Do] S. Donkin: *The q-Schur Algebra* (London Math. Soc. Lecture Note **253**), Cambridge 1998 (Cambridge Univ.)

[EGA] A. Grothendieck: Éléments de géométrie algébrique IV, Étude locale des schémas et des morphismes de schémas IV, *Publ. Math. Inst. Hautes Études Sci.* **32** (1967), 1–361

[F] J. Fogarty: *Invariant Theory*, New York etc. 1969 (Benjamin)

[G] R. Godement: *Topologie algébrique et théorie des faisceaux* (Actualités Sci. Ind. **1252**), Paris 1958 (Hermann)

[Ha] R. Hartshorne: *Algebraic Geometry* (Graduate Texts in Math. **52**), New York etc. 1977 (Springer)

[HS] P. J. Hilton, U. Stammbach: *A Course in Homological Algebra* (Graduate Texts in Math. **4**), New York etc. 1971 (Springer)

[Ho1] G. Hochschild: *Introduction to Affine Algebraic Groups*, San Francisco etc. 1971 (Holden-Day)

[Ho2] G. Hochschild: *Basic Theory of Algebraic Groups and Lie Algebras* (Graduate Texts in Math. **75**), New York etc. 1981 (Springer)

[Hu1] J. Humphreys: *Introduction to Lie Algebras and Representation Theory* (Graduate Texts in Math. **9**), New York etc. 1972 (Springer)

[Hu2] J. Humphreys: *Linear Algebraic Groups* (Graduate Texts in Math. **21**), New York etc. 1975 (Springer)

[Hu3] J. Humphreys: *Reflection Groups and Coxeter Groups* (Cambridge Studies in Advanced Math. **29**), Cambridge etc. 1990 (Cambridge Univ.)

[JK] G. James, A. Kerber: *The Representation Theory of the Symmetric Group* (Encyclopedia of Math. and its Appl. **16**), Reading, Mass. etc. 1981 (Addison-Wesley)

[J1] J. C. Jantzen: *Moduln mit einem höchsten Gewicht* (Lecture Notes in Math. **750**), Berlin etc. 1979 (Springer)

[J2] J. C. Jantzen: *Lectures on Quantum Groups* (Graduate Studies in Math. **6**), Providence, R. I. 1996 (Amer. Math. Soc.)

[K] V. G. Kac: *Infinite Dimensional Lie Algebras* (Progress in Math. **44**), Boston 1983 (Birkhäuser)

[Mac] S. Mac Lane: *Homology* (Grundlehren der mathematischen Wissenschaften **114**), Berlin etc. 1963 (Springer)

[M1] D. Mumford: *Abelian Varieties* (Tata Studies in Math. **5**), London etc. 1970 (Oxford Univ.)

[M2] D. Mumford: *Introduction to Algebraic Geometry* (preliminary version of first 3 chapters), Cambridge, Mass. (Harvard Univ.)

[MF] D. Mumford, J. Fogarty: *Geometric Invariant Theory* (Ergebnisse der Mathematik und ihrer Grenzgebiete **34**, 2nd ed.) Berlin etc. 1982 (Springer)

[N] D. G. Northcott: *Affine Sets and Affine Groups* (London Math. Soc. Lecture Note **39**), Cambridge etc. 1980 (Cambridge Univ.)

[Ra] M. Raynaud: *Faisceaux amples sur les schémas en groupes et les espaces homogènes* (Lecture Notes in Math. **119**), Berlin etc. 1970 (Springer)

[Ro] J. J. Rotman: *An Introduction to Homological Algebra* (Pure and Applied Math. **85**), New York etc. 1979 (Academic Press)

[Ru] D. E. Rutherford: *Modular Invariants*, London 1932 (Cambridge Univ.)

[Sa] I. Satake: *Classification Theory of Semi-Simple Algebraic Groups*, Chicago 1967 (Univ. of Illinois, Chicago Circle)

[Se] G. Seligman: *Algebraic Groups*, New Haven, Conn. 1964 (Yale Univ.)

[SC] Séminaire C. Chevalley, 1956–1958: *Classification des groupes de Lie algébriques*, Paris 1958 (Secr. math.)

[SHS] Séminaire Heidelberg-Strasbourg 1965-66: *Groupes algébriques linéaires*, Strasbourg 1958 (Inst. Rech. Math. Avanc.)

[Sp1] T. A. Springer: *Invariant Theory* (Lecture Notes in Math. **585**), Berlin etc. 1977 (Springer)

[Sp2] T. A. Springer: *Linear Algebraic Groups* (Progress in Math. **9**), 2nd ed., Boston etc. 1998 (Birkhäuser)

[St1] R. Steinberg: *Lectures on Chevalley Groups*, New Haven, Conn. 1968 (Yale Univ.)

[St2] R. Steinberg: *Conjugacy Classes in Algebraic Groups* (Lecture Notes in Math. **366**), Berlin etc. 1974 (Springer)

[Sw] M. Sweedler: *Hopf Algebras*, New York 1969 (Benjamin)

[Ta] M. Takeuchi: Tangent coalgebras and hyperalgebras I, *Japan. J. Math.* **42** (1974), 1–143

[Ti] J. Tits: *Lectures on Algebraic Groups, 1st Part*, New Haven, Conn. 1968 (Yale Univ.)

[Wa] W. C. Waterhouse: *Introduction to Affine Group Schemes* (Graduate Texts in Math. **66**), New York etc. 1979 (Springer)

[We] H. Weyl: *The Classical Groups, Their Invariants and Representations* (Princeton Math. Series **1**), Princeton 1946 (Princeton Univ.)

[Y] H. Yanagihara: *Theory of Hopf Algebras Attached to Group Schemes* (Lecture Notes in Math. **614**), Berlin etc. 1977 (Springer)

[P1] A. Borel et al.: *Seminar on Algebraic Groups and Related Finite Groups* (Lecture Notes in Math. **131**), Berlin etc. 1970 (Springer)

[P2] M. Collins (ed.), *Finite Simple Groups II*, Proc. Durham 1978, London etc. 1980 (Academic Press)

[P3] B. Cooperstein, G. Mason (eds.), *The Santa Cruz Conference on Finite Groups*, Proc. 1979 (Proc. Symp. Pure Math. **37**), Providence, R. I. 1980 (Amer. Math. Soc.)

[P4] O. Lehto (ed.), *Proceedings of the International Congress of Mathematicians.* Proc. Helsinki 1978, Helsinki 1980 (Acad. Sci. Fennica)

[P5] W. R. Gross, F. Gross (eds.), *Proceedings of the Conference on Finite Groups*, Proc. Park City, Utah 1975, New York etc. 1976 (Academic Press)

[P6] *Tableaux de Young et Foncteurs de Schur en Algèbre et Géométrie*, Proc. Toruń 1980 (Astérisque **87-88**), Paris 1981 (Soc. Math. France)

[P7] Tuan Hsio-Fu (ed.), *Group Theory, Beijing 1984*, Proc. (Lecture Notes in Math. **1185**), Berlin etc. 1986 (Springer)

[P8] P. Fong (ed.), *The Arcata Conference on Representations of Finite Groups*, Part 1, Proc. 1986 (Proc. Symp. Pure Math. **47:1**), Providence, R. I. 1987 (Amer. Math. Soc.)

[P9] P. Fong (ed.), *The Arcata Conference on Representations of Finite Groups*, Part 2, Proc. 1986 (Proc. Symp. Pure Math. **47:2**), Providence, R. I. 1987 (Amer. Math. Soc.)

[P10] R. Fossum et al. (eds.), *Invariant Theory*, Proc. Denton, Tex. 1986 (Contemp. Math. **88**), Providence, R. I. 1989 (Amer. Math. Soc.)

[P11] A. J. Hahn et al. (eds.), *Classical Groups and Related Topics*, Proc. Beijing 1987 (Contemp. Math. **82**), Providence, R. I. 1989 (Amer. Math. Soc.)

[P12] S. Ramanan et al. (eds.), *Proceedings of the Hyderabad Conference on Algebraic Groups*, Proc. 1989, Madras 1991 (Manoj Prakashan)

[P13] W. F. Haboush, B. J. Parshall (eds.), *Algebraic Groups and their Generalizations: Classical Methods*, Proc. University Park, Penn. 1991 (Proc. Sympos. Pure Math. **56:1**), Providence, R. I. 1994 (Amer. Math. Soc.)

[P14] W. F. Haboush, B. J. Parshall (eds.), *Algebraic Groups and their Generalizations: Quantum and Infinite-dimensional Methods*, Proc. University Park, Penn. 1991 (Proc. Sympos. Pure Math. **56:2**), Providence, R. I. 1994 (Amer. Math. Soc.)

[P15] A. Joseph, S. Shnider (eds.), *Quantum Deformations of Algebras and Their Representations*, Proc. Ramat-Gan and Rehovot 1991/1992 (Israel Math. Conf. Proc. **7**), Ramat Gan 1993 (Bar-Ilan Univ.)

[P16] V. Dlab, L. L. Scott (eds.), *Finite-Dimensional Algebras and Related Topics*, Proc. Ottawa 1992 (NATO Adv. Sci. Inst. **C 424**), Dordrecht etc. 1994 (Kluwer)

[P17] B. Allison, G. Cliff (eds.), *Representations of Groups*, Proc. Banff 1994 (CMS Conf. Proc. **16**), Providence, R. I. 1995 (Amer. Math. Soc.)

[P18] S. D. Chatterji (ed.), *Proceedings of the International Congress of Mathematicians*, Vol. 2, Proc. Zürich 1994, Basel 1995 (Birkhäuser)

[P19] G. Lehrer et al. (eds.), *Algebraic Groups and Lie Groups*, a volume in honour of R. W. Richardson (Austral. Math. Soc. Lecture Ser. **9**), Cambridge 1997 (Cambridge Univ.)

[P20] A. Adem et al. (eds.), *Group Representations: Cohomology, Group Actions and Topology*, Proc. Seattle 1996 (Proc. Sympos. Pure Math. **63**), Providence, R. I. 1998 (Amer. Math. Soc.)

[P21] R. W. Carter, J. Saxl (eds.), *Algebraic Groups and their Representations*, Proc. Cambridge 1997 (NATO Adv. Sci. Inst. **C 517**), Dordrecht etc. 1998 (Kluwer)

[P22] M. J. Collins et al. (eds.), *Modular Representation Theory of Finite Groups*, Proc. Charlottesville 1998, Berlin 2001 (de Gruyter)

[P23] J. Wang, Z. Lin (eds.), *Representations and Quantizations*, Proc. Shanghai 1998, Beijing 2000 (China High. Educ. Press)

Part B

A. M. Adamovich

1) Analogues of spaces of primitive forms over a field of positive characteristic, *Moscow Univ. Math. Bull.* **39:1** (1984), 53–56, translated from: Аналог пространства примитивных форм над полем положительной характеристики, Вестн. Моск. Ун-та. (Матем. Механ.) **1984:1**, 64–66

2) The submodule lattice for Weyl modules of symplectic groups with fundamental highest weights, *Moscow Univ. Math. Bull.* **41:2** (1986), 6–9, translated from: Структура подмодулей модулей Вейля симплектических групп с фундаментальными старшими весами, Вестн. Моск. Ун-та. (Матем. Механ.) **1986:2**, 7–10

3) Structure of cohomology modules of linear bundles over G/B, *Math. Notes* **62** (1997), 8–14, translated from: Матем. Заметки **62** (1997), 10–17

4) Structure of cohomology modules of linear bundles over G/B, and weak ordering on the Weyl group, *Math. Notes* **62** (1997), 135–140, translated from: Матем. Заметки **62** (1997), 163–168

A. M. Adamovich, G. L. Rybnikov

Tilting modules for classical groups and Howe duality in positive characteristic, *Transform. Groups* **1** (1996), 1–34

K. Akin, D. Buchsbaum

1) Characteristic-free representation theory of the general linear group, *Adv. in Math.* **58** (1985), 149–200

2) Representations, resolutions and intertwining numbers, pp. 1–19 in M. Hochster et al. (eds.): *Commutative Algebra*, Proc. Berkeley 1987 (Math. Sci. Res. Inst. Publ. **15**), New York 1989 (Springer)

3) Characteristic-free representation theory of the general linear group II: Homological considerations, *Adv. in Math.* **72** (1988), 171–210

K. Akin, D. Buchsbaum, J. Weyman

Schur functors and Schur complexes, *Adv. in Math.* **44** (1982), 207–278

J. L. Alperin

Projective modules for $SL(2, 2^n)$, *J. Pure Appl. Algebra* **15** (1979), 219–234

J. L. Alperin, L. G. Kovacs

Periodicity of Weyl modules for $SL(2, q)$, *J. Algebra* **74** (1982), 52–54

H. H. Andersen

1) Cohomology of line bundles on G/B, *Ann. scient. Éc. Norm. Sup.* (4) **12** (1979), 85–100

2) The first cohomology group of a line bundle on G/B, *Invent. math.* **51** (1979), 287–296

3) Vanishing theorems and induced representations, *J. Algebra* **62** (1980), 86–100

4) The strong linkage principle, *J. reine angew. Math.* **315** (1980), 53–59

5) The Frobenius morphism on the cohomology of homogeneous vector bundles on G/B, *Ann. of Math.* (2) **112** (1980), 113–121

6) On the structure of Weyl modules, *Math. Z.* **170** (1980), 1–14

7) Representations of algebraic groups via cohomology of line bundles, pp. 171–175 in: E. Balslev (ed.): *18th Scandinavian Congress of Mathematicians*, Proc. Aarhus 1980, Boston etc. 1981 (Birkhäuser)

8) Line bundles on flag manifolds, pp. 21–42 in [P6]

9) On the structure of the cohomology of line bundles on G/B, *J. Algebra* **71** (1981), 245–258

10) Extensions of modules for algebraic groups, *Amer. J. Math.* **106** (1984), 489–504

11) An inversion formula for the Kazhdan-Lusztig polynomials for affine Weyl groups, *Adv. in Math.* **60** (1986), 125–153

12) Filtrations of cohomology modules for Chevalley groups, *Ann. scient. Éc. Norm. Sup.* (4) **16** (1983), 495–528
13) Schubert varieties and Demazure's character formula, *Invent. math.* **79** (1985), 611-618
14) Torsion in the cohomology of line bundles on homogeneous spaces for Chevalley groups, *Proc. Amer. Math. Soc.* **96** (1986), 537–544
15) On the generic structure of cohomology modules for semisimple algebraic groups, *Trans. Amer. Math. Soc.* **295** (1986), 397–415
16) Jantzen's filtrations of Weyl modules, *Math. Z.* **194** (1987), 127–142
17) Extensions of simple modules for finite Chevalley groups, *J. Algebra* **111** (1987), 388–403
18) A new proof of old character formulas, pp. 193–207 in [P10]
19) Modular representations of algebraic groups, pp. 23–36 in [P8]
20) Finite-dimensional representations of quantum groups, pp. 1–18 in [P14]
21) Tensor products of quantized tilting modules, *Comm. Math. Phys.* **149** (1992), 149–159
22) Quantum groups, invariants of 3-manifolds and semisimple tensor categories, pp. 1–12 in [P15]
23) Modular representations of algebraic groups and relations to quantum groups, pp. 1–51 in: B. Ørsted, H. Schlichtkrull (eds.), *Algebraic and Analytic Methods in Representation Theory*, Proc. Sønderborg 1994 (Perspect. Math. **17**), San Diego 1997 (Academic)
24) The irreducible characters for semi-simple algebraic groups and for quantum groups, pp. 732–743 in [P18]
25) Quantum groups at roots of ± 1, *Commun. Algebra* **24** (1996), 3269–3282
26) Filtrations and tilting modules, *Ann. scient. Éc. Norm. Sup.* (4) **30** (1997), 353–366
27) Tilting modules for algebraic groups, pp. 25–42 in [P21]
28) A sum formula for tilting filtrations, *J. Pure Appl. Algebra* **152** (2000), 17–40
29) Tilting modules for algebraic and quantum groups, pp. 1–21 in: K. W. Roggenkamp, M. Stefănescu (eds.), *Algebra—Representation Theory*, Proc. Constanta 2000 (NATO Sci. Ser. **II 28**), Dordrecht etc. 2001 (Kluwer)
30) p-filtrations and the Steinberg module, *J. Algebra* **244** (2001), 664–683

H. H. Andersen, J. C. Jantzen

Cohomology of induced representations for algebraic groups, *Math. Ann.* **269** (1984), 487–525

H. H. Andersen, J. C. Jantzen, W. Soergel

Representations of quantum groups at a pth root of unity and of semisimple groups in characteristic p: independence of p, *Astérisque* **220** (1994), 1–321

H. H. Andersen, J. Jørgensen, P. Landrock

The projective indecomposable modules of $\mathrm{SL}(2, p^n)$, *Proc. London Math. Soc.* (3) **46** (1983), 38–52

H. H. Andersen, M. Kaneda

1) Loewy series of modules for the first Frobenius kernel in a reductive algebraic group, *Proc. London Math. Soc.* (3) **59** (1989), 74–98
2) On the D-affinity of the flag variety in type B_2, *Manuscripta Math.* **103** (2000), 393–399
3) Filtrations on G_1T-modules, *Proc. London Math. Soc.* (3) **82** (2001), 614–646

H. H. Andersen, G. Papadopoulo

Liftings of quantum tilting modules, pp. 1–8 in [P23]

H. H. Andersen, J. Paradowski

Fusion categories arising from semisimple Lie algebras, *Comm. Math. Phys.* **169** (1995), 563–588

H. H. Andersen, P. Polo, K. Wen

1) Representations of quantum algebras, *Invent. math.* **104** (1991), 1–59 and **120** (1995), 409–410

2) Injective modules for quantum algebras, *Amer. J. Math.* **114** (1992), 571–604

H. H. Andersen, K. Wen

Representations of quantum algebras. The mixed case, *J. reine angew. Math.* **427** (1992), 35–50

J. Archer

Principal indecomposable modules for some three-dimensional special linear groups, *Bull. Austral. Math. Soc.* **22** (1980), 439–455

G. Avrunin

1) 2–cohomology of some unitary groups, *Illinois J. Math.* **24** (1980), 317–332

2) Generic cohomology for twisted groups. *Trans. Amer. Math. Soc.* **268** (1981), 247–253

K. Baclawski, J. Towber

The shape-algebra and standard bases for G_2, *Amer. J. Math.* **106** (1984), 1107–1134

Y. H. Bai, J. P. Wang, K. X. Wen

Translation and cancellation of socle series patterns, *Tôhoku Math. J.* (2) **40** (1988), 633–643

J. W. Ballard

1) Some generalized characters of finite Chevalley groups, *Math. Z.* **147** (1976), 163–174

2) Projective modules for finite Chevalley groups, *Trans. Amer. Math. Soc.* **245** (1978), 221–249

3) Injective modules for restricted enveloping algebras, *Math. Z.* **163** (1978), 57–63

4) Clifford's theorem for algebraic groups and Lie algebras, *Pacific J. Math.* **106** (1983), 1–15

A. A. Baranov, I. D. Suprunenko

Branching rules for modular fundamental representations of symplectic groups, *Bull. London Math. Soc.* **32** (2000), 409–420

M. Barnabei

Schur modules, Weyl modules, and Capelli operators, *Adv. in Math.* **151** (2000), 1–35

M. Barnabei, V. Frontini, F. Sgallari

An algorithm for Weyl module irreducibility, *Rend. Sem. Mat. Univ. Politec. Torino* **49** (1991), 217–232

A. Beilinson, J. Bernstein

A proof of Jantzen conjectures, pp. 1–50 in: S. Gelfand, S. Gindikin (eds.), *I. M. Gelfand Seminar* (Adv. Soviet Math. **16:1**), Providence, R. I. 1993 (Amer. Math. Soc.)

G. Bell

1) On the cohomology of the finite special linear groups. I, II, *J. Algebra* **54** (1978), 216–238 and 239–259
2) Cohomology of degree 1 and 2 of the Suzuki groups, *Pacific J. Math.* **75** (1978), 319–329

C. Bendel
1) Projectivity of modules for infinitesimal unipotent group schemes, *Proc. Amer. Math. Soc.* **129** (2001), 671–676
2) Cohomology and projectivity of modules for finite group schemes, *Math. Proc. Cambridge Philos. Soc.* **131** (2001), 405–425

C. Bendel, D. Nakano
Complexes and vanishing of cohomology for group schemes, *J. Algebra* **214** (1999), 668–713

C. Bendel, D. Nakano, C. Pillen
On comparing the cohomology of algebraic groups, finite Chevalley groups and Frobenius kernels, *J. Pure Appl. Algebra* **163** (2001), 119–146

D. Benson
1) Projective modules for the group of twenty-seven lines on a cubic surface, *Commun. Algebra* **17** (1989), 1017–1068
2) Some remarks on the decomposition numbers for the symmetric groups, pp. 381–394 in [P8]

B. D. Boe
Geometry of the Jantzen region in Lusztig's conjecture, *Math. Comp.* **70** (2001), 1265–1280

A. Borel
1) Properties and linear representations of Chevalley groups, pp. 1–55 in [P1]
2) Linear representations of semi-simple algebraic groups, pp. 421–440 in: R. Hartshorne (ed.), *Algebraic geometry, Arcata 1974*, Proc. (Proc. Sympos. Pure Math. **29**), Providence, R. I. 1975 (Amer. Math. Soc.)

B. Braden
Restricted representations of classical Lie algebras of types A_2 and B_2, *Bull. Amer. Math. Soc.* **73** (1967), 482–486

M. Brion, P. Polo
Generic singularities of certain Schubert varieties, *Math. Z.* **231** (1999), 301–324

J. Brundan
1) Double coset density in reductive algebraic groups, *J. Algebra* **177** (1995), 755–767
2) Lowering operators for $\mathrm{GL}(n)$ and quantum $\mathrm{GL}(n)$, pp. 95–114 in [P20]
3) Multiplicity-free subgroups of reductive algebraic groups, *J. Algebra* **188** (1997), 310–330
4) Dense orbits and double cosets, pp. 259–274 in [P21]
5) Double coset density in exceptional algebraic groups, *J. London Math. Soc.* (2) **58** (1998), 63–83
6) Double coset density in classical algebraic groups, *Trans. Amer. Math. Soc.* **352** (2000), 1405–1436

J. Brundan, A. Kleshchev
1) Some remarks on branching rules and tensor products for algebraic groups, *J. Algebra* **217** (1999), 335–351

2) Modular Littlewood-Richardson coefficients, *Math. Z.* **232** (1999), 287–320
3) Tensor products and restrictions in type A, pp. 67–99 in [P22]
4) On translation functors for general linear and symmetric groups, *Proc. London Math. Soc.* (3) **80** (2000), 75–106

J. Brundan, A. Kleshchev, I. Suprunenko
Semisimple restrictions from $\mathrm{GL}(n)$ to $\mathrm{GL}(n-1)$, *J. reine angew. Math.* **500** (1998), 83–112

A. Buch, J. F. Thomsen, N. Lauritzen, V. Mehta
1) Frobenius morphisms modulo p^2, *C. R. Acad. Sci. Paris* (I) **322** (1996), 69–72
2) The Frobenius morphism on a toric variety, *Tôhoku Math. J.* (2) **49** (1997), 355–366

D. A. Buchsbaum
Aspects of characteristic-free representation theory of GL_n, and some applications to intertwining numbers, *Acta Appl. Math.* **21** (1990), 247–261

D. A. Buchsbaum, D. Flores de Chela
Intertwining numbers: the three-rowed case, *J. Algebra* **183** (1996), 605–635

D. A. Buchsbaum, R. Sánchez
On lifting maps between Weyl modules: can bad shapes be resolved by better shapes? *Adv. in Math.* **105** (1994), 59–75

N. Burgoyne
Modular representations of some finite groups, pp. 13–17 in: I. Reiner (ed.), *Representation Theory of Finite Groups and Related Topics*, Proc. Madison, Wisc., 1970 (Proc. Sympos. Pure Math. **21**), Providence, R. I. 1971 (Amer. Math. Soc.)

R. Burkhardt
1) Die Zerlegungsmatrizen der Gruppen $PSL(2, p^f)$, *J. Algebra* **40** (1976), 75–96
2) Über ein kombinatorisches Problem aus der modularen Darstellungstheorie, *J. Combinatorial Theory* (A) **21** (1976), 68–79
3) Über die Zerlegungszahlen der Suzukigruppen $Sz(q)$, *J. Algebra* **59** (1979), 421–433
4) Über die Zerlegungszahlen der unitären Gruppen $PSU(3, 2^{2f})$, *J. Algebra* **61** (1979), 548–581

M. Cabanes
Irreducible modules and Levi supplements, *J. Algebra* **90** (1984), 84–97

J. F. Carlson
The cohomology of irreducible modules over $\mathrm{SL}(2, p^n)$, *Proc. London Math. Soc.* (3) **47** (1983), 480–492

R. W. Carter
1) The relation between characteristic 0 representations and characteristic p representations of finite groups of Lie type, pp. 301–311 in [P3]
2) Representation theory of the 0-Hecke algebra, *J. Algebra* **104** (1986), 89–103
3) Raising and lowering operators for $\mathfrak{sl}_n$, with applications to orthogonal bases of $\mathfrak{sl}_n$–modules, pp. 351–366 in [P9]

R. Carter, E. Cline
The submodule structure of Weyl modules for groups of type A_1, pp. 303–311 in [P5]

R. W. Carter, G. Lusztig

1) On the modular representations of the general linear and symmetric groups, *Math. Z.* **136** (1974), 193–242
2) Modular representations of finite groups of Lie type, *Proc. London Math. Soc.* (3) **32** (1976), 347–384

R. W. Carter, M. T. J. Payne
On homomorphisms between Weyl modules and Specht modules, *Math. Proc. Camb. Phil. Soc.* **87** (1980), 419–425

P. Cartier
Représentations lineaires des groupes algébriques en caractéristique non nulle, exp. 255 in: *Séminaire Bourbaki 1962/63* (2e éd.), Paris 1964 (Secr. math.)

L. Chastkofsky
1) Characters of projective indecompsable modules for finite Chevalley groups, pp. 359–362 in [P3]
2) Projective characters for finite Chevalley groups, *J. Algebra* **69** (1981), 347–357
3) Variations on Hulsurkar's matrix with applications to representations of algebraic Chevalley groups, *J. Algebra* **82** (1983), 255–274
4) Rationality of certain zeta functions associated with modular representation theory, pp. 41–50 in: J. Mackay (ed.), *Finite Groups — Coming of Age*, Proc. Montreal 1982 (Contemp. Math. **45**), Providence, R. I. 1985 (Amer. Math. Soc.)
5) Generic Cartan invariants for Chevalley groups, *J. Algebra* **103** (1986), 466–478
6) On the Cartan invariants of a Chevalley group over $GF(p^n)$, *J. Algebra* **150** (1992), 388–401

L. Chastkofsky, W. Feit
1) On the projective characters in characteristic 2 of the groups $\mathrm{Suz}(2^m)$ and $\mathrm{Sp}_4(2^n)$, *Publ. Math. I. H. E. S.* **51** (1980), 9–35
2) On the projective characters in characteristic 2 of the groups $SL_3(2^m)$ and $SU_3(2^m)$, *J. Algebra* **63** (1980), 124–142

Y. Cheng
1) On the first Cartan invariants in characteristic 2 of the groups $SL_3(2^m)$ and $SU_3(2^m)$, *J. Algebra* **82** (1983), 194–244
2) On the Cartan invariants of $\mathrm{SL}(2, p^m)$. *Commun. Algebra* **14** (1986), 507–515

C. Chevalley
Certains schémas de groupes semi-simples, exp. 219 in *Séminaire Bourbaki 1960/61* (2e éd.), Paris 1961 (Secr. math.)

G. Cliff
A tensor product theorem for quantum linear groups at even roots of unity, *J. Algebra* **165** (1994), 566–575

E. Cline
1) Ext^1 for SL_2, *Commun. Algebra* **7** (1979), 107–111
2) A second look at Weyl modules for SL_2, preprint
3) On injective modules for infinitesimal algebraic groups II, *J. Algebra* **134** (1990), 271–297
4) Simulating algebraic geometry with algebra III: The Lusztig conjecture as a TG_1-problem, pp. 149–161 in [P9]

E. Cline, B. Parshall, L. Scott

1) Cohomology of finite groups of Lie type I, *Publ. Math. Inst. Hautes Études Sci.* **45** (1975), 169–191
2) Cohomology of finite groups of Lie type II, *J. Algebra* **45** (1977), 182–198
3) Induced modules and affine quotients, *Math. Ann.* **230** (1977), 1–14
4) Induced modules and extensions of representations, *Invent. math.* **47** (1978), 41–51
5) Induced modules and extensions of representations II, *J. London Math. Soc.* (2) **20** (1978), 403–414
6) Cohomology, hyperalgebras, and representations, *J. Algebra* **63** (1980), 98–123
7) On the tensor product theorem for algebraic groups, *J. Algebra* **63** (1980), 264–267
8) Detecting rational cohomology of algebraic groups, *J. London Math. Soc.* (2) **28** (1983), 293–300
9) A Mackey imprimitivity theory for algebraic groups, *Math. Z.* **182** (1983), 447–471
10) On injective modules for infinitesimal algebraic groups I, *J. London Math. Soc.* (2) **31** (1985), 277–291
11) Derived categories and Morita theory, *J. Algebra* **104** (1986), 397–409
12) Algebraic stratification in representation categories, *J. Algebra* **117** (1988), 504–521
13) Finite-dimensional algebras and highest weight categories, *J. reine angew. Math.* **391** (1988), 85–99
14) Duality in highest weight categories, pp. 7–22 in [P11]
15) Integral and graded quasi-hereditary algebras I, *J. Algebra* **131** (1990), 126–160
16) Infinitesimal Kazhdan-Lusztig theories, pp. 43–73 in: V. Deodhar (ed.), *Kazhdan-Lusztig Theory and Related Topics*, Proc. Chicago 1989 (Contemp. Math. **139**), Providence, R. I. 1992 (Amer. Math. Soc.)
17) Abstract Kazhdan-Lusztig theories, *Tôhoku Math. J.* (2) **45** (1993), 511–534
18) The homological dual of a highest weight category, *Proc. London Math. Soc.* (3) **68** (1994), 294–316
19) Simulating perverse sheaves in modular representation theory, pp. 63–104 in [P13]
20) Stratifying endomorphism algebras, *Mem. Amer. Math. Soc.* **124:591** (1996), 1–119
21) Graded and non-graded Kazhdan-Lusztig theories, pp. 105–125 in [P19]
22) Endomorphism algebras and representation theory, pp. 131–149 in [P21]

E. Cline, B. Parshall, L. Scott, W. van der Kallen

Rational and generic cohomology, *Invent. math.* **39** (1977), 143–163

D. H. Collingwood, R. S. Irving

A decomposition theorem for certain self-dual modules in the category $\mathcal{O}$, *Duke Math. J.* **58** (1989), 89–102

A. Cox

1) Ext^1 for Weyl modules for q-GL$(2, k)$, *Math. Proc. Cambridge Philos. Soc.* **124** (1998), 231–251
2) The blocks of the q-Schur algebra, *J. Algebra* **207** (1998), 306–325
3) On the blocks of the infinitesimal Schur algebras, *Quart. J. Math.* **51** (2000), 39–56

4) Decomposition numbers for distant Weyl modules, *J. Algebra* **243** (2001), 448–472

A. Cox, K. Erdmann

On Ext^2 between Weyl modules for quantum GL_n, *Math. Proc. Cambridge Philos. Soc.* **128** (2000), 441–463

C. W. Curtis

1) Representations of Lie algebras of classical type with applications to linear groups, *J. Math. Mech.* **9** (1960), 307–326
2) On the dimensions of the irreducible modules of the Lie algebras of classical type, *Trans. Amer. Math. Soc.* **96** (1960), 135–142
3) On projective representations of certain finite groups, *Proc. Amer. Math. Soc.* **11** (1960), 852–860
4) Irreducible representations of finite groups of Lie type, *J. reine angew. Math.* **219** (1965), 180–199
5) Modular representations of finite groups with split (B, N)-pairs, pp. 57–95 in [P1]

S. W. Dagger

On the blocks of the Chevalley groups, *J. London Math. Soc.* (2) **3** (1971), 21–29

C. De Concini

1) Symplectic standard tableaux, *Adv. in Math.* **34** (1979), 1–27
2) Characteristic free "decomposition" of the coordinate ring of the symplectic group, pp. 121–128 in: A. de Luca (ed.), *Noncommutative structures in algebra and geometric combinatorics*, Proc. Napoli 1978, Roma 1981 (Cons. Naz. delle Ricerc.)

C. De Concini, D. Eisenbud, C. Procesi

1) Young diagrams and determinantal varieties, *Invent. math.* **56** (1980), 129–165
2) *Hodge algebras* (Astérisque **91**), Paris 1982 (Soc. Math. France)

C. De Concini, D. Kazhdan

Special bases for S_N and $\mathrm{GL}(n)$, *Israel J. Math.* **40** (1981), 275–290

C. De Concini, V. Lakshmibai

Arithmetic Cohen-Macaulayness and arithmetic normality for Schubert varieties, *Amer. J. Math.* **103** (1981), 835–850

C. De Concini, C. Procesi

1) A characteristic free approach to invariant theory, *Adv. in Math.* **21** (1976), 330–354
2) Symmetric functions, conjugacy classes, and the flag variety, *Invent. math.* **64** (1981), 203–219
3) Hodge algebras, a survey, pp. 79–83 in [P6]
4) Complete symmetric varieties, pp. 1–44 in: F. Gherardelli (ed.), *Invariant theory*, Proc. Montecatini 1982, (Lecture Notes in Math. **996**), Berlin etc. 1983 (Springer)
5) Quantum Groups, pp. 31–140 in: L. Boutet de Monvel et al., *D–modules, Representation Theory, and Quantum Groups*, Proc. Venezia 1992 (Lecture Notes in Mathematics **1565**), Berlin etc. 1993 (Springer)

C. De Concini, E. Strickland

1) Traceless tensors and the symmetric group, *J. Algebra* **61** (1979), 112–128
2) On the variety of complexes, *Adv. in Math.* **41** (1981), 57–77

M. Demazure

1) Sur la formule des caractères de H. Weyl, *Invent. math.* **9** (1970) 249–252
2) Invariants symétriques entiers des groupes de Weyl et torsion, *Invent. math.* **21** (1973), 287–301
3) Désingularisation des variétés de Schubert généralisées, *Ann. scient. Éc. Norm. Sup.* (4) **7** (1974), 53–88
4) Une nouvelle formule des caractères, *Bull. Sci. Math.* (2) **98** (1974), 163–172
5) A very simple proof of Bott's theorem, *Invent. math.* **33** (1976), 271–272
6) A Moebius-like formula in the Schubert calculus, *Invent. math.* **35** (1976), 317–319
7) Démonstration de la conjecture de Mumford (d'après W. Haboush), exp. 462 in: *Séminaire Bourbaki 1974/1975* (Lecture Notes in Math. **514**), Berlin etc. 1976 (Springer)

D. I. Deriziotis

1) The Brauer complex of a Chevalley group, *J. Algebra* **70** (1981), 261–269
2) The submodule structure of Weyl modules for groups of type A_1, *Commun. Algebra* **9** (1981), 247–265
3) A proof of a Carter-Cline theorem on A_1–Weyl modules, *Bull. Acad. Polon. Sci.* (Ser. Sci. Math.) **30** (1982), 485–491
4) The matrix of an A_1-Weyl module, *Bull. Acad. Polon. Sci.* (Ser. Sci. Math.) **32** (1984), 19–22

K. Diethelm-Nüssli

Zur Darstellungstheorie von endlich dimensionalen restringierten Lie-Algebren, Diss. ETH Zürich, Nr. 7431 (1984)

J. Dieudonné

Les algèbres de Lie simples associées aux groupes simple algébriques sur un corps de caractéristique $p > 0$, *Rend. Circ. Mat. Palermo* (2) **6** (1957), 198–206

R. Dipper

1) Vertices of irreducible representations of finite Chevalley groups in the describing characteristic, *Math. Z.* **175** (1980), 143–159
2) On irreducible modules over finite Chevalley groups and parabolic subgroups, *Commun. Algebra* **10** (1982), 1073–1088
3) On irreducible modules of twisted groups of Lie type, *J. Algebra* **81** (1983), 370–389

S. Donkin

1) Hopf complements and injective comodules for algebraic groups, *Proc. London Math. Soc.* (3) **40** (1980), 298–319
2) Rationally injective modules for semisimple algebraic groups as direct limits, *Bull. London Math. Soc.* **12** (1980), 99–102
3) On a question of Verma, *J. London Math. Soc.* (2) **21** (1980), 445–455
4) A filtration for rational modules, *Math. Z.* **177** (1981), 1–8
5) The blocks of a semisimple algebraic group, *J. Algebra* **67** (1980), 36–53
6) A note on the characters of the projective modules for the infinitesimal subgroups of a semisimple algebraic group, *Math. Scand.* **51** (1982), 142–150
7) On Ext^1 for semisimple groups and infinitesimal subgroups, *Math. Proc. Camb. Phil. Soc.* **92** (1982), 231–238

8) A note on decomposition numbers of reductive algebraic groups, *J. Algebra* **80** (1983), 226–234
9) *Rational Representations of Algebraic Groups* (Lecture Notes in Math. **1140**), Berlin etc. 1985 (Springer)
10) A note on decomposition numbers for general linear groups and symmetric groups, *Math. Proc. Camb. Phil. Soc.* **97** (1985), 57–62
11) Finite resolutions of modules for reductive algebraic groups, *J. Algebra* **101** (1986), 473–488
12) On Schur algebras and related algebras I, *J. Algebra* **104** (1986), 310–328
13) Skew modules for reductive groups, *J. Algebra* **113** (1988), 465–479
14) On Schur algebras and related algebras II, *J. Algebra* **111** (1987), 354–364
15) Good filtrations of rational modules for reductive groups, pp. 69–80 in [P8]
16) Invariants of unipotent radicals, *Math. Z.* **198** (1988), 117–125
17) On conjugating representations and adjoint representations of semisimple groups, *Invent. math.* **91** (1988), 137–145
18) The normality of closures of conjugacy classes of matrices, *Invent. math.* **101** (1990), 717–736
19) Representations of symplectic groups and the symplectic tableaux of R. C. King, *Linear and Multilinear Algebra* **29** (1991), 113–124
20) Infinitesimal invariants of algebraic groups, *J. London Math. Soc.* (2) **45** (1992), 481–490
21) Invariants of several matrices, *Invent. math.* **110** (1992), 389–401
22) Invariant functions on matrices, *Math. Proc. Cambridge Philos. Soc.* **113** (1993), 23–43
23) On tilting modules for algebraic groups, *Math. Z.* **212** (1993), 39–60
24) On tilting modules and invariants for algebraic groups, pp. 59–77 in [P16]
25) On Schur algebras and related algebras III: Integral representations, *Math. Proc. Cambridge Philos. Soc.* **116** (1994), 37–55
26) On Schur algebras and related algebras IV: The blocks of the Schur algebras, *J. Algebra* **168** (1994), 400–429
27) On projective modules for algebraic groups, *J. London Math. Soc.* (2) **54** (1996), 75–88
28) On free modules for finite subgroups of algebraic groups, *J. London Math. Soc.* (2) **55** (1997), 287–296
29) An introduction to the Lusztig conjecture, pp. 173–187 in: R. W. Carter, M. Geck (eds.): *Representations of Reductive Groups* (Publ. Newton Inst.), Cambridge 1998 (Cambridge Univ.)
30) On the existence of Auslander-Reiten sequences of group representations I, *Algebr. Represent. Theory* **1** (1998), 97–127
31) On the existence of Auslander-Reiten sequences of group representations II, *Algebr. Represent. Theory* **1** (1998), 215–253
32) On the existence of Auslander-Reiten sequences of group representations III, *Algebr. Represent. Theory* **1** (1998), 399–412
33) Symmetric and exterior powers, linear source modules and representations of Schur superalgebras, *Proc. London Math. Soc.* (3) **83** (2001), 647–680

S. Donkin, K. Erdmann

Tilting modules, symmetric functions, and the module structure of the free Lie algebra, *J. Algebra* **203** (1998), 69–90

S. Donkin, I. Reiten

On Schur algebras and related algebras V: Some quasi-hereditary algebras of finite type, *J. Pure Appl. Algebra* **97** (1994), 117–134

S. Doty

1) The submodule structure of certain Weyl modules for groups of type A_n, *J. Algebra* **95** (1985), 373–383
2) Submodules of symmetric powers of the natural module for GL_n, pp. 185–191 in [P10]
3) Composition factors of induced modules for algebraic groups and their Frobenius subgroups, pp. 163–170 in [P9]
4) Character formulas and Frobenius subgroups of algebraic groups, *J. Algebra* **125** (1989), 331–347
5) The strong linkage principle, *Amer. J. Math.* **111** (1989), 135–141
6) Filtrations of rational representations of reductive groups of semisimple rank 1, *Proc. Amer. Math. Soc.* **109** (1990), 9–22
7) The symmetric algebra and representations of general linear groups, pp. 123–150 in [P12]
8) Resolutions of B modules, *Indag. Math.* (N.S.) **5** (1994), 267–283
9) Polynomial representations, algebraic monoids, and Schur algebras of classical type, *J. Pure Appl. Algebra* **123** (1998), 165–199
10) Representation theory of reductive normal algebraic monoids, *Trans. Amer. Math. Soc.* **351** (1999), 2539–2551

S. Doty, K. Erdmann, S. Martin, D. Nakano

Representation type of Schur algebras, *Math. Z.* **232** (1999), 137–182

S. Doty, D. Nakano

1) Semisimple Schur algebras, *Math. Proc. Cambridge Philos. Soc.* **124** (1998), 15–20
2) Relating the cohomology of general linear groups and symmetric groups, pp. 175–187 in [P22]

S. Doty, D. Nakano, K. Peters

1) On infinitesimal Schur algebras, *Proc. London Math. Soc.* (3) **72** (1996), 588–612
2) Polynomial representations of Frobenius kernels of GL_2, pp. 57–67 in: S.-J. Kang et al. (eds.), *Lie Algebras and Their Representations*, Proc. Seoul 1995 (Contemp. Math. **194**), Providence, R. I. 1996 (Amer. Math. Soc.)
3) Infinitesimal Schur algebras of finite representation type, *Quart. J. Math. Oxford* (2) **48** (1997), 323–345

S. Doty, J. B. Sullivan

1) The submodule structure of Weyl modules for SL_3, *J. Algebra* **96** (1985), 78–93
2) On the structure of the higher cohomology modules of line bundles on G/B, *J. Algebra* **114** (1988), 286–332
3) On the geometry of extensions of irreducible modules for simple algebraic groups, *Pacific J. Math.* **130** (1987), 253–273
4) On the intertwinings between higher cohomology modules of line bundles on G/B, *J. Algebra* **115** (1988), 289–296
5) Filtration patterns for representations of algebraic groups and their Frobenius kernels, *Math. Z.* **195** (1987), 391–407

S. Doty, G. Walker

1) The composition factors of $\mathbf{F}_p[x_1, x_2, x_3]$ as a $GL(3,p)$-module, *J. Algebra* **147** (1992), 411–441
2) Modular symmetric functions and irreducible modular representations of general linear groups, *J. Pure Appl. Algebra* **82** (1992), 1–26
3) Truncated symmetric powers and modular representations of GL_n, *Math. Proc. Cambridge Philos. Soc.* **119** (1996), 231–242

M. Dowd, P. Sin

On representations of algebraic groups in characteristic two, *Commun. Algebra* **24** (1996), 2597–2686

J. Du

A note on quantized Weyl reciprocity at roots of unity, *Algebra Colloq.* **2** (1995), 363–372

J. Du, B. Parshall, L. Scott

Quantum Weyl reciprocity and tilting modules. *Comm. Math. Phys.* **195** (1998), 321–352

J. Du, L. Scott

Lusztig conjectures, old and new I, *J. reine angew. Math.* **455** (1994), 141–182

K. Erdmann

1) Schur algebras of finite type, *Quart. J. Math. Oxford* (2) **44** (1993), 17–41
2) Symmetric groups and quasi-hereditary algebras, pp. 123–161 in [P16]
3) Ext^1 for Weyl modules of $\mathrm{SL}_2(K)$, *Math. Z.* **218** (1995), 447–459
4) Tensor products and dimensions of simple modules for symmetric groups, *Manuscripta Math.* **88** (1995), 357–386
5) Decomposition numbers for symmetric groups and composition factors of Weyl modules, *J. Algebra* **180** (1996), 316–320
6) Representations of $\mathrm{GL}_n(K)$ and symmetric groups, pp. 67–84 in: R. Solomon (ed.), *Representation Theory of Finite Groups*, Proc. Columbus, Ohio 1995 (Ohio State Univ. Math. Res. Inst. Publ. **6**), Berlin 1997 (de Gruyter)

K. Erdmann, A. Henke

1) On Ringel duality for Schur algebras, *Math. Proc. Cambridge Philos. Soc.* **132** (2002), 96–115
2) On Schur algebras, Ringel duality and symmetric groups, *J. Pure Appl. Algebra* **169** (2002), 175–199

K. Erdmann, D. Nakano

1) Representation type of q-Schur algebras, *Trans. Amer. Math. Soc.* **353** (2001), 4729–4756
2) Representation type of Hecke algebras of type A, *Trans. Amer. Math. Soc.* **354** (2002), 275–285

R. Farnsteiner, G. Röhrle

Almost split sequences of Verma modules, *Math. Ann.* **322** (2002), 701–743

R. Farnsteiner, G. Röhrle, D. Voigt

Infinitesimal unipotent group schemes of complexity 1, *Colloq. Math.* **89** (2001), 179–192

R. Farnsteiner, D. Voigt

1) On cocommutative Hopf algebras of finite representation type, *Adv. in Math.* **155** (2000), 1–22

2) Modules of solvable infinitesimal groups and the structure of representation-finite cocommutative Hopf algebras, *Math. Proc. Cambridge Philos. Soc.* **127** (1999), 441–459

3) Schemes of tori and the structure of tame restricted Lie algebras, *J. London Math. Soc.* (2) **63** (2001), 553–570

4) Representations of infinitesimal groups of characteristic 2, *Arch. Math.* **78** (2002), 184–188

J. Feldvoss, H. Strade

Restricted Lie algebras with bounded cohomology and related classes of algebras, *Manuscripta Math.* **74** (1992), 47–67

W. R. Ferrer Santos

Cohomology of comodules, *Pacific J. Math.* **109** (1983), 179–213

G. Fischer

Darstellungstheorie des ersten Frobeniuskerns der SL_2, Diss. Bielefed 1982

P. Fleischmann

Periodic simple modules for $SU_3(q^2)$ in the describing characteristic $p \neq 2$, *Math. Z.* **198** (1988), 555–568

E. Formanek, C. Procesi

Mumford's conjecture for the general linear group, *Adv. in Math.* **19** (1976), 292–305

V. Franjou, E. Friedlander, A. Scorichenko, A. Suslin

General linear and functor cohomology over finite fields, *Ann. of Math.* (2) **150** (1999), 663–728

J. Franklin

Homomorphisms between Verma modules in characteristic p, *J. Algebra* **112** (1988), 58–85

E. Friedlander

1) A canonical filtration for certain rational modules, *Math. Z.* **188** (1985), 433–438

2) Geometry of infinitesimal group schemes, pp. 55–65 in: *Proceedings of the International Congress of Mathematicians* , *Vol. II*, Proc. Berlin 1998, Doc. Math. (Extra Vol. **II**) 1998

E. Friedlander, B. Parshall

1) On the cohomology of algebraic and related finite groups, *Invent. math.* **74** (1983), 85–117

2) Cohomology of Lie algebras and algebraic groups, *Amer. J. Math.* **108** (1986), 235–253

3) Limits of infinitesimal group cohomology, pp. 523–538 in: W. Browder (ed.), *Algebraic Topology and Algebraic K–Theory*, Proc. Princeton 1983 (Ann. of Math. Studies **113**), Princeton 1987 (Princeton Univ.)

4) Cohomology of infinitesimal and discrete groups, *Math. Ann.* **273** (1986), 353–374

5) Geometry of p–unipotent Lie algebras, *J. Algebra* **109** (1987), 25–45

6) Support varieties for restricted Lie algebras, *Invent. math.* **86** (1986), 553–562

E. Friedlander, A. Suslin

Cohomology of finite group schemes over a field, *Invent. math.* **127** (1997), 209–270

O. Gabber, A. Joseph
Towards the Kazhdan-Lusztig conjecture *Ann. scient. Éc. Norm. Sup.* (4) **14** (1981), 261–302

S. Gelfand, D. Kazhdan
Examples of tensor categories, *Invent. math.* **109** (1992), 595–617

G. Georgiev, O. Mathieu
1) Catégorie de fusion pour les groupes de Chevalley, *C. R. Acad. Sci. Paris* (I) **315** (1992), 659–662
2) Fusion rings for modular representations of Chevalley groups, pp. 89–100 in: P. Sally et al. (eds.), *Mathematical Aspects of Conformal and Topological Field Theories and Quantum Groups*, Proc. South Hadley 1992 (Contemp. Math. **175**) Providence, R. I. 1994 (Amer. Math. Soc.)

P. Gilkey, G. Seitz
Some representations of exceptional Lie algebras, *Geom. dedicata* **25** (1988), 407–416

V. Ginzburg, S. Kumar
Cohomology of quantum groups at roots of unity, *Duke Math. J.* **69** (1993), 179–198

D. J. Glover
A study of certain modular representations, *J. Algebra* **51** (1978), 425–475

J. Grabmeier
On the socle of Weyl modules, *Arch. Math.* **50** (1988), 319–322

J. A. Green
1) Locally finite representations, *J. Algebra* **41** (1976), 137–171
2) *Polynomial representations of* GL_n (Lecture Notes in Math. **830**), Berlin etc. 1980 (Springer)

W. L. Griffith
1) Cohomology of flag varieties in characteristic p, *Illinois J. Math.* **24** (1980), 452–461
2) The indecomposability of $H^*(G/B, L)$, *Colloquium Math.* **47** (1982), 59–63
3) $H^*(G/B, L)$ for G of semisimple rank 2, *Fund. Math.* **115** (1983), 33–41

F. D. Grosshans
The invariants of unipotent radicals of parabolic subgroups, *Invent. math.* **73** (1983), 1–9

B. Haastert
Über Differentialoperatoren und **D**-Moduln in positiver Charakteristik, *Manuscripta Math.* **58** (1987), 385–415

W. J. Haboush
1) Reductive groups are geometrically reductive: A proof of the Mumford conjecture, *Ann. of Math.* (2) **102** (1975), 67–83
2) Homogeneous vector bundles and reductive subgroups of reductive algebraic groups, *Amer. J. Math.* **100** (1977), 1123–1137
3) Central differential operators on split semi-simple groups over fields of positive characteristic, pp. 35–85 in: M.-P. Malliavin (ed.), *Séminaire d'Algèbre Paul Dubreil et Marie-Paule Malliavin*, Proc. Paris 1979 (Lecture Notes in Math. **795**), Berlin etc. 1980 (Springer)
4) A short proof of the Kempf vanishing theorem, *Invent. math.* **56** (1980), 109–112

W. Haboush, N. Lauritzen

Varieties of unseparated flags, pp. 35–57 in: R. S. Elman et al. (eds.), *Linear Algebraic Groups and Their Representations*, Proc. Los Angeles 1992 (Contemp. Math. **153**), Providence, R. I. 1993 (Amer. Math. Soc.)

M. Hall, Jr.

The Theory of Groups, 2nd edn., New York 1976 (Chelsea)

P. W. A. M. van Ham, T. A. Springer, M. van der Wel

On the Cartan invariants of $\mathrm{SL}_2(\mathbf{F}_q)$, *Commun. Algebra* **10** (1982), 1565–1588

H. C. Hansen

On cycles in flag manifolds, *Math. Scand.* **33** (1973), 269–274

A. Henke

1) Schur subalgebras and an application to the symmetric group, *J. Algebra* **233** (2000), 342–362
2) The Cartan matrix of the Schur algebra $S(2,r)$, *Arch. Math.* **76** (2001), 416–425

G. Hiß

Die adjungierten Darstellungen der Chevalley-Gruppen, *Arch. Math.* **42** (1984), 408–416

G. Hochschild

1) Cohomology of restricted Lie algebras, *Amer. Math. J.* **76** (1954), 555–580
2) Cohomology of algebraic linear groups, *Illinois J. Math.* **5** (1961), 492–519
3) Rationally injective modules for algebraic linear groups, *Proc. Amer. Math. Soc.* **14** (1963), 880–883

M. Hochster, J. L. Roberts

The purity of the Frobenius and local cohomology, *Adv. in Math.* **21** (1976), 117–172

G. M. D. Hogeweij

Almost-classical Lie algebras I, II, *Indag. Math.* **44** (1982), 441–452 and 453–460

J. Hu

1) A combinatorial approach to representations of quantum linear groups, *Commun. Algebra* **26** (1998), 2591–2621
2) Cohomology of quantum general linear groups, *J. Algebra* **213** (1999), 513–548

S. G. Hulsurkar

Proof of Verma's conjecture on Weyl's dimension polynomial, *Invent. math.* **27** (1974), 45–52

J. E. Humphreys

1) Modular representations of classical Lie algebras and semisimple groups, *J. Algebra* **19** (1971), 51–79
2) Defect groups for finite groups of Lie type, *Math. Z.* **119** (1971), 149–152
3) Projective modules for $SL(2,q)$, *J. Algebra* **25** (1973), 513–518
4) Some computations of Cartan invariants for finite groups of Lie type, *Commun. Pure Appl. Math.* **26** (1973), 745–755
5) Weyl groups, deformations of linkage classes, and character degrees for Chevalley groups, *Commun. Algebra* **1** (1974), 475–490
6) Representations of $\mathrm{SL}(2,p)$, *Amer. Math. Monthly* **82** (1975), 21–59
7) *Ordinary and modular representations of Chevalley groups* (Lecture Notes in Math. **528**), Berlin etc. 1976 (Springer)

8) On the hyperalgebra of a semisimple algebraic group, pp. 203–210 in: H. Bass et al. (eds.), *Contributions to Algebra*, a collection of papers dedicated to Ellis Kolchin, New York etc. 1977 (Academic)

9) Symmetry for finite dimensional Hopf algebras, *Proc. Amer. Math. Soc.* **68** (1978), 143–146

10) Weyl modules and Bott's theorem in characteristic p, pp. 474–483 in: W. Rossmann (ed.), *Lie Theories and Their Applications*, Proc. Kingston, Ont. 1977 (Queen's Papers in Pure and Applied Math. **48**), Kingston, Ont. 1978 (Queen's Univ.)

11) Modular representations of finite groups of Lie type, pp. 259–290 in [P2]

12) Cartan invariants and decomposition numbers of Chevalley groups, pp. 347–351 in [P3]

13) Deligne-Lusztig characters and principal indecomposable modules, *J. Algebra* **62** (1980), 299–303

14) Ordinary and modular characters of $SL(3,p)$, *J. Algebra* **72** (1981), 8–16

15) Restricted Lie algebras (and beyond), pp. 91–98 in: S. Amitsur et al. (eds.), *Algebraists Homage*, Proc. (Conf. in honor of N. Jacobson) New Haven, Conn. 1981 (Contemp. Math. **13**), Providence, R. I. 1982 (Amer. Math. Soc.)

16) Cohomology of G/B in characteristic p, *Adv. in Math.* **59** (1986), 170–183

17) On the structure of Weyl modules, *Commun. Algebra* **12** (1984), 2665–2677

18) Cartan invariants, *Bull. London Math. Soc.* **17** (1985), 1–14

19) Nonzero Ext^1 for Chevalley groups (via algebraic groups), *J. London Math. Soc.* (2) **31** (1985), 463–467

20) Induced modules for semisimple groups and Lie algebras, pp. 341–349 in: D. J. Britten et al. (eds.), *Lie Algebras and Related Topics*, Proc. Windsor, Ont. 1984 (Canadian Math. Soc. Conf. Proc. **5**), Providence, R. I. 1986 (Amer. Math. Soc.)

21) Cohomology of line bundles on G/B for the exceptional group G_2, *J. Pure Appl. Algebra* **44** (1987), 227–239

22) Projective modules for $Sp(4,p)$ in characteristic p, *J. Algebra* **104** (1986), 80–88

23) The Steinberg representation, *Bull. Amer. Math. Soc.* (N. S.) **16** (1987), 247–263

24) Cohomology of line bundles on G/B for the exceptional group G_2, *J. Pure Appl. Algebra* **44** (1987), 227–239

25) Generic Cartan invariants for Frobenius kernels and Chevalley groups, *J. Algebra* **122** (1989), 345–352

26) Cohomology of line bundles on flag varieties in prime characteristic, pp. 193–204 in [P12]

27) Extremal composition factors for groups of Lie type, pp. 303–310 in [P13]

28) Comparing modular representations of semisimple groups and their Lie algebras, pp. 69–80 in: V. Chari, I. B. Penkov (eds.), *Modular Interfaces*, Proc. Riverside, Cal. 1995 (AMS/IP Stud. Adv. Math. **4**), Providence, R. I. / Cambridge, Mass. 1997 (Amer. Math. Soc. / Intl. Press)

J. E. Humphreys, J. C. Jantzen

Blocks and indecomposable modules for semisimple algebraic groups, *J. Algebra* **54** (1978), 494–503

J. E. Humphreys, D.-n. Verma

Projective modules for finite Chevalley groups, *Bull. Amer. Math. Soc.* **79** (1973), 467–468

S. P. Inamdar

A note on Frobenius splitting of Schubert varieties and linear syzygies, *Amer. J. Math.* **116** (1994), 1587–1590

S. P. Inamdar, V. B. Mehta

Frobenius splitting of Schubert varieties and linear syzygies, *Amer. J. Math.* **116** (1994), 1569–1586

R. S. Irving

1) The structure of certain highest weight modules for SL_3, *J. Algebra* **99** (1986), 438–457
2) The socle filtration of a Verma module, *Ann. scient. Éc. Norm. Sup.* (4) **21** (1988), 47–65

B. Iversen

The geometry of algebraic groups, *Adv. in Math.* **20** (1976), 57–85

G. D. James

1) The decomposition of tensors over fields of prime characteristic, *Math. Z.* **172** (1980), 161–178
2) Trivial source modules for symmetric groups, *Arch. Math.* **41** (1983), 294–300
3) The decomposition matrices of $\mathrm{GL}_n(q)$ for $n \leq 10$, *Proc. London Math. Soc.* (3) **60** (1990), 225–265
4) Symmetric groups and Schur algebras, pp. 91–102 in [P21]

G. D. James, G. E. Murphy

The determinant of the Gram matrix for a Specht module, *J. Algebra* **59** (1979), 222–235

I. Janiszczak

Irreducible periodic modules over $\mathrm{SL}(m, q)$ in the describing characteristic, *Commun. Algebra* **15** (1987), 1375–1391

J. C. Jantzen

1) Darstellungen halbeinfacher algebraischer Gruppen und zugeordnete kontravariante Formen, *Bonner math. Schr.* **67** (1973)
2) Zur Charakterformel gewisser Darstellungen halbeinfacher Gruppen und Lie–Algebren, *Math. Z.* **140** (1974), 127–149
3) Darstellungen halbeinfacher Gruppen und kontravariante Formen, *J. reine angew. Math.* **290** (1977), 117–141
4) Über das Dekompositionsverhalten gewisser modularer Darstellungen halbeinfacher Gruppen und ihrer Lie–Algebren, *J. Algebra* **49** (1977), 441–469
5) Weyl modules for groups of Lie type, pp. 291–300 in [P2]
6) Über Darstellungen höherer Frobenius-Kerne halbeinfacher algebraischer Gruppen, *Math. Z.* **164** (1979), 271–292
7) Darstellungen halbeinfacher Gruppen und ihrer Frobenius–Kerne, *J. reine angew. Math.* **317** (1980), 157–199
8) Zur Reduktion modulo p der Charaktere von Deligne und Lusztig, *J. Algebra* **70** (1981), 452–474
9) Zur Reduktion modulo p unipotenter Charaktere endlicher Chevalley–Gruppen, *Math. Z.* **181** (1982), 97–128

10) Filtrierungen der Darstellungen in der Hauptserie endlicher Chevalley–Gruppen, *Proc. London Math. Soc.* (3) **49** (1984), 445–482
11) Modular representations of reductive groups, pp. 118–154 in [P7]
12) Kohomologie von p–Lie–Algebren und nilpotente Elemente, *Abh. Math. Sem. Univ. Hamburg* **56** (1986), 191–219
13) Restricted Lie algebra cohomology, pp. 91–108 in: A. M. Cohen et al. (eds.), *Algebraic Groups, Utrecht 1986*, Proc. (Lecture Notes in Math. **1271**), Berlin etc. 1987 (Springer)
14) Support varieties of Weyl modules, *Bull. London Math. Soc.* **19** (1987), 238–244
15) First cohomology groups for classical Lie algebras, pp. 289–315 in: G. O. Michler, C. M. Ringel (eds.), *Representation Theory of Finite Groups and Finite–Dimensional Algebras*, Proc. Bielefeld 1991 (Progress in Math. **95**), Basel etc. 1991 (Birkhäuser)
16) Low dimensional representations of reductive groups are semisimple, pp. 255–266 in [P19]

J. C. Jantzen, B. Haastert
1) Filtrations of the discrete series of $SL_2(q)$ via crystalline cohomology, *J. Algebra* **132** (1990), 77–103
2) Filtrations of symmetric powers via crystalline cohomology, *Geom. dedicata* **37** (1991), 45–63

J. C. Jantzen, G. Seitz
On the representation theory of the symmetric groups, *Proc. London Math. Soc.* (3) **65** (1992), 475–504

A. V. Jeyakumar
1) Principal indecomposable representations for the group $SL(2,p)$, *J. Algebra* **30** (1974), 444–458
2) Periodic modules for the group $\mathrm{SL}(2,q)$, *Commun. Algebra* **8** (1980), 1721–1735

W. Jones
An algorithm for generating dominant weights, *Commun. Algebra* **5** (1977), 759–771

W. Jones, B. Parshall
On the 1-cohomology of finite groups of Lie type, pp. 313–328 in [P5]

V. G. Kac
О неприводимых представлениях алгебр Ли классического типа (Irreducible representations of Lie algebras of classical type), Успехи Мат. Наук **27:5(167)** (1972), 237–238

V. G. Kac, B. Weisfeiler
Coadjoint action of a semi-simple algebraic group and the center of the enveloping algebra in characteristic p, *Indag. Math.* **38** (1976), 136–151

W. L. J. van der Kallen
1) *Infinitesimally Central Extensions of Chevalley Groups* (Lecture Notes in Math. **356**), Berlin etc. 1973 (Springer)
2) Infinitesimally central extensions of Spin_7 in characteristic 2, *J. Pure Appl. Algebra* **14** (1979), 39–49
3) Longest weight vectors and excellent filtrations, *Math. Z.* **201** (1989), 19–31

4) Infinitesimal fixed points in modules with good filtration, *Math. Z.* **212** (1993), 157–159

5) *Lectures on Frobenius splittings and B-modules*, Notes by S. P. Inamdar (Tata Institute of Fundamental Research Lectures **84**), Berlin 1993 (Springer)

6) Steinberg modules and Donkin pairs, *Transform. Groups* **6** (2001), 87–98

M. Kaneda

1) A truncation formula for certain principal indecomposable modules of finite Chevalley groups, *Commun. Algebra* **14** (1986), 911–922

2) On the symmetry of inverse Kazhdan-Lusztig polynomials for affine Weyl groups, pp. 201–205 in [P9]

3) On the alcove identification operator of semisimple algebraic groups, *Commun. Algebra* **15** (1987), 1157–1171

4) On the inverse Kazhdan-Lusztig polynomials for affine Weyl groups, *J. reine angew. Math.* **381** (1987), 116–135

5) Extensions of modules for infinitesimal algebraic groups, *J. Algebra* **122** (1989), 188–210

6) A note on the Grothendieck-Cousin complex on the flag variety in positive characteristic, *Nihonkai Math. J.* **1** (1990), 229–251

7) On the Frobenius morphism of flag schemes, *Pacific J. Math.* **163** (1994), 315–336

8) On a theorem of O. Mathieu, *Nihonkai Math. J.* **5** (1994), 149–186

9) The Frobenius morphism of Schubert schemes, *J. Algebra* **174** (1995), 473–488

10) A survey of H. H. Andersen, J. C. Jantzen and W. Soergel, "Representations of quantum groups at a pth root of unity and of semisimple groups in characteristic p: independence of p", pp. 80–119 in: *Lusztig Program*, Proc. Kyoto 1996 (Sūrikaisekikenkyūsho Kōkyūroku **954**), Kyoto 1996 (Kyoto Univ.)

11) Some generalities on $\mathcal{D}$-modules in positive characteristic, *Pacific J. Math.* **183** (1998), 103–141

12) Based modules and good filtrations in algebraic groups, *Hiroshima Math. J.* **28** (1998), 337–344

13) Cohomology of infinitesimal quantum algebras, *J. Algebra* **226** (2000), 250–282

M. Kaneda, N. Shimada, M. Tezuka, N. Yagita

Cohomology of infinitesimal algebraic groups, *Math. Z.* **205** (1990), 61–95

M. Kashiwara, N. Lauritzen

Local cohomology and $\mathcal{D}$–affinity in positive characteristic, *C. R. Acad. Sci. Paris* (I) **335** (2002), 993–996

M. Kashiwara, T. Tanisaki

1) Kazhdan-Lusztig conjecture for affine Lie algebras with negative level, *Duke Math. J.* **77** (1995), 21–62

2) On Kazhdan-Lusztig conjectures, *Sugaku Expositions* **11** (1998), 177–195

3) Kazhdan-Lusztig conjecture for affine Lie algebras with negative level II. Nonintegral case, *Duke Math. J.* **84** (1996), 771–813

4) Characters of irreducible modules with non-critical highest weights for affine Lie algebras, pp. 275–296 in [P23]

5) Parabolic Kazhdan-Lusztig polynomials and Schubert varieties, *J. Algebra* **249** (2002), 306–325

S. Kato

1) Spherical functions and a q–analogue of Kostant's weight multiplicity formula, *Invent. math.* **66** (1982), 461–468
2) On the Kazhdan-Lusztig polynomials for affine Weyl groups, *Adv. in Math.* **55** (1982), 103–130
3) Hecke algebras and quantum general linear groups, *J. Math. Kyoto Univ.* **37** (1997), 241–249

D. Kazhdan, G. Lusztig

1) Representations of Coxeter groups and Hecke algebras, *Invent. math.* **53** (1979), 165–184
2) Schubert varieties and Poincaré duality, pp. 185–203 in: R. Osserman, A. Weinstein (eds.), *Geometry of the Laplace Operator,* Proc. Honolulu 1979 (Proc. Symp. Pure Math. **36**), Providence, R. I. 1980 (Amer. Math. Soc.)
3) Tensor structures arising from affine Lie algebras I, *J. Amer. Math. Soc.* **6** (1993), 905–947
4) Tensor structures arising from affine Lie algebras II, *J. Amer. Math. Soc.* **6** (1993), 949–1011
5) Tensor structures arising from affine Lie algebras III, *J. Amer. Math. Soc.* **7** (1994), 335–381
6) Tensor structures arising from affine Lie algebras IV, *J. Amer. Math. Soc.* **7** (1994), 383–453

D. Kazhdan, M. Verbitsky

Cohomology of restricted quantized universal enveloping algebras, pp. 107–115 in [P15]

G. R. Kempf

1) Linear systems on homogeneous spaces, *Ann. of Math.* (2) **103** (1976), 557–591
2) The geometry of homogeneous space versus induced representations of reductive groups, pp. 1–5 in: J. I. Igusa (ed.), *Algebraic Geometry, The Johns Hopkins Centenary Lectures*, Proc. 1976, Baltimore, Md., 1977 (Johns Hopkins Univ. Press)
3) The Grothendieck–Cousin complex of an induced representation, *Adv. in Math.* **29** (1978), 310–396
4) Algebraic representations of reductive groups, pp. 575–577 in [P4]
5) Representations of algebraic groups in prime characteristic, *Ann. scient. Éc. Norm. Sup.* (4) **14** (1981), 61–76
6) Tensor products of representations of the general linear group, *Amer. J. Math.* **109** (1987), 395–400
7) Tensor products of representations, *Amer. J. Math.* **109** (1987), 401–415
8) A decomposition formula for representations, *Nagoya Math. J.* **107** (1987), 63–68
9) A remark on integral representations of $\mathrm{GL}_{\mathbf{Z}}(n)$, *J. Algebra* **115** (1988), 340–341

G. R. Kempf, A. Ramanathan

Multicones over Schubert varieties, *Invent. math.* **87** (1987), 353–363

W. J. Kohler

The submodule structure of induced modules for infinitesimal groups of type B_2, Ph.D. thesis, Clark Univ. 1986

M. Koppinen

1) On the composition factors of Weyl modules, *Math. Scand.* **51** (1982), 212–216
2) On the translation functors for a semisimple algebraic group, *Math. Scand.* **51** (1982), 217–226
3) Good bimodule filtrations for coordinate rings, *J. London Math. Soc.* (2) **30** (1984), 244–250
4) Decomposing and lifting hyperalgebras, *Ann. Univ. Turku* (A.I) **184** (1983)
5) Computation of simple characters of a Chevalley group, *BIT* **26** (1986), 333–338
6) Homomorphisms between neighbouring Weyl modules, *J. Algebra* **103** (1986), 302–319
7) On stable decomposition patterns of Weyl modules, *Commun. Algebra* **15** (1987), 1649–1666
8) A note on Cabanes' decompositions for rational modules, *J. Algebra* **106** (1987), 510–516

M. Koppinen, T. Neuvonen

An imprimitivity theorem for Hopf algebras, *Math. Scand.* **41** (1977), 193–198

S. Koshitani

The Loewy structure of the projective indecomposable modules for $SL(3,3)$ and its automorphism group in characteristic 3, *Commun. Algebra* **15** (1987), 1215–1253

B. Kostant

1) Lie algebra cohomology and the generalized Borel-Weil theorem, *Ann. of Math.* (2) **74** (1961), 329–387
2) Lie group representations on polynomial rings, *Amer. Math. J.* **85** (1963), 327–404
3) Groups over **Z**, pp. 90–98 in: A. Borel, G. D. Mostow (eds.), *Algebraic Groups and Their Discontinuous Subgroups*, Proc. Boulder, Col. 1965 (Proc. Symp. Pure Math. **9**), Providence, R. I. 1966 (Amer. Math. Soc.)

F. M. Kouwenhoven

1) Schur and Weyl functors, *Adv. in Math.* **90** (1991), 77–113
2) Schur and Weyl functors II, *Commun. Algebra* **18** (1990), 2885–2941

20) Infinitesimal invariants of algebraic groups, B. Krcmar

Irreduzible rationale Darstellungen endlicher Chevalley-Gruppen, Diss. Tübingen 1979

P. Krasoń, N. J. Kuhn

On embedding polynomial functors in symmetric powers, *J. Algebra* **163** (1994), 281–294

N. J. Kuhn

1) The modular Hecke algebra and Steinberg representations of finite Chevalley groups (with an appendix by P. Landrock), *J. Algebra* **91** (1984), 125–141

20) Infinitesimal invariants of algebraic groups,

2) Rational cohomology and cohomological stability in generic representation theory, *Amer. J. Math.* **120** (1998), 1317–1341
3) The generic representation theory of finite fields: a survey of basic structure, pp. 193–212 in: H. Krause, C. M. Ringel (eds.), *Infinite Length Modules*, Proc. Bielefeld 1998 (Trends Math.), Basel 2000 (Birkhäuser)

4) A stratification of generic representation theory and generalized Schur algebras, *K-Theory* **26** (2002), 15–49

K. Kühne-Hausmann

Zur Untermodulstruktur der Weylmoduln für Sl_3, *Bonner math. Schr.* **162** (1985)

U. Kulkarni

Skew Weyl modules for GL_n and degree reduction for Schur algebras, *J. Algebra* **224** (2000), 248–262

S. Kumar

1) Proof of the Parthasarathy-Ranga Rao-Varadarajan conjecture, *Invent. math.* **93** (1988), 117–130
2) A refinement of the PRV conjecture, *Invent. math.* **97** (1989), 305–311
3) The nil Hecke ring and singularity of Schubert varieties, pp. 497–507 in: J.-L. Brylinski et al. (eds.), *Lie Theory and Geometry*, In Honor of Bertram Kostant (Progress in Math. **123**), Boston etc. 1994 (Birkhäuser)
4) The nil Hecke ring and singularity of Schubert varieties, *Invent. math.* **123** (1996), 471–506

S. Kumar, N. Lauritzen, J. F. Thomsen

Frobenius splitting of cotangent bundles of flag varieties, *Invent. math.* **136** (1999), 603–621

S. Kumar, G. Letzter

Shapovalov determinant for restricted and quantized restricted enveloping algebras, *Pacific J. Math.* **179** (1997), 123–161

S. Kumar, P. Littelmann

1) Frobenius splitting in characteristic zero and the quantum Frobenius map, *J. Pure Appl. Algebra* **152** (2000), 201–216
2) Algebraization of Frobenius splitting via quantum groups, *Ann. of Math.* (2) **155** (2002), 491–551

J. Lahtonen

1) On the submodules and composition factors of certain induced modules for groups of type C_n, *J. Algebra* **140** (1991), 415–425
2) Lower bounds for the multiplicities of simple composition factors of the induced modules, *Commun. Algebra* **20** (1992), 1873–1895
3) The reduced polynomial algebra as a module for $\mathrm{SO}_{2n}(k)$, *J. Algebra* **155** (1993), 344–368

V. Lakshmibai

1) Kempf varieties, *J. Indian Math. Soc.* (N.S.) **40** (1976), 299–349
2) Standard monomial theory for G_2, *J. Algebra* **98** (1986), 281–318
3) Singular loci of Schubert varieties for classical groups, *Bull. Amer. Math. Soc.* (N.S.) **16** (1987), 83–90
4) Bases pour les représentations fondamentales des groupes classiques I, *C. R. Acad. Sc. Paris* (I) **302** (1986), 387–390
5) Bases pour les représentations fondamentales des groupes classiques II, *C. R. Acad. Sc. Paris* (I) **302** (1986), 419–422
6) Schubert varieties and standard monomial theory, pp. 365–378 in: S. Balcerzyk et al. (eds.), *Topics in Algebra*, Part 2, Proc. Warsaw 1988 (Banach Center Publ. **26 : 2**), Warsaw 1990 (PWN)

V. Lakshmibai, P. Littelmann, P. Magyar

1) Standard monomial theory and applications, pp. 319–364 in: A. Broer (ed.), *Representation Theories and Algebraic Geometry*, Proc. Montreal 1997 (NATO Adv. Sci. Inst. **C 514**), Dordrecht etc. 1998 (Kluwer)
2) Standard monomial theory for Bott-Samelson varieties, *Compositio Math.* **130** (2002), 293–318

V. Lakshmibai, P. Magyar

1) Standard monomial theory for Bott-Samelson varieties, *C. R. Acad. Sci. Paris* (I) **324** (1997), 1211–1215
2) Standard monomial theory for Bott-Samelson varieties of $\mathrm{GL}(n)$, *Publ. Res. Inst. Math. Sci.* **34** (1998), 229–248

V. Lakshmibai, V. B. Mehta, A. J. Parameswaran

Frobenius splittings and blow-ups, *J. Algebra* **208** (1998), 101–128

V. Lakshmibai, C. Musili, C. S. Seshadri

1) Cohomology of line bundles on G/B, *Ann. scient. Éc. Norm. Sup.* (4) **7** (1974), 89–137
2) Geometry of G/P III: Standard monomial theory for a quasi-minuscule P, *Proc. Indian Acad. Sci.* (A) **88** (1979), 93–177
3) Geometry of G/P IV: Standard monomial theory for classical types, *Proc. Indian Acad. Sci.* (A) **88** (1979), 279–362
4) Geometry of G/P, *Bull. Amer. Math. Soc.* (N.S.) **1** (1979), 432–435

V. Lakshmibai, K. N. Rajeswari

1) Bases pour les représentations fondamentales des groupes exceptionels $\mathbf{E}_6$ and $\mathbf{F}_4$, *C. R. Acad. Sc. Paris* (I) **302** (1986), 575–577
2) Towards a standard monomial theory for exceptional groups, pp. 449–578 in [P10]

V. Lakshmibai, C. S. Seshadri

1) Geometry of G/P II: The work of De Concini and Procesi and the basic conjectures, *Proc. Indian Acad. Sci.* (A) **87** (1978), 1–54
2) Singular locus of a Schubert variety. *Bull. Amer. Math. Soc.* (N.S.) **11** (1984), 363–366
3) Geometry of G/P V, *J. Algebra* **100** (1986), 462–557
4) Standard monomial theory, pp. 279–322 in [P12]

M. Larsen

On the semisimplicity of low-dimensional representations of semisimple groups in characteristic p, *J. Algebra* **173** (1995), 219–236

N. Lauritzen

1) Splitting properties of complete homogeneous spaces, *J. Algebra* **162** (1993), 178–193
2) Schubert cycles, differential forms and $\mathcal{D}$-modules on varieties of unseparated flags, *Compositio Math.* **109** (1997), 1–12

N. Lauritzen, V. Mehta

Differential operators and the Frobenius pull back of the tangent bundle on G/P, *Manuscripta Math.* **98** (1999), 507–510

N. Lauritzen, J. F. Thomsen

Frobenius splitting and hyperplane sections of flag manifolds, *Invent. math.* **128** (1997), 437–442

P. Littelmann

1) Good filtrations, decomposition rules and standard monomial theory, pp. 89–106 in: A. M. Cohen (ed.), *Computational Aspects of Lie Group Representations and Related Topics*, Proc. Amsterdam 1990 (CWI Tract **84**), Amsterdam 1991 (Math. Centrum, Centrum Wisk. Inform.)
2) Good filtrations and decomposition rules for representations with standard monomial theory, *J. reine angew. Math.* **433** (1992), 161–180

J. Liu, J. Ye

Extensions of simple modules for the algebraic group of type G_2, *Commun. Algebra* **21** (1993), 1909–1946

G. Lusztig

1) Divisibility of projective modules of finite Chevalley groups by the Steinberg module, *Bull. London Math. Soc.* **8** (1976), 130–134
2) Hecke algebras and Jantzen's generic decomposition patterns, *Adv. in Math.* **37** (1980), 121–164
3) Some problems in the representation theory of finite Chevalley groups, pp. 313–317 in [P3]
4) Singularities, character formulas, and a q–analogue of weight multiplicities, pp. 208–229 in: *Analyse et topologie sur les espaces singuliers II–III*, Proc. Luminy 1981 (Astérisque **101–102**), Paris 1983 (Soc. Math. France)
5) Modular representations and quantum groups, pp. 59–77 in [P11]
6) On quantum groups, *J. Algebra* **131** (1990), 466–475
7) Finite dimensional Hopf algebras arising from quantized universal enveloping algebras, *J. Amer. Math. Soc.* **3** (1990), 257–296
8) Quantum groups at roots of 1, *Geom. Dedicata* **35** (1990), 89–114
9) Monodromic systems on affine flag manifolds, *Proc. Roy. Soc. London* (A) **445** (1994), 231–246 and **450** (1995), 731–732

A. R. Magid

On the imprimitivity theorem for algebraic groups, *Rocky Mountain J. Math.* **14** (1984), 655–660

R. Marlin

Cohomologie de de Rham des variétés de drapeaux, *Bull. Soc. math. France* **105** (1977), 89–96

S. Martin

Schur Algebras and Representation Theory (Cambridge Tracts in Math. **112**), Cambridge 1993 (Cambridge Univ.)

R. P. Martineau

On 2-modular representations of the Suzuki groups, *Amer. J. Math.* **94** (1972), 55–72

O. Mathieu

1) Filtrations of B-modules. *Duke Math. J.* **59** (1989), 421–442
2) Frobenius action on the B-cohomology, pp. 39–51 in: V. G. Kac (ed.), *Infinite-dimensional Lie Algebras and Groups*, Proc. Luminy-Marseille 1988 (Adv. Ser. Math. Phys. **7**), Teaneck, N. J. 1989 (World Scientific)
3) Filtrations of G-modules, *Ann. scient. Éc. Norm. Sup.* (4) **23** (1990), 625–644
4) On the dimension of some modular irreducible representations of the symmetric group, *Lett. Math. Phys.* **38** (1996), 23–32

5) Tilting modules and their applications, pp. 145–212 in: T. Kobayashi et al. (eds.), *Analysis on Homogeneous Spaces and Representation Theory of Lie Groups*, Proc. Okayama–Kyoto 1997 (Adv. Stud. Pure Math. **26**), Tokyo 2000 (Math. Soc. Japan)

O. Mathieu, G. Papadopoulo

1) A combinatorial character formula for some highest weight modules, *Compositio Math.* **117** (1999), 153–159
2) A character formula for a family of simple modular representations of GL_n, *Comment. Math. Helv.* **74** (1999), 280–296

G. McNinch

1) Dimensional criteria for semisimplicity of representations, *Proc. London Math. Soc.* (3) **76** (1998), 95–149
2) Semisimplicity in positive characteristic, pp. 43–52 in [P21]
3) Semisimplicity of exterior powers of semisimple representations of groups, *J. Algebra* **225** (2000), 646–666
4) Filtrations and positive characteristic Howe duality, *Math. Z.* **235** (2000), 651–685

V. B. Mehta, W. van der Kallen

1) A simultaneous Frobenius splitting for closures of conjugacy classes of nilpotent matrices, *Compositio Math.* **84** (1992), 211–221
2) On a Grauert-Riemenschneider vanishing theorem for Frobenius split varieties in characteristic p, *Invent. math.* **108** (1992), 11–13

V. B. Mehta, T. R. Ramadas

Frobenius splitting and invariant theory, *Transform. Groups* **2** (1997), 183–195

V. B. Mehta, A. Ramanathan

1) Frobenius splitting and cohomology vanishing for Schubert varieties, *Ann. of Math.* (2) **122** (1985), 27–40
2) Schubert varieties in $G/B \times G/B$, *Compositio Math.* **67** (1988), 355–358

V. B. Mehta, V. Srinivas

1) Normality of Schubert varieties. *Amer. J. Math.* **109** (1987), 987–989
2) A note on Schubert varieties in G/B, *Math. Ann.* **284** (1989), 1–5
3) A characterization of rational singularities, *Asian J. Math.* **1** (1997), 249–271

V. B. Mehta, V. Trivedi

1) Variety of complexes and F–splitting, *J. Algebra* **215** (1999), 352–365
2) The variety of circular complexes and F-splitting. *Invent. math.* **137** (1999), 449–460

V. B. Mehta, T. N. Venkataramana

A note on Steinberg modules and Frobenius splitting, *Invent. math.* **123** (1996), 467–469

A. A. Mil′ner

Irreducible representations of modular Lie algebras, *Math. USSR Izvestiya* **9** (1975), 1169–1187, translated from: Неприводимые представелия модулярных алгебр Ли, Изв. АН СССР (Серия матем.) **39** (1975), 1240–1259

C. Musili,

1) Postulation formula for Schubert varieties, *J. Indian Math. Soc.* (N.S.) **36** (1972), 143–171
2) Some properties of Schubert varieties, *J. Indian Math. Soc.* (N.S.) **38** (1974), 131–145

C. Musili, C. S. Seshadri

1) Standard monomial theory, pp. 441–476 in: M.-P. Malliavin (ed.), *Séminaire d'Algèbre Paul Dubreil et Marie-Paule Malliavin*, Proc. Paris 1980 (Lecture Notes in Math. **867**), Berlin etc. 1981 (Springer)
2) Schubert varieties and the variety of complexes, pp. 329–359 in: M. Artin, J. Tate (eds.), *Arithmetic and Geometry, Papers Dedicated to I. R. Shafarevich, Vol. II : Geometry*, (Progress in Math. **36**), Boston etc. 1983 (Birkhäuser)
3) Applications of standard monomial theory, pp. 381–406 in [P12]

D. Nakano

1) Varieties for G_rT-modules, pp. 441–452 in [P20]
2) Some recent developments in the representation theory of general linear and symmetric groups, pp. 357–373 in [P23]

D. Nakano, B. J. Parshall, D. C. Vella

Support varieties for algebraic groups, *J. reine angew. Math.* **547** (2002), 15–49

T. Neuvonen

1) On the structure of produced and induced indecomposable Lie modules, *Ann. Acad. Sci. Fenn.* (AI) **1** (1975), 199–206
2) A criterion for complete reducibility of restricted Lie modules, *Arch. Math.* **28** (1977), 149–156
3) A complete reducibility criterion with an application to representations of semisimple groups, *J. Algebra* **60** (1979), 282–288

H. Niemi

On the construction of the irreducible representations of the hyperalgebra of a universal Chevalley group, *Ann. Acad. Sci. Fenn.* (AI) **5** (1980), 17–25

P. N. Norton

0-Hecke algebras, *J. Austral. Math. Soc.* (A) **27** (1979), 337–357

U. Oberst

Affine Quotientenschemata nach affinen, algebraischen Gruppen und induzierte Darstellungen, *J. Algebra* **44** (1977), 504–538

U. Oberst, H.-J. Schneider

Über Untergruppen endlicher algebraischer Gruppen, *Manuscripta Math.* **8** (1973), 217–241

J. O'Halloran

1) Weyl modules and the cohomology of Chevalley groups, *Amer. J. Math.* **103** (1981), 399–410
2) Representation theory of an extension of a Chevalley group by a vector group, *J. Algebra* **72** (1981), 335–341
3) A vanishing theorem for the cohomology of Borel subgroups, *Commun. Algebra* **11** (1983), 1603–1606
4) One-cohomology of infinitesimal subgroups of a Chevalley group, preprint
5) Cohomology of a Borel subgroup of a Chevalley group, preprint

V. Ostrik

1) Tensor ideals in the category of tilting modules, *Transform. Groups* **2** (1997), 279–287
2) Support varieties for quantum groups, *Funct. Anal. Appl.* **32** (1998/99), 237–246, translated from: Функц. анализ и его прил. **32:4** (1998), 22–34
3) Dimensions of quantized tilting modules, *Mosc. Math. J.* **1** (2001), 65–71

4) Cohomology of subregular tilting modules for small quantum groups, *Compositio Math.* **132** (2002), 283–287

V. V. Panyukov

Representations of Lie algebras in positive characteristic, *Moscow Univ. Math. Bull.* **38:2** (1983), 64–70, translated from: О представлениях алгебр Ли в положительной характеристике, Вестн. Москов. Ун-та. (Матем. Механ.) **1983:2**, 53–58

J. Paradowski

Filtrations of modules over the quantum algebra, pp. 93–108 in [P14]

A. E. Parker

1) The global dimension of Schur algebras for GL_2 and GL_3, *J. Algebra* **241** (2001), 340–378

2) On the good filtration dimension of Weyl modules for a linear algebraic group, preprint

B. Parshall

1) Modular representations of algebraic groups, pp. 101–134 in: J. Carrell et al., *Topics in the Theory of Algebraic Groups* (Notre Dame Math. Lectures **10**), South Bend, Ind. etc. 1982 (Univ. of Notre Dame)

2) Simulating algebraic geometry with algebra II, Stratifying representation categories, pp. 263–269 in [P9]

3) The Ext algebra of a highest weight category, pp. 213–222 in [P16]

4) Koszul algebras and duality, pp. 277–285 in [P17]

B. Parshall, L. Scott

1) An imprimitivity theorem for algebraic groups, *Indag. Math.* **42** (1980), 39–47

2) Koszul algebras and the Frobenius automorphism, *Quart. J. Math. Oxford* (2) **46** (1995), 345–384

B. Parshall, J. P. Wang

1) Quantum linear groups, *Mem. Amer. Math. Soc.* **89:439** (1991), 1–157

2) Cohomology of infinitesimal quantum groups I, *Tôhoku Math. J.* (2) **44** (1992), 395–423

3) Cohomology of quantum groups: the quantum dimension, *Canad. J. Math.* **45** (1993), 1276–1298

W. Pfautsch

1) *Die Köcher der Frobeniuskerne in der* SL_2, Diss. Bielefeld 1983

2) Ein Satz über Blöcke von Frobeniuskernen einer halbeinfachen algebraischen Gruppe, manuscript 1983

3) Ext^1 for the Frobenius kernels of SL_2, *Commun. Algebra* **13** (1985), 169–179

W. Pfautsch, D. Voigt

The representation-finite algebraic groups of dimension zero, *C. R. Acad. Sci. Paris* (I) **306** (1988), 685–689

R. Pfetzing

Der Darstellungstyp der Frobeniuskerne in der SL_3, Diss. Bielefeld 1983

C. Pillen

Tensor products of modules with restricted highest weight, *Commun. Algebra* **21** (1993), 3647–3661

H. Pollatsek

First cohomology groups of some linear groups over fields of characteristic two, *Illinois J. Math.* **15** (1971), 393–417

P. Polo

1) Un critère d'existence d'une filtration de Schubert, *C. R. Acad. Sci. Paris* (I) **307** (1988), 791–794
2) Variétés de Schubert et excellentes filtrations, pp. 281–311 in: M. Andler (ed.), *Orbites unipotentes et représentations III*, Proc. Paris/Luminy 1987 (Astérisque **173-174**), Paris 1989 (Soc. Math. France)
3) Modules associés aux variétés de Schubert, *C. R. Acad. Sci. Paris* (I) **308** (1989), 123–126
4) Modules associés aux variétés de Schubert, pp. 155–171 in: S. Ramanan, A. Beauville (eds.), *Proceedings of the Indo-French Conference on Geometry*, Proc. Bombay 1989, Delhi 1993 (Hindustan Book Agency)
5) On Zariski tangent spaces of Schubert varieties, and a proof of a conjecture of Deodhar, *Indag. Math.* (N.S.) **5** (1994), 483–493
6) On Cohen-Macaulay posets, Koszul algebras and certain modules associated to Schubert varieties, *Bull. London Math. Soc.* **27** (1995), 425–434

A. A. Premet

Weights of infinitesimally irreducible representations of Chevalley groups over a field of prime characteristic, *Math. USSR Sbornik* **61** (1988), 167–183, translated from: Веса инфинитезимально неприводимых представлений групп Шевалле над полем простой характеристики, Матем. Сб. АН СССР (Нов. Сер.) **133/175** (1987), 167–183

A. A. Premet, I. D. Suprunenko

1) The Weyl modules and the irreducible representations of the symplectic group with the fundamental highest weights, *Commun. Algebra* **11** (1983), 1309–1342
2) Quadratic modules for Chevalley groups over fields of odd characteristics, *Math. Nachr.* **110** (1983), 65–96

C. Procesi

1) Les bases de Hodge dans la théorie des invariants, pp. 128–144 in: M.-P. Malliavin (ed.), *Séminaire d'Algèbre Paul Dubreil*, Proc. Paris 1976-1977 (Lecture Notes in Math. **641**), Berlin etc. 1978 (Springer)
2) Young diagrams, standard monomials and invariant theory, pp. 537–542 in [P4]

K. N. Raghavan, P. Sankaran

A new approach to standard monomial theory for classical groups, *Transform. Groups* **3** (1998), 57–73

S. Ramanan, A. Ramanathan

Projective normality of flag varieties and Schubert varieties, *Invent. math.* **79** (1985), 217–224

A. Ramanathan

1) Schubert varieties are arithmetically Cohen-Macaulay, *Invent. math.* **80** (1985), 283–294
2) Equations defining Schubert varieties and Frobenius splitting of diagonals, *Publ. Math. Inst. Hautes Études Sci.* **65** (1987), 61–90
3) Frobenius splitting and Schubert varieties, pp. 497–508 in [P12]

T. E. Rasmussen

Second cell tilting modules, Ph.D. thesis, Aarhus Univ. 2002

R. W. Richardson

The conjugating representation of a semisimple group, *Invent. math.* **54** (1979), 229–245

F. Richen

1) Modular representations of split BN pairs, *Trans. Amer. Math. Soc.* **140** (1969) 435–460

2) Blocks of defect zero of split (B, N) pairs, *J. Algebra* **21** (1972), 275–279

C. M. Ringel

The category of modules with good filtrations over a quasi-hereditary algebra has almost split sequences, *Math. Z.* **208** (1991), 209–223

M. A. Ronan

Duality for presheaves on chamber systems, and a related chain complex, *J. Algebra* **121** (1989), 263–274

M. A. Ronan, S. D. Smith

1) Sheaves on buildings and modular representations of Chevalley groups, *J. Algebra* **96** (1985), 319–346

2) On computation of sheaf homology on finite geometries, preprint 1984

3) Universal presheaves on group geometries and modular representations, *J. Algebra* **102** (1986), 135–154

J. E. Roos

Locally Noetherian categories and generalized strictly linearly compact rings, Applications, pp. 197–277 in: *Category Theory, Homology Theory and their Applications II*, Proc. Seattle 1968 (Lecture Notes in Math. **92**), Berlin 1969 (Springer)

A. N. Rudakov

1) On representations of classical semisimple Lie algebras of characteristic p, *Math. USSR Izvestiya* **4** (1970), 741–749, translated from: О представлении классических алгебр Ли в характеристике p, Изв. АН СССР (Серия матем.) **34** (1970), 735–743

2) Dimensions of certain irreducible representations of semisimple Lie algebras of classical type over fields of finite characteristic (russ.), Труды Сем. Петровск. **3** (1978), 147–160

3) A condition for the complete reducibility of representations of a Lie algebra of finite characteristic (russ.), pp. 77-82 in: A. I. Kostrikin (ed.), *Algebra, Collection*, Moskva 1980 (Izdat. Mosk. Univ.)

4) Reducible p-representations of a simple three-dimensional Lie p-algebra, *Moscow Univ. Math. Bull.* **37:6** (1982), 51–56, translated from: Приводимые p–представления простой трехмерной p–алгебры Ли, Вестн. Москов. Ун-та. (Матем. Механ.) **1982:6**, 45–49

A. N. Rudakov, I. R. Shafarevich

Irreducible representations of a simple three-dimensional Lie algebra over a field of finite characteristic, *Math. Notes* **2** (1967), 760–767, translated from: Неприводимые представления простой трехмерной алгебры Ли над полем конечной характеристики, Матем. Заметки **2** (1967), 439–454

H. Sawada

1) A characterization of the modular representations of finite groups with split (B, N)–pairs, *Math. Z.* **155** (1977), 29–41

2) Some infinite-dimensional indecomposable modules of Chevalley groups, *Arch. Math.* **36** (1981), 414–422

K. D. Schaper

Charakterformeln für Weyl-Moduln und Specht-Moduln in Primcharakteristik, Diplomarbeit, Bonn 1981

M.-Th. Schmidt

Beziehungen zwischen Homologie-Darstellungen und der Hauptserie endlicher Chevalley-Gruppen, *Bonner math. Schr.* **171** 1986

L. Scott

1) Representations in characteristic p, pp. 319–331 in [P3]
2) Simulating algebraic geometry with algebra I, The algebraic theory of derived categories, pp. 271–281 in [P9]
3) Derived categories and algebraic groups, pp. 127–142 in [P11]
4) Quasihereditary algebras and Kazhdan-Lusztig theory, pp. 293–308 in [P16]
5) Linear and nonlinear group actions, and the Newton Institute program, pp. 1–23 in [P21]

J.-P. Serre

Groupes de Grothendieck des schémas en groupes réductifs déployés, *Publ. Math. Inst. Hautes Études Sci.* **34** (1968), 37–52

C. S. Seshadri

1) Geometric reductivity over arbitrary base, *Adv. in Math.* **26** (1975), 225–274
2) Geometry of G/P I: Theory of standard monomials for minuscule representations, pp. 207–239 in: K. G. Ramanathan (ed.), *C. P. Ramanujam — A Tribute* (Tata Inst. Fund. Res. Studies in Math. **8**), Berlin etc. 1978 (Springer)
3) Standard monomial theory and the work of Demazure, pp. 355–384 in: S. Iitaka (ed.), *Algebraic Varieties and Analytic Varieties*, Proc. Tokyo 1981 (Advanced Studies in Pure Math. **1**), Tokyo 1983 (North-Holland)
4) Line bundles on Schubert varieties, pp. 499–528 in: M. Atiyah et al., *Vector Bundles on Algebraic Varieties* (Tata Inst. Fund. Res. Studies in Math. **11**), Proc. Bombay 1984, Bombay 1987 (Tata Inst.)
5) Standard monomial theory and geometry of Schubert varieties, *Proc. Indian Nat. Sci. Acad.* (A) **52** (1986), 435–441
6) The work of P. Littelmann and standard monomial theory, pp. 178–197 in: S. D. Adhikari (ed.), *Current Trends in Mathematics and Physics*, a tribute to Harish-Chandra, New Delhi 1995 (Narosa)

S. D. Smith

1) Irreducible modules and parabolic subgroups, *J. Algebra* **75** (1982), 286–289
2) Spin modules in characteristic 2, *J. Algebra* **77** (1982), 392–401
3) Sheaf homology and complete reducibility, *J. Algebra* **95** (1985), 72–80

W. Soergel

1) Roots of unity and positive characteristic, pp. 315–338 in [P17]
2) Gradings on representation categories, pp. 800–806 in [P18]
3) Conjectures de Lusztig, pp. 75–85 [exp. 793] in: *Séminaire Bourbaki 1994/95* (Astérisque **237**), Paris 1996 (Soc. Math. France)
4) Kazhdan-Lusztig polynomials and a combinatoric[s] for tilting modules, *Represent. Theory* **1** (1997), 83–114
5) Character formulas for tilting modules over quantum groups at roots of one, pp. 161–172 in: R. Bott et al. (eds.), *Current Developments in Mathematics 1997*, Proc. Cambridge, Mass., Boston 1999 (Int. Press)

6) On the relation between intersection cohomology and representation theory in positive characteristic, *J. Pure Appl. Algebra* **152** (2000), 311–335

T. A. Springer

Weyl's character formula for algebraic groups, *Invent. math.* **5** (1968), 85–105

B. Srinivasan

On the modular characters of the special linear group $SL(2, p^n)$, *Proc. London Math. Soc.* (3) **14** (1964), 101–114

R. Steinberg

1) Prime power representations of finite linear groups II, *Canad. J. Math.* **9** (1957), 347–351
2) Representations of algebraic groups, *Nagoya Math. J.* **22** (1963), 33–56

J. B. Sullivan

1) Some representation theory for the modular general linear groups, *J. Algebra* **45** (1977), 516–535
2) Representations of the hyperalgebra of an algebraic group, *Amer. J. Math.* **100** (1978), 643–652
3) Relations between the cohomology of an algebraic group and its infinitesimal subgroups, *Amer. J. Math.* **100** (1978), 995–1014
4) Simply connected groups, the hyperalgebra, and Verma's conjecture, *Amer. J. Math.* **100** (1978), 1015–1019
5) Lie algebra cohomology at irreducible modules, *Illinois J. Math.* **23** (1979), 363–373
6) The second Lie algebra cohomology group and Weyl modules, *Pacific J. Math.* **86** (1980), 321–326
7) Frobenius operations on Hochschild cohomology, *Amer. J. Math.* **102** (1980), 765–780
8) On the second socle level of induced modules for algebraic groups, *J. Algebra* **125** (1989), 400–407
9) The Euler character and cancellation theorems for Weyl modules, *Pacific J. Math.* **135** (1988), 189–208

I. D. Suprunenko

1) Сохранение систем весов неприводимых представлений алгебраической группы и алгебры Ли типа A_l с ограниченными старшими весами при редукции по модулю p (Preservation of systems of weights of irreducible representations of an algebraic group and a Lie algebra of type A_l with bounded higher weights in reduction modulo p), Весці АН БССР (Сер. фіз.-мат. навук) **1983:2**, 18–22
2) Об ограничениях фундаментальных представлений группы $A_n(K)$ на связные алгебраические подгруппы (On the restriction of the fundamental representations of the groups $A_n(K)$ to connected algebraic subgroups), preprint, Minsk, July 1983
3) Об ограничениях неприводимых представлений группы $SL(m, K)$ на $SO(m, K)$ (Constraints of irreducible representations of the group SL(m, K) on SO(m, K)), Весці АН БССР (Сер. фіз.-мат. навук) **1985:6**, 41–46
4) Условия неприводимости ограничений неприводимых представлений группы $SL(n, K)$ на связные алгебраические подгруппы (Conditions for the irreducibility of the restrictions of irreducible representations of the group

SL(n, K) to connected algebraic subgroups), Докл. АН БССР **30** (1986), 204–207

5) Restrictions of large irreducible representations of the classical groups to naturally embedded small subgroups cannot be semisimple, *Commun. Algebra* **29** (2001), 3747–3757

A. Suslin, E. Friedlander, C. Bendel

1) Infinitesimal 1-parameter subgroups and cohomology, *J. Amer. Math. Soc.* **10** (1997), 693–728
2) Support varieties for infinitesimal group schemes, *J. Amer. Math. Soc.* **10** (1997), 729–759

M. Takeuchi

A hyperalgebraic proof of the isomorphism and isogeny theorems for reductive groups, *J. Algebra* **85** (1983), 179–196

L. Thams

1) Two classical results in the quantum mixed case, *J. reine angew. Math.* **436** (1993), 129–153
2) The blocks of a quantum algebra, *Commun. Algebra* **22** (1994), 1617–1628

N. B. Tinberg

1) Some indecomposable modules of groups with split $(B,\, N)$-pairs, *J. Algebra* **61** (1979), 508–526
2) Some indecomposable modules of groups with split $(B,\, N)$-pairs, pp. 363–366 in [P3]
3) Modular representations of finite groups with unsaturated split $(B,\, N)$-pairs, *Canad. J. Math.* **32** (1980), 714–733

B. Totaro

Projective resolutions of representations of GL(n), *J. reine angew. Math.* **482** (1997), 1–13

J. Towber

1) Two new functors from modules to algebras, *J. Algebra* **47** (1977), 80–104
2) Young symmetry, the flag manifold, and representations of $GL(n)$, *J. Algebra* **61** (1979), 414–462

B. S. Upadhyaya

1) Composition factors of the principal indecomposable modules for the special linear group SL$(2, q)$, *J. London Math. Soc.* (2) **17** (1978), 437–445
2) Filtrations with Weyl module quotients of the principal indecomposable modules for the group SL$(2, q)$, *Commun. Algebra* **7** (1979), 1469–1488

B. Yu. Vejsfejler, V. G. Kats

Irreducible representations of Lie p–algebras, *Funct. Anal. Appl.* **5** (1971), 111–117, translated from: О неприводимых представлениях p–алгебр Ли, Функц. анализ и его прил. **5:2** (1971), 28–36

F. Veldkamp

1) Representations of algebraic groups of type F_4 in characteristic 2, *J. Algebra* **16** (1970), 326–339
2) The center of the universal enveloping algebra of a Lie algebra in characteristic p, *Ann. scient. Éc. Norm. Sup.* (4) **5** (1972), 217–240

D. C. Vella

1) A cohomological characterization of parabolic subgroups of reductive algebraic groups, *J. Algebra* **121** (1989), 281–300

2) Another characterization of parabolic subgroups, *J. Algebra* **137** (1991), 214–232

3) A character formula for B-modules, *Commun. Algebra* **20** (1992), 665–679

D.-n. Verma

Rôle of affine Weyl groups in the representation theory of algebraic Chevalley groups and their Lie algebras, pp. 653–705 in: I. M. Gel'fand (ed.), *Lie Groups and Their Representations*, Proc. Budapest 1971, London 1975 (A. Hilger)

D. Voigt

1) Endliche Hopfalgebren, *Math. Z.* **134** (1973), 189–203

2) *Induzierte Darstellungen in der Theorie der endlichen, algebraischen Gruppen* (Lecture Notes in Math. **592**), Berlin etc. 1977 (Springer)

3) The algebraic infinitesimal groups of tame representation type, *C. R. Acad. Sci. Paris* (I) **311** (1990), 757–760

J. Wang

1) Sheaf cohomology on G/B and tensor products of Weyl modules, *J. Algebra* **77** (1982), 162–185

2) Coinduced representations and injective modules for hyperalgebra b_r, *Chin. Ann. of Math.* (B) **4** (1983), 357–364

3) Quasi-rational modules and generic cohomology, *Northeastern Math. J.* **1** (1985), 90–100

4) Inverse limits of affine group schemes, *Chin. Ann. of Math.* (B) **9** (1988), 418–428

5) On the cyclicity and cocyclicity of G–modules, pp.231–233 in [P11]

6) Notes on some topics in the representation theory of linear algebraic groups, *Commun. Algebra* **18** (1990), 347–355

P. W. Winter

On the modular representation theory of the two-dimensional special linear group over an algebraically closed field, *J. London Math. Soc.* (2) **16** (1977), 237–252

W. J. Wong

1) Representations of Chevalley groups in characteristic p, *Nagoya Math. J.* **45** (1971), 39–78

2) Irreducible modular representations of finite Chevalley groups, *J. Algebra* **20** (1972), 355–367

3) Very strong linkage for cohomology groups of line bundles on G/B, *J. Algebra* **113** (1988), 71–80

4) Weyl modules for p–singular weights, *J. Algebra* **114** (1988), 357–368

5) A filtration of Weyl modules for large weights, *J. Austral. Math. Soc.* (A) **45** (1988), 187–194

N. Xi

Some irreducible modules of $\mathrm{Sp}(2n)$, pp. 403–409 in [P23]

B. Xu, J. Ye

Irreducible characters for algebraic groups in characteristic two I, *Algebra Colloq.* **4** (1997), 281–290

J. Ye

1) Filtrations of principal indecomposable modules of Frobenius kernels of reductive groups, *Math. Z.* **189** (1985), 515–527

2) On the first Cartan invariant of the groups SL$(3, p^n)$ and SU$(3, p^n)$, pp. 388–400 in [P7]
3) Extensions of simple modules for the group Sp$(4, K)$, *J. London Math. Soc.* (2) **41** (1990), 51–62
4) Extensions of simple modules for the group Sp$(4, K)$ II, *Chinese Sci. Bull.* **35** (1990), 450–454
5) Some results on irreducible characters for algebraic groups in characteristic two, pp. 411–420 in [P23]

J. Ye, Z. Zhou
1) Irreducible characters for algebraic groups in characteristic two II, *Algebra Colloq.* **6** (1999), 401–411
2) Irreducible characters for algebraic groups in characteristic two III, *Commun. Algebra* **28** (2000), 4227–4247
3) Irreducible characters for algebraic groups in characteristic three, *Commun. Algebra* **29** (2001), 201–223
4) Irreducible characters for algebraic groups in characteristic three II, *Commun. Algebra* **30** (2002), 273–306

A. E. Zalesskiĭ, I. D. Suprunenko
1) Представления размерности $(p^n \mp 1)/2$ симплектической группы степени $2n$ над полем характеристики p (Representations of dimension $(p^n \pm 1)/2$ of the symplectic group of degree $2n$ over a field of characteristic p), Весці АН БССР (Сер. фіз.-мат. навук) **1987:6**, 9–15
2) Truncated symmetric powers of natural realizations of the groups $SL_m(P)$ and $Sp_m(P)$ and their constraints on subgroups, *Siberian Math. J.* **31** (1990), 555–566, translated from: Срезанные симметрические степени естественных реализаций групп $SL_m(P)$ и $Sp_m(P)$ и их ограничения на подгруппы, Сиб. Мат. Журн. **31:4** (1990), 33–46

List of Notations

Part I

$\mathrm{Mor}(X, X')$	set of morphisms between two k–functors X and X', 1.2
D_X	diagonal subfunctor of $X \times X$, 1.2
$\mathbf{A}^n$	affine n–space, 1.3
$Sp_k R$	spectrum of the k–algebra R, 1.3
$k[X]$	$\mathrm{Mor}(X, \mathbf{A}^1)$ for a k–functor X, 1.3
$V(I)$	closed subfunctor defined by $I \subset k[X]$, 1.4
$D(I)$	open subfunctor defined by $I \subset k[X]$, 1.5
X_f	open subfunctor defined by $f \in k[X]$, 1.5
$\mathbf{P}^n$	projective n–space, 1.3
$\mathfrak{Mor}(X, Y)$	k–functor of morphisms between two k–functors X and Y, 1.15
$\mathrm{Hom}(G, H)$	set of homomorphisms between two k–group functors G and H, 2.1
$\mathrm{Aut}(G)$	group of automorphisms of a k–group functor G, 2.1
G_a	additive group, 2.2
M_a	additive group of a k–module M, 2.2
G_m	multiplicative group, 2.2
$GL(M)$	general linear group of a k–module M, 2.2
GL_n	$= GL(k^n)$, 2.2
$SL(M)$	special linear group of a k–module M, 2.2
SL_n	$= SL(k^n)$, 2.2
$\mu_{(n)}$	nth roots of unity, 2.2
m_G	multiplication morphism $G \times G \to G$, $(g, h) \mapsto gh$, 2.3
i_G	morphism $G \to G$, $g \mapsto g^{-1}$, 2.3
Δ_G	comultiplication on $k[G]$, i.e., comorphism of m_G, 2.3
σ_G	antipode on $k[G]$, i.e., comorphism of i_G, 2.3
ε_G	augmentation on $k[G]$, i.e., $k[G] \to k$, $f \mapsto f(1)$, 2.3
$X(G)$	group of characters of a k–group functor G, 2.4
$\mathrm{Diag}(\Lambda)$	diagonalisable group scheme associated to a commutative group Λ, 2.5
X^G	fixed point functor, 2.6
$\mathrm{Stab}_G(Y)$	stabiliser of a subfunctor Y, 2.6
$N_G(Y)$	normaliser of a subgroup functor Y, 2.6
$C_G(Y)$	centraliser of a subgroup functor Y, 2.6
$Z(G)$	centre of G, 2.6
$H \rtimes G$	semi-direct product of G and H such that H is normal in $H \rtimes G$, 2.6
k_λ	k regarded as a G–nodule via $\lambda \in X(G)$, 2.7
$\mathrm{Hom}_G(M, M')$	space of homomorphisms between two G–modules M and M', 2.7
ρ_l	left regular representation, 2.7
ρ_r	right regular representation, 2.7
Δ_M	comodule map for a G–module M, 2.8

M^G	fixed points submodule, 2.10
M_λ	weight space of weight λ, 2.10
$(e(\lambda) \mid \lambda \in \Lambda)$	canonical basis of $\mathbf{Z}[\Lambda]$, 2.11
$\mathrm{ch}(M)$	formal character of M, 2.11
$Z_G(S)$	centraliser of a subset S of a G–module, 2.12
$\mathrm{Stab}_G(N)$	stabiliser of a k–submodule N of a G–module, 2.12
$\mathrm{soc}_G M$	socle of a G–module M, 2.14
$(\mathrm{soc}_G M)_E$	isotypic component of $\mathrm{soc}_G M$ of type E, 2.14
$\mathrm{rad}_G M$	radical of a G–module M, 2.14
$[M : E]_G$	multiplicity of a simple G–module E as a composition factor of a G–module M, 2.14
${}^\alpha M$	the G–module M twisted by $\alpha \in \mathrm{Aut}(G)$, 2.15
${}^h M$	the G–module M twisted by $\mathrm{Int}(h)$, 2.15
$\mathrm{res}_H^G M$	the G–module M restricted to H, 3.1
$\mathrm{ind}_H^G M$	the G–module induced by the H–module M, 3.3
ε_M	canonical map $\mathrm{ind}_H^G M \to M$, 3.4
Q_E	injective hull of a simple G–module E, 3.17
$H^n(G, M)$	nth (rational) cohomology group of a G–module M, 4.2
$\mathrm{Ext}_G^n(M, M')$	nth Ext-group of two G–modules M and M', 4.2
$R^n \mathrm{ind}_H^G$	nth derived functor of ind_H^G, 4.2
$C^n(G, M)$	nth term of the Hochschild complex of M, 4.14
$f(X) = \mathrm{im}(f)$	image faisceau of a morphism $f : X \to Y$, 5.5
X/G	quotient faisceau of X by G, 5.5
$\mathcal{O}_X$	sheaf of regular functions on X, 5.8
$\mathcal{L}_{X/G}(M)$	sheaf associated to a G–module M, 5.8
$X \times^G Y$	bundle associated to a k–faisceau Y with G–action, 5.14
G/N	factor group of G by N, 6.1
NH	product subgroup of two subgroup faisceaux with H normalising N, 6.2
I_x	$\{ f \in k[X] \mid f(x) = 0 \}$ for any $x \in X(k)$, 7.1
$T_x X$	tangent space to X at x, 7.1
$\mathrm{Dist}(X, x)$	module of distributions on X with support in x, 7.1
$\mathcal{O}_{X,x}$	local ring of x, 7.1
$\mathfrak{m}_x$	maximal ideal of $\mathcal{O}_{X,x}$, 7.1
$(d\varphi)_x$	tangent map at x of a morphism φ, 7.2
δ_X	diagonal morphism $X \to X \times X$, 7.4
$\mathrm{Dist}(G)$	algebra of distributions on G with support in 1, 7.7
$\mathrm{Lie}(G)$	Lie algebra of G, 7.7
$d\alpha$	tangent map of a homomorphism of group schemes, 7.9
$U(\mathfrak{g})$	enveloping algebra of a Lie algebra $\mathfrak{g}$, 7.10
$U^{[p]}(\mathfrak{g})$	restricted enveloping algebra of a p–Lie algebra $\mathfrak{g}$, 7.10
Ad	adjoint action of G on $\mathrm{Dist}(G)$ or on $\mathrm{Lie}(G)$, 7.18
$M(G)$	algebra of all measures on G, 8.4
δ_G	modular function on G, 8.8
$\mathrm{coind}_H^G M$	G–module coinduced by an H–module M, 8.14
$A^{(m)}$	a k–algebra A twisted m times by the Frobenius endomorphism, 9.2
$X^{(r)}$	a k–functor X twisted r times by the Frobenius endomorphism, 9.2
F_X^r	the rth Frobenius morphism $X \to X^{(r)}$, 9.2
G_r	the rth Frobenius kernel of G, 9.4

$H^\bullet(\mathfrak{g}, M)$	Lie algebra cohomology of a $\mathfrak{g}$–module M, 9.17

Part II

$G_{\mathbf{Z}}$	a split and connected reductive $\mathbf{Z}$-group, 1.1
G	$= (G_{\mathbf{Z}})_k$, 1.1
$T_{\mathbf{Z}}$	a split maximal torus of $G_{\mathbf{Z}}$, 1.1
T	$= (T_{\mathbf{Z}})_k$, 1.1
R	root system of G with respect to T, 1.1
x_α	root homomorphism corresponding to α, 1.2
U_α	root subgroup corresponding to α, 1.2
$Y(T)$	$= \mathrm{Hom}(G_m, T)$, 1.3
$\alpha^\vee$	coroot corresponding to α, 1.3
G_α	Levi subgroup corresponding to α, 1.3
s_α	reflection with respect to α, 1.4
W	Weyl group of R, 1.4
$\dot{w}$	representative in $N_G(T)(k)$ for $w \in W$, 1.4
R^+	positive system in R, 1.5
S	set of simple roots with respect to R^+, 1.5
$\leq$	order relation on $X(T) \otimes_{\mathbf{Z}} \mathbf{R}$ determined by R^+, 1.5
$l(w)$	length of $w \in W$ with respect to the system $\{s_\alpha \mid \alpha \in S\}$ of generators of W, 1.5
w_0	longest element in W, 1.5
ρ	half sum of all positive roots, 1.5
$w \bullet \lambda$	$= w(\lambda + \rho) - \rho$, 1.5
ϖ_α	fundamental weight corresponding to $\alpha \in S$, 1.6
$U(R')$	subgroup generated by all U_α with $\alpha \in R'$, 1.7
$G(R')$	subgroup generated by all G_α with $\alpha \in R'$, 1.7
R_I	$= \mathbf{Z}I \cap R$ for $I \subset S$, 1.7
L_I	$= G(R_I)$, 1.7
W_I	$= \langle s_\alpha \mid \alpha \in I \rangle$, 1.7
U^+	$= U(R^+)$, 1.8
U	$= U(-R^+)$, 1.8
B^+	$= U^+T$, 1.8
B	$= UT$, 1.8
U_I^+	$= U(R^+ \setminus R_I)$, 1.8
U_I	$= U((-R^+) \setminus R_I)$, 1.8
P_I^+	$= U_I^+ L_I$, 1.8
P_I	$= U_I L_I$, 1.8
X_α	basis of $(\mathrm{Lie}\, G_{\mathbf{Z}})_\alpha$, 1.11
H_α	$= (d\alpha^\vee)(1) \in \mathrm{Lie}\, T_{\mathbf{Z}}$, 1.11
$X_{\alpha,n}$	$= X_\alpha^n/(n!) \otimes 1 \in \mathrm{Dist}(U_\alpha)$, 1.12
$H^i(M)$	$= R^i \operatorname{ind}_H^G(M)$, 2.1
$H^i(\lambda)$	$= H^i(k_\lambda)$ for $\lambda \in X(T)$, 2.1
$L(\lambda)$	simple G–module with highest weight λ, 2.4
$X(T)_+$	set of dominant weights in $X(T)$, 2.6
$V(\lambda)$	Weyl module with highest weight λ, 2.13
$Z_r(\lambda)$	$= \operatorname{coind}_{B_r^+}^{G_r} \lambda$, 3.7
$Z_r'(\lambda)$	$= \operatorname{ind}_{B_r}^{G_r} \lambda$, 3.7

$L_r(\lambda)$	simple G_r–module with "highest weight" λ, 3.9
$X_r(T)$	$= \{\lambda \in X(T) \mid 0 \leq \langle \lambda, \alpha^\vee \rangle < p^r$ for all $\alpha \in S\}$, 3.15
$M^{[r]}$	a G–module twisted by the rth power of the Frobenius endomorphism of G, 3.16
St_r	rth Steinberg module, 3.18
$P(\alpha)$	$= P_{\{\alpha\}}$ for $\alpha \in S$, 5.1
$\overline{C}_{\mathbf{Z}}$	$= \{\lambda \in X(T) \mid 0 \leq \langle \lambda + \rho, \beta^\vee \rangle \leq p$ for all $\beta \in R^+\}$ where $p = \infty$ if $\operatorname{char}(k) = 0$, and $p = \operatorname{char}(k)$ otherwise, 5.5
$\chi(M)$	$= \sum_{i \geq 0} (-1)^i \operatorname{ch} H^i(M)$ for a B–module M, 5.7
$\chi(\lambda)$	$= \chi(k_\lambda)$ for $\lambda \in X(T)$, 5.7
$H^i_I(\lambda)$	the analogue to $H^i(\lambda)$ for L_I, 5.21
$L_I(\lambda)$	the analogue to $L(\lambda)$ for L_I, 5.21
$s_{\beta,r}$	affine reflection $\lambda \mapsto s_\beta(\lambda) + r\beta$ for $r \in \mathbf{Z}$, $\beta \in R$, 6.1
W_p	affine Weyl group generated by all $s_{\beta,rp}$, 6.1
$\widehat{F}$	upper closure of a facet F, 6.2
C	$= \{\lambda \in X(T) \otimes_{\mathbf{Z}} \mathbf{R} \mid 0 < \langle \lambda + \rho, \beta^\vee \rangle < p$ for all $\beta \in R^+\}$, 6.2
h	Coxeter number of R, 6.2
s_F	reflection with respect to a wall F, 6.3
$\Sigma(C')$	set of all s_F with F a wall of C' (for an alcove C'), 6.3
$W^0_p(\lambda)$	stabiliser of $\lambda \in X(T)$ in W_p, 6.3
$\Sigma^0(\lambda, C')$	$= \{s \in \Sigma(C') \mid s \bullet \lambda = \lambda\}$, 6.3
$\uparrow$	order relation on $X(T)$ or on the set of alcoves, 6.4/5
$W^0_p(F)$	stabiliser of a facet F in W_p, 6.11
$\mathcal{B}(H)$	set of blocks of H, 7.1
$\operatorname{pr}_\lambda$	projection functor for $\lambda \in X(T)$, 7.3
T^μ_λ	translation functor for $\lambda, \mu \in \overline{C}_{\mathbf{Z}}$, 7.6
$V(\lambda)_A$	A–form of $V(\lambda)$, 8.3
$H^i_A(M)$	$= R^i \operatorname{ind}^{G_A}_{B_A}(M)$ for a B_A–module M, 8.6
W^+_p	if $p > h$ equal to $\{w \in W_p \mid w \bullet 0 \in X(T)_+\}$, 8.22
$\widehat{Z}'_r(\lambda)$	$= \operatorname{ind}^{G_rB}_B \lambda$ for $\lambda \in X(T)$, 9.1
$\widehat{Z}_r(\lambda)$	$= \operatorname{coind}^{G_rB^+}_{B^+} \lambda$ for $\lambda \in X(T)$, 9.1
$\widehat{L}_r(\lambda)$	simple G_rB–module with highest weight λ, 9.6
$\widehat{Q}_r(\lambda)$	injective hull of $\widehat{L}_r(\lambda)$ as a G_rT–module, 11.3
$Q_r(\lambda)$	injective hull of the G_r–module $L_r(\lambda)$, 11.3
w_I	longest element in W_I for $I \subset S$, 13.2
$X(w)$	Schubert scheme corresponding to $w \in W$, 13.3
$\leq$	Bruhat(-Chevalley) order on W, 13.7
$X(w)_P$	image of $X(w)$ in G/P, 13.8
$\mathcal{C}(\pi)$	truncated category associated to $\pi \subset X(T)_+$, A.1
O_π	truncation functor to $\mathcal{C}(\pi)$, A.1
$S_G(\pi)$	generalised Schur algebra associated to π, A.16
$T(\lambda)$	indecomposable tilting module with highest weight λ, E.4

Index

Titles in This Series

107 **Jens Carsten Jantzen,** Representations of algebraic groups, 2003

106 **Hiroyuki Yoshida,** Absolute CM-periods, 2003

105 **Charalambos D. Aliprantis and Owen Burkinshaw,** Locally solid Riesz spaces with applications to economics, second edition, 2003

104 **Graham Everest, Alf van der Poorten, Igor Shparlinski, and Thomas Ward,** Recurrence sequences, 2003

103 **Octav Cornea, Gregory Lupton, John Oprea, and Daniel Tanré,** Lusternik-Schnirelmann category, 2003

102 **Linda Rass and John Radcliffe,** Spatial deterministic epidemics, 2003

101 **Eli Glasner,** Ergodic theory via joinings, 2003

99 **Philip S. Hirschhorn,** Model categories and their localizations, 2003

98 **Victor Guillemin, Viktor Ginzburg, and Yael Karshon,** Moment maps, cobordisms, and Hamiltonian group actions, 2002

97 **V. A. Vassiliev,** Applied Picard-Lefschetz theory, 2002

96 **Martin Markl, Steve Shnider, and Jim Stasheff,** Operads in algebra, topology and physics, 2002

95 **Seiichi Kamada,** Braid and knot theory in dimension four, 2002

94 **Mara D. Neusel and Larry Smith,** Invariant theory of finite groups, 2002

93 **Nikolai K. Nikolski,** Operators, functions, and systems: An easy reading. Volume 2: Model operators and systems, 2002

92 **Nikolai K. Nikolski,** Operators, functions, and systems: An easy reading. Volume 1: Hardy, Hankel, and Toeplitz, 2002

91 **Richard Montgomery,** A tour of subriemannian geometries, their geodesics and applications, 2002

90 **Christian Gérard and Izabella Łaba,** Multiparticle quantum scattering in constant magnetic fields, 2002

89 **Michel Ledoux,** The concentration of measure phenomenon, 2001

88 **Edward Frenkel and David Ben-Zvi,** Vertex algebras and algebraic curves, 2001

87 **Bruno Poizat,** Stable groups, 2001

86 **Stanley N. Burris,** Number theoretic density and logical limit laws, 2001

85 **V. A. Kozlov, V. G. Maz′ya, and J. Rossmann,** Spectral problems associated with corner singularities of solutions to elliptic equations, 2001

84 **László Fuchs and Luigi Salce,** Modules over non-Noetherian domains, 2001

83 **Sigurdur Helgason,** Groups and geometric analysis: Integral geometry, invariant differential operators, and spherical functions, 2000

82 **Goro Shimura,** Arithmeticity in the theory of automorphic forms, 2000

81 **Michael E. Taylor,** Tools for PDE: Pseudodifferential operators, paradifferential operators, and layer potentials, 2000

80 **Lindsay N. Childs,** Taming wild extensions: Hopf algebras and local Galois module theory, 2000

79 **Joseph A. Cima and William T. Ross,** The backward shift on the Hardy space, 2000

78 **Boris A. Kupershmidt,** KP or mKP: Noncommutative mathematics of Lagrangian, Hamiltonian, and integrable systems, 2000

77 **Fumio Hiai and Dénes Petz,** The semicircle law, free random variables and entropy, 2000

76 **Frederick P. Gardiner and Nikola Lakic,** Quasiconformal Teichmüller theory, 2000

75 **Greg Hjorth,** Classification and orbit equivalence relations, 2000

74 **Daniel W. Stroock,** An introduction to the analysis of paths on a Riemannian manifold, 2000

73 **John Locker,** Spectral theory of non-self-adjoint two-point differential operators, 2000
72 **Gerald Teschl,** Jacobi operators and completely integrable nonlinear lattices, 1999
71 **Lajos Pukánszky,** Characters of connected Lie groups, 1999
70 **Carmen Chicone and Yuri Latushkin,** Evolution semigroups in dynamical systems and differential equations, 1999
69 **C. T. C. Wall (A. A. Ranicki, Editor),** Surgery on compact manifolds, second edition, 1999
68 **David A. Cox and Sheldon Katz,** Mirror symmetry and algebraic geometry, 1999
67 **A. Borel and N. Wallach,** Continuous cohomology, discrete subgroups, and representations of reductive groups, second edition, 2000
66 **Yu. Ilyashenko and Weigu Li,** Nonlocal bifurcations, 1999
65 **Carl Faith,** Rings and things and a fine array of twentieth century associative algebra, 1999
64 **Rene A. Carmona and Boris Rozovskii, Editors,** Stochastic partial differential equations: Six perspectives, 1999
63 **Mark Hovey,** Model categories, 1999
62 **Vladimir I. Bogachev,** Gaussian measures, 1998
61 **W. Norrie Everitt and Lawrence Markus,** Boundary value problems and symplectic algebra for ordinary differential and quasi-differential operators, 1999
60 **Iain Raeburn and Dana P. Williams,** Morita equivalence and continuous-trace C^*-algebras, 1998
59 **Paul Howard and Jean E. Rubin,** Consequences of the axiom of choice, 1998
58 **Pavel I. Etingof, Igor B. Frenkel, and Alexander A. Kirillov, Jr.,** Lectures on representation theory and Knizhnik-Zamolodchikov equations, 1998
57 **Marc Levine,** Mixed motives, 1998
56 **Leonid I. Korogodski and Yan S. Soibelman,** Algebras of functions on quantum groups: Part I, 1998
55 **J. Scott Carter and Masahico Saito,** Knotted surfaces and their diagrams, 1998
54 **Casper Goffman, Togo Nishiura, and Daniel Waterman,** Homeomorphisms in analysis, 1997
53 **Andreas Kriegl and Peter W. Michor,** The convenient setting of global analysis, 1997
52 **V. A. Kozlov, V. G. Maz′ya, and J. Rossmann,** Elliptic boundary value problems in domains with point singularities, 1997
51 **Jan Malý and William P. Ziemer,** Fine regularity of solutions of elliptic partial differential equations, 1997
50 **Jon Aaronson,** An introduction to infinite ergodic theory, 1997
49 **R. E. Showalter,** Monotone operators in Banach space and nonlinear partial differential equations, 1997
48 **Paul-Jean Cahen and Jean-Luc Chabert,** Integer-valued polynomials, 1997
47 **A. D. Elmendorf, I. Kriz, M. A. Mandell, and J. P. May (with an appendix by M. Cole),** Rings, modules, and algebras in stable homotopy theory, 1997
46 **Stephen Lipscomb,** Symmetric inverse semigroups, 1996
45 **George M. Bergman and Adam O. Hausknecht,** Cogroups and co-rings in categories of associative rings, 1996

For a complete list of titles in this series, visit the AMS Bookstore at **www.ams.org/bookstore/**.